APOPTOSIS

FROM SIGNALING PATHWAYS TO THERAPEUTIC TOOLS

ANNALS OF THE NEW YORK ACADEMY OF SCIENCES

Volume 1010

APOPTOSIS

FROM SIGNALING PATHWAYS TO THERAPEUTIC TOOLS

Edited by Marc Diederich

The New York Academy of Sciences
New York, New York
2003

∞ *The paper used in this publication meets the minimum requirements of the American National Standard for Information Sciences—Permanence of Paper for Printed Library Materials, ANSI Z39.48-1984.*

Library of Congress Cataloging-in-Publication Data

Apoptosis: from signaling pathways to therapeutic tools / edited by Marc Diederich.
p. ; cm. — (Annals of the New York Academy of Sciences ; v. 1010)
Includes bibliographical references and index.
ISBN 1-57331-474-9 (cloth : alk. paper) — ISBN 1-57331-475-7 (pbk. : alk. paper)
1. Apoptosis—Congresses.
[DNLM: 1. Apoptosis—Congresses. QH 671 A64375 2003] I. Diederich, Marc. II. Series.
Q11.N5 vol. 1010
[QH671]
500 s—dc22
[571.9/

2003028198

GYAT/PCP
Printed in the United States of America
ISBN 1-57331-474-9 (cloth)
ISBN 1-57331-475-7 (paper)
ISSN 0077-8923

ANNALS OF THE NEW YORK ACADEMY OF SCIENCES

Volume1010
December 2003

APOPTOSIS

FROM SIGNALING PATHWAYS TO THERAPEUTIC TOOLS

Editor
MARC DIEDERICH

This volume is the result of a conference entitled **Apoptosis 2003: From Signaling Pathways to Therapeutic Tools** held on January 29 to February 1, 2003 in Luxembourg.

CONTENTS

Part III. Mitochondria as Regulators of Apoptosis

Part IV. Caspases and Apoptosis-Related Proteases

Part V. Transcriptional and Posttranscriptional Regulation of Apoptosis

Part VI. Inhibitors and Activators of Apoptosis

Part VII. Chemopreventive Agents and Apoptosis

Part VIII. Oxidative Stress and Apoptosis

Part IX. Glutathione and Apoptosis

Part X. Apoptosis and Tissular Functions

Part XI. Apoptosis and Senescence

Part XII. Apoptosis in Human Pathologies

Part XIII. Infection-Induced Apoptosis

Part XIV. Cell Death and Neurodegenerative Diseases

Part XV. Cell Death and Cardiovasular Diseases

Part XVI. Apoptosis: From Bench to Bedside

Financial assistance was received from:

- CITY OF LUXEMBOURG
- THE NATIONAL RESEARCH FUND, LUXEMBOURG
- FONDATION DE RECHERCHE CANCER ET SANG

Preface

MARC DIEDERICH

Laboratoire de Recherche sur le Cancer et les Maladies du Sang (RCMS), Centre Universitaire de Luxembourg, L-1511 Luxembourg

Apoptosis 2003: From Signaling Pathways to Therapeutic Tools, the fifth in our series of international conferences, was held from January 29 to February 1, 2003 at the European Conference Centre in Luxembourg. The meeting was organized by the RCMS (Recherche sur le Cancer et les Maladies du Sang) research group in collaboration with the University of Luxembourg, the University of Nancy, the Free University of Brussels, the University of Liège, the University of Pisa, and the Weizmann Institute of Science. Over 1150 participants gathered for an exceptional meeting with 135 oral presentations, 550 posters, and a trade show with 63 booths.

The presentations covered fundamental and applied aspects of apoptosis, including cell death and development, mitochondria as regulators of apoptosis, apoptotic signal transduction pathways, inhibitors and activators of apoptosis, receptors as cell death mediators, apoptosis in human pathologic disorders, caspases, transcriptional and posttranscriptional regulation of apoptosis, cell death and cardiovascular diseases, glutathione and apoptosis, apoptosis and chromatin structure, p53, cell death and cell cycle, apoptosis and tissular functions, cell death and blood diseases, cell death and neurodegenerative diseases, and the pharmacology of apoptosis.

In January 2004, fundamental and clinical researchers will gather again in Luxembourg at the European Conference Centre for the sixth and seventh meetings: Signal Transduction 2004: Signal Transduction Pathways as Therapeutic Targets, and Chromatin 2004: Chromatin Structure and Gene Expression Mechanisms as Therapeutic Targets (http://www.transduction-meeting.lu). These meetings will again introduce material that will provide new insights and advance our understanding of the pathophysiology of cell signaling and gene expression at the molecular level.

I thank the Editorial Department at the New York Academy of Sciences for so expertly seeing the present volume through the press. Finally, I would like to extend my special gratitude to the City of Luxembourg, the National Research Fund in Luxembourg, as well as to the Fondation de Recherche Cancer et Sang (FRCS) for their generous contributions in support of the fifth conference.

Ann. N.Y. Acad. Sci. 1010: xix (2003).
doi: 10.1196/annals.1299.144

An Introduction to the Molecular Mechanisms of Apoptosis

SYLVIE DELHALLE, ANNELYSE DUVOIX, MICHAËL SCHNEKENBURGER, FRANCK MORCEAU, MARIO DICATO, AND MARC DIEDERICH

Laboratoire de Recherche sur le Cancer et les Maladies du Sang (RCMS), Centre Universitaire de Luxembourg, L-1511 Luxembourg, Luxembourg

ABSTRACT: Apoptosis is a type of cell death that has been observed and studied for more than a century. The process of apoptosis was described as "programmed cell death" in 1964, and the term *apoptosis*, from a Greek word meaning "to fall away from" and describing the fall of dead leaves from trees in autumn, was only coined in 1972. During the last 30 years, this type of cell death has been extensively investigated and the molecular mechanisms underlying this cell suicide well characterized. Apoptosis is a physiological phenomenon necessary to tissue and body genesis and homeostasis, but defects in its regulation may cause numerous diseases, including cancer. Investigating the mechanisms of apoptosis is thus important to discover new cellular regulators that could be potential targets for new death-inducing drugs.

KEYWORDS: apoptosis; caspases; death receptors; mitochondria

INTRODUCTION

Apoptosis, a type of cell death, has been observed and studied for more than a century.[1–3] It is now clear that commitment to apoptosis is the result of interconnecting cascades of events, involving various families of proteins. These groups of proteins are highly conserved throughout evolution, since factors initially discovered in the nematode *Caenorhabditis elegans* have human counterparts. Apoptosis not only happens after cell and DNA injury, but is also important in embryology (ontogenesis) or to maintain bodily homeostasis. In an average human adult, 50 to 70 billion cells a day undergo apoptosis,[4] which has to occur so that new cells can have their place (for example, in self-renewing tissues). Defects in the regulation of the apoptotic program may cause the accumulation of virtually immortal cells, and a lack of apoptosis can lead to limb abnormalities, autoimmune diseases, and neoplasia, while an excess of apoptotic cells may participate in AIDS or neurodegenerative and neuromuscular diseases.

Address for correspondence: Marc Diederich, Laboratoire de Recherche sur le Cancer et les Maladies du Sang (RCMS), Centre Universitaire de Luxembourg, 162A, avenue de la Faïencerie, L-1511 Luxembourg, Luxembourg. Voice: +352 46 66 44 434; fax : +352 46 66 44 438.
diederic@cu.lu

Ann. N.Y. Acad. Sci. 1010: 1–8 (2003). © 2003 New York Academy of Sciences.
doi: 10.1196/annals.1299.001

Caspases

Apoptosis is characterized by morphologic changes such as cell shrinkage, plasma membrane blebbing, mitochondrial cytochrome *c* release, fragmentation of cell DNA into multiples of 180 bp, and the ultimate breakage of cells into small apoptotic bodies which will be cleared through phagocytosis by neighboring cells.

These morphologic changes are initiated by the activation of a family of proteases called caspases (cysteine aspartate-specific proteases) cleaving their substrates after a conserved aspartate residue (reviewed in Refs. 5 and 6). They exist in cells as inactive zymogens and must be activated by proteolytic cleavage into a heterodimer consisting of a large (20-kDa) and a small (10-kDa) subunit, liberating an inactive N-terminal prodomain. Two heterodimers then associate to form a heterotetrameric active enzyme with two active sites. To date, 14 caspases have been identified in humans and mice, but their functional role has not been fully characterized yet. Though caspases 2, 3, 6, 7, 8, 9 and 10 are involved in cell death, a subgroup (caspases 1, 4 and 5 in humans and 11 to 14 in mice) is responsible for cytokine maturation. The NH_2 terminal prodomain of caspases has variable length, causing them to be classified as upstream initiator (long prodomain) and downstream effector (short prodomain) enzymes.[7]

Two canonical pathways of caspase activation have been elucidated— the death receptor (DR) or extrinsic pathway and the mitochondrial or intrinsic pathway (FIG. 1). Induction of apoptosis through extrinsic or intrinsic pathways results in activation of initiator caspases, leading to apoptosis commitment. The extrinsic pathway is triggered by activation of tumor necrosis factor (TNF) receptors' superfamily members and it recruits initiator pro-caspases 8 and 10. The intrinsic pathway involves the release by mitochondria of factors activating pro-caspase 9. Initiator caspases are then able to cleave pro-caspases and thus to turn on downstream effector pro-caspases 3, 6 and 7, which in turn activate or inhibit target proteins, leading to apoptotic cell death.

Apoptosis by Death Receptors

Death receptors are a family of cell-surface receptors connecting death-promoting extracellular signals to apoptosis execution. This connection happens through the recruitment of adaptor proteins containing peptidic domains, allowing their physical interaction with receptors and initiator caspases. Members of the DR family include TNFR1, Fas/CD95, DR3, DR4, DR5, and DR6. These receptors are responsive to cytokines belonging to the TNF family, namely TNF-α, lymphotoxin (LTα), FasL, Apo3L and TRAIL (TNF-related apoptosis-inducing ligand) (reviewed in Ref. 8). Binding of a ligand to its receptor induces receptor trimerization and recruitment of adaptor proteins through homophilic interactions of the death domains (DD) of receptors and adaptor proteins. Adaptor proteins such as FADD (Fas-associated protein with death domain) also contain a dead effector domain (DED), which can interact in a homophilic manner with the DED of initiator caspases -8 and -10 and thus recruit them to the receptor complex. Recruitment of pro-caspase-2 is mediated through interaction by a caspase activation and recruitment domain (CARD) contained in the adaptor protein CRADD (caspase and RIP adaptor with death domain).[9,10] The complex formed by receptors, adaptor proteins, and initiator

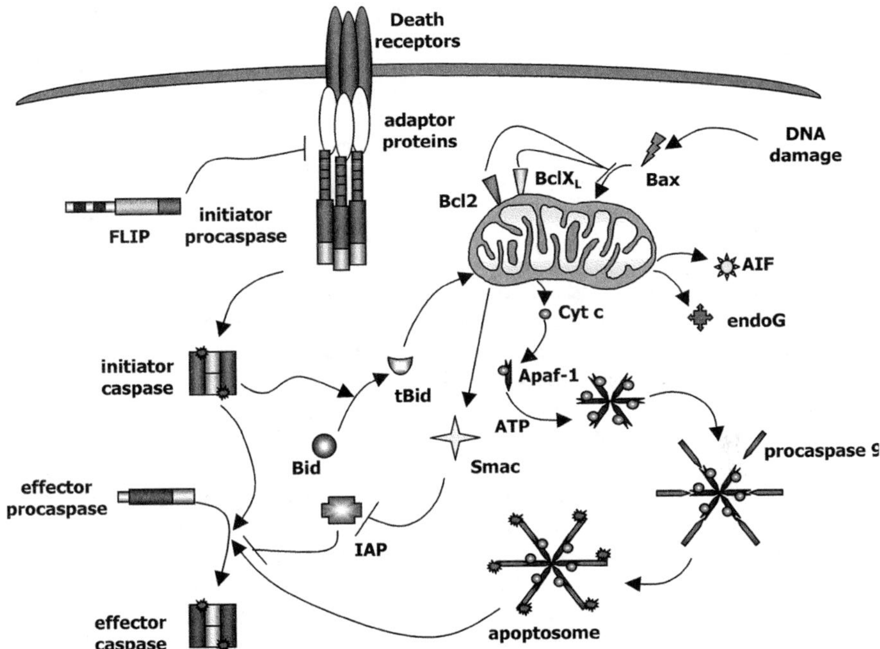

FIGURE 1. Apoptosis can be initiated by two pathways. The death-receptor pathway (extrinsic pathway) is induced by ligand binding to TNFR superfamily members. Receptors then recruit adaptor proteins through DD homophilic interactions. Adaptor proteins in turn recruit initiator pro-caspases by DED interaction, leading to DISC formation. Initiator pro-caspases are then converted in active caspases able to cleave substrates such as Bid or effector pro-caspases. The death receptor-induced triggering of apoptosis is impaired by recruitment of FLIP, an enzymatically inactive initiator caspase homologue. The second pathway (intrinsic pathway) is triggered by mitochondria in response to intracellular injuries such as DNA damage. Pro-apoptotic members of the Bcl-2 family of proteins induce mitochondrial release of various molecules such as cytochrome *c*, which can be counteracted by anti-apoptotic Bcl-2 family members. Cytochrome *c* binds to Apaf-1, which upon ATP-dependent conformational change and oligomerization associates with pro-caspase 9, forming the apoptosome, which is able to cleave effector pro-caspases. Effector caspase cleavage can be inhibited through activity of IAP proteins, which are themselves antagonized by the Smac/DIABLO protein released from mitochondria. Cleavage of Bid ensures the cross-talk between both apoptosis signaling pathways, since truncated Bid inserts itself in the mitochondrial membranes and induces cytochrome *c* release.

caspases is called death-inducing signaling complex (DISC). The inactive zymogen is then processed to give rise to an enzymatically active caspase able to further activate downstream effector caspases.

Extrinsic apoptosis can be modulated by an impaired recruitment of initiator caspases to the DISC complex. The DED-containing protein FLIP undergoes alternative splicing to give birth to $FLIP_S$ and $FLIP_L$. $FLIP_L$ contains two DED domains, but no functional protease domain, while $FLIP_S$ contains only the DED domains.[11,12] FLIP proteins compete with caspases 8 and 10 for binding to adaptor proteins, thus preventing apoptosis since they cannot transmit death-promoting signals.[13,14]

A second mechanism of DR apoptosis regulation happens through the expression of decoy death receptors. Decoy receptors DcR1 and DcR2 compete with DR4 and DR5 for TRAIL binding, but are respectively devoid of transmembrane and intracellular regions or of DD and cannot transmit death signals.[15–18] DcR3 is a soluble receptor expressed in colon and lung cancer and competing with Fas.[19]

Apoptosis by Intrinsic Pathway

Apoptosis may also be induced by intracellular signals, and mitochondria are in this case responsible for its execution. After DNA damage induced by stresses such as growth factor withdrawal, ultraviolet irradiation, or cell exposure to chemotherapeutic drugs, mitochondria release cytochrome *c* from their intermembrane space. This cytochrome *c* leakage can be initiated by the action of pro-apoptotic members of the Bcl-2 family proteins.

Cytochrome *c* then binds to the apoptotic protease-activating factor Apaf-1, which undergoes an ATP-dependent conformational change and oligomerization and then binds pro-caspase 9 by CARD domain interaction, forming the apoptosome.[20–23] Initiator caspase 9 is then activated but remains part of the apoptosome to cleave its cellular targets.

Interconnection exists between extrinsic and intrinsic pathways, since one of the caspase 8 substrate is Bid, a Bcl-2 related protein. Once it is cleaved, Bid inserts itself in the mitochondrial membrane and triggers cytochrome *c* leakage, which can in turn lead to apoptosome formation.[24]

Other soluble molecules are freed from the mitochondria intermembrane space upon induction of apoptosis.[25] In addition to releasssing cytochrome *c*, mitochondria may also release endonuclease G, implicated in apoptotic DNA degradation[26,27] and AIF (apoptosis-inducing factor), which may induce chromatin condensation and DNA fragmentation.[28] Smac/DIABLO (second mitochondria-derivated activator of caspases/direct IAP-binding protein with low pI) is also released from the mitochondria during apoptosis (see below).[29,30]

MODULATORS OF APOPTOSIS

Caspase activity is modulated by several groups of activators or inhibitors of these cell death proteases.

Inhibitors of Apoptosis Proteins (IAP)

The IAP family is a family of evolutionarily conserved apoptosis suppressors which may be found in mammals, insects, and some viruses (reviewed in Ref. 31). All six members of this family (XIAP, cIAP1, cIAP2, NAIP, survivin and apollon) contain at least one copy of a 70-amino-acid sequence called the BIR (baculovirus IAP repeat) domain, important for their anti-apoptotic behavior. Effector caspases 3, 6 and 7, as initiator caspase 9, are inhibited by human IAP survivin, XIAP (X-linked IAP), cIAP1, and cIAP2.[32–35] IAP are selective inhibitors and bind to their target caspases with a high specificity, unlike the baculoviral protein crm35, which shows a broad spectrum of caspase inhibition. Human IAP can block extrinsic and intrinsic apoptosis, probably because they target effector caspases common to both death pathways. The survival role of IAP can be inhibited by Smac/DIABLO, a-25 kDa protein released from mitochondria during apoptosis. Smac/DIABLO binds to the BIR domain of IAP, thus competing with caspases and preventing their inhibition.[29,30,36]

Bcl-2 Family Proteins

The Bcl-2 protein family includes 20 members exhibiting pro- or anti-apoptotic properties (reviewed in Refs. 37–39). Those proteins are characterized by the presence of one or several BH (Bcl-2 homology) domains. Four BH domains have been described: BH1, BH2, BH3, and BH4. Bcl-2 and Bcl-X_L, both containing four BH domains, are the most representative members of the anti-apoptotic part of the family. Among the death-inducing members, two subfamilies have been identified: a group lacking the BH4 domain and a BH3-only group. Among the proteins lacking the BH4 domain are Bax and Bak, strongly pro-apoptotic molecules whose insertion in mitochondrial membrane facilitates cytochrome *c* release. Many of those polypeptides physically interact, forming a complex network of homo- and heterodimers. The ratio of anti- and pro-apoptotic proteins determines the sensitivity or resistance of cells to various apoptotic stimuli, with pro-apoptotic members antagonizing the activity of anti-apoptotic ones. Some members of the Bcl-2 family protein are constitutively part of mitochondrial membranes, while other members locate in those membranes after specific stimuli. The exact mechanism of action is still uncertain, but Bcl-2-related proteins seem to act upstream of mitochondrial signs of apoptosis such as cytochrome *c* release, with pro-apoptotic members forming pores into mitochondrial membranes and triggering intermembrane protein leakage.[40]

Apoptosis and Cancer

Apoptosis is a physiological process to regulate tissue homeostasis, and a defect in its execution may lead to cancer. Moreover, evading apoptosis is considered as one of the six steps leading to malignant growth.[41] This resistance to apoptotic death can be due to overexpression of anti-apoptotic proteins of the Bcl-2 or IAP family. For instance, survivin, an IAP protein, is overexpressed in a majority of human cancers, which predicts a severe clinical outcome,[42–45] and Bcl-2 has been identified as overexpressed in B-cell lymphomas due to a t(14:18) translocation of its gene. Defects in the expression of death-promoting factors can also be involved in extended survey of cells, as Bax is mutated in some gastrointestinal cancers and leukemias.[46,47] Antisense oligonucleotides disrupting gene expression of Bcl-2,[48,49] Bcl-XL,[50,51] or FLIP[52] have been used in anticancer strategies, and re-expression of death-inducing proteins such as Bax sensitizes cells to cytotoxic compounds.[53,54]

Apoptosis resistance can also be conferred on cells by the activation of transcription factors such as NF-κB and AP-1. NF-κB induces transcription of various anti-apoptotic proteins (e.g., IAP, BCL-2 family proteins, and FLIP), and its overexpression has been observed in many tumors (reviewed in Ref. 55). AP-1 transcription factor is involved in oncogenesis, and its inhibition is considered as a promising anticancer therapy, as for NF-κB.[56,57] Therapies aiming to inhibit the anti-apoptotic activities of these factors might be useful in cancer treatment, and natural chemopreventive compounds acting at the stage of cancer promotion are widely studied. For instance, anti-oxidant and anti-inflammatory ingredients in spices such as curcumin, capsaicin, and gingerol can suppress or reduce NF-κB and AP-1 activation and inhibit tumor promotion (reviewed in Ref. 58).

Proteasome inhibitors sensitizing cells to chemotherapeutic drugs (bortezomib, PS-341) are used in clinical trials and seem to display convincing anticancer activities (reviewed in Ref. 59). As a conclusion, apoptosis is a complex mechanism interconnecting pathways involving several families of proteins. Though much is known about the large steps of this process, its regulation by death-promoting stimuli, cell or tissue type, and mode of execution are far from understood. Further study of fine regulation of apoptosis will be helpful in management of numerous diseases resulting from defects in apoptosis.

ACKNOWLEDGMENTS

This work was supported by the “Fondation de Recherche Cancer et Sang” and the “Recherches scientifiques Luxembourg.” S.D. is supported by a Télévie grant 7.4577.02 (FNRS, Belgium). A.D. and M.S. were supported by fellowships from the Government of Luxembourg.

REFERENCES

1. VAUX, D.L. & S.J. KORSMEYER. 1999. Cell death in development. Cell **96:** 245–254.
2. LOCKSHIN, R.A.W. & C.M. WILLIAMS. 1964. Programmed cell death.II. Endocrine potentiation of the breakdown of the intersegmental muscles of silkmoths. J. Insect Physiol. **10:** 643–649.
3. KERR, J.F., A.H. WYLLIE & A.R. CURRIE. 1972. Apoptosis: a basic biological phenomenon with wide-ranging implications in tissue kinetics. Br. J. Cancer **26:** 239–257.
4. REED, J. C. 1999. Dysregulation of apoptosis in cancer. J. Clin. Oncol. **17:** 2941–253.
5. EARNSHAW, W.C., L.M. MARTINS & S.H. KAUFMANN. 1999. Mammalian caspases: structure, activation, substrates, and functions during apoptosis. Annu. Rev. Biochem. **68:** 383–424.
6. STRASSER, A., L. O'CONNOR & V.M. DIXIT. 2000. Apoptosis signaling. Annu. Rev. Biochem. **69:** 217–245.
7. SALVESEN, G.S. & V.M. DIXIT. 1997. Caspases: intracellular signaling by proteolysis. Cell **91:** 443–446.
8. LOCKSLEY, R.M., N. KILLEEN & M.J. LENARDO. 2001. The TNF and TNF receptor superfamilies: integrating mammalian biology. Cell **104:** 487–501.
9. AHMAD, M., *et al.* 1997. CRADD, a novel human apoptotic adaptor molecule for caspase-2, and FasL/tumor necrosis factor receptor-interacting protein RIP. Cancer Res. **57:** 615–619.
10. DUAN, H. & V.M. DIXIT. 1997. RAIDD is a new “death” adaptor molecule. Nature **385:** 86–89.
11. TSCHOPP, J., M. IRMLER & M. THOME. 1998. Inhibition of fas death signals by FLIPs. Curr. Opin. Immunol. **10:** 552–558.
12. WALLACH, D. 1997. Apoptosis: placing death under control. Nature **388:** 123, 125–126.
13. GOLTSEV, Y.V. *et al.* 1997. CASH, a novel caspase homologue with death effector domains. J. Biol. Chem. **272:** 19641–19644.

14. IRMLER, M., *et al.* 1997. Inhibition of death receptor signals by cellular FLIP. Nature **388:** 190–5.
15. DEGLI-ESPOSTI, M.A. *et al.* 1997. The novel receptor TRAIL-R4 induces NF-kappaB and protects against TRAIL-mediated apoptosis, yet retains an incomplete death domain. Immunity **7:** 813–820.
16. MARSTERS, S.A. *et al.* 1997. A novel receptor for Apo2L/TRAIL contains a truncated death domain. Curr. Biol. **7:** 1003–1006.
17. PAN, G. *et al.* 1997. An antagonist decoy receptor and a death domain-containing receptor for TRAIL. Science **277:** 815–818.
18. SHERIDAN, J.P. *et al.* 1997. Control of TRAIL-induced apoptosis by a family of signaling and decoy receptors. Science **277:** 818–821.
19. PITTI, R.M. *et al.* 1998. Genomic amplification of a decoy receptor for Fas ligand in lung and colon cancer. Nature **396:** 699–703.
20. SRINIVASULA, S.M. *et al.* 1998. Autoactivation of procaspase-9 by Apaf-1-mediated oligomerization. Mol. Cell. **1:** 949–957.
21. ZOU, H. *et al.* 1997. Apaf-1, a human protein homologous to C. elegans CED-4, participates in cytochrome c-dependent activation of caspase-3. Cell **90:** 405–413.
22. ZOU, H. *et al.* 1999. An APAF-1.cytochrome c multimeric complex is a functional apoptosome that activates procaspase-9. J Biol Chem. **274:** 11549–11556.
23. LI, P. *et al.* 1997. Cytochrome c and dATP-dependent formation of Apaf-1/caspase-9 complex initiates an apoptotic protease cascade. Cell **91:** 479–89.
24. GROSS, A. *et al.* 1999. Caspase cleaved BID targets mitochondria and is required for cytochrome c release, while BCL-XL prevents this release but not tumor necrosis factor-R1/Fas death. J. Biol. Chem. **274:** 1156–1163.
25. LORENZO, H.K. *et al.* 1999. Apoptosis inducing factor (AIF): a phylogenetically old, caspase-independent effector of cell death. Cell Death Differ. **6:** 516–524.
26. LI, L.Y., X. LUO & X. WANG. 2001. Endonuclease G is an apoptotic DNase when released from mitochondria. Nature. **412:** 95–9.
27. PARRISH, J. *et al.* 2001. Mitochondrial endonuclease G is important for apoptosis in C. elegans. Nature. **412:** 90–4.
28. SUSIN, S.A. *et al.* 1999. Molecular characterization of mitochondrial apoptosis-inducing factor. Nature. **397:** 441–6.
29. DU, C. *et al.* 2000. Smac, a mitochondrial protein that promotes cytochrome c-dependent caspase activation by eliminating IAP inhibition. Cell **102:** 33–42.
30. VERHAGEN, A.M., *et al.* 2000. Identification of DIABLO, a mammalian protein that promotes apoptosis by binding to and antagonizing IAP proteins. Cell. **102:** 43–53.
31. DEVERAUX, Q.L. & J.C. REED. 1999. IAP family proteins: suppressors of apoptosis. Genes Dev. **13:** 239–52.
32. TAMM, I. *et al.* 1998. IAP-family protein survivin inhibits caspase activity and apoptosis induced by Fas (CD95), Bax, caspases, and anticancer drugs. Cancer Res. **58:** 5315–5320.
33. ROY, N. *et al.* 1997. The c-IAP-1 and c-IAP-2 proteins are direct inhibitors of specific caspases. EMBO J. **16:** 6914–6925.
34. DEVERAUX, Q.L. *et al.* 1997. X-linked IAP is a direct inhibitor of cell-death proteases. Nature. **388:** 300–304.
35. DEVERAUX, Q.L., *et al.* 1999. Cleavage of human inhibitor of apoptosis protein XIAP results in fragments with distinct specificities for caspases. EMBO J. **18:** 5242–5251.
36. GOYAL, L. 2001. Cell death inhibition: keeping caspases in check. Cell **104:** 805–808.
37. ADAMS, J.M. & S. CORY. 1998. The Bcl-2 protein family: arbiters of cell survival. Science **281:** 1322–1326.
38. GROSS, A., J.M. MCDONNELL & S.J. KORSMEYER. 1999. BCL-2 family members and the mitochondria in apoptosis. Genes Dev. **13:** 1899–1911.
39. REED, J.C. 1998. Bcl-2 family proteins. Oncogene **17:** 3225–36.
40. HENGARTNER, M. 2000. The biochemistry of apoptosis. Nature **407:** 770–776.
41. HANAHAN, D. & R.A. WEINBERG. 2000. The hallmarks of cancer. Cell **100:** 57–70.
42. AMBROSINI, G., C. ADIDA & D.C. ALTIERI. 1997. A novel anti-apoptosis gene, survivin, expressed in cancer and lymphoma. Nat. Med. **3:** 917–921.

43. AMBROSINI, G. *et al.* 1998. Induction of apoptosis and inhibition of cell proliferation by survivin gene targeting. J. Biol. Chem. **273:** 11177–11182.
44. TAMM, I. *et al.* 2000. Expression and prognostic significance of IAP-family genes in human cancers and myeloid leukemias. Clin. Cancer Res. **6:** 1796–1803.
45. VELCULESCU, V.E. *et al.* 1999. Analysis of human transcriptomes. Nat Genet. **23:** 387–388.
46. RAMPINO, N. *et al.* 1997. Somatic frameshift mutations in the BAX gene in colon cancers of the microsatellite mutator phenotype. Science **275:** 967–969.
47. YAMAMOTO, H., H. SAWAI & M. PERUCHO. 1997. Frameshift somatic mutations in gastrointestinal cancer of the microsatellite mutator phenotype. Cancer Res. **57:** 4420–4426.
48. JANSEN, B. *et al.* 1998. bcl-2 antisense therapy chemosensitizes human melanoma in SCID mice. Nat. Med. **4:** 232–234.
49. JANSEN, B. *et al.* 2000. Chemosensitisation of malignant melanoma by BCL2 antisense therapy. Lancet **356:** 1728–33.
50. BUREAU, F. *et al.* 2002. Constitutive nuclear factor-kappaB activity preserves homeostasis of quiescent mature lymphocytes and granulocytes by controlling the expression of distinct Bcl-2 family proteins. Blood **99:** 3683–3691.
51. REED, J.C. 1999. Splicing and dicing apoptosis genes. Nat. Biotechnol. **17:** 1064–1065.
52. QUE, F.G. *et al.* 1999. Cholangiocarcinomas express Fas ligand and disable the Fas receptor. Hepatology **30:** 1398–404.
53. BARGOU, R.C. *et al.* 1995. Induction of Bax-alpha precedes apoptosis in a human B lymphoma cell line: potential role of the bcl-2 gene family in surface IgM-mediated apoptosis. Eur. J. Immunol. **25:** 770–775.
54. BARGOU, R.C. *et al.* 1996. Overexpression of the death-promoting gene bax-alpha which is downregulated in breast cancer restores sensitivity to different apoptotic stimuli and reduces tumor growth in SCID mice. J. Clin. Invest. **97:** 2651–2659.
55. LIN, A. & M. KARIN. 2003. NF-kappaB in cancer: a marked target. Semin. Cancer Biol. **13:** 107–114.
56. MAEDA, S. & M. KARIN. 2003. Oncogene at last-c-Jun promotes liver cancer in mice. Cancer Cell **3:** 102–104.
57. YOUNG, M.R., H.S. YANG & N.H. COLBURN. 2003. Promising molecular targets for cancer prevention: AP-1, NF-kappa B and Pdcd4. Trends Mol. Med. **9:** 36–41.
58. SURH, Y.J. 2002. Anti-tumor promoting potential of selected spice ingredients with antioxidative and anti-inflammatory activities: a short review. Food Chem. Toxicol. **40:** 1091–1097.
59. ADAMS, J. 2003. Potential for proteasome inhibition in the treatment of cancer. Drug Discov. Today **8:** 307–315.

p73, the "Assistant" Guardian of the Genome?

GERRY MELINO[a,b]

[a]*Biochemistry Laboratory, IDI-IRCCS, Department of Experimental Medicine, University of Rome "Tor Vergata," Rome, Italy*

[b]*Medical Research Council, Toxicology Unit, Leicester, UK*

Abstract: **Although p53 is clearly involved in the salvage pathway to DNA damage, its frequent mutations do not explain the efficacy of radiotherapy and chemotherapy. Indeed, around 50% of all human cancers show mutations in p53, and a further fraction show a functional inactivation of the protein. Nevertheless, patients seem to respond to therapy that would otherwise require a functional p53. At least in part, these responses could be explained by the pathway mediated by p73. This mechanism is parallel to, but independent of the p53 pathway. Several pieces of evidence show a significant interaction between these two proteins. Therefore, while p53 can be rightly defined as the guardian of the genome, we could think of p73 as the "assistant" guardian of the genome!**

Keywords: **apoptosis; cell death; DNA damage; cell cycle**

INTRODUCTION

Eukaryotic cells have developed very sophisticated mechanisms to recognize and repair defects in their genomes.[1–4] Figure 1 contains a simplified scheme of this complex array of biochemical reactions. Despite tremendous advancements in understanding these mechanisms, the sensors and signaling mechanisms are still quite elusive. Elucidation of the role exerted by ATM is an important recent advance.[5] For example, extremely few double-strand breaks (DSB) are sufficient to trigger within minutes the autophosphorylation of ATM with dissociation of its homodimer[5]; this activates the kinase activity of ATM on other substrates, such as p53.[6,7] These and other mechanisms guarantee the correct cell cycle progression in healthy cells and prevent the occurrence of cancer.[8,9]

Abbreviations: TA-, full-length transactivating isoforms; ΔN-, amino-terminal deleted isoforms, lacking the transactivation domain; TA, transactivation domain; DBD, DNA binding domain; OD, oligomerization domain; PR, proline-rich domain; SAM, sterile alpha motif; TI, trans-inhibitory domain; NES, nuclear exporting signal; SUMO, small ubiquitin-like modifier; PML, promyelocytic leukemia protein; SAPK, stress-activated protein kinases; JNK, Jun kinase

Address for correspondence: Gerry Melino, Medical Research Council, Toxicology Unit, Hodgking Building, Leicester University, Lancaster Road, P.O. Box 138, Leicester LE1 9HN, UK.

gm89@le.ac.uk

Ann. N.Y. Acad. Sci. 1010: 9–15 (2003).
doi: 10.1196/annals.1299.002

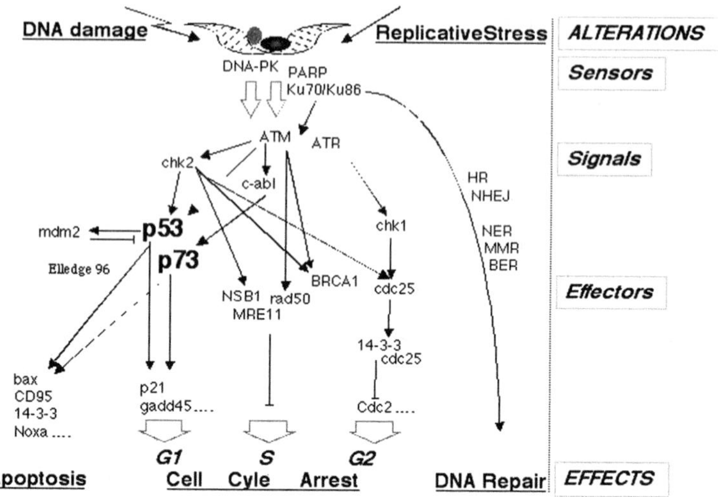

FIGURE 1. DNA damage pathways. A complex redundant pathway of biochemical reactions is required to efficiently sense and signal DNA damage to effector molecules designed to delay cell cycle progression in order to arrest potential transmission to the progeny and allow entrance inside the repair machinery. Failing that, the same systems switch the response to an apoptotic pathway, forcing the cell to die. Several pieces of this pathway are still not known. Nonetheless, it is evident that a major role is played by p53, and the novel pathway exerted by its homologue p73.

p53

The p53 protein is activated just below ATM and chk2 in response to cellular stress, such as irradiation or chemical agents able to introduce DSB or oncogene activation that cause cell-cycle arrest or apoptosis[10] (FIGS. 1 and 2). The role of p53 is so important that a vast majority of human cancers show evidence of loss of p53 function.[10,11] A set of recent papers review these aspects.[12–16] While the categorization outlined in FIGURE 2 is somewhat semantic, it helps to describe the degree of regulation of this extremely important gene.

The first level of regulation of p53 function is the regulation of the steady-state level of the protein.[16] The best example of this is the negative dominant loop exerted by mdm2. Although mdm2 causes the ubiquitination of p53, and thus its degradation, HAUSP is able to de-ubiquitinate the proteins, increasing its steady-state protein level.[17,18] Further regulations rise, for example, from the *S*-nitroxylation by NO derivatives either on p53 itself or on mdm2.[19–21]

The second type of regulation is at the subcellular localization of the protein. Interestingly, nitric oxide can affect the nuclear or cytosolic location of the protein.[22] More recently, a novel protein, Parc, has been found to regulate the subcellular localization of p53.[23] This protein seems to be responsible for the abnormal "parcking" of p53 in the cytoplasm in forms of cancer such as neuroblastoma.[23,24]

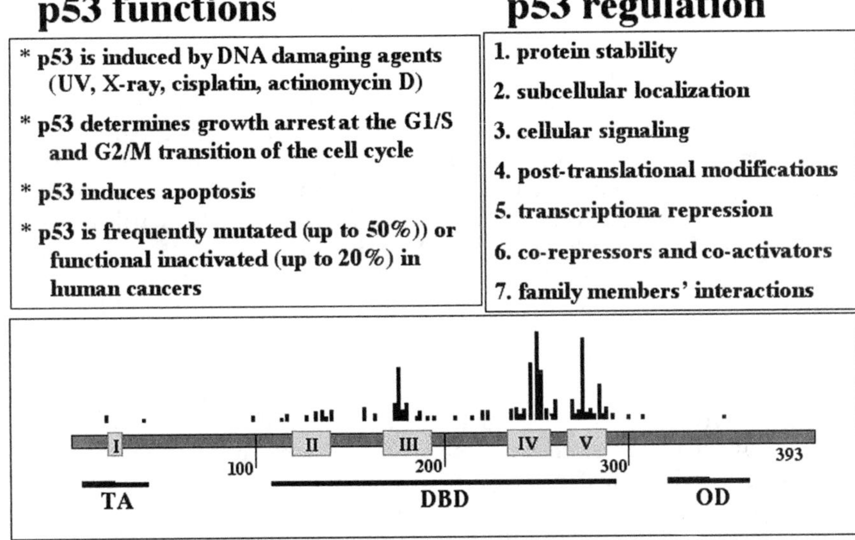

FIGURE 2. p53 is a nuclear protein. The p53 protein is normally absent or present at extremely low levels in normal cells. Upon DNA damage, though, p53 is strongly induced and activated. The major effects are the arrest of the cell cycle at the G1/S and G2/M checkpoints, and the induction of apoptosis. Clearly these effects must be finely tuned to guarantee cell integrity. These functions are so important that p53 is mutated in significant fractions of all human cancer, and it is functionally inactivated in a further significant fraction of cancers. The *right panel* shows the levels of regulation of p53. Mutations occur quite often in the DNA binding domain, as schematically reported in the frequency diagram. TA = transactivation domain; DBD = DNA binding domain; OD = oligomerization domain.

The third level of regulation is at the level of cellular signaling,[15] where different cell stresses allow activation of specific pathways (cytokines, β-catenin, PTEN, AKT), finely tuning the biological effect.[15]

The fourth level of regulation is at the post-translational modifications.[13] During DNA damage, p53 is phosphorylated by several kinases, namely ATM, p38, HIPK2, TAFII250, JNK, CK2-FACT at several Ser or Thr in the TA, Pro-rich, NLS, and COOH-terminal domains.[13] The fact that 12–15 residues can be modified shows the degree of regulation that controls the activity of the protein. When phosphorylated, p53 could become target of a further post-translational modification: the *cis-trans* isomerization of prolines catalyzed by Pin-1.[25,26]

The fifth regulation balances the level of transcriptional activation and repression.[12] Transcriptional repression is, unfortunately, less studied for obvious reasons. It occurs by interference with the function of DNA-binding transcriptional activators. Alternatively it occurs by interference with the basal transcriptional machinery. Finally, it occurs via the alteration of chromatin structure at the promoters of target genes by recruiting proteins such as histone deacetylases.

The sixth level of regulation is at level of co-factors able to interfere with the transcriptional activity. ASPP is a transcription factor able to differentially regulate the activity of specific target promoters.[27] Similarly, Brn3a is also able to modify the transcriptional activity of p53 itself.[28]

The seventh level of regulation is at the level of interference with its own family members, namely p63 and p73. Experiments made in double p63/p73 knockout mice show the inability of p53 to trigger apoptosis, even though the protein is normally induced, expressed, and functional, at least for certain promoters.[29] A dominant negative regulatory loop is also activated by p53 and TAp73, with ΔNp73 being very actively transcribed by p53/TAp73. In turn, ΔNp73 is able to inhibit the functions of p53, as well as those of TAp73.[30]

p73

In the last five years, two p53 homologues, p73 and p63, have been identified.[31,32] In this short review the involvement of p73 or p63 in development will not be discussed. It is important, however, to remember that the knockout mice of p73 presented no spontaneous tumors, only defects in neuronal development[33]and neural defects,[34] and that p73 is sufficient to trigger neuronal differentiation.[35] A general review in this area is reported by Yang *et al.* [36] and Melino *et al.*[37] FIGURE 3 shows an extremely simplified drawing of the similarity between p53 and p73. p73 is present in several isoforms in normal[1,38,39] and in cancer cells,[40] where it can regulate apoptosis.[41–43] Recent evidence indicates complex interactions among p53 and p73.[29,30,43,44] Moreover, the two isoforms TAp73 and ΔNp73 show pro-apoptotic and anti-apoptotic properties, respectively; ΔNp73 is controlled at promoter level by p53 and by TAp73. This indicates that the tumor-suppressor function of TAp73 is very finely tuned by a dominant, negative regulatory loop exerted by ΔNp73.[30] In-

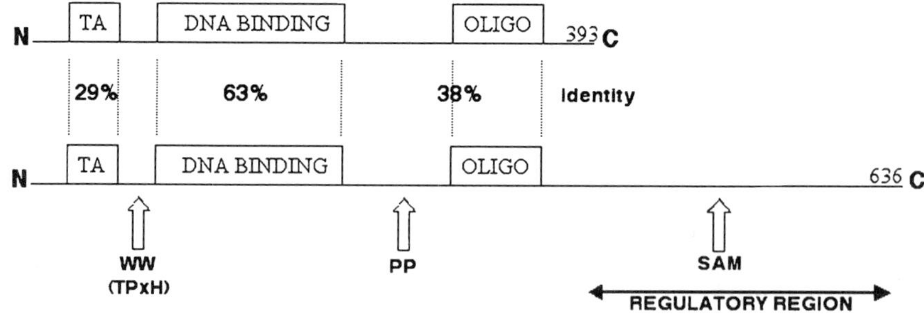

FIGURE 3. Structure of p53 and p73. The schematic structure of the p53 and p73 proteins is indicated. The overall structural domains are conserved with a high degree of residue identity. The p73 is significantly longer due to the presence of a long COOH– terminal tail which contains several regulatory components. The most relevant one is the sterile alpha motif (SAM), a protein–protein binding domain for which partners have not yet been identified. Abbreviations are identical to those of FIGURE 2.

terestingly, this regulatory loop is somehow reminiscent of that of mdm2. The relevance of ΔNp73 is indicated by its role as a prognostic factor in cancer.[45,46]

At the molecular level, the pathway involving p73 in DNA damage is totally independent from, and parallel to, p53. FIGURE 1 shows the relationship between the two. p73 is downstream of the mismatch repair gene MLH1 and c-Abl.[47–49] c-Abl is a non-receptor tyrosine kinase that is linked functionally to several key molecules involved in response to DNA damage, including ATM, p53, and p73. In fact, after a cisplatin induction, p73 is stabilized and induces apoptosis in a c-Abl-dependent manner.[47–49] p73 and c-Abl interact at the PxxP domain of p73, between the DBD and the OD, and at the SH3 domain of c-Abl. c-Abl is activated, after γ-radiation, presumably by the stress-induced protein kinase ATM. Therefore, we can consider ATM as part of this pathway.

In addition to this nuclear interaction, p73 and c-Abl can interact in the cytosol. c-Abl activates a group of stress-activated protein kinases (SAPK), such as JNK and p38MAP kinase. They are proline-targeted Ser-Thr protein kinases, associated with DNA damage due to UV light or γ radiation. After cisplatin treatment, p38 is also involved and its activation is sufficient to enhance p73 activity; p73 can be stabilized through a tyrosine 99 phosphorylation by c-Abl and a threonine phosphorylation by p38.[50,51]

p73, THE "ASSISTANT" GUARDIAN OF THE GENOME

p53 is a major player in DNA damage responses and plays a crucial role in tumorigenesis. For this, p53 has rightly been named the guardian of the genome. In normal conditions p53 is extremely more efficient than p73, but this is not true in cancer. Because p73 is very rarely mutated in cancer,[37] it is responsible for the efficacy of radiotherapy and chemotherapy in cancer treatment. For this "little help from a friend," p73 can now be regarded as the "assistant" guardian of the genome!

ACKNOWLEDGMENTS

The work was supported by grants from the Medical Research Council (to G.M.), and by Telethon (E872, E1224), AIRC, the EU (QLG1-1999-00739 and QLK-CT-2002-01956), MIUR, and MinSan (to G.M.). We apologize for the large number of authors whose relevant work could not be cited due to space limitations.

REFERENCES

1. HOEIJMAKERS, J.H. 2001. Genome maintenance mechanisms for preventing cancer. Nature **411:** 366–374.
2. LINDAHL, T. & R.D. WOOD. 1999. Quality control by DNA repair. Science **286:** 1897–1905.
3. WOOD, R.D., M. MITCHELL, J. SGOUROS & T. LINDAHL. 2001. Human DNA repair genes. Science **291:** 1284–1289.
4. BARTEK, J. & J. LUKAS. 2003. Damage alert. Nature **421:** 486–488.
5. BAKKENIST, C.J. & M.B. KASTAN. 2003. DNA damage activates ATM through intermolecular autophosphorylation and dimer dissociation. Nature **421:** 499–506.

6. BANIN, S., L. MOYAL, S. SHIEH, *et al.* 1998. Enhanced phosphorylation of p53 by ATM in response to DNA damage. Science **281:** 1674–1677.
7. CANMAN, C.E., D.S. LIM, K.A. CIMPRICH, *et al.* 1998. Activation of the ATM kinase by ionizing radiation and phosphorylation of p53. Science **281:** 1677–1679.
8. HARTWELL, L.H. & M.B. KASTAN. 1994. Cell cycle control and cancer. Science **266:** 1821–1828.
9. ELLEDGE, S.J. 1996. Cell cycle checkpoints: preventing an identity crisis. Science **274:** 1664–1672.
10. VOGELSTEIN, B., D. LANE & A.J. LEVINE. 2000. Surfing the p53 network. Nature **408:** 307–310.
11. LU, X. & K.H. VOUSDEN. 2002. Live or let die: the cell's response to p53. Nat. Rev. Cancer **2:** 594–604.
12. HO, J. & S. BENCHIMOL. 2003. Transcriptional repression mediated by the p53 tumour suppressor. Cell Death Differ. **10:** in press.
13. XU, Y. 2003. Regulation of p53 responses by post-translational modifications. Cell Death Differ. **10:** in press.
14. CLARKE, A.R. & M. HOLLSTEIN. 2003. Mouse models with modified p53 sequences to study cancer and ageing. Cell Death Differ. **10:** in press.
15. OREN, M. 2003. Decision making by p53: life, death and cancer. Cell Death Differ. **10**(4): in press.
16. WEBER, J.D. & G.P. ZAMBETTI. 2003. Revewing the debate over the p53 apoptotic response. Cell Death Differ. **10:** in press.
17. LI, M., D. CHEN, A. SHILOH, *et al.* 2002. Deubiquitination of p53 by HAUSP is an important pathway for p53 stabilization. Nature **416:** 648–653.
18. ITO, A.,Y. KAWAGUCHI, C.H. LAI, *et al.* 2002. MDM2-HDAC1-mediated deacetylation of p53 is required for its degradation. EMBO J. **21:** 6236–6245.
19. SCHONHOFF, C.M., M.C. DAOU, S.N. JONES, *et al.* 2002. Nitric oxide-mediated inhibition of Hdm2-p53 binding. Biochemistry **41:** 13570–13574.
20. HOFSETH, L.J., S. SAITO, S.P. HUSSAIN, *et al.* 2003. Nitric oxide-induced cellular stress and p53 activation in chronic inflammation. Proc. Natl. Acad. Sci. USA **100:** 143–148.
21. WANG, X., D. MICHAEL, G. DE MURCIA & M. OREN. 2002. p53 Activation by nitric oxide involves down-regulation of Mdm2. J. Biol. Chem. **277:** 15697–15702.
22. WANG, A., X. ZALCENSTEIN & M. OREN. 2003. Nitric oxide promotes p53 nuclear retention and sensitizes neuroblastoma cells to apoptosis by ionizing radiation. Cell Death Differ. **10:** in press.
23. NIKOLAEV, A.Y., M. LI, N. PUSKAS, *et al.* 2003. Parc: a cytoplasmic anchor for p53. Cell **112:** 29–40.
24. KASTAN, M.B. & G.P. ZAMBETTI. 2003. Parc-ing p53 in the cytoplasm. Cell **112:** 1–5.
25. ZACCHI, P., M. GOSTISSA & T. UCHIDA, *et al.* 2002. The prolyl isomerase Pin1 reveals a mechanism to control p53 functions after genotoxic insults. Nature **419:** 853–857.
26. ZHENG, H., H. YOU, X.Z. ZHOU, *et al.* 2002. The prolyl isomerase Pin1 is a regulator of p53 in genotoxic response. Nature **419:** 849–853.
27. SAMUELS-LEV, Y., D.J. O'CONNOR, D. BERGAMASCHI, *et al.* 2001. ASPP proteins specifically stimulate the apoptotic function of p53. Mol. Cell **8:** 781–794.
28. BUDRAM-MAHADEO, V., P.J. MORRIS & D.S. LATCHMAN. 2002. The Brn-3a transcription factor inhibits the pro-apoptotic effect of p53 and enhances cell cycle arrest by differentially regulating the activity of the p53 target genes encoding Bax and p21(CIP1/Waf1). Oncogene **21:** 6123–31.
29. FLORES, E.R., K.Y. TSAI, D. CROWLEY, *et al.* 2002. p63 and p73 are required for p53-dependent apoptosis in response to DNA damage. Nature **416:** 560–564.
30. GROB, T.J., U. NOVAK, C. MAISSE, *et al.* 2001. Human ΔNp73 regulates a dominant negative feedback loop for TA p73 and p53. Cell Death Differ. **8:** 1213–1223.
31. KAGHAD, M., H. BONNET, A. YANG, *et al.* 1997. Monoallelically expressed gene related to p53 at 1p36, a region frequently deleted in neuroblastoma and other human cancers. Cell **90:** 809–819.
32. YANG, A., M. KAGHAD, Y. WANG, *et al.* 1998. p63, a p53 homolog at 3q27–29, encodes multiple products with transactivating, death-inducing, and dominant-negative activities. Mol. Cell **2:** 305–316.

33. YANG, A., N. WALKER, R. BRONSON, *et al.* 2000. p73-deficient mice have neurological, pheromonal and inflammatory defects but lack spontaneous tumours. Nature **404:** 99–103.
34. POZNIAK, C.D., F. BARNABE'-HEIDER, V.V. RYMAR, *et al.* 2002. p73 is required for survival and maintenance of CNS neurons. J. Neurosci. **22:** 9800–9809.
35. DE LAURENZI, V., G. RASCHELLA, D. BARCAROLI, *et al.* 2000. Induction of neuronal differentiation by p73, in a neuroblastoma cell line. J. Biol. Chem. **275:** 15226–15231.
36. YANG, A., M. KAGHAD, D. CAPUT & F. MCKEON. 2002. On the shoulders of giants: p63, p73 and the rise of p53. Trends Genet. **18:** 90–95.
37. MELINO, G., V. DE LAURENZI & K.H. VOUSDEN. 2002. p73: friend or foe in tumorigenesis. Nature Rev. Cancer **2:** 605–615.
38. DE LAURENZI, V., A. COSTANZO, D. BARCAROLI, *et al.* 1998. Two new p73 splice variants, gamma and delta, with different transcriptional activity. J. Exp. Med. **188:** 1763–1768.
39. DE LAURENZI, V., M.V. CATANI, A. COSTANZO, *et al.* 1999. Additional complexity in p73: induction by mitogens in lymphoid cells and identification of two new splicing variants e and z. Cell Death Differ **6:** 389–390.
40. FILLIPPOVICH, I., N. SOROKINA, M. GATEI, *et al.* 2001. Transactivation-deficient p73alpha (p73Deltaexon2) inhibits apoptosis and competes with p53. Oncogene **20:** 514–522.
41. IRWIN, M., M.C. MARIN & A.C. PHILLIPS, *et al.* 2000. Role for the p53 homologue p73 in E2F-1-induced apoptosis. Nature **407:** 645–648.
42. BASU, S., N.F. TOTTY, M.S. IRWIN, *et al.* 2003. Akt phosphorylates the yes-associated protein, YAP, to induce interaction with 14-3-3 and attenuation of p73-mediated apoptosis. Mol. Cell **11:** 11–23.
43. MARIN, M.C., C.A. JOST, L.A. BROOKS, *et al.* 2000. A common polymorphism acts as an intragenic modifier of mutant p53 behaviour. Nat. Genet. **25:** 47–54.
44. POZNIAK, C.D., S. RADINOVIC, A. YANG, *et al.* 2000. An anti-apoptotic role for the p53 family member, p73, during developmental neuron death. Science **289:** 304–306.
45. CASCIANO, I., K. MAZZOCCO, L. BONI, *et al.* 2002. Expression of ΔNp73 is a molecular marker for adverse outcome in neuroblastoma patients. Cell Death Differ. **9:** 246–251.
46. ZAIKA, A.I., N. SLADE & S.H. ERSTER. 2002. ΔNp73, a dominant-negative inhibitor of wild-type p53 and TAp73, is up-regulated in human tumors. J. Exp. Med. **196:** 765–780.
47. GONG, J.G., A. COSTANZO, H.Q. YANG, *et al.* 1999. Regulation of the p53 homolog p73 by c-Abl tyrosine kinase in cell death response to cisplatin. Nature **399:** 806–809.
48. AGAMI, R., G. BLANDINO, M. OREN & Y. SHAUL. 1999. Interaction of c-Abl and p73alpha and their collaboration to induce apoptosis. Nature **399:** 809–813.
49. YUAN, Z.M., H. SHIOYA, T. ISHIKO, *et al.* 1999. p73 is regulated by tyrosine kinase c-Abl in the apoptotic response to DNA damage. Nature **399:** 814–817.
50. SANCHEZ-PRIETRO, R., V.J. SANCHEZ-AREVALO, J.M. SERVITA & J.S. GUTKIND. 2000. Regulation of p73 though the p38 MAP kinase pathway. Oncogene **21:** 974–979.
51. MELINO, G., X. LU, M. GASCO, *et al.* 2003. Functional regulation of p73 and p63: development and cancer. Trends Biochem. Sci **28.** In press.

Apoptosis Gene Network

Description in the GeneNet and TRRD Databases

IRINA STEPANENKO AND NIKOLAY KOLCHANOV

Institute of Cytology and Genetics, SB RAS, Lavrentieva 10, 630090, Novosibirsk, Russia

ABSTRACT: The large volume of experimental data on apoptosis requires computer processing to systematize and provide an integrated "description of operation" of the gene network that regulates apoptosis in various cell types under varying effects. The database GeneNet contains the description of the apoptosis gene network: (http://wwwmgs.bionet.nsc.ru/systems/mgl/genenet), and the Transcription Regulatory Regions Database (TRRD) contains experimental data on transcription regulation of various genes, including apoptosis genes: (http://wwwmgs.bionet.nsc.ru/mgs/gnw/trrd/).

KEYWORDS: gene networks; apoptosis; signal transduction

In the last decade, molecular biologists have shown an increasing interest in programmed cell death, morphologically classified as apoptosis, and have collected much experimental data in this area. GeneNet, including the apoptosis gene database (http://wwwmgs.bionet.nsc.ru/systems/mgl/genenet), allows for diverse experimental data on the gene network's structural and function organization to be accumulated and then visualized as interactive graphical layouts.[1] GeneNet divides the database components of the gene network into two basic classes: (1) elementary structures (genes, RNAs, proteins, signal molecules, and various types of metabolites) and (2) elementary events (metabolic and regulatory).

The elementary events comprise of basic genetic processes (i.e., replication, transcription, splicing, translation, posttranslational degradation of proteins, formation and degradation of protein complexes, biochemical reactions, transport processes, etc.,) and their regulation. Thus, gene networks are described and analyzed at the level of elementary molecular and genetic molecular structures, and their interactions are shown using gene network graphs.

A specialized tool, GeneNet Viewer, was developed for visualizing gene networks. The SRS version of GeneNet[1] can form complete lists of elementary structures (reactions, and regulatory events) specific to each network. It also gives information on the interaction of a particular protein within a network and/or on the genes whose transcription is regulated by a transcription factor of interest. GeneNet

Address for correspondence: Dr. I. Stepanenko, Institute of Cytology and Genetics, SB RAS, Lavrentieva 10, 630090, Novosibirsk, Russia. Voice: +7 3832 332971; fax +7 3832 331278. stepan@bionet.nsc.ru

Ann. N.Y. Acad. Sci. 1010: 16–18 (2003). © 2003 New York Academy of Sciences.
doi: 10.1196/annals.1299.003

contains only experimental data accumulated through annotating scientific publications by using a specialized editor, GeneNet Input, which contains a large number of controlled vocabularies.

GeneNet now includes 57 genes, 158 proteins, and 308 relationships among components of the network. The central gene network regulators are p53, NF-kappaB, and AP-1 transcription factors. An increase in concentration of each factor, or several in the cell nucleus, triggers a cassette transcription activation of many genes through interaction of this factor with its binding sites, in promoters of the corresponding genes. Activation of the gene network is controlled by apoptosis inhibitors and the heat-shock proteins HSP70, HSP90, and HSP27.[2] Regulatory circuits with a positive feedback, shift the parameters from the current state and thereby switch the gene network to a new state; as for negative feedback, it stabilizes the parameters of the gene network at a certain level. Specific to the apoptosis gene network is the prevalence of positive feedback, which ensures signal amplification and transduction of the system into a terminal state (cell death). The cascade principle of signal amplification and integration of signal transduction pathways underlie the organization of this gene network.

Transcription Regulatory Regions Database (TRRD) is used for describing regulatory regions of the genes operating in gene networks, including the apoptosis gene network.[3] TRRD contains formalized experimental data on the structure–function organization of regulatory units controlling transcription of eukaryotic genes (promoters, enhancers, and silencers), transcription factor binding sites, expression patterns, and so forth. The major transcription factors regulating the genes of the apoptosis gene network, described in Apoptosis-TRRD (http://wwwmgs.bionet.nsc.ru/mgs/papers/stepanenko/ap-trrd/), are p53, NF-κB, AP-1, HSF1, and Sp1. Characteristic of these genes are diversity of their expression patterns depending on cell type, tissue, and effects of the apoptosis gene network activators and repressors.

Accumulation of the relevant information on operation of apoptosis gene network and transcription regulation of the genes involved allows this gene network to be described in an integrative manner and the most important nodes in the network, which can be used to regulate the process, to be detected. Data formalization and systematization in the GeneNet and TRRD databases form the background for computer analysis and construction of mathematical models.

ACKNOWLEDGMENTS

This work was supported, in part, by the Russian Foundation for Basic Research (Grant Nos. 01-07-90376, 01-07-90084, 02-07-90355, and 02-07-90359); the Ministry of Industry, Science, and Technology of the Russian Federation (Grant No. 43.073.1.1.1501); and INTAS (Grant No. 21-2382). The authors thank I.V. Lokhova for her assistance with the bibliography.

REFERENCES

1. Ananko, E.A., N.L. Podkolodny, I.L. Stepanenko, *et al.* 2002. GeneNet: a database on structure and functional organization of gene networks. Nucleic Acids Res. **30:** 398–401.

2. STEPANENKO, I.L. 2001. Interference of apoptosis and heat shock response gene networks. Mol. Biol. (Mosk.) **35:** 1063–1071.
3. KOLCHANOV, N.A., E.V. IGNATIEVA, E.A. ANANKO, *et al.* 2002. Transcription Regulatory Regions Databases (TRRD): its status in 2002. Nucleic Acids Res. **30:** 312–317.

Mitochondrial Apoptosis Induced by the HIV-1 Envelope

MARIA CASTEDO,[a] JEAN-LUC PERFETTINI,[a] KARINE ANDREAU,[a] THOMAS ROUMIER,[a] MAURO PIACENTINI,[b] AND GUIDO KROEMER[a]

[a]*Centre National de la Recherche Scientifique, UMR 8125, Institut Gustave Roussy, F-94805 Villejuif, France*

[b]*Istituto Nazionale Malattie Infettive "L. Spallanzani," Rome 00149, Italy*

ABSTRACT: The envelope glycoprotein complex (Env), encoded by the human immunodeficiency virus (HIV-1), kills uninfected cells expressing CD4 and/or the chemokine receptor CXCR4 or CCR5, via at least three independent mechanisms. First, the soluble Env product gp120 can induce the apoptotic cell death of lymphocytes, neurons, and myocardiocytes, via interaction with surface receptors. Second, Env present on the surface of HIV-1 infected cells can transiently interact with cells expressing CD4 and CXCR4/CCR5, thereby provoking a hemifusion event that results in the death of the uninfected cell. Third, the interaction between Env on infected cells and its receptors on uninfected cells can result in syncytium formation. Such syncytia undergo apoptosis after a phase of latency. In several models of Env-induced apoptosis, early signs of mitochondrial membrane permeabilization (MMP) become manifest. Such signs include a loss of the mitochondrial transmembrane potential and the release of cytochrome *c* and AIF. The mechanisms of Env-triggered apoptotic MMP may involve an elevation of cytosolic Ca^{2+}, reactive oxygen species and/or the transcriptional activation of p53, with the consequent expression of proapoptotic proteins such as Bax, which permeabilizes mitochondrial membranes. The implications of these findings for the pathophysiology of HIV-1 infection is discussed.

KEYWORDS: AIDS; human immunodeficiency virus; mitochondrial transmembrane potential; p53; programmed cell death

ABBREVIATIONS: AIF, apoptosis-inducing factor; Cdk1, cyclin dependent kinase-1; $\Delta\Psi m$, mitochondrial transmembrane potential; Env, envelope glycoprotein complex; HIV, human immunodeficiency virus; mTOR, mammalian target of rapamycin; MMP, mitochondrial membrane permeabilization; p53S15, serine 15 of p53; p53S15P, p53 with phosphorylated serine 15; Z-VAD.fmk, *N*-benzyloxycarbonyl-Val-Ala-Asp-fluoromethylketone.

Address for correspondence: Dr. Guido Kroeme, CNRS-UMR 8125, Institut Gustave Roussy, Pavillon de Recherche 1, 39 rue Camille-Desmoulins, F-94805 Villejuif, France. Voice: +33-1-42-11-60-46; fax: +33-1-42-11-60-47.

kroemer@igr.fr

Ann. N.Y. Acad. Sci. 1010: 19–28 (2003). © 2003 New York Academy of Sciences.
doi: 10.1196/annals.1299.004

INTRODUCTION

Apoptosis involves the ordered dismantling of cellular structuring following a stereotyped sequence of biochemical events, at least at the end of the process. According to current understanding, at least two major pathways for activating the latent cell death machinery exist.[1,2] In the "extrinsic" pathway, the ligation of cell death receptor, such as CD95, results in the receptor-proximal activation of caspases (mainly caspases-8 and -10) and caspase-independent death effectors. In the "intrinsic" pathway, mitochondrial membrane permeabilization (MMP) occurs as a primary event.

Under normal circumstances, apoptosis is blocked due to the strict compartimentalization of catabolic hydrolases and their activators on both sides of the outer mitochondrial membrane.[3] Thus, cytochrome *c* is confined to the mitochondrial intermembrane space and therefore cannot bind to Apaf-1, a cytosolic protein. Upon permeabilization, or rupture of the outer mitochondrial membrane, cytochrome *c* suddenly interacts with Apaf-1, which then becomes an allosteric activator of procaspase-9, which in turn proteolytically activates caspase-3, one of the principal proteases participating in the apoptotic execution. Similarly, Smac/DIABLO and Omi/HtrA2, two intermembrane proteins, are normally physically separated from cytosolic inhibitors of apoptosis proteins (IAPs). Upon mitochondrial membrane permeabilization (MMP), Smac/DIABLO and Omi/HtrA2 neutralize IAPs, and thus negate the IAP-dependent inhibition of caspases-3 and -9.[4]

Two DNA-destroying enzymes (AIF and endonuclease G) are also normally confined to the mitochondrial intermembrane space, which is DNA-free. When MMP has occurred, these two proteins move to the nucleus and participate in apoptotic chromatinolysis.[5] MMP is regulated by Bcl-2-like proteins,[6,7] meaning that Bcl-2 and its analogues can inhibit apoptosis whenever MMP is the *principium movens* of the death cascade.

Peripheral blood lymphocytes from HIV-1$^+$ donors manifest a loss of the mitochondrial transmembrane potential ($\Delta\Psi m$), as well as an increase in the production of reactive oxygen species (ROS) by mitochondria, correlating with an increased level of spontaneous apoptosis.[8,9] HIV-1 carriers who not develop any signs of pathology (long term non progressors, LTNP) fail to manifest such a tendency to $\Delta\Psi m$ loss, suggesting that this laboratory parameter might have a predictive value.[10] Indeed, the level of spontaneous apoptosis (which apparently is initiated by MMP, accompanied by an $\Delta\Psi m$ dissipation) is well known to correlate with disease progression.[11,12] It is important to note that the large majority (>98%) of peripheral CD4$^+$ T cells are not infected by HIV-1, although a significant proportion (10–50%) appears to be primed to apoptosis.[11–14] Therefore, HIV-1-mediated killing must be mediated by indirect rather than direct effects.

One of the principal apoptogenic products encoded by HIV-1 is the envelope glycoprotein complex (Env).[15] The primary product of the Env gene (gp160) proteolytically matures to form to plasma membrane-anchored complex, composed by gp120 and gp41. This complex, which is present at the surface of virions and of HIV-1-infected cells, as well as soluble gp120, can interact with a suitable combination of the Env receptor (CD4) and the co-receptor, a member of the chemokine receptor family, in most cases either CXCR4 (in the case of lymphotropic Env variants) or CCR5 (in the case of monocytotropic Env variants). On the surface of HIV particles, Env forc-

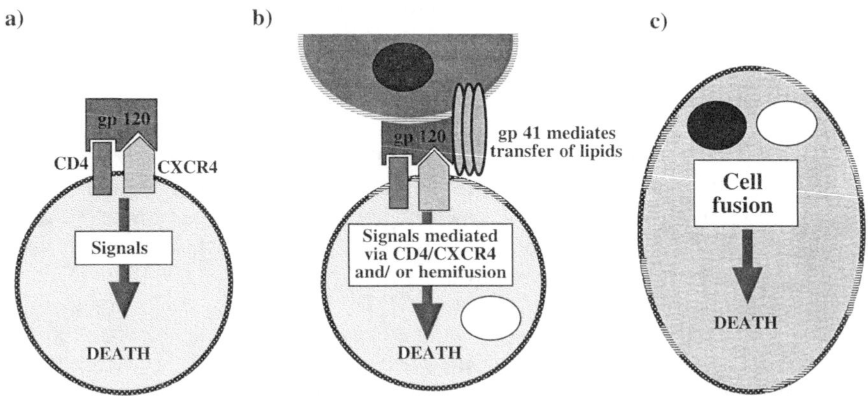

FIGURE 1. Three pathways of Env-induced apoptosis: **(a)** Soluble gp120 may engage CD4 or chemokine receptor that transmit lethal signals. **(b)** An Env (gp120/gp41) expressing cell interact with a CD4/CXCR4-expressing cell in a transient fashion that causes apoptosis after an exchange of lipids (and other membrane components?) between the two cells without an exchange of cytosol. This results in single-cell killing, affecting only the CD4/CXCR4-expressing cell. **(c)** An Env (gp120/gp41) positive cell fuses with a CD4/CXCR4-expressing cell, which results in syncytium formation and later cell death.

es the viral and the cellular membranes to fuse in a multistep process that requires the interactions of gp120 with its receptor and co-receptors. These interactions of gp120 then trigger the exposure of the transmembrane subunit of Env, gp41, and the insertion into the cellular membrane of a hydrophobic peptide (the fusion peptide), followed by the rearrangement of gp41 into a six-helix hairpin structure that tethers together viral and cellular membranes, causing them to mix and fuse.[16] Env and its subproducts can induce mitochondrial apoptosis of uninfected cells via a variety of different mechanisms (FIG.1) that will be discussed in this review.

INDUCTION OF MITOCHONDRIAL APOPTOSIS BY SOLUBLE GP120

Soluble gp120 can be found in the body fluids of HIV-1 carriers. When added to lymphocytes,[17] neurons,[18] or myocardiocytes,[19] it exerts potent cytotoxic effects that have been implicated in the HIV-1-induced lymphodepletion, AIDS dementia (also called HIV encephalitis or neuroAIDS), and AIDS cardiomyopathy, respectively.

In lymphocytes, the effect of gp120 is thought to be mediated by interactions with CD4 and CXCR4. Antibody-mediated crosslinking of CD4 in the absence of CD3 stimulation has been shown to sensitize lymphocytes to apoptosis induction,[20] and similar results have been reported for the gp120-mediated activation of T cells.[21] Such a CD4 stimulation may either activate the CD95/CD95L-triggered extrinsic pathway or activate Bax.[22,23] In Jurkat T cells, CD4 engagement by the Leu3a mAb results in: a rapid and strong increase of Lck kinase activity, its association with TCR, upregulation of Bax, subsequent alterations of the $\Delta\Psi m$, and apoptosis.[24] This effect can be counteracted by Vav, a signaling molecule that cooperates with CD28 to boost TCR signals.[24] This may explain why CD28 can inhibit gp120-induced cell death.[25] A monoclonal antibody against CXCR4 also can cause rapid lymphocyte death preceded by $\Delta\Psi m$ dissipation. This cell death occurs in a caspase-independent fashion.[20]

gp120 induces apoptosis of cultured neurons and sensitizes them to excitotoxicity and oxidative stress. Gp120 also causes neuronal dysfunction and death in rodents *in vivo*. gp120 disrupts neuronal calcium homeostasis by perturbing calcium-regulating systems in the plasma membrane (e.g., voltage-dependent calcium channels, glutamate receptor channels) and endoplasmic reticulum.[18] Ca^{2+} is indeed well known to induce or to facilitate apoptotic MMP. Accordingly, drugs that stabilize cellular calcium homeostasis can protect neurons against the toxic effects of gp120. The mitochondrial release of cytochrome *c* has been observed in cerebrocortical cultures exposed to the HIV coat protein gp120.[26] Specific inhibitors of both the Fas/tumor necrosis factor-alpha/death receptor pathway and the mitochondrial caspase pathway can prevent the gp120-induced neuronal apoptosis.[26] Mixed lineage kinase 3 (MLK3) has also been involved in the gp120-induced death of hippocampal neurons, because a chemical inhibitor of MLK3 and transfection with a dominant-negative MLK3 mutant can abolish gp120-induced killing.[27]

gp120 expressed by inflammatory cells is found, *in situ*, in the hearts of patients with HIV-1 cardiomyopathy, in which cardiomyocyte apoptosis is increased. In such hearts, active caspase 9, was expressed strongly on macrophages and weakly on cardiomyocytes. HIV-1 entered cultured neonatal rat ventricular myocytes by macropynocytosis, but did not replicate. HIV-1 or gp120 induced apoptosis of rat myocytes through a mitochondrion-controlled pathway, which was inhibited by the chemokine receptor-blocking agent AOP-RANTES, or by pertussis toxin, which interferes with signals transmitted via chemokine receptors.[19]

SINGLE-CELL KILLING BY ENV

Bystander cells expressing CD4 and CXCR4 receptor die after coculture with cells engineered to express surface-bound Env (gp120/gp41). This bystander cell killing may be secondary to syncytium formation, but it may also concern unfused cells.[28–31] Apparently, the "decision" whether or not cell fusion occurs is governed at the level of the CD4/CXCR4 expressing cell, through yet-to-be-elucidated mechanisms, since different cell lines expressing such coreceptors (e.g., Jurkat cells and U937 cells) behave in a differential fashion when exposed to the same Env-expressing cell type.[30] Agents that inhibit the interaction between gp120 and CD4 or CXCR4 inhibit both syncytium formation and single-cell killing by surface-exposed Env.[30,31] In addition, C34, a peptide derived from gp,[41] which is known to inhibit syncytium formation, also inhibits the death of single target cells, presumably by specifically acting on gp41 function. Moreover, Env-induced single cell death is associated with the gp41-mediated transfer of lipids from the membrane of Env-expressing cells to the target cell, a phenomenon which occurs without detectable cytoplasm mixing, and thus involves a hemifusion-like event.[31]

Using cell lines expressing wild-type CXCR4 and a truncated form of CD4 that binds gp120, yet lacks the ability to transduce signals, the interaction of cell-associated gp120 with CXCR4-expressing target cells, triggers signs of MMP ($\Delta\Psi m$ loss and mitochondrial release of cytochrome *c*); without which, this signal would involve a CD4-dependent signal or CD95 activation.[32] Importantly, following coculture with cells expressing Env, a CD95-independent apoptosis involving MMP is also observed in primary umbilical cord blood $CD4^+$ T lymphocytes that express high levels of CXCR4.[32] The exact mechanisms causing single-cell killing by Env-expressing cells remains largely obscure.

ENV-ELICITED SYNCYTIAL APOPTOSIS

Syncytia elicited by HIV-1 Env progress to apoptosis in a stepwise fashion. After an initial stage during which nuclei are separated by intact envelopes, cyclin B accumulates in the cytoplasm, mTOR remains a predominantly cytoplasmic protein and p53S15 is non-phosphorylated. Later, syncytia adopt a different phenotype, with nuclear fusion (karyogamy), dismantled nuclear envelopes, down-regulated cyclin B, mTOR in the nucleus, and phosphorylation of p53 on serine 15 (FIG. 2). After a latency phase, then syncytia manifest the translocation of Bax to mitochondria and signs of MMP.

Molecular ordering of the process has been achieved using a battery of chemical inhibitors (of Cdk1, mTOR, p53, caspases), micro-injection of specific inhibitors (of Bax, VDAC, AIF, cytochrome *c*), dominant negative constructs (of Cdk1 and p53),

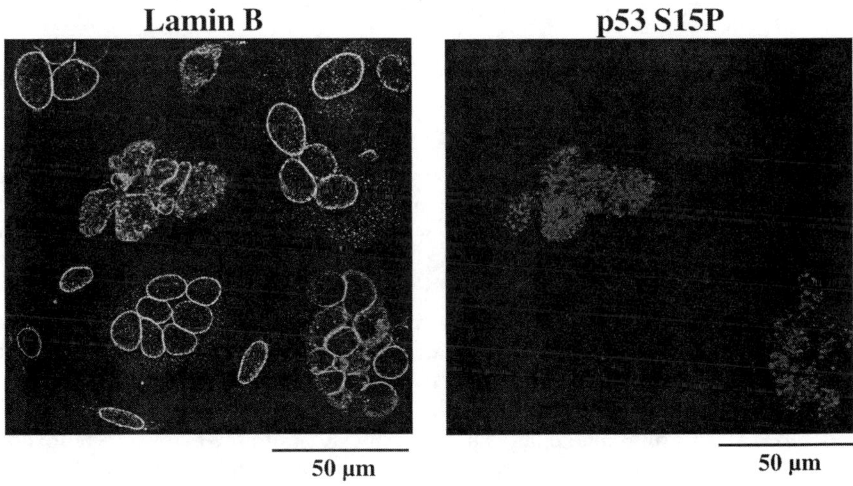

FIGURE 2. Phosphorylation of p53 on serine 15 in Env-elicited syncytia: Syncytia were generated by co-culturing HeLa cells expressing CD4 and CXCR4 with HeLa cells stably transfected with the Env gene from a lymphotropic HIV-1 isolate. Cells were stained with antibodies specific for serine 15-phosphorylated p53 (p53S15P) and lamin, a constituent of the nuclear envelope. Note that p53 is only phosphorylated on serine 15 in those syncytia that exhibit at least partial dismantling of the nuclear envelope. This indicates that the activation of p53 is secondary to entry into the prophase of mitosis.

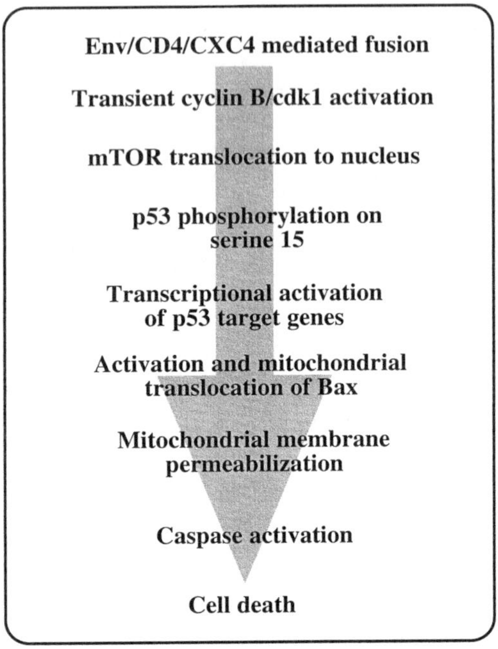

FIGURE 3. Schematic overview of the signal transduction cascade involved in Env-elicited syncytial apoptosis. For details and references see text.

as well as positive manipulations (microinjection of the cyclin B-dependent kinase-1, transfection with mTOR, microinjection of AIF, cytochrome *c*). Upon syncytium formation, cyclin B upregulated in a transient fashion at the protein level, thereby triggering cyclin B-dependent kinase-1-dependent activation of lamin disassembly (which facilitates karyogamy) and nuclear accumulation of the mammalian target of rapamycin (mTOR), a serine/threonine kinase of the phosphatidyl inositol kinase family. mTOR then directly phosphorylates p53S15, resulting in the transcriptional activation of apoptogenic p53 target genes (FIG. 3). p53 then participates in the up-regulation/activation of the pro-apoptotic Bcl-2 protein family member Bax, which undergoes a conformational change and translocates from the cytosol to mitochondria. Bax finally causes MMP with reduction of the $\Delta\Psi m$ and the release of apoptogenic proteins from mitochondria, in particular apoptosis inducing factor (AIF) and cytochrome *c*, culminating in cell death.[30,33–36]

A number of important questions, however, remain without answer. For what reason is cyclin B upregulated, and why is this upregulation transient? How is mTOR activated? Is mTOR the sole kinase to activate p53, or are there other p53 kinases implicated in the system? Do other transcription factors participate in HIV-1-induced apoptosis? Is Bax the sole pro-apoptotic target gene whose expression is responsible for apoptotic MMP? It will be important to resolve these problems in the future.

PATHOPHYSIOLOGICAL IMPLICATIONS

In the past, we have concentrated our efforts on the elucidation of apoptosis induction by Vpr, an accessory protein encoded by HIV-1 that can interact with mitochondria and kill cells through the direct induction of MMP.[37–41] In the course of these studies, we identified an α-helical dodecapeptide domain (residues 71–82) of Vpr, characterized by the presence of three critical arginine residues, which is sufficient and necessary to induce MMP.[37,39] Mutations that disrupt the function of this domain, either due to stop codons[42] or due to a selective point mutation in critical arginines (e.g., R77Q),[43] can be found among long-term nonprogressors, correlating with a decrease in apoptosis induction by the corresponding HIV-1 isolates.[42] This indicates that the study of mechanisms through which HIV-1 induces apoptosis can furnish clinically useful information.

What then, is the evidence from the results obtained with recombinant gp120, or the cell lines engineered to overexpress the gp120/gp41 complex in HIV-1-free systems, that can actually be extrapolated to the pathophysiology of HIV-1 infection? A number of observations obtained *in vitro* or *in vivo*, in HIV-1 infection, show the following: Infection of primary human lymphoblasts *in vitro*, with lymphotropic HIV-1 isolates, causes cell death mainly by syncytium formation, associated with a series of alterations that resemble those induced by Env *in vitro*, namely $\Delta\Psi m$ loss,[30,35] mitochondrial release of cytochrome *c* and AIF,[35,44] increased ROS production,[45] phosphorylation of p53 on serine 15,[35,36] and the upregulation of Bax at the mRNA and protein levels.[35] The amount of Bax is found to be increased in the mitochondrion-enriched heavy membrane fraction of HIV-infected CD4$^+$ T cells as compared to uninfected controls.[30,35] *In vitro*, pharmacological inhibition of Cdk1 (with roscovitine), mTOR (with rapamycin), or p53 (with cyclic pifithrin-α) inhibits the apoptosis induced by HIV-1 infection, while caspase inhibition (with Z-VAD.fmk) has no such effects.[30,36,44,46] Altogether, these data suggest that the cascade of events delineated above (cell fusion → activation of Cdk1/cyclin B → activation of mTOR → p53 phosphorylation on serine 15 → transcriptional activation of p53 target genes → Bax translocation to mitochondria with consequent MMP → caspase-independent cell death) is actually induced by HIV-1 infection *in vitro*.

A few observations also suggest that the above pathway is activated in HIV-1 infected individuals, *in vivo*. Thus, cyclin B reportedly is overexpressed in lymphocytes from HIV-1 carriers, and this upregulation disappears upon successful antiretroviral therapy.[47,48] Similarly, mTOR and the phosphoneoepitope indicating phosphorylation of p53 on serine 15 are overexpressed among a fraction of peripheral blood mononuclear cells, correlating with the frequency of cells expressing a marker of pre-apoptosis (tissue transglutaminase-2[49]), as well as with viral load.[36,46] Longitudinal studies indicate that highly active antiretroviral therapy cause the disappearance of both mTOR overexpression and increased p53 serine 15 phosphorylation in the patients.[46] Altogether, these data underscore that the exploration of HIV-1-induced killing yields valuable information on the pathophysiology of AIDS, as well as new laboratory parameters for disease assessment. The future will tell whether this avenue of research will also furnish new therapeutic targets for the treatment of HIV-1 infection.

ACKNOWLEDGMENTS

We thank the NIH AIDS reagents program (Bethesda, MD) for cell lines. This work has been supported by a special grant from LNC, as well as grants from ANRS, FRM, the European Commission (QLG1-CT-1999-00739) (to G.K.), and the Italian Ministry of Public Health (to M.P.).

REFERENCES

1. SCAFFIDI, C., *et al.* 1999. Apoptosis signaling in lymphocytes. Curr. Opin. Immunol. **11:** 277–285.
2. FERRI, K.F. & G.K. KROEMER. 2001. Organelle-specific initiation of cell death pathways. Nature Cell Biol. **3:** E255–E263.
3. KROEMER, G. & J.C. REED. 2000. Mitochondrial control of cell death. Nature Med. **6:** 513–519.
4. WANG, X. 2002. The expanding role of mitochondria in apoptosis. Genes Dev. **15:** 2922–2933.
5. PENNINGER, J.M. & G. KROEMER. 2003. Caspases, AIF, and mitochondria: rivaling for cell death execution. Nature Cell Biol. **5:** 97–99.
6. KROEMER, G. 1997. The proto-oncogene Bcl-2 and its role in regulating apoptosis. Nature Med. **3:** 614–620.
7. GROSS, A., J.M. MCDONNELL & S.J. KORSMEYER. 1999. Bcl-2 family members and the mitochondria in apoptosis. Genes Dev. **13:** 1988–1911.
8. MACHO, A., *et al.* 1995. Mitochondrial dysfunctions in circulating T lymphocytes from human immunodeficiency virus-1 carriers. Blood **86:** 2481–2487.
9. CASTEDO, M., *et al.* 1995. Mitochondrial perturbations define lymphocytes undergoing apoptotic depletion in vivo. Eur. J. Immunol. **25:** 3277–3284.
10. MORETTI, S., *et al.* 2000. Apoptosis and apoptosis-associated perturbations of peripheral blood lymphocytes during HIV infection: comparision between AIDS patients and asymptomatic long-term non-progressors. Clin. Exp. Immunol. **122:** 364–373.
11. GOUGEON, M.L. & L. MONTAGNIER. 1999. Programmed cell death as a mechanism of CD4 and CD8 T cell depletion in AIDS—molecular control and effect of highly active anti-retroviral therapy. Ann. N.Y. Acad. Sci. **887:** 199–212.
12. BADLEY, A.D., *et al.* 2000. Mechanisms of HIV-associated lymphocyte apoptosis. Blood **96:** 295–2964.
13. MEYAARD, L., *et al.* 1992. Programmed cell death in HIV-1 infection. Science **257:** 217–219.
14. GOUGEON, M.L., *et al.* 1993. Programmed cell death of T lymphocytes in AIDS related HIV and SIV infections. AIDS Res. Hum. Retrov. **9:** 553–563.
15. CHIRMULE, N. & S. PAHWA. 1996. Envelope glycoproteins of human immunodeficiency virus type **1:** profound influences on immune function. Microbiol. Rev. **60:** 386–406.
16. ECKERT, D.M. & P.S. KIM. 2001. Design of potent inhibitors of HIV-1 entry from the fp41 N-peptide region. Proc. Natl. Acad. Sci. USA **98:** 11187–11192.
17. CICALA, C., *et al.* 2000. HIV-1 envelope induces activation of caspase-3 and cleavage of focal adhesion kinase in primary human $CD4^+$ T cells. Proc. Natl. Acad. Sci. USA **97:** 1178–1183.
18. HAUGHEY, N.J. & M.P. MATTSON. 2002. Calcium dysregulation and neuronal apoptosis by the HIV-1 proteins Tat and gp120. J. Acquir. Immune Defic. Syndr. **31(**Suppl. 2): S55–61.

19. TWU, C., *et al.* 2002. Cardiomyocytes undergo apoptosis in human immunodeficiency virus cardiomyopathy through mitochondrion-and death receptor-controlled pathways. Proc. Natl. Acad. Sci. USA **99:** 14386–14391.
20. BERNDT, C., *et al.* 1998. CXCR4 and CD4 mediate a rapid CD95-independent cell death in CD4+ cells. Proc. Natl. Acad. Sci. USA **95:** 1255–12561.
21. FINCO, O., *et al.* 1997. Induction of CD4+ T cell depletion in mice doubly transgenic for HIV gp120 and human CD4. Eur. J. Immunol. **27:** 1319–1324.
22. HASHIMOTO, F., *et al.* 1997. Modulation of Bcl-2 protein by CD4 cross-linking: A possible mechanism for lymphocyte apoptosis in human immunodeficiency virus infection and for rescue of apoptosis by interleukin-2. Blood **1997:** 745–753.
23. SOMMA, F., *et al.* 2000. Engagement of CD4 before TCR triggering regulates both Bax- and Fas (CD95)-mediated apoptosis. J. Immunol. **164:** 5078–50787.
24. TUOSTO, L., A. MARINARI & E. PICCOLELLA. 2002. CD4-Lck through TCR and in the absence of Vav exchange factor induces Bax increase and mitochondrial damage. J. Immunol. **168:** 6106–6112.
25. TUOSTO, L., *et al.* 1995. Ligation of either CD2 or CD28 rescues CD4+ T cells from HIV-gp120-induced apoptosis. Eur. J. Immunol. **25:** 2917–2922.
26. GARDEN, G.A., *et al.* 2002. Caspase cascades in human immunodeficiency virus- associated neurodegeneration. J. Neurosci. **22:** 4015–4024.
27. BODNER, A., *et al.* 2002. Mixed lineage kinase 3 mediates gp120IIIB-induced neurotoxicit. J. Neurochem. **82:** 1424–1434.
28. KOLESNITCHENKO, V., *et al.* 1995. Human immunodeficiency virus 1 envelope-initiated G2-phase programmed cell death. Proc. Natl. Acad. Sci. USA **92:** 11889–11893.
29. BLANCO, J., *et al.* 1999. The implication of the chemokine receptor CXCR4 in HIV-1 envelope protein-induced apoptosis is independent of the G protein-mediated signalling. AIDS **13:** 909–917.
30. FERRI, K.F., *et al.* 2000. Apoptosis control in syncytia induced by the HIV-1-envelope glycoprotein complex. Role of mitochondria and caspases. J. Exp. Med. **192:** 1081–1092.
31. BLANCO, J., *et al.* 2003. Cell-surface-expressed HIV-1 envelope induces the death of CD4 T cells during GP41-mediated hemifusion-like events. Virology **305:** 318–329.
32. ROGGERO, R., *et al.* 2001. Binding of human immunodeficiency virus type 1 gp120 to CXCR4 induces mitochndrial transmembrane depolarization and cytochrome *c*-mediated apoptosis independently of Fas signaling. J. Virol. **75:** 7637–7650.
33. FERRI, K.F., *et al.* 2000. Apoptosis of syncytia induced by HIV-1-envelope glycoprotein complex. Influence of cell shape and size. Exp. Cell Sci. **261:** 119–126.
34. FERRI, K.F., *et al.* 2000. Apoptosis and karyogamy in syncytia induced by HIV-1-ENV/CD4 interaction. Cell Death Differ. **7:** 1137–1139.
35. GENINI, D., *et al.* 2001. HIV induced lymphocyte apoptosis by a p53-iniated, mitochondrion-mediated mechanism. FASEB J. **15:** 5 6.
36. CASTEDO, M., *et al.* 2001. Human immunodeficiency virus 1 envelope glycoprotein complex-induced apoptosis involves mammalian target of rapamycin/FKBP12-apamycin-associated protein-mediated p53 phosphorylation. J. Exp. Med. **194:** 1097–1110.
37. JACOTOT, E., *et al.* 2000. The HIV-1 viral protein R induces apoptosis via a direct effect on the mitochondrial permeability transition pore. J. Exp. Med. **191:** 33–45.
38. FERRI, F.K., *et al.* 2000. Mitochondrial control of cell death induced by proteins encoded by HIV-1. Ann. N.Y. Acad. Sci. **926:** 149–164.
39. JACOTOT, E., *et al.* 2001. Control of mitochondrial membrane permeabilization by adenine nucleotide translocator interacting with HIV-1 Vpr and Bcl-2. J. Exp. Med. **193:** 509–520.
40. BOYA, P., B. ROQUES & G. KROEMER. 2001. Bacterial and viral proteins regulating apoptosis at the mitochondrial level. EMBO J. **20:** 4325–4331.
41. ROUMIER, T., *et al.* 2002. The C-terminal moiety of HIV-1 Vpr induces cell death via a caspase-independent mitochondrial pathway. Cell Death Differ. **9:** 1212–1219.
42. SOMASUNDARAN, M., *et al.* 2002. Evidence for a cytopathogenicity determinant in HIV-1 Vpr. Proc. Natl. Acad. Sci. USA **99:** 9503–9508.
43. ZHANG, L., *et al.* 1997. Genetic characterization of vif, vpr, and vpu sequences from long-term survivors of human immunodeficiency virus type 1 infection. Virology **228:** 340–349.

44. Petit, F., *et al.* 2002. Productive HIV-1 infection of primary CD4+ T cells induces mitochondrial membrane permeabilization leading to caspase-independent cell death. J. Biol. Chem. **277:** 1477–1487.
45. Banki, K., *et al.* 1998. Molecular ordering in HIV-induced apoptosis—oxidative stress, activation of caspases, and cell survival are regulated by transaldolase. J. Biol. Chem. **273:** 11944–11953.
46. Castedo, M., *et al.* 2002. Sequential involvement of Cdk1, mTOR and p53 in apoptosis induced by the human immunodeficiency virus-1 envelope. EMBO J. **15:** 4070–4080.
47. Piedimonte, G., *et al.* 1999. Unscheduled cyclin B expression and p34 cdc2 activation in T lymphocytes from HIV-infected patients. AIDS **13:** 1159–1164.
48. Cannavo, G., *et al.* 2001. Abnormal intracellular kinetics of cell-cycle-dependent proteins in lymphocytes from patients infected with human immunodeficiency virus: a novel biologic link between immune activation, accelerated T-cell turnover, and high levels of apoptosis. Blood **97:** 1756–1764.
49. Amendola, A., *et al.* 1996. Induction of "tissue" transglutaminase in HIV pathogenesis: evidence for high rate of apoptosis of $CD4^+$ T lymphocytes and accessory cells in lymphoid tissues. Proc. Natl. Acad. Sci. USA **93:** 11057– 11062.

On the Role of Interferon Regulatory Factors in HIV-1 Replication

GIULIA MARSILI, ALESSANDRA BORSETTI, MARCO SGARBANTI, ANNA LISA REMOLI, BARBARA RIDOLFI, EMILIA STELLACCI, BARBARA ENSOLI, AND ANGELA BATTISTINI

Laboratory of Virology, Istituto Superiore di Sanità, Viale Regina Elena, 299, 00161 Rome, Italy

ABSTRACT: Interferons (IFNs) are pleiotropic cytokines that possess several biological activities and play a central role in basic and applied research as mediators of antiviral and antigrowth responses, modulators of the immune system, and therapeutic agents against viral diseases and cancer. Interferon regulatory factors (IRFs) have been identified together with signal transducers and activators of transcription (STAT) from studies on the type I IFN as well as IFN-stimulated (ISG) gene regulation and signaling. IRFs constitute a family of transcriptional activators and repressors implicated in multiple biological processes including regulation of immune responses and host defence, cytokine signaling, cell growth regulation, and hematopoietic development. All members share a well-conserved DNA binding domain at the NH_2-terminal region that recognizes similar DNA sequences, termed IRF element (IRF-E)/interferon-stimulated response element (ISRE), present on the promoter of target genes. Recently, a sequence homologous to the ISRE has been identified downstream from the 5′ human immunodeficiency virus type 1 (HIV-1) long terminal repeat (LTR). This sequence is a binding site for IRF-1 and IRF-2. Here we briefly summarize the role of IRFs in the regulation of HIV-1 LTR transcriptional activity and virus replication. The overall effect of IRFs on HIV-1 replication will also be discussed in the context of strategies carried out by the virus to counteract the IFN-mediated host defences both in active replication and during the establishment of viral latency.

KEYWORDS: interferons; transcription factors; IRFs; gene expression; virus replication; tat; apoptosis; viral escape

THE IRF FAMILY

Interferon regulatory factors (IRFs) constitute a family of transcriptional activators and repressors that are implicated in multiple biological processes, including regulation of immune responses and host defence, cytokine signalling, cell growth regulation, and hematopoietic development (for review, see Refs. 1,2). All members share homology in the first 115aa encompassing the DNA binding domain (DBD)

Address for correspondence: Dr. A. Battistini, Laboratory of Virology, Istituto Superiore di Sanità, Viale Regina Elena 299, 00161 Rome, Italy. Phone: +39-06-49903266; fax: +39-06-49902082.
battist@iss.it

Ann. N.Y. Acad. Sci. 1010: 29–42 (2003). © 2003 New York Academy of Sciences.
doi: 10.1196/annals.1299.005

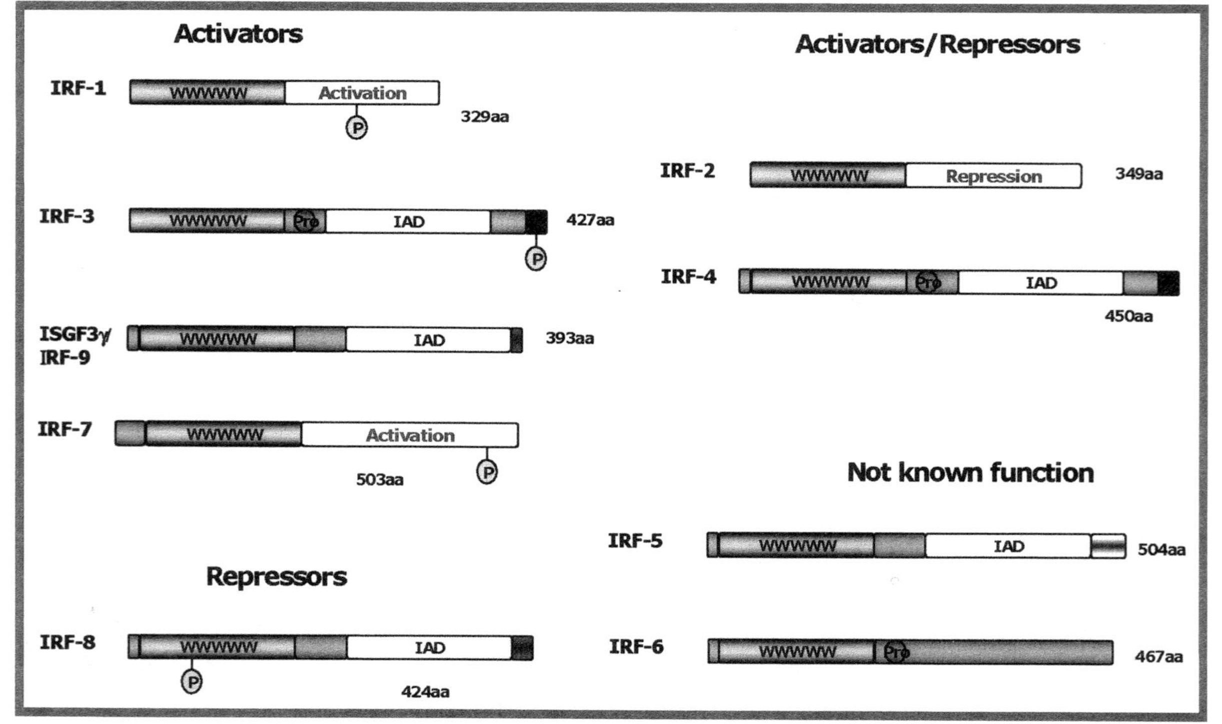

FIGURE 1. Functional domains of the IRF (interferon regulatory factor) family members. The DNA binding domain (DBD) with the five conserved repeats of tryptophan and the IRF activation domain (IAD) are indicated. Some sites of phosphorylation are also reported.

that contains a characteristic repeat of five tryptophan residues (FIG. 1). Through this DNA binding domain, the IRF family members bind to similar DNA sequences termed IRF-Element (IRF-E) whose consensus sequence G(A)AAAg/ct/cGAAAg/ct/c is present within the promoters of IFN-β and of most IFN-inducible genes (ISGs). This sequence is almost overlapping the interferon-stimulated response element (ISRE) (a/gNGAAANNGAAACT) targeted by the IFN signaling. Nine members of this family have been identified to date, based on the homologous DBD. The less conserved COOH-terminal domain acts as a regulatory region and classifies IRFs into three groups: those that activate (IRF-1, IRF-3, IRF-7 and IRF-9/ISGF-3γ), those that repress (IRF-2, IRF-8/ICSBP), and those (IRF-2, IRF-4/LSIRF/Pip, IRF-8) that are able to both activate and repress gene transcription depending on the target gene. IRFs can be expressed constitutively or upon treatment with IFN or other cytokines or in response to viral infection, in a variety of cell types, with the exception of IRF-8 and IRF-4, which are expressed exclusively in immune cells. Studies in IRF-expressing cells and knockout mice indicate that each member of the family exerts distinct roles in several biological processes as pathogen response, cytokine signaling, cell growth regulation, hematopoietic differentiation and immune regulation. The fate of cellular response is determined by a complex gene transcription network, and the distinct and not overlapping roles of IRFs are thought to be the result of slightly different DNA binding specificity, pattern of expression, posttranscriptional modifications of and/or association with other regulators (including other IRFs, STATs, Ets), and with coactivators and components of the basal transcriptional machinery.[2–5] Interactions with these partners can modify both ISRE-binding activities and the formation of initiation transcription complexes.[6]

The recent identification in the human herpesvirus-8 (HHV8) genome of virally encoded version of IRF (vIRF)[7] highlights the possibility that strategies targeting the IRF may be used by viruses to evade the antiviral effects of IFN.

REGULATION OF HIV-1 TRANSCRIPTION BY IRF

Human immunodeficiency virus-1 (HIV-1), is the pathologic agent of acquired immunodeficiency syndrome (AIDS).The syndrome is characterized by a long latency period and a gradual development of a dramatic immunodeficiency, increased susceptibility to opportunistic infections, malignancies, and neurological diseases.[8] The development of AIDS and related diseases is correlated with virus replication and spread.

HIV-1 is genetically related to the lentivirus genus. Like that of the other lentiviruses, the HIV-1 genome is complex and encodes a number of regulatory and accessory proteins. This complexity is reflected in the replication cycle, which requires, after integration in the host cell genome, intricate regulatory pathways and complex mechanisms for viral latency and persistence. All the knowledge gained since the first genetic analysis of HIV-1, 20 years ago, clearly indicates that replication of the HIV-1 provirus depends on the intracellular environment into which the virus integrates and is mainly controlled at the transcriptional level through a complex interplay between viral and cellular trans-regulatory proteins with the viral LTR.[9–12] In particular, sequences involved in viral gene expression (FIG. 2) are contained within the U3 and R regions of the HIV-1 LTR and in the intragenic enhancer in the pol

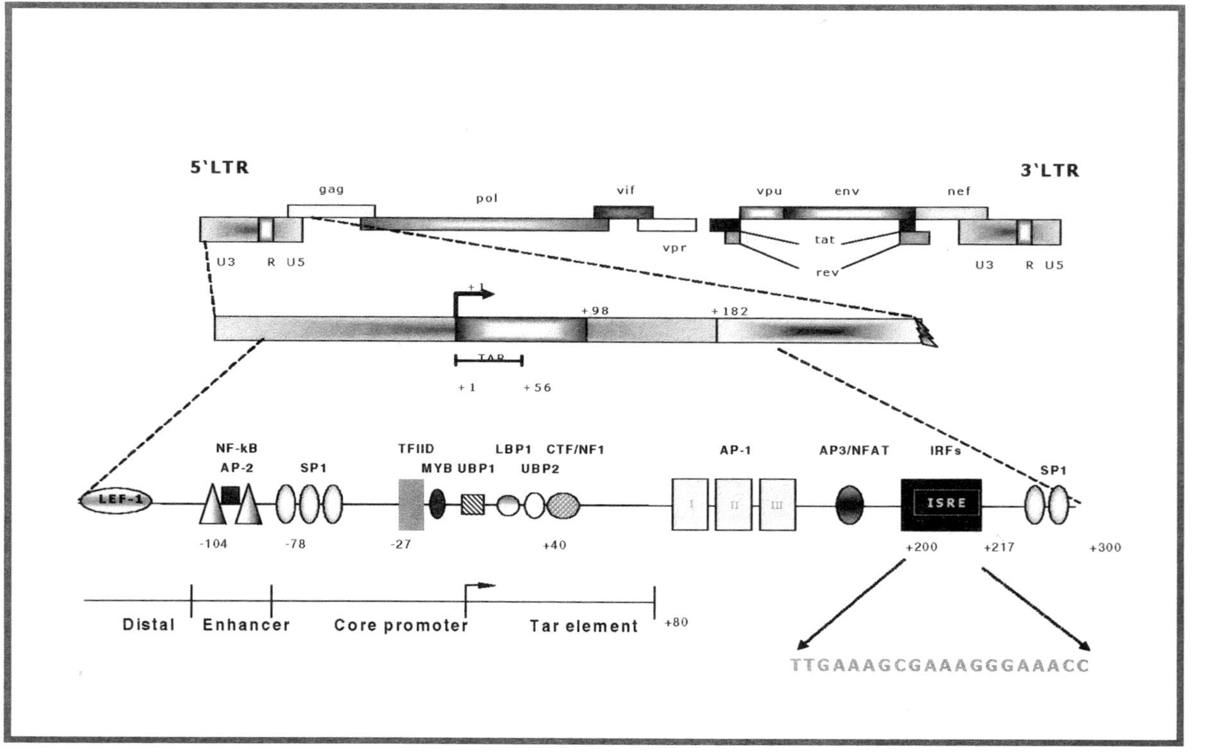

FIGURE 2. Schematic representation of the HIV-1 genome. Zoom is showed in the LTR. Several cis-elements for basal and induced cellular transcription factors are shown upstream and downstream of the transcription start site (+1). The ISRE sequence is indicated.

gene.[12–14] The U3 and R regions contain four functional domains: the modulatory and negative regulatory elements spanning nt –454 to nt –104, the enhancer from nt –109 to nt –79, the basal or core promoter from nt –78 to nt –1, and the Tat-responsive element (TAR) from nt +1 to nt +59.

Several cellular transcription factors able to bind these regions have been recognized, some of them being ubiquitous whereas others are specific and/or modulated in response to cellular activation, differentiation, cytokines and mitogens.[10] A distinctive feature of the HIV-1 LTR is the presence of transcriptional regulatory elements also downstream from the transcription start site located at the junction of the U3 and R regions. Using *in vivo* footprinting and electrophoretic mobility shift assays, binding sites for constitutive and inducible transcription factors were mapped in the untranslated region of the HIV-1 genome.[14–16] These sites include three AP-1 sites, an NF-AT site, two juxtaposed Sp-1 sites and a sequence, spanning nt +200 to +217 that is homologous to the ISRE present in the promoter of ISGs.[17] Comparison of these sequences in different HIV-1 isolates showed that all these sites are strongly conserved. Studies of mutant viruses altered in each of these motifs revealed that these regulatory elements play a critical role in HIV-1 transcription and replication, leading to the definition of a new positive transcriptional regulatory element in the HIV-1 provirus.[18,19] In particular, the sequence encompassing nt +200 +217, homologous to ISRE, is located in a DNAse I–hypersensitive site[20] that by footprinting analysis has been shown to be occupied *in vivo*;[14] it is thus reasonable that these sequences modulate viral gene expression at the transcriptional level through binding to regulatory cellular factors. It has been shown[14,21] that this sequence is a binding site for members of the interferon regulatory factor (IRF) protein family in particular IRF-1 and IRF-2.

IRF-1 and its functional antagonist, IRF-2, are the best-characterized members of the IRF family. They were originally identified as transcription factors that play a role in the regulation of the IFN-β gene.[22–25] Subsequently IRF-1 and IRF-2 were recognized to be involved in the regulation of genes expressed during inflammation, immune responses, hematopoiesis, and cell proliferation and differentiation (for a review, see Refs. 1,2).

IRF-1 is expressed at low levels in a variety of cell types, but its expression is greatly upregulated by virus infection, double-stranded (ds)RNA, IFNs, and by several cytokines such as tumor necrosis factor (TNF), interleukin-1 (IL-1), interleukin-6 (IL-6), prolactin, and leukocyte-inhibitory factor (LIF)[25–27] (FIG. 3). IRF-2 expression is, on the contrary, constitutive, but it is also inducible by IFNs.[17,24] We recently showed[21] that HIV-1 induces IRF-1 early upon infection and before expression of Tat in both Jurkat T cell line and primary CD4$^+$ T lymphocytes. Cells infected with the HIV-1 IIIB strain at a low multiplicity, to better mimic the natural infection, show an increase in both IRF-1 mRNA and protein soon after infection. The mRNA was already detectable at 3 h p.i. and returned to basal levels within 24 h. Interestingly, at the moment of the highest IRF-1 expression no doubly/spliced HIV tat/rev RNA transcripts were detected, whereas at 24 h p.i., as expected, the tat/rev mRNA was clearly detectable. Thus, HIV-1 induces IRF-1 expression early upon infection and prior to production of Tat. Stimulation of IRF-1 upon HIV-1 infection is associated with the presence of an IRF-1-specific binding activity, on its target sequence on the HIV-1 LTR and finally results in activation of transcription of the HIV-1 LTR. This indicates that upon HIV-1 infection IRF-1 can indeed activate transcrip-

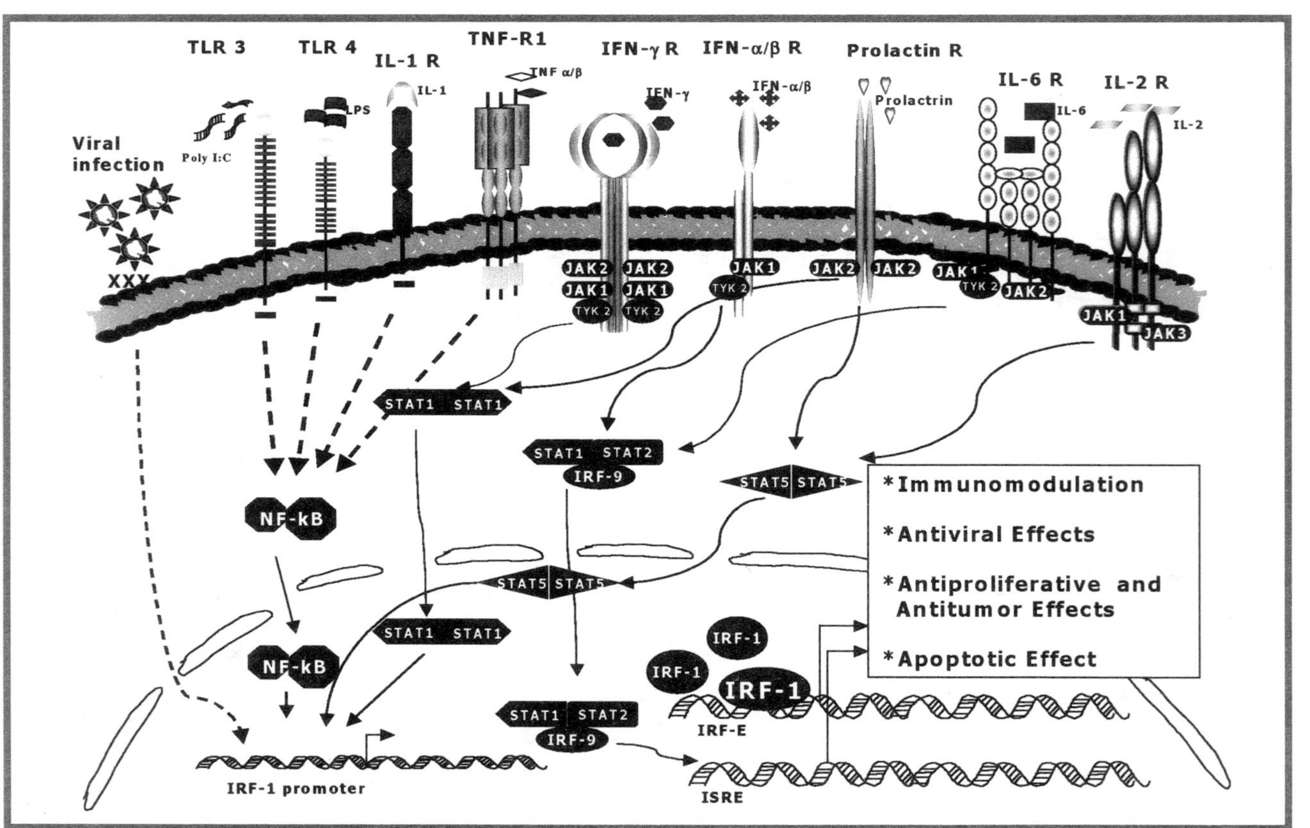

FIGURE 3. Overview of some IRF-1 inducers and their activation pathway. IRF-1 activities are indicated in the box.

tion of Tat which, in turn, thereby amplifies LTR-directed gene expression. IRF-1 also binds to Tat and cooperates with suboptimal doses of the viral transactivator to activate transcription. Both Tat and IRF-1 have been shown to functionally interact with general transcription factors such as TFIIB[3,28] and co-activators or adaptors, such as the histone acethyltransferases p300/CBP and pCAF.[4,29] These interactions occur through different domains of the proteins thus, IRF-1 might modulate the HIV-1 LTR promoter activity also by acting as a bridge between Tat and components of the basal transcriptional machinery.

In addition to its role in LTR transcription initiation during *de novo* infection, the data reported above, also suggest a potential role of IRF-1 during viral reactivation from latency. A stable reservoir of HIV-1 is latently infected resting $CD4^+$ T cells,[30] which are resting cells, whereas reactivation occurs only in activated T cells and is greatly dependent upon host transcription factors.[31–33] IRF-1 that is present at discrete levels in activated, but not in resting, T cells[34] can thus contribute to viral reactivation, even in the absence of Tat. Consistent with this, proinflammatory cytokines such as IFN-γ, IL-6, and TNF-α, which lead to cell activation and drive HIV-1 replication,[35,36] strongly activate IRF-1.[25,27]

In addition to activating HIV gene expression and replication, the IRF system can also repress HIV gene expression specifically. As previously shown,[21] in fact, another member of the IRF family, IRF-8, is able to drastically reduce HIV-1 replication *in vivo*. IRF-8, originally identified as protein binding the ISRE motif within the MHC class I gene promoter, is expressed exclusively in cells of the immune system, at low levels which are greatly enhanced by treatments with IFN-γ. It possesses an almost undetectable DNA binding activity dramatically increased after binding to IRF-1 or IRF-2. Because IRF-8 does not contain an activation/repression domain, in HIV-infected cells, an excess of IRF-8 complexing IRF-1 can inhibit both the IRF-1-induced LTR transcription and the binding of IRF-1 to Tat, further impairing HIV-1 replication and ultimately leading to a block of viral replication. These data, therefore, suggest that an increase in IRF-8 expression can be involved in the establishment of latency, whereas activation of IRF-1 expression functions as a positive regulator of HIV-1 transcription and replication.

LTR-ISRE: FUNCTIONAL AND/OR STRUCTURAL *CIS*-ELEMENT?

Results by our group and others[18,37] indicate that deletion or mutation in the ISRE sequence present on the HIV-1-LTR, also leads to a general defect in the viral promoter activity in the presence of Tat. A significant reduction in CAT activity (80%) was, in fact, obtained when a mutated construct was co-transfected with a Tat expression vector. Moreover, the replacement of the LTR-ISRE with another functional ISRE from the ISG54 gene similarly leads to a decrease in both LTR promoter activity and proviral DNA gene expression,[18] despite the fact that the ISG54-ISRE can compete for the binding of cellular factors. It is thus possible that the +200 +217 element in the LTR, is not functioning as an ISRE in the regulation of HIV-1 gene expression.

The mechanisms whereby the ISRE-like element modulates the HIV-1 LTR promoter activity are still unclear; however, this sequence might act cooperatively with other elements to position the nucleosomal structure around the LTR to regulate promoter activity in the context of chromatin as suggested.[15,18] In this respect, in addition to the positive role of IRF-1 reported above, the role of IRF-2 binding to HIV-1 LTR remains to be established. Surprisingly, in fact, overexpression of IRF-2 was not able to counteract the positive effect of IRF-1 on LTR,[21] as occurs for cellular promoters.[17,38] Recent data obtained by gene expression profiling in cells expressing Nef that indicate that IRF-2 is specifically upregulated together with other genes clearly implicated in the positive regulation of HIV,[39] are in accord with our unpublished results showing that constitutive IRF-2 expression in Jurkat T cells greatly enhances HIV-1 replication. Additional work is needed to unravel the functional significance of the IRF-2 binding to the LTR-ISRE. However the observation that IRF-2 is constitutively present in the cells and the HIV-1 ISRE is occupied *in vivo* and is necessary for efficient transcription in infected cells allows for speculation that, in the context of HIV-1 LTR, IRF-2 may act as a functional agonist rather than an antagonist of IRF-1.

EVASION OF THE IFN SYSTEM BY VIRUSES

The innate immune response is crucial during the early phase of the host defence against infections, before an antigen-specific adaptive immune response is induced. Studies of deficient cells and animals derived by gene targeting have demonstrated the essential role of IFN-mediated innate immunity against viral infections. Therefore, it is not surprising that most viruses have acquired mechanisms in order to inhibit or impair the host IFN system. Different steps in IFN signaling and antiviral pathways can be affected by different viruses and also by the same virus (for reviews, see Refs. 40–42) So some viruses shut off cellular protein synthesis, also impairing IFN production; others encode dsRNA-binding proteins that inhibit type I IFN production triggered by the dsRNAs generated during viral infection; sequestering of dsRNA by these proteins not only inhibits IFN production, but also inhibits the antiviral action of IFN by preventing activation of the dsRNA-activated enzymes PKR and OASE. IFN production and IFN signaling can also be blocked by inhibiting transcription factors involved in IFN and ISGs synthesis. The STAT and IRF family are both examples of transcription factors whose action can be directly or indirectly thwarted by viruses. The HHV-8 encodes four homologues of IRF that inhibit IRF-1 and probably IRF-3 and IRF-7 through a dominant negative effect, whereas the E6 and E7 proteins of papilloma virus bind directly to IRF-3 and IRF-1, which prevents IRF-3 activation and IRF-1 binding to the IFN-β promoter, respectively. Competition with both IRF and STAT for the binding to CBP/p300 is also exerted by the E6 protein of papilloma virus and by the E1A protein of adenovirus. Inhibition of both tyrosine and serine phosphorylation of STAT1 during Sendai virus infection has been reported. In addition, the E1A protein of adenovirus downregulates both STAT1 and IRF-9 expression, and also directly binds to STAT1 preventing its transcriptional activity. Both decoy receptors and ligands for IFN and its receptor secreted by viruses such as vaccinia virus have been described. For other viruses the mechanisms involved in the inhibition of type I IFN expression are still unknown. Generally each virus is able to affect the IFN system at different steps and by different mechanisms.[40–42]

Despite the different mechanisms and viral products that interfere with IFN action, little is still known about the exact interplay between viruses and the IFN system and about the outcome of it on viral pathogenicity, clearance, and/or establishment of a persistent infection. In the case of HIV-1, the complexity of the virus–host interactions and the profound disregulation of the host cytokine network exerted by the virus at different stages during the infection have made these studies extremely difficult. Several cytokines have been implicated, including IFNs and TNFs, in the severe immunodeficiency described in patients with HIV infection and characterized by loss of $CD4^+$ cells.[35,36] Contradictory results, however, have been so far obtained on the exact contribution of different cytokines to the AIDS progression and pathogenesis. Some cytokines and chemokines can either inhibit or enhance HIV-1 expression, and the ultimate biological effect is dependent on the phase of HIV infection, presence of other regulatory factors, time of cytokine exposure, type and maturational status of the infected cells. Indeed a significant part of the overall modulation of HIV-1 replication and progression of AIDS occurs via a balance of host factors (reviewed in Refs. 36 and 43).

Strategies used by HIV-1 for evasion from the IFN system so far described include the induction of a cellular inhibitor of Rnase-L, RLI,[44] whose activity results in the inhibition of the OASE/RNase-L system. Direct inhibition of PKR by the HIV-TAR RNA has also been described[45–47] as well as a competition of Tat with eIF-2a for binding to the activated PKR[48] that results in inhibition of PKR activity.

The involvement of IRFs in HIV-1 replication recently shown[21] allows us to speculate that targeting IRFs can be also regarded as a mechanism utilized by the HIV-1 to evade the antiviral effect of IFN. The kinetic of IRF-1 induction by HIV-1, in fact, closely resembles that described in cells infected by the vesicular stomatitis virus or Newcastle disease virus, where IRF-1 expression precedes IFN type I production.[17] Activation of IRFs and of the IFN system by virus infection usually results in a potent antiviral state in the host and bystander cells. Our results suggest that HIV-1 may, effectively, have evolved a strategy to turn this activation to its own advantage before massive IFN- and IFN-stimulated genes activation. Attractively, IRF-1 recruitment on HIV-1 LTR could thus be regarded as a strategy used by the virus to counteract host defences. This strategy seems also to involve the targeting of other members of the IRF family. The recent identification of IRF-2 as a gene upmodulated by nef,[39] and our unpublished results indicating a positive effect of IRF-2 on HIV-1 replication indicate that during the course of infection also IRF-2 can be involved in the escape of HIV-1 from the IFN system. IRF-2, in fact, not only inhibits virus-induced IFN-β transcription by preventing enhanceosome-dependent recruitment of the CBP-PolII holoenzyme complex[49] but by competing with IRF-1 on the promoter of the target ISGs can also shut off the IFN-stimulated host response. At the same time its recruitment on the LTR can contribute to LTR transcriptional activity.

Finally, the exact contribution of IRF-8, a physiological repressor of IRF-1 transcriptional activity, in HIV-1 pathogenesis, remains to be elucidated. However the clear inhibitory effect on virus replication exerted by its overexpression in T cells[21] suggests that IRF-8-increased expression triggered by cytokines could be involved in the establishment of the HIV-1 latency, in contrast to the IRF-1 activation that serves as positive regulator. Thus the differential expression of these IRFs in activated versus nonactivated cells and in different cell types may be used by the virus to determine productive infection versus virus latency at different time and/or at different tissue sites.

ROLE OF IRF-1 IN APOPTOSIS

Due to its action as inhibitor of cell growth on several cell types, the expression of IRF-1 is tightly regulated and its half-life is very short (30 min). Therefore, in general, IRF-1 activity in the cell does not lead to cell death, but under particular physiological or pathological situations, IRF-1 has been demonstrated to be a critical mediator of apoptosis. The introduction of the activated c-H-ras in IRF-$1^{-/-}$ fibroblasts alone was sufficient to oncogenically transform the cells, and these fibroblasts, in contrast with wild-type cells, survive low-serum starvation and treatment with anticancer drugs or ionizing radiations without undergoing apoptosis.[50] These results, together with the observation that DNA-damage-induced apoptosis in mitogen-activated mature T lymphocytes is dependent on IRF-1,[51] stressed the role of IRF-1 as a tumor suppressor and apoptotic gene. The induction of caspase-1 by IRF-1 has been shown to be essential for induction of apoptosis and recently a transcriptional activation of caspase-7 by IRF-1 has also been demonstrated.[2] Moreover, the involvement of IRF-1 in the control of the machinery that T cells use to induce apoptosis in their target cells is demonstrated by the requirement for IRF-1 in the transcriptional induction of the FASL.[52]

Despite the wealth of studies demonstrating IRF-1 as essential mediator of apoptotic mechanisms in several cell systems, more recent observations indicate that its role in apoptosis has now to be regarded as dual: pro- or anti-apoptotic, depending on the system analyzed.[50]

The role of apoptosis in HIV-1 infection is still debated, and the types of pro- and anti-apoptotic stimuli that have been associated with HIV-1 are multiple and often overlapping or even contradictory.[53] Nevertheless, depletion of $CD4^+$ cells is a hallmark in HIV infection and AIDS progression, and among the immunological mechanisms that may contribute to death of $CD4^+$ lymphocytes are killing by specific cytotoxic T lymphocytes and signaling through the CD4 receptor leading to apoptosis through the activity of multiple viral genes and diverse signaling pathways. In addition to direct killing of T cells by the virus indirect killing or growth inhibition of bystander cells is also surely involved.[53,54] Indeed, induction of apoptosis by HIV-1 proteins has been considered an important mechanism in the progression of immune system destruction. Specifically, there have been recent reviews on the role of viral Env and the accessory proteins Tat, nef, and vpr.[55–57]

The possible involvement of IRF-1 in this tangled picture has not yet been investigated. However, since IRF-1 is clearly involved in apoptosis of activated T lymphocytes and in the context of the machinery that T cells use to induce apoptosis in their target cells, it is an intriguing possibility that the HIV-1-induced IRF-1 may play a role in apoptosis of infected or bystander T cells.

CONCLUSIONS

Since the first identification and characterization of IRF-1 and IRF-2, several other members of the family have been identified, and their functional analyses have provided new insights in the mechanisms regulating host defence. Nevertheless, the mechanistic complexity of the IRF in terms of their multifaceted roles in pathogen response, cell growth control, and immune regulation suggests that much remains to be discovered. Future studies will bring new insights in tissue-specific, inducible, and developmental gene regulation contributing to unraveling events leading to viral, immune and tumoral diseases.

Both innate and acquired host immune systems are hit by evasive viral mechanisms. The physiological role of the IRF factors in innate immunity is supported by the fact that a number of viruses target their functions; however, IRF action seems not entirely limited to innate immunity, but it has been shown to mediate connections between various host defences. These include interference with the major histocompatibility complex antigen presentation, block of apoptosis, and modulation of multiple cytokine networks.

Studies on the role of IRF in HIV-1 pathogenesis are only beginning. Extensive work is still needed to fully delineate their induction by the virus and to understand the ultimate effect of their activity on the output of the infection in the different cell targets and/or during the different stages of the viral life cycle. Because of the broad range of effects of the IRF system in immune regulation, these studies will surely open interesting new avenues of investigation in the elucidation of HIV-1 pathogenesis.

ACKNOWLEDGMENTS

We thank members of the Laboratory of Virology at the Istituto Superiore di Sanità for helpful discussions, R. Gilardi for artwork, and S. Tocchio for editorial assistance. This work was supported by institutional grants and by the Italian AIDS Projects to A. Battistini and to B. Ensoli. Due to space limitations, not all relevant papers could be appropriately cited. We apologize to authors whose studies have not been mentioned.

REFERENCES

1. Nguyen, H., J. Hiscott & P.M. Pitha. 1997. The growing family of interferon regulatory factors. Cytokine Growth Factor Rev. **8:** 293–312.
2. Taniguchi, T., K. Ogasawara, A. Takaoka & N. Tanaka. 2001. IRF family of transcription factors as regulators of host defense. Annu. Rev. Immunol. **19:** 623–655.
3. Wang, I.M., J.C. Blanco, M.J. Tsai & K. Ozato. 1996. Interferon regulatory factors and TFIIB cooperatively regulate interferon-responsive promoter activity in vivo and in vitro. Mol. Cell. Biol. **16:** 6313–6324.
4. Merika, M., A.J. Williams, G. Chen, *et al.* 1998. Recruitment of CBP/p300 by the IFN beta enhanceosome is required for synergistic activation of transcription. Mol. Cell **1:** 277–287.
5. Hiscott, J., P. Pitha, P. Genin, *et al.* 1999. Triggering the interferon response: the role of IRF-3 transcription factor. J. Interferon Cytokine Res. **19:** 1–13.
6. Bovolenta, C., P.H. Driggers, M.S. Marks, *et al.* 1994. Molecular interactions between interferon consensus sequence binding protein and members of the interferon regulatory factor family. Proc. Natl. Acad Sci USA **91:** 5046–5050.
7. Moore, P.S., C. Boshoff, R.A. Weiss & Y. Chang. 1996. Molecular mimicry of human cytokine and cytokine response pathway genes by KSHV. Science **274:** 1739–1744.
8. Fauci, A.S. & H.C. Lane. The acquired immunodeficiency syndrome (AIDS). 1991. *In* Principles of Internal Medicine, 12th ed. J.D. Wilson, E. Graunwald, K.J. Isselbacher, *et al.*, Eds.: 1402–1410. McGraw-Hill. New York.
9. Antoni, B.A., S.B. Stein & A.B. Rabson. 1994. Regulation of human immunodeficiency virus infection: implications for pathogenesis. Adv. Virus Res. **43:** 53–145.

10. Roulston, A., R. Lin, P. Beauparlant, *et al.* 1995. Regulation of human immunodeficiency virus type 1 and cytokine gene expression in myeloid cells by NF-κB/Rel transcription factors. Microbiol. Rev. **59:** 481–505.
11. Cullen, B.R. 1991. Regulation of HIV-1 gene expression. FASEB J. **5:** 2361–2368.
12. Tang, H., K.L. Kuhen & F. Wong-Staal. 1999. Lentivirus replication and regulation. Annu. Rev. Genet. **33:** 133–170.
13. Van Lint, C., J. Ghysdael, P. Paras, Jr., *et al.* 1994. A transcriptional regulatory element is associated with a nuclease-hypersensitive site in the *pol* gene of human immunodeficiency virus type. J. Virol. **68:** 2632–2648.
14. El Kharroubi, A. & E. Verdin. 1994. Protein–DNA interactions within DNase I-hypersensitive sites located downstream of the HIV-1 promoter. J. Biol. Chem. **269:** 19916–19924.
15. El Kharroubi, A. & M.A. Martin. 1996. Cis-acting sequences located downstream of the human immunodeficiency virus type 1 promoter affect its chromatin structure and transcriptional activity. Mol. Cell. Biol. **16:** 2958–2966.
16. Roebuck, K.A., D.A. Brenner & M.F. Kagnoff. 1993. Identification of c-fos-responsive elements downstream of TAR in the long terminal repeat of human immunodeficiency virus type-1. J. Clin. Invest. **92:** 1336–1349.
17. Harada, H., T. Fujita, M. Miyamoto, *et al.* 1989. Structurally similar but functionally distinct factors, IRF-1 and IRF-2, bind to the same regulatory elements of IFN and IFN-inducible genes. Cell **58:** 729–739.
18. Liang, C., X. Li, Y. Quan, *et al.* 1997. Sequence elements downstream of the human immunodeficiency virus type 1 long terminal repeat are required for efficient viral gene transcription. J. Mol. Biol. **272:** 167–177.
19. Van Lint, C., C.A. Amella, S. Emiliani, *et al.* 1997. Transcription factor binding sites downstream of the human immunodeficiency virus type 1 transcription start site are important for virus infectivity. J. Virol. **71:** 6113–6127.
20. Verdin, E. 1991. DNase I-hypersensitive sites are associated with both long terminal repeats and with the intrangenic enhancer of integrated human immunodeficiency virus type 1. J. Virol. **65:** 6790–6799.
21. Sgarbanti, M., A. Borsetti, N. Moscufo, *et al.* 2002. Modulation of human immunodeficiency virus 1 replication by interferon regulatory factors. J. Exp. Med. **195:** 1359–1370.
22. Fujita, T., J. Sakakibara, Y. Sudo, *et al.* 1988. Evidence for a nuclear factor(s), IRF-1, mediating induction and silencing properties to human IFN-β gene regulatory elements. EMBO J. **7:** 3397–3405.
23. Fujita, T., Y. Kimura, M. Miyamoto, *et al.* 1989. Induction of endogenous IFN-α and IFN-β genes by a regulatory transcription factor IRF-1. Nature **337:** 270–272.
24. Miyamoto, M., T. Fujita, Y. Kimura, *et al.* 1988. Regulated expression of a gene encoding a nuclear factor, IRF-1, that specifically binds to IFN-β gene regulatory elements. Cell **54:** 903–913.
25. Fujita, T., L.F. Reis, N. Watanabe, *et al.* 1989. Induction of the transcription factor IRF-1 and interferon-β mRNAs by cytokines and activators of second-messenger pathways. Proc. Natl. Acad. Sci. USA **86:** 9936–9940.
26. Harada, H., E. Takahashi, S. Itoh, *et al.* 1994. Structure and regulation of the human interferon regulatory factor 1 (IRF-1) and IRF-2 genes: implications for a gene network in the interferon system. Mol. Cell. Biol. **14:** 1500–1509.
27. Abdollahi, A., K.A. Lord, B. Hoffman-Liebermann & D.A. Liebermann. 1991. Interferon regulatory factor 1 is a myeloid differentiation primary response gene induced by interleukin-6 and leukemia inhibitory factor: role in growth inhibition. Cell Growth Differ. **2:** 401–407.
28. Veschambre, P., A. Roisin & P. Jalinot. 1997. Biochemical and functional interaction of the human immunodeficiency virus type 1 Tat transactivator with the general transcription factor TFIIB. J. Gen. Virol. **78:** 2235–2245.
29. Hottiger, M.O. & G.J. Nabel. 1998. Interaction of human immunodeficiency virus type 1 Tat with the transcriptional coactivators p300 and CREB binding protein. J. Virol. **72:** 8252–8256.
30. Siciliano, R.F. 1999. Latency and reservoirs for HIV-1. AIDS **13:** 49–58.

31. NABEL, G. & D. BALTIMORE. 1987. An inducible transcription factor activates expression of human immunodeficiency virus in T cells. Nature **326:** 711–713.
32. TONG-STARKSEN, S.E., P.A. LUCIW & B.M. PETERLIN. 1987. Human immunodeficiency virus long terminal repeat responds to T-cell activation signals. Proc. Natl. Acad. Sci. USA **84:** 6845–6849.
33. GREENE, W.C. 1990. Regulation of HIV-1 gene expression. Annu. Rev. Immunol. **8:** 453–475.
34. NELSON, N., Y. KANNO, C. HONG, *et al.* 1996. Expression of IFN regulatory factor family proteins in lymphocytes. Induction of Stat-1 and IFN consensus sequence binding protein expression by T cell activation. J. Immunol. **156:** 3711–3720.
35. FAUCI, A.S. 1996. Host factors and the pathogenesis of HIV-induced disease. Nature **384:** 529–534.
36. POLI, G. & A.S. FAUCI. 1995. Role of cytokine in pathogenesis of human immunodeficiency virus infection. *In* Human Cytokines: Their Role in Disease and Therapy. B.B. Aggarwal & R.K. Pun, Eds.: 421. B. Blackwell Sciences. Oxford, England.
37. BATTISTINI, A., G. MARSILI, M. SGARBANTI, *et al.* 2002. IRF regulation of HIV-1 long terminal repeat activity. J. Interferon Cytok. Res. **22:** 27–37.
38. COCCIA, E.M., N. DEL RUSSO, E. STELLACCI, *et al.* 1999. Activation and repression of the 2-5A synthetase and p21 gene promoters by IRF-1 and IRF-2. Oncogene **18:** 2129–2137.
39. SIMMONS, A., V. ALUVIHARE & A. MCMICHAEL. 2001. Nef triggers a transcriptional program in T cells imitating single-signal T cell activation and inducing HIV virulence mediators. Immunity **14:** 763–777.
40. LEVY, D.E. & A. GARCIA-SASTRE. 2001. The virus battles: IFN induction of the antiviral state and mechanisms of viral evasion. Cytokine Growth Factor Rev. **12:** 143–156.
41. LORENZO, M.E., H.L. PLOEGH & R.S. TIRABASSI. 2001. Viral immune evasion strategies and the underlying cell biology. Seminars in Immunol. **13:** 1–9.
42. PLOEGH, H.L. 1998. Viral strategies of immune evasion. Science **280:** 248–253
43. LEVY, J.A. 1998. HIV and Pathogenesis of AIDS. 2nd ed. ASM Press. Washington DC.
44. MARTINAND, C., C. MONTAVON, T. SALEHZADA, *et al.* 1999. RNase L inhibitor is induced during human immunodeficiency virus type 1 infection and down regulates the 2-5A/RNase L pathway in human T cells. J. Virol. **73:** 290–296.
45. GUNNERY, S., A.P. RICE, H.D. ROBERTSON & M.B. MATHEWS. 1990. Tat-responsive region RNA of human immunodeficiency virus 1 can prevent activation of the double-stranded-RNA-activated protein kinase. Proc. Natl. Acad. Sci. USA **87:** 8687–8691.
46. MAITRA, R.K., N.A.J. MCMILLAN, A. DESAI, *et al.* 1994. HIV-1 TAR RNA has an intrinsic ability to activate interferon-inducible enzymes. Virology **204:** 823–827.
47. ROY, S., M. AGY, A.G. HOVANESSIAN, *et al.* 1991. The integrity of the stem structure of humans immunodeficiency virus type Tat-responsive sequence of RNA is required for interaction with the interferon-induced 68,000-Mr protein kinase. J. Virol. **65:** 632–640.
48. CAI, R., B. CARPICK, R.F. CHUN, *et al.* 2000. HIV-1 TAT inhibits PKR activity by both RNA-dependent and RNA-independent mechanisms. Arch. Biochem. Biophys. **373:** 361–367.
49. SENGER, K., M. MERIKA, T. AGALIOTI, *et al.* 2000. Gene repression by coactivator repulsion. Mol. Cell. **6:** 931–937.
50. KROGER, A., M. KOSTER, K. SCHROEDER, *et al.* 2002. Activities of IRF-1. J. Interferon Cytokine Res. **22:** 5–14.
51. TAMURA, T., M. ISHIHARA, M.S. LAMPHIER, *et al.* 1995. An IRF-1-dependent pathway of DNA damage-induced apoptosis in mitogen-activated T lymphocytes. Nature **376:** 596–599.
52. CHOW, W.A., J.J. FANG & J.K. YEE. 2000. The IFN regulatory factor family participates in regulation of Fas ligand gene expression in T cells. J. Immunol. **164:** 3512–3518.
53. ROSHAL, M., Y. ZHU & V. PLANELLES. 2001. Apoptosis in AIDS. Apoptosis **6:** 103–106.
54. KAPLAN, D. & S. SCOTT. 1998. Role of the Fas/Fas ligand apoptotic pathways in human immunodeficiency virus type 1 disease. J. Virol. **72:** 6279–6282.

55. CULLEN, B.R. 1999. HIV-1 Nef protein: an invitation to a kill. Nat Med. **5:** 985–986.
56. BADLEY, A.D., A.A. PILON, A. LANDAY & D.H LYNCH. 2000. Mechanisms of HIV-associated lymphocyte apoptosis. Blood **96:** 2951–2964.
57. AZAD, A.A. 2000. Could Nef and Vpr proteins contribute to disease progression by promoting depletion of bystander cells and prolonged survival of HIV-infected cells? Biochem. Biophys Res. Commun. **267:** 677–685.

PEA-15 Modulates TNFα Intracellular Signaling in Astrocytes

ARIANE SHARIF, BRIGITTE CANTON, MARIE-PIERRE JUNIER, AND HERVÉ CHNEIWEISS

INSERM U114, Department de Neuropharmacologie, Collège de France, 75231 Paris Cedex 05, France

ABSTRACT: PEA-15 is a small protein (15 kDa) that was first identified as an abundant phosphoprotein in brain astrocytes and subsequently shown to be widely expressed in different tissues and highly conserved among mammals. It is composed of an N-terminal death effector domain (DED) and a C-terminal tail of irregular structure. PEA-15 is regulated by multiple calcium-dependent phosphorylation pathways. PEA-15 is ideally positioned to play a major role in signal integration. Accordingly, it has been demonstrated that PEA-15 diverts astrocytes from TNFα-triggered apoptosis and regulates the actions of the ERK MAP kinase cascade by binding to ERK and altering its subcellular localization. Expression of PEA-15 directs TNFα outcomes toward survival, whereas its absence allows the development of the cytokine-induced cell death.

KEYWORDS: PEA-15; TNFalpha; death effector domain; NF-κB; ERK MAP kinase

INTRODUCTION

During the last two decades, multiple functions have been assigned to astrocytes, among which is their capacity to initiate dynamic responses when stimulated *in vivo* by a wide variety of extracellular signals. An important cellular response consists in increases in intracellular calcium that influence many astrocytic functions, ranging from cytoskeletal rearrangement to intercellular communication through calcium waves. Intracellular phosphoproteins can be considered as targets for extracellular signals received by the cell. Their phosphorylation modifies their function and consequently some of the cell properties.

PEA-15 (phosphoprotein enriched in astrocytes) is a 15 kDa acidic (pI 5.2–5.4) serine-phosphorylated protein highly expressed in the central nervous system (CNS), and particularly enriched in astrocytes, but it can be detected at lower levels in other cell types and out of the CNS.[1–4] The first 80 amino acids of PEA-15 correspond to the death effector domain (DED) sequence found in proteins that regulate apoptotic signaling pathways (FIG. 1). PEA-15 is phosphorylated on two different

Address for correspondence: Hervé Chneiweiss, INSERM U114, Chaire de Neuropharmacologie, Collège de France, 11 Place M. Berthelot, 75231 Paris Cedex 05, France. Voice: 33-1 44 27 12 19; fax: 33-1 44 27 12 60.
herve.chneiweiss@college-de-france.fr

**Ann. N.Y. Acad. Sci. 1010: 43–50 (2003). © 2003 New York Academy of Sciences.
doi: 10.1196/annals.1299.006**

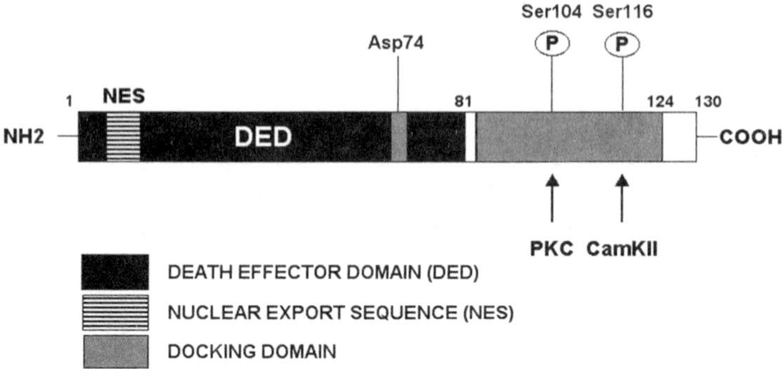

FIGURE 1. Schematic representation of PEA-15. The death effector domain is shown by the *black box*. The two phosphorylation sites are located in the C-terminal part of the molecule. Shadowed in *grey* is the double domain of ERK binding. The *hatched box* corresponds to the nuclear export signal.

seryl residues, which are located out of the DED, within the remaining 51 amino acids. Ser104 was identified as a site of regulation for protein kinase C (PKC),[1,2] whereas Ser116 is regulated by calcium-calmodulin kinase II (CamKII).[5]

The presence of a DED domain in PEA-15 associated with its high expression in cells known to be relatively resistant to apoptosis, led us to postulate an involvement of PEA-15 in apoptosis-related pathways.

ASTROCYTES ARE RESISTANT TO CYTOKINE-INDUCED APOPTOSIS

Programmed cell death of neurons and oligodendrocytes has been extensively documented in several pathological conditions. On the contrary, the astrocytes appear to be relatively resistant to CNS damage and are at the core of tissue repair and regeneration at the site of injury.[6]

Astrocytes respond to any brain injury in a process called reactive astrogliosis. Astrogliosis is characterized by profound changes in astrocytes, which pass from quiescent to reactive morphologies and metabolic profiles.[7,8] Reactive astrocytes display a higher expression level of adhesion molecules compared with resting astrocytes, as well as an increase in the production of a variety of chemokines and cytokines. This seems to engage astrocytes in repair process and immune response. The pro-apoptotic cytokine tumor necrosis factor α (TNF) expression is one the cytokines most frequently and robustly upregulated in reactive astrocytes (reviewed in Ref. 9).

Astrocytes respond to TNF exposure in several ways including cell proliferation, upregulation of TNF mRNAs, and production of interleukin 8 and macrophage-, granulocyte- and granulocyte-macrophage colony-stimulating factors. In addition, TNF primes astrocytes to render them immunocompetent through the induction of the expression of several surface molecules including MHC class II molecules, ICAM-1, and VLA-1 and -2, allowing recruitment of lymphocytes and monocytes.[10]

It is worth noting within this framework that astrocytes appear to favor the development of Th2 (i.e., non-cytolytic) immune responses through the inhibition of interleukin-12 microglia production,[11] a mechanism that could limit the ability of these cells to present antigens and generate Th1 cell-mediated diseases. These data suggest that astrocytes limit the inflammatory reaction and promote tissue repair.

TNF is thus capable of triggering multiple metabolic changes in astrocytes during the course of astrogliosis. Its main known action—apoptosis induction—does not, however, take place in astrocytes, although these cells are endowed with the whole machinery necessary for apoptosis to occur (i.e., TNF receptors, DED-containing cytosolic adaptor molecules, and caspases). Considering the central role played by TNF in the acquisition by reactive astrocytes of functions relevant to tissue repair, it appears reasonable that these cells are protected against TNF-induced apoptosis. Thus a harmonious TNF response should promote a specific differentiation of astrocytes to reactive astrocytes without a risk for apoptosis, and PEA-15 seems to play an important role in this process.

PEA-15 COUNTERACTS TNF-INDUCED APOPTOSIS WITHIN THE EARLIEST STEPS OF THE APOPTOTIC PATHWAY

Membrane receptor-induced apoptosis results from an orderly cascade of cellular events. It involves the sequential activation of aspartate-specific cysteine proteases named caspases.[12] Extracellular signals such as FasL or TNF are well known to trigger such cascades. Upon binding to their respective ligands, Fas and TNFR1 receptors initiate apoptosis by recruiting the cytosolic adaptor molecule FADD to the plasma membrane to form a multiprotein complex named death-inducing signaling complex (DISC) (FIG. 2) (reviewed in Ref. 13). The N-terminal part of FADD contains a death effector domain (DED) that binds to homologous domains located in the N-terminal part of caspase-8, allowing the activation of the caspase and its autocleavage.[14,15] Mutants of FADD lacking DED, or mutants of caspase-8 with only its DEDs, can act as dominant-negative inhibitors, suggesting that endogenous inhibitors of early steps of apoptosis could exist.

As previously stated, PEA-15 also contains a DED. We demonstrated that PEA-15 interacts *in vitro* with two other DED domain-containing proteins, FADD and caspase-8 (FIG. 2).[16] PEA-15 molecules in the vicinity of caspase-8 and caspase-10 in the DISC may prevent further cleavage of the caspases according to the induced proximity model. Indeed, PEA-15 can be recruited to the DISC.[17,18] In agreement with these biochemical findings, we found that PEA-15 expression is required to divert astrocytes from the deleterious effects of TNF. This was demonstrated using astrocytes from wild-type versus PEA-15 null mutant mice.[16] Astrocytes lacking PEA-15 and exposed to TNF rapidly exhibited the classical signs of apoptosis including inversion of membrane lipids, evidenced by annexin V labeling, and nuclear fragmentation leading to the formation of DNA ladders.[16] Re-expression of PEA-15 after transfection restored protection and survival.

The *in vitro* interaction of recombinant PEA-15 with FADD appeared weaker than that observed between FADD and caspase-8.[16] This may result from a low affinity for interaction, and correspond to an adaptation of astrocytes to their high level of expression of PEA-15 whereas FADD and caspase-8 are expressed at very low

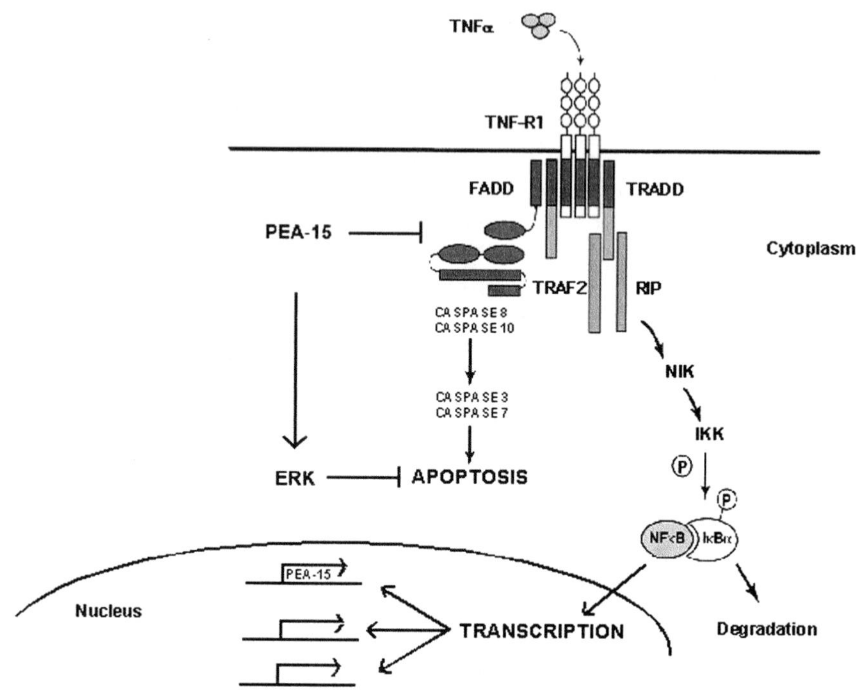

FIGURE 2. PEA-15 modulates TNFα outcomes: Binding of TNF to its receptor induces the trimerization of the receptor. This allows the binding of adaptor molecules to form the death-inducing signaling complex (DISC) through interactions of the dead domains (*dark grey boxes*). FADD then recruits procaspase-8 through binding of their death effector domains (*grey ellipses*). PEA-15 may bind FADD and/or pro-caspase-8 and inhibits further activation of the apoptotic cascade. Another signaling pathway is initiated through the oligomerization of TRAF2 homology domains (*light grey boxes*) leading to the activation of NF-κB and gene transcription, including PEA-15. Finally, PEA-15 exports ERK from the nucleus. This may enhance the effect of ERK on its cytoplasmic targets, and thus further block apoptosis.

levels. But the low affinity between PEA-15 and FADD or caspase-8 may also indicate that post-translational modifications are important. Indeed, in astrocytes PEA-15 is essentially present under its phosphorylated forms.[1,3] The importance of PEA-15 phosphorylation to prevent apoptosis has been demonstrated in other cellular models. NIH3T3 cells transfected with wtPEA-15 were protected from Fas-induced apoptosis whereas a double mutant S104A/S116A was inactive.[19] Furthermore, it was recently reported that PEA-15 is recruited to the DISC after TRAIL stimulation in astrocytoma cell lines.[17] It appears that CamKII- and PKC-dependent phosphorylations modulate PEA-15 antiapoptotic functions. PEA-15 expression renders the cells resistant to TRAIL and a doubly phosphorylated form of PEA-15 is recruited to the DISC. PKC inhibition as well as treatment with KN-93 (a CaMK inhibitor) largely rescued the glioma cell sensitivity to TRAIL.[17,18,20]

A two-hybrid search also evidenced an interaction of PEA-15 with phospholipase D1 (PLD1) that was further confirmed in intact cells.[21] PLD1 has been proposed to inhibit apoptosis[22] through the activation of PI3K and Akt stimulation.[23] PEA-15 affects the accumulation or degradation rates of PLD1 rather than its activity per se. Formation of a functional PLD1-interacting site on PEA-15 requires the C-terminus of PEA-15 and part of its DED. Thus, PEA-15 might recruit PLD1 into the complexes that initiate apoptosis through simultaneous interaction with FADD/caspase-8 and PLD1, and add the PI-3K/Akt cascade to inhibit the death-signaling machinery of the DISC. The existence of such an event remains to be precisely explored in the astrocytes, which are the main sources of PLD1 response in the CNS.[24]

A DIRECT CONTROL OF PEA-15 EXPRESSION BY TNF?

NF-κB activation is most likely the more precisely characterized of TNF-triggered intracellular cascades. Most of the pleiotropic biological actions of TNF can be attributed to its ability to activate the transcription of a startling variety of genes through the nuclear translocation of NF-κB. Initiation of NF-κB activation by TNFR1 involves primarily TRADD. Further events in the NF-κB-inducing cascade involve a member of another adaptor protein family, TRAF2, which binds to the region upstream of the DED in TRADD, as well as to the region upstream of the DED in an additional adaptor molecule, RIP, through the motif shared by this family, the TRAF domain.[25] NF-κB consists of a homo- or heterodimer of DNA-binding proteins related to the proto-oncogene c-Rel. In most cells, it exists in a latent, cytoplasmatically localized state, bound to inhibitory proteins (collectively called IkB) that mask its nuclear localization signal. Cytokines that activate NF-κB, such as TNF, do so by inducing phosphorylation of IkB. This phosphorylation targets IkB for degradation by the proteasome. One of the main biological outcomes of NF-κB translocation to the nucleus is an enhanced cell survival.[26] Accordingly, NF-κB stimulation has been found accompanied by a major increase in PEA-15 expression in B lymphocytes.[27] The activation of the TNF receptors may thus result in enhanced expression of PEA-15, which could counterbalance the mobilization of the apoptotic pathways by TNF.

INVOLVEMENT OF ERK-RELATED PATHWAYS IN PEA-15 INHIBITORY ACTION ON APOPTOSIS ?

Evidences exist for another mechanism underlying PEA-15 inhibition of TNF-induced apoptosis beyond its interaction with FADD and caspase-8. Indeed PEA-15 may have a central role in the regulation of the ERK MAP kinase signaling pathway. The first evidence for a link between PEA-15 and the MAP kinase signaling pathway came from the reversal of the H-Ras inhibition of integrin signaling by PEA-15 expression, evidenced by means of expression cloning.[28] Surprisingly, PEA-15 blocked Ras effect on integrin without affecting its stimulation of ERK activity. We subsequently demonstrated a direct interaction between PEA-15 and the MAP kinase cascade after a yeast two-hybrid screen using PEA-15 as a bait, allowing characterization of ERK1 and ERK2 as the PEA-15 partners.[29]

PEA-15 is not a substrate for ERK. Indeed, PEA-15 lacks the canonical proline-directed ERK phosphorylation sites[30] and is not phosphorylated by ERK *in vitro*. PEA-15 alters the output of ERK signaling without blocking ERK activation. Indeed, we observed that PEA-15 does not interfere with ERK activation or activity. PEA-15 does not affect the phosphorylation of ERK cytosolic substrates, such as stathmin or p90RSK, but it blocks the phosphorylation of ERK nuclear substrates, such as Elk-1, with a consequent inhibition of ERK-dependent transcription.

PEA-15 modifies ERK signaling by excluding ERK from the nucleus. High expression of PEA-15 results in export of ERK from the nucleus in astrocytes, and in PEA-15-transfected CHO and NIH3T3 cells. Moreover, we found that physiological levels of PEA-15, expressed in cultured astrocytes, are sufficient to restrict ERK to the cytoplasm and to block ERK-dependent cFos transcription and cell proliferation. PEA-15 contains a nuclear export signal (NES) that is required for ERK localization to the cytosol. Leptomycin B, a specific inhibitor of CRM1 (exportine-1/chromosome region maintenance 1)-mediated nuclear export, caused PEA-15 accumulation in the nucleus. The NES within PEA-15 is located to I15L17 and a mutation of this sequence results in both PEA-15 and ERK nuclear localization.[29] The ERKs may be inactivated by exposure to phosphatases, such as MKP-1, in the nucleus. Thus, a rapid export from the nucleus could contribute to the capacity of PEA-15 to potentiate ERK activation.[31] Up to now, we have found that the ERK–PEA-15 interaction modulates cell proliferation, but an additional modulation of TNF-induced apoptosis may also exist.

TNF induces activation of ERKs as well as the two other MAP kinases, p38 and JNK, in multiple cells including astrocytes (reviewed in Ref. 9). In human reactive astrocytes, ERK/MAP kinase is chronically activated,[32] and ERK is involved in TNFα-induced overexpression of GFAP.[21] The final outcome of this activation depends on the fine-tuning of multiple pathways, and develops temporally through several autocrine loops. For example, it was proposed, on the basis of data obtained on astrocytes, that the stromal-derived cell factor SDF-1α binds to its receptor CXCR4 and elicits intracellular signaling including the ERK MAPK cascade, which initiates activation of TNF transcription.[33] Consequently, TNF protein is synthesized and secreted to the extracellular space, where it binds to its receptors, leading to activation of NF-κB and further stimulation of ERK.[33] Activated NF-κB and ERK may in turn translocate to the nucleus and activate IL-1 and RANTES expression, leading to a new cascade of receptor activation and signal transduction. How does a given cell choose the appropriate fate upon TNF signaling? Beside the activation of deleterious or protective genes, modulation of the phosphorylation status of critical players must also be considered. For example, phosphorylation of the Bcl-2 family protein Bad may represent an important switch between survival signaling and apoptosis. Bad is phosphorylated on serine-112 in a PI-3K and/or MEK-ERK-dependent manner,[34] promoting its dissociation from Bcl-xL, and from the mitochondria, finally resulting in cell protection from apoptosis. PEA-15 expression limits ERK presence in the nucleus and enhances its activity in the cytosol. Thus, according to the level of expression of PEA-15, the outcomes of TNF effects may greatly change from cell death to cell protection. Interestingly, preliminary results in our group suggest that the levels of PEA-15 expression are variable among astrocytes according to brain regions. It remains to analyze the time course of its expression during infectious or inflammatory processes.

CONCLUSION

TNFα is now recognized to exert a major role in inflammatory processes that are elicited in the CNS after acute insults or infectious diseases and in a number of chronic neurodegenerative disorders. TNF, as the prototype of inflammatory cytokine, is often seen as a "death-only" mediator. This is not systematically the case, TNF being capable to trigger several signaling pathways, which result in the promotion of cell survival in response to a given cellular stress. Intracellular adaptor molecules such as PEA-15 participate in molecular aggregates that form signaling complexes. Their levels of expression, and the regulation of their functions through post-transcriptional processes such as phosphorylation, tune the final outcomes of a given intercellular communication.

ACKNOWLEDGMENTS

We thank all the members of the group for helpful discussions, and are particularly grateful to Drs. Mike Frohman, Joe Ramos and Mark Ginsberg for sharing unpublished data.

REFERENCES

1. Araujo, H. *et al.* 1993. Characterization of PEA-15, a major substrate for protein kinase C in astrocytes. J. Biol. Chem. **268:** 5911–5920.
2. Estelles, A. *et al.* 1996. The major astrocytic phosphoprotein PEA-15 is encoded by two mRNAs conserved on their full length in mouse and human. J. Biol. Chem. **271:** 14800–14806.
3. Danziger, N. *et al.* 1995. Cellular expression, developmental regulation, and phylogenic conservation of PEA-15, the astrocytic major phosphoprotein and protein kinase C substrate. J. Neurochem. **64:** 1016–1025.
4. Condorelli, G. *et al.* 1998. PED/PEA-15 gene controls glucose transport and is overexpressed in type 2 diabetes mellitus. EMBO J. 1998. **17:** 3858–3866.
5. Kubes, M. *et al.* 1998. Endothelin induces a calcium-dependent phosphorylation of PEA-15 in intact astrocytes: identification of Ser104 and Ser116 phosphorylated, respectively, by protein kinase C and calcium/calmodulin kinase II in vitro. J. Neurochem. **71:** 1307–1314.
6. Rzigalinski, B.A. *et al.* 1997. Effect of Ca^{2+} on in vitro astrocyte injury. J. Neurochem. **68:** 289–296.
7. Eddleston, M. & L. Mucke. 1993. Molecular profile of reactive astrocytes: implications for their role in neurologic disease. Neuroscience **54:** 15–36.
8. Ridet, J.L. *et al.* 1997. Reactive astrocytes: cellular and molecular cues to biological function. Trends Neurosci. **20:** 570–577.
9. Zvalova, D. *et al.* 2001. Keeping TNF-induced apoptosis under control in astrocytes: PEA-15 as a "double key" on caspase-dependent and MAP-kinase-dependent pathways. Prog. Brain Res. **132:** 455–467.
10. Merrill, J.E. & E.N. Benveniste. 1996. Cytokines in inflammatory brain lesions: helpful and harmful. Trends Neurosci. **19:** 331–338.
11. Dietrich, P.Y. P.R. Walker & P. Saas. 2003. Death receptors on reactive astrocytes: a key role in the fine tuning of brain inflammation? Neurology **60:** 548–554.
12. Alnemri, E.S. *et al.* 1996. Human ICE/CED-3 protease nomenclature. Cell **87:** 171.
13. French, L.E. & J. Tschopp. 2002. Defective death receptor signaling as a cause of tumor immune escape. Semin. Cancer Biol. **12:** 51–55.

14. BOLDIN, M.P. *et al.* 1996. Involvement of MACH, a novel MORT1/FADD-interacting protease, in Fas/APO-1- and TNF receptor-induced cell death. Cell **85:** 803–815.
15. MUZIO, M. *et al.* 1996. FLICE, a novel FADD-homologous ICE/CED-3-like protease, is recruited to the CD95 (Fas/APO-1) death-inducing signaling complex. Cell **85:** 817–827.
16. KITSBERG, D. *et al.* 1999. Knock-out of the neural death effector domain protein PEA-15 demonstrates that its expression protects astrocytes from TNFalpha-induced apoptosis. J. Neurosci. **19:** 8244–8251.
17. XIAO, C. *et al.* 2002. Tumor necrosis factor-related apoptosis-inducing ligand-induced death-inducing signaling complex and its modulation by c-FLIP and PED/PEA-15 in glioma cells. J. Biol. Chem. **277:** 25020–25025.
18. YANG, B.F. *et al.* 2003. Calcium/calmodulin-dependent protein kinase II regulation of c-FLIP Expression and Phosphorylation in Modulation of Fas-mediated signaling in malignant glioma cells. J. Biol. Chem. **278:** 7043–7050.
19. ESTELLES, A., C.A. CHARLTON & H.M. BLAU. 1999. The phosphoprotein protein PEA-15 inhibits Fas- but increases TNF-R1-mediated caspase-8 activity and apoptosis. Dev. Biol. **216:** 16–28.
20. HAO, C. *et al.* 2001. Induction and intracellular regulation of tumor necrosis factor-related apoptosis-inducing ligand (TRAIL) mediated apotosis in human malignant glioma cells. Cancer Res. **61:** 1162–1170.
21. ZHANG, Y. *et al.* 2000. Regulation of expression of phospholipase D1 and D2 by PEA-15, a novel protein that interacts with them. J. Biol. Chem. **275:** 35224–35232.
22. NAKASHIMA, S. & Y. NOZAWA. 1999. Possible role of phospholipase D in cellular differentiation and apoptosis. Chem. Phys. Lipids **98:** 153–164.
23. BANNO, Y. *et al.* 2001. Involvement of phospholipase D in sphingosine 1-phosphate-induced activation of phosphatidylinositol 3-kinase and Akt in Chinese hamster ovary cells overexpressing EDG3. J. Biol. Chem. **276:** 35622–35628.
24. SERVITJA, J.M. *et al.* 1998. Involvement of ET(A) and ET(B) receptors in the activation of phospholipase D by endothelins in cultured rat cortical astrocytes. Br. J. Pharmacol. **124:** 1728–1734.
25. HSU, H., J. XIONG & D.V. GOEDDEL. 1995. The TNF receptor 1-associated protein TRADD signals cell death and NF-kappa B activation. Cell **81:** 495–504.
26. GARG, A. & B.B. AGGARWAL. 2002. Nuclear transcription factor-kappaB as a target for cancer drug development. Leukemia **16:** 1053–1068.
27. LI, J. *et al.* 2001. Novel NEMO/IkappaB kinase and NF-kappa B target genes at the pre-B to immature B cell transition. J. Biol. Chem. **276:** 18579–18590.
28. RAMOS, J.W. *et al.* 1998. The death effector domain of PEA-15 is involved in its regulation of integrin activation. J. Biol. Chem. **273:** 33897–33900.
29. FORMSTECHER, E. *et al.* 2001. PEA-15 mediates cytoplasmic sequestration of ERK MAP kinase. Dev. Cell. **1:** 239–250.
30. SONGYANG, Z. *et al.* 1996. A structural basis for substrate specificities of protein Ser/Thr kinases: primary sequence preference of casein kinases I and II, NIMA, phosphorylase kinase, calmodulin-dependent kinase II, CDK5, and Erk1. Mol. Cell Biol. **16:** 6486–6493.
31. RAMOS, J.W. *et al.* 2000. Death effector domain protein PEA-15 potentiates Ras activation of extracellular signal receptor-activated kinase by an adhesion-independent mechanism. Mol. Biol. Cell **11:** 2863–2872.
32. MANDELL, J.W. & S.R. VANDENBERG. 1999. ERK/MAP kinase is chronically activated in human reactive astrocytes. Neuroreport **10:** 3567–3572.
33. HAN, Y. *et al.* 2001. TNF-alpha mediates SDF-1 alpha-induced NF-kappa B activation and cytotoxic effects in primary astrocytes. J. Clin. Invest. **108:** 425–435.
34. SCHEID, M.P., K.M. SCHUBERT & V. DURONIO. 1999. Regulation of bad phosphorylation and association with Bcl-x(L) by the MAPK/Erk kinase. J. Biol. Chem. **274:** 31108–31113.

Regulation of *Drosophila* MKP-3 by *Drosophila* ERK

SUNG-EUN KIM, SUN-HONG KIM, AND KANG-YELL CHOI

Department of Biotechnology, Protein Research Center, Yonsei University College of Engineering, Seoul 120-752, South Korea

ABSTRACT: DMKP-3 is a *Drosophila* dual-specificity phosphatase, which has high substrate specificity for *Drosophila* extracellular signal-regulated kinases (DERK). By *in vitro* reconstitution experiments, we found that DERK activates DMKP-3. Moreover, DMKP-3 was specifically activated by the addition of DERK but not by DJNK, Dp38, or *Sevenmaker* DERK D334N, a DMKP-3-binding mutant. The phosphatase activity of DMKP-3-R56A/R57A, a DERK-binding mutant, was not increased by DERK. Significantly, mammalian MKP-3 was also found to be activated by DERK. This cross-reactivity suggests a high level of conservation of the activation mechanism of ERK-specific phosphatases in *Drosophila* and mammals. When DMKP-3 was co-expressed with DERK in *Drosophila* Schneider cells, DMKP-3 protein levels increased, but this was not observed for the co-expressions of DJNK or Dp38. The stabilizations of the DERK binding mutants (DMKP-3-RR and DMKP-3-CA-RR) were not increased by DERK co-expression. Our results suggest that DERK specifically regulates DMKP-3 in terms of its enzyme activity and protein stability, and that direct protein–protein interaction is an essential aspect of this regulation.

KEYWORDS: *Drosophila*; MAP kinase; dual-specific phosphatase; signal transduction; ERK; MKP-3

INTRODUCTION

Mitogen-activated protein kinases (MAPKs) are a family of serine/threonine protein kinases, which are well conserved among eukaryotic organisms.[1,2] Moreover, the MAPKs (ERK, JNK, and p38) regulate important physiological responses such as proliferation, differentiation, apoptosis, and immune responses.[3,4] The accurate and specific regulation of MAPKs by tyrosine phosphatases and dual-specificity MAPK phosphatases (MKPs) is essential for normal cell behavior.[5–7] MKPs that inhibit several MAPKs have been identified,[8–11] and an ERK-specific dual-specificity

ABBREVIATIONS: MAPK, mitogen-activated protein kinase; MKP, MAPK phosphatase; DMKP, *Drosophila* MKP; ERK, extracellular signal–regulated kinase; DERK, *Drosophila* ERK; JNK, c-Jun NH_2-terminal kinase; DJNK, *Drosophila* JNK; Dp38, *Drosophila* p38; LPS, lipopolysaccharide; bp, base pair(s); PAGE, polyacrylamide gel electrophoresis; FBS, fetal bovine serum; PBS, phosphate-buffered saline; GST, glutathione S-transferase.

Address for correspondence: Kang-Yell Choi, Department of Biotechnology, Yonsei University College of Engineering, Seoul 120-752, Korea. Voice: +82 2-2123-2887; fax: +82 2-362-7265.

kychoi@yonsei.ac.kr

Ann. N.Y. Acad. Sci. 1010: 51–61 (2003). © 2003 New York Academy of Sciences.
doi: 10.1196/annals.1299.007

MKP has also been identified.[12,13] As mammals, three different types of MAPKs (DERK, DJNK, and Dp38) have been identified in *Drosophila*. However, the MKPs, which regulate MAPKs, are relatively less understood in *Drosophila*.

Puckered is a *Drosophila* JNK (DJNK)-specific phosphatase which is involved in embryonic dorsal closure.[14] In addition, a *Drosophila* MAPK phosphatase, DMKP, involved in the inhibition of both DERK and Dp38, has been identified, but its *in vivo* function is unknown.[15] A *Drosophila* homologue of MKP-3, DMKP-3, which shows high substrate specificity for DERK was recently identified,[16] and was found to be associated with the anti-proliferation of insulin-stimulated *Drosophila* Schneider cells.[17] DMKP-3 is localized in the cytoplasm of Schneider cells, as is mammalian MKP-3, and this cytoplasmic DMKP-3 was found to inhibit the nuclear localization of DERK.[17] One of the unique characteristics of mammalian MKP-3 is that its phosphatase activity is activated by a specific substrate.

MATERIALS AND METHODS

Plasmids

GST-DMKP-3 expression vector, pGST-DMKP-3, was described in a previous study.[16] The pGST-DMKP-3-R56A/R57A was obtained by subcloning an *Eco*RI-*Xho*I cleaved 1.2-kb DNA fragment, which was obtained by PCR against pOT2-DMKP-3-R56A/R57A using primers, 5′-CGGCACGAATTCATGCCAGAAACG-GAGCACG-3′ and 5′-GCCACTCTCGAGTCATTTAAGACCCGTGTCCG-3′, into the *Eco*RI-*Xho*I site of pGEX4T1 from Amersham Pharmacia (Uppsala, Sweden). pGST-DERK was obtained by subcloning an *Eco*RI-digested 1.1-kb DNA fragment which was obtained by PCR against pPac-His-DERK16 using 5′-GAAACGGAAT-TCATGGAGGAATTTAATTCGAGCG-3′ and 5′-TACAGCTAAT TCTTAAGGCG-CATTGTCTGGTTGTC-3′ primers, into the *Eco*RI site of pGEX4T1. To generate pBS-SK-DERK, pPac-His-DERK18 was amplified by PCR using 5′-GAAACG-GAATTCATGGAGGAATTTAATTCGAGCG-3′ and 5′-GACAGCGAATTCT-TAAGGCGCATTGTCTGGTTGTC-3′. A 1.1-kb PCR product was cleaved with an *Eco*RI, and subcloned into the *Eco*RI site of pBluescript SK from Stratagene (La Jolla, CA). pBS-SK-DERK D334N was generated by site-directed mutagenesis (Stratagene) of pBS-SK-DERK by using the primers, 5′-CAATATTATGATCCTG-GAAATGAGCCTGTCGCTG-3′ and 5′-CAGCGACAGGCTCATTTCCAGGAT-CATAATATTG-3′. To construct pGST-DERK D334N, a 1.1-kb DERK fragment was obtained by PCR against pBS-SK-DERK D334N by using the primers, 5′-GAAACGGAATTCATGGAGGAATTTAATTCGAGCG-3′ and 5′-TACAGC-GAATTCTTAAGGCGCATTGTCTGGTTGTC-3′, and the PCR product was cleaved with *Eco*RI and inserted into the *Eco*RI site of pGEX4T1. To construct pGST-DJNK, pLexA-DJNK13 was cleaved with *Sma*I and *Xho*I, and the 1.1-kb DJNK cDNA fragment obtained was inserted into the *Sma*I-*Xho*I site of pGEX4T1. To construct pGST-Dp38, a full-sized Dp38 DNA fragment was obtained by PCR against pPacPL-His-Dp38 using 5′-TCAAGCGAATTCATGTCAGTGTCCATTA-CAAAAAAG-3′ and 5′-GATGGTCTCGAGTCACTTTACATCCTTTAGAACC-3′ primers. The PCR product obtained was cleaved with *Eco*RI and *Xho*I restriction enzymes, and inserted into the *Eco*RI-*Xho*I site of pGEX4T1. The GST-MKP-3 expression vector, pGEX4T3-GST-MKP-313, was kindly provided by Dr. M. Camps (Pharmaceutical Research Institute, Geneva, Switzerland). pPacPL-DMKP-3-Myc, pPacPL-DMKP-3-CA-Myc, pPacPL-DMKP-3-RR-Myc, pPacPL-DMKP-3-RR-CA-Myc, pPacPL-His-DERK, pPacPL-His-DJNK, pPacPL-His-Dp38, and pPacPL-His-DERK D334N have been described in previous studies.[16,18]

Expression and Purification of GST Fusion Proteins

DMKP-3 and DMKP-3-R56A/R57A phosphatases, and DERK, DJNK, Dp38, and DERK D334N kinases were produced as GST-fused forms in *E. coli* BL21 (DE3) pLysS as described previously.[16] GST-fusion proteins were purified by binding them to glutathione-agarose beads, as described previously.[16]

Phosphatase Assay

Phosphatase assays of GST-DMKP-3, GST-DMKP-3-R56A/R57A, and GST-MKP-3 were performed, as previous described, by using chromogenic substrate *p*-nitrophenyl phosphate (*p*NPP) from Sigma (St. Louis, MO) as a substrate.[15,16] Reactions were performed for 2 h in 200 μl of phosphatase assay buffer (20 mM *p*NPP, 50 mM imidazole, pH 7.5, and 5 mM dithiothreitol) containing increasing amounts of purified GST-DMKP-3 (0–25 μg). In required cases, equal amounts of purified GST-fused DERK, DJNK, Dp38, or DERK D334N were added to reaction mixtures containing GST-DMKP-3 or GST-MKP-3. Similar experiments were performed by incubating 10 μg of GST-DMKP-3 or GST-MKP-3 with different quantities (0–25 μg) of a GST-fused MAPK (DERK, DJNK, or Dp38). Reactions were quenched by adding 1 M NaOH after 2 h, and phosphatase activities were measured by reading absorbance at 405 nm in a 96-well ELISA plate using a Spectra MAX250 spectrophotometer from Spectra (Sunnyvale, CA).

Cell Culture and Transient Transfection

Schneider cells were grown in Schneider medium (Sigma) supplemented with heat-inactivated 10% fetal bovine serum (FBS), 100 units of penicillin and 100 μg/ml of streptomycin at 23°C. Five micrograms of the plasmids, pPacPL-DMKP-3-Myc, pPacPL-DMKP-3-CA-Myc, or pPacPL-DMKP-3-RR-Myc were co-transfected with 0–5 μg of pPacPL-His-DERK, pPacPL-His-DJNK, or pPacPL-His-Dp38, as previously described.[17] After 48 h, cell extracts were prepared and quantified as described previously.[15] When required, transfected cells were treated with human insulin (10 μg/ml) for 12 h before being harvested for extract preparation. MAPK activation was confirmed by Western blotting using anti-phospho-ERK antibody.[15]

Western Blot

Lysates containing 30–50 μg of protein were separated by 10 % SDS-PAGE (acrylamide: bis-acrylamide at a ratio of 39:1) and transferred onto a nitrocellulose membrane. Western blotting was performed, as described previously[15] with anti-mouse-RGS His from Qiagen (Hilden, Germany) for the His-tagged MAPK proteins, and anti-Myc polyclonal antibody (sc-789) from Santa Cruz biotechnology (Santacruz, CA) for the Myc-tagged DMKP-3 proteins. Mammalian JNK antibody from New England BioLabs (Beverly, MA) recognized both His-tagged-DJNK (His-DJNK) and His-tagged-Dp38 (His-DERK) protein, and this was used to detect both His-DJNK and His-Dp38 proteins. The blots were probed with horseradish-peroxidase (HRP)-conjugated goat anti-mouse IgG secondary antibody from Bio-Rad Laboratories (Richmond, CA), and visualized by enhanced chemiluminescence from Genepia (Seoul, Korea).

RESULTS

Recombinant DMKP-3 Phosphatase Activity was Increased by DERK in Vitro

DMKP-3 most resembles mammalian MKP-3 among all the known MAPK phosphatase family members, both in terms of amino acid identity and gene structure (data not shown). In the mammalian system, MKP-3 is activated by a specific substrate ERK *in vitro*,[12] and this activation is acquired through substrate binding. To identify similarities in the regulatory mechanism of the *Drosophila* system, we experimented

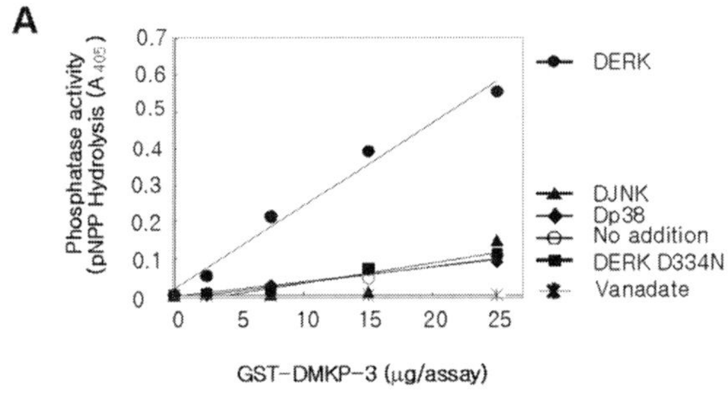

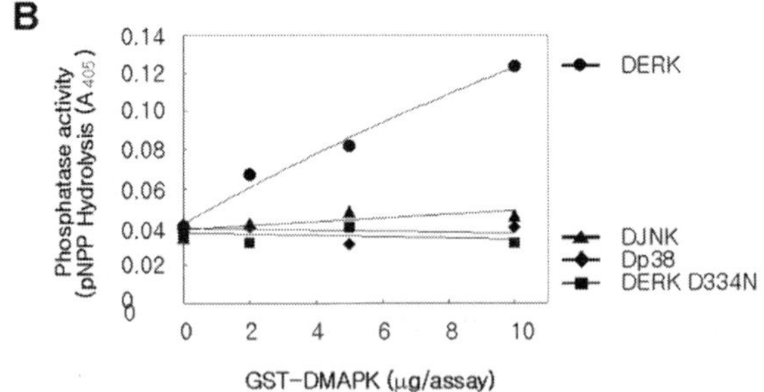

FIGURE 1. DMKP-3 phosphatase activity is specifically increased by DERK. (**A**) Purified GST-DMKP-3 protein (0–25 μg) in 200 μl of phosphatase buffer (see MATERIALS AND METHODS) was assayed in the absence or in the presence of equal amounts of GST-DERK, GST-DJNK, GST-Dp38, or GST-DERK D334N. Where required 0.5 mM sodium vanadate was added instead of the purified MAPK protein. (**B**) Phosphatase activities of 10 μg of purified GST-DMKP-3 were measured as described in (**A**) in the absence or in the presence of increasing amounts (0–10 μg) of GST-DERK, GST-DJNK, GST-Dp38 or GST-DERK D334N.

in vitro with purified proteins[12] to investigate DMKP-3 activation by DERK. Full-length MKPs (DMKP-3, DMKP-3-R56AR57A, and MKP-3) and the DMAPKs (DERK, DJNK, and Dp38) proteins were found to be overexpressed as GST fusion forms in *E. coli*, and were purified using glutathione agarose beads to purities exceeding 90% (unpublished data). The phosphatase activity of purified GST-DMKP-3 proteins was assayed directly by adding pNPP as substrate *in vitro*.

Purified GST-DMKP-3 protein was found to retain its *p*NPP hydrolyzing activity, and the enzyme activity was found to be completely abolished by the addition of the tyrosine phosphatase inhibitor, sodium vanadate (FIG. 1A). By adding recombinant GST-DERK proteins, GST-DMKP-3 phosphatase activity was linearly increased *in vitro* (FIG. 1A). On the other hand, neither GST-DJNK nor GST-Dp38 proteins significantly increased GST-DMKP-3 phosphatase activity (FIG. 1A). To better understand DMKP-3 activation by DERK, we investigated whether the DMKP-3 binding mutant *Sevenmaker*(rl sevenmaker),[8,12] DERK D334N, could activate DMKP-3. Unlike wild-type GST-DERK, GST-DERK D334N protein did not interact with DMKP-3 by two-hybrid analysis (data not shown), and did not further increase the *p*NPP hydrolyzing activity of GST-DMKP-3 (FIG. 1A).

To confirm DMKP-3 activation by DERK, we fixed the amount of GST-DMKP-3 protein, and added increasing amounts of GST-DERK, -DJNK, or -Dp38 before measuring the phosphatase activity. As shown in FIGURE 1B, GST-DMKP-3 activity

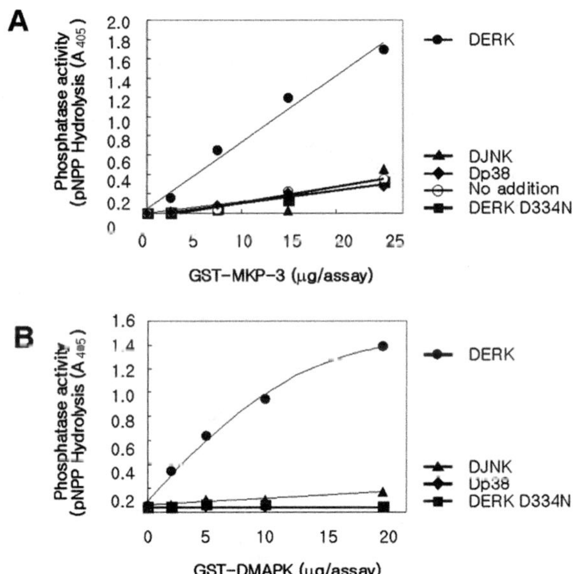

FIGURE 2. Mammalian MKP-3 activity was increased by *Drosophila* ERK. (**A**) Phosphatase activities of GST-MKP-3 protein (0–25 μg) were measured, as described in FIGURE 1A, in the presence or absence of equal amounts of DERK, DJNK, or Dp38. (**B**) The phosphatase activity of 10 μg of purified GST-DMKP-3 was measured, as described in FIGURE 1, in the presence or in the absence of increasing amounts (0–20 μg) of GST-DERK, GST-DJNK, GST-Dp38, or GST-DERK D334N.

was also increased by adding GST-DERK protein. On the other hand, DMKP-3 activity did not increase on adding GST-DJNK, -Dp38, or -DERK D334N. Therefore, DERK specifically activated DMKP-3 phosphatase activity by direct protein–protein interaction.

DERK Activates Mammalian MKP-3 in Vitro

Although *Drosophila* and mammals have similar MKPs and MAPKs, reactivity between ERK and MKP-3 has not been reported. We previously found that mammalian ERK is dephosphorylated by recombinant GST-DMKP-3 *in vitro*.[16] To identify similarities in the regulation of MKP-3 by ERK in mammals and *Drosophila*, we investigated whether DERK activates mammalian MKP-3. We found that GST-MKP-3 phosphatase activity was increased by GST-DERK (FIG. 2A). Moreover, the fold activation of GST-MKP-3 induced by GST-DERK was found to be higher than GST-DMKP-3 phosphatase activity increase by GST-DERK (compare FIGURES 1A and 2A). Moreover, GST-MKP-3 activity was not increased by adding GST-DJNK, -Dp38 or -DERK D334N (FIG. 2A). In addition, GST-MKP-3 phosphatase activity increased hyperbolically by adding more GST-DERK, when maintaining a fixed amount of GST-MKP-3 (FIG. 2B). The GST-MKP-3 activities were similarly not increased by adding GST-DJNK, -Dp38 or by -DERK D334N. Therefore, mammalian MKP-3 phosphatase activity was also specifically increased by DERK, but not by DJNK or Dp38. These results suggest that MKP-3 activation by ERK is highly conserved.

To further confirm the importance of the direct binding of DERK to DMKP-3 in DMKP-3 activation, we used a DERK binding mutant DMKP-3 protein, GST-DMKP-3-R56A/R57A, which is defective in terms of DERK binding ability on account of the mutation of two arginine residues (Arg-56 and Arg-57) within the pentapeptide sequence "IVLRR," which are involved in DERK binding (data not shown).[16] GST-DMKP-3-R56A/R57A mutant protein retained the intrinsic phosphatase activity of wild-type DMKP-3 (FIG. 3). However, GST-DMKP-3-R56A/

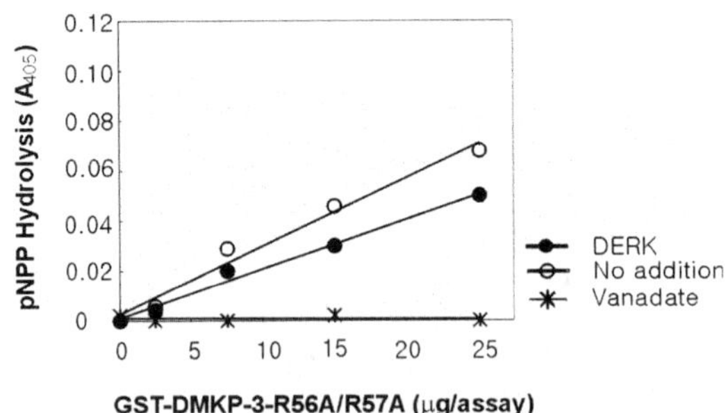

FIGURE 3. Phosphatase activity of the DERK binding mutant, DMKP-3-R56A/R57A, was not increased by DERK. Phosphatase activities of 10 μg of purified GST-DMKP-3-R56A/R57A were measured, as described in FIGURE 1A either with or without the addition of equal amounts of DERK proteins. When required, sodium vanadate (to 0.5 mM) was also added to reactions including GST-DMKP-3-R56A/R57A.

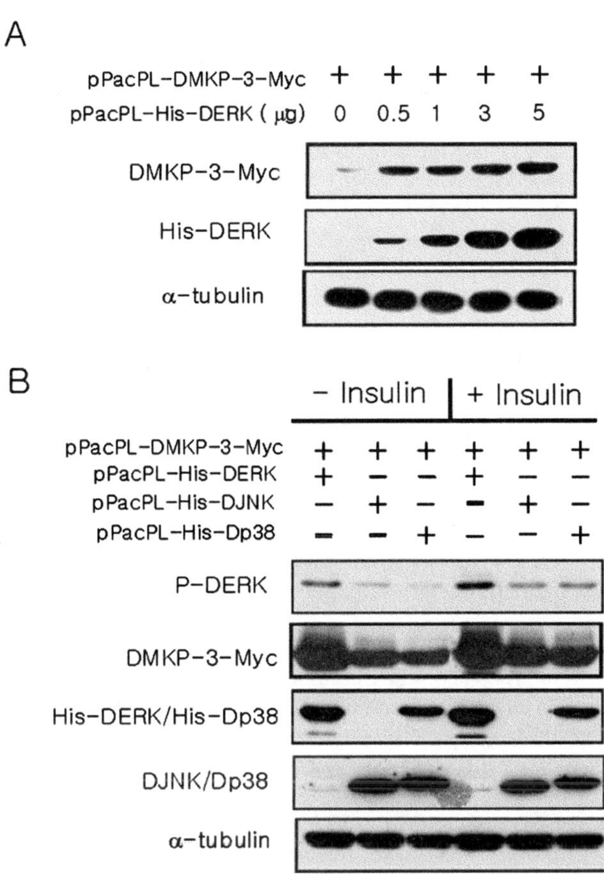

FIGURE 4. DMKP-3 protein levels were increased regardless of insulin treatment when co-expressed with DERK. (**A**) Schneider cells were grown in Schnedier medium containing 10% (v/v) FBS, and cells were transiently transfected with 5 μg of pPacPL-DMKP-3 with 5 μg of pPacPL or increasing amounts (0–5 μg) of pPacPL-DERK-His. (**B**) Schneider cells were grown and co-transfected with 5 μg of pPacPL-DMKP-3-Myc with 5 μg of pPacPL-His-DERK, pPacPL-His–DJNK or pPacPL-His-Dp38 as described in FIGURE 4A. If required, human insulin (10 μg/ml) was treated for 12 h before the cells were harvested. After 48 h, the cell extracts were prepared and Western blot analysis was performed[16] with anti-Myc polyclonal antibody, anti-(mouse RGS His) or anti-α-tubulin antibody to detect Myc-DMKP-3, His-DERK and α-tubulin, respectively. The His-DJNK proteins were detected by using anti-SAPK/JNK antibody, and His-Dp38 proteins also detected by anti-SAPK/JNK antibody (see MATERIALS AND METHODS).

R57A protein phosphatase activity was not increased; in fact, it was slightly decreased by the GST-DERK addition (FIG. 3). Similarly, MKP-3-R64A/R65A, which is known to be defective in terms of ERK binding ability, is also insensitive to phosphatase by ERK2 in the mammalian system.[19]

DMKP-3 Protein Stability was Increased by DERK Co-expression

Transient transfection studies showed that the DMKP-3 protein level was often increased by co-expressed DERK. When His-tagged DERK protein (His-DERK) was co-transfected with Myc-taaged DMKP-3 (DMKP-3-Myc) into *Drosophila* Schneider cells, DMKP-3-Myc protein level was increased (FIG. 4A). On the other hand, the DMKP-3-Myc protein level was not significantly increased when it was co-expressed with DJNK or Dp38 (FIG. 4B). Therefore, the protein level of DMKP-3-Myc was specifically increased by DERK. Because the production of both His-DERK and DMKP-3-Myc proteins in our transfection system is controlled by the actin 5C promoter in pPacPL vector,[20] the upregulation of DMKP-3 levels may be caused by protein stabilization rather than differential gene expression.

To investigate the dependency of the signal transduction of DERK-induced DMKP-3 stabilization, we treated DMKP-3-Myc-transfected Schneider cells with a DERK activator, human insulin,[16,17] and measured any changes in the stabilization of DMKP-3-Myc by DERK. As previously observed, phospho-ERK protein was increased by human insulin treatment (FIG. 4B), but DMKP-3-Myc protein stabilization by DERK did not change upon treating with insulin (FIG. 4B). In addition, insulin treatment did not change the stability of DMKP-3-Myc proteins in cells expressing His-DJNK or His-Dp38 (FIG. 4B). The level of DMKP-3 was also not changed by treating with DJNK or Dp38 activators, lipopolysaccharides, or NaCl,[18] respectively (data not shown).

To understand importance of DERK binding in the regulation of DMKP-3 stabilization, we used the DERK binding mutants (DMKP-3-RR-Myc and DMKP-3-CA-RR-Myc) to measure DMKP-3 protein stabilization by DERK. Levels of DMKP-3-RR-Myc and DMKP-3-CA/RR-Myc mutants[17] did not increase when DERK-His proteins were co-expressed with the mutant DMKP-3 protein (FIG. 5). On the other hand, the protein expression of a catalytic mutant, DMKP-3-CA-Myc,[17] was increased when co-expressed with His-DERK (FIG. 5). Stabilization of DMKP-3-RR-Myc was not increase by His-DERK regardless of insulin treatment (FIG. 5B).

DISCUSSION

The MAPKs are negatively regulated either by tyrosine phosphatases or by dual specificity MKPs,[5,21] and these regulatory mechanisms are known to be conserved among eukaryotic organisms ranging from yeast to human.[16] DMKP-3 is a *Drosophila* homologue of MKP-3, which retains high substrate specificity toward DERK. Moreover, DMKP-3 did not show significant cross-reactivity with DJNK or Dp38 in Schneider cells.[16] The DMKP-3 is involved in the G1 to S phase cell-cycle progression of *Drosophila* Schneider cells,[17] as is the MKP-3 analogue in the mammalian system.[22] DMKP-3 is most similar to mammalian MKP-3 in terms of the arrangement and sequences of the amino acids at the catalytic C-terminal and at the N-terminal domain involving substrate interaction.

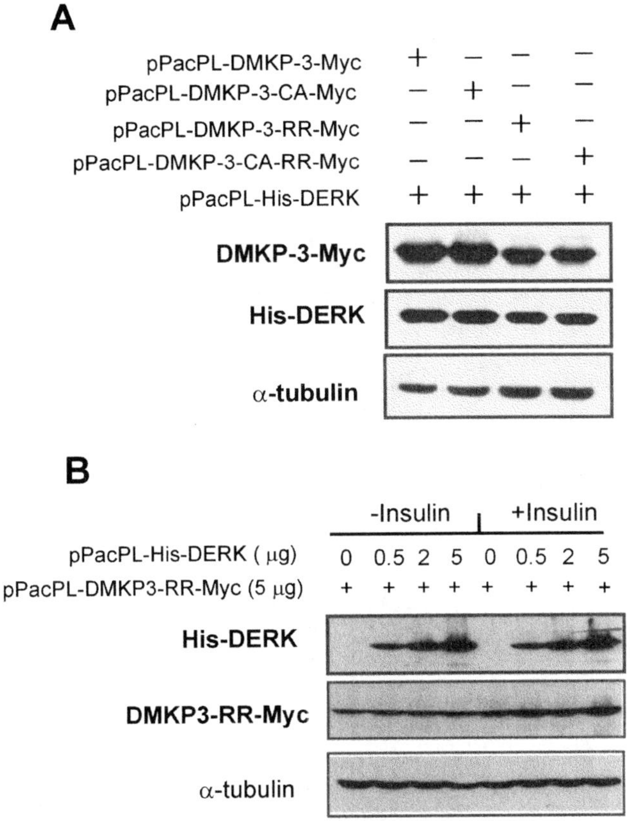

FIGURE 5. DMKP-3 protein stabilization by DERK did not occur in the DERK binding mutants, DMKP-3-RR and DMKP-3-CA-RR. (**A**) Schneider cells were grown and transfected with 5 μg of pPacPL-DMKP-3-Myc, pPacPL-DMKP-3-CA-Myc, pPacPL-DMKP-3-RR-Myc, or pPacPL-DMKP-3-CA-RR-Myc in combination with 5 μg of pPacPL-DERK. DMKP-3, His-DERK, and α-tubulin proteins were detected by Western blot as described in FIGURE 5B. Protein levels of DERK binding DMKP-3 mutants (DMKP-3-RR-Myc and DMKP-3-CA-RR-Myc) were not increased by DERK. (**B**) Schneider cells were grown and transfected with 5 μg of pPacPL-DMKP-3-RR-Myc with 0–5 μg of pPacPL-His-DERK, and treated or not treated with human insulin (10 μg/ml) for 12 h before being harvested.

In this study, we identified the specific regulation of the *Drosophila* homologue of MKP-3, DMKP-3, by DERK. Our results suggest that DERK specifically regulates DMKP-3 both in terms of its enzymatic activity and protein stabilization, and that this regulation involves direct protein–protein interaction. Moreover, MKP-3 was specifically activated by DERK, which suggests conservation of the MKP-3 regulatory mechanism between *Drosophila* and mammals. The level of DMKP-3 activation by DERK was found to be much lower than the level of MKP-3 activation by DERK, but the physiological reason for this is not known. The GST-fused DMKP-3 protein structure may not be in an optimal form for activation by GST-DERK. It is also possible that the activation level of DMKP-3 by DERK may be lower in *Drosophila* than in mammals. Activation of mammalian MKP-3 by *Drosophila* ERK suggests the conservation of the regulatory partner components (MKP-3 and ERK) between *Drosophila* and mammals. The GST-DMKP-3-R56A/R57A mutant, which does not interact with DERK, did not show increased phosphatase activity in the presence of DERK. In addition, GST-DMKP-3 phosphatase activity was not significantly increased by *Sevenmaker* DERK-D334N, which again does not interact with DMKP-3 (data not shown). These results suggest that DERK involvement is essential for DMKP-3 activation. In conclusion, we found that DMKP-3 phosphatase is activated by a specific substrate, DERK, and that this activation mechanism is highly conserved between *Drosophila* and mammal. In addition, the enzymatic regulation of DMKP-3 by DERK requires specific binding of the two proteins.

In this study, we also found that DMKP-3 is stabilized by DERK. However, such effects were not observed in DMKP-3 mutants that can not bind DERK. These results suggest that substrate binding is also essential for DMKP-3 stabilization. The null effect of insulin treatment upon the stabilization of DMKP-3 suggests that DERK tightly interacts with DMKP-3, regardless of signal transduction. The physiological significances of the regulation of MKP-3 by ERK are unknown at this time.[12] Significantly, strict regulation of DERK activity by DMKP-3 is recognized as important in many physiological processes, such as eye development, the specification of terminal structures in the embryo, the formation of wing veins, and in cellular proliferation.[17,23–25] Therefore, the specific regulation of DMKP-3 by DERK may also potentially play an important role in regulation of several important physiological processes.

ACKNOWLEDGMENTS

We thank Dr. M. Camps for providing GST-MKP-3 plasmid. This work was supported by the Molecular Cellular BioDiscovery Research Program from the Ministry of Science and Technology, the Basic Research Grant from Korea Research Foundation (KFR-2003-05-C00400), a grant from the Hallym Academy of Science, and by a Yonsei University Faculty Grant for 2002.

REFERENCES

1. Blenis, J. 1993. The human CL100 gene encodes a Tyr/Thr-protein phosphatase which potently and specifically inactivates MAP kinase and suppresses its activation by oncogenic ras in Xenopus oocyte extracts. Proc. Natl. Acad. Sci. USA **90:** 5889–5892.
2. Cobb, M.H. & E.J. Goldsmith. 1995. How MAP kinases are regulated. J. Biol. Chem. **270:** 14843–14846.
3. Schaeffer, H.J. & M.J. Weber. 1999. Mitogen-activated protein kinases: specific messages from ubiquitous messengers. Mol. Cell. Biol. **19:** 2435–2444.
4. Garrington, T.P. & G.L. Johnson. 1999. Organization and regulation of mitogen-activated protein kinase signaling pathways. Curr. Opin. Cell Biol. **11:** 211–218.
5. Keyse, S.M. 1995. An emerging family of dual specificity MAP kinase phosphatase. Biochem. Biophy. Acta **1265:** 152–160.
6. Camps, M., A. Nichols, A. & S. Arkinstall. 2000. Dual specificity phosphatases: a gene family for control of MAP kinase function. FASEB J. **14:** 6–16.

7. Andersen, J.N., O.H. Mortensen, G.H. Peters, *et al.* 2001. Structural and evolutionary relationships among protein tyrosine phosphatase domains. Mol. Cell. Biol. **21:** 7117–7136.
8. Chu, Y., P.A. Solski, R. Khosravi-Far, *et al.* 1996. The mitogen-activated protein kinase phosphatases PAC1, MKP-1, and MKP-2 have unique substrate specificities and reduced activity in vivo toward the ERK2 seven-maker mutation. J. Biol. Chem. **271:** 6497–6501.
9. Keyse, S.M. & E.A. Emslie. 1992. Oxidative stress and heat shock induce a human gene encoding a protein-tyrosine phosphatase. Nature **359:** 644–647.
10. Muda, M., U. Boschert, A. Smith, *et al.* 1997. Molecular cloning and functional haracterization of a novel mitogen-activated protein kinase phosphatase, MKP-4. J. Biol Chem. **272:** 5141–5151.
11. Rohan, P.J., P. Davis, C.A. Moskaluk, *et al.* 1993. PAC-1: a mitogen-induced nuclear protein tyrosine phosphatase. Science **259:** 1763–1766.
12. Camps, M., A. Nichols, C. Gillieron, *et al.* 1998. Catalytic activation of the phosphatase MKP-3 by ERK2 mitogen-activated protein kinase. Science **280:** 1262–1265.
13. Muda, M., U. Boschert, R. Dickinson & J.C. Martinou. 1996. MKP-3, a novel cytosolic protein-tyrosine phosphatase that exemplifies a new class of mitogen-activated protein kinase phosphatase. J. Biol. Chem. **271:** 4319–4326.
14. Martin-Blanco, E., A. Gampel, J. Ring, *et al.* 1998. A *puckered* encodes a phosphatase that mediates a feedback loop regulating JNK activity during dorsal closure in *Drosophila*. Genes Dev. **12:** 557–570.
15. Lee, W.J., S.H. Kim, Y.S. Kim, *et al.* 2000. Inhibition of mitogen-activated protein kinase by a *Drosophila* dual-specific phosphatase. Biochem. J. **349:** 821–828.
16. Kim, S.H., H.B. Kwon, Y.S. Kim, *et al.* 2002. Isolation and characterization of a *Drosophila* homologue of mitogen-activated protein kinase phosphatase-3 which has a high substrate specificity towards extracellular-signal-regulated kinase. Biochem. J. **361:** 143–151.
17. Kwon, H.B., S.H. Kim, S.E. Kim, *et al.* 2002. *Drosophila* extracellular signal-regulated kinase involves the insulin-mediated proliferation of Schneider cells. J. Biol. Chem. **277:** 14853–14858.
18. Han, S.J., K.Y. Choi, P.T. Brey & W.J. Lee. 1998. Molecular cloning and characterization of a *Drosophila* p38 mitogen-activated protein kinase. J. Biol. Chem. **273:** 369–374.
19. Nicholas, A., M. Camps, C. Gillieron, *et al.* 2000. Substrate recognition domains within extracellular signal-regulated kinase mediate binding and catalytic activation of mitogen-activated protein kinase phosphatase-3. J. Biol. Chem. **275:** 24613–24621.
20. Petersen, U.M., G. Bjorklund, Y.T. Ip & Y. Engstrom. 1995. The dorsal-related immunity factor, Dif, is a sequence-specific trans-activator of *Drosophila* Cecropin gene expression. EMBO J. **14:** 3146–3158.
21. Haneda, M., T. Sugimoto & R. Kikkawa. 1999. Mitogen-activated protein kinase phosphatase: a negative regulator of the mitogen-activated protein kinase cascade. Eur. J. Pharmacol. **365:** 1–7.
22. Brunet, A., D. Roux, P. Lenormand, *et al.* 1999. Nuclear translocation of p42/p44 mitogen-activated protein kinase is required for growth factor-induced gene expression and cell cycle entry. EMBO J. **18:** 664–674.
23. Biggs, W.H., 3rd, K.H. Zavitz, B. Dickson, B., *et al.* 1994. The *Drosophila* rolled locus encodes a MAP kinase required in the sevenless signal transduction pathway. EMBO J. **13:** 1628–1635.
24. Brunner, D., N. Oellers, J. Szabad, *et al.* 1994. A gain-of-function mutation in *Drosophila* MAP kinase activates multiple receptor tyrosine kinase signaling pathways. Cell **76:** 875–888.
25. O'Neill, E.M., I. Rebay, R. Tjian & G.M. Rubin. 1994. The activities of two Ets-related transcription factors required for *Drosophila* eye development are modulated by the Ras/MAPK pathway. Cell **78:** 137–147.

Role of p38 MAP Kinase Signal Transduction in Apoptosis and Survival of Renal Epithelial Cells

HARI K. KOUL

Signal Transduction and Molecular Biology Laboratory, Division of Urology, Department of Surgery, University of Colorado Health Sciences Center, Denver, Colorado 80262, USA

ABSTRACT: The p38 MAP kinase signal transduction pathway is activated by various forms of cellular stress. While, in many tissues, activation of p38 MAP kinase is associated with apoptosis, in some tissues p38 activation is critical for survival. Oxalate deposits are associated with several pathological conditions that involve aberrant proliferation and cellular apoptosis. Studies from our laboratory demonstrated that exposure of renal epithelial cells to oxalate and COM-crystals resulted in activation of the p38 MAP kinase pathway. Moreover, the inhibition of the p38 MAO kinase pathway resulted in the inhibition of oxalate as well as COM-crystal–induced reinitiation of the DNA synthesis. These results suggest a critical role for the p38 MAP kinase pathway in pathological conditions associated with cellular proliferation and apoptosis following deposition of calcium oxalate.

KEYWORDS: oxalate; COM-crystals; renal epithelial cells; nephrolithiasis; hyperoxaluria

INTRODUCTION

Signal transduction via MAP kinases plays a key role in a variety of cellular responses including proliferation, differentiation, and cell death. More than a dozen MAP kinase pathways have been identified so far. The three major pathways are p42/p44-MAP kinase (ERK), p38 MAP kinase, and JNK. ERK is primarily stimulated by growth factors, while p38 MAP kinase and JNK pathways are stimulated primarily by cellular stress.[1]

Oxalate and calcium oxalate deposits are associated with several pathologic conditions, including kidney stone disease, benign neoplasms of the breast, and thyroid hypertrophy. Many of these conditions are associated with aberrant proliferation and cellular apoptosis. Previously, we and others have shown that exposure of the renal epithelial cells to oxalate and calcium oxalate crystals (COM-crystals) is associated with re-initiation of the DNA synthesis, alterations in gene expression, and cellular apoptosis.[2,3] We also evaluated the effects of oxalate and COM-crystals on MAP ki-

Address for correspondence: Dr. Hari Koul, M.Sc., Ph.D., FACN, Director of Research and Professor, University of Colorado Health Sciences Center, 4200 E. 9th Ave., Campus Box C319, Denver, CO 80262. Voice: 303-315-2383; fax: 303-315-7611.

Hari.Koul@UCHSC.edu

Ann. N.Y. Acad. Sci. 1010: 62–65 (2003).
doi: 10.1196/annals.1299.008

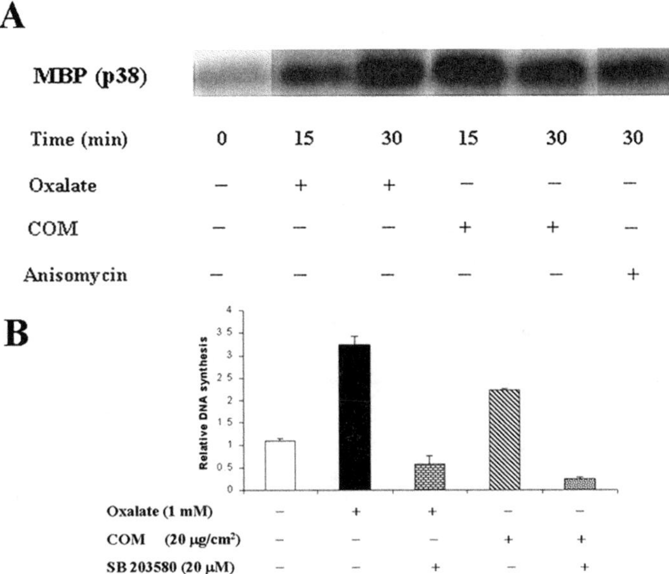

FIGURE 1. Oxalate and COM-crystal interactions with renal epithelial cells: For these studies, serum-starved, growth-arrested LLC-PK1 cells were exposed to oxalate (1 mM) or COM-crystals (20 μg/cm). Where indicated, cells were exposed to SB 203580 (20 μM) 2 h prior to other treatments. **(A)** Modulation of p38 MAP kinase by oxalate and COM-crystals, **(B)** Inhibition of oxalate and COM-crystal stimulated DNA synthesis by SB203580.

nase signal transduction pathways.[4,5] Our results demonstrated selective activation of p38 MAPK and JNK by oxalate. We also show that p38 MAP kinase is activated by COM-crystals, and activated p38 MAPK is primarily localized to nuclear compartment and is required for re-initiation of the DNA synthesis in response to COM-crystals as well as oxalate.

METHODS

Cell culture: For these studies, LLC-PK1 cells,(a line of porcine renal epithelial cells) and HK-2 cells (a line of human renal epithelial cells) were obtained from the American Type Culture Collection (Bethesda, MD). Cells were serially passaged in T-75 flasks and plated at high density in 6-well plates. Then they were grown in DMEM supplemented with serum (10%) and antibiotics as described previously.[4] Cell cultures were used at least 2–3 days post confluence.

DNA synthesis: For these studies, cells were grown to confluence in 6-well plates. Confluent cell monolayers were exposed to oxalate or COM-crystals for 24 hours. During last six hours of incubation 1–3 μCi of ^{3}H-thymidine was added to the cells. Thymidine incorporation into TCA-precipitable material was used as an index of the DNA synthesis.[2]

Western blot analysis: For these studies, cells were exposed to oxalate or COM-crystals for various time points (5–60 min). At the end of the experimental period, the cells were lysed in denaturing lysis buffer. Cell lysates were run on 10% PAGE and transferred to polyvinylidene difluoride (PVDF) membrane and probed for antibodies specific for MAP kinases as described previously.[4]

RESULTS

Oxalate and COM-crystals stimulate p38 MAP kinase: Results presented in FIGURE 1A revealed that exposure of renal epithelial cells to oxalate as well as COM-crystals resulted in rapid and robust activation of p38 MAP kinase. While exposure to oxalate required 30 min for maximal stimulation of p38 MAP kinase, it took only 15 min to achieve maximal activation with COM-crystals.

p38 MAP kinase is required for oxalate- and COM-crystal–stimulated DNA synthesis: Results presented in FIGURE 1B demonstrate that exposure of cells to oxalate as well as COM-crystals resulted in re-initiation of the DNA synthesis. Moreover, pretreatment of the cells with SB203580 (a selective inhibitor of p38 MAP kinase) abolished both oxalate-induced as well as COM-crystal-induced re-initiation of the DNA synthesis.

Additional studies[4] also demonstrated that oxalate exposure to the renal cells stimulated JNK signaling, but had no effect on ERK signaling. We observed that in addition to the re-initiation of the DNA synthesis, oxalate exposure also resulted in cytoplasmic vacuolization, nuclear blebbing and pyknosis, and DNA double-strand breaks.

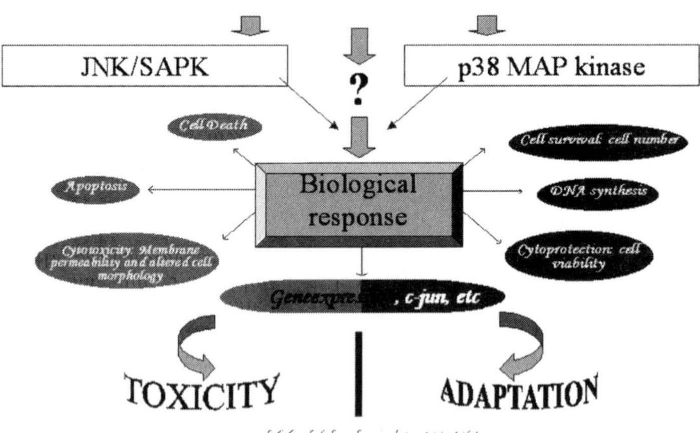

FIGURE 2. Proposed model for the role of p38 MAP kinase and JNK signal transduction in modulating effects of oxalate and COM-crystals: Oxalate and COM-crystals lead to the activation of stress kinase pathways. These pathways mediate the expression of various genes that are critical for modulating cellular adaptations and/or toxicity in response to these agents.

DISCUSSION

The p38 MAP kinase signaling pathway has been shown to be activated in many cell types by cellular stressors, like UV radiation, osmotic stress, and oxidative stress. Activation of p38 MAP kinase have been shown to result in changes in transcription, protein synthesis, cell surface receptor expression, and cytoskeletal structure, ultimately affecting cell survival or leading to the programmed cell death. Our results demonstrate that oxalate and COM-crystals are toxic to the renal epithelial cells.[2,3] We also demonstrate activation ofA p38 MAP kinase signal transduction pathway in renal cells after exposure to oxalate or to the COM-crystals.

Previous studies have shown that exposure of renal epithelial cells to oxalate as well as COM-crystals resulted in re-initiation of the DNA synthesis and alterations in gene expression.[2] In present studies we confirmed these findings. Moreover, we demonstrate inhibition of oxalate as well as COM-stimulated DNA synthesis by SB203580, a specific inhibitor of p38 MAP kinase signaling pathway. These data suggest critical requirement for p38 MAP kinase signal transduction in oxalate as well as COM-crystal-induced re-initiation of the DNA synthesis. Whether or not these pathways mediate oxalate-induced apoptosis remains to be determined. It is important to point out here that the renal epithelial cells are constantly exposed to varying concentrations of nephrotoxins, including oxalate and crystalline waste products, and undergo apoptosis and renewal on a regular basis. These considerations lead us to hypothesize that p38 MAPK and JNK signal transduction pathways play a central role in the apoptosis and renewal of renal epithelium. Studies currently in progress in our laboratory are focused in evaluating the relative contribution of these two signal transduction pathways in mediating oxalate and COM-crystal-induced changes in gene expression in the renal epithelial cells.

ACKNOWLEDGMENTS

This work was supported in part by Grant DK-RO1-54084 from the National Institutes of Health.

REFERENCES

1. JOHNSON, G.L. & R. LAPADAT. 2002. Mitogen-activated protein kinase pathways mediated by ERK, JNK, and p38 protein kinases. Science **298:** 1911–1912.
2. KOUL, H., L. RENZULLI, G. NAIR, *et al.* 1994. Oxalate-induced initiation of DNA synthesis in LLC-PK1 cells, a line of renal epithelial cells. Biochem. Biophys. Res. Commun. **205:** 1632–1637.
3. KOUL, H., M. MENON & C. SCHEID. Oxalate and renal tubular cells: a complex interaction. Ital. J. Electrol. Metab. **10:** 67–74.
4. CHATURVEDI, L.S., S. KOUL, A. SEKHON, *et al.* 2002. Oxalate selectively activates p38 mitogen-activated protein kinase and c-Jun N-terminal kinase signal transduction pathways in renal epithelial cells. J. Biol. Chem. **277:** 13321–13330.
5. KOUL, H.K., M. MENON, L. CHATURVEDI, *et al.* 2002. COM-crystals activate p38 mitogen activated protein kinase signal transduction pathway in renal epithelial cells. J. Biol. Chem. **277:** 36845–36852.

Calcium-Dependent Apoptotic Gene Expression in Cerulein-Treated AR42J Cells

JI HOON YU, HYEYOUNG KIM, AND KYUNG HWAN KIM

Department of Pharmacology and Institute of Gastroenterology, Brain Korea 21 Project for Medical Science, Yonsei University College of Medicine, Seoul 120-752, Korea

ABSTRACT: Elevated Ca^{2+} concentrations within the pancreatic acinar cells represent a risk factor for the development of acute pancreatitis. Apoptosis is an important characteristic of pancreatitis, with induction of apoptotic genes and intraceullar increase of calcium, endonucleases, and protease. The present study, which aims to investigate whether (1) cerulein induces apoptotic gene expression (bax, bid, p53) in pancreatic acinar AR42J cells and (2) cerulein-induced gene expression is mediated by intracellular Ca^{2+}, monitored the gene expression profile in the cells treated with the Ca^{2+} chelator BAPTA-AM. Results showed that cerulein (10^{-7} M) evoked an initial peak Ca^{2+} signal; a further Ca^{2+} signal was induced with second treatment of cerulein. Cerulein-induced Ca^{2+} signal could not be detected in the cells treated with the Ca^{2+} chelator BAPTA-AM. Cerulein dose-dependently induced apoptosis, determined by DNA fragmentation and pro-apoptotic bid expression in AR42J cells. Cerulein induced bid, bax, and p53 mRNA expression, which was inhibited in the cells treated with cerulein and cultured in the presence of BAPTA-AM. The present results suggest that increase in the free cytosolic Ca^{2+} may be the upstream event of apoptotic gene (bax, bid, p53) expression, which contribute to cerulein-induced apoptosis in pancreatic acinar cells.

KEYWORDS: calcium; apoptotic gene; cerulein; AR42J cells

INTRODUCTION

Pancreatic autodigestion is the underlying pathophysiologic mechanism of acute pancreatitis. This autodigestion would require a premature and intracellular activation of pancreatic proteases. Recent investigation suggested that elevated Ca^{2+} concentrations within the pancreatic acinar cells represent a risk factor for the development of acute pancreatitis and the premature activation of the protease precursor trypsinogen.[1] In addition, apoptosis has been observed in human pancreatitis and experimental models such as cerulein pancreatitis in mice.[2] Apoptosis is associated with induction of apoptotic genes and intracellular increase of calcium, endonucleases, and protease. There is rapid elevation of cytosolic Ca^{2+}, resulting in activation of proteolytic enzymes and mitochondrial damage, eventually leading to cell membrane disruption and cell death. Mitochondrial Ca^{2+} accumulation may lead

Address for correspondence: Hyeyoung Kim, Department of Pharmacology, Yonsei University College of Medicine, Seoul 120-752, Korea. Voice: +82-2-361-5232; fax: +82-2-313-1894. kim626@yumc.yonsei.ac.kr

Ann. N.Y. Acad. Sci. 1010: 66–69 (2003).
doi: 10.1196/annals.1299.009

to opening of the mitochondrial permeability transition pore with release of pro-apoptotic factor from the intermembrane space followed by apoptosis.[3] Bax and bid have been reported as potent inducers of apoptosis.[4,5] p53 activates both cell-cycle arrest and apoptosis in most types of cells.[6] Recent report showed that these pro-apoptotic genes bax and bid are transcriptionally regulated by p53.[7] Present study aims to investigate whether cerulein induces apoptotic gene expression (bax, bid, p53) in pancreatic acinar AR42J cells, and whether cerulein-induced gene expression is mediated by intracellular Ca^{2+}. This was accomplished by monitoring the gene expression profile in the cells treated with Ca^{2+} chelator BAPTA-AM.

METHODS

AR42J cells (pancreatoma, ATCC CRL 1492) were cultured in Dulbecco's modified Eagle's medium supplemented with 10% fetal bovine serum and antibiotics.

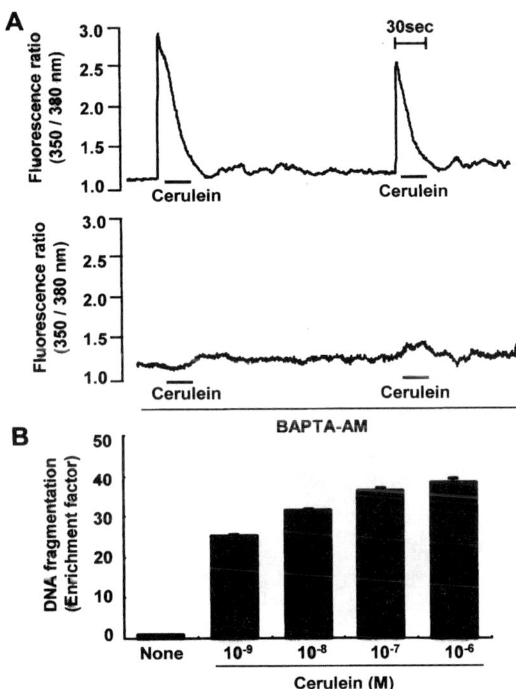

FIGURE 1. Cerulein-induced Ca^{2+} signaling and DNA fragmentation in AR42J cells. The cells were perfused with cerulein (10^{-7} M). The free cytosolic Ca^{2+} was determined by the Fura-2 fluorescences excited at 350 and 380 nm. To remove cytosolic Ca^{2+} from the cells, Ca^{2+} chelator BAPTA-AM (10 μM) was perfused before and during cerulein addition into the perfusate buffer. Ca^{2+} signaling was monitored and expressed as the fluorescence ratio of 350/380 nm (**A**). For an apoptotic index, the cells were treated with various concentrations of cerulein (10^{-9}–10^{-6} M) for 24 h. DNA fragmentation as the content of nucleosome-bound DNA was quantified by ELISA (**B**). None = the cells treated without cerulein.

For the functional assay of AR42J cells, the cells were perfused with cerulein (10^{-7} M). The free cytosolic Ca^{2+} was determined by the Fura-2 fluorescences excited at 350 and 380 nm. To remove cytosolic Ca^{2+} from the cells, the Ca^{2+} chelator BAPTA-AM (10 μM, 1,2-bis(*o*-aminophenoxy)ethane-N,N,N′,N′-tetraacetic acid tetra (acetoxymethyl) ester) was perfused before and during cerulein addition into the perfusate buffer. Ca^{2+} signaling was monitored and expressed as the fluorescence ratio of 350/380 nm. For the apoptotic indices, the cells were treated with various concentrations of cerulein (10^{-9}–10^{-6} M) for 24 h (DNA fragmentation) or 8 h (bid mRNA expression). DNA fragmentation as the content of nucleosome-bound DNA was quantified by enzyme-linked immunosorbant assay (ELISA). For mRNA expression of apoptotic genes (Bid, Bax, p53), the total RNA was isolated from the cells treated with or without cerulein (10^{-7} M) and cultured in the presence or absence of BAPTA-AM for 8 hours. mRNA expression was determined by RT-PCR and standardized by co-amplification with the housekeeping gene GAPDH as an internal control.

RESULTS AND DISCUSSION

A relatively high concentration of cerulein (10^{-7} M) evoked an initial peak Ca^{2+} signal and a further Ca^{2+} signal was induced by a second treatment of cerulein (FIG. 1A, upper panel). Cerulein-induced Ca^{2+} signal could not be detected in the

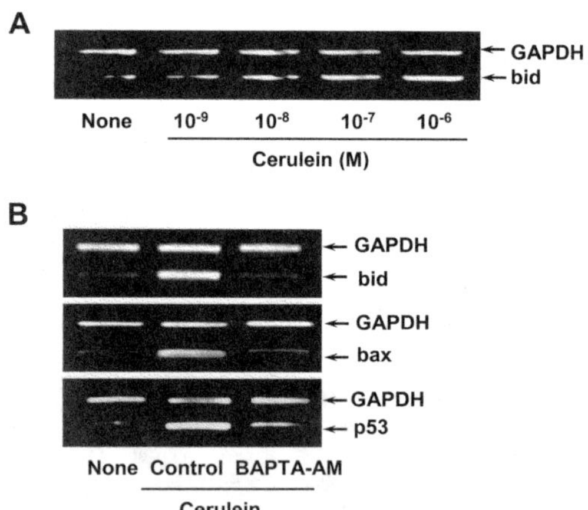

FIGURE 2. Cerulein-induced bid expression and the effect of BATA-AM on apoptotic gene expression in AR42J cells. The cells were treated with various concentrations of cerulein (10^{-9}–10^{-6} M) for 8 h. Bid mRNA expression was determined by RT-PCR (**A**). For the effect of BAPTA-AM on mRNA expression of apoptotic genes (bid, bax, p53), the cells were treated with or without cerulein (10^{-7} M) and cultured in the presence or absence of BAPTA-AM (10 μM) for 8 h. mRNA expression was determined by RT-PCR and standardized by co-amplification with the housekeeping gene GAPDH as an internal control (**B**). None = the cells treated without cerulein; control = the cells treated with cerulein and cultured in the absence of BAPTA-AM.

cells treated with the Ca^{2+} chelator BAPTA-AM (FIG. 1A, lower panel). These results confirm that AR42J cells have normal Ca^{2+}-releasing function with cerulein treatment. Cerulein dose-dependently induced apoptosis, determined by DNA fragmentation (FIG. 1B), and pro-apoptotic bid expression (FIG. 2A) in AR42J cells. Cerulein (10^{-7} M) upregulated bid, bax, and p53 mRNA expression in AR42Jcells at 8 hours. BAPTA-AM inhibited upregulation of bid, bax, p53 mRNA expression. This result demonstrates that apoptotic gene expression by cerulein may be mediated by intracellular Ca^{2+} in AR42J cells. At present it is not clear whether pro-apoptotic gene bax and bid are transcriptionally regulated by p53. Even though increased mitochondrial Ca^{2+} may open mitochondrial permeability transition pore and release pro-apoptotic proteins from the intermembrane space,[3] the present study suggests the possibility that free cytosolic Ca^{2+} may directly induce apoptotic gene expression in pancreatic acinar cells in response to the secretagogues. Our previous study showed the relation between nuclear transcription factor NF-κB and cytokine gene expression in freshly isolated rat pancreatic acinar cells.[8] Since inflammation and apoptosis are hallmarks of pancreatitis, further study should be performed to determine the time-course of apoptotic gene expression and its relation to the activation of transcription factors in pancreatic acinar cells treated with cerulein with the aim of investigating the transcriptional regulation in apoptotic gene expression.

ACKNOWLEDGMENT

This study was supported by a grant (to H. Kim) from the Korea Ministry of Health and Welfare.

REFERENCES

1. KRUGER, B., E. ALBRECHT & M.M. LERCH. 2000. The role of intracellular calcium signaling in premature protease activation and the onset of pancreatitis. Am. J. Pathol. **157:** 43–50.
2. NIEDERAU, C., L.D. FERRELL & J.H. GRENDELL. 1985. Cerulein-induced acute necrotizing pancreatitis in mice: protective effects of proglumide, benzotript, and secretin. Gastroenterology **88:** 1192–1204.
3. CROMPTON, M. 1999. The mitochodriil permeability transition pore and its role in the cell death. Biochem. J. **341:** 233–249.
4. WILDHABER, B.E., K.N. LYNN, H. YANG & D.H. TEITELBAUM. 2002. Total parenteral nutrition-induced apoptosis in mouse intestinal epithelium: regulation by the Bcl-2 protein family. Pediatr. Surg. Int. **18:** 570–575.
5. ESPOSTI, M.D. 2002. The roles of Bid. Apoptosis **7:** 433–440.
6. RYAN, K.M., A.C. PHILLIPS & K.H. VOUSDEN. 2001. Regulation and function of the p53 tumor suppressor protein. Curr. Opin. Cell Biol. **13:** 332–337.
7. SAX, J.K., P. FEI, M.E. MURPHY, *et al.* 2002. Bid regulation by p53 contributes to chemosensitivity. Nature Cell Biol. **4:** 842–849.
8. YU, J.H, J.W. LIM, W. NAMKUNG, *et al.* 2002. Suppression of cerulein-induced cytokine expression by antioxidants in pancreatic acinar cells. Lab. Invest. **82:** 1359–1368.

Effects of Intracellular Calcium on Cell Survival and the MAPK Pathway in a Human Hormone-Dependent Leukemia Cell Line (TF-1)

ÁGOTA APÁTI,[a] JUDIT JÁNOSSY,[c] ANNA BRÓZIK,[b] AND MÁRIA MAGÓCSI[b]

[a]*Membrane Research Group of the Hungarian Academy of Sciences, Nádor u. 7, H-1051 Budapest, Hungary*

[b]*Department of Cell Metabolism, National Medical Center, Budapest, Hungary*

[c]*Institute of Enzymology, Hungarian Academy of Sciences, Budapest, Hungary*

ABSTRACT: Changes in the cytoplasmic calcium concentration ($[Ca^{2+}]_i$) regulate a wide variety of cellular processes. Here we demonstrate that increased $[Ca^{2+}]_i$ was able to induce hormone-independent survival and proliferation, as well as to evoke apoptosis in human myelo-erythroid GM-CSF/IL-3 dependent leukemia cells (TF-1). Cellular responses induced by elevated $[Ca^{2+}]_i$ depended on the duration and amplitude of the calcium-signal. Moderate or high, but transient, elevation of $[Ca^{2+}]_i$ caused a transient, biphasic activation of ERK1/2 and protected cells from hormone withdrawal–induced apoptosis.[1] In contrast, high and long-lasting elevation of $[Ca^{2+}]_i$ led to sustained activation of the ERK1/2 kinases and apoptosis of TF-1 cells. Our data suggest that a time-dependent action of the MAPK pathway works as a decision-point between cell proliferation and apoptosis.

KEYWORDS: cytoplasmic calcium concentration; ERK1/2; c-Fos

INTRODUCTION

TF-1 cells have been reported to undergo apoptosis upon hormone deprivation. Activation of the MEK/ERK/MAP kinase-signaling pathway by GM-CSF rescues the TF-1 cells from hormone withdrawal–induced cell death.[2] The Ras/Raf/MEK/ERK pathway activation induces a variety of cellular responses that depend on the effector, the intensity, and amplitude of its activation and the cell type. This pathway is activated not only by specific receptor activation but also modulated by $[Ca^{2+}]_i$.[3] In order to examine the effect of calcium-signals on TF-1 cell survival and proliferation we have applied various $[Ca^{2+}]_i$ mobilizing agents, such as ER/SR Ca^{2+}-ATPase inhibitors and Ca^{2+} ionophores.

Address for correspondence: Ágota Apáti, Membrane Research Group of the Hungarian Academy of Sciences, Nádor u. 7, H-1051 Budapest, Hungary.
apati@biomembrane.hu

Ann. N.Y. Acad. Sci. 1010: 70–73 (2003). © 2003 New York Academy of Sciences.
doi: 10.1196/annals.1299.010

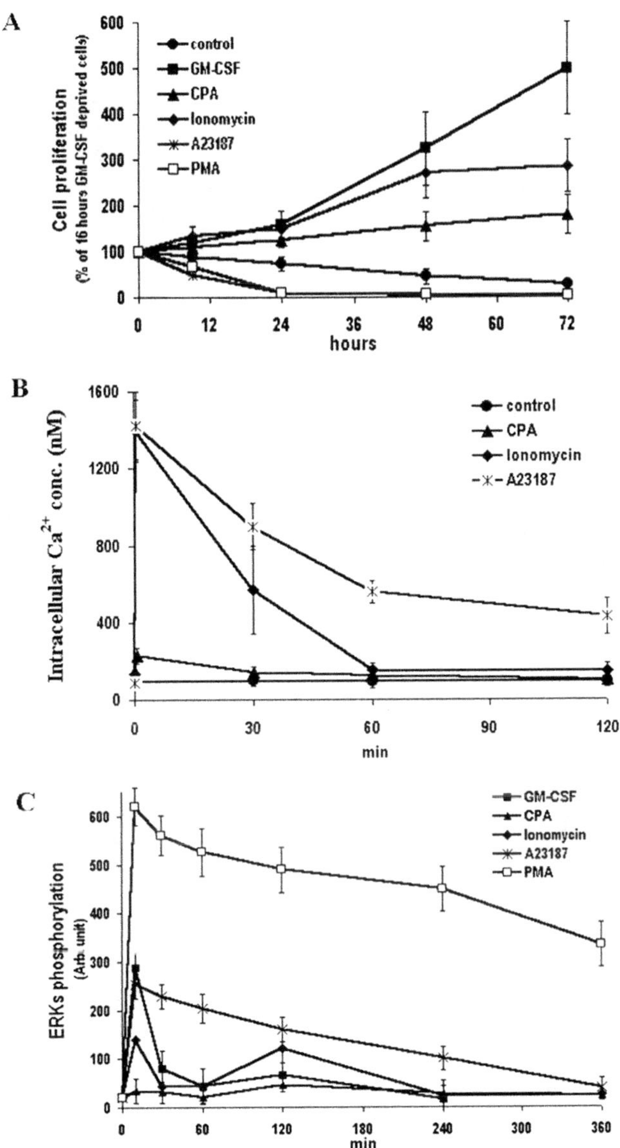

FIGURE 1. Changes in cell proliferation, Ca^{2+} concentration, and ERK1/2 activation in TF-1 cells.

RESULTS AND DISCUSSION

TF-1 cells were synchronized by GM-CSF hormone deprivation and thereafter treated with different amounts of cyclopiasonic acid (CPA), ionomycin, or A23187. We could induce cell survival or death depending on the concentration of these reagents. In FIGURE 1A relative cell growth was determined by trypan blue exclusion in control and in hormone-deprived cells treated with CPA (7.5 μM), ionomycin (1 μM), A23187 (1 μM), GM-CSF (2.5 ng/mL), or PMA (10 nM). Control cells stopped growing and lost their viability within 72 hours. Treatment of the cells with A23187 or PMA produced rapid cell death. In contrast, ionomycin and CPA protected against cell death and ionomycin (similarly to GM-CSF) even produced a significant cell growth. In order to elucidate the processes behind the rapidly decreasing cell viability of A23187- and PMA-treated cells the signs of apoptosis were examined. Poly-ADP-ribose-polymerase (PARP) cleavage studies revealed that PMA as well as A23187 treatment triggered apoptosis. In contrast, no cleavage of PARP was detected in cells treated with GM-CSF, CPA, and ionomycin for up to 48 hours. Alterations of $[Ca^{2+}]_i$ were followed by measuring the intracellular calcium–dependent fluorescence of Fura-2 for 2 hours after the addition of the above reagents (FIG. 1B). Hormone-deprived GM-CSF- and PMA-treated cells showed no significant changes in $[Ca^{2+}]_i$. Both CPA and ionomycin induced a temporary increase, whereas A23187 induced a sustained increase in $[Ca^{2+}]_i$.

The phosphorylation of ERK1/2 kinases was monitored using phospho-specific antibodies. As shown in FIGURE 1(C), two phases of ERK1/2 phosphorylation were observed in the case of GM-CSF, ionomycin, and CPA, all returning to control levels

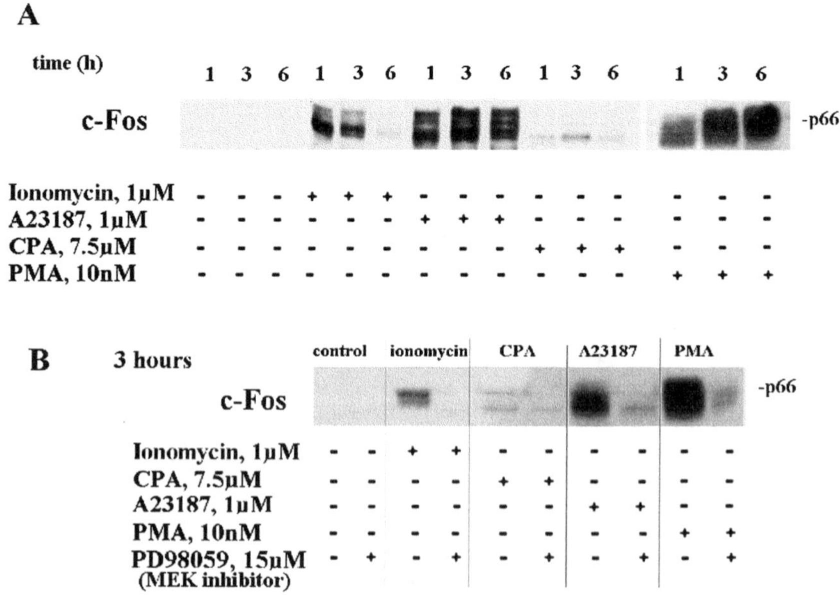

FIGURE 2. Induction and inhibition of c-Fos expression in TF-1 cells.

within 4 hours. In contrast, A23187 and PMA caused strong and sustained ERK1/2 kinase activation, which did not return to control level within the first 6 hours. ERK1/2 phosphorylation could be inhibited by specific MEK inhibitors in all cases, while in the case of $[Ca^{2+}]_i$ mobilizing agents pretreatment with EGTA alone was effective. Thus, CPA, ionomycin, and A23187 induced a Ca- and MEK-dependent ERK1/2 activation. Moreover, the transient $[Ca^{2+}]_i$ elevation accompanied with transient ERK1/2 activation and the sustained $[Ca^{2+}]_i$ elevation caused sustained activation of ERK1/2. We have examined the expression of c-Fos transcription factor, which could be stimulated in an ERK1/2-dependent manner and is thought to be involved in survival and proliferation of hematopoietic cells.[4] According to our experiments, in all cases c-Fos protein level (FIG. 2A) was increased in an ERK-dependent manner (FIG. 2B) following similar kinetics as ERK1/2 activation. This observation correlates well with the recent findings of Murphy and colleagues[5] that c-Fos functions as a sensor for ERK1/2 signal duration. Here we showed within the same cell line that $[Ca^{2+}]_i$ elevation was able to promote different cellular responses via the MEK/ERK/c-Fos pathway depending on the duration and amplitude of these signals.

ACKNOWLEDGMENTS

This work was supported by the Hungarian Research Fund (OTKA T029291), by the Scientific Research Council, Ministry of Health (ETT 206/2001), and by NRDP 1/024, Ministry of Education, Hungary.

REFERENCES

1. APATI, A., J. JANOSSY, A. BROZIK, *et al.* 2003. J. Biol. Chem. **278:** 9235–9243.
2. KOLONICS, A., A. APATI, J. JANOSSY, *et al.* 2001. Cell Signal. **13:** 743–754.
3. AGELL, N., O. BACHS, N. ROCAMORA & P. VILLALONGA. 2002. Cell Signal. **14:** 649–654.
4. YUAN, P.X., D.L. HUANG, Y.M. JIANG, *et al.* 2001. J. Biol. Chem. **276:** 31674–31683.
5. MURPHY, L.O., S. SMITH, R. CHEN, *et al.* 2002. Nat. Cell. Biol. **4:** 556–564.

Cytosolic and Endoplasmic Reticulum Ca^{2+} Concentrations Determine the Extent and the Morphological Type of Apoptosis, Respectively

CLAUDIA CERELLA,[a] MARIA D'ALESSIO,[a] MILENA DE NICOLA,[a] ANDREA MAGRINI,[b] ANTONIO BERGAMASCHI,[b] AND LINA GHIBELLI[a]

[a]*Dipartimento di Biologia, Università di Roma Tor Vergata, via della Ricerca Scientifica, 00133 Rome, Italy*

[b]*Cattedra di Medicina del Lavoro, Università di Roma Tor Vergata, 00133 Rome, Italy*

ABSTRACT: During apoptosis, an increase in cytosolic Ca^{2+} concentration ($[Ca^{2+}]_c$) accompanies the depletion of endoplasmic reticulum (ER). The actual roles of each of the two events in apoptosis are difficult to understand. In this work, we have modulated the basal $[Ca^{2+}]_c$ and the thapsigargin (THG)-dependent reticular flux (i.e., by chelating extracellular Ca^{2+} or by modulating intracellular Ca^{2+} by 3-aminobenzamide [3-ABA]). We have found that these treatments alter these Ca^{2+} parameters in a differential way and, accordingly, affect apoptosis differentially. We have found that the increase in $[Ca^{2+}]_c$ is related to the extent of apoptosis, whereas the ER depletion affects the apoptotic nuclear morphology by shifting it towards the cleavage mode.

KEYWORDS: apoptosis; calcium; endoplasmic reticulum; apoptotic morphology

INTRODUCTION

Alterations in intracellular Ca^{2+} homeostasis are known to be involved in promoting apoptosis[1]: a sustained increase in cytosolic Ca^{2+} concentration ($[Ca^{2+}]_c$) accompanies apoptosis in several cell types. Also, ER Ca^{2+} pool depletion has been reported to be involved in apoptosis.[2] Since the two events occur simultaneously, it is quite difficult to understand the precise role of each of the two events. Here, we try to uncouple them in order to analyze their role in apoptosis

MATERIAL AND METHODS

Cell Culture and Treatments

U937 cells, human tumoral monocytes, were cultured as described.[1,3] Apoptosis was induced with 10 μg/mL puromycin (PMC) and 100 μg/mL etoposide (VP16).[3]

Address for correspondence: Claudia Cerella, Dipartimento di Biologia, Università di Roma Tor Vergata, via della Ricerca Scientifica, 00133 Rome, Italy. Fax: +39062023500. sanclacer@libero.it

Ann. N.Y. Acad. Sci. 1010: 74–77 (2003). © 2003 New York Academy of Sciences. doi: 10.1196/annals.1299.011

Apoptosis was analyzed and evaluated as described.[1,3] Ten minutes before the induction of apoptosis 650 μM EGTA or 5 mM 3-ABA were added. Five micromolar BAPTA-AM were added to the cells for 20 min at 37°C. Cells were washed, resuspended in fresh medium, and apoptosis was induced.

Fluo-3 Staining and Flow Cytometric Analysis

U937 cells were stained as described.[1] $[Ca^{2+}]_c$ measurements were performed as described.[1] Fluorescence values were converted in $[Ca^{2+}]$ values as described.[4]

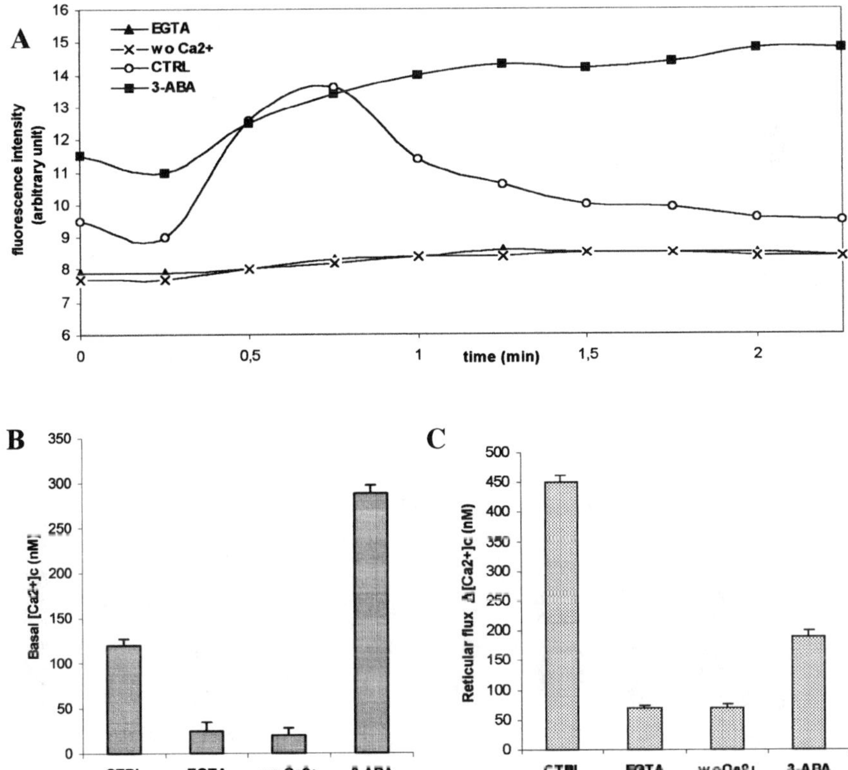

FIGURE 1. Modulations of Ca^{2+} parameters by the unavailability of extracellular Ca^{2+} and 3-ABA. U937 cells were stained with Fluo3-AM.[1] Then, they were incubated for 30 min with 650 μM EGTA, 1.5 h with 5 mM 3-ABA, or depleted of Ca^{2+} for 30 min. (**A**) Time course of Ca^{2+} fluxes of treated (10 nM THG) and untreated U937 cells to elicit store emptying (reticular flux). (**B** and **C**) Values for the basal $[Ca^{2+}]_c$ and the extent of reticular flux (Δ) induced by THG in the same time-course samples. Reticular Δ was calculated as the difference between the basal $[Ca^{2+}]_c$ and the maximum $[Ca^{2+}]_c$ measured after THG addition. The fluorescence values were converted in $[Ca^{2+}]_c$ values according to the equation of Grynkiewicz.[4] The basal $[Ca^{2+}]_c$ values reported are the average of more than 10 independent experiments ± SEM.

RESULTS

Modulations of Ca^{2+} Parameters by the Unavailability of Extracellular Ca^{2+} and 3-ABA

We have analyzed the alteration of Ca^{2+} parameters (the basal $[Ca^{2+}]_c$ and the thapsigargin [THG]-dependent reticular flux; FIG. 1) by flow cytometry, upon chelation of extracellular Ca^{2+} with EGTA or in Ca^{2+}-free medium.

The unavailability of extracellular Ca^{2+} caused a decrease in the basal $[Ca^{2+}]_c$ (FIG. 1). Accordingly, also intraluminal ER Ca^{2+} was altered since THG was able to mobilize only weak fluxes. We have observed that the presence of 3-ABA, an ADP-ribosylation inhibitor, is able to modulate intracellular Ca^{2+} with a mechanism under investigation (Cerella and colleagues, submitted for publication). Here, we show that

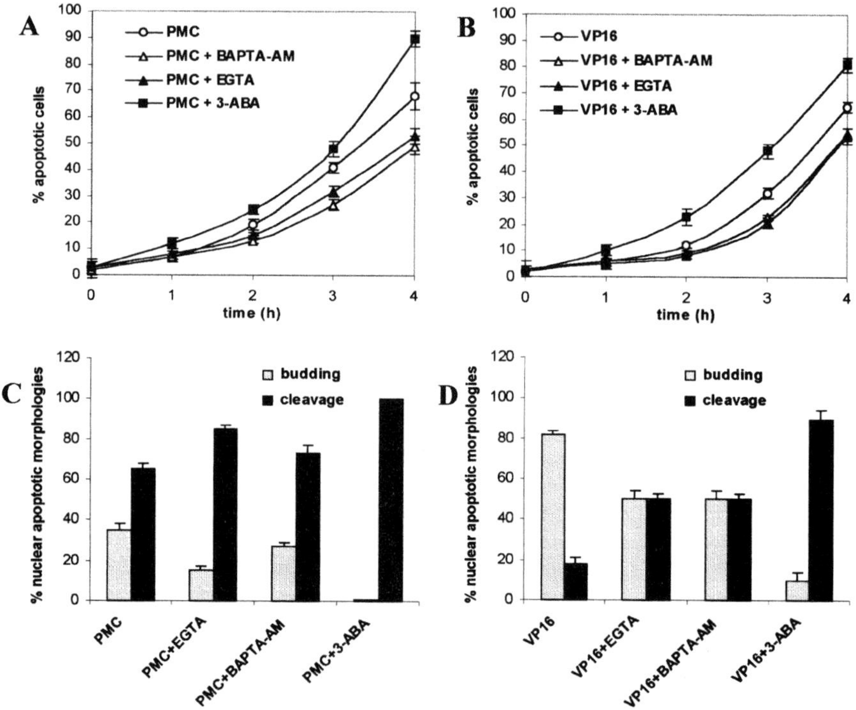

FIGURE 2. Ca^{2+} homeostasis modulators affect the extent and the nuclear morphologies of apoptosis. U937 cells were induced in apoptosis with 10 μg/mL puromycin (PMC; **A** and **C**) or 100 mM etoposide (VP16; **B** and **D**) and in presence of 650 mM EGTA, 5 μM BAPTA-AM, or 5 mM 3-aminobenzamide (3-ABA). **A** and **B** show the kinetics of apoptosis: 3-ABA increases the extent of apoptosis, whereas EGTA and BAPTA-AM decrease it. **C** and **D** show the analysis of the apoptotic nuclear morphologies at 3 h of apoptogenic treatments: all the co-treatments shift the nuclear morphologies to cleavage. Apoptosis and nuclear morphologies were evaluated according to methods described earlier.[1,3] Values reported are the average of more than 10 independent experiments ± SEM.

U937 incubated with 3-ABA for 1.5 h had a very high basal $[Ca^{2+}]_c$. The analysis of the intensity of reticular flux (FIG. 1) in the same samples shows that in cells incubated with 3-ABA, the THG-dependent increase in $[Ca^{2+}]_c$ is much lower.

Since the unavailability of extracellular Ca^{2+} and 3-ABA affects the basal $[Ca^{2+}]_c$ in the same way, but in an opposite way for the intralumen ER Ca^{2+}, all these treatments can be used to discriminate the effective role of $[Ca^{2+}]_c$ increase versus ER Ca^{2+} pool depletion in apoptosis.

Ca^{2+} Homeostasis Modulators Affect the Extent and the Nuclear Morphologies of Apoptosis

We induced apoptosis in the presence of Ca^{2+} chelating agents (EGTA and BAPTA-AM) and 3-ABA. EGTA and BAPTA-AM exert a protective effect on PMC- and VP16-induced apoptosis, whereas 3ABA has the opposite effect of increasing apoptosis (FIG. 2A and B). In contrast, the analysis of the apoptotic morphologies (FIG. 2C and D) of these samples shows that chelators exert an effect similar to 3-ABA: indeed, all agents shift the apoptotic morphologies to the cleavage mode. Thus, the ER depletion seems to be important for determining apoptotic morphologies, whereas the $[Ca^{2+}]_c$ changes affect the extent of apoptosis.

DISCUSSION

We have shown that the increase in $[Ca^{2+}]_c$ has a pro-apoptotic meaning in U937, whereas its decrease has an anti-apoptotic role. Instead, ER depletion is responsible for the shift to cleavage nuclear morphology. On the one hand, the $[Ca^{2+}]_c$ increase may cause the activation of Ca^{2+}-sensitive targets involved in apoptosis (i.e., proteases and endonucleases). On the other hand, an abnormal release of Ca^{2+} from ER, with the consequent ER Ca^{2+} pool depletion, may compromise ER structural and functional integrity, leading to ER vesiculation. Since cleavage is always accompanied by the apoptotic blebs formation, it is likely that the loss of ER integrity may direct (via cytoskeleton rearrangements?) the cleavage/blebbing pathway.

REFERENCES

1. FANELLI, C., S. COPPOLA, R. BARONE, *et al.* 1999. Magnetic fields increase cell survival by inhibiting apoptosis via modulation of Ca^{2+} influx. FASEB J. **13:** 95–102.
2. LAM, M., G. DUBYAK, L. KEN, *et al.* 1994. Evidence that Bcl2 represses apoptosis by regulating endoplasmic reticulum-associated Ca^{2+} fluxes. Proc. Natl. Acad. Sci. USA **91:** 6569–6573.
3. NOSSERI, C., S. COPPOLA & L. GHIBELLI. 1994. Possible involvement of poly-ADP-ribosyl transferase in triggering stress-induced apoptosis on U937 cells. Exp. Cell Res. **212:** 367–373.
4. GRYNKIEWICZ, G., M. POENIE & R.Y. TSIEN. 1985. A new generation of Ca^{2+} indicators with greatly improved fluorescence properties. J. Biol. Chem. **260:** 3440–3450.

G-CSF Modulates LPS-Induced Apoptosis and IL-8 in Human Microvascular Endothelial Cells

Involvement of Calcium Signaling

E.M. SCHNEIDER,[a] I. LORENZ,[a] X. MA,[a] AND M. WEISS[b]

[a]*Sektion Experimentelle Anaesthesiologie and* [b]*Department of Clinical Anaesthesiology, University Clinic, 89075 Ulm, Germany*

ABSTRACT: Microvascular endothelial cells (mECs) circulate at higher numbers in patients with severe sepsis and hemophagocytic syndromes. Although these blood mECs might stem from damaged microvasculature, they are perfectly viable and lead to the establishment of cell lines. Such mECs were cultured in low-dose human serum pools (0.5%) and MEM-alpha medium. Antigenic profiling revealed the expression of CD36, factor VIIIa, CD95-ligand, and CD44, but also CD146. We studied the antioxidative effect of the hematopoietic growth factor G-CSF[1] after *in vitro* stimulation with LPS from *E. coli* 0111:B4; the growth factor appeared to exhibit a protective effect on organ function in patients with SIRS. mECs were stimulated with 1 μg/mL of LPS for 24 h and 48 h with and without G-CSF (3×10^3 U/mL) preincubation. After 24 h, supernatants of the stimulated mEC were tested for IL-8 by ELISA, and cells were tested for hemoxygenase-1 (HO-1, Hsp32) by immunohistochemistry and flow cytometry using OSA110 (mAb, Stressgene). Stimulation with LPS upregulated IL-8 by a factor of 2 to 10 in mEC. Preincubation with G-CSF markedly downregulated the LPS-induced IL-8 secretion (20–50%), but IL-6 production was not affected. Upon 48 h of LPS stimulation, mECs developed massive signs of apoptosis and concomitant caspase 3 activation. Caspase 3 activity induced by LPS (24 h) or by staurosporin (6 h) was found to be dramatically downregulated by the G-CSF preincubation protocol.

KEYWORDS: G-CSF; microvasculature; sepsis, endothelial cells; systemic inflammatory response syndrome

INTRODUCTION

The disruption of the endothelial cell barrier is clinically manifest when severe sepsis and septic shock culminate in multiorgan dysfunction (MODS) and failure (MOF). A systemic inflammatory response syndrome (SIRS) is the first sign of the subsequent hyperinflammatory syndrome, which is not only characteristic of bacterial sepsis but is also present in a disease called hemophagocytic lymphohistiocyto-

Address for correspondence: E.M. Schneider, Sektion Experimentelle Anaesthesiologie, Department of Clinical Anaesthesiology, University Clinic, 89075 Ulm, Germany.
marion.schneider@medizin.uni-ulm.de

**Ann. N.Y. Acad. Sci. 1010: 78–85 (2003). © 2003 New York Academy of Sciences.
doi: 10.1196/annals.1299.012**

sis,[1] when hyperinflammation is related to overactivation of the phagocytes. Severe sepsis is accompanied with the systemic effects by TNF-α–secreting macrophages leading to vasodilation, loss of vascular integrity and barrier dysfunction. Finally, activation[2] and death of the endothelial cells in vital organs manifests in catecholamine resistance during septic shock.

MATERIALS AND METHODS

Microvascular endothelial cell lines (mECs) were enriched by 24 h of adherence using Ficoll-isolated mononuclear cells routinely analyzed for immune cell function from patients with hemophagocytic lymphohistiocytosis[1] and septic shock who are admitted to our university hospital intensive care unit. This study was approved by the Ethics Committee of the Ulm University Medical School and conducted according to the principles of the Declaration of Helsinki. Confluent endothelial cell layers were obtained after 3–4 weeks of culture in MEM-α without nucleotides, but with additional HEPES buffer, 25 mM, a stable glutamine compound (Glutamax, 2 mM, Gibco-BRL, Karlsruhe, Germany) and 0.5% pre-screened and heat-inactivated human serum. Antibiotics were either gentamycin (60 μg/mL) or penicillin/streptomycin (10,000 U/mL and 10 ng/mL), respectively. Following the establishment of endothelial cell islands, the medium was changed every 72 hours. Endothelial cells were detached by either 0.5% trypsin and 0.02% EDTA (Sigma, Deisenhofen, Germany) for 2 min, or by incubating the cell layer in 0.005% sodium EDTA at 37°C for 30 minutes. The latter treatment was especially useful to determine surface marker receptors as well as adhesion molecules by flow cytometry.

The following monoclonal antibodies were applied: anti-factor VIIIa, CD146, CD36, CD49d, CD49e, CD49f, PAL-E, G-CSF receptor (CD114), GM-CSF receptor-alpha chain (CD116) (Coulter/Immunotech, Krefeld, Germany). Antibodies recognizing β2-microglobulin (β2-m) were from ICN (Lisle, France), CD44 (Serotec, Hamburg, Germany), CD54, ELAM-1 from Becton-Dickinson (Heidelberg, Germany). All antibodies were diluted in either 0.002% thimerosal (Merck, Darmstadt, Germany) or 0.1% NaN_3 in 0.1% HSA (human serum albumin, Troponwerke, Köln, Germany) in PBS to obtain the requested concentration of antibodies for labeling. Following incubation at 4°C on ice, cells were washed and the secondary FITC-labeled rabbit anti-mouse F(ab′)2 recognizing heavy and light chains of mouse IgG (Cappel Lab, Rockland, purchased via Biomol, Hamburg, Germany, and pre-absorbed with human serum proteins) were applied with 2 μg/mL for 30 min followed by two washes with 0.1% HSA, 0.1% NaN_3 in PBS. To determine cell viability, the cell suspension was stained with 25 μg/mL of propidium iodide (Sigma, Deisenhofen, Germany) before analysis by flow cytometry using a FACScalibur and CellQuest software (Becton-Dickinson Europe, Heidelberg, Germany).

Cytokine Secretion Patterns

Constitutive expression of profiles of IL-6 and IL-8 by ECs were determined by ELISA (R&D Systems, Heidelberg, Germany). Confluent EC layers were washed with fresh medium and factors released within the next 24 h were determined. In G-CSF pre-incubation experiments monolayers were washed, incubated with 3000 U/

mL of rhG-CSF (Neupogen). Autochtonous G-CSF was determined by a G-CSF–specific ELISA(Amersham, Braunschweig, Germany).

CPP 32 Activity

Caspase 3 (CPP32 activity was determined by the ApoAlert Assay (Clontech, BD-Europe, Heidelberg, Germany) using caspase 3–specific AMC (7-amino-4-methylcoumarin)–labeled Z-DEVD-(Asp-Glu-Val-Asp) peptides. Upon caspase 3–specific cleavage, the substrate yields a fluorescent blue product.

TABLE 1. Representative surface marker phenotype of a microvascular endothelial cell line from a patient with HLH

Surface Antigen	Expression Density (MFI)[a]	
Control	0	
MHC-class I associated β2-microglobulin	73,018	Transplantation antigen class I
MHC-class II	1,890	Transplantation antigen class II
Pro-collagen type I	19,614	Matrix
Collagen type IV	414	Matrix specific for basal endothelium
CD49d	134	Very late function antigen D (integrin a4) binds to VCAM, MAdCAM
CD49f	0	Very late function antigen F (integrin α6, GpIc) binds to platelets
CD62 E (ELAM-1)	17,321	Endothelial leukocyte adhesion molecule 1 on activated endothelial cells binds to sialyl Lewis x, a and CD162
EN7/44	0	Antigen on migrating endothelial cells
Pal-E	853	
CD34	4,136	Stem cell antigen expressed on endothelial cells, binds to L-selectin
CD44	32,800	Adhesion molecule (heparan sulphate proteoglucan)
GM-CSF-R	0	Receptor alpha-chain for GM-CSF (CD116)
G-CSF-R	459	Receptor for G-CSF (CD114)
Ulex europaeus lectin	2,123	Lectin from *U. europaeus* specifically expressed on mEC
Ac-dil-LDL	53,836	Acetylated low density lipoproteins labeled with FITC, receptors are specific for microvascular endothelial cells

[a]MFI ([mean fluorescence intensity] corresponds to the value determined from flow cytometric analysis by multiplying the increment of the mean channel of the positive cells minus the mean channel of the negative cells with the percent of cells in the positive gate.

RESULTS

TABLE 1 demonstrates the expression pattern of one representative endothelial cell isolate determined between passage 4 to 18 of *in vitro* culture. According to this analysis, all ECs expressed β2-microglobulin associated with MHC-class I, factor VIIIa, pro-collagen I, very little collagen type IV, and were positive for CD49d and G-CSF receptors, and lacked detectable amounts of CD49f the antigen specific for

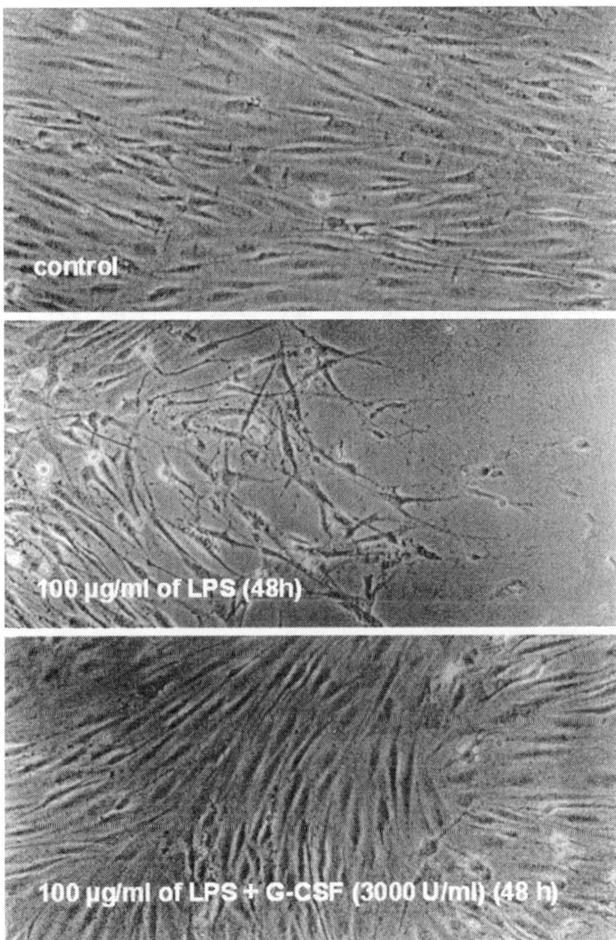

FIGURE 1. (**A**) Phase contrast image of the confluent mEC cell layer 48 h after change of the culture supernatant (0.5% human serum in MEM-α medium). Viability of the layer is perfect. (**B**) The same confluent mEC cell layer 48 h after the addition of 100 ng/mL of endotoxin in the presence of 0.5% human serum in MEM-α. Adherent cell layer is damaged and mEC are largely detached and apoptotic. (**C**) The same confluent but G-CSF pretreated (2,000 U/mL of recombinant human G-CSF Neupogen, Amgen for 24 hours) mEC cell layer treated with LPS 100 ng/mL for 48h as before. Viability is excellent.

migratory endothelial cells (EN7/44[3]) and GM-CSF receptors. The positive binding *Ulex europaeus* lectin and strong binding of the homing receptor CD44 were unequivocal characteristics of these endothelial cells. PAL-E was weakly positive, as well as the CD34 antigen and HLA-class II (syn. HLA-DR). Most importantly, the FITC-labeled ac-dil-LDL (low density lipoprotein) was even more positive than CD44 (TABLE 1).

In order to simulate the effects by high concentrations of LPS released into the circulation *in vivo*, we stimulated confluent endothelial cell layers with LPS and followed the viability by morphological examination. Interestingly, the pre-incubation of these endothelial cells with G-CSF before addition of highly stimulatory amounts of LPS (100 ng/mL–1 μg/mL), significantly inhibited cell death in these monolayers which was most prominent after 48 h of stimulation (FIG. 1).

In order to figure out whether the detachment and cell death of endothelial layers was due to apoptosis we performed caspase 3 determinations in these cells. FIGURE 2 (A) shows the representative results of one endothelial cell culture stimulated with and without G-CSF and after activation with LPS. Clearly, G-CSF pretreatment downregulated caspase 3 activity.

For control caspase 3 activation, we also stimulated endothelial cell layers with 4 μM staurosporine and determined the enzyme activity in endothelial cell lysates after 6 hours. Unexpectedly, G-CSF preincubation caused an even stronger downmodulation of caspase 3 induction than in the LPS stimulated cultures (FIG. 2B).

The cytoprotective effect by G-CSF was further substantiated by the modulation of IL-8 secretion upon *in vitro* stimulation with high dose LPS. FIGURE 2C shows that the constitutive secretion of IL-8 in confluent layers was only marginally affected by pre-incubation with 2×10^3 U/mL of G-CSF, however the increase of IL-8 secretion induced by LPS was attenuated in endothelial cell cultures with G-CSF pretreatment.

DISCUSSION

In the current study we examined the inflammatory response of endothelial cell lines established from peripheral blood mononuclear cells. These monolayer cultures and the phenotypes conformed to the expression pattern of similar cell lines described previously[4–6] (TABLE 1), although we found that CD146 and CD36 were not mutually exclusively expressed in these lines (data not shown). Nevertheless, endothelial cells circulate at higher numbers in patients suffering from acute and severe sepsis as well as in patients with hemophagocytic syndromes similar to patients with myocardial infarction.[6] Cell lines established from such isolates display high sensitivity to LPS as well as TNF-α to undergo cell death *in vitro*. Disruption of the confluent cell layer occurs within 48 h and can be prevented by exogenous G-CSF (FIG. 1). The mechanism of protection is due to the inhibition of caspase 3 activation (FIG. 2A and B).

Apoptotic cell death may occur via the receptors expressed on the cell surface and formation of the death complex but also by activation of mitochondria-dependent pathways.[7] Historically the activation of caspase 3 by the universal protein kinase inhibitor staurosporine (STS) induces apoptosis very rapidly. The maximum effect by 4 μM staurosporine appears to be around 6 hours. Here the preincubation with G-

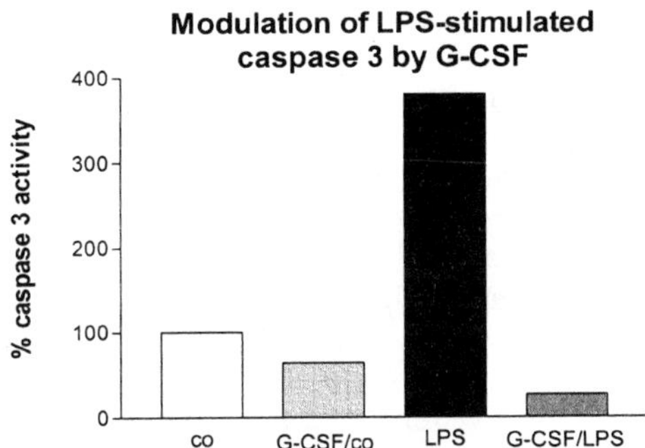

A

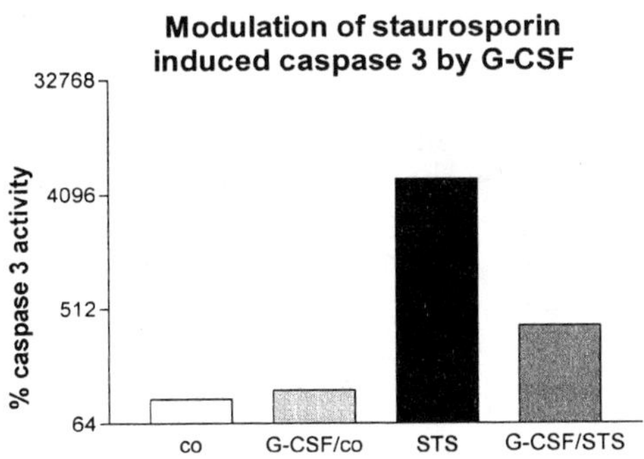

B

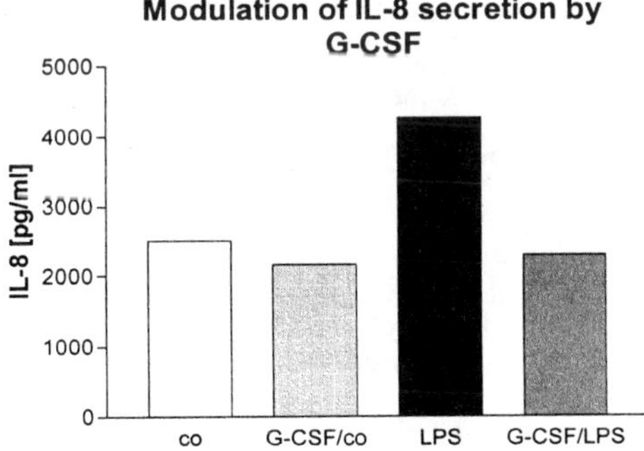

C

FIGURE 2. (**A**) Influence of G-CSF on LPS (100 ng) induced caspase 3 function from adherent mECs cultured in 6-well plates (Costar) for 24 hours. (**B**) As above, but the cell layer had been stimulated with 4 μM staurosporin for 6 hours. (**C**) As above, soluble IL-8 has been determined in supernatants of mEC cell layers.

CSF had a major effect on caspase 3 activity in the representative cell line shown in FIGURE 2. We therefore suggest that G-CSF appears to interfere with the apoptosis pathway induced via the mitochondrial pathway. Nevertheless, sepsis and hemophagocytic syndromes are associated with an increased plasma concentration of CD178 (sFAS-ligand)[1] suggesting that membrane-associated death complexes are involved as well. On the other hand, the anti-apoptotic effect by G-CSF may be mediated not only by its effect on granulopoiesis, proliferation, and recruitment of myeloid cells from bone marrow precursors, but more likely by its direct effect on mitochondria-dependent caspase 3 activation in neutrophils.[8] This effect appears to be independent of CD95 and its ligand CD178.[9] Nevertheless the significantly reduced plasma concentration of IL-8 has been previously demonstrated in patients with SIRS who were prophylactically treated with G-CSF.[10,11]

In summary these observations suggest a significant anti-apoptotic effect on microvascular endothelial cells by exogenous G-CSF, which may explain a reduced incidence of multiorgan dysfunction *in vivo* when systemic inflammatory syndromes (SIRS) are associated with major trauma and LPS released into the circulation. Although the G-CSF-receptor expression is fairly weak in endothelial cells (TABLE 1), a prominent Ca^{2+} signal is induced upon G-CSF stimulation *in vitro* (data not shown). G-CSF signaling in such endothelial cells appears to interfere with the pro-apoptotic pathway induced by LPS as well as the universal protein kinase C inhibitor staurosporine. Studies by others suggest that JNK activation may play a role in interference of LPS-induced apoptosis in endothelial cells[12] and defined regions in the G-CSF receptor could be involved.[13] Thus one of the G-CSF–induced pathways described in myeloid cells and being protein kinase C–dependent may be also functioning in endothelial layers.[8]

REFERENCES

1. SCHNEIDER, E.M. *et al.* 2002. Hemophagocytic lymphohistiocytosis is associated with deficiencies of cellular cytolysis but normal expression of transcripts relevant to killer-cell-induced apoptosis. Blood **100:** 2891–2898.
2. LEONE, M. *et al.* 2002. Systemic endothelial activation is greater in septic than in traumatic hemorrhagic shock but does not correlate with endothelial activation in skin biopsies. Crit. Care Med. **30:** 808–814.
3. HAGEMEIER, H.H. *et al.* 1986. A monoclonal antibody reacting with endothelial cells of budding vessels in tumors and inflammatory tissues, and non-reactive with normal adult tissues. Int. J. Cancer **38:** 481–488.
4. SWERLICK, R.A. *et al.* 1992. Human dermal microvascular endothelial but not human umbilical vein endothelial cells express CD36 in vivo and in vitro. J. Immunol. **148:** 78–83.
5. GERRITSEN, M.E. *et al.* 1993. Cytokine activation of human macro- and microvessel-derived endothelial cells. Blood Cells **19:** 325–339.
6. MUTIN, M. *et al.* 1999. Direct evidence of endothelial injury in acute myocardial infarction and unstable angina by demonstration of circulating endothelial cells. Blood **93:** 2951–2958.
7. HENKART, P.A. & S. GRINSTEIN. 1996. Apoptosis: mitochondria resurrected? J. Exp. Med. **183:** 1293–1295.
8. MAIANSKI, N.A. *et al.* 2002. Granulocyte colony-stimulating factor inhibits the mitochondria- dependent activation of caspase-3 in neutrophils. Blood **99:** 672–679.
9. VILLUNGER, A. *et al.* 2000. Fas ligand, Bcl-2, granulocyte colony-stimulating factor, and p38 mitogen-activated protein kinase: Regulators of distinct cell death and survival pathways in granulocytes. J. Exp. Med. **192:** 647–658.

10. WEISS, M. *et al.* 1996. Filgrastim (RHG-CSF) related modulation of the inflammatory response in patients at risk of sepsis or with sepsis. Cytokine **8:** 260–265.
11. WEISS, M., L.L. MOLDAWER & E.M. SCHNEIDER. 1999. Granulocyte colony-stimulating factor to prevent the progression of systemic nonresponsiveness in systemic inflammatory response syndrome and sepsis. Blood **93:** 425–439.
12. HULL, C. *et al.* 2002. Lipopolysaccharide signals an endothelial apoptosis pathway through TNF receptor-associated factor 6-mediated activation of c-Jun NH2-terminal kinase. J. Immunol. **169:** 2611–2618.
13. RAUSCH, O. & C.J. MARSHALL. 1999. Cooperation of p38 and extracellular signal-regulated kinase mitogen-activated protein kinase pathways during granulocyte colony-stimulating factor-induced hemopoietic cell proliferation. J. Biol. Chem. **274:** 4096–4105.

An Investigation of the MEK/ERK Inhibitor U0126 in Acute Myeloid Leukemia

A. H. J. KERR, J. A. JAMES, M. A. SMITH, C. WILLSON, E. L. COURT, AND J. G. SMITH

Centre for Research in Biomedicine, Faculty of Applied Sciences, University of the West of England, Bristol BS16 1QY, UK

ABSTRACT: Blockade of mitogen-activated protein kinase kinase (MEK1/2), part of the extracellular signal-regulated kinase (ERK) or p44/42 mitogen-activated protein kinase (MAPK) pathway, has been shown in some instances to cause apoptosis in leukemic blast cells. This investigation examined the effect of the potent MEK/ERK inhibitor U0126 on apoptosis in acute myeloblastic leukemia (AML) cell lines, and acute leukemic and non-leukemic patient samples. The pro-apoptotic effect of the inhibitor varied across the five cell lines tested (KG1a, HEL, TF-1, MO7e, and THP-1) from highly significant induction of apoptosis to no apparent response. The pro-apoptotic effect of U0126 in the most sensitive cell line, KG1a, appeared to be related to its CD34 positivity. Three of five leukemic bone marrow samples showed considerable sensitivity to the inhibitor and a similar association with CD34 expression was evident. Interestingly, control marrow cells from six non-leukemic patients did not show a significant effect when exposed to U0126. These results suggest that this agent may offer a potential alternative to standard chemotherapy with a particular role in the most primitive types of leukemia, these often being the most resistant to standard chemotherapy.

KEYWORDS: acute leukemia; apoptosis; MEK inhibitor; U0126; CD34 expression

INTRODUCTION

The Ras/Raf/MEK/ERK pathway is an important signaling cascade involved in the control of hematopoietic cell proliferation and differentiation. It has been shown that the interruption of this proliferative pathway allows pro-apoptotic signals to predominate in AML.[1,2] This study has assessed the pro-apoptotic potential of U0126 in AML versus non-AML cells. During the course of the study, it became apparent that the level of maturity of malignant cells influenced their apoptotic response to U0126. therefore an investigation of the expression of the CD34 cell surface antigen, which is expressed on very primitive progenitors, in relation to U0126 sensitivity was incorporated.

Address for correspondence: A.H.J. Kerr, Centre for Research in Biomedicine, Faculty of Applied Sciences, University of the West of England, Bristol BS16 1QY, UK. Voice: 0117 344 3812; fax: 0117 344 3609.
Alastair.Kerr@uwe.ac.uk

**Ann. N.Y. Acad. Sci. 1010: 86–89 (2003). © 2003 New York Academy of Sciences.
doi: 10.1196/annals.1299.013**

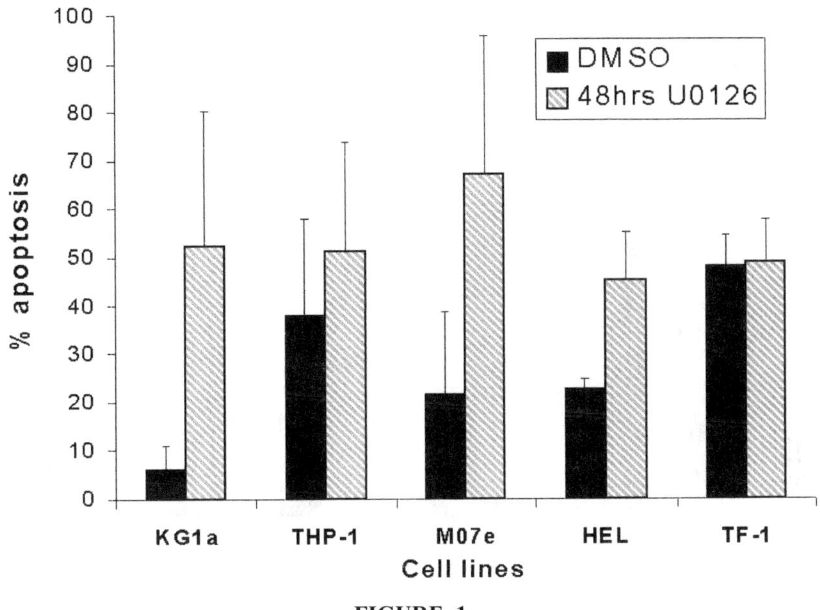

FIGURE 1.

METHODS

Prior to exposure to U0126, AML cell lines were placed in serum-free medium (SFM) for 16 hours. Marrow mononuclear cell fractions were derived from fresh AML (N=5) and non-leukemic (N=3) patient samples. Samples from three non-leukemic peripheral blood stem cell harvests (PBSCHs) were obtained from cryopreserved stocks. Cells were exposed for 48 hours to U0126 dissolved in DMSO (0.1% vol:vol) at the optimal concentration of 50 mM. DMSO and SFM controls were also included. Apo2.7-RPE, a monoclonal antibody that recognizes a mitochondrial protein 7A6 (a marker for early but committed apoptosis) was used for flow cytometric assessment of levels of apoptosis. CD34 positivity was detected using a CD34-FITC monoclonal antibody. Appropriate negative controls were used in all cases. Results of five repeat assays were analyzed using unpaired t-tests.

RESULTS

DMSO alone had no significant effect on apoptosis compared to SFM controls (data not shown). FIGURE 1 demonstrates that amongst cell lines variable sensitivity to U0126 was noted with KG1a, the most primitive cell line which is highly CD34 positive, showing the most significant levels, compared to relevant DMSO controls. FIGURE 2 illustrates that U0126 induced varying degrees of apoptosis in AML patient samples (A1–A5), with higher levels of CD34 positivity correlating with increased

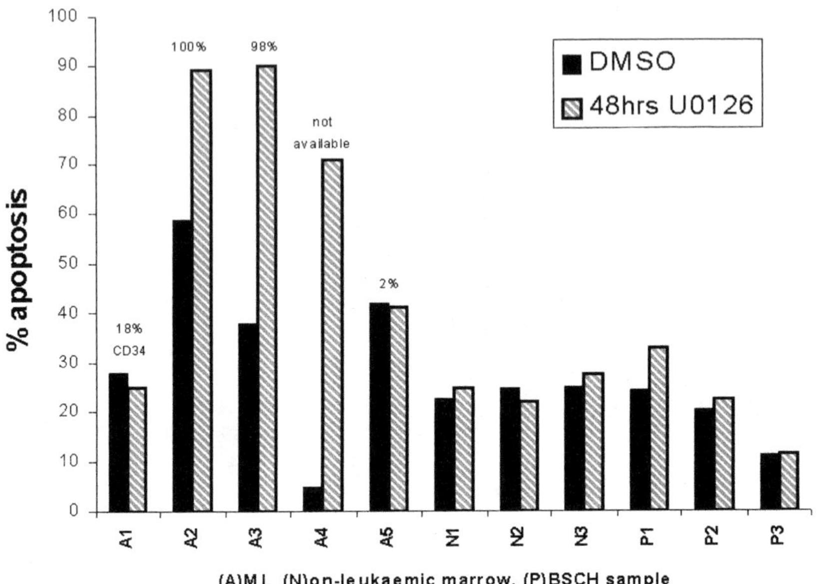

FIGURE 2.

sensitivity to U0126. Overall, non-leukemic and PBSCH samples were significantly less sensitive to the pro-apoptotic effects of the inhibitor compared to AML patient cells ($p = 0.021$).

DISCUSSION

KG1a, a primitive M0 cell line, expressing high levels of CD34 positivity, showed greatest sensitivity to U0126 compared to other cell lines derived from more mature leukemic samples (M5–M7). Repeated cell culturing of KG1a cells caused diminished CD34 expression and concurrent reduction in their sensitivity to apoptosis induced by the MEK/ERK inhibitor. Amongst AML patient samples, those exhibiting high CD34 positivity were the most sensitive to U0126. In non-leukemic and PBSCH samples, no significant increase in the level of apoptosis was noted with the inhibitor. These data suggest that the MEK/ERK inhibitor U0126 may have therapeutic potential with a particular role in the most primitive leukemias. These forms of AML are often resistant to standard treatment, therefore further work is indicated to explore the possible use of U0126 to augment/replace chemotherapy.

ACKNOWLEDGMENTS

This work was supported by the Bath Cancer Support Group and the Bath Leukaemia Research and Transplant Fund.

REFERENCES

1. MILELLA, M. 2001. J. Clin. Invest. **108:** 851.
2. MORGAN, M.A. 2001. Blood **97:** 1823.

The Effect of p38 Mitogen-Activated Protein Kinase on Mucin Gene Expression and Apoptosis in *Helicobacter pylori*–Infected Gastric Epithelial Cells

HYEYOUNG KIM, JI HYE SEO, AND KYUNG HWAN KIM

Department of Pharmacology and Institute of Gastroenterology, Brain Korea 21 Project for Medical Science, Yonsei University College of Medicine, Seoul 120-752, Korea

ABSTRACT: **The loss of mucus coat continuity and apoptosis have been shown in *Helicobacter pylori* (*H. pylori*)–infected gastric tissues. Blockade of p38 mitogen-activated kinase (MAPK) produced reversal in the LPS-induced reduction in mucin synthesis and apoptosis in gastric epithelial cells. This study investigates whether *H. pylori* induces apoptosis, alterations in mucin gene (MUC) expression, and p38 MAPK activation in human gastric epithelial AGS cells. After treatment of AGS cells with *H. pylori* at the ratio of 1:300, apoptosis was determined by DNA fragmentation and DNA laddering. MUC expression was assessed by RT-PCR. p38 MAPK activation and mucin protein level, using antimucin antibody for MUC5/6, were determined by Western blot analysis. As a result, *H. pylori* induced apoptosis and loss of mucin, which was supported by reduced mRNA expression of MUC5AC by *H. pylori* in AGS cells. MUC7/8 expression and p38 MAPK activation were induced in *H. pylori*–infected AGS cells. In conclusion, *H. pylori* induces p38 MAPK activation, wh3.ich may be the underlying mechanism of alterations in MUC expression and apoptosis in gastric epithelial cells.**

KEYWORDS: **p38 MAPK; mucin gene; apoptosis; *Helicobacter pylori***

INTRODUCTION

It has been suggested that the first line of gastric mucosal defense may reside in the mucous gel. The changes in the structural characteristics of this gel layer could be important in understanding the pathogenesis of gastric ulceration and carcinogenesis. The mucus layer acts as a protective layer for gastric epithelium against luminal attack by acid and pepsin and provides antioxidant protection to the underlying gastric epithelial cells. A decrease in the synthesis of sulfated mucus glycoprotein and an increase in mucin degradation are important features in the pathogenesis of peptic ulcer. The loss of coat continuity and the disturbances in mucin synthesis are the prominent markers in the etiology of gastric disease associated with *Helicobacter*

Address for correspondence: Hyeyoung Kim, Department of Pharmacology and Institute of Gastroenterology, Brain Korea 21 Project for Medical Science, Yonsei University College of Medicine, Seoul 120-752, Korea. Voice: 82-2-361-5232; fax: 82-2-313-1894.
kim626@yumc.yonsei.ac.kr

Ann. N.Y. Acad. Sci. 1010: 90–94 (2003). © 2003 New York Academy of Sciences.
doi: 10.1196/annals.1299.014

pylori (*H. pylori*) infection.[1] Apoptosis, programmed cell death, has been characterized morphologically by cell shrinkage and chromatin condensation, and shown in *H. pylori*–infected gastric tissues.[2] Activation of p38 mitogen–activated kinase (MAPK) has been reported to be related to apoptosis in human chondrocytes.[3] Inhibition of MAPK produced reversal in the LPS-induced reduction in mucin synthesis and apoptosis in gastric epithelial cells.[4] Present study was purposed to investigate whether *H. pylori* induces apoptosis, alterations in mucin gene (MUC) expression, and p38 MAPK activation in human gastric epithelial AGS cells.

METHODS

AGS cells (gastric adenocarcinoma, ATCC CRL 1739), 1–5$\times10^5$ cells/well, were incubated in the presence of *H. pylori* (NCTC 11637 obtained from ATCC) at a bacterium/cell ratio of 300:1 for 3 h (p38 MAPK), 12 h (MUC expression), or 24 h (DNA fragmentation, DNA laddering, mucin level). DNA fragmentation was determined by the amount of oligonucleosome-bound DNA. The relative increase in nucleosomes in the cell lysate, determined at 405 nm, was expressed as an enrichment factor. Enrichment factor of none was considered as 1. DNA laddering was visualized with ethidium bromide. Mucin level was determined by Western blotting analysis, using anti-mucin antibody for MUC5/6. Time course for p38 MAPK activation was determined by Western blotting analysis for phosphorylated p38 MAPK and total p38 MAPK. Human mucin gene (MUC) expression was assessed by RT-PCR using the primer as follows. MUC1 (5′-TCTCACCTCCTCCAATCAC-3′ as forward primer, 5′-GAAATGGCACATCACTCAC-3′ as reverse primer, giving 368 bp PCR product); MUC2 (5′-TGCCTGGCCCTGTCTTTG-3′ as forward primer, 5′-CAGCTCCAGCATGAGTGC-3′ as reverse primer, giving 440 bp PCR product); MUC3 (5′-ATAACCACCTCTGAGACCC-3′ as forward primer, 5′-TGATTGAAGAAGTGAAGCTG-3′ as reverse primer, giving 328 bp PCR product); MUC4 (5′-TTCTAAGAACCACCAGACTCAGAGC-3′ as forward primer, 5′-GAGACACACCTGGAGAGA ATGAGC-3′ as reverse primer, giving 467 bp PCR product); MUC5AC (5′-TCCGGCCTCATCTTCTCC-3′ as forward primer, 5′-ACTTGGGCACTGGTGCTG-3′as reverse primer, giving 680 bp PCR product); MUC5B (5′-ACTCCAGAGACTGTCCACAC-3′as forward primer, 5′-TACCACTGGTCTGTGTGCTA-3′ as reverse primer, giving 349 bp PCR product; MUC6 (5′-TCACCTATCACCACACAAC-3′ as forward primer, 5′-GGAGAAGAAGGAAAAAGAG-3′ as reverse primer, giving 293 bp PCR product); MUC7 (5′-CCACACCTAATTCTTCCC-3′ as forward primer, 5′-CTATTGCTCCACCATGTC-3′ as reverse primer, giving 209 bp PCR product); MUC8 (5′-ACAGGGTTTCTCCTCATTG-3′ as forward primer, 5′-CGTTTATTCCAGCACTGTTC-3′ as reverse primer, giving 239 bp PCR product); β-actin (5′-ACCAACTGGGACGACATGGAG-3′ as forward primer, 5′-GTGAGGATCTTCATGAGGTAGTC-3′ as reverse primer, giving a 349 bp PCR product).

RESULTS AND DISCUSSION

H. pylori induced apoptosis, as determined by DNA fragmentation (FIG. 1A) and DNA laddering (FIG. 1B). Mucin protein level was determined using anti-mucin antibody for MUC5/6 (FIG. 1C) since MUC5 and MUC6 were reported to be highly expressed in stomach.[5] *H. pylori* caused a loss of mucin, which was supported by reduced mRNA expression of MUC5AC, not MUC6, by *H. pylori* in AGS cells (FIG. 2A). MUC7 and MUC 8 expression were increased in *H. pylori*–infected AGS cells. MUC 7 is mainly expressed in salivary gland while MUC8 is highly expressed in tracheobronchus.[5] There have been no reports showing the expression of MUC7 or MUC8 in gastric epithelial cells. *H. pylori* may affect normal balance between cell proliferation and cell death in gastric mucosa. Previously we reported that *H. pylori*–induced apoptosis is mediated by NF-κB in human gastric epithelial cells.[6] In present study, *H. pylori*–induced apoptosis was in parallel with the activation of p38 MAPK in AGS cells (FIG. 2B). The result is in agreement with the study showing that blockade of p38 MAPK produced reversal in the LPS-induced reduction in mu-

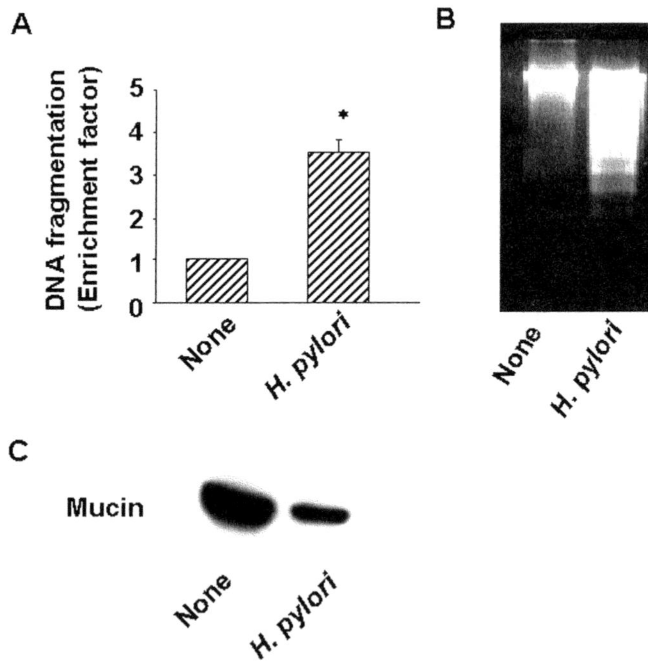

FIGURE 1. Quantification of DNA fragmentation, DNA laddering, and mucin level of AGS cells cultured in the presence of *H. pylori* (at a bacterium/cell ratio, 300:1) for 24 hours. Cell lysates were isolated from 1×10^5 cells. DNA fragmentation was determined by the amount of oligonucleosome-bound DNA in the cell lysate (**A**). The relative increase in nucleosomes in the cell lysate, determined at 405 nm, was expressed as an enrichment factor. Enrichment factor of none was considered as 1. Data represent means ± SE of four different experiments. $*P<.05$ compared to the cells cultured in the absence of *H. pylori* (none). Fragmented DNA was visualized with ethidium bromide (**B**). Mucin level was determined by Western blotting analysis using anti-mucin antibody for MUC5/6 (**C**).

cin synthesis and apoptosis in gastric epithelial cells.[4] In conclusion, *H. pylori* induces alteration of mucin gene expression and disruption of the mucus layer, which may be caused by reduced expression of MUC5AC in gastric epithelial cells. p38 MAPK activation may mediate *H. pylori*–induced apoptotic cell death and disturbance of mucin gene expression in gastric epithelial cells.

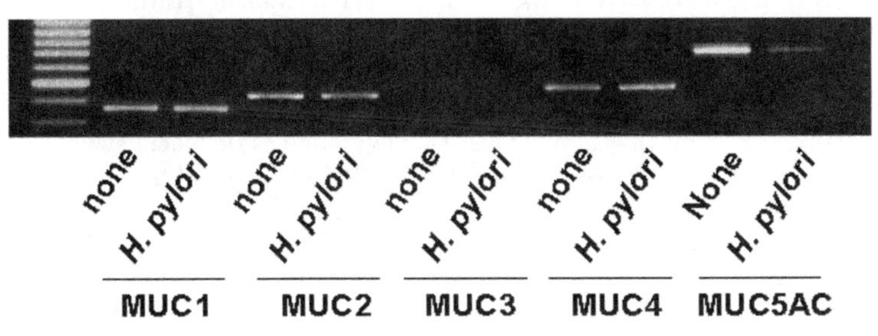

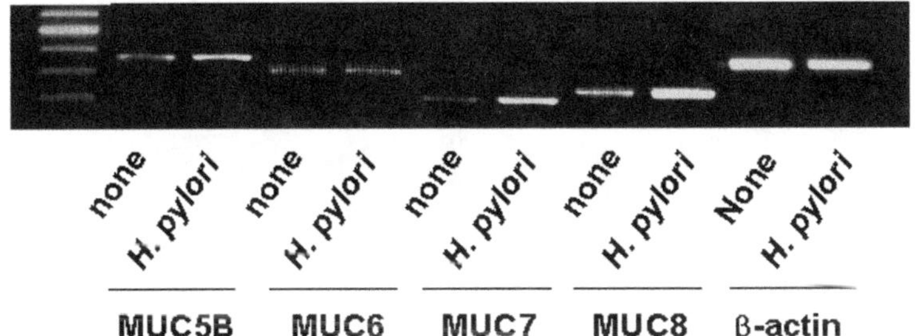

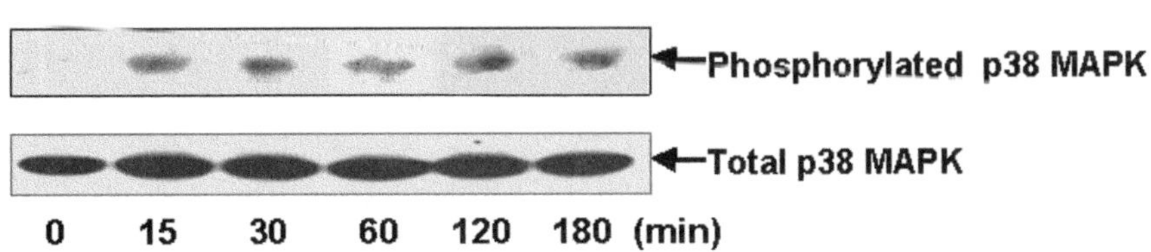

FIGURE 2. Mucin gene (MUC) expression and p38 mitogen-activated protein kinase (MAPK) activation of AGS cells cultured in the presence of *H. pylori* (at a bacterium/cell ratio, 300:1) for 12 h (MUC expression) or 3 h (p38 MAPK). mRNA level of MUC (MUC1, MUC2, MUC3, MUC4, MUC5AC, MUC5B, MUC6, MUC7, MUC8) was determined by RT-PCR (**A**). Time course of p38 MAPK activation was assessed by Western blot analysis (**B**).

ACKNOWLEDGMENT

This study was supported by a grant (to H. Kim) from Korea Science and Engineering Foundation (KOSEF) made in the program year of 2002.

REFERENCES

1. Beil, W., M.L. Enss, S. Muller, *et al.* 2000. Role of *vacA* and *cagA* in *Helicobacter pylori* inhibition of mucin synthesis in gastric mucous cells. J. Clin. Microbiol. **38:** 2215–2218.
2. Moss, S.F., J. Calam, B. Agarwal, *et al.* 1996. Induction of gastric epithelial apoptosis by *Helicobacter pylori*. Gut **38:** 498–501.
3. Shakibaei, M., G. Schulze-Tanzil, P. Desouza, *et al.* 2001. Inhibition of mitogen-activated protein kinase induces apoptosis of human chondrocytes. J. Biol. Chem. **276:** 1389–1394.
4. Slomiany, B.L. & A. Slomiany. 2002. Disruption in gastric mucin synthesis by *Helicobacter pylori* lipopolysaccharide involves ERK and p38 mitogen-activated protein kinase participation. Biochem. Biophys. Res. Commun. **294:** 220–224.
5. Kim, Y.S., J. Gum & I. Brockhausen. 1996. Mucin glycoproteins in neoplasia. Glycoconjugate J. **13:** 693–707.
6. Lim, J.W., H. Kim & K.H. Kim. 2001. NF-κB, inducible nitric oxide synthase and apoptosis by *Helicobacter pylori* infection. Free Rad. Biol. Med. **31:** 355–366.

Activation of JNK in Sensory Neurons Protects against Sensory Neuron Cell Death in Diabetes and on Exposure to Glucose/Oxidative Stress *in Vitro*

SALLY A. PRICE, LUKE HOUNSOM, TERTIA D. PURVES-TYSON,[a]
PAUL FERNYHOUGH, AND DAVID R. TOMLINSON

Division of Neuroscience, 1.124 Stopford Building, University of Manchester, Manchester, M13 9PT, England, United Kingdom

ABSTRACT: Diabetes activates all three groups of MAP kinases in sensory ganglia. Inhibition of this activation for the ERK and p38 groups prevents nerve damage, and agents that improve neuronal function in diabetic rats—antioxidants and aldose reductase inhibitors—also inhibit activation of ERK and p38 in dorsal root ganglia (DRG). However, these same treatments consistently increase activation of JNK. Thus, in DRG from rats with streptozotocin (STZ)-induced diabetes of 12-week duration, the p54/56 isoforms of JNK were activated by 2.75 compared to controls ($P < .05$). In DRG from diabetic rats treated with a gamma-linolenic acid and alpha-lipoic acid diester (GLA^^LA), the activity of the p54/56 isoform was 3.75 that of controls and the p46 isoform was also increased to 1.75 that of controls (both $P < .05$ compared to both controls and untreated diabetics). We therefore tested the hypothesis that JNK activation is protective. Exposure of rats to diabetes increased activation of JNK in DRG, but treatment with GLA^^LA increased this effect ($P < .05$). Specific inhibition of JNK in primary cultures of DRG neurons using a peptide inhibitor of JNK (JNKi1, 159-600-R100, 7.5 μM, Alexis Biochemicals) increased the release of LDH and reduced MTT staining; both findings indicate an increase in neuronal damage. Taken together these findings indicate that multiple isoforms of JNK were activated in sensory neurons of diabetic rats, probably by a combination of raised glucose and oxidative stress, and that this activation of JNK serves to protect the neurons from damage.

KEYWORDS: diabetes; dorsal root ganglia; sensory neuron; MAP kinase; JNK; c-Jun; neuroprotective

INTRODUCTION

Neuropathy is a major chronic complication of diabetes. One of the key functional markers of neuropathy is a deficit in nerve conduction velocity (NCV). In exper-

Address for correspondence: Sally A. Price, Division of Neuroscience, 1.124 Stopford Building, University of Manchester, Manchester, M13 9PT, England, UK.
sally.a.price@man.ac.uk
[a]Current address: Prince of Wales Medical Research Institute, Barker Street, Sydney, Australia.

Ann. N.Y. Acad. Sci. 1010: 95–99 (2003).
doi: 10.1196/annals.1299.015

imental diabetes NCV can be corrected by several pharmacological interventions, including antioxidants. Despite this, the cellular basis of the deficit is largely unknown. Activation of all three MAP kinase families has been demonstrated in dorsal root ganglion (DRG) neurons following both experimental diabetes and high glucose/oxidative stress *in vitro*.[1] Consequently, it has been postulated that MAP kinase signaling pathways may transduce changes resulting from high glucose/oxidative stress into changes in cellular phenotype. This is supported by experiments demonstrating that inhibitors of p38 and ERK MAP kinases prevent DRG neuronal death induced by high glucose/oxidative stress.[2] Inhibition of p38 also prevents the diabetes-induced deficits in NCV.[3] Despite convincing evidence for a detrimental role of p38 and ERK signaling following high glucose/oxidative stress, the role of JNK signaling remains elusive. This study investigated the role of JNK signaling by comparing JNK activation in DRG from untreated diabetic rats and those treated with an antioxidant (GLA^^LA—a diester of γ-linolenic acid and α-lipoic acid) known to correct NCV deficits. The role of JNK in DRG neurons was further investigated *in vitro* using a peptide inhibitor of JNK (JNKi1, 159-600-R100, Alexis Biochemicals).

METHODS

Experimental Diabetes

Diabetes was induced in male Wistar rats (300–320 g) by intraperitoneal injection of streptozotocin (50 mg/kg STZ; Sigma). Diabetes was confirmed in all STZ-treated rats 2 days later by measuring glucose concentration (Boehringer Ingelheim, Reflochek test strips) of tail-vein blood. Diabetic rats were maintained on a 12-h light/dark cycle with free access to food and water for 12 weeks. Rats treated with GLA^^LA (a gift from Scotia Pharmaceuticals) received the drug as a 5% dietary supplement for the last 4 weeks of the 12-week protocol. After 12 weeks, rats were anesthetized with halothane, the L_4 and L_5 DRG dissected out of both sides and the rats killed by exsanguination without recovery from anesthesia. All procedures were in accordance with animal usage regulations of the UK Home Office.

Primary DRG Cultures

DRG were dissociated from adult male Wistar rats (250 g). Cells were seeded onto poly ornithine-laminin–coated 35-mm dishes, 48-well plates, or 8-well chamber slides and maintained in serum-free F12 medium with N2 supplements (0.1 mg/mL transferrin, 20 nM progesterone, 100 µM putrescine, 30 nM sodium selenite, and 1 mg/mL BSA), cytosine arabinoside (0.01 mM), and 10 pM insulin at 37°C in a 95% air/5% CO_2 humidified incubator (all additives were from Sigma, Poole, UK; culture medium was from Life Technologies, Paisley, UK).

Cytotoxicity Assays

Cell viability following treatment with JNKi1 (0.2–20 µM) was examined using the CytoTox 96® nonradioactive cytotoxicity assay (lactate dehydrogenase (LDH); Promega) and MTT (3-(4,5-dimethylthiazol-2-yl)-2,5-diphenyl tetrazolium bromide; Sigma) assays as described previously.[4]

Western Blot Analysis

Tissues were homogenized in ice-cold lysis buffer (50 mM Tris pH 8.0, 150 mM NaCl, 1 mM EGTA pH 8.0, 100 mM NaF, 1.5 mM $MgCl_2$, 10% vol/vol glycerol, 1% vol/vol Triton-X100) containing 1 mM PMSF, 1 mM $NaVO_4$, and a protease inhibitor mixture. Proteins were separated by SDS-PAGE, transferred onto nitrocellulose membrane, and probed with antibodies specific to either phosphorylated JNK (Santa Cruz; 1:200), total JNK (Santa-Cruz; 1:1000), phosphorylated c-Jun, or total c-Jun (both New England Biolabs; 1:1000). The activation status was derived from the ratio of phosphorylated-to-total forms.

Immunocytochemistry

Cells were fixed in 4% paraformaldehyde and permeabilized in methanol before incubating with primary antibody (phosphorylated c-Jun, New England Biolabs; 1:100). Non-specific binding was blocked using goat serum. Immunoreactivity was detected using a biotinylated anti-rabbit secondary antibody (1:200, Vectorlabs) and fluorescein-labeled avidin (1:100, Vectorlabs).

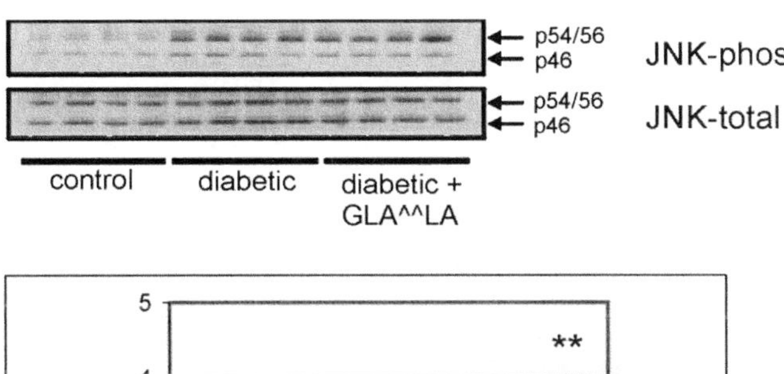

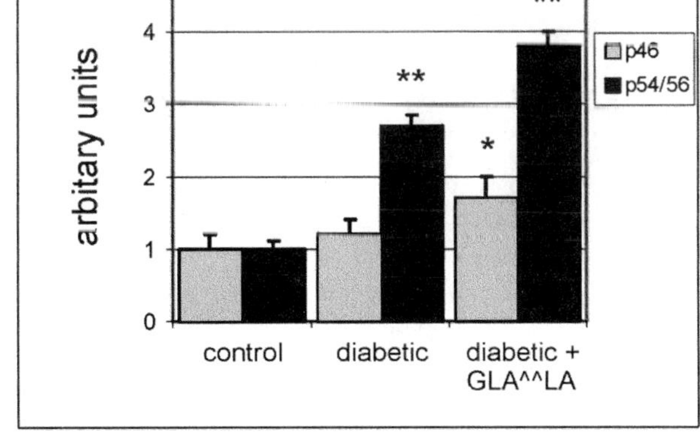

FIGURE 1. Immunoblots showing an increase in the levels of phosphorylated JNK relative to total JNK in DRG samples from untreated and GLA^^LA-treated diabetic rats compared with control rats. The bar chart shows the ratio of phosphorylated to total forms of JNK. Values are means ± 1 SD. *P<.05, **P<.01 vs. control.

RESULTS

Effect of Diabetes on JNK Activation

After 12 weeks of diabetes the p54/56 isoform of JNK was markedly activated in DRG from diabetic rats without any significant change in the total amount of JNK. In DRG from diabetic rats treated with GLA^^LA there was a further increase in the amount of activation of the p54/56 isoform of JNK and also a significant increase in the activation of the p46 isoform of JNK (FIG. 1).

Effect of JNK Inhibition on Primary Cultures of DRG

Inhibition of JNK in primary DRG cultures using the JNK inhibitor JNKi1 (0.2–20 μM) resulted in a concentration-dependent decrease in cell viability, indicated by both LDH and MTT assays. The two toxicity assays produced perfect mirror-image curves, showing identical concentration/effect relationships. The cell death induced by JNKi1 became evident 12–16 h after treatment and was caspase-independent, as indicated by the lack of effect of the broad-spectrum caspase inhibitor zVAD-fmk (100 μM). To determine the specificity of JNKi1, immunocytochemistry was carried out using an antibody specific for phosphorylated (Ser 63) c-Jun. Intense nuclear staining was present in control cells that was completely absent in cells treated with 20 μM JNKi1 (FIG. 2). Western blot analysis was also carried out on cells treated for 6 h with JNKi1 (a duration of treatment that does not induce cell death). There was a concentration-dependent decrease in immunoreactivity for phosphorylated c-Jun in cells treated with JNKi1. JNKi1 however did not have any effect on other MAP

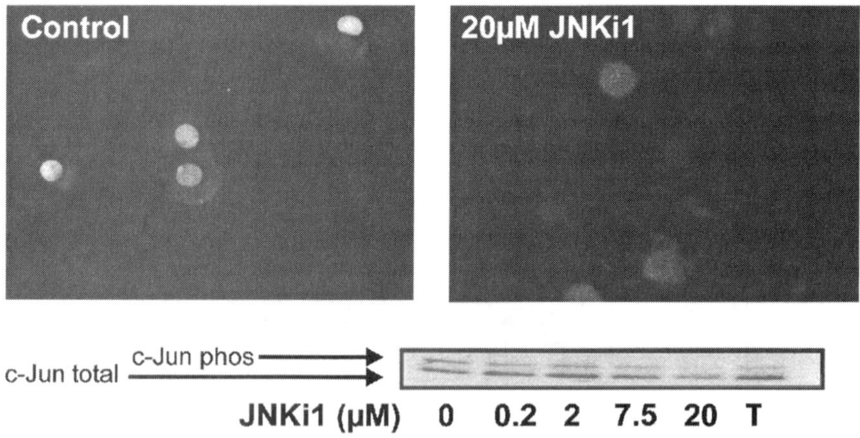

FIGURE 2. Immunocytochemistry and Western blot analysis on cultured DRG neurons treated with JNKi1 (0.2–20 μM) for 6 h. Immunocytochemistry using an antibody that recognizes only phosphorylated (Ser 63) c-Jun indicates that 20 μM JNKi1 completely inhibits c-Jun activation. Western blotting revealed that JNKi1 causes a concentration-dependent reduction in both phosphorylated and total c-Jun. T = TAT peptide.

kinases; indeed the inhibitor caused marked exaggeration of activation of ERK and a slight increase in phosphorylated p38 levels. The TAT peptide, a peptide containing the same amino acids as JNKi1, but with a scrambled sequence, was without effect on the levels of phosphorylated c-Jun and did not cause toxicity.

DISCUSSION

Experimental diabetes in rats causes a massive increase in JNK activation. Treatment with the antioxidant GLA^^LA, known to normalize NCV deficits in diabetic rats, further increased the amount of JNK activation. This suggests that JNK activation in diabetes is not a detrimental signaling pathway. If activation of JNK is protective, then beneficial treatments, such as GLA^^LA, further increase the protective effect. This finding is supported by the *in vitro* results demonstrating that inhibition of JNK with JNKi1 is toxic to DRG neurons. These results are similar to those obtained from cardiac myocytes where JNK signaling has been shown to be neuroprotective.[5] Therefore, we suggest that JNK signaling may play a neuroprotective role in sensory neurons, both under normal conditions and following experimental diabetes.

REFERENCES

1. PURVES, T.D. & D.R. TOMLINSON. 2002. Int. Rev. Neurobiol. **50:** 83–114.
2. PURVES, T.D., A. MIDDLEMAS, S. AGTHONG, *et al.* 2001. FASEB J. **15:** 2508–2514.
3. AGTHONG, S. & D.R. TOMLINSON. 2002. Ann. NY Acad. Sci. **973:** 359–362.
4. MOSMANN, T. 1983. J. Immunol. Methods **65:** 55–63.
5. Dougherty, C.J., L.A. KUBASIAK, H. PRENTICE, *et al.* 2002. Biochem. J. **15**(362)**:** 561–571.

Inhibition of JNK by HGF/SF Prevents Apoptosis Induced by TNF-α

SYLVIE REVENEAU,[a] RÉJANE PAUMELLE,[b] JULIEN DEHEUNINCK,[b] CATHERINE LEROY,[b] YVAN DE LAUNOIT,[b] AND VÉRONIQUE FAFEUR[b]

[a]*Ecole Pratique des Hautes Etudes, Institut National de la Santé et de la Recherche Médicale U517, BP 87900, 21079 Dijon, France*

[b]*Centre National de la Recherche Scientifique, Unité Mixte de Recherche 8117, Institut de Biologie de Lille, Institut Pasteur de Lille, B.P.447, 59021 Lille, France*

ABSTRACT: We investigated whether repression of JNK by hepatocyte growth factor/scatter factor (HGF/SF) in MDCK epithelial cells is linked to its ability to protect cells from apoptosis. To this purpose, cells were treated by TNF-α, a well-known inducer of JNK and of cell death, and the effects of HGF/SF were investigated under these conditions. We identified repression of JNK as a signaling target of HGF/SF for protection against TNF-α–induced cell death. This effect of HGF/SF occurs via the activation of the PI3K and MEK1 pathways.

KEYWORDS: hepatocyte growth factor; scatter factor; JNK; tumor necrosis factor-α

Hepatocyte growth factor/scatter factor (HGF/SF) is a mesenchymal-derived cytokine. It has multiple biological effects on epithelial cells, including induction of cell proliferation, migration, invasion, and morphogenesis. More recently, HGF/SF have been shown to induce cell survival, although some papers also reported a pro-apoptotic effect.[1,2] The effects of HGF/SF are mediated by activation of its receptor, the proto-oncogene c-met, which bears an intrinsic kinase activity. Binding of HGF/SF causes dimerization and autophosphorylation of c-met, which in turn activates various signaling molecules, in particular PI3K and the MAPK ERK1/2.[3] We previously found that HGF/SF can also induce weakly and then repress the activity of the MAPK JNK.[4]

It is known that JNK is primarily activated by various environmental stresses, including osmotic shock, UV radiation, heat shock, protein synthesis inhibitors, oxidative stress, and pro-inflammatory cytokines, such as TNF-α and IL1. In this study, we wished to investigate whether repression of JNK by HGF/SF in MDCK epithelial cells is linked to its ability to protect cells from apoptosis. To this purpose, cells were treated by TNF-α, a well-known inducer of JNK and of cell death and the effects of HGF/SF were investigated under these conditions.

Address for correspondence: Sylvie Reveneau, EPHE-INSERM U517, BP 87900, 21079 Dijon, France.

Sylvie.Reveneau@u-bourgogne.fr

Ann. N.Y. Acad. Sci. 1010: 100–103 (2003). © 2003 New York Academy of Sciences. doi: 10.1196/annals.1299.016

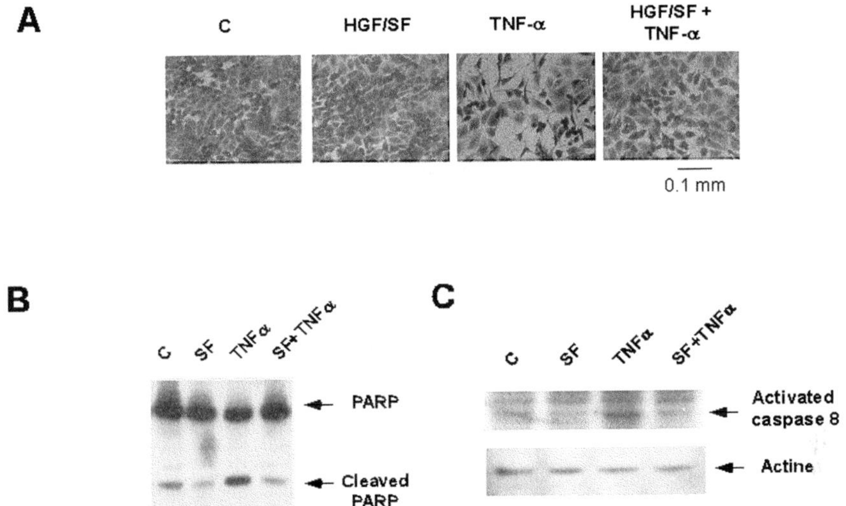

FIGURE 1. Effect of HGF/SF on TNF-α–induced cell death in MDCK cells. (**A**) Cells were incubated with HGF/SF (60 ng/mL) and/or TNF-α (30 ng/mL) in serum-free medium for 48 hours. (**B** and **C**) Cells were incubated in serum-free medium and treated by HGF/SF (30 ng/mL) and/or TNF-α (30 ng/mL) for 32 hours. Whole cell extracts were collected and cell extracts (30 μg) were resolved on 10% SDS-PAGE. Immunoblots were performed using anti-PARP antibody (**B**) and anti-caspase 8 antibody and anti-actin (**C**) antibody, which indicate comparable loading. Full and cleaved form of PARP are indicated by *arrows*.

HGF/SF PREVENTS TNF-α–INDUCED CELL DEATH

As shown in FIGURE 1A, treatment of MDCK epithelial cells by HGF/SF prevented a decrease of cell viability by TNF-α. This decrease of cell viability by TNF α corresponds to an increase in cell death, since TNF-α was found to induce DNA fragmentation, a conventional observation during the apoptotic process. In addition, this effect of TNF-α was caspase dependent, since it was prevented by Z-VAD-fmk, a pan-caspase inhibitor. Using this DNA fragmentation assay, we found that HGF/SF prevented apoptosis by TNF-α (data not shown). In agreement with these findings, HGF/SF also prevented other characteristics of TNF-α-induced cell death, such as activation of caspase-8 and cleavage of the poly-ADP-ribose polymerase (PARP), a substrate of caspase-3 (FIG. 1B and C).

HGF/SF INHIBITS TNF-α–INDUCED JNK PHOSPHORYLATION AND CELL DEATH VIA THE MEK1 AND PI3K PATHWAYS

We first established signaling pathways induced by TNF-α in MDCK cells (data not shown). TNF-α induced strongly JNK phosphorylation within 10 minutes and in a sustained manner, since it was still observed after 6 hours of treatment. TNF-α also

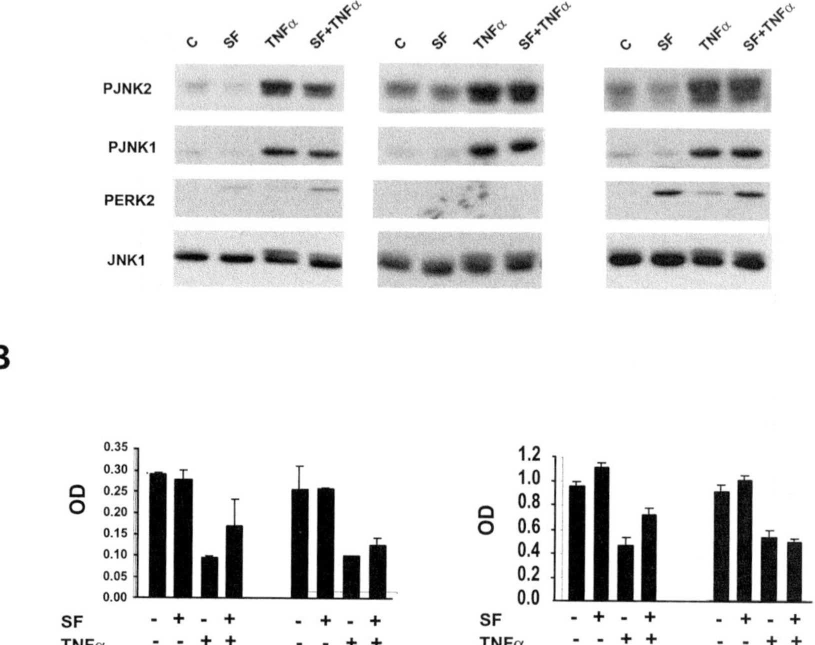

FIGURE 2. HGF/SF inhibits TNF-α–induced JNK phosphorylation via the MEKK1 and PI3K pathways. (**A**) Cells were incubated in serum-free medium (0.5%). The next day, cells were pretreated with U0126 (25 μM) or LY294002 (25 μM) for 30 minutes and then treated by HGF/SF (30 ng/mL) with or without TNF-α (30 ng/mL) for 8 hours. Phosphorylation and expression of JNK were determined by immunoblot analysis as described above. (**B**) Cells were incubated in serum-free medium and pretreated by U0126 (5 μM) or LY294002 (25 μM) inhibitors for 30 minutes and treated by HGF/SF (60 ng/mL) and/or TNF-α (30 ng/mL) for 24 hours. Cell viability was determined using a crystal violet colorimetric assay.

induced a transient degradation of IκB within 10 minutes, in agreement with the known ability of TNF-α to induce NFκB activation. In addition, TNF-α induced a weak phosphorylation of ERK and of AKT, a target of PI3K, within several minutes, which contrasts the strong induction of ERK and AKT by HGF/SF, which we previously reported.[4]

We then investigated whether these signaling targets of TNF-α were regulated by HGF/SF. As shown in FIGURE 2A, the induction of JNK phosphorylation by TNF-α was lowered when cells were pretreated by HGF/SF. In contrast, IκB degradation induced after TNF-α stimulation was not altered by pretreatment by HGF/SF (data not shown). We then used U0126 and LY294002, which are, respectively, selective inhibitors of MEK1 and PI3K. Both inhibitors were able to block the repression of JNK phosphorylation by HGF/SF (FIG. 2A). Finally, treatment of MDCK epithelial cells by HGF/SF prevented a decrease of cell viability by TNF-α and addition of U0126 and LY294002 rendered HGF/SF less efficient to prevent cell death by TNF-α (FIG. 2B).

In conclusion, we identified repression of JNK as a signaling target of HGF/SF for protection against TNF-α–induced cell death. This effect of HGF/SF occurs via the activation of the PI3K and MEK1 pathways.

REFERENCES

1. GAO, M., S. FAN, I.D. GOLDBERG, *et al.* 2001. Hepatocyte growth factor/scatter factor blocks the mitochondrial pathway of apoptosis signalling in breast cancer cells. J. Biol. Chem. **276:** 47257–47265.
2. GOHDA, E., H. OKAUCHI, M. IWAO, *et al.* 1998. Induction of apoptosis by hepatocyte growth factor/scatter factor and its augmentation by phorbol esters in Meth A cells. Biochem. Biophys Res. Commun. **245:** 278–283.
3. PAUMELLE, R., D. TULASNE, Z. KHERROUCHE, *et al.* 2002. Hepatocyte growth factor/scatter factor activates the ETS1 transcription factor by a RAS-RAF-MEK-ERK signalling pathway. Oncogene **21:** 2309–2319.
4. PAUMELLE, R., D. TULASNE, C. LEROY, *et al.* 2000. Sequential activation of ERK and repression of JNK by scatter factor/hepatocyte growth factor in Madin-Darby canine kidney epithelial cells. Mol. Biol. Cell. **11:** 3751–3763.

Signal Transduction of Cerulein-Induced Cytokine Expression and Apoptosis in Pancreatic Acinar Cells

JANGWON LEE,[a] JEONGHUN SEO,[b] HYEYOUNG KIM,[a] JAE BOCK CHUNG,[b] AND KYUNG HWAN KIM[a]

Departments of [a]Pharmacology and [b]Internal Medicine, Brain Korea 21 Project for Medical Science, Yonsei University College of Medicine, Seoul 120-752, Korea

ABSTRACT: The signaling pathways mediating cytokine production and apoptosis in pancreatic acinar cells have not been fully understood. We investigated the signal transduction of cerulein-induced cytokine IL-6 expression and apoptosis in pancreatic acinar AR42J cells. The wild-type cells and the cells transfected with mutant genes for signaling molecules (such as transcription factor NF-κB and AP-1) and ras, the upstream signal for mitogen-activated protein kinase (MAPK), were used. IL-6 level was determined in the medium by enzyme-linked immunosorbent assay (ELISA). Apoptosis was determined by viable cell counts and DNA fragmentation. Wild-type cells and transfected cells with control vector (pcDNA), IκB mutant gene (IκB mt), H-ras mutant gene (Ras N17), or c-jun dominant negative gene (TAM-67) were treated with cerulein for 24 hours. As a result, cerulein induced IL-6 expression time-dependently at 10^{-8} M, and apoptosis dose-dependently at 24 h in the wild-type cells. Cerulein-induced IL-6 expression (at 10^{-8} M) and apoptosis (at 10^{-7} M) were inhibited in the transfected cells with mutant gene (IκB mt, Ras N17, TAM-67) as compared to pcDNA cells and the wild-type cells. In conclusion, cerulein induces cytokine expression and apoptotic cell death, which may be regulated by NF-κB, AP-1, and possibly MAPK in pancreatic acinar cells.

KEYWORDS: cerulein; cytokine; apoptosis; pancreatic acinar cells

INTRODUCTION

Despite considerable progress in understanding pathophysiology of pancreatitis, the mechanism of pancreatitis remains obscure. One of the best characterized experimental models for pancreatitis involves administration of high-dose cerulein, a cholecystokinin (CCK) analogue.[1,2] Cerulein treatment provides rapid induction, mild and highly reproducible course, and easily detects changes of acute interstitial pancreatitis, which makes this secretagogue-induced model a favorite for investigations of pathophysiological events in pancreatitis. Supraphysiological doses of cerulein induced apoptosis in rat pancreatic acinar cell line AR42J cells[3] and NF-κB-

Address for correspondence: Hyeyoung Kim, Department of Pharmacology, Yonsei University College of Medicine, Seoul 120-752, Korea. Voice: 82-2-361-5232; fax: 82-2-313-1894. kim626@yumc.yonsei.ac.kr

Ann. N.Y. Acad. Sci. 1010: 104–108 (2003). © 2003 New York Academy of Sciences.
doi: 10.1196/annals.1299.017

mediated cytokine expression in freshly isolated pancreatic acinar cells.[4] The major role for p38 mitogen-activated protein kinase (MAPK) and the involvement of NF-κB in cytokine expression were suggested in dispersed pancreatic acini.[5] Reactive oxygen species (ROS) has been suggested as the signaling molecule in the pathogenesis and development of pancreatitis.[6] ROS are recognized to control signal transduction via activation of MAPK or are implicated in the regulation of transcription factors NF-κB and AP-1.[7] Therefore, it is hypothesized that NF-κB, AP-1, and MAPK may be key regulators in cytokine expression and apoptosis. We conducted the present study to investigate the role of NF-κB and AP-1, and ras, the upstream signal for MAPK on cerulein-induced IL-6 expression and apoptosis in pancreatic acinar AR42J cells.

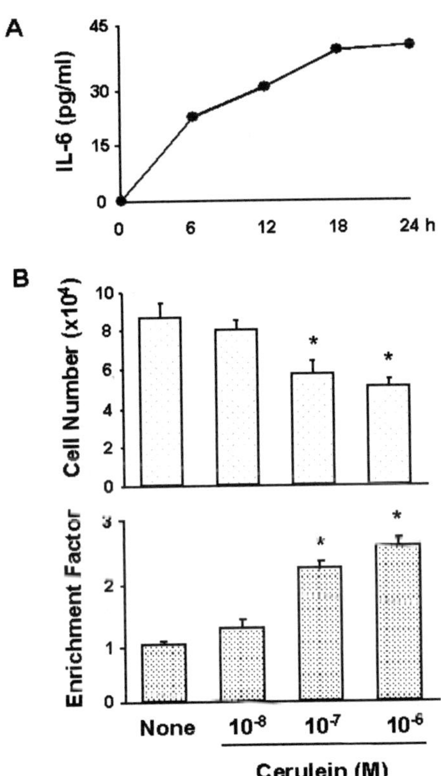

FIGURE 1. IL-6 expression and apoptosis in the wild-type cells treated with cerulein. (**A**) The wild-type cells were treated with cerulein (10^{-8} M) for 24 hours. IL-6 protein level was measured in the medium by ELISA at an indicated time point. (**B**) The wild-type cells were treated with various concentrations of cerulein (10^{-8}–10^{-6} M) for 24 hours. Viable cells were determined by trypan-blue exclusion test (*upper panel*). DNA fragmentation was assessed as nucleosome-bound DNA in the whole cell extract and expressed as enrichment factor (*lower panel*). Enrichment factor of the cells treated without cerulein (None) was considered as 1. *P<.05 compared with None (the cells without cerulein treatment).

METHODS

Rat pancreatic acinar AR42J cells (pancreatoma, ATCC CRL 1492) were incubated in DMEM with 10% FBS. The wild-type cells were treated with cerulein (10^{-8} M) for 24 h for time-course of IL-6 expression. IL-6 protein level was measured in the medium by enzyme-linked immunosorbent assay (ELISA) at an indicated time point. For the apoptotic indices, the wild-type cells were treated with cerulein (10^{-8} –10^{-6} M) for

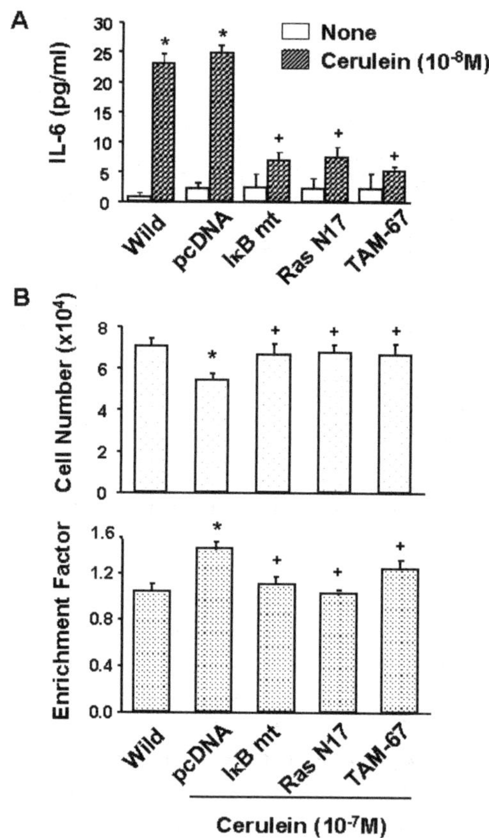

FIGURE 2. IL-6 expression and apoptosis in the wild-type cells and the transfected cells with control vector (pcDNA), IκB mutant gene (IκB mt), H-ras mutant gene (Ras N17), or c-jun dominant negative gene (TAM-67), treated with cerulein. (**A**) The cells were treated with or without cerulein (10^{-8} M) for 24 hours. IL-6 protein level was measured in the medium by ELISA at an indicated time point. *P<.05 compared with the corresponding None (the cells without cerulein treatment). (**B**) The cells were treated with or without cerulein (10^{-7} M) for 24 hours. Viable cells were determined by trypan-blue exclusion test (*upper panel*). DNA fragmentation was assessed as nucleosome-bound DNA in the whole cell extract and expressed as enrichment factor (*lower panel*). Enrichment factor of the cells treated without cerulein (None) was considered as 1. *P<0.05 compared with None (the cells without cerulein treatment), +P<.05 compared with pcDNA cells treated with cerulein.

24 hours. Viable cells were determined by trypan-blue exclusion test. DNA fragmentation was assessed as nucleosome-bound DNA in the whole cell extract and expressed as enrichment factor. Enrichment factor of the cells treated without cerulein was considered as 1. To investigate the role of NF-κB, AP-1 and ras on IL-6 expression and apoptosis, the wild-type cells and the transfected cells with control vector (pcDNA), IκB mutant gene (IκB mt), H-ras mutant gene (Ras N17), or c-jun dominant negative gene (TAM-67) were treated with cerulein (10^{-8} M for IL-6, 10^{-7} M for apoptosis) for 24 hours.

RESULTS AND DISCUSSION

Cerulein, at the concentration of 10^{-8} M, induced IL-6 expression in the wild-type cells time dependently (FIG. 1A). Cerulein-induced IL-6 expression was inhibited in the cells transfected with IκB mutant gene (IκB mt), H-ras mutant gene (Ras N17), or c-jun dominant negative gene (TAM-67) as compared with the cells transfected with control vector (pcDNA) and wild-type cells (Wild) (FIG. 2A). The wild-type cells were treated with various concentrations of cerulein (10^{-8}–10^{-6} M) for 24 h (FIG. 1B). At the concentration of 10^{-7} M and 10^{-6} M, cerulein induced apoptosis, determined by the decrease in viable cell number and DNA fragmentation, assessed as nucleosome-bound DNA. Cerulein-induced apoptosis was inhibited in the transfected cells with the mutant gene (IκB mt, Ras N17, TAM-67) as compared with pcDNA cells and wild cells (FIG. 2B). Present results are supported by previous studies showing that supraphysiological doses of cerulein induced apoptosis in AR42J cells[3] and cytokine expression in freshly isolated pancreatic acinar cells.[4] Since ROS has been considered to be a signaling molecule for MAPK activation and NF-κB and AP-1 activation,[7] present results suggest the possible involvement of ROS in cerulein-induced cytokine expression and apoptosis in pancreatic acinar cells. In conclusion, cerulein induces cytokine expression and apoptotic cell death, which may be regulated by NF-κB, AP-1, and possibly MAPK in pancreatic acinar cells.

ACKNOWLEDGMENTS

This study was supported by a grant (to H. Kim) from the Korean Ministry of Health and Welfare.

REFERENCES

1. WILLEMER, S., H.P. ELSASSER & G. ADLER. 1992. Hormone-induced pancreatitis. Eur. Surg. Res. **24**(Suppl. 1)**:** 29–49.
2. STEINLE, A.U., H. WEIDENBACH, M. WAGNER, *et al.* 1999. NF-κB/Rel activation in cerulein pancreatitis. Gastroenterology **116:** 420–430.
3. SATA, N., H. KLONOWSKI-STUMPE, B. HAN, *et al.* 1999. Supraphysiologic concentrations of cerulein induce apoptosis in the rat pancreatic acinar cell line AR42J. Pancreas **19:** 76–82.
4. YU, J.H., J.W. LIM, W. NAMKUNG, *et al.* 2002. Suppression of cerulein-induced cytokine expression by antioxidants in pancreatic acinar cells. Lab. Invest. **82:** 1359–1368.

5. BLINMAN, T.A., I. GUKOVSKI, M. MOURIA, *et al.* 2000. Activation of pancreatic acinar cells on isolation from tissue: cytokine upregulation via p38 MAP kinase. Am J. Physiol. Cell Physiol. **279:** C1993–C2003.
6. SCHOENBERG, M.H., M. BUCHLER & M. GASPER. 1990. The involvement of oxygen radicals in acute pancreatitis. Gut **31:** 1138–1143.
7. ZACKARA, R.M., W. ZHOU, Y. ZHANG, *et al.* 1998. Redox gene therapy for ischemia/reperfusion injury of the liver reduces AP-1 and NF-κB activation. Nat. Med. **4:** 698–708.

Pro-Apoptotic and Anti-Apoptotic Molecules Affecting Pathways of Signal Transduction

G. KÉRI,[a] G. RÁCZ,[b] K. MAGYAR,[c] L. ÖRFI,[d] A. HORVÁTH,[a] R. SCHWAB,[d] B.B. HEGYMEGI,[a] AND B. SZENDE[c]

[a]*Research Group of Peptide Biochemistry of Hungarian Academy of Sciences in the Department of Medical Chemistry, Molecular Biology and Pathobiochemistry, Semmelweis University, Budapest, Hungary*

[b]*First Department of Pathology and Experimental Cancer Research, Semmelweis University and Research Group of Molecular Pathology, Hungarian Academy of Sciences and Semmelweis University, Budapest, Hungary*

[c]*Department of Pharmacodynamics and Research Group of Neurochemistry, Hungarian Academy Sciences and Semmelweis University, Budapest, Hungary*

[d]*Cooperation Research Center of the Semmelweis University, Budapest, Hungary*

ABSTRACT: Selective inhibition of the "false" proliferative signals via targeting tyrosine kinases resulting in the induction of apoptosis by depletion of the "survival factors" is one of the most studied and widely accepted concepts of modern chemotherapy. We have synthesized a series of potent tyrosine kinase inhibitors and tested these compounds for apoptosis induction. Some of the tyrosine kinase inhibitors caused either apoptotic or cytoplasmic vacuolar cell death in various tumor cell cultures. The somatostatin analogue oligopeptide TT-232, which indirectly inhibits tyrosine kinases, exerted a dose-dependent apoptosis-inducing effect. The tumor growth–inhibitory effect of TT-232 and some tyrosine kinase inhibitors has also been proven by *in vivo* experiments, using human tumor xenografts. On the other hand, a dose-dependent pro- or anti-apoptotic activity of (–)-deprenyl has been shown in melanoma cell cultures, the lower doses inhibiting and the higher doses inducing apoptosis. Various metabolites of (–)-deprenyl are responsible for these actions. The effect of (–)-deprenyl is connected with depolarization of mitochondrial membranes. The kinase inhibitors act on the growth factor receptor signaling pathways (survival factor pathways) and initiate the caspase cascade. The key enzyme for the action of both pro-apoptotic and anti-apoptotic compounds is caspase 3.

KEYWORDS: apoptosis; TT-232; AG-213; deprenyl; signal transduction

INTRODUCTION

Selective inhibition of the "false" proliferative signals via targeting tyrosine-kinases resulting in the induction of apoptosis by depletion of the "survival factors" is

Address for correspondence: Dr. György Kéri, Department of Medical Chemistry, Molecular Biology and Pathobiochemistry, Semmelweis University, Puskin u. 9., Budapest H-1088, Hungary. Voice: 36-1-266-2755/4092; fax: 36-1-266-7480.
keri@puskin.sote.hu

**Ann. N.Y. Acad. Sci. 1010: 109–112 (2003). © 2003 New York Academy of Sciences.
doi: 10.1196/annals.1299.018**

one of the most studied and widely accepted concept of modern chemotherapy. We have synthesized a series of very potent tyrosine-kinase inhibitors[1–3] and tested these compounds for apoptosis induction.

The somatostatin analog oligopeptide TT-232 and the tyrphostin AG-213[4] were investigated using the HT-29 colon carcinoma cell line (ATCC = HTB38). The M-1 human melanoma cell line was used to investigate the anti-apoptotic effect of (–)-deprenyl.[5]

MATERIALS AND METHODS

Culture conditions and treatment were described earlier.[3] Cell count were taken using a Buerker's chamber and apoptotic index was determined in cover-slip cultures. Apoptotic cells were identified on the basis of morphologic signs. The effect of (–)-deprenyl was studied by the use of serum-deprived M-1 culture flow cytometric determination of apoptotic index was performed as described earlier.[5]

In Vivo *Studies*

Inbred, 8-week-old (20 g body mass) male CBA mice were immunodeprived and inoculated with HT-29 tumor. Groups of 10 mice were treated daily with 0.015, 0.15, 1.5, 15, or 30 mg/kg/i.p TT-232, respectively, and with 10 mg/kg AG-213 started 8–10 days after tumor inoculation and finished 8–10 days after the first injection. The animals were sacrificed 24 h after the last injection and the tumor mass was measured. Apoptotic and mitotic ratios of the tumors were determined by histological investigation.

RESULTS

A series of kinase inhibitors has been synthesized in our laboratory and were tested for the induction of apoptotic or Clarke III–type programmed cell death (TABLE 1).

Effects of TT-232

The somatostatin analog oligopeptide TT-232 exerted dose-dependent apoptosis-inducing effect on the HT-29 cell culture. The doses above 15 μg/mL were significantly effective. FIGURE 1 shows an example of increased apoptosis measured by flow cytometry.

In vivo, TT-232 treatment resulted in a significant decrease in tumor mass, the effective dose being 30 mg/kg/day. Histologically significant increase of apoptotic index and decrease of mitotic index were observed after this dose of TT-232.

Effect of AG-213

The tyrphostin AG-213 causes cytoplasmic non-lysosomal vacuolar cell death. This effect was reflected in a dose-dependent decrease in the number of *in vitro* cultured HT-29 cells. Our *in vivo* experiments showed retardation of the growth of HT-29 tumor xenografts in mice after the daily infections of 30 mg/kg AG-213.

TABLE 1. Type of programmed cell death caused by various kinase inhibitors

Compound	Clarke III	Apoptosis	Compound	Clarke III	Apoptosis
AG213	+++	–	HDL1322	–	+
AG17	–	+	HDL624	–	+
AG490	–	–	HDL633	–	–
AG555	++	–	HDL451	–	+
AG556	–	–	HDL622	+	–
AG879	–	–	HDL2722	–	+
AG1007	–	++	HDL2232	–	–
AG1112	++	–	HDL1735	–	+
AG1379	+	+	HDL2434	–	+
AG1317	–	++	HDL2735	–	–
AG1393	+++	–	OBF1422	–	++
AG1406	–	++	OBF1622	–	++
GO06	–	++	OBF1625	–	++
ÖL163	–	++	OBF1635	–	++
HDL1122	–	–	OBF1834	–	++

NOTE: + > 30%, ++ > 60%, +++ >90%; – = no effect.

Effect of (–)-Deprenyl

The apoptosis-inducing effect of high dose (10^{-3} M) of (–)-deprenyl and the transient apoptosis inhibitory effect of very low doses (10^{-7}-10^{-13} M) of this compound were found in cultures of M-1 melanoma cells, after serum deprivation for 5–7 days, when (–)-deprenyl was administered 3 days after serum deprivation. Caspase 3 activity was increased by 10^{-3} M (–)-deprenyl, whereas lower doses did not change the activity of this enzyme compared to the control.

DISCUSSION

The EGF receptor–specific tyrphostin causes a special type of programmed cell death, which can be defined as cytoplasmic vacuolar, non-lysosomal cell death (Clarke III).[4] The signal pathway of this type of programmed cell death is not clarified yet, but it may be induced by initiating and subsequently inhibiting apoptosis probably by disrupting the caspase cascade.[4] TT232 induces cell cycle arrest via a signaling cascade,[1–3] but this mechanism can not fully explain the strong apoptotic effect of TT232.

The MAO-B inhibitor (–)-deprenyl has been reported to exert neuroprotective effect by preventing apoptosis of the neurons in low doses.[5] The mode of action is probably related to stabilizing the mitochondrial membrane potential.[5]

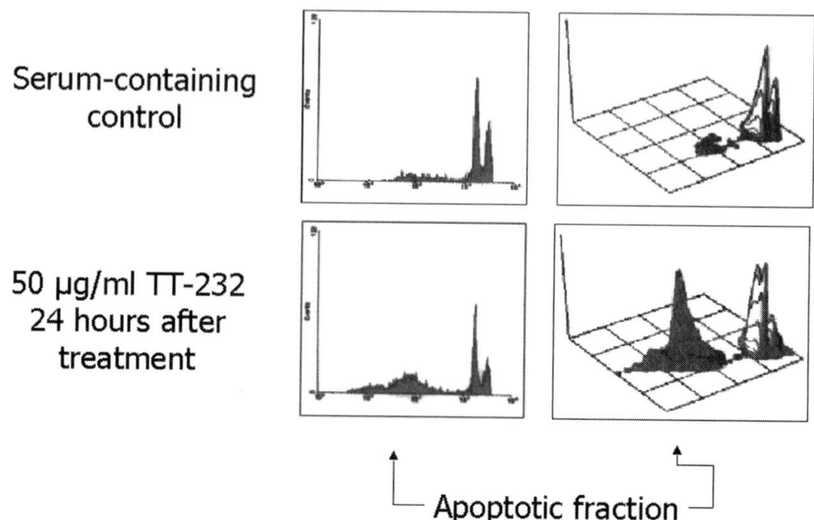

FIGURE 1. The apoptosis-inducing effect of TT-232 treatment on HT-29 human colon carcinoma cells (measured by flow cytometry).

ACKNOWLEDGMENT

This work was supported by the following grants: OTKA 32415, and NKFP-1/041 and NKFP 1/A/0020/2002.

REFERENCES

1. Õrfi, L. *et al.* 1996. Heterocondensed quinazolones: synthesis and protein-tyrosine kinase inhibitory activity of 3,4-dihydro-*1H,6H*-[1,4]-oxazino[3,4-b]quinazolin-6-one derivatives. Bioorg. Med. Chem. **4**(4)**:** 547–551.
2. Aviv, G. *et al.* 1996. Tyrphostins. 6. dimeric benzylidenemalonitrile tyrphostins: potent inhibitors of EGF receptor tyrosine kinase in vitro. J. Med. Chem. **39:** 4905–4911.
3. Kéri, G. *et al.* 1996. Tumor-selective somatostatin analog (TT-232) with strong in vitro and in vivo antitumor activity. Proc. Natl. Acad. Sci. USA **93:** 12513–12518.
4. Amin, F. *et al.* 2000. Apoptotic and non-apoptotic modes of programmed cell death in MCF-7 human breast carcinoma cells. Cell Biol. Int. **24:** 253–260.
5. Magyar, K. & B. Szende. 2000. The neuroprotective and neuronal rescue effect of (–)-deprenyl. Handb. Exp. Pharmacol. **142:** 457–472.

A Recombinant Fragment of Human Surfactant Protein D Reduces Alveolar Macrophage Apoptosis and Pro-Inflammatory Cytokines in Mice Developing Pulmonary Emphysema

HOWARD CLARK,[a] NADES PALANIYAR,[a] SAMUEL HAWGOOD,[b] AND KENNETH B. M. REID[a]

[a]*MRC Immunochemistry Unit, Department of Biochemistry, University of Oxford, Oxford, United Kingdom*

[b]*Department of Pediatrics, Cardiovascular Research Institute, University of California, San Francisco, California, USA*

ABSTRACT: Rapid removal of apoptotic cells is an important mechanism for immune homeostasis and the resolution of inflammation. Delayed clearance of apoptotic alveolar macrophages may cause activation of healthy bystander macrophages and contribute to high macrophage number and emphysema in surfactant protein D (SP-D) knock-out mice. Using flow cytometry and Annexin V and propidium iodide as markers for apoptosis and necrosis, respectively, SP-D–deficient mice were found to have a 5- to 10-fold increase in the number of apoptotic and necrotic alveolar macrophages in the lungs. SP-D–deficient mice accumulate apoptotic macrophages in the lung, and this accumulation can be reduced by treatment with recombinant SP-D (but not SP-A). The recombinant SP-D binds preferentially to apoptotic cells. The data are consistent with a specific role *in vivo* for SP-D in promoting apoptotic cell clearance in the lungs to limit macrophage-mediated inflammation and reveal a potential new mechanism for therapeutic targeting in the prevention of emphysema.

KEYWORDS: collectins; surfactant; emphysema; SP-A; SP-D

INTRODUCTION

Pulmonary emphysema is a multifactorial condition characterized by dilated airspaces and the onset of pulmonary fibrosis. By the time fibrosis has developed the changes are irreversible, so therapeutics must be directed at prevention of the inflammation causing fibrosis. The development of emphysema in smokers has been linked both to the presence of high numbers of apoptotic alveolar macrophages and to low

Address for correspondence: Dr. Howard Clark, MRC Immunochemistry Unit, Department of Biochemistry, University of Oxford, South Parks Road, Oxford OX1 3QU, United Kingdom. Voice: 01865-275795; fax: 01865-275729.

howard.clark@bioch.ox.ac.uk

Ann. N.Y. Acad. Sci. 1010: 113–116 (2003).
doi: 10.1196/annals.1299.019

levels of surfactant protein D (SP-D) in bronchoalveolar lavage (BAL).[1,2] Surfactant protein D deficiency in mice causes emphysema characterized by chronic low-grade inflammation and high numbers of alveolar macrophages with upregulated expression of reactive oxygen species and matrix metalloproteinases. We hypothesized that delayed clearance of apoptotic alveolar macrophages caused activation of healthy bystander macrophages and contributed to high macrophage number and emphysema in SP-D knock-out mice.

METHODS AND RESULTS

TUNEL assay was carried out on alveolar macrophages isolated by BAL of wild-type and SP-D knock-out mice by nick-end labeling of DNA with FITC-labeled dUTP. Direct visualiZation of macrophage cytospins by confocal microscopy showed abundant apoptotic macrophages in SP-D knock-out mice, but not in wild-type mice. Using flow cytometry and Annexin V and propidium iodide as markers for apoptosis and necrosis, respectively, we found that SP-D–deficient mice have a five- to tenfold increase in the number of apoptotic and necrotic alveolar macrophages in the lungs. Intrapulmonary replacement of a truncated 60 kDa fragment of human recombinant SP-D (rfhSP-D) to SP-D knock-out mice reduced the percentage of apoptotic and necrotic alveolar macrophages in the alveolar space, also reducing the absolute macrophage number and levels of mRNA for the pro-inflammatory cytokines MCP-1 and MIP-1alpha in the lungs (FIGS. 1 and 2).

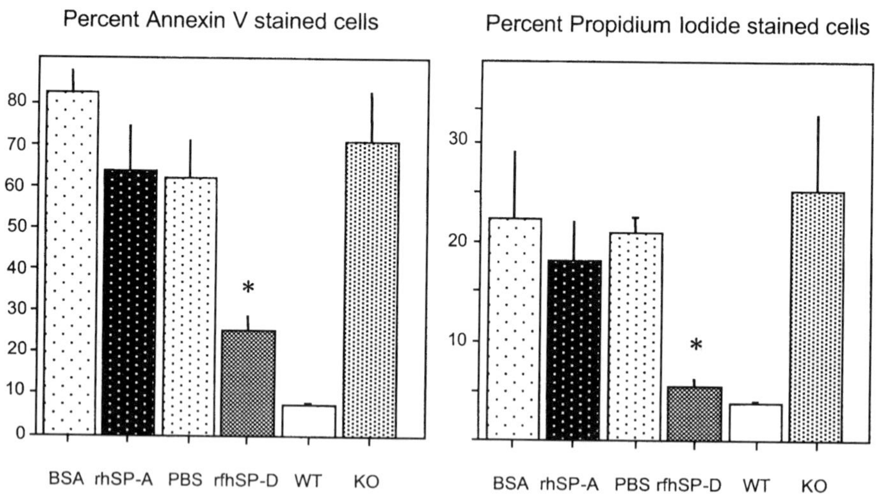

FIGURE 1. The percentage of alveolar macrophages staining with annexin V and propidium iodide from BAL of WT (wild-type) and SP-D knock-out (DKO) mice treated with saline (PBS), bovine serum albumin (BSA), surfactant protein A (SP-A), or recombinant surfactant protein D (rfhSP-D) as calculated from flow cytometry dot plots (N = 6–8 in each group, *P < .001). (Reprinted from Clark *et al.*[5] with permission.)

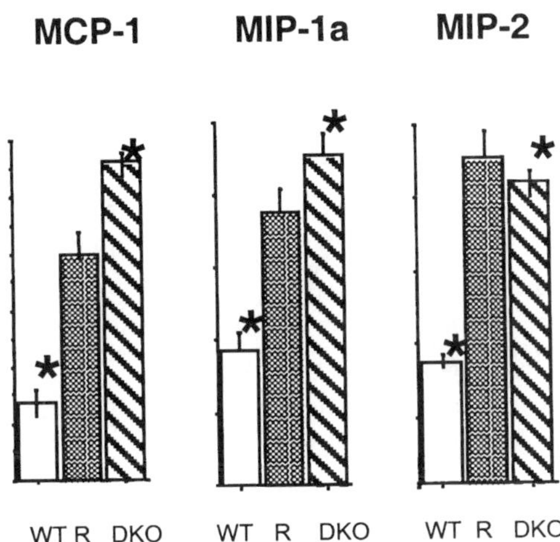

FIGURE 2. The levels of chemokine RNA for MIP-1alpha, MCP-1, and MIP-2 in total lung RNA measured in arbitrary units reflecting incorporated ^{32}P in protected RNA fragments quantified by phosphoimager and Image Quant software. Levels are corrected for unequal loading by reference to L32 and GAPDH. The effect of treatment with rfhSP-D is shown (N=3–5 mice per group, in duplicate $P < .05$). WT, wild-type mice; DKO, SP-D knock-out mice; and R, rfhSP-D–treated mice.

Double labeling of macrophages for reactive oxygen species (ROS) and annexin V demonstrated two populations of cells in knock-out mice, indicating that healthy bystander cells were activated and producing increased ROS. FITC-labeled recombinant SP-D preferentially bound to apoptotic rather than healthy alveolar macrophages, consistent with a role for SP-D in targeting the apoptotic cells for phagocytosis.

DISCUSSION AND CONCLUSIONS

Rapid removal of apoptotic cells is recognized as a centrally important mechanism for maintenance of immune homeostasis and the resolution of inflammation. If recognition and clearance of apoptotic cells by healthy macrophages is not efficient, tissue damage and prolonged low-grade inflammation could result. A number of authors have implicated lung collectins (surfactant proteins A and D) in promoting clearance of apoptotic cells. SP-A and SP-D promote clearance of apoptotic human neutrophils by alveolar macrophages *in vitro*[3] and *in vivo*.[4] The precise mechanisms involved are uncertain but appear to involve preferential recognition of the apoptotic cell surface by the collectins.[3] We have demonstrated that SP-D–deficient mice accumulate apoptotic macrophages in the lung[5] and that the number can be reduced by treatment with recombinant SP-D (but not SP-A). The recombinant SP-D binds preferentially to apoptotic cells. The data are consistent with a specific role *in vivo* for SP-D in promoting apoptotic cell clearance in the lungs to limit macrophage-mediated inflammation and reveal a potential new mechanism for therapeutic targeting in the prevention of emphysema.

REFERENCES

1. HONDA, Y., H. TAKAHASHI, Y. KUROKI, *et al.* 1996. Decreased contents of surfactant proteins A and D in BAL fluid of healthy smokers. Chest **109**(4)**:** 1006–1009.
2. MAJO, J., H. GHEZZO & M.G. COSIO. 2001. Lymphocyte population and apoptosis in the lungs of smokers and their relation to emphysema. Eur. Respir. J. **17**(5)**:** 946–953.
3. SCHAGAT, T.L., J.A. WOFFORD & J.R. WRIGHT. 2001. Surfactant protein A enhances alveolar macrophage phagocytosis of apoptotic neutrophils. J. Immunol. **166**(4)**:** 2727–2733.
4. VANDIVIER, R.W., C.A. OGDEN, V.A. FADOK, *et al.* 2002. Role of surfactant proteins A, D, and C1q in the clearance of apoptotic cells in vivo and in vitro: Calreticulin and CD91 as a common collectin receptor complex. J. Immunol. **169**(7)**:** 3978–3986.
5. CLARK, H., N. PALANIYAR, P. STRONG, *et al.* 2002. Surfactant protein d reduces alveolar macrophage apoptosis in vivo. J. Immunol. **169**(6)**:** 2892–2899.

Involvement of TNF-Related Apoptosis-Inducing Ligand (TRAIL) Induction in Interferon γ–Mediated Apoptosis in Ewing Tumor Cells

ANNIE ABADIE AND JUANA WIETZERBIN

INSERM U365, Institut Curie, Section Recherche, 26 rue d'Ulm, 75248 Paris Cedex 05 France

ABSTRACT: We investigated the effect of IFN γ or IFN γ/IFN α and IFN γ/TNF treatment on the apoptosis of Ewing tumor cells (SK-N-MC). The results show that treatment of cells with IFN γ resulted in 17% of cell death while no effect was observed when cells were treated with IFN α or TNF alone. Percentage cell death increased markedly in IFN γ/IFN α– and IFN γ/TNF–treated cells (42% and 67%, respectively) as compared to IFN γ–treated cells. These results suggest that IFN α and TNF amplified the apoptotic signal(s) induced by IFN γ. It was shown recently that Ewing cells undergo apoptosis upon treatment with recombinant TRAIL (TNF-related apoptosis inducing ligand), a member of the TNF family. Thus, since cytokines were reported to be able to induce TRAIL in other cell systems, we were prompted to investigate whether TRAIL induction was involved in the mechanism responsible for IFN γ–mediated cell death in Ewing cells. The results reported here are consistent with the notion that cell-associated and secreted TRAIL contribute in an autocrine or paracrine manner to the triggering of Ewing cell apoptosis induced by IFN γ alone or combined with IFN α or TNF. The observations reported here might contribute to the development of alternative new approaches to the treatment of Ewing tumors resistant to conventional therapy.

KEYWORDS: Ewing tumor; TRAIL; interferons; apoptosis

Ewing tumor is the second most common human bone tumor that predominantly affects children It is characterized by a t (11:22) chromosomal translocation and by the expression of the fusion protein EWS/FLI-1.[1] Although established therapeutic protocols, including local surgery and/or radiotherapy, lead to cure or remission in non-metastatic disease, conventional protocols are much less efficient in metastatic patients. Previous studies have shown that interferons and TNF are able to inhibit the growth of Ewing tumor cells and Ewing tumor xenografts.[2,3] Since induction of apoptosis is known to mediate antitumor effects, we investigated the possible trigger-

Address for correspondence: J. Wietzerbin, INSERM U365, Institut Curie, Section Recherche, 26 rue d'Ulm, 75248 Paris Cedex 05, France. Voice: +33-1-43-25-82-67; fax: +33-1-44-0- 07-85. jwietzer@curie.fr

Ann. N.Y. Acad. Sci. 1010: 117–120 (2003). © 2003 New York Academy of Sciences.
doi: 10.1196/annals.1299.020

TABLE 1. Induction of apoptosis and TRAIL expression in cytokine-stimulated Ewing cells

Treatment	% Cell Death	TRAIL mRNA 12 h	TRAIL mRNA 24 h	TRAIL protein 30 h
Control	0	–	–	–
IFN γ	17±2.25	<+	–	<+
IFN α + IFN γ	42±4.45	+	++	++
TNF α + IFN γ	67±7	++	+++	+++

NOTE: Ewing Cells (SK-N-MC from ATCC) were seeded at 2×10^5/mL. After 20-h culture to allow attachment, the cells were stimulated for 6 to 30 h with 1,000 U/mL of IFN α, IFN γ, or TNF α alone or in combination. Cell death was determined by the WST1 method (Roche) and expressed in % dead cells. Results correspond to the means of at least three independent experiments ± SEM. TRAIL mRNA was assessed by RT-PCR using specific primers for TRAIL reported previously[6]: forward 5′-CAACTCCGTCAGCTCGTTAGAAAG-3′ and reverse 5′-TTAGACCAACAACTATTTCTAGCACT-3′. TRAIL protein was evaluated in Western blotting with a goat polyclonal anti-TRAIL antibody (AF375 from R&D) and revealed by ECL.

ing by interferons and TNF of the apoptosis of Ewing tumors to contribute to the development of new pharmacological approaches.

RESULTS AND DISCUSSION

Induction of Apoptosis, TRAIL mRNA, and TRAIL Protein

We investigated the effect of IFN γ or IFN γ/IFN α and IFN γ/TNF treatment on the apoptosis of Ewing tumor cells (SK-N-MC). The results in TABLE 1 show that treatment of cells with IFN γ resulted in 17% of cell death while no effect was observed when cells were treated with IFN α or TNF alone. Percentage cell death increased markedly in IFN γ/IFN α- and IFN γ/TNF–treated cells (42% and 67%, respectively) as compared to IFN γ–treated cells. These results suggest that IFN α and TNF amplified the apoptotic signal(s) induced by IFN γ. Apoptosis was confirmed by APO 2-7 flow cytometry analysis and fragmentation of cell nuclei (not shown).

It was shown recently that Ewing cells undergo apoptosis upon treatment with recombinant TRAIL (TNF-related apoptosis inducing ligand), a member of the TNF family.[4,5] Thus, since cytokines were reported to be able to induce TRAIL in other cell systems,[6] we were prompted to investigate whether TRAIL induction was involved in the mechanism responsible for IFN γ–mediated cell death in Ewing cells. As summarized in TABLE 1, TRAIL mRNA was undetectable in control cells and in cells stimulated with TNF or IFN α alone. Treatment of cells with IFN γ alone induced transient and weakly detectable TRAIL mRNA expression from 6 to 12 h. Combined IFN γ/IFN α or IFN γ/TNF treatment markedly enhanced the level of TRAIL mRNA, which was already detectable after 6 h stimulation, with further increase up to 24 hours.

Western blot analysis of lysates of stimulated cell using a specific polyclonal antibody (AF 375 from R & D) allowed detection of TRAIL protein. Flow cytometry analysis using a different anti-TRAIL monoclonal antibody (2 E5 from Alexis) confirmed that TRAIL protein was synergistically induced in cells stimulated for 30 h with IFN γ/IFN α and IFN γ/TNF.

Secretion of TRAIL by Ewing Cells

To examine whether TRAIL protein was secreted following stimulation with cytokines, TRAIL concentrations were determined in cell culture supernatants using

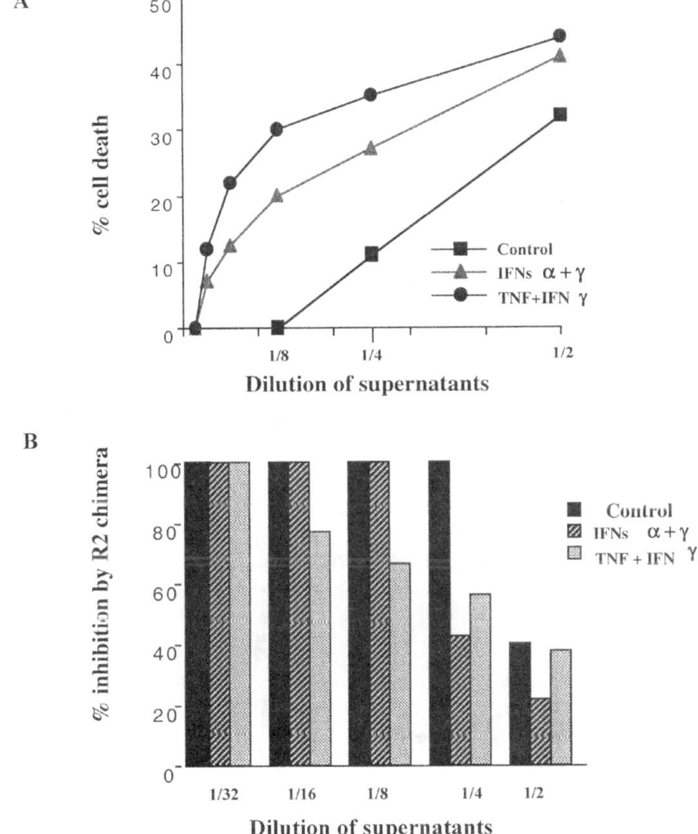

FIGURE 1. Dilution dependent cytotoxicity of sypernatants. Ewing cells (SK-N-MC) were seeded at 10^5/0.5 mL in RPMI 2% FCS in 24-well Costar plates and allowed to attach overnight. The medium was then replaced by (**A**) twofold serial dilutions of inducer-free supernatants; (**B**) the same dilutions preincubated for 1 h at room temperature with 10 μg/mL of TRAIL/ R2Fc chimera (R&D). After 24 h of incubation, cell death was determined by the WST1 method (Roche) and expressed in % dead cells (**A**) as the mean of three independent experiments, (**B**) % cytotoxicity inhibition of one representative experiment out of three.

a specific ELISA test (Function ELISA TRAIL from Active Motif). Interestingly, supernatants from stimulated cells contained considerable amounts of soluble TRAIL since around 0.6 to 0.7 ng of TRAIL was released per 10^6 cells stimulated by either IFN γ/IFN α or IFN γ/TNF combinations for 30 hours. Low amounts of TRAIL (less than 0.3 ng/10^6 cells) were also spontaneously secreted by unstimulated cells.

In order to investigate the apoptotic properties of supernatants containing secreted TRAIL without the presence of the inducing cytokines, cells were stimulated for 6 h with IFN γ/IFN α and IFN γ/TNF, carefully washed and resuspended in fresh medium. The "inducer-free" supernatants recovered after 2 days of culture were shown to induce the death of Ewing cells in a dilution-dependent manner as expected for a dose-dependent effect (FIG. 1A). Control supernatants showed a cytotoxic effect at low dilutions. The recombinant TRAIL-R2/Fc chimera (631-T2 from R&D), which neutralizes recombinant TRAIL activity (not shown), partially inhibited the cell death triggered by the supernatants (FIG. 1B), suggesting that secreted TRAIL present in the supernatants contributes to the apoptotic effect observed.

Taken together, the results reported here are consistent with the notion that cell-associated and secreted TRAIL contribute in an autocrine or paracrine manner to the triggering of Ewing cell apoptosis induced by IFN γ alone or combined with IFN α or TNF. Although more work needs to be done, particularly in animal models, the observations reported here might contribute to the development of alternative new approaches to the treatment of Ewing tumors resistant to conventional therapy.

ACKNOWLEDGMENT

This work was supported by the Institut National de la Santé et la Recherche Medicale (INSERM), the Association pour la Recherche sur le Cancer (ARC), and BioSidus, Argentine.

REFERENCES

1. DELATTRE, O., J. ZUCMAN, B. PLOUGASTEL, *et al.* 1992. Nature **359:** 162–165.
2. VAN VALEN, F., W. WINKELMANN, S. BURDACH, *et al.* 1993. J. Cancer Res. Clin. Onc. **119:** 615–621.
3. SANCÉAU, J., M.F. POUPON, O. DELATTRE, *et al.* 2002. Oncogene **21:** 7700–7709.
4. WILEY, S.R., K. SCHOOLEY, P.J. SMOLAK, *et al.* 1995. Immunity **l3:** 673–682.
5. KONTNY, H.U., K. HÄMMERLE, R. KLEIN, *et al.* 2001. Cell Death Differ. **8:** 506–514.
6. GRIFFITH, T.S., S.R. WILEY, M.Z. KUBIN, *et al.* 1999. J. Exp. Med. **189:** 1343–1351.

Cell Death in Protists without Mitochondria

OLIVIER CHOSE,[a] CLAUDE-OLIVIER SARDE,[b] CHRISTOPHE NOËL,[c] DELPHINE GERBOD,[c] JUAN-CARLOS JIMENEZ,[c] CATHERINE BRENNER,[d] MONIQUE CAPRON,[c] ERIC VISCOGLIOSI,[c] AND ALBERTO ROSETO[a]

[a]*Laboratoire de Génie Enzymatique et Cellulaire, UMR CNRS 6022, Université de Technologie de Compiègne, B.P. 20529, 60205 Compiègne Cedex, France*

[b]*Département de Génie Biologique, Université de Technologie de Compiègne, B.P. 20529, 60205 Compiègne Cedex, France*

[c]*Institut Pasteur de Lille, Unité Mixte INSERM-IPL U547, 1 Rue du Professeur Calmette, B.P. 245, 59019 Lille Cedex, France*

[d]*CNRS FRE 2445, Biochimie de l'Apoptose, Université de Versailles/St. Quentin, Bâtiment Buffon, 45 Avenue des Etats-Unis, 78035 Versailles, France*

ABSTRACT: Some protozoans, such as Trichomonad species, do not possess mitochondria. Most of the time, they harbor another type of membrane-bounded organelle, called hydrogenosome from its capacity to produce H_2. This is the case for the human parasite *Trichomonas vaginalis*. Some other parasites, such as the protist *Giardia lamblia*, do not harbor any of these organelles. From this observation arises naturally a naive question: How do cells die when the mitochondrion, the cornerstone of apoptotic process, is absent? Data strongly suggest that the mitochondrion and the hydrogenosome arose from a common ancestral endosymbiont. But hydrogenosomes do not appear to directly substitute for mitochondria in apoptotic functions. Thus, it appears judicious to examine more closely the genome of unicellular cells, which do not harbor mitochondria, and search for new molecules that could participate in the apoptotic process in these microorganisms.

KEYWORDS: programmed cell death; protist, *T. vaginalis*, *G. lamblia*

The programmed way in which most eukaryotic cells die is highly conserved from worm to mammal. Programmed cell death (PCD) is also unambiguously present in numerous protozoans harboring mitochondria. In this process, the central role of mitochondrion is widely recognized. Indeed, mitochondrion supplies or processes various essential molecules for death program execution, acting as an "integrator box" of the death signal.[1] Two redundant parallel pathways hinge on the mitochondrial core: the caspase-dependent pathway that relies on the proteolytic activity of caspases and the caspase-independent pathway in which the apoptosis-inducing factor (AIF) plays a crucial role[2] (FIG. 1). Whatever pathway is used, the engaged death

Address for correspondence: Alberto Roseto, Université de Technologie de Compiègne, Laboratoire Génie Enzymatique et Cellulaire, UMR CNRS 6022, 1 Rue Personne de Roberval, B.P. 20529, 60205 Compiègne Cedex France. Fax: +33 (0)3 44 20 39 10.
alberto.roseto@utc.fr

Ann. N.Y. Acad. Sci. 1010: 121–125 (2003). © 2003 New York Academy of Sciences.
doi: 10.1196/annals.1299.021

TABLE 1. Comparative analysis of features present in various cell death forms

	Necrosis	Apoptosis	Paraptosis	*T. vaginalis* PCD	*G. lamblia* PCD
Morphology					
Nuclear fragmentation	–	+	–	+	+
Chromatin condensation	–	+	+/–	+	+
Apoptotic bodies	–	+	–	+	+
Cytoplasmic vacuolation	+	–	+	+	+
Mitochondrial bubbling	+	sometimes	late	ND	ND
Genomic effects					
TUNEL	–	+	–	+	ND
Internucleosomal fragmentation	–	+	–	?	ND
Inhibition with		+			
ZVAD.fmk	–	+	–	+	ND
Induction with					
Staurosporin	–	+	ND	+	+
Etoposide	-	+	ND	+	+

program triggers the formation of apoptotic bodies, a phenomenon that includes chromatin condensation and simultaneous nucleosomal fragmentation of DNA. Dying cells also undergo protein cleavages that provoke genuine morphological changes, such as nuclear membrane breakdown and externalization of phosphatidylserine residues. Some protozoa, such as Trichomonad species, do not possess mitochondria. Most of the time, they harbor another type of membrane-bounded organelle, called hydrogenosome from its capacity to produce H_2.[3] This is the case for the human parasite *Trichomonas vaginalis*. Some other parasites, such as the protist *Giardia lamblia*, do not harbor any of these organelles. From this observation arises naturally a naive question: How do cells die when the mitochondrion, the cornerstone of apoptotic process, is absent?

To answer this intriguing question, *T. vaginalis* cells were treated with pro-apoptotic drugs or stressed by nutriment depletion.[4] A form of cell death with some but not all features of canonical PCD could be observed. Indeed, when epifluorescence microscopy was performed, the various aspects of chromatin condensation described in PCD for unicellular and multicellular organisms harboring mitochondria were present. Similar results were obtain when *G. lamblia* cells were treated in the same way (in preparation). Observed morphologies are common to apoptosis and paraptosis, a recently reported PCD alternative form[5] (TABLE 1). In *T. vaginalis*–treated cells, cellular membrane distension and exposure of annexin V–associated phosphatidyl serine residues on the outer leaflet of the plasma membrane were interpreted as additional common denominators with canonical PCD. However, contrary to what is described for dead cells harboring mitochondria, *T. vaginalis* dead cells exhibited a TUNEL-positive DNA fragmentation without any specific scaling pattern. In canonical PCD, DNA is known to undergo a recursive sharing process that in turn

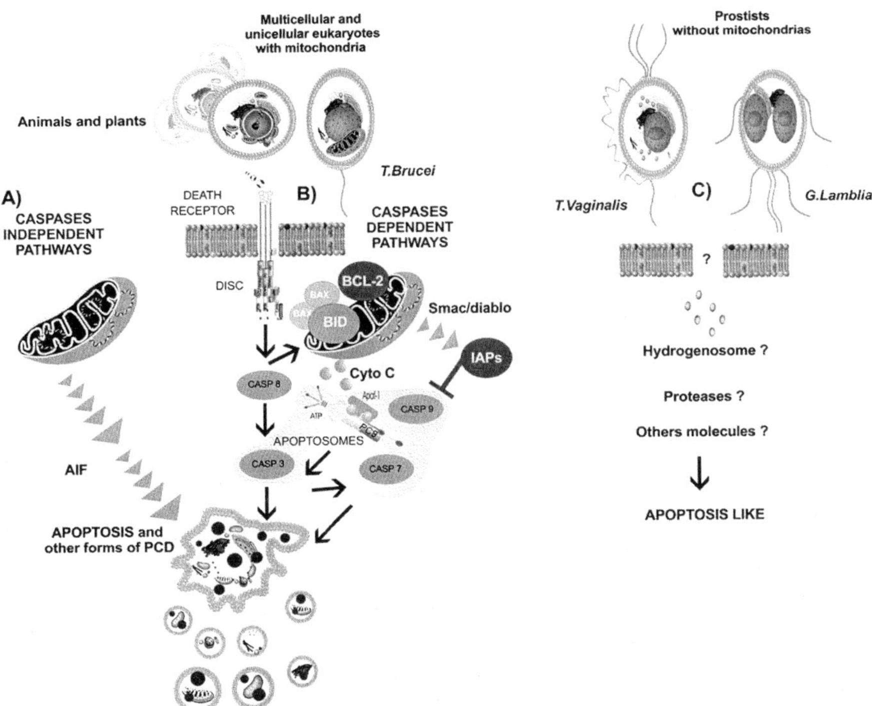

FIGURE 1. Programmed cell death in eukaryotic cells with or without mitochondria. (**A**) Endonuclease G and AIF trigger directly the nuclear DNA and represent pre-eminently molecules involved in the caspase-independent pathway. (**B**) The caspase-dependent death process is activated by two major pathways: (1) the extrinsic pathway initiated with the recruitment of procaspase 8 by ligand-activated death receptor inside the death-inducing signaling complex (DISC) and (2) the intrinsic pathway initiated by cytochrome *c* release from the mitochondria that may either, as a part of the apoptosome, activate caspase 8 and 9 or directly activate caspase 3 and 7. After the disruption of the mitochondrial membrane, various mitochondrial proteins can trigger the execution of PCD (or other apoptosis-like processes): cytochrome *c* (which belongs with Apaf1, ATP, and procaspase 9 to the apoptosome), Smac/Diablo and Omi/HtrA2. These molecules are members of different loops of the caspase-dependent pathway. Finally, active caspases can be counteracted by members of the inhibitor of apoptosis proteins (the IAP family). Smac/Diablo and Omi/HtrA2 can, in turn, neutralize the apoptotic activity of IAPs. (**C**) In protists without mitochondria, most (if not all) of known death proteins are absent. (Modified from Chose *et al.* Trends Parasitol. In press.)

generates fragments of defined length, as seen from the typical electrophoretic ladder profile. When the same analysis was performed on *T. vaginalis* dead cells, a low molecular weight persistent smear devoid of any banding pattern was observed recurrently. This feature unambiguously differentiates *T. vaginalis* PCD from other described cell death forms and, moreover, suggests the existence of a different underlying DNA fragmentation mechanism in organisms devoid of mitochondria.

It is generally admitted that caspases do not exist outside of the animal kingdom, although caspases-like proteases were recently described in plants and unicellular eukaryotes. When tested, an effective action of caspases activators and inhibitors was ascertained on both *T. vaginalis* and *G. lamblia* cells. This result suggests the involvement of caspase-like proteases in PCD in these protists. The presence of a potential caspase-dependent death pathway, the reality of which remains to be investigated, does not exclude however the co-existence of an apoptotic-like caspase-independent pathway. Recent advances in *G. lamblia* genome project may provide valuable information about the molecules involved in what appears to us to be "a new way of dying." Until now, however, *G. lamblia* sequence analysis performed at protein and DNA levels always failed to detect an equivalent to any PCD-involved molecules tested, including AIF and caspase-related compounds.

When they are present, the contribution of hydrogenosomes to cell energy supply through ATP synthesis is around 30%, most of the energy being issued from cytoplasmic anaerobic metabolism. This leads to the opinion that hydrogenosomes do not play a central role in energy production of protists devoid of mitochondria, a radically different situation from energy supply in unstressed animal cells which is essentially due to the mitochondrial contribution. This brought the role of hydrogenosomes in the apoptotic process into question. Recently, phylogenetic analyses of genes present in *T. vaginalis* indicate that some of them have a mitochondrial origin.[6] In addition, it has been shown that the mechanism whereby proteins are targeted to hydrogenosomes is similar to mitochondrial protein targeting. In other respects, an homolog of the mitochondrial carrier family has been identified in hydrogenosomes. All these data strongly suggest that the mitochondrion and the hydrogenosome arose from a common ancestral endosymbiont. Therefore, it is tempting to believe that hydrogenosomes may in some way substitute for mitochondria in apoptotic functions. But, so far, none of the proteins directly implicated in either caspase-dependent and caspase-independent pathways have been described in *T. vaginalis*, suggesting that this protist could undergo PCD by a different molecular pathway. Moreover, the lack of detection of any equivalent for most of the genes known to be required for PCD in *G. lamblia*'s almost-complete genomic sequence is somehow perplexing. Last but not least, the fact that hydrogenosomes are not present in *G. lamblia* appears to be a major impediment if one figures that *T. vaginalis* and *G. lamblia* may undergo PCD through closely related pathways.

Mitochondrion and hydrogenosome are conceptually derived from endosymbiotic alpha-proteobacteria that were early acquired by cells during eukaryogenesis. This evolutionary event was obviously after the emergence life and death and to our knowledge, most if not all proteins implicated so far in PCD are encoded by nuclear DNA. Thus, it appears judicious to examine more closely the genome of unicellular cells, which do not harbor mitochondria, and search for new molecules that could participate to PCD process in these microorganisms. We are convinced that forthcoming results from presumed early-diverging protists such as *T. vaginalis* and *G. lamblia* should be helpful in understanding how PCD emerged in eukaryotes.

REFERENCES

1. BRENNER, C. & G. KROEMER. 2000. Apoptosis: mitochondria—the death signal integrators. Science **289:** 1150–1151.
2. SUSIN, S.A *et al.* 2000. Two distinct pathways leading to nuclear apoptosis. J. Exp. Med. **192:** 571–579.

3. MÜLLER, M. 1997. The hydrogenosome. Parasitol. Today **13:** 166–167.
4. CHOSE, O. *et al.* 2002. A form of cell death with some features resembling apoptosis in the amitochondrial unicellular organism *Trichomonas vaginalis*. Exp. Cell. Res. **276:** 32–39.
5. MCARTHUR, A.G. *et al.* 2000. The *Giardia* genome project database. FEMS Microbiol. Lett. **189:** 271–273.
6. EMBLEY, T.M. & R.P. HIRT. 1998. Early branching eukaryotes? Curr. Opin. Genet. Dev. **8:** 624–629.

Study of PTPC Composition during Apoptosis for Identification of Viral Protein Target

FLORENCE VERRIER,[a] BERNARD MIGNOTTE,[a] GWENAËL JAN,[b] AND CATHERINE BRENNER[a]

[a]*CNRS FRE 2445, Université de Versailles/St. Quentin, 45, avenue des Etats-Unis, 78035 Versailles, France*

[b]*Institut National de la Recherche Agronomique, UR 121, Laboratoire de Recherches de Technologie Laitière, 35042 Rennes Cedex, France*

ABSTRACT: The permeability transition pore complex (PTPC), a mitochondrial polyprotein complex, has been previously described to be involved in the control of mitochondrial membrane permeabilization (MMP) during chemotherapy-induced apoptosis. PTPC may contain proteins from both mitochondrial membranes [e.g., voltage-dependent anion channel (VDAC), PRAX-1, peripheral benzodiazepine receptor (PBR), adenine nucleotide translocator (ANT)], from cytosol (e.g., hexokinase II, glycerol kinase), from matrix [e.g., cyclophilin D (CypD)], and from intermembrane space (e.g., creatine kinase). PTPC may also interact with tumor suppressor proteins (i.e., Bax and Bid), oncoprotein homologues of Bcl-2 and some viral proteins, which can regulate apoptosis induced by pore opening. ANT and VDAC are the target of numerous pro-apoptotic MMP inducers. However, the precise composition of PTPC as well as the respective role of each PTPC component represent major issues in the understanding MMP process. Using several experimental strategiesthat combine co-immunoprecipitation, proteomics, and functional tests with proteoliposomes, we and others have been able to characterize some of the intra/inter-PTPC protein interactions leading to a better understanding of the process of MMP. In addition, this approach could identify new putative members and regulators of PTPC pro-apoptotic function and new targets of viral protein involved in the modulation of apoptosis during infection.

KEYWORDS: mitochondrial membrane permeabilization; tumor suppressor proteins; viral proteins

Mitochondria are well known as sites of electron transport and generators of cellular ATP, but they have also been described as the site of the regulation of cell survival. Mitochondrial membrane permeabilization (MMP) is a key event leading to physiological or therapy-induced apoptosis.[1,2] Some evidence proposes that this

Address for correspondence: Catherine Brenner, CNRS FRE 2445, Université de Versailles/St. Quentin, 45, avenue des Etats-Unis, 78035 Versailles, France. Voice: 33 1 39 25 45 52; fax: 33 1 39 25 45 72.

cbrenner@genetique.uvsq.fr

Ann. N.Y. Acad. Sci. 1010: 126–142 (2003). © 2003 New York Academy of Sciences.
doi: 10.1196/annals.1299.022

event can be caused by the opening of the mitochondrial megachannel, also called the permeability transition pore complex (PTPC), and is controlled by Bcl-2 family members[3,4] and by a variety of apoptosis-inducing stimuli, such as pro-oxidant agents,[5,6] adenine nucleotide translocator ligands,[5] viral proteins,[7–9] bacterial metabolites (propionate and acetate[10]), as well as chemotherapeutic agents.[11,12] This review will focus on characterization of PTPC members and modulator proteins that can physically and functionally interact with these proteins during apoptosis.

METHODOLOGY

In order to identify new members associated with and/or regulating the PTPC and thus associated with the regulation of the mitochondrial phase of apoptosis as well as proteins that are the target of viral proteins, we used the methodology described in FIGURE 1. Mitochondria were first purified from healthy rat organs (heart, liver, and brain) or from human tumoral cell lines and solubilized with Triton X-100 detergent. Proteins interacting with PTPC were isolated by immunoprecipitation using a polyclonal rabbit serum specific to adenine nucleotide translocator (ANT), the major mitochondrial inner membrane (IM) protein. Immunocomplexes were then characterized by one-dimensional electrophoresis (1D-SDS-PAGE) and western-blot[13] or by two-dimensional electrophoresis (2D-SDS-PAGE) followed by proteomic.[14] To study the functional role of these proteins in the regulation of the pore opening, two tests have been used. PTPC, purified from rat brain mitochondria,[15,16] or ANT, purified from rat heart mitochondria,[17,18] were reconstituted in liposomes or planar lipid bilayers[4] (FIG. 1). They have been shown to exhibit a functional behavior comparable with that of the PTPC present in intact mitochondria. In these models, PTPC or ANT form non-specific pores. In the ADP/ATP translocase assay, translocation function of ANT was measured by the release of ATP preloaded in ANT-liposome induced by exogenous addition of ADP. Quantification of ATP released was measured by emitted luminescence after addition of luciferase.[11] In the pore-opening assay, 4-methylumbelliferyl phosphate (MUP) was encapsulated into ANT-liposome or PTPC-liposome. The conversion into a non-specific pore was measured by the release of MUP, which reacts with alkaline phosphatase and is converted into fluorochrome 4-methylumbelliferone (MU).[5] The electrophysiological assay, as previously described,[4,19] was used to investigate pore-forming activities of ANT, Bax, Bcl-2, and mutants of Bax and Bcl-2, at both macroscopic (black membrane) and single-channel (patch clamp) levels.[4] By this technique it has been shown that ANT, after addition of atractyloside (Atr) or Ca^{2+}, mediated a significant conductance in planar lipid bilayers and that the channel formed by ANT is cation specific.[4] Furthermore, Bax and Bcl-2 capacity to interact with ANT and to form new channels has been characterized and is described later in this review. Using the proteoliposome method, PTPC has been shown to be the target for apoptosis regulation by caspases and Bcl-2–related proteins. Although Ca^{2+}, pro-oxidants, and several recombinant caspases enhance the permeability of PTPC-containing liposomes, recombinant Bcl-2 or Bcl-XL increase the resistance of the reconstituted PTPC to pore opening.[20] Moreover, the same methodology led us to demonstrate that ANT possesses two opposite functions, ADP/ATP translocator in physiological conditions and lethal pore in apoptosis.[13] This assay has also been used to identify molecules

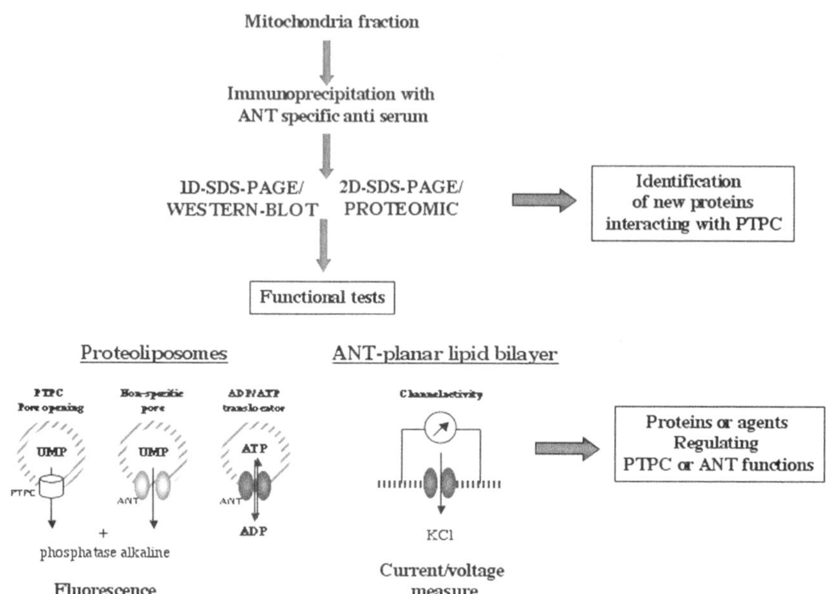

FIGURE 1. Experimental methods to identify physical and functional interactions of proteins and agents with the PTPC members. ANT, purified from rat heart mitochondria, or PTPC, purified from rat heart mitochondria, is reconstituted into phosphatidyl/cardiolopin liposomes where 4-methylumbelliferyl phosphate (MUP) was encapsulated. The conversion into a non-specific pore was measured by the release of MUP, which reacts with alkaline phosphatase to generate a fluorescent signal. Moreover, ANT-dependent translocase activity can be measured using ATP-preloaded ANT-proteoliposomes. These proteoliposomes are incubated with ADP and the ATP released was quantitatively measured by a luciferase chemoluminescence assay. ANT or PTPC were incorporated in planar lipid bilayers to measure their electrophysiological properties, such as single channel activity or macroscopic conductance, opening frequency, and ionic specificity.

modulating the pro-apoptotic function of ANT, such as viral proteins and chemotherapeutic agents.[21] Thus, atractyloside, Ca^{2+}, Bax, thiol cross-linkers (diazenedicarboxylic acid-*bis*-5*N*,*N*-dimethylamide, dithiodipyridine, *bis*-maleimido-hexane, tert-butylhydroperoxide), nitric oxide,[5,6] Vpr from HIV-1,[7,8] short-chain fatty acids,[10] or chemotherapeutic agents (such as lonidamine, arsenic trioxide, CD437, or verteporfin) can convert ANT into a non-specific pore.[11,12] In contrast bongkrekic acid (BA), Bcl-2, vMIA, a cytomegalovirus-encoded protein,[9] peptide ANT104-116, ATP, and ADP inhibit the pore formation. Finally, using proteoliposome, we showed that Bax, co-reconstituted with ANT increased the Atr-dependent pore opening since Bcl-2 decreased this effect.[3] In contrast Bax inhibits translocase activity of ANT since Bcl-2 facilitates ATP/ADP translocation.[13]

Viral proteins have been described to regulate apoptosis at a mitochondrial level, as described later in this review. The methodology described above (proteoliposome) could be used to characterize functional interaction between viral protein and mitochondrial proteins, as it has been previously described with Vpr and vMIA. Indeed, identification of new mitochondrial protein targets for viral proteins could be achieved by co-immunoprecipitation, in infected cells or in cells overexpressing viral proteins. Furthermore, knowing the mitochondrial target, cytoprotector peptides derived from the target sequence could be used to inhibit viral/mitochondrial proteins interaction and to prevent virus-induced apoptosis.

PTPC COMPOSITION EVOLUTION

Classical Members of PTPC

The PTPC is a polyprotein complex formed in the contact site between the inner (IM) and the outer mitochondrial membranes (OM), which participate in the mitochondrial matrix homeostasis and the regulation of MMP, a process leading either to apoptosis, necrosis, or autophagy.[21] In physiological conditions, PTPC controls mitochondrial Ca^{2+} homeostasis via the regulation of its conductance, by the mitochondrial pH, the inner transmembrane potential $\Delta\Psi m$, NAD/NAD(P)H redox equilibrium, and matrix protein thiol oxidation.[22-28] As measured by electrophysiology, PTPC exhibits multiple conductance states, suggesting that the channel is composed of several cooperating subunits.[29] Its opening as a large unselective channel results in an increase in the permeability of the IM, permitting the influx or efflux of any molecule of MM$\leq$1500 Da. By contrast, in lethal conditions, the extended opening of PTPC would lead to an irreversible MMP, a rupture of the OM and the release of apoptogenic proteins into the cytosol, which triggers the ultimate degradation phase of apoptosis and the cell death.[2,21,30] Depending on the conditions of apoptosis induction, PTPC opening has been shown to precede Bax translocation from cytosol to mitochondria[31] or to occur concomitantly with Bax translocation, but before cytochrome *c* or apoptosis-inducing factor (AIF) release.[2] Thus, because of its capacity to control MMP, alterations of PTPC function may lead to excess or lack of apoptosis and to numerous pathologies, such as ischemia-reperfusion,[32] cardiomyopathy,[33] neurodegeneration,[34] lupus erythematosus,[35] or cancer. The precise composition of PTPC has been a matter of debate. Based on just co-purification experiments, several non-mutually exclusive models have been proposed. The first model described the PTPC as composed by three proteins, ANT (the major protein of the IM), VDAC (the most abundant protein of the OM), both enriched at mitochondrial contact site, and the peripheral benzodiazepin receptor (PBR, in the OM),[36] implicating this protein complex in lipid metabolism by facilitating the transfer of cholesterol to the IM in steroidogenic cells.[37] Others have described PTPC as composed of ANT, VDAC, and cyclophilin D (CypD, in the matrix)[38] or as a channel unit formed by ANT and VDAC, this channel being regulated by CypD.[39] Finally, a model has been proposed in which PTPC contains, at least, hexokinase (HK, in the cytosol), PBR, VDAC, ANT, creatine kinase (CK, in the intermembrane space), and CypD (in the matrix) (FIG. 2).[15,20,21,40] The association of proteins to ANT/VDAC complexes may be related to cellular energy metabolism. Indeed, creatine kinase (mitochondrial intermembrane space) has been shown to be associated with the ANT/VDAC complex.[41] Thus ATP can translocate via ANT to generate creatine phosphate, which is then exported via VDAC to the cytosol where it is highly diffusible. In the search for the precise composition of PTPC, we found—by coim-

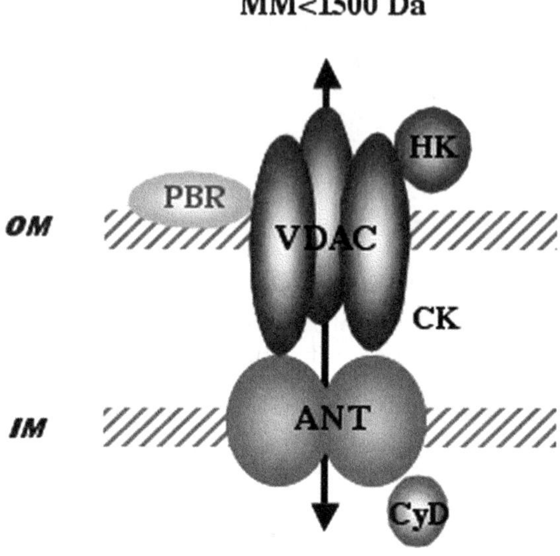

FIGURE 2. Scheme of the putative PTPC members.

munoprecipitation of solubilized mitochondrial membrane with an anti-ANT serum, followed by 1D-SDS-PAGE and western-blot analysis—that ANT associated with VDAC, HK, Bcl-2, PBR, CypD, in mitochondria isolated from various normal tissues (rat brain, heart, and liver) and three human cancer cell lines (HT29, HeLa, and MCF-7).[14] This observation is consistent with the last model previously described. Furthermore by 2D-SDS-PAGE analysis, we showed that almost 10 major proteins are co-immunoprecipitated with ANT and that some interactions seem to be favored during apoptosis induction, while others seem to be disrupted.[14]

Moreover, we and other groups have demonstrated, that PTPC function could be modulated by physical interaction of some of its members with onco-and anti-oncoproteins from the Bax/Bcl-2 family. Thus, ANT and VDAC were identified as ligands of Bax, Bcl-2, Bid, Bcl-XL,[18,23,42,43] as regulators of channel activity,[4,44] and as modulators of the ADP/ATP translocase activity of ANT.[13]

Bax/Bcl-2 Family Members

Bcl-2 family members have been described to regulate and control the mitochondrial phase of apoptosis.[45,46] They can be grouped into three categories: the anti-apoptotic members, including Bcl-2, Bcl-xL, Bcl-w, Mcl-1; the multi-domain proapoptotic members, such as Bax and Bak; and the BH3 domain–only proteins, including Bim, Bid, Bad, and Bik.[47] Generally, proapoptotic members are localized in the cytosol and move to mitochondria upon induction of apoptosis, whereas anti-apoptotic members are localized in intracellular membranes, being predominantly enriched in the contact sites of mitochondrial membranes where the PTPC is also localized.[48] Different roles have been conferred to Bcl-2–like proteins: autonomous

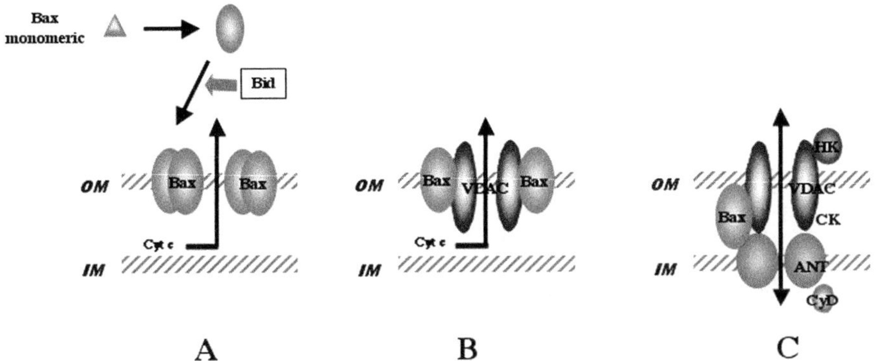

FIGURE 3. Putative mechanisms of Bax to induce outer mitochondrial rupture. (**A**) During apoptose, cytosolic Bax conformationally changes and under Bid effect can translocate and multimerize in the outer membrane where it forms large channels that can lead to the release of cytochrome *c*. (**B**) Bax and VDAC together form a novel channel that permits cytochrome *c* release. (**C**) Bax may regulate opening of PTPC via an interaction with ANT.

channel formation, mitochondrial channel regulation, increase of cell resistance to pro-oxidant stress or interaction with members of the apoptosome.[49–54] However, their exact apoptosis mode of action is still elusive. Various experimental models have been used, such as artificial membrane, isolated mitochondrial membrane, extracted mitochondrial proteins, or recombinant protein, and several hypothetical mechanisms have been proposed to explain their regulatory actions on apoptosis. Two major mechanisms for permeabilization of the OM have been suggested and summarized in FIGURE 3: (1) the formation of specific channels into the OM (Bax, Bak, Bcl-xL, and Bid form channels; Bax destabilized the mitochondrial membrane thus inducing lipidic holes,[55] Bax forms chimerics channels with VDAC or ANT) and (2) opening of the PTPC leads to mitochondrial matrix swelling and rupture of the OM.

Under physiological cellular conditions, Bax has been described to be monomeric and cytosolic.[56,57] During apoptosis, Bax changes its conformation, maybe by change in cytosolic pH,[56] by direct interaction with Bid,[58,59] or as the result of changes in mitochondrial membrane potential.[31] Bax also translocates to the mitochondria and oligomerizes, causing cytochrome *c* release from mitochondria.[58,60,61] Multimerization has been shown to be sufficient for Bax translocation[57] and oligomeric Bax is able to induce cytochrome *c* release from isolated mitochondria.[60] It has been suggested that anti-apoptotic Bcl-2 family members could inhibit Bax conformational change by direct interaction.[62] Moreover, anti-apoptotic Bcl-2 members can inhibit cytochrome *c* release from mitochondria by preventing translocation and/or activation of Bax on the mitochondria.[63,64] Bcl-2 and Bcl-xL can also inhibit apoptosis in a Bax-independent manner.[65,66] On the other hand, if BH3 proteins can activate pro-apoptotic Bcl-2 members, they have been described as inducing apoptosis by binding anti-apoptotic Bcl-2 members and inhibiting their activity.[67,68]

One mechanism for the release of cytochrome *c* and other pro-apoptotic proteins from mitochondria intermembrane space is the formation of specific channels in the OM. Some Bcl-2 family members possess structural similarities to pore-forming bacterial toxins,[69-71] and Bcl-xL, Bid, Bcl-2, and Bax have been shown to form protein artificial membranes.[50,72] However, it has also been shown that Bid, on its own, has no effect on cytochrome *c* release *in vivo*.[58] Some experimental evidence supports the idea than Bax or Bak can form ion channels in OM and trigger cytochrome *c* release. Indeed, Bax and Bak could form oligomers and tetramers of Bax could form a pore large enough to release cytochrome *c* and dextrans.[73] In addition, recombinant Bax is able to form channels in artificial membrane without any additional proteins,[49] which would permeabilize lipids membranes to cytochrome *c*.[74] Thus, it has been shown that Bax forms large oligomers in the OM of cells undergoing apoptosis.[63,75,76] However, it remains unclear whether these oligomers are exclusively composed of Bax or contain others proteins. Indeed, Bak was found to be associated with these large clusters of Bax.[76] Some experimental evidence also supports the idea that Bax or Bak can form ion channel in OM and trigger cytochrome *c* release without mitochondrial swelling.[77,78] Furthermore, in another study, cytochrome *c* depletion of isolated mitochondria by addition of Bax or Bid was not accompanied of the loss of mitochondrial transmembrane potential, suggesting that the IM remained intact.[79] Bid-induced cytochrome *c* release is Bak-dependent in mouse liver mitochondria and is not inhibited by the PTP inhibitor cyclosporin A.[58] Finally, a recent study shows that in a cell-free system, Bid (its BH3 domain) activates monomeric Bax to produce supramolecular membrane opening, allowing the passage of very large proteins (2 megadalton). This process is inhibited by Bcl-XL.[80]

The intracellular redox state has been shown to play an important role in many types of apoptosis induced by TNF, UV, growth factor withdrawal, and ceramide.[81,82] Bcl-2 might act as an antioxidant partner to block a putative reactive oxygen species (ROS)–mediated step in the cascade of event required for apoptosis. Thus in some studies Bcl-2 seems to stimulate the generation of ROS,[83,84] whereas in others Bcl-2 will prevent oxidative damage.[85-87] However, the way by which Bcl-2 protects cells from ROS is unclear and so far no direct evidence has been shown to prove this function.

The involvement of VDAC in Bax-induced apoptosis has also been supported. Using immunoprecipitation, it has been shown that Bax interacts with VDAC during apoptosis[23] and electrophysiological experiments have demonstrated that Bax and VDAC reconstituted in liposomes form a channel larger than either VDAC or Bax alone.[44] Furthermore, Bax and Bak enhanced the permeability of VDAC-liposome to cytochrome *c*. In another study, antibodies against VDAC were shown to prevent Bax-induced apoptosis while having no effect on Bid- or Bik-triggered apoptosis. Furthermore, it has been reported that VDAC was necessary to the binding of Bax and Bak to red blood cells.[88] On the other hand, it has been reported that $O_2^{\bullet}$-induced cytochrome *c* release could be mediated by a pore containing VDAC.[89] Furthermore, a recent study shows that Bim can interact and directly activate VDAC and that this interaction is enhanced during apoptosis.[90]

Interactions between Bax and ANT have also been demonstrated by co-immunoprecipitation. Additionally, Bax was shown to induce cell death when expressed in wild-type yeast but not in an ANT-deficient yeast strain.[3] In response to Atr or Ca^{2+}, ANT can form channels in lipid membranes.[17,91] Bax has been shown to enhance the

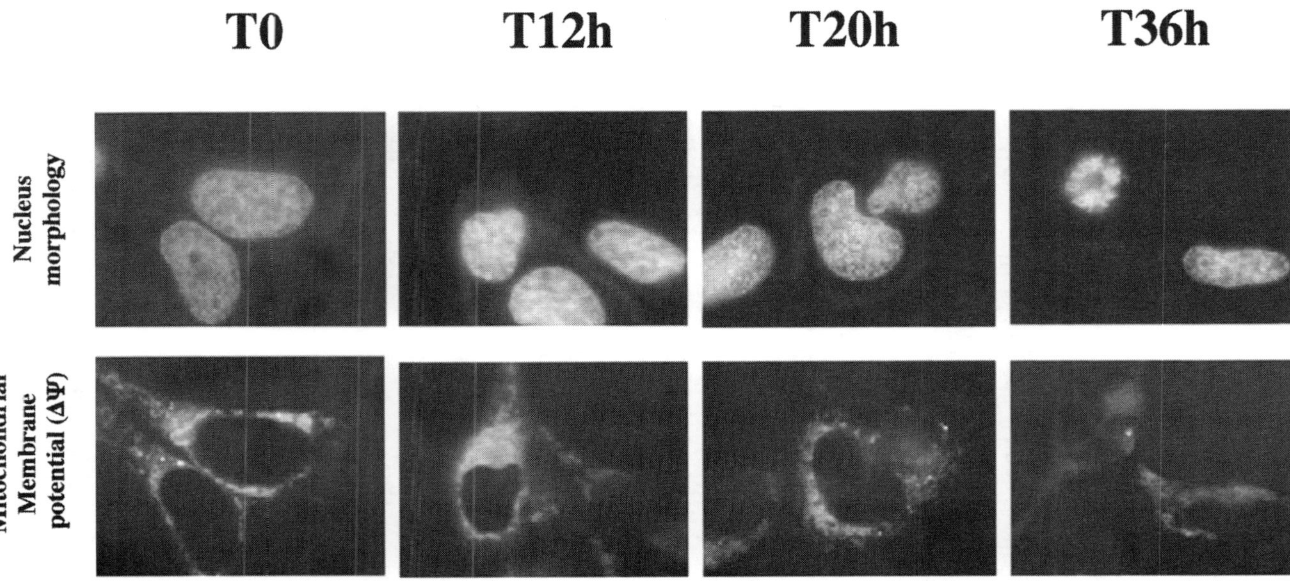

FIGURE 4. Time course effect of etoposide on HeLa cells. HeLa cells were treated with etoposide 100 μM and stained with Hoechst and JC-1. The loss of mitochondrial membrane potential is indicated by the progressive loss of thered-J-aggregate and cytoplasmic diffusion of green monomer fluorescence following exposure to etoposide. Twelve hours after treatment, etoposide induces apoptosis in HeLa cells as shown by depolarization of mitochondrial membrane, nuclear condensation (after 20 h of treatment), and apoptotic body formation (after 36 h of treatment).

ANT channel activity while endogenous ligand ATP inhibits this activity, suggesting that conformational change in ANT is necessary to this channel[4] In contrast, ANT and Bcl-2 do not form channels, Bcl-2 prevents formation of Atr-treated–ANT channel and inhibits the cooperative Bax/ANT interaction to form a channel. It has further been demonstrated that ANT/Bax channel was cationic as Bax channel was anionic, suggesting that Bax and ANT form a cooperative channel. In a recent study, it has been shown that Bcl-2 enhances the ATP/ADP exchange in proteoliposomes containing ANT, in isolated mitochondria and mitoplasts, as well as in intact cells, while Bax, which displaces Bcl-2 from ANT in apoptotic cells, inhibits ATP/ADP exchange through a direct action on ANT. Bax-mediated inhibition of ATP/ADP exchange can be separated from Bax-stimulated formation of non-specific pores by ANT, indicating a double functional interaction with ANT.[13] In agreement with the model of the PTPC opening regulation by Bax via an interaction with ANT, we have shown a dynamic interaction between PTPC and Bax and Bcl-2 during apoptosis. Thus, in mitochondria of cells treated with etoposide, Bax association with PTPC (proteins co-immunoprecipitated with ANT) displays a dramatic increase since the Bcl-2/PTPC interaction decreases[13,14] concomitantly with the loss of $\Delta\Psi m$ (FIG. 4). Furthermore, while the interaction of Bcl-2 family members with PTPC changes soon after the treatment, other interactions between ANT, VDAC, and hexokinase were maintained until chromatin condensation and nuclear degradation occurred, suggesting that they are critical for apoptosis.[14]

VIRAL PROTEINS TARGETING PTPC COMPONENTS AND REGULATING APOPTOSIS

Viruses are known to negatively or positively modulate apoptotic response.[92] Viral proteins targeting mitochondria provide a molecular link between virus infection and induction or suppression of apoptosis. A large number of viral apoptotic modulators that influence mitochondrial cell death pathways are being identified and are summarized in TABLE 1. Some of these viral proteins have been described to target mitochondria, and others bind directly to PTPC component.

Bcl-2 Homologues

The genomes of several viruses code for death-inhibitory Bcl-2 analogs, which localize to mitochondria and may interact with pro-aoptotic Bax homologues. Some proteins present sequence homology with Bcl-2 and Bcl-XL, yet their role in regulation of apoptosis has not been demonstrated so far. Viral anti-apoptotic Bcl-2 homologues include LMW5-HL of African swine fever virus[93,94] and the murine gamma herpes virus protein, M11,[95] the Bcl-2 homolog of bovine herpes virus 4,[96] BALF-1 of Epstein-Barr virus,[97] and BHRF-1 of herpes virus papio.[98] In addition, KsBcl-2 of human herpes virus 8 has been shown to be anti-apoptotic and localized to cytoplasmic membrane structure.[99] Other viral Bcl-2 homologues have also been described to display a direct, proximal effect on mitochondrial function when expressed in cells independently of other viral proteins. Thus the BHRF-1 protein of Epstein-Barr virus displays both anti-apoptotic function[100] and mitochondrial localization.[101] This viral protein, as well as E1B19K protein of adenovirus, are both able

TABLE 1. Viral proteins implicated in mitochondrial apoptosis

Protein	Virus	Mechanism/Action	Reference
Mitochondrial modulators			
Apoptin	Chicken anemia virus	Pro-apoptotic	109
E6	Human papilloma virus	Cell death sensitizer	110
Tat	Human immunodeficiency virus	Pro-apoptotic	111, 112
p13II	HTLV-1	Pro-apoptotic	107–109
Bcl-2 homologues			
BHRHF1	Epstein-Barr virus	Anti-apoptotic Bcl-2 homologue	100–102
E1B19K	Adenovirus	Anti-apoptotic Bcl-2 homologue	102, 103
KVS-Bcl-2	Herpes virus saimir	Anti-apoptotic Bcl-2 homologue	105, 106
Binding to PTPC members			
vMIA	Cytomegalovirus	Anti-apoptotic	115
M11L	Myxoma virus	Anti-apoptotic	138, 139
HBx	Hepatitis B virus	Pro-apoptotic	133, 134
Vpr	Human immunodeficiency virus	Pro-apototic	7, 8

to abolish apoptosis induced by ROS, which disrupt mitochondrial integrity.[102] Furthermore, E1B19K protein can interact with Bax, providing a protective effect against Bax-induced mitochondrial $\Delta\Psi$m loss and apoptosis.[103] Furthermore, E1B19K inhibits Bax oligomerization and blocks TNF-α–induced cell death.[104] HVS-Bcl-2 of herpes virus saimiri is also anti-apoptotic,[105] protecting cells from apoptosis induced by Fas ligation, dexamethasone treatment, DNA damage, or $O_2^{\bullet}$ production, and on the mitochondria preventing of $\Delta\Psi$m loss, cytochrome *c* translocation into cytosol, and caspase-3 activation.[106]

Mitochondrial Modulatory Proteins

Some viral proteins have been described as pro-apoptotic. p13II protein of human T cell leukemia lymphotrophic virus type 1 has been reported to accumulate in the IM and induce an alteration in mitochondrial morphology as well as $\Delta\Psi$m loss.[107,108] Nevertheless, increased levels of apoptosis in p13II-expressing cells have so far not been demonstrated.[107] Furthermore, the mitochondrial effects of the synthetic peptide spanning residues 9–41 of p13II, including the amphipathic mitochondrial targeting sequence, were insensitive to cyclosporin A, suggesting a permeability transitory pore–independent mechanism.[108] Apoptin, from chicken anemia virus, induces cell death in a caspase-dependent manner. However, in the presence of caspase inhibitors, this protein is able to induce $\Delta\Psi$m loss and cytochrome *c* release.[109] The E6 protein of human papillomavirus type 16 sensitizes cells to apoptosis induced by Atr. This effect is p53- and caspase-dependent and is abolished by cyclosporin A.[110] The transactivating protein Tat of human immunodeficiency virus type I (HIV-1), has been implicated in triggering apoptosis by a variety of mechanisms including the mitochondrial cell death pathway.[111] Stable expression of Tat in hippocampal neurons induces caspase-dependent cell death associated with an in-

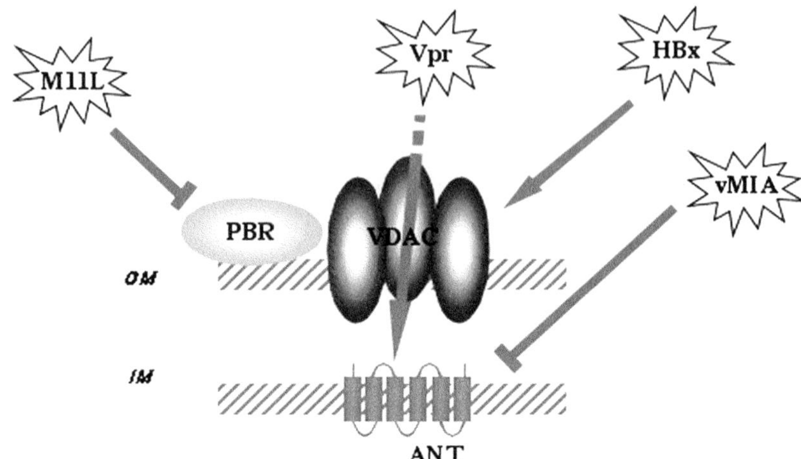

FIGURE 5. Possible mechanisms of MMP induced by viral proteins targeting the PTPC compounds VDAC, ANT, and PBR. HBx translocates to mitochondria interacting with VDAC3. Vpr interacts with ANT and induces MMP. vMIA interacts with ANT (but not with VDAC) and prevents MMP. M11L is physically associated with the mitochondrial PBR component of the PTPC.

crease in mitochondrial Ca^{2+} level and an accumulation of ROS.[112] Involvement of the mitochondrial membrane permeability transition is suggested by the finding that apoptosis is prevented by cyclosporin A.

Direct Interaction between Viral Proteins and PTPC

The importance of the PT pore in the context of apoptosis modulation in infected cells has been emphasized by the identification of several viral proteins that physically associate with PTPC members (FIG. 5). Some viral proteins exert proapoptotic effects (Vpr and Tat of HIV-1 and HBx of HBV) while others inhibit apoptosis (v-MIA of CMV, M11L of myxoma virus, and BHRF1 of EBV).

Vpr of HIV

Vpr, a small accessory protein encoded by HIV-1, has been shown to induce MMP via a specific interaction with PTPC.[7] A synthetic peptide, corresponding to the C-terminal moiety of Vpr (aa 52–96), uncouples the respiratory chain and induces a rapid inner MMP to protons and NADH. In isolated mitochondria, this peptide induces matrix swelling and inner MMP, both prevented by pre-incubation with Bcl-2 protein.[8] Furthermore it has been shown that Vpr52–96 can interact with ANT and that purified ANT in association with Vpr form large conductance channels in artificial membranes.[8] This cooperative channel formation relies on a direct protein-protein interaction, since it is abolished by the addition of a peptide corresponding to the Vpr binding site of ANT,[104–116] the same domain that interacts with Bcl-2.[3] Bax and Bcl-2 have been shown to modulate the functional and physical Vpr/ANT interaction. Indeed, while Bcl-2 inhibits this interaction, Bax increases the conductance of the Vpr/ANT channel.[8] Furthermore, the interaction of Vpr and ANT involves an α-helical dodecapeptide domain (aa 71–82) of Vpr,[113] characterized by the presence of three critical arginine residues, crucial for the 71–82 peptide to induce MMP on purified mitochondria.[8,114]

vMIA of CMV

Human cytomegalovirus (CMV) is a herpes virus that causes opportunistic infections in immunocompromised individuals. CMV inhibits apoptosis mediated by death receptors and encodes a viral mitochondria-localized inhibitor of apoptosis, namely vMIA. This viral protein is capable of suppressing apoptosis induced by diverse stimuli, as ligation of the death receptors Fas/Apo-1/CD95, TNF receptor I, TRAIL, straurosporine and maytansine, and pro-oxidants.[6,115] vMIA is predominantly localized in mitochondria, where it has been shown to be associated with ANT. These functional properties resemble those ascribed to Bcl-2. However, the absence of sequence similarity to Bcl-2 or any other known cell death suppressors suggests that vMIA defines a new class of anti-apoptotic proteins preventing cell death by a direct interaction with ANT. Unlike Bcl-2 family members,[23] vMIA does not bind VDAC.[6,115] Also, in contrast to Bcl-2, vMIA does not bind Bax.[9] Mutagenesis studies have shown that two regions of vMIA are essential for its anti-apoptotic action. The N terminal domain (aa 5–34) is required for mitochondrial localization, and the second domain (118–147) is conserved among different cytomegalovirus isolates, but whose mechanistic function is not yet known. vMIA inhibits Fas-mediated apoptosis at a point downstream of caspase-8 activation and Bid cleavage but upstream of cytochrome *c* release. Mitochondria from cells overexpressing vMIA are resistant to cytochrome *c* release induced by truncated Bid (tBid).[116] It has been proposed that vMIA could interfere with tBid inhibiting the permeabilization of the OM induced by the latter.[116]

HBx of HBV

Hepatitis B virus X protein (HBx) is implicated in the development of hepatitis B virus–induced hepatocellular carcinoma. HBx possesses multiple functions that can lead to either cellular proliferation or to apoptosis (transcriptional activation, disruption of p53 activity, DNA repair, stimulation of kinase signaling pathway activating mitotic Raf, Ras, and MAP cellular.[117–121] HBx is a potent apoptosis inducer, whose effects are neutralized by Bcl-2 or not.[122,123] The C-terminal region of HBx is involved in the inactivation of p53 inhibiting apoptosis.[124,125] On the other hand, HBx possesses two domains that inhibit kinase/serine[126] and caspases.[127] Thus expression of HBx in rat fibroblast and hepatoma cells induces cellular resistance at several apoptotic stimuli (Fas antibodies and growth factor withdrawal) which engage caspase 3.[127] Hbx can also induce spontaneous apoptosis in primary hepatocyte and in liver of HBx transgenic mice.[128–130] Furthermore, HBx can render human hepatocyte cells and rat fibroblasts sensitive to the apoptosis induced by TNF-α.[131,132] Interestingly, HBx has been localized to mitochondria and shown to promote the ΔΨm loss when expressed independently of other viral proteins.[133,134] A more recent study has shown that the transmembrane region (aa 54–70) is required in the mitochondrial targeting of HBx and that the putative α-helical regions (aa 75–88 and aa 109–131) is involved in this targeting.[135] In addition to its mitochondrial localization, HBx interacts with VDAC.[134] These protein sequences possess significant homology (HVDAC3 and HBx possess 31% amino acids similarity). Furthermore, a C-terminal HBx deletion mutant, HBx Δ99, failed to bind VDAC and activate STAT-3 and NF-kB.[135] As described above, HBx possesses both pro- or anti-apoptotic activity. It has been proposed that the pro-apoptotic activity of HBx facilitates viral propagation without immune system activation.[136] Controversial effects of HBx on apoptosis could be due to different modes of action according to HBx concentration at several stages of natural HBV infection. That is, HBx could inhibit apoptosis at an early stage of infection while at a later stage of infection, it could induce apoptosis favoring HBV replication and propagation. Pro- and anti-apoptotic effect could also result in the malignant transformation of hepatocytes. Apoptosis inhibition could then permit transforming mutations. Stimulation of hepatocyte proliferation in response to pro-apoptotic effects could lead to the selection of pre-maligned cells in the liver. This effect could increase gene activation and regulation of cellular growth.[137]

M11L of Myxoma Virus

The myxomapoxvirus protein, M11L prevents apoptosis during virus infection.[138] *In vitro*, M111 knockout viruses cause accelerated apoptosis in infected rabbit lymphocytes or monocytes.[138,139] M11L has also been reported to target the mitochondrial and inhibit ΔΨm loss and apoptotic changes (caspase-3 activation, DNA fragmentation), induced by staurosporine.[138] A recent study has shown that M11L is physically associated with the mitochondrial PBR (component of the PT-PC) and, expressed in stable cell line, inhibits the release of cytochrome *c* and ΔΨm loss induced by staurosporine, alone or in combination with pro-apoptotic ligands of PBR. This indicates that M11L functionally prevents transmission of apoptotic signals that proceed via mitochondria.[140] This represents the first evidence that the PBR may physically associate with an anti-apoptotic protein able to modulate mitochondrial function. Interestingly, expression of the non-targeted M11L truncation mutant protein in a stable cell line failed to protect it from apoptosis. Furthermore, it has been described in the same study that M111/PBR functional interaction was not disturbed by the binding of PBR ligand, FGIN-1-27, suggesting that M11L acts more as a modulator of PBR than as a competitive molecule.[140,141] The mechanism by which M11L could inhibit apoptosis is not fully understood, but some hypotheses have been proposed. PBR is an OM protein, possessing five transmembrane structures and is involved in several functions including cholesterol transport from the OM to the IM, regulation of steroidogenesis, porphyrin transport, sensing of ROS species, and regulation of apoptosis.[142] Although previous studies have shown that PBR could be the target of several anti- or pro-apoptotic drugs (rasagiline, platelet-activating factor, porphyrins), indicating the role of PBR as a modulator of mitochondrial apoptosis.[143–145] PBR could regulate apoptosis through indirect effects on mitochondrial metabolism. PBR also interacts with several mitochondrial protein—VDAC and ANT and members of PTPC.[146] Thus, by this direct interaction or by indirect interaction with other members, it could modify the molecular dynamics of intra-PTPC protein interactions (inhibiting, for example, the interaction of pro-apoptotic proteins such as cyclophilin D/ANT or Bax/VADC/ANT).

PERSPECTIVES AND OPEN QUESTIONS

In apoptotic cascades, mitochondria has been described as a checkpoint for relaying and amplifying many of the cell death signals.[147] Although some of the mechanistic details remain unclear, the finding that a pathogenic virus targets mitochondrial proteins to inhibit or induce apoptosis, as reviewed before, provides valuable tools for a better understanding of the participation of mitochondria in the regulation of cellular death.

Furthermore, increased knowledge of physiopathological mechanisms of viral infection and apoptosis modulation will help in the design and development of new therapies for human disease. Viral mitochondrial proteins have been shown to either protect from (vMIA) or induce (Vpr, Hbx, M11L) apoptosis. Indeed, interaction of M11L and PBR is required for the lethal development of myxomatosis. Moreover, vpr mutations that abolish the apoptogenic Vpr/ANT interaction can be found among long-term nonprogressors, correlating with decreases in apoptosis induction by HIV-1.[148,149] For instance, drugs or peptides inhibiting the interaction of M11L, HBx, or Vpr with PTPC members could be used to neutralize their effect on apoptosis and limit viral infection. CMV is a herpes virus that can be the cause of mortality in immunodeficient patients, such as individuals with AIDS or organ- transplant recipients.[116] As described in this review, vMIA can target mitochondria where it has been associated with ANT and possesses an anti-apoptotic function. Moreover, infection of cells with vMIA-deficient CMV is not productive and induces apoptosis in these cells. Taken together these observations establish the possibility that drugs inhibiting ANT-vMIA interaction could restore apoptosis, killing the infected cells and affecting the infection.

REFERENCES

1. Costantini, P., E. Jacotot, D. Decaudin & G. Kroemer. 2000. J. Natl. Cancer Inst. **92:** 1042–1053.
2. Kroemer, G. & J. Reed. 2000. Nat. Med. 6. 513–519.
3. Marzo, I., C. Brenner, N. Zamzami, *et al.* 1998. Science **281:** 2027–2031.
4. Brenner, C., H. Cadiou, H.L. Vieira, *et al.* 2000. Oncogene **19:** 329–336.
5. Costantini, P., A.S. Belzacq, H.L. Vieira, *et al.* 2000. Oncogene **19:** 307–314.
6. Vieira, H.L., A.S. Belzacq, D. Haouzi, *et al.* 2001. Oncogene **20:** 4305–4316.
7. Jacotot, E., L. Ravagnan, M. Loeffler, *et al.* 2000. J. Exp. Med. **191:** 33–46.
8. Jacotot, E., K.F. Ferri, C. El Hamel, *et al.* 2001. J. Exp. Med. **193:** 509–520.
9. Goldmacher, V.S., L.M. Bartle, A. Skaletskaya, *et al.* 1999. Proc. Natl. Acad. Sci. USA **96:** 12536–12541.
10. Jan, G., A.S. Belzacq, D. Haouzi, *et al.* 2002. Cell Death Differ. **9:** 179–188.
11. Belzacq, A.S., E. Jacotot, H.L. Vieira, *et al.* 2001. Cancer Res. **61:** 1260–1264.
12. Belzacq, A.S., C. El Hamel, H.L. Vieira, *et al.* 2001. Oncogene **20:** 7579–7587.
13. Belzacq, A.S., H.L. Vieira, F. Verrier, *et al.* 2003. Cancer Res. **63:** 541–546.
14. Verrier, F., A. Deniaud, V. Rincheval, *et al.* 1996. FEBS Lett. **396:** 189–195.
15. Marzo, I., C. Brenner, N. Zamzami, *et al.* 1998. J. Exp. Med. **187:** 1261–1271.

16. Ruck, A., M. Dolder, T. Wallimann & D. Brdiczka. 1998. FEBS Lett. **426:** 97–101.
17. Marzo, I., C. Brenner, N. Zamzami, *et al.* 1998. Science **281:** 2027–20231.
18. Brullemans, M., O. Helluin, J.Y. Dugast, *et al.* 1994. Eur. Biophys. J. **23:** 39–49.
19. Marzo, I., C. Brenner, N. Zamzami, *et al.* 1998. J. Exp. Med. **187:** 1261–1271.
20. Belzacq, A.S., H.L. Vieira, G. Kroemer & C. Brenner. 2002. Biochimie **84:** 167–176.
21. Zoratti, M. & I. Szabo. 1995. Biochim. Biophys. Acta **1241:** 139–176.
22. Shimizu, S., M. Narita & Y. Tsujimoto. 1999. Nature **399:** 483–487.
23. Crompton, M. 1999. Biochem. J. **341:** 233–249.
24. Woodfield, K., A. Ruck, D. Brdiczka & A.P. Halestrap. 1998. Biochem. J. **336:** 287–290.
25. Bernardi, P., K.M. Broekemeier & D.R. Pfeiffer. 1994. J. Bioenerg. Biomembr. **26:** 509–517.
26. Ichas, F., L. Jouaville & J. Mazat. 1997. Cell **89:** 1145–1153.
27. Bernardi, P., R. Colonna, P. Costantini, *et al.* 1998. Biofactors **8:** 273–281.
28. Szabo, I., P. Bernardi & M. Zoratti. 1992. J. Biol. Chem. **267:** 2940–2946.
29. Zamzami, N., P. Marchetti, M. Castedo, *et al.* 1995. J. Exp. Med. **181:** 1661–1672.
30. De Giorgi, F., L. Lartigue, M.K. Bauer, *et al.* 2002. FASEB J. **16:** 607–609.
31. Lemasters, J., A. Nieminen, T. Qian, *et al.* 1997. Mol. Cell. Biochem. **174:** 159–165.
32. Dorner, A., K.M. Schulze, U. Rauch & H.P. Schultheiss. 1997. Mol. Cell. Biochem. **174:** 261–269.
33. Friberg, H. & T. Wieloch. 2002. Biochimie **84:** 241–250.
34. Bribes, E., S. Galiegue, B. Bourrie & P. Casellas. 2003. Immunol. Lett. **85:** 13–18.
35. McEnery, M. *et al.* 1992. Proc. Natl. Acad. Sci. USA **89:** 3170–3174.
36. Papadopoulos, V., A.M. Dharmarajan, H. Li, *et al.* 1999. Biochem. Pharmacol. **58:** 1389–1393.
37. Crompton, M., E. Barksby, N. Johnson & M. Capano. 2002. Biochimie **84:** 143–152.
38. Halestrap, A.P., G.P. McStay & S.J. Clarke. 2002. Biochimie **84:** 153–166.
39. Vieira, H.L., D. Haouzi, C.E. Hamel, *et al.* 2000. Cell Death Differ. **7:** 1146–1154.
40. O'Gorman, E., G. Beutner, M. Dolder, *et al.* 1997. FEBS Lett. **414:** 253–257.
41. Cao, G., M. Minami, W. Pei, *et al.* 2001. J. Cereb. Blood Flow Metab. **21:** 321–333.
42. Capano, M. & M. Crompton. 2002. Biochem. J. **367:** 169–178.
43. Shimizu, S., T. Ide, T. Yanagida & Y. Tsujimoto. 2000. J. Biol. Chem. **275:** 12321–12325.
44. Kroemer, G., N. Zamzami & S.A. Susin. 1997. Immunol. Today **18:** 44–51.
45. Kroemer, G. 1997. Nat. Med. **3:** 614–620.
46. Adams, J.M. & S. Cory. 1998. Science **281:** 1322–1326.
47. Vander Heiden, M.G. & C.B. Thompson. 1999. Nat. Cell. Biol. **1:** E209–E216.
48. Antonsson, B., F. Conti, A. Ciavatta, *et al.* 1997. Science **277:** 370–372.
49. Schendel, S.L., Z. Xie, M.O. Montal, *et al.* 1997. Proc. Natl. Acad. Sci. USA **94:** 5113–5118.
50. Spector, M.S., S. Desnoyers, D.J. Hoeppner & M.O. Hengartner. 1997. Nature **385:** 653–656.
51. Pan, GH., E.W. Humke & V.M. Dixit. 1998. FEBS Lett. **426:** 151–154.
52. Hengartner, M. 1997. Nature **388:** 714–715.
53. Chinnaiyan, A., D. Chaudhary, K. O'Rourke, *et al.* 1997. Nature **388:** 728–729.
54. Basanez, G., A. Nechushtan, O. Drozhinin, *et al.* 1999. Proc. Natl. Acad. Sci. USA **96:** 5492–5497.
55. Khaled, A.R., K. Kim, R. Hofmeister, *et al.* 1999. Proc. Natl. Acad. Sci. USA **96:** 14476–14481.
56. Gross, A., J. Jockel, M.C. Wei & S.J. Korsmeyer. 1998. EMBO J. **17:** 3878–3885.
57. Wei, M.C., T. Lindsten, V.K. Mootha, *et al.* 2000. Genes Dev. **14:** 2060–2071.
58. Eskes, R., S. Desagher, B. Antonsson & J.C. Martinou. 2000. Mol. Cell. Biol. **20:** 929–935.
59. Antonsson, B., S. Montessuit, S. Lauper, *et al.* 2000. Biochem. J. **345:** 271–278.
60. Jurgensmeier, J.M., Z.H. Xie, Q. Deveraux, *et al.* 1998. Proc. Natl. Acad Sci USA **95:** 4997–5002.

61. OLTVAI, Z.N., C.L. MILLIMAN & S.J. KORSMEYER. 1993. Cell **74:** 609–619.
62. MIKHAILOV, V., M. MIKHAILOVA, D.J. PULKRABEK, *et al.* 2001. J. Biol. Chem. 20.
63. MURPHY, K.M., U.N. STREIPS & R.B. LOCK. 2000. J. Biol. Chem. **275:** 17225–17228.
64. CHENG, E.H., B.LEVINE, L.H. BOISE, *et al.* 1996. Nature **379:** 554–556.
65. ZHA, H. & J.C. REED. 1997. J. Biol. Chem. **272:** 31482–31488.
66. O'CONNOR, L., A. STRASSER, L.A. O'REILLY, *et al.* 1998. EMBO J. **17:** 384–395.
67. HUANG, D.C. & A. STRASSER. 2000. Cell **103:** 839–842.
68. MUCHMORE, S.W., M. SATTLER, H. LIANG, *et al.* 1996. Nature **381:** 335–341.
69. CHOU, J.J., H. LI, G.S. SALVESEN, *et al.* 1999. Cell **96:** 615–624.
70. MCDONNELL, J.M., D. FUSHMAN, C.L. MILLIMAN, *et al.* 1999. Cell **96:** 625–634.
71. SCHENDEL, S.L., M. MONTAL & J.C. REED. 1998. Cell Death Differ. **5:** 372–380.
72. KORSMEYER, S.J., M.C. WEI, M. SAITO, *et al.* 2000. Cell Death Differ. **7:** 1166–1173.
73. SAITO, M., S.J. KORSMEYER & P.H. SCHLESINGER. 2000. Nat. Cell Biol. **2:** 553–555.
74. ANTONSSON, B., S. MONTESSUIT, B. SANCHEZ & J.C. MARTINOU. 2001. J. Biol. Chem. **276:** 11615–11623.
75. NECHUSHTAN, A., C.L. SMITH, I. LAMENSDORF, *et al.* 2001. J. Cell Biol. **153:** 1265–1276.
76. MARTINOU, I., S. DESAGHER, R. ESKES, *et al.* 1999. J. Cell Biol. **144:** 883–889.
77. FINUCANE, D.M., N.J. WATERHOUSE, G.P. AMARANTE-MENDES, *et al.* 1999. Exp. Cell Res. **251:** 166–174.
78. VON AHSEN, O., C. RENKEN, G. PERKINS, *et al.* 2000. J. Cell Biol. **150:** 1027–1036.
79. KUWANA, T., M.R. MACKEY, G. PERKINS, *et al.* 2002. Cell **111:** 331–342.
80. MIGNOTTE, B. & J.L. VAYSSIERE. 1998. Eur. J. Biochem. **252:** 1–15.
81. FLEURY, C., B. MIGNOTTE & J.L.VAYSSIERE. 2002. Biochimie **84:** 131–141.
82. GOTTLIEB, E., M.G. VANDER HEIDEN & C.B. THOMPSON. 2000. Mol. Cell. Biol. **20:** 5680–5689.
83. KANE, D.J., T.A. SARAFIAN, R. ANTON, *et al.* 1993. Science **262:** 1274–1277.
84. HOCKENBERY, D.M., Z.N. OLTVAI, X.M. YIN, *et al.* 1993. Cell **75:** 241–251.
85. TYURINA, Y.Y., V.A. TYURIN, G. CARTA, *et al.* 1997. Arch. Biochem. Biophys. **344:** 413–423.
86. SATOH, T., Y. ENOKIDO, H. AOSHIMA, *et al.* 1997. J. Neurosci. Res. **50:** 413–420.
87. SHIMIZU, S., Y. MATSUOKA, Y. SHINOHARA, *et al.* 2001. J. Cell Biol. **152:** 237–250.
88. MADESH, M. & G. HAJNOCZKY. 2001. J. Cell Biol. **155:** 1003–1015.
89. SUGIYAMA, T., S. SHIMIZU, Y. MATSUOKA, *et al.* 2002. Oncogene **21:** 4944–4956.
90. BRUSTOVETSKY, N. & M. KLINGENBERG. 1994. J. Biol. Chem. **269:** 27329–27336.
91. TEODORO, J.G. & P.E. BRANTON. 1997. J. Virol. **71:** 1739–1746.
92. AFONSO, C.L., J.G. NEILAN, G.F. KUTISH & D.L. ROCK. 1996. J. Virol. **70:** 4858–4863.
93. BRUN, A., C. RIVAS, M. ESTEBAN, *et al.* 1996. Virology **225:** 227–230.
94. WANG, G.H., T.L. GARVEY & J.I. COHEN. 1999. J. Gen. Virol. **80:** 2737–2740.
95. BELLOWS, D.S., B.N. CHAU, P. LEE, *et al.* 2000. J. Virol. **74:** 5024–5031.
96. MARSHALL, W.L., C. YIM, E. GUSTAFSON, *et al.* 1999. J. Virol. **73:** 5181–5185.
97. MESEDA, C.A., J.R. ARRAND & M. MACKETT. 2000. J. Gen. Virol. **81:** 1801–1805.
98. CHENG, E.H., J. NICHOLAS, D.S. BELLOWS, *et al.* 1997. Proc. Natl. Acad. Sci. USA **94:** 690–694.
99. HENDERSON, S., D. HUEN, M. ROWE, *et al.* 1993. Proc. Natl. Acad. Sci. USA **90:** 8479–8483.
100. HICKISH, T., D. ROBERTSON, P. CLARKE, *et al.* 1994. Cancer Res. **54:** 2808–2811.
101. DUMONT, A., S.P. HEHNER, T.G. HOFMANN, *et al.* 1999. Oncogene **18:** 747–757.
102. HAN, J., D. MODHA & E. WHITE. 1998. Oncogene **17:** 2993–3005.
103. SUNDARARAJAN, R. & E. WHITE. 2001. J. Virol. **75:** 7506–7516.
104. NAVA, V.E., E.H. CHENG, M. VELIUONA, *et al.* 1997. J. Virol. **71:** 4118–4122.
105. DERFUSS, T., H. FICKENSCHER, M.S. KRAFT, *et al.* 1998. J. Virol. **72:** 5897–5904.
106. CIMINALE, V., L. ZOTTI, D.M. D'AGOSTINO, *et al.* 1999. Oncogene **18:** 4505–4514.
107. D'AGOSTINO, D.M., L. RANZATO, G. ARRIGONI, *et al.* 2002. J. Biol. Chem. **277:** 34424–34433.
108. DANEN-VAN OORSCHOT, A.A., A.J. VAN DER EB & M.H. NOTEBORN. 2000. J. Virol. **74:** 7072–7078.

109. BROWN, J., H. HIGO, A. MCKALIP & B. HERMAN. 1997. J. Cell. Biochem. **66:** 245–255.
110. GOLDBERG, B. & R.B. STRICKER. 1999. Immunol. Lett. **70:** 5–8.
111. KRUMAN, II, NATH, A. & M.P. MATTSON. 1998. Exp. Neurol. **154:** 276–288.
112. SCHULER, W., K. WECKER, H. DE ROCQUIGNY, *et al.* 1999. J. Mol. Biol. **285:** 2105–2117.
113. JACOTOT, E., L. RAVAGNAN, M. LOEFFLER, *et al.* 2000. J. Exp. Med. **191:** 33–46.
114. GOLDMACHER, V.S., L.M. BARTLE, A. SKALETSKAYA, *et al.* 1999. Proc. Natl. Acad. Sci. USA **96:** 12536–12541.
115. GOLDMACHER, V.S. 2002. Biochimie **84:** 177–185.
116. BENN, J. & R.J. SCHNEIDER. 1994. Proc. Natl. Acad. Sci. USA **91:** 10350–10354.
117. BERGAMETTI, F., S. PRIGENT, B. LUBER, *et al.* 1999. Oncogene **18:** 28660–28671.
118. QADRI, I., M.E. FERRARI & A. SIDDIQUI. 1996. J. Biol. Chem. **271:** 15443–15450.
119. QADRI, I., J.W. CONAWAY, R.C. CONAWAY, *et al.* 1996. Proc. Natl. Acad. Sci. USA **93:** 10578–10583.
120. ROULSTON, A., R.C. MARCELLUS & P.E. BRANTON. 1999. Annu. Rev. Microbiol. **53:** 577–628.
121. TERRADILLOS, O., A. DE LA COSTE, T. POLLICINO, *et al.* 2002. Oncogene **21:** 377–386.
122. SCHUSTER, R., W.H. GERLICH & S. SCHAEFER. 2000. Oncogene **19:** 1173–1180.
123. WANG, X.W., M.K. GIBSON, W. VERMEULEN, *et al.* 1995. Cancer Res. **55:** 6012–6016.
124. ELMORE, L.W., A.R. HANCOCK, S.F. CHANG, *et al.* 1997. Proc. Natl. Acad. Sci. USA **94:** 14707–14712.
125. TAKADA, S. & K. KOIKE. 1990. Jpn. J. Cancer Res. **81:** 1191–1194.
126. GOTTLOB, K., M. FULCO, M. LEVRERO & A. GRAESSMANN. 1998. J. Biol. Chem. **273:** 33347–3353.
127. POLLICINO, T., O. TERRADILLOS, H. LECOEUR, *et al.* 1998. Biomed Pharmacother **52:** 363–368.
128. TERRADILLOS, O., T. POLLICINO, H., LECOEUR, *et al.* 1998. Oncogene **17:** 2115–2123.
129. KOIKE, K., K. MORIYA, H. YOTSUYANAGI, *et al.* 1998. Cancer Lett **134:** 181–186.
130. SU, F. & R.J. SCHNEIDER. 1997. Proc. Natl. Acad. Sci. USA **94:** 8744–8749.
131. SHERON, N., J. LAU, H. DANIELS, *et al.* 1991. J. Hepatol. **12:** 241–245.
132. TAKADA, S., Y. SHIRAKATA, N. KANENIWA & K. KOIKE. 1999. Oncogene **18:** 6965–6973.
133. RAHMANI, Z., K.W. HUH, R. LASHER & A. SIDDIQUI. 2000. J. Virol. **74:** 2840–2846.
134. WARIS, G., K.W. HUH & A. SIDDIQUI. 2001. Mol. Cell. Biol. **21:** 7721–7730.
135. LAU, J.Y., X. XIE, M.M. LAI & P.C. WU. 1998. Semin. Liver Dis. **18:** 169–176.
136. TERRADILLOS, O., O. BILLET, C.A. RENARD, *et al.* 1997. Oncogene **14:** 395–404.
137. EVERETT, H., M. BARRY, S.F. LEE, *et al.* 2000. J. Exp. Med. **191:** 1487–1498.
138. MACEN, J.L., K.A. GRAHAM, S.F. LEE, *et al.* 1996. Virology **218:** 232–237.
139. EVERETT, H., M. BARRY, X. SUN, *et al.* 2002. J. Exp. Med. **196:** 1127–1139.
140. CASTEDO, M., J.L. PERFETTINI & G. KROEMER. 2002. J. Exp. Med. **196:** 1121–1125.
141. CASELLAS, P., S. GALIEGUE & A.S. BASILE. 2002. Neurochem. Int. **40:** 475–486.
142. YOUDIM, M.B. & M. WEINSTOCK. 2001. Cell. Mol. Neurobiol. **21:** 555–573.
143. VERMA, A., S.L. FACCHINA, D.J. HIRSCH, *et al.* 1998. Molec. Med. **4:** 40–45.
144. PARKER, M.A., H.E. BAZAN, V. MARCHESELLI, *et al.* 2002. J. Neurosci. Res. **69:** 39–50.
145. MCENERY, M., A. SNOWMAN, R. TRIFILETTI & S. SNYDER. 1992. Proc. Natl. Acad. Sci. USA **89:** 3170–3174.
146. BRENNER, C. & G. KROEMER. 2000. Science **289:** 1150–1151.
147. SOMASUNDARAN, M., M. SHARKEY, B. BRICHACEK, *et al.* 2002. Proc. Natl. Acad. Sci. USA **99:** 9503–9508.
148. ZHANG, L., Y. HUANG, H. YUAN, *et al.* 1997. Virology **228:** 340–349.

Antagonism between Apoptotic (Bax/Bcl-2) and Anti-Apoptotic (IAP) Signals in Human Osteoblastic Cells under Vector-Averaged Gravity Condition

HIROSHI NAKAMURA, YASUHIRO KUMEI, SADAO MORITA, HITOYATA SHIMOKAWA, KEIICHI OHYA, AND KENICHI SHINOMIYA

Section of Orthopedic Spinal Surgery, Department of Frontier Surgical Therapeutics, Division of Advanced Therapeutical Sciences, Graduate School of Tokyo Medical and Dental University, Tokyo 113-8519, Japan

ABSTRACT: A functional disorder associated with weightlessness is well documented in osteoblasts. The apototic features of this disorder are poorly understood. Harmful stress induces apoptosis in cells via mitochondria and/or Fas. The Bax triggers cytochrome *c* release from mitochondria, which can be blocked by the Bcl-2. Released cytochrome *c* then activates the initiator caspase, caspase-9, which can be blocked by the anti-apototic (IAP) family of molecules. The effector caspase, caspase-3, finally exerts DNA fragmentation. We conducted this study to examine the apoptotic effects of vector-averaged gravity on normal human osteoblastic cells. Cell culture flasks were incubated on the clinostat, which generated vector-averaged gravity condition (simulated microgravity) for 12, 24, 48, and 96 hours. Upon termination of clinostat cultures, the cell number and cell viability were assessed. DNA fragmentation was analyzed on the agarose-gel electrophoresis. The mRNA levels for Bax, Bcl-2, XIAP, and caspase-3 genes were analyzed by semi-quantitative RT-PCR. Twenty-four hours after starting clinostat rotation, the ratios of Bax/Bcl-2 mRNA levels (indicator of apoptosis) were significantly increased to 136% of the 1G static controls. However, the XIAP mRNA levels (anti-apoptotic molecule) were increased concomitantly to 138% of the 1G static controls. Thus, cell proliferation or cell viability was not affected by vector-averaged gravity. DNA fragmentation was not observed in clinostat group as well as in control group. Finally, the caspase-3 mRNA levels were not affected by vector-averaged gravity. Simulated microgravity might modulate some apoptotic signals upstream the mitochondrial pathway.

KEYWORDS: weightlessness; gravity; osteoblast; Bax; Bcl-2; caspase-3; XIAP

Address for correspondence: Dr. Hiroshi Nakamura, Section of Orthopedic Spinal Surgery, Department of Frontier Surgical Therapeutics, Division of Advanced Therapeutical Sciences, Graduate School of Tokyo Medical and Dental University, Tokyo 113-8519, Japan. Voice: +81-3-5803-5279, fax: +81-3-5803-5273.

nakamura.orth@tmd.ac.jp

Ann. N.Y. Acad. Sci. 1010: 143–147 (2003). © 2003 New York Academy of Sciences.

doi: 10.1196/annals.1299.023

INTRODUCTION

Apoptotic signals are transmitted via mitochondria- and/or Fas-mediated pathways.[1] The Bax triggers apoptotic signals by enhancing cytochrome *c* release from mitochondria. The Bcl-2 blocks cytochrome *c* release by stabilizing the mitochondria membrane. The released cytochrome *c* then activates the caspase cascade: the initiator caspase, caspase-9 through the effector caspase, caspase-3 downstream in the apoptotic pathway. Thus, Bax, Bcl-2, and caspase-9 exert pro-apoptotic and/or anti-apoptotic effects in the mitochondrial pathway. The caspase-3 finally activates DNase, resulting in DNA fragmentation. The XIAP inhibits activation of caspase-3.

It is well documented that weightlessness induces bone mass loss and functional disorder of osteoblasts, but the mechanisms remain unknown. It was recently reported that vector-averaged gravity induced apoptosis of rat osteosarcoma-derived ROS17/2.8 cells, without affecting cell proliferation,[2] However, the pro- and anti-apoptotic molecules were not examined. The increased ratio of Bax/Bcl-2 indicates the overall surplus of cytochrome *c* release from mitochondria.[1,3] XIAP is an anti-apoptotic molecule that inhibits procaspase-3 binding to apoptosome.[4] The purpose of this study is to examine the effects of weightlessness on induction of apoptotic changes in human osteoblasts that were cultured under vector-averaged gravity condition. We examine the mRNA levels of Bax, Bcl-2, XIAP, and caspase-3.

MATERIALS AND METHODS

Human osteoblastic (HOB) cells were obtained from trabecular bone taken from a tibia of a 19-year-old male patient undergoing corrective surgery after traumatic fracture. The patient had no history of infective and metabolic diseases. This study was approved by the Declaration of Helsinki and informed consent was obtained from the subject prior to the experiment. HOB cells were maintained in α-modified Eagle minimum essential medium supplemented with 10% heat-treated fetal bovine serum, 0.2 mM L-ascorbic acid phosphate, 2 mM β-glycerophosphate, and 10 nM dexamethasone. The fifth passage of the cells was used throughout the present study. Vector-averaged gravity was generated by a three-dimensionally rotating clinostat (Mitsubishi Heavy Industry, Co. Ltd.) to simulate weightlessness. The 1G static control cultures were placed on the clinostat device in the same incubator to undergo the identical condition of vibration and temperature except clinostat rotation. After cells were inoculated and attached onto 12.5 cm^2 flasks, the medium was replaced by fresh medium with 1 nM vitamin D_3. Clinostat rotation was performed for 12, 24, 48, and 96 hours.

Upon termination of clinostat culture, both floating attached cells were collected separately for proliferation and cell viability assays. Cells were also fixed to prepare genomic DNA (QIAmp™ DNA Mini kit, QIAGEN), which was applied to 2% agarose gel for electrophoresis. DNA fragments were visualized by ethidium bromide staining and UV irradiation. Cells were also fixed by guanidine solution to analyze the mRNA levels by semi-quantitative RT-PCR.[5] Statistical significance was evaluated by Student's *t*-test.

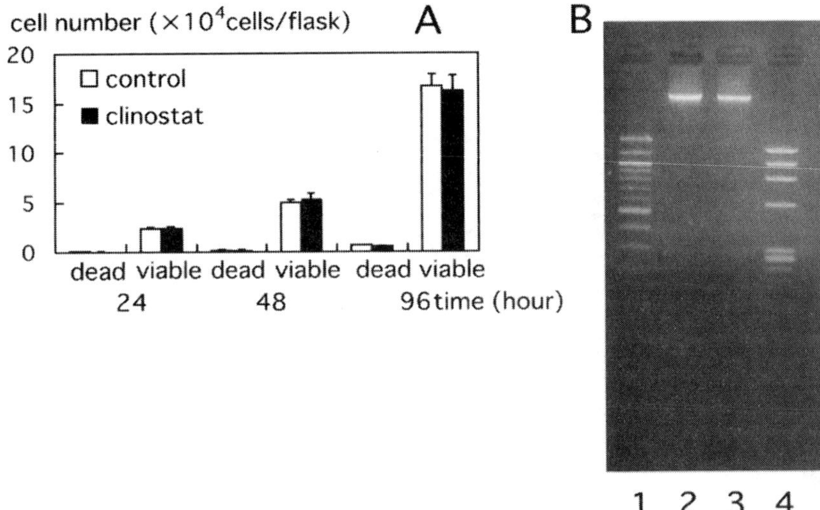

FIGURE 1. (**A**) HOB cells were inoculated at a density of 1,500 cells/cm^2 into 12.5 cm^2 culture flasks. After a one-day culture, the conditioned medium was discarded and the flasks were filled with fresh medium. At 24, 48, and 96 h after initiation of clinostat rotation, the cells were counted under a microscope using a hemocytometer ($N = 3$). (**B**) DNA fragmentation of HOB cells under vector-averaged gravity condition (clinostat) and static condition (control). Lane1, 100 bp ladder; Lane2, control; Lane 3, clinostat; Lane4, maker.

RESULTS AND DISCUSSION

In this study, the population doubling time of HOB cells under vector-averaged gravity was 26 h, which was equal to that of the static control. Approximately 96% of the cells were viable throughout the culture (FIG. 1, A). There was no difference in the percentage of non-viable cells between clinostat group and static control. After the agarose-gel electrophoresis, DNA fragmentation was not observed either in the clinostat group or in the control group (FIG. 1, B). Data strongly suggest that vector-averaged gravity did not affect proliferation nor did it induce apoptosis of HOB cells.

On the mitochondrial pathway of apoptosis, Bax releases cytochrome *c* and promotes apoptosis, whereas Bcl-2 blocks the apoptotic process by stabilizing the mitochondrial membrane and inhibiting the release of cytochrome *c*. The increased ratio of Bax/Bcl-2 has been used as the apoptotic sign via mitochondrial pathway.[1,3] On the first day of clinostat rotation, the ratio of mRNA levels for Bax/Bcl-2 was increased to 136% as compared to the static controls (FIG. 2, A), suggesting an apoptotic trend by the condition of vector-averaged gravity. However, clinostat rotation also increased the mRNA levels of XIAP, which inhibits caspase-9, to 138% of the controls on the first day (FIG. 2, B).

Ultimately, vector-averaged gravity did not increase the expression of the effector caspase caspase-3 (FIG. 2, C) nor did it induce DNA fragmentation. Data suggested

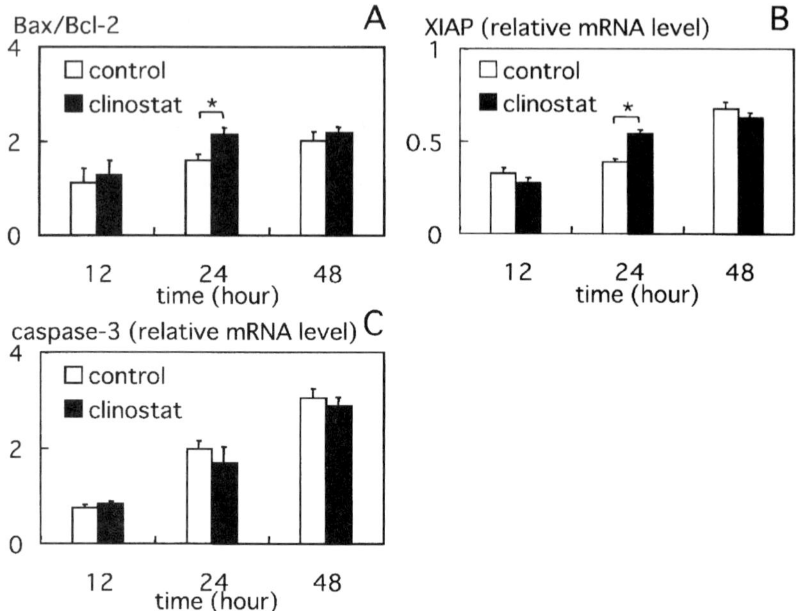

FIGURE 2. Quantitative RT-PCR analyses of mRNA levels. (**A**) Bax/Bcl-2, (**B**) XIAP/GAPDH, (**C**) caspase-3/GAPDH. Results are shown as the mean ± SD ($N = 6$). The asterisk denotes a significant difference ($P < .05$, unpaired t-test) between control and clinostat group.

that the vector-averaged gravity condition induced an early pro-apoptotic trend of which signaling was blocked by the concurrent upregulation of an anti-apoptotic molecule, XIAP, at the later phase of the caspase cascade. Weightlessness might not induce final apoptotic features of human osteoblasts, but it might modulate both pro- and anti-apoptotic molecules in apoptotic signaling.

ACKNOWLEDGMENTS

We thank Professor Masamichi Yamashita (Institute of Space and Astronautical Science, Japan) and Masaru Uemura (Mitsubishi Heavy Industry, Co. Ltd., Kobe, Japan).

REFERENCES

1. Green, D.R. 2000. Apoptotic pathways: Paper wraps stone blunt scissors. Cell **102:** 1–4.
2. Sarker, D., T. Nagata, K. Koga, *et al.* 2000. Culture in vector-averaged gravity under clinostat rotation results in apoptosis of osteoblastic ROS 17/2.8 cells. J. Bone Miner. Res. **15:** 489–498.

3. URAYAMA, S., A. KAWAKAMI, T. NAKASHIMA, *et al.* 2000. Effect of vitamin K2 on osteoblast apoptosis: vitamin K2 inhibits apoptotic cell death of human osteoblasts induced by Fas, proteasome, inhibitor, etoposide, and staurosporine. J. Lab. Clin. Med. **136:** 181–193.
4. DEVERAUX, Q.L., N. ROY, H.R. STENNICKE, *et al.* 1998. IAPs block apoptotic events induced by caspase-8 and cytochrome *c* by direct inhibition of distinct caspases. EMBO J. **17:** 2215–2223.
5. KUMEI, Y., H. SHIMOKAWA, H. KATANO, *et al.* 1996. Microgravity induces prostaglandin E2 and interleukin-6 production in normal rat osteoblasts: Role in bone demineralization. J. Biotechnol. **47:** 313–324.

Bcl-2 Prevents Loss of Cell Viability and Caspase Activation Induced by 3-Nitropropionic Acid in GT1-7 Cells

OLGA BRITO, SANDRA ALMEIDA, CATARINA R. OLIVEIRA, AND A. CRISTINA REGO

Institute of Biochemistry, Faculty of Medicine, and Center for Neuroscience and Cell Biology of Coimbra, University of Coimbra, 3004-504 Coimbra, Portugal

ABSTRACT: In this study, we analyzed the effect of Bcl-2 overexpression, an important anti-apoptotic protein, by using two hypothalamic cell lines, GT1-7*puro* or GT1-7*bcl-2*. 3-Nitropropionic acid (3-NP) mediated a dose-dependent decrease in cell viability in GT1-7*puro* cells, as determined by following the MTT (3-(4,5-dimethylthiazol-2-yl)-2,5-diphenyltetrazolium bromide) reduction, which was significantly prevented in GT1-7*bcl-2* cells. In addition, activation of caspases-2, -3, and -6 induced by 3-NP was prevented by Bcl-2 overexpression. The data suggest that irreversible inhibition of mitochondrial complex II induces apoptotic features of cell death in a process prevented by Bcl-2.

KEYWORDS: apoptosis; Bcl-2; GT1-7 cells; mitochondria; 3-nitropropionic acid

INTRODUCTION

Selective neuropathology associated with Huntington's disease (HD) affects mainly the striatum and the cortex, and has been reported to involve mitochondrial dysfunction, namely through the inhibition of respiratory chain complexes II and III. 3-NP, an irreversible inhibitor of complex II (succinate dehydrogenase), appears to cause neural degeneration in the striatum, mimicking the neuropathological, biochemical, and behavioral alterations observed in HD.[1] In addition, cell death with features of apoptosis has been observed after metabolic dysfunction induced by inhibitors of mitochondrial function, such as 3-NP. In fact, 3-NP–mediated cell death was reported to occur through the release of cytochrome *c*, activation of caspase-3, alterations in the levels of the apoptotic related proteins, Bax and Bcl-2, DNA fragmentation, as well as the cleavage of poly(ADP-ribose)polymerase, in a process that seems to be related to both apoptosis and necrosis.[5,6] Nevertheless, analysis of Bcl-2 protection in cell death processes induced by 3-NP has been less explored. Therefore, in the present study, we analyzed the role of Bcl-2 in 3-NP–mediated changes

Address for correspondence: Ana Cristina Rego, Ph.D., Institute of Biochemistry, Faculty of Medicine, and Center for Neuroscience and Cell Biology of Coimbra, University of Coimbra, 3004-504 Coimbra, Portugal. Voice: +351 239 820 190; fax: +351 239 822 776.
acrego@cnc.cj.uc.pt

**Ann. N.Y. Acad. Sci. 1010: 148–152 (2003). © 2003 New York Academy of Sciences.
doi: 10.1196/annals.1299.024**

in cell viability and caspases-2, -3, and -6 activation, by comparing the effects on cultured GT1-7*puro* and GT1-7*bcl-2* cells, a hypothalamic neural cell line overexpressing this anti-apoptotic protein.[2,3]

METHODS

Culture of GT1-7 Cells

GT1-7*puro* and GT1-7*bcl2* cells[2,3] were kindly donated by Dr. D. Bredesen and Dr. L. Ellerby (Buck Institute for Research in Aging, CA, USA). The cells were maintained as previously described.[4] On the day prior to the experiments, the cells were plated at a density of 0.2×10^6 cells/cm^2. The cells were further incubated with 3-NP (0.1 to 7.5 mM) for 24 h, in culture medium.

Analysis of Caspase-like Activity

The activity of caspases was determined after extraction with lysis buffer containing 0.1 mM PMSF, 2 mM DTT, and 1:1000 cocktail of protease inhibitors (antipain, leupeptin, pepstatin, and chymostatin), followed by centrifugation at 15,800*g* for 10 min, at 4°C. The supernatants were analyzed for protein content (BioRad reagent). Caspase activity was determined by adding the reaction buffer (10% sucrose, 0.1% CHAPS, 25 mM HEPES, pH 7.4, containing 10 mM DTT), 25 μg protein for caspase-3, or 40 μg protein for caspases-2 and -6 activities, and the following colo-

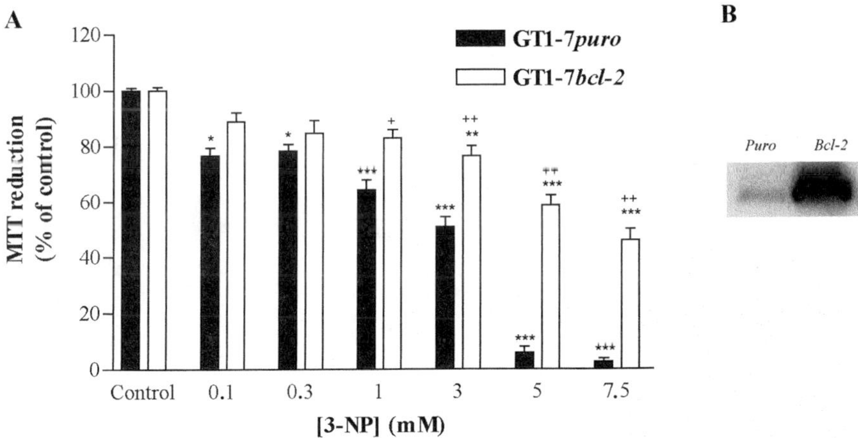

FIGURE 1. Overexpression of Bcl-2 correlates with 3-NP–mediated changes in cell viability. (**A**) Viability of GT1-7*puro* or GT1-7*bcl-2* cells was analyzed as the capacity to reduce the tetrazolium salt MTT to formazan. The results, expressed as the percentage of optical density observed in the control (non-treated cells), are represented as the means ± SEM of at least 3 experiments, run in duplicate. Statistical analysis: $^*P < .05$, $^{**}P < .01$ or $^{***}P < .001$ as compared to the control; $^+P < .05$ or $^{++}P < .001$ as compared to GT1-7*puro* cells. (**B**) Overexpression of Bcl-2 is shown in GT1-7*bcl-2* cells by western blotting.

rimetric substrates (100 μM): Ac-VDVAD-pNa (acetyl-Val-Asp-Val-Ala-Asp-p-nitroanilide), Ac-DEVD-pNa (acetyl-Asp-Glu-Val-Asp-*p*-nitroanilide) or Ac-VEID-pNa (acetyl-Val-Glu-Ile-Asp-p-nitroanilide), respectively, for analysis of caspases-2, -3, or -6. After 2 h at 37°C, the absorbance was measured at 405 nm (ELISA SLT.Spectra).

MTT Assay

The conversion of MTT to formazan, indicative of cell viability, was measured at 570 nm, after 1 h incubation with 0.5 mg/mL MTT, followed by the addition of an equal volume of 0.04 M HCl-isopropanol.

Western Blot Analysis of Bcl-2

The cells were homogenized and total cellular protein (without nuclei) was obtained according to Rego and colleagues.[4] The samples were analyzed by western blotting, after protein (40 μg protein/lane) separation by SDS-PAGE (12%) and transferred to a PVDF membrane (Amersham). After blocking with 5% fat-free milk in TBS-Tween (0.1%), the membranes were incubated with anti-Bcl-2 (1:200; Santa Cruz Biotechnology, Inc) overnight, at 4°C. The membranes were further incubated with a secondary antibody bound to alkaline phosphatase (1:20,000). The bands were revealed with the ECF substrate (Amersham) and further visualized using a STORM-860 apparatus and the program ImageQuant 5.0 (Molecular Dynamics).

RESULTS AND DISCUSSION

In FIGURE 1(A), we studied the role of Bcl-2 on 3-NP-induced neurotoxicity. Loss of cell viability occurred in a dose-dependent manner in GT1-7 cells upon incubation with 3-NP. In the presence of 0.1–0.3 mM 3-NP, a significant ($P<.05$) decrease (by about 20%) in cell viability was observed in GT1-7*puro* cells (FIG. 1, A). A decrease of 49% in cell viability was determined in the presence of 3 mM 3-NP ($P<.001$), whereas for higher 3-NP concentrations (5 mM or 7.5 mM), there was a large decrease in the percentage of viable GT1-7*puro* cells (FIG. 1, A). Interestingly, Bcl-2 overexpression in GT1-7*bcl-2* cells (FIG. 1, B) could prevent the loss of cell viability even in the presence of higher 3-NP concentrations (FIG. 1, A). As an example, in the presence of 5 mM 3-NP, GT1-7*puro* cells showed 5.84±2.19% MTT reduction, as compared to the control, whereas in GT1-7*bcl-2* cells a value of 58.80±3.41% ($P<.01$ compared to GT1-7*puro* cells) was obtained. These data highly suggested that Bcl-2 protected from 3-NP–induced neural cell degeneration.

Because the activation of caspases is a crucial event during apoptosis, we have further evaluated the protective role of Bcl-2 by measuring the activity of caspases-2, -3, and -6, an initiator and two effector caspases, in GT1-7*puro* and GT1-7*bcl-2* cells (FIG. 2). In the presence of 0.3 mM 3-NP, a concentration of 3-NP not reflecting an extensive cell death process (FIG. 1, A), a significant increase in caspases activation was observed in GT1-7*puro* cells, as determined for caspase-2 (5.07±2.27 fold), caspase-3 (2.97±1.00 fold), or caspase-6 (1.55±0.01 fold) (FIG. 2), suggesting that, similarly to what has been reported,[5,6] 3-NP cytotoxicity is associated with features of apoptosis.

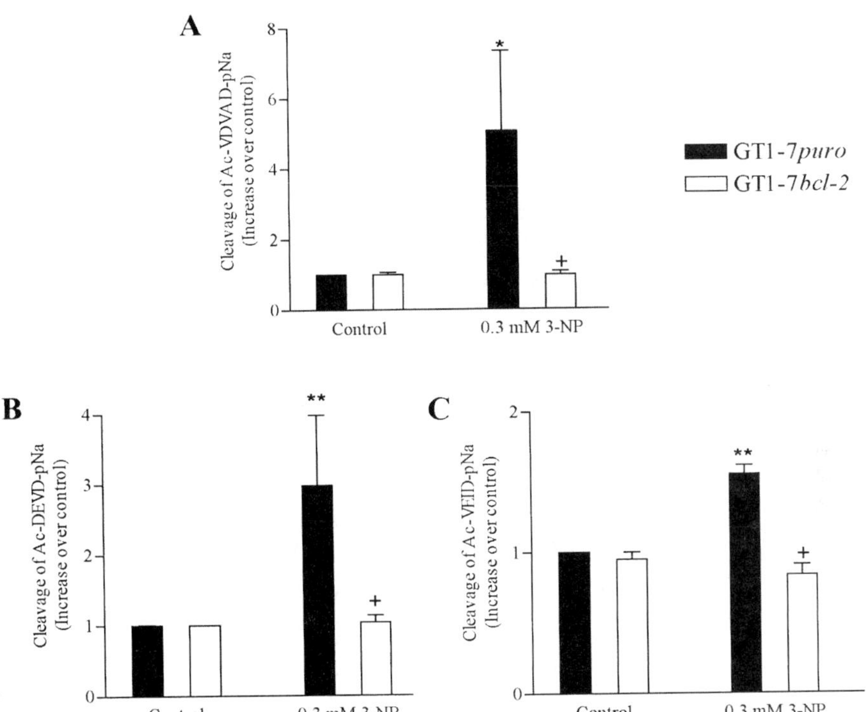

FIGURE 2. Effect of Bcl-2 on 3-NP–induced activation of caspases-2 (**A**), -3 (**B**), or -6 (**C**). GT1-7*puro* or GT1-7*bcl-2* cells were incubated in the presence of 0.3 mM 3-NP. The activity of caspases was determined through the cleavage of the colorimetric substrates: Ac-VDVAD-pNa, Ac-DEVD-pNa, or Ac-VEID-pNa, respectively, for the analysis of the activity of caspase-2, -3 or -6. The results were expressed as an increase in optical density compared to the control (cells incubated in the absence of 3-NP), and represented as the means ± SEM of 3–6 independent experiments. Statistical analysis: *$P < .05$ or **$P < .01$ as compared to the control, +$P < .01$ as compared to GT1-7*puro* cells.

Furthermore, 3-NP–exposed organotypic rat corticostriatal slice cultures were previously reported to highly express the proapoptotic protein Bax, whereas Bcl-2 was not expressed in the striatal neurons,[6] largely implicating the occurrence of apoptosis. In addition, exposure of GT1-7*bcl-2* cells to 3-NP significantly reduced the activation of these caspases to control levels (FIG. 2), highly supporting a protective effect of Bcl-2 against activation of caspases and cell death induced by complex II inhibition with 3-NP. Previously, the anti-apoptotic activity of Bcl-2 was implicated not only in the prevention of mitochondrial cytochrome *c* release,[4] but also in decreasing the generation of intracellular reactive oxygen species.[3]

REFERENCES

1. Beal, M.F., E. Brouillet, B.G. Jenkins, *et al.* 1993. Neurochemical and histologic characterization of striatal excitotoxic lesions produced by the mitochondrial toxin 3-nitropropionic acid J. Neurosci. **13:** 4181–4192.
2. Mellon, P.L., J.J. Windle, P.C. Goldsmith, *et al.* 1990. Immortalization of hypothalamic GnRH neurons by genetically targeted tumorigenesis Neuron **5:** 1–10.
3. Kane, D.J., T.A. Sarafian, R. Anton, *et al.* 1993. Bcl-2 inhibition of neural death: decreased generation of reactive oxygen species. Science **262:** 1274–1277.
4. Rego, A.C., S. Vesce & D.G. Nicholls. 2001. The mechanism of mitochondrial membrane potential retention following release of cytochrome c in apoptotic GT1-7 neural cells. Cell Death Differ. **8:** 995–1003.
5. Rodrigues, C.M., C.L. Stieers, C.D. Keene, *et al.* 2000. Tauroursodeoxycholic acid partially prevents apoptosis induced by 3-nitropropionic acid: evidence for a mitochondrial pathway independent of the permeability transition. J. Neurochem. **75:** 2368–2379.
6. Vis, J.C., R.T. de Boer-van Huizen, M.M. Verbeek, *et al.* 2002. 3-Nitropropionic acid induces cell death and mitochondrial dysfunction in rat corticostriatal slice cultures. Neurosci. Lett. **329:** 86–90.

Cisplatin-Induced Apoptosis in HL-60 Human Promyelocytic Leukemia Cells

Differential Expression of BCL2 and Novel Apoptosis-Related Gene BCL2L12

KOSTAS V. FLOROS,[a] HELLINIDA THOMADAKI,[a] GEORGE LALLAS,[a] NIKOS KATSAROS,[a] MAROULIO TALIERI,[b] AND ANDREAS SCORILAS[a]

[a]*National Center for Scientific Research "Demokritos," Athens 15310, Greece*

[b]*"G. Papanicolaou" Research Center of Oncology, "Saint Savas" Hospital, 171 Alexandras Avenue, Athens 11522, Greece*

ABSTRACT: The anti-apoptotic molecule BCL2 delays cell-cycle entry from quiescence. We have recently cloned a new member of the BCL2 family of apoptosis-related genes, BCL2L12. In the present study, the expression of BCL2 and BCL2L12 genes during cisplatin-induced apoptosis in HL-60 leukemic cells was investigated. The kinetics of apoptosis induction and cell toxicity were evaluated by DNA laddering and the MTT method, respectively. BCL2 and BCL2L12 expression was analyzed by RT-PCR using gene-specific primers. The ratio of apoptotic cells increased with increasing concentrations of cisplatin and exposure time of cell culture to the drug. Gradual, time-dependent downregulation of BCL2 gene was observed during cisplatin treatment. Upregulation of BCL2L12 was observed 3 h (no DNA fragmentation) and 6 h (initiation of the DNA fragmentation) after treatment with cisplatin, followed by a decrease of its expression after 12-h continuous treatment with the drug. It is known that the main anti-carcinogenic effect of cisplatin is due to the induction of cell apoptosis. Present results indicate that downregulation of the BCL2 and upregulation of the BCL2L12 gene may be the underlying mechanisms.

KEYWORDS: apoptosis; BCL2L12; Bcl2 gene family; promyelocytic cell lines; HL-60; cisplatin

INTRODUCTION

Apoptosis is a highly regulated physiological process. Apoptotic events are regulated by a number of proteins that exert either a positive (pro-apoptotic) or a negative (anti-apoptotic) effect. Proteins participating in these events include members of the BCL2 family, which are characterized by the presence of at least one of the BH1, BH2, BH3, or BH4 domains. BCL2 is perhaps the best characterized member of the

Address for correspondence: Dr. Andreas Scorilas, National Center of Scientific Research "Demokritos," IPC, Athens, Greece. Voice: +30210-650-3640; fax: +30210-681-5034. scorilas@netscape.net

Ann. N.Y. Acad. Sci. 1010: 153–158 (2003). © 2003 New York Academy of Sciences. doi: 10.1196/annals.1299.025

BCL2 family and one of the pro-survival proteins that prevents the release of cytochrome *c* from mitochondria and interacts with several proteins participating in cell death regulation.[1] BCL2L12 is a novel gene that encodes a BCL2-like, proline-rich protein and contains one BH2 homology domain.[2]

Cisplatin is an effective antineoplastic agent for metastatic testicular, breast and ovarian cancer. Cisplatin, after entering the cell, undergoes strong hydration to form positively charged species that can interact with DNA to yield a variety of adducts with inter-strand and intra-strand DNA crosslinks as well as DNA-protein crosslinks. The mechanism by which DNA-adducts kill cells is not fully understood.[3] It has been previously reported that cells treated with cisplatin were arrested in the G_2-phase of the cell cycle before inter-nucleosomal degradation of genomic DNA and loss of membrane integrity, presenting evidence that cisplatin kills by the mechanism of apoptosis.[4]

It was recently shown that cisplatin alone is able to cause cell death through the induction of apoptosis in a variety of cell lines, including the human leukemic cell line HL-60.[4] In the present study, we investigated the alterations in the mRNA levels of BCL2 and BCL2L12 genes under cisplatin treatment.

MATERIALS AND METHODS

Human promyelotic leukemia HL-60 cell line was maintained in RPMI 1640 medium, supplemented with 10% fetal bovine serum, 200,000 U/L benzopenicillin, 0.1 g/L streptomycin, 0.3 g/L L-glutamine, and 0.85 g/L $NaHCO_3$. Cells were plated at 4×10^5 cells/mL and incubated at 37°C for 48 h before treatment. Cell viability was tested using a standard MTT (3-[4,5-dimethylthiazol-2-yl]-2,5-diphenyltetrazolium bromide thiazolyl blue indicator dye) chemosensitivity assay. Apoptosis was induced after treatment with 5, 15, 30 μg/mL of cisplatin (Faulding Pharmaceuticals Plc; U.K.), for 3, 6, and 12 hours. Inter-nucleosomal DNA fragmentation was detected by electrophoresis of cells (10^6 cells/well), using 2% agarose gel.[5] The gel was ethidium bromide (10 μg/mL) stained and photographed. Total RNA was extracted using the Trizol method (Gibco BRL, USA). Two μg of total RNA were reverse-transcribed into first-strand cDNA using the Superscript™ preamplification system (Gibco BRL). The final volume was 20 μL. Gene-specific primers were designed (GGAGACCGCAAGTTGAGTGG and GTCATCCCGGCTACAGAACA for BCL2L12, TTTGAGTTCGGTGGGGTCAT and TGACTTCACTTGTGGCCCAG for BCL2, ATCTGGCACCACACCTTCTA and CGTCATACTCCTGCTTGCTG for actin), and PCR was carried out in a reaction mixture containing 1 μL of cDNA, 10 mM Tris-HCl (pH 8.3), 50 mM KCl, 1.5 mM $MgCl_2$, 200 mM dNTPs (deoxynucleoside triphosphates), 150 ng of primers, and 2.5 U of HotStar™ DNA polymerase (Qiagen Inc., Valencia, CA, USA) on a thermal cycler (Eppendorf Mastercycler Gradient). The cycling conditions contained: a denaturation step at 95°C for 15 min, followed by 30 cycles of 94°C for 30 sec, 62°C for 30 sec, 72°C for 1 min, and a final extension step at 72°C for 10 min. Equal amounts of PCR products were electrophoresed on 2% agarose gels and visualized by ethidium bromide staining. Actin was used as internal control for the integrity of the mRNA.

(A)

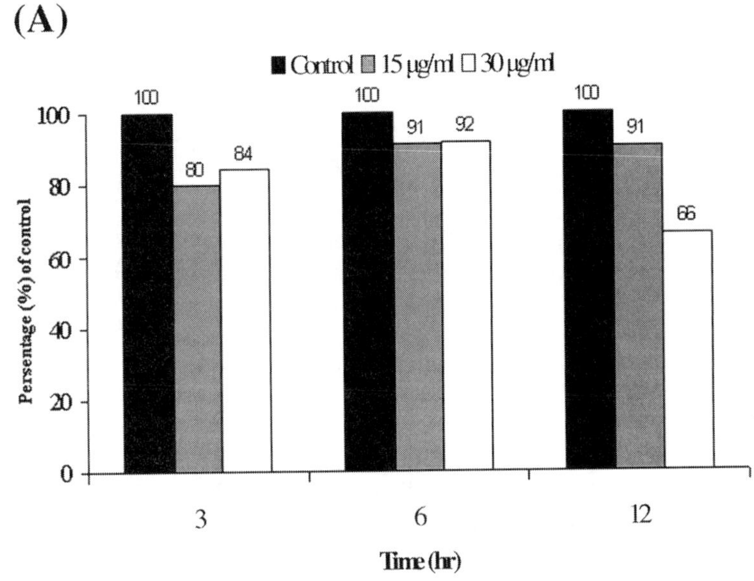

(B)

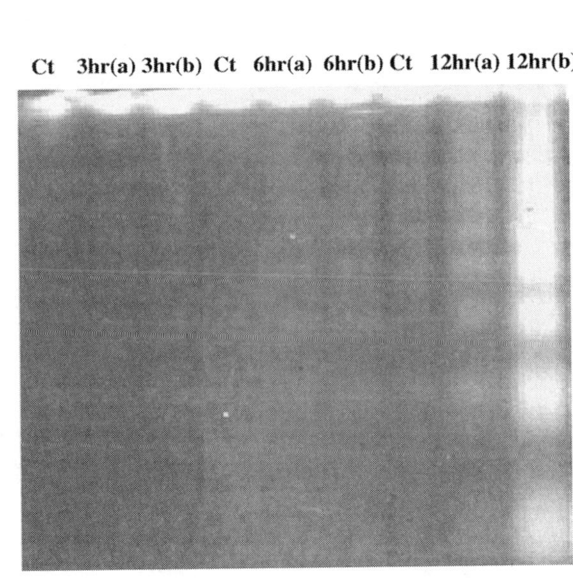

FIGURE 1. (**A**) Viability of HL-60 cells after treatment with different concentrations of cisplatin (MTT assay). (**B**) DNA ladder formation after treatment with cisplatin. Ct, control cells without cisplatin treatment; 3hr(a), cells treated with 15 µg/mL cisplatin for 3 h; 3hr(b), cells treated with 30 µg/mL cisplatin for 3 h; 6hr(a), cells treated with 15 µg/mL cisplatin for 6 h; 6hr(b), cells treated with 30 µg/mL cisplatin for 6 h; 12hr(a), cells treated with 15 µg/mL cisplatin for 12 h; 12hr(b), cells treated with 30 µg/mL cisplatin for 12 hours.

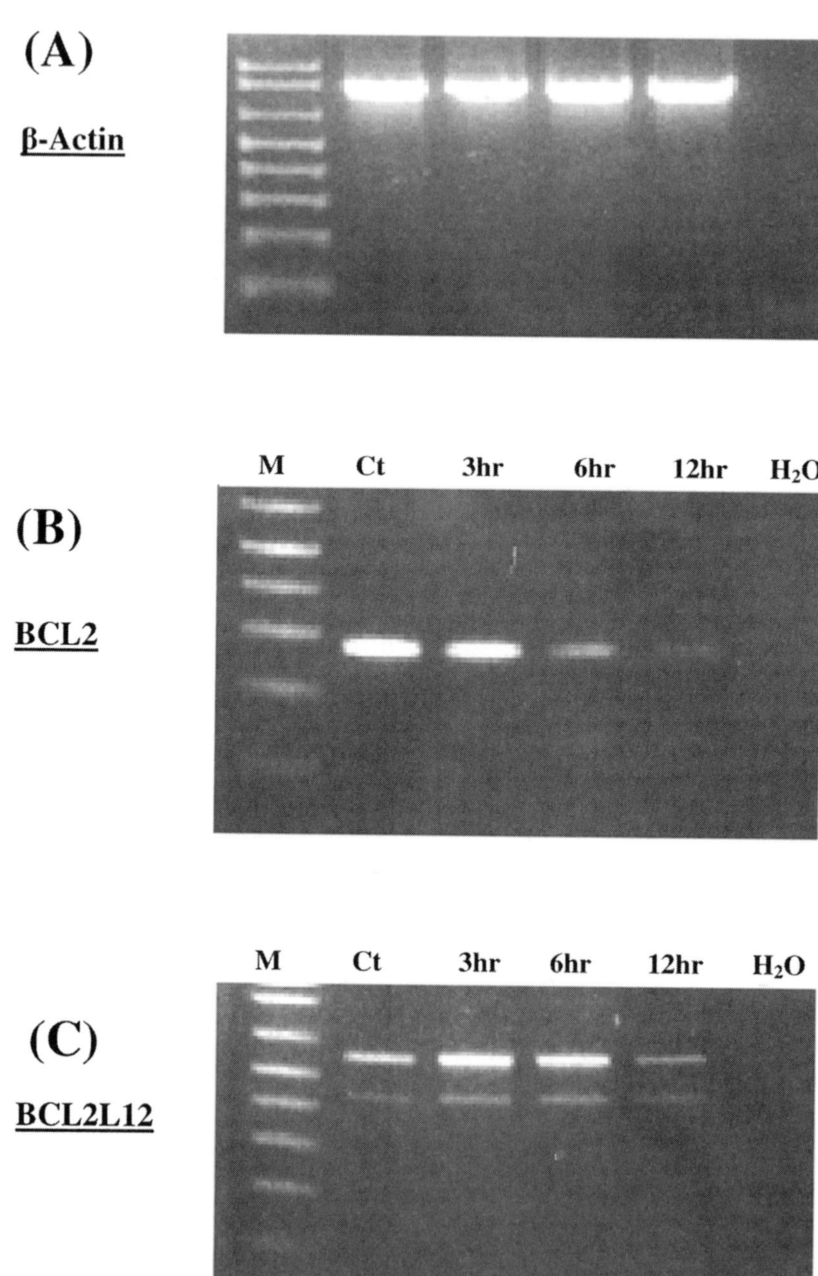

FIGURE 2. mRNA expression of β-actin (**A**), BCL2 (**B**), and BCL2L12 (**C**). Ct, control cells without cisplatin treatment; 3hr, cells 3 hours after treatment with cisplatin (30 μg/mL); 6hr, cells 6 h after treatment with cisplatin (30 μg/mL); 12hr, cells 12 h after treatment with cisplatin (30 μg/mL); H_2O, negative control.

RESULTS AND DISCUSSION

Using the MTT assay, we evaluated the cytotoxic effect of cisplatin on HL-60 cells and we determined the time points at which the percentage of cells with low mitochondrial metabolism was less than 50%. We observed constant decrease of mitochondrial metabolism at cisplatin concentration of 15 μg/mL of cell suspension. Results obtained at cisplatin concentration of 15 μg/mL were similar to those obtained at 30 μg/mL, 3 and 6 h after treatment, where we also observed a steady decrease of mitochondrial metabolism. Finally, the cytotoxic effect of cisplatin was higher after 12 h treatment (FIG. 1).

Apoptosis was confirmed in each case by endonucleotic cleavage of DNA (laddering). DNA fragments were not evident at 3 hours. Low percent of DNA fragmentation was observed when HL60 cells were treated with 15 or 30 μg cisplatin/mL of cell suspension for 6 h and it was clearly visible 12 h after cell exposure to cisplatin. At the same time a gradual increase of DNA was noticed (FIG. 1). Expression of BCL2 and BCL2L12 gene was investigated at this concentration.

Gradual downregulation of BCL2 mRNA expression was observed (using RT-PCR technique), which almost disappeared 12 h after drug treatment (FIG. 2). Present results are in agreement with other studies concerning BCL2 expression in MCF-7 cells treated with cisplatin.[6] Upregulation of BCL2L12 gene was observed 3 h (no DNA fragmentation) and 6 h (initiation of the DNA fragmentation) after cisplatin treatment, followed by decrease of its mRNA expression after 12 h treatment (FIG. 2). All experiments were performed three times with the same results. These findings indicate the possibility that the alterations in the mRNA expression of BCL2L12 are due to a pro-apoptotic effect.

The main anti-carcinogenic effect of cisplatin is due to the induction of cell apoptosis. Downregulation of the BCL2 and upregulation of the BCL2L12 gene may be an underlying mechanism. To the best of our knowledge, this is the first study examining the expression of BCL2L12 in relation to BCL2 gene under cisplatin-induced apoptosis.

ACKNOWLEDGMENTS

We would like to thank Dr. M. Havredaki, Dr. G. Voutsinas, A. Panagiotopoulou, MSc., and M. Katsarou, MSc., for their valuable assistance.

REFERENCES

1. REED, J.C. 1997. Double identity for proteins of the Bcl-2 family. Nature **387:** 773–776.
2. SCORILAS, A. *et al.* 2001. Molecular cloning, physical mapping, and expression analysis of a novel gene, BCL2L12, encoding a proline-rich protein with a highly conserved BH2 domain of the Bcl-2 family. Genomics **72:** 217–221.
3. GONZALEZ, V.M. *et al.* 2001. Is cisplatin-induced cell death always produced by apoptosis? Molec. Pharmacol. **59:** 657–663.
4. BARRY, M.A. *et al.* 1990. Activation of programmed cell death (apoptosis) by cisplatin, other anticancer drugs, toxins and hyperthermia. Biochem. Pharmacol. **40:** 2353–2362.

5. EASTMAN, A. 1995. Assays for DNA fragmentation, endonucleases, and intracellular pH and Ca^{2+} associated with apoptosis. Methods Cell Biol. **46:** 41–55.
6. POLISENO, L. *et al.* 2002. Bcl2-negative MCF7 cells overexpress p53: implications for the cell cycle and sensitivity to cytotoxic drugs. Cancer Chemother. Pharmacol. **50:** 127–130.

TNFα Potentiates 2-Methoxyestradiol-Induced Mitochondrial Death Pathway

MOJGAN DJAVAHERI-MERGNY, JUANA WIETZERBIN, DANY ROUILLARD, AND FRANÇOISE BESANÇON

INSERM U365, Institut Curie, 26 rue d'Ulm, 75248 Cedex 05 Paris, France

ABSTRACT: Ewing sarcoma cells are resistant to TNFα-induced cell death and this resistance results from the activation of the transcription factor NF-κB. Here, we investigated whether NF-κB activation interferes with 2-Me–induced cell death signaling in Ewing sarcoma cells and we examined the effect of treatment of these cells with 2-Me either alone or in combination with TNFα. Our results show that TNFα cooperates with 2-Me to induce apoptosis in Ewing tumor cells through mitochondrial cell death signaling. These results suggest that the use of TNFα in combination with 2-Me may be beneficial for Ewing tumor treatment.

KEYWORDS: 2-methoxyestradiol; TNFα; apoptosis; NF-κB; mitochondria

INTRODUCTION

Ewing sarcoma is the second most common bone tumor in children and young adults.[1] Despite advances in therapy, the five-year survival rate for patients with metastatic disease is poor, indicating the need for alternative treatment. 2-Methoxyestradiol (2-Me), a physiological metabolite of 17 β-estradiol, has recently emerged as a promising anti-cancer agent.[2] We have reported that 2-Me induces apoptosis of Ewing sarcoma–derived cells through the mitochondrial death pathway as evidenced by reduction of the mitochondrial transmembrane potential, cytochrome *c* release, and caspase-9 activation.[3] Many chemotherapeutic drugs have found limited use as anti-cancer agents due to their ability to induce anti-apoptotic responses.[4] We have previously shown that Ewing sarcoma cells are resistant to TNFα-induced cell death and that this resistance results from the activation of the transcription factor NF-κB.[5] Here, we investigated whether NF-κB activation interferes with 2-Me induced cell death signaling in Ewing sarcoma cells and examined the effect of treatment of these cells with 2-Me either alone or in combination with TNFα.

Address for correspondence: Mojgan Djavaheri-Mergny, INSERM U504, Bât Inserm, 16, avenue P.V. courturier 94807 Villejuif Cedex, France. Voice: +33-1-45595041; fax: +33-1-46770233.

Mergny@vjf.inserm.fr

Ann. N.Y. Acad. Sci. 1010: 159–162 (2003). © 2003 New York Academy of Sciences.
doi: 10.1196/annals.1299.026

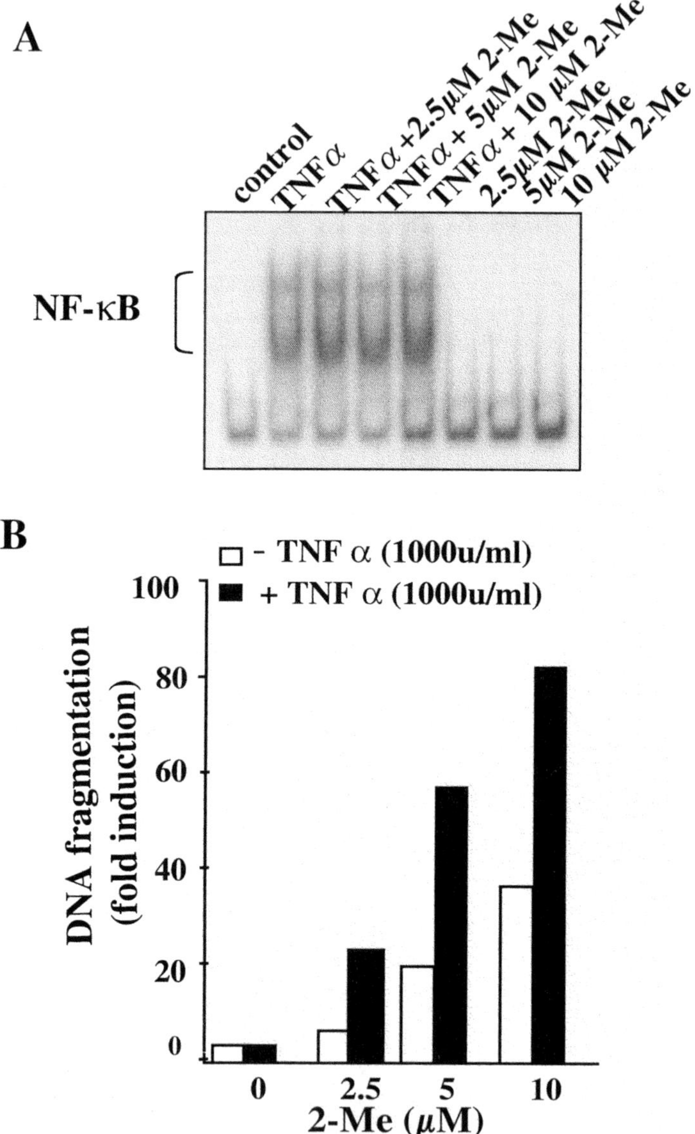

FIGURE 1. 2-Me-induced apoptosis in Ewing tumor cells is enhanced by TNFα. EW7 cells were treated with the indicated concentrations of 2-Me and TNFα (1000 U/mL) for 2 hours. (**A**) Nuclear extracts were then prepared and assayed for NF-κB DNA-binding activity. (**B**) Apoptosis was measured after 16 h of treatment by nucleosomal DNA fragmentation assay (Roche Diagnostics GmbH, Mannheim, Germany). The ratio of nucleosomal DNA fragmentation of treated cells to DNA fragmentation in control cells was determined.

RESULTS

2-Me–Induced Apoptosis in Ewing Tumor Cells Is Enhanced by TNFα

We first examined the effect of treatment of EW7 cells with either 2-Me, TNFα, or 2-Me+TNFα on NF-κB activity. This activity was measured by electrophoretic mobility shift assay (EMSA) analysis of nuclear extracts using a ^{32}P-labeled oligonucleotide containing the NF-κB consensus sequence. As shown in FIGURE 1(A), 2-Me treatment alone did not induce NF-κB activity whereas, as previously shown, TNFα induced NF-κB DNA binding activity. Moreover, treatment of EW7 cells with 2-Me in combination with TNFα did not modify TNFα-induced NF-κB activity. As NF-κB induces anti-apoptotic responses, we further examined whether activation of NF-κB by TNFα inhibits apoptosis induced by 2-Me. As shown in FIGURE 1(B), 2-Me induced apoptosis in a dose-dependent manner in EW7 cells as measured by nucleosomal DNA fragmentation assay. Surprisingly, combination of 2-Me treatment

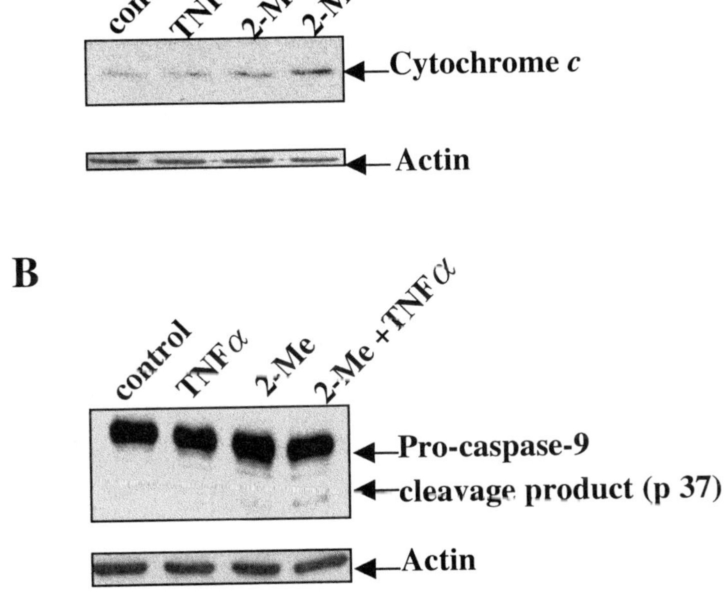

FIGURE 2. TNFα enhances 2-Me-induced activation of the mitochondrial death pathway. Ewing tumor cells (EW7)were treated with 5 μM 2-Me and with TNFα (1000 U/mL). (**A**) Cytosolic fractions were prepared after 8 h of treatment and subjected to SDS-PAGE followed by immunoblotting with anti–cytochrome *c* (BD Biosciences, Heidelberg, Germany). Actin was used as a control for protein loading. (**B**) Cellular extracts were prepared after 16 h of treatment and analyzed by Western blot using anti–caspase-9 antibody (New England BioLabs, UK). Actin was used as a control for protein loading.

with TNFα enhanced (by about two to threefold) cell death induced by 2-Me treatment alone (FIG. 1, B).

TNFα Enhances Activation of the Mitochondrial Death Pathway Induced by 2-Me

As we have previously shown that 2-Me induced apoptosis through the mitochondrial death pathway,[3] we examined whether the increase in the killing effect of 2-Me by TNFα was accompanied by an increase in the activity of this pathway. Mitochondrial release of cytochrome *c* is an early step in mitochondrial death pathway activation that subsequently leads to activation of caspase-9.[6] To investigate the effect of a combined treatment with 2-Me and TNFα on the mitochondrial death pathway, we measured the level of cytosolic cytochrome *c* and caspase–9 activation by immunoblot analysis. As shown in FIGURE 2A, 2-Me (5 μM) induced the release of cytochrome *c* after 8 h of treatment and the combination of 2-Me with TNFα increased the level of cytosolic cytochrome *c* release. Caspase-9 activation revealed by the presence of pro-caspase-9 cleavage products was higher in cells treated with both compounds than in cells treated with 2-Me or TNFα alone (FIG. 2, B).

CONCLUSION

Our results show that TNFα cooperates with 2-Me to induce apoptosis in Ewing tumor cells through mitochondrial cell death signaling. More investigations are needed to elucidate the precise molecular mechanisms involved in the activation of this pathway by both compounds. These results suggest that the use of TNFα in combination with 2-Me may be beneficial for Ewing tumor treatment.

REFERENCES

1. KOVAR, H. 1998. Ewing's sarcoma and peripheral primitive neuroectodermal tumors after their genetic union. Curr. Opin. Oncol. **10:** 334–342.
2. PRIBLUDA, V.S., E.R. GUBISH, T.M. LAVALLEE, *et al.* 2000. 2-Methoxyestradiol: an endogenous antiangiogenic and antiproliferative drug candidate. Cancer Metastasis Rev. **19:** 173–179.
3. DJAVAHERI-MERGNY, M., J. WIETZERBIN & F. BESANÇON. 2003. 2-Methoxyestradiol induces apoptosis in Ewing sarcoma cells through mitochondrial hydrogen peroxide production. Oncogene **22:** 2558–2567.
4. PAHL, H.L. 1999. Activators and target genes of Rel/NF-kappa B transcription factors. Oncogene **18:** 6853–6866.
5. JAVELAUD, D., J. WIETZERBIN, O. DELATTRE & F. BESANCON. 2000. Induction of p21Waf1/Cip1 by TNFalpha requires NF-kappaB activity and antagonizes apoptosis in Ewing tumor cells. Oncogene **19:** 61–68.
6. REED, J.C. & D.R. GREEN. 2002. Remodeling for demolition: changes in mitochondrial ultrastructure during apoptosis. Mol. Cell. **9:** 1–3.

Induction of Apoptosis by Ultrasound Application in Human Malignant Lymphoid Cells

Role of Mitochondria-Caspase Pathway Activation

F. FIRESTEIN,[a] L.A. ROZENSZAJN,[a] L. SHEMESH-DARVISH,[a] R. ELIMELECH,[a] J. RADNAY,[a] AND U. ROSENSCHEIN[b]

[a]*Faculty of Life Sciences, Bar-Ilan University, Ramat-Gan, Israel*

[b]*Department of Cardiology, Tel Aviv Medical Center, Tel Aviv University, Tel Aviv, Israel*

ABSTRACT: In the present study, we have focused on the specific question of whether ultrasound application (ULS) delivered with optimized parameters for cavitation generation can stimulate apoptosis in lymphoid cell lines. Suspended T and B lymphoid cell lines (Jurkat and Raji, respectively) were exposed to low frequency ULS (750 KHz) at an intensity level of 54.6 W/cm^2 spatial peak temporal average (SPTA) at focal area, which was found to be the optimal physical parameter to induce apoptosis in these malignant cell lines. Unsonicated cells and cells exposed to γ-radiation (20 Gy) using ^{137}Cs source were used as control. Apoptosis was evaluated by cell morphology changes, cell-cycle analysis, and phosphatidylserine exposure. Fraction of cells with low mitochondria membrane potential was observed 1 h after sonication, accompanied by cytochrome *c* release from mitochondria to the cytosol and caspase-3 activation. Here we present evidence that ULS exposure with cavitation formation on malignant lymphoid cell lines differs from γ-radiation and is associated with time-dependent apoptosis, which is mitochondria-caspase dependent.

KEYWORDS: ultrasound; cavitation; apoptosis

INTRODUCTION

We have recently demonstrated for the first time that selected physical parameters utilized in the application of therapeutic low-frequency (750 kHz), high-power, focused and pulsed wave ultrasound (ULS), which results in a cavitation process, induces time-dependent apoptotic cell death in human myeloid leukemia cell lines.[1] Resent observations confirmed our findings in human histiocytic lymphoma U937 cells[2] and in leukemic cell lines.[3] Apoptosis is a crucial pathway for a variety of biological events, including physiology of mammalian development and surveillance against tumors or other malfunctioning cells. An important and even decisive role in

Address for correspondence: Frida Firestein, Faculty of Life Sciences, Bar-Ilan University, Ramat-Gan 52900, Israel. Voice: 972-3-5318383; fax: 972-3-5351824.
fridabox@hotmail.com

Ann. N.Y. Acad. Sci. 1010: 163–166 (2003).
doi: 10.1196/annals.1299.027

the initiation and accomplishment of the apoptotic process has been ascribed to mitochondria.[4] In an attempt to explore the mechanism of ultrasound-mediated apoptosis, we exposed lymphoid cell lines, apoptosis-sensitive Jurkat (T cell leukemia), and apoptosis-resistant cell line Raji (Burkitt B cell lymphoma),[5] to therapeutic ultrasound, with the focus on the role of mitochondria in the apoptotic process. Our results show that after ULS exposure of 54.6 W/cm^2 spatial peak temporal-average intensity (SPTA) on malignant lymphoid cells, a release of cytochrome *c* to the cytosol accompanied by early disruption of mitochondrial transmembrane potential and caspase activation was displayed.

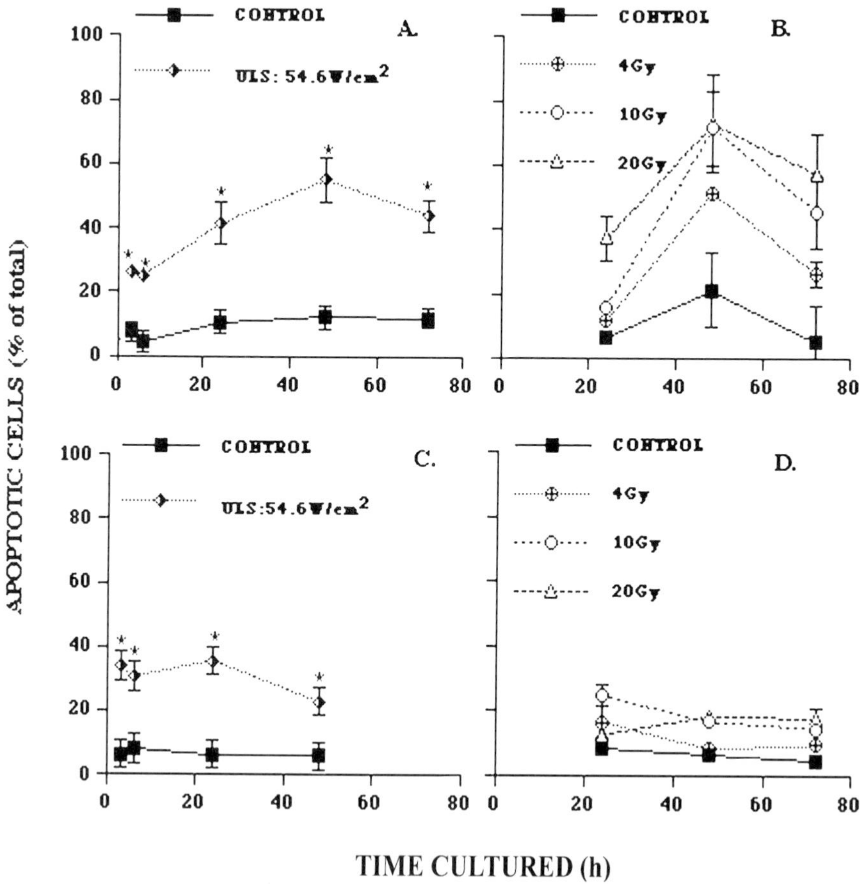

FIGURE 1. Ultrasound induces apoptotic cell death as a function of time, determined by fluorescent microscopy. Lymphatic cell line suspensions Jurkat and Raji were sonicated at duty cycle 1:50, pulsed duration 100 μsec at intensity of 54.6 W/cm^2 SPTA. Apoptotic cells were determined with nuclei acid binding dyes acridine orange and ethidium bromide. Values were calculated from three independent experiments. $^*P < .05$. (**A** and **B**) Jurkat cells. (**C** and **D**) Raji cells.

RESULTS AND DISCUSSION

Ultrasound Dose, Cell Viability, and Free Radicals

We have previously established the desirable parameters for our ULS system and chose conditions to deliver high-power, low-frequency (750 KHz), focused and pulsed ULS at intensity level of 54.6 W/cm^2 SPTA at focal area to result in inertial cavitation effects and cell damage. The number of cells surviving ULS under these conditions was approximately 50% as detected by trypan blue exclusion. No significant temperature changes (1–2°C) and no free hydrogen radicals, nitric oxide radicals, or reactive oxygen species formation (ROS) were observed (data not shown). These results may exclude free radical–dependent mechanisms to mediate apoptosis in our system.

Detection and Evaluation of Apoptotic Cells

The evaluation of apoptotic cells using ethidium bromide and acridine orange staining is a rapid and objective method to study the effect of ULS on a signal-cell suspension. ULS treatment results in apoptotic cell death, started 3 h post-treatment and increased during incubation time (FIG. 1). Further positive identification of apoptotic cells was established based on morphological changes, cell-cycle analysis, and annexin V/propidium iodide double staining (data not shown). These results im-

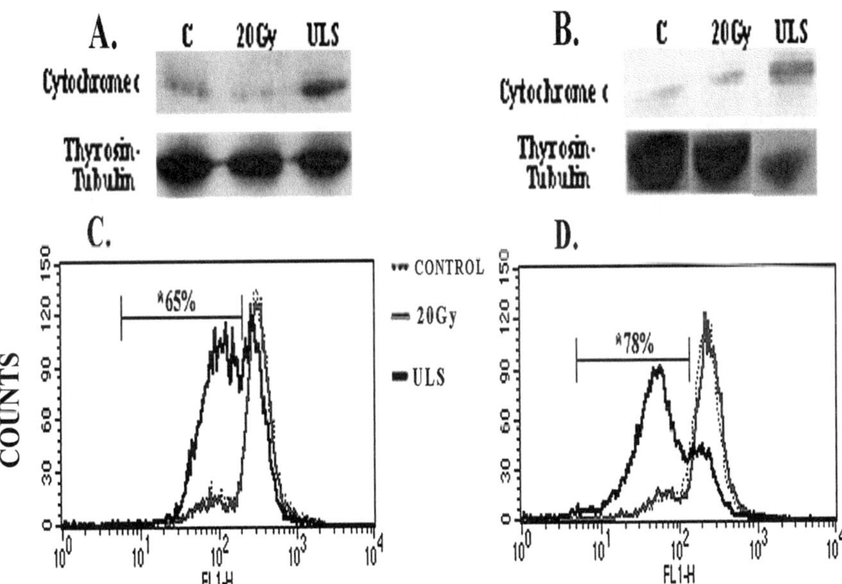

FIGURE 2. Effect of ultrasound on mitochondria. (**A** and **B**) Western blot analysis of cytochrome *c* release from the mitochondria to cytosol 1 h after sonication. (**C** and **D**) Reduction of the mitochondrial transmembrane potential was determined by rhodamine-123 uptake.

ply that apoptosis induced by ULS is an early event that increases with time, and may be independent of cell-cycle checkpoints, in contrast to γ-rays which exhibited cell cycle G2/M phase block (data not shown). Apoptosis also occurred in γ-ray-resistant Raji cells (FIG. 1C and D).

Mechanism of Apoptosis Induced by ULS

In order to investigate the molecular signaling pathway of apoptosis induced by ULS, disruption of mitochondrial transmembrane potential ($\Delta\Psi m$), cytochrome *c* release, and caspase activation, were explored. The release of cytochrome *c* from mitochondria acts as a central gate in turning on/off apoptosis by the formation of "apoptosome." As shown in FIGURE 2, release of cytochrome *c* to the cytosol was already observed 1 h post-treatment, accompanied by early disruption of mitochondrial transmembrane potential ($\Delta\Psi m$) and caspase activation (data not shown). This observation implies that ULS differs from a γ-irradiation apoptotic mechanism by its early events of apoptotic cascade, probably due to early activation of the mitochondria to induce apoptotic death. Further studies of ULS response should be performed to increase understanding of the ULS distinct pathway. The effect of ULS on lymphoid cells may offer a noninvasive, localized method of apoptosis activation and can be utilized in the future for therapeutic application in cancer.

REFERENCES

1. ASHUSH, H. *et al.* 2000. Apoptosis induction of human myeloid leukemic cells by ultrasound exposure. Cancer Res. **60:** 1014–1020.
2. HONDA, H. *et al.* 2002. Effect of dissolved gases and an echo contrast agent on apoptosis induced by ultrasound and its mechanism via the mitochondria-caspase pathway. Ultrasound Med. Biol. **28:** 673–682.
3. LAGNEAUX, L. *et al.* 2002. Ultrasonic low energy treatment: a novel approach to induce apoptosis in human leukemic cells. Exp. Hematol **30:** 1293–1301.
4. GREEN, D.R. & G. KROEMER. 1998. The central executioners of apoptosis: Caspase or mitochondria? Trends Cell Biol. **8:** 267–271.
5. KAWABATA, Y. *et al.* 1999. Defective apoptotic signal transduction pathway downstream of caspase-3 in human B-lymphoma cells: A novel mechanism of nuclear apoptosis resistance. Blood **94:** 3523–3530.

Apoptotic Mitochondrial Dysfunction Induced by Benzo(*a*)pyrene in Liver Epithelial Cells

Role of p53 and pH_i Changes

LAURENCE HUC, DAVID GILOT, CLAIRE GARDYN, MARY RISSEL, MARIE-THÉRÈSE DIMANCHE-BOITREL, ANDRÉ GUILLOUZO, OLIVIER FARDEL, AND DOMINIQUE LAGADIC-GOSSMANN

INSERM U456, Faculté de Pharmacie, Université Rennes I, 2 avenue Prof. Léon Bernard, 35043 Rennes cedex, France

ABSTRACT: How pH_i changes, more specifically alkalinization, affect the apoptotic cascade has yet to be determined. The aim of the present work was to test the involvement of mitochondria in the apoptotic cascade triggered by benzo(*a*)pyrene [B(a)P] and to determine the role of pH_i changes and p53 relative to mitochondria. Our results indicate that B(a)P-induced apoptosis might rely upon a p53-dependent and a pH-sensitive mitochondrial dysfunction.

KEYWORDS: benzo(*a*)pyrene; rat liver epithelial cell; apoptosis; mitochondria; intracellular pH; p53; reactive oxygen species

INTRODUCTION

Polycyclic aromatic hydrocarbons (PAHs), such as benzo(*a*)pyrene [B(a)P], are responsible for important carcinogenic and apoptotic effects, whose mechanisms are still poorly understood. With respect to PAH-induced apoptosis, this phenomenon has been shown to require caspase[1] as well as p53 activation.[2] An important role for reactive oxygen species (ROS) production has been demonstrated in cell death induced by various chemicals.[3] Regarding this latter point, although PAHs, via metabolism through cytochrome 1A1, are well known to induce ROS production, a role for such species in the apoptotic cascade triggered by these compounds remains to be established. Besides ROS, cytoplasmic acidification has also been proposed as an important factor in apoptosis.[4] In this context, it is worth noting that we have shown previously that biphasic perturbations of H^+ homeostasis (i.e., early alkalinization followed by late acidification) were elicited during B(a)P-induced cell death in the rat liver F258 epithelial cell line and that inhibition of those changes afforded a significant cell protection against B(a)P-induced toxicity.[5] However, how pH_i changes, more specifically alkalinization, affect the apoptotic cascade has yet to be determined. The aim of the present work was to test the involvement of mitochondria in

Address for correspondence: Dominique Lagadic-Gossmann, INSERM U456, Faculté de Pharmacie, Université Rennes I, 2 av Prof Léon Bernard, 35043 Rennes cedex, France. Voice: (33)2.23.23.48.37; fax: (33)2.23.23.47.94.
dominique.lagadic@rennes.inserm.fr

**Ann. N.Y. Acad. Sci. 1010: 167–170 (2003). © 2003 New York Academy of Sciences.
doi: 10.1196/annals.1299.028**

TABLE 1. Apoptotic effects of benzo(*a*)pyrene (72h) in F258 rat liver epithelial cell line

Cell Treatment	DNA Fragmentation Detected by Hoechst 33342 Staining of Cells (% control population)	Caspase-3–like Activity (% control population)
B(a)P (50 nM)	274.0 ± 1.7%	692.2 ± 69.9%
B(a)P + α-Naphthoflavone (10 μM)	118.3 ± 3.4%[a]	85.6 ± 0.1%[a]
B(a)P + Cariporide (30 μM)	181.2 ± 2.7%[a]	276.6 ± 21.7%[a]
B(a)P + Bongkrekic acid (10 μM)	128.9 ± 2.8%[a]	378.2 ± 26.2%[a]
B(a)P + Pifithrin α (10 μM)	121.1 ± 2.1%[a]	79.6 ± 0.1%[a]

Results were quoted as mean±SEM of three independent experiments. They were expressed as percentage of effects compared to values obtained from control, untreated cells. Inhibitor, B(a)P-treated cells versus control, B(a)P-treated cells, [a]$P<.05$ (ANOVA followed by Student-Newman-Keuls test).

the apoptotic cascade triggered by B(a)P and to determine the role of pH_i changes and p53 relative to mitochondria.

MATERIAL AND METHODS

The F258 rat liver epithelial cell line, cultured in Williams' medium E supplemented with 10% fetal calf serum at 37°C under a 5% CO_2 atmosphere, were treated during the exponential phase with 50 nM B(a)P and/or the different inhibitors tested (applied to cells 2 h prior to PAH exposure) for various treatment times.

Microscopical detection of apoptosis was performed in both floating and adherent cells, using Hoechst 33342 (0.5 μg/mL) labeling. Caspase activity was measured in cell lysates using the fluorogenic substrate DEVD-AMC.

Mitochondrial potential and ROS production were measured by flow cytometry using $DiOC_6(3)$ (50 nM) and dihydroethidium (DHE; 5 μM), respectively.

RESULTS AND DISCUSSION

Results quoted in TABLE 1 show that treatment of F258 cells by B(a)P (50 nM, 72 h) led to a significant apoptotic effect (as detected by Hoechst 33342 staining and caspase-3–like activity) compared to untreated counterparts. Using α-naphthoflavone (10 μM; a CYP1A1 inhibitor) and cariporide (30 μM; an inhibitor of Na^+/H^+ exchange which has been shown to inhibit PAH-induced pH_i change[5]), we demonstrated the involvement of B(a)P metabolism and associated pH_i changes in the apoptotic cascade elicited by B(a)P. Furthermore, a role for mitochondria and the tumor suppressor p53 was suggested since B(a)P-induced apoptosis was markedly inhibited by bongkrekic acid (10 μM) and pifithrin α (10 μM), respectively. In order to gain more insight into a putative effect of B(a)P on mitochondrial function, we decided to measure mitochondrial membrane potential ($\Delta\Psi m$) using $DiOC_6(3)$ staining of cells. As illustrated in FIGURE 1, a mitochondrial hyperpolarization was

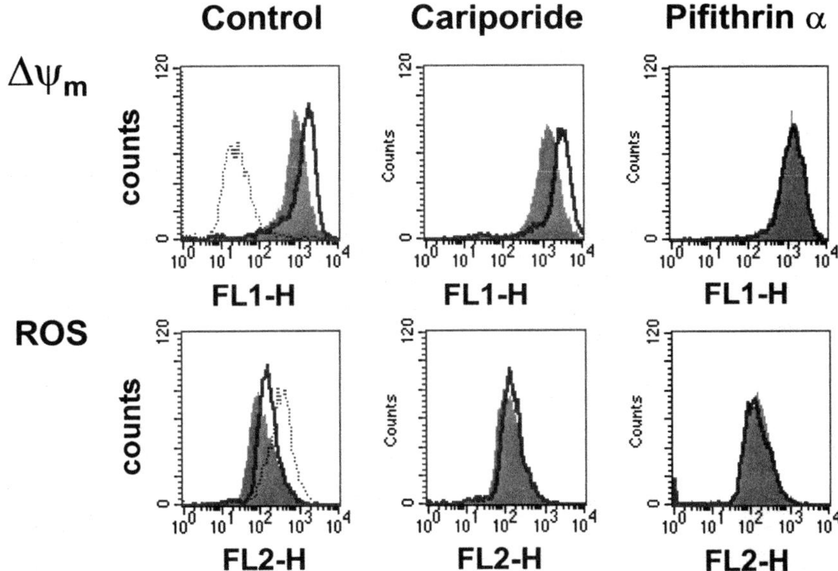

FIGURE 1. Effects of B(a)P (50 nM, 48 h) on mitochondrial membrane potential and ROS production in F258 cells. Mitochondrial membrane potential (ΔΨm) and ROS production were measured using $DiOC_6(3)$ and DHE staining of the cells, respectively. Illustrated histograms were representative of three independent experiments. (*Filled, grey peak*) Control, untreated cells; (*open, black peak*) B(a)P-treated cells; (*open, dotted peak*) positive control cells (i.e., in the presence of FCCP [50 μM] and menadione [100 μM] when $DiOC_6(3)$ and DHE were used, respectively).

detected following a 48-h treatment, in contrast to the well described depolarization observed upon FCCP application. We next tested the effects of B(a)P upon ROS production. FIGURE 1 shows that B(a)P induced a marked production of ROS in F258 cells, as pointed out by a rightward shift of the fluorescent peak, similar to that observed with menadione, a known pro oxidant molecule. This production was found to be inhibited by bongkrekic acid (not shown). In order to test the involvement of p53 and of the previously observed B(a)P-induced alkalinization in the changes described above, we evaluated the effects of pifithrin α (10 μM) and cariporide (30 μM). Our data show that whereas pifithrin α prevented both hyperpolarization and ROS production, cariporide was effective only upon ROS production (FIG. 1). Altogether, our results indicate that B(a)P-induced apoptosis might rely upon a p53-dependent and a pH-sensitive mitochondrial dysfunction.

ACKNOWLEDGMENTS

This work was supported by grants from the Institut National de Recherche et de Sécurité (INRS) and the Ministère de l'Aménagement du Territoire et de l'Environnement.

REFERENCES

1. LEI, W., R. YU, S. MANDLEKAR & A.-N.T. KONG. 1998. Induction of apoptosis and activation of interleukin 1β-converting enzyme/Ced-3 protease (caspase-3) and cJun NH_2-terminal kinase 1 by benzo(a)pyrene. Cancer Res. **58:** 2102–2106.
2. KWON, Y.-W., S. UEDA, M. UENO, *et al.* 2002. Mechanism of p53-dependent apoptosis induced by 3-methylcholanthrene. Involvement of p53 phosphorylation and p38 MAPK. J. Biol. Chem. **277:** 1837–1844.
3. FLEURY, C., B. MIGNOTTE & J.L. VAYSSIERE. 2002. Mitochondrial reactive oxygen species in cell death signaling. Biochimie **84:** 131–141.
4. PERVAIZ, S. & M.-V. CLÉMENT. 2002. A permissive apoptotic environment: Function of a decrease in intracellular superoxide anion and cytosolic acidification. Biochem. Biophys. Res. Commun. **290:** 1145–1150.
5. HUC, L., L. SPARFEL, A. GUILLOUZO, *et al.* 2002. Apoptosis-related pH_i changes in rat liver cells treated with benzo(a)pyrene. Pharmacologist **44**(suppl. 1)**:** A124 (abstract).

Orphan Nuclear Receptor Nur77 Translocates to Mitochondria in the Early Phase of Apoptosis Induced by Synthetic Chenodeoxycholic Acid Derivatives in Human Stomach Cancer Cell Line SNU-1

JIN HEE JEONG,[a] JOO-SUNG PARK,[a] BONGKYUNG MOON,[a] MIN CHAN KIM,[a] JAE-KON KIM,[b] SUNGEUN LEE,[b] HONGSUK SUH,[b] NAM DEUK KIM,[c] JONG-MIN KIM,[a] YOUNG CHUL PARK,[a] AND YOUNG HYUN YOO[a]

[a]*Department of Anatomy and Cell Biology (BK21 program), Dong-A University College of Medicine and Institute of Science, Busan, South Korea*

[b]*Department of Chemistry, Pusan National University College of Natural Science, Busan, South Korea*

[c]*Department of Pharmacy, Pusan National University College of Pharmacy, Busan, South Korea*

ABSTRACT: Apoptosis-inducing activity of synthetic CDCA derivatives, HS-1199 and HS-1200, on gastric cancer cell line SNU-1 cells was explored. CDCA derivatives demonstrated various apoptosis hallmarks, such as mitochondrial changes, activation of caspase, DNA fragmentation, and nuclear condensation. Importantly, the orphan receptor Nur77 (TR3) was shown to translocate from the nucleus to mitochondria at the early time points after CDCA derivatives treatment. These data support the theory that CDCA derivatives-induced apoptosis of SNU-1 gastric cancer cell lines is mediated by mitochondria and caspase, and, at least in part, by Nur77.

KEYWORDS: apoptosis; CDCA derivatives; SNU-1; orphan receptor; Nur77

Address for correspondence: Young Hyun Yoo, Department of Anatomy and Cell Biology, Dong-A University College of Medicine, 3-1 Dongdaesin-Dong, Seo-Gu, Busan, South Korea 602-714. Fax: 82-51-241-3767.

yhyoo@daunet.donga.ac.kr

Current addresses: Joo-Sung Park and Bongkyung Moon at Department of Family Medicine; Min Chan Kim and Young Hyun Yoo at Department of General Surgery, Dong-A University College of Medicine, Busan, South Korea; and Sungeun Lee at LG Electronics Institute of Technology, 16 Woomyeon-Dong, Seocho-Gu, Seoul 137-724.

Ann. N.Y. Acad. Sci. 1010: 171–177 (2003). © 2003 New York Academy of Sciences.
doi: 10.1196/annals.1299.029

INTRODUCTION

Recently, we have developed synthetic CDCA derivatives and have been investigating their antiproliferating activity. L-phenyl alanine benzyl ester conjugate (HS-1199) and alanine benzyl ester conjugate (HS-1200) of CDCA induced apoptosis via a p53-independent pathway in human breast carcinoma cells. HS-1199 and HS-1200 also showed apoptotic activity in human leukemic T cells through the activation of caspases.[1,2]

A prior study showed that mitochondrial targeting of nuclear orphan receptor Nur77 (TR3) induced cytochrome *c* release and resulted in apoptosis.[3] Furthermore, several laboratories identified farnesoid X receptor (FXR), an orphan nuclear receptor, as a CDCA receptor and PKC was suggested as a signaling mediator of CDCA. Nur77 was also demonstrated to play a role in regulation of apoptosis in gastric cancer cells.[4] Although understanding how orphan receptors are involved in the apoptotic regulatory machinery is necessary, many important issues remain to be elucidated.

In this study we, for the first time, investigated their apoptosis-inducing activity in gastric cancer cells as it pertains to the fundamental mechanism via which synthetic CDCA derivatives regulate apoptosis and potential therapeutics in gastric cancer cells. In particular, we here focused on the involvement of an orphan receptor Nur77.

MATERIAL AND METHODS

Cell Culture

SNU-1 cells were maintained at 37°C with 5% CO_2 in air atmosphere in RPMI 1640 with 300 mg/L L-glutamine, 2 g/L sodium bicarbonate, and 2 g/L glucose supplemented with 10% FBS.

Synthetic CDCA Derivatives Treatment

CDCA derivatives were synthesized as described previously.[2] Twenty-four hours after cells were subcultured, the original medium was removed. Cells were washed with PBS and then incubated in the same fresh medium. CDCA, HS-1199 or HS-1200 from a stock solution was added to the medium to obtain 50 μM dilution of the drug, which was the concentration required for approximately half-maximal inhibition of viability 5 h after treatment. The concentration of ethanol used in this study, both as a vehicle for bile acids, had no effect on cell viability in the preliminary studies.

Assessment of Characteristic Apoptotic Manifestations at a Single Time Point (5 h)

Nuclear Stain

The samples were stained in 4 μg/mL Hoechst 33342 at 37°C for 30 min and fixed for 10 min in 4% paraformaldehyde.

TUNEL Technique

Cells were labeled in TUNEL reaction mixture for 60 min at 37°C.

Immunofluorescent Staining

The cytocentrifuged cells were fixed in 4% paraformaldehyde for 10 min, incubated with anti–cytochrome *c* antibody for 1 h and then with FITC-conjugated secondary antibody for 1 h at room temperature.

Cell Counts and Photomicrography

Cells showing nuclear changes, positive TUNEL reaction, or having the punctate staining pattern for cytochrome *c*, were determined by a blinded observer from a random sampling of 300 cells per experiment.

Assay of Mitochondrial Membrane Potential (MMP)

JC-1 was added directly to the cell culture medium and incubated for 15 min. Cells were quantified for J-aggregated fluorescence intensity in a modular fluorimetric system using excitation and emission filters of 492 and 590 nm, respectively.

Western Blot Analysis

Protein concentrations of cell lysates were determined by the method of Bradford (Bio-Rad protein assay). Equal amounts of protein were subjected to 15% or 7.5% SDS/PAGE for caspase-3 and PARP, respectively, and transferred to a nitrocellulose membrane. Western blot analyses were carried out by using appropriate antibody and immunostaining with antibodies was performed using Supersignal West Pico Chemiluminescent Substrate.

Determination of Involvement of Orphan Receptor

Confocal Microscopy

The cytocentrifuged cells were fixed for 10 min in 4% paraformaldehyde, incubated first with anti-Nur77 IgG antibody, and then reacted with corresponding FITC-conjugated anti-IgG as the secondary antibody. To show mitochondria, cells were incubated with anti-Hsp60 mouse IgG antibody and then reacted with Texas red–conjugated anti-mouse IgG as secondary antibody. Fluorescent images were observed and analyzed under a laser-scanning confocal microscope.

Mitochondrial Fractionation and Western Blotting

Mitochondria were prepared by sucrose density gradients. Cells were dounce-homogenized, and nuclear contaminants were removed. The supernatant was layered over a 1–2 M sucrose step gradient and spun at 4°C for 30 min at 22,000 rpm. Mitochondria were collected at the 1–1.5 M interphase by lateral suction through the tube, washed (in 4 volumes of MS buffer, 15,000 rpm), resuspended in a final volume of 200 μL of MS buffer, and used for Western blot as above.

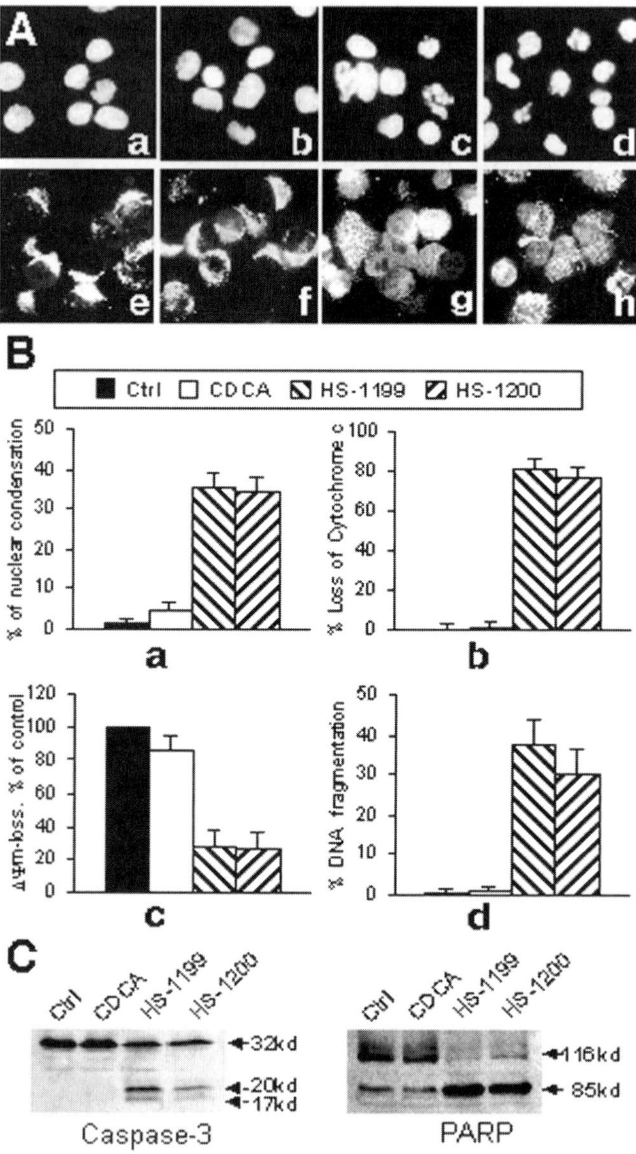

FIGURE 1. Demonstration of CDCA derivatives-induced apoptosis in SNU-1 cells. (**A**) Nuclear morphology of Hoechst stainings (**a-d**) and immunofluorescent staining to cytochrome *c* (**e-h**). (**B**) Quantification data of apoptosis as determined by Hoechst staining (**a**), immunofluorescent staining to cytochrome *c* (**b**), reduction of MMP measured employing JC-1 (**c**) and TUNEL assay (**d**). (**C**) Western blot assay showing that caspase activation and PARP degradation were produced in CDCA derivatives-treated cells. 32-kDa precursor of caspase-3 was degraded, and 20-kDa and 17-kDa cleavage products were produced. The PARP 85-kDa cleavage products were shown to increase in CDCA derivatives–treated cells.

RESULTS

Induction of Apoptosis in SNU-1 Cells by CDCA Derivatives Treatment

Hoechst stainings demonstrated that CDCA derivatives induced a change in nuclear morphology. Compared to the control or CDCA-treated cells having the typical round nuclei (FIG. 1, Aa and Ab), HS-1199 and HS-1200-treated cells displayed condensed and fragmented nuclei (FIG. 1, Ac and Ad). Immunofluorescent staining to cytochrome *c* showed cytochrome *c* release to cytosol (FIG. 1, Ag and Ah), whereas cytochrome *c* in the control and CDCA-treated cells was found in a punctate pattern, in keeping with its normal mitochondrial location (FIG. 1, Ae and Af). Quantification data of above two assays are displayed in FIGURE 1 (Ba and Bb). MMP was measured employing JC-1, which was significantly reduced in CDCA derivative–treated cells (FIG. 1, Bc). To prove that treatment of SNU-1 cells with CDCA derivatives resulted

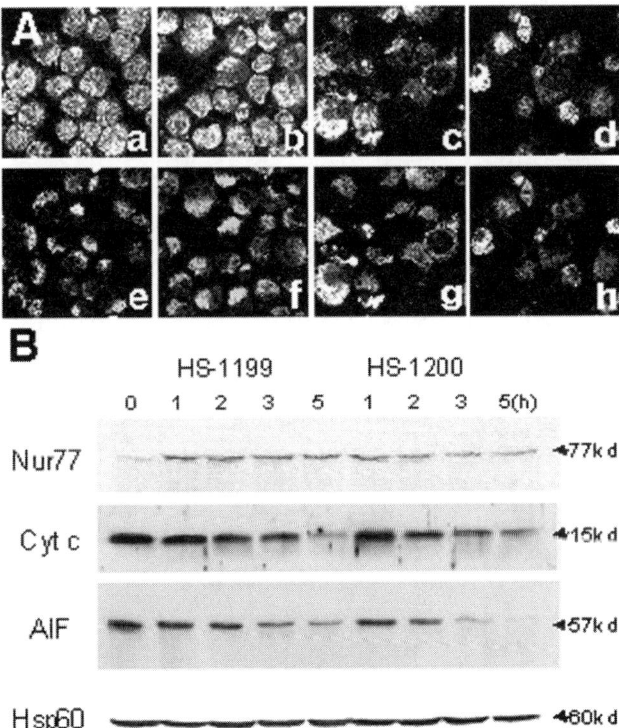

FIGURE 2. Involvement of Nur77 in CDCA derivatives-induced apoptosis in SNU-1 cells. Confocal microscopical observation displayed that Nur77, which was localized in the nucleus of the control and CDCA-treated cells (**Aa** and **Ab**), translocated into mitochondria in CDCA derivatives–treated cells (**Ac** and **Ad**). Translocation of Nur77 was confirmed by a double staining of mitochondrial protein HSP-60 in the same sample (**Ae-h**). Western blot assay of mitochondrial fraction showed that Nur77 translocated 1 h after treatment, with consequent release of cytochrome *c* and AIF. Hsp60 used as a control did not show alteration (**B**).

in DNA fragmentation, TUNEL assay was undertaken. As shown in FIGURE 1 (Bd), CDCA derivatives–treated cells showed the positive reaction to TUNEL assay. Western blot assay showed that caspase activation and PARP degradation were produced in CDCA derivatives–treated cells. 32-kDa precursor of caspase-3 was degraded, and 20-kDa and 17-kDa cleavage products were produced. The PARP 85-kDa cleavage products were shown to increase in CDCA derivatives–treated cells (FIG. 1C).

The Implication of Orphan Receptor Nur77

We examined the involvement of Nur77 in this type of apoptosis. Confocal microscopic observation displayed that Nur77, which seemed to have localized in the nucleus of the control and CDCA-treated cells (FIG. 2, Aa and Ab), was translocated into mitochondria in CDCA derivatives–treated cells (FIG. 2, Ac and Ad). Translocation of Nur77 was confirmed by a double staining of mitochondrial protein Hsp60 in the same sample (FIG. 2, Ae–h). Western blot assay of the mitochondrial fraction showed that although a trace amount of Nur77 was observed in the mitochondria of control cells, its amount began to increase 1 h after CDCA-derivatives treatment, and then release of cytochrome *c* and AIF followed. Hsp60 was used as a control showing no alteration of amount in the mitochondria even after the translocation of Nur77 (FIG. 2B).

DISCUSSION

Gastric cancer remains an important cause of death worldwide despite the declining overall incidence of gastric cancer. The curative treatment of gastric cancer requires gastric resection. However, most patients are not cured by this surgery. The failure following resection occurs in 40–65% of patients after gastric resection with curative intent. The frequency of such relapses makes postoperative adjuvant chemotherapy an attractive possibility. Therefore, novel chemicals inducing apoptosis of gastric cancer cells may contribute potential therapeutics to the management of gastric cancer. In this study, we demonstrated for the first time that synthetic CDCA derivatives HS-1199 and HS-1200 induced apoptosis of gastric cancer cells. The reduction of MMP, cytochrome *c* release into cytosol, caspase-3 activation, PARP cleavage, nuclear condensation, and DNA fragmentation were demonstrated. Taken together, HS-1199 and HS-1200 induce apoptosis of SNU-1 gastric cancer cell line via caspase and mitochondrial pathway. Furthermore we observed the early translocation of Nur77 into mitochondria and consequent release of mitochondrial apoptogenic factors. These data support that CDCA derivatives-induced apoptosis of SNU-1 cells is in part mediated by an orphan receptor Nur77.

Nur77 (also known as TR3 or NGFI-B), known as an orphan member of the steroid/thyroid/retinoid receptor superfamily, is an immediate early response gene whose expression is rapidly induced by a variety of growth stimuli. Nur77 is involved in the regulation of apoptosis in different cells.[4] Interestingly, Nur77 may also induce mitochondrial membrane permeabilization. Indeed, in cells undergoing apoptosis, Nur77 was demonstrated to translocate to mitochondrial membranes and recombinant Nur77 induces the release of cytochrome *c* when added to purified mitochondria *in vitro*.[3] Nur77 was also implicated in gastric cancer cells. A study showed that Nur77 translocated from the nucleus to the mitochondria and the subsequent cytochrome *c* release from the mitochondria to the cytosol are required for apoptosis of stomach cells.[5] Our data are also consistent with the report.

Although many important issues for their therapeutic application remain to be elucidated, CDCA derivatives HS-1199 and HS-1200 produce apoptosis of gastric cancer cells via caspase-3, mitochondria, and, at least in part, Nur77.

ACKNOWLEDGMENTS

This work was supported by Grant No R01-2001-00145 from the Korea Science & Engineering Foundation.

REFERENCES

1. Choi, Y.H. *et al.* 2001. Apoptotic activity of novel bile acid derivatives in human leukemic T cells through the activation of caspases. Int. J. Oncol. **18:** 979–984.
2. Im, E.O. *et al.* 2001. Novel bile acid derivatives induce apoptosis via a p53-independent pathway in human breast carcinoma cells. Cancer Lett. **163:** 83–93.
3. Li, H. *et al.* 2000. Cytochrome c release and apoptosis induced by mitochondrial targeting of nuclear orphan receptor TR3. Science **289:**1159–1164.
4. Uemura, H. & C. Chang. 1998. Antisense TR3 orphan receptor can increase prostate cancer cell viability with etoposide treatment. Endocrinology **139:** 2329–2334.
5. Wu, Q. *et al.* 2002. Dual roles of Nur77 in selective regulation of apoptosis and cell cycle by TPA and ATRA in gastric cancer cells. Carcinogenesis **23:** 1583–1592.

Prodigiosin Induces Apoptosis by Acting on Mitochondria in Human Lung Cancer Cells

ESTHER LLAGOSTERA, VANESSA SOTO-CERRATO, BEATRIZ MONTANER, AND RICARDO PÉREZ-TOMÁS

Departament de Biologia Cellular i Anatomia Patològica, Cancer Cell Biology Research Group, Universitat de Barcelona, Barcelona, Spain

ABSTRACT: Prodigiosin (PG) is a secondary metabolite, isolated from a culture of *Serratia marcescens*, which has shown potent cytotoxicity against various human cancer cell lines as well as immunosuppressive activity. The purpose of this study was to evaluate the role of mitochondria in PG-induced apoptosis. Therefore, we evaluated the apoptotic action of PG in GLC4 small cell lung cancer cell line by Hoechst 33342 staining. In these cells, we examined mitochondrial apoptosis-inducing factor (AIF) and cytochrome *c* (cyt *c*) release to the cytosol in PG time-response studies. These findings suggest that PG induces apoptosis in both caspase-dependent and caspase-independent pathways.

KEYWORDS: AIF; apoptosis; cytochrome *c*; mitochondria; prodigiosin

INTRODUCTION

Prodigiosin (PG) belongs to a family of tripyrrole red pigments produced by a restricted group of *Streptomyces* and *Serratia* strains, which exhibit promising antitumor and immunosuppressive activities. PG has shown useful cytotoxicity against various human cancer cell lines[1,2] as well as immunosuppressive activity.[3]

The caspase family of proteins is one of the main effectors of apoptosis. However, they are not required for cell death to occur in many systems. For example, apoptosis-inducing factor (AIF) is a caspase-independent death effector which, upon apoptosis induction, translocates from mitochondrial intermembrane to the nucleus and causes chromatin condensation and fragmentation of DNA into 50 kb fragments.[4] In contrast to cytochrome *c*, AIF does not appear to require the presence of further cytosolic factors to induce apoptotic features in nuclei.

On the basis of previous results that showed the activation of caspase-9, -8, and -3 in PG-treated Jurkat cells,[5] we pursued the characterization of the role of mitochondria in PG-induced apoptosis in GLC4 cells.

Address for correspondence: Ricardo Pérez-Tomás, Departament de Biologia Cellular i Anatomia Patològica, Cancer Cell Biology Research Group, Universitat de Barcelona, Ed. Pavelló Central, LR 5101, C/ Feixa Llarga s/n, E-08907 L'Hospitalet (Barcelona), Spain. Voice: 34-93 4024288; fax: 34-93 4029082.

rperez@bell.ub.es

Ann. N.Y. Acad. Sci. 1010: 178–181 (2003). © 2003 New York Academy of Sciences. doi: 10.1196/annals.1299.030

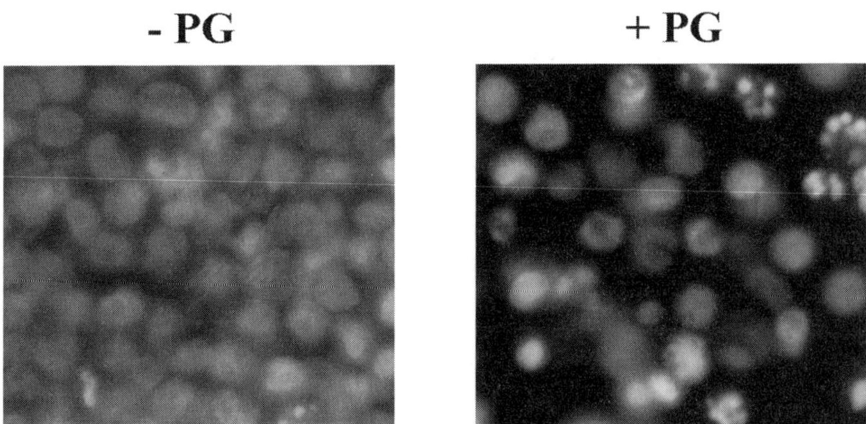

FIGURE 1. Fluorescence microscopic analysis of GLC4 nuclei with Hoechst 33342 staining. Cells were untreated (–PG) or treated (+PG) with 200 nM PG for 16 hours.

MATERIAL AND METHODS

Cell Culture

GLC4 small cell lung cancer cell lines were cultured in RPMI 1640 medium supplemented with 10% fetal calf serum, 100 U/mL penicillin, 100 μg/mL streptomycin, and 4 mM L-glutamine, at 37°C, 5% CO_2 in air.

Isolation and Purification of PG

PG was isolated from a culture of *S. marcescens* 2170 as described previously.[1]

Hoechst Staining

For cell morphology evaluation, cells (5×10^5/mL) were untreated or treated with 200 nM PG for 16 h, washed with PBS, and incubated with 2 μg/mL Hoechst 33342 (Sigma, USA) for 30 min, at 37°C in the dark. The sections were examined with a Leitz Diaplan microscope and photographed with a Wild MPS 45 Photoautomat system.

AIF and Cyt c *Release Analysis*

AIF and cyt *c* release from mitochondria to cytosol were measured by Western blot. After cells (5×10^6) were treated with 200 nM PG for 15 to 720 min, they were lysed in ice-cold lysis buffer (250 mM sucrose, 1 mM EDTA, 0.05% digitonin, 25 mM Tris pH 6.8, 1 mM DTT, 1 μg/mL aprotinin, 1 μg/mL leupeptin, 1 μg/mL pepstatin, and 0.1 mM PMSF), for 30 sec, then centrifuged at 13,000×*g* at 4°C for 3 min. Supernatants (cytosolic extracts) were electrophoresed on 12% polyacrylamide gels and transferred to Immobilon-P membranes (Millipore, USA). Then they were incubated overnight with a polyclonal antibody against AIF (Oncogen Research prod-

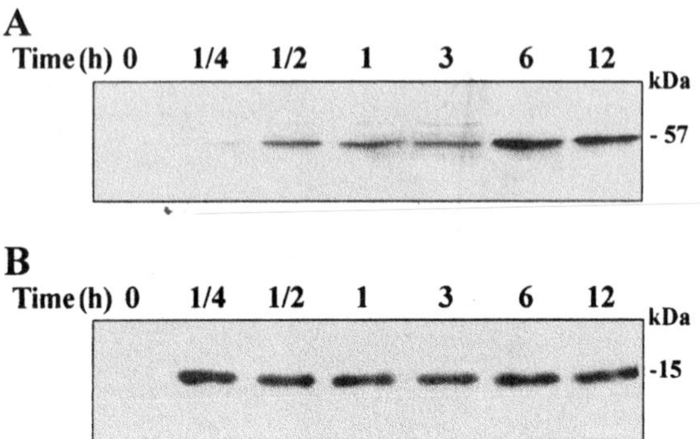

FIGURE 2. Effect of PG on mitochondrial AIF and cyt *c* release. (**A**) Analysis of AIF in cytosolic extracts. Time-dependent release of AIF into the cytosol upon exposure to PG (200 nM) is observed. (**B**) Time-course of cyt *c* release to the cytosol in response to PG treatment (200 nM).

ucts, USA) or with a monoclonal antibody against cyt *c* (Pharmingen International). Secondary antibodies conjugated to HRP were goat anti-rabbit IgG or goat anti-mouse IgG (BioRad, UK). Peroxidase was developed using the enhanced chemiluminescence (ECL) detection kit (Amersham, UK).

RESULTS AND DISCUSSION

The exceptional cytotoxic potency of PG in GLC4 cells (IC_{50} 129 and 143 nM at 16 and 24 h, respectively) prompted us to determine the apoptotic effect of PG in GLC4 cells. We characterized several morphological nuclear changes at microscopic level using Hoechst 33342 staining (FIG. 1). In contrast to untreated cells, chromatin condensation and nuclear fragmentation were observed in PG-treated cells, which are the hallmarks of apoptosis. These changes were triggered by apoptogenic factors harbored by mitochondria. To determine the role of mitochondria in PG-induced apoptosis, we studied its effect on some of these molecules. We found that PG induced mitochondrial release of AIF (FIG. 2A) and also demonstrated the release of cyt *c* into the cytosol (FIG. 2B). Both phenomena occur promptly upon exposure to PG.

These findings suggest that PG causes apoptosis to use the mitochondrial pathway and that PG apoptosis induction occurs in both caspase-dependent and -independent manner to ensure efficient DNA breakdown. Altogether, PG induces the release of cyt *c* and AIF into the cytoplasm. Cyt *c* binds to the adaptor Apaf-1 and the complex recruits and activates caspase 9. This initiates a caspase cascade responsible for the hydrolysis of key cytoplasmic proteins, for cleavage among genomic DNA nucleosomes into 180 bp fragments via the caspase-activated DNase (CAD). On the other hand, AIF migrates to the nucleus and induces high-molecular-mass DNA fragmentation and marginal chromatin condensation. The combined action of caspases and AIF causes the apoptotic phenotype observed in GLC4 cells and cell death.

ACKNOWLEDGMENTS

This work was supported by a grant from the Ministry of Science and Technology and the European Union (SAF2001-3545), and by a "Marató de TV3" grant (Ref. # 001510).

REFERENCES

1. MONTANER, B., S. NAVARRO, M. PIQUÉ, *et al.* 2000. Prodigiosin from the supernatant of *Serratia marcescens* induces apoptosis in haematopoietic cancer cell lines. Br. J. Pharmacol. **131:** 585–593.
2. MONTANER, B. & R. PÉREZ-TOMÁS. 2001. Prodigiosin-induced apoptosis in human colon cancer cells. Life Sci. **68:** 2025–2036.
3. CAMPÀS, C., M. DALMAU, B. MONTANER, *et al.* 2003. Prodigiosin induces apoptosis of B and T cells from B-cell chronic lymphocytic leukemia. Leukemia **17:** in press.
4. LORENZO, H.K., S.A. SUSIN, J. PENNINGER & G. KROEMER. 1999. Apoptosis inducing factor (AIF): a phylogenetically old, caspase-independent effector of cell death. Cell Death Differ. **6:** 516–524.
5. MONTANER, B. & R. PÉREZ-TOMÁS. 2002. Prodigiosin induces caspase-9 and caspase-8 activation and cytochrome *c* release in Jurkat T cells. Ann. N.Y. Acad. Sci. **973:** 246–249.

Molecular Interaction between Cyclophilin D and Adenine Nucleotide Translocase in Cytochrome *c* Release

Does It Determine Whether Cytochrome *c* Release Is Dependent on Permeability Transition or Not?

KIYOTAKA MACHIDA AND HIROYUKI OSADA

Antibiotics Laboratory, Riken, Hirosawa 2-1, Saitama 351-0198, Japan

ABSTRACT: We propose a model for the mechanism of permeability transition (PT) related cytochrome *c* release. It is likely that the Ca^{2+} requirement for the induction of the mitochondrial permeability transition pore (MPTP) opening might be due to the Ca^{2+}-dependent interaction between cyclophilin D and ANT. We show here that the modification of adenine nucleotide translocase (ANT), which is one of the components of MPTP, can induce two different types of the cytochrome *c* release. One is dependent on classical PT, resulting in mitochondrial swelling, and is inhibited by cyclosporin A. The other is PT-independent, without swelling, and is insensitive to cyclosporin A.

KEYWORDS: mitochondrial permeability transition; cytochrome *c* release; cyclophilin D

In the early steps of apoptotic process, apoptogenic factors such as cytochrome *c* and apoptosis-inducing factor are released from mitochondrial intermembrane space into cytosol. Once cytochrome *c* is released from mitochondria, it binds to Apaf-1 and promotes the assembly of a multiprotein complex that induces proteolytic processing and activation of the caspases.[1] However, the mechanism underlying cytochrome *c* release from mitochondria is not fully understood. One hypothesis is that a mitochondrial protein complex called the mitochondrial permeability transition pore (MPTP) mediates cytochrome *c* release. The components of this pore include the voltage-dependent anion channel (VDAC), the adenine nucleotide translocase (ANT), cyclophilin D, and kinases (such as hexokinase and creatine kinase).[2] Opening of MPTP induces alteration of inner membrane permeability, called permeability transition (PT), resulting in mitochondrial swelling, rupture of outer membrane, and the release of proteins from the intermembrane space. However, the pro-apoptotic Bcl-2 family protein, Bax, causes cytochrome *c* release without altering inner membrane permeability and a cyclophilin inhibitor, cyclosporin A, did not inhibit this

Address for correspondence: Hiroyuki Osada, Antibiotics Laboratory, Riken, Hirosawa 2-1, Saitama 351-0198, Japan.Voice: +81-48-467-9541; fax +81-48-462-4669.
antibiot@postman.riken.go.jp

Ann. N.Y. Acad. Sci. 1010: 182–185 (2003). © 2003 New York Academy of Sciences.
doi: 10.1196/annals.1299.031

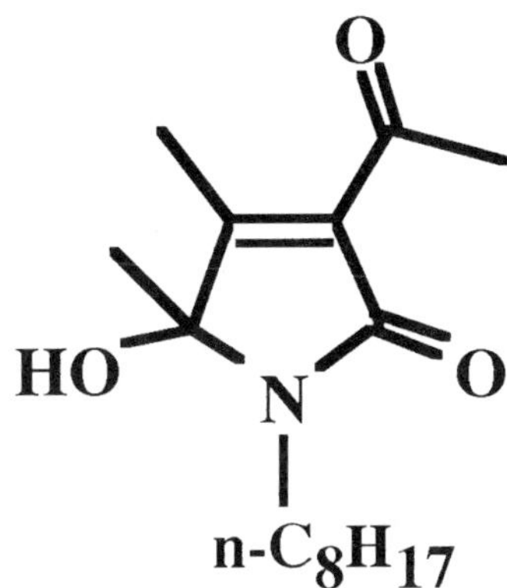

FIGURE 1. Structure of MT-21.

event.[3] Interestingly, other investigators reported that the Bax-mediated cytochrome *c* release is PT-dependent and inhibited by cyclosporin A.[4] Therefore, even in a simple assay system using isolated mitochondria, the involvement of the MPTP complex in cytochrome *c* release remains controversial.

We reported previously[5] that a synthetic compound, MT-21 (FIG. 1), induced apoptotic cell death in human promyelocytic leukemia HL-60 cells through induction of cytochrome *c* release from mitochondria. Furthermore, we demonstrated recently that the molecular target of MT-21 was ANT and that MT-21 could dissociate the ANT-cyclophilin D complex *in vitro*.[6]

Interestingly, we found that MT-21 and atractyloside, specific inhibitors of ANT, could induce cytochrome *c* release without induction of PT and that this event was highly dependent on the presence of Mg^{2+}. These results suggest that modification of ANT in the mitochondrial inner membrane can induce cytochrome *c* release without having an effect on inner membrane permeability. It has been suggested that divalent cations, such as Mg^{2+} and Ca^{2+}, bind VDAC or ANT and affect their conformation.[7] Thus, some alteration of the state of MPTP complex that resides in mitochondrial membrane contact site might occur in the presence of Mg^{2+}. To investigate whether MPTP complex is altered in this condition or not, we analyzed the translocation of cyclophilin D from matrix to the membrane. As expected, we observed that cyclophilin D binding to the mitochondrial membrane was significantly inhibited by the addition of Mg^{2+} or the depletion of Ca^{2+}. It was strongly suggested that the association of cyclophilin D with MPTP complex is highly dependent on the presence of Ca^{2+}. We also found that a cyclophilin inhibitor, cyclosporin A, did not inhibit the PT-independent cytochrome *c* release. A hypothetical model of our proposed mechanism of cytochrome *c* release is shown in FIGURE 2. In the presence of Mg^{2+} or EGTA, cyclophilin D is not associated with MPTP complex, thus cyclosporin A could not inhibit the PT-independent cytochrome *c* release. It is likely that the Ca^{2+} requirement for the induction of the MPTP opening might be due to the Ca^{2+}-dependent interaction between cyclophilin D and ANT. Taken together, we conclude that MPTP opening is highly dependent on the molecular interaction between cyclo-

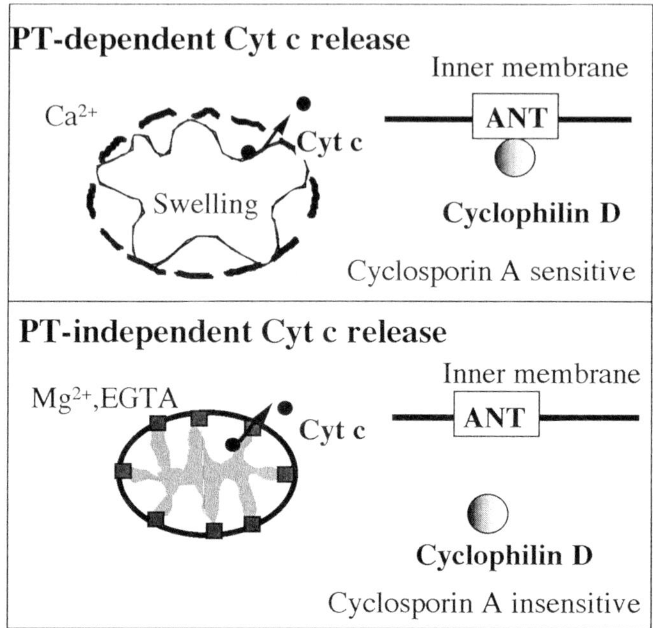

FIGURE 2. A proposed model for the mechanism of PT-dependent or PT-independent cytochrome *c* release.

philin D and ANT, however alteration of inner membrane permeability is not necessary for the cytochrome *c* release. Therefore, we show here that the modification of ANT, which is one of the components of MPTP, can induce two different types of the cytochrome *c* release. One is dependent on classical PT, resulting in mitochondrial swelling, and is inhibited by cyclosporin A, and the other is PT-independent, without swelling, and is insensitive to cyclosporin A.

REFERENCES

1. Li, P. *et al.* 1997. Cytochrome *c* and dATP-dependent formation of Apaf-1/caspase-9 complex initiates an apoptotic protease cascade. Cell **91:** 479–489.
2. Beutner, G. *et al.* 1998. Complexes between porin, hexokinase, mitochondrial creatine kinase and adenylate translocator display properties of the permeability transition pore. Implication for regulation of permeability transition by the kinases. Biochim. Biophys. Acta **1368:** 7–18.
3. Jurgenmeier, J.M. *et al.* 1998. Bax directly induces release of cytochrome *c* from isolated mitochondria. Proc. Natl. Acad. Sci. USA **95:** 4997-5002.
4. Pastorino, J.G. *et al.* 1998. The overexpression of Bax produces cell death upon induction of the mitochondrial permeability transition. J. Biol. Chem. **273:** 7770–7775.
5. Watabe, M. *et al.* 2000. MT-21 is a synthetic apoptosis inducer that directly induces cytochrome *c* release from mitochondria. Cancer Res. **60:** 5214–5222.

6. Machida, K. *et al.* 2002. A novel adenine nucleotide translocase inhibitor, MT-21, induces cytochrome *c* release by a mitochondrial permeability transition-independent mechanism. J. Biol. Chem. **277:** 31243–31248.
7. Gincel, D. *et al.* 2001. Calcium binding and translocation by the voltage-dependent anion channel: a possible regulatory mechanism in mitochondrial function. Biochem. J. **358:** 147–155.

Caspase-12 and ER-Stress–Mediated Apoptosis

The Story So Far

EVA SZEGEZDI, UNA FITZGERALD, AND AFSHIN SAMALI

Cell Stress and Apoptosis Research Group, Department of Biochemistry, National Centre for Biomedical Engineering Science, National University of Ireland, Galway, Ireland

ABSTRACT: **The labyrinth of the endoplasmic reticulum (ER) interweaves the cytosol and connects to the nucleus, mitochondria, and the plasma membrane. In the lumen of the ER, the essential function of lipid synthesis, Ca^{2+} storage, folding, and maturation of proteins take place. Therefore, the tight regulation and maintenance of ER homeostasis is vital. Disturbance of the Ca^{2+} homeostasis during hypoxia, or imbalance between the demand and capacity of the protein-folding apparatus, initiates an adaptive response of the cell, termed the unfolded protein response (UPR, ER stress response). As a result, ER-localized chaperones are induced, protein synthesis is slowed down, and a protein degrading system is initiated. However, if the ER stress cannot be alleviated, it culminates in apoptosis. This paper reviews the newly outlined signaling pathways of the unfolded protein response and describes the central role of caspase-12 in the initiation of cell death. The complex role of the ER and its signaling pathways provides a novel angle on apoptosis research and may offer a key to apoptosis-associated diseases.**

KEYWORDS: **caspase-12; ER-stress; unfolded protein response (UPR); apoptosis**

THE ENDOPLASMIC RETICULUM AND APOPTOSIS

In the thirty years since Kerr and Wyllie coined the term *apoptosis* to describe the active participation of a cell in its own demise, innumerate reports have been published, in which apoptosis has been directly implicated in the pathology of human disease.[1] In an effort to understand the underlying mechanisms, attention was initially focused on the cellular response to the engagement of death receptors, or on the induction of apoptosis by DNA damaging agents. Later, the release of cytochrome c from mitochondria during apoptosis caused by a variety of treatments was an early indication of its role as a potential modulator of cell death. With the cloning of the mammalian homologues of CED-3, the central role of caspases in apoptosis was defined. More recently, the endoplasmic reticulum (ER) has been proposed as a critical regulator of cell death, with the discovery of an ER-anchored caspase, together

Address for correspondence: Afshin Samali, Cell Stress and Apoptosis Research Group, Department of Biochemistry, NUI, Galway, Ireland. Voice: +353-91-750393; fax: +353-91-512504.
afshin.samali@nuigalway.ie

**Ann. N.Y. Acad. Sci. 1010: 186–194 (2003). © 2003 New York Academy of Sciences.
doi: 10.1196/annals.1299.032**

with its activation by agents that specifically target the ER.[2] The ER is a subcellular organelle, whose luminal volume constitutes at least one-tenth the volume of the cell and where the vast majority of secreted, glycosylated, and lipid proteins are folded into their tertiary and quaternary structure. A variety of conditions, such as the loss of the luminal oxidizing environment or imbalance of the Ca^{2+} homeostasis, hypoxia/ischemia, or protein overload in the ER, lead to the accumulation of unfolded proteins. These proteins tend to form aggregates through hydrophobic interactions that may go on to activate an evolutionarily conserved compensatory mechanism, termed the ER stress response (or unfolded protein response: UPR).[3–5] Several signals, such as viral infection, that initiate an UPR also trigger an ER overload response (EOR). The cellular response to EOR differs slightly, in that it activates the NFκB transcription factor. However, as our interest lies chiefly in ER-dependent caspase activation, EOR is not covered by this review.

The UPR is an adaptive cellular response and it involves (1) cell-cycle arrest in G1/S phase; (2) attenuation of protein synthesis in order to prevent further protein aggregation and accumulation in the ER; (3) induction of ER-localized chaperone proteins and folding catalysts, such as Grp78/BiP, Grp94, calnexin, and calreticulin; and (4) activation of ER-associated protein degradation (ERAD) that enhances elimination of unwanted aggregates.[6–9] This concerted and complex response of the cell is mediated by three molecules: Ire1(α and β), ATF6, and PERK/PEK.[10,11] All are localized to the ER membrane and are able to sense protein aggregation and trigger UPR. In resting cells these ER stress sensors are thought to be associated with the Grp78/Bip chaperone protein, which prevents their activation. This concept follows from reports that overexpression of Grp78/BiP prevented ionophore-induced inhibition of protein synthesis and ER stress-specific transcription.[12,13] PERK/PEK and the two Ire1 receptor molecules share homology in their luminal, N-terminal regions, and it is these domains that are thought to sense the occurrence of protein aggregation. The ligands of PERK/PEK and Ire1 are not known, although their kinase activity may be induced when they oligomerize following interaction with the hydrophobic regions of unfolded proteins.

Ire1, in addition to its stress-sensing and transmembrane domains, contains a kinase domain that activates in *cis*, an endoribonuclease domain.[10] In response to ER stress, the ribonuclease activity cleaves the mRNA of the transcription factor XBP1 and so generates a frame-shift splice variant with a functional transactivation domain. The expressed XBP1 protein translocates to the nucleus, and, as a heterodimer with NF-Y, induces the expression of itself and other ER stress-specific genes that will alleviate the stress. The ATF6 transcription factor, which possesses a luminal sensing domain and a cytosolic transactivation domain, is constitutively expressed and activated by proteolysis.[14] Upon ER stress, Grp78/BiP dissociates from ATF6, allowing ATF6 to translocate to the Golgi where the S1P and S2P proteases liberate its cytosolic transactivation domain.[15] The soluble ATF6 transactivation domain then enters the nucleus and induces UPR target genes, such as chaperone proteins and the transcription factor XBP1. In resting cells the XBP1 mRNA level is hardly detectable; therefore a fast XBP1 induction in response to ER stress is vital. The coordinated action of ATF6 and XBP1 constitutes a very efficient system for stimulating XBP1 expression and enhancing protein folding and maturation in general. A second line of defence, which halts protein synthesis and thus relieves the ER from further protein loading, is mediated by the PERK/PEK kinase.

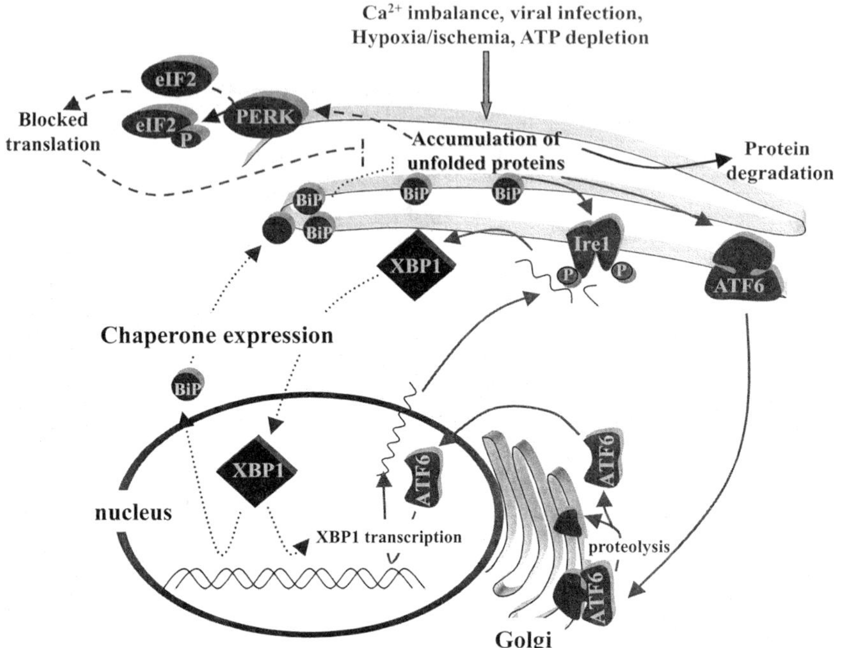

FIGURE 1. The adaptive phase of the unfolded protein response. The coordinated action of three systems is required for the elimination of protein aggregates accumulating in ER stress. (1) Grp78, which inhibits translocation of ATF6 to the Golgi apparatus, dissociates from ATF6 and associates with aggregating proteins. The free ATF6 translocates to the Golgi, where it is activated by proteolysis. The active ATF6 enters the nucleus, where it initiates the expression of ER-localized chaperones and XBP1 transcription factor. The transcribed XBP1 mRNA is processed in the cytosol by the endoribonuclease activity of Ire1. The resulting mRNA is translated into an XBP1 protein with an active transactivation domain, which translocates to the nucleus and induces its own expression as well as other chaperones. (2) The PERK/PEK system—upon protein aggregation—attenuates protein synthesis by phosphorylating the eIF2 translational initiation factor. (3) The ERAD (ER Associated Protein Degradation) mediates ubiquitination and elimination of denatured proteins from the ER.

During ER stress, PERK/PEK phosphorylates the α subunit of eIF2, thus stabilizing its inactive, GDP-bound form, thereby blocking the initiation of protein translation and attenuating protein synthesis (FIG. 1).[16]

In summary, stress conditions that interfere with the homeostasis of the ER initiate diverse signaling responses, resulting in a decreased rate of protein translation to prevent further accumulation of unfolded proteins. Simultaneously, transcription factors are activated in order to induce the expression of ER-resident chaperones to deal with accumulated protein aggregates. Furthermore, the ER-specific protein-degrading apparatus also becomes activated and eliminates denatured proteins. This coordinated response halts the buildup of proteins and allows time for the elimination of unfolded proteins and reestablishes cellular homeostasis. However, if the stress cannot be resolved, the cell dies by apoptosis.

ER STRESS AND THE TRIGGERING OF APOPTOSIS

At present, the exact signaling mechanism underlying ER stress-induced apoptosis is poorly understood. However, it is clear that two main pathways, a transcription factor- and a caspase-dependent one, are activated. In the former, Ire1 is thought to serve a pro-apoptotic function, possibly through upregulation of the transcription factor, GADD153/CHOP, which may, in turn, amplify the pro-apoptotic signal. The mechanism that is achieved is brought about through altering the balance between Bcl-2 and Bax, as indicated when neuronal cells were treated with amyloid beta (Aβ) peptide.[17,18] Alternatively, Ire1 may activate proximal components of the c-jun N-terminal kinase (JNK) pathway by recruiting the TRAF2 adaptor molecule.[19] What is the connection (if there is one) between CHOP and the Ire1/PERK/ATF6 ER stress-sensing molecules? What specific cellular alterations initiate CHOP activation? Does ER stress-mediated downregulation of Bcl-2 lead to loss of mitochondrial membrane potential and cytochrome c release? These issues remain unresolved.

ER can also initiate cell death through activation of caspases. This pathway is independent of mitochondria and death receptors and thought to be mediated by caspase-12, as discussed in detail below.

CASPASE-12, THE CENTRAL PLAYER IN ER STRESS-INDUCED APOPTOSIS

Caspase-12 belongs to the group I family of caspases (also known as ICE-like caspases), which also includes caspases-1, -4, -5, -11, and -13.[20] These caspases contain a long pro-domain responsible for interaction with various adaptor proteins (CARD domain) and two catalytic subunits: p20 and p10. So far, functional caspase-12 has been cloned from the mouse and rat, but the question of whether or not a human isoform of caspase-12 exists remains controversial. Alignment of the murine caspase-12 cDNA with the human genome sequence revealed a human homologue within the I/ICE gene cluster on chromosome 11, being 68% homologous to murine caspase-12 and 57% homologous to human caspase-4.[21] However, analysis of this gene disclosed that the nine different, alternatively spliced mRNA species expressed from the gene have premature stop codons or frame-shift mutations that prevent its expression or lead to expression of a truncated version.

Further studies are essential in order to clarify whether a different locus in the human genome carries a functional caspase-12 gene, or if its function is carried out by other caspases. Also, it cannot be excluded that mRNA editing, or other mechanisms existing in the cell, could override stop codons. The reported activation of caspase-12 in human cells following the induction of ER stress supports this view (TABLE 1).[22–24]

Activation of caspase-12 during apoptosis has been reported in mouse, rat, rabbit, cow, and human cells (see TABLE 1 for summary). In murine cells several studies examined the activation and function of caspase-12 in response to ER stress-mediated apoptosis. For example, caspase-12$^{-/-}$ mice were found to be resistant to Aβ peptide-induced apoptosis in an Alzheimer's disease model.[2] The report of Morishima *et al.* described the caspase cascade initiated in UPR. This cascade is novel and does not depend on either mitochondria or death receptor activation. Upon activation,

TABLE 1. Caspase-12 activation in disease-linked forms of apoptosis

Associated Disorder	Species	Cell/Tissue Type	Inducing Signal
Alzheimer's disease	mouse	Cortical neurons[2] Renal epithelial cells[36]	Tm, Tg, Brefeldin A, A23187, amyloid β
	rat	PC12 cells and primary hippocampal neurons[37,38]	DNA damage, Brefeldin A
	rabbit	hippocampal tissue[18]	aluminum maltolate
Parkinson's disease	mouse	SN4741 neuronal cells[39]	Manganese
	mouse	NIH3T3[40]	manganese (II)
	mouse	C2C12 myoblasts[41]	tunicamycin
Huntington's disease	mouse	C2C5[42]	polyglutamine aggregates
White matter diseases	mouse	oligodendrocytes[43]	PLP mutation
Cancer	human	melanoma[22]	cisplatin
	human	colon cancer cells[44]	PPME photosensitizer
	human	CCRF-CEM, Jurkat[23]	TC-A
Ischemic heart disease	rat	primary cardiomyocytes and H9C2 cells[49,50]	serum/glucose/oxygen deprivation
Viral infection	human	A549 ling epithelial[45]	RSV
	cow	MDBK[46]	CpBVDV
Liver disease	human	Huh7 liver cells[24]	thapsigargin
Diabetes	rat	BRIN-BD11 tumoral cells[47]	thapsigargin
General apoptosis–associated disorders	mouse	ARK-2B fibroblasts[48]	serum withdrawal
		myoblasts[25]	recombinant caspase-12
		Sak2 cell line[26]	thapsigargin, brefeldin A
	human	HEK 293T[30]	thapsigargin, brefeldin A

ABBREVIATIONS: CpBVDV, cytopathic bovine diarrhea virus; PLP, proteolytic protein; PPME, pyropheophorbide-a methylester; RSV, respiratory syncytial virus; TC-A, tetracarcin-A; Tg, thapsigargin; Tm, tunicamycin.

caspase-12 translocates from the ER to the cytosol, where it directly cleaves pro-caspase-9, which, in turn, activates the effector caspase, caspase-3.[25] The existence of a mitochondrion- and apoptosome-independent activation of caspase-9 by caspase-12 was further supported by an examination of Apaf1$^{-/-}$ mouse embryonic fibroblast cells, where the ER stress-mediated apoptotic process remained intact.[26] In these cells inhibition of either caspase-12 or caspase-9 rescued the cells from thapsigargin-induced cell death. By contrast, Michalak's lab showed that the ER and the mitochondria are linked closely and that Ca^{2+} released from the ER eventually accumulates in mitochondria.[27] In addition, ER stress causes oxidative stress and mitochondrial changes that can be blocked by overexpressing Bcl-2.[28] Beside the transcription- and caspase-mediated cell death pathways, disruption of the Ca^{2+}-balance leads to calpain activation, which through the cleavage of Bid and pro-caspase-12 contributes to caspase-9 activation.

Recent studies investigating the mechanism of caspase-12 activation suggest several models. Pro-caspase-12 could interact with Ire1α through the adaptor molecule TRAF2: upon ER stress, pro-caspase-12 would be released from TRAF2, homodimerize and undergo auto-processing upon ER stress.[19] Another proposed mechanism of pro-caspase-12 activation is that Ca^{2+} released from the ER during ER stress, activates m-calpain, which, in turn, would translocate from the cytosol to the ER and cleave off the CARD pro-domain of caspase-12. The resultant p30 subunit of caspase-12 is then auto-processed into active p20 and p10 subunits.[29] Besides these suggested mechanisms, the processing of pro-caspase-12 by caspase-7 has also been reported. According to this, caspase-7 cleaves pro-caspase-12 in the middle of the pro-domain, which triggers auto-processing between the pro-domain and the p20 subunit, and between the p20 and p10 subunits of caspase-12 (FIG. 2).[30] Thus, the

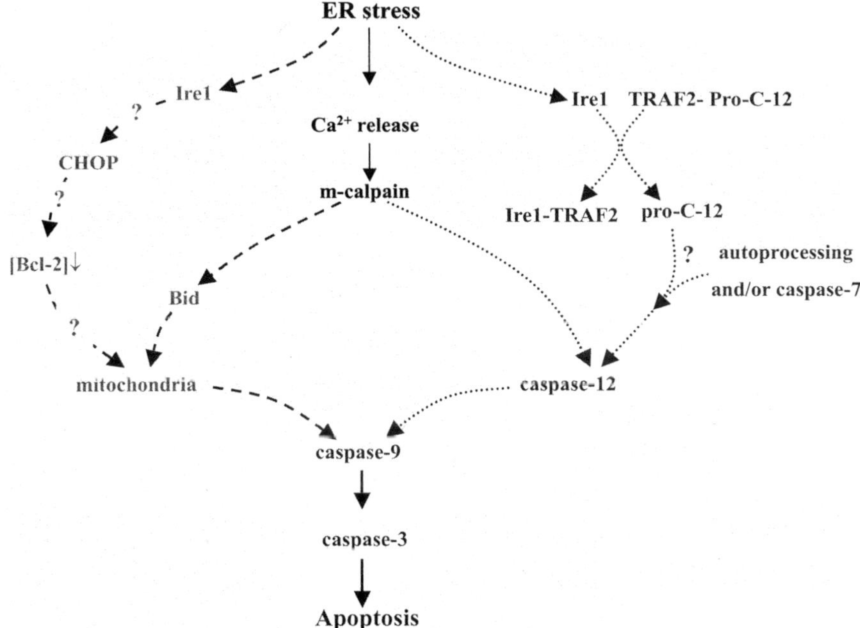

FIGURE 2. The postadaptive phase of UPR. Upon prolonged ER stress the proapoptotic CHOP/GADD153 transcription factor is induced and activated. CHOP/GADD153 and/or other processes may be responsible for the downregulation of Bcl-2, which results in mitochondrial cytochrome c release. A parallel pathway involving caspase-12 activation converges with the transcription-dependent pathway, both leading to caspase-9 and then caspase-3 activation, and finally cell death. Furthermore, disruption of the calcium homeostasis can contribute to the activation of both pathways.

activation of pro-caspase-12 is complex and may depend on the cell type and the nature of the cellular insult.

CONCLUSIONS

The pivotal role of ER in the life of the cell has always been known, but the part it plays in the progression of human disease has only recently been appreciated.[31] Many neurodegenerative disorders, including Alzheimer's, Huntington's, or Parkinson's diseases, are associated with pathological accumulation of protein aggregates.[32] For example, Alzheimer's disease (AD) has been linked to the pathological accumulation of amyloid-β protein aggregates, and mutations in ER-associated presenilin 1 have been implicated in the development of familial AD.[33,34]

The importance of ER stress-induced apoptosis during hypoxia/ischemia-induced brain and heart damage has also been realized. During hypoxia/ischemia, intracellular ATP and pH drop, causing Ca^{2+} imbalance and switching on the UPR.[35] If the hypoxia/ischemia is prolonged, the cell enters an apoptotic program, which leads to permanent tissue loss; being terminally differentiated, cardiomyocytes and neurons are unable to regenerate lost tissue.

There is mounting evidence that impairment of the ER stress response machinery itself can cause severe disease. In the β cells of the pancreas, PERK is essential in regulating the capacity of the ER to fold and process insulin. Upon increased insulin requirement in the fed state, PERK, by slowing insulin synthesis, resolves the temporary imbalance between insulin load and the folding capacity of the ER. Mutation of PERK leads to insulin overload, which over time causes chronic ER stress, for β cell loss and diabetes.[16] Finally, the relevance of ER stress to the development of cancer therapies has recently been emphasized, by the reported activation of caspase-12 by cisplatin and tetrocarcin A, two chemotherapeutic drugs.[22,23]

The list of diseases where ER stress significantly contributes to or accelerates disease progression is growing. The ER's importance in many aspects of apoptosis means that it could be a key element in the design of successful therapies targeting diseases hitherto believed incurable.

ACKNOWLEDGMENTS

Our work is supported by the Higher Education Authority of Ireland (PRTLI), the Irish Heart Foundation, Enterprise Ireland, and the Millenium Research Fund of NUI, Galway.

REFERENCES

1. Kerr, J.F., A.H. Wyllie & A.R. Currie. 1972. Apoptosis: a basic biological phenomenon with wide-ranging implications in tissue kinetics. Br. J. Cancer **26:** 239–257.
2. Nakagawa, T. *et al.* 2000. Caspase-12 mediates endoplasmic-reticulum-specific apoptosis and cytotoxicity by amyloid-beta. Nature **403:** 98–103.
3. Gething, M.J. & J. Sambrook. 1992. Protein folding in the cell. Nature **355:** 33–45.

4. PAHL, H.L. & P.A. BAEUERLE. 1995. A novel signal transduction pathway from the endoplasmic reticulum to the nucleus is mediated by transcription factor NF-kappa B. Embo J. **14:** 2580–2588.
5. PRICE, B.D. & S.K. CALDERWOOD. 1992. Gadd45 and Gadd153 messenger RNA levels are increased during hypoxia and after exposure of cells to agents which elevate the levels of the glucose-regulated proteins. Cancer Res. **52:** 3814–3817.
6. BREWER, J.W. & J.A. DIEHL. 2000. PERK mediates cell-cycle exit during the mammalian unfolded protein response. Proc. Natl. Acad. Sci. USA **97:** 12625–12630.
7. PROSTKO, C.R. *et al.* 1992. Phosphorylation of eukaryotic initiation factor (eIF) 2 alpha and inhibition of eIF-2B in GH3 pituitary cells by perturbants of early protein processing that induce GRP78. J. Biol. Chem. **267:** 16751–16754.
8. KOZUTSUMI, Y. *et al.* 1988. The presence of malfolded proteins in the endoplasmic reticulum signals the induction of glucose-regulated proteins. Nature **332:** 462–464.
9. VAN LAAR, T., A.J. VAN DER EB & C. TERLETH. 2001. Mifl: a missing link between the unfolded protein response pathway and ER-associated protein degradation? Curr. Protein Pept. Sci. **2:** 169–190.
10. TIRASOPHON, W., A.A. WELIHINDA & R.J. KAUFMAN. 1998. A stress response pathway from the endoplasmic reticulum to the nucleus requires a novel bifunctional protein kinase/endoribonuclease (Ire1p) in mammalian cells. Genes Dev. **12:** 1812–1824.
11. SHI, Y. *et al.* 1998. Identification and characterization of pancreatic eukaryotic initiation factor 2 alpha-subunit kinase, PEK, involved in translational control. Mol. Cell. Biol. **18:** 7499–7509.
12. MORRIS, J.A. *et al.* 1997. Immunoglobulin binding protein (BiP) function is required to protect cells from endoplasmic reticulum stress but is not required for the secretion of selective proteins. J. Biol. Chem. **272:** 4327–4334.
13. BERTOLOTTI, A. *et al.* 2000. Dynamic interaction of BiP and ER stress transducers in the unfolded-protein response. Nat. Cell Biol. **2:** 326–332.
14. HAZE, K. *et al.* 1999. Mammalian transcription factor ATF6 is synthesized as a transmembrane protein and activated by proteolysis in response to endoplasmic reticulum stress. Mol. Biol. Cell **10:** 3787–3799.
15. SHEN, J. *et al.* 2002. ER stress regulation of ATF6 localization by dissociation of BiP/GRP78 binding and unmasking of Golgi localization signals. Dev. Cell **3:** 99–111.
16. HARDING, H.P. *et al.* 2001. Diabetes mellitus and exocrine pancreatic dysfunction in perk$^{-/-}$ mice reveals a role for translational control in secretory cell survival. Mol. Cell. **7:** 1153–1163.
17. WANG, X.Z. *et al.* 1998. Cloning of mammalian Ire1 reveals diversity in the ER stress responses. EMBO J. **17:** 5708–5717.
18. GHRIBI, O. *et al.* 2001. Abeta(1-42) and aluminum induce stress in the endoplasmic reticulum in rabbit hippocampus, involving nuclear translocation of gadd 153 and NF-kappaB. Brain Res. Mol. Brain Res. **96:** 30–38.
19. YONEDA, T. *et al.* 2001. Activation of caspase-12, an endoplastic reticulum (ER) resident caspase, through tumor necrosis factor receptor-associated factor 2-dependent mechanism in response to the ER stress. J. Biol. Chem. **276:** 13935–13940.
20. VAN DE CRAEN, M. *et al.* 1997. Characterization of seven murine caspase family members. FEBS Lett. **403:** 61–69.
21. FISCHER, H. *et al.* 2002. Human caspase 12 has acquired deleterious mutations. Biochem. Biophys. Res. Commun. **293:** 722–726.
22. MANDIC, A. *et al.* 2003. Cisplatin induces ER stress and nucleus-independent apoptotic signaling. J. Biol. Chem. **278:** 9100–9106.
23. TINHOFER, I. *et al.* 2002. Stressful death of T-ALL tumor cells after treatment with the anti-tumor agent Tetrocarcin-A. FASEB J. **16:** 1295–1297.
24. XIE, Q. *et al.* 2002. Effect of tauroursodeoxycholic acid on endoplasmic reticulum stress-induced caspase-12 activation. Hepatology **36:** 592–601.
25. MORISHIMA, N. *et al.* 2002. An endoplasmic reticulum stress-specific caspase cascade in apoptosis. Cytochrome c-independent activation of caspase-9 by caspase-12. J. Biol. Chem. **277:** 34287–34294.
26. RAO, R.V. *et al.* 2002. Coupling endoplasmic reticulum stress to the cell death program. An Apaf-1-independent intrinsic pathway. J. Biol. Chem. **277:** 21836–21842.

27. NAKAMURA, K. *et al.* 2000. Changes in endoplasmic reticulum luminal environment affect cell sensitivity to apoptosis. J. Cell Biol. **150:** 731–740.
28. HACKI, J. *et al.* 2000. Apoptotic crosstalk between the endoplasmic reticulum and mitochondria controlled by Bcl-2. Oncogene **19:** 2286–2295.
29. NAKAGAWA, T. & J. YUAN. 2000. Cross-talk between two cysteine protease families. Activation of caspase-12 by calpain in apoptosis. J. Cell Biol. **150:** 887–894.
30. RAO, R.V. *et al.* 2001. Coupling endoplasmic reticulum stress to the cell death program. Mechanism of caspase activation. J. Biol. Chem. **276:** 33869–33874.
31. RUTISHAUSER, J. & M. SPIESS. 2002. Endoplasmic reticulum storage diseases. Swiss Med. Wkly. **132:** 211–222.
32. FORLONI, G. *et al.* 2002. Protein misfolding in Alzheimer's and Parkinson's disease: genetics and molecular mechanisms. Neurobiol. Aging **23:** 957–976.
33. IMAIZUMI, K. *et al.* 2001. The unfolded protein response and Alzheimer's disease. Biochim. Biophys. Acta **1536:** 85–96.
34. KUDO, T. *et al.* 2002. The unfolded protein response is involved in the pathology of Alzheimer's disease. Ann. N. Y. Acad. Sci. **977:** 349–355.
35. CHEN, M. *et al.* 2002. Calpain and mitochondria in ischemia/reperfusion injury. J. Biol. Chem. **277:** 29181–29186.
36. SIMAN, R. *et al.* 2001. Endoplasmic reticulum stress-induced cysteine protease activation in cortical neurons: effect of an Alzheimer's disease-linked presenilin-1 knock-in mutation. J. Biol. Chem. **276:** 44736–44743.
37. CHAN, S.L. *et al.* 2002. Presenilin-1 mutations sensitize neurons to DNA damage-induced death by a mechanism involving perturbed calcium homeostasis and activation of calpains and caspase-12. Neurobiol. Dis. **11:** 2–19.
38. CHEN, L. & X. GAO. 2002. Neuronal apoptosis induced by endoplasmic reticulum stress. Neurochem. Res. **27:** 891–898.
39. CHUN, H.S., H. LEE & J.H. SON. 2001. Manganese induces endoplasmic reticulum (ER) stress and activates multiple caspases in nigral dopaminergic neuronal cells, SN4741. Neurosci. Lett. **316:** 5–8.
40. OUBRAHIM, H., P.B. CHOCK & E.R. STADTMAN. 2002. Manganese(II) induces apoptotic cell death in NIH3T3 cells via a caspase-12-dependent pathway. J. Biol. Chem. **277:** 20135–20138.
41. FUJITA, E. *et al.* 2002. Caspase-12 processing and fragment translocation into nuclei of tunicamycin-treated cells. Cell Death Differ. **9:** 1108–1114.
42. KOUROKU, Y. *et al.* 2002. Polyglutamine aggregates stimulate ER stress signals and caspase-12 activation. Hum. Mol. Genet. **11:** 1505–1515.
43. CERGHET, M. *et al.* 2001. Differential expression of apoptotic markers in jimpy and in Plp overexpressors: evidence for different apoptotic pathways. J. Neurocytol. **30:** 841–855.
44. MATROULE, J.Y. *et al.* 2001. Mechanism of colon cancer cell apoptosis mediated by pyropheophorbide-a methylester photosensitization. Oncogene **20:** 4070–4084.
45. BITKO, V. & S. BARIK. 2001. An endoplasmic reticulum-specific stress-activated caspase (caspase-12) is implicated in the apoptosis of A549 epithelial cells by respiratory syncytial virus. J. Cell. Biochem. **80:** 441–454.
46. JORDAN, R. *et al.* 2002. Replication of a cytopathic strain of bovine viral diarrhea virus activates PERK and induces endoplasmic reticulum stress-mediated apoptosis of MDBK cells. J. Virol. **76:** 9588–9599.
47. DIAZ-HORTA, O. *et al.* 2002. Na/Ca exchanger overexpression induces endoplasmic reticulum-related apoptosis and caspase-12 activation in insulin-releasing BRIN-BD11 cells. Diabetes **51:** 1815–1824.
48. KILIC, M. *et al.* 2002. Formation of noncanonical high molecular weight caspase-3 and -6 complexes and activation of caspase-12 during serum starvation induced apoptosis in AKR-2B mouse fibroblasts. Cell Death Differ. **9:** 125–137.
49. LOGUE *et al.* 2003. Ordering of caspases activated during ER stress: caspase-12-dependent activation of caspase-3, -7, -9 and -2. J. Biol. Chem. In preparation.
50. O'MAHONEY, M. *et al.* 2003. Hypoxia and ischemia induce nuclear condensation and caspase activation in cardiomyocytes. Ann. N.Y. Acad. Sci. This volume.

Restoration of TRAIL-Induced Apoptosis in a Caspase-8–Deficient Neuroblastoma Cell Line by Stable Re-expression of Caspase-8

ANNICK MÜHLETHALER-MOTTET, KATIA BALMAS, KATYA AUDERSET, JEAN-MARC JOSEPH, AND NICOLE GROSS

Onco-Hematology Research Unit, Department of Medical and Surgical Pediatrics, Centre Hospitalier Universitaire Vaudois, CH-1011 Lausanne, Switzerland

ABSTRACT: Tumor necrosis factor–related apoptosis-inducing ligand (TRAIL) selectively induces apoptosis in most tumor cells, a process sometimes potentiated by chemotherapeutic drugs or cycloheximide (CHX). Childhood neuroblastoma (NB) is a clinically and biologically heterogeneous neoplasm whose behavior can be explained by differential regulation of apoptosis. The non-invasive S-type NB cell lines are sensitive to TRAIL, whereas the invasive N-type NB cell lines are resistant. We have reported the silencing of caspase-8 expression in N-type cells as a possible mechanism of death receptor–mediated resistance to apoptosis in NB. The recently observed deregulation of caspase-10 in these cells prompted us to investigate the particular contribution of caspase-8 silencing in the resistance to TRAIL in N-type cells. Stable caspase-8 expression was therefore restored in the IGR-N91 cell line by retroviral infection. The IGR-N91-C8 cells became sensitive to TRAIL-mediated apoptosis, whereas the control vector-infected IGR-N91-M cells remained resistant. Interestingly, the apoptotic response to TRAIL was enhanced by co-treatment of SH-EP and IGR-N91-C8 cells with CHX or with sub-toxic concentration of doxorubicin (DOX) in a caspase-dependent manner, as cells could be protected from death by specific caspase-8 or pan-caspase inhibitors. CHX or DOX was shown to enhance TRAIL-induced caspase-8 activation and loss of mitochondrial transmembrane potential. In conclusion, restoration of active caspase-8 expression in caspase-8– and caspase-10–deficient IGR-N-91 cell line is necessary and sufficient to fully restore TRAIL-mediated cell death. Moreover, DOX and CHX were able to sensitize NB cell lines to TRAIL-induced apoptosis in a caspase-8–dependent manner by engaging death receptor and mitochondrial signaling pathways.

KEYWORDS: apoptosis; TRAIL; caspase-8; neuroblastoma; doxorubicin; cycloheximide

Address for correspondence: Nicole Gross, Onco-Hematology Research Unit, Department of Medical and Surgical Pediatrics, CHUV, CH-1011 Lausanne, Switzerland. Voice: ++41-21-3143997; fax: ++41-21-3142664.
Nicole.Gross@chuv.hospvd.ch

**Ann. N.Y. Acad. Sci. 1010: 195–199 (2003). © 2003 New York Academy of Sciences.
doi: 10.1196/annals.1299.033**

INTRODUCTION

Caspase-8 is an upstream and initiator protease essential for TRAIL-mediated cell death. It is also involved in amplification of the apoptotic process by activation of the mitochondrial signaling pathway. Once recruited to the death inducing signaling complex (DISC), activated caspase-8 can signal the mitochondrial pathway through cleavage of the BCL2 family member Bid.[1] The *CASP8* gene is frequently inactivated by a combination of methylation and allelic deletion in several cancers, including aggressive and *MYC*N-amplified neuroblastoma, a clinically and biologically heterogeneous childhood neoplasm. We have proposed that caspase-8 silencing in NB could be responsible for the observed resistance to TRAIL-mediated apoptosis.[2] However, the involvement of deregulation of additional signaling molecules is suspected in resistance to TRAIL-mediated cell death, namely inactivation of *CASP10* gene, downregulation of TRAIL-receptors expression, overexpression of inhibitory molecules (such as c-FLIP), or yet-unknown mechanisms.[3] Cycloheximide (CHX) or sub-toxic concentration of chemotherapeutic drugs were shown to sensitize various tumor cell types to TRAIL-induced cell death.[4]

The main objective of this study was to identify the precise contribution of caspase-8 expression and activation in the resistance of aggressive neuroblastoma cells to TRAIL, a potential candidate molecule for innovative therapy of highly resistant malignancies. We further examined the cytotoxic activity of TRAIL in combination with CHX or the chemotherapeutic drug DOX in NB cell lines expressing endogenous or exogenous caspase-8.

RESULTS AND DISCUSSION

To analyze the contribution of caspase-8 silencing in the resistance to TRAIL-induced apoptosis in N-type NB cells, we restored a stable caspase-8 expression in the caspase-8 and caspase-10 negative cell line IGR-N91. Therefore, the IGR-N91 cells were infected with a bicistronic retroviral vector encoding for the GFP and the caspase-8 genes (IGR-N91-C8) or with the GFP control vector (IGR-N91-M). The GFP-positive cells were selected as bulk populations and several clones of IGR-N91-C8 cells were isolated. The expression level of caspase-8 protein was then analyzed by Western blot in the different clones and compared with the caspase-8–positive cell line SH-EP. The IGR-N91-C8 cells and clones expressed a level of caspase-8 equivalent to that of SH-EP control cells, whereas the IGR-N91-M cells remained deficient in caspase-8 (FIG. 1A). All cells expressed similar levels of caspase-3. Note that the SH-EP cell line expresses both caspase-8 and caspase-10, whereas the IGR-N91-C8 cells are still deficient in caspase-10 (data not shown).

The caspase-8–complemented IGR-N91-C8 cells were then analyzed for their sensitivity to TRAIL-induced cell death (FIG. 1B). The IGR-N91-C8 cells were resensitized to TRAIL-mediated apoptosis as the viability of the bulk IGR-N91-C8 cells after TRAIL induction is similar to that of the TRAIL-sensitive SH-EP cell line (79% and 77%, respectively), whereas the control IGR-N91-M cells remain resistant (FIG. 1B). The IGR-N91-C8 clones were even more sensitive to TRAIL than SH-EP cells or the bulk IGR-N91-C8 cells, with two clones displaying high response to TRAIL (11% and 5% of viability) (FIG. 1B). These results show that the caspase-8

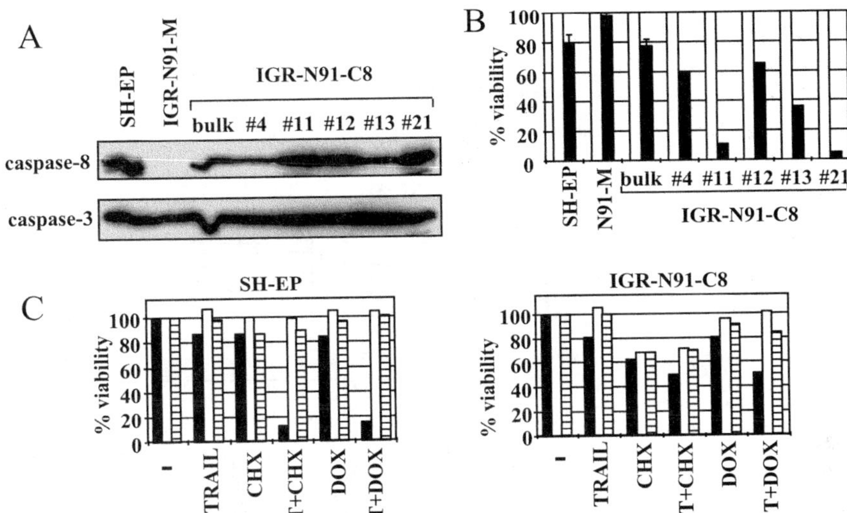

FIGURE 1. (**A**) Restoration of caspase-8 expression in IGR-N91-C8 cells. IGR-N91 cells were infected with retroviral supernatant produced as previously described.[5] Fifty micrograms of whole cell protein extracts were loaded on a 12% SDS-PAGE gel and the expression level of caspase-8 and caspase-3 proteins were analyzed by immunoblotting using mouse mAbs raised against caspase-8 (MBL) and caspase-3 (BD Transduction Lab). (**B**) Restoration of TRAIL sensitivity of IGR-N91-C8 cells. NB cell lines were unstimulated (*white*) or treated with 100 ng/mL of TRAIL and 1 μg/mL of enhancer for 48 h (*black*). MTS cell proliferation assay (Promega) was performed to measure cell viability. The mean of three representative experiments are shown for SH-EP, IGR-N91-M (N91-M), and IGR-N91-C8 bulk cells, and one assay for clones #4, 11, 12, 13, and 21. (**C**) Sensitization of SH-EP and N91-C8 bulk cells to TRAIL-induced cytotoxicity by DOX and CHX. Cells were induced during 48 h by TRAIL (T) in the presence or absence of DOX or CHX and either unprotected (*black columns*) or inhibited by zIETD-fmk (*white*) or zVAD-fmk (*striated*). Cell viability was measured by MTS.

expressed in the complemented cells is functional and sufficient to fully re-sensitize NB cells to TRAIL-induced cell death.

Next, the ability of CHX or the DNA-damaging agent doxorubicin (DOX) to increase TRAIL-mediated cytotoxicity was tested both in the caspase-8–complemented cells IGR-N91-C8 and in the SH-EP cell line. The results show that cell death induced by TRAIL was enhanced by co-treatment of SH-EP cells or IGR-N91-C8 cells with CHX or with sub-toxic concentration of DOX (FIG. 1C). The IGR-N91-M control cells could not be sensitized to TRAIL (data not shown). The enhanced response was caspases dependent, as cell death is inhibited by the caspase-8 protease inhibitor zIETD-fmk or by the pan-caspase inhibitor zVAD-fmk (FIG. 1C).

The effect of sensitization to TRAIL by CHX on activation of apoptotic proteins was then analyzed by western blot. Co-treatment of SH-EP or IGR-N91-C8 cells with TRAIL and CHX increased the activation of caspase-8 and the effector caspases-3, -7, and -9, when compared to induction by TRAIL alone (FIG. 2A). However caspase-3 was not cleaved in the IGR-N91-C8 cells. The cleavage of caspases sub-

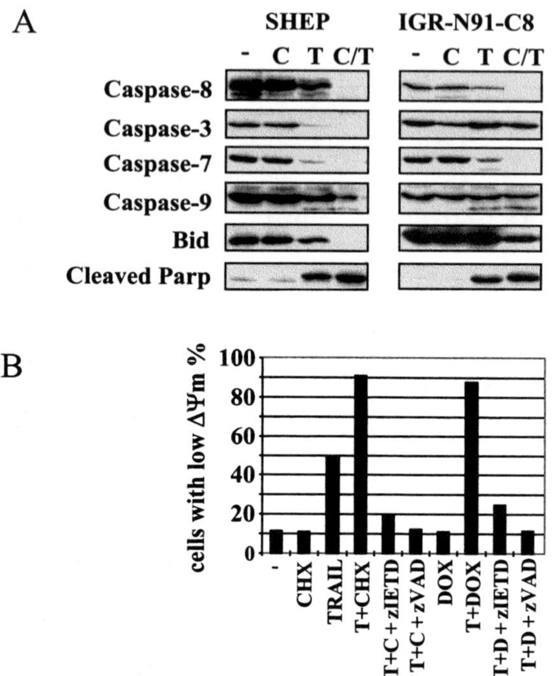

FIGURE 2. (**A**) CHX increases TRAIL-induced caspase activation. Cells were treated with TRAIL (T) and/or CHX (C) or left unstimulated (–) during 8 hours. Cell lysates were prepared and 50 μg of proteins were loaded on 12% SDS-PAGE gels. Cleavage of apoptotic proteins was analyzed by immunoblotting using mouse mAbs raised against caspase-8 (MBL), caspase-3 (BD Transduction Lab), caspase-7 (BD Pharmingen), and PARP (BD Pharmingen), and rabbit polyclonal Abs against caspase-9 (Biolabs) and Bid (R&D). (**B**) CHX and DOX increased TRAIL-induced activation of the mitochondrial signaling pathway. The loss of ΔΨm was measured with the fluorescent dye JC-1 (Calbiochem) according to manufacturer's protocol. SH-EP cells were uninduced (–) or induced 15 h with TRAIL (T, 100 ng/mL + enhancer 1 μg/mL), Cycloheximide (CHX, 1 μg/mL) or doxorubicin (DOX, 0.1 μg/mL). Cells were pre-treated 30 min with the indicated caspase inhibitor. Loss of ΔΨm was measured by flow cytometry using the propidium iodide channel (FL2) for reduction of red aggregates. The percentage of cells with low ΔΨm is indicated.

strates such as Bid and PARP was also increased by combination of TRAIL and CHX when compared to TRAIL alone (FIG. 2A). It should be noted that CHX alone did not affect caspases activation nor the expression level of the analyzed proteins.

As Bid and caspase-9 cleavage was increased by co-treatment with TRAIL and CHX, we analyzed if this sensitization also affects the activation of the mitochondrial pathway. Therefore, disruption of the mitochondrial transmembrane potential (ΔΨm) was measured by flow cytometry in SH-EP cell line with the fluorescent dye JC-1. The loss of ΔΨm induced by TRAIL was increased by co-treatment with CHX or DOX and was highly prevented by the caspase-8 inhibitor zIETD-fmk and the pan-caspase inhibitor zVAD-fmk (FIG. 2B).

In conclusion, restoration of active caspase-8 expression in a NB cell line deficient for caspase-8 and caspase-10 is sufficient to fully restore TRAIL sensitivity. This indicates that caspase-8 silencing is on its own responsible for the resistance to TRAIL-mediated apoptosis in the NB cell line IGR-N91. Moreover, TRAIL-induced apoptosis can be potentiated by co-treatment with CHX and DOX in neuroblastoma cell lines. This sensitization to TRAIL by CHX or DOX is caspase-8–dependent and engages both the death receptor and the mitochondrial signaling pathways.

REFERENCES

1. Chen, M. & J. Wang. 2002. Apoptosis **7:** 313-319.
2. Hopkins-Donaldson, S., J.L. Bodmer, K.B. Bourloud, *et al.* 2000. Cancer Res. **60:** 4315-4319.
3. Igney, F.H. & P.H. Krammer. 2002. Nat. Rev. Cancer **2:** 277-288.
4. Wajant, H., K. Pfizenmaier & P. Scheurich. 2002. Apoptosis **7:** 449-459.
5. Soneoka, Y., P.M. Cannon, E.E. Ramsdale, *et al.* 1995. Nucleic Acids Res. **23:** 628-633.

Cisplatin-Induced Apoptosis in Melanoma Cells

Role of Caspase-3 and Caspase-7 in Apaf-1 Proteolytic Cleavage and in Execution of the Degradative Phases

BARBARA DEL BELLO, MARTA A. VALENTINI, MARIO COMPORTI, AND EMILIA MAELLARO

Department of Pathophysiology and Experimental Medicine, University of Siena, via A. Moro, 53100 Siena, Italy

ABSTRACT: Apoptosis protease–activating factor-1 (Apaf-1), which plays a central role in the formation of the apoptosome, is absent or poorly expressed (because of a transcriptional silencing by methylation) in a substantial percentage of metastatic melanomas and melanoma cell lines, which are unable to activate caspase-9 and execute the mitochondrial pathway of apoptosis. We studied cisplatin-induced apoptosis of the Apaf-1–positive human metastatic Me665/2/21 melanoma cells. Our results indicate that caspase-7 is already processed in still-adhering cells and such activation, contrary to the common view, precedes caspase-3 processing. As expected by the cytochrome *c* release into the cytosol, caspase-9 is processed to active forms (p37 and p35), along with a yet-unidentified p28. Interestingly, we also demonstrate a remarkable loss of Apaf-1 protein, along with the appearance of a related immunoreactive fragment of ≅ 26 kDa; such proteolytic degradation proves to be a caspase-3/-7–mediated event. Our data also indicate that the inhibition afforded by ac-DEVD-CHO on several components (i.e., caspase-3/-7 and caspase-9 activities), and Apaf-1 proteolytic degradation, does not significantly abrogate either the apoptotic morphology or the cleavage of canonical targets, such as poly(ADP-ribose) polymerase (PARP) and lamin B. These results suggest that caspase-3 and caspase-7 are dispensable for the execution of apoptosis and, in our cellular model, the point of no return could be out of the mitochondrial cascade.

KEYWORDS: melanoma cells; caspase-3; caspase-7; caspase-9; apoptosome; Apaf-1; proteolytic cleavage

These data are part of a more detailed paper due to appear in *Experimental Cell Research* (2004).

Address for correspondence: Emilia Maellaro, Department of Pathophysiology and Experimental Medicine, University of Siena, via A. Moro, 53100 Siena, Italy. Voice: 0577-234006; fax 0577-234009.

maellaro@unisi.it

Ann. N.Y. Acad. Sci. 1010: 200–204 (2003). © 2003 New York Academy of Sciences. doi: 10.1196/annals.1299.034

INTRODUCTION

Metastatic melanoma, a cancer that fails to respond to conventional chemotherapy and radiotherapy, shows a low degree of both spontaneous and drug-induced apoptotic cell death. The apoptosis protease-activating factor-1 (Apaf-1), which plays a central role in the formation of apoptosome, is absent or poorly expressed (because of a transcriptional silencing by methylation) in a substantial percentage of metastatic melanomas and melanoma cell lines, which are unable to activate caspase-9 and execute the mitochondrial pathway of apoptosis.[1] In the present investigation, we studied cisplatin-induced apoptosis of the Apaf-1–positive human metastatic Me665/2/21 melanoma cells. In particular, we addressed the role played by the effector caspases-3 and -7 on Apaf-1 status and caspase-9 activation, and on the execution phases of cell death

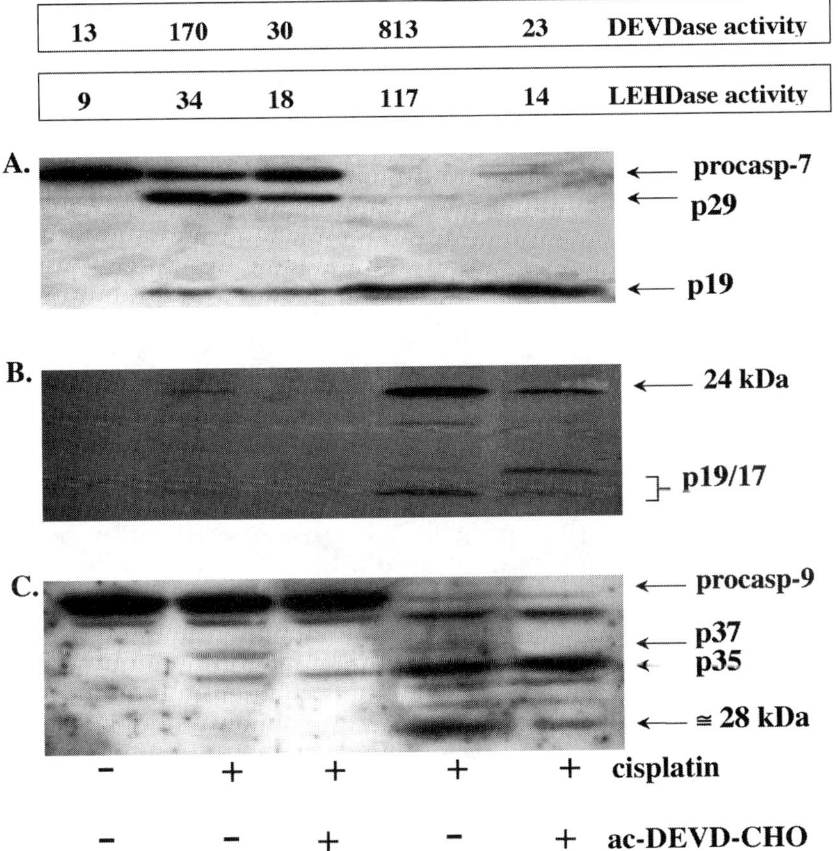

FIGURE 1. (**A**) Caspase-7 processing, (**B**) Caspase-3 processing (active forms only), (**C**) caspase-9 processing. Lanes 2 and 3, treated, adhering cells, and Lanes 4 and 5, treated, detached cells.

RESULTS AND DISCUSSION

After 48 h of cisplatin treatment (1 μg/mL), we harvested and analyzed the population of still undetached cells, proved to be committed cells at an early stage of cell damage, separately from frankly apoptotic, floating cells. Along with a remarkable caspase-3 and -7 enzymatic activity (as evaluated by DEVDase activity), we demonstrated that caspase-7 is already processed in still-adhering cells and that such activation, contrary to the common view, precedes caspase-3 processing (FIG. 1). As expected by cytochrome *c* redistribution from mitochondria to cytosol, a significant caspase-9 enzymatic activity (LEHDase activity) was detected, along with the formation of its processing forms, p37, p35, and a yet-unidentified p28, the latter possibly deriving from p37 (FIG. 1). As expected, the presence of ac-DEVD-CHO (50 μM) in intact cells strongly inhibited caspase-3 and -7 activity. Caspase-9 activity was also inhibited, this being due (1) to a direct inhibition by ac-DEVD-CHO, as demonstrated by DEVD-based inhibitors on the recombinant human protein, and/or (2) to the impairment of p37 formation, resulting from the positive activation loop from caspase-3 and -7 back to caspase-9, in spite of the major presence of p35. In fact, caspase-9 autoprocessing, while producing the large subunit p35, exposes an N-terminal ATPF motif on the linker region, available for interaction with the BIR3 domain of XIAP, thus generating a mature caspase-9 in a dormant state. The most intriguing and novel finding is the potential role of effector caspases–3 and –7 on the status of Apaf-1, the central scaffold protein of the apoptosome. Interestingly, for the first time in melanoma cells, we demonstrated a significant decrease of Apaf-1 protein in still adhering cells of cisplatin-treated sample, and the almost complete disappearance in apoptotic, floating cells, along with the formation of an Apaf-1 N-terminal fragment of ≈26 kDa. Both Apaf-1 loss and its proteolytic cleavage were significantly prevented in the presence of caspase-3 and -7 inhibitor ac-DEVD-CHO, thus suggesting that Apaf-1 is a potential target of effector caspases during apoptosis (FIG. 2). On the basis of calculated molecular weight, the Apaf-1 isoform(s)[2]

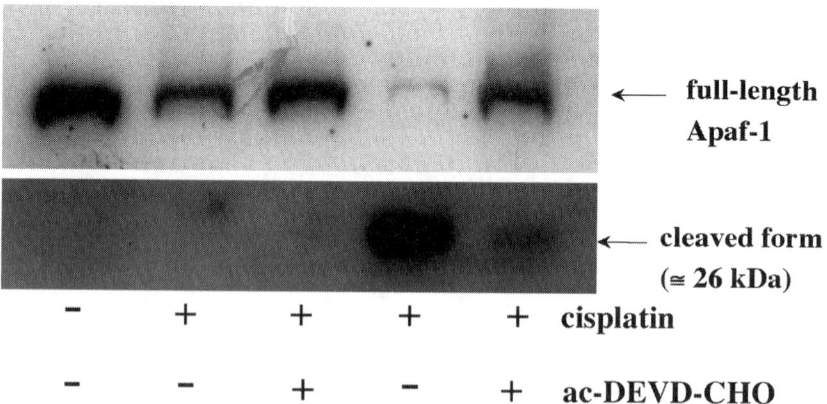

FIGURE 2. Western blot of Apaf-1 cleavage. Lanes 2 and 3, treated, adhering cells, and Lanes 4 and 5, treated, floating cells.

expressed in these melanoma cells is likely S or LN. Due to the abundance of aspartate residues in both Apaf-1 isoforms, several potential cleavage sites can be identified, nearly giving a size of ≈26 kDa. We suggest that a putative cleavage motif could be DVWD↓S at aspartate 236 or 245, in Apaf-1S or Apaf-1LN, respectively. As far as the responsible caspase is concerned, caspase-7 appears to be the unique candidate for Apaf-1 decrease in still-adhering cells, in that caspase-3 is not yet processed in such a sample, while both caspase-7 and caspase-3 are potential effectors of Apaf-1 cleavage in frankly apoptotic cells. What is the functional consequence of Apaf-1 loss on the apoptosome complex? We suggest that Apaf-1 does not necessarily serve as an allosteric activator, as proposed by some investigators, since caspase-9 is well active in apoptotic cells, in spite of a strong decrease of Apaf-1. Alternatively, only a very small moiety of Apaf-1 could be required to trigger the caspase cascade in intact cells, or different mechanisms for caspase-9 activation, alternative to the Apaf-1–dependent pathway, must be taken in account. The role played by the apoptosome and by Apaf-1 is particularly interesting in melanoma cells undergoing apoptosis. In this cell type *Apaf-1* has been considered a tumor suppressor gene,[1] its silencing being a rate-limiting factor in the execution of apoptotic cell death, and partially explaining both the progression of cancer disease and the failure of cancer cells to respond to conventional chemotherapy. In fact, Apaf-1–negative metastatic melanoma cells proved to be invariably chemoresistant, while Apaf-1 restoration markedly enhanced chemosensitivity in diverse cell types.[1,3] The above results strongly suggest a central role of Apaf-1 in determining the sensitivity to apoptosis. In this perspective, the above Apaf-1 fragment, likely comprising the CARD domain and a moiety of the CED-4-like domain, is likely able to recruit caspase-9 and to elicit its autoprocessing. Moreover, consistent with the finding that the WD-40–deleted form of Apaf-1 proved to be constitutively active for caspase-9 processing, in the absence of dATP and cytochrome *c*,[4] the Apaf-1 cleavage could paradoxically result in a peptide even more effective than the full-length protein.

For pathological conditions in which excessive apoptotic cell death occurs, caspase inhibition is considered a potential tool of therapeutic intervention. In this perspective, the present study also addressed the protection from apoptosis in conditions where both caspase-3 and -7 (considered the major effector caspases) and caspase-9 resulted to be inhibited. Unexpectedly, cell rounding and detachment from the monolayer was almost unchanged compared to cisplatin-treated cells. Furthermore, the cleavage of two key targets, i.e., PARP and lamin B, considered essential for DNA stability and chromatin structural organization, were not prevented in the presence of ac-DEVD-CHO. In accordance to this, the nuclear apoptotic morphology (i.e., chromatin condensation and fragmentation) was not significantly affected by ac-DEVD-CHO. We suggest that, besides the well known redundancy of the effector caspases with respect to substrates cleavage, compensatory activation pathways of alternative caspases, as recently documented in caspase-$3^{-/-}$ and caspase-$9^{-/-}$ mice,[5] could account for the absence of any protective effect against apoptosis during caspase-3 and –7 inhibition.

In conclusion, our results indicate that in cisplatin-induced apoptosis of Me665/2/21 melanoma cells (1) Apaf-1 proteolytic degradation is a caspase-3 and -7–mediated event, not compromising caspase-9 activity and apoptosis, and (2) caspase-3 and –7, along with caspase-9, are dispensable for the execution of apoptosis, suggesting that in our cellular model the point of no return is out of the mitochondrial cascade. In this instance, novel functions of Apaf-1, alternative to the apoptosome machinery, could be taken into account to explain the role of Apaf-1 as a critical mediator of apoptosis in diverse cell types.

REFERENCES

1. Soengas, M.S. *et al.* 2001. Inactivation of the apoptosis effector Apaf-1 in malignant melanoma. Nature **409:** 207–211.
2. Cain, K. *et al.* 2002. The Apaf-1 apoptosome: a large caspase-activating complex. Biochimie **84:** 203–214.
3. Jia, L. *et al.* 2001. Apaf-1 protein deficiency confers resistance to cytochrome-c-dependent apoptosis in human leukemic cells. Blood **98:** 414–421.
4. Hu, J.M. *et al.* 1998. WD-40 repeat region regulates Apaf-1 self-association and procaspase-9 activation. J. Biol. Chem. **273:** 33489–33494.
5. Zheng, T.S. *et al.* 2000. Deficiency in caspase-9 or caspase-3 induces compensatory caspase activation. Nature Med. **6:** 1241–1247.

A Novel Antioxidant-Inhibited Dexamethasone-Mediated and Caspase-3–Independent Muscle Cell Death

ARKADIUSZ ORZECHOWSKI,[a] MICHAL JANK,[a] BARBARA GAJKOWSKA,[b] TOMASZ SADKOWSKI,[a] AND MICHAL MAREK GODLEWSKI[a]

[a]*Department of Physiological Sciences, Faculty of Veterinary Medicine, Warsaw Agricultural University, Warsaw, Poland*

[b]*Laboratory of Cell Ultrastructure, M. R. C. Polish Academy of Sciences, Warsaw, Poland*

ABSTRACT: Dexamethasone (Dex)-mediated cell death is associated with repression of survival factors (AP-1, c-myc, NF-κB). Dex suppressed the activity of genes encoding antioxidant enzymes leading to impaired viability and apoptotic cell death. These findings suggest that reactive oxygen species inhibit protein synthesis and amplify m-calpain–dependent proteolysis. The events that led to death of L6 muscle cells were most likely triggered by Dex-mediated repression of antioxidative defenses on the genomic level.

KEYWORDS: dexamethasone; hydrogen peroxide; muscle cell death; m-calpain; caspase-3

In lymphoid tissue glucocorticoid-induced cell death was reported to occur downstream from gene regulation, including execution caspases and proteosome complex activation.[1] Recently, it was shown that dexamethasone-mediated cell death is associated with repression of survival factors (AP-1, *c*-myc, NF-κB). Additionally, dexamethasone (Dex) suppressed the activity of genes encoding antioxidant enzymes leading to impaired viability and apoptotic cell death.[2] Nevertheless, molecular mechanisms of dose-dependent action of Dex on muscle death, in contrast to other cell types, are poorly understood. Increasing evidence has accumulated concerning the role of calpains in pathogenesis of muscle diseases. The activity of m-calpain is elevated in the mutant muscular dysgenesis mouse,[3] while mutations in calpain 3 are known to cause limb-girdle muscular dystrophy type 2A.[4] Accelerated proteolysis in dystrophic myotubes is associated with calcium influx[5] and Ca^{2+} is indispensable for calpain activation. The Ca^{2+} response is also crucial for the cytoskeleton remodeling in muscle cells and myotubes mediated by calpains. Therefore, calcium-medi-

Address for correspondence: Arkadiusz Orzechowski, Department of Physiological Sciences, Faculty of Veterinary Medicine, Warsaw Agricultural University, Warsaw, Poland.
orzechowski@alpha.sggw.waw.pl

**Ann. N.Y. Acad. Sci. 1010: 205–208 (2003). © 2003 New York Academy of Sciences.
doi: 10.1196/annals.1299.035**

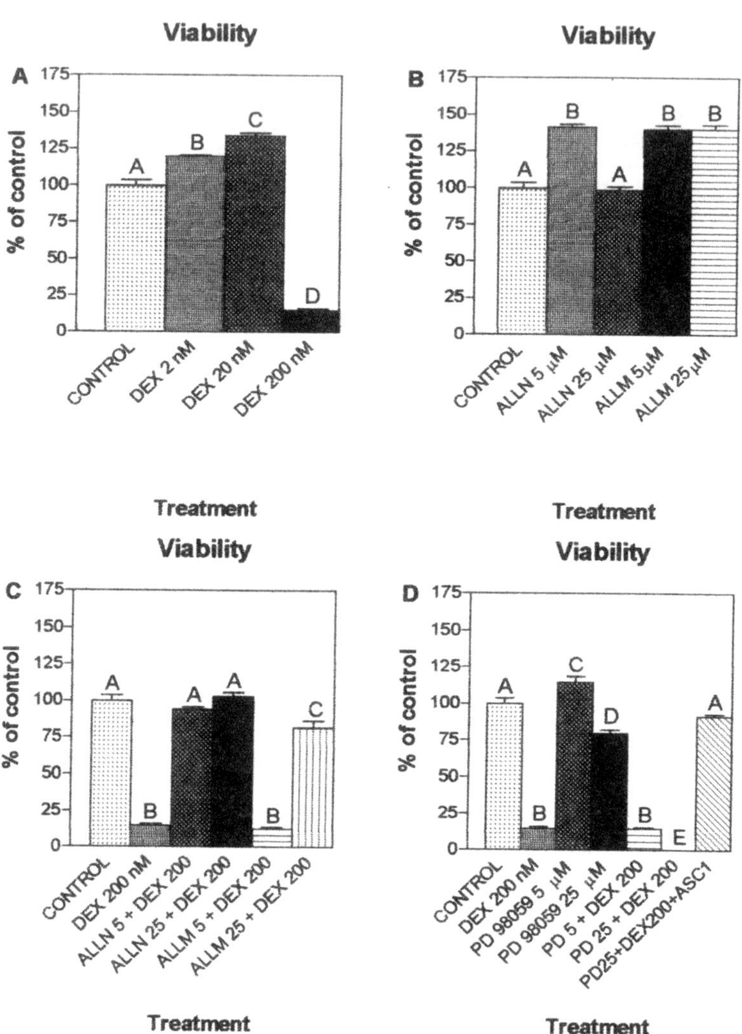

FIGURE 1. Bar charts illustrating the cell respiration (viability) measured by the conversion of soluble MTT into insoluble formazan at day 4. of chronic treatment of L6 muscle cells with some of the experimental agents. Typical results from two experiments performed in quadruplicates are shown as the means ± SEM. (**A**) The effect of Dex (2, 20, 200 nM) on cell viability; (**B**) the effect of calpain inhibitors ALLM (5 or 25 μM) for μ-calpain, or ALLN (5 or 25 μM) for m-calpain; (**C**) the effect of Dex (200 nM) without or with calpain inhibitors ALLM (5 or 25 μM) for μ-calpain, or ALLN (5 μM or 25 μM) for m-calpain; (**D**) the effect of Dex (200 nM) without or with MEK inhibitor PD 98059 (5 or 25 μM) and ASC (1 mM). The means that differ at least significantly ($P<.05$) are marked by different uppercase letters.

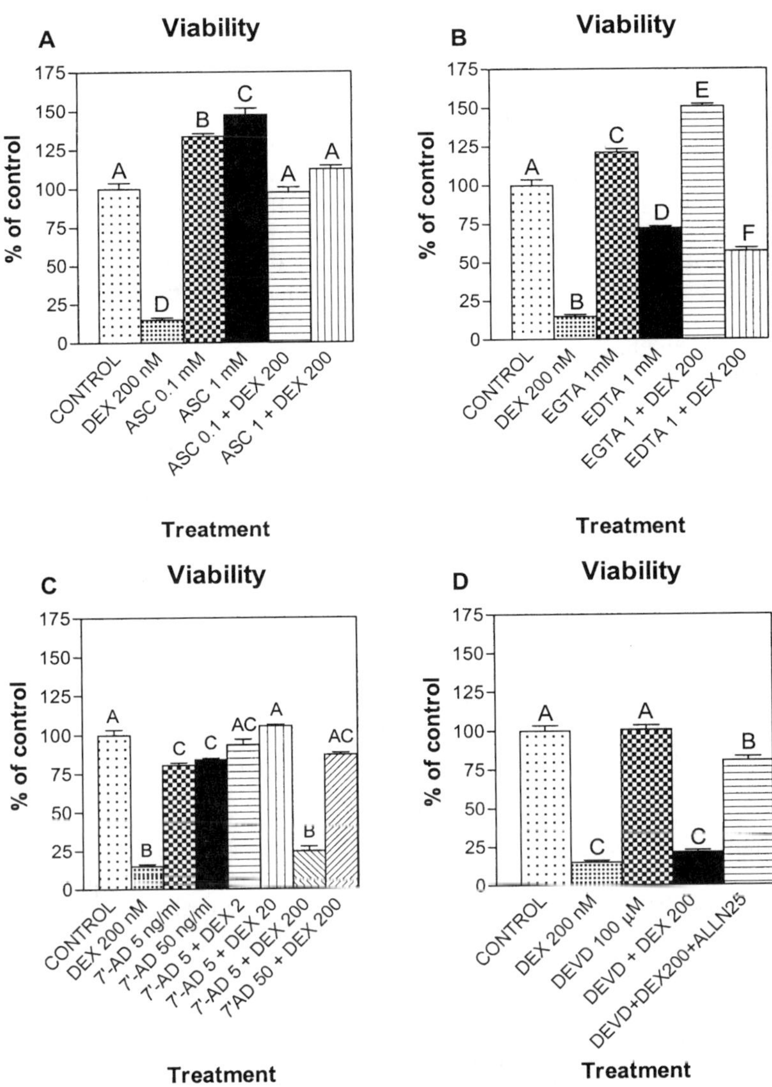

FIGURE 2. Bar charts illustrating cell respiration (viability) measured by the conversion of soluble MTT into insoluble formazan at day 4 of chronic treatment of L6 muscle cells with some of the experimental agents. Typical results from two experiments performed in quadruplicate are shown as the means ± SEM. (**A**) The effect of ASC (0.1, 1 mM) on Dex-mediated (200 nM) repression of cell viability; (**B**) the effect of EDTA (1 mM) on Dex-mediated (200 nM) repression of cell viability; (**C**) the effect of 7′-AD (5, 50 nM) on Dex-affected (2, 20, 2000 nM) cell viability; (**D**) the effect of Ac DEVD CHO (100 µM) on Dex-mediated (200 nM) repression of cell viability. The means that differ at least significantly ($P<.05$) are marked by different uppercase letters.

ated proteolytic damage is often linked with calpain activity. On the other hand, glucocorticoids are potent regulators of muscle cell differentiation and myotube formation, as well as protein synthesis.

Cell death–promoting effect of Dex was studied during myogenesis (10 days) in L6 muscle cells by making use of several indices such as cell viability (protein synthesis, mitochondrial respiration), mortality (DNA fragmentation, chromatin condensation, structural modifications), and immunocytochemical studies [hydrogen peroxide, m-calpain (calpain 2)]. Dex initially (2 nM) stimulated protein synthesis ($P<.001$), but a further increase (20 nM) did not stimulate, whereas a higher dose (200 nM) inhibited formation of cellular proteins ($P<.001$). The latter, apparently, resulted from the mitochondrial dysfunction ($P<.001$) confirmed by MTT assay (FIG. 1). From the day 4, structural changes featuring cell death were observed. Antioxidants [sodium ascorbate (ASC), catalase (CAT), or N-acetyl-L-cysteine (NAC)] as well as the inhibition of transcription and translation by actinomycin D abrogated Dex-induced cell death ($P<.001$, FIG. 2). Using a fluorescent probe (DCFH-DA), we directly corroborated the working hypothesis of the mediating role of H_2O_2 in the reduction of cell viability by the excess of glucocorticoids. We also found that PKC, PLC-gamma, and PLA_2 were required to induce Dex-dependent cell death since inactivation of PKC by H7 completely abolished cytotoxic effect of Dex, while the blockade of PLC-gamma and PLA2 by U 73122 partially abolished the effect. Cell death was triggered by Ca^{2+} influx necessary to activate m-calpain since it was reversed by the calcium chelator EGTA or m-calpain inhibitor ALLN but not EDTA nor ALLM (FIG. 2). However, cell viability impaired by ionophore A 23187 ($P<.001$) was not reversed by EGTA, EDTA, or caspase-3 blocker - Ac DEVD CHO, nor ALLN, nor antioxidants (ASC, NAC, CAT), indicating that apparently other than calcium ions contributed to the induction of muscle cell death caused by A23187. Specific caspase-3 inhibitor Ac DEVD CHO also did not rescue cells from Dex-induced cell death ($P<.001$, FIG. 2), in contrast to m-calpain inhibitor ALLN (FIG. 1). Taken together, these findings suggest that reactive oxygen species inhibit protein synthesis and amplify m-calpain–dependent proteolysis. The events that led to death of L6 muscle cells were most likely triggered by Dex-mediated repression of antioxidative defenses on the genomic level. Whether activation of m-calpain resulted in degradation of prosurvival transcription factors should be further investigated.

REFERENCES

1. DISTELHORST, C.W. 2002. Cell Death Differ. **9:** 6–19.
2. BAKER, A.F. *et al.* 1996. Cell Death Differ. **3:** 207–213.
3. JOFFROY, S. *et al.* 2000. Int. J. Dev. Biol. **44:** 421–428.
4. RICHARD, I. *et al.* 1995. Cell **81:** 27–40.
5. ALDERTON, J.M. *et al.* 2000. J. Biol. Chem. **275:** 9452–9460.

Caspase Inhibitors as a Supplement in Immune Activation Therapies to Achieve Eradication of HIV in Its Latent Reservoirs

CARSTEN SCHELLER,[a] SIEGHART SOPPER,[a] ELENI KOUTSILIERI,[a] STEPHAN LUDWIG,[b] VOLKER TER MEULEN,[a] AND CHRISTIAN JASSOY[a]

[a]*Institute for Virology and Immunobiology, Julius-Maximilians-Universität, 97078 Würzburg, Germany*

[b]*Institut für Medizinische Strahlenkunde und Zellforschung (MSZ), Julius-Maximilians-Universität, 97078 Würzburg, Germany*

Abstract: Synthetic caspase inhibitors are known to block death receptor–induced apoptosis. We have reported previously that in addition to apoptosis inhibition, caspase inhibitors induce a switch from proapoptotic to proinflammatory signaling in T cells. This switch sensitizes TNF-R1 to trigger TNF-α–induced reactivation of HIV in latently infected T cells. Here we ponder the prospects of caspase inhibitors as a supplement in immune activation therapies (IAT) to achieve eradication of HIV from the latent virus reservoirs in HIV-infected patients.

Keywords: apoptosis; caspase inhibitors; eradication; HIV; reactivation; TNF-R1

Highly active antiretroviral therapy (HAART) has dramatically improved treatment of HIV infection. Most patients treated with HAART maintain clinically undetectable virus titers and HIV-associated mortality has significantly decreased. However, patients treated with HAART cannot be cured from HIV, because antiretroviral therapy is not able to eradicate the virus from the body. The virus persists in latently infected cells from which it spreads throughout the body whenever HAART is interrupted.

New approaches are now being developed to achieve eradication of HIV from the infected patient. They are called "immune activation therapies" (IAT) and aim to stimulate HIV replication in the latent virus reservoirs in order to make these compartments vulnerable to antiretroviral therapy. A number of different substances are known to trigger reactivation of HIV in latently infected cells, such as TNF-α, IL-2, OKT3, or IFN-γ.[1] However, none of them could have achieved eradication of HIV

Address for correspondence: Carsten Scheller, Institute for Virology and Immunobiology, Versbacher Strasse 7, 97078 Würzburg, Germany. Voice: ++49 931 201-49897; fax: ++49 931 201-49553.

scheller@vim.uni-wuerzburg.de

**Ann. N.Y. Acad. Sci. 1010: 209–212 (2003). © 2003 New York Academy of Sciences.
doi: 10.1196/annals.1299.036**

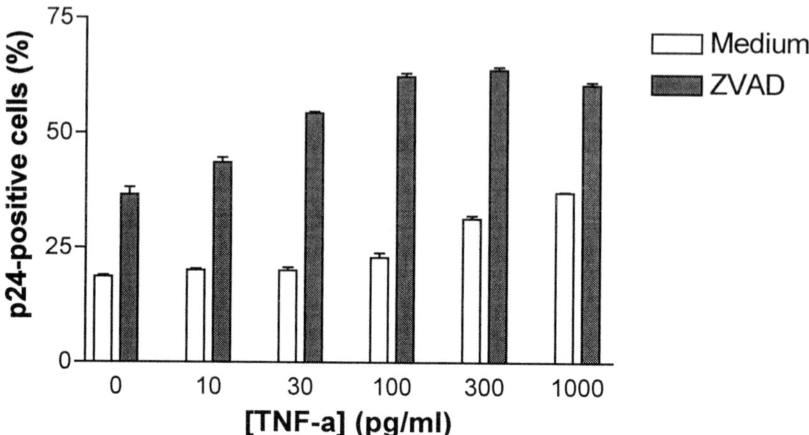

FIGURE 1. Caspase inhibition enhances HIV reactivation triggered by TNF-R1. ACH-2 T lymphoblasts were cultured for 7 h with TNF-α at the indicated concentrations. Cells were cocultured with the caspase inhibitor zVADfmk (ZVAD; 100 μM) or left untreated (Medium). HIV-p24 expression was monitored by intracellular staining and flow cytometry. All assays were adjusted to the same solvent concentration (DMSO; 0.1%).

in infected patients so far. One of the main reasons for this failure is the disadvantages associated with current IAT: administration or generation of high amounts of proinflammatory cytokines to trigger virus reactivation is associated with severe side effects which force the withdrawal of therapy.

One of the first stimuli described to trigger reactivation of HIV in latently infected T cells is TNF-α.[2] The tumor necrosis factor receptor 1 is known to activate two different signaling cascades, a proinflammatory and a proapoptotic signaling pathway.[3] TNF-α–induced HIV replication is mediated by the proinflammatory signaling cascade triggered by TNF-R1. We have further investigated the signaling underlying TNF-α–mediated virus reactivation using the latently infected T cell line ACH-2. We have found that virus reactivation was significantly increased when cells are incubated with peptidic caspase inhibitors, such as zVADfmk.[4] To obtain comparably high levels of virus-producing cells, about 100–1,000 times lower amounts of TNF-α were needed when cells were coincubated with zVADfmk compared to the situation when cells were incubated with TNF-α alone (FIG. 1). In fact, the low amounts of endogenously produced TNF-α were sufficient to induce virus reactivation in a considerably high proportion of cells. This sensitization is caused by a blockade of the proapoptotic signaling cascade by zVADfmk, which results in enhancement of the alternative signaling branch.

CD95, a receptor structurally and functionally related to TNF-R1, can similarly activate both proapoptotic and proinflammatory signaling cascades.[5] Although the proinflammatory signaling activity is almost cryptic in activated T cells, it is greatly enhanced when caspase activation is inhibited. As depicted in FIGURE 2, stimulation of CD95 results in reactivation of HIV. Similar to the situation with TNF-R1, caspase inhibition enhances the stimulatory effect (FIG. 2).

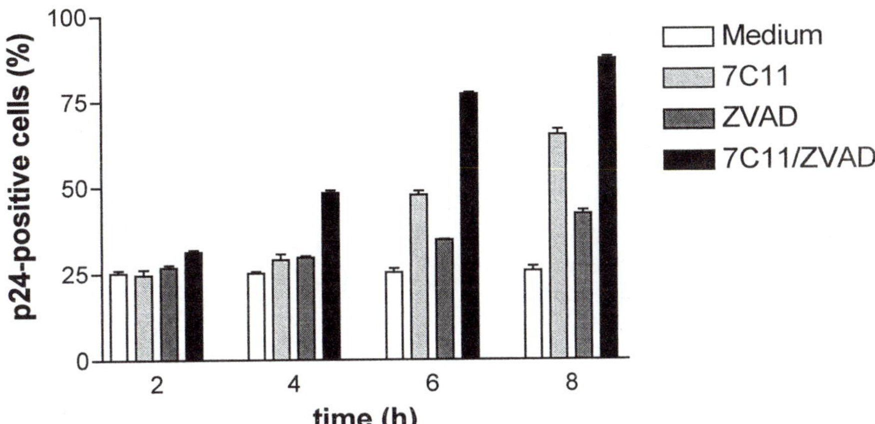

FIGURE 2. Caspase inhibition enhances HIV reactivation triggered by CD95. ACH-2 T lymphoblasts were cultured with medium alone (Medium), the agonistic anti-CD95 mAb 7C11 (7C11; 200 ng/mL), the apoptosis inhibitor zVADfmk (ZVAD; 100 μM) or both (7C11/ZVAD) for the indicated time periods. HIV-p24 expression was monitored by intracellular staining and flow cytometry. All assays were adjusted to the same solvent concentration (DMSO; 0.1%).

We have shown that caspase inhibitors potently enhance HIV reactivation induced by different members of the tumor necrosis factor receptor family. Even low levels of endogenously produced TNF-α are sufficient to trigger reactivation of HIV in cultured cells when caspase activation is inhibited. These findings raise hope that caspase inhibitors may similarly enhance HIV reactivation *in vivo* triggered by endogenously produced TNF-α. Alternatively, when applied with exogenous TNF-α, caspase inhibitors may help to reduce the amount of cytokines needed to trigger virus reactivation in HIV patients receiving IAT.

ACKNOWLEDGMENTS

The study was supported by a grant from the Bundesministerium für Bildung, Wissenschaft, Forschung und Technologie, Germany (BMBF 01 K1 0211 Competence Network HIV/AIDS).

REFERENCES

1. KULKOSKY, J. & R.J. POMERANTZ. 2002. Approaching eradication of highly active antiretroviral therapy-persistent human immunodeficiency virus type 1 reservoirs with immune activation therapy. Clin. Infect. Dis. **35:** 1520–1526.
2. FOLKS, T.M., K.A. CLOUSE, J. JUSTEMENT, *et al.* 1989. Tumor necrosis factor alpha induces expression of human immunodeficiency virus in a chronically infected T-cell clone. Proc. Natl. Acad. Sci. USA **86:** 2365–2368.

3. BAUD, V. & M. KARIN. 2001. Signal transduction by tumor necrosis factor and its relatives. Trends Cell Biol. **11:** 372–377.
4. SCHELLER, C., S. SOPPER, P. CHEN, *et al.* 2002. Caspase inhibition activates HIV in latently infected cells. Role of tumor necrosis factor receptor 1 and CD95. J. Biol. Chem. **277:** 15459–15464.
5. SCHELLER, C., S. SOPPER, C. EHRHARDT, *et al.* 2002. Caspase inhibitors induce a switch from apoptotic to proinflammatory signaling in CD95-stimulated T lymphocytes. Eur. J. Immunol. **32:** 2471–2480.

Apoptotic Cell Death of Cybrid Cells Bearing Leber's Hereditary Optic Neuropathy Mutations Is Caspase Independent

CLAUDIA ZANNA, ANNA GHELLI, ANNA MARIA PORCELLI, VALERIO CARELLI,[a] ANDREA MARTINUZZI,[b] AND MICHELA RUGOLO

Dipartimento di Biologia Evoluzionistica Sperimentale, Università di Bologna, Bologna, Italy

[a]*Dipartimento di Science Neurologiche, Università di Bologna, Bologna, Italy*

[b]*IRCCS "E.Medea," Conegliano, Treviso, Italy*

ABSTRACT: Leber's hereditary optic neuropathy (LHON) is a maternally inherited disease characterized by selective death of retinal ganglion cells. Three pathogenic mtDNA point mutations induce an impairment of oxidative phosphorylation. We have investigated whether the release of cytochrome *c* during incubation of LHON cybrids in galactose medium leads to activation of the executive caspase-3 and to alteration of the energetic status of cells. From our research, it can be concluded that apoptotic cell death induced in LHON cybrid by galactose medium is caspase independent. It remains to be explained how the significant fragmentation of intranucleosomal DNA observed in LHON cybrids could also occur in the absence of caspase activation.

KEYWORDS: LHON (Leber's hereditary optic neuropathy); mitochondrial DNA; complex I; apoptosis; cytochrome *c*; caspase-3; ATP; galactose medium

INTRODUCTION

Leber's hereditary optic neuropathy (LHON) is a maternally inherited disease characterized by a rapid loss of central vision due to a selective death of retinal ganglion cells. Three pathogenic mtDNA point mutations at positions 11778/ND4, 3460/ND1, 14484/ND6, all affecting ND subunits of complex I, are generally agreed to cause LHON. The pathogenic role of complex I dysfunction in this disease remains unclear, since it cannot be correlated with a biochemical alteration common to all the mutations. Cellular studies using transmitochondrial cell lines (cybrids) provided the confirmation that the three LHON mutations induce an impairment of oxidative phosphorylation.[1]

Recently, it has been shown that mutations 11778 and 3460 predispose cybrid cells to Fas-induced apoptosis.[2] We also reported that LHON cybrids exhibited im-

Address for correspondence: Claudia Zanna, Dipartimento di Biologia Evoluzionistica Sperimentale, Università di Bologna, Via Irnerio 42, 40126 Bologna, Italy. Voice: +39 051 2091286; fax +39 051 242576.

zanna@alma.unibo.it

Ann. N.Y. Acad. Sci. 1010: 213–217 (2003). © 2003 New York Academy of Sciences.
doi: 10.1196/annals.1299.037

paired growth and increased cell death compared to controls when incubated in glucose-free/galactose-supplemented medium. The type of cell death was apoptotic, as indicated by the occurrence of nuclear condensation, DNA laddering and cytochrome *c* release from mitochondria.[3] It is well known that cytochrome *c* in the cytosol interacts with Apaf-1, leading to the ATP-dependent formation of a macromolecular complex known as apoptosome.[4] This complex recruits and activates the caspase-9, which in turn can activate additional caspases, such as caspase-3, -6, and -7, resulting in cell death.

In the present study we have investigated whether the release of cytochrome *c* during incubation of LHON cybrids in galactose medium leads to activation of the executive caspase-3 and to alteration of the energetic status of cells.

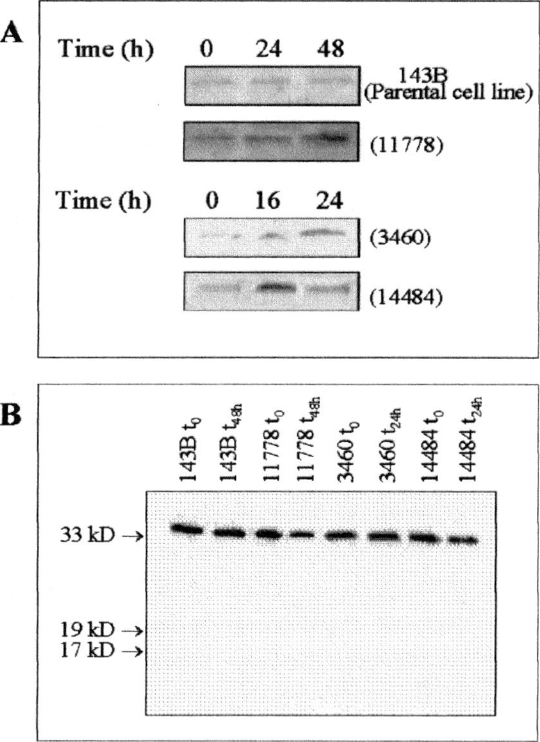

FIGURE 1. Release of cytochrome *c* in the cytosolic fractions (**A**) and cleavage of caspase-3 (**B**) in parental and cybrid cell lines with LHON mutations at different times of incubation in galactose medium. Data are representative of three similar experiments.

METHODS

Cell Lines and Culture Conditions

Cybrid cell lines were constructed using enucleated fibroblasts from six LHON probands as mitochondria donors carrying 11778, 3460, 14484 LHON primary mutations and the osteosarcoma (143B.TK$^-$)-derived 206 cell line as acceptor rho0 cell line (both 143B.TK$^-$ and rho^0 206 were kindly provided by Giuseppe Attardi and Michael King).[3] Parental and cybrid cell lines were grown in DMEM medium supplemented with 10% fetal calf serum (FCS), 2 mM L-glutamine, 100 U/mL penicillin, 100 μg/mL streptomycin, and 0.1 mg/mL bromodeoxyuridine. For the experiments, cells were seeded 4×10^5 cells/cm^2 and incubated in DMEM glucose-free medium supplemented with 5 mM galactose, 5 mM Na-pyruvate, and 5% FCS (DMEM galactose-medium) at 37°C in an incubator with a humidified atmosphere of 5% CO_2.

Western Blot Analysis of Cytochrome c and Caspase-3

Cytochrome *c* release and caspase-3 cleavage were determined in cytosolic fractions or cell lysates, respectively, obtained from cells incubated with DMEM galactose medium.[3,5] Samples of 80–100 μg protein of cytosolic fractions or cell lysates were separated by 12–15% SDS PAGE, transferred onto nitrocellulose membrane, and probed with anti–cytochrome *c* or anti–caspase-3 antibody.

ATP Assay

The amount of cellular ATP was determined by the luciferin/luciferase assay. After incubation in DMEM galactose-medium, cells were harvested in phosphate buffered solution (PBS) and disrupted with 2.5% perchloric acid for 1 min at 4°C. Subsequently, samples were neutralized with 0.1 M K_2CO_3 and 0.07 M Tris and the ATP content was measured using the BioOrbit ATP monitoring kit according to manufacturer's instruction.

RESULTS AND DISCUSSION

FIGURE 1A shows the time course of cytochrome *c* release in the cytosolic fractions obtained from subcellular fractionation of homogenates of the parental 143B.TK$^-$ cells and LHON cybrids. No release of cytochrome *c* was assessed in the parental cell line 143B, whereas in cybrids with 11778/ND4 mutation the release was apparent after 24 h, in cybrids with the 3460/ND1 and 14484/ND6 mutations after 16 h incubation in galactose medium. Caspases are synthesized as intact zymogens and then undergo proteolytic cleavage to become active. Thus activation of caspase-3 has been determined by assaying the cleavage of the procaspase-3. Surprisingly, at the times indicated, no cleavage of caspase-3 in cell lysates of both parental and LHON cybrid cell lines could be detected (FIG. 1B). Therefore, despite the significant cytochrome *c* release, LHON cybrid did not display any significant caspase-3 activation. It is noteworthy that caspase-8 activity also was not increased by incubation in galactose medium (results not shown). In addition to cytochrome *c*,

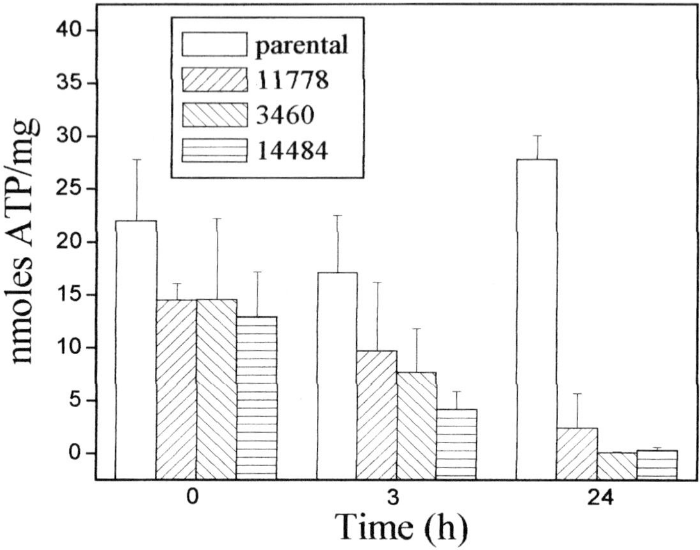

FIGURE 2. Measurement of cellular ATP amount in parental and cybrid cell lines with LHON mutations at different times of incubation in galactose medium. Data are representative of three similar experiments.

ATP is also required for the assembly of apoptosome;[4] therefore, evaluation of the energetic status of cells incubated in galactose medium has been carried out. As showed in FIGURE 2, in glucose medium (t_0), the amount of ATP of cybrids with LHON mutations was significantly reduced in comparison with that of parental 143B cells. After 3-h incubation in galactose medium, a decrease of ATP was observed in all cell lines. At 24 h, the ATP level of parental 143B cell line returned to values even higher than at t_0, whereas it further decreased in all LHON cybrids. It is therefore likely that the drop in ATP levels occurring only in LHON cybrids might be responsible for the lack of caspase-3 activation, also in the presence of a significant amount of cytochrome *c* in the cytosol.

From these results it can be concluded that apoptotic cell death induced in LHON cybrid by galactose medium is caspase independent. It remains to be explained how the significant fragmentation of intranucleosomal DNA observed in LHON cybrids could occur also in the absence of caspases activation. It is possible that apoptosis-inducing factor (AIF) and endonuclease-G might be involved. These apoptogenic factors, normally sequestered within the mitochondrial intermembrane space, might be released in the cytosol together with cytochrome *c*, and then translocated into the nucleus to contribute to the DNA fragmentation. Experiments are in progress to test this hypothesis.

ACKNOWLEDGMENTS

This work was supported by grants from PRIN 2001-2002: "Mitochondria in Cellular Pathology"; Progetto Dipartimentale "Approcci molecolari e genetici allo studio delle patologie," "Progetto Giovani Ricercatori," University of Bologna; and Telethon (n.GGP02323).

REFERENCES

1. BROWN, M.D., I.A. TROUNCE, A.S. JUN, *et al.* 2000. J. Biol. Chem. **275:** 39831–39836.
2. DANIELSON, S.R., A. WONG, V. CARELLI, *et al.* 2002. J. Biol. Chem. **277:** 5810–5815.
3. GHELLI, A., C. ZANNA, A.M. PORCELLI, *et al.* 2003. J. Biol. Chem. **278:** 4145–4150.
4. LIU, X., C.N. KIM, J. YANG, *et al.* 1996. Cell **86:** 147–157.
5. GHELLI, A., A.M. PORCELLI, C. ZANNA & M. RUGOLO. 2002. Arch. Biochem. Biophys. **402:** 208–217.

Involvement of the Change in Chromatin Structure in Thymocyte Apoptosis Induced by Phosphorylation of Histones

RIYO ENOMOTO, YUKARI YOSHIDA, TOMOE KOMAI, CHIYOKO SUGAHARA, YUKA YASUOKA, AND EIBAI LEE

Department of Pharmacology, Faculty of Pharmaceutical Sciences and High Technology Research Center, Kobe Gakuin University, Ikawadani-cho, Nishi-ku, Kobe 651-2180, Japan

ABSTRACT: The inhibitors of protein phosphatase such as calyculin A and okadaic acid induced thymocyte apoptosis. DNA fragmentation was increased in the nuclei from thymocytes treated with calyculin A, which accelerated phosphorylation of histones. Judging from the circular dichroism analysis, the structure of soluble chromatin was changed by the treatment with calyculin A. These results suggest that the change in chromatin structure may be one of the molecular mechanisms of internucleosomal DNA fragmentation.

KEYWORDS: apoptosis; calyculin A; chromatin; DNA fragmentation; histone phosphorylation

Internucleosomal DNA fragmentation is a biochemical feature of apoptotic cell death.[1] A caspase-activated DNase (CAD) is responsible for DNA fragmentation.[2] Since the packaging of genome in chromatin restricts the access of the enzymes,[3] the only activation of CAD seems to be insufficient for DNA fragmentation. To clarify the mechanism of DNA fragmentation of thymocytes undergoing apoptosis induced by calyculin A,[4,5] we studied the role of chromatin structure in apoptotic cell death.

A significant increase in the histone H1 and H2A phosphorylation induced by calyculin A and okadaic acid was detected before proceeding to DNA fragmentation. Calyculin A had no effect on the activities of caspase-3 and -8, suggesting that caspase cascade is unlikely to be involved in the apoptotic process induced by this inhibitor. FIGURE 1 shows that the incubation of isolated nuclei from thymocytes treated with calyculin A in the buffer solution resulted in an increase in DNA fragmentation in comparison with control. Thus, a constitutive DNase in the nuclei seems to catalyze DNA fragmentation. To demonstrate the change in chromatin structure in thymocytes treated with calyculin A, circular dichroism analyses of sol-

Address for correspondence: Riyo Enomoto, Department of Pharmacology, Faculty of Pharmaceutical Sciences and High Technology Research Center, Kobe Gakuin University, Ikawadani-cho, Nishi-ku, Kobe 651-2180, Japan. Voice: +81-78-974-4429; fax: +81-78-974-5689.
enomoto@pharm.kobegakuin.ac.jp

**Ann. N.Y. Acad. Sci. 1010: 218–220 (2003). © 2003 New York Academy of Sciences.
doi: 10.1196/annals.1299.038**

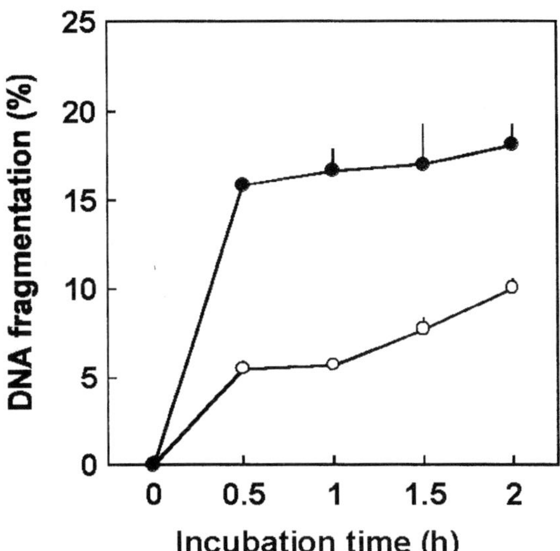

FIGURE 1. The effect of incubation time on nuclear DNA fragmentation. Thymocytes were incubated in the presence or absence of 5 nM calyculin A for 4 hours. After the treatment with calyculin A, the nuclear fractions were prepared. The resulting nuclei were incubated in the DNA fragmentation buffer for 0.5, 1, 1.5, and 2 h, and then DNA in the soluble fraction and pellet were determined using Pico Green. *open circle*, control; *closed circle*, 5 nM calyculin A. Results are means ± S.E. ($N = 4$).

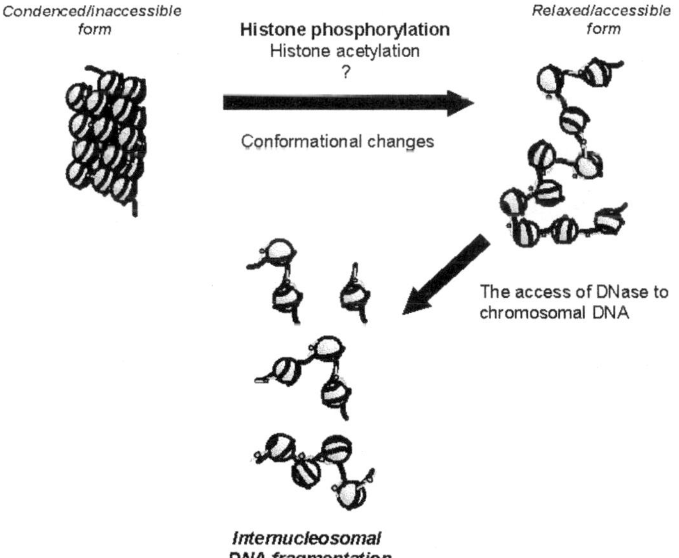

FIGURE 2. A possible mechanism of DNA fragmentation in the cells undergoing apoptosis induced by chemical modifications of histones.

uble chromatin were performed. The negative peak at 200–240 nm, showing α-helical content of chromatin, was decreased in the soluble chromatin from calyculin A–treated thymocytes compared with control. The structural alteration of chromatin was also observed in the cells treated with sodium butyrate, an inhibitor of histone deacetylase inhibitor, which causes histone hyperacetylation. Thus, chemical modifications of histones, such as phosphorylation and acetylation, seem to alter the chromatin structure. Moreover, the structural change of chromatin preceded DNA fragmentation. The present results suggest that the change of chromatin structure induced by chemical modifications of histone, especially hyperphosphorylation, allows easy accessibility of nuclear endonuclease to DNA. FIGURE 2 illustrates the possible molecular mechanism of DNA fragmentation of the cells undergoing apoptosis.

REFERENCES

1. WYLLIE, A.H. 1980. Glucocorticoid-induced thymocyte apoptosis is associated with endogenous endonuclease activation. Nature **284:** 555–556.
2. ENARI, M., H. SAKAHIRA, H. YOKOYAMA, *et al.* 1998. A caspase-activated DNase that degrades DNA during apoptosis, and its inhibitor ICAD. Nature **391:** 43–50.
3. WOLFFE, A. 1995. Chromatin: structure and function. Academic Press. London.
4. LEE, E., A. NAKATSUMA, R. HIRAOKA, *et al.* 1999. Involvement of histone phosphorylation in thymocyte apoptosis by protein phosphatase inhibitors. IUBMB Life **48:** 79–83.
5. ENOMOTO, R., R. KOYAMAZAKI, Y. MARUTA, *et al.* 2001. Phosphorylation of histones triggers DNA fragmentation in thymocyte undergoing apoptosis induced by protein phosphatase inhibitors. Mol. Cell. Biol. Res. Commun. **4:** 276–281.

Accumulation of Histones in Cell Lysates Precedes Expression of Apoptosis-Related Phagocytosis Signals in Human Lymphoblasts

CHRISTOPH GABLER, NORBERT BLANK, SILKE WINKLER, JOACHIM R. KALDEN, AND HANNS-M. LORENZ

Department of Medicine III, Institute for Clinical Immunology, Krankenhausstrasse 12, University of Erlangen-Nuremberg, 91054 Erlangen, Germany

ABSTRACT: Systemic lupus erythematosus (SLE) is characterized by the production of autoantibodies directed against several nuclear components, such as DNA and histones. Apoptosis was induced in activated human lymphoblasts ($n = 6$) by UV-B irradiation for 30 sec followed by continuous culturing. An extranuclear accumulation of the nucleosomal histones H2A, H2B, H3, and H4 in cell lysates was observed very early in the process of apoptosis, even before phosphatidylserine externalization occurred on the outer membrane surface of apoptotically dying lymphoblasts. We hypothesize that a dysregulation of apoptosis during these early phases may contribute to the induction of autoimmunity against nuclear autoantigens as seen in SLE.

KEYWORDS: systemic lupus erythematosus (SLE); apoptosis; histone

INTRODUCTION

In recent years, it has become evident that systemic lupus erythematosus (SLE) is a disease characterized by an array of autoantibodies directed against the native nucleosome, its DNA component, and/or its histone component.[1] The nucleosome is composed of a core particle with a tetramer of each two molecules histone H3 and H4 in the center flanked by two histone H2A-H2B dimers. Two superhelical turns of DNA are wound around this histone octamer. Adjacent nucleosome particles are linked like beads on a string of histone-free linker DNA, and a molecule of histone H1 is located at the point where DNA enters and exits the nucleosome. Nuclear antigens are generated and released in vivo during apoptosis.[2] A hallmark of apoptosis is the cleavage of chromatin by caspase-activated DNase. This fragmentation occurs at the internucleosomal level and leads to DNA ladder formation classically associated with apoptosis. In addition, we reported recently that activated lymphoblasts contain higher core histone content in cell lysates of early apoptotic cells.[3] This accumulation in cell lysates correlated highly with the status of early apoptotic cells

Address for correspondence: Hanns-M. Lorenz, Department of Medicine III, Institute for Clinical Immunology, Krankenhausstrasse 12, University of Erlangen-Nuremberg, 91054 Erlangen, Germany. Voice: 49-9131-853-9107; fax: 49-9131-853-4770.
Hannes.Lorenz@med3.imed.uni-erlangen.de

**Ann. N.Y. Acad. Sci. 1010: 221–224 (2003). © 2003 New York Academy of Sciences.
doi: 10.1196/annals.1299.039**

A (H2A)
OD x mm²
12
10
8
6
4
2
0
0h
30min UV
1h UV
2h UV
B (H2B)
OD x mm²
24
20
16
12
8
4
0
0h
30min UV
1h UV
2h UV
C (H3)
OD x mm²
15
12
9
6
3
0
0h
30min UV
1h UV
2h UV
D (H4)
OD x mm²
28
21
14
7
0
0h
30min UV
1h UV
2h UV

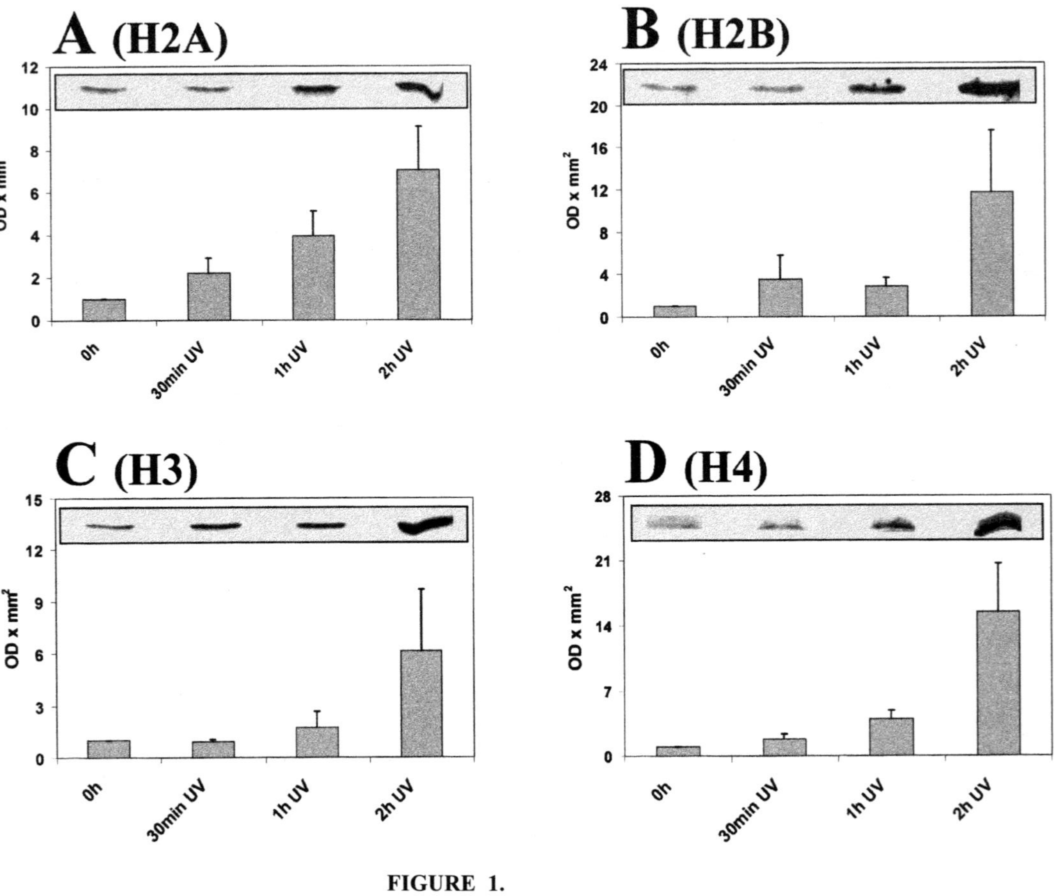

FIGURE 1.

(Annexin V (AxV) positive, propidium iodide (PI) negative), but not with late apoptotic or necrotic cells. Aim of this study was to evaluate the kinetics of accumulation of histones in cell lysates of human lymphoblasts after induced apoptosis. Moreover, we wanted to investigate the order of histone appearance in cell lysates in relation to markers of apoptosis (AxV binding and/or PI permeability).

RESULTS

Apoptosis was induced in activated human lymphoblasts (N=6) by UV-B irradiation for 30 sec followed by incubation for 30 min, 1 h, and 2 h, respectively. Cell pellets were subsequently lysed with a modified RIPA buffer, followed by centrifugation to separate the nucleus. Cell lysates (supernatants containing the cytoplasm) were carefully removed without destroying the pellet (containing the nucleus). Histones were detected in these cell lysates with specific antibodies by Western blot analysis. Simultaneously, we performed AxV/PI staining in irradiated cells for all timepoints. After UV-B irradiation, we observed a continuous increase of the relative amount of histones in the cell lysates compared to lymphoblasts before UV exposure (t0). Relative ratios of H2B or H4 content in cell lysates at the distinct timepoints to t0 were as follows: 1.3 and 1.8 (30 min), 2.7 and 3.9 (1 h), 11.7 and 15.4 (2 h), respectively (FIG. 1B and D). The increase of the H2A or H3 content in the cell lysates was much weaker and reached relative contents after 2 h compared to t0 as follows: 7.0 (H2A) and 6.1 (H3) (FIG. 1A and C). Interestingly, histone H1 content in cell lysates did not change significantly after induced apoptosis even two hours after UV light exposure. In contrast to the core histone data, the percentage of early apoptotic cells (AxV positive, PI negative) did not alter during the first hour (~2%), but

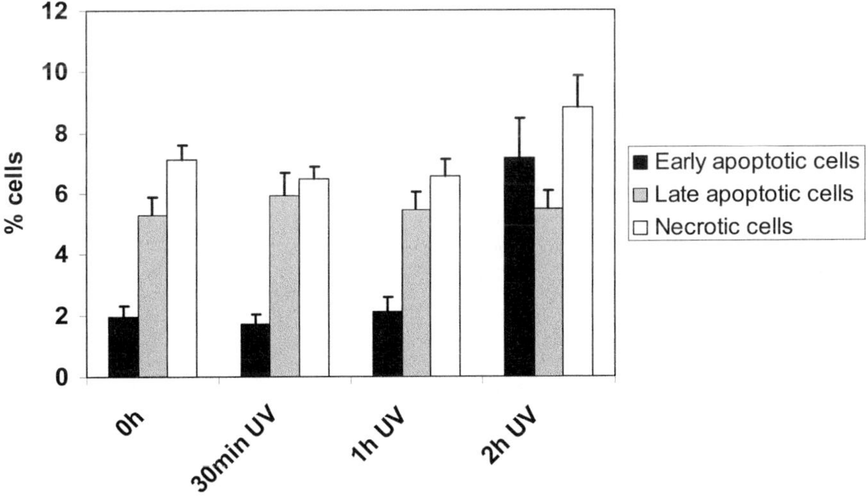

FIGURE 2.

increased after 2 h to about 7% (FIG. 2), whereas the percentage of late apoptotic cells (about 6%) or necrotic cells (about 7%) did not differ during the first 2 hours after UV-induced apoptosis.

DISCUSSION

The present study showed that extranuclear accumulation of histones in cell lysates occurs very early in the process of apoptosis. Our results revealed that the accumulation of core histones in cell lysates was detected as early as 30 min after UV irradiation, whereas phosphatidylserine externalization occurred 2 h after apoptosis induction. These results are supported by the findings that cell membranes which bind annexin V contained fragmented DNA.[4] Because H2A, H2B, H3, and H4 form the core structure of the nucleosome and are bound to DNA, they are normally insoluble in non-ionic detergent containing buffers like RIPA buffer. Consequently, their detection in these cell lysates indicates that they have been released or separated from chromatin and that the detected histone bands are likely to represent cytoplasmic histones. We showed earlier that in human lymphoblasts after IL-2 withdrawal the amount of cells with positive staining for annexin V correlated with the quantity of core histones in cell lysates.[3] In Jurkat cells, it has been shown that the appearance of the histones H2A, H2B, H3 and H4 occurred in cell lysates also very early after apoptosis induction with staurosporin, a very potent apoptosis inducer.[5] However, not all cell lines released histones from nucleosomes during DNA fragmentation and apoptosis: based on their experiments Wu and colleagues[5] showed convincingly that histone release and DNA fragmentation can be clearly separated.

In conclusion, our results suggest that accumulation of core histones in the cytoplasm is a very early event in apoptosis, preceding the externalization of phagocytosis signals on the outer membrane surface of apoptotically dying lymphoblasts. We therefore speculate that a dysregulation of apoptosis during these early phases may contribute to the induction of autoimmunity against nuclear autoantigens as seen in SLE.

REFERENCES

1. AMOURA, Z., H. CHABRE, S. KOUTOUZOV, *et al.* 1994. Nucleosome-restricted antibodies are detected before anti-dsDNA and/or antihistone antibodies in serum of MRL-Mp lpr/lpr and +/+ mice, and are present in kidney eluates of lupus mice with proteinuria. Arthritis Rheum. **37:** 1684–1688.
2. BELL, D.A, B. MORRISON & P. VANDENBYGAART. 1990. Immunogenic DNA-related factors. Nucleosomes spontaneously released from normal murine lymphoid cells stimulate proliferation and immunoglobulin synthesis of normal mouse lymphocytes. J. Clin. Invest. **85:** 1487–1496.
3. GABLER, C., M. SCHILLER, N. BLANK, *et al.* 2002. Contents of core histones in the cytoplasm of human lymphoblasts correlates with features of early apoptosis. Clin. Exp. Rheumatol. **20:** 262.
4. KOOPMAN, G., C.P. REUTELINGSPERGER, G.A. KUIJTEN, *et al.* 1994. Annexin V for flow cytometric detection of phosphatidylserine expression on B cells undergoing apoptosis. Blood **84:** 1415–1420.
5. WU, D., A. INGRAM, J.H. LAHTI, *et al.* 2002. Apoptotic release of histones from nucleosomes. J. Biol. Chem. **277:** 12001–12008.

Early Activation and Induction of Apoptosis in T Cells Is Independent of c-Fos

HANNA BIERBAUM,[a] SVEN BAUMANN,[b] INGRID HERR,[c] JAN P. TUCKERMANN,[d] JOCHEN HESS,[a] MARINA SCHORPP-KISTNER,[a] AND PETER ANGEL[a]

[a]*Division of Signal Transduction and Growth Control,* [b]*Tumor Immunology Program,* [c]*Division of Pediatric Oncology,* [d]*Division of Molecular Biology of the Cell I, Deutsches Krebsforschungszentrum Heidelberg, Im Neuenheimer Feld 280, D-69120 Heidelberg, Germany*

ABSTRACT: We used c-Fos–deficient activated T cells from the spleen and c-Fos–deficient thymocytes to address the capacity of these cells to undergo apoptosis in response to various stimuli. To determine the role of c-Fos in apoptosis regulation in thymocytes, we challenged thymocytes from wild-type and c-Fos–deficient mice with either TPA or the glucocorticoid dexamethasone. After various time points cells were stained according to the Nicoletti method and analyzed by FACS. Thymocytes from both genotypes exhibited similar efficiency of apoptosis in response to treatment with TPA or dexamethasone. Our data provide clear evidence that c-Fos is not required for apoptosis regulation in activated T cells as well as in thymocytes.

KEYWORDS: c-Fos; apoptosis; early activation; induction; T cells

The dimeric transcription factor AP-1, consisting of different members of the Fos (c-Fos, FosB, Fra-1, and Fra-2), Jun (c-Jun, JunB, and JunD), and ATF (ATF-2, ATF-3/LRF1, ATFa, and B-ATF) protein families, mediates gene regulation in response to many extracellular stimuli and thereby regulates various cellular responses, such as proliferation, differentiation, cell survival, and apoptosis.[1,2] Apoptosis plays an important role in the development and function of T cells as it is required for negative and positive selection of thymocytes and elimination of target cells as well as lymphocytes at the decline of the immune response.[3] In T cells the most important apoptosis trigger is the CD95 pathway, which is initiated by binding of the CD95 ligand (CD95-L) to its receptor CD95. In the lymphatic system, CD95 is poorly expressed on resting, naive T cells, but is rapidly upregulated during T cell activation, leading to activation-induced cell death (AICD).[3] Several reports have shown that the CD95-L is an AP-1 target gene.[4,5] Evidence for a specific function of AP-1 in apoptosis regulation was provided by the analysis of fibroblasts lacking individual AP-1 members.[2] While c-Jun was identified to act as a proapoptotic regulator in cells exposed to DNA damage via induction of CD95-L, c-Fos–deficient fibroblast

Address for correspondence: Marina Schorpp-Kistner, Division of Signal Transduction and Growth Control, Deutsches Krebsforschungszentrum Heidelberg, Im Neuenheimer Feld 280, D-69120 Heidelberg, Germany. Voice: 49-6221-42-4575; fax: 49-6221-42-4554. marina.schorpp@dkfz.de

Ann. N.Y. Acad. Sci. 1010: 225–231 (2003). © 2003 New York Academy of Sciences.
doi: 10.1196/annals.1299.040

showed enhanced apoptosis in response to UV-irradiation. In contrast, light-induced apoptosis in cells of the retina is impaired in the absence of c-Fos, suggesting that c-Fos exhibits both pro- and antiapoptotic functions, depending on the cell type and nature of the inducer.[1,2] In our hands c-Fos–deficient mice showed frequent inflammations, e.g., of the eyes, compared to age- and sex-matched wild-type mice, which may be explained by a defect in the regulation of T-cell apoptosis and/or an impaired immune system. Thus, we used c-Fos–deficient activated T cells from the spleen and c-Fos–deficient thymocytes to address the capacity of these cells to undergo apoptosis in response to various stimuli.

MATERIALS AND METHODS

T Cell Isolation

Thymus and spleen were prepared from 6-week-old mice and chopped into small pieces, processed through a 30 μm nylon net and the obtained cells were washed twice with medium.

Thymocytes were directly cultivated in a concentration of $1–2 \times 10^6$ cells/mL, while T cells from spleen were isolated by incubation for 1 h at 37°C to sediment the macrophages. Thereafter, the supernatant containing the T and B cells was transferred onto an IgM-coated tissue culture flask for 1 h to remove the B cells. The isolated T cells remaining in suspension were subsequently cultivated at a concentration of $1–2 \times 10^6$ cells/mL.

T Cell Activation

Freshly isolated splenic T cells (1×10^6 cells/mL) from wild-type and c-Fos–deficient mice were activated with either 10 ng/mL TPA and 1 μg/mL ionomycin, 5 μg/mL ConA, and 50 U/mL IL-2 or by culturing on αCD3/αCD28-coated plates. To determine T cell activation potential, 1×10^5 cells were stained for the surface activation markers CD25 and CD69 (BD Pharmingen) at various time points. Stained cells were analyzed by flow cytometry.

Induction of Apoptosis

Freshly isolated thymocytes or splenic T cells (1×10^6 cells/mL) from wild-type and c-Fos–deficient mice were plated in doublets and apoptosis was induced by different stimuli as follows: 10 ng/mL TPA, 1 μg/mL ionomycin, 1×10^{-6} M dexamethasone and recombinant CD95-L as described previously.[5] To measure DNA content (apoptotic nuclei), cells were harvested at indicated time points, washed with phosphate-buffered saline, and lysed according to the Nicoletti method in a hypotonic buffer containing (per mL) 0.1% sodium citrate, 0.1% Triton X-100, and 50 mg of propidium iodide. The fluorescence intensity of propidium iodide–stained nuclei was determined by flow cytometry (FACScan; Becton Dickinson, Heidelberg, Germany) with Cell Quest software. Segmented apoptotic nuclei were recognized by subdiploid DNA content. Early apoptotic changes were identified by staining of cells with fluorescein thiocyanate-conjugated annexin V (Becton Dickinson, Heidelberg, Germany) and analysis by flow cytometry.

RESULTS AND DISCUSSION

Primary T cell activation results in upregulation of CD95 mRNA and protein followed by the gradual acquisition of sensitivity to different apoptosis inducers.[3] This process also known as activation-induced cell death (AICD) can be mimicked *in vitro* by treatment of primary activated T cells with apoptosis-inducing stimuli.

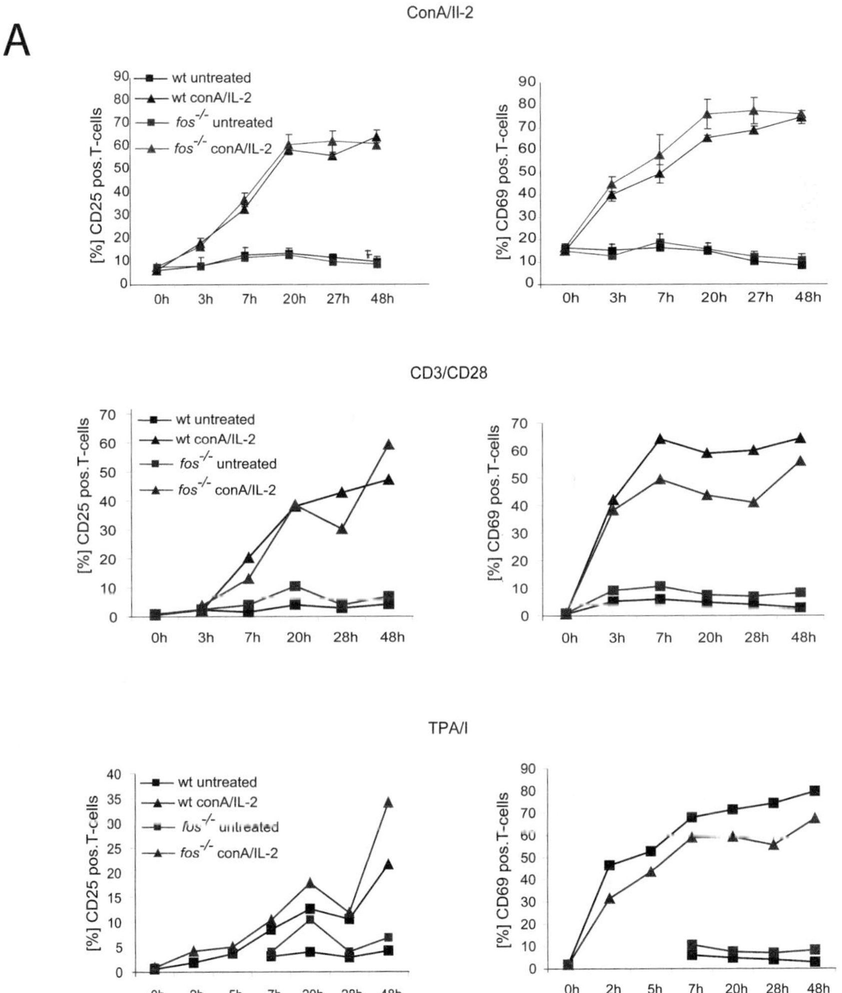

FIGURE 1. Normal activation and apoptosis response in splenic T cells lacking c-Fos. (**A**) Isolated splenic T cells from wild-type or c-Fos–deficient mice were activated with ConA/IL-2, αCD3/αCD28, or TPA/ionomycin. At the indicated time points cells were incubated with antibodies specific for CD25 (*left*) or CD69 (*right*) and analyzed by FACS.

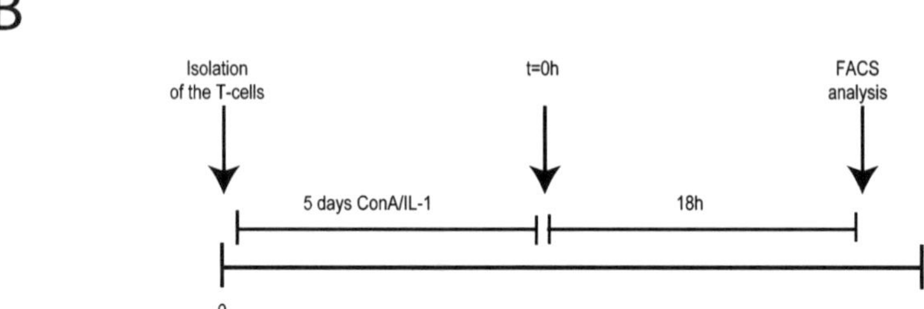

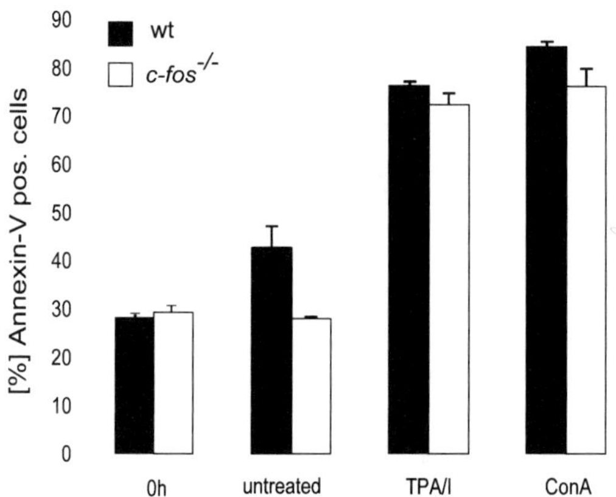

FIGURE 1. (**B**) Five days post activation of the splenic T-cells (t = 0), apoptosis was induced with TPA/I and ConA/IL2. At the indicated time points cells were stained with annexinV and measured by FACS.

T-Cell Activation in c-Fos–Deficient T Cells from the Spleen

Although it has been shown that c-Fos is required at late stages of T cell activation,[6] early T cell activation has not been analyzed by functional means employing c-Fos–deficient mice. To ensure a functional CD95 pathway, we analyzed the activation of wild-type and c-Fos–deficient splenic T cells upon activation with Concanavalin A (ConA) and IL-2 by FACS analysis for the early activation markers CD25 and CD69.

In agreement with previous studies, we could not find any differences in the amount and ratio of splenic T cells and B cells.[6] Moreover, early activation of c-Fos–deficient T cells was comparable to that of wild-type T cells (FIG. 1). To confirm these findings for other activation signals, we stimulated T cells by the physiological activation stimuli anti-CD3 and anti-CD28 or by raising the intracellular level of Ca^{2+} using TPA and ionomycin. Both treatments resulted in a similar efficiency of early activation of wild-type and c-Fos–deficient T cells (FIG. 1). Thus, early T cell activation appears Fos independent, although c-Fos is part of the AP-1 and NFAT transcription factor complex responsible for IL-2 induction.[6] Our data on the early differentiation markers CD25 and CD69 are in line with previous observations,[6] suggesting that c-Fos is not required for both early and late stages of T cell activation or that lack of c-Fos can be compensated by other Fos family members.

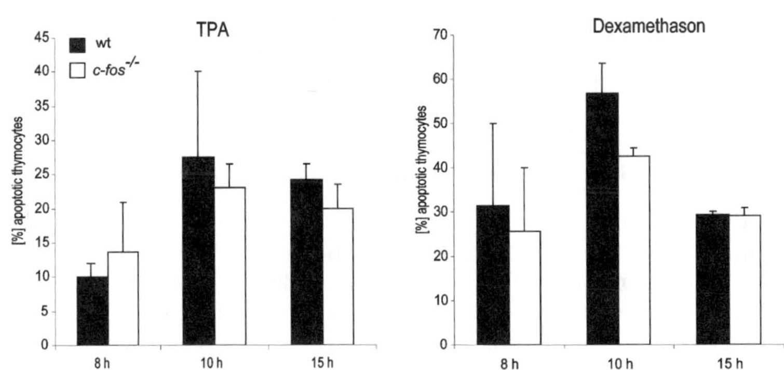

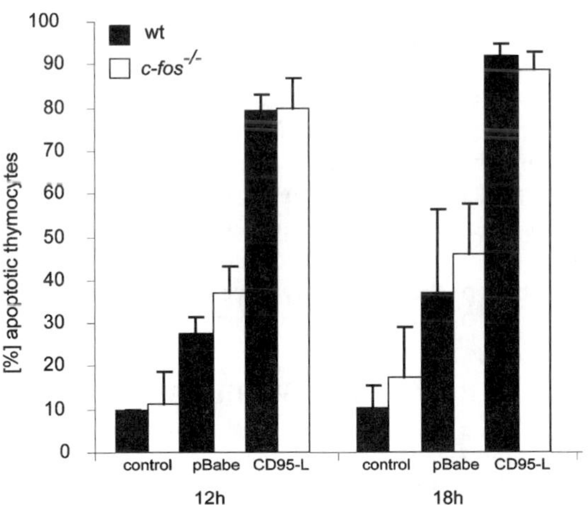

FIGURE 2. Apoptosis in thymocytes is not affected by lack of c-Fos. Apoptosis in thymocytes from wild-type and c-Fos–deficient mice was induced with (**A**) TPA or the glucocorticoid dexamethasone, or (**B**) recombinant CD95-L. At the indicated time points cells were stained by the Nicoletti method and analyzed by FACS.

Apoptosis in c-Fos–Deficient Activated T Cells

As T-cell activation is independent of c-Fos, we next used the ConA/IL-2 activated T cells and reactivated them five days post primary activation with ConA or TPA/ionomycin to check for alterations in the apoptotic response. The apoptosis rate was measured by annexin-V and propidium iodide (PI) staining and subsequent FACS analysis. No difference could be observed in spontaneous apoptosis of T cells and in apoptosis induction after either treatment among wild-type and *c-fos*$^{-/-}$ T cells (data not shown and FIG. 1B).

Although a series of cell types have been found where c-Fos exhibits either pro- or anti-apoptotic functions in non-activated T cells, our approach strongly suggests that the AICD, which is considered as one of the most physiological relevant apoptosis processes, is c-Fos independent.

Apoptosis Response in c-Fos–Deficient Thymocytes

During the maturation of T cells in the thymus, apoptosis is required for positive and negative selection. There are hints that c-Fos could be implicated in the control of this programmed cell death,[2] yet functional data on this issue are still missing. To determine the role of c-Fos in apoptosis regulation in thymocytes, we challenged thymocytes from wild-type and c-Fos–deficient mice with either TPA or the glucocorticoid dexamethasone. After various time points cells were stained according to the Nicoletti method (or annexin-V staining; data not shown) and analyzed by FACS. Thymocytes from both genotypes exhibited similar efficiency of apoptosis in response to treatment with TPA or dexamethasone (FIG. 2A). To confirm these results for a physiological stimulus, we measured apoptosis in wild-type and c-Fos–deficient thymocytes upon treatment of cells with CD95-L, which is involved in the *in vivo* apoptosis induction of thymocytes.[3] In agreement with our results from TPA- and dexamethasone-treated thymocytes, CD95-L stimulated apoptosis in wild-type and c-Fos–deficient thymocytes with similar efficiency (FIG. 2B). These data show that apoptosis induction in thymocytes in response to the physiological stimulus CD95-L and to experimental inducers, such as TPA (AP-1 inducing stimuli) and glucocorticoids (AP-1 inhibiting stimuli) is independent of c-Fos.

In summary, our data provide clear evidence that c-Fos is not required for apoptosis regulation in activated T cells as well in thymocytes. However, at this point we cannot exclude that c-Fos function can be compensated for by other Fos family members still present in *c-fos*$^{-/-}$ mice. To address this question, double or triple knock-out mice lacking multiple members of the Fos family have to be analyzed in the future.

ACKNOWLEDGMENT

This work was supported by the German–Israeli Cooperation in Cancer Research.

REFERENCES

1. SCHORPP-KISTNER, M., P. HERRLICH & P. ANGEL. 2001. The AP-1 family of transcription factors: structure, regulation and functional analysis in mice. *In* Targets for Can-

cer Chemotherapy. N.B. Cathangue and L.R. Bandara, Eds.: 29–52. Humana Press, Inc. Totowa, NJ.
2. JOCHUM, W., E. PASSEGUE & E.F. WAGNER. 2001. AP-1 in mouse development and tumorgenesis. Oncogene **20:** 2401–2412.
3. BAUMANN, S., A. KRUEGER, S. KIRCHHOFF & P.H. KRAMMER. 2002. Regulation of T cell apoptosis during the immune response. Curr. Mol. Med. **2:** 257–272.
4. EICHHORST, S.T., S. MULLER, M. LI-WEBER *et al.* 2000. A novel AP-1 element in the CD95 ligand promoter is required for induction of apoptosis in hepatocellular carcinoma cells upon treatment with anticancer drugs. Mol. Cell. Biol. **20:** 7826–7837.
5. KOLBUS, A., I. HERR, M. SCHREIBER, *et al.* 2000. c-Jun-dependent CD95-L expression is a rate limiting step in the induction of apoptosis by alkylating agents. Mol. Cell. Biol. **20:** 575–582.
6. JAIN, J., E.A. NALEFSKI, P.G. MCCAFFREY, *et al.* 1994. Normal peripheral T-cell function in c-Fos-deficient mice. Mol. Cell. Biol. **14:** 1566–1574.

NF-κB Inhibition Restores Sensitivity to Fas-Mediated Apoptosis in Lymphoma Cell Lines

MARIA MELI,[a] NATALE D'ALESSANDRO,[a] MANLIO TOLOMEO,[b] LUCIANO RAUSA,[a] MONICA NOTARBARTOLO,[a] AND LUISA DUSONCHET[a]

[a]*Department of Pharmacological Sciences, Università di Palermo, 90127 Palermo, Italy*

[b]*Department of Haematology and AIDS Center, Università di Palermo, 90127 Palermo, Italy*

ABSTRACT: Failure to perform the Fas-related apoptosis pathway can account for tumor resistance both to chemotherapeutic agents and to immunological effectors. We studied the role of NK-κB in Fas-resistance, employing the Fas-sensitive human T-lymphoma HuT78 cell line and its Fas-resistant variants HuT78B1 and HuT78G9. All these cell lines expressed high levels of constitutively activated NF-κB. Pretreatment of cells with NF-κB inhibitors (PDTC, MG132, or SN50) strongly enhanced CH11-induced apoptosis in HuT78 and Hut78G9 cells, while only MG132 showed a similar potentiating effect in HuT78B1. The described synergism was significantly inhibited by pretreatment with the anti-Fas–blocking antibody ZB4 or with the pancapsase inhibitor Z-VAD-FMK, but not by capsase-8 or -9 inhibitors. Overall, these data suggest that NF-κB inhibition may restore the Fas-pathway in Fas-resistant NF-κB–overexpressing tumors.

KEYWORDS: apoptosis; NF-κB; MG132; Fas/FasL system

INTRODUCTION

Apoptosis is the major process involved in cell death induced by both anticancer drugs and immune effectors, and its impairment may be considered a major point for cancer progression and poor response to therapies. The role of the cell death receptors in drug-induced apoptosis has been debated: it has been reported that chemotherapy-induced apoptosis may proceed via activation of Fas, and cross-resistance between drugs and cell death receptor-ligands has also been described.[1] Notably, human tumors often lose the ability to activate Fas-mediated apoptosis, but the mechanisms involved are still poorly understood. NF-κB is a transcription factor activated in cells after noxious and stressing events; it promotes cell survival by inducing the expression of various factors which inhibit apoptosis. However, it has been also re-

Address for correspondence; Dr. Maria Meli, Department of Pharmacological Sciences, University of Palermo, via del Vespro 129, 90127 Palermo, Italy. Voice: +39 091 6553261; fax: +39 091 6553249.
mmeli@unipa.it

Ann. N.Y. Acad. Sci. 1010: 232–236 (2003). © 2003 New York Academy of Sciences.
doi: 10.1196/annals.1299.041

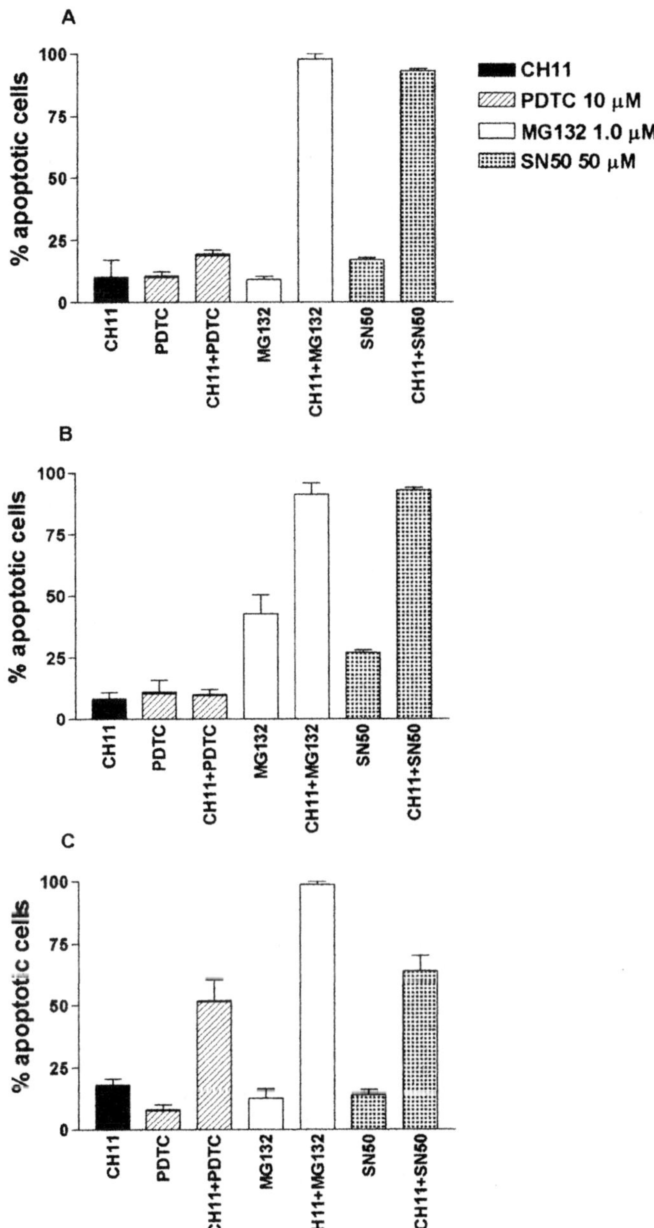

FIGURE 1. Influence of different NF-κB inhibitors on CH11-induced apoptosis in HuT78 (**A**), HuT78B1 (**B**), and HuT78G9 (**C**) cells. After 1-h exposure to the inhibitors at the concentrations indicated, the cells were treated for 24 h with CH11 0.05 μg/mL (HuT78) or 0.5 μg/mL (HuT78B1 and HuT78G9). Data are the mean±SE of at least three independent experiments.

ported that NF-κB can promote apoptosis, probably by upregulating FasL expression.[2] In this work we used a T-cell lymphoma cell line (HuT78) with high levels of constitutively activated NF-κB.[3] These cells, however, express surface Fas receptor and undergo apoptosis when challenged with an agonistic anti-Fas antibody (CH11) or with different anticancer drugs, indicating that their apoptotic machinery is preserved.[4] We asked whether NF-κB inhibition might affect CH11-sensitivity in the parental HuT78 cell line and in two variants (HuT78B1 and HUT78G9) resistant to Fas-mediated apoptosis and endowed with different degree of sensitivity to cytotoxic drug–induced cell death.

MATERIALS AND METHODS

The HuT78B1 clone was generated by *in vitro* exposure of HuT78 to the Fas-agonistic mAb CH11 and was completely resistant to CH11-induced apoptosis owing to the expression of both the wild-type and a truncated mutated form of the Fas receptor.[5] The G9 clone was similarly, but separately, selected and exhibited, in addition to CH11 resistance, a significant resistance to apoptosis by different stimuli including several anticancer drugs. All the cell lines did not express p53 protein or P-glycoprotein. Apoptosis was detected by fluorescence microscopy or by flow cytometry. Fas expression was evaluated by flow cytometry after incubation of cells with a FITC-conjugated anti-Fas DX2 mAb. The level of activated NF-kB was measured on whole cells or nuclear extracts employing the ELISA-based TransAM™ NF-κB kit (Active Motif, Rixensart, Belgium) and a specific anti-p50 mAb.

RESULTS

Initial experiments confirmed that NF-κB was constitutively activated in HuT78 cells. High levels of activation were also found in the CH11-resistant HuT78B1 and HuT78G9. To evaluate the role of NF-κB in sensitivity to CH11-induced apoptosis, we treated the cells with CH11 in combination with three different NF-κB inhibitors, namely the antioxidant PDTC, the proteasome inhibitor MG132, and the inhibitor of NF-κB nuclear translocation SN50 (Fig. 1). Overall, MG132 and SN50 strongly enhanced apoptosis induced by subcytotoxic concentrations of CH11 in all the cell lines, whereas PDTC showed a minor effect only in HuT78G9 cells. On the contrary, minimal effects were noted in other CH11-sensitive T-cell tumor cell lines (CCRF-CEM and Jurkat) that were shown to express significantly lower levels of activated NF-κB. This suggested that the overexpression of the transcription factor was required for the observed synergism. Furthermore, to confirm the role of NF-κB, we measured the influence of the inhibitors on nuclear NF-κB activation. Preliminary results suggest that MG132 employed at 1 μM is able to suppress by 40–60% the levels of activated NF-κB both in sensitive and resistant cells. On the contrary, this agent did not modify Fas expression. To identify the apoptotic pathway involved in the described synergism, different inhibitors of the apoptotic pathway were employed. In preliminary experiments, we studied the effects of the blocking anti-Fas antibody ZB4 or several caspase inhibitors on apoptosis induced by fully cytotoxic concentrations of CH11 or MG132 in HuT78 cells. Results showed that CH11- or

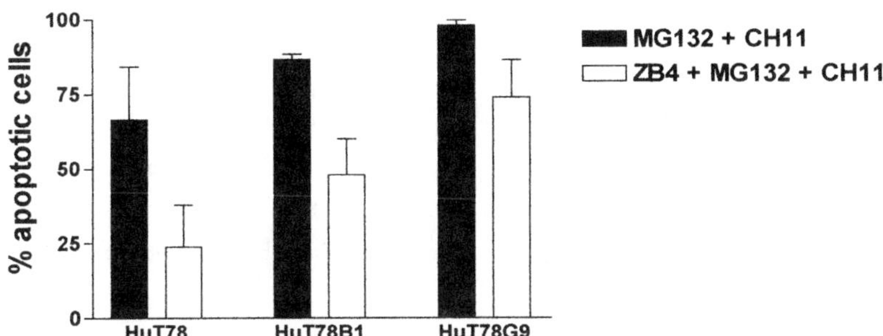

FIGURE 2. Apoptosis induced by the combination of MG132 and CH11 in the presence of the Fas-blocking antibody ZB4. After 30-min exposure to ZB4 (1.0 μg/mL), cells were treated with MG132 0.1 μM (HuT78 cells) or 1.0 μM (HuT78B1 and HuT78G9) and CH11, 0.05 μg/mL or 0.5 μg/mL, respectively, for 24 hours. Data are the mean±SE of at least three independent experiments.

MG-132-dependent apoptosis involved the activation of both caspase 8- and caspase 9-dependent pathways; however, MG132-induced apoptosis was not inhibited by ZB4, indicating that its effects were not mediated by Fas-ligation. On the contrary, when the cells were exposed to the combination of CH11 and MG132, ZB4 was able to decrease apoptosis in HuT78; it exhibited also a lower, yet significant, effect in the resistant HuT78B1 and HuT78G9 cells, suggesting that the combination had restored the Fas-triggered pathway of apoptosis (FIG 2). Surprisingly, only the pan-caspase inhibitor Z-VAD-FMK, but not the specific caspase 8- (Z-IETD-FMK) or caspase 9- (Z-LEHD-FMK) inhibitors, abrogated the synergism between CH11 and MG132, indicating that the combinations probably activate caspase 3 by a particular unidentified pattern.

CONCLUSIONS

In conclusion, our data support the possible involvement of NF-κB in the regulation of the Fas-pathway as reported also by others. Interestingly, NF-κB inhibitors and in particular MG132 were capable of restoring Fas-sensitivity in Fas-resistant cells, suggesting their possible use in the clinic, either in combination with other cancer therapies, or alone, and to potentiate the antitumor immune response. Other studies are needed to elucidate whether these results may be broadened to other NF-κB overexpressing tumors and which molecular mechanisms are actually involved in the effects of MG132 described herein.

REFERENCES

1. FRIESEN, C., I. HERR, P.H. KRAMMER & K.-M. DEBATIN. 1996. Involvement of the CD95 (APO-1/Fas) receptor/ligand system in drug-induced apoptosis in leukaemia cells. Nature Med. **2:** 574–577.

2. Kasibhatla, S., T. Brunner, L. Genestier, *et al.* 1998. DNA damaging agents induce expression of Fas ligand and subsequent apoptosis via the activation of NF-kappa B and AP-1. Mol. Cell **1:** 543–551.
3. Giri, D.K. & B.B. Aggarwal. 1998. Constitutive activation of NF-κB causes resistance to apoptosis in human cutaneous T cell lymphoma HuT-78 cells. J. Biol. Chem. **273:** 14008–14014.
4. Tolomeo, M., L. Dusonchet, M. Meli, *et al.* 1998. The CD95/CD95 ligand system is not the major effector in anticancer drug-mediated apoptosis. Cell Death Differ. **5:** 735–742.
5. Cascino, I., G. Papoff, R. De Maria, *et al.* 1996. Fas/Apo-1 (CD95) receptor lacking the intracytoplasmic signaling domain protects tumor cells from Fas-mediated apoptosis. J. Immunol. **156:** 13–17.

Convergence of the NF-κB and Interferon Signaling Pathways in the Regulation of Antiviral Defense and Apoptosis

JOHN HISCOTT, NATHALIE GRANDVAUX, SONIA SHARMA,
BENJAMIN R. TENOEVER, MARC J. SERVANT, AND RONGTUAN LIN

*Lady Davis Institute for Medical Research, Jewish General Hospital,
Departments of Microbiology and Immunology and Medicine, McGill University,
Montreal, Canada H3T 1E2*

ABSTRACT: The ubiquitously expressed interferon regulatory factor 3 (IRF-3) is directly activated following virus infection and functions as a key activator of the immediate-early Type 1 interferon (IFN) genes. Using DNA microarray analysis (8,556 genes) in Jurkat T cells inducibly expressing constitutively active IRF-3, several target genes directly regulated by IRF-3 were identified. Among the genes upregulated by IRF-3 were transcripts for a subset of known IFN-stimulated genes (ISGs), including ISG56, which functions as an inhibitor of translation initiation. Phosphorylation of C-terminal Ser/Thr residues—^{382}GGASSLENTVDLHISNSHPLSLTSDQY^{408}—is required for IRF-3 activation. Using C-terminal point mutations and a novel phosphospecific antibody, Ser396 was characterized as the minimal phosphoacceptor site required *in vivo* for IRF-3 activation following Sendai virus (SeV) infection, expression of viral nucleocapsid, or double-stranded RNA (dsRNA) treatment. The identity of the virus-activated kinase (VAK) activity that targets and activates IRF-3 and IRF-7 has remained a critical missing link in the understanding of interferon signaling. We report that the IKK-related kinases—IKKε/TBK-1—are components of VAK that mediate IRF-3 and IRF-7 phosphorylation and thus functionally link the NF-κB and IRF pathways in the development of the antiviral response.

KEYWORDS: IRF-3; interferon signaling; IKK-related kinases; antiviral responses

INTRODUCTION

The success of the host defense against viral infection is dependent on the ability of the cell to detect the presence of an invading pathogen. In response to recognition of viral and bacterial specific antigens, the cell activates a multitude of signal trans-

Address for correspondence: John Hiscott, Lady Davis Institute for Medical Research, 3755 Cote Ste. Catherine, Montreal, Quebec H3T 1E2. Voice: 514-340-8222, ext. 5265; fax: 514-340-7576.
john.hiscott@mcgill.ca

Ann. N.Y. Acad. Sci. 1010: 237–248 (2003).
doi: 10.1196/annals.1299.042

duction cascades to produce protein messengers in the form of cytokines that impede pathogen replication and spread through innate and adaptive immune mechanisms.[1–3] One of the best-characterized components of the innate host defense against virus is the family of transcriptionally activated interferon (IFN) proteins, which act in a paracrine fashion to induce gene expression in cells of the adjacent microenvironment, through engagement of cell surface IFN receptors and activation of the JAK-STAT signaling pathway.[4]

Rapid induction of type I IFNs following virus infection requires posttranslational modification of latent transcription factors involved in immunomodulation, including NF-κB, ATF-2/c-Jun, and IRF-3.[5–7] IRF-3 is a 55 kDa protein expressed constitutively in a variety of tissues and maintained in a latent conformation in the cytoplasm. Virus-induced C-terminal phosphorylation of IRF-3 represents an important post-translational modification, leading to dimerization, cytoplasmic-to-nuclear translocation, association with CBP/p300 coactivators, and stimulation of DNA binding and transcriptional activities through binding to interferon-stimulated response element (ISRE) sites.[8] Substitution of the serine and threonine residues in the aa396–aa405 region of IRF-3 with the phosphomimetic aspartic acid creates a constitutively active form of IRF-3 (IRF-3 5D) that is able to stimulate transcription in the absence of virus infection and also induces apoptosis.[9–12] IRF-3 alone is not sufficient to induce expression of endogenous human Type I IFNs,[12,13] but cooperates with NF-κB and ATF-2/c-Jun to form a transcriptionally active enhanceosome complex on the IFN-β promoter.[5,14,15]

Like IRF-3, the closely related IRF-7 is activated by phosphorylation following virus infection and contributes to the amplification of the transcriptional response through a second wave of IFN gene expression, which includes other IFNα genes not induced during the early, IRF-3–dependent stage of infection.[16–18] Gene knockout studies targeting IRF-3 and IRF-9/p48 support the concept that the early induction of antiviral genes is further amplified through the production of IFN-induced transcription factors such as IRF-7 and ISGF3.[17,19,20] Activated genes that contribute to the generation of the anti-viral state include PKR, ISG56, Mx, and 2′5′ OAS.[2,4,21]

The present review provides a summary of three recent studies describing the activation and function of IRF-3. (1) A tetracycline-inducible system expressing a constitutively active form of IRF-3 together with DNA microarray analysis was used to profile genes that are directly responsive to IRF-3. (2) Several biochemical approaches were used to identify the minimal phosphoacceptor site required for IRF-3 phosphorylation. Of the seven potential phosphoacceptor sites present in the C-terminal cluster, the single-point mutation of Ser 396 to Asp (S396D) was sufficient to generate a constitutively active form of IRF-3. Moreover, a novel phosphospecific antibody was used to demonstrate that Ser 396 is phosphorylated *in vivo* following virus infection, nucleocapsid expression, or dsRNA treatment. (3) Identification of virus activated kinase (VAK) activity responsible for IRF-3 phosphorylation is critical to an understanding of the signaling pathways involved in linking the host response to pathogens with the establishment of the antiviral state. We demonstrate that the IKK-related kinases—IKKε/TBK1—are components of VAK and specifically phosphorylate the C-terminal residues of IRF-3 and IRF-7. These studies constitute the first biochemical evidence linking the NF-κB and IRF pathways in the response to pathogens, specifically through IKKε/TBK1 activity.

TABLE 1. IRF-3 5D responsive genes identified by microarray

Accession Number	Fold Induction	Gene Name
Known interferon induced proteins		
M14660	+18.3	ISG54/IFI54
X03557	+8.4	ISG56/IFI56
AF026939	+3.4	Interferon-induced protein with TPR 4/IFI60/ISG60
D28915	+1.9	Hepatitis C associated microtubular aggregate protein 44kDa
M55542	+1.7	Guanylate binding protein 1 (GBP1)
M87284	+1.5	2′-5′ oligoadenylate synthase 2 (2′-5′OAS)
AI739106	+1.5	ISG15
Metabolism		
D86724	+3.6	Arginase type II
Phospholipases and related proteins		
X14034	+3.4	Phospholipase C-γ 2
M80899	+1.8	AHNAK nucleoprotein
Nuclear receptor		
X03225	+2	Alpha-glucocorticoid receptor
Post-translational modification		
AB004550	+1.8	UDP-Gal:betaGlcNAc beta 1,4-galactosyltransferase, polypeptide 5
Unknown function		
AF026941	+3.5	*CIG 5* /Viperin
D90070	+2.1	PMA-induced protein 1

NOTE: Genes identified by microarray analysis that are upregulated (+) or downregulated (–) in response to IRF-3 5D induction in Jurkat T cells. Genes that exhibit a >1.5-fold increase in expression level are listed.

Characterization of IRF-3 Responsive Genes by DNA Microarray

To identify novel genes whose expression is under the control of IRF-3, a Jurkat cell line inducibly expressing the constitutively active IRF-3 5D in response to doxycycline stimulation was used together with DNA microarray analysis.[21] Only eight clones of the 9,182 elements monitored had an increased expression of greater than twofold (TABLE 1). To confirm the induced gene expression, poly(A)-mRNAs isolated from rtTA-Neo- and rtTA-IRF-3 5D-Jurkat were analyzed by Northern blot. Three of the eight upregulated genes, namely ISG56, PLC-γ2, and arginase II, were strongly induced in rtTA-IRF-3 5D- versus rtTA-Neo-Jurkat. Moreover, ISG54 has already been reported to be induced in response to IRF-3 5D in IFN-unresponsive cells.[22]

ISG56 regulation by IRF-3 was further examined biochemically by immunoblot analysis. A strong induction of ISG56 was observed in rtTA-IRF-3 5D-Jurkat beginning at 12 h and was sustained throughout doxycycline (DOX) treatment, with the same kinetics profile as IRF-3 5D expression. The expression of ISG56 in response

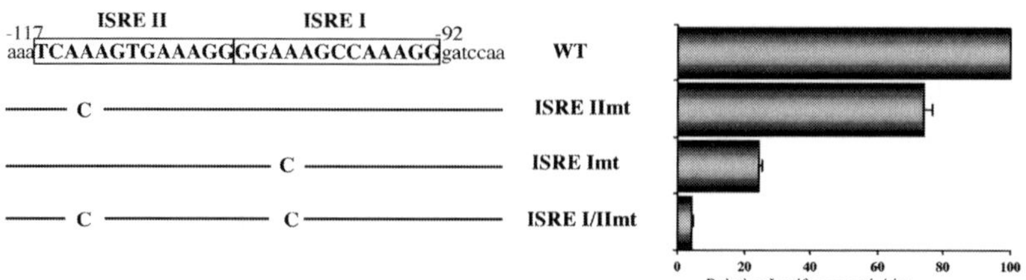

FIGURE 1. Involvement of ISRE sites in IRF-3 5D-dependent induction of the ISG56 promoter. A schematic representation of the ISG56 promoter mutants generated by overlap PCR is provided. HEC1B cells were co-transfected with the wild type (wt) or mutated 561-luciferase reporter constructs and 200 ng of IRF-3 5D expression plasmid. Relative luciferase activities were measured at 24 h post-transfection. After normalization for the basal level in the presence of the reporter construct alone and Renilla, luciferase activities were expressed as percentages of the activation observed with the wt-561-reporter. Each value represents the mean ± SE of triplicate independent samples. The data are representative of at least two different experiments with similar results.

to virus infection was further examined in HEC1B cells that do not respond to treatment with type 1 IFN.[23] ISG56 protein expression was induced in HEC1B cells either by ectopic IRF-3 5D expression or by Sendai virus stimulation. Furthermore, virus induction was inhibited in a concentration-dependent manner by expression of a dominant negative form of IRF-3, IRF-3ΔN, which lacks the N-terminal DNA binding domain.[11] These results demonstrate that IRF-3 is sufficient to induce ISG56 in response to virus infection in an IFN-independent pathway.[21]

IRF-3 co-expression directly stimulated the *ISG56* promoter[24] in HEC1B cells. Sendai virus infection of HEC1B resulted in a 17-fold induction of promoter activity and this induction was increased in a dose-dependent manner by co-transfection of an IRF-3–expressing plasmid (FIG. 1). Two ISRE consensus sites are present in the promoter of *ISG56*, at positions -92 to -104 (ISRE I) and -105 to -117 (ISRE II).[24] The ISRE I and ISRE II sites were mutated either independently (ISRE Imt and ISRE IImt constructs, respectively) or together (ISRE I/IImt) to differentiate the contribution of each site in IRF-3 activation. Analysis of the promoter activity demonstrated that mutation of the ISRE II site slightly decreased the activation of the promoter (26%), whereas mutation of the ISRE I site resulted in a stronger inhibition of the activation (76%). Mutation of both sites completely abrogated the induction of the promoter (FIG. 1). These results demonstrate that IRF-3–dependent activation of the *ISG56* promoter is mediated through both ISRE sites, with a major contribution by the ISRE I.[21]

These studies demonstrate that, *in vivo*, activation of IRF-3 was sufficient to activate only a subset of ISGs (namely ISG54, ISG56, and ISG60) that establish the antiviral state. Other interferon-inducible genes present in the microarray gene panel (including ISG15, OAS, PKR, and GBP) were not upregulated, indicating that IRF-3 is able to discriminate among ISRE-containing genes. These results are in accordance with the results of Taniguchi's group: using MEF cells with a targeted disruption of the IRF-3 gene, they showed that full level induction of ISG15 required not only

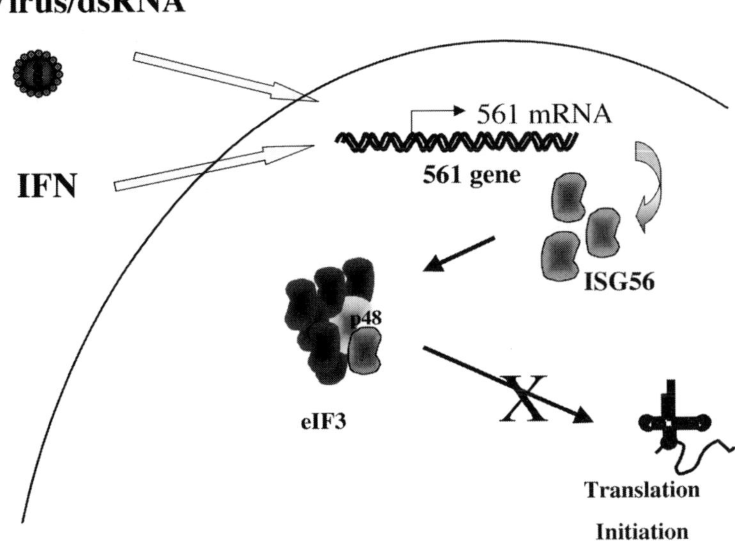

FIGURE 2. Schematic illustrating the role of ISG56 in the inhibition of translation initiation. Based on the studies of Guo and colleagues, the 561 gene is transcriptionally induced by virus infection, dsRNA and IFN treatment leading to ISG56 production. ISG56 in turn binds to the p48 subunit of the elongation initiation factor 3 (eIF3) and blocks protein translation initiation in IFN- and virus-infected cells.

IRF3, but ISGF3 as well. They also showed that IRF3 was involved in ISG54, but not OAS or PKR activation in response to Newcastle disease virus (NDV) infection.[19]

The specific functions of many IFN-induced proteins remain to be determined. Interestingly, ISG56 has recently been shown to downregulate protein synthesis through interaction with the p48 subunit of eIF-3 (FIG. 2),[25] and to act as a strong negative regulator of cell proliferation.[26] IRF-3 5D has previously been shown to induce apoptosis when overexpressed in cell lines,[12] although the exact mechanism mediating this process remains unknown. Thus, the translation inhibitory function of ISG56 (FIG. 2) may provide, at least in part, a mechanistic explanation for the ability of IRF-3 5D to mediate apoptosis. It will also be interesting to analyze the involvement of the other genes identified in this study as new IRF-3 target genes in the apoptosis process. These results demonstrate that IRF-3 is sufficient to induce a specific protein synthesis independent antiviral response. Indeed, IRF-3 acts not only by contributing to the induction of the immediate-early type I *IFN* genes, but also through direct activation of a subset of the antiviral interferon-stimulated genes.

Identification of the Minimal Phosphoacceptor Site Required for in Vivo *Activation of IRF-3*

The rate-limiting step in the activation of IRF-3 is its phosphorylation by at least one kinase activity, namely virus-activated kinase (VAK), stimulated following virus infection or dsRNA treatment. In order to delineate the minimal residues required for C-terminal IRF-3 activation, the effect of various point mutations on the transac-

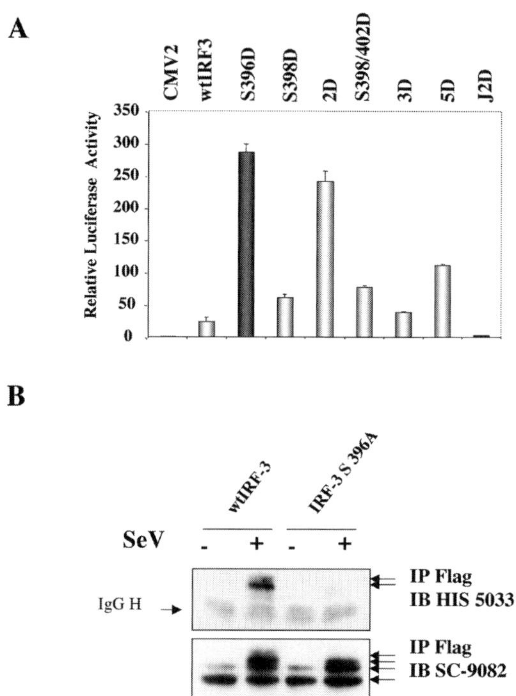

FIGURE 3. IRF-3 Ser396 phosphomimetic is a strong transactivator of the IFN-β promoter and is phosphorylated. (**A**) HEK 293 cells were transiently transfected with luciferase reporter constructs containing the IFN-β promoter and different IRF-3 point mutants as indicated above the bar graphs: CMV2, control vector; wtIRF3, wild type IRF3; S396D; S398D; 2D: S396/398D; S398/402D; 3D, S402D, T404D, S405D; 5D, S396D, S398D, S402D, T404D, S405D; J2D, S385D, S386D. At 36 h post-transfection, luciferase activity was measured. Relative luciferase activity was measured as fold activation over the transfection of pFlag-CMV2 alone. Each value represents the mean ± S.E. of triplicate determinations. The data are representative of at least four different experiments with similar results. (**B**) *In vivo* phosphorylation of Ser 396 of IRF-3 after virus infection. HEK 293 cells were transiently transfected with Flag-tagged constructs (wtIRF3 or IRF3 S396A) as indicated. At 36 h post-transfection, cells were left untreated or infected with Sev(40 HAU/106 cells) for 6 hours. WCE (200 μg) were immunoprecipitated with anti-Flag antibody linked to protein G-Sepharose beads. The immunoprecipitated proteins were resolved by 10% SDS-acrylamide gel electrophoresis, transferred to nitrocellulose and probed with HIS5033 antibody. Following stripping, the membrane was reprobed with an antibody that reacts against whole IRF-3 (Santa Cruz; SC-9082).

tivating potential, CBP association, and endogenous gene activation was analyzed. Of the seven potential phosphoacceptor sites present in the C-terminal cluster, the single-point mutation of Ser396 to Asp (S396D) was sufficient to generate a constitutively active form of IRF-3 that transactivated PRD I-III- and ISRE-containing target gene promoters in transient co-transfection assays (FIG. 3A), interacted with CBP co-activator, and induced the endogenous *ISG56* gene.[27]

To verify whether *in vivo* phosphorylation of IRF-3 occurred at Ser396, an antibody directed against a phosphopeptide spanning Ser396 (HIS5033) was raised.[27] Using this novel phosphospecific antibody, we showed for the first time that Ser 396 is phosphorylated *in vivo* following virus infection (FIG. 3B) or dsRNA treatment. HIS5033 antibody recognized only the slowly migrating form of IRF-3, previously shown to represent the activated form of IRF-3; furthermore the antibody did not recognize the Ser396Ala mutated IRF-3 (FIG. 3B). To study the kinetics of Ser 396 phosphorylation in response to virus, dsRNA or LPS, the LPS-sensitive human U373/CD14 astrocytoma cell line was used as previously described.[28] Detailed kinetics of SeV, LPS, and poly I:C induction was performed and the phosphorylation state of Ser396 was analyzed with the HIS5033 antibody. Both SeV and poly I:C induced Ser396 phosphorylation, whereas LPS treatment did not lead to Ser396 phosphorylation. Kinetic analysis of IRF-3 phosphorylation also revealed quantitative and qualitative differences in Ser396 phosphorylation by virus and dsRNA. To further analyze the LPS effect, IRF-3 Ser396 phosphorylation was also examined in the monocytic cell line U937 (FIG. 4). Infection of U937 cells with SeV resulted in IRF-

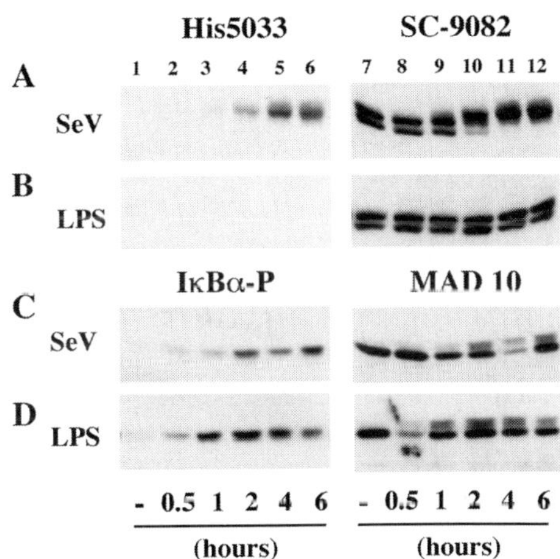

FIGURE 4. IRF-3 Ser 396 phosphorylation is detected *in vivo* after Sendai virus infection but not after LPS treatment. Whole cell extracts (45 μg) prepared from the U937 cells untreated (–) or infected with SeV (40 HAU/106 cells) (**A** and **C**) and LPS (10 μg/mL) (**B** and **D**) for the indicated times were resolved on SDS-gels. The phosphorylation of IRF-3 in U937 cells was measured with HIS 5033 antibody in immunoblot analysis (A and B, lanes 1–6) and with anti-IRF-3 antibody SC-9082 (**A** and **B**, lanes 7–12). Extracts were immunoblotted with a phosphospecific antibody against IkBa (IkBa-P) (**C** and **D**, lanes 1–6) and an antibody that recognized non-phosphorylated and phosphorylated IkBa (MAD 10) (**C** and **D**, lanes 7–12).

3 Ser396 phosphorylation beginning as early as 2h post-infection (FIG. 4A, lanes 3–6), whereas no signal with HIS5033 antibody was detected with cell extracts derived from LPS-stimulated cells (FIG. 4B). Both SeV and LPS were able to induce the phosphorylation of IκBα (FIG. 4C and D).

This study demonstrated that of the seven possible phosphoacceptor sites present in the C-terminal end of IRF-3, mutation of Ser396 to Asp was the minimal mutation required to generate a constitutively active form of IRF-3. Using a phosphospecific antibody raised against a peptide containing the phosphorylated Ser residue at position 396, we showed for the first time that this site is phosphorylated *in vivo* when cells were exposed to SeV, nucleocapsid, or poly I:C, well characterized inducers of IRF-3.[29] Our data suggest that LPS-induced activation of IRF-3 does not lead to phosphorylation of Ser 396. How LPS activates IRF-3 remains to be determined; treatment of target cells with LPS generates multiple signaling pathways in addition to IKK/JNK/p38, such as protein kinase C (PKC), Src-type tyrosine kinases, and the phosphatidylinositol 3-kinase-protein kinase B pathway.[1] Finally, the critical role of Ser 396 phosphorylation in IRF-3 activation following virus infection and the development of a specific and sensitive phosphospecific antibody against Ser396 may be useful as a research and diagnostic tool as a marker of virus infection.

IKKε/TBK1 Mediates an Antiviral Program through Activation of IRF-3 and IRF-7

Similarities in virus-inducible activation of IRF-3 and IRF-7 through C-terminal phosphorylation suggest that the same or similar kinase(s) may phosphorylate both transcription factors during the course of viral infection. To date, pharmacological and molecular studies, using well characterized kinase inhibitors or numerous dominant negative kinase mutants, have failed to identify VAK.[30,31] Interestingly however, a two-hybrid screen using IRF-3 as a bait revealed that the C-terminal end of the IKKα subunit interacted with IRF-3. This result prompted us to analyze the possibility that IKK subunits IKK α, β and the recently identify IKK-related kinases, IKKε/TBK-1 may phosphorylate the C-terminus of IRF-3.

IKK α, IKKβ, and IKKε were expressed in HEK293 cells and with these whole cell extracts, *in vitro* kinase assays using GST-IRF-3 (aa380-427) as substrate were performed. Phosphorylation of IRF-3 was observed only with extracts derived from IKKε-expressing cells. Furthermore, using *in vitro* transcribed and translated IKKα, IKKβ, and IKKε, *in vitro* kinase assays showed that only IKKε was able to directly phosphorylate the IRF-3 substrate.

To further substantiate the ability of IKKε to activate IRF-3 and IRF-7, several key events in their activation were analyzed in cells expressing IKKε. (1) Subcellular localization studies showed that co-expression of IKKε together with GFP-IRF-3 or -IRF-7 in Cos7 cells induced translocation of approximately 35% of IRF-3 and >95% of IRF-7 to the nucleus. This nuclear translocation was also observed when the related TBK-1 kinase was co-expressed. (2) EMSA experiments also showed that co-expression of IKKε induced both IRF-3 and IRF-7 DNA binding. (3) In reporter gene analysis, co-expression of IKKε together with IRF-7 resulted in a dramatic 2,000-fold stimulation of the IRF-7–responsive IFNα4 promoter. Likewise, the IRF-3–responsive promoters, IFNβ and RANTES, were also stimulated in response to IKKε co-expression, effects that were blocked by a dominant negative form of IRF-

3 (IRF-3ΔN). Similar results were obtained through co-expression of TBK-1, whereas IKKα, IKKβ, or NIK did not stimulate IRF-responsive promoters. Together these experiments demonstrate that expression of IKKε or TBK-1 is sufficient to induce phosphorylation, cytoplasmic-to-nuclear translocation, DNA binding, and transcriptional activity of IRF-3 and IRF-7.

To investigate the involvement of IKKε in the IFN response to virus infection, BJAB, a B-lymphoid cell line, was infected with vesicular stomatitis virus (VSV), a rhabdovirus that is highly sensitive to the effects of IFN, in the presence of wild-type or dominant negative IKKε. VSV infection induced both IFNβ and IFNα mRNA at 6 h and the level of endogenous IFNα and IFNβ expression was augmented ~fivefold with co-expression of IKKε. Interestingly, the utilization of BJAB cells expressing an IκB super-repressor which blocks NF-κB activity,[32] together with IKKε (K38A) completely ablated expression of IFNα and IFNβ mRNA, demonstrating the requirement for NF-κB and IRF cooperation.[33]

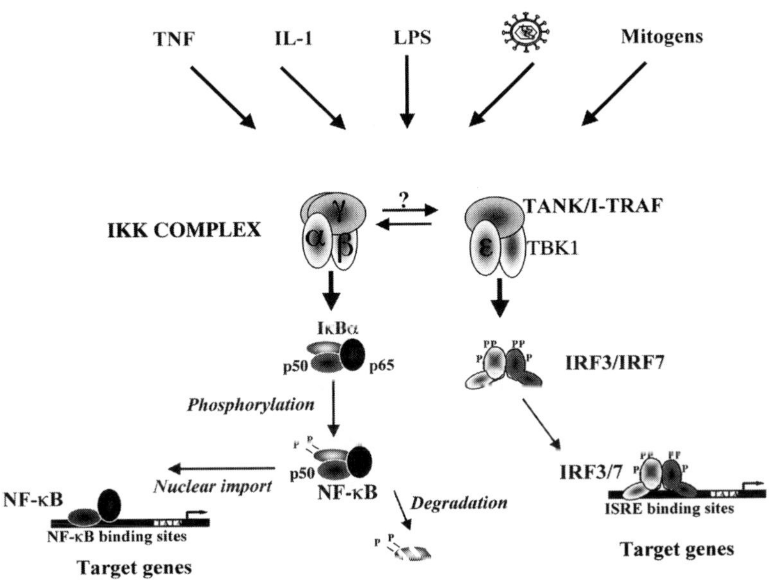

FIGURE 5. Schematic representation of IKK complexes and signaling to NF-κB and IRF3/7. Multiple inducers of the NF-κB pathway converge on the IKK complexes and phosphorylate downstream substrates including IkBa and IRF3/7. The clasical IKK complex is composed of IKKα, IKKβ and IKKγ while a novel, IKK-related complex is composed of IKKε, TBK1 and TANK/I-TRAF. Crosstalk between the kinase complexes is also suggested as a component of the complex regulation of NF-κB target genes that may permit modulation of aspects of NF-κB and IRF activation. Recent studies suggest that the IKK-related kinases IKKε and TBK1 exist as a multiprotein complex in association with the regulatory adapters TANK/I-TRAF and TRAF2.[40,41]

As discussed above and elsewhere, IRF-3 plays an important role in the establishment of the antiviral state. The antiviral activity of IKKε was studied through the analysis of virus multiplication in the presence of the kinase. The yield of infectious virus was decreased by 4 logs in cells overexpressing IKKε; the virus titer was ~10^6 pfu/mL in IKKε and IRF-3 expressing cells, whereas in cells expressing IKKε (K38A) and IRF-3, the 12 h virus titer was approximately 10^{10}. Similarly, if the early phase of IFN induction is blocked by expression of the IRF-3ΔN, the VSV titer increases to 10^{10} pfu/mL, further demonstrating that the antiviral activity of IKKε is IRF-3 dependent. Thus, engagement of the antiviral program through IKKε expression is sufficient to induce interferon-stimulated genes and inhibit VSV multiplication.

The identification of IKKε/TBK-1 as a component of VAK responsible for IRF-3 and IRF-7 phosphorylation in response to virus infection highlights several important points: (1) IRF-3 and IRF-7 represent direct physiological targets for IKKε/TBK-1; (2) the signaling pathways leading to NF-κB and IRF-3/IRF-7 activation in response to virus infection not only cooperate in cytokine activation but are directly linked through IKKε/TBK-1; and (3) establishment of the antiviral response through IRF phosphorylation illustrates a new previously unrecognized function for IKKe/TBK-1 (FIG. 5). Another implication of these results is that phosphorylation by IKKε and TBK-1 may be cell-type specific, since IKKε was originally identified as a cytokine- and PMA-inducible activity in lymphoid cells; the related TBK1/NAK kinase was shown to represent a constitutive form of this IKK-related activity.[34–37] TBK1/T2K–deficient mice have also been generated, but TBK1$^{-/-}$ mice died at day 14.5 of massive liver degeneration and apoptosis, a phenotype that is similar to the phenotype of RelA$^{-/-}$ and IκBα$^{-/-}$ mice.[38,39] Interestingly, in response to either TNFα or IL-1 induction, TBK1$^{-/-}$ fibroblasts exhibited normal phosphorylation and degradation of IκB, as well as NF-κB binding activity. However, NF-κB directed transcription was dramatically reduced, suggesting that TBK-1 may possess a unique role in direct phosphorylation of the transactivation domain of RelA.[38] Recent studies indicate that TANK (TRAF family member associated NF-κB activator) can synergize with IKKε/TBK to link these kinases to IKK complexes where the two kinases may modulate aspects of NF-κB activation (FIG. 6).[40,41] The present studies identify a role for IKKε/TBK1 in virus-mediated activation of the IRFs that is completely distinct from its role in the activation of NF-κB–dependent signaling pathway.[42]

ACKNOWLEDGMENTS

The authors wish to thank Dr. Ganes Sen (Lerner Research Institute) and John Bell (Ottawa General Research Institute) for reagents, and members of the Molecular Oncology Group, Lady Davis Institute, McGill University for helpful discussions. This research was supported by grants from the Canadian Institutes of Health Research and National Cancer Institute with funds from the Canadian Cancer Society, as well as the CANVAC Network Centres of Excellence. S.S. was supported by a CIHR studentship, B.R.t.O. by an NSERC Studentship, N.G. by FRSQ Post-doctoral Fellowships, R.L. by an FRSQ Chercheur Boursier, and J.H. by a CIHR Senior Scientist award.

REFERENCES

1. AKIRA, S., K. TAKEDA & T. KAISHO. 2001. Toll-like receptors: critical proteins linking innate and acquired immunity. Nat. Immunol **2:** 675–680.
2. SAMUEL, C.E. 2001. Antiviral actions of interferons. Clin. Microbiol. Rev. **14:** 778–809.
3. HUANG, Q., D. LIU, P. MAJEWSKI, *et al.* 2001. The plasticity of dendritic cell responses to pathogens and their components. Science **294:** 870–875.
4. SEN, G.C. 2001. Viruses and interferons. Annu. Rev. Microbiol. **55:** 255–281.
5. KIM, T.K. & T. MANIATIS. 1997. The mechanism of transcriptional synergy of an in vitro assembled interferon-beta enhanceosome. Mol. Cell. **1:** 119–129.
6. FALVO, J.V., B.S. PAREKH, C.H. LIN, *et al.* 2000. Assembly of a functional beta interferon enhanceosome is dependent on ATF-2-c-jun heterodimer orientation. Mol. Cell. Biol. **20:** 4814–4825.
7. GRANDVAUX, N., B.R. TENOEVER, M.J. SERVANT, *et al.* 2002. The interferon antiviral response: from viral invasion to evasion. Curr. Opin. Infect. Dis. **15:** 259–267.
8. SERVANT, M.J., B. TEN OEVER & R. LIN. 2002. Review: Overlapping and distinct mechanisms regulating IRF-3 and IRF-7 function. J. Interferon Cytokine Res. **22:** 49–58.
9. AZIMI, N., Y. TAGAYA, J. MARINER, *et al.* 2000. Viral activation of interleukin-15 (IL-15): characterization of a virus-inducible element in the IL-15 promoter region. J. Virol. **74:** 7338–7348.
10. LIN, R., C. HEYLBROECK, P. GENIN, *et al.* 1999. Essential role of interferon regulatory factor 3 in direct activation of RANTES chemokine transcription. Mol. Cell. Biol. **19:** 959–966.
11. LIN, R., C. HEYLBROECK, P.M. PITHA, *et al.* 1998. Virus dependent phosphorylation of the IRF-3 transcription factor regulates nuclear translocation, transactivation potential and proteasome mediated degradation. Mol. Cell. Biol. **18:** 2986–2996.
12. HEYLBROECK, C., S. BALACHANDRAN, M.J. SERVANT, *et al.* 2000. The IRF-3 transcription factor mediates Sendai virus-induced apoptosis. J. Virol. **74:** 3781–3792.
13. YEOW, W.S., W.C. AU, Y.T. JUANG, *et al.* 2000. Reconstitution of virus-mediated expression of interferon alpha genes in human fibroblast cells by ectopic interferon regulatory factor-7. J. Biol. Chem. **275:** 6313–6320.
14. MERIKA, M., A.J. WILLIAMS, G. CHEN, *et al.* 1998. Recruitment of CBP/p300 by the IFN· enhanceosome is required for synergistic activation of transcription. Molec. Cell **1:** 277–287.
15. WATHELET, M.G., C.H. LIN, B.S. PAREKH, *et al.* 1998. Virus infection induces the assembly of coordinately activated transcription factors on the IFN beta enhancer in vivo. Mol. Cell **1:** 507–518.
16. MARIE, I., J.E. DURBIN & D.E. LEVY. 1998. Differential viral induction of distinct interferon-alpha genes by positive feedback through interferon regulatory factor-7. EMBO J. **17:** 6660–6669.
17. SATO, M., H. SUEMORI, N. HATA, *et al.* 2000. Distinct and essential roles of transcription factors IRF-3 and IRF-7 in response to viruses for IFN-alpha/beta gene induction. Immunity **13:** 539–548.
18. LIN, R., P. GENIN, Y. MAMANE, *et al.* 2000. Selective DNA binding and association with the CREB binding protein coactivator contribute to differential activation of alpha/beta interferon genes by interferon regulatory factors 3 and 7. Mol. Cell. Biol. **20:** 6342–6353.
19. NAKAYA, T., M. SATO, N. HATA, *et al.* 2001. Gene induction pathways mediated by distinct irfs during viral infection. Biochem. Biophys. Res Commun. **283:** 1150–1156.
20. YEOW, W.S., W.C. AU, W.J. LOWTHER, *et al.* 2001. Downregulation of IRF-3 levels by ribozyme modulates the profile of IFNA subtypes expressed in infected human cells. J. Virol. **75:** 3021–3027.
21. GRANDVAUX, N., M.J. SERVANT, B. TENOEVER, *et al.* 2002. Transcriptional profiling of interferon regulatory factor 3 target genes: direct involvement in the regulation of interferon-stimulated genes. J. Virol. **76:** 5532–5539.

22. WEAVER, B.K., O. ANDO, K.P. KUMAR, *et al.* 2001. Apoptosis is promoted by the dsRNA-activated factor (DRAF1) during viral infection independent of the action of interferon or p53. FASEB J. **15:** 501–515.
23. FUSE, A., H. ASHINO-FUSE & T. KUWATA. 1984. Binding of 125I-labeled human interferon to cell lines with low sensitivity to interferon. Gann **75:** 379–384.
24. WATHELET, M.G., I.M. CLAUSS, C.B. NOLS, *et al.* 1987. New inducers revealed by the promoter sequence analysis of two interferon-activated human genes. Eur. J. Biochem. **169:** 313–321.
25. GUO, J., D.J. HUI, W.C. MERRICK, *et al.* 2000. A new pathway of translational regulation mediated by eukaryotic initiation factor 3. EMBO J. **19:** 6891–6899.
26. GEISS, G., G. JIN, J. GUO, *et al.* 2001. A comprehensive view of regulation of gene expression by double-stranded RNA-mediated cell signaling. J. Biol. Chem. **30:** 30.
27. SERVANT, M.J., N. GRANDVAUX, B.R. TENOEVER, *et al.* 2003. Identification of the minimal phosphoacceptor site required for in vivo activation of IRF-3 in response to virus and dsRNA. J. Biol. Chem. **10:** 10.
28. ORR, S.L. & P. TOBIAS. 2000. LPS and LAM activation of the U373 astrocytoma cell line: differential requirement for CD14. J. Endotoxin Res. **6:** 215–222.
29. TENOEVER, B.R., M.J. SERVANT, N. GRANDVAUX, *et al.* 2002. Recognition of the Measles viral nucleocapsid as a mechanism of IRF-3 activation. J. Virol. **76:** 3659–3669.
30. SERVANT, M.J., B. TEN OEVER, C. LEPAGE, *et al.* 2001. Identification of distinct signaling pathways leading to the phosphorylation of interferon regulatory factor 3. J. Biol. Chem. **276:** 355–363.
31. SMITH, E.J., I. MARIE, A. PRAKASH, *et al.* 2001. IRF3 and IRF7 phosphorylation in virus-infected cells does not require double-stranded RNA-dependent protein kinase R or Ikappa B kinase but is blocked by Vaccinia virus E3L protein. J. Biol. Chem. **276:** 8951–8957.
32. KWON, H., N. PELLETIER, C. DE LUCA, *et al.* 1998. Inducible expression of IkappaBalpha repressor mutants interferes with NF-kappaB activity and HIV-1 replication in Jurkat T cells. J. Biol. Chem. **273:** 7431–7440.
33. GENIN, P., M. ALGARTE, P. ROOF, *et al.* 2000. Regulation of RANTES chemokine gene expression requires cooperativity between NF-kappa B and IFN-regulatory factor transcription factors. J. Immunol. **164:** 5352–5361.
34. SHIMADA, T., T. KAWAI, K. TAKEDA, *et al.* 1999. IKK-i, a novel lipopolysaccharide-inducible kinase that is related to IkappaB kinases. Int. Immunol. **11:** 1357–1362.
35. POMERANTZ, J.L. & D. BALTIMORE. 1999. NF-kappaB activation by a signaling complex containing TRAF2, TANK and TBK1, a novel IKK-related kinase. EMBO J. **18:** 6694–6704.
36. PETERS, R.T., S.M. LIAO & T. MANIATIS. 2000. IKKepsilon is part of a novel PMA-inducible IkappaB kinase complex. Mol. Cell. **5:** 513–522.
37. PETERS, R.T. & T. MANIATIS. 2001. A new family of IKK-related kinases may function as I kappa B kinase kinases. Biochim. Biophys. Acta **1471:** M57–62.
38. BONNARD, M., C. MIRTSOS, S. SUZUKI, *et al.* 2000. Deficiency of T2K leads to apoptotic liver degeneration and impaired NF-kappaB-dependent gene transcription. EMBO J. **19:** 4976–4985.
39. TOJIMA, Y., A. FUJIMOTO, M. DELHASE, *et al.* 2000. NAK is an IkappaB kinase-activating kinase. Nature **404:** 778–782.
40. NOMURA, F., T. KAWAI, K. NAKANISHI, *et al.* 2000. NF-kappaB activation through IKK-i-dependent I-TRAF/TANK phosphorylation. Genes Cells **5:** 191–202.
41. CHARIOT, A., A. LEONARDI, J. MULLER, *et al.* 2002. Association of the adaptor TANK with the I kappa B kinase (IKK) regulator NEMO connects IKK complexes with IKK epsilon and TBK1 kinases. J. Biol. Chem. **277:** 37029–37036.
42. SHARMA, S., B. TENOEVER, N. GRANDVAUX, *et al.* 2003. Triggering the interferon antiviral response through an IKK-related pathway. Science **300:** 1148–1151.

Translational Upregulation of the X-Linked Inhibitor of Apoptosis

MARTIN HOLCIK

Solange Gauthier Karsh Molecular Genetics Laboratory, Department of Pediatrics, Children's Hospital of Eastern Ontario Research Institute, Ottawa, Ontario, K1H 8L1 Canada

ABSTRACT: The X-linked inhibitor of apoptosis protein (XIAP) is the most potent and best studied intrinsic regulator of programmed cell death. The critical role XIAP plays in the control of apoptosis is also reflected in the complex ways the activity of XIAP is regulated. In addition to regulating the function of the protein, the synthesis of XIAP is also selectively regulated. XIAP is translated by a cap-independent mechanism of translation initiation that is mediated by a unique internal ribosome entry site (IRES) sequence element located in its 5′ untranslated region. This allows XIAP mRNA to be actively translated during conditions of cellular stress when the majority of cellular protein synthesis is inhibited. The IRES regulation of XIAP translation points to an important mechanism in the control and regulation of apoptosis.

KEYWORDS: IRES; translational control; cellular stress; apoptosis; caspase inhibition; IAP

INTRODUCTION

Many, and perhaps most, apoptotic processes are ultimately channeled through the caspases. Caspases are cysteine proteases that cleave their substrates after a conserved aspartate residue in their substrate target sequence and that facilitate apoptosis by cleaving a growing number of cellular targets.[1] Caspases are synthesized as inactive zymogens that are proteolytically processed to produce mature, active proteases. In addition, caspases are arranged into a self-amplifying cascade such that the upstream caspases (initiator caspases, such as caspase-9) activate downstream caspases (effector caspases, such as caspase-3). Tight control of caspase activity is thus critical for cellular homeostasis. Opposing the cellular destruction by caspases are two classes of cellular apoptotic inhibitors—members of the Bcl-2 and IAP gene families. Whereas the Bcl-2 proteins can block only the mitochondrial branch of apoptosis by preventing the release of cytochrome *c*, the IAPs block both the mitochondria- and death-receptor–mediated pathways of apoptosis by directly binding to and inhibiting both the initiator and effector caspases.[2] For this reason the IAPs are con-

Address for correspondence: Martin Holcik, Solange Gauthier Karsh Molecular Genetics Laboratory, Department of Pediatrics, Children's Hospital of Eastern Ontario Research Institute, 401 Smyth Road, Ottawa, Ontario, K1H 8L1 Canada. Voice: 613-738-3207; fax: 613-738-4833.
martin@mgcheo.med.uottawa.ca

Ann. N.Y. Acad. Sci. 1010: 249–258 (2003). © 2003 New York Academy of Sciences.
doi: 10.1196/annals.1299.043

sidered to be the key regulators of apoptosis. IAPs were first described in baculoviruses where their role is to block the apoptotic response initiated in response to viral infection.[3] The homologues of viral IAPs were soon discovered in other metazoans including *Caenorhabditis elegans*, yeast, *Drosophila*, and vertebrates.[2] There are at present eight mammalian IAPs that inhibit apoptosis induced by a variety of triggers both *in vitro* and *in vivo*. Significantly, IAPs are the only endogenous inhibitors that block both the initiator and effector caspases.

The IAP proteins are typified by a ~80 amino acid zinc finger BIR (baculoviral IAP repeat) motif. While the number of BIR domains in a given IAP is variable (ranging from one to three) they are invariably present in the N-terminus of the protein and mediate the interaction with caspases. Several, but not all, IAPs also contain a C-terminal RING (really interesting new gene) zinc finger domain, which possesses an E3 ubiquitin ligase activity that catalyzes its own, as well as caspase, ubiquitination. The modular structure of IAPs is reflected in their multifunctionality. In addition to regulating apoptosis, IAPs also play a role in other important cellular processes including receptor-mediated signaling, cell cycle, and ubiquitination.[4]

XIAP

The X-chromosome linked inhibitor of apoptosis, XIAP, is likely the most potent cellular inhibitor of caspases and is the prototype member of the IAP family.[4] XIAP contains three BIR domains that are essential and sufficient for the binding to and inhibition of distinct caspases. While the BIR2 domain inhibits the effector caspases 3 and 7, the BIR3 domain blocks the initiator caspase 9. The role of the BIR1 domain is unclear at this time. In addition to the BIR domains, XIAP also contains a RING zinc finger that catalyzes its own ubiquitination *in vitro* and in cultured cells. Similar to other IAPs, XIAP is involved in several important cellular processes in addition to the control of apoptosis. XIAP was found to interact with the BMP (bone morphogenic protein) receptor (type I) and with its signaling molecule TAB1.[5] While the interaction with TAB1 is mediated by the BIR domain region of XIAP, the binding to the BMP receptor is facilitated by the RING zinc finger domain. XIAP was suggested to act as an activator of TAB1, which in turn activates the downstream MAP kinase kinase kinase family member TAK1. In addition, XIAP has been found to selectively activate JNK1 (c-jun N-terminal kinase 1), another downstream member of the MAP kinase signaling pathway.[6] Interestingly, there exists a possibility that JNK1 and XIAP are part of the same positive-feedback loop since protection of cells by XIAP in caspase-1–induced apoptosis is dependent on activated JNK1.[6]

Given the critical roles XIAP plays in the cell's biology, it is not surprising that it is regulated at multiple levels. Two negative regulators of XIAP have recently been identified, termed Smac/DIABLO[7,8] and XAF1.[9] Smac/DIABLO is a mitochondrial protein that is processed and released from mitochondria following an apoptotic insult and prevents XIAP from binding to caspases. In contrast, XAF1 is a nuclear protein that also inhibits XIAP from inhibiting caspases although this mechanism is not understood. In addition, XIAP is subject to transcriptional control by the stress-inducible transcriptional activator NF-κB.[10] While these modes of regulation are shared between XIAP and other IAP genes, XIAP is also uniquely regulated at the level of protein synthesis.

TRANSLATIONAL CONTROL OF XIAP EXPRESSION

Analysis of the genomic organization of both the human and murine XIAP loci was instrumental in the discovery of the translational regulation of XIAP expression (FIG. 1A). The human XIAP contains an unusually large 5′ untranslated region (5′ UTR) (~1.7 kb) that also contains a large number of out-of-frame initiation codons and short open reading frames.[11] The murine XIAP has a similar 5′ genomic organization with a 5′ UTR in excess of 5 kb.[12] The majority of eukaryotic mRNAs have short 5′ UTRs with an average length of 300 nt[13] and it has been proposed that they are translated by linear ribosomal scanning.[14] In this model, the ribosome is recruited, with the aid of several initiation factors, to the m^7G cap-structure at the 5′-terminus of the mRNA and is then thought to proceed in the 5′→3′ direction until an initiation AUG codon in a favorable context is encountered and the translation is initiated.[15] However, the extremely long 5′ UTR of XIAP, the high degree of its predicted secondary structure and the large number of potential initiation start sites would present a significant obstacle to efficient translation. Yet, the XIAP protein is found abundantly in all tissues examined.[16] We have also noticed that while the human and mouse XIAP 5′ UTR do not share any significant sequence similarity throughout most of their length (<25%) the region of approximately 350 nt located just upstream of the initiation codon is more conserved (76% similarity), suggesting that this part of the UTR may contain regulatory element(s) (FIG. 1A).

An alternative mechanism of translation initiation by internal initiation which is not hindered by a long 5′ UTR and multiple upstream AUG codons was identified and is mediated by IRES (internal ribosome entry site).[17] The IRES element is a regulatory motif that exists in the 5' untranslated region of some viral and cellular gene transcripts and allows translation of these transcripts without the complete cell machinery used for normal translation of m^7G-capped mRNAs. We have therefore tested whether the 5′ UTR of XIAP contains an IRES element that would mediate translation of XIAP by internal initiation.[18] Indeed, by using bi-cistronic mRNA constructs, both human and murine XIAP 5′ UTRs were shown to contain functional IRES elements. Deletional analysis of the human XIAP 5′ UTR delineated the XIAP IRES element to be located within a 162 nt region of the 5′ UTR, placed just upstream of the initiation codon AUG. Furthermore, mutational analysis identified a critical 12 nt polypyrimidine tract that is essential, but not sufficient, for IRES activity (FIG. 1B).[18]

XIAP IRES BINDING PROTEINS

Sequence characterization of the XIAP IRES element enabled us to begin to search for specific RNA-binding proteins that bind XIAP IRES and thus may be involved in the regulation of XIAP expression. A sequence-specific RNA-protein (RNP) complex was identified that assembles on the 162 nt XIAP IRES.[19] The boundaries of this complex were defined, with the aid of antisense oligonucleotides, to be between the UTR coordinates –34 and –62, thus overlapping the critical polypyrimidine tract (FIG. 1C). Analysis of this complex by UV-crosslinking identified four specific cellular proteins (p42, p50, p75, p100) as part of the XIAP RNP complex. Further characterization of XIAP RNP identified two of these proteins as the autoantigen La[19] and the heterogeneous nuclear ribonucleoprotein C (hnRNPC),[20]

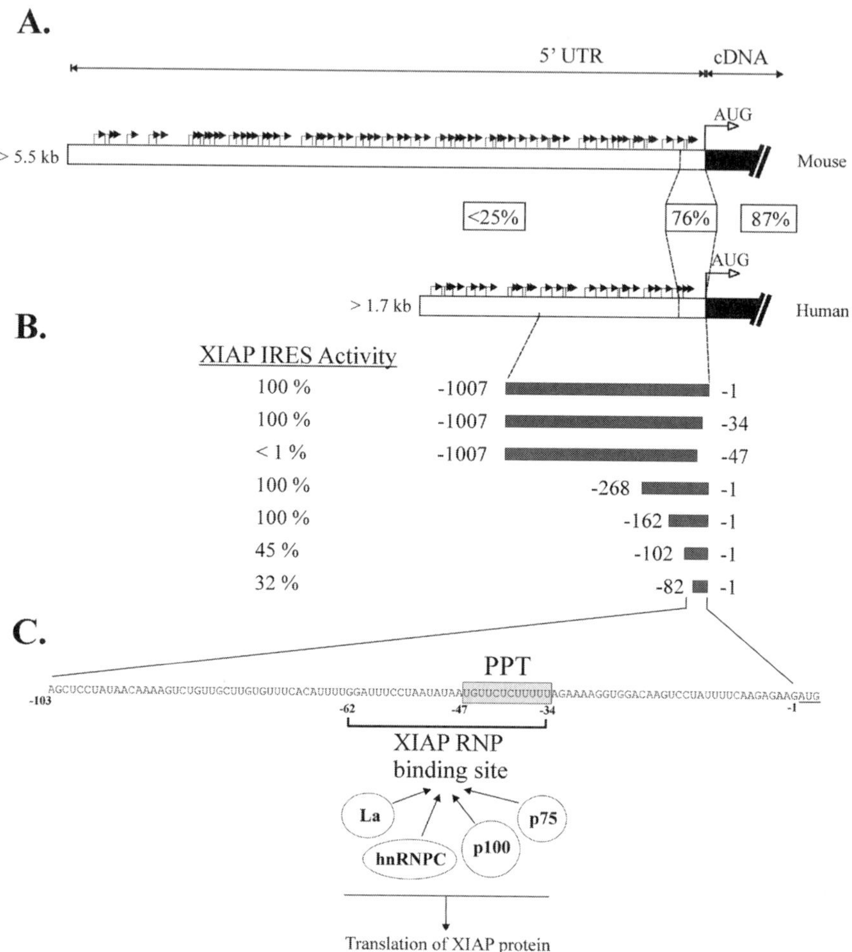

FIGURE 1. XIAP IRES element. (**A**) Schematic organization of the 5′ region of mouse (*top*) and human (*bottom*) XIAP. *Solid rectangles* represent the cDNA of human and mouse XIAP while the *white rectangles* represent the 5′ untranslated regions. *Black arrows* indicate potential initiation AUG codons; the *white arrows* indicate the authentic XIAP initiation codons. The sequence similarity between mouse and human sequences is shown in *boxes*. (**B**) Deletion analysis of XIAP IRES. Indicated segments of human XIAP 5′ UTR (*right*) were cloned into a bi-cistronic plasmid and the activity of IRES was determined (*left*). These experiments delineated the XIAP IRES to be within the –34 to –162 segment of the 5′ UTR. (**C**) UV-crosslinking identified the –34 to –62 segment of XIAP 5′ UTR as the RNP complex binding site with at least four cellular proteins binding here. The RNP complex binding site overlaps with the polypyrimidine tract (–34 to –47) that is critical for XIAP function. The concerted action of XIAP IRES binding proteins regulates XIAP translation and hence apoptosis.

and excluded two known IRES binding factors PTB and PCBP.[19] Importantly, the binding of La and hnRNPC to the XIAP IRES element was confirmed both *in vitro* and *in vivo*, underlining the biological significance of this observation.[19,20] The autoantigen La is a general RNA binding protein that participates in diverse aspects of RNA metabolism including the regulation of RNA polymerase III transcription, the processing of tRNA precursors, and the translation of several viral and cellular RNAs.[21] La exists as a dimer and this dimerization is essential for its functions. We have therefore used the dominant negative mutant of La, which prevents dimerization of the endogenous La, to demonstrate that La is specifically required for the translation of XIAP IRES-dependent but not cap-dependent translation.[19] In addition, purified recombinant La was sufficient to bind XIAP IRES RNA although the resultant RNP complex was smaller than that formed with cellular extracts. These data suggested that the La autoantigen is a central component of the XIAP IRES RNP complex that may facilitate the binding of additional cellular factors to the XIAP IRES sequence and is essential for the modulation of XIAP mRNA translation.

The second cellular protein that was shown to bind the XIAP IRES element is hnRNPC. hnRNPC is the most abundant nuclear RNA binding protein. Similar to La, hnRNPC was implicated in diverse roles of RNA biogenesis including polyadenylation, RNA turnover, splicing, and translation.[22] hnRNPC exists in two splice forms, hnRNPC1 and hnRNPC2, which display almost identical biochemical and RNA-binding properties. We have shown that both forms were capable of binding the XIAP IRES element.[20] Importantly, we observed that the cellular levels of hnRNPC1/C2 correlated with the XIAP IRES activity.[20] In SKOV3 carcinoma cells the levels of hnRNPC1/C2 were significantly lower than in other cells examined. At the same time, the activity of XIAP IRES in SKOV3 cells was also reduced. In contrast, transient overexpression of either isoform of hnRNPC, alone or in combination, resulted in enhanced translation of XIAP IRES. Importantly, this induction of IRES translation was specific to the XIAP IRES since the overexpression of hnRNPC1/C2 had no effect on the translation of the IRES element of an unrelated apoptotic gene Apaf-1.

Many cellular proteins are specifically cleaved during apoptosis. Recently, identification of proteins that are modified during Fas-induced apoptosis in Jurkat cells revealed that a large number of the apoptosis-modified proteins are, in fact, RNA-binding proteins.[23] More specifically, both autoantigen La and hnRNPC1/C2 were shown to be cleaved by caspases in cells induced to undergo apoptosis by a variety of triggers.[24–26] The advantage of regulating apoptosis by modulating RNA metabolism is not clear. Since regulation of apoptosis requires rapid responses, this could be achieved by regulation of RNA metabolism (e.g., splicing, turnover, translation) more efficiently than by *de novo* mRNA synthesis.[23] XIAP is a potent inhibitor of apoptosis by virtue of blocking caspase activation. It is tempting to speculate that the proteolytic cleavage of La, hnRNPC1/C2, and perhaps other IRES binding proteins at the onset of apoptosis is targeted to attenuate the synthesis of XIAP.

WHY IRES-MEDIATED CONTROL FOR XIAP?

A distinguishing mark of IRES-mediated translation is that it allows for enhanced or continued expression under conditions where normal, cap-dependent translation

TABLE 1. Distinct regulation of XIAP IRES under physiological stress

Conditions	Cell Line	XIAP IRES Activity[a]
Serum withdrawal	HeLa, 293T, HEL299, SH-SY5Y	Increased
γ-Irradiation	H661, H520	Increased
	H460, SKOV3	Reduced
Anoxia	293T	Increased
Heat shock	293T	Reduced
Etoposide-induced apoptosis	293T	Reduced

[a]When compared to untreated controls.

is shut off or compromised, such as during heat shock, hypoxia, growth arrest or position in the cell cycle.[17] It was demonstrated by several laboratories that cellular protein synthesis is severely compromised following the induction of apoptosis. This is accomplished by at least two different mechanisms: (1) the phosphorylation of some initiation factors and/or their regulators[27] or (2) by the proteolytic cleavage of several initiation factors.[28] The rapid inhibition of protein synthesis is believed to function as a protective homeostatic mechanism. We have therefore hypothesized that the IRES-mediated translation of XIAP will be resistant to the repression of protein synthesis that precedes the execution of apoptosis and could be critical for cell survival.[29] Indeed, the XIAP IRES element is actively translated during various conditions of cellular stress such as serum starvation and low-dose γ-irradiation (TABLE 1).[18,30] In addition, overexpression of the XIAP cDNA containing the IRES element protected cells from serum starvation–induced apoptosis more efficiently than did the coding region alone. In a different model of cellular stress, endogenous levels of XIAP protein were found to be selectively upregulated following low-dose g-irradiation of some non-small cell carcinoma cell lines.[30] The mechanism of this upregulation was shown to be via IRES-mediated translation of XIAP. Significantly, those cell lines able to upregulate XIAP protein displayed enhanced resistance to low-dose γ-irradiation.

ARE THERE OTHER APOPTOSIS-INVOLVED PROTEINS REGULATED BY IRES?

In addition to viral RNA, IRES elements are found in a limited but growing number of cellular mRNAs. Interestingly, these mRNAs include several growth factors, oncogenes, and genes that are critically involved in the regulation of apoptosis (TABLE 2). There is no structural or sequence homology among cellular IRES elements so their discovery remains largely experimental. Also, the regulation of various cellular IRES elements is not fully understood, although the evidence to date suggests that the IRES elements are differentially regulated. For instance, the IRES element of the oncogene c-Myc is upregulated following genotoxic stress,[31] while the IRES of VEGF is enhanced by hypoxia.[32] Investigation of a range of cellular stresses revealed that XIAP IRES is active under conditions of acute, but transient stress (such as anoxia), but is compromised under severe apoptotic conditions of

TABLE 2. Cellular mRNAs with Internal Ribosome Entry Sites

Gene product	Regulation (if known)
Transcription factors	
Antennapadia	–
Ultrabithorax	–
MYT2	–
NRF	–
AML1/RUNX1	Development
Gtx homeodomain protein	–
Hypoxia inducible factor 1α	Hypoxia
Mnt	–
Protooncogenes	
c-Myc	Genotoxic stress, apoptosis
N-Myc	–
Pim1	–
Protein kinase $p58^{PITSLRE}$	Cell cycle
c-Jun	–
Ornithine decarboxylase	Cell cycle
Blk	–
Tumor suppressors	
APC	–
Proteins involved in apoptosis	
Apaf-1	Apoptosis, anoxia
XIAP	Apoptosis, anoxia, radiation
Bcl2	–
Bag 1	–
Growth factors	
FGF2	Heat shock, oxidative stress
PDGF/c-sis	Differentiation
VEGF	Hypoxia
Cyr61	–
IGFII	–
Translation factors	
eIF4G	–
p97/DAP5/NAT1	Apoptosis
Transporters/receptors	
CAT-1	Amino acid starvation, ER stress
Notch 2	–
Estrogen receptor α	–
IFG-1 receptor	–
Dendritically localized proteins	
ARC	–
MAP2	–
RC3	–
α-Subunit of calcium calmodulin-dependent kinase II dendrin	–
Others	
BiP	ER stress
La autoantigen	–
β-Subunit of mitochondrial H^+-ATP synthase	–
p27Kip1	–
Hairless	Cell cycle, development
Protein kinase C delta	Apoptosis
Connexin 43	–
Connexin 32	–
FMR1	–

etoposide-induced cell death.[33] In contrast, however, IRES elements of the pro-apoptotic genes Apaf-1 and DAP5 were enhanced in etoposide-treated cells.[33,34] Furthermore, this activation was caspase dependent and was mediated by caspase-cleaved fragments of the initiation factors eIF4GI and p97/DAP5/NAT1.[33]

These data support the hypothesis that IRES-mediated translation escapes the control mechanisms that regulate cap-dependent translation during conditions of cellular stress. In addition, conditions of transient cellular stress, such as anoxia, favor translation of pro-survival IRES elements, such as XIAP, while severe apoptotic conditions result in the activation of the pro-death IRES elements of Apaf-1 and DAP5. While the exact molecular pathways that regulate IRES translation in transient cellular stress need to be further investigated, these results indicate that the apoptotic fragments of the eIF4G translation initiation factor family mediate upregulation of pro-death IRES elements during apoptotic stress presumably to enforce the commitment of cells to die.

CONCLUSIONS

XIAP is a critical regulator of cell death by virtue of its ability to inhibit both initiator and effector caspases. Among the IAPs, XIAP translation is uniquely regulated by a rare cap-independent mechanism that allows XIAP protein to be synthesized during conditions of global protein synthesis shut-off. Although we have previously postulated that the IRES-mediated translation of XIAP would be critical to cell survival under all stress conditions,[29] recent evidence suggests that only conditions of transient cellular stress support XIAP IRES translation. This mechanism may assure that XIAP levels are increased in cells undergoing transient stress to allow them to recover and survive once normal conditions are restored. On the other hand, high levels of XIAP protein would be detrimental in cells with irreparable damage that are committed to die since it would prevent, or significantly delay, their orderly apoptotic disassembly. The XIAP IRES is thus not very efficient under these conditions. This model of IRES regulation of XIAP suggests that translational control contributes to the fine-tuning of cell fate. Investigations of proteins and cellular pathways that control IRES translation of XIAP, and other apoptotic proteins, will no doubt produce exciting discoveries.

REFERENCES

1. NICHOLSON, D.W. 1999. Caspase structure, proteolytic substrates, and function during apoptotic cell death. Cell Death Differ. **6**(11)**:** 1028–1042.
2. SALVESEN, G.S. & C.S. DUCKETT. 2002. Apoptosis: IAP proteins: blocking the road to death's door. Nat. Rev. Mol. Cell. Biol. **3**(6)**:** 401–410.
3. CROOK, N.E., R.J. CLEM & L.K. MILLER. 1993. An apoptosis-inhibiting baculovirus gene with a zinc finger-like motif. J. Virol. **67**(4)**:** 2168–2174.
4. HOLCIK, M. & R.G. KORNELUK. 2001. XIAP, the guardian angel. Nat. Rev. Mol. Cell. Biol. **2**(7)**:** 550–556.
5. YAMAGUCHI, K. *et al.* 1999. XIAP, a cellular member of the inhibitor of apoptosis protein family, links the receptors to TAB1-TAK1 in the BMP signaling pathway. EMBO J. **18**(1)**:** 179–187.
6. SANNA, M.G. *et al.* 1998. Selective activation of JNK1 is necessary for the anti-apoptotic activity of hILP. Proc. Natl. Acad. Sci. USA **95**(11)**:** 6015–6020.

7. Du, C. *et al.* 2000. Smac, a mitochondrial protein that promotes cytochrome c-dependent caspase activation by eliminating IAP inhibition. Cell **102**(1)**:** 33–42.
8. Verhagen, A.M. *et al.* 2000. Identification of DIABLO, a mammalian protein that promotes apoptosis by binding to and antagonizing IAP proteins. Cell **102**(1)**:** 43–53.
9. Liston, P. *et al.* 2001. Identification of XAF1 as an antagonist of XIAP anti-caspase activity. Nat. Cell Biol. **3**(2)**:** 128–133.
10. Stehlik, C. *et al.* 1998. Nuclear factor (NF)-kappa B-regulated X-chromosome-linked IAP gene expression protects endothelial cells from tumor necrosis factor alpha–induced apoptosis. J. Exp. Med. **188**(1)**:** 211–216.
11. Lagace, M. *et al.* 2001. Genomic organization of the x-linked inhibitor of apoptosis and identification of a novel testis-specific transcript. Genomics **77**(3)**:** 181–188.
12. Farahani, R. *et al.* 1997. Genomic organization and primary characterization of MIAP-3: the murine homologue of human X-linked IAP. Genomics **42**(3)**:** 514–518.
13. Lander, E.S. *et al.* 2001. Initial sequencing and analysis of the human genome. Nature **409**(6822)**:** 860–921.
14. Kozak, M. 1978. How do eucaryotic ribosomes select initiation regions in messenger RNA? Cell **15**(4)**:** 1109–1123.
15. Hershey, J.W.B. & W.C. Merrick. 2000. Pathway and mechanism of initiation of protein synthesis. *In* Translation Control of Gene Expression. N. Sonenberg, J.W.B. Hershey, & M.B. Mathews, Eds.: 33–88. Cold Spring Harbor Laboratory Press. Cold Spring Harbor, NY.
16. Holcik, M. *et al.* 2002. Cloning and characterization of the rat homologues of the inhibitor of apoptosis protein 1, 2, and 3 genes. BMC Genomics **3**(1)**:** 5–10.
17. Hellen, C.U. & P. Sarnow. 2001. Internal ribosome entry sites in eukaryotic mRNA molecules. Genes Dev. **15**(13)**:** 1593–1612.
18. Holcik, M. *et al.* 1999. A new internal-ribosome-entry-site motif potentiates XIAP-mediated cytoprotection. Nat. Cell Biol. **1**(3)**:** 190–192.
19. Holcik, M. & R.G. Korneluk. 2000. Functional characterization of the X-linked inhibitor of apoptosis (XIAP) internal ribosome entry site element: Role of La autoantigen in XIAP translation. Mol. Cell Biol. **20**(13)**:** 4648–4657.
20. Holcik, M., B.W. Gordon & R.G. Korneluk. 2003. The internal ribosome entry site-mediated translation of antiapoptotic protein XIAP is modulated by the heterogeneous nuclear ribonucleoproteins C1 and C2. Mol. Cell. Biol. **23**(1)**:** 280–288.
21. Wolin, S.L. & T. Cedervall. 2002. The La protein. Annu. Rev. Biochem. **71:** 375–403.
22. Krecic, A.M. & M.S. Swanson. 1999. hnRNP complexes: composition, structure, and function. Curr. Opin. Cell Biol. **11**(3)**:** 363–371.
23. Thiede, B. *et al.* 2001. Predominant identification of RNA-binding proteins in Fas-induced apoptosis by proteosome analysis. J. Biol. Chem. **276**(28)**:** 26044–26050.
24. Waterhouse, N. *et al.* 1996. Heteronuclear ribonucleoproteins C1 and C2, components of the spliceosome, are specific targets of interleukin 1 beta-converting enzyme-like proteases in apoptosis. J. Biol. Chem. **271**(46)**:** 29335–29341.
25. Rutjes, S.A. *et al.* 1999. The La (SS-B) autoantigen, a key protein in RNA biogenesis, is dephosphorylated and cleaved early during apoptosis. Cell Death Differ. **6**(10)**:** 976–986.
26. Brockstedt, E. *et al.* 1998. Identification of apoptosis-associated proteins in a human Burkitt lymphoma cell line. Cleavage of heterogeneous nuclear ribonucleoprotein A1 by caspase 3. J. Biol. Chem. **273**(43)**:** 28057–28064.
27. Brostrom, C.O. & M.A. Brostrom. 1998. Regulation of translational initiation during cellular responses to stress. Prog. Nucl. Acid Res. Mol. Biol. **58:** 79–125.
28. Clemens, M.J. *et al.* 2000. Translation initiation factor modifications and the regulation of protein synthesis in apoptotic cells. Cell Death Differ. **7**(7)**:** 603–615.
29. Holcik, M., N. Sonenberg & R.G. Korneluk. 2000. Internal ribosome initiation of translation and the control of cell death. Trends Genetics **16**(10)**:** 469–473.
30. Holcik, M. *et al.* 2000. Translational upregulation of X-linked inhibitor of apoptosis (XIAP) increases resistance to radiation induced cell death. Oncogene **19**(36)**:** 4174–4177.
31. Subkhankulova, T., S.A. Mitchell & A.E. Willis. 2001. Internal ribosome entry segment-mediated initiation of c-Myc protein synthesis following genotoxic stress. Biochem. J. **359**(Pt 1)**:** 183–192.

32. STEIN, I. *et al.* 1998. Translation of vascular endothelial growth factor mRNA by internal ribosome entry: implications for translation under hypoxia. Mol. Cell. Biol. **18**(6): 3112–3119.
33. NEVINS, T.A. *et al.* 2003. Distinct regulation of IRES-mediated translation following cellular stress is mediated by apoptotic fragments of eIF4G translation initiation factor family members eIF4GI and p97/DAP5/NAT1. J. Biol. Chem. **278**(6): 3572–3579.
34. HENIS-KORENBLIT, S. *et al.* 2002. The caspase-cleaved DAP5 protein supports internal ribosome entry site-mediated translation of death proteins. Proc. Natl. Acad. Sci. USA **99**(8): 5400–5405.
35. BONNAL, S. *et al.* 2003. IRESdb: the Internal Ribosome Entry Site database. Nucl. Acids Res. **31**(1): 427–428.

Role of NF-κB and DNA Repair Protein Ku on Apoptosis in Pancreatic Acinar Cells

JI YEON SONG, JOO WEON LIM, HYEYOUNG KIM, AND KYUNG HWAN KIM

Department of Pharmacology and Institute of Gastroenterology, Brain Korea 21 Project for Medical Science, Yonsei University College of Medicine, Seoul 120-752, Korea

ABSTRACT: **Reactive oxygen species have been known to cause DNA damage and induce apoptosis. During DNA damage, DNA repair proteins Ku70 and Ku80 prevent cell death, but severe DNA damage beyond the repair capacity of the DNA repair proteins triggers necrosis or apoptosis. Recent reports have shown that NF-κB plays a critical role in protecting the cells from apoptosis. We investigated whether glucose oxidase acting on β-D-glucose (G/GO), which continuously produces H_2O_2, induces apoptosis, and whether NF-κB and Ku are involved in G/GO-induced apoptosis in pancreatic acinar AR42J cells. Electron microscopic observation showed that apoptotic cells with characteristic nuclear condensation and shrinkage as well as large vacuoles were detected after G/GO treatment. G/GO treatment induced apoptotic cell death, as determined by viable cell count and DNA fragmentation. G/GO-induced apoptosis was increased in the cells transfected with the Ku-dominant negative mutant (Ku D/N) and mutated IκBα gene (IκB mt) as compared to the wild-type cells (Wild) and the cells transfected with the control pcDNA3 vector (pcN-3). G/GO treatment caused nuclear loss of both Ku70 and Ku80 in Wild cells and pcN-3 cells. Even without G/GO treatment, nuclear loss of Ku proteins was observed in IκB mt cells. These results suggest that oxidative stress–induced reduction of nuclear Ku proteins may cause loss of defense against DNA damage and thus induce apoptosis in pancreatic acinar cells. The novel finding is that nuclear translocation of Ku proteins may be mediated by NF-κB.**

KEYWORDS: **NF-κB; Ku; apoptosis; pancreatic acinar cells**

INTRODUCTION

Oxidative stress is regarded as a major pathogenic factor in acute pancreatitis.[1] Acinar cell death linked to oxidative DNA damage occurs in the development of acute pancreatitis. Recent studies have revealed that apoptosis is a major mechanism of acinar cell death in various experimental models of acute pancreatitis, suggesting a role for apoptosis in the pathophysiology of the disease.[2,3] However, the molecular mechanism of apoptosis in acute pancreatitis has not been clearly elucidated. Ku70 and Ku80 proteins as regulatory parts of the DNA-dependent protein kinase (DNA-PK) initiate the repair process of DNA double-strand breaks, which produce DNA

Address for correspondence: Hyeyoung Kim, Department of Pharmacology, Yonsei University College of Medicine, Seoul 120-752, Korea. Voice: 82-2-361-5232; fax: 82-2-313-1894. kim626@yumc.yonsei.ac.kr

Ann. N.Y. Acad. Sci. 1010: 259–263 (2003).
doi: 10.1196/annals.1299.044

fragmentation.[4,5] Treatment of embryo fibroblast (MEF) derived from Ku80-null (Ku80$^{-/-}$) mice with hydrogen peroxide (H_2O_2) induced more severe DNA damage than those from wild-type mice.[6] Therefore, we assume that Ku may play a role in cell death mechanisms, especially DNA fragmentation after oxidative stress. It has been known that an inducible transcription factor, NF-κB plays an anti-apoptotic role. Um and colleagues[7] showed that Ku activity positively correlates with NF-κB activity in multidrug-resistant leukemia cells. Our previous study showed that Ku70 and Ku80 expression is mediated by NF-κB–dependent mechanism.[8] The present

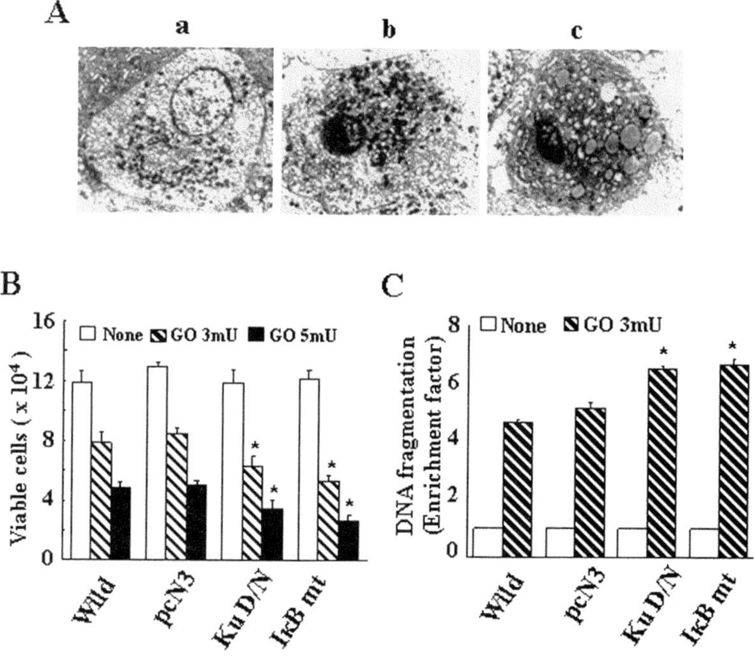

FIGURE 1. Electron-microscopic appearance (**A**), cell viability (**B**), and DNA fragmentation (**C**) of wild-type cells and the transfected cells treated with G/GO. (**A**) Wild-type cells were cultured in the absence or presence of 10 mM glucose/5 mU glucose oxidase (G/GO). The cells without G/GO treatment showed normal morphological appearance with zymogen granules. The plasma and nuclear membranes were defined and chromatin was homogeneously distributed in the nucleus (**a**). After G/GO treatment for 12 h and 24 h, apoptotic cells with characteristic nuclear condensation and shrinkage (**b**, **c**) and large vacuoles (**c**) were detected (**b**, 12 h; **c**, 24 h) (×8,000). (**B**) Wild-type cells (Wild) and the cells transfected with the control pcDNA3 vector (pcN-3), the Ku-dominant negative mutant (Ku D/N), or mutated IκBα gene (IκB mt) were treated with or without G/GO (GO 3 and 5 mU/mL) for 24 hours. Cell viability was determined by trypan blue exclusion test. (**C**) DNA fragmentation was assessed as the content of nucleosome-bound DNA by ELISA. The relative increase in nucleosomes in the cell lysate, determined at 405 nm, was expressed as an enrichment factor. Enrichment factor of wild-type cell treated without G/GO was considered as 1. Each bar represents means ± standard error of five separate experiments. *$P<.05$ versus the corresponding pcN-3.

study aims to investigate the role of Ku proteins and the involvement of NF-κB and Ku on apoptotic cell death induced by oxidative stress in pancreatic acinar AR42J cells.

METHODS

Rat pancreatic acinar AR42J cells (pancreatoma, ATCC CRL 1492) were incubated without or with glucose oxidase (GO) (3 and 5 mU/mL) and β-D-glucose (10 mM) (G/GO). For the determination of morphological appearance by electron microscopy, AR42J cells were incubated with G/GO (5 mU/mL) for 12 h and 24 hours. To investigate the role of NF-κB and Ku on apoptosis of AR42J cells, the cells were transiently transfected with control pcDNA3 vector (pcN-3), mutated I-κBα gene (IκB mt), or Ku-dominant negative mutant gene (Ku D/N). Cell viability and DNA fragmentation were determined by viable cell counting using trypan blue exclusion and quantities of nucleosome-bound DNA, respectively. Localization of Ku after G/GO treatment was determined using FITC-conjugated secondary antibody for anti-Ku70 and anti-Ku80 antibody and examined by immunofluorescence.

RESULTS AND DISCUSSION

In the present study, we showed evidence of apoptosis in G/GO-treated AR42J cells by electron microscopic observation, viable cell counts, and quantitative DNA fragmentation. Wild-type cells without G/GO treatment showed normal morphological appearance with zymogen granules. The plasma and nuclear membranes were defined and chromatin was homogeneously distributed in the nucleus (FIG. 1A, a). After G/GO (5 mU/mL) treatment for 12 h, apoptotic cells with condensed and fragmented nuclei, but intact plasma integrity were observed (FIG. 1A, b). At 24-h culture with G/GO, the cells had severe nuclear shrinkage and large vacuoles, demonstrating progressive apoptosis (FIG. 1A, c). In Wild cells and pcN-3 cells, apoptotic cell death was induced by 24-h treatment of G/GO (3 and 5 mU/mL), determined by viable cell count and DNA fragmentation (FIG. 1B and C). G/GO-induced apoptotic cell death was accelerated in Ku D/N cells and IκB mt cells. This result suggests that NF-κB and Ku proteins may provide protection against oxidative stress–induced apoptosis. For the determination of Ku localization changes after G/GO treatment, immunostaining with anti-Ku70 and anti-Ku80 monoclonal antibody, followed by FITC-conjugated secondary anti-goat IgG antibody were used (FIG. 2). G/GO treatment caused nuclear loss of both Ku70 and Ku80 in Wild cells and pcN-3 cells. Even without G/GO treatment, nuclear loss of Ku proteins was observed in IκB mt cells. The present study suggests that NF-κB regulates nuclear localization of Ku proteins. Our results are supported by the report showing that Ku70 and Ku80 expression is mediated by NF-κB–dependent mechanism.[8] In conclusion, nuclear loss of Ku70 and Ku80 may cause the loss of defense against oxidative DNA damage and may be responsible for apoptotic cell death in pancreatic acinar cells. In addition, nuclear location of Ku proteins may be mediated by NF-κB.

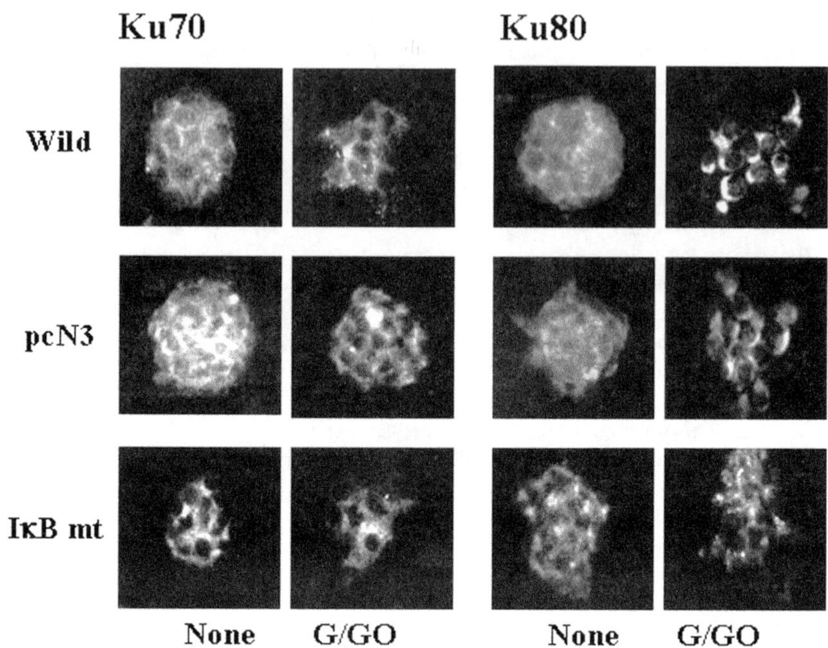

FIGURE 2. The localization of Ku in wild-type cells and the transfected cells treated with G/GO. Wild-type cells (Wild) and the cells transfected with the control pcDNA3 vector (pcN-3) or mutated IκBα gene (IκB mt) were treated with or without 10 mM glucose/5 mU/mL glucose oxidase (G/GO) for 12 hours. The cells were fixed and stained with anti-Ku70 and anti-Ku80 monoclonal antibody, followed by FITC-conjugated secondary anti-goat IgG antibody. Localization of Ku70 and Ku80 was determined by immunofluorescence (×400).

ACKNOWLEDGMENT

This study was supported by a grant from Brain Korea 21 Project for Medical Science, Yonsei University College of Medicine, Seoul, Korea.

REFERENCES

1. SANFEY, H., G.B. BULKELEY & J.L. CAMERON. 1985. The pathogenesis of acute pancreatitis: the source and role of oxygen-derived free radicals in three different experimental models. Ann. Surg. **201:** 63–69.
2. KAISER, A.M., A.K. SALUJA, A. SENGUPTA, *et al.* 1995. Relationship between severity, necrosis, and apoptosis in five models of experimental acute pancreatitis. Am. J. Physiol. **269:** C1295–C1304.
3. SANDOVAL, D., A. GUKOVSKAYA, P. REAVEY, *et al. 1996.* The role of neutrophils and platelet-activating factor in mediating experimental pancreatitis. Gastroenterology **111:** 1081–1091.
4. FEATHERSTONE, C. & S.P. JACKSON. 1999. Ku, a DNA repair protein with multiple cellular functions. Mutation Res. **434:** 3–15.

5. Kim, G.W., N. Noshita, T. Sugawara & P.H. Chan. 2001. Early decrease in DNA repair proteins, Ku70 and Ku86, and subsequent DNA fragmentation after transient focal cerebral ischemia in mice. Stroke **32:** 1401–1407.
6. Arrington, E.D., M.C. Caldwell, T.S. Kumaravel, *et al.* 2000. Enhanced sensitivity and long-term G2 arrest in hydrogen peroxide-treated Ku80-null cells are unrelated to DNA repair defects. Free Rad. Biol. Med. **29:** 1166–1176.
7. Um, J.H., C.D. Kang, B.G. Lee, *et al.* 2001. Increased and correlated nuclear factor-kappa B and Ku autoantigen activities are associated with development of multidrug resistance. Oncogene **20:** 6048–6056.
8. Lim, J.W., H. Kim & K.H. Kim. 2002. Expression of Ku70 and Ku80 mediated by NF-κB and cyclooxygenase-2 is related to proliferation of human gastric cancer cells. J. Biol. Chem. **277:** 46093–46100.

PLAG Proteins

How They Influence Apoptosis and Cell Proliferation

JEROEN DECLERCQ, KAREN HENSEN, WIM J. VAN DE VEN, AND MARCELA CHAVEZ

Department of Human Genetics (K. U. Leuven) and Flanders Interuniversity Institute for Biotechnology, Leuven B-3000, Belgium

ABSTRACT: Pleomorphic adenoma gene 1 product (PLAG1) a transcription factor with zinc finger domains, is upregulated in several tumors, such as pleomorphic adenomas and lipoblastomas. PLAGL2 is a second member of the same subfamily of zinc finger proteins. Cells overexpressing either of these two genes display the typical markers of neoplastic transformation. We investigated whether PLAG1 and PLAGL2 could prevent cells from undergoing apoptosis, leading them to grow uncontrollably.

KEYWORDS: pleomorphic adenoma gene 1; PLAG1; PLAG2

Pleomorphic adenoma gene 1 product (PLAG1) a transcription factor with zinc finger domains, is upregulated in several tumors, such as pleomorphic adenomas and lipoblastomas. PLAGL2 is a second member of the same subfamily of zinc finger proteins. Cells overexpressing either of these two genes display the typical markers of neoplastic transformation: (1) the cells lose cell-cell contact inhibition; (2) show anchorage-independent growth; and (3) cause tumor growth when injected in nude mice.[1] Tumor cells often lose several controls regulating their cell cycle leading to an impairment of progression towards controlled cell death by apoptosis. We are particularly interested in investigating whether PLAG1 and PLAGL2 could prevent cells from undergoing apoptosis, leading them to uncontrolled growth.

Two non-exclusive ways can lead to an inhibition of apoptosis by PLAG proteins. One or several of the anti-apoptotic proteins can be upregulated and/or some of the apoptotic proteins can be downregulated by these transcription factors. PLAG1 immunoreactive cells in pleomorphic adenomas of the salivary gland express the anti-apoptotic factor, Bcl-2.[2] In line with this result, the expression of the anti-apoptotic insulin-like growth factor 2 (IGF-2) gene is highly upregulated in salivary gland adenomas overexpressing PLAG1. In contrast, IGF-2 is not detectable in adenomas without abnormal PLAG1 expression or in normal salivary gland tissues. Potential PLAG1 binding sites were also found in the promotor 3 of the IGF-2 gene.[3] On the

Address for correspondence: Marcela Chavez, Department of Human Genetics (K. U. Leuven) and Flanders Interuniversity Institute for Biotechnology, Leuven B-3000, Belgium.
marcela.chavez@med.kuleuven.ac.be

Ann. N.Y. Acad. Sci. 1010: 264–265 (2003). © 2003 New York Academy of Sciences.
doi: 10.1196/annals.1299.045

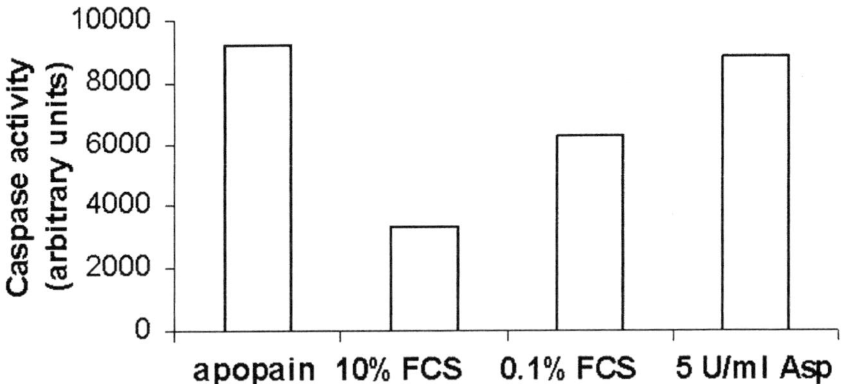

FIGURE 1. Caspase 3 activity. NIH3T3 cells (1.5×10^6) were incubated in low serum medium (0.1% FCS) or with 5 U/mL L-asparaginase for 48 hours. Caspase 3 activity in cell lysates was monitorited *in vitro* with the fluorogenic caspase substrate (Ac-DEVD-AFC) for 2 hours. The fluorometric intensity was measured at 520 nm after excitation at 390 nm.

basis of these results we sought to investigate the mechanisms through which PLAG1 and PLAGL2 may be involved in anti-apoptosis. For that, we generated stable NIH3T3 cell lines overexpressing PLAG1 or PLAGL2 in order to direct them to enter apoptosis. Several studies have shown that growth factors promote survival in various experimental cell systems. In the case of IGF-2, the major pathway that protects cells from apoptosis is dependent on PI3 kinase and Akt. The latter phosphorylates and inhibits Bad, a pro-apoptotic member of the Bcl-2 family and caspase 9.[3,4] A future report will show whether these factors or any others are modulated in PLAG-overexpressing cells, but not in the Mock cells. These results will give us the first clue of PLAG anti apoptotic mechanism. We will also investigate pathways independent from IGF-2 in fibroblasts derived from IGF-2 KO mice. At present, we succeeded in inducing apoptosis in NIH3T3 cells by asparaginase (5 U/mL) treatment or 0.1% FCS starvation. Indeed high levels of caspase 3 are induced after both treatments (FIG. 1).

REFERENCES

1. HENSEN, K. *et al.* 2002. The tumorigenic diversity of the three PLAG family members is associated with different DNA binding capacities. Cancer Res. **62:** 1510–1517.
2. DEBIEC-RYCHTER, M. *et al.* 2001. Histologic localization of PLAG1 (pleomorphic adenoma gene 1) in pleomorphic adenoma of the salivary gland: cytogenetic evidence of common origin of phenotypically diverse cells. Lab. Invest. **81:** 1289–1297.
3. VOZ, M.L. *et al.* 2000. PLAG1, the main translocation target in pleomorphic adenoma of the salivary glands, is a positive regulator of IGF-II. Cancer Res. **60:** 106–113.
4. KULIK, G., A. KLIPPEL & M.J. WEBER. 1997. Antiapoptotic signalling by the insulin-like growth factor I receptor, phosphatidylinositol 3-kinase, and Akt. Mol. Cell Biol. **17:** 1595–1606.
5. SCHEID, M.P., K.M. SCHUBERT & V. DURONIO. 1999. Regulation of bad phosphorylation and association with Bcl-x(L) by the MAPK/Erk kinase. J. Biol. Chem. **274:** 31108–31113.

How the Nucleolar Sequestration of p53 Protein or Its Interplayers Contributes to Its (Re)-Activation

JÓZEFA WĘSIERSKA-GĄDEK AND MARCEL HORKY[a]

Cell Cycle Regulation Group, Institute of Cancer Research, Faculty of Medicine, University of Vienna, Vienna, Austria

[a]*Department of Pathological Physiology, Faculty of Medicine, Masaryk University, Brno, Czech Republic*

ABSTRACT: The tumor suppressor p53 is a short-lived protein that under normal conditions is reduced to a barely detectable level. The stability of p53 protein is primarily regulated in normal non-transformed cells by two interplayers: Mdm2 and p14^{ARF}. Relocation of p53, Mdm2, and p14^{ARF} to the nucleolus seems to regulate, at least partially, the steady-state of p53. Moreover, there are alternative pathways of the regulation of p53 stability in unstressed cells. Jun-N(amino)-terminal kinase (JNK) and poly(ADP-ribose) polymerase-1 (PARP-1) are involved in the regulation of the steady-state of wild-type (wt) p53 protein. However, in most human cervical carcinomas, which express the high-risk human papilloma viruses (HPVs) E6 protein, a complete switch from Mdm2 to HPV E6-mediated degradation of p53 occurs. Virally encoded E6 protein utilizes the cellular ubiquitin-protein ligase termed E6-associated protein (E6-AP) to target p53 protein for proteolytic degradation. We recently addressed the question of whether p53 protein can be generally reactivated by chemotherapy in HeLa cells despite the E6 activity. We observed an increase of cellular p53 after cisplatin (CP) treatment. p53 protein accumulated preferentially in the nucleoli. We checked the cellular level of E6 during CP therapy. Six hours after application of CP the expression of E6 protein was markedly reduced. This coincided with the increase of cellular p53 level and preceded the nucleolar accumulation of p53 protein, thereby indicating that repression of virally coded E6 protein by CP contributes to the restoration of p53 expression.

KEYWORDS: apoptosis; HPV; JNK; Mdm2; p14^{ARF}; PARP-1; p53 stability; repression of E6

REGULATION OF P53 STABILITY IN UNSTRESSED CELLS

From its initial discovery in 1979 as an oncogene to the recognition a decade later that the p53 protein is a critical regulator of cell-cycle progression and apoptosis in

Address for correspondence: Prof. Dr. Jozefa Gadek-Wesierski, Cell Cycle Regulation Group, Institute of Cancer Research, Borschkegasse 8 a, A-1090 Vienna, Austria. Voice: +(43-1)-4277-65247; fax: +(43-1)-4277-65194.
Jozefa.Antonia.Gadek-Wesierski@univie.ac.at

**Ann. N.Y. Acad. Sci. 1010: 266–272 (2003). © 2003 New York Academy of Sciences.
doi: 10.1196/annals.1299.046**

the cellular response to stress stimuli, the p53 tumor suppressor protein was promoted to "the molecule of the year" and became a superstar.[1–3] The diversity of p53 functions is based on the fact that its normal control mission can be directly inactivated by mutations that result in the loss or the gain of function, or indirectly by interaction with other cellular or virally encoded factors.[1–3]

The wild-type (wt) p53 phosphoprotein is a very potent sensor of DNA damage and of cellular stresses. Its fundamental function is to maintain genomic stability by the control and regulation of the cell-cycle progression or by apoptosis. p53 tumor suppressor is a short-lived protein. The low basal level of p53 is exactly regulated and its expression fluctuates during the cell cycle. p53 phosphoprotein reaches the highest concentration at G_1 phase of the cell cycle.[4] Mdm2 protein (human Hdm2), the product of a p53-dependent gene, plays a critical role in regulating p53 level and p53 activity as transcription factor.[5,6] Mdm2 (mouse double minute 2), possesses intrinsic E3 ubiquitin-ligase activity and efficiently targets p53 for polyubiquitination and proteosome-dependent degradation.[7,8] Moreover, Mdm2, through specific interaction with transactivation domain of p53 protein, is able to inhibit its transcriptional activity.[5] The Mdm2 shuttling between the nucleus and the cytoplasm is essential for Mdm2 activity to target p53 for degradation.[9] The importance of Mdm2 in the negative regulation of the p53 functions is reflected in the fact that inactivation of Mdm2 by gene disruption was lethal for early embryos, whereas the simultaneous disruption of the *p53* gene rescued the mice from lethality.[10,11] Thus, an autoregulatory feedback loop has been established between p53 and Mdm2, in which p53 induces expression of its own antagonist. The negative regulation of p53 transcription factor can be abrogated by human alternative reading frame p14 protein ($p14^{ARF}$).[12–14] The INK4a-ARF locus encodes two distinct tumor suppressors, $p16^{INK4a}$ and $p14^{ARF}$ (murine $p19^{ARF}$), which are generated by alternative splicing. $p14^{ARF}$ mRNA is transcribed from exons 1β, 2, and 3.[12,13] Whereas $p16^{INK4a}$ restrains cell proliferation through preventing phosphorylation of the retinoblastoma protein,[15] $p14^{ARF}$ acts by attenuating Mdm2-mediated degradation of p53. $p14^{ARF}$ binds to Mdm2 in a region distinct from the p53 binding site.

It is becoming evident that alternative mechanisms supervising the stability of p53 protein exist in unstressed cells. Recently the role of nonactive Jun-N (amino)-terminal kinase (JNK) in regulating p53 stability was reported.[16] JNK has been found to complex with distinct nuclear protein ATF2, c-jun, and p53 and to target the associated proteins for the ubiquitination and degradation.[16,17] JNK-p53 complexes were preferentially found in G_0/G_1, whereas Mdm2-p53 complexes preponderated in S/G_2 phases of the cell cycle. These data indicate that JNK is an Mdm2-independent regulator of p53 steady state in nonstressed cells. Moreover, the role of PARP-1 in the regulation of the basal expression of wt p53 protein was demonstrated in two different cell systems.[18–20] First, the inactivation of the PARP-1 by gene disruption resulted in the shortening of the half-life of the p53 protein.[18,19] The reconstitution of PARP-1–deficient cells with the human counterpart abrogated the reduced stability of p53 protein. A second line of evidence revealed that co-expression of PARP-1 with temperature-sensitive (ts) $p53^{135Val}$ mutant in primary rat cells increased the stability of wt p53.[20] The feature of this ts $p53^{135Val}$ mutant is temperature-dependent switching between mutant and wild-type phenotype. At basal or elevated temperature (37°C and 39°C, respectively) p53 has an oncogenic character and is localized in the cytoplasm. After temperature shift to 32°C p53 translocates into the nucleus and acts as tumor suppressor. The co-expression of PARP-1 did not affect the stability of mutant form of $p53^{135Val}$ protein.[20] However, PARP-1 complexed with wt p53 prevented its degradation even after temperature shift up to basal conditions and delayed the recovery of G_1-arrested cells in S-phase after elevation of temperature.[20] Thus, it shows that normal cells possess a few independent mechanisms responsible for regulation of the steady states of p53 protein.

ROLE OF p53 IN CELLULAR RESPONSE TO STRESS CONDITIONS

In response to various stressors the level of p53 protein rises and p53 becomes activated. An increase in the concentration of p53 protein is to a large extent regulated through control of p53 stability. The enhanced half-life is the result of post-translational modifications[21] that alter the interaction of p53 with its negative regulator and render p53 resistant to destruction via the ubiquitin-mediated proteolysis pathway. The activation of p53 is due to its conversion from latent to active form, which is capable of sequence-specific DNA binding and transactivation of p53-responsive genes. Activated p53 protein plays a pivotal role in the regulation of the cell cycle.[1–3] Wt p53 can induce cell cycle in late G_1 via upregulation of $p21^{waf1}$ or at G_2 through growth arrest DNA damage (GADD)-45. p53 acts also as a mitotic checkpoint factor and controls centromere duplication. Moreover, p53 is also capable of inducing of apoptosis as well as of promoting discrete apoptotic steps.[1–3]

INACTIVATION OF WT PROTEIN IN TUMOR CELLS

Wt p53 protein has received a great deal of attention for its key role in the prevention of malignant transformation. Inactivation of wt p53 and disruption of the p53 signaling occurs in a wide range of human cancers.[22,23] The normal function of p53 is very frequently inactivated by mutations. About half of the major forms of cancer contain p53 missense mutations. The impact of altered or lost wt p53 function on malignancy may be even greater than initially predicted because mechanisms other than mutations can functionally inactivate p53 protein. Mdm2 amplification in human sarcomas and leukemias,[24] expression of E6 protein in HPV high risk-positive cervical carcinomas,[25,26] or inappropriate p53 localization in some breast cancer tumors and in a large majority of undifferentiated neuroblastomas[27,28] represent such other mechanisms, by which cancer cells can overcome p53-mediated growth control. Whereas the localization of wt p53 in the cytoplasm precludes its ability to act as transcription factor, the expression of E6 oncoprotein or the amplification of Mdm2, which in turn targets p53 for accelerated degradation, eliminates p53 functions.[29,30]

NUCLEOLAR SEQUESTRATION OF P53 INTERPLAYERS

$p14^{ARF}$ is thought to activate p53 by abrogation of the Mdm2 activities. It binds to and inhibits the E3 ubiquitin ligase activity of Mdm2 towards p53 *in vitro*[31,32] and *in vivo*[32,33] and prevents Mdm2-dependent p53 nucleo-cytoplasmic shuttling which leads to stabilization and activation of p53. Current models propose that p14^{ARF} sequesters Mdm2 in the nucleoli, thus physically separating Mdm2 and p53 in different subcellular compartments.[34] Whereas p53 and Mdm2 are primarily nucleoplasmic proteins, p14^{ARF} resides within nucleoli. Human and mouse ARF are highly basic proteins (~20% Arg residues) that possess within the amino-terminus the amino acid sequence RRPR representing the nucleolar localization signal (NoLS). When p14^{ARF} protein binds to Mdm2, its NoLS becomes masked. On the other hand the complex formation between p14^{ARF} and Mdm2 leads to the exposure of cryptic NoLS within the RING domain of Mdm2 and the retention of the p14^{ARF}-Mdm2 complex in the nucleolus is attributable to the exposed NoLS of Mdm2.[34,35] However, recent reports call into question the requirement of these topological interactions[36] and describe situations in which stabilization of p53 by p14^{ARF} does not require relocation of Mdm2 to the nucleolus.[33,37] Moreover, forms of p14^{ARF} that do not accumulate in the nucleolus retain the capacity to stabilize p53.[33] Furthermore, a lack of correlation between p14^{ARF}-mediated stabilization of p53 and induction of transcriptional activation was reported. These data suggest that additional events other than Mdm2 import are required for p14^{ARF} function. Indeed, recently CARF, a novel protein cooperating with p14^{ARF} in activating p53 was described.[38] The newly discovered collaborator of ARF (CARF) co-localizing with p14^{ARF} in the nucleolus is co-regulated with p14^{ARF} and thus acts as a novel component of the p14^{ARF}-p53 pathway.

ESCAPE OF P53 PROTEIN FROM ACCELERATED DEGRADATION BY NUCLEOLAR TARGETING

Interestingly, not only factors regulating p53 functions but also p53 protein itself accumulates in some situations in the nucleoli. After inhibition of proteosome activity in normal human fibroblasts, upregulated p53 protein accumulated in nucleolar structures and co-localized with nucleolin.[39] p53 seems to possess affinity for a few nucleolar components. The specific binding to nucleolin, nucleophosmin, poly(ADP-ribose)polymerase-1 (PARP-1) and topoisomerase I was reported.[40-43] Nucleophosmin continuously shuttling between the nucleus and cytoplasm interacts directly with p53 and regulates the increase in stability and transcriptional activation of p53 after different types of stress.[40] Under some stress conditions such as heat shock or ionizing radiation, translocation of nucleolin from the nucleolus to the nucleoplasm was observed. The relocation was mediated by formation of p53-nucleolin complex.[41]

Interestingly, the nucleolar sequestration of p53 is one of the mechanisms contributing to the reactivation of p53 in HeLa cells after chemotherapy with cisplatin.[44,45] HeLa cell line derived from a malignant cervical carcinoma harbors wild-type p53 and retinoblastoma tumor suppressor genes. However, the high-risk HPV E6 and E7 oncoproteins expressed in HeLa cells exert profound effects on tumor suppressor genes and inactivate them.[46-48] Therefore, we raised the question whether p53 can be reactivated by chemotherapy in HeLa cells despite of the presence HPV encoded E6 activity. We have observed that cellular levels of p53 protein increased after CP treatment reaching a maximum after 6 hours. p53 protein accumulated preferentially in the nucleoli with a peak after 15 hours.[44] CP-induced nucleolar targeting of p53 seems to be selective because p73, another member of p53 gene family accumulated primarily in nuclei in response to CP. Monitoring of the intranuclear distribution of Hdm2, a negative regulator of p53, revealed that this protein present in the nucleoli of untreated control translocated into chromatin during the therapy by CP. Interestingly, quite inverse intranuclear redistribution showed p14^{ARF}. This observation indicating that Hdm2 is not involved in the regulation of p53 function in HeLa cells is consistent with a report demonstrating a switch from Hdm2 to E6-mediated degradation of p53 in cervical cancer cells.[49] Proteosome inhibitors were not able to mimic the effect of CP on p53 level. Since the reduced stability of wild-type p53 proteins in HeLa cells is a consequence of its enhanced ubiquitination by virally encoded E6 protein resulting in its an accelerated degradation, we checked the cellular level of E6 during the CP therapy. Six hours after application of CP the expression of E6 protein was markedly reduced. This coincided with the increase of cellular p53 level and preceded the nucleolar accumulation of p53 protein, thereby indicating that repression of virally coded E6 protein by CP contributes to the restoration of p53 expression. The reactivation of p53 protein in HeLa cells by CP could be important for the progression of apoptotic process. Indeed, double-immunostaining revealed nuclear accumulation of p53 protein in almost all cells in which caspase-3 was activated. In this context it is reasonable that reactivation of p53 expression sensitizes HeLa cells to CP-induced apoptosis.

REFERENCES

1. Mowat, M.R. 1998. p53 in tumor progression: life, death, and everything. Adv. Cancer Res. **74:** 25–48.
2. Prives, C. & P.A. Hall. 1999. The p53 pathway. J. Pathol. **187:** 112–126.
3. Bargonetti, J. & J.J. Manfredi. 2002. Multiple roles of the tumor suppressor p53. Curr. Opin. Oncol. **4:** 86–91.
4. Shaulsky, G., A. Ben-Ze'ev & V. Rotter. 1990. Subcellular distribution of the p53 protein during the cell cycle of Baalb/c 3T3 cells. Oncogene **5:** 1707–1711.
5. Momand, J., G.P. Zambetti, D.C. Olson, *et al.* 1992. The Mdm2 oncogene product forms a complex with the p53 protein and inhibits p53-mediated transactivation. Cell **69:** 1237–1245.
6. Finlay, C.A. 1993. The Mdm2 oncogene can overcome wild-type p53 suppression of transformed cell growth. Mol. Cell. Biol. **13:** 301–306.
7. Haupt, Y., R. Maya, A. Kazaz & M. Oren. 1997. Mdm2 promotes the rapid degradation of p53. Nature **387:** 296–299.
8. Kubbutat, M.H.G., S.N. Jones & K.H. Vousden. 1997. Regulation of p53 stability by Mdm2. Nature **387:** 299–303.
9. Tao, W. & A.L. Levine. 1999. Nucleocytoplasmic shuttling of oncoprotein Hdm2 is required for Hdm2-mediated degradation of p53. Proc. Natl. Acad. Sci. USA **96:** 3077–3080.
10. Jones, S.N., A.E. Roe, L.A. Donehower & A. Bradley. 1995. Rescue of embryonic lethality in Mdm2-deficient mice by absence of p53. Nature **378:** 206–208.
11. Montes de Oca Luna, R., D.S. Wagner & G. Lozano. 1995. Rescue of early embryonic lethality in mdm2-deficient mice by deletion of p53. Nature **378:** 203–206.
12. Mao, L., A. Merlo, G. Bedi, *et al.* 1995. A novel p16INK4A transcript. Cancer Res. **55:** 2995–2997.
13. Kamijo T, F. Zindy, M.F. Roussel, *et al.* 1997. Tumor suppression at the mouse INK4a locus mediated by the alternative reading frame product p19ARF. Cell **91:** 649–659.
14. Quelle, D.E., M. Cheng, R.A. Ashmun & C.J. Sherr. 1997. Cancer-associated mutations at the INK4a locus cancel cell cycle arrest by p16INK4a but not by the alternative reading frame protein p19ARF. Proc. Natl. Acad. Sci. USA **94:** 669–673.

15. LUKAS, J., D. PARRY, I. AAGAARD, *et al.* 1995. Retinoblastoma-protein-dependent cell cycle inhibition by the tumour suppressor p16. Nature **375:** 503–506.
16. FUCHS, S., V. ADLER, T. BUSCHMANN, *et al.* 1998. JNK targets p53 ubiquitination and degradation in nonstressed cells. Genes Dev. **12:** 2658–2663.
17. FUCHS, S.Y., V. ADLER, M.R. PINCUS & Z. RONAI. 1998. MEKK1/JNK signaling stabilizes and activates p53: Proc. Natl. Acad. Sci. USA **95:** 10541–10546.
18. WESIERSKA-GADEK, J., Z-Q. WANG & G. SCHMID. 1999. Reduced stability of regularly spliced but not alternatively spliced p53 protein in PARP-deficient mouse fibroblasts. Cancer Res. **59:** 28–34.
19. WESIERSKA-GADEK, J., E. BOHRN, Z. HERCEG, *et al.* 2000. Differential susceptibility of normal and PARP knock-out mouse fibroblasts to proteosome inhibitors. J. Cell. Biochem. **78:** 681–696.
20. WESIERSKA-GADEK, J. & G. SCHMID. 2000. Overexpressed poly(ADP-ribose)polymerase delays the release of rat cells from p53-mediated G1 checkpoint. J. Cell. Biochem. **80:** 85–103.
21. APPELLA, E. & C.W. ANDERSON. 2001. Post-translational modifications and activation of p53 by genotoxic stresses. Eur. J. Biochem. **268:** 2764–2772.
22. HOLLSTEIN, M., D. SIDRANSKY, B. VOGELSTEIN & C.C. HARRIS. 1991. p53 mutations in human cancers. Science **253:** 49–53.
23. HAINAUT, P. & M. HOLLSTEIN. 2000. p53 and human cancer: the first ten thousand mutations. Adv. Cancer Res. **77:** 81–137.
24. BUESO-RAMOS, C.E., Y. YANG, E. DELEON, *et al.* 1993. The human MDM2 oncogene is overexpressed in leukemias. Blood **82:** 2617–2623.
25. SCHWARZ, E., U.K. FREESE, L. GISSMANN, *et al.* 1985. Structure and transcription of human papillomavirus sequences in cervical carcinoma cells. Nature **314:** 111-114.
26. DURST, M., C.M. CROCE, L. GISSMANN, *et al.* 1987. Papillomavirus sequences integrate near cellular oncogenes in some cervical carcinomas. Proc. Natl. Acad. Sci. USA **84:** 1070–1074.
27. MOLL, U.M., G. RIOU & A.J. LEVINE. 1992. Two distinct mechanisms alter p53 in breast cancer: mutation and nuclear exclusion. Proc. Natl. Acad. Sci. USA **89:** 7262–7266.
28. MOLL, U.M., M. LAQUAGLIA, J. BERNARD & G. RIOU. 1995. Wild-type p53 protein undergoes cytoplasmic sequestration in undifferentiated neuroblastomas but not in differentiated tumors. Proc. Natl. Acad. Sci. USA **92:** 4407–4411.
29. SCHEFFNER, M., B.A. WERNESS, J.M. HUIBREGTSE, *et al.* 1990. The E6 oncoprotein encoded by human papillomavirus types 16 and 18 promotes the degradation of p53. Cell **63:** 1129–1136.
30. SCHEFFNER, M., J.M. HUIBREGTSE, R.D. VIERSTRA & P.M. HOWLEY. 1993. The HPV-16 E6 and E6-AP complex functions as a ubiquitin-protein ligase in the ubiquitination of p53. Cell **75:** 495–505.
31. HONDA, R. & H. YASUDA. 1999. Association pf p19(ARF) with Mdm2 inhibits ubiquitin ligase activity of Mdm2 for tumour suppressor p53 EMBO J. **18:** 22–27.
32. MIDGLEY, C.A., J.M. DESTERRO, M.K. SAVILLE, *et al.* 2000. An N-terminal p14ARF peptide blocks Mdm2-dependent ubiquitination in vitro and can activate p53 in vivo. Oncogene **19:** 2312–2323.
33. LIANOS, S., A.P. CLARK, J. ROWE & G. PETERS. 2001. Stabilization of p53 by p14ARF without relocation of MDM 2 to the nucleolus: Nat. Cell Biol. **3:** 445-451.
34. WEBER, D.J., J.L. TAYLOR, F.M. ROUSSEL, *et al.* 1999. Nucleolar Arf sequesters Mdm2 and activates p 53. Nat. Cell Biol. **1:** 20–26.
35. LORUM, M.A., M. ASHKROFT, M.H. KUBBUTAT & K.H. VOUSDEN. 2000. Identification of a cryptic nucleolar-localization signal in MDM2. Nat. Cell Biol. **2:** 179–181.
36. BOTHNER, B., W.S. LEWIS, L.E DI GIAMMMARINO, *et al.* 2001. Defining the molecular basis of Arf and Hdm 2 interactions, J. Mol. Biol. **314:** 263–277.
37. KORGAONKAR, CH., L. ZHAO, M. MODESTOU & E.D. QUELLE. 2002. ARF function does not require p53 stabilisation or Mdm2 relocalization. Mol. Cell. Biol. **22:** 196–206.
38. HASANT, M.K., T. YAGUCHI, T. SUGIHARA, *et al.* 2002. CARF Is a novel protein that cooperates with mouse p19ARF(Human p14ARF) in activating p53. J. Biol. Chem. **40:** 37765–37770.

39. Klibanov, A.S., M.H. O'Hagan & M. Ljungman. 2001. Accumulation of soluble and nucleolar-associated p53 proteins following cellular stress. J. Cell Sci. **114:** 1867–1873.
40. Colombo, E., J.C. Marine, D. Danovi, *et al.* 2002. Nucleophosmin regulates the stability and transcriptional activity of p53. Nat. Cell Biol. **4:** 529–533.
41. Daniely, Y., D. Dimitrova & A.J. Boroweiec. 2002. Stress-dependent nucleolin mobilisation mediated by p53–nucleolin complex formation. Mol. Cell. Biol. **22:** 6014–6022.
42. Karayan, L., J-F. Riou, P. Seite, *et al.* 2001. Human ARF protein interacts with Topoisomerase I and stimulates its activity. Oncogene **20:** 836–848.
43. Wesierska-Gadek, J., A. Bugajski & C. Cerni. 1996. ADP-ribosylation of p53 tumor suppressor protein: mutant but not wild-type p53 is modified. J. Cell Biochem. **62:** 90–101.
44. Wesierska-Gadek, J., D. Schloffer, V. Kotala & M. Horky. 2002. Escape of p53 protein from E6-mediated degradation in HeLa cells after cisplatin therapy. Int. J. Cancer **101:** 128–136.
45. Horky, M., G Wurzer, V. Kotala, *et al.* 2001. Segregation of nucleolar components coincides with caspase-3 activation in cisplatin-treated HeLa cells. J. Cell Sci. **114:** 663–670.
46. Heck, D.V., C.L. Yee, P.M. Howley & K. Munger. 1992. Efficiency of binding the retinoblastoma protein correlates with the transforming capacity of the E7 oncoproteins of the human papillomaviruses. Proc. Natl. Acad. Sci. USA **89:** 4442–4446.
47. Slebos, R.J., M.H. Lee, B.S. Plunkett, *et al.* 1994. p53-dependent G1 arrest involves pRB-related proteins and is disrupted by the human papillomavirus 16 E7 oncoprotein. Proc. Natl. Acad. Sci. USA **91:** 5320–5324.
48. Kessis, T.D., R.J. Slebos, W.G. Nelson, *et al.* 1993. Human papillomavirus 16 E6 expression disrupts the p53-mediated cellular response to DNA damage. Proc. Natl. Acad. Sci. USA **90:** 3988–3992.
49. Hengstermann, A., L.K. Linares, A. Ciechanover, *et al.* 2001. Complete switch from Mdm2 to human papillomavirus E6-mediated degradation of p53 in cervical cancer cells. Proc. Natl. Acad. Sci. USA **98:** 1218–1223.

Combination of Tumor Necrosis Factor-α with Sulindac in Human Carcinoma Cells *in Vivo*

HIROSHI YASUI, MASAAKI ADACHI, AND KOHZOH IMAI

First Department of Internal Medicine, Sapporo Medical University School of Medicine, S-1, W-16, Chuo-ku, Sapporo, 060-8543, Japan

ABSTRACT: Transcription factor NF-κB plays a pivotal role in cancer cells in the resistance to apoptosis, since NF-κB is frequently activated in many primary carcinoma cells. Indeed, several NF-κB inhibitors are found to be promising anti-cancer agents. However, some anti-cancer agents activate NF-κB signals and may reduce their potential, including tumor necrosis factor (TNF)-α. Recently, the nonsteroidal anti-inflammatory drug (NSAID) sulindac and its metabolites have been shown to inhibit the NF-κB–mediated survival signals through inhibition of IKK-β by their direct interaction. We thus investigate whether sulindac and its metabolite can augment TNF-α–mediated apoptosis in human carcinoma cells and be applicable for *in vivo* clinical usage. We here demonstrate that sulindac inhibited TNF-α–mediated NF-κB activation and greatly enhanced TNF-α–induced apoptosis in human gastric MKN45 and cervical HeLa carcinoma cell lines. The *in vivo* tumor growth of MKN45 cells was most strongly inhibited by a combination of TNF-α with sulindac compared with TNF-α or sulindac alone. Moreover, we demonstrate that sulindac sulfide further augmented TNF-α–mediated apoptosis. Our data strongly suggest that combination therapy of TNF-α with sulindac and its metabolites may sensitize cancer cells to TNF-α and augment its pro-apoptotic potential. Therefore, in combination with sulindac or its metabolites, TNF-α may become a potentially useful anti-cancer agent to suppress tumor.

KEYWORDS: TNF-α; sulindac; NF-κB; apoptosis; tumor growth suppression

Resistance to apoptosis appears to be responsible for a principal mechanism by which cancer cells are able to overcome anti-cancer therapies.[1] Although tumor necrosis factor (TNF)-α was considered to be a strong anti-cancer reagent, it cannot induce apoptosis in many cancer cells. This may be due to its adverse effect, i.e., it activates NF-κB.[2] Recently, several nonsteroidal anti-inflammatory drugs (NSAID), sulindac and its metabolites, have been shown to inhibit the NF-κB–mediated signals through inhibition of IKK-β by their direct interaction.[3,4] We thus investigate whether sulindac and its metabolite can augment TNF-α–mediated apoptosis.

Address for correspondence: Hiroshi Yasui, First Department of Internal Medicine, Sapporo Medical University School of Medicine, S-1, W-16, Chuo-ku, Sapporo, 060-8543, Japan. Voice: 81-11-611-2111; fax: 81-11-611-2282.
hiroyasu@sapmed.ac.jp

Ann. N.Y. Acad. Sci. 1010: 273–277 (2003).
doi: 10.1196/annals.1299.047

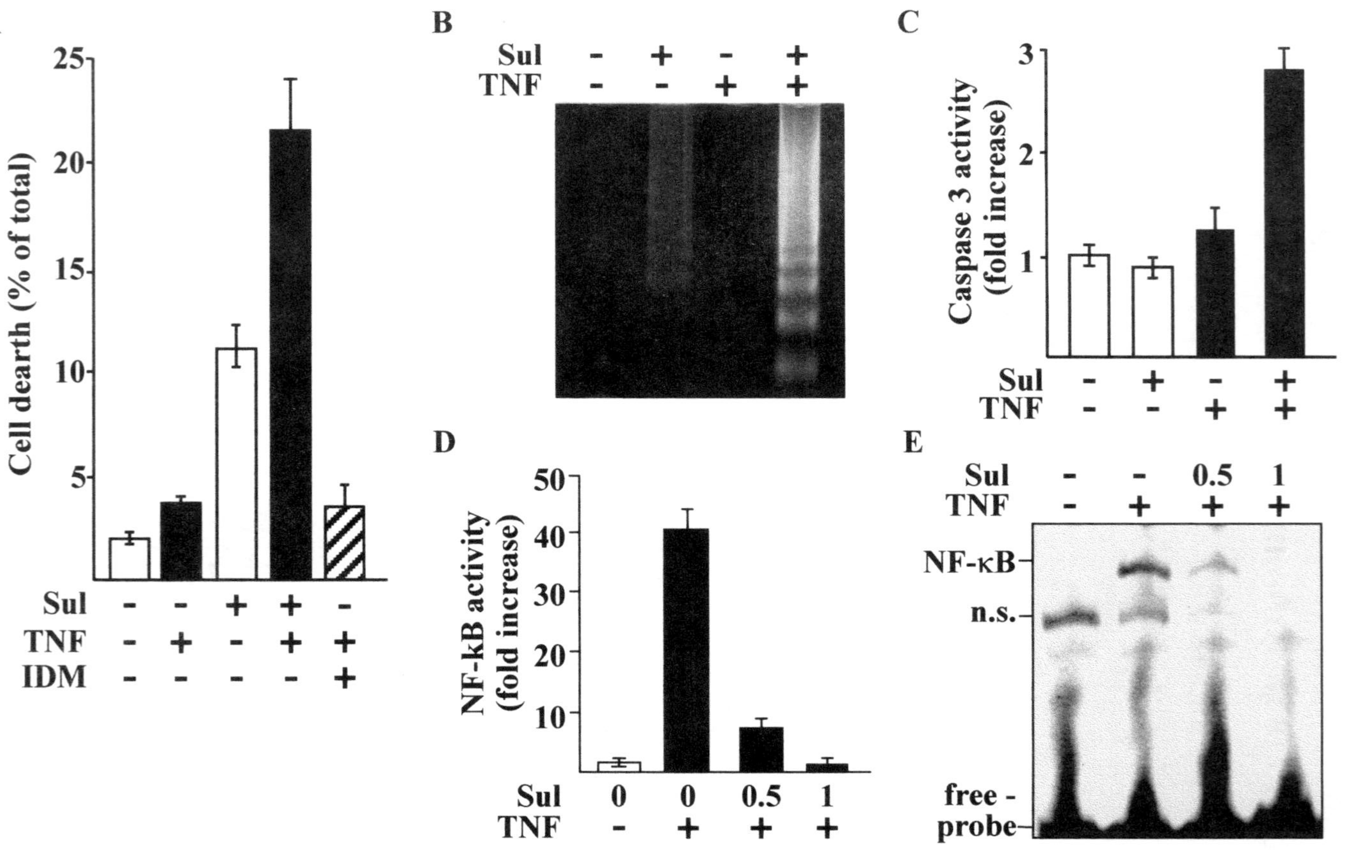

FIGURE 1. *See following page for legend.*

First, MKN45 cells were treated with 1 mM sulindac (Sul) and/or 20 ng/mL recombinant human TNF-α (TNF). At 24 h after incubation with 20 ng/mL TNF-α alone, MKN45 cells minimally reduced cell viability. The 1 mM sulindac treatment slightly induced cell death (FIG. 1, A). Importantly, TNF-α plus sulindac treatment strongly augmented cell death, DNA fragmentation, and caspase-3 activity (FIG. 1 A–C). However, a structurally related indomethacin (IDM), which can not inhibit NF-κB activation,[4] showed no distinct augmentation of TNF-α–mediated cell death (FIG. 1A).

Sulindac strongly inhibited TNF-α–mediated NF-κB activation evaluated by reporter-luciferase assay (FIG. 1D) and gel-shift assays (FIG. 1E). In contrast, sulindac did not affect HSE-mediated transcriptional activation (data not shown). These results suggest that the augmented effect of sulindac on anti-tumor activity of TNF-α may depend upon its inhibitory effect on NF-κB activation. Moreover, similar effects of sulindac on TNF-α–mediated NF-κB signals and apoptosis were observed in HeLa cells (data not shown). These results suggest that sulindac can sensitize wide range of carcinoma cells to TNF-α through its inhibitory effect on NF-κB signals.

We further investigated the effect of 20 ng TNF-α plus 1 μmol sulindac intra-tumor injections (day 0, 3, 6, 9, 12, and 15) on progression of the MKN45-generated subcutaneous tumors. The *in vivo* tumor growth was most strongly inhibited by the combination of TNF-α with sulindac compared with TNF-α or sulindac alone (FIG. 2A). Finally, we demonstrated that a sulindac metabolite sulindac sulfide augmented TNF-α–mediated apoptosis more strongly than sulindac in MKN45 cells. While

---

FIGURE 1. Sulindac enhanced TNF-α–mediated apoptosis and inhibited TNF-α–mediated NF-κB activation in MKN45 cells. (**A**) Augmentation of TNF-α–mediated cell death by sulindac. MKN45 cells were treated with 1 mM sulindac (Sul) or 25 mM indomethacin (IDM) for 2 h prior to incubation with or without 20 ng/mL TNF-α for 24 hours. Cell death was evaluated by trypan blue exclusion assay. (**B**) Augmented DNA fragmentation by the combination therapy. MKN45 cells were treated with or without 1 mM Sul for 2 h prior to incubation with or without 20 ng/mL of TNF-α for 24 hours. Low-molecular weight DNAs were separated by 1.2% agarose gel and stained with ethidium bromide. (**C**) Augmented caspase-3 activity by the combination therapy. MKN45 cells were treated with or without 1 mM Sul for 2 h prior to incubation with the indicated amounts of TNF-α for 12 hours. Caspase-3 activity was measured with a Caspase-3 Colorimetric Protease Assay Kit. The numbers represent the fold increase in absorbance at 405 nm relative to the mock transfectants. (**D**) Sulindac inhibits TNF-α–induced NF-κB activation. MKN45 cells were transiently transfected with both pRL-TK and pNF-κB-Luc using the Effectene Transfection Regent. After 24-h transfection, the cells were incubated with the indicated amounts of Sul (mM) for 2 h prior to incubation with or without 20 ng/mL TNF for 8 hours. Cells were harvested to detect luciferase activity using Dual-Luciferase reporter assay system, and luciferase activity was measured using a luminometer and normalized to RL activity. (**E**) Sulindac inhibits TNF-α–induced DNA binding of NF-κB. MKN45 cells were treated with the indicated amounts of sulindac for 2 h prior to incubation with or without 20 ng/mL of TNF-α for 30 min, and their total cell extracts were prepared. DNA binding activity of the extracts was analyzed by gel-shift assay. Equal amounts of ^{32}P-labeled NF-κB oligo were run in each lane, as shown in free-probe. Non-specific binding was not inhibited by addition of excess cold oligo probe and is shown as (n.s.). In **A**, **C**, and **D**, each column displays the mean ± standard deviation (S.D.) of data from three separate experiments. Three independent experiments exhibited similar results.

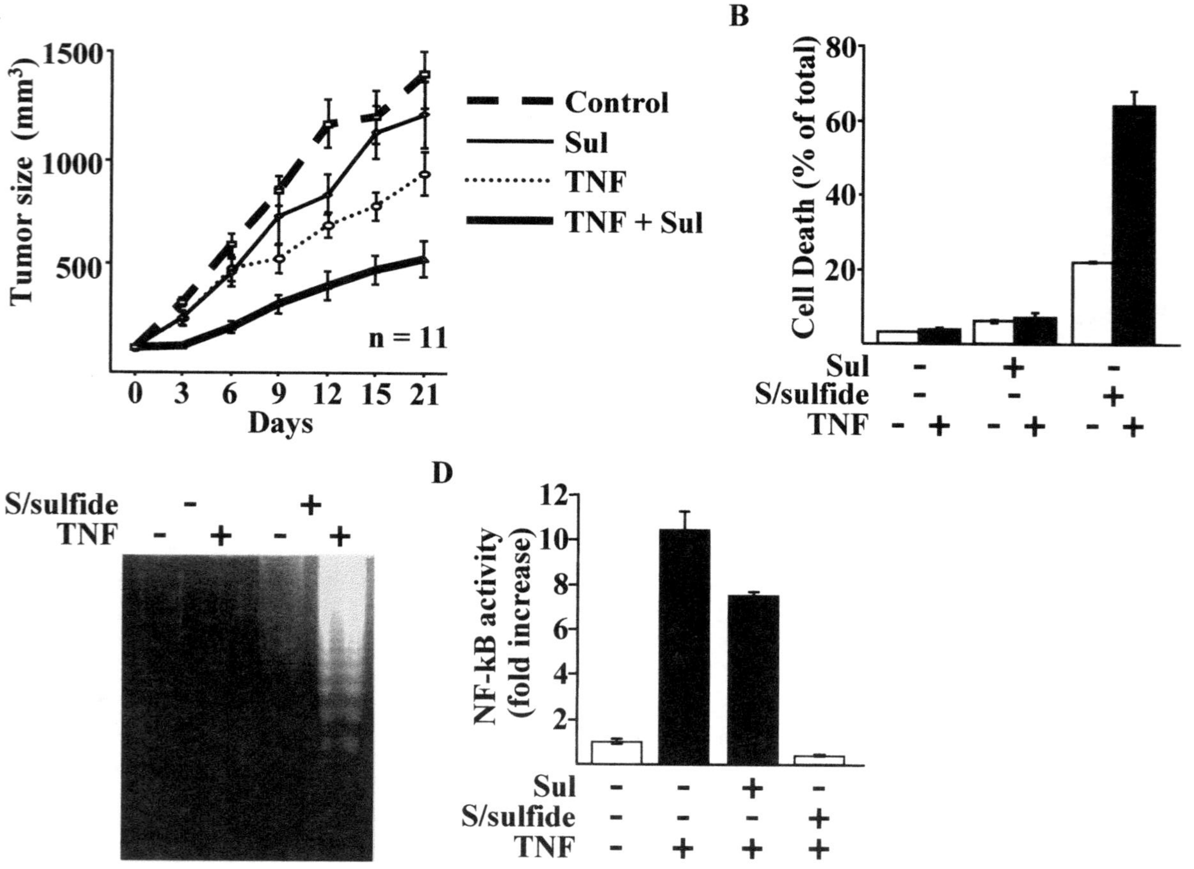

FIGURE 2. *See following page for legend.*

0.2 mM sulindac minimally affected TNF-α–mediated cell death, 0.2 mM sulindac sulfide (S/sulfide) strongly augmented TNF-α–induced cell death and DNA fragmentation (FIG. 2B and C). Sulindac sulfide (0.2 mM) more strongly inhibited TNF-α–mediated NF-κB activation than 0.2 mM sulindac evaluated by luciferase-reporter assay (FIG. 2D).

A recent report also shows that the combination of TNF-α with sulindac enhances TNF-α–mediated apoptosis in two non-small cell lung carcinoma cell lines *in vitro*.[5] Although we demonstrated more potent pro-apoptotic activity of the combination therapy (TNF-α and sulindac) only in human gastric and cervical adenocarcinoma cell lines, these data suggest that the combination enables TNF-α to be applicable for therapies against a wide range of carcinomas. In addition, inhibition of NF-κB is a pivotal role of sulindac for its augmentation of pro-apoptotic functions of TNF-α, allowing us to consider that NF-κB inhibitors are promising anticancer agents as a partner of TNF-α.

REFERENCES

1. JOHNSTONE, R.W., A.A. RUEFLI & S.W. LOWE. 2002. Apoptosis: A link between cancer genetics and chemotherapy. Cell **108:** 153–164.
2. KARIN, M. & A. LIN. 2002. NF-kappaB at the crossroads of life and death. Nat. Immunol. **3:** 221–227.
3. YIN, M-J., Y. YAMAMOTO & R.B. GAYNOR. 1998. The anti-inflammatory agents aspirin and salicylate inhibit the activity of I (kappa)B kinase-beta. Nature **396:** 77–80.
4. YAMAMOTO, Y., M-J. YIN, K.M. LIN & R.B. GAYNOR. 1999. Sulindac inhibits activation of the NF-kappaB pathway. J. Biol. Chem. **274:** 27307–27314.
5. BERMAN, K.S., U.N. VERMA, G. HARBURG, *et al.* 2002. Sulindac enhances tumor necrosis factor-α-mediated apoptosis of lung cancer cell lines by inhibition of nuclear factor-κB. Clin. Cancer Res. **8:** 354–360.

FIGURE 2. *In vivo* effect of sulindac and further augmentation of sulindac sulfide on TNF-α mediated apoptosis. (**A**) Combination of TNF-α with sulindac strongly reduced in vivo tumor progression. Viable MKN45 cells (1×10^6 cells) were subcutaneously inoculated into the mice. At approximately 20 days after inoculation, subcutaneous tumors became visible and were then treated or untreated with Sul (1 μmol/each injection) and/or TNF-α (20 ng/each injection) on days 0, 3, 6, 9, 12, and 15. Calculated tumor sizes of MKN45 cells treated with Sul (*thin line with open squares*), TNF-α (*broken line with open circles*) both (*thick line with closed triangles*), or untreated (*thick line with closed squares*) are shown. Bars are the means ± S.E. of calculated tumor sizes of subcutaneous tumors. (N=11 per each group). (**B**) Sulindac sulfide (S/sulfide) augmented TNF-α–mediated apoptosis more strongly in MKN45 cells. MKN45 cells were treated with 0.2 mM Sul or 0.2 mM S/sulfide for 2 h prior to incubation with or without 20 ng/mL TNF-α for 24 hours. (**C**) Augmented DNA fragmentation by TNF-α plus sulindac sulfide. MKN45 cells were treated with or without 0.2 mM S/sulfide for 2 h prior to incubation with or without 20 ng/mL of TNF-α for 24 hours. Low-molecular weight DNAs were separated by 1.2% agarose-gel and stained with ethidium bromide. (**D**) Sulindac sulfide inhibited TNF-α–induced NF-κB activation. MKN45 cells were transiently transfected with both pRL-TK and pNF-κB-Luc using the Effectene Transfection Regent. After 24-h transfection, the cells were incubated with the indicated amounts of 0.2 mM sulindac or 0.2 mM S/sulfide for 2 h prior to incubation with or without 20 ng/mL TNF for 8 hours. Cells were harvested to detect luciferase activity using Dual-Luciferase reporter assay system, and luciferase activity was measured using a luminometer and normalized to RL activity. In **B** and **D**, each column displays the mean ± standard deviation (S.D.) of data from three separate experiments.

MNNG Induces Dramatic DNA Damage and Non-Apoptotic Changes in Cervical Carcinoma HeLa Cells

JÓZEFA WĘSIERSKA-GĄDEK, MARIETA GUEORGUIEVA, AND JACEK WOJCIECHOWSKI

Cell Cycle Regulation Group, Institute of Cancer Research, Faculty of Medicine, University of Vienna, Vienna, Austria

ABSTRACT: It has been previously reported that a short treatment of human cervix carcinoma HeLa cells with N-methyl-N′-nitro-N-nitrosoguanidine (MNNG) induced apoptosis. We examined the action of MNNG on HeLa cells and compared it with that of cisplatin. MNNG damaged the integrity of the cell membrane and killed the cells within 3 hours. During this period no changes characteristic for apoptosis, such as cell shrinkage, condensation of nuclei, chromatin fragmentation or activation of caspases, could be detected. However, the exposure of HeLa cells to 50 μM MNNG for 1 h resulted in dramatic DNA damage. The MNNG-induced disruption of cell membrane associated with cell death indicates that HeLa cells die by necrosis.

KEYWORDS: apoptosis; DNA breaks; necrosis; Comet assay; single-cell gel electrophoresis

INTRODUCTION

Cell death is a crucial phenomenon in normal development of organisms and in prevention of diseases. Abnormalities in the regulation of cell death lead to various pathologies. There is more than one way to die.[1] It is widely acknowledged that depending on the type of stress stimuli, cells undergo necrosis or apoptosis. Necrosis occurs in response to harmful insults such as hyperthermia, physical damage, or chemical injury. Apoptosis is implicated in cell turnover in normal adult tissues, such as intestinal crypts; in the immune and hematopoietic system; or after cell injury due to a variety of agents, such as radiation, viruses, and chemotherapeutic agents.[1] The two types of cell death are manifested by different morphological changes and distinct patterns of cell resolution. The earliest morphological changes that occur during necrosis are swelling of the cytoplasm and organelles. This leads to the loss of selective permeability of the plasma membrane followed by total dissolution of the organelles. In contrast to necrosis, apoptosis shows a morphologically

Address for correspondence: Prof. Dr. Jozefa Gadek-Wesierski, Cell Cycle Regulation Group, Institute of Cancer Research, Borschkegasse 8 a, A-1090 Vienna, Austria. Voice: +(43-1)-4277-65247; fax +(43-1)-4277-65194.

Jozefa.Antonia.Gadek-Wesierski@univie.ac.at

Ann. N.Y. Acad. Sci. 1010: 278–282 (2003). © 2003 New York Academy of Sciences.
doi: 10.1196/annals.1299.048

distinct pattern of cell resolution.[1] The onset of apoptosis includes the loss of cell junctions and condensation of the cytoplasm leading to cell shrinkage. The nucleus condenses and coalesces into several large structures, which then break up into fragments. The mitochondria initially remain apparently intact. We recently extensively investigated distinct steps of apoptosis in HeLa cervical carcinoma cells induced by cisplatin or other anti-cancer agents.[2,3] We observed that apoptosis is a slow biphasic process executed within 24 hours. Therefore, we wondered about research reporting that a short treatment of HeLa cells with N-methyl-N′-nitro-N-nitrosoguanidine

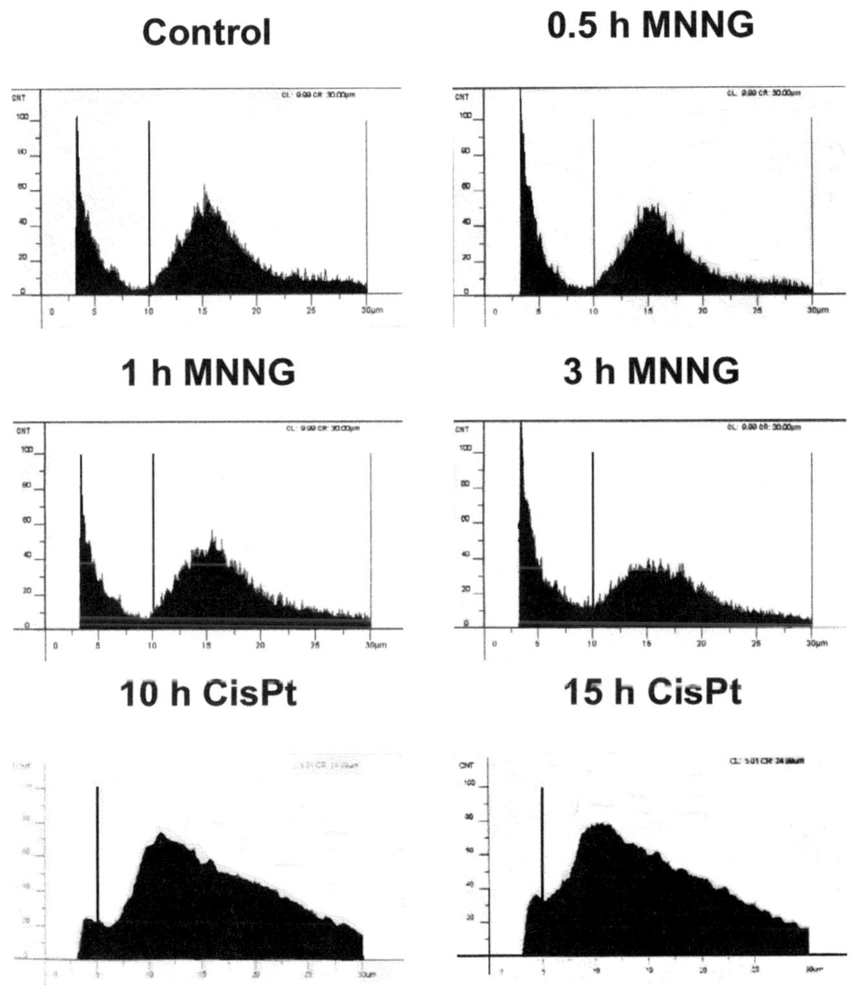

FIGURE 1. Swelling of HeLa cells after treatment with 50 μM MNNG. The volume of exponentially growing HeLa cells treated with 50 μM MNNG for 0.5 h, 1 h, and for 3 h or with 40 μM CP for 10 h and 15 h was measured. Each measurement was repeated three times using at least 4,000 cells.

(MNNG)–induced apoptosis.[4] It has been reported that apoptosis was induced within 10 min upon exposure of HeLa cells to 50 μM MNNG. However, the authors did not show any evidence of apoptotic changes.[4] MNNG is a strong alkylating agent resulting in DNA double-strand breaks.

Therefore, we examined the action of MNNG on HeLa cells and compared it with that of cisplatin (CP). MNNG damaged the integrity of the cell membrane and killed the cells within 3 hours. During this period no changes characteristic for apoptosis could be detected. However, the exposure of HeLa cells to 50 μM MNNG for 1 h resulted in dramatic DNA damage. The MNNG-induced disruption of cell membrane associated with cell death indicates that HeLa cells die by necrosis.

MATERIAL AND METHODS

Cell Culture and CASY

The human cervical carcinoma cell line $HeLaS_3$ was cultured in Dulbecco's Modified Eagle Medium (DMEM) supplemented with 10% fetal calf serum. Cells were grown up to 60–70% confluence and then treated with 5 μM and 50 μM MNNG or with 40 μM cisplatin for indicated periods of time. Cell volume was determined by CASY.

Comet Assay

The generation of DNA-strand breaks was assessed by the single cell gel electrophoresis assay performed under alkaline conditions.[5] The experiments were carried out according to the guidelines published by Tice and colleagues.[5] Cell suspensions (1×10^5 cells) were mixed with LMP agarose and spread on agarose-precoated slides. After electrophoresis, slides were neutralized and stained with propidium iodide (20 μg/mL). Comet tail length (μm) and tail moment were measured under a fluorescence microscope (Nikon Model 027012) using an automated image analysis system based on a public domain NIH image program.

RESULTS AND DISCUSSION

We examined the action of MNNG on HeLa cells and compared it with that of CP. In the first step we determined the direct cytotoxicity of MNNG on HeLa cells by dye exclusion. We observed that exposure of cells to 50 μM MNNG resulted in accumulation of Trypan blue. After 1 h, about 15% of cells were positive and after 3 h the majority of HeLa cells accumulated the dye. It shows that the treatment with MNNG impaired the integrity of the plasma membrane. Determination of cell volume revealed slight cell swelling observed after 1 h and 3 h treatment with MNNG (FIG. 1). During this period no changes characteristic for apoptosis such as cell shrinkage assessed by CASY (FIG. 1) or condensation of nuclei or chromatin fragmentation determined by inspection of Hoechst-stained cells could be detected (not shown). During this time no activation of caspases was observed either (not shown). However, the exposure of HeLa cells to 50 μM MNNG for 1 h resulted in dramatic

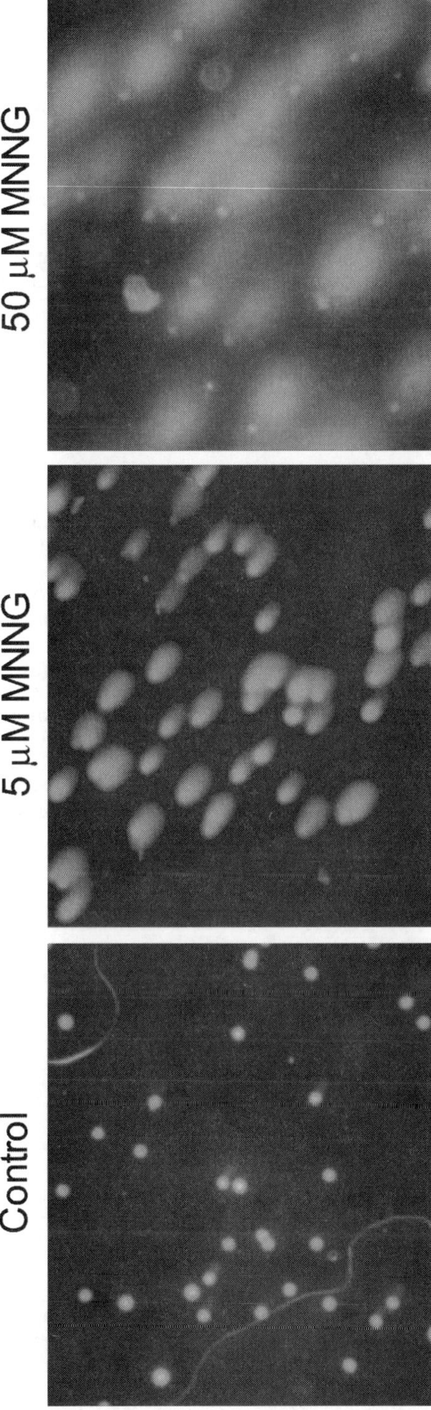

FIGURE 2. Strong DNA damage in HeLa cells after treatment with MNNG. Untreated HeLa cells and cells treated with 5 µM or with 50 µM MNNG for 1 h were subjected to the single-cell gel electrophoresis assay. The DNA was stained with propidium iodide. Three independent experiments were performed; in each experiment cells harvested from three distinct petri dishes were resolved by electrophoresis. Fifty cells per petri dish were evaluated.

DNA damage as determined by single cell electrophoresis (FIG. 2). The Comet assay was not evaluable by the image analysis program. Damaged DNA migrated almost completely out of the nuclei and there remained only residual DNA anchored to the nuclear envelope or in the nucleoli. Even treatment of cells with a tenfold lower MNNG concentration resulted in strong DNA strand breaks (FIG. 2).

These results evidence that MNNG-treated HeLa cells did not induce apoptosis despite strong DNA damage. In response to genotoxic stimuli cells initiate two different pathways of cell death: apoptosis and necrosis. The lack of cell shrinkage, condensation of nuclei, and activation of caspases accompanied by impairment of plasma membrane and cell swelling indicate that MNNG is directly cytotoxic and kills cells by necrosis. It seems that the severe DNA damage generated by MNNG is responsible for induction of rapid cell death.

REFERENCES

1. VERMES, I. & C. HAANEN. 1994. Apoptosis and programmed cell death in health and disease. Adv. Clin. Chem. **31:** 177–246.
2. HORKY, M., G. WURZER, V. KOTALA, *et al.* 2001. Segregation of nucleolar components coincides with caspase-3 activation in cisplatin-treated HeLa cells. J. Cell Sci. **114:** 663–670.
3. WESIERSKA-GADEK, J., D. SCHLOFFER, V. KOTALA & M. HORKY. 2002. Escape of p53 protein from E6-mediated degradation in HeLa cells after cisplatin therapy. Int. J. Cancer **101:** 128–136.
4. KUMARI, S.R., H. MENDOZA-ALVAREZ & R. ALVAREZ-GONZALEZ. 1998. Functional interaction of p53 with poly(ADP-ribose) polymerase (PARP) during apoptosis following DNA damage: covalent poly(ADP-ribosyl)ation of p53 by exogenous PARP and noncovalent binding of p53 to the Mr 85,000 proteolytic fragment. Cancer Res. **58:** 5075–5078.
5. TICE, R.R., E. AGURELL, D. ANDERSON, *et al.* 2000. Single cell gel/comet assay: guidelines for in vitro and in vivo genetic toxicology testing. Environ. Mol. Mutagen. **35:** 206–221.

Thallium Induces Apoptosis in Jurkat Cells

MARCANTONIO BRAGADIN,[a] ANTONIO TONINELLO,[b] ALBERTO BINDOLI,[c] MARIA PIA RIGOBELLO,[b] AND MARCELLA CANTON[b]

[a]*Dipartimento di Scienze Ambientali, Università di Venezia, DD2137 30123 Venezia, Italy*

[b]*Dipartimento di Scienze Biologiche, Università di Padova, Viale G. Colombo 3, 35121 Padova, Italy*

[c]*Istituto di Neuroscienze (CNR), Sezione di Biomembrane, Università di Padova, iale G. Colombo 3, 35121 Padova, Italy*

ABSTRACT: Thallium induces swelling in mitochondria and apoptosis in Jurkat cells. Both the swelling and the apoptosis are inhibited by means of cyclosporine A (CsA), thus suggesting that these phenomena result from the opening of the membrane transition pore (MTP) in mitochondria. Therefore, apoptosis can be explained as due to the opening of the MTP in mitochondria.

KEYWORDS: mitochondria; thallium; apoptosis

INTRODUCTION

The transport of Tl^+ across biological membranes has already been studied, either as a particular study regarding the transport of monovalent cations, or as a biological study in order to explain the toxicity of Tl^+ at a molecular level.[1–7] It has already been evidenced that Tl^+ penetrates the membranes of erythrocytes and bacteria by means of an electrophoretic mechanism[3,4] and that, once inside, Tl^+ induces a swelling in the mitochondria. In isolated mitochondria, Tl^+ induces a swelling, a stimulation of the respiratory rate, a collapse in the membrane potential, the release of endogenous Ca^{2+} and an uncoupling effect.[8–12]

All these phenomena have been up to now interpreted as due to an electrophoretic uptake of Tl^+ accompanied by an uncoupling effect, which collapses the potential, $\Delta\Phi$, induces the release of Ca^{2+}, and stimulates the respiratory chain.

In this paper we have reconsidered the interactions of Tl^+ with cells and mitochondria. In particular, we have further investigated the previously analyzed swelling mechanism, taking into account the fact that recent observations have suggested that many metals induce the opening of a large size pore (MTP), with a consequent swelling of the mitochondria and inhibition of the oxidative phosphorylation.[13,14] The results of this paper do, in fact indicate that Tl^+ induces swelling both in isolated mitochondria and in cells. However, this swelling is inhibited by cyclosporine A

Address for correspondence: Marcantonio Bragadin, Dipartimento di Scienze Ambientali, Venezia, Italy. Voice: +39 41 234 8507; fax: +39 41 234 8584.
bragadin@unive.it

Ann. N.Y. Acad. Sci. 1010: 283–291 (2003).
doi: 10.1196/annals.1299.049

(CsA), thus suggesting that the phenomenon which causes cell death by apoptosis, results from the opening of the MTP in mitochondria. This interpretation can not only explain all previously observed phenomena, but also those relating to cell death from apoptosis and the toxicity of Tl^+.

MATERIALS AND METHODS

Cell Culture

The human T-lymphoblastoid Jurkat cell line was cultured in an RPMI 1640 medium supplemented with 10% heat-inactivated fetal bovine serum, 2 mM of L-glutamine, 100 IU/mL of penicillin, and 100 mg/mL of streptomycin (Life Technologies, Paisley, Scotland) in 5% CO_2-95% air at 37°C in a humidified atmosphere.

Assessment of Cell Viability

To assess the changes in nuclear morphology typical in apoptosis, the Jurkat cells (10^6 cells/mL), resuspended in a serum-free medium, were stained with Hoechst 33258 solution and 1 μM propidium iodide for 5 min following Tl^+ treatment. The cells were then washed with Hanks balanced salt solution and visualized with the fluorescence microscope using excitation/emission cubes of 340/440±25 nm and 568/585±25 nm longpass filter for Hoechst 33258 and propidium iodide, respectively. Three randomly selected fields were acquired from each treatment. The corresponding bright-field images were also acquired, and three channels were overlaid using the appropriate function of the Metamorph software. The images were stored and the cells in each field were counted and illustrated as detailed in FIGURE 1.

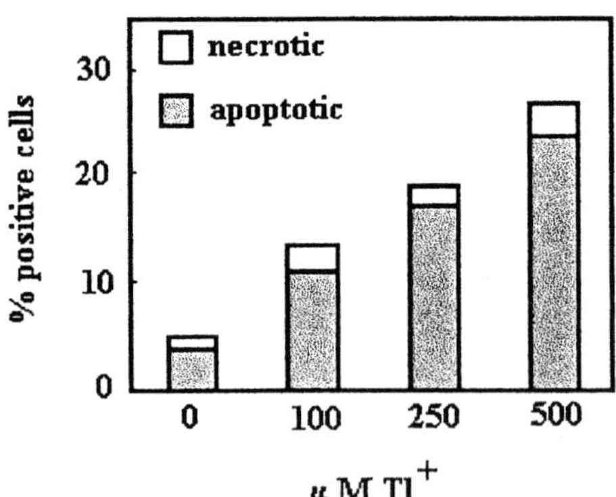

FIGURE 1. Loss of viability induced by Tl^+ in Jurkat cells. Jurkat cells (10^6/mL) were incubated with increasing concentrations of thallium acetate (0.25–1 mM) at 37°C for 16 hours. Necrosis was assessed by the staining of nuclei with PI (propidium iodide) whereas apoptosis was visualized by the appropriate changes of nuclei stained with Hoechst 33258. Values are the mean value of at least three experiments.

TMRM Staining and Imaging

After treatment with increasing concentration of Tl^+, Jurkat cells were washed with Hank's balanced salt solution supplemented with 10 nM Hepes, pH 7.4 and then incubated for 30 min at 37°C with 25 nM TMRM. CsA (1.6 μM) was also added to inhibit multidrug-resistance pumps, which can affect TMRM loading. Cellular fluorescence images were acquired with an Olympus IMT-2 inverted microscope.

For the detection of TMRM fluorescence, 568 ± 25 nm excitation and 585 nm longpass emission filter settings were used. Data were acquired and analyzed using Metamorph software (Universal Imaging). Mitochondria were identified as regions of interest (ROI) and at least 30 ROIs were considered for each experiment. The decrease in fluorescence intensities induced by FCCP was expressed as the difference of the values obtained before and after the addition of the uncoupler. The values of these differences obtained in the treated cells were normalized to the difference values of the untreated cells (control).

Mitochondrial and Lysosomal Preparation

The rat liver mitochondria and lysosomes were prepared using the conventional methods.[15,16] The protein concentration was determined by means of the Lowry method.[17]

The spectrophotometric measurements were performed using a Jenway 6400 Spectrophotometer. The oxygen uptake was determined using a Clark oxygen electrode placed in a thermostated (20°C) chamber, equipped with a magnetic stirring mechanism. All the reagents were of an analytical grade. The *p*-trifluoro-methoxycarbonyl cyanide phenylhydrazone (FCCP), acridine orange, and thallium acetate were supplied by FLUKA. The cytochrome *c* and CsA were supplied by Sigma, all the other reagents were supplied by Molecular Probes.

RESULTS AND DISCUSSION

Tl^+ induces cell death by means of an apoptosis mechanism in Jurkat cells. As the extent of the phenomenon is dependent on the concentration of Tl^+, FIGURE 1 shows the percentage of cell death by apoptosis produced by increasing amounts of Tl^+.

Since the apoptosis can result from the opening of the CsA-sensitive pore (MTP) in mitochondria, the experiments in FIGURE 1 were performed with addition of CsA. The results indicate (FIG. 2) that CsA inhibits the Tl^+ induced apoptosis, thus suggesting that the apoptosis caused by Tl^+ is effectively due to the opening of the MTP in mitochondria.

This behavior was further confirmed by the experiments shown in FIGURE 3. Since the opening of the MTP necessarily induces a collapse in the membrane potential, this figure shows the decrease in mitochondrial potential, $\Delta\Phi$, in Jurkat cells produced by increasing amounts of Tl^+. Further experiments were performed on isolated mitochondria in order to confirm the hypothesis, that Tl^+ is an inducer of the opening of the MTP in mitochondria.

This phenomenon is evidenced by means of swelling experiments. The mitochondria are resuspended in a sucrose medium and the opening of a large size pore (the

MTP allows for the transport of solutes with molecular masses of about 1,500 Da[18]) allows for the entrance of sucrose into the mitochondrial matrix from the resuspending medium. The entrance of sucrose is accompanied by that of water and the result is mitochondrial swelling. The swelling is monitored by means of an absorbance decrease at 540 nm and, if it results from the opening of the MTP, it is inhibited by CsA.[13,14]

FIGURE 4 shows the swelling induced by phosphate (Pi) in the presence of Ca^{2+} in the resuspending medium. Pi is a typical inducer of the opening of the MTP since the swelling is inhibited by CsA (FIG. 4). Analogously, Tl^+ induces swelling in isolated mitochondria (FIG. 5) and, also in this case, the swelling is inhibited by CsA,

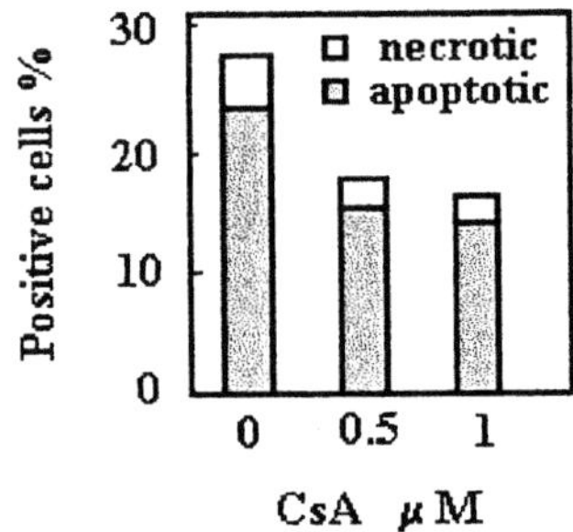

FIGURE 2. CsA protection on the apoptosis induced by Tl^+. Jurkat cells (10^6/mL) were preincubated in the absence or presence of CsA (0.5–1 mM) for 1 h and were subsequently treated with 1 mM thallium acetate for 16 hours. Necrosis was assessed by the staining of nuclei with PI, whereas apoptosis was visualized by the appropriate changes of nuclei stained with Hoechst 33258. Values are the mean of at least three experiments.

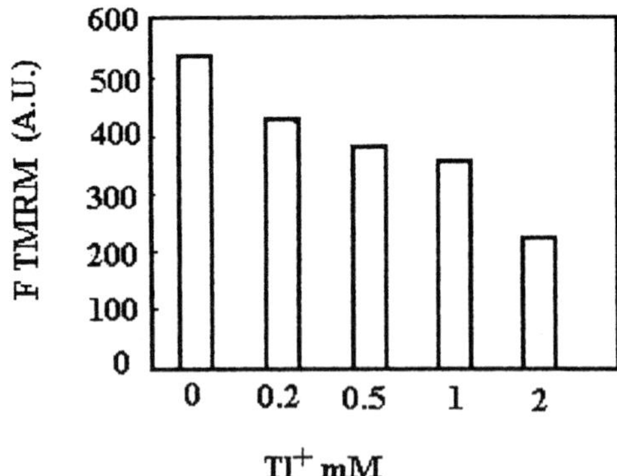

FIGURE 3. $\Delta\Phi$ decrease induced by Tl^+ collapse in Jurkat cells. Jurkat cells (10^6/mL) were treated with thallium acetate as described in FIGURE 1. After 16 h of incubation, cells were stained with 25 nM TMRM and 1.6 mM CsA as detailed in MATERIALS AND METHODS.

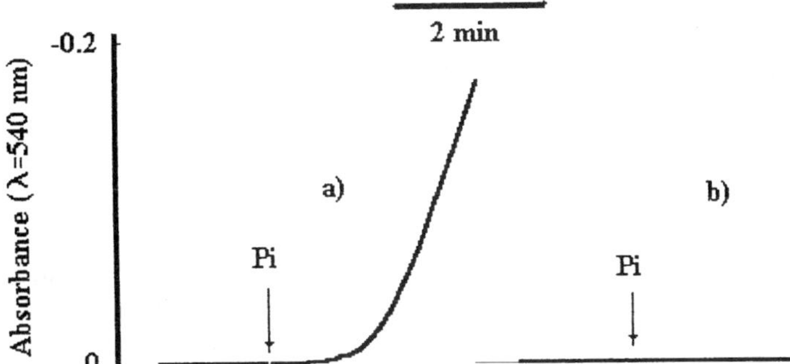

FIGURE 4. Swelling of mitochondria induced by phosphate (Pi) and its inhibition by CsA. The mitochondria (0.5 mg/mL) were resuspended in 2.5 mL of the operating medium (medium composition: 0.25 M sucrose, 2 mM succinate, 0.1 mM Ca^{2+}, 10 mM Hepes-Mops pH 7.4). The spectrophotometer was then adjusted to zero absorbance (λ=540 nm) before the addition of 2 mM phosphate (Pi) (**a**). In (**b**), the same experiment was performed as in (**a**), but the medium contained 2 γ CsA.

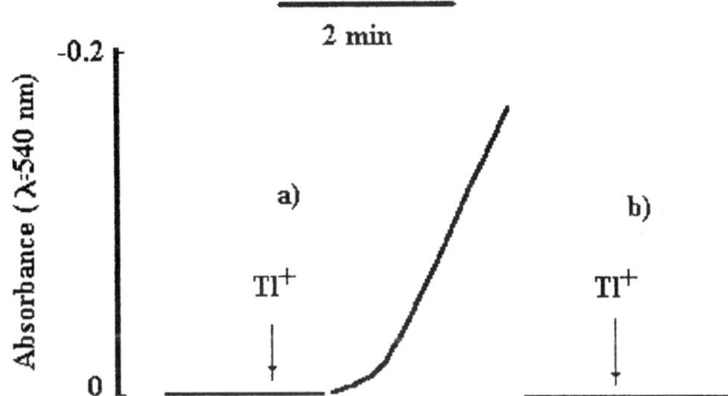

FIGURE 5. Swelling of mitochondria induced by thallium and its inhibition by CsA. Rat liver mitochondria were incubated as in FIGURE 4. Tl^+ 200 μM (in the form of thallium acetate) was added when indicated. In (**a**), no CsA was present, in (**b**), the medium contained 2 γ of CsA. The dose of Tl^+ necessary to induce swelling can depend on the preparation and above all on the aging of mitochondria. Therefore, the dose of 100 μM Tl^+ is indicative of the concentration range, but in all tested experiments, the swelling is CsA dependent, thus indicating that it is due to the opening of the MPT.

thus suggesting that the swelling is due to the opening of the MTP. This sensitivity to CsA indicates that the possibility that the swelling is due to other causes (such as the presence of native or artificial antiporters, transport of solutes[13]) can be excluded from now on, since only the swelling induced by the opening of the MTP is CsA sensitive.

The above interpretation is different from the one already proposed in order to explain the mitochondrial swelling induced by Tl^+.[9,10] In fact, the interpretation in Skulskii and colleagues[10] indicated that Tl^+ enters the mitochondria by means of an electrophoretic mechanism. This Tl^+ uptake was said to be followed by a Tl^+ release, in all probability resulting from an "increased proton permeability" (caused by Tl^+) or from a "release of endogenous Ca^{2+} followed by a reuptake."

Both of these effects can now be easily explained as resulting from the opening of the MTP (which occurs, as is true in the case of many other metals,[14] at low Tl^+ concentrations). The opening of the pore explains the Tl^+ uptake and subsequent release, the uncoupling effect, the collapse of the potential and the release of the endogenous Ca^{2+}.[13,14,18]

In Melnick and colleagues,[9] the authors observed that, after the addition of Tl^+ to the mitochondria, not only a swelling, but also a stimulation of the State 4 mitochondrial respiration took place. This last phenomenon was interpreted as due to an uncoupling effect induced by the Tl^+, since it is well known that an uncoupler stimulates the respiratory rate. We are now able to confirm here that Tl^+ induces a respiratory rate stimulation (FIG. 6, a). In this figure it can be seen that, after some minutes, the stimulation is followed by a respiratory rate inhibition.

Our interpretation is that the stimulation is due to a $\Delta\Phi$ collapse consequent to the opening of the pore. The subsequent inhibition results from the cytochrome *c* release which occurs after the swelling.

FIGURE 6(b) shows that, in the presence of cytochrome *c* in the resuspending medium, the respiratory rate stimulation by Tl^+ was not followed by inhibition, while in FIGURE 6(c), where CsA is present, no stimulation of the respiratory rate by Tl^+ was observed. FIGURE 6(d) shows that, in the presence of CsA, the FCCP uncoupler induces a respiratory rate stimulation, thus demonstrating that Tl^+ is not an uncoupler.

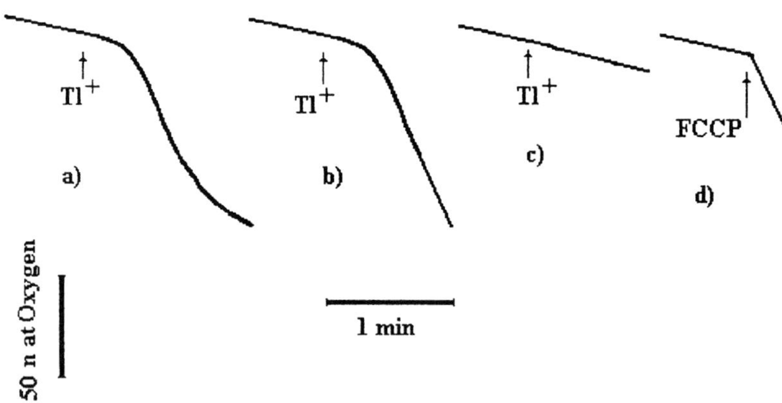

FIGURE 6. Respiratory rate of mitochondria: the effects of the thallium and FCCP. Medium composition: 0.25 M sucrose, 10 mM Hepes-Mops pH 7.4, 2 mM succinate, 2 μM rotenone. The addition of 200 μM Tl^+ to the medium (2 mL) containing the mitochondria (0.5 mg/mL) induces an enhancement of the respiratory rate (**a**), which is then followed by an inhibition. In (**b**), the conditions were as in (**a**), but the medium contained 200 μM of cytochrome *c*. In (**c**), the medium composition was as in (**a**), but contained in addition, 1γ of CsA. In (**d**), the addition of FCCP (50 nmoles/mg protein), an uncoupler, to the medium containing 1 γ CsA, induces a rapid increase in the respiratory rate.

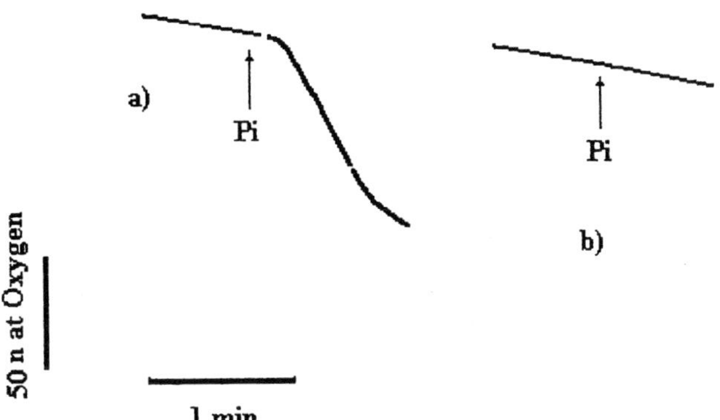

FIGURE 7. The effect of phosphate (Pi) on the respiratory rate of mitochondria. Both the medium composition and the mitochondrial concentration were as in FIGURE 3. The addition of 2 mM of phosphate (Pi) to the respiring mitochondria induced a fast increase in the respiratory rate in (**a**), followed by an inhibition. In (**b**), the medium contained 2γ of CsA and the addition of 2 mM of Pi did not induce any increase in the respiratory rate.

The respiratory rate stimulation obtained by means of Tl^+ is similar to that observed when Pi is added (FIG. 7, a), since Pi also induces the opening of the MTP. Once again the stimulation was not observed in the presence of CsA (FIG.7, b) since, under these conditions, the opening of the pore does not occur. It should be noted that the effect of Tl^+ on the respiratory rate is not instantaneous, but takes place after about one minute. This period of time is of the same order of magnitude as that required to induce the opening of the MTP (FIG. 5).

In order to further exclude the possibility that Tl^+ is an uncoupler, we made use of the response given by energized lysosomes. The internal pH in energized lysosomes is acid resulting from two factors: the ATP-driven proton pump pumps the protons into the matrix and a selective channel allows for the transport of the chloride ions into the matrix.[20–22] The entry of the chloride ions (partially) neutralizes the positive potential created by the proton pump, thus favoring the acidification process and consequently the internal pH reaches a value of about 4.2 (the lysosomal membrane being impermeable to the protons). The pH is measured using an acridine orange (AO) probe. AO is a weak permeant base and, following the mechanism of the permeant bases,[20–22] AO accumulates inside the lysosomes, whose driving force is the acid pH inside. The accumulation of this dye is followed by an absorbance quenching. Following the absorbance quenching at 492 nm, it is subsequently possible to measure the internal pH in the lysosomes. FIGURE 8 shows a typical experiment concerning the measurement of the internal pH in lysosomes under energized conditions. Some minutes after the addition of ATP, the steady state indicates an internal pH of about 4.2. The addition of an uncoupler which permeates the membrane to the protons induces a fast internal alkalinization while the addition of Tl^+ (5 mM!) does not induce any alkalinization in energized lysosomes. This experiment demonstrates that Tl^+ does not alter the membrane permeability to the protons.

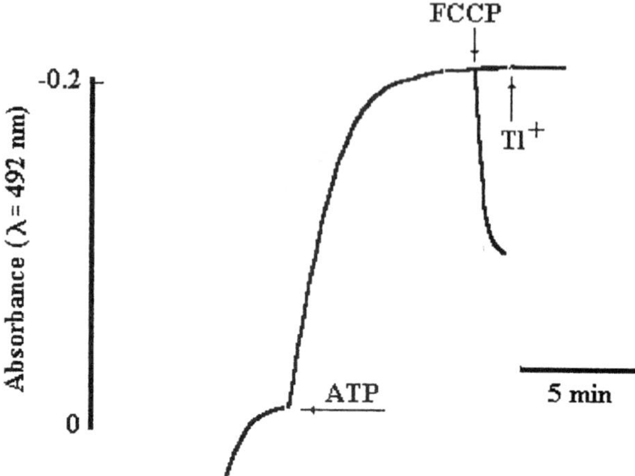

FIGURE 8. The effects of FCCP and Tl^+ on energized lysosomes. The lysosomes (0.2 mg/mL) were resuspended in 2.5 mL of the incubating medium (medium composition: 125 mM KCl, 10 mM $MgCl_2$, 10 mM Hepes pH 7.4). After the addition of 5 μM of acridine orange (AO), the spectrophotometer (λ=482 nm) was adjusted to zero absorbance. The addition of 2 mM of ATP induced an absorbance quenching. After about 15 min, the addition of 1 μM FCCP in steady state, induced a fast internal alkalinization, while the addition of 5 mM Tl^+ (as acetate) did not induce any effect on the energized lysosomes.

CONCLUSIONS

In conclusion, our experiments indicate that: (1) Tl^+ induces apoptosis in cells; (2) the apoptosis is inhibited by CsA; and (3) Tl^+ induces a $\Delta\Phi$ collapse in cells. This behavior suggests that Tl^+ induces apoptosis since it induces the opening of the MTP in mitochondria. In order to confirm this hypothesis, experiments were performed on isolated mitochondria. The results indicate that: (1) Tl^+ induces the opening of the MTP, since the swelling is inhibited by CsA; (2) as a consequence of the swelling, a release of cytochrome *c* occurs, which causes a respiratory rate inhibition; (3) the respiratory rate stimulation and the release of cytochrome *c* are inhibited by CsA.

The experiments performed on isolated mitochondria thus confirmed the starting hypothesis regarding the mechanism that causes apoptosis. In addition, the mechanism outlined in this paper explains the previously observed phenomena, in particular the swelling[9,10] and the release of cytochrome *c*, which accompany the apoptosis.

Our interpretation excludes the possibility that Tl^+ is an uncoupler as suggested by other researchers.[9–11] The opening of the MTP can explain the toxicity of Tl^+.

REFERENCES

1. Gehring, P.J. & P.B. Hammond. 1964. the uptake of thallium by rabbit erythrocytes. J. Pharm. Exp. Ther. **145:** 215–221.
2. Spencer, P.S., E.R. Peterson R.A. Madrid & C.S. Raine. 1973. Effects of thallium salts on neuronal mitochondria in organotypic cord-ganglia-muscle combination cultures, J. Cell Biol. **58:** 79–75.

3. SKULSKII, I.A., V. MANNINEN & J. JARNEFELT. 1976. Factors affecting the relative magnitudes of the ouabain-insensitive fluxes of thallium ion in erythrocytes, Biochim. Biophys. Acta **506:** 233–241.
4. SKULSKII, I.A., V.V. GLASUNOV, I.D. RJABOVA & G.A. GORNEVA. 1977. Selective permeability of bacterial membranes for monovalent thallium ions, Biochimija **42:** 1637–1656.
5. CAVIERAS, J.D. & J.C. ELLORY. 1974. Thallium and the sodium pump in human red cells, J. Physiol. (London) **243:** 243–266.
6. CORNELIUS, G., W. GARTNER & D.H. HAYNES. 1974. Cation complexation by valinomycin and nigerycin-type ionophores registered by the fluorescence signal of Tl^+. Biochemistry **13:** 3052–3057.
7. NEHER, E. 1975. Ion specificity of the gramicidin channel and the thallous ion, Biochim. Biophys. Acta **401:** 540–544.
8. SKULSKII, I.A., L.N.E. SARIS, M. SAVINA & V.V. GLASUNOV. 1981. Uptake of thallous ions by mitochondria is stimulated by nonactin but not by respiration alone. Eur. J. Biochem. **120:** 263–266.
9. MELNICK, R.L., L.G. MONTI & S.M. MOTZKIN. 1976. Uncoupling of mitochondrial oxidative phosphorylation by thallium. Biochem. Biophys. Res. Commun. **69:** 68–73.
10. SKULSKII, L.A., M. SAVINA, V.V. GLASUNOV & N.E. SARIS. Electrophoretic transport of Tl^+ in mitochondria. J. Membr. Biol. **44:** 187–194.
11. SARIS, N.E., I.A. SKULSKII, M.V. SAVINA & V.V. GLASUNOV. 1981. Mechanism of mitochondrial transport of thallous ions. J. Bioenerg. Biomembr. **13:** 51–59.
12. DIWAN, J.J., R. PALIWAL E. KAFTAN & R. BAWA. 1990. A mitochondrial protein fraction catalyzing transport of the K+ analog Tl^+. FEBS Lett. **273:** 215–218.
13. BERNARDI, P. 1999. Mitochondrial transport of cations: Channels, exchangers, and permeability transitions. Physiol. Rev. **79:** 1127–1155.
14. ZORATTI, M. & I. SZABÒ. 1995. The mitochondrial permeability transition Biochim. Biophjs. Acta **1241:** 139–176.
15. BRAGADIN, M., T. POZZAN & G.F. AZZONE. 1980 Kinetics of Ca^{++} carrier in rat liver mitochondria. Biochemistry **18:** 5972–5978.
16. DELL'ANTONE, P. 1979. Evidence for an ATP-driven "proton pump" in rat liver lysosomes by basic dyes uptake. Biochem. Biophys. Res. Commun. **86:** 180–189.
17. LOWRY, O.H., N.J. ROSENBROUGH, A.L. FARR & R.J. RANDALL. 1951. Protein measurement with folin phenol reagent. J. Biol. Chem. **193:** 265–275.
18. BERNARDI, P. & V. PETRONILLI. 1996. The permeability transition pore as a mitochondrial calcium release channel: A critical appraisal. J. Bioenerg. Biomem. **28:** 129–136.
19. JACOBS, E.E. & D.R. SANADI. 1960. The reversible removal of cytochrome c from mitochondria. J. Biol. Chem. **235:** 531–534
20. DELL'ANTONE, P. 1979 Evidence for an ATP-driven "Proton Pump" in rat liver lysosomes by basic dyes uptake. Biochem. Biophys. Res. Commun. **86:** 180–189.
21. MELLMAN, I., R. FUCHS & A. HELENIUS. 1986. Acidification of the endocytic pathways. Ann. Rev. Biochem. **55:** 663–700.
22. DE CAMILLI, P. 1990. Pathways to regulated exocytosis in neurons. Annu. Rev. Physiol. **52:** 625–645.

Oxalate Exposure Promotes Reinitiation of the DNA Synthesis and Apoptosis of HK-2 Cells, a Line of Human Renal Epithelial Cells

SWEATY KOUL,[a] SHERRY FU,[a] AND HARI KOUL

Signal Transduction and Molecular Biology Laboratory, Division of Urology, Department of Surgery, University of Colorado Health Sciences Center, Denver, Colorado 80262, USA

ABSTRACT: Oxalate, a metabolic end product, is a major constituent of kidney stones. Previously, we and others have demonstrated that oxalate is toxic to renal epithelial cells. In the present study, we characterized oxalate-induced cell death in HK2 cells, a line of renal epithelial cells from the human kidney. For these studies, HK2 cells were exposed to oxalate for various time points. Cells were examined for nuclear morphology, DNA fragmentation, and expression of various apoptosis-related proteins. Apoptotic mode of cell death was also confirmed by TUNEL assay. Results from these studies revealed that oxalate exposure resulted in time-dependent increase in DNA fragmentation. Maximum DNA fragmentation was observed at 2–24 hours following oxalate exposure. Results from Western blot analysis demonstrated an increased expression of the FAS ligand. Taken together, these data reveal that oxalate-associated nephrotoxicity results from oxalate-induced apoptosis of renal epithelial cells.

KEYWORDS: oxalate; renal epithelial cells; apoptosis; nephrolothiasis; hyperoxaluria

INTRODUCTION

Oxalate is an organic dicarboxylate and is primarily excreted by the kidney. It is a major constituent of a majority of urinary tract stones. Oxalate-induced alterations in renal epithelial cells are the critical events in crystal retention and clinical stone disease. We demonstrated that oxalate is toxic to renal epithelial cells and that exposure of renal epithelial cells to oxalate resulted in selective activation of p38 MAP kinase.[1–3] Moreover, we observed a net cell loss at high levels of oxalate, despite increased DNA synthesis. In the present studies, we characterized oxalate-induced cell death in HK-2 cells (a line of human proximal tubular epithelial cells). The results from these studies reveal that oxalate exposure results in apoptosis of the renal epithelial cells.

Address for correspondence: Dr. Hari Koul, M.Sc., Ph.D., FACN, Sr. Director of Research and Professor, University of Colorado Health Sciences Center, 4200 E. 9th Ave., Denver, CO 80262. Voice: 303-315-2383; fax: 303-315-7611.

Hari.Koul@UCHSC.edu

[a]Equal contributors.

Ann. N.Y. Acad. Sci. 1010: 292–295 (2003).
doi: 10.1196/annals.1299.050

METHODS

Cell Culture

For these studies, HK-2 cells, (a line of human renal epithelial cells) were obtained from the American Type Culture Collection (Bethesda, MD). Cells were serially passaged in T-75 flasks, plated at high density in six well plates, and grown in DMEM supplemented with serum (10%) and antibiotics, as described previously.[4] Cell cultures were used at leaset 2–3 days postconfluence.

Quantitative DNA Fragmentation

For these studies, cells were subcultured in 6 well plates and exposed to oxalate (0.4 mM or 1.0 mM) for various intervals of time (1, 2, 4, and 24 h). Cells were washed in PBS, and DNA fragmentation was measured by Boehringer Mannheim cell death detection ELISA, as described by the manufacturer (Cat. No. 1544675).

FACS Analysis

For these studies, cells were grown to confluence in 6 well plates. Confluent cell monolayers were exposed to oxalate (0.4 mM or 1.0 mM) for various intervals of time (24 h). At the end of the experimental period, cells were trypsinized and washed with ice-cold PBS, pH 7.4. After treatments with trypsin and RNase A, cells were stained with propidium iodide and analyzed by flow cytometry. Apoptotic cells were determined by their hypochromic, sub-G1, and staining profiles.

Western Blot Analysis

Apoptosis-related proteins were analyzed by Western blot. For these studies, cells were treated as described for FACS analysis, except that at the end of the experimental period, cells were lysed in denaturing lysis buffer. Cell lysates were run on 10% PAGE and transferred to polyvinylidene difluoride (PVDF) membrane and probed

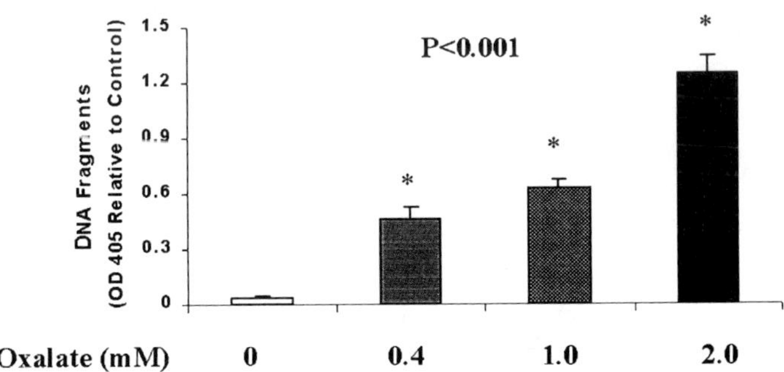

FIGURE 1. Effect of oxalate on the DNA fragments in HK-2 cells: HK-2 cells were exposed to various concentrations (0.4–2.0 mM) of oxalate for 2 h. DNA fragmentation was determined by a Boehringer Mannheim cell death detection ELISA, as described by the manufacturer.

for various apoptosis-related proteins (BCL-2, CAS, Fas, Fas ligand, p53, ICH-1_L, RIP, and TIAR), using monoclonal antibodies, as described previously.[3]

RESULTS

Oxalate Exposure Promotes DNA Fragmentation

Results presented in FIGURE 1 revealed that cytoplasmic DNA fragments were seen within 2 h of oxalate exposure. Oxalate exposure increased DNA fragmentation in a dose-dependent manner. Accumulation of DNA fragments in cytoplasm is an early indicator of apoptosis.

Oxalate Exposure Increases Number of Hypochromic, Sub-G1 Cells

We further characterized this process of cell death following oxalate exposure by flow cytometry. Results presented in FIGURE 2 demonstrated that oxalate exposure increased the number of hypochromic, sub-G1 cells.

We also evaluated the effects of oxalate on cell and nuclear morphology, *in situ* terminal deoxynucleotidyl transferase-mediated dUTP-biotin nick end labeling (TUNEL) method, and the expression of apoptosis-related proteins. Results of these studies (presented at the meeting) demonstrated that oxalate exposure resulted in cytoplasmic vacuolization, nuclear blebbing and pyknosis, and DNA double-strand breaks. Moreover, oxalate exposure did not have any significant effect on the protein levels of any of the apoptotic proteins BCL-2, CAS, p53, ICH-1_L, RIP, and TIAR, but significantly increased the protein levels of Fas ligand.

DISCUSSION

Two distinct forms of eukaryotic cell death can be characterized by morphological and biochemical criteria, necrosis, and apoptosis. Cell swelling and plasma membrane rupture accompany necrosis within minutes of exposure to a necrotic agent. Apoptosis is characterized by membrane blebbing, nuclear and cytoplasmic condensation, and DNA fragmentation. The DNA fragments accumulate in the cytoplasm and can be detected several hours before the plasma membrane breakdown. Apoptosis is therefore also referred to as a programmed mode of cell death. It involves a pro-

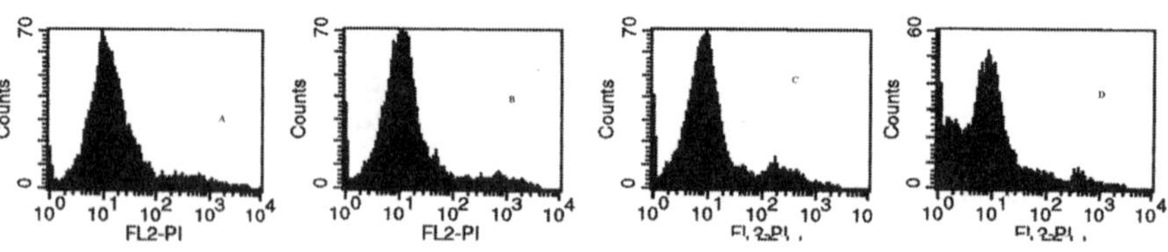

FIGURE 2. FACS analysis of oxalate-treated cells: HK-2 cells were exposed to oxalate for 1 h, harvested and fixed, stained with propidium iodide, and analyzed for DNA staining profiles by FACS analysis.

gram of events rather than physical or chemical cell membrane lysis. Results of our studies presented in brief demonstrate that exposure of proximal tubular epithelial cells to oxalate results in DNA fragmentation, hypochromic nuclei, and increased DNA double-strand breaks, features consistent with cell apoptosis.

The exact mechanism of how oxalate exposure promotes apoptosis is not clearly understood at this time. Understanding the processes that mediate oxalate nephrotoxicity is clinically relevant, since renal tubular damage and dysfunction are observed in patients with hyperoxaluria and nephrolithiasis, and at present no treatment regiments are available to prevent these disorders. Our results demonstrate increased expression of Fas ligand in oxalate-treated cells. Moreover, recent studies from our laboratory demonstrated selective activation of p38 MAP kinase and a mild activation of Jun N-terminal kinase in renal epithelial cells following oxalate exposure.[3] Whether or not these pathways mediate oxalate-induced apoptosis remains to be determined. Studies currently in progress in our laboratory are focused on evaluating the relative contribution of these two signal transduction pathways in mediating oxalate-induced apoptosis of the renal epithelial cells.

ACKNOWLEDGMENT

This work was supported in part by NIH–DK-RO1-54084 to Hari Koul.

REFERENCES

1. Koul, H., L. Renzulli, G. Nair, *et al.* 1994. Oxalate-induced initiation of DNA synthesis in LLC-PK1 cells, a line of renal epithelial cells. Biochem. Biophys. Res. Commun. **205:** 1632–1637.
2. Koul, H., M. Menon & C. Scheid. 1996. Oxalate and renal tubular cells: a complex interaction, Ital. J. Electrol. Metab. **10:** 67–74.
3. Chaturvedi,* L.S., S. Koul,* A. Sekhon, *et al.* 2002. Oxalate selectively activates p38 mitogen-activated protein kinase and c-Jun N-terminal kinase signal transduction pathways in renal epithelial cells. J. Biol. Chem. **277:** 13321–13330. (*Equal contributors.)
4. Bhandari, A., S. Koul, A. Sekhon, *et al.* 2002. Effects of oxalate on HK-2 cells, a line of proximal tubular epithelial cells from normal human kidney. J. Urol. **168:** 253–259.

Triggering of T-Cell Apoptosis by Toxin-Related Ecto-ADP-Ribosyltransferase ART2

FELIX SCHEUPLEIN,[a] SAHAHIL ADRIOUCH,[b] GUSTAVO GLOWACKI,[a] FRIEDRICH HAAG,[a] MICHEL SEMAN,[b] AND FRIEDRICH KOCH-NOLTE[a]

[a]*Institute of Immunology, University Hospital, D-20246 Hamburg, Germany*

[b]*Laboratoire d'Immunodifferenciation, Université Denis Diderot, Paris, France*

ABSTRACT: Cytotoxicity induced by protein ADP-ribosylation is a common theme of certain bacterial toxins and of the mammalian ectoenzyme ART2. Exposure of T cells to NAD, the substrate for ART2-catalyzed ADP-ribosylation, induces exposure of phosphatidylserine, uptake of propidium iodide, and fragmentation of DNA. ART2-specific antibodies raised by gene gun immunization block NAD-induced apoptosis. ART2 catalyzed ADP-ribosylation of cell membrane proteins induces formation of cytolytic membrane pores by activating the P2X7 purinoceptor. This alternative pathway to T cell apoptosis could be triggered upon the release of NAD from intracellular stores, for example, during inflammatory tissue damage.

KEYWORDS: ADP-ribosylation; FACS; phosphatidyl-serine exposure; NAD

The maintenance of homeostasis in the immune system and the focusing of immune reactions on appropriate targets require the coordinated interplay of regulatory mechanisms, many of which involve programmed cell death.[1] Consequently, throughout their life span, T lymphocytes are continually subject to processes that may result in their death.

ADP-ribosyltransferases (ARTs) are NAD-dependent enzymes that post-translationally modify proteins by transferring an ADP-ribose moiety from NAD to specific amino acids (e.g., arginine residues) of target proteins. Several prokaryotic ARTs function as secretory toxins that exert cytotoxic effects on mammalian cells by inactivating key proteins in their target cells. In mammals, a family of toxin-related extracellular ARTs have been identified that are expressed either as GPI-anchored or secreted molecules by different cell types.[2]

Exposure of phosphatidylserine (PS), as evidenced by AnnexinV-binding, is an early indicator of apoptosis in many cell types.[3] Treatment of T lymphocytes with the ART substrate NAD induces rapid T-cell death by apoptosis, as first evidenced by exposure of PS on the outer leaflet of the plasma membrane (FIG. 1A) and ultimately by failure to exclude propidium iodide (PI) (FIG. 1B).

[a]Address for correspondence: Friedrich Koch-Nolte, Institute of Immunology, University Hospital, Martinistr. 52, D-20246 Hamburg, Germany. Voice: +49-40-42803-3612; fax: +49-40 42803-4243.

nolte@uke.uni-hamburg.de

Ann. N.Y. Acad. Sci. 1010: 296–299 (2003). © 2003 New York Academy of Sciences.
doi: 10.1196/annals.1299.051

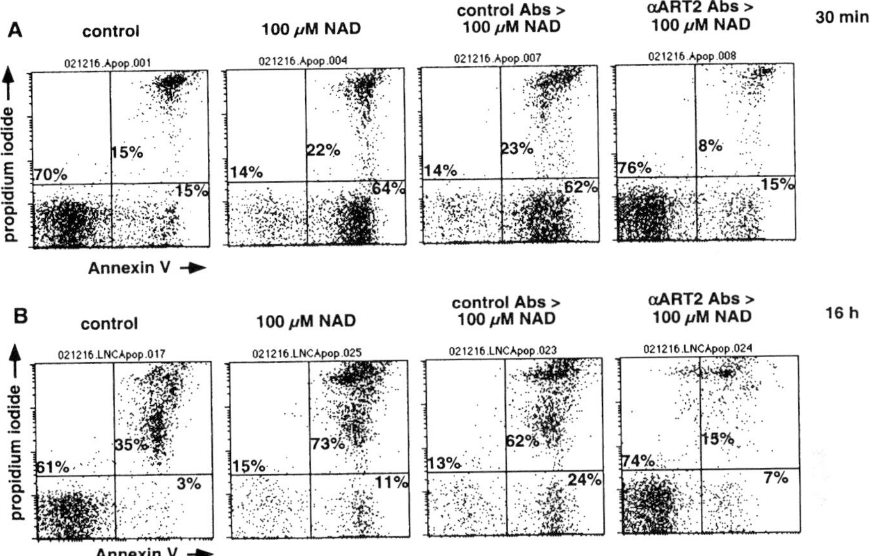

FIGURE 1. Extracellular NAD induces cell death in T lymphocytes. Purified lymph node T cells were preincubated for 30 minutes with preimmune (*panel 3*) or with ART2.2-specific immune serum (*panel 4*) (diluted 1:200 in RPMI-medium). NAD was then added to a final concentration of 100 μM (*panels* 2–4) and cells were incubated further for 30 minutes **(A)** or 16 hours **(B)** before staining with annexing-FITC and propidium iodide. Controls (*panel 1*) were incubated in medium without NAD or antibodies.

Gene-gun immunization with expression constructs encoding ART2.2 induce potent ART2.2-specific antibodies in both, rats and rabbits.[4] To examine whether such antibodies could protect T cells against NAD-induced apoptosis, we incubated T cells with ART2.2-specific or control antibodies for 30 minutes before treatment with NAD. Indeed, pretreatment of T cells with polyclonal immune sera (FIG. 1, panel 3) or a cocktail of anti-ART2.2 monoclonal antibodies (not shown) caused a marked reduction in the level of PS-exposure compared to T cells treated with preimmune sera or isotype-matched control antibodies. Similarly, treatment of cells with phosphatidylinositol-specific phospholipase C (PI-PLC), which removes glycosyl-phosphatidylinositol-anchored ART2.2 from the cell surface, rendered cells resistant to NAD-induced apoptosis.[5]

Naturally occurring deficiencies of ART2 (formerly designated Rt6) have been observed in several mouse and rat models for autoimmune diseases. It is thus an attractive hypothesis that ecto-NAD-induced cell death contributes to the prevention of autoimmunity. We propose that this occurs by eliminating unwanted bystander cells that might otherwise become activated in the course of an immune reaction.

Under normal conditions the concentration of extracellular NAD is low, that is, in the submicromolar range. Micromolar concentrations of extracellular NAD, which are sufficient to trigger NAD-induced cell death, may occur either as a result

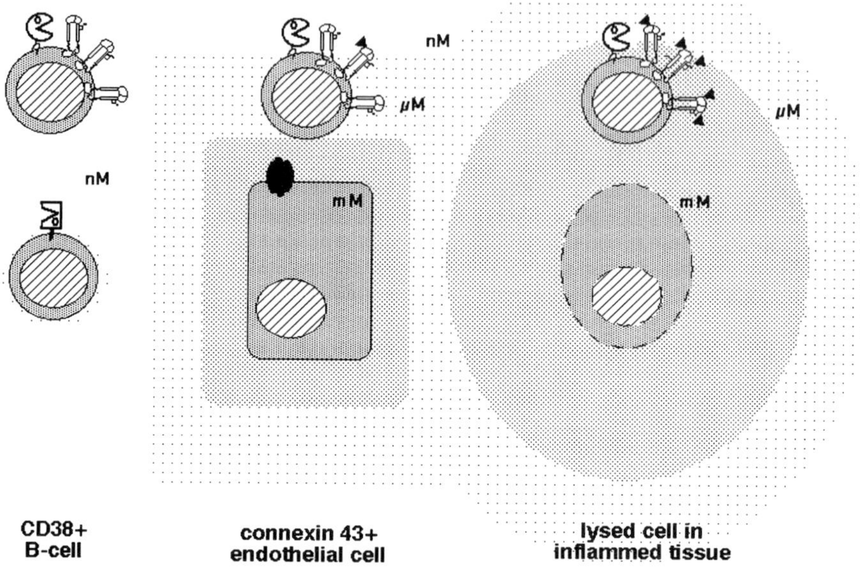

FIGURE 2. NAD may be released from lysed cells during inflammatory reactions or via controlled release mechanisms. NAD concentration is indicated by degree of shading. ADP-ribosylation of cell surface target proteins by black triangles.

of local tissue injury, as happens during inflammation, or as the result of regulated secretion mechanisms (FIG. 2). Ecto-NAD-induced cell death provides a novel mechanism of focusing an apoptotic signal to a selected population of cells that express appropriate "receptor" molecules that are able to process the signal. These molecules include the ARTs, but also one or more as yet unknown effector molecules that are targets for ADP-ribosylation.

EXPERIMENTAL PROCEDURES

ART2-specific antibodies were induced by repeated gene gun immunization of rats and rabbits.[4] Lymph nodes were isolated from sacrificed BALBc/ByJ mice, and single cell suspensions were prepared and processed for flow cytometry on a FACS-Calibur (Becton Dickinson). T cells were enriched by depletion of B cells using magnetic cell separation with Dynabead-immobilized goat anti-mouse IgG (Dynal, 4–6 beads/cell). Cells were incubated for 30 minutes at 37°C with or without antiserum (diluted 1:200 in RPMI medium), washed, and incubated further in RPMI containing b-NAD^+ (Sigma) as indicated for 0.5–16 hours. Cells were washed in RPMI medium supplemented with 2 mM $CaCl_2$, and stained in this medium for 20 min on ice with FITC-conjugated AnnexinV (1 μg/mL) and propidium iodide (10 μg/mL).

ACKNOWLEDGMENTS

We thank Dunja Freese, Gudrun Dubberke, and Fenja Braasch for technical assistance. This work was supported by grants from the Deutsche Forschungsgemeinschaft, the Bundesministerium für Wirtschaft und Technologie, and the Ministère de la Recherche et de la Technologie.

REFERENCES

1. VAN PARIJS, L. & A.K. ABBAS. 1998. Homeostasis and self-tolerance in the immune system: turning lymphocytes off. Science **280:** 243–248.
2. GLOWACKI, G. *et al.* 2002. The family of toxin-related ecto-ADP-ribosyltransferases in humans and the mouse. Protein Sci. **11:** 1657–1670.
3. BOSSY-WETZEL, E. & D.R. GREEN. 2000. Detection of apoptosis by annexin V labeling. Methods Enzymol. **322:** 15–18.
4. KOCH-NOLTE, F. *et al.* 1999. A new monoclonal antibody detects a developmentally regulated mouse ecto ADP-ribosyltransferase on T cells: subset distribution, inbred strain variation, and modulation upon T cell activation. J. Immunol. **163:** 6014–6022.
5. ADRIOUCH, S. *et al.* 2001. Rapid induction of naive T cell apoptosis by ecto-nicotinamide adenine dinucleotide: requirement for mono(ADP-ribosyl)transferase 2 and a downstream effector. J. Immunol. **167:** 196–203.

ICBP90 Expression Is Downregulated in Apoptosis-Induced Jurkat Cells

ABDUL-QADER ABBADY,[a] CHRISTIAN BRONNER,[a] MARIE-ALINE TROTZIER,[a] RAPHAEL HOPFNER,[a] KAWTAR BATHAMI,[a] CHRISTIAN D. MULLER,[b] MICHAEL JEANBLANC,[a] AND MARC MOUSLI[a]

[a]*Institut National de la Santé et de la Recherche Médicale, UMRS 392, and* [b]*Centre National de la Recherche Scientifique, UMR7034, Faculté de Pharmacie, B.P. 60024, 74, route du Rhin, 67401 Illkirch Cedex, France*

ABSTRACT: **Inhibition of apoptosis, resulting from an increase in anti-apoptotic protein, plays a fundamental role in carcinogenesis. Because ICBP90 gene expression is deregulated in cancer cells, we studied its expression in Jurkat cells under apoptotic conditions to see whether ICBP90 is involved in the regulation of apoptosis. We found that ICBP90 expression and the percentage of living cells were dose-dependently decreased in PHA and ionophore A23187–stimulated Jurkat cells, but not in THP-1 cells. These results suggest that apoptosis is dependent upon ICBP90 expression downregulation and that ICBP90 exhibits its anti-apoptotic properties.**

KEYWORDS: **apoptosis; calcium; cancer; cell proliferation; ICBP90; Jurkat cells**

INTRODUCTION

Inhibition of apoptosis plays an important role in the carcinogenic process.[1] Indeed, anti-apoptotic actors, including the Bcl-2 family, are often over-expressed in cancer cells.[1] Identifying the transcription factors that are responsible for this deregulation will help to understand how apoptosis is inhibited in carcinogenesis.

We recently described a human protein named ICBP90 (Inverted CCAAT box Binding Protein of 90 kDa) that is able to regulate the topoisomerase IIα gene expression, a target for chemotherapy essential for chromosome segregation.[2] ICBP90 expression is cell-cycle-regulated in non-cancer cells, but has a permanent expression in cancer cells even at confluence.[3] We proposed that ICBP90 belongs to a new family of proteins, including Np95 and NIRF (Np95/ICBP90 ring finger) whose expressions are deregulated in cancer cell lines.[3]

This study investigates whether ICBP90 is involved in the regulation of the signaling transduction of apoptosis in Jurkat cells. Apoptosis was induced in the Jurkat cell line by phytohemagglutinin (PHA) or by the calcium ionophore A23187. The

Address for correspondence: Dr. Marc Mousli, Inserm UMRS 392, Faculté de Pharmacie, 74 route du Rhin, B.P. 60024, 67401 Illkirch, France. Voice: +33-390-24-41-95; fax: +33-390-24-43-08

Marc.Mousli@pharma.u-strasbg.fr

Ann. N.Y. Acad. Sci. 1010: 300–303 (2003).
doi: 10.1196/annals.1299.052

human myeloid leukemia cell line THP-1 was used as a negative control, because apoptosis cannot be induced by PHA or by ionophore A23187 in this cell line.

MATERIALS AND METHODS

The living Jurkat or THP-1 cell number was evaluated in 96-well micro-plates by a colorometric assay using the XTT kit (Roche Diagnostics, Meylan, France). The percentage of cell death was determined using Annexin-V-FLUOS (Beckman

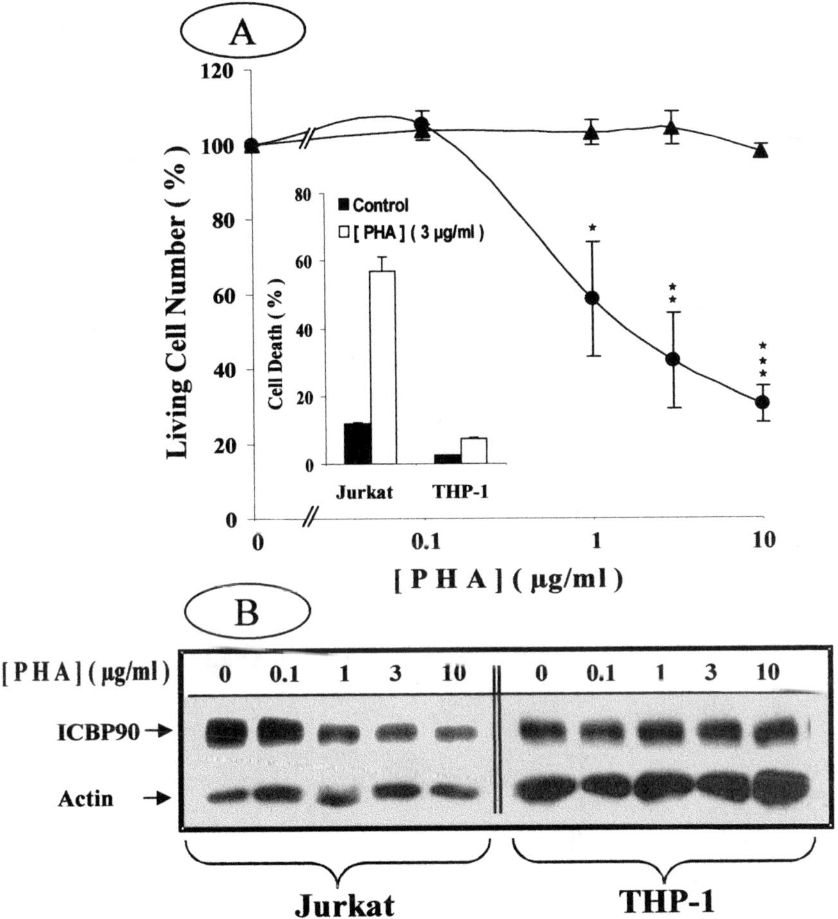

FIGURE 1. Effect of PHA on Jurkat or THP-1 living cell number and ICBP90 expression. **(A)** Jurkat (●) or THP-1 cells (▲) were plated (4×10^5cells/mL) for 24 h and stimulated by the indicated concentrations of PHA for 24 h. Living cell number was expressed as percentage of control (i.e., unstimulated cells). **(B)** ICBP90 expression was analyzed by Western blot after 24 h of stimulation. Data are means ± SEM of three separate experiments performed in triplicate and analyzed using Student's *t* test. $^{*}P < .05$, $^{**}P < .01$, $^{***}P < .001$.

Coulter Company, Marseille, France). ICBP90 expression was analyzed by Western blot as described elsewhere.[2,3] The ICBP90 protein was detected using the mAb 1RC1C10 (IGBMC, Illkirch, France).

RESULTS

FIGURE 1 shows the percentage of living cells (Jurkat or THP-1 cells), actin, and ICBP90 expressions after 24 h of stimulation by various concentrations of PHA. Jurkat living-cell number dose-dependently decreased between 0.1 and 10 µg/mL of PHA, whereas no effect was observed on the THP-1 cell line (FIG. 1A). FACS analysis showed that PHA (3 µg/mL) induced a strong apoptosis in Jurkat cells, but

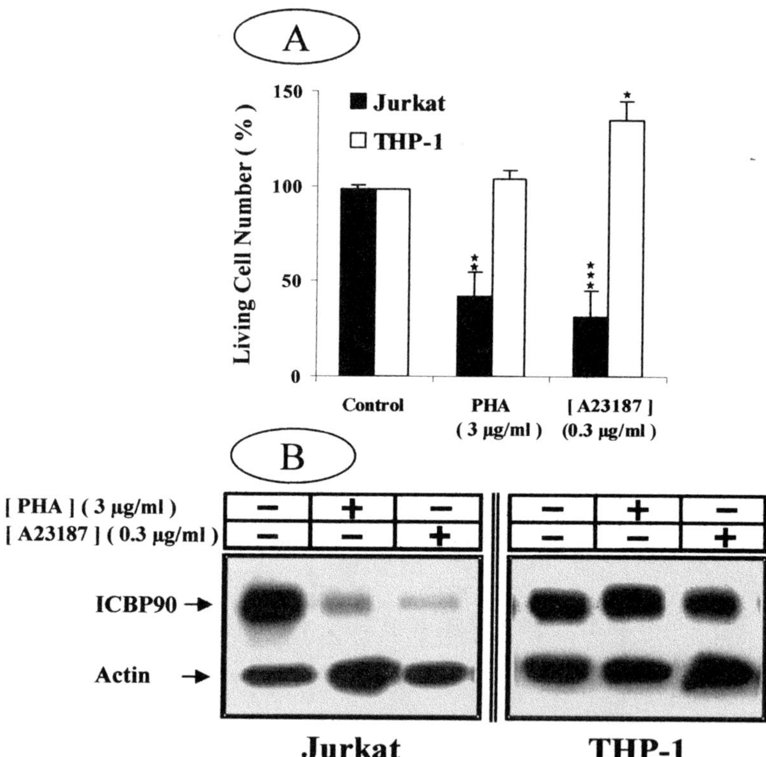

FIGURE 2. Effect of calcium ionophore A23187 and PHA on Jurkat or THP-1 living cell number and ICBP90 expression. **(A)** Jurkat or THP-1 cells were plated (4×10^5cells/mL) for 24 h and stimulated by the indicated concentrations of ionophore A23187 or PHA for 24 h. Living cell number was expressed as percentage of control (i.e., unstimulated cells). **(B)** ICBP90 expression was analyzed by Western blot after 24 h of stimulation. Data are means ± SEM of three separate experiments performed in triplicate and analyzed using Student's t test. $^{*}P < .05$; $^{**}P < .01$; $^{***}P < .001$.

exerted only a slight effect in THP-1 cells (insert of FIG. 1A). Western blot analysis showed that ICBP90 expression is dose-dependently decreased in PHA treated Jurkat cells from 1 to 10 μg/ml, but not in THP-1 cells (FIG. 1B).

FIGURE 2 shows the effect of calcium ionophore A23187 on the percentage of living Jurkat and THP-1 cells. Elevating intracellular calcium by the ionophore A23187, induced a strong decrease of the percentage of living cells (FIG. 2A) that was due to apoptosis (data not shown). This effect was similar to that obtained with 3 μg/ml of PHA (FIG. 2A). Parallel to the decrease of living cell number, ICBP90 expression was downregulated in A23187-stimulated Jurkat cells. In contrast, in THP-1 cells, ionophore A23187 induced a significant increase in living cell number but did not significantly affect ICBP90 expression (FIG. 2B).

DISCUSSION

The expression profile of ICBP90 in Jurkat cells was investigated under apoptotic conditions in consideration that deregulated expression of ICBP90 occurs in several cancer cell lines,[2,3] and that overexpression of ICBP90 overcomes cell contact inhibition in normal cells.[4] In PHA-stimulated Jurkat cells, a strong dose-dependent correlation was found between ICBP90 down-regulation and the decrease of living cell number. Ionophore A23187-induced apoptosis shows that downregulation of ICBP90 occurs after calcium mobilization. Interestingly, neither downregulation of ICBP90 expression nor apoptosis occurs in THP-1 cells. This suggests a relationship between the lack of ICBP90 downregulation and the failure of apoptosis induction.

We propose that ICBP90 permanent expression in cancer cells participates in the inhibition of apoptosis in carcinogenesis because ICBP90 expression peaks during G1/S transition in non-cancer cells[3] and the critical regulation of apoptosis occurs during this transition. Recently, we proposed that Np95 and ICBP90 belongs to the same family of transcription factors with altered expression in cancer cells.[3] The adenovirus E1A oncoprotein was shown to activate a new pathway involving Np95 that is important in the transition of G1/S phases.[5] Therefore, ICBP90 as well as Np95 may be involved in a new transduction-signaling pathway exhibiting anti-apoptotic properties.

REFERENCES

1. KAUFMANN, S.H. & G.J. GORES. 2000. Apoptosis in cancer: cause and cure. Bioessays **22:** 1007–1017.
2. HOPFNER, R., M. MOUSLI, J.M. JELTSCH, *et al.* 2000. ICBP90, a novel human CCAAT binding protein, involved in the regulation of topoisomerase IIα expression. Cancer Res. **60:** 121–128
3. MOUSLI, M., R. HOPFNER, A.Q. ABBADY, *et al.* 2003. ICBP90 belongs to a new family of proteins with an expression that is deregulated in cancer cells. Br. J. Cancer **89:** 120–127.
4. HOPFNER, R., M. MOUSLI, P. OUDET & C. BRONNER. 2002. Overexpression of ICBP90, a novel CCAAT-binding protein, overcomes cell contact inhibition by forcing topoisomerase IIα expression. Anticancer Res. **22:** 3165–3170
5. BONAPACE, I.M., L. LATELLA, R. PAPAIT, *et al.* 2002. Np95 is regulated by E1A during mitotic reactivation of terminally differentiated cells and is essential for S phase entry. J. Cell Biol. **157:** 909–914.

1-Trichloromethyl-1,2,3,4-tetrahydro-β-carboline (TaClo) Induces Apoptosis in Human Neuroblastoma Cell Lines

RAVI SHANKAR AKUNDI, MICHAEL HÜLL, HANS WILLI CLEMENT, AND BERND L. FIEBICH

Department of Psychiatry, University of Freiburg Medical School, Hauptstrasse 5, Freiburg 79104, Germany

ABSTRACT: The former popular hypnotic chloral hydrate has been known to spontaneously condense with tryptamine in the body to give rise to a highly unpolar 1-trichloromethyl-1,2,3,4-tetrahydro-β-carboline (TaClo) derivative. Earlier studies have revealed the relative permeability of the molecule through the body, and its ability to induce Parkinson-like symptoms in rats. In this study, we report that TaClo is a highly cytotoxic agent that leads to apoptosis in human neuroblastoma cells involving caspase-3 activation.

KEYWORDS: TaClo; Parkinson's disease; caspase-3; neuroblastoma; apoptosis

INTRODUCTION

Chloral hydrate is used for the treatment of insomnia, especially for children and elderly patients. Chloraldehyde (Clo), which is widely distributed throughout the body within 30 minutes of application, is a highly reactive aldehyde that spontaneously condenses with the biogenic amine tryptamine (Ta) by a Pictet-Spengler type cyclization, resulting in the highly unpolar harman derivative 1-trichloromethyl-1,2,3,4-tetrahydro-β-carboline (TaClo).[1] TaClo, structurally similar to the dopaminergic neurotoxin 1-methy-4-phenyl-1,2,3,6-tetrahydropyridine (MPTP), was found to easily penetrate the blood–brain barrier of rats. Peripheral application of TaClo in rats has been able to initiate a slowly progressing neurodegeneration with a late onset of the first Parkinson-like symptoms.[2] Recent studies also detected TaClo in the serum and urine of patients treated regularly with chloral hydrate for various symptoms.[3] In this study, we show that TaClo is a pro-apoptotic agent for the human neuroblastoma cell lines SK-N-SH and SH-SY5Y.

ABBREVIATIONS: TaClo, 1-trichloromethyl-1,2,3,4-tetrahydro-β-carboline; MPTP, 1-methyl-4-phenyl-1,2,3,6-tetrahydropyridine; ROS, reactive oxygen species; PARP, poly(ADP-ribose) polymerase; DEVD-pNA, Asp-Glu-Val-Asp-paranitroaniline; z-VAD.fmk, z-Val-Ala-Asp-fluoromethylketone.

Address for correspondence: Bernd L. Fiebich, Neurochemistry Lab, Department of Psychiatry, University of Freiburg Medical School, Hauptstrasse 5, Freiburg 79104, Germany. Voice: +49-761-2706501; fax. +49-761-2706917.

bernd_fiebich@psyallg.ukl.uni-freiburg.de

Ann. N.Y. Acad. Sci. 1010: 304–306 (2003). © 2003 New York Academy of Sciences.
doi: 10.1196/annals.1299.053

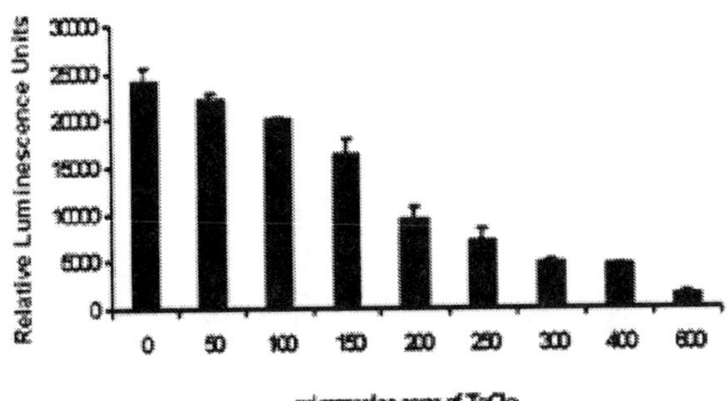

FIGURE 1. TaClo was chemically synthesized and kindly provided by Dr. Bringmann (University of Würzburg, Germany). Cell viability was tested based on ATP levels as a function of metabolically active cells using a luminescent assay (Promega).

RESULTS AND DISCUSSION

Cell viability experiments based on ATP levels of actively metabolizing cells revealed that TaClo is cytotoxic to SK-N-SH cells at an LD_{50} of 150 μM concentration (FIG. 1). TaClo-mediated cell-death involves the activation of caspase 3, whose activity increases threefold after 24 h of exposure with 150 μM TaClo (FIG. 2). The process of cell death is apoptotic at low concentrations of TaClo (below or equal to 300 μM) involving activation of caspase 3 and cleavage of PARP. At higher concentrations of TaClo (equal or more than 600 μM) cell death occurred without cleavage of PARP. This indicates a change towards the activation of other processes of cell toxicity. The activity of caspase 3 was diminished by the pan-caspase inhibitor z-VAD-fmk (FIG. 2). However, cell death could not be prevented, indicating that caspase 3 was a point of no return in the process of apoptosis induced by TaClo.

TaClo-induced caspase 3 activity is supposed to be dependent upon the mitochondrial cytochrome *c* release into cytosol. Earlier reports show that TaClo inhibits the mitochondrial complex I[1] and increases the concentration of reactive oxygen species.[4] This process might be responsible for changes in mitochondrial membrane permeability that result in leakage of cytochrome *c*.

Our *in vivo* experiments in rats show that TaClo-induced reduction of striatal dopamine metabolism could be significantly ameliorated through pretreatment of the animals with the glutamate antagonist MK-801 (data not shown). This suggests an involvement of glutamate release and NMDA receptors. SK-N-SH cells have been shown to lack ionotropic glutamate receptors,[5] which indicate apoptotic cell death in SK-N-SH cells is independent of NMDA receptor. To investigate an influence of NMDA receptor blockade on TaClo-induced cell death, we performed parallel experiments on a different neuroblastoma cell line, SH-SY5Y, which contains both ionotropic and metabotropic glutamate receptors. SH-SY5Y cells were also observed to undergo cell death in the presence of TaClo (not shown). Further, TaClo-induced cell death could not be prevented in either SK-N-SH or SH-SY5Y cells by MK-801. This conclusively shows that the cytotoxicity of TaClo *in vitro* is indepen-

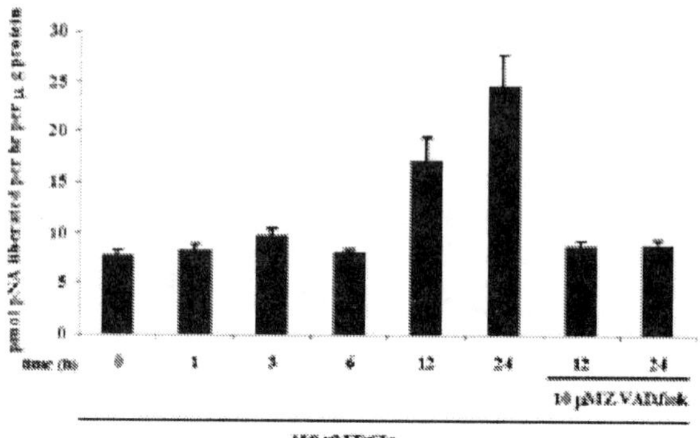

FIGURE 2. Cell lysates for caspase assays were prepared by freeze–thaw lysis of the cells in 0.1 M HEPES (pH 7.4) buffer comprising 2 mM DTT, 0.1% CHAPS, and 1% sucrose, followed by clarification at 15,000 *g*. 50 μg of protein was incubated with caspase-3 substrate, DEVD-pNA (Calbiochem) at 37°C. Cleavage of the substrate was colorimetrically measured at 405 nm.

dent of NMDA receptors. However, besides the demonstrated direct neurotoxicity of TaClo *in vitro*, there may also be an indirect neurotoxicity involving TaClo-induced excitotoxicity *in vivo*.

REFERENCES

1. Janetzky, B., R. God, G. Bringmann & H. Reichmann. 1995. J. Neural Trans. Suppl. **46:** 265–273.
2. Heim, C. & K.H. Sontag. 1997. J. Neural Trans. Suppl. **50:** 107–111.
3. Bringmann, G., M. Münchbach, D. Feineis, *et al.* 2002. J. Chromatogr. B. **767:** 321–332
4. Gerlach, M., A.Y. Xiao, C. Heim, *et al.* 1998. Neurosci. Lett. **257:** 17–20.
5. Zaulyanov, L.L., P.S. Green & J.W. Simpkins. 1999. Cell. Mol. Neurobiol. **19:** 705–718.

Induction of Apoptosis by Erucylphospho-*N,N,N*-trimethylammonium Is Associated with Changes in Signal Molecule Expression and Location

MARTIN R. BERGER,[a] IANA TSONEVA,[b] SPIRO M. KONSTANTINOV,[a,c] AND HANSJÖRG EIBL[d]

[a]*Unit of Toxicology and Chemotherapy, German Cancer Research Center, Im Neuenheimer Feld 280, 69120 Heidelberg, Germany*

[b]*Institute of Biophysics, BAS, Acad. G. Bonchev St., bl 21, 1113 Sofia, Bulgaria*

[c]*Lab for Experimental Chemotherapy, Department of Pharmacology, Medical University of Sofia, 2 Dunav St., 1000 Sofia, Bulgaria*

[d]*Max Planck Institute of Biophysical Chemistry, Am Faßberg, 37077 Göttingen, Germany*

ABSTRACT: At concentrations effecting apoptosis, the alkylphosphocholine ErPC3 induced increased expression of the Rb protein in breast cancer (MCF-7) and leukemia (SKW-3, AR-230) cell lines as well as hypophosphorylation (K-562, CMLT-1, DOHH-2) and fragmentation of Rb (BV-173, SKW-3) in leukemia cell lines. ErPC3 exerts at least part of its antineoplastic activity by apoptosis, and this chain of events comprises early changes in the lipid raft fraction of the cellular membrane as well as modulation of different signal molecules, such as Abl, Bcr-Abl (fusion protein), and Rb.

KEYWORDS: alkylphosphocholine; mechanism of action; modulation of signal molecules; lipid rafts

INTRODUCTION

Alkylphosphocholines are membrane-seeking antineoplastic agents that act as signal transduction modulators. Erucylphosphocholine and its congener erucylphospho-*N,N,N*-trimethylammonium (ErPC3) belong to a new subgroup of i.v. injectable agents[1] that elicit apoptosis in a series of leukemia cell lines, but cause proliferation of normal hematopoietic stem cells.[2] The intention of these experiments was to contribute to understanding the underlying mechanism of action of ErPC3.

Address for correspondence: Prof. M.R. Berger, Unit of Toxicology and Chemotherapy, German Cancer Research Center, Im Neuenheimer Feld 280, 69120 Heidelberg, Germany. Voice: +49 6221 423310; fax: +49 6221 423313.
M.Berger@Dkfz-Heidelberg.de

Ann. N.Y. Acad. Sci. 1010: 307–310 (2003). © 2003 New York Academy of Sciences.
doi: 10.1196/annals.1299.054

MATERIALS AND METHODS

ErPC3 was synthesized as published previously.[3] Methyl-β-cyclodextrin (MCD), cholera toxin subunit B conjugated with HRP or FITS were purchased from Sigma (Munich, Germany). For analysis, cells were exposed to appropriate concentrations of ErPC3 and lysed (whole cell lysates) or nuclei were separated and lysed (nuclei). Thereafter, proteins were detected by Western blot with primary antibodies against Abl and Rb and appropriate secondary antibodies (from Santa Cruz, CA and Heidelberg, Germany). In addition, DNA was isolated and separated by agarose gel electrophoresis to detect ladder formation. Lipid rafts were isolated by sucrose gradient centrifugation technique and fractions of the gradient were analyzed by Western blot as well with subsequent detection of membrane associated proteins (Abl) and the ganglioside G_{M1} for detection of lipid rafts. In addition, lipid rafts were visualized by staining of cells placed onto glass slides by cytospin with cholera toxin subunit B conjugated with FITS. Stained cells were inspected by using a fluorescence microscope.

RESULTS AND DISCUSSION

In line with our hypothesis that ErPC3 induces activation of signal cascades that transmit messages into the nucleus, the following investigations were carried out. Rb protein expression was checked in various tumor cell lines. As evidenced by Western blot, breast cancer (MCF-7) and leukemia cells (SKW-3 and AR-230) showed induction of Rb protein expression in response to treatment with the alkylphosphocholine ErPC3 (FIG. 1). Another type of response, hypophosphorylation of Rb, was found in

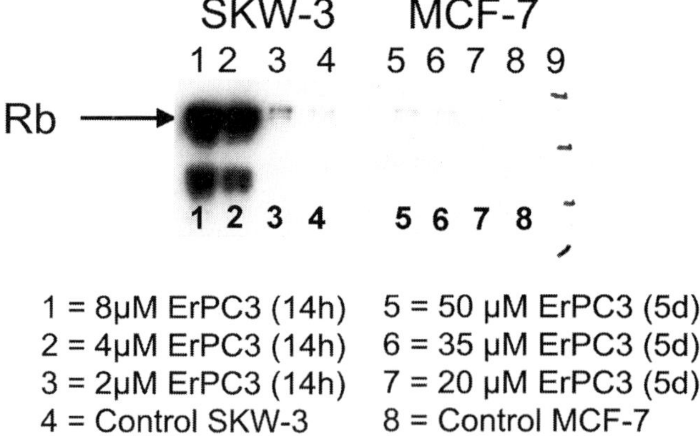

FIGURE 1. The influence of treating SKW-3 (*lanes 1–4*) and MCF-7 cells (*lanes 5–8*) with ErPC3 on Rb protein expression. *Lane 1*: 8 μM, 14 h; *lane 2*: 4 μM, 14 h; *lane 3*: 2 μM, 14 h; *lane 4*: SKW-3 control; *lane 5*: 50 μM, 5d; *lane 6*: 35 μM, 5d; *lane 7*: 20 μM, 5d; *lane 8*: MCF-7 control.

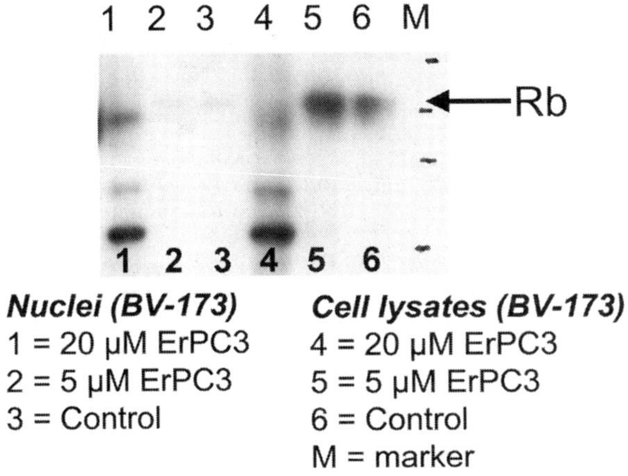

FIGURE 2. The influence of treating BV-173 cells with ErPC3 on Rb protein expression in nuclei (*lanes 1–3*) and whole cell lysates (*lanes 4–6*). *Lane 1*: 20 μM; *lane 2*: 5 μM; *lane 3*: control; *lane 4*: 20 μM; *lane 5*: 5 μM; *lane 6*: control.

K-562, CML-T1 and DOHH-2 leukemic cells. In addition, at concentrations inducing apoptosis, ErPC3 caused dephosphorylation and fragmentation of the Rb protein in BV-173 and SKW-3 lymphoid cells as visualized by fragment sizes in the order of 70 and 50 kDa (FIG. 2). Pre-treatment of SKW-3 cells with methyl-beta-cyclodextrin was able to reduce, but not to prevent oligonucleosomal DNA fragmentation caused by ErPC3 in SKW-3 cells. Leukemia cells exposed to ErPC3 changed the distribution of lipid rafts, as visualized by fluorescence microscopy. One hour of exposure to ErPC3 caused the lipid rafts to form large condensed areas in K-562 and HL-60 cells. After five hours of exposure, these areas had decreased in size; and at 24 hours of exposure, they had almost returned to the state before exposure to ErPC3. Exposure to ErPC3 also changed the location of Abl in AML-derived HL-60 cells, as evidenced by Western blotting of isolated lipid rafts (sucrose gradient centrifugation technique). Reorganization of rafts containing Abl and Bcr-Abl was detected also in K-562 cells, which are relatively resistant towards ErPC3. These data may help to explain why an antisense oligonucleotide directed against Bcr-Abl is synergistically active with ErPC3 in suppressing the growth of colonies derived from leukemia cell lines containing the protein Bcr-Abl.[4,5] Taken together, our data indicate that the alkylphosphocholine ErPC3 exerts at least part of its antineoplastic activity by apoptosis; and that this chain of events comprises early changes in the lipid-raft fraction of the cellular membrane and modulation of different signal molecules such as Abl, Bcr-Abl (the fusion protein), and Rb.

REFERENCES

1. BERGER, M.R., S. SOBOTTKA, S.M. KONSTANTINOV & H. EIBL. 1998. Erucylphosphocholine is the prototype of i.v. injectable alkylphosphocholines. Drugs Today **34**(Suppl.): F73–81.

2. KONSTANTINOV, S.M., M. TOPASHKA-ANCHEVA, A. BENNER & M.R. BERGER. 1998. Alkylphosphocholines: effects on human leukemic cell lines and normal bone marrow cells. Int. J. Cancer **77:** 778–786.
3. EIBL, H. & J. ENGEL. 1992. Synthesis of hexadecylphosphocholine (Miltefosine). *In* New Drugs in Cancer Therapy: Progress in Experimental Tumor Research, Vol. 34. H. Eibl, P. Hilgard & C. Unger, Eds.: 1–5. Karger. Basel.
4. KONSTANTINOV, S.M., M.C. GEORGIEVA, M. TOPASHKA-ANCHEVA, *et al.* 2002. Combination with an antisense oligonucleotide synergistically improves the antileukemic efficacy of erucylphospho-*N*,*N*,*N*-trimethylpropyl-ammonium in CML cell lines. Mol. Cancer Ther. **1:** 877–884.
5. KONSTANTINOV, S.M., H. EIBL & M.R. BERGER. 1999. BCR-ABL influences the antileukaemic efficacy of alkylphosphocholines. Br. J. Haematol. **107:** 365–374.

Antiproliferative and Apoptosis-Inducing Effects of Hemin in Hepatoma Cells

CARMEN-CRISTINA DIACONU,[a] MARGIT SZATHMÁRI,[b] AND ANIKÓ VENETIANER[b]

[a] *"Stefan S. Nicolau" Institute of Virology, Romanian Academy, Bucharest, Romania*

[b] *Institute of Genetics, Biological Research Center, Hungarian Academy of Sciences, Szeged, Hungary*

ABSTRACT: Hemin is an extremely versatile molecule that may have cytotoxic or cytoprotective effects on certain cells. We investigated the effect of hemin on the growth of hepatoma cells, including the multidrug-resistant ones. Searching for new tools that interfere with the growth of hepatomas is an important area of clinical research. Cell viability and proliferation of drug-sensitive and multidrug-resistant hepatoma cell lines was determined using the trypan-blue exclusion test XTT/PMS and colony-forming ability assays. Apoptosis was assessed by confocal microscopy and DNA ladder assay. Hemin inhibited the proliferation and induced apoptosis in both drug-sensitive and multidrug-resistant hepatoma cells overexpressing functional *P*-glycoprotein. zVAD-fmk inhibited the hemin-induced decrease in cell viability, pointing to a role of caspases in hemin-induced apoptosis. The antiproliferative and apoptosis-inducing effects of hemin might be considered in the design of treatment for patients with hepatoma.

KEYWORDS: hemin; antiproliferative; apoptosis; hepatoma; multidrug-resistant

INTRODUCTION

Heme, the prosthetic group of hemoproteins, exerts multiple effects in a variety of *in vitro* and *in vivo* systems, although the mechanisms of action are not yet fully understood.[1] Hemin (oxidized form of heme) may have cytotoxic and cytoprotective effects and can increase or decrease the sensitivity of certain normal cells as well as tumor cells to chemotherapeutic agents. This led us to speculate whether hemin itself has any effect on the growth of hepatoma cells, including the multidrug-resistant ones. Hepatocellular carcinomas are essentially refractory to chemotherapy. *P*-glycoprotein (P-gp) plays an important role in the development of simultaneous resistance to multiple cytotoxic drugs in cancer cells. The failure or suppression of apoptosis may contribute to the appearance of tumor cells resistant to cytotoxic therapy. This paper reports that hemin inhibits the proliferation and induces apoptosis in both drug-sensitive and multidrug-resistant hepatoma cells.

Address for correspondence: Dr. Carmen-Cristina Diaconu, Institute of Virology, Romanian Academy, 285 Mihai Bravu, Bucharest, Romania. Voice: 401-324-2590; fax: 401-324-14-71. ccdiaconu@yahoo.com

Ann. N.Y. Acad. Sci. 1010: 311–315 (2003). © 2003 New York Academy of Sciences. doi: 10.1196/annals.1299.055

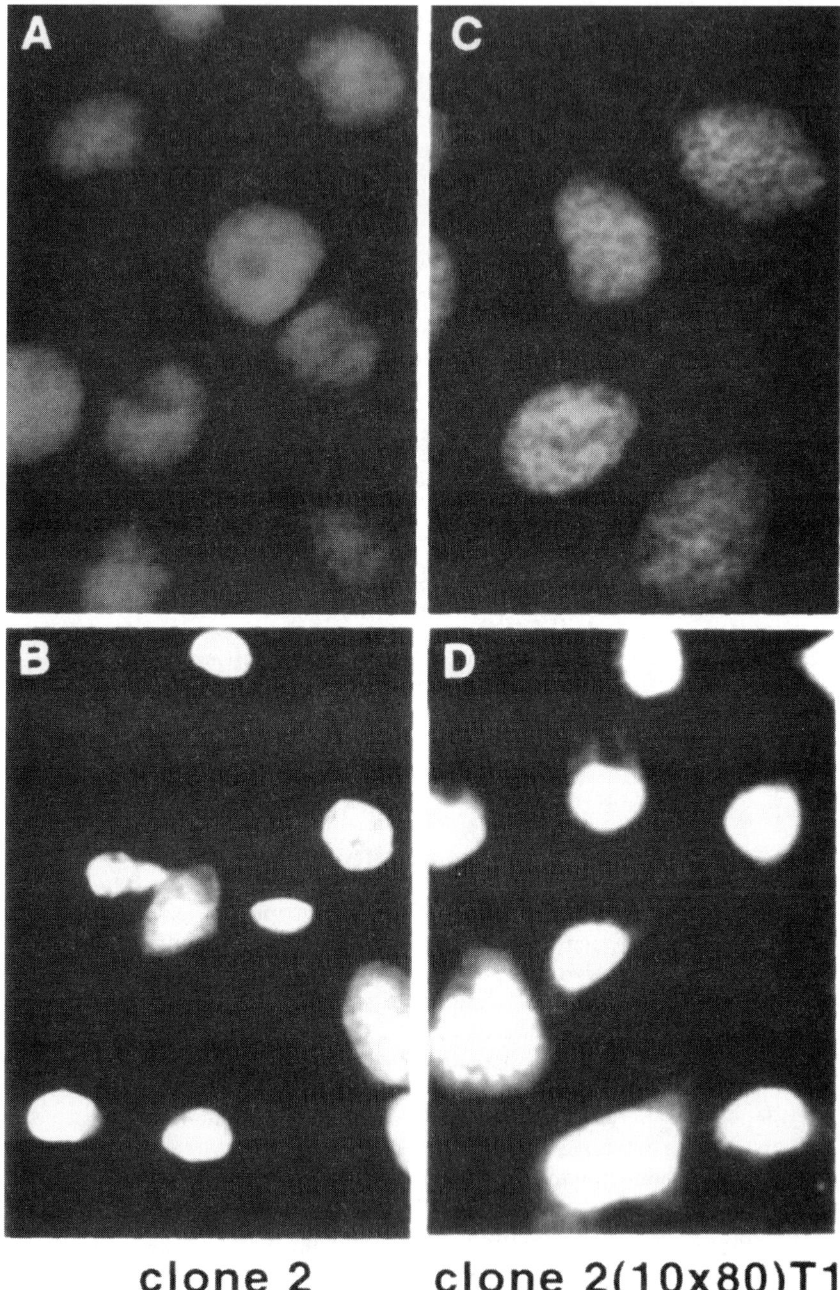

FIGURE 1. Hemin induces apoptosis in hepatoma variants: Morphological appearance of apoptotic cells. Drug-sensitive **(B)** and multidrug-resistant **(D)** cells were treated with 40 μM hemin for 1 day and stained with PI. **(A)** and **(C)** represent control cells.

MATERIALS AND METHODS

Cell Lines

The culture conditions and characterization of rat hepatoma clone 2 cells have been described earlier.[2] The moderately multidrug-resistant clone 2(10×80)T1 was isolated from clone 2 cells and the highly multidrug-resistant clone 2(10×80)T1_{c1000} was isolated from clone 2(10×80)T1 by using stepwise selection with colchicines.

Cell Viability and Proliferation Tests

The numbers of viable cells were determined 24, 48, or 72 hours after the cells were treated with different concentrations of hemin (Sigma), Benzyloxycarbonyl-Val-Ala-Asp(*O*-methyl)-fluoromethylketone (zVAD-fmk, Enzyme Systems Products) or simultaneously hemin and zVAD-fmk, by trypan blue (TB) or XTT/PMS assays. Colony-forming ability after hemin exposure was determined as described earlier.[3]

Assay for Induction of Apoptosis

After treatment with different concentrations of hemin, zVAD-fmk or simultaneously with hemin and zVAD-fmk, cells seeded on coverslips were stained with propidium iodide (PI). The percentage of necrotic cells was determined on a May-Grünwald-Giemsa-stained cytospin preparation.[3] For quantitative evaluation of apoptosis and necrosis, at least 400 cells were counted in each preparation. Nuclear area measurements for PI-stained cells were performed with the LSM software on a Zeiss Axiovert 135 M, LSM 410 Confocal Microscope and a 100× (N.A. = 1.5, oil) objective.[3]

DNA Ladder Assay

Cells were treated with 40 or 60 μM hemin for 21 h. As internal control for ladder formation, clone 2(10×80)T1_{c1000} and clone 2 cells were treated with 600 and 60 ng/mL actinomycin D for 21 h, respectively. Fragmented DNA was separated from intact DNA and then extracted and loaded onto agarose gels.[3]

RESULTS

In Vitro *Antiproliferative Effect of Hemin on Hepatoma Cells*

Treatment with increasing concentrations of hemin (20–100 μM) for 1, 2, or 3 days reduced the viability of both the drug-sensitive clone 2 and the multidrug-resistant 2(10×80)T1 clone cells in a time- and dose-dependent manner. We did not observe a growth-inhibitory effect on primary human embryo fibroblasts. Treatment with 40 μM hemin for 2 weeks almost totally inhibited colony formation by clone 2 and the multidrug-resistant variants.

Hemin Induces Apoptosis in Drug-Sensitive and Multidrug-Resistant Cells

Hemin induced typical apoptotic morphology in both drug-sensitive and multidrug-resistant cells (FIG. 1). Treatment of the variants with increasing concentrations of hemin for 24 h induced apoptosis in a dose-dependent manner, necrosis being below 10%. Treatment of hepatoma cells with 80 μM hemin for 1, 2, or 3 days induced a time-dependent decrease in the areas of the nuclei. While the areas of the nuclei of the control cells were mainly in the range 200–360 μm^2 (90%), they were most often (65%) found to be around 140 μm^2, following treatment with 80 μM hemin for 3 days. Reductions in the sizes of the treated nuclei were likewise observed for the drug-sensitive and drug-resistant hepatoma cells. A characteristic pattern of DNA fragments was observed in both drug-sensitive and highly multidrug-resistant cells after treatment with 40–60 μM hemin for 21 h (FIG. 2).

For comparison, the cells were also treated with actinomycin D, which induces apoptosis and DNA ladder formation in these cells.[3] In order to determine whether hemin-induced apoptosis in hepatoma cells requires interleukin-1β-converting enzyme (ICE) family proteases, we used zVAD-fmk, an irreversible tripeptide inhibitor.[4] Whereas the treatment of clone 2 cells with 40 or 80 μM hemin in the absence of zVAD-fmk for 24 h induced 27% and 46% of apoptosis; in the presence of 100 μM zVAD-fmk, the levels were reduced to 13% and 12%, respectively. In the cells treated only with 100 μM zVAD-fmk, the percentage of apoptotic cells was 7%.

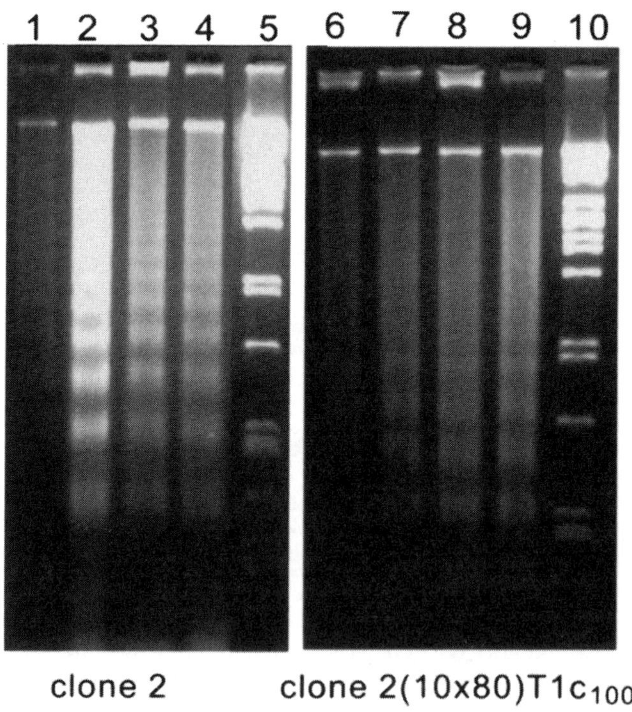

FIGURE 2. DNA fragmentation in hemin-treated hepatoma cells: Clone 2 cells treated with 60 ng/mL actinomycin D **(2)**, or with 40 or 60 μM hemin **(3** and **4)** for 21 h; control cells **(1)**. Clone 2(10×80)$T1_{c1000}$ cells treated with 600 ng/mL actinomycin D **(7)**, or with 40 or 60 μM hemin **(8** and **9)** for 21 h; control cells **(6)**. MW markers: lambda/Pst1 **(5** and **10)**.

DISCUSSION

Hemin is an extremely versatile molecule that exerts a large number of effects in a variety of *in vitro* and *in vivo* systems.[1] This report provides the first evidence that hemin exerts a pronounced antiproliferative effect, and it is able to induce apoptosis in both drug-sensitive and multidrug-resistant hepatomas in a dose- and time-dependent manner. Apoptosis induced by hemin was associated with typical morphological and biochemical changes, such as condensation of the nucleus and cytoplasm, decreases in the areas of the nuclei, and a characteristic pattern of DNA fragmentation. An important conclusion of our experiments is that hemin-induced apoptosis is caspase-dependent and occurs even in multidrug-resistant hepatoma cells. When a cell is provoked to commit "suicide," cytochrome *c* is relocalized from the mitochondria to the cytosol, where it helps to activate the death proteases, known as caspases. Heme is known to be within the prosthetic group of cytochrome *c*. Our results suggest that not only an increased level of cytosolic cytochrome *c*, but also elevated levels of hemin can induce programmed cell death. The finding that zVAD-fmk prevents hemin-induced apoptosis in hepatoma cells suggests that Ced-3/ICE-family proteases are required.[4] Earlier data showed that hemin, which can be safely administered to patients,[5] is able to enhance the interleukin-2-mediated anti-tumor effect.[6] On the basis of these observations, we propose that hemin be considered in the design of treatment for patients with hepatoma.

ACKNOWLEDGMENTS

We thank Prof. Dr. C. Cernescu, Dr. G. Tardei, D. Petrusca, Dr. L. Menczel, and M. Tóth for their helpful suggestions. This work was supported by grants from the Dr. János Bástyai Holczer Fund, the National Scientific Research Fund (OTKA T025008), and the National Research and Development Project Fund (PNCD VIASAN 051/2001-2004).

REFERENCES

1. PONKA, P. 1999. Cell biology of heme. Am. J. Med. Sci. **318:** 241–256.
2. VENETIANER, A. & Z. BŐSZE. 1983. Expression of differentiated functions in dexamethasone-resistant hepatoma cells. Differentiation **2:** 270–278.
3. DIACONU, C.C., M. SZATHMÁRI, G. KÉRI & A. VENETIANER. 1999. Apoptosis is induced in both drug-sensitive and multidrug-resistant hepatoma cell by somatostatın analogue TT-232. Br. J. Cancer **80:** 1197–1203.
4. JACOBSON, D.M., M. WEIL & C.M. RAFF. 1996. Role of Ced-3/ICE-family proteases in staurosporine-induced programmed cell death. J. Cell Biol. **133:** 1041–1051.
5. RUND, D. & E. RACHMILEWITZ. 2000. New trends in the treatment of beta-thalassemia. Crit. Rev. Oncol. Hematol. **33:** 105–118.
6. TSUJI, A., J. WANG, K.H. STENZEL & A. NOVGRODSKY. 1993. Immune stimulatory and anti-tumour properties of haemin. Clin. Exp. Immunol. **93:** 308–312.

Effect of Zinc on Prostatic Tumorigenicity in Nude Mice

PEI FENG, TIE LUO LI, ZHI XIN GUAN, RENTY B. FRANKLIN, AND LESLIE C. COSTELLO

Department of Biomedical Sciences, University of Maryland Dental School, Baltimore, Maryland 21201, USA

ABSTRACT: Prostate epithelial cells accumulate the highest zinc levels of any cells in the body. Evidence indicates that zinc plays critical roles in the normal function and pathology of the prostate gland. We have identified two important effects of zinc in the prostate epithelial cells: the inhibition of m-aconitase and the induction of mitochondrial apoptogenesis. However, at the present time, the effects of zinc on prostatic cells in *in vivo* conditions have not yet been reported. The objectives of this *in vivo* study were to investigate the effect of zinc on: tumorogenicity in nude mice, zinc accumulation in tumor tissues, and the levels of mitochondrial membrane permeability related proteins, Bax/Bcl-2. A tumorigenicity animal model was established using male nude mice (4–6 weeks old) with inoculation of PC-3 cells (5–10$\times$10^6/mL) prepared in 10% Matrigel. The mice were treated with zinc by ALZET osmotic pumps (Durect Corporation), with a releasing rate of 0.25 μl/h for 28 days. Zinc concentrations of the tumor tissues were determined by Atomic Absorption Spectrophotometer method. Frozen sections of tumor tissues were prepared for TUNEL assay. The levels of Bax and Bcl-2 in the tumor tissues were determined by Western blot analyses. Our study demonstrated that *in vivo* treatment of zinc increased zinc accumulation and citrate production in PC-3 cell induced tumor tissues and inhibited tumor growth. The inhibitory effect of zinc appears to result from zinc-induced apoptosis by regulation of mitochondrial membrane permeability-related Bax/Bcl-2 proteins.

KEYWORDS: prostate; zinc; tumor; apoptosis; Bcl-2/Bax

Prostate epithelial cells accumulate the highest zinc levels of any cells in the body. However, malignant prostate cells have lost this ability; and evidence indicates that zinc plays critical roles in the normal function and pathology of the prostate gland. We have identified two important effects of zinc in the prostate epithelial cells: the inhibition of *m*-aconitase and the induction of mitochondrial apoptogenesis.[1,2,3] In a normal prostate, zinc inhibition of *m*-aconitase activity results in suppressing citrate oxidation, and increases citrate production. Our recent studies demonstrate that zinc induces mitochondrial apoptogenesis in cultured PC-3 (human prostatic malignant cell line) and BPH (benign prostatic hyperplasia) cells. Furthermore, exposure of mi-

Address for correspondence: Pei Feng, Department of Biomedical Sciences, University of Maryland Dental School, Baltimore, MD 21201. Voice and fax: 410-706-7340.
pfeng@umaryland.edu

Ann. N.Y. Acad. Sci. 1010: 316–320 (2003).
doi: 10.1196/annals.1299.056

tochondria isolated from these cells to zinc results in the release of cytochrome *c* and redistribution of Bax/Bcl-2 on mitochondrial membranes (Bax/Bcl-2 data unpublished). However, the effects of zinc on prostatic cells in *in vivo* conditions have not yet been reported.

The objectives of this *in vivo* study were to investigate the effect of zinc on tumorogenicity in nude mice, zinc accumulation in tumor tissues, and the levels of mitochondrial membrane permeability related proteins, Bax/Bcl-2.

The experiments were carried out using male nude mice (4-6 weeks old) as a tumorigenicity animal model. The mice were housed in a pathogen-free environment under controlled light and humidity. Tumors were established by inoculation of PC-3 cells (5–10×10^6/mL) prepared in 10% Matrigel and using 0.1 mL for each injection (s.c.) at both flanks of the animals. The size of tumors was measured weekly. Zinc treatment was given by ALZET osmotic pumps (Durect Corporation), with a releasing rate of 0.25 μL/h for 28 days. The pumps were filled with PBS (control), zinc sulfate 5 mg/mL (low dose) and 7.5 mg/mL (high dose), respectively, and were implanted s.c. at the lower back of the animals. The operations of implanting pumps and inoculating PC-3 cells were carried out simultaneously.

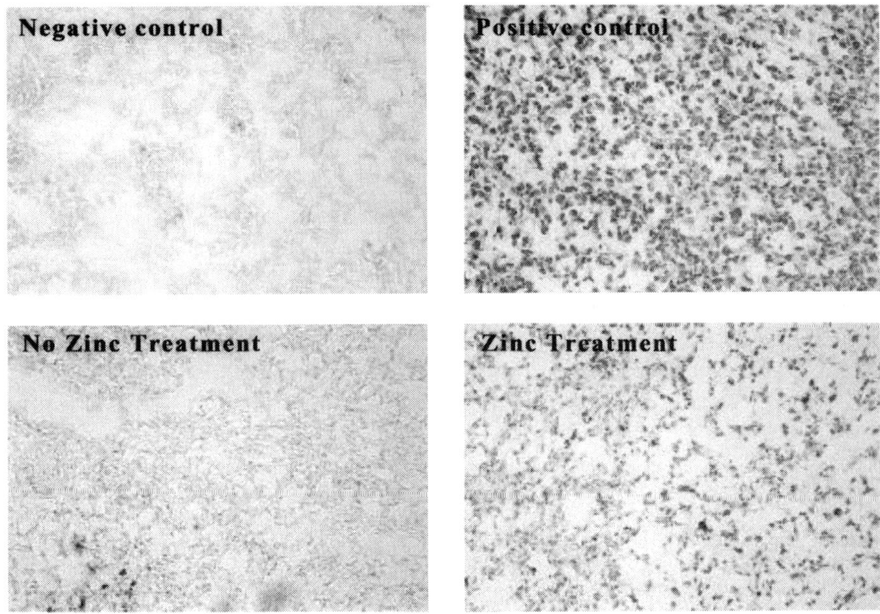

FIGURE 1. Zinc treatment induces apoptosis in tumor tissues: Frozen sections of the tumor tissues (i.m.) were stained for apoptotic DNA fragmentation with an *in situ* cell death detection kit (TUNEL), followed by DAB staining. The positive control was treated with DNase I, and the negative control was assayed without TdT. The results showed that *in vivo* treatment with zinc induced tumor cell apoptosis significantly compared to tumors without zinc treatment.

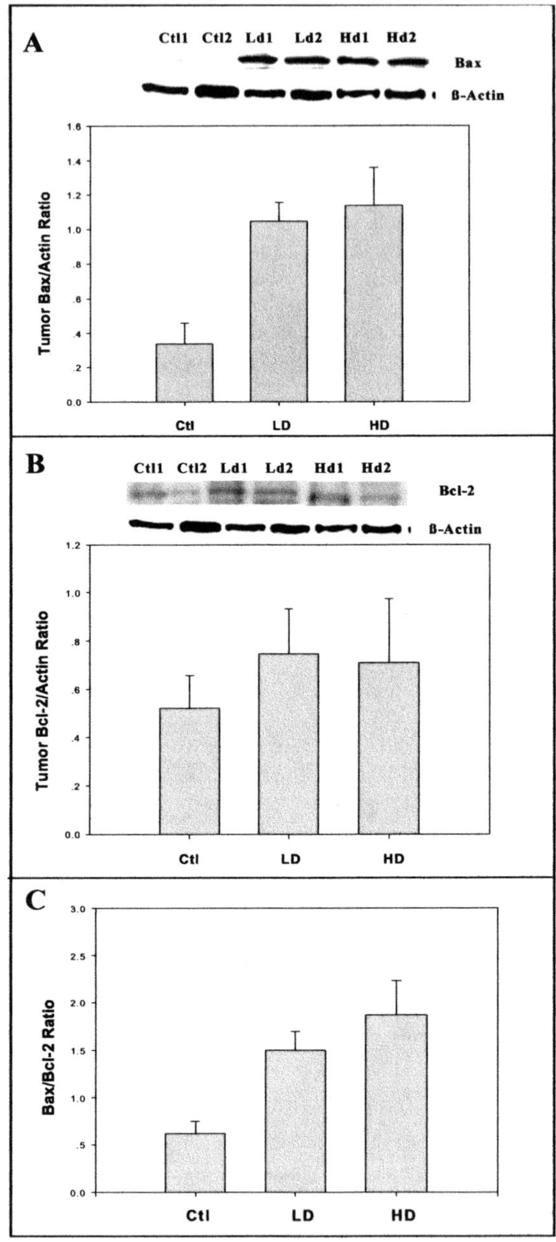

FIGURE 2. The effect of zinc on the ratio of Bax/Bcl-2 levels: **(A)** After 28 days of treatment, the levels of Bax in tumor tissues were detected by Western blot. Significantly increased Bax levels were identified in both zinc-treated groups compared to controls. **(B)** In contrast, Bcl-2 levels increased slightly compared to controls. **(C)** The Bax/Bcl-2 ratio increased significantly in zinc-treated groups, increasing mitochondrial outer membrane permeability and leading to cytochrome *c* release. (Data shown as representatives of the analyzed samples.)

The animal body weights were examined weekly, and their general health conditions were monitored closely. After 28 days of treatment, the mice were sacrificed, and the samples were collected for further studies. Zinc concentrations were determined by atomic absorption spectrophotometer method. Frozen sections of tumor tissues (8 μm) were prepared for TUNEL assay (according to the manufactory protocol) and stained by DAB. The results were examined and recorded by light microscopy.

The levels of Bax and Bcl-2 in the tumor tissues were determined by Western blot analyses with specific antibodies, and the amount of protein loaded for each sample was justified by β actin as an internal control. The citrate level in the tumor tissues was assayed using a previously established fluoroenzymatic method.[3]

The results showed evidence of zinc-induced inhibition of tumor growth in zinc-treated animals by the relatively smaller size and lighter weight of tumors compared with those of the controls. However, there was no significant effect of zinc on animal body weight. The incidence of tumor occurrence was 12/12 (control), 11/12 (low dose), and 9/12 (high dose). A higher zinc accumulation of tumor tissues was observed in zinc treated animals (16.3–16.4 ng/mg protein) than that of controls (12.5 ng/mg protein), leading to about 20–40% increase of citrate production in zinc-treated tumor tissues. In order to investigate the inhibitory effect of zinc on PC-3 cell–induced tumors, frozen sections of tumor tissues were examined by a TUNEL assay (FIG. 1). The results showed that extensive cell apoptosis was observed in zinc treated tumor tissues compared to the controls, in which only a few apoptotic cells were detected. Most recently, we have identified that zinc induces PC-3 cell apoptosis through regulating the mitochondrial outer membrane permeability-related proteins, Bax/Bcl-2 (data not shown) in accordance with previous findings.[4,5] Thus, to further ascertain the mechanism of zinc induced apoptosis *in vivo*, Bax and Bcl-2 levels in the tumor tissues were determined by Western blot (FIG. 2). The results showed that Bax levels significantly increased in zinc-treated tumor tissues compared with those of the controls (about threefold), and that Bcl-2 levels only increased slightly. The ratio of Bax/Bcl-2 in zinc-treated groups was significantly elevated compared to that of the controls.

Our study demonstrates that *in vivo* treatment of zinc increases zinc accumulation and citrate production in PC-3 cell–induced tumor tissues and inhibits tumor growth. The inhibitory effect of zinc appears to result from zinc-induced apoptosis by regulation of mitochondrial membrane permeability-related Bax/Bcl-2 proteins.

ACKNOWLEDGMENTS

We thank Dr. JuRen He for her expert help with TUNEL assay and Ms. Maggie Pierce for her assistance with *in vivo* studies. This project is supported by DOD Grant No. DAMD17-01-1-0072.

REFERENCES

1. FENG, P., J.Y. LIANG, T.L. LI, *et al.* 2000. Zinc induces mitochondria apoptogenesis in prostate cells. Mol. Urol. **4**(1): 31–36.
2. FENG, P., T.L. LI, Z.X. GAUN, *et al.* 2002. Direct effect of zinc on mitochondrial apoptogenesis in prostate cells. Prostate **52:** 311–318.

3. Costello, L.C. & R.B. Franklin. 1998. The novel role of zinc in the intermediary metabolism of prostate epithelial cells and its implications in prostate malignancy. Prostate **35:** 285–296.
4. Kroemer, G. & J.C. Reed. 2000. Mitochondrial control of cell death. Nature Med. **6**(6): 513–519.
5. Heiden, M.G.V., N.S. Chandel, X.X. Li *et al.* 2000. Outer mitochondrial membrane permeability can regulate coupled respiration and cell survival. PNAS **97**(9): 4666–4671.

The Selective Estrogen Receptor Modulator 4-Hydroxy Tamoxifen Induces G1 Arrest and Apoptosis of Multiple Myeloma Cell Lines

JULIETTE GAUDUCHON,[a] FABRICE GOUILLEUX,[b] SÉBASTIEN MAILLARD,[c] VÉRONIQUE MARSAUD,[c] MICHEL J. RENOIR,[c] AND BRIGITTE SOLA[a]

[a]*UPRES-EA 2128, UFR de Médecine, Université de Caen, Caen, France*

[b]*Laboratoire d'Immunologie, UFR de Médecine, Université Jules Verne, Amiens, France*

[c]*Laboratoire de Pharmacologie Cellulaire et Moléculaire des Anti-cancéreux, CNRS UMR 8612, UFR de Pharmacie, Châtenay-Malabry, France*

ABSTRACT: Multiple myeloma (MM) is an incurable hematological malignancy for which new therapeutic strategies should be envisaged. The selective estrogen receptor modulator (SERM), 4-hydroxy tamoxifen (4-OHTam), in the range of 1 to 10 μM, was able to impair the cell proliferation of MM cell lines. This was achieved by blocking cells at the G1 phase of the cell cycle and by inducing apoptosis. This cellular response was observed in five out of six tested cell lines, all five expressing both α and β estrogen receptor forms. No modifications of Bcl-2, Bcl-X, and Bax levels were observed, as well as no changes in Pi3K/Akt and JAK/STAT pathways that are often constitutively active in these cells. The signalization of 4-OHTam-induced cell death needs further investigation.

KEYWORDS: multiple myeloma; cell cycle; apoptosis; Bcl-2 family; p53; survival pathway

INTRODUCTION

Multiple myeloma (MM) is a B-cell malignancy characterized by the clonal expansion of plasma cells in the bone marrow and the occurrence of bone destruction.[1] Accounting for approximately 1–2% of all human cancers and 10% of hematological malignancies, MM is a relatively frequent disease.[1] Although high-dose chemotherapy associated with autologous hematopoietic stem cell transplantation has improved overall patient survival, the prognosis is still poor and new therapeutic strategies have to be designed.[2] In two recent reports, tamoxifen and toremifene, active as efficient antiestrogens (AE) in the treatment of estrogen-dependent breast cancer, induced growth inhibition and apoptosis of MM cell lines and MM patients.[3,4] However, the apoptosis-related molecular events are unclear. In order to get new insights, we analyzed the response of six MM cell lines towards 4-hydroxy tamoxifen (4-OHTam) treatment.

Address for correspondence: Brigitte Sola, UPRES-EA 2128, UFR de Médecine, CHU Côte de Nacre, 14032 Caen Cedex, France. Voice: +33-2-31-06-82-25; fax: +33-2-31-47-40-84. sola@medecine.unicaen.fr

Ann. N.Y. Acad. Sci. 1010: 321–325 (2003).
doi: 10.1196/annals.1299.057

MATERIALS AND METHODS

LP1, U266, NCI H929, and Karpas 620 MM cell lines have been described previously.[5] CAC-2 and CAC-6 have been derived in our laboratory and exhibited key features of MM cell lines (data not shown). Cells were maintained in RPMI 1640 and supplemented with 10% heat-inactivated fetal calf serum (FCS), 100 U/mL penicillin, 100 μg/mL streptomycin, and 2 mM L-glutamine. In experiments using 4-OHTam, cells were cultured in phenol red–free RPMI 1640 medium supplemented with charcoal-treated, inactivated FCS in order to eliminate estrogen agonistic or anti-agonistic activities. For cell proliferation and viability determinations, cells excluding trypan blue were counted in a hemocytometer at different time intervals. Cell viability was assayed using CellTiter 96 Aqueous One Solution® (Promega, France) as recommended by the supplier. Cell cycle analysis and apoptosis determination were realized after propidium iodide staining and fluorescent cell sorter examination. Western blots were obtained after electrophoresis of whole-cell extract proteins, transfer onto nylon membranes, and immunodetection.

RESULTS

We first analyzed the effects of concentrations of 4-OHTam varying from 0.01 to 100 μM on the six cell lines (LP1, CAC-2, CAC-6, NCI H929, U266, and Karpas 620) using an XTT assay (see MATERIALS AND METHODS). The tests determined that concentrations higher than 10 μM were highly toxic and led to cell death after a 24- or 48-h treatment (data not shown). We then studied the effects of 4-OHTam (0–10 μM) on cell proliferation and found in all cases, except Karpas 620, that the 4-OHTam impaired cell proliferation. As exemplified for the LP1 and NCI H929 cell lines (FIG. 1A), this inhibition correlated with the AE concentration used.

As demonstrated by cytofluorometric analysis, the decrease of cell proliferation was the result of a block at the G1 phase of the cell cycle, as well as the induction of apoptotic cell death (exemplified for LP1 and NCI H929; FIG. 1B). Again, both the percentage of cells accumulated in G1 and the percentage of apoptotic cells correlated with the AE concentration (TABLE 1).

TABLE 1. Results obtained with each MM cell line

	Increase in sub-G1 fraction			Increase in G1 fraction		
MM cell line	1 μM	5 μM	10 μM	1 μM	5 μM	10 μM
LP1	1	–	7	2	10	19
CAC-2	–	1	8	–	9	20
CAC-6	–	6	nd	6	9	nd
NCI H929	1	1	19	–	9	7
U299	–	4	9	–	9	18
Karpas 620	–	–	–	–	–	–

NOTE: The increase in sub-G1 and G1 fractions was calculated as the difference between the percentage of treated vs. nontreated cells in each fraction; –, no increase; nd, not done.

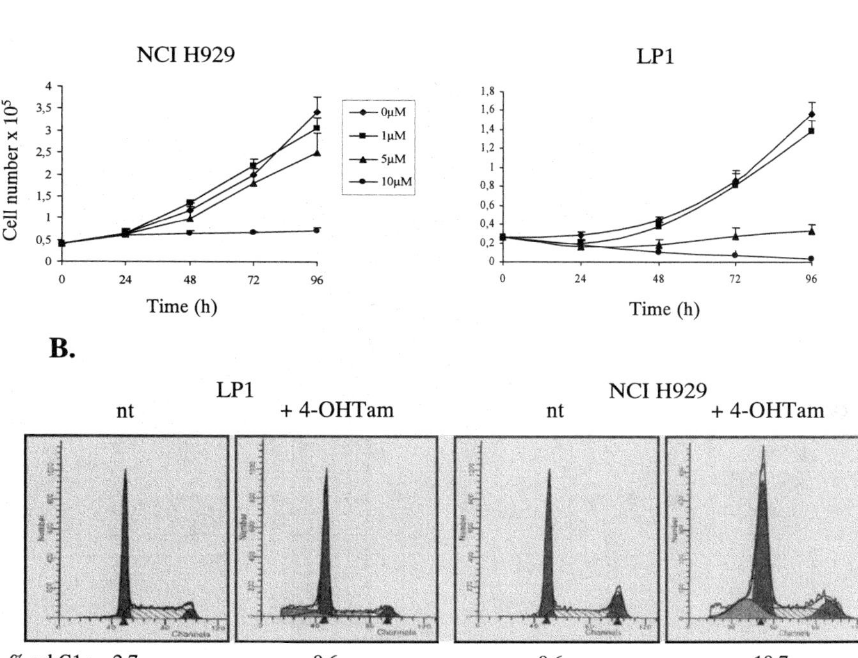

FIGURE 1. Effects of 4-OHTam on MM cell line proliferation and apoptosis. **(A)** Exponentially growing cells were treated, or not for controls, with various concentrations of 4-OHTam (1–10 µM). At various times (24–96 h), viable cells, excluding trypan blue, were enumerated. Experiments were done in triplicate. Representative proliferation curves of NCI H929 and LP1 cell lines are presented. **(B)** MM cells were treated for 48 h with various concentrations of 4-OHTam (0–10 µM). They were then stained with propidium iodide and sorted with the FACScalibur cytometer. Ten thousand events were acquired for each sample, and the results were analyzed with CellQuest v.1.1.2 and ModFit LT v.1.01 softwares. Representative profiles obtained with LP1 and NCI H929 cells are presented. The percentages of cells in the sub-G1 and G1 fractions are noted under the graphs.

The cell lines responding to 4-OHTam treatment possessed both α and β estrogen receptors (ER), which were absent on Karpas 620 cells (FIG. 2A). The apoptotic cell death elicited by the 4-OHTam did not require Bcl-2, Bcl-X, or Bax proteins (FIG. 2B). The Pi3K/Akt pathway constitutively activated in MM cell lines, despite a normal level of PTEN, was not modified by 4-OHTam treatment, nor was the JAK/STAT3 pathway constitutively activated in CAC-2, LP1, and U266 (data not shown; FIG. 2C).

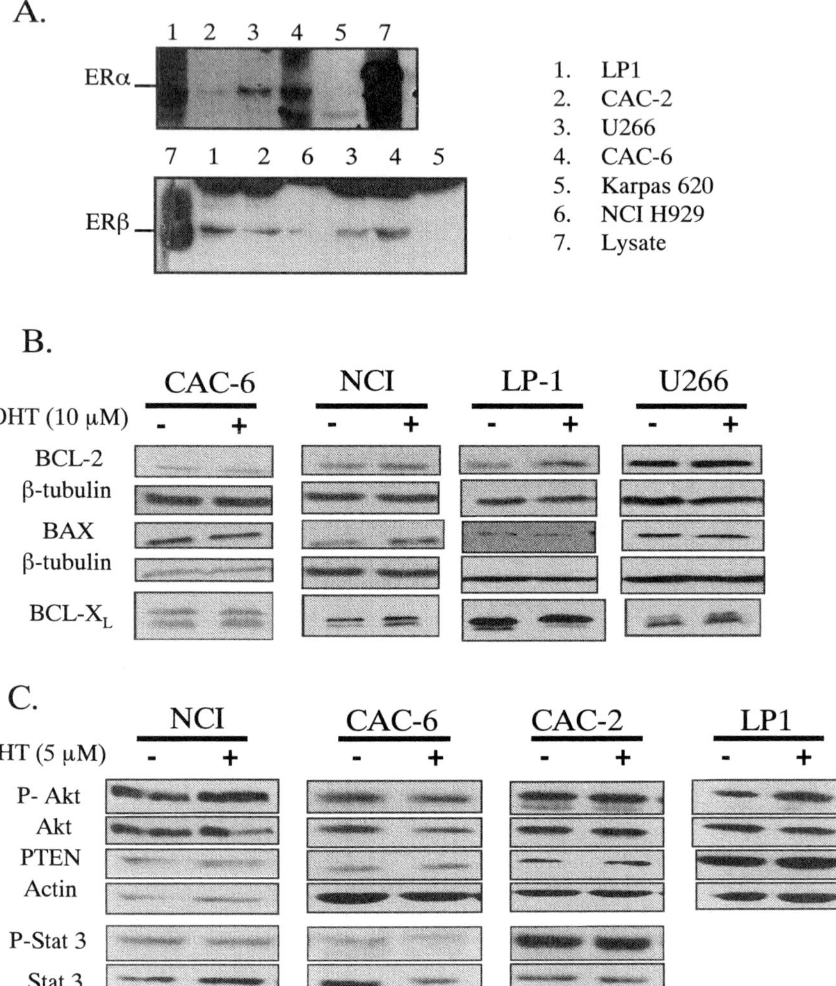

FIGURE 2. Expression of estrogen receptors and modulation of signaling molecules by the 4-OHTam. **(A)** Total cell extracts were prepared from MM cells that were grown in phenol red–free medium containing stripped FCS for 48 h. Aliquots (100 μg of proteins) were separated on an 8% SDS-PAGE, transferred onto a PVDF membrane, and blotted with D12 anti-ER mouse antibody (Santa Cruz Lab.) or with E8 anti-ERβ rabbit antibody (a gift of G. Greene). As a control, 80 ng of ERα and ERβ translated in rabbit reticulocyte lysates were used (*lane 7*). **(B)** Western blots, realized as in part A, were performed with the following antibodies: anti-Bcl-2 from Dako; anti-Bcl-X, anti-Bax, and antitubulin from Santa Cruz Biotech. **(C)** Western blots were performed with the following antibodies: antiphospho-Akt, anti-Akt, and anti-PTEN from Cell Signaling; anti-phospho-STAT3, anti-STAT3, and anti-actin from Santa Cruz Biotech.

DISCUSSION

The 4-OHTam inhibits cell proliferation of ER-expressing myeloma cells by blocking the cell cycle at G1 and inducing apoptosis. The 4-OHTam does not modify the phosphorylation state of Akt and STAT3, known to be important regulators of proliferation and survival of MM cells (FIG. 2C). The induction of apoptotic cell death is independent of Bcl-2, Bcl-X, and Bax proteins, for which the levels remain unmodified after 4-OHTam treatment. It has been previously proposed that the pro-apoptotic effect of tamoxifen was due to a positive regulation of the Fas receptor.[4] The expression of this molecule will be studied, as well as the role of the caspase family in the induction and execution of apoptosis.

ACKNOWLEDGMENTS

This work was supported by grants from ARERS–Verre Espoir, Fondation de France (No. 2002004328), and Ligue contre le cancer–Comité de la Manche. J. Gauduchon received a Scholar Award from the Académie de Médecine. We thank A. Barbaras and A. Régner for their technical help, and F. Freymuth (Laboratoire de Virologie, CHU de Caen), X. Troussard (Laboratoire d'Hématologie, CHU de Caen), and E. Lebrun (Etablissement Français du Sang, Caen) for their help in the characterization of MM cell lines.

REFERENCES

1. HALLEK, M., P.L. BERGSAGEL & K.C. ANDERSON. 1998. Multiple myeloma: increasing evidence for a multistep transformation process. Blood **91:** 3–21.
2. HIDESHIMA, T. & K.C. ANDERSON. 2002. Molecular mechanisms of novel therapeutic approaches for multiple myeloma. Nat. Rev. Cancer **2:** 927–937.
3. TREON, S.P., G. TEOH, M. URASHIMA *et al.* 1998. Anti-estrogens induce apoptosis of multiple myeloma cells. Blood **92:** 1749–1757.
4. OTSUKI, T., O. YAMADA, J. KUREBAYASHI *et al.* 2000. Estrogen receptors in myeloma cells. Cancer Res. **60:** 1434–1441.
5. TROUSSARD, X., H. AVET-LOISEAU, M. MACRO *et al.* 2000. Cyclin D1 expression in patients with multiple myeloma. Hematol. J. **1:** 181–185.

Ultraviolet Ray Induces Chromosomal Giant DNA Fragmentation Followed by Internucleosomal DNA Fragmentation Associated with Apoptosis in Rat Glioma Cells

YOSHIRO HIGUCHI,[a] YUJI MIZUKAMI,[b] AND TANIHIRO YOSHIMOTO[a]

[a]*Department of Molecular Pharmacology, Kanazawa University Graduate School of Medical Science, Kanazawa 920-8640, Japan*

[b]*Department of Radiological Technology, Faculty of Medicine, Kanazawa University, Kanazawa 920-8640, Japan*

ABSTRACT: Giant DNA fragments (1–2 Mbp) were found in C6 rat glioma cells irradiated by a lethal dose of ultraviolet-C (UV-C, 254 nm) at 50 J/m^2. After irradiation, the fragments mutated into high-molecular-weight (100–800 kbp) DNA fragments and then into ladder-formed internucleosomal DNA fragments. Poly-ADP-ribose polymerase (PARP) activity and NAD levels were reduced during DNA fragmentation. Some inhibitors of caspase and protease inhibited DNA ladder formation, but not giant DNA fragmentation, whereas antioxidants did not inhibit DNA fragmentation. These results suggest that a lethal dose of UV radiation induces giant DNA fragmentation and leads to internucleosomal DNA fragmentation associated with apoptosis through some caspases and non-reactive oxygen species in cells.

KEYWORDS: DNA fragmentation; UV-C; giant DNA; apoptosis

INTRODUCTION

The biological effects of UV with 200–290 nm of wavelength (UV-C) are multiple and include the release of soluble mediators that induce apoptosis. UV-C must first be absorbed by a chromophore within the cell, which then transduces energy into a biochemical signal.[1] We have demonstrated that the size distribution of X-ray irradiation induced double-strand breaks in DNA fragments with a range of 0.1 to 10 Mbp in mammalian cells.[2] However, it remains to be elucidated whether UV-C can induce such giant DNA fragmentation in mammalian cells. We investigated DNA fragmentation during UV-induced cell death in C6 rat glioma and considered the mechanism of giant DNA fragmentation that implicates internucleosomal DNA fragmentation.

Address for correspondence: Yoshiro Higuchi, Department of Molecular Pharmacology, Kanazawa University Graduate School of Medical Science, Kanazawa 920-8640, Japan.
higuchiy@med.kanazawa-u.ac.jp

Ann. N.Y. Acad. Sci. 1010: 326–330 (2003). © 2003 New York Academy of Sciences.
doi: 10.1196/annals.1299.058

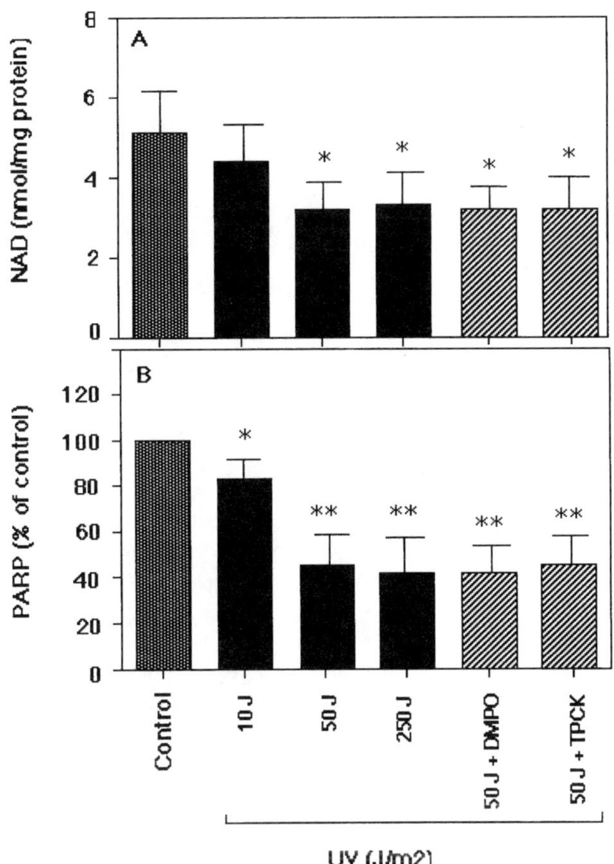

FIGURE 1. Intracellular NAD levels and poly(ADP-ribose) polymerase activity in UV-C-irradiated C6 cells. C6 rat glioma cells were grown in Dulbecco's modified Eagle's medium, supplemented with 5% fetal calf serum (medium DF-5) at 37°C, in a humidified atmosphere, containing 5% CO_2. **(A)** The cell pellet (2×10^6) prepared from C6 cells irradiated with UV-C and incubated for 12 h was suspended in 25 μL of 0.5 M $HClO_4$ and then vortexed for 5 min. The homogenate was centrifuged at 12,000*g* for 10 min, and the supernatant was applied to the HPLC separating system for NAD, which consisted of a UV detector (260 nm) equipped with a TSK-GEL ODS-80T column (4.6 × 250 mm, Tosoh, Tokyo). For the PARP assay **(B)**, C6 cells (5×10^6) were incubated for 12 h after irradiation with UV-C. The cells were washed once with PBS and suspended in 5–10 pellet volume with a buffer of 50 mM Tris-HCl, pH 8.0, and 25 mM $MgCl_2$, containing 0.1 mM phenylmethyl sulfonyl fluoride. The cells were sonicated and then centrifuged at 3000*g* for 5 min at 4°C. The supernatant was used for the PARP assay. The method was adopted from an assay protocol provided by Trevigen (USA) following the method of Kaufmann.[3] The values are means ± SD of three independent experiments; * and ** indicate a significant difference from control at $p < 0.05$ and $p < 0.01$, respectively.

RESULTS AND DISCUSSION

The C6 cell death rate by UV-C irradiation was dose-dependent, and the colony-forming ability of cells irradiated at 50 J/m^2 was less than 0.1% of nonirradiated control cells (data not shown).

To evaluate the contribution of PARP enzyme in repairing UV-induced DNA damage, PARP catabolism in C6 cells was monitored at 6, 12, and 24 h after UV irradiation. Giant DNA fragments, located in approximately the 2-Mbp range, were insignificant at 6 h after irradiation at 50 J/m^2 with UV. Twelve hours after UV irradiation, PARP activity in C6 cells irradiated with UV at both 50 and 250 J/m^2 decreased compared to cells with nonirradiated C6 cells (FIG. 1).

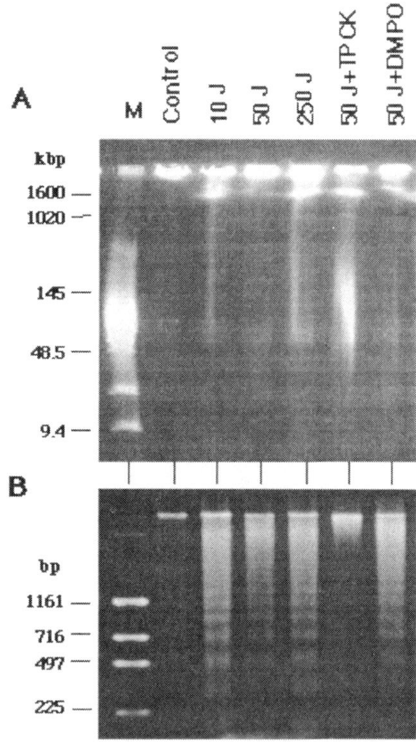

FIGURE 2. DNA fragmentation analysis by pulsed-field gel electrophoresis (PFGE) and agarose gel electrophoresis (AGE). C6 cells were irradiated in a culture disk with UV-C using UV lump (254 nm), replaced in complete media (DF-5), and then incubated at 37°C. PFGE (**A**) and AGE (**B**) analyses of the C6 cells were carried out according to the method described previously.[4] PFGE was performed at 14°C for 28 h (constant 150 volts; pulse time, 90 s for 20 h and 120 s for 8 h) by using the Pharmacia Pulsaphor PFGE system. AGE was performed at 15 volts/cm for 90 min at room temperature. The gel was stained with ethidium bromide, visualized on a UV transilluminator, and photographed with Polaroid 667 film. Chromosomal DNA from a mixture of lambda DNA, its concatemers, and *BSP*I-digested pUC-19 plasmid DNA were used as DNA size markers.

After 24 h, the amount of giant DNA fragments had increased. Smaller (30–700 kbp) DNA fragments were observed after a 24-h incubation period (FIG. 2A). There was no observation of either 2-Mbp giant DNA fragments or 30- to 700-kbp high-molecular-weight fragments during a 24-h period in UV-irradiated cells at 10 J/m^2.

At 6 h, internucleosomal DNA fragments were weakly observed in cells irradiated with UV at 50 J/m^2, and cell fragmentation continued to increase with further incubation for 24 h (FIG. 2B). Tosyl phenylamine chloromethyl ketone (TPCK), a serine protease inhibitor and an agent inhibiting apoptosis, inhibited DNA ladder fragmentation, but not 1- to 2-Mbp giant DNA fragmentation. 5,5′-Dimethyl-1-pyrroline-*N*-oxide (DMPO), an oxygen radical scavenger, did not inhibit either UV-C-induced 1- to 2-Mbp giant DNA or internucleosomal DNA fragmentation.

In this study, lethal doses of UV-C damaged target cells by a biochemical process that accompanied the formation of giant DNA fragments (1–2 Mbp). Some DNA fragments may be repaired to sizes comparable to chromosomes, as seen in the sample-loading wells, by a cellular repair system consuming NAD. PARP is involved in DNA excision repair in mammalian cells and is decayed by caspase-3 in apoptosis. PARP activation accompanies a decrease of NAD levels followed by depletion of ATP. This is a suicide mechanism of cell death to prevent the perpetuation of damaged DNA. Thereafter, the residual parts of giant DNA fragments were found to be degraded with time and to accumulate ladder-forming DNA fragments that are characteristic of apoptosis.[5]

Where and how does UV attack chromatin DNA? There appear to be preferential cleavage sites for UV-C radiation in the hinge domains that connect the individual chromatin domains. These domains can be attacked by reactive oxygen species or by secondary substances produced in cells stimulated by UV radiation. Moreover, chromosomes may have several "hot spots" that are sensitive to DNA breakage by UV radiation.

The mechanisms behind UV-induced DNA damage remain to be clarified, but it may result from UV-induced activation of unknown factors. After giant DNA fragmentation, repair systems (such as PARP and NAD) are suppressed, and the cell death signal may proceed into fragmented DNA modification. Ladder-formed internucleosomal DNA fragments (approximately 200 to 2000 bp) were detected only after UV irradiation. This suggests that UV-C may induce apoptosis accompanying such internucleosomal DNA fragments that are produced by some nuclear endonucleases at 200-bp increments.[6,7] Giant (1–2 Mbp) DNA fragmentation has been observed in HeLa carcinoma cells treated with H_2O_2[7] and in HT-29 carcinoma cells irradiated with γ rays.[8] Thus, 1- to 2-Mbp giant cell fragmentation seems to be a ubiquitous phenomenon occurring in the early stages of mammalian cell death that is induced by active oxygen-producing agents and UV irradiation. This event leads to a process resulting in ladderlike internucleosomal DNA fragmentation, which is characteristic of apoptosis.

REFERENCES

1. MOUNT, D.W. 1996. Nature **383:** 763–764.
2. HIGUCHI, Y. & S. MATSUKAWA. 1997. Free Radical Biol. Med. **23:** 90–99.
3. KAUFMANN, S.H., S. DESNOYERS, Y. OTTAVIANO *et al.* 1993. Cancer Res. **53:** 3976–3985.
4. HIGUCHI, Y. 2002. Methods Mol. Biol. **186:** 161–170.

5. CARSON, D.A. & J.M. RIBEIRO. 1993. Lancet **341:** 1251–1254.
6. STEWART, B.W. 1994. J. Natl. Cancer Inst. **86:** 286–1295.
7. HIGUCHI, Y. & T. YOSHIMOTO. 2001. Recent Res. Dev. Biophys. Biochem. Res. **Signpost:** 161–171.
8. DUSENBURY, C.E., M.A. DAVIS, T.S. LAWRENCE & J. MAYBAUM. 1990. Mol. Pharmacol. **39:** 285–289.

Apoptosis Induced by the Alkaloid Sampangine in HL-60 Leukemia Cells

Correlation between the Effects on the Cell Cycle Progression and Changes of Mitochondrial Potential

JÉRÔME KLUZA,[a] ALICE M. CLARK,[b] AND CHRISTIAN BAILLY[a]

[a]*INSERM U-524 and Laboratoire de Pharmacologie Antitumorale du Centre Oscar Lambret, IRCL, 59045 Lille Cedex, France*

[b]*National Center for Natural Products Research and Department of Pharmacognosy, Research Institute of Pharmaceutical Sciences, School of Pharmacy, University of Mississippi, Oxford, Mississippi 38677, USA*

ABSTRACT: Sampangine, a plant-derived copyrine alkaloid extracted from the stem bark of *Cananga odorata*, primarily exhibits antifungal and antimycobacterial activities, but it also displays *in vitro* antimalarial activity against *Plasmodium falciparum* and is cytotoxic to human malignant melanoma cells. It inhibits cell aggregation, but no molecular target has yet been identified. We investigated the biochemical pathway involved in sampangine-induced cytotoxicity toward HL-60 cells. These leukemia cells are prone to enter apoptosis after treatment with various stimuli, including genotoxic compounds structurally close to sampangine, such as ascididemin.

KEYWORDS: sampangine; apoptosis; mitochondrial membrane potential; cell cycle; cytotoxicity

INTRODUCTION

Sampangine is a plant-derived copyrine alkaloid extracted from the stem bark of *Cananga odorata*. This azaoxoaporphine alkaloid primarily exhibits antifungal and antimycobacterial activities, but it also displays *in vitro* antimalarial activity against *Plasmodium falciparum* and is cytotoxic to human malignant melanoma cells.[1] This alkaloid inhibits cell aggregation, although no molecular target has yet been identified. We have investigated the biochemical pathway involved in sampangine-induced cytotoxicity toward HL-60 cells. These leukemia cells are prone to enter apoptosis after treatment with various stimuli, including genotoxic compounds structurally close to sampangine, such as ascididemin.[2]

Address for correspondence: Ch. Bailly, Laboratoire de Pharmacologie Antitumorale du Centre Oscar Lambret, IRCL, Place de Verdun, 59045 Lille Cedex, France.
bailly@lille.inserm.fr

Ann. N.Y. Acad. Sci. 1010: 331–334 (2003).
doi: 10.1196/annals.1299.059

SAMPANGINE IS A POTENT INHIBITOR OF HL-60 CELL PROLIFERATION

We evaluated the activity of sampangine by studying the effects of the alkaloid on cell proliferation and viability of HL-60 leukemia, using two complementary assays: CellTiter 96 and Cytotox 96 (Promega). The first test measured a reduction of activity of mitochondrial enzymes by live cells, whereas the second measured the enzymatic activity from lactate dehydrogenase (LDH) released upon necrotic cell death. After 48 h of treatment, a concentration of 2.65 μM (IC_{50}) is required to reduce the proliferation of HL-60 by 50% and a concentration of 24.47 μM is necessary to induce 50% cell death (DL_{50}). The ratio IC_{50}/DL_{50} gives information on the type of cell death induced by this alkaloid. A ratio of 1 is expected for a drug inducing cell necrosis, whereas a ratio of <1 would reflect a marked contribution of apoptosis. With a ratio of 0.11, we hypothesize that apoptosis plays a role in the cytotoxic action of sampangine.

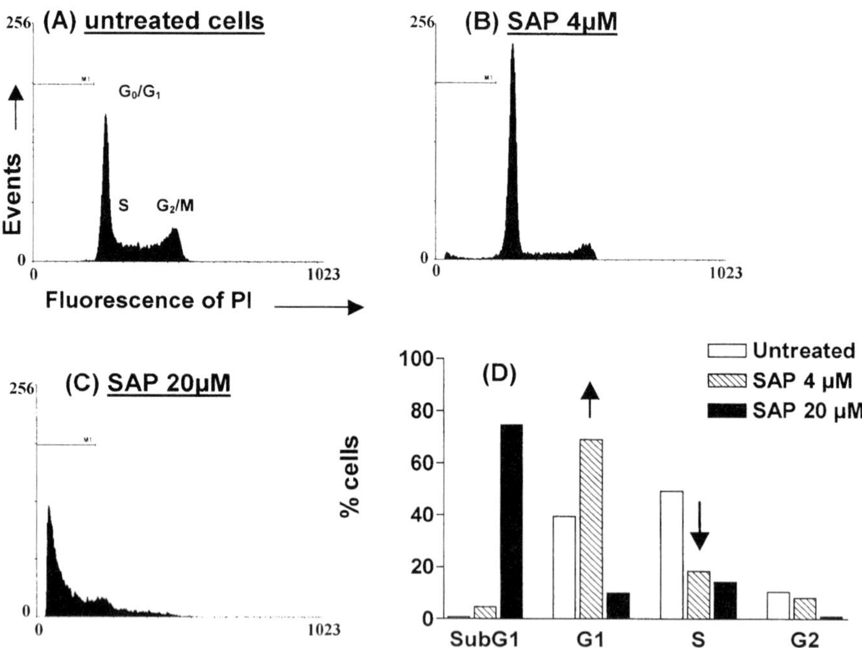

FIGURE 1. Cell cycle distribution of HL-60 in the presence or absence of sampangine. Panels A, B, and C show cytofluorimetric profiles of cells staining with propidium iodide (PI). Cells were treated with the drug for 48 h before PI labeling. The software, Cellquest, was used to determine the percentage of cells in the different phases of the cell cycle (panel D).

SAMPANGINE INDUCES G_1 ARREST AND APOPTOSIS

A flow cytometry analysis was performed to better characterize the mechanism of action of sampangine in HL-60. Propidium iodide (PI) was used to monitor sampangine-induced cell cycle effects (FIG. 1). We found that cells treated for 48 h with 4 μM sampangine induced an accumulation of cells in the G_0/G_1 phase from 40% in controls to 69% in the drug-treated samples, whereas the S-phase cell population decreased from 49% to 18%. In contrast, a hypodiploid DNA content peak (75% sub-G_1) was detected with 20 μM sampangine. This cell population, characterized by a loss of DNA content, is typical of apoptotic cells.

We next measured the effect of sampangine on the activity of caspase-3 by using the specific probe, Phiphilux G_1D_2, because caspases are very often key participants of drug-induced apoptosis. We found that HL-60 cells treated with 20 μM sampangine showed a massive activation of caspase-3 that is consistent with the induction of apoptosis (data not shown).

SAMPANGINE MODULATES THE POTENTIAL OF MITOCHONDRIAL MEMBRANES

Apoptotic stimuli generally alter the mitochondrial transmembrane potential ($\Delta\Psi_m$). We used the fluorochrome TMRM (tetramethyl-rhodamine ester) to monitor the changes in $\Delta\Psi_m$ induced by sampangine (FIG. 2). This probe is useful to probe either an increase (as induced by the mitochondrial ATP synthetase inhibitor, oligomycin) or a decrease (as induced by the mitochondrial uncoupling agent, carbonyl cyanide *m*-chlorophenylhydrazone, mClCCP) of the mitochondrial membrane potential. After 48 h of treatment, distinct effects were observed on mitochondria of HL-60: a marked hyperpolarization was measured after incubation with 4 μM sampangine and a depolarization with 20 μM. Similar results were obtained with the fluorescent probe JC-1. Moreover, preliminary observations suggest a contribution of reactive oxygen species (ROS) (data not shown). These ROS may arise from the

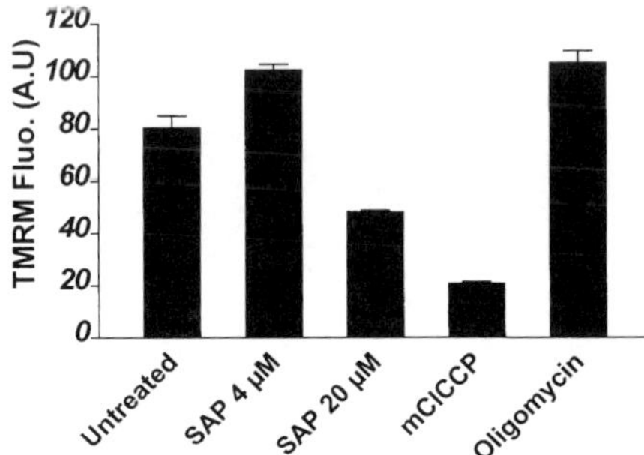

FIGURE 2. Mitochondrial changes during apoptosis. Cells were incubated with sampangine for 48 h or with oligomycin (1.25 μg/mL, 10 min at 37°C) or mClCCP (50 μM, 10 min at 37°C). Cells were stained with tetramethyl-rhodamine ester probes (TMRM, 150 nM, 20 min at 37°C), and fluorescence was quantified by flow cytometry.

accumulation of intracellular hydrogen peroxide correlated with an abnormal activity of the mitochondrial respiratory chain.[3]

CONCLUSIONS

We conclude that sampangine triggers apoptosis in HL-60 cells. Two distinct events were observed: (1) At low concentrations, sampangine induces a marked G_1 arrest and concomitantly provokes a hyperpolarization of mitochondria. On the basis of the literature data, the increased membrane potential may be correlated with a priming phase of apoptosis mechanism, followed by a mitochondrial swelling and a release of cytochrome *c*.[4] (2) In contrast with higher drug concentrations, nuclear alteration and depolarization of mitochondria were observed.

At present, it is difficult to conclude whether the observed modifications reflect different kinetics of drug action or two different biochemical pathways. The present observation confirms the relationship between cell cycle arrest (G_0/G_1 or G_2/M) induced by cytotoxic compounds[5] and hyperpolarization of mitochondria in the HL-60 cell model.

REFERENCES

1. ORABI, K., E. LI, A. CLARK & C. HUFFORD. 1999. Microbial transformation of sampangine. J. Nat. Prod. **62:** 988–992.
2. DASSONNEVILLE, L., N. WATTEZ, B. BALDEYROU *et al.* 2000. Inhibition of topoisomerase II by the marine alkaloid ascididemin and induction of apoptosis in leukemia cells. Biochem. Pharmacol. **60:** 527–537.
3. JOSHI, B., L. LI, B.G. TAFFE *et al.* 1999. Apoptosis induction by a novel anti-prostate cancer compound, BMD188 (a fatty acid–containing hydroxamic acid), requires the mitochondrial respiratory chain. Cancer Res. **59:** 4343–4355.
4. SCARLETT, J.L., P.W. SHEARD, G. HUGHES *et al.* 2000. Changes in mitochondrial membrane potential during staurosporine-induced apoptosis in Jurkat cells. FEBS Lett. **475:** 267–277.
5. KLUZA, J., A. LANSIAUX, N. WATTEZ *et al.* 2000. Apoptotic response of HL-60 human leukaemia cells to the antitumor drug TAS-103. Cancer Res. **60:** 4077–4084.

The Antiproliferative Alkylphospholipid S-1-*O*-Phosphocholine-2-*N*-Acetyl-Octadecane Induces Apoptosis in Leukemia Cell Lines

H. F. KRUG,[a] C. OBERLE,[a] A. MATZKE,[a] AND U. MASSING[b]

[a]*Forschungszentrum Karlsruhe GmbH, Institute of Toxicology and Genetics, 76021 Karlsruhe, Germany*

[b]*Tumor Biology Center, Breisacher Str. 117, 79106 Freiburg, Germany*

ABSTRACT: Lipids are involved in a multitude of important cellular functions. They act as signaling molecules and can even provoke apoptosis. In this context we investigated the efficacy of synthetic alkylphosphocholines (APCs) as potential anti-cancer membrane-affecting drugs. Leading to novel therapeutic strategies for cancer treatment, the new agents interact with the cell membrane and do not affect the DNA. The data presented here show a cell death–inducing capacity for 1-*O*-phosphocholine-2[S]-*O*-acetyl-octadecane and 1-*O*-phosphocholin-2[S]-*N*-acetyl-octadecane in Jurkat T cells as well as in BJAB cells. The activation of caspases is generally required for the induction of apoptosis as shown by experiments with specific caspase inhibitors. The results point on the one hand to the formation of a functional DISC after APC-treatment as indicated by the clustering of receptor molecules and on the other hand to the dependency on the instrinsic apoptotic machinery and the downstream of mitochondria-activated apoptosome.

KEYWORDS: leukemia cell lines; apoptosis; lipids

Among other important functions, lipids are involved in the formation of cellular structures, cell protection, and energy storage: moreover, they act as signaling molecules that regulate many biological functions. Some lipids can even provoke apoptosis. In this context, we investigated the efficiency of synthetic alkylphosphocholines (APCs) as potential anti-cancer membrane-affecting drugs.

APCs represent a new and promising class of antitumor agents. They induce apoptosis and do not interact with DNA like common antitumor medication.

APCs are derived from 2-lysophosphatidylcholine and are based on natural components of the cell membrane. It was shown, that the glycerol backbone is not an essential structural element and that APCs exert similar antitumor effects. A great

Address for correspondence: H. F. Krug, Forschungszentrum Karlsruhe GmbH, Institute of Toxicology and Genetics, P.O. Box 3540, 7621 Karlsruhe, Germany. Voice: +49-7247-82-3262; fax: +49-7247-82-3557.
krug@itg.fzk.de

**Ann. N.Y. Acad. Sci. 1010: 335–338 (2003). © 2003 New York Academy of Sciences.
doi: 10.1196/annals.1299.060**

disadvantage of the majority of anticancer drugs is the destruction of normal tissue with a high proliverative rate such as hematopoetic bone marrow. There are only a few antineoplastic substances with no haematological cytotoxicity, such as hexadecylphosphocholine. However, their exact mode of action is still unknown.

Until now there are two prominent examples of APCs: 1-*O*-octadecyl-2-*O*-methyl-sn-glycero-3-phosphocholine (Edelfosine®, Et-18-OCH_3) and hexadecylphosphocholine (Miltefosine®, HePC) which have passed through clinical trials. Edelfosine is used in experiments in purging leukemic bone marrow before transplantation and Miltefosine is used in the topical treatment of skin metastasis in breast cancer patients.

Our research presents newly synthesized, enantiomerically pure APC compounds.[1] It was our aim to develop APCs with high antineoplastic activity and lower side effects as conventional APCs. The new compounds are structurally similar to HePC, but contain additional functional groups: acetic ester, hydroxyl, acetamid and amino adjacent to phosphocholine moiety (2 position) and synthesized in the R- and S-configuration. In order to achieve the same effective length as HePC, the new compounds are based on an octadecyl chain, instead of a hexadecyl chain.[2]

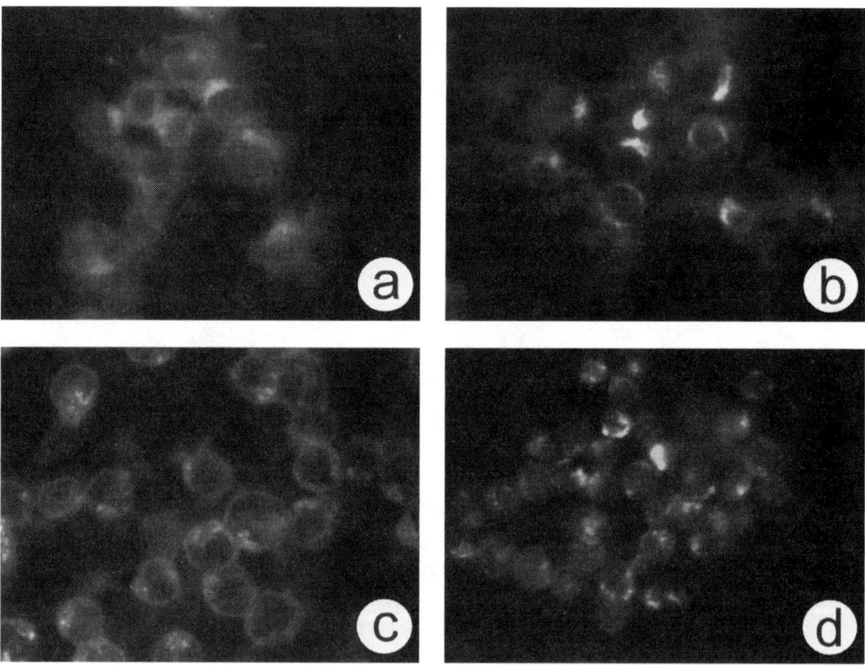

FIGURE 1. CD95-Receptor capping after treatment of Jurkat cells with S-NC-2. After fixation, cells were specifically labeled by use of CD95 antibody and a secondary antibody conjugated to FITC. Cells were treated with ethanol as negative control (**a:** 4 h), CD95 ligand as positive control (**b:** 30 min) and with S-NC-2 for two different time points (**c:** 1 h; **d:** 4 h). Fluorescence was determined in an Axiovert microscope at identical exposure times and offset/gain adjustment. Original magnification: 630×.

The treatment of cancer cell lines with the different compounds revealed characteristic biological and morphological changes of apoptosis. The externalization of phosphatidylserin, activation of proteases and endonucleases, membrane blebbing, chromatin condensation and cleavage of specific apoptotic markers, such as PARP could be observed.

The molecular target of these compounds is possibly situated upstream of caspases, presumably at the receptor level. Experiments with pro- and anti-apoptotic molecules, as well as with different cell lines (e.g., a FADD dominant negative version of BJAB cells) indicated the participation of the CD95 death receptor.[1]

Plasma membranes contain microdomains rich in cholesterol and sphingolipids, so-called lipid rafts. More and more evidence arises that lipid rafts are involved in signal transduction. Investigations by Gajate and Mollinedo[3] demonstrated that the ether lipid Et-18-OCH_3 induces a co-capping of Fas and lipid rafts. They also demonstrated that this Fas-co-aggregation with lipid rafts is essential for the induction of apoptosis by Et-18-OCH_3. By using immune histochemistry, we showed that R-NC-2 along with S-NC-2 induce a capping of Fas as well (FIG. 1). This provides further evidence of the participation of the CD95-receptor in apoptosis signaling induced by S-NC-2.

Experiments with Jurkat A3, FADD-deficient, and caspase-8-deficient cell lines point to a receptor dependent component within APC-induced apoptosis. This effect was clearly reduced in the deficient cell lines, but to a different extent than in BJAB cells.

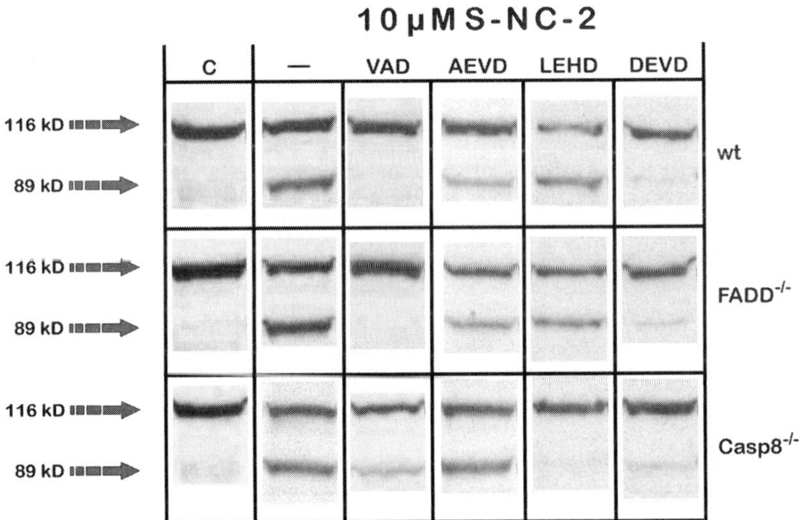

FIGURE 2. Prevention of S-NC-2 induced apoptosis by various caspase inhibitors. Cells were pretreated without (–) or with the indicated caspase inhibitors and then incubated with S-NC-2 for 16 h. Control cells were incubated with ethanol only (C). After solubilization of the cells, proteins were separated by SDS-PAGE, and PARP cleavage was analyzed by Western blotting. The antibody used recognizes the uncleaved PARP (116 kDa) and the large fragment (89 kDa).

A closer examination of the possible target molecules within the apoptotic machinery revealed that caspases, in general, play an important role. Preventing the activation of caspases by the general caspase-inhibitor zVADfmk, results in a strong reduction of PARP-cleavage in all cell lines investigated (FIG. 2). A reduction of apoptotic cells could also be observed via FACS-analysis. A systematically inhibition of various caspases gave an overview about the importance of single caspases in this context. The inhibition of the initiator-caspase-10 by zAEVDfmk had almost no effect in these Jurkat cell lines (FIG. 2). This leads to the conclusion that the initiator-caspase-10 plays no role in the induction of apoptosis by S-NC-2. On the other hand, the specific inhibition of caspase-9 by zLEHDfmk provokes a moderate, and the inhibition of caspase-3 by zDEVDfmk a complete, reduction of PARP-cleavage, which indicates a stronger dependency on these two caspases (FIG. 2).

These results point to a dual apoptosis-inducing mechanism by S-NC-2. First, the death receptors might be involved, as shown by the use of signaling-deficient cell lines. Receptor capping possibly within stabilized lipid rafts could be demonstrated by immune histochemistry. Second, a more distinct activation of the intrinsic mitochondrial pathway could be discussed, because of the failure of a total inhibition of apoptosis within the $FADD^{-/-}$ Jurkat cells. There might be a caspase-independent stimulation of cytochrom *c* release and formation of downstream apoptosome leading to enhanced activation of caspase-9 and strong activation of caspase-3. We therefore investigated the signaling elements upstream of the mitochondria in more detail.

REFERENCES

1. MATZKE, A., U. MASSING & H.F. KRUG. 2001. Killing tumour cells by alkylphosphocholines: evidence for involvement of CD95. Eur. J. Cell. Biol. **80:** 1–10.
2. MASSING, U. & H. EIBL. 1994. Synthesis of enantiomerically pure 1-*O*-phosphocholine-2-*O*-acyl-octadecane and 1-*O*-phosphocholine-2-*N*-acyl-octadecane. Chem. Phys. Lipids **69:** 105–120.
3. GAJATE, C. & F. MOLLINEDO. 2001. The antitumor ether lipid Et-18-OCH(3) induces apoptosis through translocation and capping of Fas/CD95 into membrane rafts in human leucemic cells. Blood **98:** 3860–3863.

Gene Expression Induced by X-ray Irradiation in Human Blood Cell Lines

M. MORI,[a] M. A. BENOTMANE,[a] J. VERHEYDE,[a] P. VAN HUMMELEN,[b] E. L. HOOGHE-PETERS,[c] AND C. DESAINTES[a]

[a]*Laboratory of Radiobiology, Belgian Nuclear Research Center (SCK-CEN), B-2400 Mol, Belgium*

[b]*VIB MicroArray Facility, Flanders Interuniversity Institute for Biotechnology, B-3000 Leuven, Belgium*

[c]*Department of Pharmacology, Medical School, Free University of Brussels (VUB), B-1090 Brussels, Belgium*

ABSTRACT: The response to X-ray irradiation of three different human hematopoietic cell lines originating from T (Jurkat), B (Raji), and promyelocytic (HL60) leukemia was analyzed. The survival after irradiation differed among the three cell lines, with Jurkat cells being the most vulnerable and HL60 being the least sensitive. The profile of gene expression was studied with the microarray technique in both Jurkat and HL60 cell lines. Out of the 13,800 different genes spotted on microarrays, very few genes (<0.5%) appeared to be induced more than 2-fold or repressed more than 2.5-fold in both cell lines.

KEYWORDS: X rays; irradiation; Jurkat; Raji; HL60

INTRODUCTION

The energy transferred to a cell by ionizing radiation causes damage to all the cellular compartments. Random lesions are induced by direct and indirect ionization of the cellular molecules. Various repair pathways could be triggered to allow the damage to be efficiently repaired. However, the response depends on the severity and complexity of the damage, which are related to the type of radiation, the dose, the dose rate, the phase of the cell cycle, the cell type, and the microenvironment. At a cellular level, different responses could include apoptosis, chromosome aberrations, and cell cycle deregulation. Exposure of cells to X rays delays cell cycle progression as a consequence of an arrest of the cells in G_1, S, or G_2. The block in G_1 requires the expression of wild-type p53.[1] Many tumor cells lacking functional p53 fail to arrest in G_1 following exposure to ionizing radiation, but show arrest of the cell cycle in G_2.[2,3] Cell death can be induced via activation of proapoptotic pathways or reproductive cell death that occurs after several mitotic cycles. At a molecular level, the

Address for correspondence: Marcella Mori, Radiobiology Laboratory, SCK-CEN, Studiecentrum voor kernenergie–Centre d'étude de l'énergie nucléaire, Boeretang 200, B-2400 Mol, Belgium. Voice: +32 14 33 27 32; fax: +32 14 31 47 93.
mmori@sckcen.be

**Ann. N.Y. Acad. Sci. 1010: 339–341 (2003). © 2003 New York Academy of Sciences.
doi: 10.1196/annals.1299.061**

response could include mutations in the DNA, subtle modifications of proteins, or modulation of gene expression.

In this study, we compared the response to X-ray irradiation of three different human hematopoietic cell lines originating from T (Jurkat), B (Raji), and promyelocytic (HL60) leukemia. The survival after irradiation differed among the three cell lines, with Jurkat cells being the most vulnerable (LD_{50} of 0.5 Gy) and HL60 being the least sensitive (LD_{50} of 2 Gy). The profile of gene expression was studied with the microarray technique in both Jurkat and HL60 cell lines. Out of the 13,800 different genes spotted in duplicate on 3-glass microarrays, very few genes (<0.5%) appeared to be induced more than 2-fold or repressed more than 2.5-fold in both cell lines.

MATERIALS AND METHODS

Jurkat, Raji, and HL60 cells were grown in RPMI 1640 supplemented with 10% FCS. Cells were irradiated with X rays at 0.317 Gy/min (250 kV, 15 mA, filtered with 1 mm Cu). For the clonogenic survival assay, cells were irradiated at different doses and plated on semisolid medium (GIBCO). The survived clones were counted after 2 weeks.

Total RNA was isolated from treated and sham controls using Trizol extraction. The quality of the RNA was checked with Agilent bioanalyzer. cDNA was generated by reverse transcription with oligo-dT linked to a T7 promoter. T7 was used to generate antisense RNA (10-fold amplification), further retrotranscribed, and labeled with Cy3 (treated) and Cy5 (control) fluorochromes. Equimolar concentrations of control and treated labeled cDNA were mixed and hybridized on different glass microarrays containing ±13,800 different genes (VIB MicroArray Facility, Flanders Interuniversity Institute for Biotechnology).

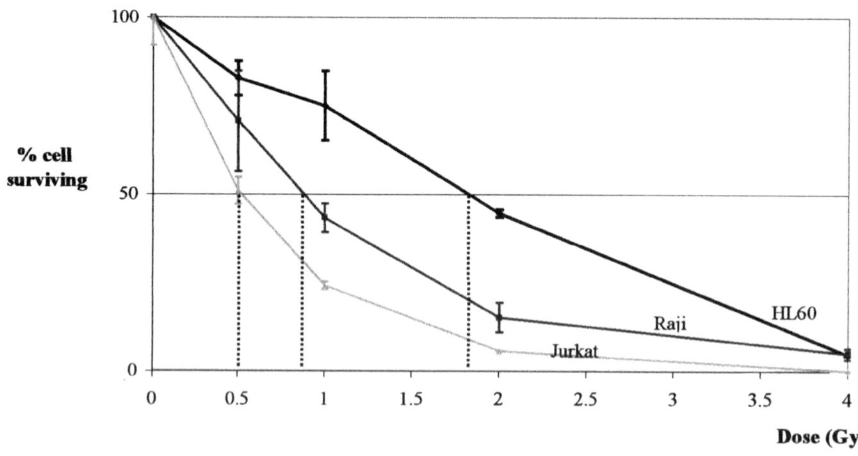

FIGURE 1. Comparison of survival curves for three different cell lines: Jurkat, Raji, and HL60. Large differences in radiosensitivity are observed. The doses corresponding to 50% survival are 0.5, 0.83, and 1.8 Gy, respectively.

TABLE 1. Microarray results of X-ray-irradiated Jurkat and HL60 cells collected at various times after different doses (0.5 and 2 Gy for Jurkat; 2 Gy for HL60)

	Jurkat		HL60	
Tested genes	13,826	100%	4,609	100%
Controls	800	5.8%	315	6.8%
Relevant genes	13,026	94.2%	4,294	93.2%
Activated genes	42	0.3%	64	1.4%
Known genes	30	0.22%	46	0.9%
Unknown genes	12	0.08%	18	0.5%
Fold changes				
Between 2 and 3	35	83.4%	38	60%
>3-fold	7	16.6%	27	40%
Repressed genes	41	0.3%	13	0.3%
Known genes	32	0.23%	8	0.2%
Unknown genes	9	0.07%	5	0.1%
Fold changes				
Between 2.5 and 3.5	33	80.5%	5	39%
>3.5-fold	8	19.5%	8	61%

RESULTS AND DISCUSSION

The three cell lines showed differential radiosensitivity. The assessment of the radiosensitivity was determined by calculating the LD_{50} dose (50% of survived cells). Jurkat cells were the most sensitive and HL60 showed less vulnerability (FIG. 1).

Data from the microarray experiments were analyzed (1) according to the level of modulation (>2-fold for activation and <2.5-fold for repression), (2) in terms of the average of background red/green (irradiated/control), higher than 1 in at least one condition, and (3) according to the coefficient of variation of the duplicate spot intensity (CV < 0.7).

TABLE 1 summarizes the results from the first analysis. The modulation induced by radiation concerns few of the tested genes: 0.3% for Jurkat and 1.4% for HL60.

REFERENCES

1. XIE, G., R.C. HABBERSETT, Y. JIA *et al.* 1998. Requirements for p53 and the ATM gene product in the regulation of G1/S and S phase checkpoints. Oncogene **16:** 721–736.
2. TAYLOR, W.R. & G.R. STARK. 2001. Regulation of the G2/M transition by p53. Oncogene **20:** 1803–1815.
3. DESIMONE, J.N. *et al.* 2003. Complexity of the mechanism of initiation and maintenance of DNA damage-induced G2-phase arrest and subsequent G1-phase arrest: TP53-dependent and TP53-independent roles. Radiat. Res. **159:** 72–85.

Staurosporine Induces Apoptotic Volume Decrease (AVD) in ECV304 Cells

A. M. PORCELLI,[a] A. GHELLI,[a] C. ZANNA,[a] P. VALENTE,[b] S. FERRONI,[b] AND M. RUGOLO[a]

[a]*Dipartimento di Biologia Ev. Sp.*, [b]*Dipartimento di Fisiologia Umana e Generale, Università di Bologna, 40126 Bologna, Italy*

ABSTRACT: Incubation of ECV304 cells with 1 μM staurosporine (STS) causes apoptotic cell death. In the present study, we investigate whether a significant apoptotic volume decrease (AVD) was apparent during the very early times (1 h) of the apoptotic process. Our data suggest that upregulation of Cl^- (and possibly K^+) channels by STS may be a very early primary event required for the subsequent onset of AVD, which results in apoptosis.

KEYWORDS: apoptotic volume decrease (AVD); cell shrinkage; Cl^- currents; patch clamp; glibenclamide; NPPB

INTRODUCTION

Apoptosis is a unique physiological process where the activation of specific biochemical and morphological events results in cellular suicide. Although the current model supports a cross-talk of mitochondria and caspases during the commitment and execution of the death process, the early events triggering apoptosis are still poorly understood. A number of recent studies have shown that a dramatic decrease of intracellular ions, primarily K^+ and Na^+, occurs during apoptosis and that prevention of this efflux results in protection from cell death in a number of model systems. The efflux of K^+, accompanied by the electroneutral efflux of Cl^-, permits the concomitant movement of osmotically obliged water out of the cell, generating the characteristic loss of cell volume (apoptotic volume decrease or AVD).[1,2] The ion transport pathways mediating KCl efflux during AVD have not been identified, although indirect evidence suggested the involvement of K^+ and Cl^- channels.[3]

We have previously described that incubation of ECV304 cells with 1 μM staurosporine (STS) caused apoptotic cell death, characterized by cytochrome c release from mitochondria (already detectable after 4–5 h), subsequent activation of caspase-3, and finally nuclear DNA fragmentation (starting at 16 h and maximal after 24 h).[4] In the present study, we have investigated whether a significant AVD was apparent during the very early times (1 h) of the apoptotic process. Furthermore, some of the

Address for correspondence: A. M. Porcelli, Dipartimento di Biologia Ev. Sp., Università di Bologna, Via Irnerio 42, 40126 Bologna, Italy. Voice: +39 051 2091286; fax: +39 051 242576. porcelli@alma.unibo.it

Ann. N.Y. Acad. Sci. 1010: 342–346 (2003).
doi: 10.1196/annals.1299.062

ion channels that might be involved in STS-induced cell shrinkage have been identified by patch-clamp analysis.

MATERIALS AND METHODS

Real-Time Dynamic Changes of Cell Morphology

Cells were seeded onto 24-mm coverslips and transfected with 8 μg of DNA of green fluorescent protein (GFP), using the Ca^{2+}-phosphate technique. Experiments were performed 36 h after transfection. The coverslips with the cells were transferred to a thermostated stage of a digital imaging system composed of an inverted epifluorescence microscope (Nikon Eclipse 300) with a back-illuminated CCD camera (Princeton Instruments, AZ) and Metamorph acquisition/analysis software (Universal Imaging Corporation, PA), as described in reference 4. Images of a single cell were acquired, the coverslip was then perfused with a medium containing STS, and images were captured every 10 min.

Electrophysiological Measurements

Low-density seeded cells were analyzed with the perforated-patch configuration of the patch-clamp technique.[5] This configuration allows the measurements of transmembrane ionic currents without perturbing the cytoplasmic pool of enzymes and metabolites that may be essential for channel modulation. To this purpose, the antibiotic amphotericin B, which forms nonselective ion-permeable pores, was added into the pipette solution. Stock solutions of amphotericin B were prepared in DMSO (60 mg/mL). The pipette was back-filled with amphotericin-containing intracellular saline (20 μL/5 mL); because amphotericin B gradually loses its efficacy, pipette solution containing amphotericin B was freshly prepared every 2–3 h. Bathing external solution was as follows (in mM): 140 NaCl, 4 KCl, 2 $MgCl_2$, 2 $CaCl_2$, 10 HEPES, 5 glucose, pH 7.4, with NaOH, and osmolarity adjusted to ~315 mOsm with mannitol. The intracellular (pipette) saline was as follows (in mM): 144 KCl, 2 $MgCl_2$, 10 HEPES, pH 7.2, with KOH, and osmolarity of ~300 mOsm. Membrane

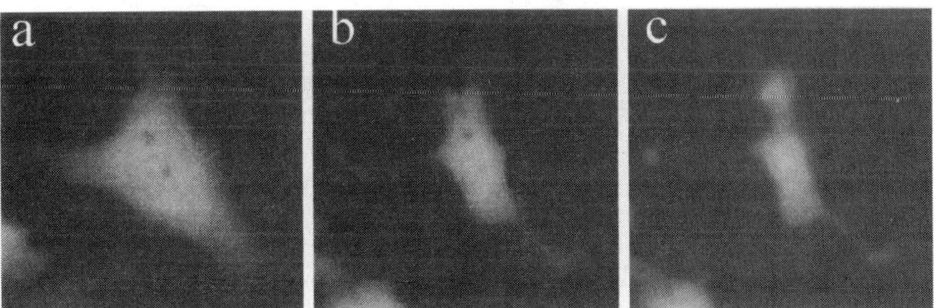

FIGURE 1. Time course of STS-induced AVD. Images of the same GFP-transfected ECV304 cell were captured after 0 min **(a)**, 30 min **(b)**, and 180 min **(c)** of incubation with 1 μM STS. This set of images is representative of five others.

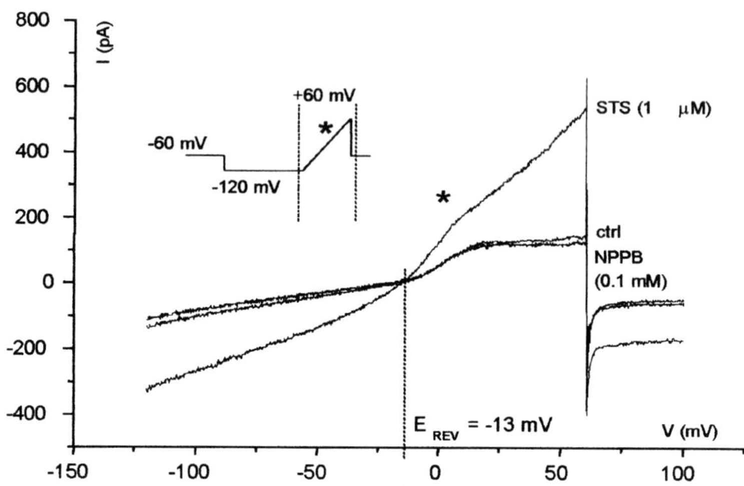

B

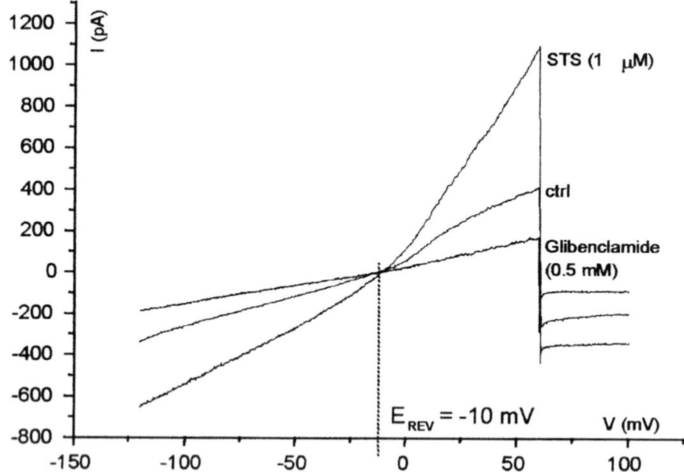

FIGURE 2. STS positively modulates Cl^- channels in ECV304 cells. Cells were voltage clamped at −60 mV with the perforated mode of the patch-clamp technique and stimulated with a voltage ramp paradigm (*inset*) from −120 to 60 mV, 180 mV/500 ms. Exposure to 1 μM STS caused an increase in ramp currents (*) at both positive and negative membrane potentials. Note that the intercept of control and STS-induced ramp currents, which depicts the reversal potential (E_{REV}) of the STS-induced conductance, was close to the equilibrium potential for Cl^- under our experimental conditions. STS-activated currents were fully inhibited by the Cl^- channel blockers, NPPB **(A)** and glibenclamide **(B)**.

currents were amplified (EPC-7, List Electronic) and stored in a microcomputer for off-line analysis (pClamp 6.0 and Origin). Ramp currents shown are those of the maximal effect under the different conditions. Currents were not leak subtracted and were filtered at 3 kHz before acquisition.

RESULTS AND DISCUSSION

Real-time changes of cell morphology were determined in single ECV304 cells transfected with GFP. The diffuse fluorescence of a cell incubated in the growth medium reveals the typical polygonal shape (FIG. 1a). The same cell was then perfused with the growth medium containing STS and, from this time, images were captured at defined times. A significant decrease of cell volume was already apparent after 30 min (FIG. 1b) and maximal after 180 min (FIG. 1c). STS-induced AVD was abolished in cells incubated in high extracellular [K^+] medium (results not shown), suggesting that osmotic water efflux is due to activation of K^+ and Cl^- efflux pathways.

Because there is accumulating evidence that channel-mediated transmembrane K^+ and Cl^- fluxes play a critical role in AVD,[3] we next explored whether STS could stimulate an ion channel activity compatible with AVD in ECV304 cells. Cells were voltage clamped at a holding potential of –60 mV and stimulated with voltage ramps from −120 to 60 mV (FIG. 2A, inset). Under control conditions, cells displayed a linear current behavior from −120 and –20 mV. More positive potentials usually elicited small outward currents that were inhibited by a blocker of K^+ channels (TEA, data not shown). Upon application of bath solution containing 1 μM STS, membrane conductance increased over the entire range of ramp potentials (FIG. 2; ~2-fold increase at 60 mV). The STS-induced currents reversed around 0 mV (–7 ± 5 mV, mean ± SD; $n = 9$), a value close to the Cl^- equilibrium potential under our experimental conditions. Moreover, STS-mediated currents were inhibited by application of the Cl^- channel blockers, 5-nitro-(3-phenylpropylamino) benzoic acid (NPPB, 0.1 mM; $n = 5$) and glibenclamide (0.5 mM; $n = 3$). The STS effect developed gradually, as first changes in membrane conductance occurred 8–10 min after starting STS challenge and currents reached the maximum values at 15–20 min. The effect was irreversible since prolonged (>10 min) washout of cells with control saline solution failed to reduce the ion currents. These results clearly indicate that STS causes an upregulation of Cl^- channel activity in ECV304 cells. Furthermore, in addition to high extracellular [K^+], incubation with quinine or glibenclamide completely prevented both AVD and nuclear DNA fragmentation. In conclusion, our data suggest that activation of Cl^- (and possibly K^+) channels by STS may be a very early primary event required for the subsequent onset of AVD, which results in apoptosis. The ability of channel blockers to promote cell survival indicates that these channels are important regulatory targets during STS-induced cell death.

ACKNOWLEDGMENTS

This work was supported by grants from PRIN 2001–2002 (Mitocondri nella fisiolopatologia cellulare) and from Progetto Dipartimentale "Biologia della riproduzione e dello sviluppo: aspetti morfologici e funzionali", University of

Bologna. We thank R. Rizzuto, University of Ferrara, for the generous gift of GFP and for making available the Telethon Cell Imaging facility.

REFERENCES

1. Bortner, C.D. & J.A. Cidlowski. 2002. Annu. Rev. Pharmacol. Toxicol. **42:** 259–281.
2. Lang, F., M. Ritter, N. Gamber *et al.* 2000. Cell. Physiol. Biochem. **10:** 417–428.
3. Maeno, E., Y. Ishizaki, T. Kanaseki *et al.* 2000. Proc. Natl. Acad. Sci. USA **97:** 9487–9492.
4. Ghelli, A., A.M. Porcelli, C. Zanna & M. Rugolo. 2002. Arch. Biochem. Biophys. **402:** 208–217.
5. Horn, R. & A. Marty. 1988. J. Gen. Physiol. **92:** 145–159.

The Role of β-Glucuronidase in Induction of Apoptosis by Genistein Combined Polysaccharide (GCP) in Xenogeneic Mice Bearing Human Mammary Cancer Cells

LAN YUAN,[a] CHIHIRO WAGATSUMA,[a] BUXIANG SUN,[a] JUNG-HWAN KIM,[b] AND YOUNG-JOON SURH[b]

[a]*Laboratory of Biochemistry, Amino Up Chemical Company, Limited, Sapporo 004-0839, Japan*

[b]*College of Pharmacy, Seoul National University, Seoul 151-742, Korea*

ABSTRACT: In order to reveal the mechanisms underlying genistein combined polysaccharide (GCP) inhibition of tumor proliferation, we investigated the effects of GCP on the growth of human breast cancer cells transplanted to athymic nude mice. We further investigated the metabolism of genistein and its association with the β-glucuronidase activity in tumor and normal tissues.

KEYWORDS: genistein combined polysaccharide (GCP); β-glucuronidase; tumor; apoptosis

INTRODUCTION

GCP (genistein combined polysaccharide) is a novel functional food obtained from the extracts of soybean isoflavone fermented with Basidiomycetes mycelia.[1] GCP contains high concentrations of isoflavone aglycons, genistein, and polysaccharides. GCP has antitumorigenic activities, including suppression of angiogenesis and induction of apoptosis.[2,3] In order to reveal the mechanisms underlying GCP inhibition of tumor proliferation, we have investigated the effects of GCP on the growth of human breast cancer cells transplanted to athymic nude mice. We have further investigated the metabolism of genistein and its association with the β-glucuronidase activity in tumor and normal tissues.

METHODS

- Transplantable tumor models: xenogeneic mouse models bearing human breast cancer cell line (MDA-MB-231).

Address for correspondence: Lan Yuan, Laboratory of Biochemistry, Amino Up Chemical Company, Limited, Sapporo 004-0839, Japan.
yuanlan@hotmail.com

Ann. N.Y. Acad. Sci. 1010: 347–349 (2003).
doi: 10.1196/annals.1299.063

- Apoptosis analysis: flow cytometry analysis for cell cycle by DNA content measurement and detection of apoptotic cells by annexin V staining; Western blot analysis for poly(ADP-ribose)polymerase (PARP), p53, p21, and cyclin.
- Detection of β-glucuronidase activities in tumor tissues and other organs.

RESULTS

Inhibition of tumor proliferation: GCP significantly inhibited the growth of human breast cancer (MDA-MB-231) cells ($p < 0.01$) in xenogeneic mice models. Induction of apoptosis: GCP significantly decreased the proportion of MDA-MB-231 cells in G1/S phase from 73.13% to 14.91%; cells arrested in G2/M phase increased from 20.21% to 45.27%. GCP induced apoptosis in 33.43% of tumor cells versus 0.23% in the control group (FIG. 1). GCP activated cleavage of PARP and also caused induction of p21 protein expression and reduction of cyclin B1 expression in the tumor tissues. A subsequent study has revealed that genistein exists as a glucuronide conjugate in systemic circulation: the conjugated genistein could not exert any physiological activity. The level of genistein aglycon in tumor tissues was much higher than that in normal tissues (FIG. 2), which might contribute to the

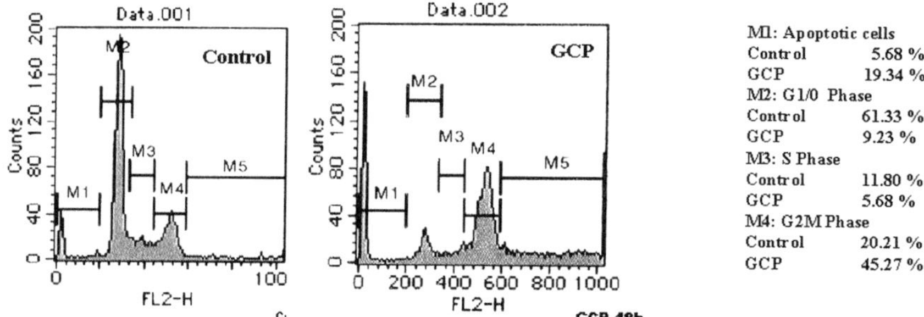

FIGURE 1. Induction of apoptosis in MDA-MB-231 cells by GCP: analysis of cell cycle by flow cytometry.

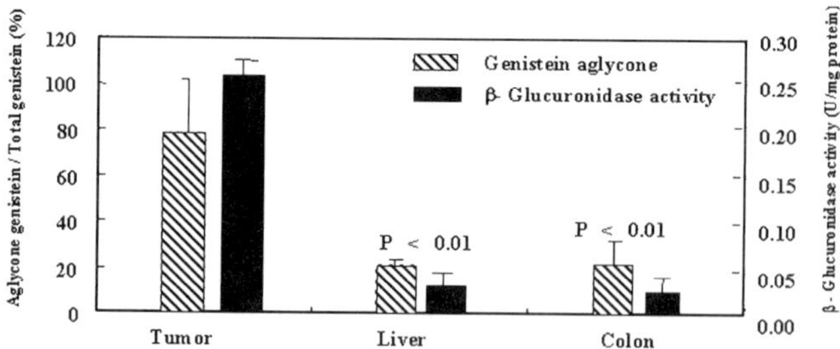

FIGURE 2. Relationship between genistein aglycon and β-glucuronidase activity in tumor tissue and normal organs.

higher susceptibility of tumor cells to GCP-induced apoptosis. These results, taken together, indicate that GCP is a potential anticarcinogenic and chemopreventive dietary supplement.

DISCUSSION

The salient features of our present study are as follows: (1) GCP inhibits tumor cell growth through multiple mechanisms, including induction of tumor apoptosis. (2) The biological activities of genistein (aglycon) are more evident in tumor tissues than in normal tissues. This might explain the reason why even high dosage of genistein administration rarely induces toxicity to normal tissues. (3) Higher levels of β-glucuronidase expression in tumor tissues results in production of more genistein aglycon, leading to tumor destruction. Because of the differential expression of β-glucuronidase in different types of tumors, clinical application of genistein may be preferentially considered for those tumors that express high levels of glucuronidase.

REFERENCES

1. MIURA, T. *et al.* 2001. Development of a natural antitimor substance GCP™. New Food Industry **43:** 17–22.
2. MIURA, T. *et al.* 2002. Isoflavone aglycon produced by culture of soybean extracts with basidiomycetes and its anti-angiogenic activity. Biosci. Biotechnol. Biochem. **66:** 2626–2631.
3. YUAN, L. *et al.* 2003. Inhibition of human breast cancer growth by GCP™ (genistein combined polyscccharide) in xenogeneic athymic mice: involvement of genistein biotransformation by beta-glucuronidase from tumor tissues. Mutat. Res. **523-524:** 55–62.

Vitamin D Protects Keratinocytes from Apoptosis Induced by Osmotic Shock, Oxidative Stress, and Tumor Necrosis Factor

TALIA DIKER-COHEN, RUTH KOREN, URI A. LIBERMAN, AND AMIRAM RAVID

Department of Physiology and Pharmacology, Felsenstein Medical Research Center, Sackler Faculty of Medicine, Tel-Aviv University, Tel-Aviv, Israel

ABSTRACT: Calcitriol, the hormonal form of vitamin D, inhibited caspase-3-like activation in HaCaT keratinocytes exposed to hyperosmotic and oxidative stresses, heat shock, and the inflammatory cytokine TNF. The hormone also protected the cells from caspase-independent cell death induced by hyperosmotic and oxidative stresses. The protection against hyperosmotic stress is not affected by inhibitors of the EGF receptor, ERK or PI13 kinase pathways, neither is it due to reduced activity of the proapoptotic p38 MAP kinase. These results are in accordance with previous *in vivo* findings that vitamin D protects epidermal keratinocytes from apoptosis due to UV radiation or chemotherapy.

KEYWORDS: calcitriol; vitamin D; keratinocytes; apoptosis; stress

INTRODUCTION

Being the barrier between the body and the exterior, epidermal keratinocytes are routinely exposed to various environmental and physiological stresses, including radiation, dehydration, changes in ambient temperature, and inflammation. Keratinocytes respond to such noxious stimuli by activation of distinct signaling pathways that lead to programmed cell death (PCD) and cellular survival, and the balance between the two will determine the ultimate cell fate.

Calcitriol, the hormonally active form of vitamin D, is synthesized from its precursor, 7-dehydrocholesterol, by a photochemical reaction, followed by two hydroxylations at C25 and C1. Epidermal keratinocytes contain the full machinery for the production and degradation of calcitriol[1] as well as the nuclear vitamin D receptors. Thus, the hormone acts in autocrine and paracrine manners and its concentrations in the skin may exceed those in the circulation.

In addition to the well-established role of calcitriol as a promoter of keratinocyte differentiation,[2] it was recently shown to have a protective effect on keratinocytes *in vivo* and *in vitro*. For instance, calcitriol decreased formation of apoptotic cells in epidermis exposed to UV[3] and reduced keratinocyte death following treatment with

Address for correspondence: Amiram Ravid, Felsenstein Medical Research Center, Beilinson Campus, Petah-Tikva 49100, Israel. Voice: +972-3-9377394; fax: +972-3-9211478.
aravid@post.tau.ac.il

Ann. N.Y. Acad. Sci. 1010: 350–353 (2003). **© 2003 New York Academy of Sciences.**
doi: 10.1196/annals.1299.064

chemotherapeutic drugs.[4] Such findings suggest a role of calcitriol in the cutaneous response to stresses as a factor shifting the life to death balance towards cell survival. We aim to examine the notion that calcitriol is an endogenous protective factor of keratinocytes from a variety of environmental and physiological stresses.

MATERIALS AND METHODS

Human immortalized nontumorigenic keratinocytes (HaCaT cells) were maintained in low calcium MEM containing 10% FCS. Twenty-four hours after plating, medium was replaced with serum-free medium containing 0.5 mg/mL albumin with or without calcitriol (100 nM). Cytotoxicity was assessed by propidium iodide staining or crystal violet staining. Caspase-3-like activity was assayed by cleavage of the fluorogenic substrate, ac-DEVD-AMC, and quantified using a fluorimeter plate reader. Cleavage of the caspase-3-like endogenous substrate, PARP, was detected by immunoblotting.

RESULTS AND DISCUSSION

Exposure of HaCaT cells to hyperosmotic shock (0.3 M sorbitol, 1 h) mimicking dehydration, heat shock (45°C, 1 h), oxidative stress (0.4 mM H_2O_2, 1 h), or the proinflammatory cytokine TNF (10 ng/mL, 24 h) brought about the activation of executioner caspases assessed by PARP and ac-DEVD-AMC cleavage. Pretreatment (24 h) with calcitriol prior to the application of all the above stresses brought about 70–90% inhibition of DEVD-ase activity (FIG. 1). The effect of calcitriol on sorbitol-treated cells was dose-dependent, already observed at the physiological concentration of 0.1 nM, and reached its maximum level after a minimum of 16-h preincubation,

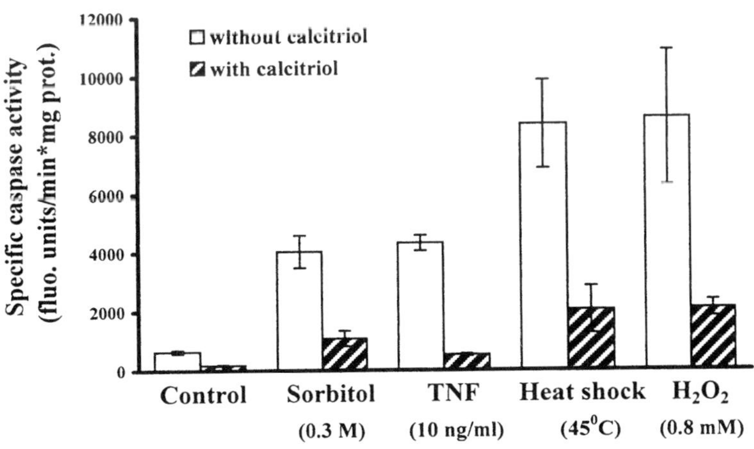

FIGURE 1. Calcitriol inhibits stress-induced caspase activation.

TABLE 1. Calcitriol protects keratinocytes from caspase-independent cell death

	Cytotoxicity (%), mean ± SD	
	Sorbitol (PI staining)	H_2O_2 (CV staining)
zD-DCB	60.5 ± 8.5	79.4 ± 5.5
zD-DCB + calcitriol	20.3 ± 4.3	49.9 ± 10.0

indicating a genomic mechanism of the hormone. Sorbitol-induced PARP cleavage was also inhibited by 24-h pretreatment with calcitriol. Treatment with the pan-caspase inhibitor zD-DCB practically abolished PARP cleavage. Calcitriol protected keratinocytes from sorbitol- and H_2O_2-induced cell death also in the presence of zD-DCB as assessed by propidium iodide and crystal violet stainings, respectively (TABLE 1), indicating that calcitriol can inhibit both caspase-dependent and caspase-independent modes of cell death.

Any defense factor against cell death can exert its protection either by enhancing cellular pathways responsible for cell survival or by inhibiting pathways leading to cell death. The EGF receptor (EGFR), the MAP kinase ERK, and PI3 kinase are key elements in cell survival. We have previously shown that calcitriol enhances signaling via the EGFR, resulting in enhanced ERK activation.[5] However, blocking this effect by the EGFR tyrosine kinase inhibitor AG1478 did not reduce the protective effect of calcitriol on DEVD-ase activity in sorbitol-treated cells. The protective effect of calcitriol was not affected by the addition of wortmannin, a PI3 kinase inhibitor, nor of SB203580, an inhibitor of the proapoptotic p38 MAP kinase, thereby excluding the involvement of these two signaling pathways in calcitriol action.

This study suggests a novel role for the epidermal vitamin D endocrine system in the cutaneous defense against stress. We conclude that calcitriol protects keratinocytes from both caspase-dependent and caspase-independent cell death induced by various stresses, probably via a mechanism involving gene expression.

ACKNOWLEDGMENTS

This research was supported by the Israel Science Foundation administered by the Israel Academy of Sciences and Humanities (Research Grant No. 601/99). This work was performed in partial fulfillment of the requirements for the Ph.D. degree of Talia Diker-Cohen, Sackler Faculty of Medicine, Tel-Aviv University.

REFERENCES

1. LEHMANN, B., T. GENEHR, P. KNUSCHKE *et al.* 2001. UVB-induced conversion of 7-dehydrocholesterol to 1α,25-dihydroxyvitamin D_3 in an *in vitro* human skin equivalent model. J. Invest. Dermatol. **117:** 1179–1185.
2. BIKLE, D.D. 1995. 1,25$(OH)_2D_3$-regulated human keratinocyte proliferation and differentiation: basic studies and their clinical application. J. Nutr. **125:** 1709S–1714S.

3. LEE, J. & J.I. YOUN. 1998. The photoprotective effect of 1,25-dihydroxyvitamin D_3 on ultraviolet light B–induced damage in keratinocytes and its mechanism of action. J. Dermatol. Sci. **18:** 11–18.
4. CHEN, G., A. BAECHLE, T.D. NEVINS *et al.* 1998. Protection against cyclophosphamide-induced alopecia and inhibition of mammary tumor growth by topical 1,25-dihydroxy-vitamin D_3 in mice. Int. J. Cancer **75:** 303–309.
5. GARACH-JEHOSHUA, O., A. RAVID, U.A. LIBERMAN & R. KOREN. 1999. 1,25-Dihydroxy-vitamin D_3 increases the growth-promoting activity of autocrine epidermal growth factor receptor ligands in keratinocytes. Endocrinology **140:** 713–721.

Inhibition of Apoptosis by Amphiregulin via an Insulin-like Growth Factor-1 Receptor–Dependent Pathway in Non-Small Cell Lung Cancer Cell Lines

AMANDINE HURBIN,[a] LAURENCE DUBREZ,[b] JEAN-LUC COLL,[a] AND MARIE C. FAVROT[a]

[a]*Groupe de Recherche sur le Cancer du Poumon, INSERM-U578, Institut Albert Bonniot, La Tronche 38706 Cedex, France*

[b]*INSERM U517, Dijon 21033 Cedex, France*

ABSTRACT: The reciprocal activation of amphiregulin (AR) and insulin-like growth factor-1 (IGF1) pathways has been shown to induce inhibition of serum deprivation apoptosis in non-small cell lung cancer (NSCLC) cell lines H358 and H322. We demonstrated that AR activated the IGF1 receptor (IGF1-R), which in turn induced the secretion of AR and IGF1. Transactivation of the IGF1-R by AR is independent of its binding to EGFR. Thus, AR can inhibit apoptosis in NSCLC cells through an IGF1-R-dependent pathway.

KEYWORDS: amphiregulin; insulin-like growth factor-1 receptor; apoptosis; serum deprivation; lung cancer

In lung cancer, well-characterized molecular abnormalities affect cell cycle control and apoptosis;[1] in addition, a variety of peptide hormones and growth factors, which bind to tyrosine kinase receptors,[2] induce autocrine or paracrine stimulation of cell survival, proliferation, migration, and angiogenesis. The major autocrine/paracrine loops for the growth and survival of non-small cell lung cancer (NSCLC) cells involve the insulin-like growth factor-1 receptor (IGF1-R) and erbB receptors such as epidermal growth factor receptor (EGFR) pathways.[2] Several ligands, including amphiregulin (AR), bind the EGFR and are produced by NSCLC cells.[3] IGF1 is a major survival factor involved in the pathogenesis of lung cancers.[4] However, the mechanisms of action and cooperation of IGF1-R and EGFR pathways in growth and survival of NSCLC are poorly understood. We showed that a reciprocal activation of AR and IGF1 pathways induced cell survival in NSCLC cells: AR activates the IGF1-R, which in turn induces the secretion of AR and IGF1.[5]

Address for correspondence: Amandine Hurbin, Groupe de Recherche sur le Cancer du Poumon, INSERM-U578, Institut Albert Bonniot, La Tronche 38706 Cedex, France. Voice: +33 47 654 9554; fax: +33 47 654 9509.
amandine.hurbin@ujf-grenoble.fr

**Ann. N.Y. Acad. Sci. 1010: 354–357 (2003). © 2003 New York Academy of Sciences.
doi: 10.1196/annals.1299.065**

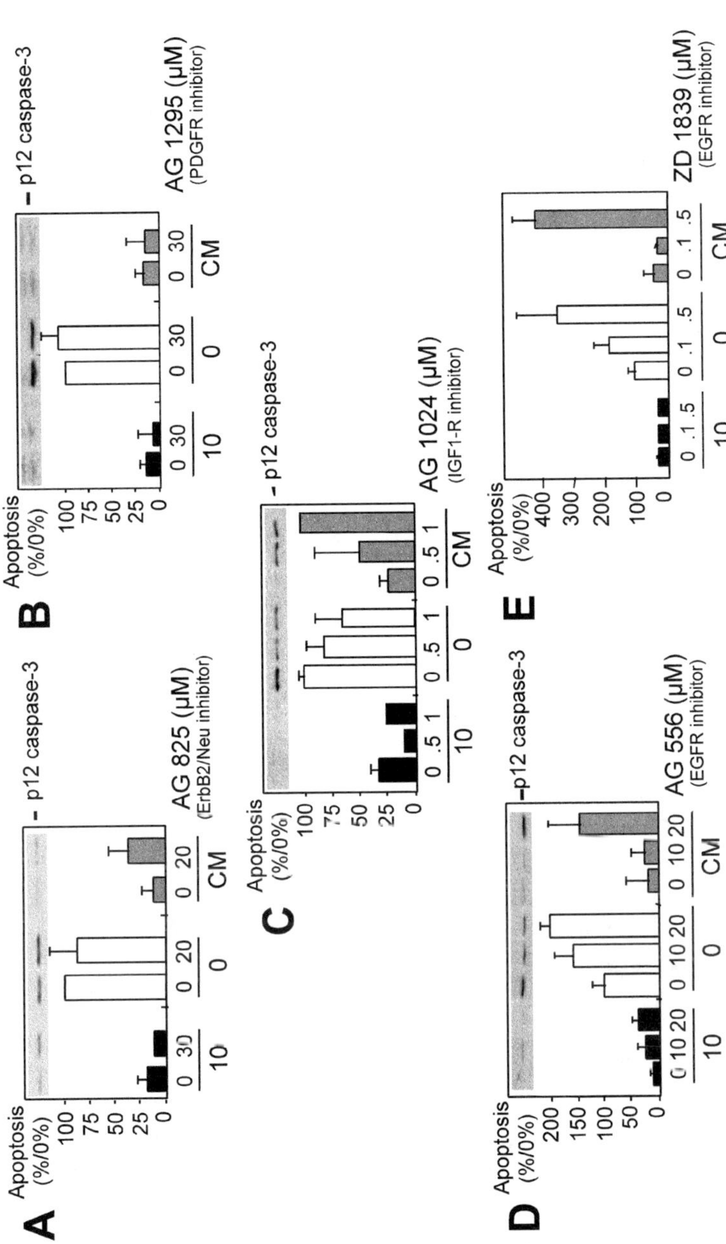

FIGURE 1. Effect of tyrosine kinase receptor inhibition on serum deprivation apoptosis in H322 cells. H322 cells were cultured in RPMI medium supplemented with 10% serum (10), serum-free medium (0), or conditioned medium from H358 cells (CM). Activity of the indicated concentrations of AG825 **(A)**, AG1295 **(B)**, AG1024 **(C)**, AG556 **(D)**, and ZD1839 **(E)** was analyzed. Apoptosis was analyzed by fluorescence microscopy after Hoechst staining of cells and by detection of the p12 active fragment of caspase-3 by immunoblotting. The percentage of apoptosis detected in the serum-free medium (0) was normalized at 100. Apoptosis was expressed as a rate of apoptosis in serum-free medium and as the mean ± SD of three independent experiments.

In the absence of serum, the NSCLC cell line H358 resisted apoptosis, and serum-free conditioned medium from H358 cells (H358 CM) protected the NSCLC H322 cells from serum deprivation apoptosis. We considered the possible involvement of several tyrosine kinase receptors and of their ligands, expressed by human lung cancer cells, in the survival of H358 and H322 cells. The IGF1-R inhibitor, tyrphostin AG1024 (1 μM), and both EGFR inhibitors, tyrphostin AG556 (20 μM) and quinazolin ZD1839 (0.5 μM), restored serum deprivation apoptosis in H358 cells and in H322 cells cultured in H358 CM (FIG. 1). Conversely, other inhibitors had no effect on H358 or H322 cell survival (FIG. 1). Altogether, these data highly suggest

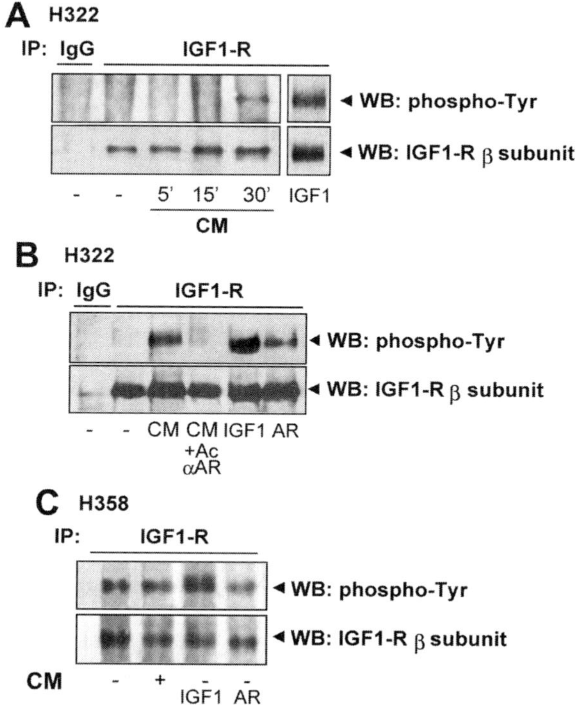

FIGURE 2. Effect of H358 conditioned medium, AR, and IGF1 on IGF1-R tyrosine phosphorylation. Serum-deprived H322 cells were incubated: **(A)** for 30 min in serum-free medium (–), in serum-free medium with 50 ng/mL of IGF1 (IGF1), or for the indicated times in serum-free H358 CM (CM); **(B)** for 30 min in serum-free medium (–), in H358 CM (CM), in H358 CM preincubated with anti-AR neutralizing antibody (CM + Ac αAR), or in serum-free medium with 50 ng/mL of IGF1 (IGF1) or AR (AR), prior to detergent lysis and immunoprecipitation of the IGF1-R β-subunit. **(C)** Serum-deprived H358 cells were incubated for 30 min in serum-free medium (–), in H358 CM (CM), or in serum-free medium with 50 ng/mL of IGF1 (IGF1) or AR (AR), prior to detergent lysis and immunoprecipitation of the IGF1-R β-subunit. Tyrosine phosphorylation was visualized by protein immunoblotting with monoclonal antiphosphotyrosine IgG as described (*upper lane*). Equal loading of immunoprecipitated proteins was confirmed by reprobing immunoblots with polyclonal anti-IGF1-R β-subunit antibody (*lower lane*). IgG: immunoglobulin control for immunoprecipitation. Data are representative of at least three separate experiments.

that H358 cells secreted autocrine/paracrine factors activating the IGF1-R and EGFR in H358 and H322 cells. The antiapoptotic activity of H358 CM is completely abolished after incubation with anti-AR neutralizing antibody and only partially with anti-IGF1 neutralizing antibody. In contrast, preincubation of H358 CM with other anti-EGF-like or IGF-like ligand antibodies did not alter antiapoptotic activity of H358 CM. The level of IGF1, detected by ELISA, secreted by serum-deprived H358 cells was not significantly different from that secreted by other cancer cell lines from lung or other origins, but the level of AR was 50 times higher. Furthermore, AR was below detectable levels in H322 cells. AR recombinant protein, in combination with IGF1, prevented serum deprivation apoptosis. These data confirmed that both AR and IGF1 are involved in the antiapoptotic activity of H358 CM and suggested a cooperation between the growth factors to inhibit serum deprivation apoptosis in the two cell lines.

We observed the phosphorylation of the IGF1-R on H322 cells after 30 min of incubation in H358 CM, which was prevented when H358 CM was preincubated with the anti-AR neutralizing antibody (FIG. 2). Moreover, incubation of H322 cells for 30 min with AR recombinant protein in serum-free medium induced the phosphorylation of the IGF1-R (FIG. 2). Thus, in H322 cells, AR activated the IGF1-R. Accordingly, H358 cells cultured in serum-free medium, which secreted IGF1 and a high amount of AR, presented a constitutive phosphorylation of the IGF1-R. Interestingly, we showed that IGF1-R phosphorylation by H358 CM or AR in H322 cells was not inhibited in the presence of the two EGFR inhibitors, AG556 (20 μM) and ZD1839 (0.5 μM). Altogether, these data demonstrated that the IGF1-R can be activated by AR independently of the activation of its own receptor, the EGFR.

In order to determine the mechanism by which a large quantity of AR is released by H358 cells, we assessed whether treatment with the IGF1-R inhibitor, AG1024, or the metalloprotease inhibitor, 1,10-phenanthroline, affected the AR release by serum-deprived H358 cells. Incubation with AG1024 (1 μM) or 1,10-phenanthroline (100 μM) markedly decreased the AR level detected in serum-free culture medium. In contrast, incubation of H358 cells with 1,10-phenanthroline had no effect on IGF1 level, but AG1024 decreased IGF1 release in serum-free culture medium. The data suggest that the IGF1-R pathway regulated AR secretion through metalloprotease activation. Furthermore, activation of the IGF1-R might also induce IGF1 release.

We have shown here that the reciprocal activation of AR and IGF1 pathways induced inhibition of serum deprivation apoptosis in NSCLC cell lines H358 and H322. We demonstrated, for the first time to our knowledge, that AR activated the IGF1-R, which in turn induced the secretion of AR and IGF1. Transactivation of the IGF1-R by AR is independent of its binding to EGFR. Thus, AR can inhibit apoptosis in NSCLC cells through an IGF1-R-dependent pathway.

REFERENCES

1. SALGIA, R. & A.T. SKARIN. 1998. J. Clin. Oncol. **16:** 1207–1217.
2. ROZENGURT, E. 1999. Curr. Opin. Oncol. **11:** 116–122.
3. RUSCH, V., D. KLIMSTRA, E. VENKATRAMA *et al.* 1997. Clin. Cancer Res. **3:** 515–522.
4. GRIMBERG, A. & P. COHEN. 2000. J. Cell. Physiol. **183:** 1–9.
5. HURBIN, A., L. DUBREZ, J-L. COLL & M.C. FAVROT. 2002. J. Biol. Chem. **277:** 49127–49133.

The Overexpression of Bcl-2 Antagonizes the Proapoptotic Function of the Kappa-Opioid Receptor

NAISUM WONG,[a] CATHERINE TEIMEI DIAO,[a] AND TAKMING WONG[b]

Departments of [a]Biochemistry and [b]Physiology, Faculty of Medicine, University of Hong Kong, Hong Kong

ABSTRACT: Opioid receptors are G-protein-coupled cell-surface receptors that are mainly expressed in neuronal cells. Stimulation of the κ-opioid receptor expressed by cultured human epithelial cancer cells promotes staurosporine-induced apoptosis. In this study, while Bcl-2 did not inhibit staurosporine-induced apoptosis, it did inhibit the κ-opioid receptor–mediated potentiation of apoptosis. The results suggest that Bcl-2 targets a step that is specific to the signaling pathway of the κ-opioid receptor.

KEYWORDS: opioid; receptor; staurosporine; epithelial

INTRODUCTION

Opioid receptors are G-protein-coupled cell-surface receptors that are mainly expressed in neuronal cells. The function of opioid receptors in nonneuronal cells has not been investigated extensively. We demonstrated previously that stimulation of the κ-opioid receptor expressed by cultured human epithelial cancer cells promoted staurosporine-induced apoptosis. The underlying mechanism is not known. In the present study, while Bcl-2 did not inhibit staurosporine-induced apoptosis, it did inhibit the κ-opioid receptor–mediated potentiation of apoptosis. The results suggest that Bcl-2 targets a step that is specific to the signaling pathway of the κ-opioid receptor.

RESULTS

When the human epithelial cancer cell line (CNE2) was treated with staurosporine for 24 h, a dose-dependent increase in the number of apoptotic cells was observed. The percentage of apoptotic cells increased substantially in the presence of a specific agonist for the κ-opioid receptor (U50488H) (FIG. 1). To investigate if the κ-opioid receptor–mediated potentiation of apoptosis was sensitive to Bcl-2, this

Address for correspondence: Dr. NaiSum Wong, Department of Biochemistry, Faculty of Medicine, University of Hong Kong, 3/F, Laboratory Block, Faculty of Medicine Building, 21, Sassoon Road, Hong Kong. Voice: +852-28199142; fax: +852-28551254.
nswong@hkucc.hku.hk

Ann. N.Y. Acad. Sci. 1010: 358–360 (2003). © 2003 New York Academy of Sciences.
doi: 10.1196/annals.1299.066

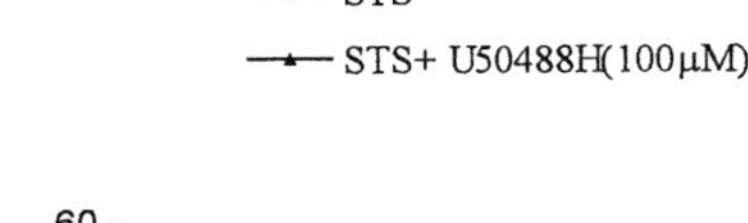

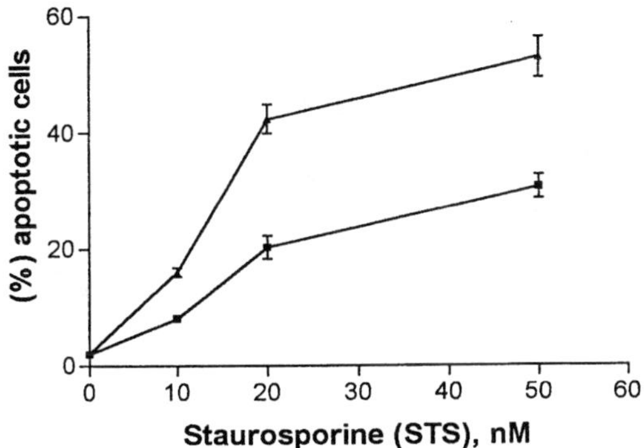

FIGURE 1. Dose-dependent induction of apoptosis by staurosporine. CNE2 cells were grown in 60-mm dishes until 80% confluence. The cells were then incubated in 0.2% serum medium for 12 h. Stock staurosporine solution (in DMSO) was added to the cells at the indicated concentrations. After 24 h, cells were harvested and analyzed for apoptosis using procedures as detailed in reference 1.

protein was overexpressed in the CNE2 cells that were then treated with staurosporine, either with or without coincubation with U50488H. In cells that were transfected with a plasmid that did not have the cDNA insert for Bcl-2, treatment with staurosporine resulted in apoptosis that was enhanced in the presence of U50488H. The enhancement in apoptosis was the same between the empty-plasmid-transfected and the wild-type cells. In cells that were transfected with the Bcl-2 cDNA and hence overexpressed the Bcl-2 protein, treatment with staurosporine caused the same amount of apoptosis as was observed in the empty-plasmid-transfected or the wild-type cells. Thus, the overexpression of Bcl-2 protein did not inhibit apoptosis caused by staurosporine. However, when the Bcl-2 overexpressing cells were treated together with staurosporine and U50488H, apoptosis was only slightly enhanced in comparison to the empty-plasmid-transfected or the wild-type cells. Results are summarized in FIGURE 2.

DISCUSSION

The present study has provided two important findings. First, the overexpression of Bcl-2 has no effect on apoptosis induced by staurosporine. The antiapoptotic function of Bcl-2 is most probably due to its ability to regulate cytochrome C release from mitochondria, an event that is crucial to the activation of caspases. The lack of effect of Bcl-2 suggests that multiple mechanisms may exist for the activation of caspases and some of these mechanisms are not inhibited by Bcl-2, as illustrated by the activation of caspase-8 by the death receptors. Further experiments will be necessary to distinguish between these possibilities. Second, the enhancement of

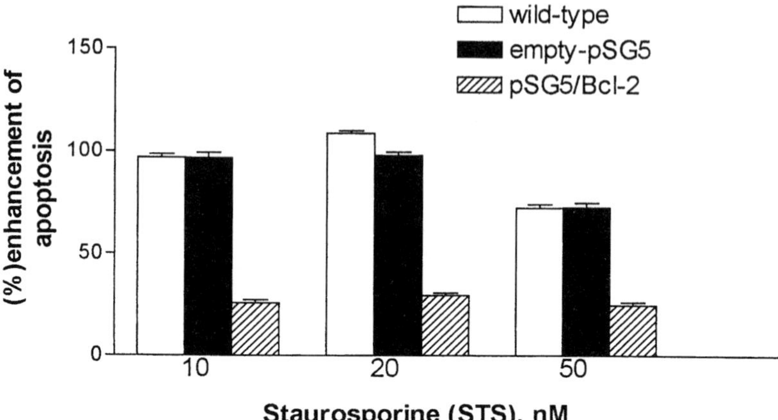

FIGURE 2. The loss of enhancement of apoptosis mediated by the κ-opioid receptor in CNE2 cells that overexpressed Bcl-2. CNE2 cells were transfected with a pSG5 plamid carrying a 910-bp cDNA encoding the full-length Bcl-2 protein. Expression of the Bcl-2 protein was verified by Western blot analysis using an anti-Bcl-2 antibody obtained from Transduction Laboratories (Lexington, KY). Transfected cells were stimulated to undergo apoptosis in the absence or presence of U50488H. The % enhancement of apoptosis was measured in the wild-type cells, in cells transfected with an empty pSG5 plasmid, and in cells transfected with the pSG5/Bcl-2 plasmid.

apoptosis mediated by the stimulation of the κ-opioid receptor was abolished by the overexpression of Bcl-2. This observation suggests that Bcl-2 is acting at a step that is specific to the activation of the κ-opioid receptor. The stimulation of the κ-opioid receptor in CNE2 cells produced the second messenger, inositol 1,4,5-trisphosphate, which is known to mobilize intracellular calcium ions. Indeed, a rise in intracellular calcium was detected when these cells were stimulated by U50488H (data not shown). Intracellular calcium ion is known to affect mitochondrial function, notably that of the PTP, as well as other organelles, resulting in apoptosis. Recently, the over-expression of Bcl-xl has been shown to inhibit the intracellular mobilization of calcium ion through the inhibition of the synthesis of the inositol 1,4,5-trisphosphate receptor. If this effect of Bcl-xl is also shared by Bcl-2, then the expression of the inositol 1,4,5-trisphosphate receptor is expected to be inhibited, hence suppressing the calcium ion signal elicited by the κ-opioid receptor. Preliminary experiments showed that the inhibition of phospholipase C activity can also abolish the enhancement of apoptosis due to the stimulation of the κ-opioid receptor. Verification of this hypothesis will require the actual measurement of the calcium ion level, as well as the expression of the inositol 1,4,5-trisphosphate receptor in the Bcl-2 overexpressing CNE2 cells.

REFERENCE

1. Diao, C.T.M., L. Li., S.Y. Lau *et al.* 2000. κ-Opioid receptor potentiates apoptosis via a phospholipase C pathway in the CNE2 human epithelial tumor cell line. Biochim. Biophys. Acta **1499:** 49–62.

Induction of Apoptosis in Human Pancreatic Cancer Cells by Docosahexaenoic Acid

N. MERENDINO,[a] R. MOLINARI,[a] B. LOPPI,[a] G. PESSINA,[a] M. D' AQUINO,[b] G. TOMASSI,[a] AND F. VELOTTI[a]

[a]*Department of Environmental Sciences, Tuscia University, Viterbo, Italy*

[b]*National Institute for Food and Nutrition Research (INRAN), Rome, Italy*

ABSTRACT: Polyunsaturated fatty acids have been indicated to induce anti-proliferative and/or apoptotic effects in various tumor cells. We showed that, at a 200-μM concentration, both alpha-linoleic (18:2 n-6; LA) or docosahexaenoic (22:6 n-3; DHA) acid inhibited cell growth, while only DHA induced apoptosis in the human Paca-44 pancreatic cancer cell line. Investigating the mechanism underlying DHA-induced apoptosis, we showed that DHA induced a rapid and dramatic (>60%) intracellular depletion of reduced glutathione (GSH), without affecting oxidized glutathione (GSSG). Moreover, using two specific inhibitors of carrier-mediated GSH extrusion, cystathionine or methionine, we observed that GSH depletion occurred via an active GSH extrusion, and that inhibition of GSH efflux completely reversed apoptosis. These results provide the first evidence for a possible causative role of GSH depletion in DHA-induced apoptosis.

KEYWORDS: polyunsaturated fatty acids (PUFAs); butyric acid (BA); alpha-linoleic acid (LA); docosahexaenoic acid (DHA); apoptosis; oxidation; glutathione; pancreatic cancer

There is evidence that some dietary polyunsaturated fatty acids (PUFAs) can inhibit the growth and/or kill a variety of tumor cells *in vitro* and *in vivo*,[1] with particular emphasis on docosahexaenoic acid (22:6 n-3; DHA) contained mainly in seafood products. PUFAs can also reverse and/or inhibit tumor cell drug resistance.[2] This has led to certain PUFAs being proposed as possible anti-cancer therapeutic agents, or potential adjuvants to radio- or chemotherapy.[3] However the molecular mechanism(s) underlying these activities are still some time from being clarified.

Several studies have indicated that DHA induces apoptosis by a protein phosphatase–mediated process and that it is able to activate the caspases 3 and 9 and belongs to a family of cysteine proteases, which are involved downstream in the process of apoptosis.[4,5] Moreover DHA decreased the antiapoptotic bcl-2 level,[5,6]

Address for correspondence: Nicolò Merendino, Laboratory of Immunology and Nutrition, Department of Environmental Sciences, Largo dell'Università, Tuscia University, 01100 Viterbo, Italy. Voice: +39-0761-357133; fax: +39-0761-357134.
merendin@unitus.it

Ann. N.Y. Acad. Sci. 1010: 361–364 (2003).
doi: 10.1196/annals.1299.143

inactive prostaglandin family of the gene, and decreased lipoxygenase activity, and it altered expression of peroxisome proliferators PARP, while PUFA-induced apoptotic cells appear to be independent of the p53 pathway.[6,7]

An assumed mechanism by which PUFAs could affect survival of cancer cells and induce apoptosis is the modulation of redox status and the involvement of the mithocondrial pathway of apoptosis induction leading to cytochrome *c* release, caspase activation, loss of mitochondrial membrane potential, and DNA fragmentation.[8,9] In the oxidative mechanism, concentration of glutathione (γ-glutamylcysteinylglicine, GSH), the most abundant intracellular radical scavenger, could play a key role, considering that intracellular GSH depletion has been reported to occur in different apoptotic systems.[10,11]

Pancreatic cancer represents the fifth leading cause of cancer death in Western countries. It is associated with severe cachexia and is almost incurable, displaying a high degree of resistance to conventional radio- and chemotherapy.[12] The identification of anti-proliferative/apoptotic agents for pancreatic cancer cells and the understanding of their mechanism of action might contribute to the development of a more effective anti-pancreatic cancer therapy.

In the present study, we investigated the ability of DHA to induce growth inhibition and apoptosis in the Paca-44 human pancreatic cancer cell line and we analyzed the mechanism(s) underlying its activity. We observed that DHA, used at a concentration (200 μM) achievable *in vivo,* induced both growth inhibition (approximately 50%) and apoptosis (more than 50%) in the Paca-44 pancreatic cancer cell line. To evaluate the effectiveness of DHA biological activity, we compared DHA anti-proliferative/apoptotic activities with those exerted by other fatty acids, such as the less unsaturated fatty acid, alpha-linoleic acid (LA), and the short-chain saturated fatty acid, butyric acid (BA), both being indicated as anti-proliferative/apoptotic compounds.[13] We observed that, although both LA and BA exerted an anti-proliferative effect on pancreatic cancer cells, the long-chain PUFAs, DHA and LA, were more effective in inducing cell growth inhibition than was the short-chain fatty acid, BA. However, both LA and BA, used at the same concentration as DHA, were not able to induce apoptosis in pancreatic cancer cells.

To understand the mechanisms underlying the apoptotic activity induced by DHA, we analyzed whether an oxidative mechanism was involved in this process. Treatment of Paca-44 cells with the antioxidant butylated hydroxytoluene (BHT), vitamin A and vitamin C, completely reversed apoptosis by DHA, indicating that an oxidative mechanism was involved in DHA-induced apoptosis in pancreatic cancer cells. To investigate a possible role of gluthatione in DHA-induced apoptosis, Paca-44 cells were treated with DHA (200 μM), and the intracellular GSH (reduced form) and GSSG (oxidized form) contents were measured by HPLC. GSH content was dramatically reduced (more than 60%) following 6 h of DHA treatment, while intracellular GSSG was not modified at both 6 and 18 h (FIG. 1). This result showed that DHA induced GSH depletion in pancreatic cancer cells undergoing apoptosis, and suggested that GSH depletion was the result of an active GSH extrusion from the cells. Using two specific inhibitors of carrier-mediated GSH extrusion, such as cystathionine and methionine, we examined whether GSH depletion was the result of an active GSH extrusion from pancreatic cancer cells. As illustrated in TABLE 1, the inhibition of GSH efflux completely reversed the apoptosis by DHA, showing that GSH was actively extruded from the cell. In this study, we showed that the PUFA

TABLE 1. Effect of GSH-carrier inhibitors on DHA-induced apoptosis

	Percentage of Apoptotic Cells
Untreated	10.2
DHA (200 μM)	33.5
DHA + cystathionine (1 mM)	17.48
DHA + methionine (1 mM)	13.52
Cystathionine (1 mM)	12.02
Methionine (1 mM)	9.87

NOTE: Paca-44 cells were pre-treated with ethanol alone (control), 1 mM of methionine, or 1 mM of cystathionine, added 1 hour before DHA (200 μM) treatment, and apoptosis was assessed at 48 hours. Paca-44 cells were treated with ethanol alone, 1 mM of methionine, or 1 mM of cystathionine for 48 hours. This experiment is representative of at least three different experiments.

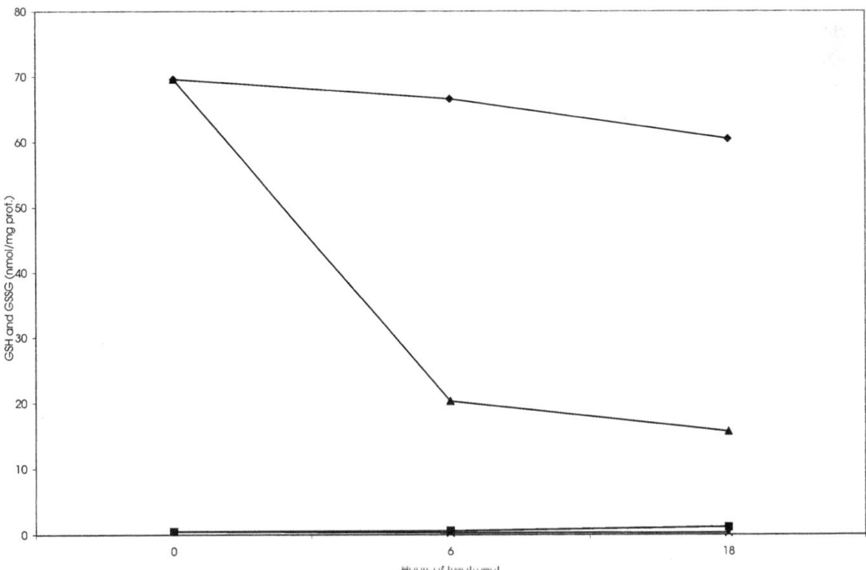

FIGURE 1. Intracellular glutathione depletion after DHA treatment of Paca-44 cells. Glutathione was determined by HPLC. GSH (*solid diamond*) and GSSG (*solid square*) content in untreated cells; GSH (*solid triangle*) and GSSG (×) content in cells treated with DHA (200 μM). This experiment is representative of three different experiments.

DHA induced both growth inhibition and apoptosis in the human Paca-44 pancreatic cancer cell line. Moreover, this result provides the first evidence for the requirement of specific GSH depletion in the commitment of apoptosis by DHA. We postulate that GSH depletion might be the cause of oxidative stress, by altering the reducing power of the cell and its ability to scavenge or detoxify the various reactive oxygen intermediates that form in cell metabolism, leading to or making cells more susceptible to apoptosis.

Overall our results suggest DHA as a possible therapeutic agent in pancreatic carcinoma.

ACKNOWLEDGMENTS

This work was supported in part by a grant from MIUR (MM06158571_003) and MIPAAF (5-C71).

REFERENCES

1. DIGGLE, C.P. 2002. In vitro studies on the relationship between polyunsaturated fatty acids and cancer: tumour or tissue specific effects? Prog. Lipid. Res. **41:** 240–253.
2. DAS, U.N. 1999. Reversal of tumor cell drug resistance by essential fatty acids. Lipids **34:** S103.
3. HAWKES, N. 1994. Hopeful newspaper articles are justified. Br. Med. J. **309:** 543–544.
4. SIDDIQUI, R.A., L.J. JENSKI, K. NEFF, *et al.* 2001. Docosahexaenoic acid induces apoptosis in Jurkat cells by a protein phosphatase-mediated process. Biochim. Biophys. Acta **1499**: 265–275.
5. NARAYANAN, B.A., N.K. NARAYANAN & B.S. REDDY. 2001. Docosahexaenoic acid regulated genes and transcription factors inducing apoptosis in human colon cancer cells. Int. J. Oncol. **19**: 1255–1262.
6. CHEN, Z.Y. & N.W. ISTFAN. 2000. Docosahexaenoic acid is a potent inducer of apoptosis in HT-29 colon cancer cells. Prostaglandins Leukot. Essent. Fatty Acids **63:** 301–308.
7. DIGGLE, C.P., E. PITT, P. ROBERTS, *et al.* 2000. N;-3 and n;-6 polyunsaturated fatty acids induce cytostasis in human urothelial cells independent of p53 gene function. J. Lipid Res. **41:** 1509–1515.
8. NARAYANAN, B.A., N.K. NARAYANAN & B.S. REDDY. 2001. Docosahexaenoic acid regulated genes and transcription factors inducing apoptosis in human colon cancer cells. Int. J. Oncol. **19**: 1255–1262.
9. ARITA, K., H. KOBUCHI, T. UTSUMI, *et al.* 2001. Mechanism of apoptosis in HL-60 cells induced by n-3 and n-6 polyunsaturated fatty acids. Biochem. Pharmacol. **62:** 821–828.
10. GHIBELLI, L., S. COPPOLA, G. ROTILIO, *et al.* 1995. Non-oxidative loss of glutathione in apoptosis via GSH extrusion. Biochem. Biophys. Res. Commun. **216:** 313–320.
11. VAN DEN DOBBELSTEEN, D.J., C.S. NOBEL, J. SCHLEGEL, *et al.* 1996. Rapid and specific efflux of reduced glutathione during apoptosis induced by anti-Fas/APO-1 antibody. J. Biol. Chem. **271:** 15420–15427.
12. MATSUNO, S., S. EGAWA, & K. ARAI. 2001. Trends in treatment for pancreatic cancer. J. Hepatobiliary Pancreat. Surg. **8:** 544–548.
13. GAMET, L., D. DAVIAUD, C. DENIS BOUXVITT, *et al.* 1992. Effects of short-chain fatty acids on growth and differentiation of the human colon-cancer cell line HT29. Int. J. Cancer **52:** 286–289.

Pro-oxidant Activity of Low Doses of Resveratrol Inhibits Hydrogen Peroxide–Induced Apoptosis

KASHIF ADIL AHMAD,[a] MARIE-VERONIQUE CLEMENT,[b] AND SHAZIB PERVAIZ[a]

Departments of [a]Physiology and [b]Biochemistry, Faculty of Medicine, National University of Singapore, Singapore 117597

ABSTRACT: We have recently shown that efficient apoptotic signaling is a function of a permissive intracellular milieu created by a decrease in the ratio of superoxide to hydrogen peroxide and cytosolic acidification. Resveratrol, a phytoalexin found in grapes and wines, triggers apoptosis in some systems and inhibits the death signal in others. In this regard, the reported inhibitory effect on hydrogen peroxide–induced apoptosis has been attributed to its antioxidant property. Here, we provide evidence that exposure of human leukemia cells to low concentrations of resveratrol (4–8 μM) inhibits caspase activation and DNA fragmentation induced by incubation with hydrogen peroxide or upon triggering apoptosis with a novel compound that kills via intracellular hydrogen peroxide production. At these concentrations, resveratrol elicits pro-oxidant properties as evidenced by an increase in intracellular superoxide concentration. This pro-oxidant effect is further supported by our observations that the drop in intracellular superoxide and cytosolic acidification induced by hydrogen peroxide is completely blocked in cells preincubated with resveratrol. Thus, the inhibitory effect of resveratrol on hydrogen peroxide–induced apoptosis is not due to its antioxidant activity, but contrarily via a pro-oxidant effect that creates an intracellular environment nonconducive for apoptotic execution.

KEYWORDS: apoptosis; hydrogen peroxide; resveratrol; reductive stress

INTRODUCTION

Apoptosis is an orderly series of intracellular events triggered by a variety of stimuli as diverse as death receptor ligation, ultraviolet irradiation, and anticancer drugs.[1,2] The effector components of this death pathway and their intricate networking have been unraveled over the past couple of decades. Consequently, it is now well established that depending upon the level of activation of the initiator caspase, such as caspase 8, the death signal can directly recruit downstream effector caspases or

Address for correspondence: Shazib Pervaiz, Associate Professor, Department of Physiology, National University of Singapore, MD9, #09-06, Singapore 117597. Voice: +65 6874 6602; fax: +65 6778 8161.

phssp@nus.edu.sg

Ann. N.Y. Acad. Sci. 1010: 365–373 (2003).
doi: 10.1196/annals.1299.067

engage the mitochondria with the resultant release of death amplification factors from the mitochondrial intermembranous space.[3–5] Death signaling by anticancer drugs is generally reliant upon the positive input from the mitochondria, which explains for the resistance of tumor cells overexpressing the death inhibitory protein Bcl-2.[6,7] Therefore, by implication, an intracellular milieu permissive for the activation of caspases is critical for an efficient execution of the death signal. To that end, we have been working on the role of cellular redox state in the regulation of death signaling.[8–11] The cellular redox state is a product of intracellular levels of reactive oxygen species (ROS), such as superoxide (O_2^-) and hydrogen peroxide (H_2O_2), and the inherent ability of the cells to maintain a nontoxic constitutive concentration of these reactive species via the antioxidant defense systems. Whereas an overwhelming accumulation of intracellular ROS could create an oxidatively stressed environment leading to necrotic death of the cells, a slight increase on the other hand is a stimulus for cellular proliferation.[12,13] Indeed, a slight pro-oxidant intracellular milieu is a hallmark of many tumor cells and is believed to endow tumor cells with a survival advantage over their normal counterparts.[12,14–16] In this regard, we previously demonstrated that a slightly elevated intracellular concentration of O_2^- was inhibitory to apoptotic signaling, irrespective of the trigger.[8,9,17,18] On the contrary, our data supported a death-promoting effect of H_2O_2.[9,19] These observations formed the basis for our hypothesis that a critical balance between intracellular H_2O_2 and O_2^- may dictate the response of cells to apoptotic stimuli.[10,20] A tilt in the ratio toward H_2O_2 is favorable for apoptotic signaling; however, if O_2^- predominates, the death signal is significantly impeded. Interestingly, the apoptosis-promoting activity of H_2O_2 is also linked to its ability to reduce the intracellular concentration of O_2^-, in addition to inducing a drop in cytosolic pH, a state that we have defined as *reductive stress–induced apoptosis.*[10,11,20] We have recently demonstrated that exposure of tumor cells to anticancer drugs can result in enhanced intracellular H_2O_2 production that renders the cytosolic milieu permissive for efficient execution of the death signal.[19] Therefore, any stimulus or signal that inhibits the ability of H_2O_2 to reduce the intracellular environment could potentially block the death signal and give rise to drug resistance.

One area of recent interest in cancer biology is the chemopreventive potential of natural products. Among the compounds being evaluated for their cancer-inhibiting activity is a phytoalexin, resveratrol (RSV), found in grapes and wines and known for its diverse biological activities, including its antioxidant property.[21–23] In this regard, we have previously shown that the chemopreventive activity of RSV could be due to its ability to induce apoptotic death in tumor cells.[24] However, depending upon the cell type and the concentration used, RSV has been shown to promote or inhibit cellular proliferation and death signaling.[21,25] Our present study was stimulated by a recent report that H_2O_2-induced apoptotic signaling was inhibited in the presence of RSV and the implication that this could be a function of its antioxidant activity.[26,27] Using human leukemia cell line HL60 that is highly sensitive to H_2O_2-induced apoptosis, we show here that preincubation of cells with low doses of RSV (4–8 μM) significantly inhibits death signaling upon exposure to 100 μM H_2O_2. Furthermore, we provide evidence that this inhibitory activity is not due to an antioxidant effect of RSV, but contrarily mediated by a significant increase in intracellular O_2^- production, thereby creating a pro-oxidant intracellular milieu nonpermissive for death execution.

MATERIALS AND METHODS

Determination of Cell Viability

The human promyelocytic leukemia (HL60) cell line is purchased from ATCC (Rockville, MD) and maintained in RPMI 1640 supplemented with 10% FBS, 1% L-glutamine, and 1% *S*-penicillin. In a typical survival assay, HL60 cells (1×10^5 cells/well) plated in 96-well plates were preincubated with RSV (4–8 µM) for 2 h and then treated with 100 µM H_2O_2 or 25 µg/mL of merocil (C2) for 18 h. Cytotoxicity was determined by MTT assay as described previously.[28] After drug exposure, 50 µL of MTT from the stock solution was added to each well and incubated for 4 h at 37°C in the dark. After 4 h of incubation with MTT, the plate was spun at 3000 rpm for 5 min. Crystals were dissolved in 200 µL DMSO + 10 µL Sorenson's glycine buffer after removing the supernatant. Viability was determined by absorbance at 570-nm wavelength using an automated ELISA reader.

Analysis of DNA Fragmentation by PI

Following treatment with RSV, H_2O_2, and/or merocil, HL60 cells were centrifuged and pellets were resuspended in 500 µL of 1× PBS + 1% FBS and immediately fixed with 70% ethanol:water. The 70% ethanol:water was added while vortexing the cells to avoid clumping. Next, the cells were incubated for 15 min on ice and spun at 1200 rpm for 5 min. Cells were washed with 1× PBS + 1% FBS once and then resuspended in 500 µL of PI:RNase A solution for 30 min at 37°C in the dark. Stained cells were analyzed by flow cytometry (Coulter EPICS Elite ESP) with excitation wavelength at 488 nm and emission wavelength at 610 nm. At least 10,000 events were analyzed by the WINMDI software.

Determination of Caspase 3 and 9 Activities

Caspase 3 and 9 activities were determined by using the Biorad Fluorescent Assay kit for the two caspases. Tumor cells (1×10^6) were preincubated for 2 h with 4 or 8 µM RSV and then incubated with 100 µM H_2O_2 or 25 µg/mL of C2 for 18 h. Cells were washed once with 1× PBS, centrifuged and resuspended in 50 µL of chilled cell lysis buffer (provided by the supplier), and incubated at 4°C. Fifty µL of 2× reaction buffer with 10 µM DTT and 5 µL of the conjugate substrate (DEVD-AFC for caspase-3 and LEHD-AFC for caspase-9) were added to each sample. Cells were incubated for 60 min at 37°C within the spectrofluorometer. Caspase activity was determined by the relative fluorescence intensity at 505 nm following excitation at 400 nm with a spectrofluorometer (Tecan Spectrofluoroplus, Austria). Five cycles of reading were taken every 15 min. Results are shown as fold increase (X-increase) in activity relative to untreated control cells.

Measurements of Intracellular O_2^- and pH

Intracellular O_2^- was assayed by a lucigenin-based chemiluminescence assay as described elsewhere.[11] Briefly, 2×10^6 cells were plated in 6-well plates, centrifuged following gentle washing with 1× PBS, and permeabilized in 400 µL of 1× releasing agent (Sigma, St. Louis, MO). Next, 100 µL of 850 µM freshly prepared lucigenin was autoinjected and chemiluminescence was detected by a luminometer (Turner designs, TD-20/20) and monitored for 60 s. Data are shown as relative light units per 60 s (RLU/60 s). The mean ± SD of three independent measurements is shown. For measurement of cytosolic pH, cells were loaded with BCECF-AM (Sigma) and the fluorescence ratio of 525 nm/610 nm was used to derive cytosolic pH using a standard pH calibration curve as described previously.[11,19]

RESULTS AND DISCUSSION

H_2O_2-Induced Apoptosis Is Blocked by Low Doses of RSV

We have previously demonstrated that H_2O_2 can trigger apoptosis or necrosis depending upon the concentration used in the assay.[9] Concentrations below 500 μM elicit a classical apoptotic response, whereas higher concentrations result in necrotic death. Here, we show that exposure of HL60 cells to H_2O_2 (50–150 μM) resulted in a significant decrease in cell survival as measured by the MTT assay and appearance of sub-G1 fraction on cell cycle analysis (FIG. 1A). That this cytotoxic activity of H_2O_2 is indeed a function of activation of the apoptotic death pathway is evidenced by the significant increases in caspase 3 and 9 activities (FIG. 1B). These data corroborate earlier findings that exposure of human tumor cells to H_2O_2 triggers activation of the apoptotic pathway mediated by caspase activation.[29–31] The observed

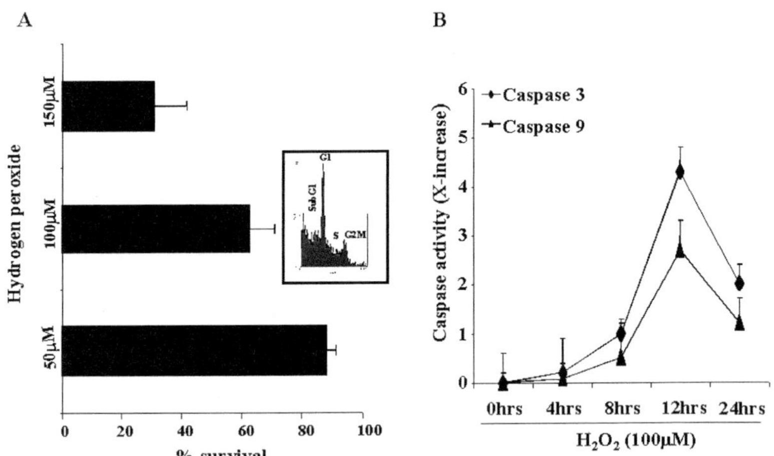

FIGURE 1. Exposure of HL60 cells to H_2O_2 results in apoptosis. **(A)** HL60 cells (1×10^6/mL) were incubated with H_2O_2 (50–100 μM) for 18 h and cell death was determined by either MTT assay or PI staining for DNA fragmentation (sub-G1 fraction). Survival data are the mean ± SD of three independent experiments. **(B)** Cells (2×10^6) were exposed to 100 μM H_2O_2 for 4–24 h, and caspase 3 and 9 activities were assessed by fluorogenic assays using DEVD-AFC (caspase 3) or LEHD-AFC (caspase 9) substrates. Data shown are the mean ± SD of three independent experiments and are presented as the fold increase (X-increase) in enzyme activity over untreated cells.

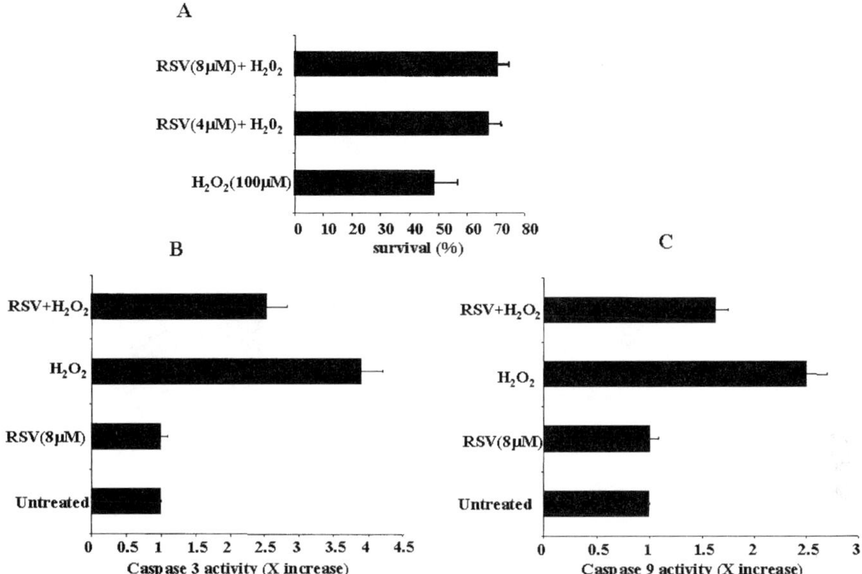

FIGURE 2. Preincubation with RSV inhibits H_2O_2-induced death signaling. HL60 cells (2×10^6) were exposed to 4 or 8 μM RSV for 2 h, followed by 18 h of incubation with 100 μM H_2O_2. **(A)** Cell survival was determined by the MTT assay. **(B)** Caspase 3 and **(C)** caspase 9 activities were assessed by fluorogenic assays using DEVD-AFC (caspase 3) or LEHD-AFC (caspase 9) substrates. Data shown are the mean ± SD of three independent experiments.

increase in caspase 9 activity indicates involvement of the mitochondrial pathway that provides amplification factors, such as cytochrome C, apoptosis-inducing factor (AIF), and Smac/Diablo.[5] Preincubation of cells with low doses of RSV (4–8 μM) prior to the addition of H_2O_2 for 18 h resulted in a significant increase in cell survival (FIG. 2A), which was a direct outcome of the inhibitory effect on caspase 3 and 9 activities (FIGS. 2B and 2C). We were intrigued by these findings, particularly in the light of our recent reports demonstrating the apoptotic activity of RSV on tumor cells. However, in those studies, the concentrations of RSV used were at least 10 fold higher than those used in this study. These results imply that, depending upon the concentration used, RSV could trigger or inhibit death signaling in tumor cells. Indeed, this divergent signaling by RSV has previously been demonstrated in other systems. What is of interest is our data showing a significant drop in H_2O_2-induced caspase 9 activity upon preincubation with RSV. These data suggest that the inhibitory activity of RSV on H_2O_2 signaling could involve a pathway (or pathways) upstream of the mitochondria or signals that engage the mitochondrial death machinery.

RSV Inhibits H_2O_2-Induced Apoptosis by Creating a Nonpermissive Milieu for Caspase Activation

We have previously shown that apoptosis induced by H_2O_2 is mediated by a decrease in intracellular concentration of O_2^- and a significant acidification of the cytosolic milieu.[9,19] Our results also provide evidence that a critical determinant of

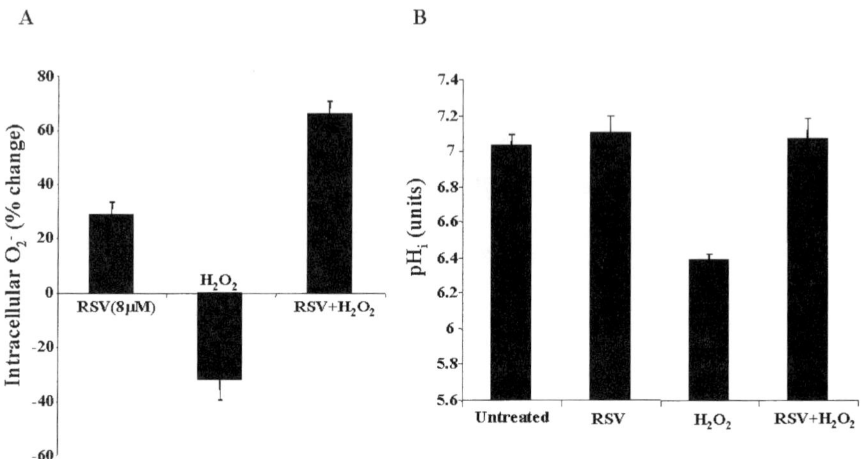

FIGURE 3. Pretreatment of cells with RSV induces a dramatic increase in intracellular O_2^- concentration and inhibits H_2O_2-induced cytosolic acidification. HL60 cells (2×10^6) were exposed to 8 μM RSV for 2 h, followed by 4 h of incubation with 100 μM H_2O_2. **(A)** Intracellular O_2^- was measured by a lucigenin-based chemiluminescence assay (shown as % change of intracellular O_2^-). **(B)** Cytosolic pH was determined by the ratio of the fluorescence (525 nm/610 nm) of BCECF-AM preloaded into the cells. Data shown are the mean ± SD of three experiments.

the efficacy of the death signal is the intracellular redox state of the cells. In that respect, maintaining a slightly elevated intracellular concentration of O_2^- promotes cellular proliferation[13] and inhibits death signaling via a direct or indirect effect on caspase activation pathways.[18] Indeed, a pro-oxidant intracellular milieu is an invariable finding in tumor cells and has been shown to provide them with a survival advantage over their normal counterparts.[16] In support of this hypothesis, we demonstrated that decreasing intracellular O_2^- via pharmacological inhibition of the NADPH oxidase complex[18] or by transfection of tumor cells with the dominant negative form of Rac (RacN17)[17] significantly increased the sensitivity of tumor cells to apoptotic stimuli. Contrarily, inhibition of superoxide dismutase (Cu/Zn SOD), a critical enzyme that dismutates O_2^- to generate H_2O_2, resulted in a significant increase in the constitutive intracellular levels of O_2^- and a reciprocal decrease in cell sensitivity to apoptosis.[18] Thus, the ability of H_2O_2 to induce a significant drop in intracellular O_2^- is a critical effector mechanism for sensitizing tumor cells to death triggers. Indeed, exposure of HL60 cells to H_2O_2 resulted in a significant drop in intracellular O_2^- and acidification of the cytosol (FIGS. 3A and 3B). Interestingly, exposure of cells to 8 μM RSV resulted in a significant increase in intracellular O_2^- (25% over the untreated cells), and addition of H_2O_2 to cells primed with RSV resulted in even higher levels of O_2^- (FIG. 3A). In addition, preincubation with RSV completely inhibited the drop in cytosolic pH triggered by H_2O_2 (FIG. 3B). These data suggest that the inhibitory activity of RSV on H_2O_2-mediated death signaling could be attributed to (a) its ability to maintain a higher intracellular O_2^- concentration and (b) blocking H_2O_2-induced cytosolic acidification, thereby creating an environment

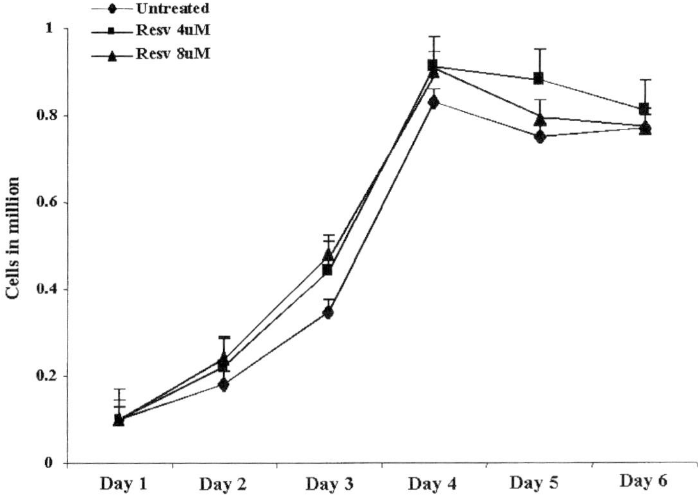

FIGURE 4. Low doses of RSV promote cell proliferation. HL60 cells (0.5×10^5) were plated in 6-well plates and cell proliferation was monitored by daily counting the number of cells using a Coulter counter. Data shown are the mean ± SD of three observations.

nonpermissive for caspase activation and efficient death execution. Furthermore, at the concentrations used in this study, RSV stimulated cellular proliferation (FIG. 4), which could in part be due to its slight pro-oxidant activity.

Low Doses of RSV Inhibit Drug-Induced Apoptosis Mediated by Intracellular H_2O_2

Compounds such as RSV are currently under clinical evaluation for potential use in combination chemotherapy regimens. Stimulated by our findings on the inhibitory effect of RSV on H_2O_2-induced death signaling in tumor cells, we next asked if this could potentially be of significance in a protocol that involves simultaneous exposure of cells to RSV and an anticancer compound. Here, we used a novel anticancer compound, the biological activity of which we have recently demonstrated,[32] and in these studies we implicated intracellular H_2O_2 as the mediator of the death signal triggered by C2.[19] Indeed, our data show that preincubation of cells with RSV resulted in a decrease in their sensitivity to apoptosis triggered by C2 (FIG. 5). Just as with H_2O_2-induced apoptosis, RSV inhibited C2-induced cell death via a pathway (or pathways) that involved activation of the caspase proteases (FIG. 5B).

Taken together, these data support a death inhibitory activity of RSV at doses that may be physiologically relevant. The use of these compounds in combination with anticancer drugs that signal through intracellular generation of H_2O_2 could be a dangerous mixture as the slight pro-oxidant effect elicited in tumor cells may indeed provide tumor cells with a survival advantage on the one hand and impede death signals on the other. This could then be an ideal environment for the propagation and proliferation of tumor cells. The precise mechanisms underlying the pro-oxidant and

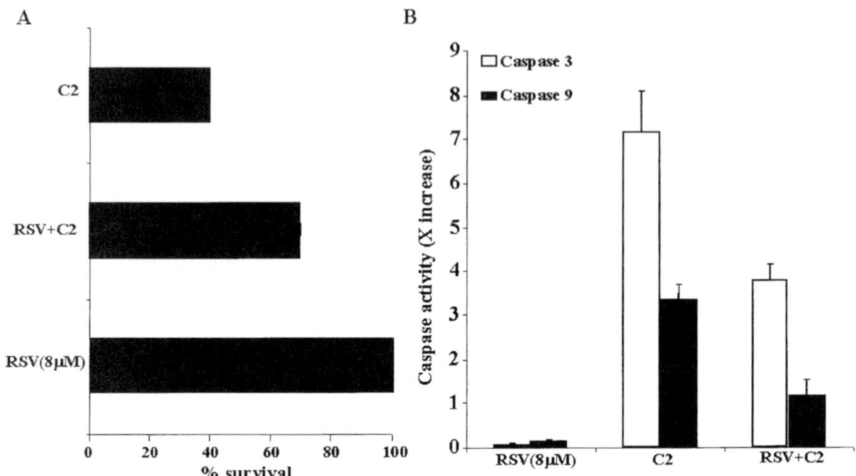

FIGURE 5. RSV pretreatment inhibits C2-induced apoptosis. HL60 cells (1×10^6/mL) were exposed to 8 μM RSV for 2 h, followed by incubation with 25 μg/mL of C2 for 18 h. **(A)** Cell survival was assessed by the MTT assay (shown as % survival). **(B)** Caspase 3 and 9 activities were determined by fluorogenic assays using specific substrates. Fold increase (X-increase) in enzyme activity over untreated control cells is shown from three independent experiments.

the death inhibitory activities of RSV are the focus of our current investigations. These studies could have tremendous implications for the use of compounds such as RSV in combination chemotherapy of leukemia cells.

ACKNOWLEDGMENTS

This work was supported by research grants from the National Medical Research Council and the Biomedical Research Council of Singapore.

REFERENCES

1. Green, D. & G. Kroemer. 1998. The central executioners of apoptosis: caspases or mitochondria? Trends Cell Biol. **8:** 267–271.
2. Strasser, A., L. O'Connor & V.M. Dixit. 2000. Apoptosis signaling. Annu. Rev. Biochem. **69:** 217–245.
3. Scaffidi, C. *et al.* 1998. Two CD95 (APO-1/Fas) signaling pathways. EMBO J. **17:** 1675–1687.
4. Reed, J.C. & D.R. Green. 2002. Remodeling for demolition: changes in mitochondrial ultrastructure during apoptosis. Mol. Cell. **9:** 1–3.
5. Kroemer, G. & J.C. Reed. 2000. Mitochondrial control of cell death. Nat. Med. **6:** 513–519.
6. Kroemer, G. 1997. The proto-oncogene Bcl-2 and its role in regulating apoptosis. Nat. Med. **3:** 614–620.
7. Nicotra, A. & S. Parvez. 2002. Apoptotic molecules and MPTP-induced cell death. Neurotoxicol. Teratol. **24:** 599–605.

8. CLEMENT, M.V. & S. PERVAIZ. 1999. Reactive oxygen intermediates regulate cellular response to apoptotic stimuli: an hypothesis. Free Radical Res. **30:** 247–252.
9. CLEMENT, M.V., A. PONTON & S. PERVAIZ. 1998. Apoptosis induced by hydrogen peroxide is mediated by decreased superoxide anion concentration and reduction of intracellular milieu. FEBS Lett. **440:** 13–18.
10. PERVAIZ, S. & M.V. CLEMENT. 2002. A permissive apoptotic environment: function of a decrease in intracellular superoxide anion and cytosolic acidification. Biochem. Biophys. Res. Commun. **290:** 1145–1150.
11. PERVAIZ, S. & M.V. CLEMENT. 2002. Hydrogen peroxide–induced apoptosis: oxidative or reductive stress? Methods Enzymol. **352:** 150–159.
12. BURDON, R.H. 1995. Superoxide and hydrogen peroxide in relation to mammalian cell proliferation. Free Radical Biol. Med. **18:** 775–794.
13. BURDON, R.H. 1996. Control of cell proliferation by reactive oxygen species. Biochem Soc. Trans. **24:** 1028–1032.
14. BURDON, R.H., V. GILL & C. RICE-EVANS. 1990. Oxidative stress and tumour cell proliferation. Free Radical Res. Commun. **11:** 65–76.
15. BURDON, R.H. & V. GILL. 1993. Cellularly generated active oxygen species and HeLa cell proliferation. Free Radical Res. Commun. **19:** 203–213.
16. CERUTTI, P.A. 1985. Prooxidant states and tumor promotion. Science **227:** 375–381.
17. PERVAIZ, S. *et al.* 2001. Activation of the RacGTPase inhibits apoptosis in human tumor cells. Oncogene **20:** 6263–6268.
18. PERVAIZ, S. *et al.* 1999. Superoxide anion inhibits drug-induced tumor cell death. FEBS Lett. **459:** 343–348.
19. HIRPARA, J.L., M.V. CLEMENT & S. PERVAIZ. 2001. Intracellular acidification triggered by mitochondrial-derived hydrogen peroxide is an effector mechanism for drug-induced apoptosis in tumor cells. J. Biol. Chem. **276:** 514–521.
20. CLEMENT, M.V. & S. PERVAIZ. 2001. Intracellular superoxide and hydrogen peroxide concentrations: a critical balance that determines survival or death. Redox Rep. **6:** 211–214.
21. BHAT, K.P.L., J.W. KOSMEDER & J.M. PEZZUTO. 2001. Biological effects of resveratrol. Antioxid. Redox Signal. **3:** 1041–1064.
22. PERVAIZ, S. 2001. Resveratrol—from the bottle to the bedside? Leuk. Lymphoma **40:** 491–498.
23. SOLEAS, G.J., E.P. DIAMANDIS & D.M. GOLDBERG. 2001. The world of resveratrol. Adv. Exp. Med. Biol. **492:** 159–182.
24. CLEMENT, M.V. *et al.* 1998. Chemopreventive agent resveratrol, a natural product derived from grapes, triggers CD95 signaling-dependent apoptosis in human tumor cells. Blood **92:** 996–1002.
25. BHAT, K.P. *et al.* 2001. Estrogenic and antiestrogenic properties of resveratrol in mammary tumor models. Cancer Res. **61:** 7456–7463.
26. JANG, J.H. & Y.J. SURH. 2001. Protective effects of resveratrol on hydrogen peroxide–induced apoptosis in rat pheochromocytoma (PC12) cells. Mutat. Res. **496:** 181–190.
27. KAMPA, M. *et al.* 2000. Wine antioxidant polyphenols inhibit the proliferation of human prostate cancer cell lines. Nutr. Cancer **37:** 223–233.
28. HIRPARA, J.L. *et al.* 2000. Induction of mitochondrial permeability transition and cytochrome C release in the absence of caspase activation is insufficient for effective apoptosis in human leukemia cells. Blood **95:** 1773–1780.
29. YAMAKAWA, H. *et al.* 2000. Activation of caspase-9 and -3 during H_2O_2-induced apoptosis of PC12 cells independent of ceramide formation. Neurol. Res. **22:** 556–564.
30. NAKAMURA, T. & K. SAKAMOTO. 2001. Reactive oxygen species up-regulates cyclooxygenase-2, p53, and Bax mRNA expression in bovine luteal cells. Biochem. Biophys. Res. Commun. **284:** 203–210.
31. SIMIZU, S. *et al.* 1998. Induction of hydrogen peroxide production and Bax expression by caspase-3(-like) proteases in tyrosine kinase inhibitor–induced apoptosis in human small cell lung carcinoma cells. Exp. Cell Res. **238:** 197–203.
32. PERVAIZ, S. *et al.* 1999. Purified photoproducts of merocyanine 540 trigger cytochrome C release and caspase 8–dependent apoptosis in human leukemia and melanoma cells. Blood **93:** 4096–4108.

The Sesame Seed Oil Constituent, Sesamol, Induces Growth Arrest and Apoptosis of Cancer and Cardiovascular Cells

ALIX JACKLIN,[a] COLIN RATLEDGE,[a] KEVIN WELHAM,[b] DENNIS BILKO,[c] AND CHRISTOPHER J. NEWTON[c,d]

[a]*Biological Sciences and* [b]*Department of Chemistry, University of Hull, Hull, United Kingdom*

[c]*Jacob's Well Medical Research, Beverley, East Yorkshire, United Kingdom*

[d]*Cytogenex Ltd. (Techfill), Road 6, South Humberside Industrial Estate, Grimsby, Lincolnshire, United Kingdom*

ABSTRACT: Sesamol is a potent inhibitor of fungal fatty acid biosynthesis. This effect is apparently due to inhibition of malic enzyme and the supply of NADPH that is required for this biosynthetic pathway. It is the ability of sesamol to reduce the synthesis of the coenzyme, NADPH, that makes it attractive for use in studying the effect of oxidants on tumor and vascular endothelial cells. By conducting preliminary studies on the effect of sesamol alone, it was clear that the compound demonstrated marked cytotoxicity. This paper describes the experiments performed.

KEYWORDS: sesamol; FACS; endothelial cells; growth

INTRODUCTION

The health-promoting properties of sesame oil have been observed for millennia. Amongst these in particular, sesame oil is attributed to having an antiaging effect.[1] More recently, attention has focused on the compounds in sesame seed oil that may be responsible for these effects, and sesamol has emerged as a strong candidate due to its antioxidant activity. Sesamol is not present in virgin oil, but it is a derivative of sesaminol and it is released during the refining of unroasted sesame oil or when the oil itself is fried as in culinary use.[2]

More recent work conducted by members of our group working at the University of Hull has demonstrated that sesamol is a potent inhibitor of fungal fatty acid biosynthesis.[3] This effect is apparently due to inhibition of malic enzyme and the supply of NADPH that is required for this biosynthetic pathway. It is the ability of sesamol to reduce the synthesis of the coenzyme, NADPH, that attracted us to its use for work on the effect of oxidants on tumor and vascular endothelial cells. By conducting preliminary studies on the effect of sesamol alone, it was clear that the compound demonstrated marked cytotoxicity. This paper describes the experiments performed.

Address for correspondence: Dr. Chris Newton, Cytogenex Ltd. (Techfill), Road 6, South Humberside Industrial Estate, Grimsby, Lincolnshire, United Kingdom.
cjnewton@compuserve.com

Ann. N.Y. Acad. Sci. 1010: 374–380 (2003).
doi: 10.1196/annals.1299.068

METHODOLOGY

Cell Culture

Breast cancer cell lines, MCF-7 and T47D, were maintained at 37°C and with 5% CO_2 in RPMI 1640 medium containing antibiotics, glutamine, and 10% FCS. The transformed human umbilical vein endothelial cell line, EA.hy 926, was maintained in DMEM under otherwise similar culture conditions. For experiments, cells were seeded into 48-well plates at densities over the range of 2000 to 10,000 per well. For fluorescent activated cell sorting (FACS) analysis, cells were seeded into 25-cm^2 or 75-cm^2 culture flasks.

Light and Fluorescence Microscopy

Prior to microscopic observations using a Leica DMIL fitted with IMC optics, cells were stained with propidium iodide (PI) and Hoechst 33342 as previously described.[4] Images under normal and UV light were obtained using a SPOT^{JUNIOR} cooled CCD camera.

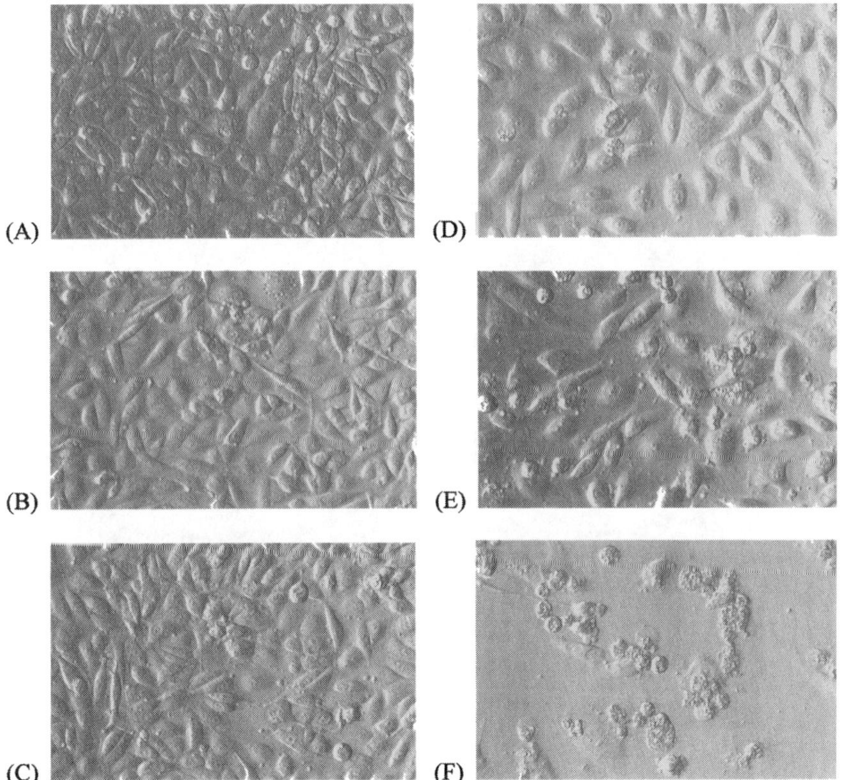

FIGURE 1. Morphological appearance of endothelial cells exposed to a dose range of sesamol: **(A)** control, **(B)** 50 μM, **(C)** 100 μM, **(D)** 250 μM, **(E)** 500 μM, and **(F)** 1 mM.

Cell Proliferation Assays

Cell numbers after treatment with sesamol were determined using a size-calibrated Coulter counter as described previously.[4]

As a sensitive method to determine the effect of sesamol on S-phase activity, tritiated thymidine (^{3}H-T) was added to cells (0.5 μCi/well) for the last 2 h of the treatment period. Thymidine uptake into DNA was determined according to a previously published method.[5]

FACS Analysis

FACS analysis was performed on cells treated for various time periods and at several concentrations of sesamol. The methodology employed has been described previously.[4]

FIGURE 2. Combined normal light and fluorescence (Hoechst 33342) microscopic appearance of endothelial cells exposed to 1 mM sesamol for 72 h.

RESULTS

Morphological Appearance of Cells Exposed to Sesamol

For a 72-h period, endothelial cells were exposed to sesamol over the range of (A) control, (B) 50 μM, (C) 100 μM, (D) 250 μM, (E) 500 μM, and (F) 1 mM. FIGURE 1 shows the appearance of these cells under normal light at a magnification of 200× [note that the figure has been reduced to 75%]. At the lower doses, there is a clear increase in cell size; in contrast, at 1 mM, cell death is apparent. To obtain more information as to the mode of death, similarly treated cells were exposed to PI and Hoechst 33342. FIGURE 2 shows that, in comparison to control cells (panel 1), those exposed to 1 mM sesamol (panel 2) for 72 h have fragmented nuclei, a hallmark characteristic of apoptotic cell death.

Growth Response to Sesamol

Although high doses of sesamol were observed to kill cells, morphological observations suggested that lower doses slowed growth. This was formally checked by observing the effect of 100 and 250 μM sesamol on the growth of the MCF-7 breast cancer cells. FIGURE 3a shows that, even after 3 days of treatment, cell numbers with 100 μM sesamol were lower than controls and this effect was maintained for the duration of the study.

Lower doses of sesamol were studied with the thymidine uptake assay. This assay is extremely sensitive to small changes in growth parameters. FIGURE 3b shows that even doses as low as 10 μM inhibited thymidine uptake (control ^{3}H-T uptake: 124,400 dpm).

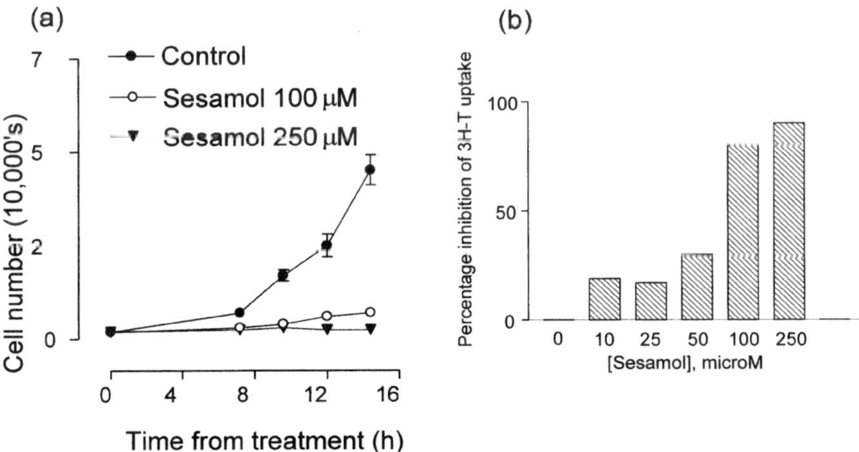

FIGURE 3. Growth response of breast cancer cells to sesamol measured **(a)** as cell numbers and **(b)** as tritiated thymidine uptake.

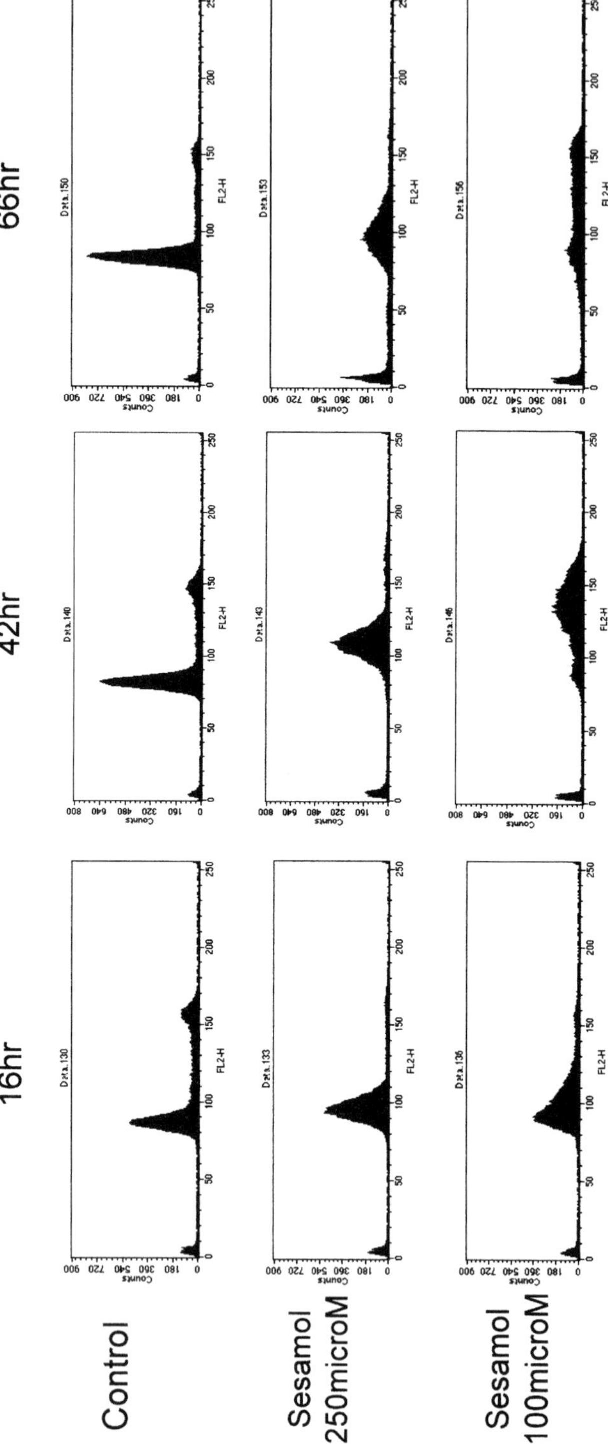

FIGURE 4. The effect of sesamol on cell cycle parameters measured by FACS analysis.

Effect of Sesamol on the Cell Cycle

As MCF-7 cells had a tendency to clump together, endothelial cells were used for cell cycle analysis by FACS. FIGURE 4 shows the analysis of cells stained with PI after 16, 42, and 66 h of exposure to sesamol at 100 and 250 μM. In comparison to control cells, where the usual G_1 (left peak), S (middle area), and G_2/M (right, smaller peak) phases are present, cells with sesamol at 250 μM are observed to accumulate in S phase after 42 h. Transit through G_2/M is completely blocked. At the lower dose of 100 μM, cells appear to transit S phase into G_2/M and there is some evidence of cells returning to G_1 (peak higher at 66 h than at 42 h).

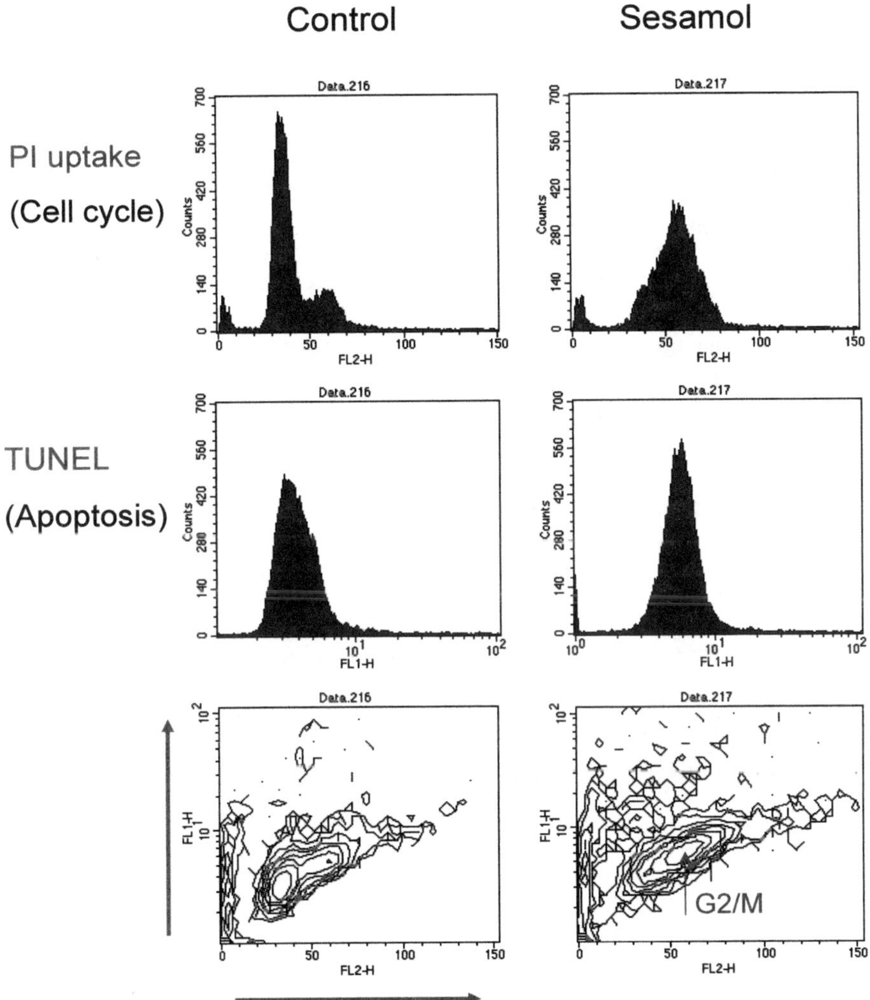

FIGURE 5. Combined cell cycle and TUNEL end-labeling analysis of fragmented DNA as determined by two-channel FACS.

Cell Death and Phase of the Cell Cycle

Although the FACS analysis showed that most cells are in S or G_2/M after 72 h with sesamol, PI labeling was combined with the TUNEL assay to obtain information as to the position in the cell cycle where cells are dying (S or G_2/M). The latter measures DNA fragmentation that occurs in apoptosis, and cells fluoresce green in addition to the red of PI. FIGURE 5 shows the two-channel FACS analysis of T47D breast cancer cells treated with 1 mM sesamol for 72 h. Sesamol reduces the G_1 peak and cells accumulate in S, G_2, and M. Green fluorescence increases with sesamol and the contour plot shows that, in parallel with the cell cycle effect, maximum fluorescence matches the distribution of cells in the cell cycle. It would therefore seem that cell death is occurring from both S and G_2/M.

DISCUSSION

Sesamol can now be added to a growing number of phenolic and polyphenolic antioxidant compounds that interfere with the cell cycle.[6] The question now arises as to why this should be the case. There are phenolics like genistein that have been reported to inhibit tyrosine kinase enzymes,[7] but equally there are reports that polyphenolics interfere with the so-called antioxidant response pathways and redox-sensitive protein regions.[8] With studies to determine whether sesamol interferes with signal transduction pathways, our work is also under way to look at the protein expression signature following sesamol treatment. For this, we are using 2D gel electrophoresis, Phoretix software, and MALDI-TOF protein analysis.

ACKNOWLEDGMENTS

This work was supported by CRODA International Limited and CYTOGENEX Limited.

REFERENCES

1. NAMIKI, M. 1995. The chemistry and physiological functions of sesame. Food Rev. Int. **11:** 281–329.
2. KAMEL-ELDIN, A. & L.A. APPELQVIST. 1994. Variations in the composition of sterols, tocopherols and lignans in seed oils from four sesamum species. J. Am. Oil Chem. Soc. **71:** 149–156.
3. WYNN, J.P., A. KENDRICK & C. RATLEDGE. 1997. Sesamol as an inhibitor of growth and lipid metabolism in *Mucor circenelloides* via its action on malic enzyme. Lipids **32:** 605–610.
4. NEWTON, C.J., G. RAN, Y-X. XIE *et al.* 2002. Statin-induced apoptosis of vascular endothelial cells is blocked by dexamethasone. J. Endocrinol. **174:** 7–16.
5. PAGOTTO, U., T. ARZBERGER, U. HOPFNER *et al.* 1995. Expression and localization of endothelin-1 and endothelin receptors in human meningiomas: evidence for a role in tumour growth. J. Clin. Invest. **96:** 2017–2025.
6. ZHANG, G., Y. MIURA & K. YAGASAKI. 2000. Induction of apoptosis and cell cycle arrest in cancer cells by *in vivo* metabolites of teas. Nutr. Cancer **38:** 265–273.
7. LI, Y., M. BHUIYAN & F.H. SARKAR. 1999. Induction of apoptosis and inhibition of c-erbB-2 in MDA-MB-435 cells by genistein. Int. J. Oncol. **15:** 525–533.
8. GOPALAKRISHNA, R. & U. GUNDIMEDA. 2002. Antioxidant regulation of protein kinase C in cancer prevention. J. Nutr. **13:** 3819S–3823S.

The Inducible NO Synthase Is Downregulated during Apoptosis of Malignant Cells from B-Cell Chronic Lymphocytic Leukemia Induced by Flavopiridol and Polyphenols

CHRISTIAN BILLARD,[a] CLAIRE QUINEY,[a] RUOPING TANG,[b] CATHERINE KERN,[a] FLORENCE AJCHENBAUM-CYMBALISTA,[b] DANIEL DAUZONNE,[c] AND JEAN-PIERRE KOLB[a]

[a]*INSERM E355, Centre Biomédicale des Cordeliers, Paris, France*

[b]*INSERM EPI 9912, Hôpital Hôtel-Dieu, Paris, France*

[c]*CNRS UMR 176, Institut Curie, Paris, France*

ABSTRACT: Downregulation of iNOS and NO is a pathway common for flavones and polyphenols, two distinct families of phytoalexins. Our data suggest that inhibition of the NO pathway could be one of the mechanisms involved in the proapoptotic properties of these phytoalexins in leukemia B-cells.

KEYWORDS: B-cells; leukemia; apoptosis; NO; iNOS; flavopiridol; polyphenols

It is now well established that B-cell chronic lymphocytic leukemia (B-CLL) is characterized by deficiency in apoptosis rather than accelerated cell proliferation.[1] We previously reported that leukemic cells from patients with B-CLL and hairy cell leukemia (HCL), another chronic B-cell malignancy, but not normal B-lymphocytes, constitutively express a nitric oxide (NO) synthase of the inducible type (iNOS) and that the NO produced contributes to the defective apoptosis.[2,3] Recently, we showed that resveratrol, a polyphenolic phytoalexin, promotes *in vitro* apoptosis of B-CLL cells and B-cell lines and that this effect is associated with inhibition of the NO pathway.[4] Furthermore, previous studies reported that flavopiridol, a semisynthetic phytoalexin belonging to the flavone family, induces *in vitro* apoptosis of B-CLL cells (reviewed in ref. 1).

We show here that flavopiridol-induced apoptosis is associated with downregulation of both iNOS expression and NO production in several leukemic B-cell lines and B-CLL patient cells. Similar results were found with a synthetic flavonoid (DD1), with acetate derivatives of resveratrol and of its dimer (viniferin), and with other grape-derived polyphenols (vineatrols).

Address for correspondence: Christian Billard, INSERM E355, Centre Biomédicale des Cordeliers, Paris, France.
christian.billard@bhdc.jussieu.fr

Ann. N.Y. Acad. Sci. 1010: 381–383 (2003). © 2003 New York Academy of Sciences.
doi: 10.1196/annals.1299.069

TABLE 1. Effects of various flavones and polyphenols on the NO pathway and on apoptosis processes in B-cell lines[a] and B-CLL cells

	FL[b]	DD1[c]	R[d]	RTA[e]	VF[e]	VFPA[e]	V10[e]	V25[e]
iNOS downregulation[f]	+++	+++	++	+	+/–	+++	+++	+
Inhibition of NO release[g]	+++	++	+	+	nt	+++	nt	nt
PS externalization[h]	+++	++	++	+	+/–	+++	+++	++
DNA fragmentation[i]	+++	+++	++	+	+/–	++	+++	++
Caspase-3 activation[j]	+++	nt	nt	++	NS	+	+++	++
Bcl-2 downregulation[k]	+	+	+	+	NS	++	+++	++
$\Delta\Psi_m$ decrease[l]	++	nt	+	+	+/–	+	++	++

NOTE: nt, not tested; NS, not significant.
[a]Cell lines: EHEB, WSU-CLL, and ESKOL.
[b]Flavopiridol: a gift from Aventis Pharmaceuticals (Bridgewater, MA).
[c]A synthetic aminoflavone prepared as described previously (ref. 5).
[d]Resveratrol: purchased from Sigma (St. Louis, MO).
[e]Viniferin (VF, resveratrol dimer), resveratrol triacetate (RTA), viniferin penta-acetate (VFPA), and the two preparations of vineatrols isolated from vine-shoots and containing 10% or 25% of resveratrol (V10 and V25) were isolated and kindly provided by Actichem (Montauban, France).
[f]Determined by immunofluorescence and FACS analysis from permeabilized cells and by Western blotting (ref. 2).
[g]Endogenous NO production was measured by *in situ* fixation of the specific fluorescent probe, DAF-2 DA (ref. 4).
[h]Assayed by FITC–annexin V fixation and FACS analysis (ref. 4).
[i]Quantified by the formation of cytoplasmic histone–associated DNA fragments using an ELISA (ref. 4).
[j]Measured by the capacity of cell lysates to cleave the specific substrate, DEVD-AMC (ref. 4).
[k]Evaluated with a quantitative ELISA (ref. 4).
[l]The dissipation of the mitochondrial transmembrane potential ($\Delta\Psi_m$) was analyzed using a mitochondrial permeability detection kit (Biomol, Plymouth Meeting, PA), as already used and described (ref. 4).

Leukemic B-cells were isolated from the peripheral blood of patients with B-CLL (Binet stage A), which were selected according to the International CLL Workshop criteria, as previously described.[2] Three B-cell lines were studied: EHEB established from B-CLL; WSU-CLL originally described as a B-CLL cell line, but presenting with phenotypic and genotypic characteristics of a pre-B acute lymphoblastic leukemia; and ESKOL, an HCL cell line with a preplasma cell phenotype.

Cells were induced to apoptosis by culturing them in the presence of various concentrations of flavopiridol (ranging from 0.05 to 5 μM), and iNOS expression was analyzed at different times, as compared to untreated controls. Treatment with flavopiridol resulted in a time- and dose-dependent inhibition of the percentage of iNOS-positive cells in B-cell lines as well as B-CLL cells. This effect was detectable at 0.1 μM and was maximal at 1 μM, generally by 24 h of culture. Western blot experiments performed with cells from 5 B-CLL patients, which were treated with flavopiridol (1 μM) for 18 h, confirmed the almost complete downregulation of iNOS protein. As measured in parallel, phosphatidylserine (PS) externalization and

DNA fragmentation were stimulated, although to different extents, revealing an inverse relation between iNOS expression and apoptosis levels during flavopiridol exposure. All the other apoptotic events analyzed, that is, caspase-3 activity, Bcl-2 downregulation, and dissipation of the mitochondrial transmembrane potential, were found to be stimulated after flavopiridol treatment (TABLE 1). The decrease of iNOS expression clearly preceded PS externalization at least in WSU-CLL cells. In addition, the endogenous NO production was strongly reduced in both B-cell lines and B-CLL cells upon flavopiridol exposure. These results show that flavopiridol-induced apoptosis was associated with inhibition of the NO pathway. The synthetic flavonoid, DD1, was also found to trigger apoptosis and to downregulate iNOS expression and function in both cell lines and B-CLL cells (TABLE 1). This indicated that, in addition to flavopiridol, another molecule of the flavone family is able to act through inhibition of the NO pathway. Similar results were observed with several polyphenols: a vineatrol was more efficient than resveratrol, while viniferin exhibited only slight effects; interestingly, strong effects were exerted by an acetate derivative of viniferin, but not by resveratrol triacetate (TABLE 1).

In conclusion, downregulation of iNOS and NO is a pathway common for flavones and polyphenols, two distinct families of phytoalexins. Our data also suggest that inhibition of the NO pathway could be one of the mechanisms involved in the proapoptotic properties of these phytoalexins in leukemia B-cells. In addition, our data could be useful for designing new clinical trials in the therapy of malignancies characterized by defective apoptosis such as B-CLL, with phytoalexins capable of *in vivo* blocking of the NO pathway.

REFERENCES

1. BYRD, J., K. RAI, E. SAUSVILLE & M. GREVER. 1998. Old and new therapies in chronic lymphocytic leukemia: now is the time for a reassessment of therapeutic goals. Semin. Oncol. **25:** 65–74.
2. ZHAO, H., N. DUGAS, C. MATHIOT *et al.* 1998. B-cell chronic lymphocytic leukemia cells express a functional inducible nitric oxide synthase displaying anti-apoptotic activity. Blood **92:** 1031–1043.
3. ROMAN, V., H. ZHAO, J.M. FOURNEAU *et al.* 2000. Expression of a functional inducible nitric oxide synthase in hairy cell leukemia and ESKOL cell line. Leukemia **14:** 696–715.
4. ROMAN, V., C. BILLARD, C. KERN *et al.* 2002. Analysis of resveratrol-induced apoptosis in human B-cell chronic leukaemia. Br. J. Haematol. **117:** 842–851.
5. DAUZONNE, D., B. FOLLÉAS, L. MARTINEZ & G.G. CHABOT. 1997. Synthesis and *in vitro* cytotoxicity of a series of 3-aminoflavones. Eur. J. Med. Chem. **32:** 71–82.

Catha edulis (Khat) Induces Cell Death by Apoptosis in Leukemia Cell Lines

E. DIMBA,[a] B. T. GJERTSEN,[b] G. W. FRANCIS,[c] A. C. JOHANNESSEN,[a] AND O. K. VINTERMYR[d]

[a]*Department of Odontology–Oral Pathology and Forensic Odontology, Faculty of Dentistry, University of Bergen, Bergen, Norway*

[b]*Institute of Medicine, Hematology Section, University of Bergen, Bergen, Norway*

[c]*Department of Chemistry, University of Bergen, Bergen, Norway*

[d]*Department of Pathology, The Gade Institute, Haukeland University Hospital, Bergen, Norway*

ABSTRACT: Khat is the *Celastraceus edulis* plant, a flowering evergreen tree or large shrub, which grows in the Horn of Africa and southwestern Arabia. Khat use has been associated with development of oral cancer, but its molecular effects remain controversial. This study describes a novel cytotoxic effect of whole khat extract on three leukemia cell lines. Cells were exposed to khat extract and harvested for analysis by fluorescent and electron microscopy, trypan blue exclusion, as well as immunoblotting to characterize the mode of cell death. In a separate series, cells were pretreated with a panel of caspase inhibitors for possible inhibitory effects. Khat induced a rapid cell death effect in HL-60, Jurkat, and NB4 cells that occurred within 2 h of exposure. The treated cells retained their ability to exclude trypan blue dye, a key feature in the apoptotic process. Exposed cells consistently developed morphological features of manifest apoptosis. Z-VAD, a pan-caspase inhibitor, completely inhibited toxic activity for up to 8 h, with partial inhibition by other caspase-specific agents. Western blot analysis showed specific cleavage of caspase-3 in khat-exposed cells. This study shows that khat induces cell death by apoptosis in a process sensitive to inhibition by caspase inhibitors, suggesting that subcellular interactions could be of particular relevance for the biological effects of khat in the cell death process and possibly carcinogenesis.

KEYWORDS: toxicology; apoptosis; human; khat; cell culture; HL-60; Jurkat; NB4

Celastraceus edulis (khat) is an alkaloid-containing plant that is used by millions of people in the Horn of Africa and southwestern Arabia for its euphoric effects. The distribution of khat has become more rapid and efficient with modern transportation, facilitating an increase in the usage of this drug. Somatic, behavioral, and neural

Address for correspondence: Elizabeth Anne Okumu Dimba, Department of Odontology–Oral Pathology and Forensic Odontology, Faculty of Dentistry, University of Bergen, N-5021 Bergen, Norway. Voice: +47 55 97 46 88; fax: +47 55 97 31 58.
elizabeth.dimba@student.uib.no

**Ann. N.Y. Acad. Sci. 1010: 384–388 (2003). © 2003 New York Academy of Sciences.
doi: 10.1196/annals.1299.070**

effects have been ascribed to specific psychoactive components, but there is limited biochemical information about the adverse effects of khat.[1] Khat chewing has been associated with the etiology of oral cancer, but its interactions with oral mucosa at the cellular level are controversial. This paucity of information may be attributed to ethical and logistic limitations of *in vivo* approaches.[2] We describe a novel cytotoxic effect of *Catha edulis* extract on a selected panel of leukemia cell lines.

MATERIALS AND METHODS

Khat

Plant samples were processed using an adaptation of the methanolic extraction protocol described by M. M. Lee, excluding acid-base purification.[3] The plant material was vacuum-dried and resuspended in dimethyl sulfoxide (Sigma, St. Louis, MO) for storage at –80°C. Caspase inhibitors were from Molecular Biology Labs (Tokyo, Japan).

Cell Culture and Handling of Cells

HL-60, Jurkat, and NB4 cells were cultured in RPMI 1640 medium (Sigma) supplemented with 10% heat-inactivated fetal bovine serum, 2 mM L-glutamine, 100 IU/mL penicillin, and 100 μM streptomycin (Gibco, Grand Island, NY) and maintained in a humidified atmosphere at 37°C and 5% CO_2. Cells were seeded at a density of 3×10^5 cells/mL, exposed to test dilutions of khat extract, and harvested at designated time points for analysis. Solvent controls were included in the experiments and dimethyl sulfoxide concentrations kept below 0.1%.

Dye Exclusion Test

The number of cells that excluded trypan blue were counted in a hemocytometer and expressed as arithmetic means ± SD of three or more separate experiments.

Nuclear Chromatin Condensation Assay

Samples were fixed in 4% formaldehyde containing 10 μg/mL Hoechst 33342 (Sigma), and nuclear morphology examined at 400-fold magnification with incident light fluorescent microscopy for features of manifest apoptosis. The percentage of apoptotic cells was determined as the fraction of intensely staining, condensed, and fragmented nuclei against the total number of cells in each field, in three or more separate experiments.

Electron Microscopy

Cells were fixed in 0.1 M Na-cacodylate buffer, pH 7.4, containing 2% glutaraldehyde. Samples were then rinsed and postfixed in 1% osmium tetroxide. The specimens were then dehydrated using graded ethanols and embedded in epoxy resin, and ultrathin sections were double-stained with uranyl acetate and lead citrate before examination with a Jeol 100CX electron microscope.

Caspase-3 Cleavage

In this study, 10×10^6 cells were washed in phosphate-buffered saline and lysed. Samples were homogenized on ice, centrifuged, and stored at –80°C. The protein extracts were resolved using a 12.5% SDS-polyacrylamide gel and electroblotted onto polyvinylfluoride membranes (Amersham Diagnostics, United Kingdom). Caspase-3 cleavage was detected using enhanced chemiluminescence (Pierce Biotechnology, Rockford, IL).

Statistical Analysis

One-way repeated-measure ANOVA with SPSS version 11.0 was used to analyze data on cell viability and apoptotic morphology.

RESULTS AND DISCUSSION

We report a novel cell death effect by khat, inducing apoptosis in leukemia cell lines, in a process sensitive to modulation by caspase inhibitors. Cytotoxic effects of the extract were rapid, occurring within 2 h. After 8 h, 99.5% of the exposed cells

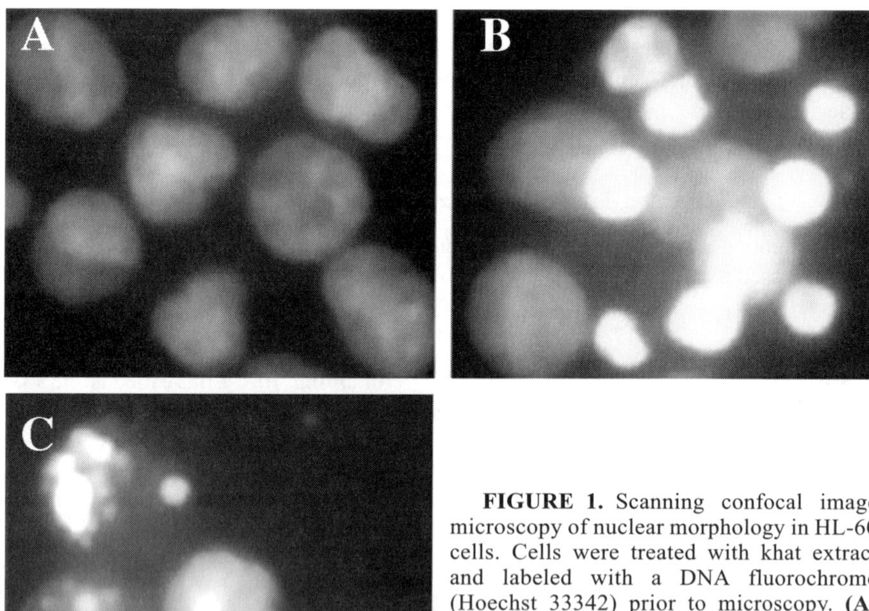

FIGURE 1. Scanning confocal image microscopy of nuclear morphology in HL-60 cells. Cells were treated with khat extract and labeled with a DNA fluorochrome (Hoechst 33342) prior to microscopy. **(A)** Fluorescent nuclear morphology of normal HL-60 cells. **(B)** Early apoptosis induced in HL-60 cells by exposure to khat extract for 2 h; chromatin condensation. **(C)** Late apoptosis induced in HL-60 cells by exposure to khat extract for 8 h; apoptotic bodies.

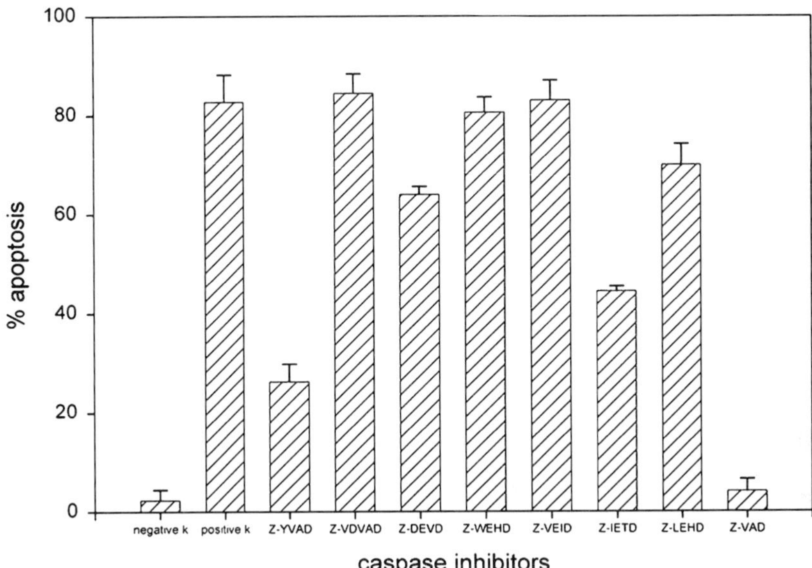

FIGURE 2. Inhibition of khat toxicity by pretreatment of leukemia cell lines by caspase inhibitors.

became apoptotic as evaluated by nuclear morphology compared to 39.4% of cells as evaluated by dye (trypan blue) exclusion test. The morphological and ultrastructural changes observed were also consistent with apoptosis and included loss of microvilli, blebbing of the cell membrane, condensation of nuclear chromatin, vacuolization of the cytosolic compartment, segregation of subcellular organelles, and formation of apoptotic bodies (FIG. 1). Pretreatment with 0.1 μM pan-caspase inhibitor Z-VAD-fmk blocked toxic activity for 8 h, while other caspase inhibitors resulted in partial inhibition (FIG. 2). Western blot analysis of cell proteins showed a specific cleavage of α-caspase-3 after khat exposure (data not shown).

Apoptosis is a major mechanism of cancer suppression and is characterized by morphological and ultrastructural changes related to a cascade of caspase-regulated biochemical events.[4,5] *Catha edulis* has a complex phytochemistry, with fractions like flavonoids that could induce apoptosis and act against cancer cells, as well as mutagenic phenylalkylamines like cathinone.[6] The overall *in vivo* effect of the khat habit may depend on the mode and duration of use, as well as synergistic reactions with other substances. Oral mucosa may be exposed to varying concentrations of khat, concomitant with other substances of abuse such as tobacco and alcohol. This study suggests that the subcellular interactions of khat are relevant in understanding the activity of antagonistic fractions in the processes of cellular cytotoxicity and possibly carcinogenesis.

ACKNOWLEDGMENTS

We thank Therese Bredholt for technical support and Daniela-Elena Costea for helpful suggestions on cell culture work. This study was supported by the Gade Legat Fund and the Norwegian Government Quota Program.

REFERENCES

1. HILL, C.M. & A. GIBSON. 1987. The oral and dental effects of khat chewing. Oral Surg. Oral Med. Oral Pathol. **63:** 433–436.
2. KASSIE, F., F. DARROUDI, M. KUNDI *et al.* 2001. Khat (*Catha edulis*) consumption causes genotoxic effects in humans. Int. J. Cancer **92:** 329–332.
3. LEE, M.M. 1995. The identification of cathinone in khat (*Catha edulis*): a time study. J. Forensic Sci. **40**(1)**:** 116–121.
4. EVAN, G.I. & K.H. VOUSDEN. 2001. Proliferation, cell cycle, and apoptosis in cancer. Nature **411:** 342–348.
5. HENGARTNER, M.O. 2000. The biochemistry of apoptosis. Nature **407:** 770–776.
6. KALIX, P. 1990. Pharmacological properties of the stimulant khat. Pharmacol. Ther. **48:** 397–416.

Curcumin-Induced Cell Death in Two Leukemia Cell Lines: K562 and Jurkat

A. DUVOIX,[a] F. MORCEAU,[a] M. SCHNEKENBURGER,[a] S. DELHALLE,[a] M.M. GALTEAU,[b] M. DICATO,[a] AND M. DIEDERICH[a]

[a]*Laboratoire RCMS, Centre Universitaire du Luxembourg, 162A Avenue de la Faïencerie, L-1511 Luxembourg, Luxembourg*

[b]*Laboratoire Thiols et Fonctions Cellulaires, Faculté de Pharmacie, Université Henri Poincaré-Nancy 1, 30 rue Lionnois, 54000 Nancy, France*

ABSTRACT: Curcumin presents strong antioxidant and anticancer properties. However, molecular mechanisms leading to curcumin-induced cell death are poorly understood. The effect of curcumin was compared in two different leukemia cell lines: K562 and Jurkat. Cell death was induced in both cell lines, and apoptosis pathways were investigated by Western blot analysis. Decreases in pro-caspase 8 and 9 levels were observed. BH_3 interacting domain death agonist (Bid) was also cleaved. Jurkat cells appeared to be more sensitive to curcumin, and apoptosis takes place earlier.

KEYWORDS: curcumin; leukemia; apoptosis

INTRODUCTION

Curcumin, the yellow pigment extracted from *Curcuma longa,* has shown strong anticancer and antioxidant properties. Its chemopreventive effect and ability to induce cell death in leukemia cell lines could become of therapeutic importance. We compare here its effect on two different leukemia cell lines: K562, a Philadelphia-positive CML and Jurkat, a T cell leukemia. K562 cells present the Bcr/Abl translocation that decreases apoptosis and upregulates cell proliferation. By Western blot analysis we show that curcumin induces apoptosis in both cell lines and that Jurkat T cells are more sensitive to curcumin than is the K562 cell line.

MATERIALS AND METHODS

Cells were cultured for 72 h with 0.1% FCS before treatment with curcumin (Sigma, Bornem, Belgium). They were then harvested and washed with PBS 1X and the total protein content was extracted with the M-Per Mammalian Protein Extraction

Address for correspondence: M. Diederich, Centre Universitaire du Luxembourg, 162A avenue de la Faiencerie, L-1511 Luxembourg, Luxembourg. Voice: +352 46 66 44 434; fax: +352 46 66 44 438.
diederic@cu.lu

**Ann. N.Y. Acad. Sci. 1010: 389–392 (2003). © 2003 New York Academy of Sciences.
doi: 10.1196/annals.1299.071**

Reagent (Pierce, Rockford, USA). Proteins were separated by SDS-PAGE, transferred on Hybond P membrane (AP-Biotech, Little Chalfont, UK) and incubated with anti-pro-caspase 8, -pro-caspase 9, BH_3 interacting domain death agonist (Bid), and β-actin antibodies. Membranes were then incubated with secondary anti-rabbit or anti-mouse antibodies; proteins of interest were visualized with ECL-plus Western blot detection reagent (AP-Biotech). Cytotoxicity was measured with Cytotox 96® Non-Radioactive Cytotoxicity Assay (Promega) by determining the quantity of lactate dehydrogenase (LDH released in the medium.

RESULTS

Cells from the two leukemia cell lines, K562 and Jurkat, were treated for 24 h with different concentrations of curcumin, and the cytotoxicity of this chemopreventive agent was measured. FIGURE 1A shows that Jurkat T cells are more sensitive to culture in 0.1% FCS, as 50% of the cells were dead after 72 h and only 20% of the K562 cells were dead after the same treatment. Curcumin treatments, then, induce more cell death, and 100% of cell death is obtained with 10 μM and 20 μM of curcumin for Jurkat and K562 cells, respectively.

Curcumin was then used at 20 μM in both K562 and Jurkat cell lines. Proteins were extracted and analyzed by Western blotting using anti-pro-caspase 8, anti-pro-caspase 9, anti-Bid antibodies and anti-β-actin antibodies. In both cell lines, pro-caspase 8 and 9 levels were reduced by curcumin treatments (FIG. 1B). In Jurkat cells, pro-caspase 8 and 9 decrease at 6 h to disappear at 24 h. Interestingly, in K562

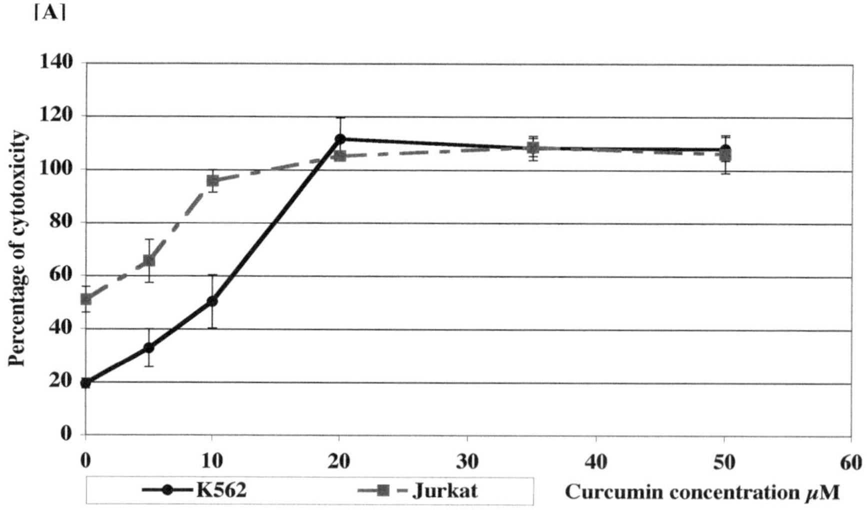

FIGURE 1A. Effect of curcumin on K562 and Jurkat cell lines: Percentage of cell death after 24-h curcumin treatment. LDH release was measured with Cytotox 96® Non-Radioactive Cytotoxicity Assay.

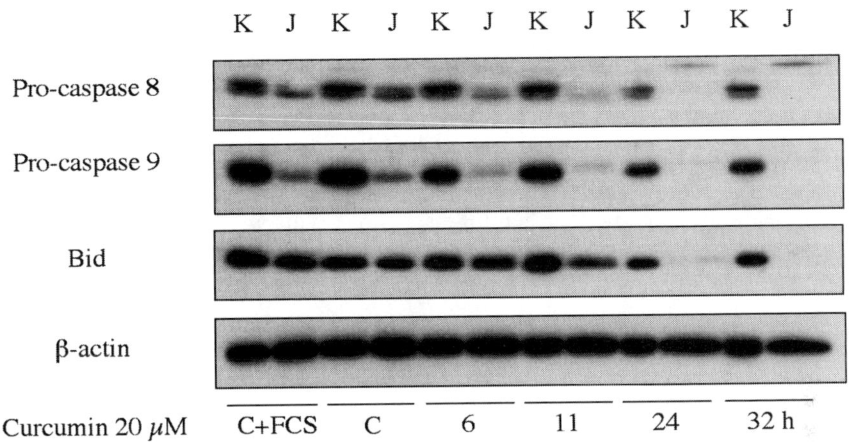

FIGURE 1B. Effect of curcumin on K562 and Jurkat cell lines: Western blotting of total protein extract after 20-μM curcumin treatments using anti-pro-caspase 8, anti-pro-caspase 9, anti-Bid and anti-β-actin antibodies. K: K562 cell line; J: Jurkat cell line; C+FCS: untreated cells cultured in 10 % FCS medium; C: untreated cells cultured in 0.1 % FCS medium.

cells, pro-caspase 8 and 9 are reduced at 24 h without disappearing, even at 32 h. These results indicate that apoptosis is induced in both leukemia cell lines by curcumin even if they react differently to the treatment. It shows that both mitochondrial and receptor apoptosis pathways are used in both cell lines. To confirm this, we used anti-Bid antibody. Indeed, mitochondrial pathway is induced through the cleavage of Bid and its insertion in the mitochondrial membrane. Western blottings reveal that Bid is cleaved in Jurkat cell line at 24 h. Levels of intact Bid are reduced in the K562 cell line at 32 h. These data confirm the fact that both apoptosis pathway are used in leukemia cell lines and that K562 cells are less sensitive to curcumin treatments than Jurkat cells. Hybridization with β actin shows a regular pattern, indicating that the same amount of protein was charged onto the gel.

DISCUSSION

Curcumin is a compound known for its anti-inflammatory[1] and anti-carcinogenic properties.[2] Indeed, it can reduce cancer incidence as well as tumor progression. One way curcumin can fight cancer is to induce apoptosis in transformed cells. It has been shown that curcumin alone induced programmed cell death in the Jurkat cell line.[3] Mukhopadhyay *et al*. showed in 2001 that curcumin also induced apoptosis in human androgen-dependent LNCaP as well as in androgen-independent DU145 human prostate cancer cell lines by downregulating the expression of Bcl-2 and Bcl-XL and the activation of the pro-caspases 3 and 8.[4] Similar results were obtained by Bush *et al*. by using a melanoma cell model.[5] Our results agree with the data showed by Bharti *et al.* demonstrating that curcumin induces caspases 7 and 9 in multiple myeloma cells.[6] We also show here that curcumin is able to induce apoptosis in the K562 cell line, which is Philadelphia-positive and thus apoptosis-resistant.

In conclusion, our work demonstrates the ability of curcumin to provoke programmed cell death in leukemia cell death, showing its anti-cancer property.

ACKNOWLEDGMENTS

We thank B. Aggarwal for helpful discussions. This work was supported by the Fondation de Recherche "Cancer et Sang." A.D. and M.S. are supported by fellowships from the Government of Luxembourg. This work was in part supported by the Télévie grant 7.4577.02.

REFERENCES

1. Gukovsky, I., C.N. Reyes, E.C. Vaquero, *et al.* 2003. Curcumin ameliorates ethanol and nonethanol experimental pancreatitis. Am. J. Physiol. Gastrointest. Liver Physiol. **284:** G85–95.
2. Ikezaki, S., A. Nishikawa, F. Furukawa, *et al.* 2001. Chemopreventive effects of curcumin on glandular stomach carcinogenesis induced by N-methyl-N′-nitro-N-nitrosoguanidine and sodium chloride in rats. Anticancer Res. **21:** 3407–3411.
3. Piwocka, K., K. Zablocki, M.R. Wieckowski, *et al.* 1999. A novel apoptosis-like pathway, independent of mitochondria and caspases, induced by curcumin in human lymphoblastoid T (Jurkat) cells. Exp. Cell Res. **249:** 299–307.
4. Mukhopadhyay, A., C. Bueso-Ramos, D. Chatterjee, *et al.* 2001. Curcumin downregulates cell survival mechanisms in human prostate cancer cell lines. Oncogene **20:** 7597–7609.
5. Bush, J.A., K.J. Cheung, Jr. & G. Li. 2001. Curcumin induces apoptosis in human melanoma cells through a Fas receptor/caspase-8 pathway independent of p53. Exp. Cell Res. **271:** 305–314.
6. Bharti, A.C., N. Donato, S. Singh & B.B. Aggarwal. 2003. Curcumin (diferuloylmethane) down-regulates the constitutive activation of nuclear factor-kappa B and Ikappa Balpha kinase in human multiple myeloma cells, leading to suppression of proliferation and induction of apoptosis. Blood **101:** 1053–1062.

Sulforaphane Modulates Cell Cycle and Apoptosis in Transformed and Non-transformed Human T Lymphocytes

CARMELA FIMOGNARI,[a] MICHAEL NÜSSE,[b] FAUSTO BERTI,[a] RENATO IORI,[c] GIORGIO CANTELLI-FORTI,[a] AND PATRIZIA HRELIA[a]

[a]*Department of Pharmacology, University of Bologna, 40126 Bologna, Italy*

[b]*GSF-Flow Cytometry Group, 85758 Neuherberg, Germany*

[c]*Research Institute for Industrial Crops, MiPA, 40129 Bologna, Italy*

ABSTRACT: Isothiocyanates exert chemopreventive effects against chemically induced tumors in animals, modulating enzymes required for carcinogens' activation/detoxification and/or the induction of cell-cycle arrest and apoptosis in tumor cell lines. To investigate the chemopreventive potential of isothiocyanates, we studied proliferation, apoptosis induction and p53, bcl-2 and bax protein expression in Jurkat T-leukemia cells by the isothiocyanate sulforaphane. Sulforaphane caused G_2/M-phase delay and increase of apoptotic cell fraction in a time- and dose-dependent manner. Necrosis was observed after prolonged exposure to elevated sulforaphane doses. Moreover, it markedly increased p53 and bax protein expression, and slightly affected bcl-2 expression. Since selective targeting and low toxicity for normal host tissues are fundamental requisites for proposed chemopreventive agents such as sulforaphane, we tested sulforaphane on non-transformed phytohemagglutinin-stimulated human T-lymphocytes. We demonstrated that sulforaphane arrested cell-cycle progression in G_1 phase by a significant down-modulation of cyclin D3. Moreover, sulforaphane induced apoptosis (and also necrosis), mediated by an increase in the expression of p53, whereas it exerted little effect on bcl-2 and bax levels. These findings indicate that sulforaphane can exert protective effects inhibiting leukemic cell growth. Moreover, sulforaphane is active not only in transformed lymphocytes but also in their normal counterpart. Although *in vitro* studies do not necessarily predict *in vivo* outcomes, our findings raise important questions regarding the suitability of sulforaphane for cancer chemoprevention.

KEYWORDS: chemoprevention; isothiocyanates; sulforaphane; apoptosis; cell cycle; cyclins; p53; bcl-2; bax; T lymphocytes; leukemia

INTRODUCTION

Glucosinolates are naturally occurring thioglucosides present in cruciferous vegetables. They are hydrolyzed by myrosinase to a number of compounds including

Address for correspondence: Carmela Fimognari, Ph.D., Department of Pharmacology, University of Bologna, Via Irnerio, 48, 40126 Bologna, Italy. Voice: +39-051-253705; fax: +39-051-253705.
fimognari@biocfarm.unibo.it

Ann. N.Y. Acad. Sci. 1010: 393–398 (2003). © 2003 New York Academy of Sciences.
doi: 10.1196/annals.1299.072

isothiocyanates (ITCs), depending on the substrate structure and the reaction conditions.[1] Isothiocyanates exert chemopreventive effects against chemically induced tumors in animals, modulating enzymes required for carcinogens' activation/detoxification and/or the induction of cell-cycle arrest and apoptosis in tumor cell lines.[1] One of the most commonly studied isothiocyanates is sulforaphane. Sulforaphane elicits high levels of mammalian phase-2 enzymes by antioxidant response element–mediated transcriptional activation.[2] Moreover, in a rat model, sulforaphane reduced the incidence of breast cancer, and also delayed the appearance and reduced the size of the tumors.[2] It exerts cytostatic and cytotoxic effects on different human cancer cell lines.[3] In order to expand our knowledge on the chemopreventive potential of ITCs, we decided to study the effects of sulforaphane on cell growth, cell-cycle progression, and apoptosis induction in Jurkat cells (i.e., human T cell leukemia). We also assessed the apoptotic response of Jurkat cells in terms of p53, bcl-2, and bax expression. Since selective targeting and low toxicity for normal host tissues are fundamental requisites for proposed chemopreventive agents such as sulforaphane, we therefore tested sulforaphane for its effects on cell-cycle progression and apoptosis on phytohemagglutinin (PHA)-stimulated, non-transformed T lymphocytes.

MATERIALS AND METHODS

Cell treatment: Exponentially growing Jurkat cells or T lymphocytes (isolated by density gradient centrifugation, depletion of adherent cells on plastic dishes, and erythrocyte rosetting) were treated with sulforaphane generated *in situ* by myrosinase-catalyzed hydrolysis of glucoraphanin.[4]

Flow cytometry: Flow cytometry was performed using a FACStar$^+$ flow cytometer (Becton Dickinson, CA) equipped with an argon laser (Innova 90; Coherent Radiation, CA) operating at 488 nm (500 mW).

Flow cytometric measurement of cell proliferation: The preparation of samples for measurements of the cell-cycle distribution of nuclei by DNA content was performed according to a two-step method reported elsewhere.[5]

Analysis of cell-cycle proteins: Cells were fixed and permeabilized by Leucoperm solution A and B (Serotec, UK). They were then incubated with 20 μL of each antibody, that is. fluorescein isothiocyanate (FITC) cyclin D2 (1 mg/mL; Serotec), FITC cyclin D3 (BD PharMingen, CA), cyclin-dependent kinase (CDK) 4 (BD PharMingen), and CDK6 (1 mg/mL; Serotec), or isotype-matched negative control (Serotec). The cells (except those stained with cyclin D2 and D3) were incubated with 20 μL of FITC-labeled secondary antibody (5 μg/mL; Serotec). The cells were then analyzed to quantitate FITC binding.

Measurement of apoptosis by flow cytometry: The cell pellet was resuspended in 100 μL of labeling solution (ANNEXIN-V-FLUOS, Boehringer Mannheim, Germany) containing 2 μL annexin V labeling reagent, and 0.1 μg propidium iodide (PI) (Sigma). Green (FITC) and red (PI uptake) fluorescence of individual cells was measured, using 488-nm excitation and a 530-nm bandpass filter for FITC detection and a filter >590 nm for PI detection.

Flow cytometric evaluation of p53, bax and bcl-2 proteins: Cells were fixed, permeabilized, and incubated with 10 μL of each antibody, that is, FITC p53 (35 μg/mL;

A

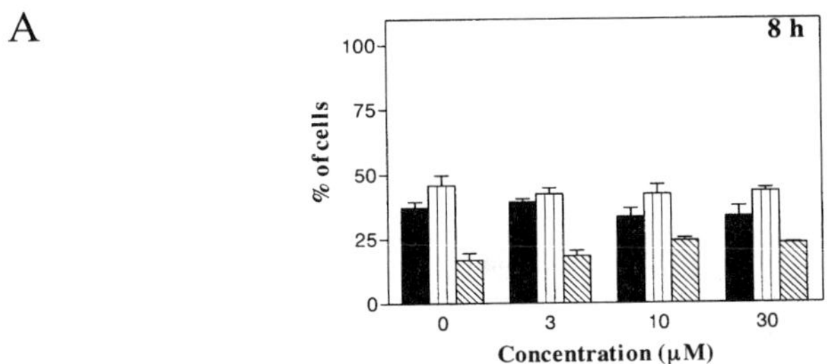

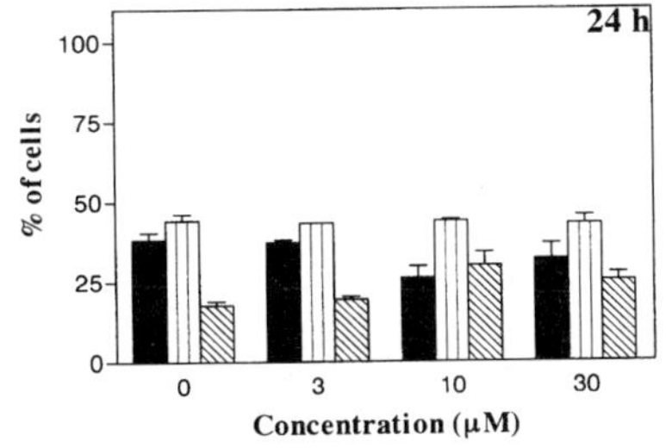

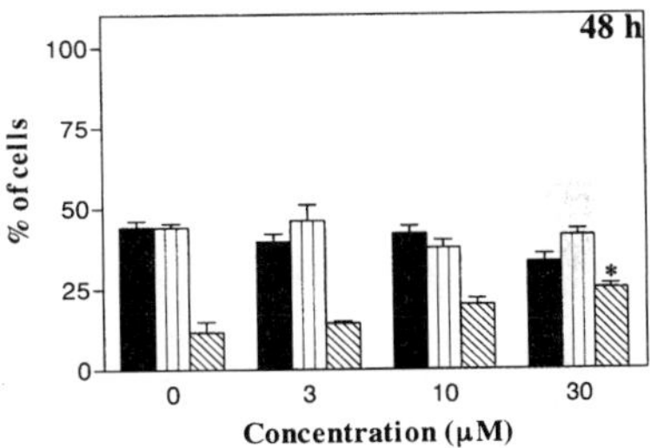

B

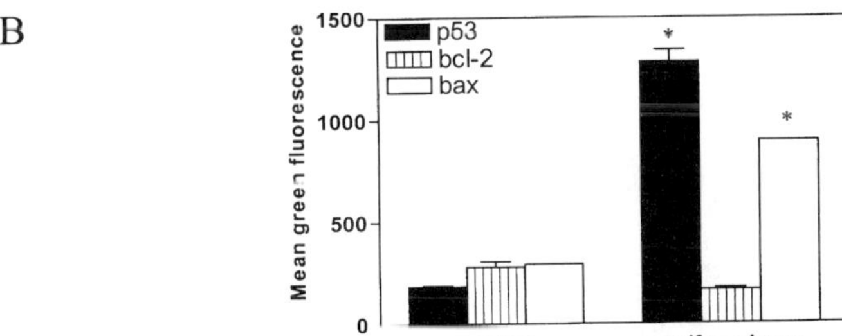

FIGURE 1. (**A**) Effects of sulforaphane on Jurkat cell cycle with respect to controls. Results are expressed as percentages of total cell counts. Data are means ± SE (deviation bars, except when smaller than the symbol size) of three independent experiments. (**B**) Representative histograms showing p53, bcl-2, and bax protein level following 48-h culture in the presence of no compound and sulforaphane (30 μM), respectively. When they do not appear, deviation bars are smaller than the symbol size.

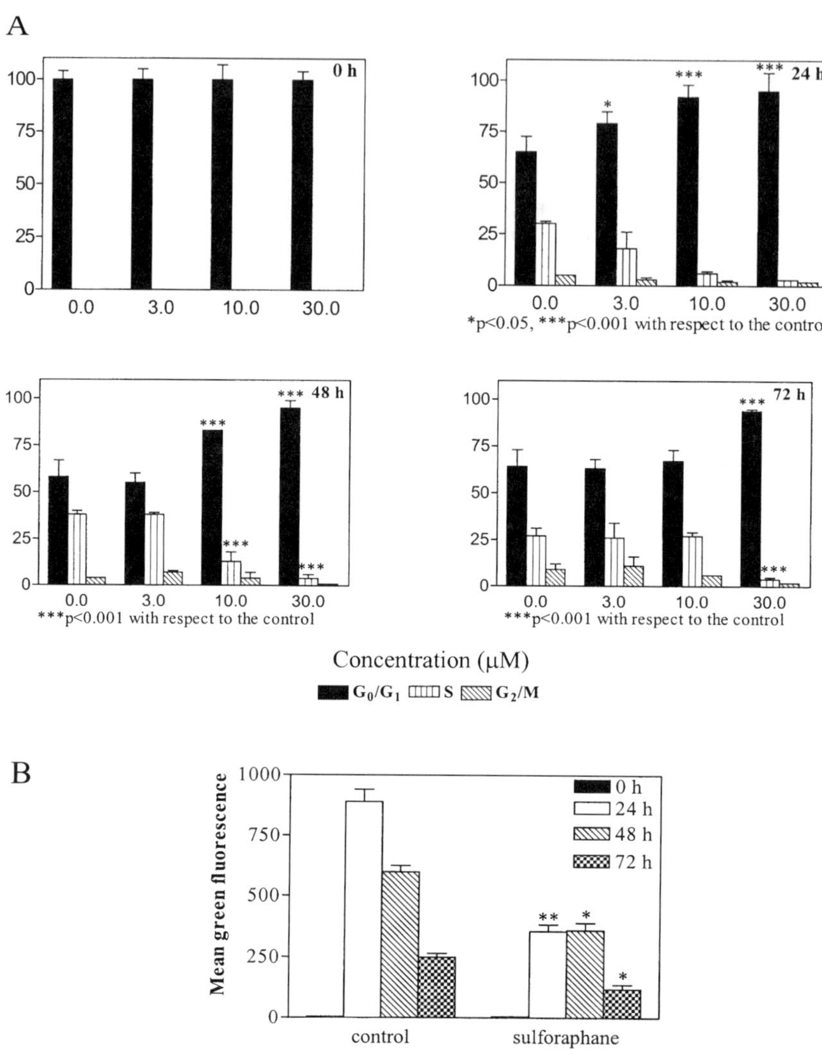

FIGURE 2. (**A**) Effects of sulforaphane on T cell cycle with respect to controls. Results are expressed as percentages of total cell counts. Data are means ± SE (deviation bars, except when smaller than the symbol size) of three independent experiments. (**B**) Representative histograms showing cyclin D3 levels following 0-, 24-, 48-, and 72-h culture in the absence or presence, respectively, of sulforaphane (30 μM). Each bar represents mean ± SE (deviation bars, except when smaller than the symbol size) of three independent experiments.

Novocastra, UK) against phosphorylated and non-phosphorylated protein, FITC bcl-2 (200 μg/mL; Serotec), and bax (200 μg/mL; Santa Cruz Biotechnology, CA), or isotype-matched negative control (Serotec). The cells were washed and incubated with 10 ml of the FITC-labeled secondary antibody for the bax-stained cells (5 μg/mL; Serotec). The cells were then analyzed to quantitate FITC binding.

Statistical analysis: All data are the mean ± SE of three experiments. Fisher's exact test was adopted for statistical evaluation of the results. *T* test was used for the analysis of protein level measurements. All *P* values are two-sided. All statistical analyses were performed in GraphPad InStat version 3.00 for Windows 95, GraphPad Software, San Diego, California, USA.

RESULTS AND DISCUSSION

In Jurkat cells, sulforaphane inhibits cell-cycle progression, with a marked increase in the proportion of G_2/M phase cells after 48 h of treatment (FIG. 1A). Apoptosis was observed at the highest dose of sulforaphane and at late stages during the treatment (at 24 and 48 h). In fact, the fraction of apoptotic cells was 16% (vs. 2% in the control) after 24 h and 32% (vs. 3.50 in the control) after 48 h (data not shown). The analysis of p53, bcl-2 and bax protein levels indicated that p53 and bax expression were found to be increased in treated cells relative to the untreated cells. (FIG. 1B). In non-transformed T lymphocytes, the most marked effect of sulforaphane was the suppression of PHA-driven proliferation. In particular, cell-cycle analysis revealed that sulforaphane inhibited the cell-cycle progression of the normal T cells in the G_1 phase. For example, after 72 h of treatment at 30 μM, the fraction of cells in the G_0/G_1 phase increased to 94% (FIG. 2A). We therefore analyzed the expression of the major D cyclins and associated kinases in human peripheral blood T lymphocytes after T-cell activation. We found that sulforaphane treatment caused a marked decrease in cyclin D3, with a maximum effect registered at 24 h, where the expression of cyclin D3 was decreased by 60% (FIG. 2B). The expression of cyclin D2, CDK4, and CDK6 was more mildly attenuated (data not shown). The weak effect of sulforaphane on the expression of CDK4 and CDK6 is not surprising, in the light of our results indicating a delay of cell-cycle progression. In fact, down-modulation of CDK4 and CDK6 is involved in blocking, rather than delaying, G1 phase after sulforaphane treatment. We also detected a progressive dose-related increase in the fraction of apoptotic cells. For example, after 72 h of treatment at 30 μM, sulforaphane increased the fraction of apoptotic cells to 50% with respect to the control (data not shown). Treated cells showed raised p53 expression (292.5 vs. 151.5 in controls, $P<0.001$), which however did not reach the 7-fold increase we previously recorded in Jurkat leukemia cells. As in Jurkat cells, bcl-2 levels showed a slight, non-significant decrease (835.0 vs. 761.5 in the control). On the other hand, bax levels rose only slightly (202.0 vs. 196.5, P = n.s.), as compared with the highly-significant 3-fold increase previously found in Jurkat cells.

These findings indicate that sulforaphane can exert protective effects inhibiting leukemia cell growth. Moreover, sulforaphane is active not only in transformed lymphocytes but also in their normal counterpart. Although *in vitro* studies do not necessarily predict *in vivo* outcomes, our findings raise important questions regarding the suitability of sulforaphane for cancer chemoprevention. Thus, any chemopreventive use of sulforaphane would have to be very carefully examined, as dietary supplementation with single, putative anticarcinogenic compounds is not warranted without extensive investigation into their possible harmful effects.

REFERENCES

1. MITHEN, R.F., M. DEKKER, R. VERKERK, *et al.* 2000. The nutritional significance, biosynthesis and bioavailability of glucosinolates in human foods. J. Sci. Food Agric. **80:** 967–984.
2. FAHEY, J.W., Y. ZHANG & P. TALALAY. 1997. Broccoli sprouts: an exceptionally rich source of inducers of enzymes that protect against chemical carcinogens. Proc. Natl. Acad. Sci. USA **94:** 10367–10372.
3. CHIAO, J.W., F.L. CHUNG, R. KANCHERLA, *et al.* 2002. Sulforaphane and its metabolite mediate growth arrest and apoptosis in human prostate cancer cells. Int. J. Oncol. **20:** 631–636.
4. VISENTIN, M., A. TAVA, R. IORI & S. PALMIERI. 1991. Isolation and identification of trans-4-(methylthio)-3-butenyl glucosinolate from radish roots (*Raphanus sativus* L.). J. Agric. Food Chem. **40:** 1687–1691.
5. FIMOGNARI, C., M. NÜSSE & P. HRELIA. 1999. Flow cytometric analysis of genetic damage, effect on cell cycle progression and apoptosis by Thiophanate-methyl in human lymphocytes. Env. Mol. Mutagen. **33:** 220–223.

Peptides from Anchovy Sauce Induce Apoptosis in a Human Lymphoma Cell (U937) through the Increase of Caspase-3 and -8 Activities

YOUNG GON LEE,[a] JI YEON KIM,[a] KI WON LEE,[a] KYOUNG HEON KIM,[b] AND HYONG JOO LEE[a]

[a]*Department of Food Science and Technology and School of Agricultural Biotechnology, Seoul National University, Seoul 151-742, South Korea*

[b]*Division of Food Science, Korea University, Seoul 136-701, South Korea*

ABSTRACT: Various biological activities of peptides have been found. In this study, the induction of apoptosis in a human lymphoma cell line (U937) by peptide fraction separated from anchovy sauce was studied using biochemical and flow cytometric methods. After U937 was treated with a hydrophobic peptide fraction of the anchovy sauce, a sub-G1 peak, which represents apoptosis, was found in the cell cycle analysis. The apoptosis in the peptide-treated U937 cell was also confirmed by apoptotic DNA fragmentation. Furthermore, an increase of caspase-3 and -8 activities was observed. Our results suggest that the peptide fraction separated from the anchovy sauce may have cancer chemopreventive effects through the induction of apoptosis in cancer cells.

KEYWORDS: fish sauce; peptide; apoptosis; caspase; lymphoma cell line

INTRODUCTION

Apoptosis has been considered as a promising key target for cancer chemoprevention. Despite the diversity of apoptosis-inducing agents, numerous experiments gave consensus that signals leading to the activation of caspases may play an important role in the initiation and execution of apoptosis.[1] A number of food peptides have been reported to show various physiological activities such as those seen in anti-hypertension, anti-thrombosis, hypocholesterolemia, and opioid and anti-carcinogenic activities.[2,3] Fermented fish sauces rich in peptides have been used as important food in Asia.[4] In this study, the anti-carcinogenic activities of hydrophobic peptide fraction separated from anchovy sauce were found by investigating the induction of apoptosis and increase of caspase activities.

Address for correspondence: Professor Hyong Joo Lee, Department of Food Science and Technology and School of Agricultural Biotechnology, Seoul National University, Seoul 151-742, South Korea. Voice: +82-2-880-4853; fax: +82-2-873-5095.
leehyjo@snu.ac.kr

Ann. N.Y. Acad. Sci. 1010: 399–404 (2003).
doi: 10.1196/annals.1299.073

MATERIALS AND METHODS

Extraction of Hydrophobic Peptides

The anchovy sauce, obtained from Daesang Co. (Seoul, South Korea), was extracted by a HP-20 nonionic polymeric adsorbent (Aldrich, USA) with 95% (v/v) methanol. The extracts were then suspended in 50% (v/v) butanol, transferred to a separatory funnel, and left standing for one day to separate into two phases: the lower phase was considered to be the water layer containing hydrophilic peptides; and the upper phase (Aob) to be the butanol layer containing hydrophobic peptides. The peptide fractions were confirmed by ninhydrin reaction after thin-layer chromatography.

Cell Culture

A human lymphoma cell line U937 was purchased from Korea Cell Lines Bank (Seoul, South Korea), and cultured in RPMI-1640 (Gibco BRL, USA) medium with 10% (v/v) fetal bovine serum (Gibco BRL, USA) and 1% (w/v) penicillin/streptomycin (10,000 units of penicillin/mL and 10 mg/mL streptomycin) at 37°C under 5% (w/w) CO_2.

Cell Cycle Analysis

Cell cycle analysis was confirmed according to the previous method.[5] In brief: the U937 cell was cultured, and treated with the hydrophobic peptide fraction of the anchovy sauce (Aob) at 500 µg/mL for 12, 24, 36, and 48 h. Subsequently, DNA content of nuclei was determined by staining nuclear DNA with propiodium iodide (50 µg/mL) (Sigma, USA) and then measuring the relative DNA content of nuclei with flow cytometry (Becton Dickinson, USA).

Analysis of DNA Fragmentation

The U937 cell, cultured and treated with Aob at 100, 200, and 500 µg/mL for 48 hours, was incubated with a lysis buffer, and treated with RNase A at 37°C for 2 h and with proteinase K at 50 °C for 2 h. DNA electrophoresis was performed in a 1.5% agarose gel. The gel was stained with 0.5 µg of ethidium bromide, and photographed by UV transillumination.

Measurement of Caspases Activities

Activities of caspase-3, and -8 were measured using a caspase-3 and -8 fluorescent assay kit (Peptron, South Korea). U937 cells, treated with Aob at 250 or 500 µg/mL for 48 h, were dissolved in a lysis buffer. The cell lysates were combined with 10 mM DTT and fluorogenic substrate. After a reaction at 37°C for 30 min, fluorescences of the supernatants were determined using a fluorescence spectrophotometer (Hitachi, Japan) with excitation and emission at 360 nm and 460 nm, respectively. The enzyme activity was expressed as a unit of nmol of AMC per hour, and actinomycin-D (1 µM) was used as a positive control. The statistical significance of differences between mean values was assessed by a Student's *t*-test. Only test values which resulted in $P<0.01$ were considered to be significant.

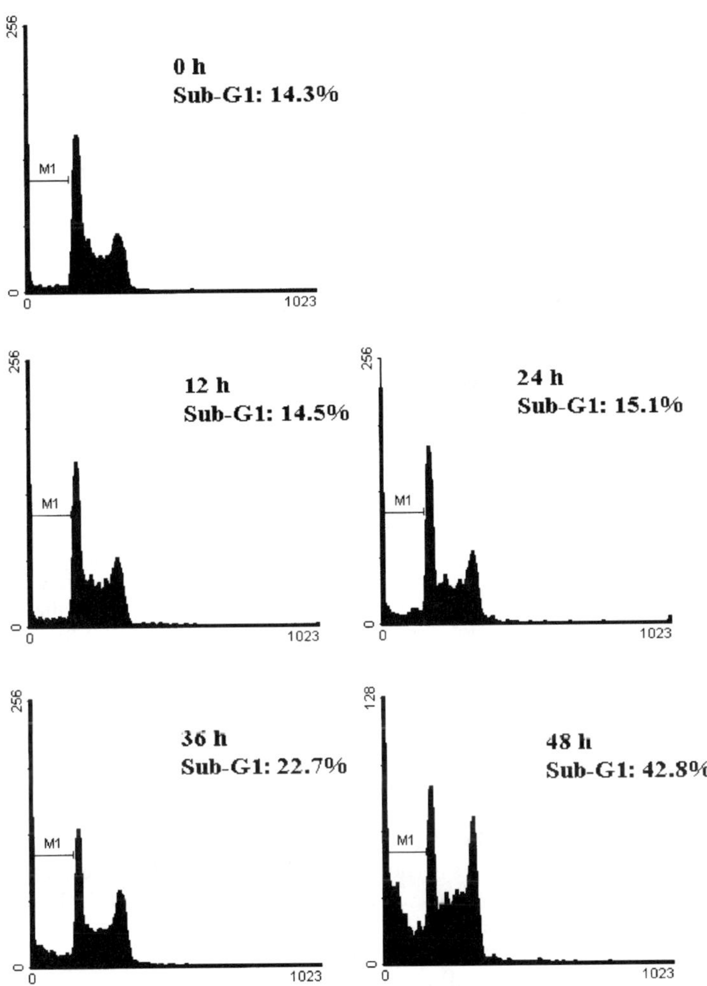

FIGURE 1. Effect of the hydrophobic peptide fraction derived from anchovy sauce (Aob) on the cell cycle of U937. Cells were cultured, and treated with Aob at 500 μg/mL. After incubating for the indicated times, cells were stained with propiodium iodide, and analyzed by flow cytometry.

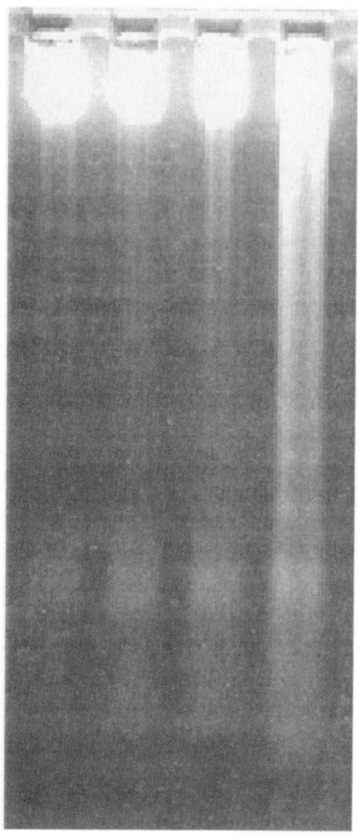

FIGURE 2. The apoptotic DNA fragmentation of U937 cells treated with the indicated concentration of Aob for 48 h. DNA electorphoresis was performed on 1.5% agarose gel, and visualized by a UV light. *Lane* 1, control; *lane* 2, 100 μg/mL; *lane* 3, 200 μg/mL; *lane* 4, 500 μg/mL.

RESULTS AND DISCUSSION

U937 cells were treated with 500 μg/mL of the hydrophobic peptide fraction of the anchovy sauce (Aob), and the cell cycle was examined after treatments for 12, 24, 36, and 48 h. The number of cells in the sub-G1 fraction, which represents apoptotic cells, was found to be higher than the control. The content of sub-G1 phases also increased from 14.3 (for the control) to 42.8 % (after 48-h exposure to Aob) (FIG. 1). This observation provides preliminary evidence that Aob induced apoptosis.[6]

One of the most important hallmarks of apoptosis is the fragmentation of chromosomal DNA into multiples of 180–200-bp DNA, which typically appears as a "DNA laddering" pattern on DNA electrophoresis gel.[6] FIGURE 2 shows the apoptotic DNA fragmentation of U937 cell treated with Aob at a final concentration of 500 μg/mL for 48 h.

Various apoptotic signals such as death factors, growth factor–deprivation, and genotoxic agents are known to activate caspase-8. The activated caspase-8 in turn activates caspase-3 in a cascade mode. The activated caspase-3 in the cascade down-

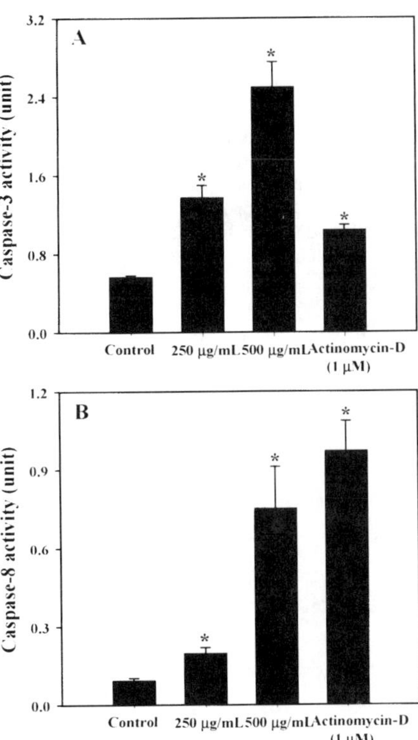

FIGURE 3. The caspase-3 (**A**) and -8 (**B**) activities of U937 cells treated with the indicated concentration of Aob for 48 h. The unit of caspase activity was expressed in nmol AMC released per hour. Concentration of AMC was obtained using a standard curve prepared with a stock solution of AMC. Actinomycin-D (1 μM) was used as a positive control. Data indicate means of the values (bar, SD) obtained from three independent experiments. *($P < 0.01$, Student's *t*-test.)

stream cleaves ICAD at two positions, consequently inactivating the inhibitory activity of ICAD.[1] To measure the enzymatic activity of caspases during the peptide fraction-induced apoptosis, we used a fluorogenic peptide substrate. The Aob was found to induce significant increases in the activities of both caspase-3 (FIG. 3A) and -8 (FIG. 3B). On the basis of the foregoing results, the DNA fragmentation of the Aob-treated U937 can be explained by the activation of capases and then release of CADs by the activated capases to degrade the chromosomal DNA of U937.

REFERENCES

1. FALEIRO, L., R. KOBAYASHI, H. FEARNHEAD & Y. LAZEBNIK. 1997. Multiple species of CPP32 and Mch2 are the major active caspases present in apoptotic cells. EMBO J. **16:** 2271–2281.

2. Shin, Z.I., R. Yu, S.A. Park, *et al.* 2001. His-His-Leu, an angiotensin I converting enzyme inhibitory peptide derived from Korean soybean paste, exerts antihypertensive activity *in vivo*. J. Agric. Food Chem. **49:** 3004–3009.
3. Kim, S.E., H.H. Kim, J.Y. Kim, *et al.* 2000. Anticancer activity of hydrophobic peptides from soy proteins. BioFactors **12:** 151–155.
4. Raksakulthai, N. & N.F. Haard. 1992. Correlation between concentration of peptide and amino acids and the flavor of fish sauces. J. Asian Food **7:** 86–90.
5. Kim, J.Y. 2002. Inhibitory effect of tumor cell proliferation and induction of G2/M cell cycle arrest by panaxytriol. Planta Med. **68:** 119–122.
6. Bortner, C.D., N.B.E. Oldenburg & J.A. Cidlowski. 1995. The role of DNA fragmentation in apoptosis. Trends Cell Biol. **5:** 21–26.

Alveolar Cell Death in Hyperoxia-Induced Lung Injury

ALESSANDRA PAGANO AND CONSTANCE BARAZZONE-ARGIROFFO

Departments of Pediatrics and Pathology, University of Geneva Medical School, Geneva, Switzerland

Abstract: Exposure to high oxygen concentration causes direct oxidative cell damage through increased production of reactive oxygen species. *In vivo* oxygen-induced lung injury is well characterized in rodents and has been used as a valuable model of human respiratory distress syndrome. Hyperoxia-induced lung injury can be considered as a bimodal process resulting (1) from direct oxygen toxicity and (2) from the accumulation of inflammatory mediators within the lungs. Both apoptosis and necrosis have been described in alveolar cells (mainly epithelial and endothelial) during hyperoxia. While the *in vitro* response to oxygen seems to be cell type–dependent in tissue cultures, it is still unclear which are the death mechanisms and pathways implicated *in vivo*. Even though it is not yet possible to distinguish unequivocally between apo-ptosis, necrosis, or other intermediate form(s) of cell death, a great variety of strategies has been shown to prevent alveolar damage and to increase animal survival during hyperoxia. In this review, we summarize the different cell death pathways leading to alveolar damage during hyperoxia, with particular attention to the pivotal role of mitochondria. In addition, we discuss the different protective mechanisms potentially interfering with alveolar cell death.

Keywords: hyperoxia; lung; mice; apoptosis; necrosis; mitochondria

INTRODUCTION

High oxygen concentrations are commonly used to treat premature babies, children, and adults with pulmonary failure. However, prolonged exposure to high oxygen tension is known to induce lung tissue damage. Hyperoxia-induced lung injury (100% oxygen exposure) has been used as a well-established model of lesions mimicking acute respiratory distress syndrome. In rodents, oxygen-induced lung injury is characterized by extensive alveolar damage leading to the disruption of the alveolo-capillary barrier, pulmonary edema, and pleural effusion.[1] Alveolar leak occurs when epithelial cells are damaged, suggesting that these cells are crucial in maintaining the barrier integrity.[2–4] It is generally assumed that lung damage results essentially from the direct action of increased intracellular reactive oxygen species (ROS), which is accompanied by a secondary inflammatory and stress response of the

Address for correspondance: Dr. Constance Barazzone Argiroffo, Department of Pathology and Department of Pediatrics, Centre Médical Universitaire, 1211 Geneva 4, Switzerland. Voice: +41 22 702 57 57; fax: +41 22 702 57 46.
constance.barazzone@hcuge.ch

Ann. N.Y. Acad. Sci. 1010: 405–416 (2003).
doi: 10.1196/annals.1299.074

host.[5–7] All these pathologic alterations converge toward a central event: alveolar cell death.

HYPEROXIA-INDUCED CELL DEATH

Apoptosis and Necrosis in Hyperoxia

Two major patterns of cell death are recognized today: apoptosis and necrosis.[8] *Apoptosis* is defined as a programmed cell death, morphologically characterized by nuclear shrinkage, chromatin condensation, and DNA fragmentation to oligonucleosome-size fragments, while intracellular organelles and plasma membrane remain intact.[9] In contrast, *necrosis* is an accidental cell death, resulting from unscheduled, acute injury. It is characterized by the acute disruption of cellular metabolism leading to ATP depletion, rupture of the plasma membrane, and mitochondrial and cytoplasma swelling.[8] In particular conditions of tissue or cell stress, it has been suggested that both or intermediate forms of cell death might occur in the same cell (*necrapoptosis*).[10] Recently, it has been proposed to replace the term *necrosis* by *oncosis*, which designates the cellular swelling, while the term *necrosis* is rather attributed to post-mortem morphological cellular alterations.[11–12] However, to simplify the nomenclature in the present review we will refer to the classical term necrosis in opposite to apoptosis.

We will review here the principal evidence that allows us to define and decipher the mechanisms of alveolar cell death occurring during hyperoxia.

The morphological changes of alveolar cells during exposure to hyperoxia have been widely described in rodents.[13,14] However, despite a vast literature resulting from nearly 30 years of research, the mechanisms that underlie cell injury and death, following exposure to hyperoxia, are not completely understood. Morevover, it remains unknown which cells are the primary targets of oxygen damage. Capillary endothelial cells are believed to be the most susceptible cells to direct oxygen poisoning, which induces the loss of cellular volume, at early times of oxygen exposure, as far as from 48 hours.[13–15] Lesions of epithelial cells become appearent only at later times (72 hours).[1–13] Aspect of apoptosis, such as condensed nuclear chromatin, has been observed mainly in endothelial cells, while epithelial cells show generally necrotic aspect, such as disruption of plasma membrane, organelle swelling, and intracellular blebbing.[3–16] Some cells can present overlapping features of apoptosis and necrosis.[3]

These morphological observations have been highlighted by studies using molecular approaches. Several reports have shown oligonuclesomal DNA fragmentation in lungs of animals exposed to hyperoxia, as assessed by *in situ* TUNEL analysis and DNA electrophoresis, thus confirming that apoptosis is induced during hyperoxia *in vivo*.[3–7,17] However, it is now recognized that TUNEL positivity is not a hallmark of apoptosis but rather of DNA damage.[9] The results on the concomitant presence of apoptosis and necrosis might reflect the fact that both pathways of cell death co-exist or that another mechanism of cell death distinct from apoptosis or necrosis might be induced in hyperoxia. Alternatively, secondary necrosis could result from terminal steps of apoptosis.[8] This last hypothesis is very unlikely, since a great amount of apoptosis preceding necrosis has never been observed nor described until now *in vivo*.

Finally, it cannot be excluded that distinct cell death pathways may be specifically induced in different cell types (i.e., endothelial, epithelial, or inflammatory cells).

This question, which is still open *in vivo,* has been tentatively addressed *in vitro* using several cell lines. Hyperoxia-exposure of cultured A549 cells, derived from alveolar type II epithelial cells, resulted in extensive cell swelling and death within few days without typical signs of apoptosis, as no nuclear chromatin condensation was evident and these cells were TUNEL-negative.[16–18] Untransformed human small airway epithelial (SAE) cells, simultaneously exposed with A549 cells to 95% oxygen, showed some extent of apoptosis, as did endothelial cell lines under similar conditions.[18–20] In addition, hyperoxic exposure has also been shown to induce apoptosis in cultured murine macrophages, PC12 and Rat-1 cells, as assessed by DNA laddering and chromatin condensation.[21–23] These results suggest that, although most of the cell lines die by apoptosis when cultured in hyperoxic conditions, pulmonary epithelial cells seem to use an alternative pathway of cell death. Nevertheless, one could keep in mind that the differences observed among cell types might also be related to different culture conditions, the duration, and the dose of oxygen exposure. For instance, according to the dose applied, the oxidant H_2O_2, can induce *in vitro* either necrosis or apoptosis within the same cells.[19–24] Supporting this idea, type II alveolar epithelial cells from rats exposed for 48 hours to sublethal doses of oxygen were shown to be committed to an apoptotic pathway when cultured *ex vivo.*[2] Therefore, even if the extrapolation from the *in vitro* to *in vivo* results should be done with precautions, it might be possible that hyperoxia could induce apoptosis and/or necrosis in endothelial or epithelial cells *in vivo*, according to the accessibility of oxygen *in situ* and to the duration of oxygen aggression.

Oxidative Stress, Source of ROS and Cell Death

Reactive oxygen species (ROS) have been recognized to act as biochemical intermediates in apoptotic signaling.[24–26] Their excessive generation mediates the damage of cellular macromolecules, leading to irreversible dysfunction and cell death.[27] Mitochondrial oxidative phosphorylation is one of the major source of ROS in eukaryotic cells.[24] During oxygen exposure the increased rate of intracellular ROS production was believed to be dependent on mitochondria. Indeed, overproduction of superoxide was specifically detected in rat lung mitochondria and submitochondrial particles.[28,29] The ROS production was associated with an increased release in H_2O_2 by lung mitochondria and microsomes.[30] Therefore, until now, it has been postulated that mitochondrially derived ROS produced during hyperoxia are directly implicated in apoptotic cell death. This concept has been recently questioned by Budinger *et al.,* who reported that Rat-1 cells depleted in mitochondrial DNA (therefore unable to perform oxidative phosphorylation and produce ROS), underwent apoptosis normally when exposed to high oxygen concentration.[23] These data therefore suggest that the high amount of ROS produced outside mitochondria during hyperoxia is sufficient to induce cell death.

Infiltrating inflammatory cells, which are trapped in the lung, may represent an important source of ROS. These cells are potent producers of superoxide radicals.[31] Hyperoxia-induced lung injury is associated with the accumulation of inflammatory cells, such as alveolar macrophages, neutrophils, and platelets.[32,33] The intensity of the inflammatory response has been shown to be dependent on the animal species and strain. A large number of infiltrating inflammatory cells was found in rabbit and rats.[33,34] In contrast, in mice, even if a certain variability in sensitivity to hyperoxic injury exists according to the strain,[35] neutrophils were present in small quantity in the lungs and appeared at later time points.[3,36] Accordingly, a lower level of myeloperoxidase (MPO) activity was measured in lung extracts compared to those of other animal species.[38] In rats and mice, studies preventing hyperoxia-induced lung injury have been often correlated with a reduced influx of inflammatory cells in broncholaveolar lavage (BAL) and lung interstitium.[39–41] More directly, depleting neutrophils and macrophages with chemical agents or stopping their influx by chemokine blockade protected the lungs of hyperoxia-exposed rabbits and rats.[34-42,43] However, in mice, neutrophil-depletion studies were not conclusive.[36–40] More recently, it has been reported that blocking neutrophil influx into the alveoli of newborn rats reduced the number of TUNEL-positive alveolar cells.[44] It is possible that the positive TUNEL staining and DNA laddering might result from inflammatory cells entrapped in lung interstitium and dying by apoptosis. Indeed, neutrophils obtained from BAL of patients affected by acute respiratory distress syndrome (ARDS) showed characteristic features of apoptosis, as determined by morphological analysis and DNA laddering.[45]

A major efforts has been made to prevent lung injury by neutralizing the high levels of ROS produced in the alveoli. Numerous strategies have been attempted to increase the local concentration of antioxidants, for instance, by delivering or overexpressing superoxide dismutase (SOD) and catalase in the animals.[40-46,49] In these studies, hyperoxia-exposed animals exhibited a lesser extent of lung damage compared with control animals, as shown by decreased pulmonary inflammation, lung edema, and hemorrhages. No direct analyses of alveolar cell death were performed in those reports, making it difficult to prove the causality between increased levels of antioxidant and cell survival. However, it can be assumed that longer survival of the animals is tightly dependent on alveolar cell damage. Of note, all these strategies were partially successful, since the mean survival of the treated animals was only slightly increased. Therefore, it can be postulated that, *in vivo,* antioxidants contribute to mitigate secondary inflammation, but they may be insufficient *per se* to overcome the direct effects of ROS-mediated oxygen toxicity.

MOLECULAR MECHANISMS OF CELL DEATH IN HYPEROXIA

Cell Death Receptor Pathway and Apoptosis-Related Genes

The apoptotic signaling pathway may be initiated either by cell surface molecules interacting with their receptors, many of which contain death domains, or alternatively by directly targeting the mitochondria or the endoplasmic reticulum.[50] Several pathways, involving the signaling by cell death receptors, have been investigated during hyperoxic acute lung injury. While Fas was upregulated in lungs exposed to hyperoxia, Fas-null mice did not show any differences in susceptibility to oxygen toxicity.[3] Blocking TNF-α-mediated cell signaling by administration of polyclonal antibodies did not ameliorate oxygen-induced lung damage.[38] Finally, the disruption of the CD40-CD40 ligand (CD40L) signaling pathway in mice deficient for CD40 or for its ligand, or in mice treated with an anti-CD40L monoclonal antibody, did not increase tolerance to oxygen.[51] All these results suggest that the cell surface receptor pathway of programmed cell death is not of central importance in hyperoxic lung cell death.

The transcription factor p53 is also known to promote apoptosis, in response to DNA damage, by increasing the transcription of the pro-apoptotic molecule Bax *in vitro.*[52] Here again, besides its increased expression in mouse lungs exposed to hyperoxia, mice deficient for p53 were not protected from hyperoxia.[3–53] The protein p21, a cyclin-dependent kinase inhibitor implicated in DNA repair, was proposed to promote apoptosis, since it was significantly increased in alveolar epithelial cells during hyperoxia.[54] However, p21-deficient mice, whose cells do not undergo cell cycle arrest, were more susceptible to oxygen, suggesting that cell cycle arrest allows DNA to repair rather than contributing to hyperoxic lung injury.[55]

Mitochondria-Dependent Cell Death Pathway

Mitochondria play a central role in directing the cell's fate. Indeed, they maintain the cellular levels of ATP and are able to release death-promoting factors, such as cytochrome *c*, when cells are exposed to pro-apoptotic signals. Although several death stimuli are able to induce the mitochondrial release of cytochrome *c*, the molecular mechanisms of this process are still under investigation.[56,57] It has been shown that the activation and translocation from the cytosol to the mitochondrial membrane of the pro-apoptotic members of the Bcl-2 family, such as Bax, leads to the formation of channels through which cytochrome *c* is released without alteration of the morphology of mitochondria.[57,58] Alternatively, strong cell death signals, such as oxidants and pathologic increases in cytosolic Ca_2, or Bax itself, can directly affect the permeability of the mitochondrial membrane, inducing the opening of the high-conductance channel permeability transition pore (PTP) to solutes of molecular mass less than 1500 Da.[57-59,60] This provokes mitochondrial swelling, outer membrane break, and subsequent release of cytochrome *c*.[61,62] In this latter case, mitochondria are irreversibly damaged. These mechanisms are counter-regulated by the anti-apoptotic members of Bcl-2 family, Bcl-2 and Bcl-x_L.[57–63]

Morphologic changes of alveolar cell mitochondria, such as swelling and cristae disorganization, have been described more than 30 years ago in hyperoxia-exposed animals.[13–64] These mitochondrial alterations have been only recently correlated with perturbation of the mitochondrial membrane permeability (MMP) *in vitro.*[65] In another study, it has been shown that a late depolarization of the mitochondrial membrane was associated with cytochrome *c* release in Rat-1 cells exposed to hyperoxia. In these cells apoptosis occurred through caspase-9 activation. The pro apoptotic proteins Bax and Bak were required for hyperoxia-induced cell death.[23] Accordingly, recent results obtained by our group suggest that alteration of mitochondria structure is associated with cytochrome *c* release in alveolar cells, when mice are exposed to high oxygen concentrations.[95]

Preventing mitochondria-dependent cell death by modulation of anti-apoptotic proteins may be a good strategy to induce protection from hyperoxia-induced cell damage. This has been suggested by several *in vitro* studies showing that Bcl-2 and Bcl-x_L overexpression can rescue cells from death during hyperoxia.[22,23] Indeed, hyperoxia-induced death of Rat-1 cells was completely prevented by the overexpression of the anti-apoptotic Bcl-x_L protein.[23] *In vivo*, there is only indirect evidence

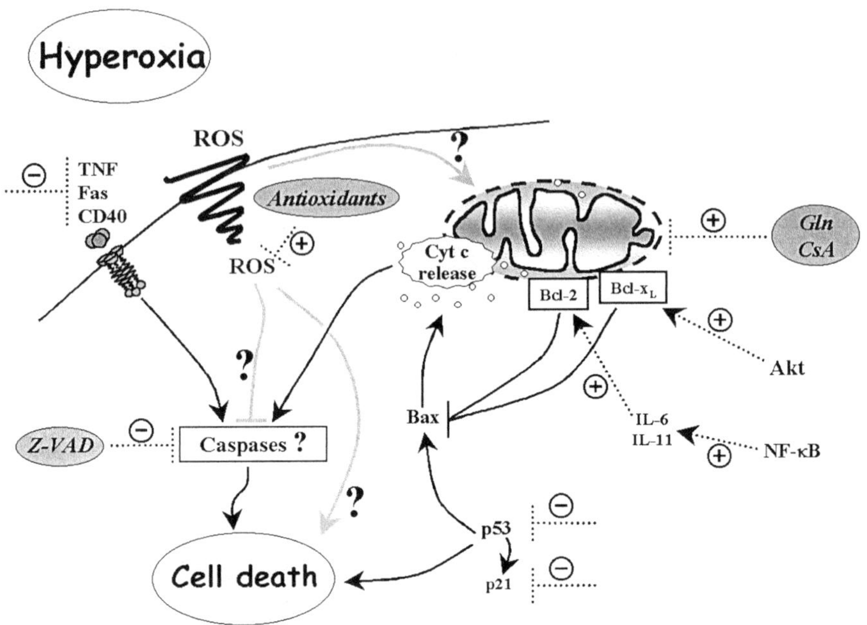

FIGURE 1. Schematic and non-exhaustive representation of the pathways implicated in hyperoxia-induced alveolar cell death. The protective strategies explored in this model are also shown. *Solid lines* represent the known signaling pathways (solid arrows and lines: activating and inhibitory pathways). *Dotted lines* represent the approaches interfering with these pathways (dotted arrows and lines: positive and negative regulation). The positive (+) and negative (–) signs are used to indicate whether the strategies applied were efficient or inefficient in inducing protection. Cyt c, cytochrome *c*; CsA, cyclosporin A; Gln, glutamine; NF-κB, nuclear factor-κB; ROS, reactive oxygen species; SOD, superoxide dismutase; TNF, tumor necrosis factor ; Z-VAD, caspase inhibitor.

that Bcl-2 upregulation can prevent pulmonary cell death. Indeed, the overexpression of IL-6-type cytokines, IL-6, and IL-11, in the lungs dramatically increased the mean survival of mice exposed to 100% oxygen and was associated with a lesser extent of inflammation, lung injury, and DNA fragmentation.[41–66] Importantly, IL-6 protection was correlated with the increased expression of the antiapoptotic protein Bcl-2 in lung extracts of transgenic mice.[66] Since overexpression of Bcl-2 and Bcl-x_L was succesfully applied *in vivo* in different models of induced cell death,[67,68] it would be interesting to evaluate their role during hyperoxia-induced cell death.

Recently, it was also reported that the modulation of another anti-apoptotic protein attenuated hyperoxia-induced lung injury. The protein kinase B/Akt is a kinase known to phosphorylate, among its different substrates, and inactivate the pro-apoptotic protein Bad, thereby promoting cell survival.[69] Intratracheal introduction of the activated form of Akt by adenoviral gene transfer techniques prevented hyperoxia-induced lung injury in mice and quantitatively reduced TUNEL-positive alveolar cells.[70] These data also highlight that direct targeting of alveolar epithelial cells is sufficient to induce an efficient protection from oxidative lung injury.

Strategies to avoid mitochondrial alterations and membrane depolarization have been also used succesfully to prevent hyperoxia-induced cell damage. Studies were recently performed on cultured epithelial pulmonary cells. Glutamine treatment of A549 cells exposed to hyperoxia prevented mitochondrial membrane depolarization, alteration of mitochondria morphology, and cell death. Similarly, overexpression in the same cell type of hexokinase II, an enzyme implicated in the glycolytic pathway, protected cells against hyperoxia-induced cell death also by preserving mitochondrial structure.[71]

Cyclosporin A, which is a specific blocker of PTP opening,[72,73] was effective in protecting mitochondria from swelling and cytochrome *c* release in several experimental conditions.[74–75] Administration of this drug ameliorated lung function in hyperoxia-exposed mice, although the mechanism of protection was not elucidated in that study.[76] Our work suggests that cyclosporin A may prevent hyperoxia-induced lung injury in mice by preserving alveolar epithelial cell mitochondria from swelling and cytochrome *c* release.[95] All these studies provide evidence that mitochondria are important targets and emphasize that maintaining a sufficient proportion of intact organelles is crucial for cell survival in this model.

Activation of Caspases

Once cytochrome *c* is released in the cytosol, one of the key events of apoptosis is the activation of the caspase family members. These enzymes are considered as essential molecules of the apoptotic pathway, since cell death can be prevented by dominant negative mutations, or pharmacological inhibitors affecting caspase activity.[77] Activation of caspase-9 after oxygen exposure has been shown only in cultured fibroblasts, while data are missing for other cell cultures.[23] Besides morphologic features of apoptosis during hyperoxia, until now there are no evidences of caspase activation *in vivo*, except for activated caspase-1 (p20).[78] We were unable to detect any caspase activity or cleaved form of effector caspase (-3, -7, -9) within hyperoxia-exposed lungs (Ref. 3 and our unpublished results). In agreement with these results, attempts to prevent lung injury by continuous infusion of the tripeptide caspase inhibitor Z-VAD were unsuccessful.[3] It cannot be excluded that the intracellular level of active caspases is too low or restricted to too small a number of alveolar cells to be detected by the current available analysis tools.

Recently, it has been postulated that the level of ATP might be critical in the activation of caspases and in directing cell death toward apoptosis or necrosis.[57–79] Apoptosis requires ATP to activate the signaling pathway of caspases, whereas necrosis follows the depletion of intracellular ATP.[80] Since mitochondria are the key regulators of ATP levels, they could be responsible for the switch between apoptosis and necrosis. It could be hypothesized that during hyperoxia the amount of functional mitochondria and ATP available may be too low to ensure caspase activation and cells would be rather committed to necrosis. Once again, this can be restricted to a cell subpopulation, such as type II epithelial cells, while apoptosis could involve less exposed (endothelial?) cells. Alternatively, it has been noted that caspases are redox-sensitive enzymes and a prolonged and excessive oxidative stress can prevent their activation and resulting apoptosis.[81] Finally, it can be also possible that hyperoxia triggers a caspase-independent apoptotic pathway, as suggested in other pathological models.[82–83] However, no data are available at the present time to confirm this hypothesis.

MODULATION OF THE CELL STRESS RESPONSE

Hyperoxia induces rapid expression of several stress-response genes.[84] In the last years efforts have been made to modulate the response of these genes in order to induce protection against hyperoxia-induced lung injury. It is beyond the scope of this review to enumerate all protective mechanisms involved in the modulation of the cell response. We will focus on heme oxygenase-1 (HO-1) and the redox-sensitive transcription factor NF-κB, which are closely related to cell survival and death.

NF-κB has been shown to prevent apoptosis and to act as a cell survival factor in different conditions.[85] Several studies have shown that this transcription factor is activated in hyperoxia-exposed cells.[86,87] Its activity is essentially repressed by steroids.[88] Work from our laboratory demonstrated that the endogenous level of corticosterone (CS) increased during hyperoxia in mice serum and that administration of hydroxycorticosterone acetate aggravated lung injury.[89] Since glucocorticoids were shown to promote apoptosis not only in thymocytes but also in cultured epithelial cells,[90] it is likely that their increased production during hyperoxia might accelerate alveolar cell death. The removal of endogenous CS production by mouse adrenalectomy decreased hyperoxia-induced lung injury and correlated with a maintained NF-κB activity.[91] These results are in agreemeent with Franek and coworkers, who found that the induction of higher basal levels of NF-κB before oxidant exposure prevented cell death in cultured pulmonary epithelial cells.[92] These data suggest that the modulation of the glucocorticoid-mediated stress response modifies the signaling pathway of cell survival.

Heme oxygenase-1 (HO1) is a well-studied stress-induced gene, which is upregulated in hyperoxic conditions.[93] Exogenous administration of HO-1 by adenoviral gene transfer increased rat survival and attenuated neutrophil influx and apoptosis of lung cells during hyperoxia.[39] Similarly, administration of a low dose of carbon monoxide is thought to prevent lung oxidative stress through HO-1 elevation.[94] However, the exact pathway used by this antioxidant enzyme to increase cell survival is not yet elucidated.

SUMMARY

In summary, the molecular and biochemical mechanisms involved in hyperoxic lung injury are very complex and include several and redundant pathways. Despite the huge amount of work performed over the last years, further attention is needed to better understand the relevance of these multiple signaling pathways. From the recent literature it can be extrapolated that alveolar epithelial cells should be the main target of therapeutic approaches. Interfering with cell death by upregulating anti-apoptotic proteins have given promising results. In particular, preserving mitochondrial integrity and function in these cells might provide encouraging future prospects.

ACKNOWLEDGMENT

This work is supported by the FNRS (Swiss National Fund) (C.B.A.), by the Wolfermann-Nägele foundation (C.B.A. and A.P.), and by the Novartis Foundation (A.P.).

REFERENCES

1. CRAPO, J.D. 1986. Morphologic changes in pulmonary oxygen toxicity. Annu. Rev. Physiol. **48:** 721–731.
2. MATALON, S. & E.A. Egan. 1984. Interstitial fluid volumes and albumin spaces in pulmonary oxygen toxicity. J. Appl. Physiol. **57:** 1767–1772.
3. BARAZZONE, C. *et al.* 1998. Oxygen toxicity in mouse lung: pathways to cell death. Am. J. Respir. Cell Mol. Biol. **19:** 573–581.
4. BARAZZONE, C. *et al.* 1999. Keratinocyte growth factor protects alveolar epithelium and endothelium from oxygen-induced injury in mice. Am. J. Pathol. **154:** 1479–1487.
5. FREEMAN, B.A. *et al.* 1982. Hyperoxia increases oxygen radical production in rat lung homogenates. Arch. Biochem. Biophys. **216:** 477–484.
6. FOX, R.B. *et al.* 1981. Pulmonary inflammation due to oxygen toxicity: Involvment of chemotactic factors and polymorphonuclear cells. Am. Rev. Respir. Dis. **123:** 521–523.
7. MANTELL, L.L. & P.J. Lee. 2000. Signal transduction pathways in hyperoxia-induced lung cell death. Mol. Genet. Metab. **71:** 359–370.
8. KROEMER, G. *et al.* 1998. The mitochondrial death/life regulator in apoptosis and necrosis. Annu. Rev. Physiol. **60:** 619–642.
9. SARASTE, A. & K. Pulkki. 2000. Morphologic and biological hallmarks of apoptosis. Cardiovasc. Res. **45:** 528–537.
10. LEMASTERS, J.J. 1999. V. Necrapoptosis and the mitochondrial permeability transition: shared pathways to necrosis and apoptosis. Am. J. Physiol. **276:** G1–6.
11. VAN CRUCHTEN, S. & W. VAN DEN BROECK. 2002. Morphological and biochemical aspects of apoptosis, oncosis and necrosis. Anat. Histol. Embryol. **31:** 214–223.
12. KANDUC, D. *et al.* 2002. Cell death: apoptosis versus necrosis (review). Int. J. Oncol. **21:** 165–170.
13. KISTLER, G.S. *et al.* 1967. Development of fine structural damage to alveolar and capillary lining cells in oxygen-poisoned rat lungs. J. Cell Biol. **32:** 605–628.
14. CRAPO, J.D. *et al.* 1980. Structural and biochemical changes in rat lungs occurring during exposures to lethal and adaptive doses of oxygen. Am Rev Respir Dis. **122:** 123–143.
15. MASSARO, D. & G.D. MASSARO. 1978. Biochemical and anatomical adaptation of the lung to oxygen-induced injury. Fed. Proc. **37:** 2485–2488.
16. KAZZAZ, J.A. *et al.* 1996. Cellular oxygen toxicity. Oxidant injury without apoptosis. J. Biol. Chem. **271:** 15182–15186.
17. OTTERBEIN, L.E. *et al.* 1998. Pulmonary apoptosis in aged and oxygen-tolerant rats exposed to hyperoxia. Am. J. Physiol. **275:** L14–20.
18. JYONOUCHI, H. *et al.* 1998. The effects of hyperoxic injury and antioxidant vitamins on death and proliferation of human small airway epithelial cells. Am. J. Respir. Cell Mol. Biol. **19:** 426–436.
19. HOGG, N. *et al.* 1999. Apoptosis in vascular endothelial cells caused by serum deprivation, oxidative stress and transforming growth factor-beta. Endothelium **7:** 35–49.
20. SHAIKH, A.Y. *et al.* 1997. Melatonin protects bovine cerebral endothelial cells from hyperoxia- induced DNA damage and death. Neurosci. Lett. **229:** 193–197.
21. PETRACHE, I. *et al.* 1999. Mitogen-activated protein kinase pathway mediates hyperoxia-induced apoptosis in cultured macrophage cells. Am. J. Physiol. **277:** L589–595.
22. KATOH, S. *et al.* 1997. The rescuing effect of nerve growth factor is the result of up-regulation of bcl-2 in hyperoxia-induced apoptosis of a subclone of pheochromocytoma cells, PC12h. Neurosci. Lett. **232:** 71–74.
23. BUDINGER, G.R. *et al.* 2002. Hyperoxia-induced apoptosis does not require mitochondrial reactive oxygen species and is regulated by Bcl-2 proteins. J. Biol. Chem. **277:** 15654–15660.
24. SLATER, A.F. *et al.* 1995. The role of intracellular oxidants in apoptosis. Biochim. Biophys. Acta **1271:** 59–62.
25. BUCKLEY, S. *et al.* 1998. Apoptosis and DNA damage in type 2 alveolar epithelial cells cultured from hyperoxic rats. Am. J. Physiol. **274:** L714–720.
26. HOIDAL, J.R. 2001. Reactive oxygen species and cell signalling. Am. J. Respir. Cell Mol. Biol. **25:** 661–663.

27. FREEMAN, B.A. & J.D. CRAPO. 1982. Biology of disease: free radicals and tissue injury. Lab. Invest. **47:** 412–426.
28. FREEMAN, B.A. & J.D. CRAPO. 1981. Hyperoxia increases oxygen radical production in rat lungs and lung mitochondria. J. Biol. Chem. **256:** 10986–10992.
29. TURRENS, J.F. *et al.* 1982. The effect of hyperoxia on superoxide production by lung submitochondrial particles. Arch. Biochem. Biophys. **217:** 401–410.
30. TURRENS, J.F. *et al.* 1982. Hyperoxia increases H2O2 release by lung mitochondria and microsomes. Arch. Biochem. Biophys. **217:** 411–421.
31. FORMAN, H.J. & M.J. THOMAS. 1986. Oxidant production and bactericidal activity of phagocytes. Annu. Rev. Physiol. **48:** 669–680.
32. HARADA, R.N. *et al.* 1984. Macrophage effector function in pulmonary oxygen toxicity: hyperoxia damages and stimulates alveolar macrophages to make and release chemotaxins for polymorphonuclear leukocytes. J Leukoc Biol. **35:** 373–383.
33. BARRY, B.E. & J.D. CRAPO. 1985. Patterns of accumulation of platelets and neutrophils in rat lungs during exposure to 100% and 85% oxygen. Am. Rev. Respir. Dis. **132:** 548–555.
34. SHASBY, D.M. *et al.* 1982. Reduction of the edema of acute hyperoxic lung injury by granulocyte depletion. J. Appl. Physiol. **52:** 1237–1244.
35. JOHNSTON, C.J. *et al.* 1998. Inflammatory and epithelial responses in mouse strains that differ in sensitivity to hyperoxic injury. Exp. Lung Res. **24:** 189–202.
36. SMITH, L.J. *et al.* 1988. Hyperoxic lung injury in mice: effect of neutrophil depletion and food deprivation. J. Lab. Clin. Med. **111:** 449–458.
37. WELTY, S.E. *et al.* 1993. Increases in lung tissue expression of intercellular adhesion molecule-1 are associated with hyperoxic lung injury and inflammation in mice. Am. J. Respir. Cell Mol. Biol. **9:** 393–400.
38. BARAZZONE, C. *et al.* 1996. Hyperoxia induces platelet activation and lung sequestration: an event dependent on tumor necrosis factor-alpha and CD11a. Am. J. Respir. Cell Mol. Biol. **15:** 107–114.
39. OTTERBEIN, L.E. *et al.* 1999. Exogenous administration of heme oxygenase-1 by gene transfer provides protection against hyperoxia-induced lung injury. J. Clin. Invest. **103:** 1047–1054.
40. FOLZ, R.J. *et al.* 1999. Extracellular superoxide dismutase in the airways of transgenic mice reduces inflammation and attenuates lung toxicity following hyperoxia. J. Clin. Invest. **103:** 1055–1066.
41. WAXMAN, A.B. *et al.* 1998. Targeted lung expression of interleukin-11 enhances murine tolerance of 100% oxygen and diminishes hyperoxia-induced DNA fragmentation. J. Clin. Invest. **101:** 1970–1982.
42. BERG, J.T. *et al.* 1995. Response of alveolar macrophage-depleted rats to hyperoxia. Exp Lung Res. **21:** 175–185.
43. AUTEN, R.L., Jr. *et al.* 2001. Anti-neutrophil chemokine preserves alveolar development in hyperoxia- exposed newborn rats. Am. J. Physiol. Lung Cell Mol. Physiol. **281:** L336–344.
44. AUTEN, R.L. *et al.* 2002. Blocking neutrophil influx reduces DNA damage in hyperoxia-exposed newborn rat lung. Am. J. Respir. Cell Mol. Biol. **26:** 391–397.
45. MATUTE-BELLO, G. *et al.* 1997. Neutrophil apoptosis in the acute respiratory distress syndrome. Am. J. Respir. Crit. Care Med. **156:** 1969–1977.
46. TURRENS, J.F. *et al.* 1984. Protection against oxygen toxicity by intravenous injection of liposome- entrapped catalase and superoxide dismutase. J. Clin. Invest. **73:** 87–95.
47. WHITE, C.W. *et al.* 1991. Transgenic mice with expression of elevated levels of copper-zinc superoxide dismutase in the lungs are resistant to pulmonary oxygen toxicity. J. Clin. Invest. **87:** 2162–2168.
48. WISPE, J.R. *et al.* 1992. Human Mn-superoxide dismutase in pulmonary epithelial cells of transgenic mice confers protection from oxygen injury. J. Biol. Chem. **267:** 23937–23941.
49. DANEL, C. *et al.* 1998. Gene therapy for oxidant injury-related diseases: adenovirus-mediated transfer of superoxide dismutase and catalase cDNAs protects against hyperoxia but not against ischemia-reperfusion lung injury. Hum. Gene Ther. **9:** 1487–1496.

50. GREEN, D.R. 2000. Apoptotic pathways: paper wraps stone blunts scissors. Cell **102:** 1–4.
51. BARAZZONE ARGIROFFO, C. *et al.* 2002. CD40-CD40 ligand disruption does not prevent hyperoxia-induced injury. Am. J. Pathol. **160:** 67–71.
52. MIYASHITA, T. & J.C. REED. 1995. Tumor suppressor p53 is a direct transcriptional activator of the human bax gene. Cell **80:** 293–299.
53. O'REILLY, M.A. *et al.* 1998. Exposure to hyperoxia induces p53 expression in mouse lung epithelium. Am. J. Respir. Cell Mol. Biol. **18:** 43–50.
54. O'REILLY, M.A. *et al.* 1998. Accumulation of p21(Cip1/WAF1) during hyperoxic lung injury in mice. Am. J. Respir. Cell Mol. Biol. **19:** 777–785.
55. O'REILLY, M.A. *et al.* 2001. The cyclin-dependent kinase inhibitor p21 protects the lung from oxidative stress. Am J Respir Cell Mol Biol. **24:** 703–710.
56. FERRI, K.F. & G. KROEMER. 2001. Organelle-specific initiation of cell death pathways. Nat Cell Biol. **3:** E255–263.
57. MARTINOU, J.C. & D.R. GREEN. 2001. Breaking the mitochondrial barrier. Nat. Rev. Mol. Cell Biol. **2:** 63–67.
58. ANTONSSON, B. *et al.* 2000. Bax oligomerization is required for channel-forming activity in liposomes and to trigger cytochrome c release from mitochondria. Biochem. J. **345 (Pt. 2):** 271–278.
59. LEMASTERS, J.J. *et al.* 1998. The mitochondrial permeability transition in cell death: a common mechanism in necrosis, apoptosis and autophagy. Biochim. Biophys. Acta **1366:** 177–196.
60. ZAMZAMI, N. & G. KROEMER. 2001. The mitochondrion in apoptosis: how Pandora's box opens. Nat Rev Mol Cell Biol. **2:** 67–71.
61. HIRSCH, T. *et al.* 1997. Role of the mitochondrial permeability transition pore in apoptosis. Biosci. Rep. **17:** 67–76.
62. PETIT, P.X. *et al.* 1998. Disruption of the outer mitochondrial membrane as a result of large amplitude swelling: the impact of irreversible permeability transition. FEBS Lett. **426:** 111–116.
63. GROSS, A. *et al.* 1999. BCL-2 family members and the mitochondria in apoptosis. Genes& Dev. **13:** 1899–1911.
64. MASSARO, G. & D. MASSARO. 1973. Pulmonary granular pneumocytes: Loss of mitochondrial granules during hyperoxia. J. Cell Biol. **59:** 246–250.
65. AHMAD, S. *et al.* 2001. Glutamine protects mitochondrial structure and function in oxygen toxicity. Am. J. Physiol. Lung Cell. Mol. Physiol. **280:** L779–791.
66. WARD, N.S. *et al.* 2000. Interleukin-6-induced protection in hyperoxic acute lung injury. Am. J. Respir. Cell Mol. Biol. **22:** 535–542.
67. MARTINOU, J.C. *et al.* 1994. Overexpression of BCL-2 in transgenic mice protects neurons from naturally occurring cell death and experimental ischemia. Neuron **13:** 1017–1030.
68. BILBAO, G. *et al.* 1999. Reduction of ischemia-reperfusion injury of the liver by in vivo adenovirus-mediated gene transfer of the antiapoptotic Bcl-2 gene. Ann. Surg. **230:** 185–193.
69. DATTA, S.R. *et al.* 1997. Akt phosphorylation of BAD couples survival signals to the cell-intrinsic death machinery. Cell **91:** 231–241.
70. LU, Y. *et al.* 2001. Activated Akt protects the lung from oxidant-induced injury and delays death of mice. J. Exp. Med. **193:** 545–549.
71. AHMAD, A. *et al.* 2002. Elevated expression of hexokinase II protects human lung epithelial- like A549 cells against oxidative injury. Am. J. Physiol. Lung Cell Mol. Physiol. **283:** L573–584.
72. PETRONILLI, V. *et al.* 1994. Regulation of the permeability transition pore, a voltage-dependent mitochondrial channel inhibited by cyclosporin A. Biochim. Biophys. Acta **1187:** 255–259.
73. ZAMZAMI, N. *et al.* 1996. Inhibitors of permeability transition interfere with the disruption of the mitochondrial transmembrane potential during apoptosis. FEBS Lett. **384:** 53–57.
74. KANTROW, S.P. & C.A. Piantadosi. 1997. Release of cytochrome c from liver mitochondria during permeability transition. Biochem. Biophys. Res. Commun. **232:** 669–671.

75. FRIBERG, H. *et al.* 1998. Cyclosporin A, but not FK 506, protects mitochondria and neurons against hypoglycemic damage and implicates the mitochondrial permeability transition in cell death. J. Neurosci. **18:** 5151–5159.
76. MATTHEW, E. *et al.* 1999. Cyclosporin A protects lung function from hyperoxic damage. Am. J. Physiol. **276:** L786–795.
77. HENGARTNER, M.O. 2000. The biochemistry of apoptosis. Nature **407:** 770–776.
78. MANTELL, L.L. *et al.* 1999. Hyperoxia-induced cell death in the lung--the correlation of apoptosis, necrosis, and inflammation. Ann. N. Y. Acad. Sci. **887:** 171–180.
79. HIRSCH, T. *et al.* 1997. The apoptosis-necrosis paradox. Apoptogenic proteases activated after mitochondrial permeability transition determine the mode of cell death. Oncogene **15:** 1573–1581.
80. LEIST, M. *et al.* 1997. Intracellular adenosine triphosphate (ATP) concentration: a switch in the decision between apoptosis and necrosis. J. Exp. Med. **185:** 1481–1486.
81. HAMPTON, M.B. *et al.* 1998. Redox regulation of the caspases during apoptosis. Ann. N.Y. Acad. Sci. **854:** 328–335.
82. SUSIN, S.A. *et al.* 2000. Two distinct pathways leading to nuclear apoptosis. J. Exp. Med. **192:** 571–580.
83. ZHANG, X. *et al.* 2002. Intranuclear localization of apoptosis-inducing factor (AIF) and large scale DNA fragmentation after traumatic brain injury in rats and in neuronal cultures exposed to peroxynitrite. J. Neurochem. **82:** 181–191.
84. CHOI, A.M. *et al.* 1995. Molecular responses to hyperoxia in vivo: relationship to increased tolerance in aged rats. Am. J. Respir. Cell Mol. Biol. **13:** 74–82.
85. BALDWIN, A.S., Jr. 1996. The NF-kappa B and I kappa B proteins: new discoveries and insights. Annu. Rev. Immunol. **14:** 649–683.
86. LI, Y. *et al.* 1997. Nuclear factor-kappaB is activated by hyperoxia but does not protect from cell death. J. Biol. Chem. **272:** 20646–20649.
87. SHEA, L.M. *et al.* 1996. Hyperoxia activates NF-kappaB and increases TNF-alpha and IFN-gamma gene expression in mouse pulmonary lymphocytes. J. Immunol. **157:** 3902–3908.
88. NEWTON, R. 2000. Molecular mechanisms of glucocorticoid action: what is important. Thorax. **55:** 603–613.
89. BARAZZONE-ARGIROFFO, C. *et al.* 2001. Hyperoxia increases leptin production: a mechanism mediated through endogenous elevation of corticosterone. Am. J. Physiol. Lung Cell. Mol. Physiol. **281:** L1150–1156.
90. DORSCHEID, D.R. *et al.* 2001. Apoptosis of airway epithelial cells induced by corticosteroids. Am. J. Respir. Crit. Care Med. **164:** 1939–1947.
91. BARAZZONE-ARGIROFFO, C. *et al.* 2003. Glucocorticoids aggravate hyperoxia-induced lung injury through decreased nuclear factor-kappa B activity. Am. J. Physiol. Lung Cell Mol. Physiol. **284:** L197–204.
92. FRANEK, W.R. *et al.* 2001. Hyperoxia inhibits oxidant-induced apoptosis in lung epithelial cells. J. Biol. Chem. **276:** 569–575.
93. LEE, P.J. *et al.* 1996. Regulation of heme oxygenase-1 expression in vivo and in vitro in hyperoxic lung injury. Am. J. Respir. Cell Mol. Biol. **14:** 556–568.
94. OTTERBEIN, L.E. *et al.* 1999. Carbon monoxide provides protection against hyperoxic lung injury. Am. J. Physiol. **276:** L688–694.
95. PAGANO, A. *et al.* 2003. Mitochondrial cytochrome c release is a key event in hyperoxia-induced lung injury: protection by Cyclosporin A. Am. J. Physiol. Lung Cell Mol. Physiol. Oct 3 [epub ahead of print].

Hypoxic Regulation of Neutrophil Apoptosis

Role of Reactive Oxygen Intermediates in Constitutive and Tumor Necrosis Factor α-Induced Cell Death

J. MURRAY, S.R. WALMSLEY, K.I. MECKLENBURGH, A.S. COWBURN, J.F. WHITE, A.G. ROSSI, AND E.R. CHILVERS

Respiratory Medicine Division, Department of Medicine, University of Cambridge School of Clinical Medicine, Addenbrooke's and Papworth Hospitals, Cambridge CB2 2QQ, UK

Rayne Laboratory, Respiratory Medicine Unit, MRC Center for Inflammation Research, University of Edinburgh Medical School, Teviot Place, Edinburgh EH8 9AG, UK

ABSTRACT: Activation of the NADPH oxidase system to generate reactive oxygen species (ROS) plays a key role in bacterial killing by human neutrophils. However, the involvement of such radicals in spontaneous and TNFα-driven neutrophil apoptosis remains uncertain. While incubation of cells under anoxic conditions attenuated the pro-apoptotic effect of TNFα, full activation of the respiratory burst using PAF followed by fMLP, or the addition of physiologically relevant concentrations of H_2O_2, had no effect on the rate of apoptosis. Furthermore, the phosphoinositide 3-kinase inhibitor, LY294002, which abolishes receptor-mediated activation of the NADPH oxidase, and five discrete anti-oxidants all failed to affect apoptotic thresholds. Thus ROS do not appear to modulate constitutive apoptosis in neutrophils or appear sufficient to mediate the pro-apoptotic effect of TNFα.

KEYWORDS: apoptosis; neutrophils; NADPH oxidase; TNFα

INTRODUCTION

Human neutrophils undergo constitutive apoptosis when aged *in vitro* and this results in hypo-responsiveness to external signals and the expression of cell surface motifs that initiate recognition by inflammatory macrophages.[1] The speed and capacity of the macrophage phagocytic response towards apoptotic neutrophils, together with the observation that engulfment excites an anti-inflammatory rather than pro-inflammatory macrophage response, predicts that this process plays a pivotal role in the safe disposal of intact but effete neutrophils from an inflamed focus. This view is supported by *in vivo* data in animal models of acute lung injury and glomerulonephritis and in the neonatal respiratory distress syndrome.[2]

Address for correspondence: Professor Edwin Chilvers, Department of Medicine, Box 157 Addenbrooke's Hospital, University of Cambridge School of Clinical Medicine, Hills Road, Cambridge CB2 2QQ, UK. Voice and fax: (44) 1223 762007.
erc24@cam.ac.uk

**Ann. N.Y. Acad. Sci. 1010: 417–425 (2003). © 2003 New York Academy of Sciences.
doi: 10.1196/annals.1299.075**

While the mechanism(s) regulating neutrophil survival and death are poorly understood, there is now considerable evidence to show that this process can be altered by physiological stimuli. For example the pro-inflammatory mediators G-CSF, GM-CSF, lipopolysaccharide, C5a, IL-1β, IFNγ, and LTB_4 all increase the survival of these cells by delaying apoptosis.[3] Moreover, compared to other myeloid cells, neutrophils are relatively resistant to agonist-stimulated apoptosis. We have previously reported that TNFα has the unique capacity to drive neutrophil apoptosis in a proportion of cells at early times (<12 h) while decreasing the overall extent of apoptosis at later times. Of note, TNFα has also been shown to "prime" neutrophils for apoptotic killing by agents such as C5a, IgG, fMLP and *E. coli*[4] and to enhance the phagocytosis of apoptotic neutrophils by monocyte-derived macrophages. These data suggest a potentially synergistic or dual mechanism whereby TNFα may drive the removal of neutrophils from an inflamed site both by accelerating receptor-mediated apoptosis and by facilitating apoptotic cell clearance.

TNFα causes a rapid rise in the levels of intracellular ROS[5] and this response has been proposed to underly the cytotoxic effects of this cytokine. Consistent with this, the anti-oxidants, thioredoxin and *N*-acetylcysteine, have been reported to inhibit TNFα-mediated apoptosis. TNFα cytotoxicity can also be affected by certain mitochondrial inhibitors,[6] and alterations in mitochondrial ultrastructure after TNFα treatment are well documented. Furthermore, cellular sensitivity or resistance to TNFα have been correlated with decreased or increased levels of mitochondrial superoxide dismutase, respectively.[7]

The aim of this study was to determine the role of ROS in constitutive and TNFα-stimulated apoptosis in human neutrophils. Neutrophils were resistant to killing by H_2O_2 or maximal activation of the NADPH oxidase, and the mitochondrial complex III inhibitor actinomycin that increases ROS release inhibited rather than increased TNFα-induced apoptosis. While TNFα-mediated killing was inhibited under anoxic conditions, the above data suggest that neutrophil apoptosis is not influenced by ROS when generated under physiological conditions.

MATERIALS AND METHODS

Neutrophil Preparation and Culture

Human neutrophils were purified from the peripheral blood of healthy human volunteers as previously detailed.[8] Cell purity was >95% neutrophils with <0.1% mononuclear cell contamination and >99% cell viability. Freshly harvested neutrophils were suspended at 5×10^6/mL in Iscoves MDM supplemented with 10% autologous serum and 50 U/mL penicillin and 50 U/mL streptomycin and cultured at 37°C in flat-bottomed 96-well Falcon flexiwell plates (Becton-Dickinson, U.K.) in the presence or absence of test reagents (TNFα, 12.5 ng/mL; PAF, 1 μM; fMLP, 100 nM; LY 294002, 10 μM; Trolox, 10 mM; SOD, 200 μg/mL; CAT, 250 μg/mL; glutathione, 5 mM; NAC, 10 mM; rotenone, 0.1 μg/mL; actinomycin A, 50 μM). Normoxic environments (PO_2 of media 19 kPa; radiometer, Copenhagen) were controlled using a humidified 5% CO_2/air incubator, and anaerobic conditions (PO_2 0 kPa) maintained with a MACS 500 Don Whitley H2 catalyst-dependent anaerobic incubator with a 5% CO_2/10% H_2/balanced N_2 gas mix.

Morphological Assessment of PMN Apoptosis

Cell morphology was examined under oil immersion light microscopy (×100 objective) and apoptotic neutrophils were defined as cells containing darkly stained condensed pyknotic nuclei.[1] A total of 300 neutrophils were counted over at least five high-power fields with the observer blinded to the assay conditions. Our previous studies have shown that morphologic assessment of apoptosis correlates closely to other apoptotic markers including annexin V binding, propidium iodide staining, CD16 shedding, and DNA cleavage, and remains the most reliable index of apoptosis in neutrophils.[9,10]

Assessment of Superoxide Anion Generation

Neutrophils were suspended at 10^6 cells in 90 μL PBS with $CaCl_2$ and $MgCl_2$ and equilibrated at 37°C in a thermomixer for 5 min. Cells were primed with platelet-

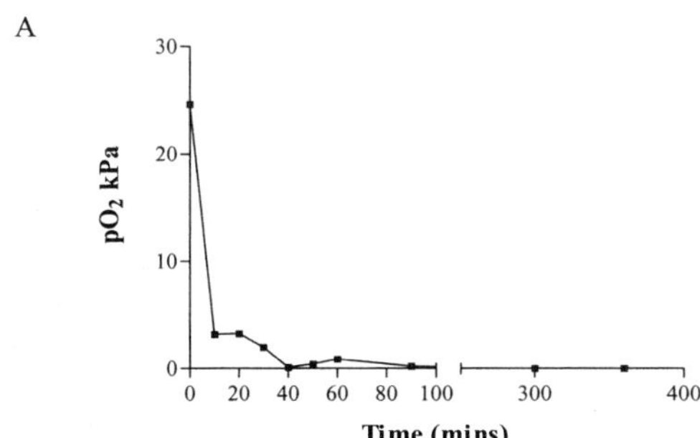

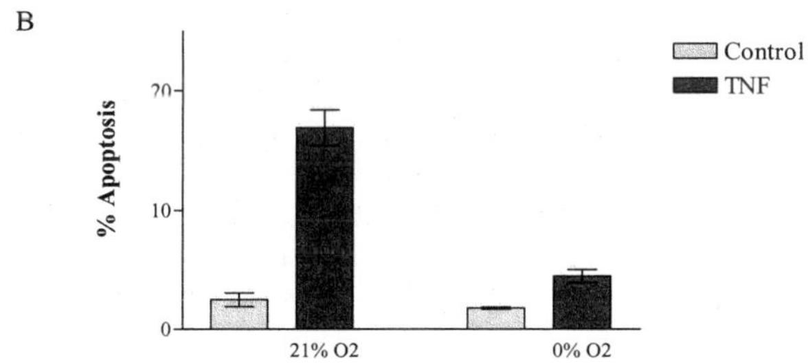

FIGURE 1. TNFa-induced apoptosis is inhibited under anoxic conditions. (**A**) Medium was pre-equilibrated to 0% by incubating for 40 min in a MACS 500 Don Whitley H_2 catalyst-dependent anaerobic incubator using 5% CO_2/10% H_2/balanced N_2 gas mix. (**B**) Neutrophils were subsequently incubated in 21% oxygen or in 0% oxygen in the pre-equilibrated medium at 5×10^6 cells/mL. TNFα (12.5 ng/mL), when present, was added at the start of each incubation, and cells harvested at 6 h. Apoptosis was assessed morphologically from cytocentrifuge preparations. (**B**) Data represent mean±SEM of n = 3 experiments each performed in triplicate.

activating factor (PAF, 1 μM) or PBS for 5 min and then stimulated with fMLP (100 nM) or PBS in the presence of pre-warmed cytochrome *c* (1.2 mg/ml in PBS) or lucigenin (0.25 mM in BSA 1 mg/mL PBS without $MgCl_2$ or $CaCl_2$). The superoxide dismutase-inhibitable reduction of cytochrome *c* was determined as previously described and expressed as nmoles O_2^- generated per 10^6 neutrophils. Lucigenin-dependant chemiluminescence (LDCL) was recorded continuously at 9-s intervals for 7 min; data were recorded on-line (Cellular Chemiluminescence, Dynatech Laboratories Ltd.) to produce mean LDCL values from triplicate wells.

Statistical Analysis

All values are presented as the mean ± SEM of (*n*) number of independent experiments. The data were evaluated using the paired or unpaired Students *t*-test: *P* values <0.05 were considered to be statistically significant.

RESULTS

TNFα-Induced Apoptosis is Inhibited under Anoxic Conditions

To address whether the pro-apoptotic effect of TNFα in neutrophils requires oxidative metabolism, the effect of absolute hypoxia was examined. Neutrophils were cultured in the presence or absence of TNFα under normoxic or anoxic conditions using medium that had been fully deoxygenated prior to the addition of TNFα (FIG. 1A). The pro-apoptotic effect of TNFα was greatly reduced under such conditions (FIG. 1B), indicating an essential role for molecular oxygen in the cytocidal effect of TNFα. In subsequent experiments, neutrophils were suspended in normoxic rather than deoxygenated medium prior to anoxic incubation. In these experiments the extent of TNFα-mediated apoptosis was not significantly different from that observed under normoxic conditions, indicating that the commitment of these cells to undergo apoptosis in response to TNFα is extremely rapid, occurring within 30 min of exposure to this cytokine (data not shown).

Role of the NADPH Oxidase in TNFα-Stimulated Neutrophil Apoptosis

Oxygen radicals have been proposed as playing an essential role in TNFα and phagocytosis-induced apoptotic cell death in neutrophils.[11] However it has yet to be determined whether activation of the NADPH oxidase under physiological conditions results in apoptosis. To address this we activated neutrophils by first incubating with a priming concentration of PAF (1 μM, 5 min) followed by the addition of a maximally effective concentration of the bacterial tripeptide fMLP (100 nM, 10 min). As shown in FIGURE 2A and B, this protocol resulted in a robust stimulation of superoxide anion release yet had no effect on the subsequent rate of constitutive apoptosis measured either at 6 or 20 hours (FIG. 2C).

Activation of the phosphoinositide 3-kinase (PI3-kinase)/Akt pathway is an important mechanism underlying growth factor–mediated survival in mesenchymal cells and cytokine-stimulated neutrophils. Gardai and colleagues[12] proposed that NADPH oxidase-derived oxidants, acting through Lyn and the SH2-containing inositol-5-phosphatase (SHIP), serve to inhibit Akt and thereby trigger neutrophil apop-

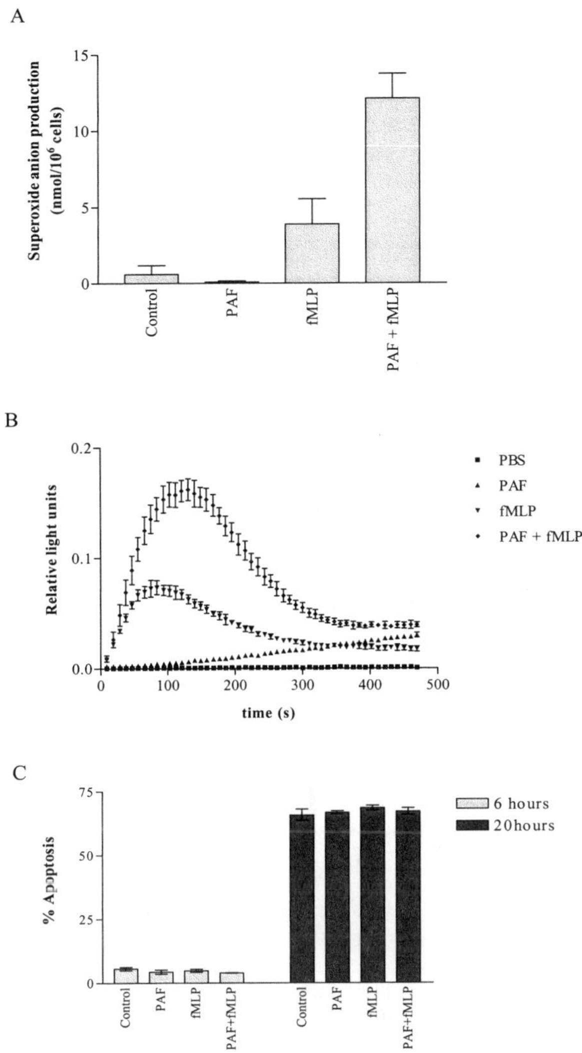

FIGURE 2. Role of NADPH oxidase in TNFα-stimulated neutrophil apoptosis. (**A**) Neutrophils were incubated at 5×10^6 cells/ml with test reagents PAF (1 μM) and/or fMLP (100 nM). Superoxide dismutase–inhibitable reduction of cytochrome *c* was measured at 35–565 nm and expressed as nmoles O_2^- generated per 10^6 neutrophils. (**B**) Kinetics of ROS release was measured by lucigenin-dependent chemiluminescence with mean data produced from triplicate wells. (**C**) At the time points indicated, cells were recovered and apoptosis assessed by morphologic examination of cytocentrifuge preparations. Data in **A** and **C** represent mean±SEM of three separate experiments, each performed in triplicate. Data in **B** are from triplicate determinations from a single experiment and representative of two further experiments.

tosis. Despite this, LY29008, a PI3-kinase inhibitor that we have shown abolishes agonist-stimulated NADPH oxidase and the survival effect of GM-CSF, had no effect on the extent of apoptosis under normoxic conditions (% apoptosis 20 h: normoxia 63±3, normoxia + LY29008 62±2, n =3). These data concur with our previous observation that the PI3-kinase inhibitor wortmannin fails to block TNFα-induced neutrophil apoptosis[13] and would support our hypothesis that activation of the NADPH oxidase does not mediate the pro-apoptotic effect of TNFα.

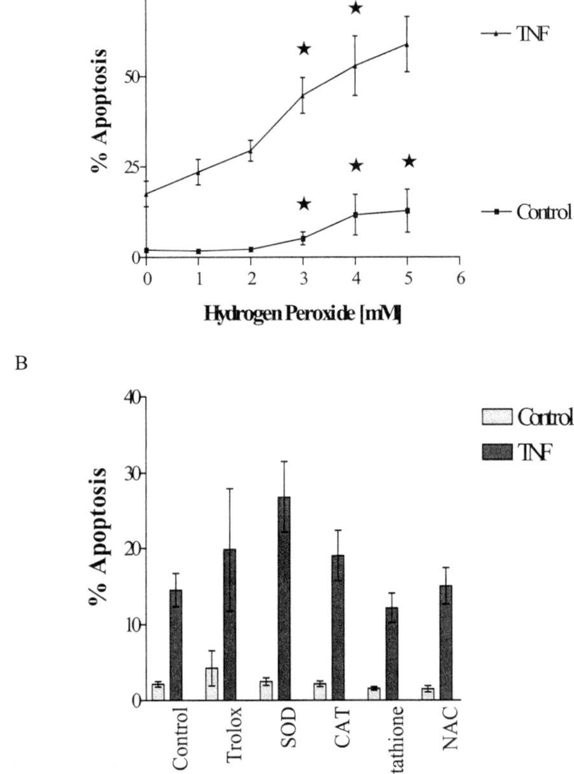

FIGURE 3. Role of reactive oxygen species and anti-oxidants on TNFα-stimulated neutrophil apoptosis. (**A**) The effect of exogenous H_2O_2 on neutrophil apoptosis was assessed by incubating cells at 5×10^6/mL with a range of H_2O_2 concentrations from 1–5 mM. Cells were harvested at 6 h and apoptosis scored morphologically. (**B**) Neutrophils were incubated at 5×10^6 cells/ml with or without TNFα (12.5 ng/mL) in the presence of the anti-oxidants trolox (10 mM), SOD (200 μg/mL), CAT (250 μg/mL), glutathione (5 mM) or NAC (10 mM). Cells were harvested at 6 h and apoptosis assessed morphologically. Data represent mean ± SEM of n = 3 experiments each performed in triplicate with significant differences to control values ($P < 0.05$) marked *.

Role of Reactive Oxygen Species in TNFα-Stimulated Neutrophil Apoptosis

To explore further the capacity of ROS to influence apoptotic thresholds in neutrophils these cells were incubated for 6 hours in the presence or absence of TNFα with increasing concentrations of H_2O_2. Detailed concentration-response studies demonstrated that H_2O_2 (up to 1 mM) had no effect on the rate of constitutive or TNFα-stimulated apoptosis (data not shown). While the use of much higher (supraphysiological) concentrations of H_2O_2 increased the extent of basal and TNFα-stimulated apoptosis at 6 h (FIG. 3A), the overall extent of apoptosis remained less than 20%, even in the presence of 5 mM H_2O_2. These data again suggest that human neutrophils are highly resistant to oxidant-induced apoptosis.

To assess whether the inhibition of TNFα-induced neutrophil apoptosis observed under hypoxic conditions reflected changes in the intracellular generation of ROS the effects of a panel of structurally discrete antioxidants were examined. In agreement with the above data, trolox (10 μM), superoxide dismutase (SOD, 200 μg/mL), catalase (CAT, 250 μg/mL), reduced glutathione (5 mM) and *N*-acetyl-L-cysteine (NAC, 10 mM) were all unable to inhibit the pro-apoptotic effect of TNFα (FIG. 3B).

The mitochondria are an important source of ROS formation through auto-oxidation of Complex I flavin mononucleotide and Complex III ubisemiquinone, and have been previously shown to modulate TNFα-induced cytotoxicity in L929 and WEH1 cells. To investigate the putative role of these organelles in TNFα-mediated neutrophil apoptosis we preincubated cells with the mitochondrial inhibitors rotenone (0.1 μg/mL), a Complex I inhibitor, or actinomycin A (50 μM), a Complex III inhibitor, prior to cytokine stimulation. While neither of these agents influenced the rate of constitutive apoptosis at 6 h (data not shown), actinomycin A caused a significant inhibition TNFα-stimulated apoptosis (% apoptosis 6 h: TNFα 17 ± 3, TNFα + actinomycin A 8 ± 1, $n = 3$, $P < 0.05$).

DISCUSSION

We have previously reported that TNFα has a bimodal effect on the survival of human neutrophils *in vitro* with the capacity to induce early apoptosis but decrease the extent of cell death at later times. This early killing effect of TNFα requires activation of the high-affinity TNFR75 in addition to TNFR55.[14] Despite many attempts to elucidate the mechanisms of TNFα–mediated cytotoxicity in hematopoietic and other cell types, there is currently no unifying theory to account for the pro-apoptotic effect of TNFα in neutrophils. In this study we have sought to identify the intracellular mechanisms underlying the induction of neutrophil apoptosis by TNFα and, in particular, assess the involvement of ROS in both constitutive and TNFα-induced apoptosis.

Gardai and co-workers[12] demonstrated that TNFα in combination with either the β2-integrin-activating antibody VIM12, GM-CSF, or LPS, promoted apoptosis via the phosphorylation and activation of the tyrosine kinase Lyn with subsequent recruitment and activation of SHIP and downregulation of Akt. This work also proposed a role for oxidants in the apoptotic response since the initial activation of Lyn was dependent on the NADPH oxidase. Oxidant-dependent induction of neutrophil apoptosis has also been proposed in models of neutrophil phagocytosis of *E. coli* or opsonized particles.[15] By contrast, in our experimental system we have been unable to modulate neutrophil apoptosis with maximal physiological stimulation of the NADPH oxidase or by the addition of exogenous H_2O_2 (up to a concentration of 1 mM).

Studies in L929 cells have identified the mitochondria as the first site of damage in TNFα-mediated cytolysis,[16] and mitochondrial inhibitors have been shown to influence TNFα-induced cytotoxicity in several cell types. It has long been recognized that the mitochondria are an important source of ROS under certain conditions and according to Boveris and Chance[17] (1973), this pathway accounts for about 1–2% of the oxygen consumed by cells during resting respiration. Two sites within the mitochondrial respiratory chain have been identified as responsible for ROS formation; the first is dependent on the auto-oxidation of the flavin mononucleotide from the NADH-dehydrogenase (Complex I), whereas the second and probably more relevant site depends on the auto-oxidation of the unstable ubisemiquinone (Complex III), which is an intermediate of the Q-cycle reaction. It is important to note that while the production of oxygen radicals at the ubiquinone site is diminished by complex I and II inhibitors the formation of ROS actually increases several-fold in the presence of actinomycin A, implicating ubisemiquinone as the main reductant site of oxygen.[18] Indeed this ability of actinomycin A to enhance mitochondrial ROS generation appears to underly its capacity to augment TNFα-stimulated apoptosis in both L929 and WEHI mouse fibrosarcoma cells.[6] Hence our finding that this compound inhibits rather than increases the pro-apoptotic effect of TNFα further questions the role of ROS in TNFα-induced killing in neutrophils. This conclusion is supported by the lack of effect of the antioxidants trolox, SOD, catalase, reduced glutathione, and NAC on the TNFα response.

While ROS generation either via activation of the NADPH oxidase or the mitochondrial respiratory burst does not appear to be involved in TNFα-killing, our data indicate that the ability of TNFα to induce neutrophil apoptosis still requires the presence of molecular oxygen. Two models exist regarding the effect of oxygen deprivation on the mitochondrial respiratory chain. Firstly, a reduction in O_2 can be associated with a decline in ROS.[19] Secondly, under hypoxic conditions the reduction of O_2 to H_2O by cytochrome *c* oxidase (Complex IV) is inhibited, resulting in release of O_2^- and electrons at Complex III.[20] In neutrophils, while it is evident that the absence of environmental oxygen, or the presence of mitochondrial complex III inhibitors, blocks TNFα killing, the lack of effect of exogenous H_2O_2, antioxidants, and complex I inhibitors indicates a regulatory step independent of ROS.

Together, our data indicate that the generation of ROS under physiological conditions does not affect the rate of constitutive apoptosis in neutrophils or underlie the pro-apoptotic effect of TNFα. This conclusion would be entirely consistent with the fact that these cells are required to produce large quantities of such intermediates to effect efficient bacterial killing.

ACKNOWLEDGMENTS

This work was supported by the Wellcome Trust (035662), the MRC (G9016491), and the National Asthma Campaign (01/042). S.R.W. and J.A.W. hold MRC Clinical Research Training Fellowships.

REFERENCES

1. SAVILL, J., J.E. HENSON, A.H. WYLLIE, *et al.* 1989. Macrophage phagocytosis of aging neutrophils in inflammation. Programmed cell death in the neutrophil leads to recognition by macrophages. J. Clin. Invest. **83:** 865–875.
2. GRIGG, J.M., J. SAVILL, C. SARRAF, *et al.* 1991. Neutrophil apoptosis and clearance by macrophages in the lungs of neonates with pulmonary inflammation Lancet **338:** 720–722.
3. LEE, A., M.K. WHYTE & C. HASLETT. 1993. Inhibition of apoptosis and prolongation of neutrophil functional longevity by inflammatory mediators. J. Leukoc. Biol. **54:** 283–288.
4. SALAMONE, G., M. GIORDANO, A.S. TREVANI, *et al.* 2001. Promotion of neutrophil apoptosis by TNF-alpha. J. Immunol. **166:** 3476–3483.
5. MATTHEWS, N., M.L. NEALE, S.K. JACKSON & J. STARK. 1987. Tumour cell killing by tumour necrosis factor: inhibition by anaerobic conditions, free-radical scavengers and inhibitors of arachidonate metabolism. Immunology **62:** 153–155.
6. SCHULZE-OSTHOFF, K., A.C. BAKKER, B. VAN HAESEBROECK, *et al.* 1992. Cytotoxic activity of tumour necrosis factor is mediated by early damage of mitochondrial function: evidence for the involvement of mitochondrial radical generation. J. Biol. Chem. **267:** 5317–5323.
7. HIROSE, K.D., D.L. LONGO, J.J. OPPENHEIM & K. MATSUSHIMA. 1993. Over expression of mitochondrial manganese superoxide dismutase promotes the survival of tumour cells exposed to interleukin-1, tumour necrosis factor, selected anticancer drugs, and ionizing radiation. FASEB J. **7:** 361–368.
8. HASLETT, C., L.A. GUTHRIE, M.M. KOPAMIAK, *et al.* 1985. Modulation of multiple neutrophil functions by preparative methods or trace concentrations of bacterial lipopolysaccaride. Am. J. Path. **119:** 101–110.
9. HOWBURN, A., K.A. CADWALLADER, B.J. REED, *et al.* 2002. Role of PI3-kinase-dependent Bad phosphorylation and altered transcription in cytokine-mediated neutrophil survival. Blood **100:** 2607–2616.
10. DRANSFIELD, I., A.M. BUCKLE, J. SAVILL, *et al.* 1994. Neutrophil apoptosis is associated with a reduction in CD16 (Fc gamma RIII) expression. J. Immunol. **153:** 1254–1263.
11. COXON, A., P. RIEU, F.J. BARKLOW, *et al.* 1996. A novel role for the β2 integrin CD11b/CD18 in neutrophil apoptosis: a homeostatic mechanism in inflammation. Immunity **5:** 653–666.
12. GARDAI, S., B.B. WHITLOCK, C. HELGASON, *et al.* 2002. Activation of SHIP by NADPH oxidase-stimulated Lyn leads to enhanced apoptosis in neutrophils. J. Biol. Chem. **277:** 5236–5246.
13. MURRAY, J., A.M. CONDLIFFE, C. HASLETT & E.R. CHILVERS. 1996. Wortmannin enhances tumour necrosis factor alpha-stimulated neutrophil apoptosis. Biochem. Soc. Trans. **24:** 80s.
14. MURRAY, J., J.A. BARBARA, S.A. DUNKLEY, *et al.* 1997. Regulation of neutrophil apoptosis by tumor necrosis factor-alpha: requirement for TNFR55 and TNFR75 for induction of apoptosis in vitro. Blood **90:** 2772–2783.
15. LOCK, R., C. DAHLGREN, M. LINDEN, *et al.* 1990. Neutrophil killing of two type 1 fimbria-bearing *Escherichia coli* strains: dependence on respiratory burst activation. Infect. Immun. **58:** 37–42.
16. MATTHEWS, N. 1983. Anti-tumour cytotoxin produced by human monocytes: studies on its mode of action. Br. J. Cancer **48:** 405–410.
17. BOVERIS, A. & B. CHANCE. 1973. The mitochondrial generation of hydrogen peroxide: general properties and effect of hyperbaric oxygen. Biochem. J. **134:** 707–716.
18. BOVERIS, A. 1976. Mitochondrial production of hydrogen peroxide in *Saccharomyces cerevisiae.* Acta Physiol. Lat. Am. **26:** 303–309.
19. EHLEBEN, W., B. BOLLING, E. MERTEN, *et al.* 1998. Cytochromes and oxygen radicals as putative members of the oxygen sensing pathway. Respir. Physiol. **114:** 25–36.
20. CHANDEL, NS., E. MALTEPE, E. GOLDWASSER, *et al.* 1998. Mitochondrial reactive oxygen species trigger hypoxia-induced transcription. Proc. Natl. Acad. Sci. USA **95:** 11715–11720.

Hypoxic Stress Stably Alters Apoptotic Parameters on U937 Cells

M. DE NICOLA, F. LIUZZI, C. CERELLA, M. D'ALESSIO, A. BERGAMASCHI,[a] A A. MAGRINI,[a] AND L. GHIBELLI

Dipartimento di Biologia, [a]Cattedra di Medicina del Lavoro, Università di Roma Tor Vergata, via della Ricerca Scientifica, 00133 Roma, Italy

ABSTRACT: **Tumor promonocytic U937 cells cultured under a low O_2/high CO_2 atmosphere display altered characteristics after restoration of normal atmosphere: increased resistance to apoptosis induced by different treatments; apoptotic morphology; lack of glutathione (GSH) extrusion in apoptosis; lack of protection by antioxidants; and lack of Ca^{2+} mobilization with thapsigargin. These alterations were stably maintained for many months of culture in normal conditions, originating the stable U937-HX variant. Since the hypoxic treatment did not produce a great selective pressure, the alterations are conceivably the result of stable adaptative response.**

KEYWORDS: **hypoxia; apoptosis; glutathione; U937**

INTRODUCTION

Growing evidence points to the fundamental pathophysiologic role of hypoxia in malignant progression due to its ability to alter cell propensity to undergo apoptosis. Apoptosis is a physiological cellular suicide program by which damaged or unnecessary cells undergo self-destruction. Apoptosis can be induced by physiological and damaging agents that activate two different pathways of apoptotic signaling characterized by typical morphological and biochemical alterations. With the goal of understanding how hypoxia affects apoptosis on U937 monocytic cells, we analyzed a variant of U937 (U937-HX) that results from a non-toxic period of culture in hypoxic conditions. In this study the comparison between the standard U937 cell line and the variant U937-HX is used as a strategic tool to explore which events in the apoptotic signaling are altered by hypoxia.

MATERIALS AND METHODS

Cell culture and treatments: U937 were cultured as described in Reference 1. U937-HX were maintained in a controlled atmosphere for 48 hours without O_2 and

Address for correspondence: Milena De Nicola, Dipartimento di Biologia, Università di Roma Tor Vergata, via della Ricerca Scientifica, 00133 Roma, Italy. Voice: +39-06-72594335; fax:+39 062023500
milden73@libero.it

Ann. N.Y. Acad. Sci. 1010: 426–429 (2003).
doi: 10.1196/annals.1299.076

with 45% CO_2 at 37°C in a complete culture medium. Apoptosis was induced with 10 μg/ml puromycin (PMC), or 100 μg/ml VP16, and was quantified by fluorescence microscopy analysis after nuclear staining with Hoechst.[1] Antioxidant mix (3 mM mannitol and 2 mM α-ketoglutarate) were added 1 h before apoptotgenic treatment and maintained throughout the experiment.

Analysis of intracellular GSH: cells were washed, resuspended in RPMI 1640 medium, stained with 10 nm of chloromethyl fluorescein diacetate (CMFDA) for 15 min at 37°C, and then analyzed using Dako Galaxy flow cytometer.

RESULTS

Hypoxic stress determines resistance to damage-induced apoptosis: First we analyzed the effect of hypoxic stress on damage-induced apoptosis. FIGURE 1 shows that hypoxia reduces the extent of apoptosis induced by PMC. Resistance to apoptosis is substantial also with VP16 (data not shown). Thus the resistance developed by hypoxia is not specific for one inducer. This resistance is stably maintained for a long time, without restoration of normal sensitivity for at least 3 months of culture.

Hypoxic stress alters apoptotic parameters: In order to investigate the mechanism(s) through which hypoxia modulates resistance to apoptosis we analyzed some peculiar morphological and biochemical features of U937 that could be altered in U937-HX.

We observed that in U937-HX the inhibition of the endoplasmatic reticulum Ca^{2+} ATPase Serca 2b with thapsigargin failed to mobilize internal Ca^{2+} fluxes (data not shown).

Another difference concerns apoptotic morphology. We described two different apoptotic morphologies on U937 apoptotic cells, budding and cleavage, which un-

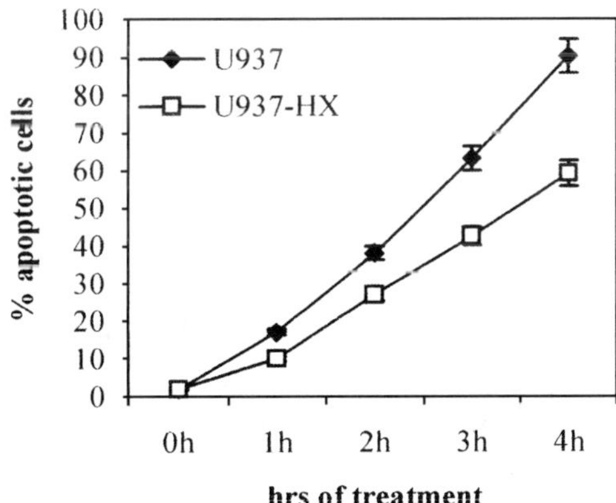

FIGURE 1. Effect of hypoxia on apoptosis extent. Time course of apoptosis on U937 and U937-HX induced with 10 μg/mL PMC. The values are the average of $n > 10$ experiments ± SD. Similar results were obtained with VP16 (not shown).

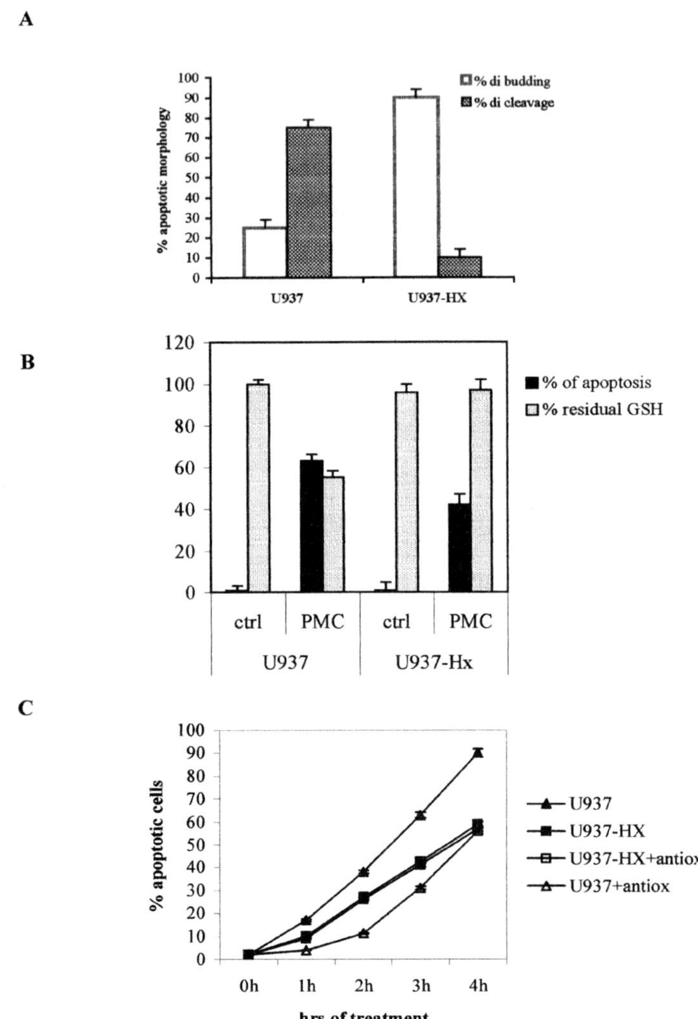

FIGURE 2. Effects of hypoxia on apoptotic parameters. **(A)** *Effect of hypoxia on apoptotic morphology.* U937 and U937-HX cells were induced to apoptosis with 10 μg/mL PMC. Apoptotic morphology was determined at 3 h of treatment. The values are the average of $n = 4$ experiments ± SD. Similar results were obtained with VP16 (not shown). **(B)** *Effect of hypoxia on apoptosis and GSH efflux.* Extent of apoptosis and GSH content were determined at 3 h of treatment with PMC as described in MATERIALS AND METHODS. The values are the average of $n = 3$ experiments ± SD. Similar results were obtained with VP16 (not shown).**(C)** *Effect of hypoxia on protection from apoptosis by antioxidants.* Time course of apoptosis on U937 and U937-HX treated with 10 μg/mL PMC in the presence or absence of antioxidant mix (3 mM mannitol and 2 mM α-ketoglutarate). Antioxidants (antiox) were added 1 h before apoptotgenic treatment and maintained throughout the experiment. The values are the average of $n = 3$ experiments ± SD.

dergo a different type of nuclear vesiculation.[2] FIGURE 2A shows a significant difference of apoptotic morphology between U937 and the HX variant: U937 show mainly cleavage, whereas U937-HX show only the budding morphology. Moreover, standard apoptotic U937 develop cytoplasmatic protrusion (blebs), whereas the variant HX do not.

Redox imbalance play a crucial role in blebbing.[3] The observation that apoptotic U937-HX cells never show blebbing led us to investigate whether hypoxia modulated the fate of GSH, the main intracellular antioxidant. FIGURE 2B compares the effects of hypoxia on apoptotic GSH efflux and on the extent of apoptosis on U937 and U937-HX. We observed that standard U937 (as expected)[1] but not U937-HX, lost GSH when induced to apoptosis.

The oxidative stress consequent to GSH loss is responsible for U937 apoptosis since antioxidants protect U937 from damage-induced apoptosis.[4] Since apoptotic U937-HX did not lose GSH, we expect that antioxidants fail to protect damage-induced apoptosis. This is what indeed occurs (FIG. 2C): the time course of apoptosis induced by PMC in the presence/absence of an antioxidant mix shows that oxidative stress is not involved in the U937-HX apoptotic pathway.

DISCUSSION

The hypoxia-induced increased resistance to apoptosis is in line with much evidence indicating that hypoxic episodes during tumor growth may influence the degree of resistance of tumor cells to antitumor therapy. These results, together with the observation that there is no free-radical generation in U937-HX apoptosis,[5] indicate that hypoxia interferes with damage-induced apoptosis. Besides, the hypoxia adaptive stress response can engender permanent modification since we observe that hypoxia-induced apoptotic alterations are stably maintained for many months. We are now investigating whether the morphological shift and the different GSH fates are linked by a cause–effect relationship.

REFERENCES

1. GHIBELLI, L., S. COPPOLA, G. ROTILIO, *et al.* 1995. Non-oxidative loss of glutathione in apoptosis via GSH extrusion. Biochem. Biophys. Res. Commun. **216:** 313–320.
2. DINI, L., S. COPPOLA, M.T. RUZITTU & L. Ghibelli. 1996. Multiple pathways for apoptotic nuclear fragmentation. Exp. Cell Res. **223:** 340–347.
3. JEWELL, S.A., G. BELLOMO, H. THOR, *et al.* 1982. Bleb formation in hepatocytes during drug metabolism is caused by disturbances in thiol and calcium ion homeostasis. Science **217:** 1257–1259.
4. LIUZZI, F., C. FANELLI, M.R. CIRIOLO, *et al.* 2003. Rescue of cells from apoptosis by antioxidants occurs downstream to GSH extrusion. Ann. N.Y. Acad. Sci. **1010:** this volume
5. D'ALESSIO, M., C. CERELLA, M. DENICOLA, *et al.* 2003. Apoptotic GSH extrusion is associated with free radical generation. Ann. N.Y. Acad. Sci. **1010:** this volume.

Na/Ca Exchanger Overexpression Induces Endoplasmic Reticulum Stress, Caspase-12 Release, and Apoptosis

OSCAR DÍAZ-HORTA,[a,b] FRANÇOISE VAN EYLEN,[a] AND ANDRÉ HERCHUELZ[a]

[a]*Laboratory of Pharmacology, Brussels University School of Medicine, Brussels, Belgium*

[b]*National Institute of Endocrinology, Havana, Cuba*

ABSTRACT: This paper describes a possible strategy to control Ca^{2+} homeostatsis as a possible approach to prevent or enhance β-cell apoptosis

KEYWORDS: apoptosis; Na/Ca exchanger overexpression; endoplasmic reticulum; caspase-12

INTRODUCTION

Ca^{2+} may trigger programmed cell death (apoptosis) and regulate death-specific enzymes.[1] Therefore, the development of strategies to control Ca^{2+} homeostasis may represent a potential approach to prevent or enhance β-cell apoptosis.

RESULTS AND DISCUSSION

To test this hypothesis, the plasma membrane Na/Ca exchanger (NCX1.7 isoform) was stably overexpressed in insulin-secreting tumoral cells. NCX1.7 overexpression increased apoptosis induced by endoplamic reticulum (ER) Ca^{2+}-ATPase inhibitors (thapsigargin and cyclopiazonic acid) but not by agents increasing cytosolic Ca^{2+} concentration ($[Ca^{2+}]_i$) through opening of plasma membrane Ca^{2+}-channels (FIG. 1A). FIGURE 1B illustrates the time course of the effect of thapsigargin and cyclopiazonic acid on cell viability. In overexpressing cells, there was a shift to the left in the time-dependence of the apoptosis induced by both inhibitors, confirming that overexpression sensitizes the cells to apoptotic cell death.

NCX1.7 overexpression reduced the rise in $[Ca^{2+}]_i$ induced by all agents, depleted ER Ca^{2+} stores, sensitized the cells to Ca^{2+}-independent proapoptotic signaling pathways, and reduced cell proliferation by about 40%. Depletion of ER Ca^{2+} stores

Address for correspondence: Dr. André Herchuelz, Université libre de Bruxelle, Faculté de Médicine, Laboratoire de Pharmacologie et de Therapeutique, Route de Lennik, 1070, Bruxelles, Belgium.

herchu@ulb.ac.be

Ann. N.Y. Acad. Sci. 1010: 430–432 (2003). © 2003 New York Academy of Sciences.
doi: 10.1196/annals.1299.077

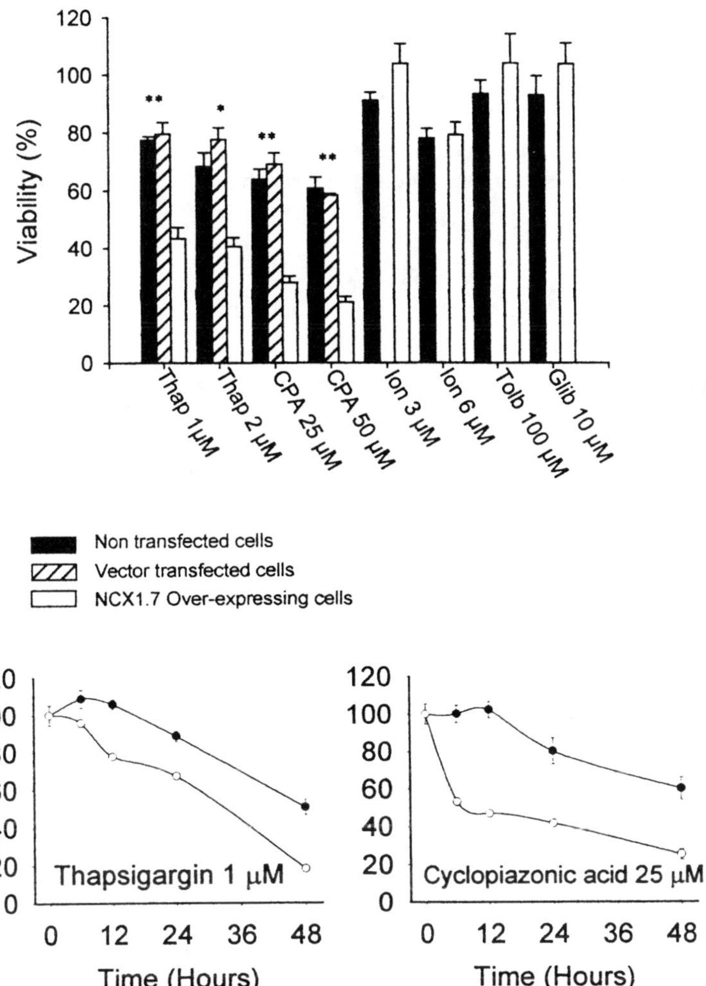

FIGURE 1. (**A**) Effect of NCX1.7 overexpression on cell viability in the presence of thapsigargin (Thap; 1 and 2 μM), cyclopiazonic acid (CPA; 25 and 50 μM), ionomycin (Ion; 3 and 6 μM), tolbutamide (Tolb; 100 μM), and glibenclamide (Glib; 10 μM). The MTT assay was used to measure cell survival. *Solid columns:* non-transfected cells; *hatched columns:* vector-only transfected cells; *open columns:* NCX1.7-transfected cells. Data are given as means ± SEM from at least 4 individual experiments comprising each at least 4 replicates. Statistical significance: *P<0.005, ** P<0.001. (**B**) Time course of the effect of thapsigargin and CPA on cell viability. Data are given as means ± SEM from at least 4 individual experiments each comprising 4 replicates.

was accompanied by the activation of the ER-specific caspase (caspase-12)[2], the activation being enhanced by ER Ca^{2+}-ATPase inhibitors.

CONCLUSIONS

Na/Ca exchanger overexpression, by depleting ER Ca^{2+} stores, induces ER stress with resulting activation of caspase-12, and increase in apoptotic cell death. By increasing apoptosis and decreasing cell proliferation, overexpression of Na/Ca exchanger may represent a new potential approach in cancer gene therapy.

REFERENCES

1. NICOTERA, P. & S. ORRENIUS. 1998. The role of calcium in apoptosis. Cell Calcium **213:** 173–180.
2. NAKAGAWA, T., H. ZHU, N. MORISHIMA, *et al.* 2000. Caspase-12 mediates endoplasmic-reticulum-specific apoptosis and cytotoxicity by amyloid-β. Nature **403:** 98–103.

A Novel Gene, *Jpk*, Induces Apoptosis in F9 Murine Teratocarcinoma Cell through ROS Generation

KYOUNG-AH KONG, SUNGDO PARK, HYOUNGWOO PARK, AND MYOUNG HEE KIM

Department of Anatomy, Embryology Laboratory, Brain Korea 21 Project for Medical Sciences, Yonsei University College of Medicine, Seoul 120-752, Korea

ABSTRACT: A novel gene *Jpk* (*Jopock*) has been originally isolated through yeast 1 hybridization technique as a *trans*-acting factor interacting with the position-specific regulatory element of a murine *Hoxa-7*. Northern analysis revealed that the *Jpk* was expressed at day 7.0 post coitum (p.c.) during early gastrulation. Previously it has been shown that a trace amount of JPK protein led bacterial cells to death. In eukaryotic F9 cells, *Jpk* also led the cell to death-generating DNA ladder: fewer than 50% of the cells survived after 72-h transfection. Flow cytometric analysis with cells stained with each Annexin V/7-amino-actinomycin D (7-AAD), MitoTracker, and hydroethidine (HE) revealed that *Jpk* induced apoptotic cell death in a time-dependent manner, reduced mitochondrial membrane potential, and increased ROS (reactive oxygen species) production, respectively. Additionally, *Jpk* seemed to regulate the Bcl family at the transcriptional level when RT-PCR was performed. Although the precise mechanism is not clear, these results altogether suggest that *Jpk* is a potent inducer of apoptosis through generation of ROS as well as concomitant reduction of mitochondrial membrane potential.

KEYWORDS: Jpk; EGFP; apoptosis; ROS; mitochondrial transmembrane potential

INTRODUCTION

The *Hox* gene plays a crucial role in pattern formation and tissue identity by expression at a specific time and position during early embryogenesis.[1] In the previous study, we have isolated a novel gene, *Jpk*, associating with the upstream regulatory element of a murine *Hoxa 7*.[2] Interestingly enough, however, *Jpk* turned out to be a potent inducer of cell death for both bacterial and mammalian cells.[3,4] Apoptosis occurs during the development of all animals[5] and has been thought to relate to oxygen free radical production and mitochondria.[6] In this study, we have examined the mechanism by which *Jpk* induces apoptosis in F9 cells, trying especially to elucidate

Address for correspondence: Myoung Hee Kim, Department of Anatomy, Embryology Lab., Brain Korea 21 Project for Medical Science, Yonsei University College of Medicine, C.P.O. Box 8044, Seoul, 120-752, Korea. Voice: 82-2-361-5173; fax: 82-2-365-0700.
mhkim1@yumc.yonsei.ac.kr

Ann. N.Y. Acad. Sci. 1010: 433–436 (2003).
doi: 10.1196/annals.1299.078

the relationship between *Jpk*-induced apoptosis and oxygen free radical production. The results show for the first time that *Jpk* induces ROS production which was followed by an apoptotic alteration such as a reduction of mitochondrial membrane potential.

METHODS

F9 cells (murine embryonic teratocarcinoma, ATCC, Rockville, MD) were maintained in Dulbecco's modified Eagle's medium (DMEM) containing 10% fetal bovine serum, 60 μg/mL penicillin, and 100 μg/mL streptomycin. Plasmids pEGFP-*Jpk* and pEGFP-C1, described previously,[4] have been transfected into the F9 cells with Lipofectamine Reagent (Invitrogen). For flow cytometric analysis, cells were stained with annexin V-PE and 7-AAD[7] according to the manufacturer's instructions. The ROS was measured by incubating the cells with 2μM HE (Sigma) for 15 min at 37°C. To evaluate mitochondrial transmembrane potential (ΔΨm), cells were incubated in PBS containing 100 nM MitoTracker (Molecular Probes, Inc.) for 30 min at 37°C, and then analyzed by flow cytometry. All flow cytometric analysis was performed with the signal gated to measure only specific GFP-mediated fluorescence. For RT-PCR analysis, total RNAs were isolated using RNAzolB (LPS industries, Inc.) and subjected to cDNA synthesis, which was followed by the amplification for 35 cycles with specific primers.

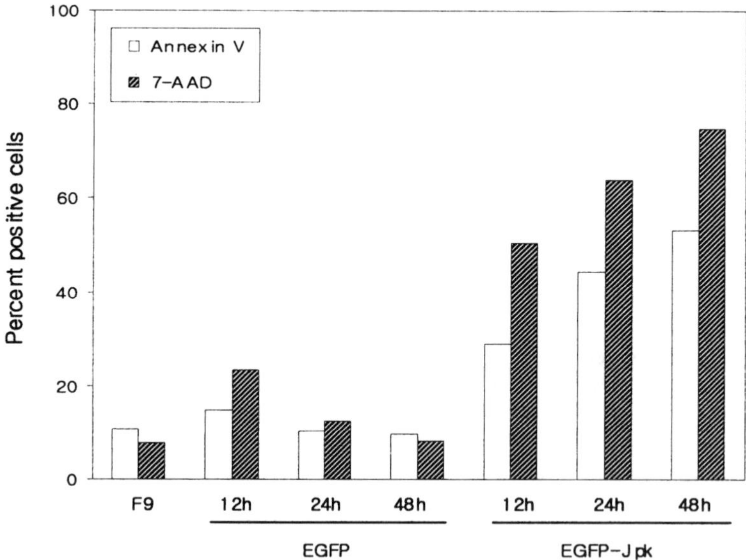

FIGURE 1. Time-dependent apoptotic cell death of F9 cells by *Jpk*. F9 cells were seeded in 6-well culture plates at 2×10^5 cells per well and transfected with either pEGFP-C1 or pEGFP-*Jpk*. The cells were incubated for the indicated time, stained by incubation with Annexin V-PE and 7-AAD, and subjected to flow cytometric analysis, gated only GFP signal.

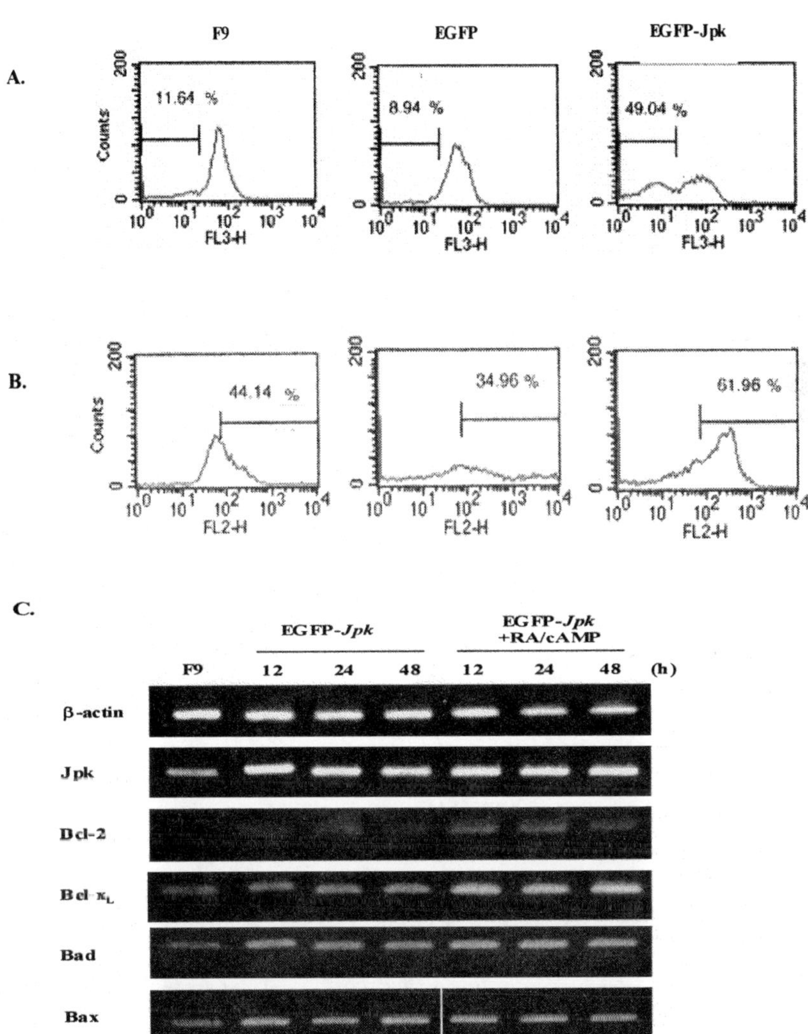

FIGURE 2. Assessment of mitochondrial transmembrane potential (ΔΨm) and ROS generation. F9 cells were transfected with either pEGFP-C1 or pEGFP-*Jpk* in 6-well culture plates at 2×10^5 cells per well and harvested at 24-h incubation. The cells were stained with either MitoTracker (**A**) or HE (**B**) and subjected to flow cytometric anaylsis, gated only GFP signal. For RT-PCR analysis, *Jpk*-transfected cells were treated with RA (100 nM) and cAMP (0.5 mM). After incubation for the indicated time, cells were harvested and RT-PCR was perforned to see the expression of pro-apoptotic (*Bad, Bax*) and anti-apoptotic (*Bcl-2, Bcl-xL*) genes. Transfection of *Jpk* was confirmed by the band intensity of *Jpk* mRNA (C).

RESULTS AND DISCUSSION

In order to examine the apoptotic effect of *Jpk*, F9 cells were incubated first with Annexin V and 7-AAD. Since Annexin V has high affinity for phosphatidylserine (PS) externalized in the earlier stages of apoptosis and 7-AAD is a DNA intercalator, the same as propidium iodide (PI),[7] flow cytometric data of double-stained cells wouldreveal apoptotic cell death. As shown in FIGURE 1, *Jpk* induced apoptotic cell death, which continuously increased up to 48 h. We then measured the mitochondrial membrane potential by staining cells with MitoTracker. The result showed that incorporation of MitoTracker was reduced in *Jpk*-transfected cells, indicating the loss of membrane potential (FIG. 2A). The intracellular ROS production was measured by utilizing HE that is oxidized by ROS to ethidium, which emits red fluorescence. As shown in FIGURE 2B, the population of ethidium-containing cells increased in *Jpk*-transfected cells compared with controls, suggesting that *Jpk* promoted the production of ROS. The anti-apoptotic proteins Bcl-2 and Bcl-x_L can bind to pro-apoptotic Bax and Bad to maintain mitochondrial membrane potential,[6] and thus the ratio of anti-/pro-apoptotic genes is very important in apoptotic signal. FIGURE 2C shows alteration of mRNA expression level of pro-apoptotic (*Bax, Bad*) and anti-apoptotic (*Bcl-2, Bcl-*x_L) genes. *Jpk* increased mRNA levels of both pro- and anti-apoptotic genes. However, the mRNA level of anti-apoptotic genes such as *Bcl-2* and *Bcl-*x_L increased when cells were treated with retinoic acid and cAMP, which leads to cell differentiation and survival. These data presented here demonstrate that *Jpk* may trigger apoptosis through reduction of mitochondrial membrane potential and induction of intracellular ROS levels, regulating Bcl family at the transcriptional level.

ACKNOWLEDGMENTS

This work was supported by a grant from the Korea Science and Engineering Foundation, R04-2001-000-00156-0 (2002), and partly by the Brain Korea 21 Project for Medical Science at Yonsei University.

REFERENCES

1. KRUMLAUF, R. 1994. Hox genes in vertebrate development. Cell **78:** 191–201
2. CHO, M., C. SHIN, W. MIN & M.H. KIM. 1997. Rapid analysis for the isolation of novel genes encoding putative effectors to the position specific regulatory element of murine Hoxa-7. Mol. Cells **7:** 220–225.
3. PARK, S.D., H.W. PARK & M.H. KIM. 2002. A novel factor associating with the upstream regulatory element of murine Hoxa-7 induces bacterial cell death. Mol. Biol. Rep. **29:** 363–368.
4. KIM, H.N., S.D. PARK, H.W. PARK & M.H. KIM. 2002. Isolation and characterization of a novel gene, Jpk, from murine embryonic cDNA library. Korean J. Genet. **24:** 197–203.
5. JACOBSON, M.D., M. WEIL & M.C. RAFF. 1997. Programmed cell death in animal development. Cell **88:** 347–354.
6. RAHA, S. & B.H. ROBINSON. 2001. Mitochondria, oxygen free radicals, and apoptosis. Am. J. Med. Genet. **106:** 62–70.
7. DERBY, E., V. REDDY, W. KOPP, E. NELSON, M. BASELER, T. SAYERS & A. MALYGUINE. 2001. Three-color flow cytometric assay for the study of the mechanisms of cell-mediated cytotoxicity. Immunol. Lett. **78:** 35–39.

Vitamin D Enhances Caspase-Dependent and Independent TNF-Induced Breast Cancer Cell Death

The Role of Reactive Oxygen Species

G.E. WEITSMAN, A. RAVID, U.A. LIBERMAN, AND R. KOREN

Felsenstein Medical Research Center, Department of Physiology and Pharmacology, Sackler Faculty of Medicine, Tel Aviv University, Tel Aviv, Israel

ABSTRACT: Calcitriol, the hormonal form of vitamin D, enhanced TNF-induced cytotoxicity in MCF-7 breast cancer cells. It increased the induction of caspase-3-like activity and TNF-induced caspase-independent cytotoxicity in the presence of a pan-caspase inhibitor. The antioxidants *N*-acetylcysteine, glutathione, lipoic acid, and ascorbic acid markedly reduced the effect of the hormone on TNF-induced caspase activation, attesting to the involvement of reactive oxygen species (ROS) in the cross-talk between the hormone and the cytokine. Calcitriol augmented the drop in mitochondrial membrane potential induced by TNF as assessed by the fluorescent probe JC-1. We postulate that the interaction of TNF and calcitriol on the level of the mitochondria underlies the enhancement of TNF-induced, ROS-mediated caspase-dependent and -independent cell death.

KEYWORDS: calcitriol; mitochondria; reactive oxygen species; TNF

INTRODUCTION

The *in vivo* and *in vitro* anticancer activity of calcitriol, the hormonal form of vitamin D, is well documented. In addition to its antiproliferative and pro-apoptotic activities as a single agent, it also potentiates the activity of some common anticancer drugs and agents of the immune system including tumor necrosis factor-α (TNF-α.).[1,2] The cytokine induces programmed cell death by two distinct pathways: caspase-dependent and caspase-independent.[3,4] There is ample evidence supporting the mediatory role of reactive oxygen species (ROS) in both modes of cell death. Impairment of mitochondrial metabolism may be a key event upstream to the bifurcation to caspase-dependent and -independent death pathways. Mitochondrial injury, associated with a decrease in mitochondrial membrane potential (ΔΨ), results in secondary excessive ROS generation and release of a battery of proteins involved in caspase activation as well as morphologic manifestations of programmed cell death that do not require caspase activation.[5]

Address for correspondence: R. Koren, Felsenstein Medical Research Center, Beilinson Campus, Petah-Tikva, 49100, Israel.
rkoren@post.tau.ac.il

**Ann. N.Y. Acad. Sci. 1010: 437–440 (2003). © 2003 New York Academy of Sciences.
doi: 10.1196/annals.1299.0079**

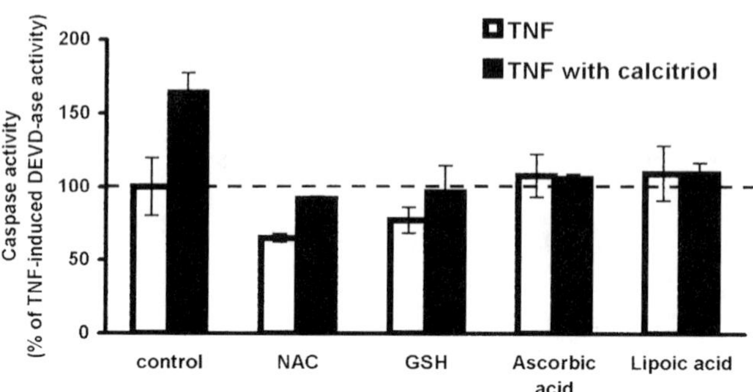

FIGURE 1. Antioxidants selectively inhibit TNF-induced caspase activity enhanced by calcitrol.

MATERIAL AND METHODS

MCF-7 human breast cancer cells were used as an experimental model. Cell number was assessed by staining with Crystal Violet. Caspase-3-like activity was monitored by cleavage of the fluorogenic substrate, Ac-DEVD-AMC, and the endogenous substrate, PARP. The changes in mitochondrial membrane potential were estimated by the use of a cationic fluorescent dye (JC-1) that undergoes potential-dependent accumulation in the mitochondria and emits red (590 nm) or green (540 nm) light; the ratio between red and green fluorescence is a measure for $\Delta\Psi$.

RESULTS AND DISCUSSION

MCF-7 cells treated with TNF-α exhibited morphologic signs of apoptosis. We observed substantial DEVD-ase activity in extracts of TNF-α-treated cells that was markedly enhanced in extracts from cells treated with calcitriol 24 h before harvesting. The enhancing effect of the hormone was dose-dependent, significant already at a concentration of 1 nM, and saturating at 10 nM. Calcitriol also increased PARP cleavage induced by TNF-α . Exposure to calcitriol alone (100 nM for 24 h) did not induce DEVD-ase activity or PARP cleavage. Another dihydroxylated metabolite of vitamin D_3, 24,25-dihydroxyvitamin D_3, which does not bind to the nuclear vitamin D receptor, had no effect on caspase activation in TNF-α-treated cells. These findings support the notion that the increase in TNF-α-induced caspase activation is mediated by the nuclear vitamin D receptor. Calcitriol also markedly increased the cytotoxicity of TNF-α in MCF-7 cells co-treated with the pan-caspase inhibitor zD-DCB in a dose-dependent manner. We conclude that the hormone potentiates both caspase-dependent and -independent modes of TNF-induced cell death in MCF-7 cells.

Next we assessed the involvement of ROS in the potentiating effect of calcitriol. Treatment of cells with the antioxidants *N*-acetylcysteine (NAC), reduced glutathione (GSH), and ascorbic and lipoic acids prior to the addition of TNF-α abol-

TABLE 1. Effect of calcitriol on TNF-α-induced drop of mitochondrial membrane potential

	Calcitrol (100 nM)	zD-DCB (50 μM)	JC-1 Fluorescence Ratio (590 nm/540 nm) mean ± SD, $n = 5$	
			None	TNF (10 ng/mL)
Exp. 1	–	–	3.0 ± 0.1	2.3 ± 0.2[a]
	+	–	2.9 ± 0.2	1.4 ± 0.1[b]
Exp. 2	–	+	3.6 ± 0.4	2.9 ± 0.3[a]
	+	+	3.2 ± 0.4	1.9 ± 0.2[b]

[a]Significant difference between TNF-α-treated and control groups ($P<0.05$).
[b]Significant difference between groups treated with TNF-α alone and in combination with calcitriol ($P<0.05$)

ished the enhancing effect of calcitriol on TNF-α-induced caspase activation (FIG. 1). It is noteworthy that NAC and GSH only marginally inhibited and ascorbic and lipoic acids had no effect on caspase activation by TNF-α alone. The thiol antioxidants NAC and GSH inhibited caspase-independent cytotoxicity induced by TNF-α and calcitriol. The implication of these findings is that ROS are involved in the enhancing effect of calcitriol on both modes of cell death.

In accordance with our findings, we hypothesized that mitochondria are the target of calcitriol action since they are involved in ROS homeostasis, and at the same time play a role in both caspase-dependent and -independent modes of cell death. Indeed, we found out that a 20-hour treatment with TNF-α brought about a marked drop in ΔΨ evidenced by the decreased prevalence of energized mitochondria (TABLE 1) and release of cytochrome *c* into the cytosol. The extent of mitochondrial injury, as manifested by these two measures, was significantly increased by co-treatment with calcitriol. Exacerbation of TNF-induced mitochondrial damage by the hormone may lead to altered ROS homeostasis and enhanced caspase-dependent and -independent cell death. It remains to be assessed whether damage to mitochondria is involved also in the reported synergistic interaction of calcitriol and its analogues with other anticancer modalities.

ACKNOWLEDGMENTS

This work was supported by the Israel Science Foundation (Grant No. 601/99) and by a travel grant from the Constantiner Institute for Molecular Genetics (to G. W.).

REFERENCES

1. ROCKER D., A. RAVID, U.A. LIBERMAN, *et al.* 1994. 1,25-dihydroxyvitamin D_3 potentiates the cytotoxic effect of TNF on human breast cancer cells. Mol. Cell. Endocrinol. **106:** 157–162.
2. KOREN R., D. ROCKER, O. KOTESTIANO, *et al.* 2000. Synergistic anticancer activity of 1,25-dihydroxyvitamin D_3 and immune cytokines: the involvement of reactive oxygen species. J. Steroid Biochem. Mol. Biol. **73:** 105–112.

3. WALLACH D., M. BOLDIN, E. VARFOLOMEEV, *et al.* 1997. Cell death induction by receptor of the TNF family: towards a molecular understanding. FEBS Lett. **410:** 96–106.
4. LEIST M. & M. JAATTELA. 2001. Four deaths and a funeral: from caspases to alternative mechanisms. Nat. Rev. Mol. Cell. Biol. **2:** 589–598
5. FLEURY C., B. MIGNOTTE & J.L. VAYSSIERE. 2002. Mitochondrial reactive oxygen species in cell death signaling. Biochimie **84:** 131–141.

Rescue of Cells from Apoptosis by Antioxidants Occurs Downstream from GSH Extrusion

FRANCESCA LIUZZI, CLAUDIA FANELLI, MARIA ROSA CIRIOLO, CLAUDIA CERELLA, MARIA D'ALESSIO, MILENA DENICOLA, ANDREA MAGRINI, ANTONIO BERGAMASCHI, AND LINA GHIBELLI

Dipartimento di Biologia, Cattedra Medicina del Lavoro, Università di Roma Tor Vergata, 00133 Rome, Italy

ABSTRACT: Antioxidants–that is, scavengers of free radicals and anaerobic conditions (5%CO_2, 95% N_2)–protect monocytic U937 cells from damage-induced apoptosis. Antioxidants rescue the cells acting on the apoptotic pathway at a step downstream from gluthatione extrusion. Reducing agents, such as DTT, also reduce stress-induced apoptosis. Thus, apoptotic GSH extrusion triggers the downstream events of apoptosis by leaving cells unprotected against thiol\s oxidation and radical production.

KEYWORDS: glutathione; antioxidants; apoptosis

INTRODUCTION

Apoptotic signaling is a set of intracellular reactions leading to the coordinated involution of cell structures. It is accepted that oxidative stress is part of apoptosis signaling. Treatments capable of inducing oxidative stress are apoptogenic agents, such as H_2O_2 that induce apoptosis in a variety of cell types.[1,2] Moreover, glutathione, the most abundant antioxidant in the cell, has been shown to play a crucial role in apoptosis.

Indeed, we showed that cells get rid of intracellular GSH by actively extruding it through specific carriers.[3] The inhibition of such extrusion rescues stressed cells from apoptosis.[4] Furthermore, the oncogene bcl-2, a protein capable of blocking the onset of apoptosis from many stimuli,[5] can act through a radical-scavenging mechanism. The role of oxidative stress in apoptosis has been questioned by reports indicating that apoptosis, induced by physiological agents, may occur in anaerobic conditions, such as in the absence of reactive oxygen species (ROS).[6]

Physiological and damaging agents activate two different pathways of apoptotic signaling. New evidence suggests that redox alterations may affect a damaging-induced pathway, but not a physiological pathway. In this study, we want to investigate the role of antioxidants and anaerobic conditions in damage-induced apoptosis.

Address for correspondence: Lina Ghibelli, Dipartimento di Biologia, Università di Roma Tor Vergata, via della Ricerca Scientifica, 00133 Roma, Italy. Fax: +39-062023500. ghibelli@uniroma2.it

Ann. N.Y. Acad. Sci. 1010: 441–445 (2003). © 2003 New York Academy of Sciences. doi: 10.1196/annals.1299.080

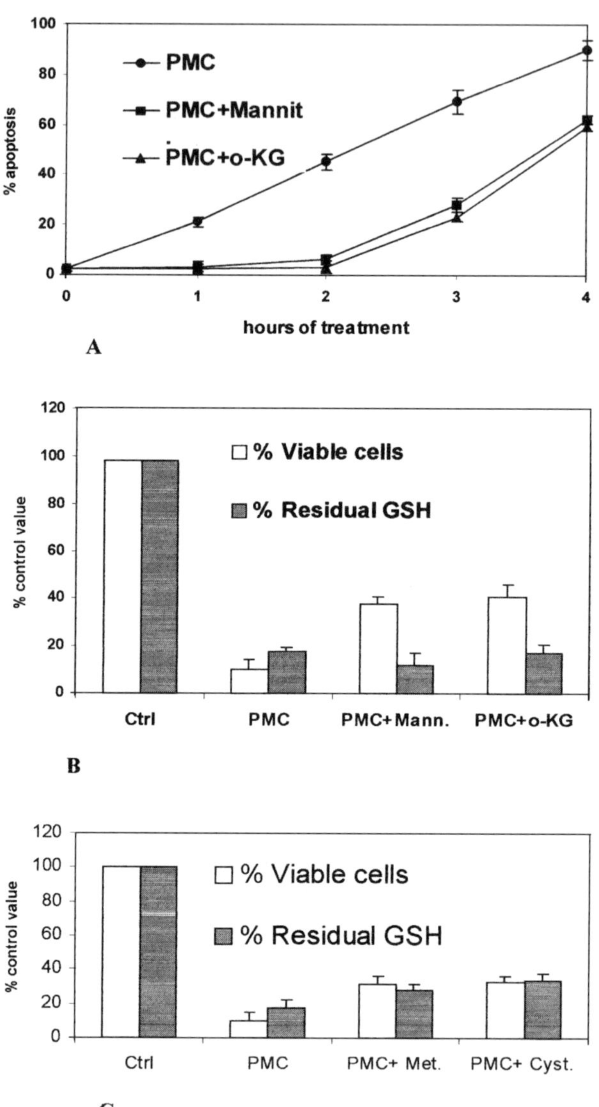

FIGURE 1. Antioxidants protect from damage-induced apoptosis downstream from GSH extrusion. 3 mM α-ketoglutarate (α–KG), 2 mM mannitol (Mann.), 1 mM cystathionine (cyst) or methionine (met) was added before apoptogenic treatment with 10 μg/mL puromycin (PMC) and maintained throughout the experiment. **(A)** Time course of PMC-induced apoptosis in the absence or presence of α-KG and Mann. (data are shown as mean ± SD, $n = 5$, $P < 0.01$ at 4 h of treatment). **(B-C)** Viable cells and GSH content at 4 h of treatment.

MATERIALS AND METHODS

Cell Culture and Treatment

U937 tumoral human pro-monocytic cells were cultured as described.[4] An anaerobic condition was obtained by maintaining cells overnight in a controlled atmosphere by placing cell flasks in airtight bags (atmosbags) filled with N_2 (95%) and CO_2 (5%) at 37°C in complete medium with 10% FCS.

Apoptosis was induced with 10 μg/mL puromycin (PMC), in either normal or anaerobic conditions, as specified and was quantified as the fraction of apoptotic nuclei, as described.[1] Antioxidant treatments were done with 2 mM α-chetoacids (α-ketoglutarate, α-ketobutirarte, α-ketoisovalerate); 3 mM mannitol or benzoate; and 50 μM dithiothreitol (DTT). The compounds were added 1 h before the apoptogenic treatment and kept throughout the experiment.

Glutathione determination was performed by HPLC, as previously described.[1]

RESULTS

Antioxidants Protect Cells from Damage-Induced Apoptosis Downstream from GSH Extrusion

It is known that redox imbalance is required for the damage-induced pathway. We wanted to verify the effect of antioxidants on apoptosis in U937 cells treated with PMC, in presence/absence of different antioxidants. We used two antioxidant categories: (1) the hydroxy radical (˙OH) scavengers mannitol and benzoate and (2) α-ketoacids, which detoxify cells of H_2O_2 (R-COCOOH + $H_2O_2 \rightarrow$ R-COOH + H_2O + CO_2). FIGURE 1a shows the time course of apoptosis induced by PMC in the presence/absence of mannitol and α-ketoglutarate. Both antioxidants are able to reduce apoptosis with the same kinetics. Other antioxidants belonging to the same categories, such as benzoate, α-ketobutirate, and α-ketoisovalerate, are equally effective in reducing apoptosis (data not shown).

These results imply that ˙OH and H_2O_2 play a role in apoptotic signaling induced by PMC. In this instance, the cells rescued by antioxidants may already have gone through the characteristic apoptotic GSH extrusion. Thus, we expect to find living cells without GSH. This is what indeed occurs, as shown in FIG. 1b, where the values of residual GSH and the viability cells are compared after 4 hours of PMC treatment. Instead, agents that inhibit GSH extrusion rescue cells with normal GSH levels (FIG. 1c). Thus, antioxidants protect from apoptosis downstream with respect to GSH extrusion.

Anaerobic Conditions Protects from PMC-Induced Apoptosis

Next, we analysed the effect of anaerobic conditions on PMC-induced apoptosis. Anaerobic conditions (FIG. 2a) were able to reduce the extent of apoptosis throughout the experiment. In this condition, antioxidants failed to exert their protective effect (FIG. 2b).

Because redox alterations also imply thiol oxidation, we investigated the role of redox alterations in apoptosis by probing PMC-induced apoptosis with the thiol-re-

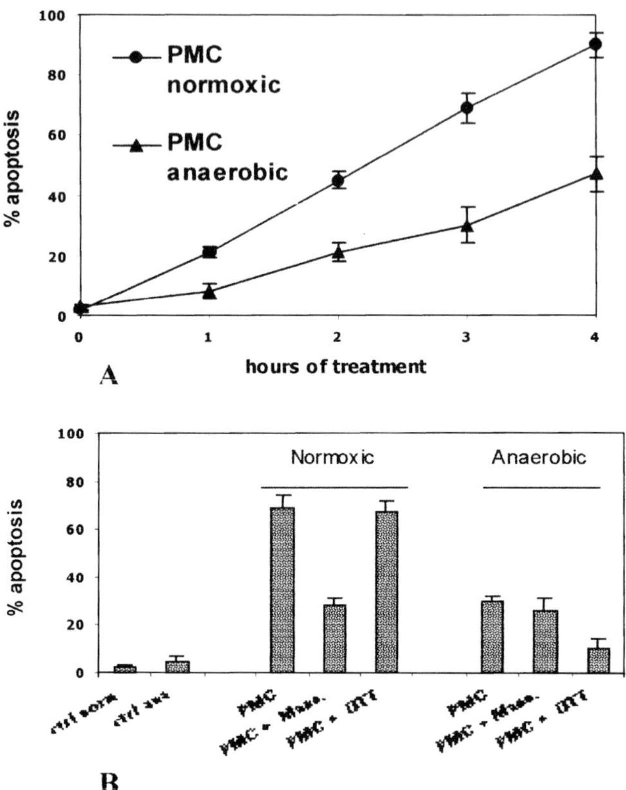

FIGURE 2. Anaerobic conditions protect from PMC-induced apoptosis. We kept cells in anaerobic condition overnight (5% CO2, 95% N2) and throughout the experiment. 2 mM Mann. or 50 μM DTT were added before apoptogenic treatment with 10 μg/mL PMC. **(A)** Time course of PMC-induced apoptosis in normoxic and anaerobic conditions (data are shown as mean ± SD, $n = 5$, $P < 0.01$ at 4 h of treatment. **(B)** Percentage (at 3 h) of PMC-induced apoptosis in the presence/absence of Mann. or DTT, in normoxic and anaerobic conditions.

ducing agent DTT. DTT induces an extra-protection in anaerobic, but not in normoxic condition (FIG. 2b). This indicates that reduction of disulfide bridges is actually able to interfere with apoptosis. However, this effect might be difficult to detect in normoxic conditions that may frustrate the action of 50 μM DTT. (We could not use higher concentrations, because they would create toxic effects in our experimental system).

DISCUSSION

The absence of any synergistic or additive protective effect of antioxidants and lack of O_2 indicate that they act on the same target: reactive oxygen species. This indicates that ROS play an important role in PMC-induced apoptosis. We previously showed that prevention of GSH extrusion with specific carrier inhibitors rescued cells from stress induced apoptosis.[4] With this study, we go a step forward by demonstrating that ROS production and thiol oxidation are the mechanism through which GSH extrusion is able to trigger apoptosis. This apparently contradicts previous reports.[6] However, that study mainly concerned physiological, rather than stress-induced, apoptosis. Thus, we may hypothesize that anaerobiosis is able to interfere with stress-induced, but not physiologically induced, apoptosis. This is in line with evidence indicating that redox alterations are required for the damage, but not for the physiological, apoptotic pathway.

REFERENCES

1. LENNON, S.V., S.J. MARTIN & T.G. COTTER. 1991. Dose-dependent induction of apoptosis in human tumour cell lines by widely diverging stimuli. Cell Prolif. **24:** 203–214.
2. NOSSERI, C., S. COPPOLA & L. GHIBELLI. 1994. Possible involvement of poly(ADP-ribosyl) polymerase in triggering stress-induced apoptosis. Exp. Cell. Res. **212:** 367–373.
3. GHIBELLI, L., S. COPPOLA, G. ROTILIO, *et al.* 1995. Non-oxidative loss of gluthatione in apoptosis via GSH extrusion. Bioch. and Bioph. Res. Comm. **216:** 313–320.
4. GHIBELLI, L., C. FANELLI, G. ROTILIO, *et al.* 1998. Rescue of cells from apoptosis by inhibition of active GSH extrusion. FASEB J. **12:** 479–486.
5. KORSMEYER, S.J. 1992. Bcl-2: a repressor of lymphocyte death. Immunol. Today **14:** 131–136.
6. JACOBSON, M.D. & M.C. RAFF. 1995. Programmed cell death and Bcl-2 protection in very low oxygen. Nature (Lond.) **374:** 814–816.

Cytochrome *c*, Glutathione, and the Possible Role of Redox Potentials in Apoptosis

JOHN T. HANCOCK, RADHIKA DESIKAN, AND STEVEN J. NEILL

Centre for Research in Plant Science, University of the West of England, Bristol, Coldharbour Lane, Bristol, BS16 1QY, United Kingdom

ABSTRACT: The redox environment of the cell is now thought to be extremely important to control the activity of many proteins. During apoptosis, the intracellular redox potential (E_h) becomes more positive, with possible consequences for the mechanisms of apoptosis. Glutathione and cytochrome *c* might both influence and be influenced by the cellular redox environment and therefore be important in the progression of apoptosis.

KEYWORDS: apoptosis; cytochrome *c*; glutathione; hydrogen peroxide; redox environment

Many factors can influence the induction and maintenance of apoptotic mechanisms, including the presence of reactive oxygen species (ROS),[1] such as hydrogen peroxide (H_2O_2) and superoxide (O_2^-). ROS can arise from several cellular sources, including electron leakage in mitochondria and dedicated enzymes at the cell surface, such as the NADPH oxidase complex. ROS may be produced by the cell that is about to undergo apoptosis, or by neighboring cells. Either way, regardless of the source, ROS must be perceived by a cell and subsequently invoke the signal transduction pathways leading to apoptosis.

A way in which ROS might act on cells is via modulation of the redox status of the cell. Many proteins contain amino acids with side chains that may be reduced or oxidised. Typically, these are cysteine residues, in which the thiol side chain can be in the –SH state, or as a cystine disulfide –S–S–. The redox status of such thiols may have considerable influence on the structure of the protein; thus, affecting the cysteine oxidation state is a potential way of modulating protein activity.

A main factor influencing the redox status of the intracellular environment is the redox status and concentration of glutathione. Typically, glutathione concentration is relatively high (1–10 mM) and mainly reduced (GSH). By measuring GSH/GSSG levels in cells, and using the Nernst equation, an estimate of the redox potential in the cell (E_h) can be calculated. Such studies have shown that intracellular E_h alters

Address for correspondence: Dr. John T. Hancock, Centre for Research in Plant Science, University of the West of England, Bristol, Coldharbour Lane, Bristol, BS16 1QY, UK. Voice: +44-117-328-2475; fax: +44-117-328-2904.
john.hancock@uwe.ac.uk

**Ann. N.Y. Acad. Sci. 1010: 446–448 (2003). © 2003 New York Academy of Sciences.
doi: 10.1196/annals.1299.081**

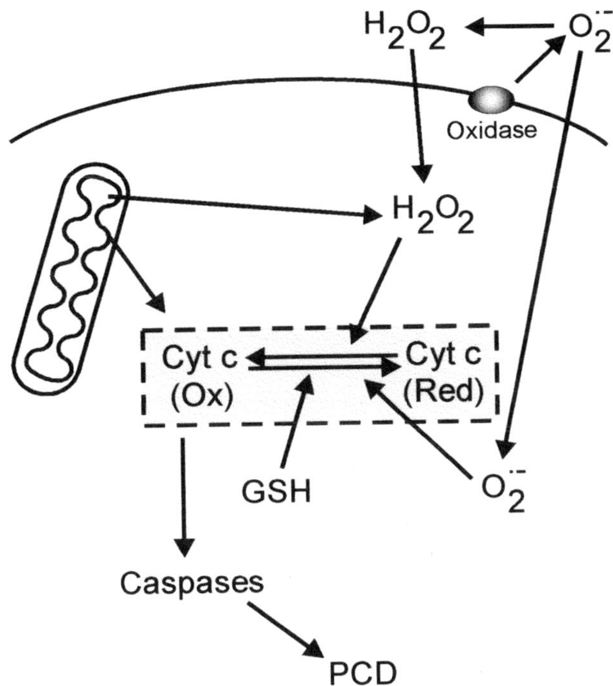

FIGURE 1. A schematic showing the possible influence of reactive oxygen species and glutathione in the redox state of cytochrome *c*. Both GSH and superoxide ions are able to reduce cytochrome *c*, whereas hydrogen peroxide can cause its oxidation.

during differentiation and apoptosis.[2,3] However, in making these calculations, it has to be appreciated that the E_h imposed by glutathione is influenced both by the pH of the solution, and by the concentration of total glutathione (GSH+GSSG), as the Nernst equation is a squared relationship with respect to the concentration of GSH.

It is not just the GSH/GSSG couple that will influence the redox status as the phrase "redox environment" of the cell suggests.[3] It has been proposed that the redox environment can be summarised by the equation: redox environment = $\sum E \times$ [reduced species] (the summation of the product of the reduction potential and the concentration of the reduced species of the couples present).

However, it can not be assumed that all parts of the cell, and organelles, have the same redox state; even the cytoplasm probably has redox "hot-spots" within cellular microdomains. Measurement of global redox couples will give a false impression of the true redox environment experienced by cellular proteins.

One of the proteins often implicated in the induction of apoptosis is cytochrome *c*, released from the mitochondria, this in turn leading to the activation of caspases and, therefore, cellular death.[4] Of particular relevance is the fact that cytochrome *c* is released and exists in the cytoplasm as a holoenzyme; and therefore, it is able to exist in two redox states: reduced or oxidized. Therefore, the redox state, and hence the structure of cytochrome *c*, may be influenced by the redox environment in which it is found.[5] Normally, under quiescent conditions, the high GSH content of the cell will keep the cytochrome *c* in a reduced state. However, if the cell has been exposed to oxidizing conditions (relatively high concentrations of hydrogen peroxide, for example) the cellular redox environment will become more oxidized, and cytochrome *c* may be oxidized, adopting a new conformation and hence triggering apoptosis. Clearly, robust experimentation is needed to verify if such a mechanism (FIG. 1) is used by cells. However, it is certain that many proteins will, in fact, be regulated by the redox environment inside the cell.

REFERENCES

1. JABS, T. 1999. Reactive oxygen intermediates as mediators of programmed cell death in plants and animals. Biochem. Pharmacol. **57:** 231–245.
2. KIRLIN, W.G, J. CAI, S.A. THOMPSON, *et al.* 1999. Glutathione redox potential in response to differentiation and enzyme inducers. Free Radic. Biol. Med. **27:** 1208–1218.
3. SCHAFER, F.Q. & G.R. BUETTNER. 2001. Redox environment of the cell as viewed through the redox state of the glutathione disulphide/glutathione couple. Free Radic. Biol. Med. **30:** 1191–1212.
4. ALNEMRI, E.S. 1999. Hidden powers of the mitochondria. Nat. Cell Biol. **1:** E40–E42.
5. HANCOCK, J.T., R. DESIKAN & S.J. NEILL. 2001. Does the redox status of cytochrome *c* act as a fail-safe mechanism in the regulation of programmed cell death? Free Radic. Biol. Med. **31:** 697–703.

Apoptotic GSH Extrusion Is Associated with Free Radical Generation

M. D'ALESSIO,[a] C. CERELLA,[a] M. DE NICOLA,[a] A. BERGAMASCHI,[b] A. MAGRINI,[b] G. GUALANDI,[c] A.M. ALFONSI,[c] AND L. GHIBELLI[a]

Departments of [a]Biology and [b]Occupational Health, University of Rome "Tor Vergata," Via della Ricerca Scientifica, 00133 Rome, Italy

[c]DABAC, University of Tuscia, Viterbo, Italy

ABSTRACT: Reactive oxygen species (ROS) are involved in many forms of apoptosis and mediate apoptosis in a number of cell types. In this paper, we use a variant of U937 monocytic cells (U937 HX) that show different biochemical features with respect to standard U937. Apoptotic standard U937 extrude reduced glutathione (GSH) and generate free radicals concomitantly with loss of mitochondria transmembrane potential (mt $\Delta\Psi$). These events are correlated with the extrusion of intracellular GSH. Conversely, apoptotic U937 HX cells retain GSH, and the loss of mt $\Delta\Psi$ is not accompanied by generation of free radicals. The perfect inverse correlation between (a) ROS generation and (b) the presence of intracellular GSH during apoptosis suggests novel mechanisms to finely tune ROS generation in apoptosis.

KEYWORDS: apoptosis; reactive oxygen species; ROS; glutathione; mitochondrial transmembrane potential

INTRODUCTION

Apoptosis can be triggered by different stimuli and can evolve through multiple pathways; that is, the physiological and the stress/mitochondria-induced pathway.[1] Reactive oxygen species (ROS) mediate apoptosis in a number of cell types.[2] The mechanism involved in the generation of ROS in apoptosis is not yet clear, although it has been hypothesized to be caused by the alteration of mitochondrial transmembrane potential (mt$\Delta\Psi$). Indeed, the literature describes apoptosis as characterized by rapid loss of mt$\Delta\Psi$ that enhances generation of superoxide anions.[3] Recent studies show redox imbalance is necessary for the mitochondrial apoptotic pathway.[2] In this instance, the extrusion of GSH, the major endogenous antioxidant, alters the intracellular redox status.[4,5] Thus, we wanted to explore whether the generation of ROS in stress-induced apoptosis may be correlated with a different redox status of the apoptotic cells.

Address for corresspondence: M. D'Alessio, Dipartimento di Biologia, Università di Roma Tor Vergata, via della Ricerca Scientifica, 00133 Rome, Italy. Fax: +39-06 2023500. dalessio.maria@libero.it

Ann. N.Y. Acad. Sci. 1010: 449–452 (2003). © 2003 New York Academy of Sciences. doi: 10.1196/annals.1299.082

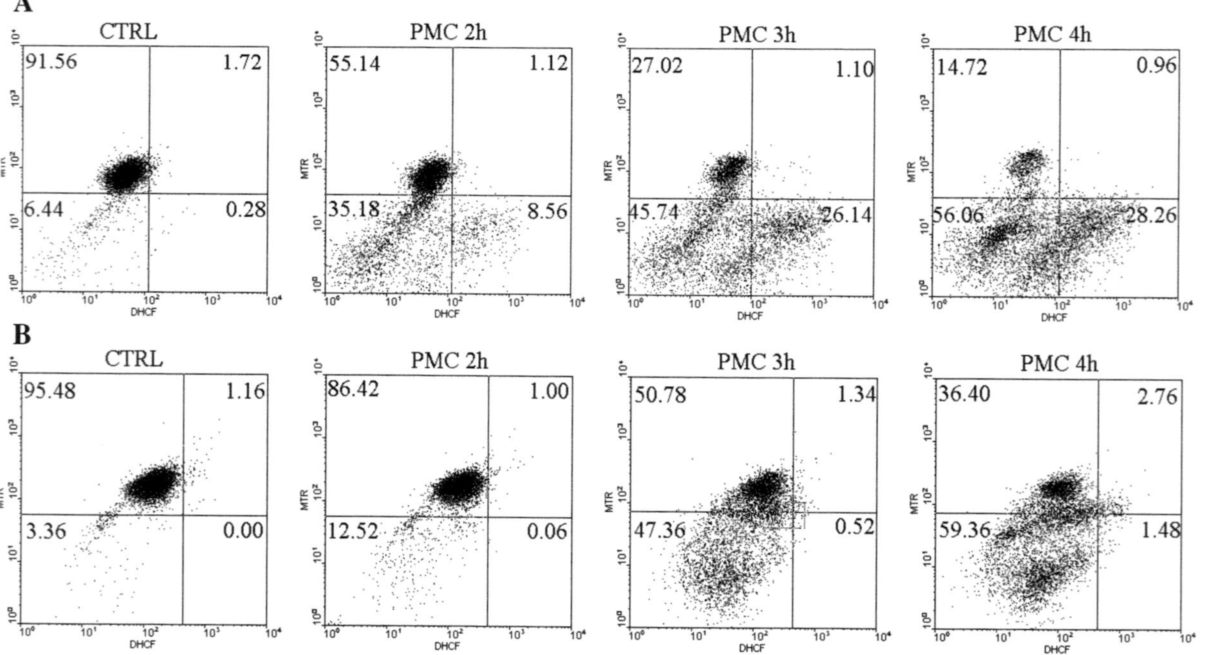

FIGURE 1. In U937 HX the loss of mtΔΨ is not accompanied by ROS generation. Panels **A** and **B** show biparametrical flow cytometric analysis of cells, double-stained with 100 nM MTR and DHCFDA. **(A)** Comparison of two apoptotic poulations in standard U937 cells: (1) unaltered and (2) those with higher DHCF fluorescence. **(B)** In U937 HX cells, the loss of mtΔΨ is not accompanied by alterations of DHCF fluorescence. (One of four independent experiments is shown.)

MATERIALS AND METHODS

Cell Culture and Treatment

U937 human pro-monocytic tumor cells were cultured as previously described.[4] U937 HX resulted from a nontoxic period of culture in hypoxic (i.e., low O_2, high CO_2) conditions.[6] Apoptosis was induced with 10 μg/mL puromycin (PMC) and was quantified as the fraction of apoptotic nuclei as previously described.[4]

Analysis of Mitochondrial ΔΨ, ROS, and Intracellular GSH

Cells were loaded respectively, with 100 nM MitotrackerRed (MTR), dichlorodihydrofluoresceindiacetate (DHCF-DA), or chloromethylfluoresceindiacetate (CMF-DA) by incubation at 37°C for 10 min. Cells were then washed and immediately analyzed using FACScan (Bekton & Dickinson); data (5000 events) were elaborated with the Cell Quest software. MTR, DHCF-DA, and CMFDA were purchased from Molecular Probes.

RESULTS

Loss of mt ΔΨ Is Accompanied by Generation of Free Radicals in Apoptotic U937 but Not in Apoptotic U937 HX

Upon induction of apoptosis by purmicin (PMC), the kinetics of mtΔΨ loss and ROS generation was followed with a biparametrical flow cytometric analysis of cells, double stained with: MTR (a dye cross-linked to the mitochondrial matrix, according to an existing mtΔΨ*);* and DHCFDA (a dye that becomes fluorescent only when it oxdidates. FIGURE 1 shows that in U937 HX the loss of mtΔΨ is not accompanied by alterations of DHCF fluorescence. Instead, in standard U937, it is possible to observe two apoptotic populations: unaltered, and those with higher DHCF fluorescence (compare panels A and B). These results show that ROS generation in apoptosis requires additional events, because it cannot be explained only on the basis of mtΔΨ loss.

Standard U937, but Not U937 HX, Extrude GSH When Induced in Apoptosis

In order to analyze possible alterations of intracellular redox balance in apoptosis, we analyzed intracellular GSH content during apoptosis by using flow cytomet-

TABLE 1. Standard U937, but not U937 HX, lose GSH when induced in apoptosis

Treatment	**(a)** Percent of apoptotic cells	**(b)** Percent of apoptotic cells with GSH loss	**(c)** Percent of apoptotic cells with higher ROS
Untreated	2 ± 1	2 ± 1	1 ± 1
PMC 3 h U937 HX	52 ± 6	2 ± 8	2 ± 4
PMC 3h U937 standard	68 ± 3	35 ± 8	28 ± 4

NOTE: The different parameters are calculated as described in MATERIALS AND METHODS. The results are reported as the average of $n = 4$ experiments ±SD. $^*P = 0.01$.

ric analysis of cells stained with CMFDA, a dye used *in situ* for glutathione analysis. In TABLE 1, we compared the fraction of cells with: (a) apoptotic morphology, (b) loss of GSH, and (c) higher ROS content; in standard U937 and in U937 HX after 3 h of PMC treatment. This way, we were able to observe that standard U937, but not U937 HX, lose GSH when induced in apoptosis and that the fraction of cells that produce ROS correspond roughly to the fraction of cells losing GSH.

DISCUSSION

In this paper, we propose a simple but novel link with GSH extrusion, a key event in stress-induced apoptosis, suggesting a possible relationship between generation of ROS and GSH extrusion in apoptosis. In this way, the generation of ROS in apoptosis may be correlated with a different redox status of the apoptotic cells, rather than with the alteration of mitochondrial transmembrane potential. Indeed, in U937 HX cells late in apoptosis, we observed one population with low fluorescence in both MTR (apoptotic cells) and DHCF. However, at the corresponding time points, there appear two populations in standard U937 cells: one with low fluorescence for MTR and DHCF, and another that maintains low fluorescence for MTR but presents a higher fluorescence for DHCF (FIG. 1). At present, the significance of these two populations is yet to be investigated.

REFERENCES

1. COPPOLA, S. & L. GHIBELLI. 2000. GSH extrusion and the mitochondrial pathway of apoptotic signalling. Biochem. Soc. Trans. **28:** 56–61.
2. SUZUKI, Y.S., H.J. FORMAN & A. SEVANIAN. 1997. Oxidants as stimulators of signal transduction. Free Radical Biol. Med. **22:** 269–285.
3. MARCHETTI, P. T. HIRSCH, N. ZAMZAMI, *et al.* 1996. Mitochondrial permeability transition triggers lymphocyte apoptosis. J. Immunol. **157**(11): 4830–4836.
4. GHIBELLI, L. C. FANELLI, G. ROTILIO, *et al.* 1998. Rescue of cells from apoptosis by inhibition of active GSH extrusion. FASEB J. **12:** 479–486.
5. GHIBELLI, L. S. COPPOLA, C. FANELLI, *et al.* 1999. Glutathione depletion causes cytochrome c release even in the absence of cell commitment to apoptosis. FASEB J. **13:** 2031–2036.
6. DE NICCOLA, F., C. LIUZZI, M. CERELLA, *et al.* 2003. Hypoxia stress stably alters apoptotic parameters on U937 cells. Ann. N.Y. Acad. Sci **1010:** this volume.

The Developing Mouse Dentition

A New Tool for Apoptosis Study

RENATA PETERKOVÁ,[a] MIROSLAV PETERKA,[a] AND HERVÉ LESOT[b]

[a]*Institute of Experimental Medicine, Academy of Sciences CR, Videnska 1083, 142 20 Prague 4, Czech Republic*

[b]*INSERM U-595, 11, rue Humann, 67085 Strasbourg, France*

ABSTRACT: Developing limb or differentiating neural and blood cells are traditional models used to study programmed cell death in mammals. The developing mouse dentition can also be an attractive model for studying apoptosis regulation. Apoptosis is most extant during early odontogenesis in mice. The embryonic tooth pattern is comprised not only of anlagen of functional teeth (incisor, molars), but also of vestiges of ancestral tooth primordia that must be suppressed. Apoptosis is involved in (a) the elimination of vestigial tooth primordia in the prospective toothless gap (diastema) between the incisor and molars and (b) the shaping of germs in functional teeth. This type of apoptosis occurs in the dental epithelium according to a characteristic temporo-spatial pattern. Where apoptosis concentrates, specific signaling is also found. We proposed a hypothesis to explain the stimulation of apoptosis in the dental epithelium by integrating two concepts: (1) The regulation of epithelial budding by positional information generated from interactions between growth-activating and growth-inhibiting signals, and (2) apoptosis stimulation by the failure of death-suppressing signals. During the budding of the dental epithelium, local excess in growth inhibitors (e.g., Bmps) might lead to the epithelial cells' failure to receive adequate growth-activating (apoptosis-suppressing) signals (e.g., Fgfs). The resulting signal imbalance leads to cell "suicide" by apoptosis. Understanding of apoptosis regulation in the vestigial tooth primordia can help to elucidate the mechanism of their suppression during evolution and to identify factors essential for tooth survival. The latter knowledge will be important for developing a technology of tooth engineering.

KEYWORDS: apoptosis; tooth; development; model; regulation; growth factor

INTRODUCTION

Programmed cell death (PCD) is a physiological phenomenon representing an essential part of normal development.[1,2] PCD is a controlled mechanism that is beneficial to the organism.[3] Apoptosis is a type of death characterized by morphological changes: shrinking of the cell, condensation of the cytoplasm, compaction and seg-

Address for correspondence: R. Peterková, Institute of Experimental Medicine, Academy of Sciences CR, Videnska 1083, 142 20 Prague 4, Czech Republic. Voice: +420-241-062-232; fax: +420-241-062-604.

repete@biomed.cas.cz

Ann. N.Y. Acad. Sci. 1010: 453–466 (2003).
doi: 10.1196/annals.1299.083

regation of the chromatin, and convolution of the cell surface. Fragments of the condensed nucleus and/or cytoplasm are then separated as apoptotic bodies of various sizes.[4] Nearly all PCD in vertebrates, show morphological features of apoptosis.[3]

In general, apoptosis is involved in the control of cell numbers and the suppression of abnormal, misplaced, nonfunctional, or harmful cells and in the renewal or regeneration of tissues. In addition, apoptosis occurs regularly at specific places and stages during individual development: In the course of deleting or shaping cell structures and in cell differentiation, it eliminates unwanted cells.[2] Apoptosis thus plays essential roles in animal homeostasis and development.[3–6] Consequently, its dysfunction leads to a range of developmental abnormalities and diseases.[5,6]

The apoptotic process is regulated by a set of conserved genes, which shares similarities from invertebrates to mammals.[5,7,8] Basic data on apoptosis regulation was achieved in the nematode *Caenorhabditis elegans*. The apoptotic machinery in *C. elegans* (e.g., Ref. 7) can be considered as an evolutionary simple prototype of the more sophisticated regulation seen in other metazoa.[5] The regulation of apoptosis reached a maximum complexity in mammals.[5,8,9]

There are two main apoptotic pathways in mammalian cells: the death-receptor pathway and the mitochondrial pathway.[9] The death-receptor pathway is triggered by death receptors after they bind appropriate ligands: members of the tumor necrosis factor (TNF) superfamily (e.g., Fas or TNF). The receptor-ligand interaction switches on, via adapter molecules, the intracellular caspase machinery resulting in cell death.[9,10] The mitochondrial pathway is initiated by mitochondrial stress in response to extracellular cues and internal insults (DNA damage). This step may include the activation of a pro-apoptotic member of the Bcl-2 family (e.g., Bax), which is antagonized by anti-apoptotic members (Bcl-2) of the same group. The stressed mitochondria release cytochrome *c*, which associates with Apaf-1. Both the death-receptor and mitochondrial pathway converge when activating caspase-3. The activation is antagonized by the IAP (inhibitor of apoptosis) proteins. The latter proteins are counteracted by the Smac/DIABLO protein from mitochondria. Caspase-3 activation is followed by the completion of cell destruction and subsequent removal.[9]

DISCUSSION

PCD during Development in Mammals

According to its role during mammalian development, Glücksmann[11] distinguished (a) "phylogenetic" cell death participating in the loss of vestigial structures, (b) "morphogenetic" cell death involved in specific shaping of structures, and (c) "histogenetic" cell death associated with cell differentiation. There are many instances of the physiological occurrence of PCD during development in mammals (e.g., Refs. 2 and 3). Most refer to PCD during cell differentiation (histogenetic cell death) in the nervous system. Differentiating cells also die in somites, gonads, kidneys, and in the hematopoietic and immune systems. Morphogenetic cell death has been reported during the formation of free spaces (e.g., ducts of the salivary gland) and the separation or fusion of embryonic structures. During separation, apoptosis is involved in the detachment of invaginated structures (e.g., splitting off the neural tube or the lens from surface ectoderm) or the individualization of structures (digits).[2,3] The elimination of surface epithelium by apoptosis is required for the subsequent fusion of structures, after they approach each other and come into contact, as observed in palatal shelves.[12]

There are several classical models to study apoptosis during mammal development. The developing limb is employed to investigate the morphogenetic effect of apoptosis and its regulation.[13] The kidney,[14] and the nervous[15,16] or immune[17] systems are used to study the regulation of differentiation-associated apoptosis.

A detailed review of data on the molecular aspects of apoptosis regulation in classical mammalian models is beyond the scope of this paper. The present aim is to summarize the current knowledge on PCD in developing dentition of mice; and because all phylogenetic, morphogenetic and differentiation-associated types of PCD occur there, it can be considered a further attractive tool for studies on apoptosis regulation.

Apoptosis during Dentition Development

PCD occurs in the dental epithelium at different stages of tooth development (odontogenesis) (FIG. 1). Cell degeneration was first observed in developing molars in humans and mice.[18,19] This type of cell death was later specified as apoptosis in mice,[20] using morphological criteria and labeling of DNA fragments by the TUNEL method.[4,21] Systematic descriptive data are now available on the apoptosis occurrence during odontogenesis in mice (see below), which is the most frequently used model in tooth development studies.

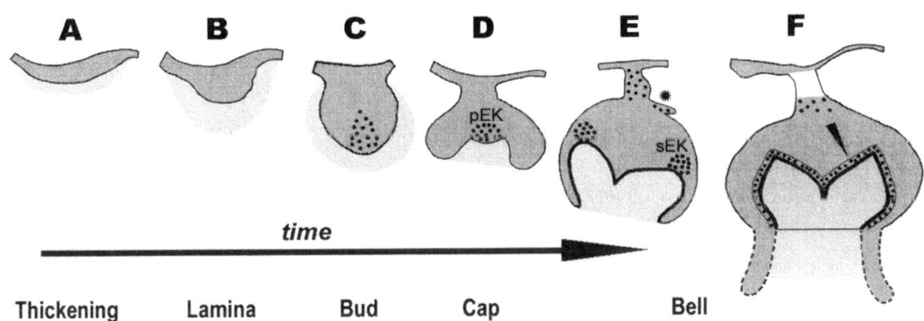

FIGURE 1. Stages of tooth development. Teeth are produced by matured tooth germs that develop through reciprocal tissue interactions between the epithelium (*dark gray*) and ecto-mesenchyme (*light gray*). Developmental morphological stages are named according to the shape of the dental epithelium on frontal sections: **(A)** dental thickening, **(B)** lamina, **(C)** bud, **(D)** cap, and **(E, F)** bell. The epithelium-mesenchyme interface (*bold line*) anticipates the prospective crown shape with the cusps on its occlusal surface. The root shape is determined later (*dashed line*). The primary and secondary enamel knot (pEK and sEK, respectively) comprise densely packed cells of the dental epithelium[26] and exhibit signaling activity.[38] The *arrowhead* points to the layer of the inner dental epithelium, which gives rise to ameloblasts producing tooth enamel. The *star* indicates the vestigial anlage of the secondary dentition.[34] The *white area* corresponds to the regressing stalk of the enamel organ. Apoptosis (*black dots*) accumulates in the infolded dental epithelium according to the specific temporo-spatial pattern.

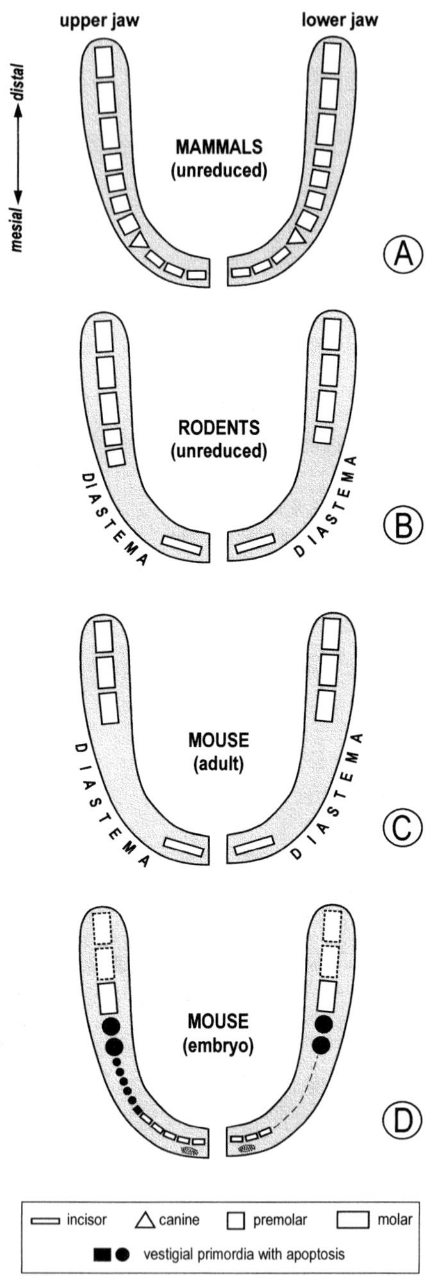

FIGURE 2. Tooth patterns in the upper and lower jaw. **(A)** The maximum (unreduced) tooth number in mammalian ancestors and some living mammals; **(B)** the maximum (unreduced) tooth number in rodent ancestors and some living rodents; **(C)** the functional dentition in adult mice; and **(D)** the tooth pattern in mouse embryos. The latter pattern includes

Apoptosis accumulates in specific areas of mouse dental epithelium according to a specific temporo-spatial pattern (FIG. 1). Apoptosis can be considered as "phylogenetic" cell death, when it suppresses vestigial tooth buds (abortive primordia of ancestral teeth[22,23]) (FIG. 1C) during the determination of the functional dental pattern and the delimitation of the toothless diastema.[24,25] Apoptosis has a "morphogenetic" role in the course of shaping the primordia of the functional teeth in mice.[24–29] (e.g., FIG. 1D). "Histogenetic" cell death occurs during late tooth development.[30,31] It is associated with tissue maturation (FIG. 1F).

Apoptosis and Determination of Dental Pattern

Apoptosis is most extent during the early stages of mouse odontogenesis. Then, the conserved embryonic tooth pattern also includes the vestigial tooth primordia[23] (FIG. 2). Such an embryonic tooth pattern, where the number of tooth primordia is high (FIG. 2D), has to be reduced to a functional one (FIG. 2C). Apoptosis acts as phylogenetic PCD in this process by suppressing the epithelium of the "supernumerary" tooth placodes/buds (FIGS. 2 and 3). These vestiges are either eliminated during subsequent development, or their remnants are integrated into the anlagen of functional teeth.[23]

In the upper jaw, vestigial tooth primordia develop in the prospective toothless gap (diastema) separating the incisor from molars in adult mice[23–25] (compare FIG. 2C and 2D). Most of these diastemal primordia are small (FIG. 2D), and their morphology is similar to that of the tearly tooth anlagen in reptiles.[23] The life span of these vestiges is only about 24 hours, during which time their epithelium undergoes apoptosis and disappears (FIG. 3). A similar structure also occurs transiently at the distal limit of the upper incisor.[23] Its suppression by apoptosis determines the incisor–diastema boundary (FIG. 2D). Such small tooth anlagen do not occur in the prospective diastema of the lower jaw. The presence of vestigial placodes/buds in the prospective maxillary diastema and their absence in the mandible (FIG. 2D) can be explained by differential rates of tooth suppression during evolution.[23,32] This fits with the general rule that tooth suppression in rodents always appears less advanced in the maxilla than in the mandible.[33]

Furthermore, large vestigial tooth anlagen are present at the mesial limit of the molar segment in both the upper and lower jaws of mouse embryos (FIGS. 2D, 3) putatively corresponding to rudiments of the premolars in primitive rodents (FIG. 2B). These vestiges are also affected by apoptosis, but not totally suppressed. They are reduced, reshaped, and incorporated into the mesial part of the first molar during further development[23,25] (FIG. 3).

Apoptosis is also involved in the regression of the mesial limit of the incisor epithelium in both the lower[29] and upper jaws. There, an abortive incisor forms at later stages[22] and resembles, by its shape, a primitive reptilian tooth.[23] Therefore,

---

both the primordia of prospective functional teeth and the transient (vestigial) primordia that, presumably, correspond to the abortive development of teeth present in the ancestors of the mouse. The *dashed line* shows molars developing during later stages. The *dotted area* shows epithelial apoptosis in the area from which the abortive incisor will later originate. The functional incisor originates from several fused vestigial primordia.

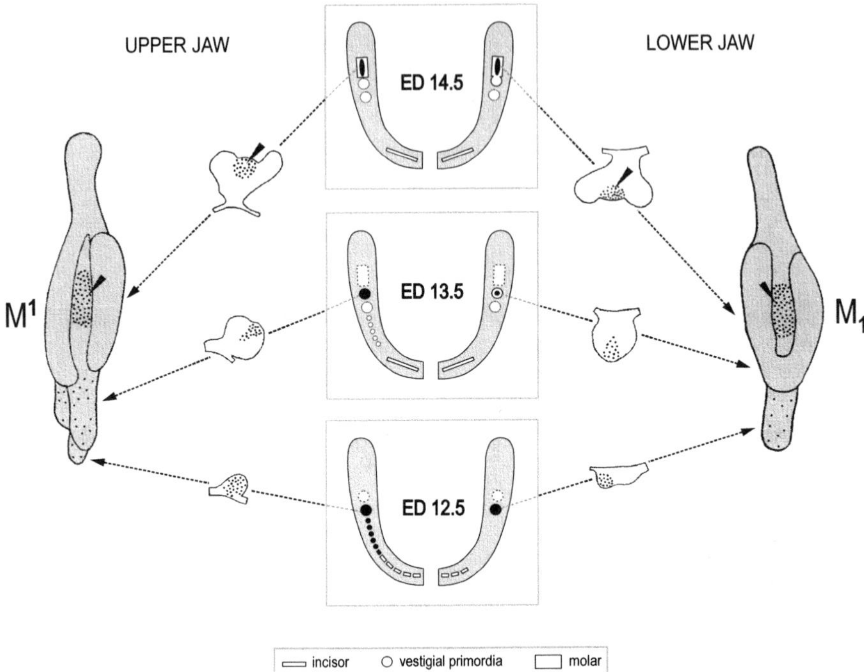

FIGURE 3. Three episodes of apoptosis (*black*) in the dental epithelium at embryonic day (ED) 12.5, 13.5, and 14.5, respectively.[32] Apoptosis participates in the formation of the diastema and in the development of the cap of the first upper (M^1) and lower (M_1) molars. The *dashed line* interconnects the most prominent tooth primordium at a specific stage with its drawing recorded from a frontal section; the *dashed arrow* shows the final location of the structure in the 3D reconstruction of the dental epithelium in mouse embryos at ED 15.0. The disappearing or emerging tooth primordia are delineated by the *dotted* and *dashed line*, respectively. The vestigial primordia develop and then regress sequentially along the mesio-distal (bottom-up) direction. Apoptosis only transiently affects the vestige, which fuses with and contributes to the mesial limit of the first molar in the mandible. As the first molar cap develops, the third episodic concentration of cell death appears in the enamel knot (*arrowhead*). Note the delay in the formation of the first molar compared to the ante-molar vestiges.

apoptosis at the mesial limit of the incisor epithelium might be considered phylogenetic PCD suppressing a tooth-forming potential inherited from ancestors. The apoptosis that has been observed[27] in the rudimental anlage of the secondary (replacing) mouse dentition[34] might also be considered phylogenetic PCD.

Apoptotic removal of the vestigial tooth primordia has been also reported in other species—the field vole[35] and *Suncus murinus*.[36]

Apoptosis and Tooth Morphogenesis

Cell death occurs during the morphogenesis of functional incisors and molars. A characteristic feature of developing mouse molars is an accumulation of cell death that typically concentrates in the enamel knots (FIG. 1D, E).[18–20,26–28,37] The enamel knots

(EKs) are considered signaling centers that regulate tooth morphogenesis (e.g., Refs. 37 and 38). The intense apoptotic process in the primary enamel knot (FIG. 1D) has a morphogenetic effect that retards growth in the center of the molar cap.[26,32,39] It might also represent the elimination of cells after they complete their signaling activity.[40]

Contrary to what happens in the molar, the EK in the incisor shows only sparse apoptosis.[29] This EK disappears, probably as a result of histological rearrangement.[29]

Although apoptosis concentrates in the EK of molars, apoptotic bodies in contact with the basement membrane are very rare.[41] Similarly, granulosa cells in the ovarian follicle undergo apoptosis if they are distal from the basement membrane.[42] However, electron microscopy documents that EK cells in contact with basement membrane in molars also show initial stages of apoptosis.[41] It must thus be assumed that the cells in contact with basement membrane also undergo apoptosis and lose contact with their neighbors[43] and with the BM itself. The localization of apoptotic bodies does not necessarily correspond to the cell's original position. A loss of laminin-5 (LN-5) initiates apoptosis induced by hypoxia in corneal epithelial cells.[44] LN-5 is present in the basement membrane underlying EK cells in incisors[45] where apoptosis is almost absent.[46] It is disappearing from basement membrane associated with EK in molar[45] where apoptotic events accumulate.[26–28] Specific epithelial cell—basement membrane interactions involving LN-5 might play a major role in controlling apoptosis in EK and offer insight as to which situations trigger such control in incisors versus molars.

Apoptosis and Cell Differentiation

Differentiation-associated cell death occurs during the late stages of tooth development (FIG. 1E, F). Apoptosis concentrates in the regressing stalk (FIG. 1E) of the molar[27] and incisor[29] enamel organs, while it is sparsely dispersed in the internal cells (stellate reticulum) of the enamel organ.[27,28] Cell death is upregulated during the differentiation of ameloblasts that secrete enamel and later is involved in the removal of nonfunctional ameloblasts[30,31] (FIG. 1F). PCD also occurs during the late phase of differentiation and following involution of odontogenous tissues in rats, hamsters,[31,47–52] and humans.[53]

Molecules Involved in the Intracellular Control of Apoptosis in the Dental Epithelium

Factors involved in the regulation of apoptosis (Fas, Fas-ligand, Bax, Bcl-2) have been investigated by immunolocalization during morphogenesis of human teeth.[54] The results show that Fas can mediate apoptosis in the dental epithelium and this effect appears to be antagonized by Bcl-2 in developing human teeth.[54]

Distribution of pro-apoptotic (Bax and Bak) and anti-apoptotic (Bcl-2 and Bcl-x) members of the Bcl-2 family of intracellular proteins have been investigated during mouse odontogenesis.[55] Contemporaneous immunoreactivity for Bcl-x and Bak or for Bcl-2 and Bax have been observed in various types of cells, mainly at places of ectoderm–mesenchyme interactions.[55]

The caspase 3 represents a step of apoptosis regulation in which both the receptor-mediated and mitochondrial pathway converge.[9] The activated caspase 3 in the mouse lower molar colocalizes with apoptosis in the epithelium,[56] whose pattern was documented in 3D reconstructions during early morphogenesis.[28]

Most data on PCD regulation relate to ameloblast differentiation during the last phase of tooth development. Bcl-2 and Bax are involved in the regulation of the mitochondrial pathway of apoptosis and show antagonistic effects.[9] Bax is increased in the ameloblasts layer, whereas the Bcl-2 shows strong immunolabeling in the whole enamel organ, except for the differentiating ameloblasts.[57,58] Both proteins have been assumed to play essential roles in the process of amelogenesis and to control the survival of dental epithelial cells.[57,58] Among molecules tested (Fas, Fas-ligand, TNF-α and TNF-receptor 1) for involvement in experimentally induced apoptosis, only Fas was detected as mediator of this process in ameloblasts.[59]

Putative Mechanisms Involved in Apoptosis Stimulation during Mouse Odontogenesis

The dental epithelium shows specific gene expression for various signaling molecules such as Fgfs, Shh, and Bmps[38,61–63] where apoptosis also concentrates. Among these molecules, the Bmp-4 and –2 appear to be involved in the stimulation of apoptosis in the dental epithelium.[21,24,37,64] This is in agreement with the similarity in the temporo-spatial distribution of apoptosis and *Bmp-2* and *Bmp-4* expressions.[64] In addition, exogenous Bmp-4 has been shown to induce cessation in both cell proliferation and survival in the dental epithelium *in vitro* through activation of the cyclin-dependent kinase inhibitor p21.[37]

Bmp-4 has been experimentally shown to upregulate apoptosis in neural crest cells[65] and developing limbs.[(e.g., 66–68)] The induction of apoptosis by Bmp-4 requires expression of *Msx-2* in neural crest cells,[69] in embryonic stem cells,[70] and in the molar EK.[37] Functional tests have demonstrated that both Msx-2 and p21 are necessary for the pro-apoptotic effect of Bmp-4 in cultured neuroblasts, but not for its anti-proliferative effect.[71] However, this is not a general rule. For example, the expression of *Msx-2* and *Msx-1* is not essential for Bmp-induced apoptosis in limbs.[68] Bmp-2 causes apoptosis and the cell cycle arrest mediated by p21 such as those in mouse hybridoma cells.[72]

On the contrary, Fgf may prevent apoptosis.[6] Fgfs are essential during tooth development where they focus on determining the toothless and tooth-bearing regions in mice[35,74] and stimulating dental epithelium growth.[38,73] Fgf and Bmp exhibit counteraction during tooth patterning: the Bmp-4 inhibits tooth-inducing activity of Fgf-8 in the prospective toothless diastema.[74] Similarly at later stages, Bmp-4 signal seems to be stronger than the Fgf-8 signal in the future diastema while the Fgf-8 is stronger in the molar region.[35] This antagonism has been also observed in the molar EK. There the apoptosis, which is presumably mediated by Bmps, can be prevented by exogenous Fgf-4.[27] Mutual antagonism between Bmp and Fgf has also been reported during the regulation of development in limbs, feathers and lungs.[68,75–80]

Bmps and Fgfs act as growth inhibitors and activators, respectively, in developing lungs and feathers.[78,80] Morphogenesis of these structures represents models of epithelial budding based on the positional information mediated by reaction-diffusion mechanisms.[a] We have applied similar principles to interpret the development of dentition.[32] The budding morphogenesis in feathers, lungs, and teeth show similari-

[a]Positional information specifies cells in a coordinate system.[81] Reaction-diffusion processes generate spatial variation depending on interactions between activator and inhibitor substances.[63,82]

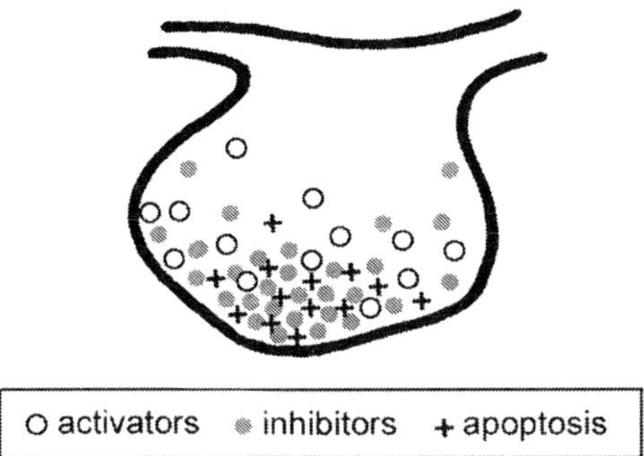

FIGURE 4. A hypothetical scheme showing the stimulation of apoptosis in the budding dental epithelium by predominance of growth inhibitors. During budding morphogenesis, both the growth activators (*circles*) and inhibitors (*gray spots*) co-express at the tip of an epithelial bud, and reaction-diffusion processes influence their effects.[78,80] Activators (e.g., Fgfs) induce and maintain bud growth, whereas inhibitors (e.g., Bmps) antagonize this effect. As bud growth proceeds, increased levels of inhibitors may finally counteract activators at the bud tip and lead to growth arrest.[78,80] The Bmps might also act as growth inhibitors during budding of the dental epithelium; and their local excess might lead to the epithelial cells' failure to receive adequate growth-activating (apoptosis-suppressing) signals (e.g., Fgf). The resulting signal imbalance leads to apoptosis (*black crosses*).

ties at early stages. In all instances, for example, Bmp-4 seems to act as inhibitor of epithelial budding. When the effects of growth inhibitors (e.g., Bmps) override those of growth activators (e.g., Fgfs), the growth of the epithelial bud is stopped.[32,78,80]

Positional information mediated by the reaction-diffusion mechanisms might also play a role in the stimulation of apoptosis. Most cells require survival (death-suppressing) signals to avoid PCD. Accordingly, the PCD is activated either by insufficient death-suppressing signals, or by death-inducing signals that override the action of survival factors.[1,2,6]

We have formulated a hypothesis about the regulation of apoptosis in the dental epithelium[32] by taking into account (a) the regulation of epithelial budding by reaction-diffusion mechanisms[78,80] and (b) the stimulation of PCD by an imbalance in extracellular factors.[1,2,6] The Bmps might function as inhibitors of further budding in the dental epithelium. Their local excess might lead to the epithelial cells' failure to receive adequate growth-activating (apoptosis-suppressing) signals (e.g., Fgf) and to cell death by apoptosis (FIG. 4).

Bmps are members of the TGF-β superfamily. The effects of Bmps are mediated through activation of specific combinations of type I and type II serine/threonine kinase receptors.[83] The activated BMP type I receptors propagate the signal through the specific receptor-regulated Smads.[83] These proteins include the Bmp-activated Smad1, Smad5, and Smad8 and the TGF-β–activated Smad2 and Smad3. Besides, there are also inhibitory Smads (Smad6 and Smad7) acting as inhibitors of TGF-β, activin and Bmp signaling.[83,84] Both Smad1/5 and apoptosis decrease in developing limbs of transgenic mice with experimental inhibition of Bmp signaling.[68] In contrast, the experimental attenuation of Smad7 results in significant inhibition of embryonic tooth development and increase of apoptosis in the dental epithelium.[85]

These data suggest that apoptosis in the specific loci of dental epithelium (e.g., FIG. 3) can be stimulated by an excess of Bmps. This excess might lead to growth inhibition and to pro-apoptotic effects in the dental epithelium via the Bmp–Smad pathway. The growth-inhibitory effect of this pathway may involve the activation of *p21*.[86] Indeed, the growth inhibition associated with *p21* expression has been documented in the EK and ameloblast layer,[37,87] where both *Bmp-2* and *Bmp-4* expression[38] and apoptosis[25,30] occur.

CONCLUSION

Mouse odontogenesis provides a tool to study the regulation of different types of PCD (phylogenetic, morphogenetic and histogenetic) in the same developing system. It will be interesting to compare the differences and similarities in the mechanisms that underlie apoptosis with different developmental roles. Understanding of apoptosis regulation in vestigial tooth primordia can help to elucidate the mechanism of their suppression during evolution and to identify factors essential for tooth survival. Such knowledge will be helpful for developing a technology of tooth engineering

ACKNOWLEDGMENTS

We thank Mr. J. Dutt for critical reading of the manuscript. Technical assistance of Miss B. Rokytová is acknowledged. Supported by Grants GACR (304/02/0448/A) and ASCR (AVOZ5039906), and by French Grant APS02005MSA for the project A02178MS GIS-"Maladies rares."

REFERENCES

1. RAFF, M.C. 1996. Size control: the regulation of cell numbers in animal development. Cell **86:** 173–175.
2. JACOBSON, M.D., M. WEIL & M.C. RAFF. 1997. Programmed cell death in animal development. Cell **88:** 347–354.
3. COUCOUVANIS, E.C., G.R. MARTIN & J.H. NADEAU. 1995. Genetic approaches for studying programmed cell death during development of the laboratory mouse. *In* Methods in Cell Biology: Cell Death. Vol. 46. L.M. Schwartz & B.A. Osborne, Eds.: 387–440. Academic Press. San Diego, CA.
4. KERR, J.F.R., G.C. GOBÉ, C.M. WINTERFORD & B.V. HARMON. 1995. Anatomical methods in cell death. *In* Methods in Cell Biology: Cell Death. Vol. 46. L.M. Schwartz & B.A. Osborne, Eds.: 1–27. Academic Press. San Diego, CA.
5. MEIER, P., A. FINCH & G. EVAN. 2000. Apoptosis in development. Nature **407:** 796–801.
6. VERMES, I. & C. HAAEN. 1994. Apoptosis and programmed cell death in health and disease. Adv. Clin. Chem. **31:** 177–246.
7. KAUFMANN, S.H. & M.O. HENGARTNER. 2001. Programmed cell death: alive and well in the new millennium. Trends Cell Biol. **11:** 526–534.
8. ARAVIND, L., V.M. DIXIT & E.V. KOONIN. 1999. The domains of death: evolution of the apoptosis machinery. TIBS **24:** 47–53.
9. HENGARTNER, M.O. 2000. The biochemistry of apoptosis. Nature **407:** 770–776.
10. BAKER, S.J. & E.P. REDDY. 1998. Modulation of life and death by the TNF receptor superfamily. Oncogene **17:** 3261–3270.

11. GLÜCKSMANN, A. 1951. Cell deaths in normal vertebrate ontogeny. Biol. Rev. **26:** 59–86.
12. MORI, C., N. NAKAMURA, Y. OKAMOTO, *et al.* 1994. Cytochemical identification of programmed cell death in the fusing fetal mouse palate by specific labeling of DNA fragmentation. Anat. Embryol. **190:** 21–28.
13. ZUZARTE-LUIS, V. & J.M. HURLE. 2002. Programmed cell death in the developing limb. Int. J. Dev. Biol. **46:** 871–876.
14. POHL, M., R.O. STUART, H. SAKURAI & S.K. NIGAM. 2000. Branching morphogenesis during kidney development. Ann. Rev. Physiol. **62:** 595–620.
15. WELLER, M., J.B. SCHULZ, U. WULLNER, *et al.* 1997. Developmental and genetic regulation of programmed neuronal death. J. Neural. Transm. Suppl. **50:** 115–123.
16. CALOF, A.L., N. HAGIWARA, J.D. HOLCOMB, *et al.* 1996. Neurogenesis and cell death in olfactory epithelium. J. Neurobiol. **30:** 67–81.
17. MARSDEN, V.S. & A. STRASSER. 2003. Control of apoptosis in the immune system: Bcl, BH3 – only proteins and more. Ann. Rev. Immunol. **21:** 71–105.
18. NOZUE, T. 1971. Specific spindle cells and globular substances in enamel knot. Okajimas Folia Anat. Jpn. **48:** 139–152.
19. KINDAICHI, K. 1980. An electron microscopic study of cell death in molar tooth germ epithelia of mouse embryos. Arch. Histol. Jpn. **43:** 289–304.
20. LESOT, H., J.L. VONESCH, M. PETERKA, *et al.* 1995. Spatial distribution of mitoses and apoptosis. Proceedings of the 10th International Symposium on Dental Morphology. R.J. Radlanski & H. Renz, Eds.: 27–32. C. & M. Brünne GbR. Berlin.
21. TUREČKOVÁ, J., H. LESOT, J-L. VONESCH, *et al.* 1996. Apoptosis is involved in disappearance of the diastemal dental primordia in mouse embryo. Int. J. Dev. Biol. **40:** 483–489.
22. MOSS-SALENTIJN, L. 1978. Vestigial teeth in rabbit, rat and mouse; their relationship to the problem of lacteal dentitions. *In* Development, Function and Evolution of Teeth. P.M. Butler & K.A. Joysey, Eds.: 13–29. Academic Press. London.
23. PETERKOVÁ, R., M. PETERKA, L. VIRIOT & H. LESOT. 2002. Development of the vestigial tooth primordia as part of mouse odontogenesis. Connect. Tissue Res. **43:** 120–128.
24. PETERKOVÁ, R., H. LESOT, J-L. VONESCH, *et al.* 1996. Mouse molar morphogenesis revisited by three dimensional reconstruction: analysis of initial stages of the first upper molar development revealed two transient buds. Int. J. Dev. Biol. **40:** 1009–1016.
25. VIRIOT, L., H. LESOT, J-L. VONESCH, *et al.* 2000. The presence of rudimentary odontogenic structures in the mouse embryonic mandible requires reinterpretation of developmental control of first lower molar histomorphogenesis. Int. J. Dev. Biol. **44:** 233–240.
26. LESOT, H., J-L. VONESCH, M. PETERKA, *et al.* 1996. Mouse molar morphogenesis revisited by three dimensional reconstruction: spatial distribution of mitoses and apoptosis in cap to bell stages first and second molar teeth. Int. J. Dev. Biol. **40:** 1017–1031.
27. VAAHTOKARI, A, T. ÅBERG, & I. THESLEFF. 1996. Apoptosis in the developing tooth: association with an embryonic signaling center and suppression by EGF and FGF-4. Development **122:** 121–129.
28. VIRIOT L, R. PETERKOVÁ, J-L. VONESCH, *et al.* 1997. Mouse molar morphogenesis revisited by three dimensional reconstruction: III) Spatial distribution of mitoses and apoptoses up to bell-staged first lower molar teeth. Int. J. Dev. Biol. **41:** 679–690.
29. KIEFFER, S., R. PETERKOVÁ, J-L. VONESCH, *et al.* 1999. Morphogenesis of the lower incisor in mouse from the bud to early bell stage. Int. J. Dev. Biol. **43:** 531–539.
30. SHIBATA, S., S. SUZUKI, T. TENGAN & Y. YAMASHITA. 1995. A histochemical study of apoptosis in the reduced ameloblasts of erupting mouse molars. Arch. Oral Biol. **40:** 677–680.
31. BRONCKERS, A.L.J.J., D.M. LYARUU, W. GOEI, *et al.* 1996. Nuclear DNA fragmentation during postnatal tooth development of mouse and hamster and during dentin repair in the rat. Eur. J. Oral Sci. **104:** 102–111.
32. PETERKOVÁ, R., M. PETERKA, L. VIRIOT & H. LESOT. 2000. Dentition development and budding morphogenesis. J. Craniofac. Genet. Dev. Biol. **20:** 158–172.

33. LUCKETT, W.P. 1985. Superordinal and intraordinal affinities of rodents: developmental evidence from the dentition and placentation. *In* Evolutionary Relationships among Rodents. W.P. Luckett & J-L. Hartenberger, Eds.: 227–276. Plenum Press. New York.
34. GAUNT, W.A. 1966. The disposition of the developing cheek teeth in the albino mouse. Acta Anat. (Basel) **64:** 572–585.
35. KERÄNEN, S.V.E., P. KETTUNEN, A. ÅBERG, *et al.* 1999. Gene expression patterns associated with suppression of odontogenesis in mouse and vole diastema regions. Dev. Genes Evol. **209:** 495–506.
36. SASAKI, C., T. SATO & Y. KOZAWA. 2001. Apoptosis in regressive deciduous tooth germs of *Suncus murinus* evaluated by the TUNEL method and electron microscopy. Arch. Oral Biol. **46:** 649–660.
37. JERNVALL, J., T. ÅBERG, P. KETUNEN, *et al.* 1998. The life history of an embryonic signaling center: BMP-4 induces *p21* and is associated with apoptosis in the mouse tooth enamel knot. Development **125:** 161–169.
38. THESLEFF, I. & M. MIKKOLA. 2002. The role of growth factors in tooth development. Int. Rev. Cytol. **217:** 93–135.
39. PETERKA, M., H. LESOT & R. PETERKOVÁ. 2002. Body weight in mouse embryos specifies staging of tooth development. Connect. Tissue Res. **43:** 186–190.
40. COIN, R., H. LESOT, J-L. VONESCH, *et al.* 1999. Aspects of cell proliferation kinetics of the dental epithelium during mouse molar and incisor morphogenesis: a reappraisal of the role of the enamel knot area. Int. J. Dev. Biol. **43:** 261–267.
41. LESOT, H., R. PETERKOVÁ, R. SCHMITT, *et al.* 1999. Initial features of the inner dental epithelium histo-morphogenesis in the first lower molar in mouse. Int. J. Dev. Biol. **43:** 245–254.
42. AHARONI, D., I. MEIR, R. ATZMON, *et al.* 1997. Differential effect of components of the extracellular matrix on differentiation and apoptosis. Curr. Biol. *7*: 43–51.
43. CLARKE, P.H.G. 1990. Developmental cell death: morphological diversity and multiple mechanisms. Anat. Embryol. **181:** 195–213.
44. ESCO, M.A., Z. WANG, M.L. MCDERMOTT & M. KURPAKUS-WHEATER. 2001. Potential role for laminin 5 in hypoxia-mediated apoptosis of human corneal epithelial cells. J. Cell Sci. **114:** 4033–4040.
45. YOSHIBA, K., N. YOSHIBA, D. ABERDAM, *et al.* 2000. Differential expression of laminin-5 subunits during incisor and molar development in the mouse. Int. J. Dev. Biol. **44:** 337–340.
46. KIEFFER-COMBEAU, S., J.M. MEYER & H. LESOT. 2001. Cell-matrix interactions and cell–cell junctions during epithelial histo-morphogenesis in the developing mouse incisor. Int. J. Dev. Biol. 2001. **45:** 733–742.
47. MOE, H. 1997. Physiological cell death of secretory ameloblasts in the rat incisors. Cell Tissue Res. **197:** 443–451.
48. JOSEPH, B.K., G.C. GOBE, N.W. SAVAGE & W.G. YOUNG. 1994. Expression and localization of sulphated glycoprotein-2 mRNA in the rat incisor tooth ameloblasts: relationships with apoptosis. Int. J. Exp. Pathol. **75:** 313–320.
49. YAMAMOTO, H., K. ISHIZEKI, J. SASAKI & T. NAWA. 1998. Ultrastructural and histochemical changes and apoptosis of inner enamel epithelium in rat enamel free area. J. Craniofac. Genet. Dev. Biol.**18:** 44–50.
50. BARATELLA, L., V.E. ARANA-CHAVEZ, E. KATCHBURIAN. 1999. Apoptosis in the early involuting stellate reticulum of rat molar tooth germs. Anat. Embryol. **200:** 49–54.
51. KANEKO, H., S. HASHIMOTO, Z. ENOKIYA, *et al.* 1999. Cell proliferation and death of Hertwig´s epithelial root sheat in the rat. Cell Tissue Res. **298:** 95–103.
52. CERRI, P.S., E. FREYMÜLLER & E. KATCHBURIAN. 2000. Apoptosis in the early developing periodontium of rat molars. Anat. Rec. **258:** 136–144.
53. FRANQUIN, J.C., M. REMUSAT, I. ABOU HASHIEH & J. DEJOU. 1998. Immunocytochemical detection of apoptosis in human odontoblasts. Eur. J. Oral Sci. **106** (Suppl. 1): 384–387.
54. HATAKEYAMA, S., N. TOMICHI, Y. OHARA-NEMOTO & M. SATOH. 2000. The immunohistochemical localization of Fas ligand in jaw bone and tooth germ of human fetuses. Calcif. Tissue Int. **66:** 330–337.
55. KRAJEWSKI, S., A. HUGGER, M. KRAJEWSKA, *et al.* 1998. Developmental expression patterns of Bcl-2, Bcl-x, Bax, and Bak in teeth. Cell Death Differ. **5:** 408–415.

56. SHIGEMURA, N., T. KIYOSHIMA, T. SAKAI, *et al.* 2001. Localization of activated caspase-3-positive and apoptotic cells in the developing tooth germ of the mouse lower first molar. Histochem. J. **33:** 253–258.
57. SLOOTWEG, P.J. & R.A. WEGER. 1994. Immunohistochemical demonstration of bcl-2 protein in human tooth germs. Arch. Oral Biol. **39:** 545–550.
58. KONDO, S., Y. TAMURA, J.W. BAWDEN & S. TANASE. 2001. The immunohistochemical localization of Bax and Bcl-2 and their relation to apoptosis during amelogenesis in developing rat molars. Arch. Oral Bio. **41:** 557–568.
59. NISHIKAWA, S. 2002. Colchicine-induced apoptosis and anti-Fas localization in rat-incisor ameloblasts. Anat. Sci. Int. **77:** 175–181.
60. TUREČKOVÁ, J., C. SAHLBERG, T. ÅBERG, *et al.* 1995. Comparison of expression of the msx-1, msx-2, BMP-2 and BMP-4 genes in the mouse upper diastemal and molar tooth primordia. Int. J. Dev. Biol. **39:** 459–468.
61. DASSULE, R.H. & A.P. MCMAHON. 1998. Analysis of epithelial-mesenchymal interactions in the initial morphogenesis of the mammalian tooth. Dev. Biol. **202:** 215–227.
62. TUCKER, A.S. & P.T. SHARPE. 1999. Molecular genetics of tooth morphogenesis and patterning: the right shape in the right place. J. Dent. Res. **78:** 826–834.
63. WEISS, K.M., D.W. STOCK & Z. ZHAO. 1998. Dynamic interactions and the evolutionaty genetics of dental patterning. Crit. Rev. Oral Biol. Med. **9:** 369–398.
64. PETERKOVÁ, R., M. PETERKA, J-L. VONESCH, *et al.* 1998. Correlation between apoptosis distribution and Bmp-2 and BMP-4 expression in vestigial tooth primordia in mice. Eur. J. Oral Sci. **106:** 667–670.
65. SMITH, A. & A. GRAHAM. 2001. Restricting Bmp-4 mediated apoptosis in hindbrain neural crest. Dev. Dyn. **220:** 276–283.
66. ZHOU, H. & L. NISWANDER. 1996. Requirement for BMP signaling in interdigital apoptosis and scale formation. Science **272:** 738–741.
67. YOKOUCHI, Y., J. SAKIYAMA, T. KAMEDA, *et al.* 1996. BMP-2/-4 mediate programmed cell death in chicken limb buds. Development **122:** 3725–3734.
68. GUHA, U., W.A. GOMES, T. KOBAYASHI, *et al.* 2002. In vivo evidence that BMP signaling is necessary for apoptosis in the mouse limb. Dev. Biol. **249:** 108–120.
69. GRAHAM, A. & A. LUMSDEN. 1996. Patterning the cranial neural crest. Biochem. Soc. Symp. **62:** 77–83.
70. MARAZZI, G., Y. WANG & D. SASSOON. 1997. Msx2 is a transcriptional regulator in the BMP-4 mediated programmed cell death pathway. Dev. Biol. 127–138.
71. GOMES, W.A. & J.A. KESSLER. 2001. Msx-2 and p21 mediate the pro-apoptotic but not the anti-proliferative effects of BMP4 on cultured sympathetic neuroblasts. Dev. Biol. **237:** 212–221.
72. YAMATO, K, S. HASHIMOTO, N. OKAHASHI, *et al.* 2000. Dissociation of bone morphogenetic protein-mediated growth arrest and apoptosis of mouse B cells by HPV-16 E6/E7. Exp. Cell Res. **257:** 198–205.
73. HARADA, H., T. TOYONO, K. TOYOSHIMA, *et al.* 2002. Fgf-10 maintains stem cell compartment in developing mouse incisors. Development **129:** 1533–1541.
74. NEUBÜSE, A., H. PETERS & R. BALLING. 1997. Antagonistic interactions between FGF and BMP signaling pathways: a mechanism for positioning the sites of tooth formation. Cell **90:** 247–255.
75. NIESWANDER, L. & G.R. MARTIN. 1993. FGF-4 and BMP 2 have opposite effects on limb growth. Nature **361:** 68–71.
76. BUCKLAND, R.A., J.M. COLLINSON, E. GRAHAM, *et al.* 1998. Antagonistic effects of FGF4 on BMP induction of apoptosis and chondrogenesis in the chick limb bud. Mech. Dev. **71:** 143–150.
77. CHUONG, C.M. & A. NOVEEN. 1999. Phenotypic determination of epithelial appendages: genes, developmental pathways, and evolution. J. Invest. Dermatol. Symp. Proc. **4:** 307–311.
78. HOGAN, B.L.M. 1999. Morphogenesis. Cell **96:** 225–233.
79. METZGER, R.J. & M.A. KRASNOW. 1999. Genetic control of branching morphogenesis. Science **284:** 1635–1639.
80. JUNG, H.S., P.H. FRANCIS-WEST, R.B. WIDELITZ, *et al.* 1998. Local inhibitory action of BMPs and their relationships with activators in feather formation: implications for periodic patterning. Dev. Biol. **196:** 11–23.

81. WOLPERT, L. 1989. Positional information revisited. Development **107**(Suppl.): 3–12.
82. TURING, A.M. 1952. The chemical basis of morphogenesis. Philos. Trans. R. Soc. London B **237:** 37–72.
83. ITOH, F., H. ASAO, K. SUGAMURA, *et al.* 2001. Promoting bone morphogenetic protein signaling through negative regulation of inhibitory Smads. EMBO J. **20:** 4132–4142.
84. LUTZ, M. & P. KNAUS. 2002. Integration of the TGF-β pathway into the cellular signalling network. Cell Signal **14:** 977–988.
85. ITO, Y., J. ZHAO, A. MOGHARELI, C.F. SHULER, *et al.* 2001. Antagonistic effects of Smad2 versus Smad7 are sensitive to their expression level during tooth development. J. Biol. Chem. **276:** 44163–44172.
86. POULIOT, F. & C. LABRIE. 2002. Role of Smad1 and Smad4 proteins in the induction of p21 WAF1, Cip1 during bone morphogenetic protein-induced growth arrest in human breast cancer cells. J. Endocrinol. **72:** 187–198.
87. BLOCH-ZUPAN, A., T. LEVEILLARD, P. GORY, *et al.* 1998. Expression of p21 WAF1/CIP1 during mouse odontogenesis. Eur. J. Oral Sci. **106** (Suppl.1): 104–111.

Innate Immune Collectins Bind Nucleic Acids and Enhance DNA Clearance *in Vitro*

NADES PALANIYAR, JEYA NADESALINGAM, AND KENNETH B. M. REID

MRC Immunochemistry Unit, Department of Biochemistry, Oxford University, Oxford OX1 3QU, UK

ABSTRACT: Collectins are a class of innate immune proteins that opsonize microbes and other ligands via their carbohydrate-recognition domains and enhance their clearance by phagocytes. We hypothesized that collectins, such as pulmonary surfactant protein (SP-) A and D and mannose-binding lectin (MBL), could bind nucleic acid, which is a pentameric sugar-based anionic polymer. We have shown that SP-D and MBL bind to both DNA and RNA effectively and also that SP-D enhances the uptake of DNA by human monocytic cells. Therefore, our results suggest that nucleic acids may be a novel class of ligands for collectins. This class of innate immune proteins could enhance the clearance of DNA that is released by necrotic cells or microbial cells. Our findings may have important biological implications, such as the alleviation of DNA-mediated tissue inflammation.

KEYWORDS: collectins; nucleic acids; pulmonary surfactant proteins; mannose-binding lectin

INTRODUCTION

Collectins such as pulmonary surfactant proteins (SP-) A and D, and mannose-binding lectin (MBL) have fibrillar collagen-like regions and globular C-type lectin domains, and are known pattern recognition proteins.[1] These collectins recognize a variety of hexose sugar moieties of the microorganisms. Nucleic acid is a pentose-based polyanionic fiber, whcih causes tissue inflammation[2] and required for chronic infection of certain pulmonary pathogens.[3] We hypothesized that collectins may bind DNA by their lectin domains and polycationic stretches present in the collagen-like regions. Our results show that SP-A, SP-D and MBL bind DNA and RNA but with different affinities. Native SP-D shows stronger interaction with DNA, compared with that of SP-A. Our results also provide evidence that both carbohydrate recognition domains (CRDs) and collagen-like regions of SP-D are involved in the interaction with DNA. In addition, SP-D enhances the uptake of DNA by U937 monocytic cell line, more effectively than SP-A and C1q. Collectively, our data suggest that collectins bind DNA and enhance their clearance *in vitro*.

Address for correspondence: Nades Palaniyar, MRC Immunochemistry Unit, Department of Biochemistry, Oxford University, South Parks Road, Oxford OX1 3QU, UK. Voice: +44-1865-275795; fax: +44-1865-275729.

Palaniyar.Nadesalingam@bioch.ox.ac.uk

Ann. N.Y. Acad. Sci. 1010: 467–470 (2003).
doi: 10.1196/annals.1299.084

MATERIALS AND METHODS

DNA–protein complexes were prepared for gel shift analyses by incubating 0.5 μg of genomic DNA with protein in a 20-μL reaction containing 20 mM Tris-HCl (pH 7.4), 150 mM NaCl, 5 mM $CaCl_2$ buffer. After 20 min at 37°C, the DNA–protein complexes were fractionated by agarose gel (1%, wt/vol) electrophoresis, and visualized by UV transilluminator. The $[^{35}]$S or Cy3-labeled plasmid or genomic DNA (2–5 μg) was mixed with different concentrations of proteins (0–10 μg/mL) in HENK balance salts (Flow laboratories, UK) containing 5 mM $CaCl_2$ or 2 mM $MgCl_2$ and added to RPMI media (Gibco BRL, UK). This mixture was added to the U937 cells (10^5/well) grown in 24-well plates (Nunc, UK) in RPMI media and incubated at 37°C for 0–60 min, washed twice with PBS, fixed in 0.5% (vol/vol) formaldehyde/PBS, and the intensity of Cy3 fluorescence in the cells was determined in a FACS station (Becton Dickinson, UK).

RESULTS

Collectins Bind DNA

Although collectins contain both collagen-like regions and CRDs, whether these proteins bind DNA is unknown. To determine this function, we purified collectins such as SP-A, SP-D, and MBL, and related proteins C1q, SAP and gp-340 (gift from Dr. Uffe Holmskov, Denmark), and tested their nucleic acid binding abilities. EMSA assays showed that SP-D, MBL, and C1q form large complexes and retard the migration of the viral DNA and RNA (FIG. 1). DNA binding ability of SP-A, SAP and SP-D(n/CRD) were detected by surface plasmon resonance analyses and electron microscopy (unpublished data). These data suggest that collectins bind DNA via both the collagen-like region and CRDs.

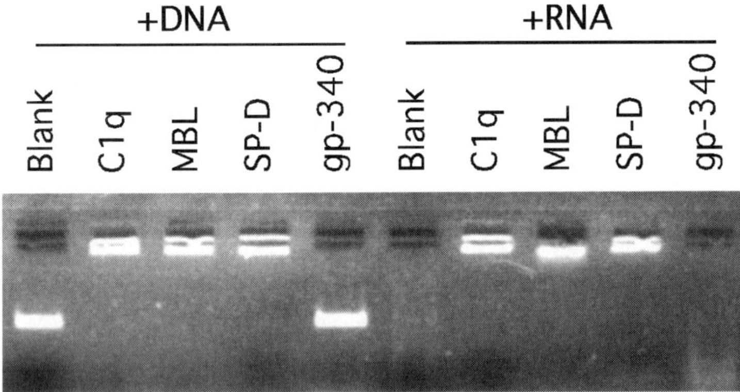

FIGURE 1. Binding of collectins and related proteins to DNA and RNA. Single-stranded M13mp19 viral DNA or mouse lung total RNA was incubated with proteins, and the mixtures were size-fractionated in an agarose gel (0.8% (wt/vol). Free RNA is quickly degraded in the binding buffer, but is protected by C1q, MBL and SP-D.

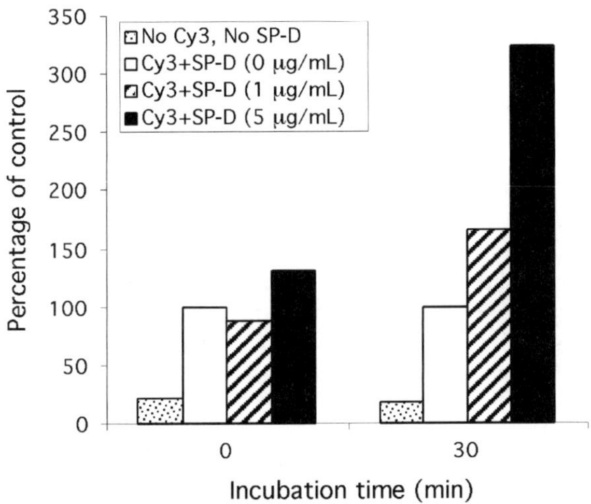

FIGURE 2. SP-D enhances the phagocytosis of genomic DNA. The U937 cells were incubated with Cy3-labeled DNA in the presence of different concentrations of SP-D, and the cell-associated label was determined by FACS analyses.

SP-D Enhances the Phagocytosis of DNA

To determine whether the binding of collectins to DNA leads to an enhanced phagocytosis of the ligand, we mixed the Cy3-labeled DNA and proteins and incubated them with U937 cells. FACS analyses of the cell-associated label showed that SP-D enhances the clearance of DNA by phagocytes (FIG. 2). Quantifying the $^{[25]}$S-labeled DNA distribution in U937 cells indicated that SP-D enhances the clearance of DNA more effectively than SP-A and C1q. Taken together, these results suggest that SP-D effectively enhances the clearance of DNA *in vitro.*

DISCUSSION

We found that collectins bind DNA and RNA. In addition we showed that SP-D enhances the clearance of DNA by phagocytes *in vitro.* The simple EMSA assay using agarose gels detects the binding of collectins to DNA and RNA, however, this experiment does not detect carbohydrate-based interactions. Certain proteins that contain collagen-like regions, including C1q, bind DNA. Our EMSA results suggest that, like C1q, both SP-D and MBL, but not SP-A, bind DNA via their collagen-like regions. It is well known that lectins, including the SAP,[4] bind agarose matrix; hence, a more sensitive assay system is necessary to detect this activity. Since the recombinant fragment that contained only the neck+CRD region binds DNA in BIAcore and electron microscopy, these domains can also directly bind DNA. Interestingly, DNA is a pentose-based polymer; hence, it is probable that these two molecules also interact with each other via the lectin–carbohydrate interactions. Further studies are necessary to understand the exact molecular nature of these interactions. Since DNA is an inflammation-causing molecule,[2] clearance of free DNA from any microenvironment is essential. SP-D deficiency in mice results in the inflammation of the lung,[5] and therefore DNA may be one of the important contributing factors in this pathophysiological process.

SUMMARY

We hypothesized that collectins may bind nucleic acid, that is, an anionic polymer of repeating pentose-based subunits, by their lectin domains as well as the polycationic stretches of collagen-like regions. Our data suggest that SP-A, SP-D, and MBL can bind DNA but with different affinities. Native SP-D, but not SP-A, showed strong interaction with DNA. Although the removal of the collagen-like region of SP-D reduced its DNA-binding ability, the trimeric lectin units of these highly oligomeric proteins bind to DNA *in vitro*. These results indicate that the interaction between collectins and DNA involves both collagen-like regions and lectin domains of these proteins. The results further suggest that both carbohydrate recognition domains and collagen-like regions of SP-D interact with DNA via the lectin activity and the polyionic ligand-binding capability, respectively. FACS and scintillation counting analyses show that SP-D enhances the uptake of Cy3 or 3H-labeled DNA by U937 monocytic cell line, more effectively than SP-A and C1q. Collectively, our data suggest that collectins bind DNA and enhance their clearance *in vitro*.

ACKNOWLEDGMENTS

This study was funded by the Wellcome Trust/Canadian Institutes of Health Research fellowships (PN), and grants from MRC-UK and EU QLK2-CT-2000-00325 (JN, KBMR).

REFERENCES

1. Palaniyar, N., J. Nadesalingam & K.B. Reid. 2002. Pulmonary innate immune proteins and receptors that interact with gram-positive bacterial ligands. Immunobiology. **205:** 575–594.
2. Schwartz, D.A. *et al.* 1997. CpG motifs in bacterial DNA cause inflammation in the lower respiratory tract. J. Clin. Invest. **100:** 68–73.
3. Whitchurch, C.B. *et al.* 2002. Extracellular DNA required for bacterial biofilm formation. Science **295:** 1487.
4. Hind, C.R. *et al.* 1984. Binding specificity of serum amyloid P component for the pyruvate acetal of galactose. J. Exp. Med. **159:** 1058–1069.
5. Clark, H. *et al.* 2002. Surfactant protein D reduces alveolar macrophage apoptosis in vivo. J. Immunol. **169:** 2892–2899.

Surfactant Protein D Binds Genomic DNA and Apoptotic Cells, and Enhances Their Clearance, *in Vivo*

NADES PALANIYAR,[a] HOWARD CLARK,[a] JEYA NADESALINGAM,[a] SAMUEL HAWGOOD,[b] AND KENNETH B.M. REID[a]

[a]*MRC Immunochemistry Unit, Department of Biochemistry, Oxford University, South Parks Road, Oxford, OX1 3QU, United Kingdom*

[b]*Department of Pediatrics and Cardiovascular Research Institute, University of California, San Francisco, California, USA*

ABSTRACT: Collectins, such as surfactant protein D (SP-D), bind apoptotic cells; however, the ligands that they recognize on these cells are unknown. We hypothesized that SP-D binds to the DNA present on these cells. We show that SP-D binds and aggregates mouse alveolar macrophage DNA effectively. Alveolar macrophages of SP-D$^{(-/-)}$ mice contained more nicked DNA than those of SP-A$^{(-/-)}$ and wild type mice. Our results also suggest that carbohydrate recognition domains of SP-D may recognize DNA present on the apoptotic cells. Therefore, cell-surface DNA could be a ligand for recognition of apoptotic cells by collectins.

KEYWORDS: surfactant protein D; pulmonary surfactant proteins; alveolar macrophages

INTRODUCTION

Collectins are well-studied innate immune proteins that contain fibrillar collagen-like regions and globular C-type lectin domains. Pulmonary surfactant proteins (SP) A and D of the lung, and the mannose-binding lectin (MBL) of the serum are some of the major innate immune collectins. Recent studies indicate that these collectins bind apoptotic cells to varying degrees;[1–3] however, the ligands for these lectins on the surface of the dying cells are unknown. We hypothesized that these lectins bind DNA present on the surface of apoptotic cells. To test this hypothesis, we used SP-A and SP-D knockout mice and studied the nature of alveolar macrophages (AM) and their ability to bind DNA and SP-D.

Address for correspondence: Nades Palaniyar, MRC Immunochemistry Unit, Department of Biochemistry, Oxford University, South Parks Road, Oxford, OX1 3QU, United Kingdom. Voice: +44-1865-275795; fax: +44-1865-275729.
Palaniyar.Nadesalingam@bioch.ox.ac.uk

Ann. N.Y. Acad. Sci. 1010: 471–475 (2003). © 2003 New York Academy of Sciences.
doi: 10.1196/annals.1299.085

METHODS AND MATERIALS

DNA-protein complexes were prepared for gel shift analyses by incubating 0.5 μg of genomic DNA with protein in a 20-μL reaction containing 20 mM Tris-HCl (pH 7.4), 150 mM NaCl, 5 mM $CaCl_2$ buffer. After 30 min at 37°C, the DNA-protein complexes were fractionated by agarose gel (1%, w/v) electrophoresis and visualized by UV transilluminator. AMs isolated from different mice were separated from the BAL fluid by centrifugation and processed for nicked DNA labeling as described by the manufacturer (Pharmingen, UK). These cells were incubated with FITC-labeled SP-D (n/CRD) either before or after the TUNEL assay. The label associated with AMs were determined by FACS analyses (Beckman Dickson, UK).

RESULTS

SP-D Binds Genomic DNA

Apoptotic and necrotic macrophages contain degraded/nicked DNA, which is a putative ligand to signal their clearance. To test whether SP-D binds to genomic DNA of AMs, we isolated the ligand from SP-D$^{(-/-)}$ and wild type mice and incubated them with purified native SP-D. When size-fractionated on agarose gels, it was apparent that SP-D, but not SP-D (n/CRD), formed large complexes with DNA (FIG. 1). The intense ethidium bromide staining of DNA detected in the SP-D (n/CRD) lane may indicate that this recombinant protein fragment alters the structure of DNA. No DNA laddering was apparent in the genomic DNA isolated from the SP-D$^{(-/-)}$, and the DNA appeared similar to the one isolated from the wild type mice (FIG. 1,

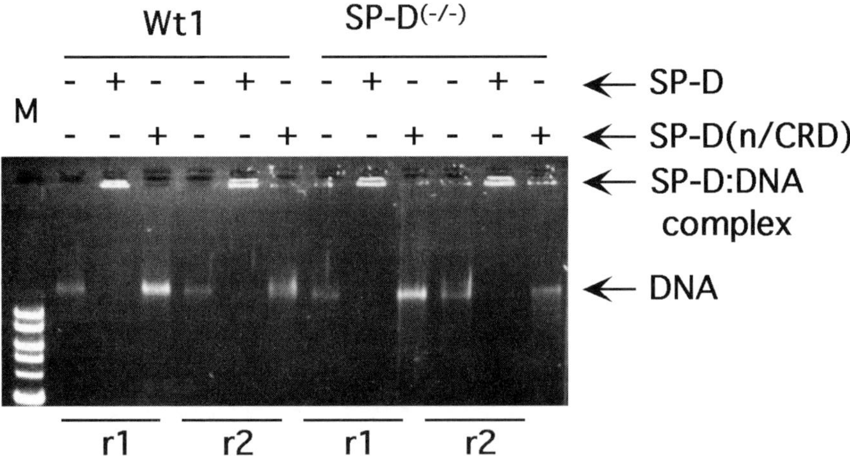

FIGURE 1. Agarose gel showing that SP-D binds genomic DNA. SPD was incubated with genomic DNA isolated from AMs of SP-D$^{(-/-)}$ and wild type mice and separated on a 0.8% (w/v) gel. Each replicate (r1, r2) consists of DNA isolated from AMs of 3–4 mice. M, 1Kb DNA ladder (10, 8, 6, 5, 4, and 3 kb; New England Biolabs).

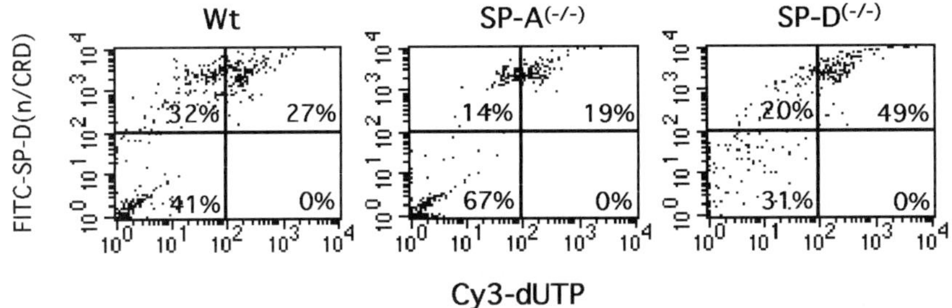

FIGURE 2. FACS analyses showing that SP-D (n/CRD) preferentially binds to nick-DNA-containing AMs. AMs were isolated from wild type, SP-A$^{(-/-)}$, and SP-D$^{(-/-)}$ mice, and the nicked DNA was labeled with Cy3-dUTP by TUNEL assay and incubated with FITC-labeled SP-D (n/CRD) and analyzed by FACS.

data not shown). SP-D formed complexes with DNA isolated from the AMs of both SP-D$^{(-/-)}$ and wild type mice. We also found that AMs isolated from the SP-D$^{(-/-)}$ mice cleared free DNA less effectively compared with that of the ones of SP-A$^{(-/-)}$ and wild type mice (unpublished data). This finding may indicate that the SP-D$^{(-/-)}$ mice is defective in clearing DNA from the lung.

SP-D (n/CRD) Binds Well to Apoptotic Alveolar Macrophages

SP-D binds to AM.[4] To determine whether SP-D preferentially binds apoptotic macrophages, and to understand which domain of the protein is necessary for the interaction with these phagocytes, we conducted double labeling experiments. We isolated AMs from SP-A$^{(-/-)}$, SP-D$^{(-/-)}$, and wild type mice, and labeled the nicked DNA with Cy3-dUTP by the TUNEL assay. A 60-kDa recombinant fragment of SP-D (n/CRD) that contains eight Gly-X-Y repeats of the proximal collagen-like region, neck, and CRDs were labeled with FITC, and allowed to bind to these AMs. Analyses of these cells showed that AMs isolated from SP-D$^{(-/-)}$ mice contained more nicked DNA than those from SP-A$^{(-/-)}$ and wild type mice (FIG. 2). SP-D (n/CRD) bound to AM of all three mice genotypes, however, showed preferential interaction with the phagocytes that contained nicked DNA.

DISCUSSION

We showed that SP-D binds genomic DNA isolated from AMs, and SP-D (n/CRD) preferentially interacts with nicked-DNA containing phagocytes. These results suggest that SP-D could bind DNA present on the surface of the apoptotic or necrotic cells, and signals their clearance. Since the intranasal application of SP-D (n/CRD) to the SP-D$^{(-/-)}$ mice results in the clearance of apoptotic macrophages from the lungs,[1] at least some of the minimal domains necessary for this function is present in the globular C-terminal portion of the protein.

Although agarose gel electrophoresis analyses of DNA isolated from AMs show no obvious sign of DNA laddering, direct labeling experiments by TUNEL assay

(FIG. 2) suggest that AMs of the SP-D$^{(-/-)}$ have nicked DNA. The presence of high levels of nicked DNA in these mice may suggest that the cells are apoptotic or accumulate fragments of DNA in the phagocyte. AMs of SP-D$^{(-/-)}$ mice are large and foamy and known to accumulate other compounds such as lipids;[5] however, whether they accumulate extranuclear DNA is unknown. In addition to the TUNEL assay, propidium iodide (PI) and Annexin V staining of these AMs show signs of apoptosis.[1] Taken together, these results may suggest that considerable number of AMs of the SP-D$^{(-/-)}$ mice undergo asynchronized apoptosis or are necrotic or accumulate DNA. Several infection experiments by others suggest that these macrophages are active and clear pathogens, effectively;[6] however, their intrinsic capacity needs to be determined clearly. Further experiments are underway to describe the detailed mechanisms involved in the defective clearance of DNA and apoptotic cells from these mice.

SUMMARY

We hypothesized that collectins bind free DNA and DNA present on the surface of apoptotic cells. To test this hypothesis, we used SP-A$^{(-/-)}$ and SP-D$^{(-/-)}$ mice, and studied the nature of AMs and their ability to bind DNA and SP-D. FACS analysis and confocal microscopy showed that a recombinant fragment of SP-D, that contains the coiled-coil neck region and the CRDs but lacks the N-terminal and most of the collagen-like regions (SP-D (n/CRD)), binds to the apoptotic AMs with high efficiency. This result suggests that the lectin activity of SP-D may be involved in the binding of this protein to the apoptotic cells; therefore, SP-D is likely to play an important role in the clearance of apoptotic cells. AMs isolated from the SP-D$^{(-/-)}$ mice are defective in clearing free DNA compared with those of SP-A$^{(-/-)}$ and wild type mice. In summary, these studies indicate that SP-D binds free DNA as well as the DNA present on the surface of apoptotic cells, and enhances their clearance, *in vivo*.

ACKNOWLEDGMENTS

This study was funded by the Wellcome Trust/Canadian Institutes for Health Research fellowships (N.P.), a Beit Fellowship (H.C.), and grants from the MRC-UK, EU QLK2-CT-2000-00325 (J.N. and K.B.M.R.) and the NIH HL-24075, HL-58047 (S.H.).

REFERENCES

1. CLARK, H. *et al.* 2002. Surfactant protein D reduces alveolar macrophage apoptosis in vivo. J. Immunol. **169:** 2892–2899.
2. OGDEN, C.A. *et al.* 2001. C1q and mannose binding lectin engagement of cell surface calreticulin and CD91 initiates macropinocytosis and uptake of apoptotic cells. J. Exp. Med. **194:** 781–795.
3. VANDIVIER, R.W. *et al.* 2002. Role of surfactant proteins A, D, and C1q in the clearance of apoptotic cells in vivo and in vitro: calreticulin and CD91 as a common collectin receptor complex. J. Immunol. **169:** 3978–3986.

4. MIYAMURA, K. *et al.* 1994. Surfactant protein D binding to alveolar macrophages. Biochem. J. **300:** 237–242.
5. BOTAS, C. *et al.* 1998. Altered surfactant homeostasis and alveolar type II cell morphology in mice lacking surfactant protein D. Proc. Natl. Acad. Sci. USA **95:** 11869–11874.
6. LEVINE, A.M. & J.A. WHITSETT. 2001. Pulmonary collectins and innate host defense of the lung. Microbes Infect. **3:** 161–166.

Inhibition of HSP70 and a Collagen-Specific Molecular Chaperone (HSP47) Expression in Rat Osteoblasts by Microgravity

YASUHIRO KUMEI,[a] SADAO MORITA,[a] HITOYATA SHIMOKAWA,[a] KEI'ICHI OHYA,[a] HIDEO AKIYAMA,[b] MASAHIKO HIRANO,[b] CLARENCE F. SAMS,[c] AND PEGGY A. WHITSON[c]

[a]*Graduate School of Tokyo Medical and Dental University, Tokyo 113-8549, Japan*

[b]*Toray Research Center, Kamakura 248-8555, Japan*

[c]*NASA Johnson Space Center, Houston, Texas 77058, USA*

ABSTRACT: Rat osteoblasts were cultured aboard a space shuttle for 4 or 5 days. Cells were exposed to 1α, 25 dihydroxyvitamin D_3 during the last 20 h and then solubilized by guanidine solution. The mRNA levels for molecular chaperones were analyzed by semi-quantitative RT-PCR. ELISA was used to quantify TGF-β1 in the conditioned medium. The HSP70 mRNA levels in the flight cultures were almost completely suppressed, as compared to the ground (1 × *g*) controls. The inducible HSP70 is known as the major heat shock protein that prevents stress-induced apoptosis. The mean mRNA levels for the constitutive HSC73 in the flight cultures were reduced to 69%, ~ 60% of the ground controls. HSC73 is reported to prevent the pathological state that is induced by disruption of microtubule network. The mean HSP47 mRNA levels in the flight cultures were decreased to 50% and 19% of the ground controls on the 4th and 5th days. Concomitantly, the concentration of TGF-β1 in the conditioned medium of the flight cultures was reduced to 37% and 19% of the ground controls on the 4th and 5th days. HSP47 is the collagen-specific molecular chaperone that controls collagen processing and quality and is regulated by TGF-β1. Microgravity differentially modulated the expression of molecular chaperones in osteoblasts, which might be involved in induction and/or prevention of osteopenia in space.

KEYWORDS: space flight; stress; osteoblast; HSP70; HSP47; HSC73; TGF-β1

INTRODUCTION

Microgravity induces bone mass loss in humans and animals. Although a number of *in vivo* and *in vitro* studies have been conducted, the mechanisms remain unknown. In response to mechanical forces, osteoblasts play an important role in bone

Address for correspondence: Yasuhiro Kumei, Biochemistry, Department of Hard Tissue Engineering, Graduate School of Tokyo Medical and Dental University, Tokyo 113-8549, Japan. Voice and fax: +81-3-5803-4555.
kumei.bch@tmd.ac.jp

Ann. N.Y. Acad. Sci. 1010: 476–480 (2003).
doi: 10.1196/annals.1299.086

formation and bone remodeling. Microgravity might be perceived as a physical stress, generating significant changes in the microenvironment of osseous tissue.

It is well documented that cells exposed to harmful stress produce heat shock proteins (HSPs) that act as molecular chaperones and play key roles in mediating protein folding, assembly, transport, and degradation. HSPs also protect cells from apoptosis by preventing denaturation and dysfunction of proteins.

However, space flight data were extremely limited concerning apoptosis and/or molecular chaperons in bone cells. The purpose of this study was to examine the expression of the major stress-inducible HSP70, the constitutive (cognate) HSC73, and the collagen-specific molecular chaperone HSP47 in rat osteoblasts that were cultured during space flight.

MATERIALS AND METHODS

Primary cultures of osteoblasts were obtained from femurs of young male Wistar rats, by using α-modified Eagle's medium supplemented with 10% fetal bovine serum and 2 mM β-glycerophosphate.[1] Cells were inoculated three days before launch, and cultured aboard a space shuttle for 4 days (Experiment I) or 5 days (Experiment II), using a treatment of 1 nM of 1α,25 dihydroxyvitamin D_3 for the last 20 hours. Cells were then solubilized with guanidine isothiocyanate solution during the spaceflight. The ground ($1 \times g$) controls and flight experiments had identical procedures and timelines. After return to earth, the mRNA levels for HSP70, HSC73, HSP47 were analyzed by semiquantitative RT-PCR after normalization to glyceraldehyde 3-phosphate dehydrogenase.[1] The gene-specific primers used were: HSP70 forward 5′-CTCGTGCGTGGGCGTGTTCC-3′; HSP70 reverse 5′-TTTCCCTTGTAGTTCACCTG-3′; HSC73 forward 5′-TGAAGAAGCACCACCAGATG-3′; HSC73 reverse 5′-TGATGAAGACAAACAGAAGA-3′; HSP47 forward 5′-CAATGTGACCTGGAAACTGG-3′; and HSP47 reverse 5′-ATGAAGCCACGGTTGTCTAC-3′. TGF-β1 in the conditioned medium was quantified by ELISA (Quantikine™, R&D Systems, Minneapolis).

RESULTS AND DISCUSSION

The HSP70 mRNA expression was completely suppressed in three of the four flight cultures, whereas significant levels for HSP70 mRNA were observed in the ground ($1 \times g$) controls (FIG. 1A). The mean HSC73 mRNA levels in the flight cultures were 69% and 60% of the ground controls on the fourth and fifth days of the mission, respectively (FIG. 1B). We observed a similar decrease of the HSP70 mRNA levels in human osteoblasts that were cultured under a simulated microgravity condition (unpublished data). Our data were conflicted with the report that microgravity increased apoptosis in human T lymphoblastoid Jurkat cells with upregulation of HSP27, but not HSP70.[2] In our study, the total amount of DNA per flight culture vessel was decreased to half of the ground controls, which suggested that apoptosis might occur during the early phase of the spaceflight.[1] However, we did not find any difference of HSP27 expression between flight cultures versus

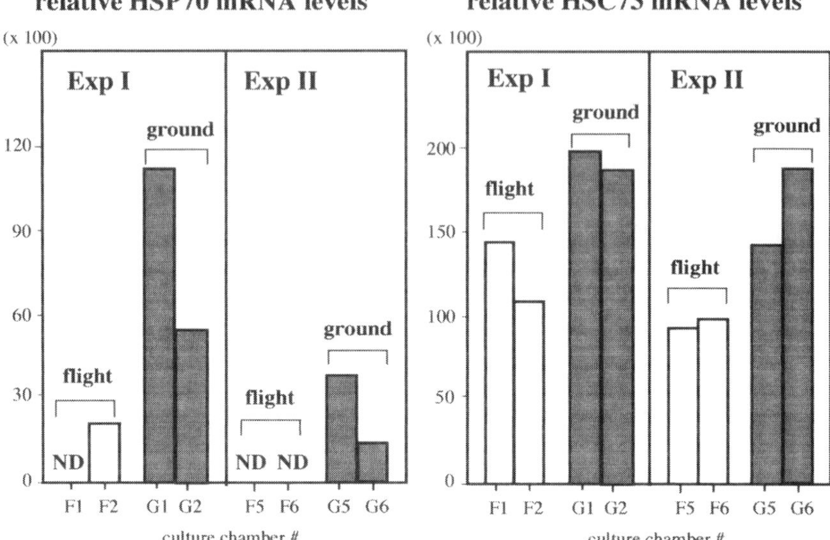

FIGURE 1. The mRNA levels of HSP70 **(A)** and HSC73 **(B)** of rat osteoblasts during space flight. ND: Gene transcripts were not detected even after RT-PCR amplification.

ground controls (unpublished data). This discrepancy might be due to the differences in experimental conditions, including cell type, timeline, and flight hardware.

The mean HSP47 mRNA levels in the flight cultures were decreased to 50% and 19% of those of the ground controls on the fourth and fifth days of the mission (FIG. 2A). On the fourth day, the concentration of TGF-β1 in the conditioned medium was 292 ± 74 pg/mL and 800 ± 108pg/mL in the flight and ground control cultures, respectively. On the fifth day, TGF-β1 was found at the concentration of 104 ± 104 pg/mL and 539 ± 171 pg/mL in the flight and ground control cultures, respectively (FIG. 2B). HSP47 binds specifically to procollagens.[3] Expression of HSP47 and procollagen a1(I) was coordinately upregulated by TGF-β1 and most sensitively around 1000 pg/mL of TGF-β1 in mouse osteoblastic cell culture.[4] A decrease such as 60% to 80% decrease of TGF-β1 in the medium might reduce HSP47 transcription in the flight cultures. However, we did not observe a decrease of the procollagen a1(I) mRNA levels in the flight cultures (unpublished data).

Statistical analyses were not performed for the mRNA levels of HSPs, because data were obtained only from duplicate cultures, due to the limited amounts of RNA. However, data from our repeated spaceflight experiments strongly suggested that microgravity decreased the mRNA levels for HSP70 (plus HSC73) and HSP47 in rat osteoblasts. HSP70 prevents stress-induced apoptosis, by inhibiting the c-Jun NH_2-terminal kinase activation[5] and caspase recruitment.[6] HSP47 is essential for collagen processing and quality control under stress, and it also prevents secretion of abnormal procollagen.[3] HSC73 directly interacts with tubulin and prevents ischemic disruption of the microtubule network.[7]

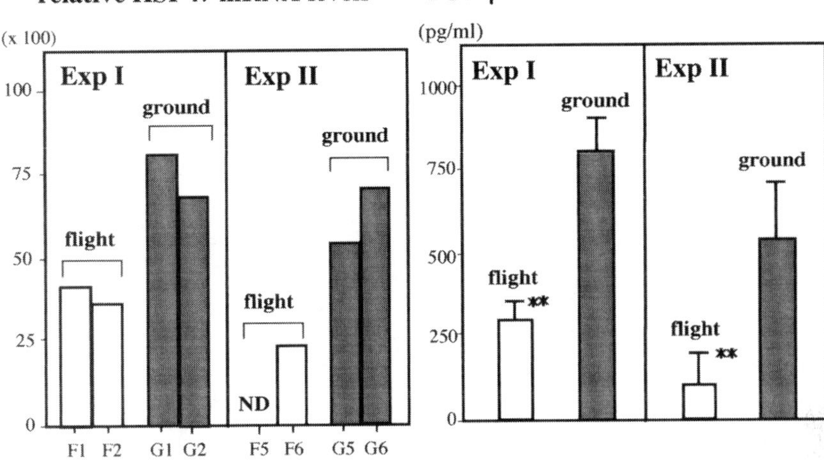

FIGURE 2. The HSP47 mRNA levels **(A)** and TGF-β1 concentration in the conditioned medium **(B)** of rat osteoblast culture during space flight. ND: Gene transcripts were not detected even after RT-PCR amplification. Each column/bar represents the mean ±SD from quadruplicate cultures. **Significantly lower than controls ($P < 0.01$).

Decrease in the expression of these HSPs might indicate an apoptotic trend in rat osteoblasts during spaceflight. Further studies are needed to investigate the regulatory mechanisms of HSPs and apoptotic signaling under microgravity.

ACKNOWLEDGMENTS

The authors thank Prof. Masaki Yanagishita of Tokyo Medical and Dental University for his helpful advice in sample preparation and for his careful reading of the manuscript.

REFERENCES

1. Kumei, Y., H. Shimokawa, H. Katano, *et al.* 1996. Microgravity induces prostaglandin E2 and interleukin-6 production in normal rat osteoblasts: role in bone demineralization. J. Biotechnol. **47:** 313–324.
2. Cubano, L.A. & M.L. Lewis. 2001. Effect of vibrational stress and spaceflight on regulation of heat shock proteins hsp70 and hsp27 in human lymphocytes (Jurkat). J. Leukoc. Biol. **69:** 755–761.
3. Nagata, K. 1996. Hsp47: a collagen-specific molecular chaperone. Trends Biol. Sci. **21:** 23–26.
4. Yamamura, I., H. Hirata, N. Hosokawa & K. Nagata. 1998. Transcriptional activation of the mouse HSP47 gene in mouse osteoblast MC3T3-E1 cells by TGF-β1. Biochem. Biophys. Res. Commun. **244:** 68–74.

5. MOSSER, D.D., A.W. CARON, L. BOURGET, *et al.* Role of the human heat shock protein hsp70 in protection against stress-induced apoptosis. Mol. Cell. Biol. **17:** 5317–5327.
6. SALEH, A., S.M. SRINIVASULA, L. BALKIR, *et al.* 2001. Negative regulation of the Apaf-1 apoptosome by Hsp70. Nature Cell Biol. **2:** 476–483.
7. DECKER, R.S., M.L. DECKER, S. NAKAMURA, *et al.* HSC73-tubulin complex formation during low-flow ischemia in the canine myocardium. Am. J. Physiol. Heart Circ. Physiol. **283:** H1322–1333.

Coinduction of GTP Cyclohydrolase I and Inducible NO Synthase in Rat Osteoblasts during Space Flight

Apoptotic and Self-protective Response?

YASUHIRO KUMEI,[a] SADAO MORITA,[a] HIROSHI NAKAMURA,[a] HIDEO AKIYAMA,[b] MASAHIKO HIRANO,[b] HITOYATA SHIMOKAWA,[a] AND KEI'ICHI OHYA[a]

[a]*Graduate School of Tokyo Medical and Dental University, Tokyo 113-8549, Japan*

[b]*Toray Research Center, Kamakura 248-8555, Japan*

ABSTRACT: The mechanism underlying space flight–induced osteopenia is unknown. In osteoblasts, the inducible nitric oxide (NO) synthase (iNOS) is involved in the early response to mechanical strain and induction of apoptosis. GTP cyclohydrolase I (GTPCH) is a key enzyme that is essential for iNOS activity. The coordinate expression of GTPCH prevents apoptosis that is induced by iNOS/NO. The purpose of this study was to investigate the effects of space flight on the expression of apoptotic/anti-apoptotic molecules iNOS and GTPCH in rat osteoblasts. Rat osteoblasts were cultured aboard a space shuttle and solubilized on the 4th and 5th days of the mission. The mRNA levels for iNOS and GTPCH in the flight cultures were increased to at least 120-fold and threefold higher than the ground (1 × g) controls, respectively. The amount of cellular DNA per flight culture vessel was 53% and 58% of the ground controls on the 4th and 5th days, respectively. However, the increasing rate of the DNA amount from the 4th to the 5th day was not different between the flight cultures and the ground controls. Morphologically, the cells grew in space as well as on the ground. Co-expression of GTPCH and iNOS may indicate a self-protective mode of action in osteoblasts against the harmful stress under microgravity.

KEYWORDS: **space flight; osteoblast; GTP cyclohydrolase; NOS; apoptosis**

INTRODUCTION

It is well documented that space flight conditions change the function of osteoblasts *in vivo* and *in vitro*. We previously showed cytokine production in rat osteoblast culture was greatly enhanced during space flight.[1] A report indicated that proinflammatory cytokines stimulated co-expression of GTP cyclohydrolase I

Address for correspondence: Dr. Yasuhiro Kumei, Biochemistry, Department of Hard Tissue Engineering, Graduate School of Tokyo Medical and Dental Univ., Tokyo 113-8549, Japan. Voice and fax: +81-3-5803-4555.
kumei.bch@tmd.ac.jp

Ann. N.Y. Acad. Sci. 1010: 481–485 (2003). © 2003 New York Academy of Sciences.
doi: 10.1196/annals.1299.087

(GTPCH) and inducible nitric oxide (NO) synthase (iNOS) in mouse osteoblastic cells.[2] GTPCH is the rate-limiting enzyme for *de novo* synthesis of tetrahydrobiopterin (BH_4) that is the co-factor for iNOS.[3] Endogenous NO plays important roles in osteoblasts, not only for cell proliferation and function,[4] but also for induction of apoptosis.[5] In the former case, induction of NO and iNOS is an early response of osteoblasts to mechanical strain.[6] In the latter case, co-expression of GTPCH with iNOS mRNAs prevented the NO-induced apoptotic changes in osteoblasts.[2] This is the first report on the co-expression of iNOS with GTPCH in rat osteoblasts that were cultured aboard a space shuttle.

MATERIALS AND METHODS

Primary culture of osteoblasts was prepared from male Wistar rat femur marrows.[1] Cells were inoculated into flight culture vessels at a density of $10^4/cm^2$ three days before launch. Cells were cultured continuously for four and five days aboard a space shuttle.[1] On the third day of the mission, cells were treated with 1 nM of 1α,25-dihydroxyvitamin D_3 for one day, then solubilized with guanidine isothiocyanate solution on the fourth day (Experiment I). The entire process of the operation was repeated one day later during the space flight (Experiment II). Morphological observation was also conducted during the spaceflight by using a phase-contrast microscope. The ground ($1 \times g$) control samples were synchronized with flight cultures under conditions identical except for microgravity: the latent hours at 22 ~ 25°C before the flight incubator was activated at 37°C and a 30-min exposure to $2 \sim 3 \times g$ to simulate the hypergravity that was generated by shuttle lift off.

After the return to Earth, cellular RNA and DNA in each culture were purified separately for examining the mRNA levels and the amount of cellular DNA (indicator of cell proliferation).[1] The mRNA levels of iNOS and GTPCH were analyzed by semiquantitative RT-PCR, using gene-specific primers[7]: GTPCH forward 5′-GGATACCAGGAGACCATCTCA-3′; GTPCH reverse 5′-TAGCATGGTGCTAGTGACAGT-3′; iNOS forward 5′-CTGCAGGTCTTTGACGCTCGG-3′; iNOS reverse 5′-CTGGAACACAGGGGTGATGCT-3′. Data were normalized against the housekeeping gene glyceraldehyde 3-phosphate dehydrogenase.[1]

RESULTS AND DISCUSSION

The GTPCH mRNA levels in the flight cultures were threefold higher than the ground controls on the fourth day of the mission (FIG. 1A). On the fifth day, the GTPCH mRNA levels in the flight cultures were further increased to six- to eightfold higher than the ground control (FIG. 1A). The iNOS gene transcripts were not detected at all in two of the four ground controls even after RT-PCR (FIG. 1B). On the contrary, the iNOS mRNA levels in the flight cultures were increased to 120-fold and 180-fold higher than the remaining two ground controls on the fourth and fifth days (FIG. 1B). As for the mechanism of coordinate increase of GTPCH and iNOS expression, positive regulation by cAMP and NF-κB might be involved at the transcriptional level.[2] We observed upregulation of cAMP/adenylate cyclase in osteoblasts under microgravity (unpublished data).

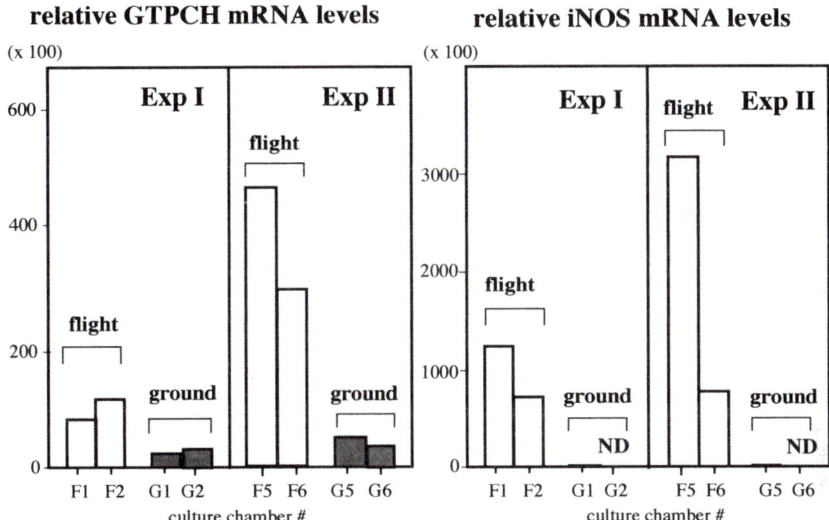

FIGURE 1. The mRNA levels of GTPCH **(A)** and iNOS **(B)** of rat osteoblasts during space flight. ND: gene transcripts not detected even after RT-PCR amplification.

Morphologically, on the fourth day of the mission, cells in the flight cultures as well as in the ground controls had grown to the preconfluent phase, showing no serious impairment or cell death (FIG. 2A). The total amount of cellular DNA in the flight cultures was 53% and 58% as small as the ground controls on the fourth and fifth days (FIG. 2B). However, the increasing rate of DNA amount per flight culture vessel (an indicator of cell proliferation) was not different from the ground controls from the fourth to the fifth day (FIG. 2B).[1] Collectively, data suggested that growth inhibition or apoptosis might be induced in the early days of space flight but recovered later during the mission.

Although endogenous NO induced apoptosis in osteoblasts,[5] coordinate expression of GTPCH with iNOS protected against apoptosis that was induced by NO in osteoblasts.[2] On the other hand, another report addressed the physiological significance of endogenous NO as an important regulator of osteoblastic cell proliferation and function.[4] In addition, iNOS/NO inhibited apoptosis by *S*-nitrosylation and inactivation of caspase.[8] We have not determined whether the upreguation of iNOS/NO is the apoptotic or anti-apoptotic feature in osteoblasts under microgravity. Data were not statistically analyzed owing to limited amounts of RNA samples for the flight cultures. However, our repeated spaceflight experiments and the reproducible results strongly suggested that microgravity markedly increased the mRNA levels for GTPCH and iNOS in rat osteoblasts. Microgravity may modulate apoptotic and/or anti-apoptotic molecules in osteoblasts. Further studies are needed to examine the possible induction of apoptosis in an early phase of spaceflight and clarify the mechanism for iNOS and GTPCH induction under microgravity.

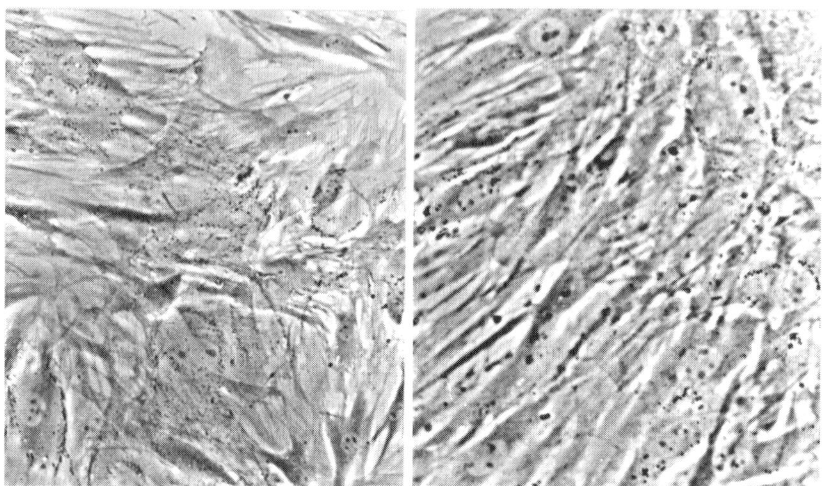

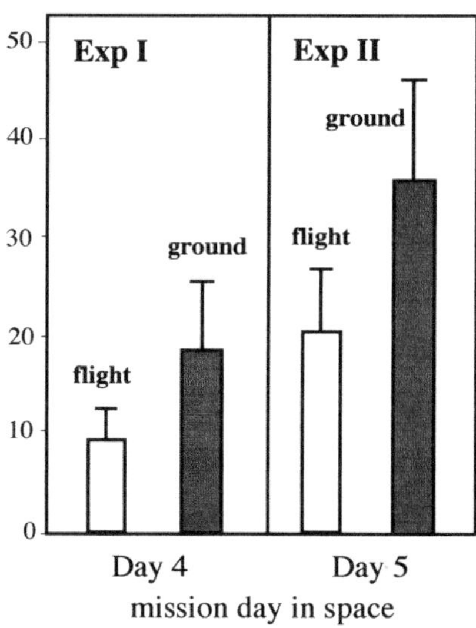

FIGURE 2. Phase-contrast microscopy of rat osteoblasts in flight (*left*) and ground (*right*) cultures (×100) **(A)** and total amount of cellular DNA per culture vessel **(B)**. Photos were taken on the fourth day of the mission. Each column/bar represents the mean ±SD from quadruplicate cultures.

REFERENCES

1. KUMEI, Y., H. SHIMOKAWA, H. KATANO, *et al.* 1996. Microgravity induces prostaglandin E_2 and interleukin-6 production in normal rat osteoblasts: role in bone demineralization. J. Biotechnol. **47:** 313–324.
2. TOGARI, A., M. ARAI, M. MOGI, *et al.* 1998. Coexpression of GTP cyclohydrolase I and inducible nitric oxide synthase mRNAs in mouse osteoblastic cells activated by proinflammatory cytokines. FEBS Lett. **428:** 212–216.
3. BAEK, K.J., B.A. THIEL, S. LUCAS & D.J. STUEHR. 1993. Macrophage nitric oxide synthase subunits. J. Biol. Chem. **268:** 21120–21129.

4. RIANCHO, J.A., E. SALAS, M.T. ZARRABEITIA, *et al.* 1995. Expression and functional role of nitric oxide synthase in osteoblast-like cells. J. Bone Miner. Res. **10:** 439–446.
5. DAMOULIS, P.D. & P.V. HAUSCHKA. 1997. Nitric oxide acts in conjunction with proinflammatory cytokines to promote cell death in osteoblasts. J. Bone Miner. Res. **12:** 412–422.
6. PITSILLIDES, A.A., S.C.F. RAWLINSON, R.F. SUSWILLO, *et al.* 1995. Mechanical strain-induced NO production by bone cells: a possible role in adaptive bone (re)modeling? FASEB J. **9:** 1614–1622.
7. HATTORI, Y., M. OKA, K. KASAI, *et al.* 1995. Lipopolysaccharide treatment in vivo induces tissue expression of GTP cyclohydrolase I mRNA. FEBS Lett. **368:** 336–338.
8. MANNICK, J.B., A. HAUSLADEN, L. LIU, *et al.* 1999. Fas-induced caspase denitrosylation. Science **284:** 651–654.

Apoptosis Signal Transduction and the Maturity Status of Human Spermatozoa

UWE PAASCH,[a] ASHOK AGARWAL,[b] AKSHAY K. GUPTA,[b] RAKESH K. SHARMA,[b] SONJA GRUNEWALD,[a] ANTHONY J. THOMAS, Jr.,[b] AND HANS-JUERGEN GLANDER[a]

[a]*EAA Center University of Leipzig, Stephanstrasse11, 04109 Leipzig, Germany*

[b]*Cleveland Clinic Foundation, Cleveland, Ohio*

Keywords: maturity; spermatozoa; human; caspases; TUNEL; mitochondrial membrane potential

INTRODUCTION

Numerous studies have shown the presence of DNA strand breaks in human ejaculated spermatozoa that might be a result of impaired spermatogenesis or degradation due to apoptosis (referred to as programmed cell death).[1] Although the pathways of apoptosis used in spermatogenesis are poorly understood,[2–4] differences in selected molecular markers involved in apoptosis, between males with normal and abnormal sperm parameters, are thought to be indicative of an abortive apoptosis.[5–7] The fertilization capacity of human sperm correlates with their maturity status. It was our aim to investigate the main elements of the apoptosis signaling cascade in mature and immature sperm: Caspase activity (aCP), disruption of mitochondrial membrane potential ($\Delta\psi_m$) and DNA fragmentation (TU+).

METHODS

Semen samples from healthy donors were pooled to create 18 pools. The liquefied semen was loaded onto a 47% and 90% discontinuous isolate gradient (Irvine Scientific, Santa Ana, CA) and centrifuged at 500 × *g* for 20 min at room temperature. The resulting interface between the 47% and 90% layers (immature spermatozoa) and the 90% pellet (mature spermatozoa) were aspirated, and transferred to separate test tubes. The pellets for both fractions were resuspended in Biggers-Whitten-Whittingham medium (BWW) and centrifuged at 500 × *g* for 7 minutes.

Active caspases 8, 9, and 3 were detected in living spermatozoa by carboxyfluorescein-labeled caspase inhibitors. The detection of activated caspases by the inhibitor was performed according to the manufacturer's instruction manual in the flu-

Address for correspondence: Uwe Paasch, EAA Center University of Leipzig, Stephanstrasse11, 04109 Leipzig, Germany.
andro@medizin.uni-leipzig.de

**Ann. N.Y. Acad. Sci. 1010: 486–488 (2003). © 2003 New York Academy of Sciences.
doi: 10.1196/annals.1299.088**

orescein caspase activity kit (CaspaTag™, S730x, Intergen Co., Oxford, England) with controls. Mitosensor™, a lipophilic cationic dye (5,5′, 6,6′-tetrachloro-1,1′,3,3′-tetraethylbenzimid-azolyl carbocyanine chloride, C25H27C13N), was used to detect disrupted or intact transmembrane potential of mitochondria in vital spermatozoa (Apo Alert™ Mitochondrial Membrane Sensor kit, Ct No. K2017-1, Chlontech, CA, USA).

Spermatozoa with intact mitochondria (IM) excite an intense red fluorescence as a result of forming aggregates. Green fluorescence of their monomers indicates disrupted $\Delta\psi_m$. The proportion of cells with DNA fragmentation (TU+) was measured in the same fractions using the TUNEL assay (APO-DIRECT™, Flow Cytometry Kit for Apoptosis, Cat. No. APT110, Chemicon, USA). The assay was used according to the manufacturer's instructions. All fluorescence signals of labeled spermatozoa were analyzed by the flow cytometer, FACscan (Becton Dickinson).

RESULTS

Compared to the immature fraction, the mature subset had a lower level of aCP 9 (31.3 ± 12.5 vs. 53.1 ± 14.7, [% ± SD], $P < 0.01$), CP8 (29.4 ± 13.7 vs. 53.6 ± 18.1, $P < 0.01$) and CP3 (30.2 ± 14.4 vs. 53.6 ± 17.4, $P < 0.01$); a higher proportion of cells with intact $\Delta\psi_m$ (56.1 ± 22.2 vs. 40.7 ± 15.2, $P < 0.05$); and a lower proportion with TU+ (31.6 ± 15.6 vs. 39.7 ± 16.7, $P < 0.01$).

There was a significant positive correlation of activity between, aCP 3 with aCP 8 and aCP 9. Intact $\Delta\psi_m$ was inversely correlated with all aCP in mature cells, and with aCP 8 and aCP 3 in immature cells ($P < 0.01$ for each aCP, respectively). The aCP 9 of an immature fraction followed the same trend ($P = 0.051$). The TU+ cells showed no significant correlation with aCP or $\Delta\psi_m$.

CONCLUSIONS

Incomplete maturation of human ejaculated spermatozoa is associated with an increase of caspase 3, 8, and 9 activity in the spermatozoa. This activity is also associated with the disruption of mitochondrial membrane potential in the immature fraction. The activated apoptotic process does not immediately affect the levels of DNA fragmentation. Reaching maturity may implicate a deactivation of the apoptosis-signaling cascade in human sperm. However, the exact mechanisms of caspase activation and disruption of the mitochondrial membrane potential in ejaculated sperm are yet to be elucidated and further research in this direction is needed.

REFERENCES

1. GORCZYCA, W., F. TRAGANOS, H. JESIONOWSKA & Z. DARZYNKIEWICZ. 1993. Exp. Cell. Res. **207:** 202–205.
2. FRANCAVILLA, S., P. D'ABRIZIO, N. RUCCI, *et al.* 2000. J. Clin. Endocrinol. Metab. **85:** 2692–2700.
3. PENTIKAINEN, V., K. ERKKILA & L. DUNKE.1999. Am. J. Physiol. **276:** E310–E316.

4. D'Alessio, A., A. Riccioli & P. Lauretti. 2000. Proc. Natl. Acad. Sci. USA **98:** 3316–3321.
5. Sakkas, D., E. Mariethoz & J.C. St. John. 1999. Exp. Cell Res. **251:** 350–355.
6. Gandini, L., F. Lombardo, D. Paoli, *et al.* 2000. Hum. Reprod. **15:** 830–839.
7. Sakkas, D., O. Moffatt, G. Manicardi, *et al.* 2002. Biol. Reprod. **66:** 1061–1067.

Opposite Phenotypes of Cancer and Aging Arise from Alternative Regulation of Common Signaling Pathways

SVETLANA V. UKRAINTSEVA AND ANATOLY I. YASHIN

Max-Planck Institute for Demographic Research, 18057 Rostock, Germany

ABSTRACT: Phenotypic features of malignant and senescent cells are in many instances opposite. Cancer cells do not "age"; their metabolic, proliferative, and growth characteristics are opposite to those observed with cellular aging (both replicative and functional). In many such characteristics cancer cells resemble embryonic cells. One can say that cancer manifests itself as a local, uncontrolled "rejuvenation" in an organism. Available evidence from human and animal studies suggests that the opposite phenotypic features of aging and cancer arise from the opposite regulation of genes participating in apoptosis/ growth arrest or growth signal transduction pathways in cells. This fact may be applicable in the development of new anti-aging treatments. Genes that are contrarily regulated in cancer and aging cells (e.g., proto-oncogenes or tumor suppressors) could be candidate targets for anti-aging interventions. Their "cancer-like" regulation, if strictly controlled, might help to rejuvenate the human organism.

KEYWORDS: cancer; aging; apoptosis; growth signal pathway

CANCER AS "REJUVENESCENCE"

A comparison of phenotypic features of cancer and aging shows that, in many instances, they are opposite. Cancer cells are potentially immortal, while aging cells normally die through apoptosis. Cancer cells may proliferate unlimited, while aging cells decline in proliferative ability. Cancer cells are autonomous from growth suppressing signals, while aging cells, if they do not die, undergo irreversible growth arrest. Cancer cells are de-differentiated, while aging cells enter terminal differentiation. Cancer cells have an increased metabolism, while aging cells decline in metabolic activity. Furthermore, cancer cells may secrete embryonic proteins and factors promoting angiogenesis, while aging cells do not.[1–3]

Many cancer features are also inherent to most "young" (embryonic) cells in the organism. Embryonic cells have a high proliferative potential, are able to migrate, secrete factors that promote angiogenesis, secrete embryonic proteins such as α-fetoprotein, and produce enzymes capable of degrading basement membranes.[3] Thus, cancer cells do not "age." They can be viewed as a local uncontrolled

Address for correspondence: Svetlana V. Ukraintseva, Max-Planck Institute for Demographic Research, Konrad Zuse Str. 1, 18057 Rostock, Germany. Voice: +1 (919)942-5672. svu@mail.ru

Ann. N.Y. Acad. Sci. 1010: 489–492 (2003). © 2003 New York Academy of Sciences. doi: 10.1196/annals.1299.089

TABLE 1. Typical features of cancer and aging cells

Cancer cells	Aging cells
potential immortality	programmed death
growth autonomy	growth arrest
nonlimited proliferation	decline in proliferation
ability to migrate	not able to migrate
increased metabolism	decline in metabolism
de-differentiation	terminal differentiation
secret embryonic proteins	no embryonic proteins
may promote angiogenesis	no angiogenesis

"rejuvenation" within an organism because their metabolic, proliferative and other characteristics are opposite to those observed with aging (summarized in TABLE 1).

Recent evidence suggests that the opposite phenotypic features of aging and cancer may arise from alternative regulation of the same genetic pathways, such as apoptosis and growth signal transduction pathways (TABLE 2).

APOPTOSIS

Potential immortality of cancer cells refers to their ability to avoid apoptosis.[2] A powerful inductor of apoptosis, such as *p53,* is downregulated in a majority of human cancers. However, expression of *p53* is elevated in senescent cells, providing irreversible growth arrest or death (see TABLE 2). Cancer cells may also avoid apoptosis with increased expression of anti-apoptotic proteins (such as Bcl-2). At the same time, expression of this protein is lower in cells from old individuals, compared with young ones. These and other examples are summarized in TABLE 2. One can see from this table that the key genes participating in apoptotic pathways seem to be alternatively regulated in cancer and aging.

GROWTH SIGNALING

Growth autonomy and high proliferative ability of cancer cells are both associated with stimulation of growth signaling pathways. Relevant genes are found to be upregulated in many human cancers.[4] At the same time, their expression was shown to decrease in senescent cells (see examples in TABLE 2). For instance, *myc* is expressed at a much higher level in cancer and in normal young proliferating cells, relative to those of terminally differentiated nondividing (senescent) cells.[4] The growth signal pathway was recently proposed to be an evolutionary-conserved aging regulatory pathway in different species, because many genes, in which mutations were found to increase longevity in experimental animals, were also involved in this pathway.[5] Some of these mutations result in a "cancer-like" type of gene expression[6] (TABLE 2).

TABLE 2. Signaling pathways alternatively regulated in cancer and aging

Gene	Protein Function	In Cancer Cells	In Aging Cells	In Aging Organism
		Apoptosis		
p53	Inductor of apoptosis or growth arrest	Downregulated in most cancers[7]	Elevated expression[11]	Upregulation is associated with early aging phenotype and low cancer risk[14]
CD95	"Death" receptor	Decreased expression[8]	Higher expression[12]	Proportion of cells expressing CD95 is higher in older individuals[15]
Bcl-2	Anti-apoptotic protein	Overexpressed in some cancers[9]	Decreased levels[4]	Lower expression in cells from older individuals[15]
		Growth Signal Transduction		
myc	Transcription factor	Overexpressed in many cancers[4]	Decreased expression[4]	
ras	Signal transducer	Activated in some cancers[10]	Decreased expression[13]	Activation extends the life span in yeast[16]
tyrosine kinase receptors	Growth factor receptors	Overexpressed in some cancers[4]	Fewer receptors with increasing donor's age[17]	Overexpression of tkr-1 increases longevity in nematodes[6]

CONCLUSION

In many instances, phenotypes of cancer and aging are opposites. Available evidence suggests that this is caused by alternative regulation of the same genetic pathways, such as apoptosis and growth signal transduction pathways. Respective genes could be candidate targets for new anti-aging interventions. More studies are needed to understand if "cancer-like" regulation of these genes may decelerate or reverse aging in organisms. Other pathways, which are alternatively manifested in cancer and aging (e.g., responsible for differentiation, angiogenesis, and rate of metabolism), may also provide researchers with new candidates for "anti-aging" genes. However, this area needs further investigation.

REFERENCES

1. RUBIN, H. 1997. Mech. Ageing Dev. **98:** 1.
2. HANAHAN, D. 2000. Cell **100: 57**.
3. MINTZ, B. 1981. Adv. Cancer Res. **34:** 211.
4. PETERS, G. 1997. Oncogenes and Tumour Suppressors: 332. Oxford University Press. New York.
5. KENYON, C. 2001. Cell **105:** 165.
6. MURAKAMI, S. & T.E. JOHNSON. 1998. Life extension and stress resistance in *Caenorhabditis elegans* modulated by the tkr-1 gene. Cur. Biol. **8:** 1091–1094.
7. SOUSSI, T. 2000. The p53 tumor suppressor gene: from molecular biology to clinical investigation. Ann. N. Y. Acad. Sci. **910:** 121–137; discussion, 137–139.
8. PINKOSKI, M. & D. GREEN. 2000. Cloak and dagger in the avoidance of immune surveillance. Curr. Opin. Genet. Dev. **10:** 114–119.
9. ROSS, D. 1998. Introduction to Oncogenes and Molecular Cancer Medicine. Springer. New York.
10. FRAME, S. & A. BALMAIN. 2000. Integration of positive and negative growth signals during ras pathway activation in vivo. Curr. Opin. Genet. Dev. **10:** 106–113.
11. KULJU, K.S. & J.M. LEHMAN. 1995. Increased p53 protein associated with aging in human diploid fibroblasts. Exp. Cell. Res. **217:** 336–345.
12. MIYAWAKI, T., T. UEHARA, R. NIBU, *et al.* 1992. Differential expression of apoptosis-related Fas antigen on lymphocyte subpopulations in human peripheral blood. J. Immunol. **149:** 3753–3758.
13. DELGADO, D.L. RAYMOND & R. DEAN. 1986. C-ras expression decreases during in vitro senescence in human fibroblasts. Biochem. Biophys. Res. Commun. **137:** 917–921.
14. DONEHOWER, L. 2002. Does p53 affect organismal aging? J. Cell. Physiol. **192:** 23–33.
15. AGGARWAL, S. & S. GUPTA. 1998. Increased apoptosis of T cell subsets in aging humans: altered expression of Fas (CD95), Fas ligand, Bcl-2, and Bax. J. Immunol. **160:** 1627–1637.
16. JAZWINSKI, S.M., J.B. CHEN & J. SUN. 1993. A single gene change can extend yeast life span: the role of Ras in cellular senescence. Adv. Exp. Med. Biol. **330:** 45–53.
17. REENSTRA, W.R., M. YAAR & B.A. GILCHREST. 1996. Aging affects epidermal growth factor receptor phosphorylation and traffic kinetics. Exp. Cell Res. **227:** 252–255.

Gelsolin for Senescence-Associated Resistance to Apoptosis

JEONG SOO AHN,[a] IK-SOON JANG,[a] JI HEON RHIM,[a] KYUNGTAE KIM,[a] EUI-JU YEO,[b] AND SANG CHUL PARK[a]

[a]*Department of Biochemistry, Seoul National University College of Medicine, Seoul, South Korea*

[b]*Department of Biochemistry, Gachon Medical School, Inchon, South Korea*

ABSTRACT: One of the characteristics of the senescent cell is apoptotic resistance. Gelsolin, a Ca^{2+}-dependent actin regulatory protein, is believed to regulate the intracellular movements which are necessary for cell growth, proliferation, and differentiation. Recently, gelsolin was suggested to play a role in apoptotic resistance, which led us to examine its involvement in the apoptotic resistance of senescent cells. We found that the protein and mRNA levels of gelsolin were increased in senescent human diploid fibroblasts (HDFs). Gelsolin was intracellularly co-localized to the actin stress fiber and distributed to the nucleus and mitochondria in old HDFs. To examine the anti-apoptotic function of gelsolin in senescent HDFs, we tried to downregulate the expression of gelsolin by using antisense oligonucleotide in old HDFs. We then treated the senescent HDFs with the apoptosis-inducing agent menadione. Downregulation of gelsolin in senescent HDFs resulted in increased sensitivity to menadione-induced apoptotic cell death. This suggests that gelsolin plays a role in the apoptotic resistance observed in senescent HDFs.

KEYWORDS: senescence; gelsolin; apoptotic resistance

INTRODUCTION

Although many studies have been carried out to understand aging-specific resistance to apoptosis,[1] the role of gelsolin had yet to be fully examined. Gelsolin is an actin regulatory protein that is isolated from rabbit lung macrophages. It is a modulator of the cytoplasmic actin gel-sol transformation,[2] and it is responsible for mechanisms of intracellular movements such as cytokinesis and those of transport vesicles, which are required for cell growth, proliferation, and differentiation. Contrasting roles of gelsolin have been suggested such as oncogenic, anti-oncogenic, pro-apoptotic, and anti-apoptotic function in different cell lines.[3] Recently, it was reported that human gelsolin prevents apoptosis by inhibiting cytochrome *c* release from the mitochondrial inner membrane by closing mitochondrial pores in gelsolin-

Address for correspondence: Sang Chul Park, M.D., Ph.D., Department of Biochemistry, Seoul National University College of Medicine, 28 Yon-gon-Dong, Chongno-Gu, Seoul, 110-799, South Korea. Voice: +82-2-740-8244; fax: +82-2-744-4534.
scpark@snu.ac.kr

**Ann. N.Y. Acad. Sci. 1010: 493–495 (2003). © 2003 New York Academy of Sciences.
doi: 10.1196/annals.1299.090**

overexpressed cell lines.[4] Therefore, we tested the role of gelsolin in senescent HDFs in resistance to apoptosis.

RESULTS AND DISCUSSION

Upregulation of Gelsolin in Senescent HDFs: Senescent cells have different phenotypes compared with young cells at the molecular, cellular, and intercellular levels. Age-related apoptotic resistance is not fully understood. When HDFs enter replicative senescence, the growth rate of the HDF decreases rapidly, SA-b-galactosidase activity is increased, and the morphologic shape of the cell becomes flattened and enlarged. When senescent cells were visualized *in vitro* by confocal microscopy, actin stress fibers were observed to have increased; we assumed that these increased actin fibers were regulated by their related protein and we measured the level of gelsolin. We found that the gelsolin levels were significantly increased, not only in RNA transcript, but also in protein (FIG. 1). This result suggests that the level of gelsolin increases significantly with aging and that this increase contributes to the senescent phenotype both morphologically and functionally.

Anti-Apoptotic Effect of Gelsolin in Senescent HDF: For apoptosis, activation of the caspase cascade by the cytochrome *c* and Apaf-1 complex has been well illustrated. But the essential release of cytochrome *c* from the mitochondria is incompletely understood.[5] It is suspected that cytochrome *c* release is strictly regulated by the voltage-dependent anion channel (VDAC) of the mitochondrial pore. Because gelsolin was recently reported to interact with VDAC, we assumed that the high level of gelsolin in senescent HDFs might play a role in apoptotic resistance. Therefore, we tried to modulate the gelsolin level of senescent HDFs by using a specific antisense oligonucleotide and monitoring the effect of gelsolin reduction in an apoptosis-inducing condition. The antisense oligonucleotide treatment reduced the gelsolin level efficiently (FIG. 2A). We treated the cells with menadione, a well-known apoptosis-inducing agent via oxidative stress. We observed that when the HDFs were

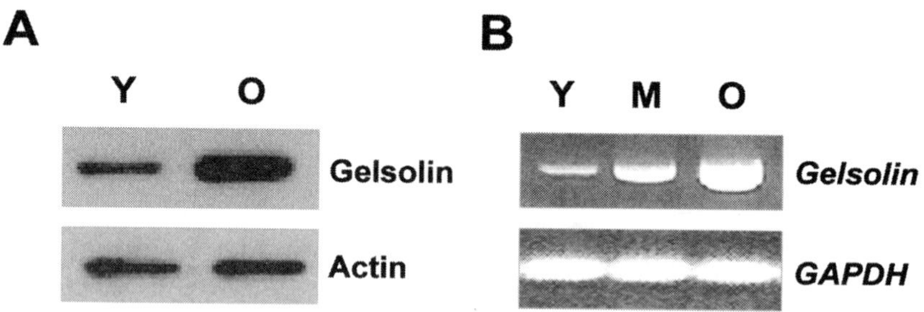

FIGURE 1. Expression level of cytoplasmic gelsolin in young and old human diploid fibroblast. **(A)** Gelsolin was analyzed by using immumoblot with monoclonal antibody to gelsolin in young (passages 10) and old (passages 28) human diploid fibroblast. **(B)** Expressed mRNA level of gelsolin with RT-PCR in young, middle, and old HDFs. Actin and GAPDH were used for internal control.

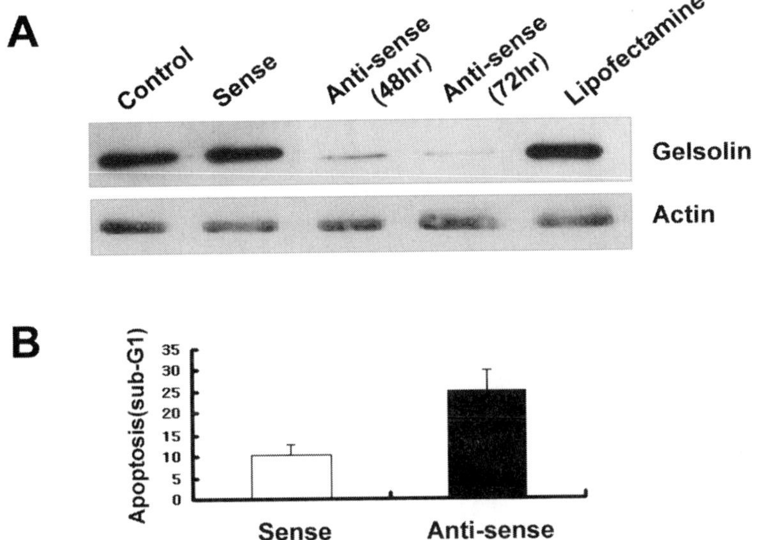

FIGURE 2. Downregulation of gelsolin and anti-apoptotic effect of cytoplasmic gelsolin in senescent HDFs. **(A)** Gelsolin-specific synthetic oligonucleotides were transfected with lipofectamine for 48 h or 72 h. Downregulated gelsolin was confirmed with immunoblot. **(B)** Sense (SS) and antisense (AS) oligonucleotides were transfected into old HDFs for 72 h to downregulate gelsolin and thereafter were treated with menadione (30 mM). After apoptotic stimuli, the subG1 fraction was analyzed.

treated with antisense oligonucleotides for gelsolin reduction, those cells showed apoptotic sensitivity (25%), compared with cells treated with sense oligonucleotides (10%) (FIG. 2B). The data suggest that the high level of gelsolin in the senescent HDFs is related to the senescence-associated apoptotic resistance in oxidative stress-dependent apoptosis induction. But it is not clear whether the high level of gelsolin in senescent HDFs is responsible for the general apoptotic resistance to other causative stresses for apoptosis. However, because the regulation of VDAC on cytochrome *c* release by gelsolin is known, it can be assumed that the apoptosis-resistant property of senescent cells is derived from their high level of gelsolin.

REFERENCES

1. WANG, E. 1995. Senescent human fibroblasts resist programmed cell death, and failure to suppress bcl 2 is involved. Cancer Res. **55:** 2284–2292.
2. YIN, H.L. & T.P. STOSSEL. 1979. Control of cytoplasmic actin gel-sol transformation by gelsolin, a calcium-dependent regulatory protein. Nature **281:** 583–586.
3. SUN, H.Q., M. YAMAMOTO, M. MEJILLANO & H.L.YIN. 1999. Gelsolin, a multifunctional actin regulatory protein. J. Biol. Chem. **274:** 33179–33182.
4. KUSANO, H., S. SHIMIZU, R.C. KOYA, *et al.* 2000. Human gelsolin prevents apoptosis by inhibiting apoptotic mitochondrial changes via closing VDAC. Oncogene **19:** 4807–4814.
5. TSUJIMOTO, Y. & S. SHIMIZU. 2000. Bcl-2 family. FEBS Lett. **466:** 6–10.

An Age-Associated Decrease in the Frequency of C4B*Q0 Indicates That Null Alleles of Complement May Affect Health or Survival

GUÐMUNDUR JÓHANN ARASON,[a] SIGURDUR BÖÐVARSSON,[b] SIGURÐUR ÞOR SIGURÐARSON,[b] GARÐAR SIGURÐSSON,[b] GUÐMUNDUR ÞORGEIRSSON,[b] SVEINN GUÐMUNDSSON,[c] JUDIT KRAMER,[d] AND GEORG FÜST[e]

[a]*Department of Immunology, Institute of Laboratory Medicine and* [b]*Department of Medicine, Landspítali University Hospital, LSH Hringbraut, 101 Reykjavík, Iceland*

[c]*National Blood Bank, 101 Reykjavík, Iceland*

[d]*Central Laboratory, St. John's Hospital, Budapest, Hungary*

[e]*Third Department of Medicine, Semmelweis University and Research Group of Metabolism and Atherosclerosis, Hungarian Academy of Sciences, Budapest, Hungary*

ABSTRACT: We studied the distribution of complement C4, C3, and factor B allotypes in 423 healthy Icelandic subjects from 17 to 89 years of age. A marked decrease was observed in the carrier frequency of variant alleles of complement C4B (C4B*Q0) and C3 (C3*F). These results confirm our previous observations on Hungarian subjects and suggest a negative effect of C4B*Q0 on health or survival.

KEYWORDS: coronary artery disease; Iceland; Hungary; C4B*Q0

INTRODUCTION

Coronary artery disease (CAD) is considered to be the most important disease in Western countries in regards to both mortality and health care cost. The etiology is unknown, but known risk factors include smoking, diabetes, hypertension, and hypercholesterolemia. In previous studies, we identified null alleles of complement C4B (C4B*Q0) as a new risk factor in the disease, because they were found to occur in raised frequency in CAD patients in Iceland as well as Hungary.[1,2] Consistent with this, the incidence of C4B*Q0 shows a marked age-associated drop in the Hungarian population, suggesting it has a negative effect on survival.[3] These results strongly implicate the C4B*Q0 allele as a candidate gene for CAD association. However, this is complicated by linkage disequilibrium of the C4 genes with neighboring

Address for correspondence: Guðmundur Jóhann Arason, Department of Immunology, Institute of Laboratory Medicine, Landspítali University Hospital, LSH Hringbraut (hus 14), 101 Reykjavík, Iceland. Voice: +354-543-5800; fax: +354-543-4828.
garason@landspitali.is

**Ann. N.Y. Acad. Sci. 1010: 496–499 (2003). © 2003 New York Academy of Sciences.
doi: 10.1196/annals.1299.091**

genes in the MHC. Our goal was to confirm and extend these findings by studying age-associated differences in alleles of C4, C3, and factor B (Bf) in a population with a different genetic background.

MATERIALS AND METHODS

The study involved 423 healthy Icelandic subjects from 17 to 89 years of age. C3 and C4 allotypes were determined by high-voltage agarose electrophoresis, and factor B was typed on cellulose acetate membranes.[4–6] Categorical variables were compared by the chi square test, and multiple regression analysis was performed using SPSS 10.0.

RESULTS

*Age-Associated Decrease in C4B*Q0 and Bf*F*

Polymorphisms of C4A, C4B, C3, and Bf alleles were determined in 423 healthy Icelanders that were divided into 5 age groups (17–24, 25–34, 35–44, 45–54, and over 54). The percentage of subjects carrying the silent C4A allele (C4A*Q0), the silent C4B allele (C4B*Q0), and a silent allele at either locus is shown in FIGURE 1A, while FIGURE 1B depicts the percentage of those carrying the variant (F) allele of C3 and Bf. A decrease in the carrier frequency of C4B*Q0, C4 null alleles, and Bf*F was found in the eldest age group.

When the group of "elderly subjects" (55–74) was compared to "young adults" (17–54) (TABLE 1), these differences proved significant at the 0.01 level. The odds ratio of elderly people to carry these alleles was 5.16, 2.96, and 6.71, respectively. The age-dependent decrease in carrier status of silent C4 was due mainly to a decrease in C4B*Q0 (TABLE 1).

TABLE 1. Distribution of the variant C4, C3, and Bf alleles in "young" (17–54 years old) and "elderly" (55–74 years old) healthy Icelanders

	C4A*Q0		C4B*Q0		C4A*Q0 or C4B*Q0		C3*F		Bf*F	
Group	yes	no	yes	no	yes	no	yes	no	yes	no
Young	83	282	87	278	159	206	80	194	106	162
Old	10	48	3	55	12	46	10	36	5	41
P value*	0.397		<0.001		0.001		0.376		<0.0001	
Odds ratio** (95% CI***)	1.41 (0.68–2.98)		5.69 (1.74–18.66)		2.96 (1.52–5.77)		1.49 (0.70–3.14)		6.71 (2.33–19.28)	

*Fisher's exact test.
**Odds ratio of young people to carry the variant allele.
***Confidence interval.

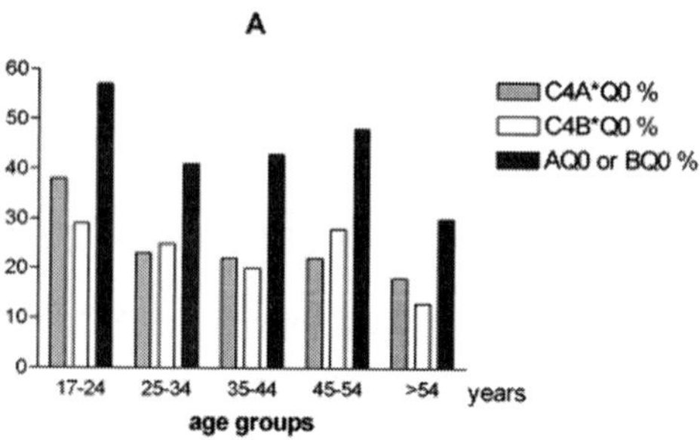

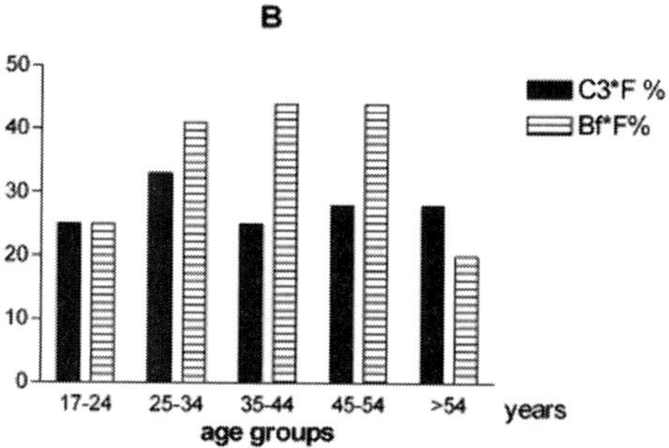

FIGURE 1. Carrier percentage of variant alleles of **(A)** C4 and **(B)** C3 and Bf, in different age groups of healthy Icelandic people.

Gene Frequencies

The gene frequencies of C4A, C4B, C3, and Bf alleles were similar to those previously described for Caucasians[3] (C4A*3 = 0.78, C4A*4 = 0.13, C4A*Q0 = 0.12, C4A*6 = 0.05, C4A*rare = 0.01; C4B*1 = 0.78, C4B*2 = 0.18, C4B*3 = 0.03, C4B*rare = 0.01; C3*S = 0.83, C3*F = 0.17); and identical for all age groups, with the exception of C4B*Q0 and factor B*F, being 0.12 resp. 0.21 in younger subjects compared to 0.02 resp. 0.07 for elderly subjects. These differences were statistically significant (P = .001 resp. P = .003).

Influence of Gender on the Decrease of C4 Null Alleles

Silent and variant alleles occurred more frequently among males than among females, but a significant carrier frequency difference (P=.013) was observed only in the case of C4 null alleles. Because the multiple regression analysis indicated there was a significant interaction (P = .002) between age and gender when C4B*Q0 carrier/non-carrier status was considered as the dependent variable, we analyzed whether the gender of the subjects influenced the age-dependent decrease in carrier frequency (TABLE 2). The decrease in null allele carrier frequency was more marked in females (P = .025), although it approached significance also in males (P = .056).

DISCUSSION

An age-associated decrease was observed in the carrier status and gene frequency of C4B*Q0 and Bf*F. This suggests a negative effect of these alleles on health or survival. In order to distinguish between mortality-associated and health-associated effects, further trans-sectional studies are warranted. Although statistically significant in both cases, the decrease was more marked when expressed in terms of carrier status, rather than gene frequencies. This indicates that the presence of these alleles in the heterozygous state may be sufficient for a negative effect to be felt. It appears likely that linkage disequilibrium accounts for the decrease in Bf*F frequency, as we previously observed an age-associated increase, but not a decrease, in this allotype in Hungarians. C4B*Q0, on the other hand, has now been shown to display an age-associated decrease in two totally unrelated Caucasian populations. These results strongly support that C4B*Q0 is a negative selection factor for health or survival.

ACKNOWLEDGMENTS

This work was supported by grants from the Icelandic Research Council (No. 011330003) and the Science Fund of Landspítali University Hospital of Iceland.

REFERENCES

1. KRAMER, J., K. RAJCZY, L. HEGYI, *et al.* 1994. C4B*Q0 allotype as risk factor for myocardial infarction. Br. Med. J. **309:** 313–314.
2. ARASON, G.J., J. KRAMER, S. BÖDVARSSON, *et al.* 1998. Increased frequency of C4B null alleles in Icelandic patients with coronary disease. Mol. Immunol. **35:** 412.
3. KRAMER, J., T. FULÖP, K. RAJCZY, *et al.* 1991. A marked drop in the incidence of the null allele of the B gene of the fourth component of complement (C4B*Q0) in elderly subjects: C4B*Q0 as a probable negative selection factor for survival. Hum. Genet. **86:** 595–598.
4. TEISBERG, P. 1970. High voltage agarose electrophoresis in the study of C3 polymorphism. Vox Sang. **19:** 47–56.
5. SIM, E. & F.J. CROSS. 1986. Phenotyping of human complement component C4, a class-III HLA antigen. Biochem. J. **239:** 763–767.
6. KRAMER, J., E. THIRY & G. FÜST. 1989. Rapid determination of the human complement factor B phenotypes. Haematologia **122:** 97–100.

Ethanol-Induced Decrease of the Expression of Glucose Transport Protein (Glut3) in the Central Nervous System as a Predisposing Condition to Apoptosis

The Effect of Age

P. FATTORETTI, C. BERTONI-FREDDARI, T. CASOLI, G. DI STEFANO, G. GIORGETTI, AND M. SOLAZZI

Neurobiology of Aging Laboratory, "N. Masera" INRCA Research Department, 60121 Ancona, Italy

ABSTRACT: We measured the effect of chronic ethanol administration on the expression of Glut3 in the cerebellum and hippocampus of adult and old rats. Glut3 expression significantly decreased in aging, in ethanol-treated rats vs. age-matched controls, and in adult- vs. old ethanol-treated rats. These findings lend consistent support to the hypothesis that disturbances of glucose metabolism due to ethanol may constitute an unfavorable condition predisposing to neuronal death.

KEYWORDS: Glut3; ethanol; cell death; aging; brain metabolism; hippocampus; cerebellum

INTRODUCTION

Neuronal death is a prominent damage in chronic ethanol intake. Although no consensus has been reached on the mechanism(s) underlying the ethanol-induced loss of neurons, it has been documented that a very specific action of this alcohol is the alteration of brain glucose utilization rates in discrete areas of the central nervous system (CNS). The decrease of glucose use due to ethanol intake is reported to affect those brain regions characterized by a high energy metabolism, thus the CNS areas most vulnerable to ethanol are also particularly sensitive to glucose deprivation.[1]

Glucose metabolism is of critical importance for nervous functions: in addition to meeting brain energy requirements, it provides ribose precursors, which are involved in the synthesis of lipids and neurotransmitters as well as in the removal of free radicals.

Glucose transport into mammalian cells is facilitated by a family of transmembrane proteins called Gluts. With specific reference to the nerve cells, the primary

Address for correspondence: Dr. Patrizia Fattoretti, Neurobiology of Aging Laboratory, "N. Masera" INRCA Research Department, Via Birarelli 8, 60121 Ancona, Italy. Voice: +39-71-8004163; fax: +39-71-206791.
p.fattoretti@inrca.it

Ann. N.Y. Acad. Sci. 1010: 500–503 (2003). © 2003 New York Academy of Sciences.
doi: 10.1196/annals.1299.092

glucose transporter is Glut3 isoform, which has high affinity for glucose, and thus determines the rate of glucose uptake by its membrane concentration.[2] With advancing age, the efficiency of energy metabolism, particularly in postmitotic cells (e.g. neurons) is reported to decline significantly, and this is widely supported to represent an unfavorable, though physiological, condition predisposing to age-related functional impairments and pathologic conditions.[3]

MATERIALS AND METHODS

Considering that a critical step for glucose use by neurons is its transport into these cells by Glut3 and that the neuronal energy providing system is significantly impaired in aging, we investigated the effect of chronic ethanol administration (20% in the drinking water for 4 weeks) on the level of expression of Glut3 in adult (12 months) and old (24 months) rats. As controls, we used isocalorically pair-fed rats. The quantitative evaluation of Glut3 was carried out by Western blot analysis in the cerebellum and hippocampus.[4]

RESULTS

Both zones of the CNS showed a significant decrease of the level of expression of Glut3 due to age (hippocampus: –24%; cerebellum: –50%). In ethanol-treated rats versus age-matched controls, Glut3 expression decreased significantly by 18% and 38.8% in the hippocampus, while in the cerebellum it decreased by 35% and

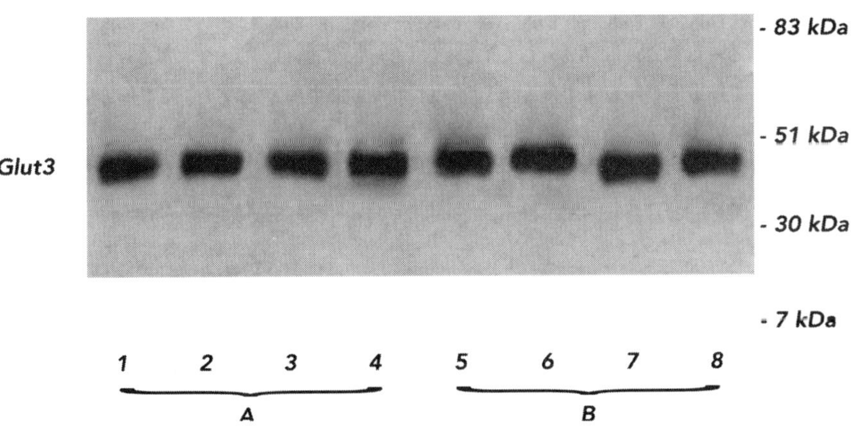

FIGURE 1. Glut3 analysis. Representative Western blot showing the 45-kDa Glut3 protein bands from hippocampal **(A)** and cerebellar **(B)** membranes of 12-month-old pair-fed (Adult); 12-month-old ethanol treated (Adult + Et.); 24-month-old pair fed (Old); and 24-month-old ethanol-treated (Old + Et.) rats. Protein blots were probed with an anti-rat brain Glut3 antibody. Right side: molecular mass markers. *Lanes 1 and 5*: adult rats; *lanes 2 and 6*: old animals; *lanes 3 and 7*: adult ethanol-fed rats; *lanes 4 and 8*: old ethanol-fed animals.

27% in adult and old rats, respectively. Comparing adult-ethanol-treated rats versus old-ethanol-treated animals, we found a 43% and 44% age-related decrease of Glut3 expression in the hippocampus and cerebellum, respectively. Although it is reported that the two zones analyzed in our study are selectively vulnerable to ethanol intoxication, our data suggest a higher sensitivity to the toxicity of this alcohol of the hippocampus of old rats (–38.8% in old vs. –18% in adult rats) and of the cerebellum of adult animals (–35% in adult vs. –27% in old rats).

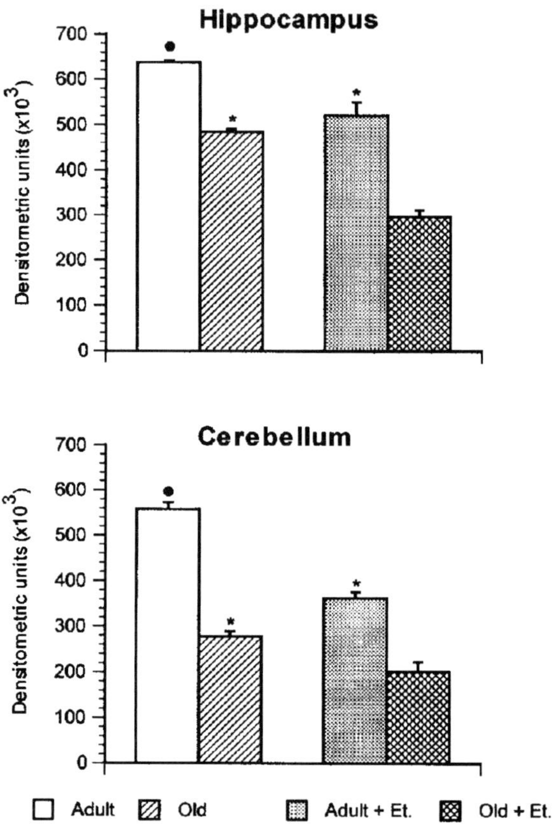

FIGURE 2. Densitometric assessments of Glut3 protein levels in adult, old, adult ethanol-fed, and old ethanol-fed rats. The densitometric quantitation of band intensity revealed a significant age-dependent decline of Glut3 both in the hippocampus (24%) and cerebellum (50%), respectively. In ethanol-fed rats versus normally fed controls, Glut3 expression significantly decreased by 18% and 38.8% in the hippocampus, while in the cerebellum it decreased by 35% and 27% in adult and old rats, respectively. Comparing adult-ethanol treated rats versus old-ethanol-treated animals, we found a 43% and 44% age-related decrease of Glut3 expression in the hippocampus and cerebellum, respectively. Statistical comparison: ANOVA and Tukey test. Data are expressed as mean ± SEM; $P < .05$ was considered to be statistically significant. •$P < .05$ versus normally fed old animals and adult ethanol-fed rats; *$P < .05$ vs. old ethanol-treated group.

DISCUSSION

These findings clearly document that chronic ethanol administration significantly impairs the expression of Glut3 both in the hippocampus and cerebellum of adult and old animals. Aging *per se* appears to affect Glut3 expression; however, old ethanol-treated rats show further reductions in both the CNS zones investigated as compared to the values of adult ethanol-fed animals. Among the different factors that may be responsible for the alterations observed in the present study, Glut3 location and function and ethanol metabolism and mechanism(s) of action are of primary importance.

Glut3 is an integral membrane glycoprotein whose membrane-spanning regions are reported to form a channel through which glucose moves. The transport of glucose into neurons is supposed to occur by association-dissociation steps. Taking into account this peculiar location of Glut3, any perturbation of neuronal membrane may be reasonably supposed to affect the physiological function of this protein. In this context, both aging and chronic ethanol intake are well reported to alter the structural integrity and chemical composition of neuronal membranes and, in turn, to impair membrane physicochemical properties, for example, fluidity. Specifically, age-related changes of neuronal membranes include an increase in the level of cholesterol and of lipid peroxidation metabolites as well as a decrease of phospholipid unsaturation index.[5]

Therefore, neuronal membranes from an old individual represent a different target for ethanol action when compared with those from an adult subject. With specific reference to the toxicity of ethanol for neuronal membranes, it is well documented that this alcohol is able to interact with different receptor systems to increase membrane fluidity and that it is responsible for interdigitations in membrane model systems.[5] On the basis of these data, the present findings lend consistent support to the hypothesis that ethanol-induced glucose disturbances may lead to interruptions of neurotrophic support and to the activation of apoptotic mechanisms.

REFERENCES

1. Hu, I.C., S.P. Singh & A.K. Snyder. 1995. Effects of ethanol on glucose transporter expression in cultured hippocampal neurons. Alcohol Clin. Exp. Res. **19:** 1398–1402.
2. Mueckler, M. 1994. Facilitative glucose transporters. Eur. J. Biochem. **219:** 713–725.
3. Meier-Ruge, W. & C. Bertoni-Freddari. 1996. The significance of glucose turnover in the brain pathogenetic mechanisms of Alzheimer's disease. Rev. Neurosci. **7:** 1–19.
4. Fattoretti, P., C. Bertoni-Freddari, T. Casoli, *et al.* 2002. Aging and vitamin E-deficiency affect the expression of glucose transport protein (Glut3) in the rat hippocampus and cerebellum. Free Rad. Res. **36:** 91–92.
5. Gibson-Wood, W., F. Schroeder, A.M. Rao, *et al.* 1996. Membranes and ethanol: lipid domains and lipid-protein interactions. *In* Pharmacological Effects of Ethanol on the Nervous System. R.A. Deitrich & V.G Erwin, Eds.: 207–212. CRC Press. Boca Raton, FL.

CUGBP2 Plays a Critical Role in Apoptosis of Breast Cancer Cells in Response to Genotoxic Injury

DEBNATH MUKHOPADHYAY,[a] JESSE JUNG,[a] NABENDU MURMU,[a] COURTNEY W. HOUCHEN,[a,b] BRIAN K. DIECKGRAEFE,[a,b] AND SHRIKANT ANANT[a,b]

[a]*Department of Internal Medicine, Washington University School of Medicine, St. Louis, Missouri 63110, USA*

[b]*Siteman Cancer Center, Washington University School of Medicine, St. Louis, Missouri 63110, USA*

ABSTRACT: Posttranscriptional control of gene expression plays a key role in regulating gene expression in cells undergoing apoptosis. Cyclooxygenase-2 (COX-2) is a crucial enzyme in the conversion of arachidonic acid to prostaglandin E2 (PGE_2) and is significantly upregulated in many types of adenocarcinomas. COX-2 overexpression leads to increased PGE_2 production, resulting in increased cellular proliferation. PGE_2 enhances the resistance of cells to ionizing radiation. Accordingly, understanding mechanisms regulating COX-2 expression may lead to important therapeutic advances. Besides transcriptional control, COX-2 expression is significantly regulated by mRNA stability and translation. We have previously demonstrated that RNA binding protein CUGBP2 binds AU-rich sequences to regulate COX-2 mRNA translation. In the current study, we have determined that expression of both COX-2 mRNA and CUGBP2 mRNA are induced in MCF-7 cells, a breast cancer cell line, following exposure to 12 Gy γ-irradiation. However, only CUGBP2 protein is induced, but COX-2 protein levels were not altered. Silencer RNA (siRNA)–mediated inhibition of CUGBP2 reversed the block in COX-2 protein expression. Furthermore, MCF-7 cells underwent apoptosis in response to radiation injury, which was also reversed by CUGBP2 siRNAs. These data suggest that CUGBP2 is a critical regulator of the apoptotic response to genotoxic injury in breast cancer cells.

KEYWORDS: radiation-induced injury; RNA binding proteins; AU-rich sequence; cyclooxygenase-2

INTRODUCTION

Cyclooxygenase (COX) is the enzyme catalyzing the rate-limiting step in prostaglandin synthesis, of which there are two forms: COX-1 and COX-2. COX-1 is

Address for correspondence: Shrikant Anant, Ph.D., Department of Medicine, Washington University School of Medicine, 660 South Euclid Ave., St. Louis, MO 63110. Voice: 314-747-4752; fax: 314-362-8959.
sanant@im.wustl.edu

**Ann. N.Y. Acad. Sci. 1010: 504–509 (2003). © 2003 New York Academy of Sciences.
doi: 10.1196/annals.1299.093**

widely expressed and plays a role in tissue homeostasis. COX -2 is induced in response to growth factors, cytokines, and tumor promoters.[1] COX-2 is upregulated in many cancers, including colon, breast, lung, pancreas, esophagus, and squamous cell carcinoma of the head and neck. The excessive production of prostaglandins by COX-2 enhances tumor cell growth. Inhibition of COX-2 by specific inhibitors suppresses proliferation of cells in a number of tumors including adenocarcinoma of the colon, gall bladder, and esophagus. Furthermore, genetic elimination of COX-2 protects decreased intestinal adenomas in mice lacking the adenomatosis polyposis coli tumor suppressor gene.[2]

Understanding the mechanisms that regulate COX-2 expression is of great interest given its role in cancer progression. COX-2 is regulated at the posttranscriptional level of mRNA stability and translation.[3] Recently, we identified an RNA-binding protein, CUGBP2, which has structural homology to the HuR family of RNA stabilizing factors. CUGBP2 binds and stabilizes COX-2 mRNA in colon cancer cells.[4] Upon binding to COX-2 mRNA, CUGBP2 inhibited COX-2 mRNA translation resulting in cells undergoing radiation induced apoptosis.[5] However, the global relevance of these findings in other tumor cells is currently unknown.

In this study, identified CUGBP2 expression is induced inMCF-7 breast cancer cells when those cells are exposed to ionizing radiation. While COX-2 m RNA is increased in the cells, the message is not translated. Silencer RNA (siRNA)-mediated downregulation of CUGBP2 led to increased COX-2 protein expression and decreased apoptosis. These data suggest that CUGBP2 can function as RNA stabilizer and a translational inhibitor in breast cancer cells.

MATERIALS AND METHODS

Cells and Radiation Treatment

The human breast cancer cell lines MCF-7 were obtained from the American Type Culture Collection (ATCC). The cells were subjected to either 12 Gy γ radiation (0.94 cGy/min) in a gammacel 40 cesium irradiator. Apoptotic cells were identified by hematoxylin and eosin staining and confirmed by dUTP fluorescein nick end labeling (TUNEL, Boehringer Mannheim: Indianapolis, IN) staining according to manufacturer's recommendations.

RNA and Protein Analyses

Total RNA from cells was isolated using Trizol (Invitrogen, Carlsbad, CA) and subjected to reverse transcription to generate first strand cDNA. This was used to determine CUGBP2 and COX-2 mRNA levels by Real Time PCR (RT-PCR) as previously published.[5] Total cell lysates from cells were subjected to SDS-PAGE electrophoresis and subjected to Western blot analyses as previously described.

siRNA

Log-phase growing MCF-7 cells was transfected with three different siRNAs for CUGBP2. Cells were treated with radiation, following which apoptosis and protein expression were determined.[5]

RESULTS

We have previously shown that CUGBP2 and COX-2 mRNA are induced in a colon cancer cell line in response to radiation, but COX-2 protein expression is inhibited at the translational level by CUGBP2.[5] To determine whether a similar inhibition of COX-2 protein occurs in breast cancer cells, we treated MCF-7 cells with 12 Gy γ irradiation.

Analysis of total RNA from these cells demonstrated that both CUGBP2 and COX-2 mRNA are rapidly induced within 1 h following radiation that continued to remain high at 12 h (FIG. 1A). However, while CUGBP2 protein expression was observed during this time, no COX-2 protein was observed after radiation treatment

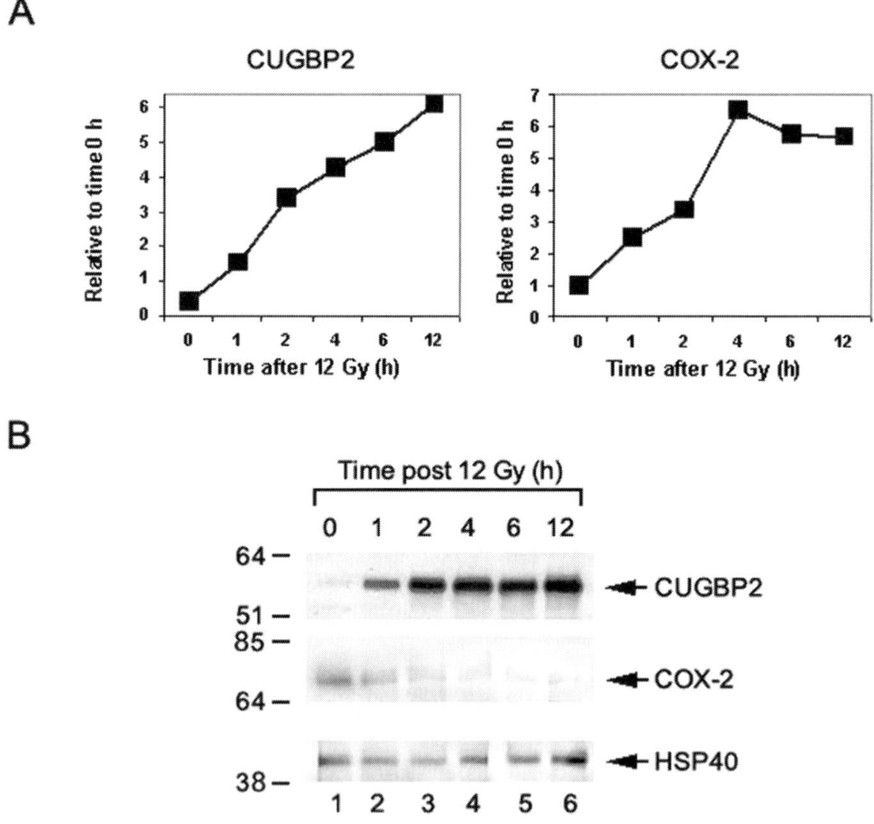

FIGURE 1. Effect of radiation exposure on CUGBP2 and COX-2 expression in MCF-7 cells. **(A)** Time course of CUGBP2 and COX-2 mRNA expression in MCF-7 cells after exposure to 12 Gy γ irradiation. Total RNA was isolated at the indicated time (0–12 h) following radiation and subjected to real time PCR analysis. The data is plotted as relative to that observed at time 0 h. **(B)** Western blot analyses for CUGBP2 and COX-2 protein expression. Total cell lysates generated at the indicated times were resolved by SDS–PAGE and sequentially probed with antibodies to CUGBP2, COX-2, and HSP40. Location of the molecular weight standards is shown to the left.

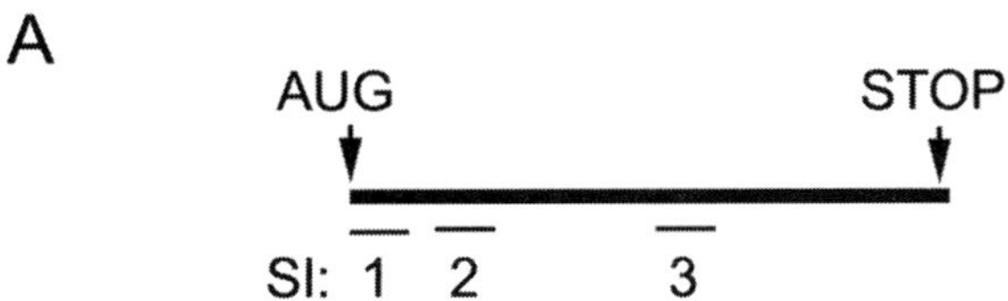

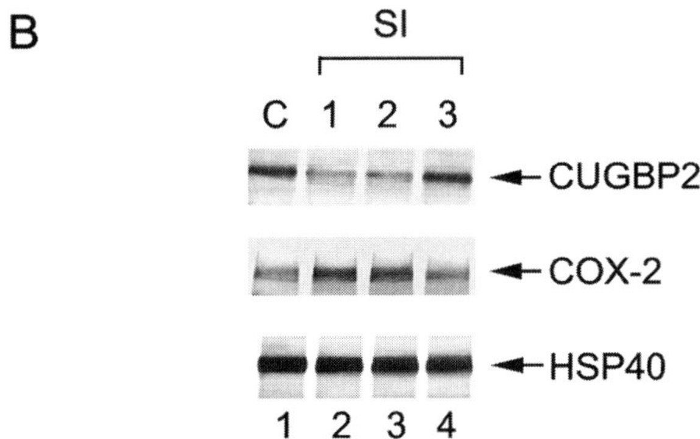

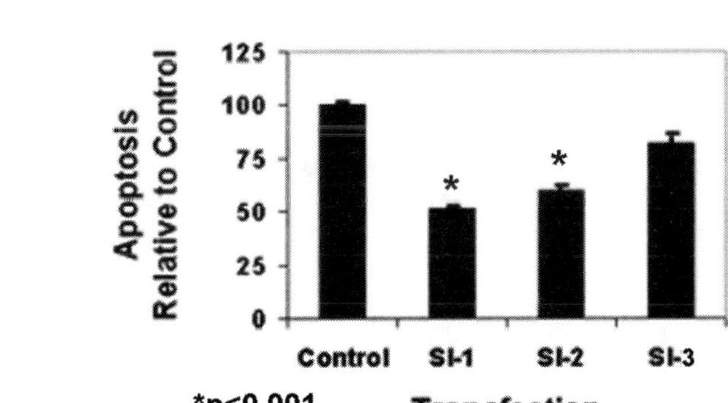

FIGURE 2. siRNA-mediated inhibition of CUGBP2 reduces radiation-induced apoptosis in MCF-7 cells. **(A)** Schematic representation of CUGBP2 cDNA and the location of the three siRNAs. **(B)** 48 h after transfection of siRNAs, cell lysates were prepared and Western blotted for CUGBP2, COX-2 and HSP40. **(C)** Cells transfected with the siRNAs were subjected to 12 Gy γ irradiation. 24 h after radiation, the number of apoptotic cells was visualized by hematoxylin and eosin and TUNEL staining procedures. The total number of apoptotic cells in each transfection is represented as relative to control. Data is mean ± standard error of the mean from three different transfections.

(FIG. 1B). Furthermore, the lower levels of COX-2 protein observed in the control cells at the 0 h time point was suppressed following radiation (FIG. 1B). These data suggest that radiation treatment induces CUGBP2 in MCF-7 cells and that CUGBP2 may play a role in inhibiting COX-2 mRNA translation.

To demonstrate that CUGBP2 regulates translation of COX-2 mRNA in MCF-7 cells, we performed siRNA-mediated knockdown of CUGBP2 and determined its effect on COX-2 protein expression and radiation induced apoptosis. MCF-7 cells were transfected with three different silencer RNA located in different regions of the CUGBP2 coding region, and then were treated by 8-Gy γ irradiation (FIG. 2A). CUGBP2 expression was inhibited by two siRNAs located near the 5′ end of the CUGBP2 cDNA. In contrast, COX-2 protein was induced in the cells when CUGBP2 expression was inhibited (FIG. 2B).

To determine the effect of CUGBP2 inhibition on radiation mediated apoptosis, cells were stained, and the apoptosis levels determined. Transfection of SI-1, SI-2, and SI-3 RNAs resulted in 15, 18, and 25% apoptotic cells, respectively, while control cells demonstrated 30% apoptotic cells in response to 8 Gy γ irradiation.

These data demonstrate that cells with decreased expression of CUGBP2 (SI-1, SI-2) demonstrated a significant decrease in apoptosis compared to control transfected cells (FIG. 2C). On the other hand, transfection of SI-3 siRNA that did not reduce CUGBP2 expression also did not have significant differences in apoptosis. These data suggest that CUGBP2 mediated suppression of COX-2 mRNA enhances susceptibility to radiation-induced cellular apoptosis.

DISCUSSION

COX-2 is a potent inhibitor of apoptosis. This is due to the activity of prostaglandins that are generated by COX-2 from arachidonic acids.[1] The reduction of apoptosis in cells expressing COX-2 was also associated with an elevated expression of anti-apoptotic factors, such as Bcl-2; and the reduced expression of proapoptotic factors, such as transforming growth factor receptor.[6] Increased expression of Bcl-2, along with reduced expression of the pro-apoptotic protein Bax, have also been observed in mammary tumors derived from COX-2 overexpressing transgenic mice.[7] Together, these studies suggest that COX-2 is an important factor in the development of breast neoplasias. Hence, mechanisms that inhibit COX-2 may be beneficial in treatment of breast cancer. Indeed, COX-2 selective nonsteroidal anti-inflammatory drugs (NSAIDs), which inhibit COX-2 function, have been studied in mouse models of colon cancer and in clinical trials. Results have shown that COX-2 may be a useful target for the prevention and treatment of colon cancer.[8]

COX-2 is regulated at the posttranscriptional level of mRNA stability and translation.[3] This is mediated by AU-rich element (ARE) in the 3′ untranslated region (3′UTR) of COX-2 mRNA.[3] RNA stability factor HuR, which is overexpressed in colon cancer cells, was initially identified to bind the COX-2 ARE and to stabilize COX-2 mRNA, resulting in increased COX-2 protein expression.[4]

We have demonstrated that CUGBP2 also interacts with COX-2 3′UTR, but that this interaction results in the inhibition of COX-2 translation. This results in radiation-induced apoptosis of HT-29 cells, a colon cancer cell line.[5] These data suggest that CUGBP2 plays a critical role in regulating COX-2 mRNA translation in the cells. Knockdown of CUGBP2 in breast cancer cells results in increased expression, further demonstrating that CUGBP2 dependent inhibition of COX-2 is not restricted to colon cancer.

The molecular basis of radiation sensitivity in breast cancer cells has not been well understood. Breast tumors that have mutations in one of the breast cancer susceptibility genes, BRCA1 and BRCA2, demonstrate marked sensitivity to ionizing radiation.[9] However, the extent to which these mutations occur in sporadic cancers is currently unknown. Another gene that is a contributing factor to breast cancer[9] predisposition is the ataxia telangiectasia (A-T) gene (ATM), which regulates cellular sensitivity to radiation.

However, the extent to which the ATM gene is disrupted in breast cancer is not clearly understood. Taken together these data imply that the therapeutic responses in breast cancer cells may be complex. We have observed that CUGBP2 is significantly induced in cells after radiation treatment. Furthermore, inhibition of CUGBP2 decreased the number of cells undergoing radiation-induced apoptosis, suggesting that CUGBP2 may be essential for radiation sensitivity of the breast cancer cell. This will be the focus of further research in our laboratory.

ACKNOWLEDGMENTS

This work is supported by National Institutes of Health Grants DK-02822 (C.W.H), DK-60106 (B.K.D., DK-52574, and DK-62265 (S.A.). S.A. is a Research Scholar of the American Gastroenterology Association.

REFERENCES

1. WILLIAMS, C.S., M. MANN & R.N. DUBOIS. 1999. The role of cyclooxygenases in inflammation, cancer, and development. Oncogene **18:** 7908–7916.
2. OSHIMA, M. *et al.* 1996. Suppression of intestinal polyposis in Apc delta716 knockout mice by inhibition of cyclooxygenase 2 (COX-2). Cell **87:** 803–809.
3. COK, S.J. & A.R. MORRISON. 2001. The 3′-untranslated region of murine cyclooxygenase-2 contains multiple regulatory elements that alter message stability and translational efficiency. J. Biol. Chem. **276:** 23179–23185.
4. DIXON, D.A. *et al.* 2000. Post-transcriptional control of cyclooxygenase-2 gene expression. The role of the 3′-untranslated region. J. Biol. Chem. **275:** 11750–11757.
5. MUKHOPADHYAY, D., *et al.* 2003. Coupled mRNA stabilization and translational silencing of cyclooxygenase-2 by a novel RNA binding protein, CUGBP2. Mol. Cell **11:** 113–126.
6. SHENG, H. *et al.* 1998. Modulation of apoptosis and Bcl-2 expression by prostaglandin E2 in human colon cancer cells. Cancer Res. **58:** 362–366.
7. LIU, C.H. *et al.* 2001. Overexpression of cyclooxygenase-2 is sufficient to induce tumorigenesis in transgenic mice. J. Biol. Chem. **276:** 18563–18569.
8. GUPTA, R.A. & R.N. DUBOIS. 2001. Colorectal cancer prevention and treatment by inhibition of cyclooxygenase-2. Nat. Rev. Cancer **1:** 11–21.
9. XIA, F. & S.N. POWELL. 2002. The molecular basis of radiosensitivity and chemosensitivity in the treatment of breast cancer. Semin. Radiat. Oncol. **12:** 296–304.

Concerted Deregulations of Multiple Apoptosis-Controlling Genes in Pancreatic Carcinoma Cells

ANNA TRAUZOLD, STEFAN SCHMIEDEL, STEPHANIE OESTERN, MATTHIAS CHRISTGEN, SABINE WESTPHAL, CHRISTIAN ROEDERM, AND HOLGER KALTHOFF

Molecular Oncology, Clinic for General Surgery, Christian Albrechts University, 24 105 Kiel, Germany

KEYWORDS: **apoptosis resistance; death receptors; pancreatic cancer; CD95; TRAIL**

INTRODUCTION

CD95, TRAIL-R1, and TRAIL-R2 are members of the TNF-receptor family of transmembrane proteins that are capable of inducing apoptosis, thus named "death receptors." Several intracellular proteins, such as caspases, FADD, FLIP, Bid, Bcl-xL, and IAPs participate in, or interfere with, the signal transduction of the death receptors.[1] Perturbations in the expression of these molecules result in an increase or a decrease in the apoptotic response of cells to CD95-L or TRAIL. The majority of pancreatic adenocarcinoma cell lines is resistant to anti-CD95 and TRAIL-mediated apoptosis.[2–4] Recently, we revealed an anti-apoptotic role of FAP1 and Bcl-x_L in these cells.[3,5] We also demonstrated that PKC and NF-κB protected pancreatic tumor cells from anti-CD95 and TRAIL-mediated apoptosis.[4]

Because pancreatic adenocarcinoma is one of the most aggressive cancer types with an extremely poor prognosis, it is crucial to characterize the common molecular changes in pancreatic tumor cells to develop new therapeutic strategies. Thus, we determined the expression levels of several apoptosis-relevant proteins in five pancreatic tumor cell lines with different potential to undergo apoptosis. We then correlated our data with the selected protein's sensitivity to anti-CD95- or TRAIL-induced cell death.

MATERIALS AND METHODS

Human pancreatic adenocarcinoma cell lines Capan1, Colo357, PancTuI, Panc89, and Panc1 were cultured in RPMI 1640 supplemented with 10% FCS, 2 mM

Address for correspondence: Anna Trauzold, Molecular Oncology, Clinic for General Surgery Christian-Albrechts-University, Arnold-Heller Str. 7, 24 105 Kiel, Germany. Voice: +49-431-597-1962; fax: +49-431-597-1939.
atrauzold@1med.uni-kiel.de

**Ann. N.Y. Acad. Sci. 1010: 510–513 (2003). © 2003 New York Academy of Sciences.
doi: 10.1196/annals.1299.094**

glutamine, and 1 mM sodium pyruvate. For induction of apoptosis, agonistic monoclonal anti-CD95-antibodies clone CH11 (100 ng/mL; Coulter Immunotech) or recombinant TRAIL (100 ng/mL; R&D Systems) were added to the culture medium for 24 hours. The extent of apoptosis was measured with the JAM assay as previously described.[2] Cell surface expression of CD95 and TRAIL was determined by FACS analysis. The expression of apoptosis-relevant proteins was detected by Western blot.

RESULTS AND CONCLUSIONS

Pancreatic adenocarcinoma cell lines PancTuI, Panc1, and Panc89 were almost completely resistant to anti-CD95; whereas Colo357 cells were moderately sensitive and Capan1 cells highly sensitive to those cell lines (FIG. 1). PancTuI cells were also completely refractory to TRAIL, while about 30% of Panc1; 40% of Panc89; 60% of Colo357; and more than 90% of Capan1 cells died due to the TRAIL treatment. Determination of death receptor expression on the cell surface by FACS analysis revealed that all cells expressed TRAIL-R1 and TRAIL-R2, and all but Panc1 cells also expressed CD95 (data not shown).

These data suggested that the block in CD95- (except for Panc1 cells) and TRAIL-mediated apoptosis is located downstream of the receptor level. Consequently, we systematically analyzed the expression of intracellular proteins involved in transmission or inhibition of the death signal in these cells. We found that all cells expressed proteins known to be necessary for apoptosis such as caspases-8; -9; -3; -

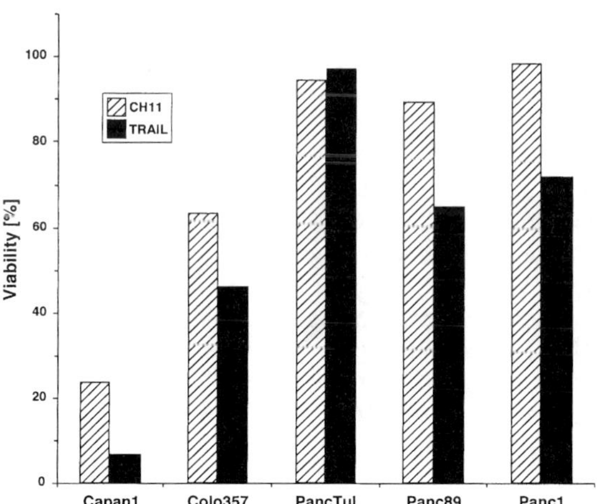

FIGURE 1. Sensitivity of pancreatic tumor cells to anti-CD95 and TRAIL-mediated apoptosis. Cells were treated with an agonistic anti-CD95 antibody CH11 (100 ng/mL) or with TRAIL (100 ng/mL) for 24 h. Viability was determined by JAM-DNA fragmentation assay. Data shown are the results of one representative experiment (n = 6) out of three performed. Standard deviations were smaller that 10%.

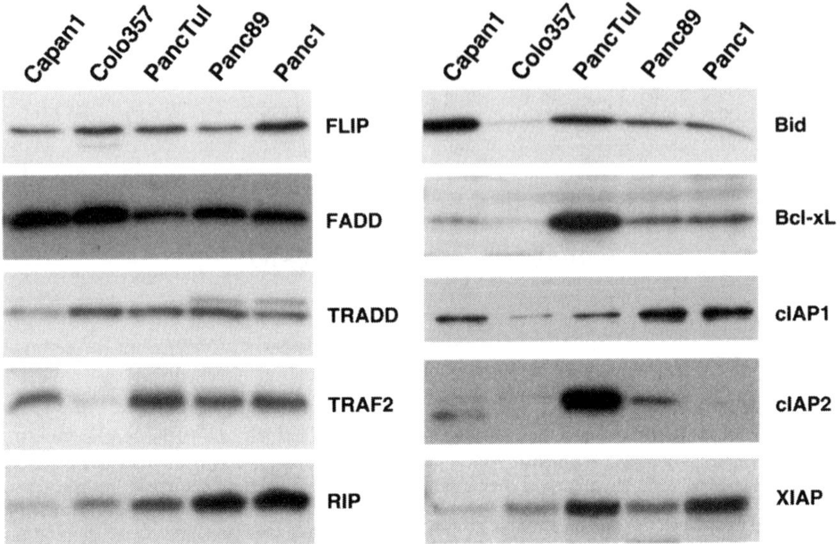

FIGURE 2. Expression of apoptosis relevant proteins in pancreatic tumor cells. Thirty micrograms of whole-cell lysates were analyzed using Western blot for constitutive expression of proteins as shown in the figure. Equal gel loading was verified by assessing β-actin levels (data not shown).

7; -2 as well as FADD, Bid, APAF1, and Smac/DIABLO (data not shown). Interestingly, we found that DISC interacting proteins FLIP, TRADD, TRAF2, and RIP as well as Bcl-x_L, XIAP, cIAP1 and cIAP2 were overexpressed, whereas FADD and Bid were clearly downregulated in resistant cells (FIG. 2).

These changes in expression levels of anti- and pro-apoptotic proteins might contribute to resistance towards CD95- and TRAIL-mediated apoptosis. To confirm this hypothesis, we generated dominant negative mutants of FADD and overexpressed XIAP and Bcl-x_L in sensitive cell lines. All mutants showed significantly reduced sensitivity towards death receptor-mediated apoptosis. Additionally, overexpression of pro-apoptotic Bid in resistant cell lines rendered them sensitive (data not shown).

The results revealed multiple changes of the expression of anti-apoptotic proteins in resistant pancreatic tumor cells. Concurring in all resistant cells, elevated levels of FLIP, TRADD, TRAF2, and RIP were expressed. This suggested that protection from apoptosis already occurs at the receptor level, either by inhibition of the DISC formation or by receptor-mediated NF-κB activation. Additionally, a decrease of Bid and increase of Bcl-x_L expression impede the mitochondrial amplification loop essential for apoptosis in type II cells. The expression of caspase inhibitors XIAP, cIAP1, and cIAP2 may provide additional protection. In conclusion, multiple changes of death receptor signaling seem to be responsible for the resistance of pancreatic adenocarcinoma cells to anti-CD95- and TRAIL-mediated apoptosis.

REFERENCES

1. IGNEY, F.H.& P.H. KRAMMER. Nat. Rev. Cancer. 2002. **2:** 277–288.
2. UNGEFROREN, H., M. VOSS, M. JANSEN, *et al.* 1998. Cancer Res. **58:** 1741–1749.
3. HINZ, S., A. TRAUZOLD, L. BOENICKE, *et al.* 2000. Oncogene **19:** 5477–5486.
4. TRAUZOLD, A., H. WERMANN, A. ARLT, *et al.* 2001. **20:** 4258–4269.
5. UNGEFROREN, H., M-L. KRUSE, A. TRAUZOLD, *et al.* 2001. J. Cell Sci. **114:** 2735–2746.

Nuclear Translocation of a Clusterin Isoform Is Associated with Induction of Anoikis in SV40-Immortalized Human Prostate Epithelial Cells

A. E. CACCAMO,[a] M. SCALTRITI,[b] A. CAPORALI,[a]
D. D'ARCA,[b] F. SCORCIONI,[b] G. CANDIANO,[c]
M. MANGIOLA,[d] AND S. BETTUZZI[a]

[a]*Department of Medicina Sperimentale, University of Parma, 43100 Parma, Italy*

[b]*Department of Scienze Biomediche, University of Modena, Modena, Italy*

[c]*Fisiopatologia dell'Uremia, Institute G. Gaslini, Genova, Italy*

[d]*Immunologia Clinica, Institute G. Gaslini, Genova, Italy*

ABSTRACT: **Clusterin gene expression is potently induced in experimental models in which apoptosis is activated, such as rat prostate involution following castration. Nevertheless, its precise physiological role has not yet been established, and both anti-apoptotic and pro-apoptotic functions have been suggested for this gene. Clusterin expression level depends on cell proliferation state, and we recently showed that its over-expression inhibited cell cycle progression of SV40-immortalized human prostate epithelial cells PNT2 and PNT1a. Here we studied clusterin expression in PNT1a cells subjected to serum-starvation with the aim of defining clusterin early molecular changes following apoptosis induction. Under serum-starvation conditions, decreased growth rate, slow rounding-up of cells, cell detachment, and formation of apoptotic bodies indicative of anoikis (detachment-induced apoptosis) were preceded by significant downregulation of 70 kDa clusterin precursor and upregulation of 45–40 kDa isoforms. On the 8th day of serum-free culturing, only the higher molecular-weight protein-band of about 45 kDa was clearly induced and accumulated in detached cells and apoptotic bodies in which PARP was activated. Anoikis was preceded by induction and transloction of a 45-kDa clusterin isoform to the nucleus. Thus, nuclear targeting of a specific 45-kDa isoform of clusterin appeared to be an early and specific molecular signal triggering anoikis-death. Considering also that clusterin is downregulated during prostate cancer onset and progression, and that its upregulation has inhibited DNA synthesis and cell cycle progression of immortalized human prostate epithelial cells, we suggest that clusterin might be a new anti-oncogene in the prostate.**

KEYWORDS: **clusterin; prostate epithelial cells; nuclear translocation; anoikis**

Address for correspondence: S. Bettuzzi, Dept. of Medicina Sperimentale, Univ. of Parma, Via Volturno 39, 43100 Parma, Italy. Voice: +0039-0521-903803; fax: +0039-0521-903802.
saverio.bettuzzi@unipr.it

**Ann. N.Y. Acad. Sci. 1010: 514–519 (2003). © 2003 New York Academy of Sciences.
doi: 10.1196/annals.1299.095**

INTRODUCTION

Clusterin gene expression is potently induced during rat prostate involution following castration.[1] Its precise physiological role has not been established,[2] although many pathophysiological functions have been proposed, among which are an anti-apoptotic role[3,4] and pro-apoptotic functions.[5] Clusterin is upregulated in quiescent human fibroblasts,[6] but downregulated after induction of cell proliferation.[7] Thus its expression level depends on the proliferation state.

Recently, clusterin was shown to slow down cell cycle progression of SV40-immortalized human prostate epithelial cells PNT2 and PNT1a.[8] Because apoptosis can be induced upon serum-starvation in many cell types, including PNT1a cells, we studied its expression in serum-starved PNT1a cells to better understand the possible role of this gene in apoptotic processes.

MATERIALS AND METHODS

Cell culturing. PNT1a cells were cultured in standard serum-fed or in serum-free conditions with BSA 0.1%, wt-vol.

Cell growth analysis. Adhering cells were harvested, fixed with 4% paraformaldehyde in PBS and stained with 0.1% crystal violet. Color was extracted with Na-citrate 2M, pH 4.5, and absorbance was measured at 540 nm.

Western blotting. Total protein extracts in SDS-PAGE loading buffer (equivalent to 4×10^5 cells) were loaded, resolved on 10% polyacrylamide gel and blotted on a PVDF membranes. Immunoreactive bands were detected using mouse monoclonal anti-human clusterin antibody, clone 41d from Upstate Biotechnology (Lake Placid, NY).[8]

Nuclear extracts. Cells were lysed with 50 mM Tris-HCl pH 7.4, 75 mM NaCl, 25 mM KCl, 5 mM $MgCl_2$, 0.5% Triton X-100, and 0.25% deoxycolate. Nuclear extracts obtained by centrifugation at 13,000 rcf for 10 min were dissolved in SDS-PAGE loading buffer and used for Western blot analysis.

Detection of apoptotic bodies. Apoptotic bodies were detected by DAPI and Giemsa nuclear staining.

DNA fragmentation and cell cycle progression analysis. Ipo-diploid cells and cell cycle progression were detected by FACS analysis.[8]

PARP activation. PARP was detected by Western blot using a polyclonal anti-human PARP (cleavage site 214/215) specific antibody (Biosource, Int. Inc., CA).

Immunocytochemistry analysis. Cells were fixed, permeabilized in 0.3% Triton X-100 in PBS for 15 min at room temperature, and subjected to immunocytochemistry[8] using monoclonal anti-human clusterin plus anti-mouse Alexa Fluor™ 568-red fluorescence as secondary antibody (Eugene-Leiden, the Netherlands).

Fluorescence microscopy analysis. Digital images were acquired with a color CCD camera: red fluorescence, clusterin immunostaining; blue fluorescence, DAPI nuclear staining.

RESULTS AND DISCUSSION

PNT1a cells that were grown in serum-fed standard condition produced clusterin as an intracellular precursor of about 70 kDa that is progressively downregulated until the eighth day of culturing, while mature low-molecular weight isoforms accumulate as a triplet of about 45-kDa protein bands (FIG. 1H). The same cells grown

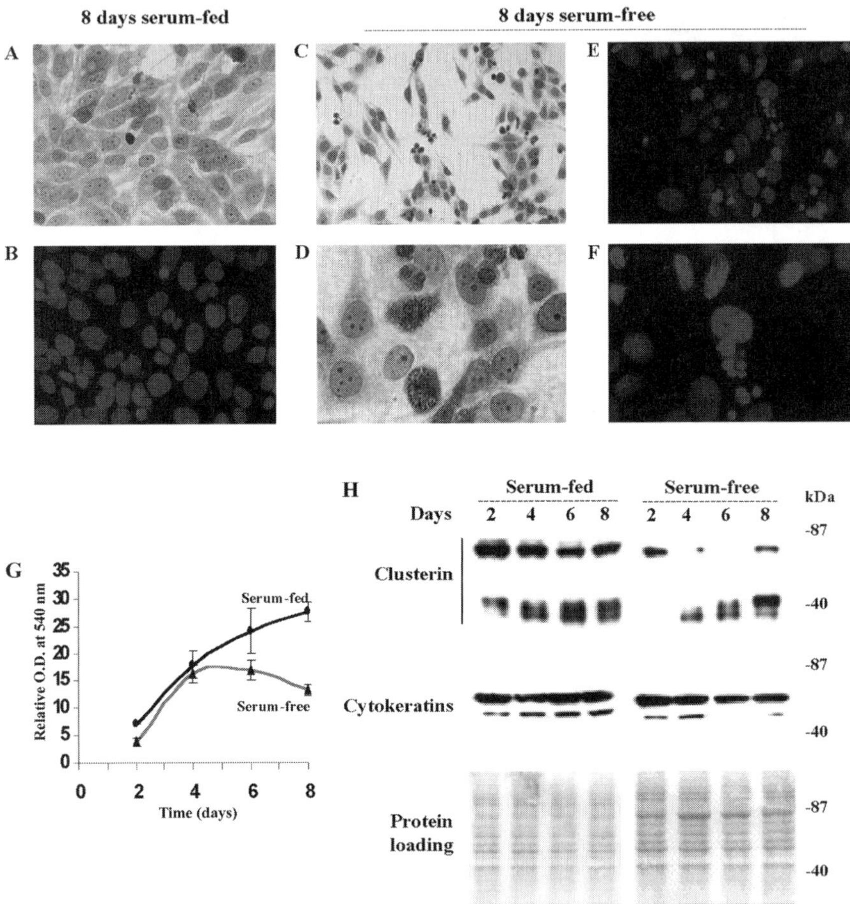

FIGURE 1. Detection of anoikis and clusterin induction in PNT1a cells following serum starvation. Cells were cultured in serum-fed or in serum-free medium **(A–F)**, harvested after 8 days, fixed and stained with Giemsa **(A, C, D)** or DAPI (blue fluorescence**) (B, E, F)**; cell detachment and anoikis induction, chromatin condensation and formation of apoptotic bodies are shown in **(C–F)**; magnification: ×10 **(C)**; ×20 **(A, B, E)**; ×40 **(D, F)**; cell growth time course of PNT1a cells under serum-fed or serum-free conditions **(G)**. Representative Western blot analysis of time-course clusterin expression in PNT1a cells grown under serum-fed or serum-free condition and harvested 2, 4, 6, or 8 days after plating as indicated **(H)**. Cytokeratins expression and red-ponceau staining of the blots (*protein loading*) are reported to show equal loading of the samples.

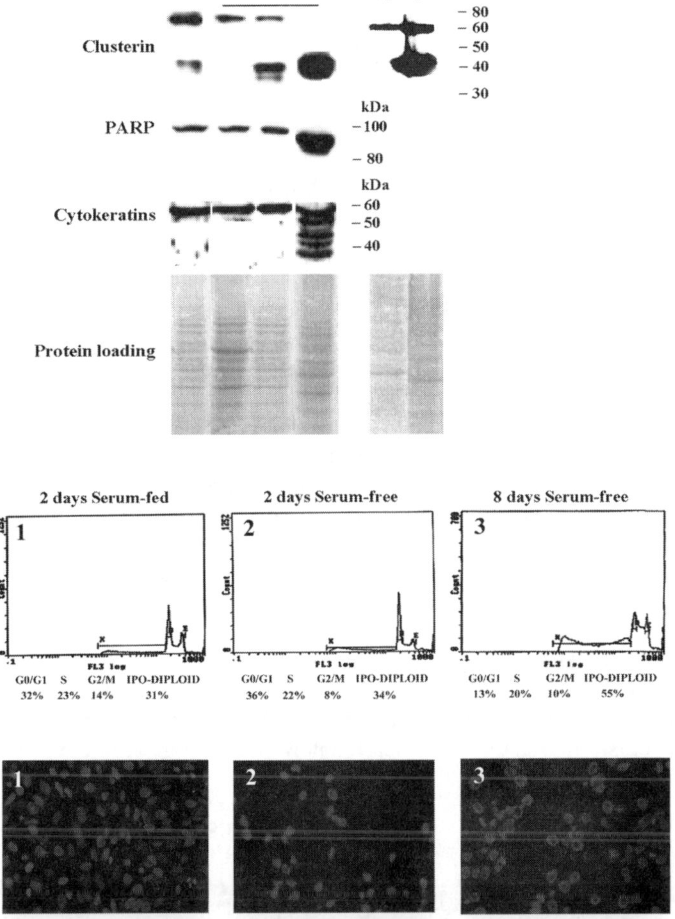

FIGURE 2. Western blot and immunocytochemistry analysis of clusterin accumulation in nucleus and apoptotic bodies during anoikis induction in PNT1a cells. **(A)** Clusterin protein expression profile: Cytocheratins and PARP activation were studied in adhering cells (samples 1–3: 2 days serum-fed, 2 days serum-free and 8 days serum-free, respectively); in detached cells (sample 4: 8 days serum-free; note that cytocheratins profile is altered by cell detachment); and in nuclear extracts (sample 1, 2 days serum-fed; sample 3, 8 days serum-free). Red-ponceau staining of the blots (*protein loading*) is reported to show equal loading

up to 8 days under serum-free condition showed decreased growth rate (FIG. 1G), cell detachment, and formation of apoptotic bodies (FIG. 1, C–F). Under these conditions, significant downregulation of clusterin precursor and upregulation of the 45-kDa isoforms were evident up to the 6th day (FIG. 1H). On the eighth day of serum-free culturing, only the higher molecular mass protein-band of about 45 kDa was clearly induced (FIG. 1H; FIG. 2A). Accumulation of this clusterin isoform was more evident in detached cells and apoptotic bodies (FIG. 2A), associated with anoikis (detachment-induced apoptosis) as demonstrated by slowly rounding up cells followed by detachment (FIG.1, C–F), chromatin condensation (FIG. 1D, E), formation of apoptotic bodies (FIG. 1E, F; FIG. 2C), caspase-3, -8, -9 activation (not shown), PARP activation (FIG. 2A), and induction of DNA fragmentation (ipo-diploid peak) by FACS analysis (FIG. 2B). Anoikis was preceded by translocation of a 45-kDa clusterin isoform to the nucleus (FIG. 2A).

Nuclear clusterin translocation was an early event, because clusterin was associated with condensed chromatin already two days after switching to a serum-free condition, and only later was it accumulated into apoptotic bodies (FIG. 2C). Thus, clusterin upregulation and nuclear translocation were a specific responses preceding morphological changes leading to cell detachment and decreased proliferation. In fact, clusterin translocation was associated with S phase arrest at the G0/G1-S phase checkpoint (FIG. 2B) probably because of incomplete DNA synthesis similar to what we have previously found by transient over-expression of clusterin in the same cells.[8] Nuclear targeting of a specific 45-kDa isoform of clusterin, probably due to posttranslational modification (protein truncation?), appears to be an early and specific molecular signal activated by environmental/hormonal stress. Our data are consistent with the finding that clusterin expression was recently shown repressed by anchorage by DNA microarray analysis.[9] Considering that clusterin is downregulated during prostate cancer onset and progression,[10] and that its upregulation inhibits DNA synthesis and cell cycle progression of immortalized human prostate epithelial cells,[8] clusterin must be considered a new anti-oncogene in the prostate.

REFERENCES

1. BETTUZZI, S., R.A. HIIPAKKA, P. GILNA & S.T. LIAO. 1989. Identification of an androgen-repressed mRNA in rat ventral prostate as coding for sulphated glycoprotein 2 By CDNA cloning and sequence analysis. Biochem. J. **257:** 293–296.
2. ROSENBERG, M.E. & J. SILKENSEN. 1995. Clusterin: physiologic and pathophysiologic considerations. Int. J. Biochem. Cell Biol. **27:** 633–645.
3. KOCH BRANDT, C. & C. MORGANS. 1996. Clusterin: a role in cell survival in the face of apoptosis? Prog. Mol. Subcell. Biol. **16:** 130–149.
4. MIYAKE, H., C. NELSON, P.S. RENNIE & M.E. GLEAVE. 2000. Testosterone-repressed prostate message-2 is an antiapoptotic gene involved in progression to androgen independence in prostate cancer. Cancer Res. **60:** 170–176.

---

of the samples. **(B)** Representative histograms and quantification of the relative percentage of total cell population in G_0/G_1, S and G_2/M phases of the cell cycle and of the ipo-diploid peak (DNA fragmentation). Sample labels are as in panel A. **(C)** Detection of clusterin accumulation in nuclei and apoptotic bodies (red fluorescence) by immunocytochemistry. Nuclear staining was obtained with DAPI (blue fluorecence). Sample labels are as in panel A. Magnification: ×10.

5. YANG, C.R., K. LESKOV, K. HOSLEY-EBERLEIN, *et al.* 2000. Nuclear clusterin/Xip8, an x-ray-induced Ku70-binding protein that signals cell death. Proc. Natl. Acad. Sci. USA **97:** 5907–5912.
6. BETTUZZI, S., S. ASTANCOLLE, G. GUIDETTI, *et al.* 1999. (SGP-2) Gene Expression Is Cell Cycle Dependent In Normal Human Dermal Fibroblasts. FEBS Lett. **448:** 297–300.
7. GRASSILLI, E., S. BETTUZZI, D. MONTI, *et al.* 1991. Studies on the relationship between cell proliferation and cell death: opposite patterns of SGP-2 and ornithine decarboxylase mRNA accumulation in Pha-stimulated human lymphocytes. Biochem. Biophys. Res. Commun. **180:** 59–63.
8. BETTUZZI, S., F. SCORCIONI, S. ASTANCOLLE, *et al.* 2002. Clusterin (SGP- 2) transient overexpression decreases proliferation rate of SV40-immortalized human prostate epithelial cells by slowing down cell cycle progression. Oncogene **21:** 4328–4334.
9. GOLDBERG, G.S., Z. JIN, H. ICHIKAWA, *et al.* 2001. Global effects of anchorage on gene expression during mammary carcinoma cell growth reveal role of tumor necrosis factor-related apoptosis-inducing ligand in anoikis. Cancer Res. **61:** 1334–1337.
10. BETTUZZI, S., P. DAVALLI, S. ASTANCOLLE, *et al.* 2000. Tumor progression is accompanied by significant changes in the levels of expression of polyamine metabolism regulatory genes and clusterin (sulfated glycoprotein 2) in human prostate cancer specimens. Cancer Res. **60:** 28–34.

The Role of Apoptosis in Pituitary Adenomas in the Field of Conventionally Used Therapeutic Approaches

JAROSLAV CERMAN,[a] JAN CAP,[b] MARTINA MAREKOVA,[a] STANISLAV NEMECEK,[c] JOSEF MAREK,[e] EMIL RUDOLF,[d] AND MIROSLAV CERVINKA[d]

[a]Department of Medical Biochemistry, [b]2nd Department of Internal Medicine, [c]Department of Neurosurgery, [d]Department of Medical Biology and Genetics, Charles University in Prague, Faculty of Medicine in Hradec Kralove, 500 38 Hradec Kralove, Czech Republic

[e]3rd Department of Internal Medicine in Prague, Charles University in Prague, Prague, Czech Republic

ABSTRACT: The aim of the study was to investigate the mechanism of action of somatostatin analogues (SSA), ionizing radiation, and their combination on pituitary adenoma cells with special emphasis on proliferative and apoptotic activity. In the 14 GH-secreting adenomas pretreated with SSA before surgery, more prominent regressive changes were found accompanied by compensatory increase in perivascular fibrosis than in the reference group of 17 unpretreated adenomas. The proliferative Ki-67 labeling index was significantly lower in the treated group (median 1.6 per 1000) than in the untreated patients (median 5.0 per 1000). Apoptosis was detected in only 2 of the 14 pretreated adenomas, and it was more frequent (9/17) and more prominent in the untreated group. In cell lines, the SSA had minimal antiproliferative effect, and they were unable to induce apoptosis. Ionizing radiation at doses of 5–20 Gy induced apoptosis in the corticotroph cell line AtT20 with no cell-cycle block. In the somatotroph GH3 cell line, the early (premitotic) apoptosis was detectable using only a high dose of 200 Gy; after irradiation with doses of 20–50 Gy, apoptosis appeared with the latency of 48–72 hours, and was preceded by cell-cycle arrest in the G_2/M phase. The treatment with somatostatin-14 during irradiation increased the percentage of apoptotic cells in culture 10 days after irradiation (11% versus 3% using 20 Gy).

KEYWORDS: apoptosis; pituitary adenomas; acromegaly; somatostatin analogues; irradiation; cell lines

Address for correspondence: Jaroslav Cerman, Department of Medical Biochemistry, Charles University in Prague, Faculty of Medicine in Hradec Kralove, P.O. Box 38, 500 38 Hradec Kralove, Czech Republic. Voice: +420-495816222; fax: +420-495513597.
cerman@lfhk.cuni.cz

**Ann. N.Y. Acad. Sci. 1010: 520–524 (2003). © 2003 New York Academy of Sciences.
doi: 10.1196/annals.1299.096**

INTRODUCTION

Apart from surgery, certain drugs and ionizing radiation are used in the treatment of pituitary adenomas. Somatostatin analogues (SSA) are used for acromegaly treatment both after the operation[1] and as a primary therapy.[2] They are able not only to inhibit GH secretion, but also to reduce tumor size in 30–40% of acromegalic patients.[3] The detailed mechanism of the observed adenoma shrinkage is not known. The GH-secreting adenomas are less sensitive to radiotherapy than ACTH-secreting ones in Cushing's disease,[4] and the results were even worse when applied during octreotide treatment.[5] The aim of the study was to investigate the mechanism of action of SSA, ionizing radiation, and their combination on pituitary adenoma cells with special emphasis on proliferative and apoptotic activity.

PATIENTS AND METHODS

Fourteen patients were treated with SSA for three months before surgery. Tumor shrinkage was observed in seven patients. In two cases the shrinkage was about half of the starting volume. No medical therapy of acromegaly was applied to the reference group of 17 patients.

Adenoma tissue specimens obtained at surgery were investigated using both routine histopathology and immunocytochemistry (ICC) with specific antibodies (Ab). Besides evaluation of all pituitary hormones, markers pertinent to (a) apoptosis and (b) proliferation were detected by ICC: (a) Cytokeratin-18 (Ab-clone C-04, EXBIO Prague), apoptosis-specific fragment M30, that is, caspase-cleaved formalin-resistant neo-epitope of cytokeratin-18 (monoclonal Ab, Roche Molecular Biochemicals), and cleaved 17-kDa form of caspase-3 (rabbit polyclonal Ab, Cell Signaling Technology Inc.); (b) proliferation marker–the Ki-67 antigen (Ab-clone MIB-1, DAKO). The Ki-67 labeling index was calculated from at least 1000 nuclei and expressed as median, 25th, and 75th percentile.

The Mann-Whitney Rank Sum Test was used to estimate the statistical significance of the difference between the two groups. The presence of fibrosis and apoptosis was evaluated by the same pathologist (S.N.), who was unaware of the clinical characteristics and treatment assignments of the patients. The cell lines GH3 (rat somatotrophs) and AtT20 (subclone D16v-F2 of mouse corticotrophs) were obtained from the American Type Culture Collection (Manassas, VA). Cells were irradiated at room temperature using a ^{60}Co γ-ray source with a dose rate of 3 Gy/min. For cell-cycle analysis the cells were washed, fixed by 70% ethanol, and then stained with propidium iodide. Fluorescence (DNA content) was measured with a Coulter Electronic (Hialeah, FL) flow cytometric apparatus. The percentage of cells showing morphologic characteristics of apoptosis in Diff-Quick stained cytospin preparations was calculated as well.

RESULTS

Histological examination revealed regressive changes in somatostatin analogue-treated adenomas accompanied by hemorrhage and compensatory fibrosis. Considerable fibrosis was found in twelve out of fourteen treated adenomas and in six out of seventeen untreated tumors. This difference was statistically significant using the Fisher Exact Test ($P = .009$). The apoptotic marker M30 was observed in only two out of the fourteen treated adenomas and in nine out of the seventeen untreated ones. This difference was close to statistical significance ($P = .057$). For ICC detection of the M30 and Ki-67 antigens, see FIG. 1. The Ki-67 labeling index was 1.6 per mille (median) in

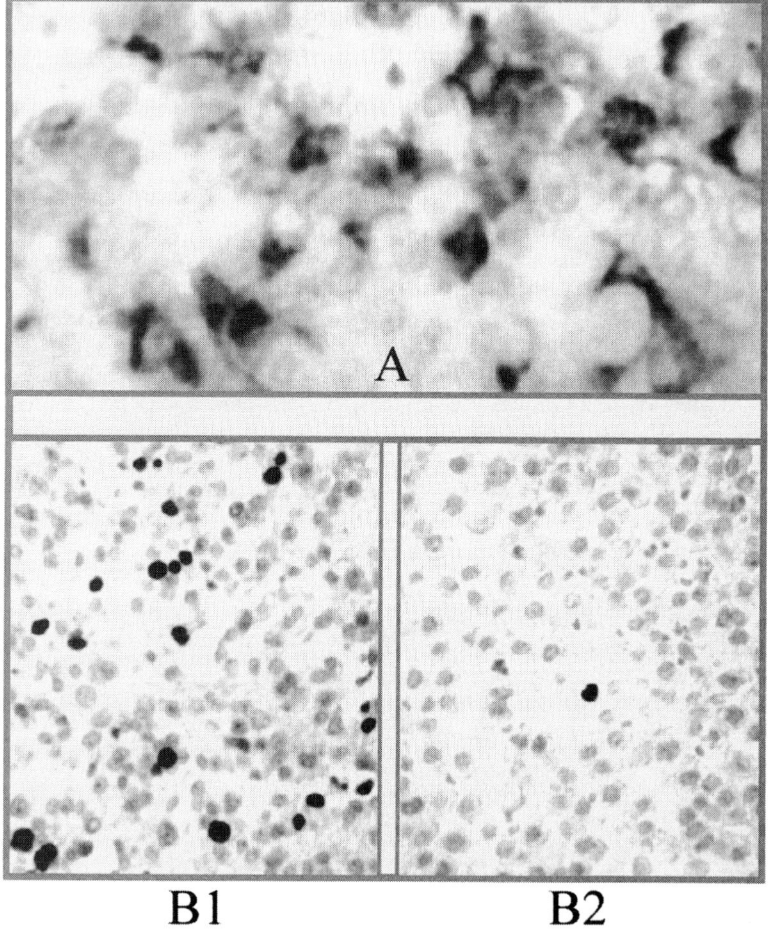

FIGURE 1. ICC detection of antigens M30 and Ki-67 in formalin-fixed paraffin embedded sections from GH-secreting pituitary adenomas. The system of biotinylated second antibody and peroxidase-labeled streptavidin (Px-L-SA-B) with 3,3′-diaminobenzidine (DAB; as a chromogen) was used for visualization. Cell nuclei were weakly counterstained with hematoxylin. (Original magnification: A: ×400, B1,2: ×100.) **(A)** The M30 antigen in an adenoma untreated by SSA. A case with a high incidence of apoptotic cells. **(B1)** Relatively numerous finding of the Ki-67 antigen in an untreated adenoma. **(B2)** An example of sporadic occurrence of the Ki-67 positive nuclei in one of the adenomas pretreated with SSA.

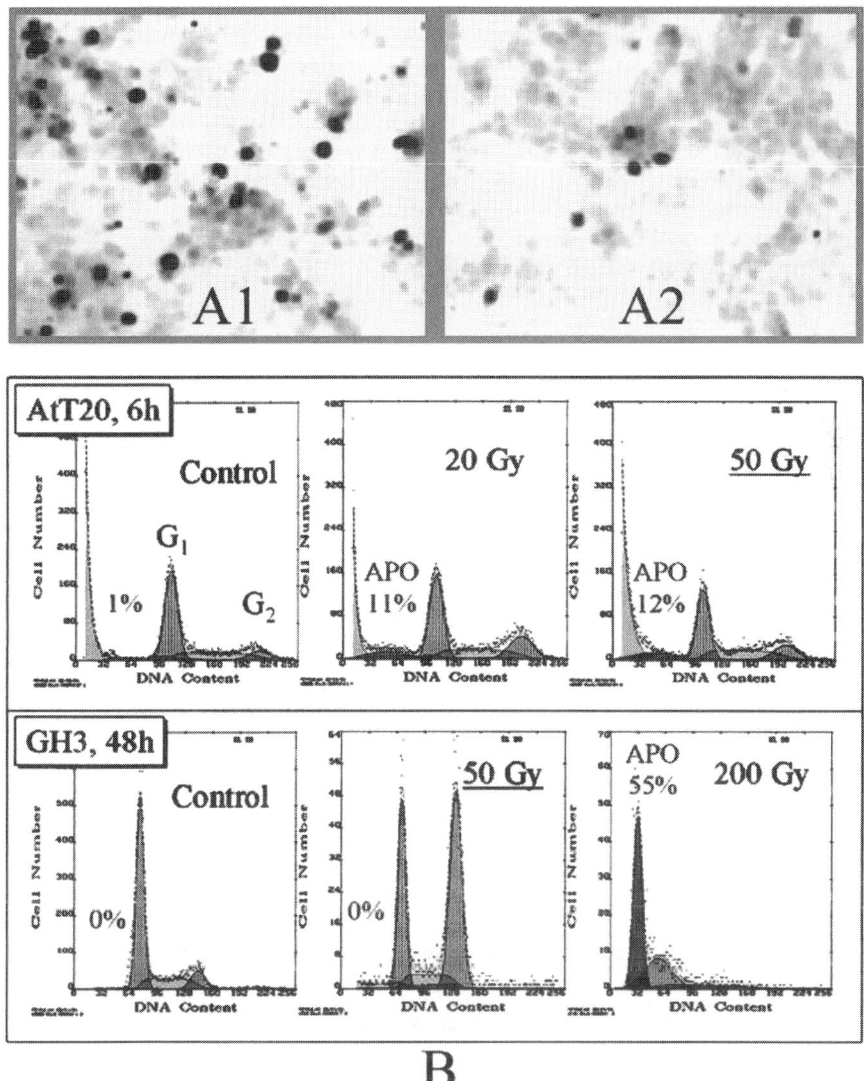

FIGURE 2. The influence of ionizing radiation on pituitary cell lines. **(A1,2)** ICC detection of caspase-3 using the system Px-L-SA-B (see FIG. 1) for visualization in a cytospin preparation of the corticotroph AtT20 cell line. **(A)** The AtT20 cells 6 h after the 20-Gy dose of irradiation. **(A2)** The control slide without irradiation. This pair of preparations and the mutual ratio of positive cells are representative of common experiments. **(B)** Flow-cytometric analysis of DNA content of AtT20 (above) and GH3 cell lines after irradiation (6 and 48 h, respectively). G_1: cells in G_0/G_1 phase; G_2: cells in G_2/M phase (duplex amount of DNA); APO: apoptotic cells (subG_1 peak). Early apoptosis has been detected in the AtT20 line. In the GH3 line irradiation produced the arrest in the G_2/M phase. Early (premitotic) apoptosis was detectable only using a high radiation dose of 200 Gy.

treated adenomas. It was significantly higher in the untreated group (5.0 per mille). This difference reached statistical significance ($P = .049$, Mann-Whitney test).

In cell lines, the SSA had minimal antiproliferative effect and they were unable to induce apoptosis. Ionizing radiation (FIG. 2) induced apoptosis in the corticotroph cell line AtT20 using doses of 5–20 Gy with no cell-cycle block. In the somatotroph GH3 cell line, the early (premitotic) apoptosis was detectable using only a high dose of 200 Gy; after irradiation with doses of 20–50 Gy it appeared with the latency of 48–72 hours and was preceded by cell-cycle arrest in the G_2/M phase. The treatment with somatostatin-14 during irradiation increased the percentage of apoptotic cells in culture 10 days after irradiation (11% versus 3% using 20 Gy).

CONCLUSIONS

Somatostatin analogues (SSA) have an antiproliferative effect on GH-secreting adenomas that is mainly mediated by a drop in the number of replicating cells. No induction of apoptosis by SSA has been demonstrated either in patients or in cell lines. The shrinkage in the size of some adenomas seems to be caused by the loss of functional parenchyma with compensatory fibrosis. Ionizing radiation in doses comparable to those used during gamma knife radiosurgery induced apoptosis in the AtT20 cell line after 6 hours, while the cell-cycle arrest in the G_2 phase dominated in the GH3 line. Somatostatin augmented radiation-induced apoptosis in the somatotrophs line.

ACKNOWLEDGMENTS

This study was supported by project MSM 111500001 from Ministry of Education (CR).

REFERENCES

1. FREDA, P.U. 2002. Somatostatin analogs in acromegaly. J. Clin. Endocrinol. Metab. **87:** 3013–3018.
2. NEWMAN, C.B., S. MELMED, A. GEORGE, *et al.* 1998. Octreotide as primary therapy for acromegaly. J. Clin. Endocrinol. Metab. **83:** 3034–3040.
3. AMATO, G., G. MAZZIOTTI, M. ROTONDI, *et al.* 2002. Long-term effects of lanreotide SR and octreotide LAR on tumour shrinkage and GH hypersecretion in patients with previously untreated acromegaly. Clin. Endocrinol. (Oxford) **56:** 65–71.
4. VLADYKA, V., R. LIŞCAK, G. SIMONOVA, *et al.* 2000. Gamma Knife (GK) radiosurgery for pituitary adenomas: evaluation of a series of 163 patients. J. Radiosurgery **3:** 113–131.
5. LANDOLT, A.M., D. HALLER, N. LOMAX, *et al.* 2000. Octreotide may act as a radioprotective agent in acromegaly. J. Clin. Endocrinol. Metab. **85**(3): 1287–1289.

In Vivo Imaging of Chemotherapy-Induced Apoptosis in Human Cancers

TARIK BELHOCINE,[a] NEIL STEINMETZ,[b] ALLAN GREEN,[b] AND PIERRE RIGO[a]

[a]*University Hospital of Liège, Division of Nuclear Medicine, Sart Tilman, 4000 Liège, Belgium*

[b]*Theseus Imaging Corporation, 101 Arch Street Suite 1762, Boston, Massachusetts 02110, USA*

ABSTRACT: *Rationale.* Induction of apoptosis in sensitive tumor cells is the main mechanism of action of chemotherapy agents in human cancers. Also, the assessment of drug-induced apoptosis soon after chemotherapy may be an early predictor of treatment efficacy. *Patients and Methods.* A phase I/II study was prospectively conducted in 15 patients presenting with proven lung cancers ($n = 10$), breast cancers ($n = 2$), and lymphomas ($n = 3$) to assess the value of the ^{99m}Tc-radiolabeled recombinant human (rh) Annexin V for imaging apoptosis immediately after completion of the first course of chemotherapy. Early Annexin V findings post-chemotherapy (day+1, day+2) were also compared to the tumor status at 6 to 12 weeks post-treatment. *Results.* All lung and lymphoma patients with an increased tracer uptake post-treatment ($n = 8$) had either partial or complete tumor response. Five patients with no tracer uptake had progressive disease. However, two breast cancers had a response to treatment, although no significant tracer uptake was observed. Tumor response and survival time were significantly correlated with the ^{99m}Tc-labeled Annexin V uptake. No serious events related to tracer administration were noted. *Conclusion.* Preliminary results of this pilot study demonstrate the feasibility of the ^{99m}Tc-labeled Annexin V uptake for the *in vivo* imaging of apoptosis after one course of chemotherapy. If confirmed on larger series, these promising results may open new perspectives in the management of oncology patients.

KEYWORDS: apoptosis; cancer; chemotherapy; ^{99m}Tc-labeled Annexin V

INTRODUCTION

The role of programmed cell death, also called *apoptosis*, in many vital functions such as embryogenesis, immune system, and homeostasis, is now well established.[1] On the other hand, a dysregulation of apoptosis may be observed in a number of diseases with various aspects. For instance, neurodegenerative or immune diseases, as well as ischemic pathologies or viral infections are usually related to excessive apoptosis. Conversely, cancer cells most often exhibit a lack of apoptosis so that most

Address for correspondence: Tarik Belhocine, University Hospital of Liège, Division of Nuclear Medicine, Sart Tilman, 4000 Liège, Belgium. Voice: +324-366-7199; fax: +324-366-7933.

tarik.bel@swing.be

Ann. N.Y. Acad. Sci. 1010: 525–529 (2003). © 2003 New York Academy of Sciences.
doi: 10.1196/annals.1299.097

chemotherapy regimens are intended to restore or to initiate apoptosis in sensitive tumor cells.[2] In other words, the early demonstration of drug-induced apoptosis in target cancer cells may be indicative of potential treatment efficacy. We studied the feasibility of a new nuclear imaging agent for the *in vivo* imaging of the apoptosis and/or necrosis after a single course of chemotherapy.

METHODS

A prospective study was conducted in 15 patients (11 male, 4 female; mean age of 60 years) presenting with histologically confirmed non–small cell lung cancer (NSCLC; $n = 7$), small cell lung cancer (SCLC; $n = 3$), non-Hodgkin's lymphoma (NHL; $n = 2$); Hodgkin's lymphoma (HL; $n = 1$), and breast cancer (BC; $n = 2$). All patients scheduled for at least one course of chemotherapy underwent ^{99m}Tc-labeled rh-Annexin V imaging immediately before (day–2, day–1) and after (day+1, day+2) treatment. Annexin V images were qualitatively interpreted in blinded fashion. Any tumor uptake post-chemotherapy was quantitatively measured using a routine computer method. Post-treatment assessment of objective tumor responses to chemotherapy was based on the response criteria in solid tumors.[3] Accordingly, all subjects were classified as having either complete (CR) or partial response (PR) or progressive disease (PD). In addition, all patients had [^{18}F]fluorodeoxyglucose positron emission tomography (^{18}FDG PET) in their workup. Survival curves in patients with and without tracer uptake were obtained using the Kaplan-Meier method. The log-rank test was used for statistical comparison of the curves' equality. Similarly, tumor responses were correlated to the degree of tracer uptake. Tracer safety was evaluated in terms of clinical and/or biological adverse events. Median follow-up was 117 days (47–356).

RESULTS

In all patients, pre-treatment ^{99m}Tc-labeled rh-Annexin V images showed only a physiological distribution of tracer to the naso-pharynx, spleen, liver, bowels, and bone marrow. No tumor uptake of the apoptosis agent was observed immediately before initiation of chemotherapy.

After treatment, seven patients (one NHL, one HL, two SCLC, three NSCLC) presented with a variable degree of ^{99m}Tc-labeled Annexin V uptake in at least one tumor site (FIG. 1). All these patients had either complete or partial response to treatment. In the present series, the optimal timing for detecting an apoptotic signal was the 24th hour after tracer injection corresponding to the 48th hour after completion of the first course of chemotherapy. In the remaining eight patients, no tracer uptake was observed at day+1 or day+2 post-treatment. Among them, six patients had progressive disease (four NSCLC, one SCLC, one NHL), whereas two breast cancer patients showed partial or complete response to treatment (TABLE 1). Overall, tumor responses to chemotherapy (CR, PR, PD) were significantly correlated to the degree of ^{99m}Tc-labeled rh-Annexin V uptake ($P < .01$). Similarly, survival times in lung cancer and lymphoma patients were correlated to tumor uptake ($P < .01$). No serious adverse events related to tracer administration were noted.

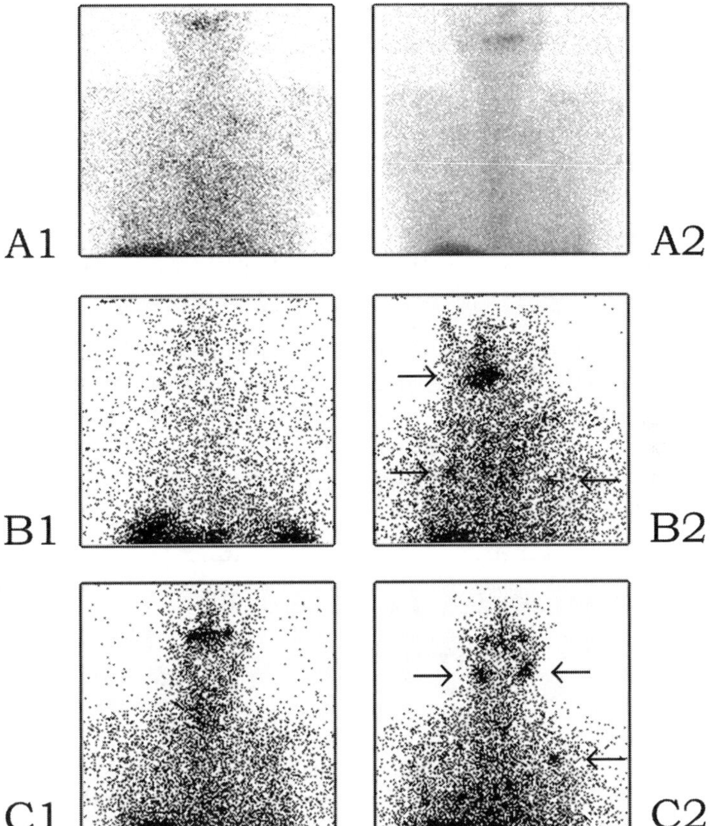

FIGURE 1. (A) A case of non–small cell lung cancer. **(A1)** Pretreatment image with no tracer uptake. **(A2)** Post-treatment image with no tracer uptake detected at tumor sites. The patient had progressive disease. **(B)** A case of non–small cell lung cancer. **(B1)** Pretreatment image with no tracer uptake. **(B2)** Post-treatment image showed an intense tracer uptake located at the cervical tumor site and bilateral hilary nodes (*arrows*). The patient had complete response to chemotherapy. **(C)** A case of non-Hodgkin's lymphoma. **(C1)** Pre-treatment image with no tracer uptake. **(C2)** Post-treatment image showed a clear tracer uptake located at the cervical nodes and the left axillary node (*arrows*). The patient had complete response to chemotherapy.

COMMENTS

Many chemotherapy agents exert their cytotoxic or cytostatic effects by inducing apoptosis in sensitive tumor cells.[2] In clinical routine, several methods may be used for assessing drugs-induced programmed cell death.[4] For instance, morphological features of apoptotic cells are readily and accurately described in microscopic analysis. However, these techniques tend to underestimate the rate of apoptosis. *In vitro* assays assessing the DNA degradation of apoptotic cells by means of electrophoresis

TABLE 1. Outcome of chemotherapy treatment

Patient	Tumor histology	Annexin V imaging 1	Annexin V imaging 2	Tumor response
1	SCLC	Negative	Negative	PD
2	NSCLC	Negative	Positive	CR
3	NHL	Negative	Positive	CR
4	NSCLC	Negative	Negative	PD
5	NSCLC	Negative	Negative	PD
6	BC	Negative	Negative	CR
7	NSCLC	Negative	Positive	PR
8	BC	Negative	Negative	PR
9	NSCLC	Negative	Positive	PR
10	NSCLC	Negative	Negative	PD
11	NSCLC	Negative	Negative	PD
12	NHL	Negative	Negative	PD
13	HL	Negative	Positive	CR
14	SCLC	Negative	Positive	PR
15	SCLC	Negative	Positive	CR

ABBREVIATIONS: Annexin V imaging 1: pretreatment imaging (J–2, J–1); Annexin V imaging 2: post-treatment imaging (J+1, J+2); CR: complete response; PR: partial response; PD: progressive disease; SCLC: small cell lung cancer; NSCLC: non–small cell cancer; NHL: non-Hodgkin's lymphoma; HL: Hodgkin's lymphoma; BC: breast cancer.

methods are simple and easily available. Nonetheless, such methods are usually poorly sensitive for the detection of nucleosomal apoptotic fragments with high molecular weights, while detecting small random necrotic fragments. Furthermore, these methods are not quantitative. The TUNEL technique or terminal deoxyribonucletidyl transferase-mediated dUTP nick end labelling, which measures an apoptotic signal in flow cytometry or in fluorescence microscopy, has been shown to increase the sensitivity for the detection of apoptotic cells. This method, however, cannot distinguish between apoptosis and necrosis. In addition, the DNA degradation is a late and a transient event in the apoptotic cascade. Owing to its high affinity for the externalized phosphatidylserine (PS), an early membrane event in tumor cells committed to apoptosis, fluorescein-labeled Annexin V has been used for the detection and the quantification of apoptotic cells.[5] Nonetheless, this technique is invasive, which is difficult to apply in oncology practice, especially soon after treatment. Additionally, tumor sites are not always accessible to biopsies.

In this pilot study, we assessed the feasibility of ^{99m}Tc-labeled Annexin V for the noninvasive detection of apoptosis immediately after completion of the first course of chemotherapy. Preliminary results obtained in oncology patients confirm those reported in preclinical and clinical studies demonstrating the accurate localization of this new imaging agent at histologically proven sites of apoptosis.[6] Although the radiolabeled Annexin V may not distinguish between apoptosis and necrosis, this *in vivo* tracer could offer a number of valuable advantages.[7] In oncology patients, Annexin V imaging may help to select the most efficient therapies on a patient-by-patient basis, whereas chemotherapy regimens are usually based on overall statistical data. Furthermore, after chemotherapy, several weeks or months are commonly required to demonstrate significant tumor changes in morphological imaging. Similarly, in the first weeks following the treatment, ^{18}FDG PET images may be flawed either by a "flare phenomenon" or by a "cellular stunning," which may lead to false positive findings or false negative results, respectively. This warrants the need for a molecular imaging technique such as ^{99m}Tc-labeled rh-Annexin V designed for the early assessment of the *in situ* tumor response to therapies inducing apoptosis.

CONCLUSION

In vivo imaging of cell death after a single course of chemotherapy is feasible in human cancers by using the radiolabeled Annexin V as a molecular probe of the externalized phosphatidylserine.

If these preliminary results are confirmed on larger homogenous series, ^{99m}Tc-labeled Annexin V may play a determinant role in the management of oncology patients.

ACKNOWLEDGMENTS

Part of this work was presented for the 47th annual meeting of the Society for Nuclear Medicine (SNM, St. Louis, MO, June 2000), and for the 93rd annual meeting of the American Association for Cancer Research (AACR, San Francisco, CA, April 2002).

REFERENCES

1. KERR, J.F.R., A.H. WYLLIE & A.R. CURRIE. 1972. Apoptosis: a basic biological phenomenon with wide-ranging implications in tissue kinetics. Br. J. Cancer **26:** 239–257.
2. KAUFMANN, S.H. & W.C. EARNSHAW. 2000. Induction of apoptosis by cancer chemotherapy. Exp. Cell Res. **256:** 42–49.
3. THERASSE, P., S. ARBUCK, E.A. EISENHAUER, *et al.* 2000. New guidelines to evaluate the response to treatment in solid tumors. J. Natl. Cancer Inst. **92:** 205–216.
4. MESNER, P.W. & S.H. KAUFMANN. 1997. Methods utilized in the study of apoptosis. Adv. Pharmacol. **41:** 57–87.
5. MARTIN, S.J., C.P.M. REUTELINGSPERGER, A.J. MCGAHON, *et al.* 1995. Early redistribution of plasma membrane phosphatidylserine is a general feature of apoptosis regardless of the initiating stimulus: inhibition by overexpression of Bcl-2 and Abl. J. Exp. Med. **182:** 1545–1556.
6. TAKAFUMI MOCHIZUKI, YUJI KUGE, SONGJI ZHAO, *et al.* 2003. Detection of apoptotic tumor response in vivo after a single dose of chemotherapy with ^{99m}Tc-Annexin V. J. Nucl. Med. **44:** 92–97.
7. VAN DE WIELE, C., C. LAHORTE, H. VERMEERSCH, *et al.* 2003. Quantitative tumor apoptosis imaging using technetium-99m–HYNIC annexin V single photon emission computed tomography. J. Clin. Oncol. **21:** 3483–3487.

Effects of Membrane Attack Complex of Complement on Apoptosis in Experimental Autoimmune Encephalomyelitis

TEODORA NICULESCU,[a] SUSANNA WEERTH,[b] LUCIAN SOANE,[a] FLORIN NICULESCU,[a] VIOLETA RUS,[a] CEDRIC S. RAINE,[b] MOON L. SHIN,[a] AND HOREA RUS[a]

[a]*University of Maryland, School of Medicine, Baltimore, Maryland 21201, USA*

[b]*Albert Einstein College of Medicine, Bronx, New York, USA*

ABSTRACT: Complement activation is involved in the initiation of inflammation and antibody-mediated demyelination in experimental autoimmune encephalomyelitis (EAE). We investigated the role of MAC in apoptosis in myelin-induced EAE in complement C5-deficient (C5-d) and C5-sufficient (C5-s) mice. The number of apoptotic cells assessed by TUNEL assay was significantly increased in C5-d mice during clinical recovery as compared with C5-s mice. Most of the apoptotic cells were lymphocytes, monocytes, and oligodendrocytes. DNA microarray was performed using total RNA extracted from spinal cords. Genes expressed higher in C5-s included members of the caspase (caspase 6, 7), TNF and TNFR families (CD27, FasL, lymphotoxin-beta R) and survivin. These results indicate that C5 and possibly MAC may be required for the limitation of inflammatory response within the central nervous system.

KEYWORDS: experimental autoimmune encephalomyelitis; membrane attack complex; complement system; apoptosis; oligodendrocyte

INTRODUCTION

Experimental autoimmune encephalomyelitis (EAE) is the animal model for multiple sclerosis. With EAE the inflammatory demyelination is affected by a complement system, especially C5b-9, the membrane attack complex (MAC).[1,2] MAC is assembled when complement C5 is cleaved to C5b by C5 convertases generated by either the classical, alternative, or lectin pathways, and C5b interacts with C6-C9.[3] MAC has a dose-dependent dual activity on oligodendrocytes (OLG): A sublytic MAC enhances OLG survival by rescuing cells from apoptosis, while a lytic dose causes cell death.[4]

Recently, Weerth *et al.* studied the possible role of MAC in EAE using C5deficient (C5-d) and C5 sufficient (C5-s) mice.[5] Dramatic differences between

Address for correspondence: Horea Rus, M.D., Ph.D., University of Maryland, School of Medicine, Dept. of Pathology, 10. S. Pine St., Baltimore, MD 21201. Voice: 410-706-3170; fax: 410-706-7706.

hrus@umaryland.edu

Ann. N.Y. Acad. Sci. 1010: 530–533 (2003). © 2003 New York Academy of Sciences.
doi: 10.1196/annals.1299.098

C5-d and C5-s were observed during chronic EAE: demyelination and Wallerian degeneration in the acute phase was followed by axonal loss and glial scarring in C5-d, whereas active remyelintion is associated with axon sparing in C5-s mice.

In this study, inflammation and demyelination were more restricted in C5-d than in C5-s mice in acute EAE.[5] Also, lower clinical scores and much less severe demyelintion were found in C6-deficient than in C6-sufficient rats using an antibody-mediated EAE model.[6] In this study, we investigated the role of complement C5 and MAC in apoptosis in myelin-induced EAE in C5-d and C5-s mice.

MATERIALS AND METHODS

EAE Induction and Immunohistochemistry

Adult female mice from a congenic outbred strain, deficient in C5 (D10.D2/oSnJ) and C5-sufficient controls (B10.D2/nSnJ) (Jackson Lab, Bar Harbor, ME), were injected with purified guinea pig myelin to induce chronic EAE.[5] Mice were observed daily for signs of EAE and scored for neurological deficits. Representative animals were taken from the acute (day 10 post-immunization) and recovery (day 24 post-immunization), phases of the disease. Apoptosis detection was performed on paraffin-embedded cervical spinal cord sections, using an ApopTag Peroxidase Kit (Intergen, Purchase, NY) as previously described.[7] Apoptotic OLG were defined by TUNEL-positive nuclei in cells stained by a monoclonal antibody against myelin/oligodendrocyte specific protein (MAB 328) (Chemicon, Temecula, CA), and the Vector M.O.M. immunodetection kit (Vector Lab, Burlingame, CA). Quantitative evaluation of apoptosis was performed on three consecutive sections per animal and expressed as mean ± SEM. Statistical analysis was performed using paired Student's *t* test.

Apoptosis Gene Expression

The expression of apoptosis genes was evaluated by DNA gene array. A total RNA extraction from the spinal cords of frozen mice and DNA array was performed using a GEArray Q series array kit (Super Array Inc., Bethesda, MD), according to the manufacturer's instructions. RT PCR was performed for differentially expressed genes using specific primers and standard protocols.

RESULTS AND DISCUSSION

All mice developed EAE with an acute phase, a brief remission, and a stable chronic phase. In an acute EAE (day 10 p.i.), a similar number of apoptotic cells were found in C5-s (372 ± 90) and C5-d (398 ± 140) mice respectively (FIG. 1). In the early recovery phase (day 24 p.i.), C5-d animals (395 ± 15) showed significantly more apoptotic cells in association with persistent inflammation as compared to C5-s mice (255 ± 48) (FIG. 2).

These data suggest a decrease in ongoing apoptosis during recovery in C5-s mice. Apoptotic cells were mainly monocytes, lymphocytes and OLG. Among TUNEL-

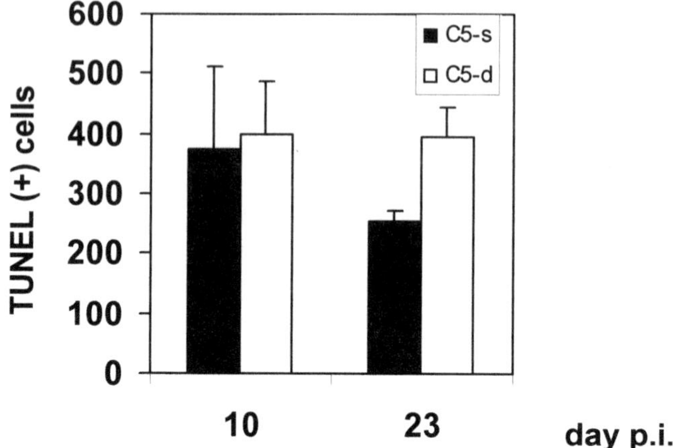

FIGURE 1. Quantitaive evaluation of apoptosis in C5-s and C5-d mice with EAE. A quantitative evaluation of apoptosis was performed on three consecutive cervical spinal cord sections per animal and expressed as mean ±SEM. The statistical analysis was performed using paired Student's t test. A statistically significant higher number of apoptotic cells were found in C5-d during remission (day 24) when compared with C5-s mice ($P < .04$).

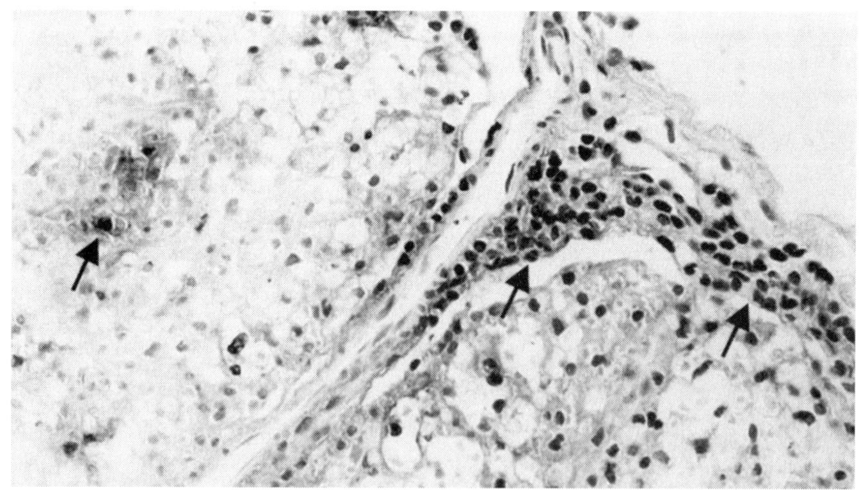

FIGURE 2. TUNEL analysis of *in situ* apoptosis in C5-d mice with EAE. Apoptosis was determined by TUNEL assay day 24 p.i. in a C5-d mice. A large number of apoptotic cells were found in the cervical spinal cord in C5-d mice during remission (*arrows*) as part of the infiltrate and deep in the white matter.

positive cells, 33% of those cells also stained for the MAB328 OLG marker; this indicated that OLG apoptosis is an integral part of acute EAE.

We performed gene array of the 96 mouse genes involved in regulation of apoptosis. Differences in expression over 2.5-fold were found in 23 genes. Of seven genes that expressed higher in C5-s mice, five encoded pro-apoptotic proteins. These included members of the caspases (caspase 6, 7), TNF and TNFR families (CD27, FasL, lymphotoxin-beta R), and survivin.

Differentially expressed genes were confirmed by RT-PCR. These results indicate that the absence of C5, which prevents C5b-9 assembly, is associated with a deficiency in the clearance of apoptotic cells and decreased expression of proapoptotic genes during acute EAE. We conclude that C5 and membrane attack complex are required for the limitation of inflammatory response and tissue damage, which increases the chances of an efficient recovery in EAE.

ACKNOWLEDGMENTS

This work was supported in part by National Institutes of Health Grants NS 42011 (HR), NS15662 (MLS), NS07098, 08952, and 11920 (CSR).

REFERENCES

1. RAINE, C.S. 1997. The Norton Lecture: a review of the oligodendrocyte in the multiple sclerosis lesion. J. Neuroimmunol. **77:** 135.
2. SHIN, M.L., H. RUS & F. NICULESCU. 1998. Complement System in Central Nervous System Disorders. *In* The Human Complement System in Health and Disease. J.E. Volanakis & M.M. Frank, Eds.: 499. Marcel Dekker. New York.
3. MULLER-EBERHARD, H.J. 1988. Molecular organization and function of the complement system. Annu. Rev. Biochem. **57:** 321.
4. SOANE, L., H.J. CHO, F. NICULESCU, *et al.* 2001. C5b-9 terminal complement complex protects oligodendrocytes from death by regulating Bad through phosphatidylinositol 3-kinase/Akt pathway. J. Immunol. **167:** 2305.
5. WEERTH, S., H. RUS, M.L. SHIN & C.S. RAINE. 2003. Complement C5 in experimental autoimmune encephalomyelitis (EAE) facilitates remyelination and prevents gliosis. Am. J. Pathol. **163:** 1069.
6. MEAD, R.J., S.K. SINGHRAO, J.W. NEAL, *et al.* 2002. The membrane attack complex of complement causes severe demyelination associated with acute axonal injury. J. Immunol. **168:** 458.
7. RUS, H.G., F. NICULESCU & M.L. SHIN. 1996. Sublytic complement attack induces cell cycle in oligodendrocytes. J. Immunol. **156:** 4892.

Vaccinia Virus Complement Control Protein Modulates Inflammation Following Spinal Cord Injury

D. N. REYNOLDS,[a] S. A. SMITH,[a] Y.-P. ZHANG,[b] D. K. LAHIRI,[c] D. J. MORASSUTTI,[b] C. B. SHIELDS,[b] AND G. J. KOTWAL[a,d]

[a]*Department of Microbiology and Immunology, University of Louisville School of Medicine, Louisville, Kentucky 40202, USA*

[b]*Department of Neurological Surgery, University of Louisville School of Medicine, Louisville, Kentucky 40202, USA*

[c]*Department of Psychiatry, Indiana University School of Medicine, Indianapolis, Indiana 46202, USA*

[d]*Division of Medical Virology, University of Cape Town Medical School, Cape Town, South Africa*

ABSTRACT: The vaccinia virus complement control protein (VCP) possesses multiple modulatory functions. Functioning as a complement inhibitory protein, VCP reduces production of proinflammatory chemotactic factors produced during complement activation. Additionally, VCP binds heparin and heparan sulfate proteoglycans, resulting in added functions shown to block monocyte chemotaxis *in vitro*. Using an *in vivo* spinal cord contusive injury model in rats, the inflammation-modulating abilities of VCP were evaluated. The results of both myeloperoxidase assaying and H&E stained section counts of spinal tissue reveal that neutrophil infiltration to the area of the lesion was reduced in animals that received VCP as compared to saline-injected controls.

KEYWORDS: spinal cord injury; vaccinia virus complement control protein; inflammation

INTRODUCTION

Acute spinal cord injury (SCI) resulting from automobile accidents and other accidents is a devastating and complex disease to understand and manage. In part, this is a result of the many interactions involved. These include the direct injury, secondary and delayed cell death, and loss of long axons and local neural circuits. The immune system is believed to play a major role in promoting secondary injury mechanisms. In particular, neutrophils and moncytes are thought to be the initial me-

Address for correspondence: Girish J. Kotwal, Division of Medical Virology, University of Cape Town Medical School, Anzio Road, Observatory, Cape Town, South Africa. Voice: +27-21-406-6676; fax: +27-21-448- 4110.
gjkotw01@yahoo.com

Ann. N.Y. Acad. Sci. 1010: 534–539 (2003). © 2003 New York Academy of Sciences.
doi: 10.1196/annals.1299.099

diators of secondary damage caused by the influx of peripheral immune cells into the injured spinal cord.[1] Restoring function in this milieu is a difficult challenge. Successful treatment of this disease will likely require a combination of interventions, aimed at minimizing secondary injury, rebuilding circuitry, and restoring dysfunctional neurons.

VCP, first identified in 1988, is a 35-kDa (244 amino acid) nonglycosylated soluble protein encoded by vaccinia virus that is secreted from vaccinia virus–infected cells.[2] It is functionally similar to the family of human complement control proteins. VCP binds to C4b and C3b, blocking both the classical and alternative pathways of complement. The results of this activity include the downregulation of proinflammatory chemotactic factors (C3a, C4a, and C5a), resulting in reduced cellular influx and inflammation. VCP also has been shown to have heparin-binding capabilities. Through this additional activity, it has been shown to block MIP-1α activity, inhibiting monocyte attachment and egression through endothelial cells *in vitro.*[3] It is postulated that VCP binds to heparan sulfate proteoglycans on the surface of endothelial cells, directly blocking MIP-1α binding, thereby inhibiting formation of a chemokine gradient.

MATERIALS AND METHODS

Spinal Cord Injury Model

Adult Sprague-Dawley rats weighing between 200 and 250 g were anesthetized with pentobarbital (50 mg/kg, i.p.). With the NYU Impactor, the height of the drop was set at mild (12.5 g/cm) severity, which results in reproducible complete hindlimb paralysis. The dura remained intact during injury induction. Animals designated to receive intraparenchymal injections were immediately subjected to the injection protocol described below, and wounds closed with sutures and clips, as described previously.

Spinal Cord Intraparenchymal Injections

Borosilicate pipettes (1.2 μm O.D.) were pulled on a Sutter Electrode Puller (P-87) and beveled to 25 μm (WPI Beveler).[4] Animals received two bilateral 5 μL injections (each given over 5 min), for a total of 10 μL per animal of the desired solution, at the visible T9 injury epicenter. The injection pipettes were placed 1.6 mm deep in the spinal cord tissue and 0.7 mm from the midline of the spinal cord, which results in delivery of the injected solution to the gray matter.

Tissue Sectioning and Neutrophil Counting

Animals designated for perfusion were deeply anesthetized with 0.3 mL–0.4 mL pentobarbital, perfused with 400–500 mL of 4% paraformaldehyde (Sigma, St. Louis, MO) in 0.01 M PBS, and their spinal cord segments were removed and placed in 4% paraformaldehyde overnight. The section thickness was set to 20 μm and the tissue cut in cross section or longitudinally. For each longitudinal tissue section, 30 fields were counted under a magnification of 100×. A series of 10 fields were counted along either edge and through the center of the tissue section for a total of 30. A total of 5,040 fields were counted for this experiment.

Myeloperoxidase Microassay

From each animal, three 8-mm segments of spinal cord were immediately dissected, placed in a freezing vial, snap-frozen in liquid nitrogen, and then transferred to −80°C. MPO was assayed photometrically on a plate reader at 630 nm (Biotek Instruments, Elx-800, Winooski, VT), as described previously.[5] One unit of MPO activity was defined as the amount of enzyme reducing 1 μmol of peroxide/min. Results were expressed as units of MPO activity per gram of tissue.

RESULTS AND DISCUSSION

To evaluate the diffusion rate of VCP in the spinal cord, six uninjured female Sprague-Dawley rats received injection of biotinylated VCP. Of the six rats, two were des-

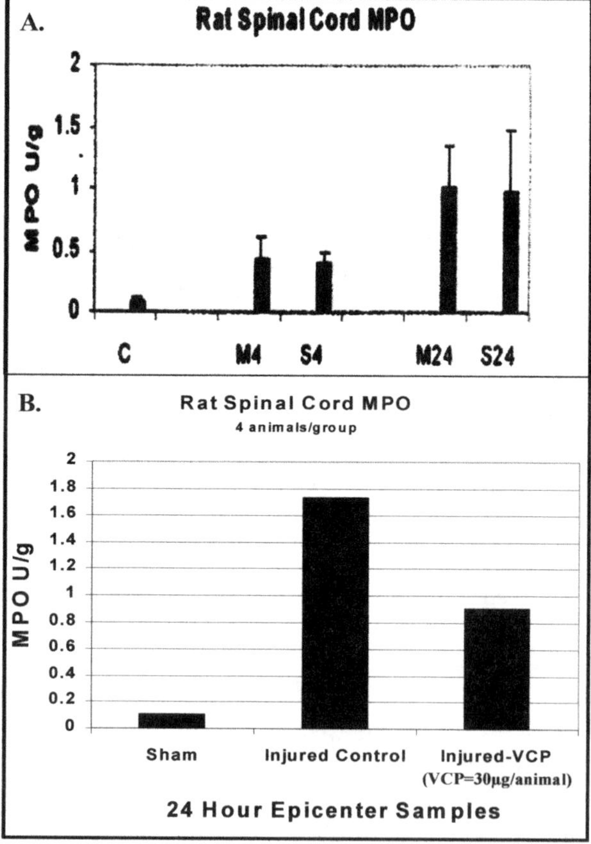

FIGURE 1. (A) Rat spinal cord MPO following SCI at T8 level. Tissue was harvested at 4 h mild (M4) and severe (S4) injury and at 24 hours (M24, S24). **(B)** Rat spinal cord MPO, 24 h following SCI with and without VCP injection. Results are shown as an average of four animals for each group.

ignated for a survival time of 2 h, two for 4 h, and two for 6 h; a dose of either 2 μL (10 μL/μg) or 8 μL (10 μg/μL) of biotinylated VCP was injected in one rat for each survival time point. Following injection, rats were sacrificed at the appropriate time points, spinal cords were sectioned and the biotinylated VCP was detected using streptavidin-horseradish peroxidase. VCP staining was clearly visible within the spinal cord parenchyma 2 h after receiving either the low or high dose of VCP. Staining was most intense within the central gray matter, and less intense in the surrounding white matter. After 6 h, VCP staining was less intense but still visible, and again, staining was most intense in the central gray matter (data not shown). These results confirmed that VCP injected directly into spinal cord parenchyma remains present for at least 6 h and is found primarily within the gray matter into which it is initially injected.

To determine whether VCP could reduce the degree of neutrophil influx following SCI, rats were subjected to a contusive SCI. First, as a control to evaluate the model, the level of neutrophil infiltration was determined using the MPO assay at

Injured Control **Injured w/VCP**

A.

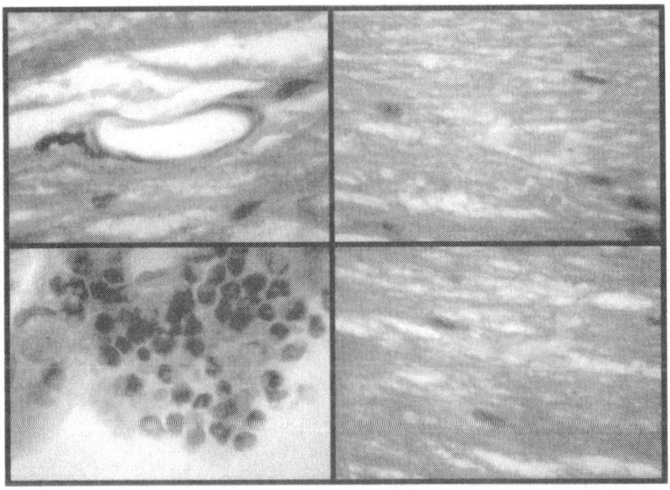

B.

	Total	Average/Section
Sham-6h (8 sections)	8	1
Sham-24h (9 sections)	2	0.2
Injured Control- 6h (8 sections)	37	4.6
Injured Control-24h (8 sections)	405	50.6
Injured w/VCP-6h (4 sections)	33	8.25
Injured w/VCP- 24h (4 sections)	47	11.75

FIGURE 2. **(A, L. Col.)** H&E stains of injured control and injured–VCP-treated. **(A, R. Col.)** spinal cords 24 h after injury. **(B)** Neutrophil infiltration of sham (6 h and 24 h), injured control (6 h and 24 h), and injured–VCP-treated (6 h and 24 h) spinal cord epicenters.

4 h and 24 h post-injury (FIG. 1A). An increase in neutrophil levels, as indicated by elevated MPO, was consistently achieved between the 4 h and 24 h time points with both the severe and mild injuries. As can be seen, the MPO levels present following mild injury were quite similar to those observed following severe injury. For this reason the mild injury was chosen for use in evaluating VCP. Next, to test VCP's effectiveness, an injection of either saline or VCP (21 μg/μL) was administered immediately following a mild contusion injury and the degree of neutrophil infiltration was evaluated using the MPO assay. The results indicate the influx of neutrophils in spinal tissue was considerably lower in rats receiving VCP than in those receiving saline injections post-injury (FIG. 1B).

To help confirm these findings, neutrophils were counted in H&E stained spinal cord sections (FIG. 2A). Little difference was observed in the number of neutrophils between the injured control, and the injured–VCP-treated animals at the 6-h time point. However, a substantial difference is evident 24 h post-injury (FIG. 2B). These data are very similar to that obtained using the myeloperoxidase assay, confirming the reduced number of neutrophils infiltrating the injury site of the VCP-treated animals 24 h following SCI.

In a recent report on immune responses following acute SCI, the MPO assay was used to show that neutrophil recruitment peaks at 24 h.[1] It has also been known for some time that complement levels are elevated (66% of patients) following SCI in humans.[6] This acute timing of the inflammatory response in SCI should be a target for acute therapeutic intervention. Prevention of secondary injury and its resultant axonal loss and cell death are important goals in spinal cord injury therapy. The multiple functions exhibited by VCP, especially its ability to inhibit complement and modulated cellular influx, may prove beneficial in reducing damage caused by secondary injury following spinal cord injury. The therapeutic effects of VCP administered following traumatic SCI could greatly diminish the early tissue damage and thus enhance the opportunity for future regenerative repair. Further studies are currently being conducted to evaluate VCP's effectiveness in promoting functional recovery following contusive SCI.

ACKNOWLEDGMENTS

The authors gratefully acknowledge the support provided by the Kentucky Spinal Cord and Head Injury Research Trust (KSCHIRT). We would also like to thank Melissa M. Hiser for her role in counting neutrophils. G. J. Kotwal is a Senior Wellcome Trust International Fellow for Biomedical Science in South Africa.

REFERENCES

1. CARLSON, S.L., M.E. PARRISH, J.E. SPRINGER, *et al.* 1998. Acute inflammatory response in spinal cord following impact injury. Exp. Neurol. **151:** 77–88.
2. KOTWAL, G.J. & B. MOSS. 1988. Vaccinia virus encodes a secretory polypeptide structurally related to complement control proteins. Nature **335:** 176–178.
3. REYNOLDS, D.N., K.L. KEELING, R. MOLESTINA, *et al.* 2000. Heparin binding activity of vaccinia virus complement control protein confers additional properties of uptake by mast cells and attachment to endothelial cells. Advances in Animal Virology. S. Jamee, Ed.: 337–342. Oxford and IBH. New York.

4. Magnuson, D.S., T.C. Trinder, Y.P. Zhang, *et al.* 1999. Comparing defects following excitotoxic and contusion injuries in the thoracic and lumbar spinal cord of the adult rat. Exp. Neurol. **156:** 191–204.
5. Keeling, K.L., R.R. Hicks, J. Mahesh, *et al.* 2000. Local neutrophil influx following lateral fluid-percussion brain injury in rats is associated with accumulation of complement activation fragments of the third component (C3) of the complement system. J. Neuroimmunol. **105:** 20–30.
6. Rebhun, J. & J. Botvin. 1980. Complement elevation in spinal cord injury. Ann. Allergy **44:** 287–288.

Blood Cell Apoptosis/Necrosis

Some Clinical and Laboratory Aspects

JERARD SEGHATCHIAN[a] AND GRACINDA DE SOUSA[b]

[a]*Blood Component Technology Consultancy London, England*

[b]*Lisbon Regional Blood Centre, Lisbon, Portugal*

ABSTRACT: Exposure of phosphatidyl-serine on blood cell membrane surface and microvesiculation as the hallmarks of apoptosis/necrosis were investigated. The effect of leukofiltration on the retention/generation of microvesicules, leukocyte subsets and major biological response modifiers were evaluated in a like study. It is concluded that apoptotic cells potentially contribute to transfusion reactions in donor/recipient-specific ways.

KEYWORDS: TGFβ; transfusion complications; apoptosis; microvesiculation; DNA fragmentation

BACKGROUND

Apoptosis (controlled cell death) and signal transduction (interaction of extracellular signaling molecules with membrane-associated receptors) are two interrelated physiological phenomena that play essential roles in maintenance of hemostasis and immune regulation.

While leukocytes are the most vulnerable blood cells, because they contain the constituent apoptotic machinery to undergo apoptosis, there is some evidence that platelets, as enucleated blood cells, also contain most constituents of apoptotic cells. This enables them to display similar apoptotic morphological changes and transduction signaling.

Transduction signaling, leading to mitochondrial DNA fragmentation and exposure of phosphotidyl-serine (PS) on their surfaces, occurs during the conventional five-day storage of platelet concentrates (PC). Even the organnelle lacking red cells display distinct morphologic characteristics that enable their recognition and clearance by the spleen and bone marrow. In fact about 1% of erythrocytes die in circulation everyday by apoptotic process and are replaced by an equal number of newly differentiated cells.

From the transfusion standpoint, it is unclear to what extent blood collection, processing, and storage in different plastic packs could modify the cellular viability.

It is clinically established that about 30–70% of transfused red cells disappear from circulation after 1 h and 72 h, respectively, because of PS exposure. Moreover

Address for correspondence: Jerard Seghatchian, Blood Component Technology Consultancy, 50 Primrose Hill Road, London, England. Voice/fax: +44-207-722-9596.
jseghatchian@btopenworld.com

Ann. N.Y. Acad. Sci. 1010: 540–547 (2003). © 2003 New York Academy of Sciences.
doi: 10.1196/annals.1299.100

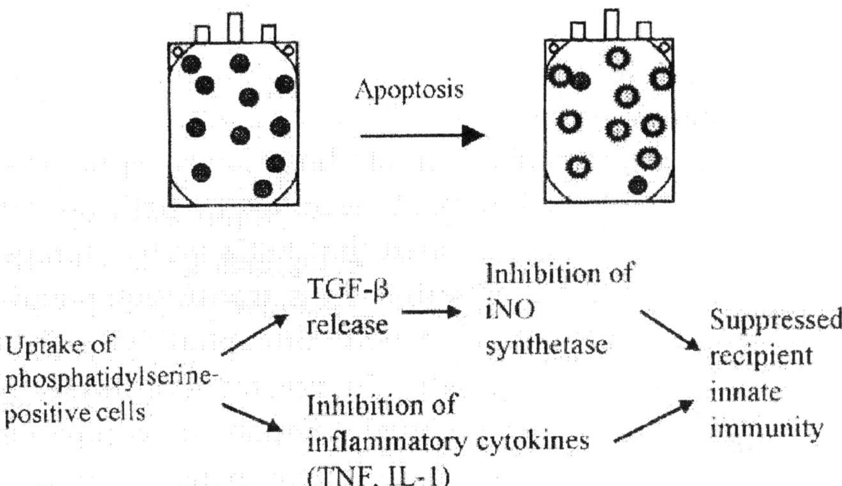

FIGURE 1. A possible mechanism for transfusion-related immunosuppression.

the presence of leukocytes and platelets, because of the cellular lysis within 24–72 h, accelerate red cell apoptosis/necrosis, as identified by progressive microvesiculation during storage and by the generation of cells of a smaller size, an increased density, and which express PS on their surface as a hallmark of apoptosis.

The current hypothesis linking apoptosis to hemostasis and immunomodulation is based on the development of the PS positive cells and their microvesicule in supernatant plasma. This is facilitated by TGFβ as shown below (FIG. 1).

OBJECTIVES

Our duel objectives were: (1) to establish the degree to which blood component collection and processing technologies, in particular those in which cells come into contact with various artificial filter surfaces, influence apoptosis, and/or the generation/retention of microvesicles; and (2) to assess the development of some biological response modifiers (BRM), in particular TGFβ and RANTES, which possibly account for residual donor-specific transfusion reactions in blood component therapy.

MATERIALS AND METHODS

Blood components and the following related tests were used to assess each test's functional activity, in accordance with established and validated methods:

- Flow cytometry was used to assess DNA with two different protocols: (1) PS exposure using Annexin V–FITC, platelet activation markers, red cell, and platelet microvesicle (MV) and (2) leukocyte subpopulation by immunophenotyping.

- ELISA was used to estimate released cytosolic protein, a new global marker of cellular injury. Soluble *p*-selectin, platelet, and leukocyte-derived cytokines were used.
- ICAM 3 using Western blotting and platelet pro/anticoagulant activities, in purified system, with chromogenic substrates.

When a comparative analysis was carried out, the pooling and splitting protocol was used for eliminating the donor variability. The filter back-flush procedure was used for enrichment of the retained cells for leukocyte subset analysis.

RESULTS

Considerable apoptotic events occur during the 8 h of blood storage at an ambient temperature or during 24 h at 4°C, before leukofiltration, as measured by ICAM 3 or the release of Annexin V (not shown).

A donor-specific leukofitration failure was observed in 0.3–0.5% of donations. These donations showed a considerably higher amount of apoptotic cells resembling DNA laddering. Furthermore, the presence of smaller cells was found with a lower propidium iodite (PI) fluorescence intensity (FIG. 2). A similar pattern was observed

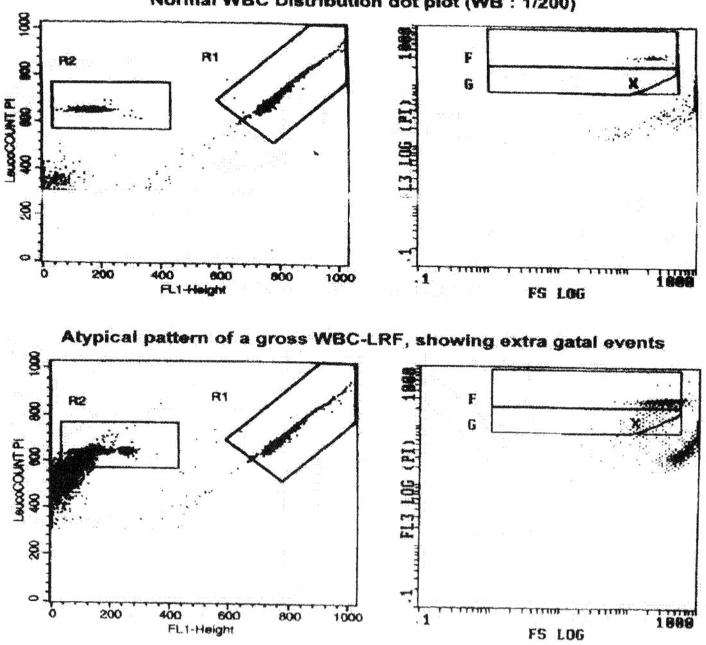

FIGURE 2. Comparison of whole blood and leukofiltration failure measured by Becton Dickinson leukocount and DNA prep methods. NOTE: DNA fragmentation as measured by propidium iodide.

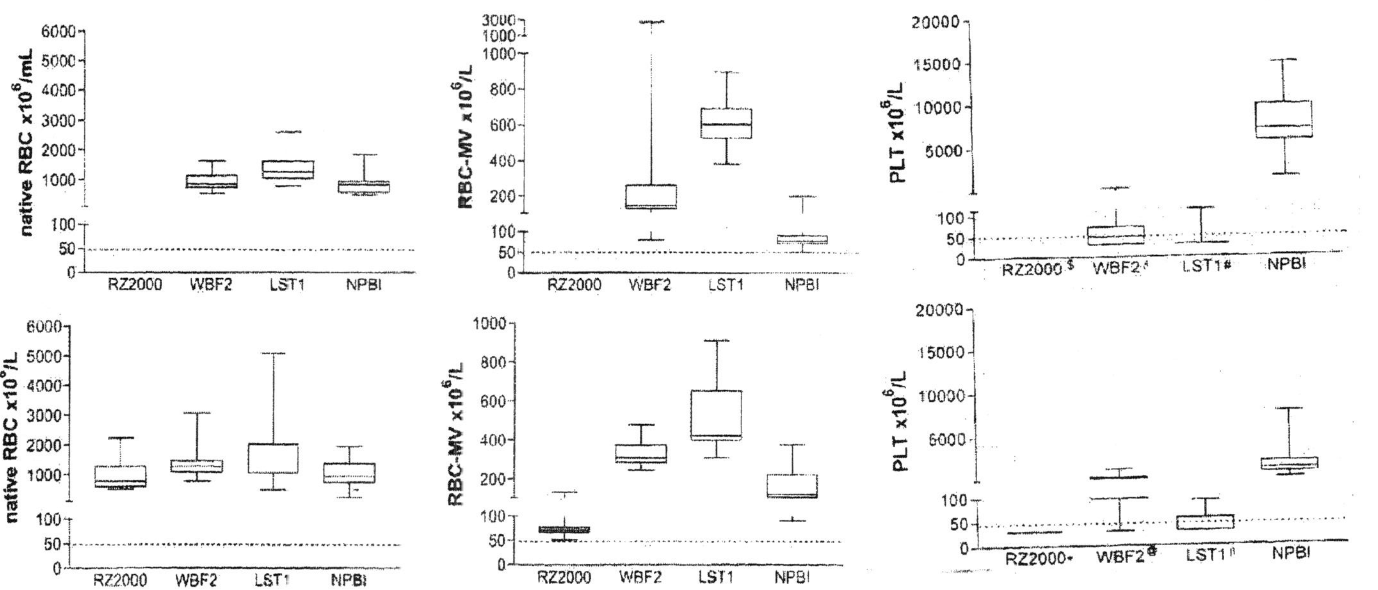

FIGURE 3. Native red blood cell (RBC) and RBC microvesicle (MV) in WBC reduced plasma derived from filtered whole blood day 0 (*left panel*) and day 1 (*central panel*). Residual platelet in WBC reduced plasma day 0 (*right upper panel*), day 1 (*right bottom panel*). *Box*, 25 and 75 percentile; *horizontal line*, median; *bars*, lowest and highest values. Day 0, RBC-MV, LST1 was higher than NPBI (P<.001). Day 1, RBC-MV, RZ2000 was lower than WBF2 and LST1 (P<.001), NPBI was lower than LST1 (P<.01). Day 0 (*top right*): half of WBF2 samples were below 50×10^6/L. Only 2 of LST1 samples were above the detection limit. Day 1 (*bottom right*): all were below 50×10^6/L.

in red cell (RCC) and platelet (PC) concentrates that failed leukodepletion. Whole blood filters have different capacities in removing apoptotic cells and red cell or platelet microvesicules (FIG. 3).

Changes in both red cell and platelet cytoskeleton structure, which are associated with cell shrinkage or microvesiculation and which resemble apoptotic bodies occurred during storage (not shown).

Leukocyte subpopulations in pre (PF), post (S), and filter content (F) differed, and leukocytes appeared to be activated upon contact with filter (FIG. 4).

The levels of PS exposure were different in various PC prepared by different technologies, and a significant correlation was found between PS exposure and microvesiculation, and between *p*-selectin expression on platelet and soluble *p*-selectin in plasma supernatants (FIG. 5).

In purified system, the procoagulant activity of the isolated microvesicule was increased during storage, whereas the anticoagulant activity decreased. This finding supports the notion that MV are potentially procoagulant (FIG. 6).

Leukodepleted products differed in terms of both leukocyte and platelet-derived cytokines during storage, as exemplified by the distribution of TGFβ and RANTES (FIG. 7). This suggests various platelet products have different immunomodulatory potential.

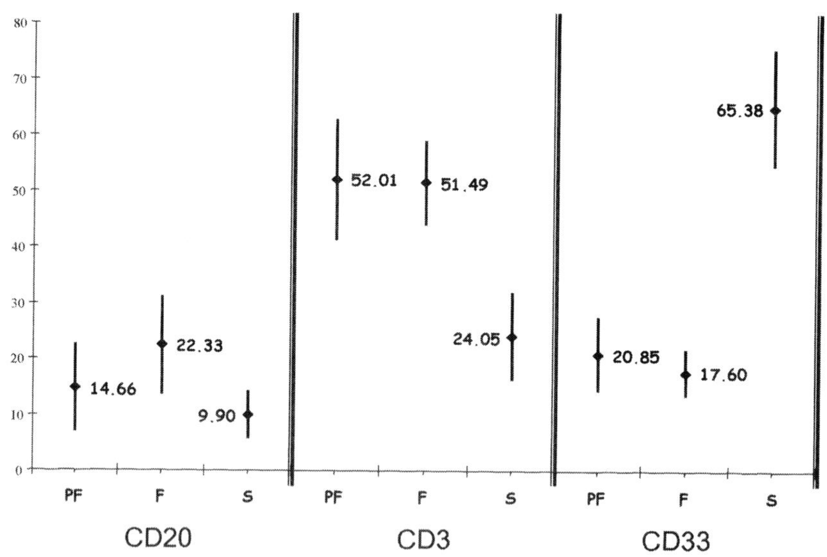

FIGURE 4. Leukocyte subsets prefiltration (PF), reverse filter washing (F), and postfiltration (S), mean and 95% CI. Paired samples (PF-S) test for CD20 and CD3 show a decrease (P = .0899 and $P < .005$, respectively), while CD33 shows an increase (P = .000).

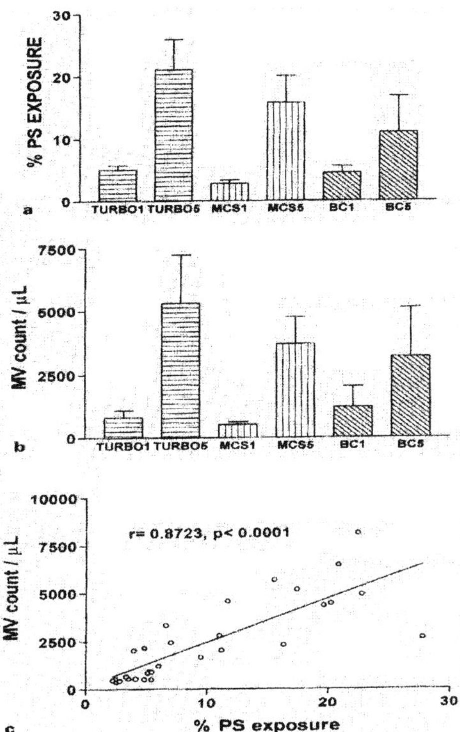

FIGURE 5. PS exposure, MV count, and the correlation between both markers. (**a**) Both Cobe Turbo and buffy coat platelet concentrations (BC-PC) showed significantly higher degrees of PS exposure than Haemonetics (P=.007 and .003), but on day 5, significance was found only between Cobe Turbo and BC-PC. (**b, c**) MV counts showed no significance.

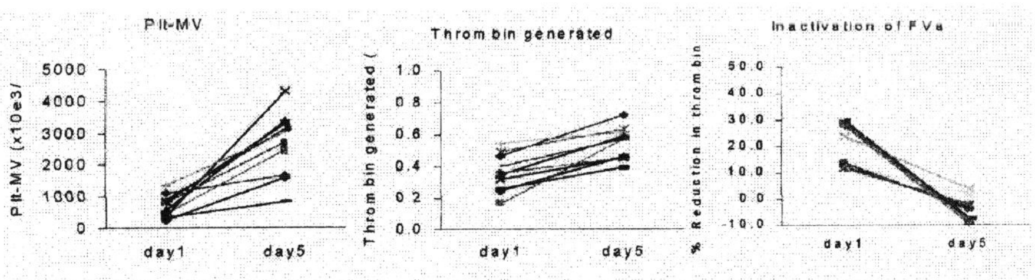

FIGURE 6. The procoagulant and anticoagulant activities of platelet microvesicule (Plt-MV) during storage.

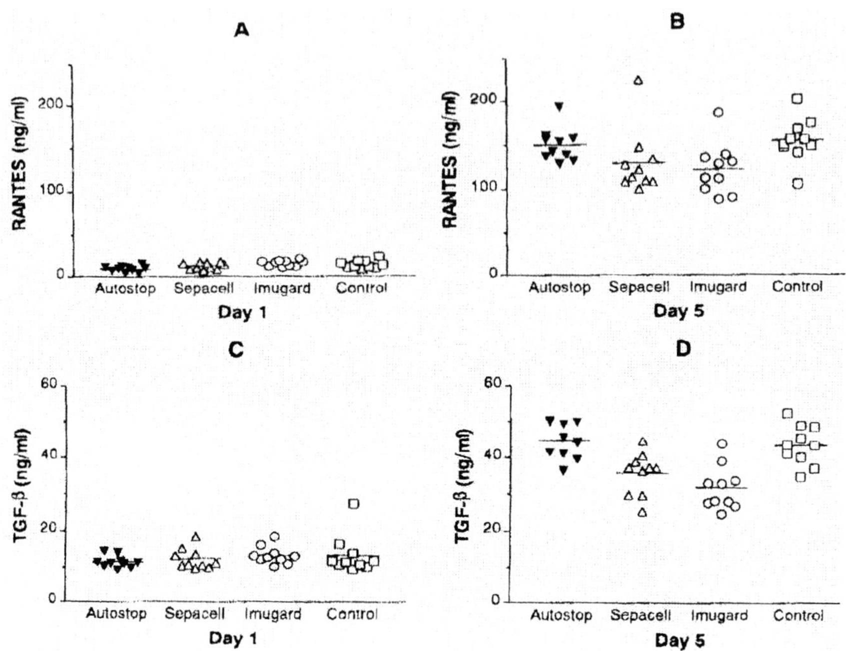

FIGURE 7A. Levels of RANTES in buffy coat platelet concentrates.

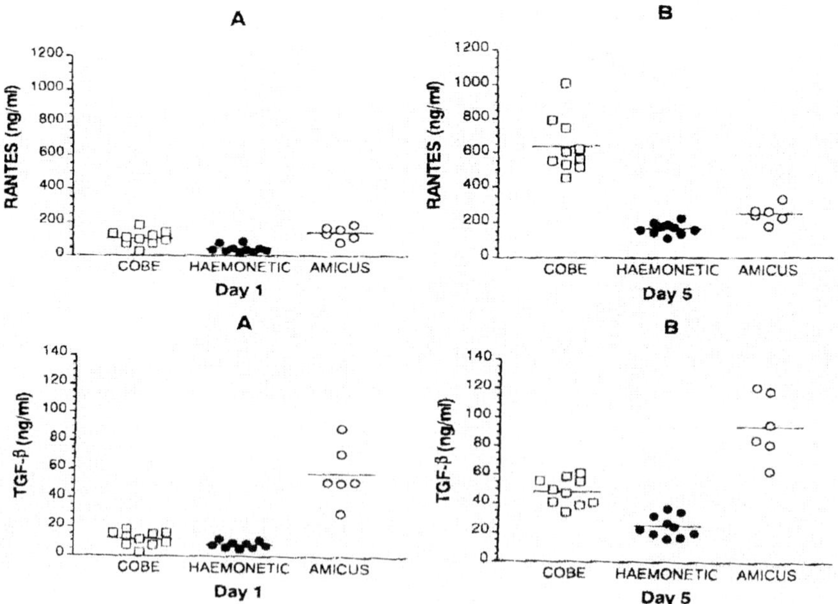

FIGURE 7B. Levels of RANTES in apheresis platelet concentrates.

CONCLUSION

The conventional protocol for blood collection, processing, and storage is associated with considerable time and temperature-dependent apoptosis/necrosis, as identified by a donor-specific DNA fragmentation, microvesiculation, and the release of activation/secretion markers.

The filters and blood pack systems in current use differ in terms of generation/retention of microvesicule and apoptotic cells. Blood components derived from different technologies have different rates of PS exposure, microvesiculation, generation of cytokines, and other biological response modifiers, even when the products are stored in the same type of container. The cellular lesion appears to be greatest in the nonfiltered apheresis process.

These data support the notion that microvesiculation, PS exposure (as the hallmark of apoptosis/necrosis), and TGFβ play important roles in both hemostasis and immunomodulation.

Clinically high levels of TGFβ are seen in some patients receiving platelet concentrates; and excessive PS exposure is associated with systemic lupus anticoagulant and other immunomodulatory conditions. Clearly more clinical data are required with respect to the effective dose of apoptotic cells and TGFβ to exert transfusion complications and the donor/recipient-specific interrelationship.

REFERENCES

1. Dzik, S., M. Minchef & F. Puppo. 2002. Apoptosis, transforming growth factor-β, and the immunosuppressive effect of transfusion. Transfusion **42:** 1221–1223.
2. Seghatchian, J., P. Krailadsiri, M. Beard, *et al.* 2002. Studies on the characterization of the cause of leucoreduction failures, with particular reference to extra gatal events. Transfusion Apheresis Sci. **26:** 47–60.
3. Krailadsiri, P. & J. Seghatchian. 2001. Residual red cell and platelet content in WBC-reduced plasma measured by a novel flow cytometry method. Transfusion Apheresis Sci. **24:** 279–286.
4. Krailadsiri, P. & J. Seghatchian. 2000. Are all leucodepleted platelet concentrates equivalent? Vox Sang. **78:** 171–175.
5. Trindade, H., H. Carvalho, G. Sousa, *et al.* Filtration induces changes in activity states and leucocyte populations. Transfusion Apheresis Sci. **28:** 319–327.

T-Cell Survival Regulator LKLF Is Not Involved in Inappropriate Apoptosis of Diabetes-Prone BBDP Rat T Cells

PHILIP DIESSENBACHER, KATRIN BARTELS, FRIEDRICH KOCH-NOLTE, AND FRIEDRICH HAAG

Institute of Immunology, University Hospital, D-20246 Hamburg, Germany

ABSTRACT: Diabetes-prone BB (dpBB) rats develop autoimmune insulin-dependent diabetes mellitus (IDDM) at high frequency as a consequence of a defect in T cell development, caused by a mutation in a single gene locus on rat chromosome 4 (*lyp*) which has recently been identified as immune-associated nucleotide 4 (*ian4*). A phenotype similar to dpBB rat lymphopenia has recently been described in the mouse as the result of the targeted inactivation of the gene for the transcription factor LKLF (Lung *Krüppel*-like factor, KLF2) in the immune system. We cloned the LKLF gene of the rat and screened a panel of rat/hamster radiation hybrid cell lines to determine its chromosomal localization. We conclude that the LKLF gene is not defective in dpBB rats and that its expression is not compromised by the *lyp* mutation.

KEYWORDS: LKLF; apoptosis; T cells; diabetes-prone BB rats

Diabetes-prone BB (dpBB) rats develop autoimmune insulin-dependent diabetes mellitus (IDDM) at high frequency as a consequence of a defect in T cell development, caused by a mutation in a single gene locus on rat chromosome 4 (*lyp*) which has recently been identified as immune-associated nucleotide 4 (*ian4*).[1] These rats display profound T cell lymphopenia in the periphery, while the distribution of T cell subsets in the thymus appears grossly normal. The few T cells found in the periphery are short-lived, display an activated phenotype,[2] and are characterized by a lack of expression of the T cell differentiation marker RT6.[3] The closely related diabetes-resistant drBB strain shows normal peripheral T cell numbers and normal RT6 expression.

A phenotype similar to dpBB rat lymphopenia has recently been described in the mouse as the result of the targeted inactivation of the gene for the transcription factor LKLF (Lung *Krüppel*-like factor, KLF2) in the immune system.[4] Like dpBB rats, these mice showed a paucity of mature T cells in the periphery without significant

Address for correspondence: Friedrich Haag, Institute of Immunology, University Hospital, Martinistr. 52, D-20246 Hamburg, Germany. Voice: +49-40-42803 4595; fax: +49-40-42803 4243.
haag@uke.uni-hamburg.de

Ann. N.Y. Acad. Sci. 1010: 548–551 (2003). © 2003 New York Academy of Sciences.
doi: 10.1196/annals.1299.101

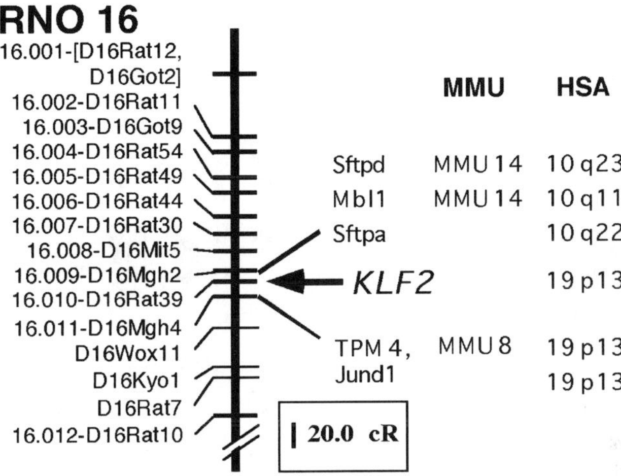

FIGURE 1. Position of the *KLF2* gene on rat chromosome 16. The linkage of KLF2 and neighboring genes to markers on rat chromosome 16 is shown, using the rat radiation hybrid map obtained from the Wellcome Trust Centre for Human Genetics (www.well.ox.ac.uk). The positions of orthologous genes of the mouse and the human were obtained from the genome-wide comparative map of the rat available on the web at Otsuka GEN Research Institute (www.otsuka.com).

perturbation of thymocyte subsets. KLFs are a family of zinc finger transcription factors characterized by C-terminal DNA-binding domains with three Cys2-His2 zinc fingers. While most family members are broadly expressed, a small subfamily, including erythroid, gut, and lung KLFs (EKLF, GKLF, and LKLF; or KLFs 1, 4, and 2, respectively) is expressed in a restricted, tissue-specific manner. Its members are associated with late stage differentiation and/or cell cycle control in their respective tissues. These observations prompted us to investigate whether defective expression of LKLF might contribute to the lymphopenia observed in BB rats.

We cloned the LKLF gene of the rat and screened a panel of rat/hamster radiation hybrid cell lines to determine its chromosomal localization. The results indicated that rat *LKLF* lies on RNO16. The tightest linkage was to the anonymous marker *D16Rat39* (LOD=20,040, theta=0,054), followed by *D16Mgh2* (LOD=18,264, theta=0,083) and *D16Mgh4* (LOD=17,832, theta=0,104), which are markers for the genes *Sftpa* (surfactant associated protein A) and *TPM4* (tropomyosin 4), respectively (FIG. 1).

We compared lymphocyte subpopulations from lymphopenic dpBB and non-lymphopenic drBB rats with respect to expression of LKLF, RT6, and the anti-apoptotic gene Bcl-2 by RT-PCR (FIG. 2). As expected, RT6 transcripts were not detectable in B cells or thymocytes, and were reduced in dpBB as compared to drBB rats. Only low levels of LKLF transcripts were detectable in the thymus, consistent with restriction of LKLF expression to the most mature thymocyte subset in the mouse.[4] No differences with regard to expression levels of LKLF in thymocytes, T cells, or

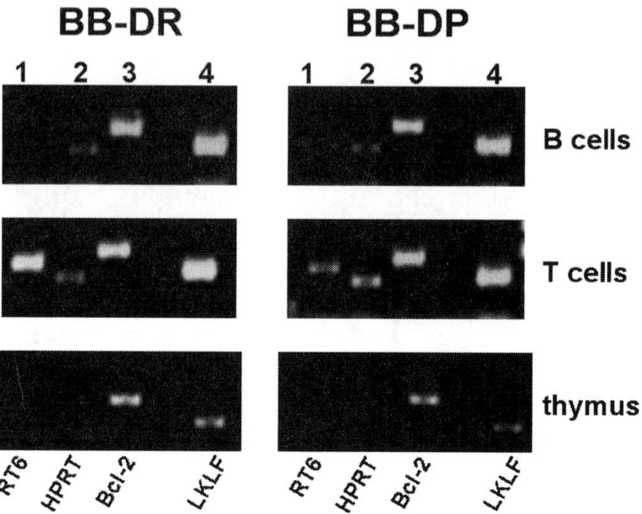

FIGURE 2. Expression of LKLF, RT6, and Bcl-2 in lymphopenic dpBB and non-lymphopenic drBB rats. RNA from thymocytes and purified lymph node B and T cells was analyzed by RT-PCR for expression of RT6, HPRT, Bcl-2, and LKLF.

B cells were found between normal and lymphopenic rats. We thus conclude that the LKLF gene is not defective in dpBB rats and that its expression is not compromised by the *lyp* mutation. It remains possible, however, that the *lyp* mutation affects a gene further downstream in the genetic program controlled by LKLF.

EXPERIMENTAL PROCEDURES

A panel of 106 DNAs from rat/hamster radiation hybrid cell lines (Research Genetics, Huntsville, AL) was subjected to PCR amplification using primers LKLF-3utrF (5′-CAC GGC CCC CTT GCA ACA A-3′) and LKLF-3utrR (5′-GTC ATA CAG ATT GCC ACG GTC A-3′). PCR amplification was performed using 30 cycles: 30 sec at 94°C, 30 sec at 58°C, 30 sec at 72°C, and 5 min final extension time at 72°C. T and B cells were separated using MACS beads (Miltenyi Biotech) specific for rat Ig. For reverse transcription PCR analysis, the following primers were used: LKLF, TA1F (5′-TGC GAG CGC GGC CTC CAG GAG MGN TGG-3′) and NLS1R (5′-CCA GGA GCG GCG GCC GCG TTT GGG CT-3′); RT6, VX1 (5′-CAG TCT ACC GAG GCA CTA AGA CC-3′) and CRE (5′-GAC CGA GGA GAA CCA CAA GGA ACA G-3′); Bcl-2, BCL-F (5′-CCC TCC GCC GGG CTG GGG ATG-3′) and BCL-R (5′-CCT CAC TTG TGG CCC AGG TAT G-3′); HPRT, HPRT-F (5′-GTT GGA TAC AGG CCA GAC TTT GTT G-3′) and HPRT-R (5′-GAT TCA ACT TGC GCT CAT CTT AGG C-3′).

REFERENCES

1. HORNUM, L., J. ROMER & H. MARKHOLST. 2002. The diabetes-prone BB rat carries a frameshift mutation in Ian4, a positional candidate of Iddm1. Diabetes **51:** 1972–1979.
2. HERNANDEZ-HOYOS, G., S. JOSEPH, N.G. MILLER & G.W BUTCHER. 1999. The lymphopenia mutation of the BB rat causes inappropriate apoptosis of mature thymocytes. Eur. J. Immunol. **29:** 1832–1841.
3. GREINER, D.L., E.S. HANDLER, K. NAKANKO, *et al.* 1986. Absence of the RT6 T cell subset in diabetes-prone BB/W rats. J. Immunol. **136:** 148–151.
4. KUO, C.T., M.L. VESELITS & J.M. LEIDEN. 1997. LKLF: A transcriptional regulator of single-positive T cell quiescence and survival [see comments]. Science **277:** 1986–1990.

Early Activation of Antioxidant Mechanisms in Muscle of Mutant Cu/Zn-Superoxide Dismutase-Linked Amyotrophic Lateral Sclerosis Mice

NATASA JOKIC, FRANCK DI SCALA, LUC DUPUIS, FREDERIQUE RENE, ANDRE MULLER, JOSE-LUIS GONZALEZ DE AGUILAR, AND JEAN-PHILIPPE LOEFFLER

Laboratoire de Signalisations Moléculaires et Neurodégénérescence, Université Louis Pasteur, Faculté de Médecine, 11 rue Humann, 67085 Strasbourg, France

ABSTRACT: A subset of familial ALS cases is associated with missense mutations in the gene encoding Cu/Zn-superoxide dismutase (SOD1), a free radical scavenging enzyme that protects cells against oxidative stress. Overexpression of these ALS-linked mutations confers an unidentified gain of function to the enzyme that triggers a series of neurological disorders characteristic of human ALS. To understand how skeletal muscle may counteract the progression of the disease, we explored the expression of different molecular effectors involved in antioxidant pathways. Our results are strongly indicative of the early and long-lasting activation of a series of molecular effectors thought to act coordinately in preventing the increased oxidative stress characteristic of ALS.

KEYWORDS: amyotrophic lateral sclerosis; axotomy; c-Jun; gastrocnemius muscle; γ-glutamylcysteine synthetase; glutathione peroxidase; Nrf-2; oxidative stress; SOD1

Amyotrophic lateral sclerosis (ALS) is a fatal neurological disorder affecting middle-aged individuals. It is characterized by progressive muscle atrophy and selective loss of both lower and upper motor neurons in the spinal cord, brainstem, and motor cortex. A subset of familial ALS cases is associated with missense mutations in the gene encoding Cu/Zn-superoxide dismutase (SOD1), a free radical scavenging enzyme that protects cells against oxidative stress.[1] Overexpression of these ALS-linked mutations confers an unidentified gain of function to the enzyme that triggers a series of neurological disorders characteristic of human ALS. Although the precise mechanisms underlying the pathogenesis of the disease are still poorly understood, many studies agree that increased oxidative stress plays a central role.[2] To under-

Address for correspondence: Jean-Phillipe Loeffler, Laboratoire de Signalisations Moléculaires et Neurodégénérescence, EA 3433, Université Louis Pasteur, Faculté de Médecine, 8e étage, bâtiment 3, 11 rue Humann, 67085 Strasbourg, France. Voice: +33-3-90-24-30-91; fax: +33-3-90-24-30-65.

loeffler@neurochem.u-strasbg.fr

Ann. N.Y. Acad. Sci. 1010: 552–556 (2003). © 2003 New York Academy of Sciences.
doi: 10.1196/annals.1299.102

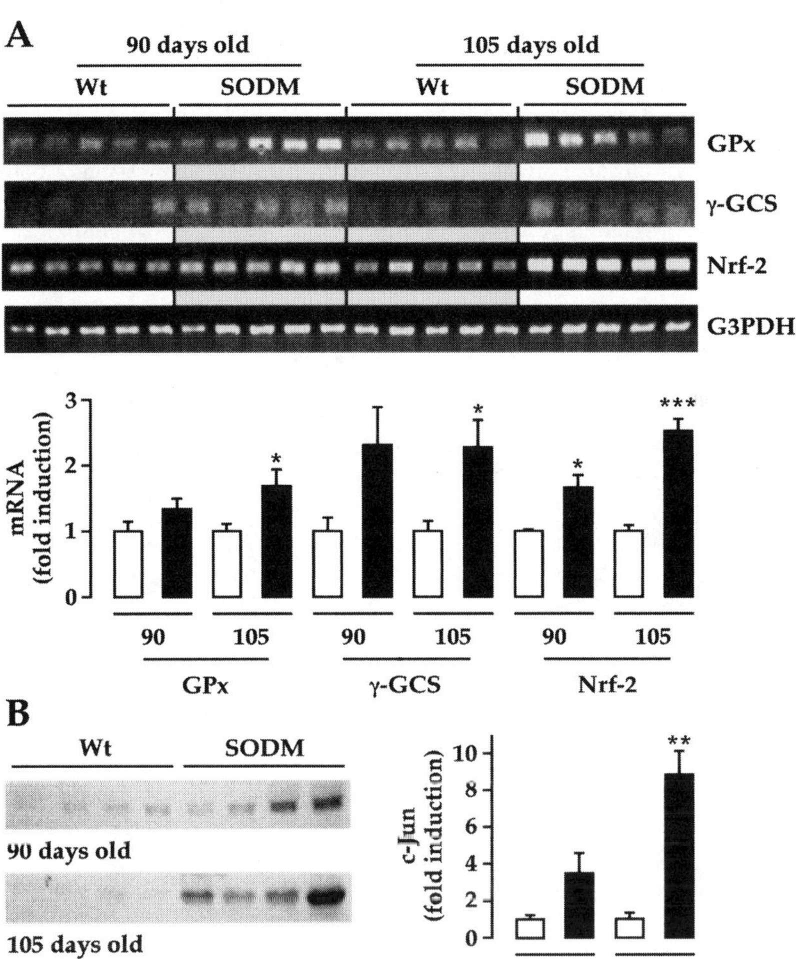

FIGURE 1. (**A**) Representative RT-PCR showing GPx, γ-GCS, and Nrf-2 mRNA levels in the gastrocnemius muscle of wild-type (Wt, *open bars*) and G86R (SODM, *closed bars*) mice of 90 and 105 days old. RT-PCR was performed as described previously.[3] To normalize quantification (as shown on the graph), glyceraldehyde 3-phosphate dehydrogenase (G3PDH) mRNA levels were used as internal control. The data represent means ± SE (N = 5). (**B**) Representative Western blots show c-Jun protein levels in the gastrocnemius muscle of mice as above. A quantitative analysis is shown on the graph. Protein extracts were prepared and electrophoresed as described previously.[3] To ensure equal loading, extracts were assayed for protein quantification and blots were pre-visualized using Ponceau Red staining. The data represent means ± SE (N = 4). $^{*}P < .05$, $^{**}P < .01$, $^{***}P < .001$ vs. corresponding Wt (Student's t test).

stand how skeletal muscle may counteract the progression of the disease, we explored here the expression of different molecular effectors involved in antioxidant pathways. To this aim, we used the gastrocnemius muscle of G86R mice, a transgenic line developing ALS-like pathology. In these animals, histological modifications and motor dysfunctions (i.e., hind limb paralysis) start from approximately 105 days of age, no apparent symptoms being detected at earlier stages.[3]

Using RT-PCR analysis, we showed that glutathione peroxidase (GPx) and γ-glutamylcysteine synthetase (γ-GCS) mRNA levels appeared higher in asymptomatic 90-day-old G86R mice than in wild-type littermates. The expression of both enzymes still increased significantly at 105 days of age, when mice start to exhibit the first ALS-like symptoms (FIG. 1A). Transcriptional regulation of antioxidant en-

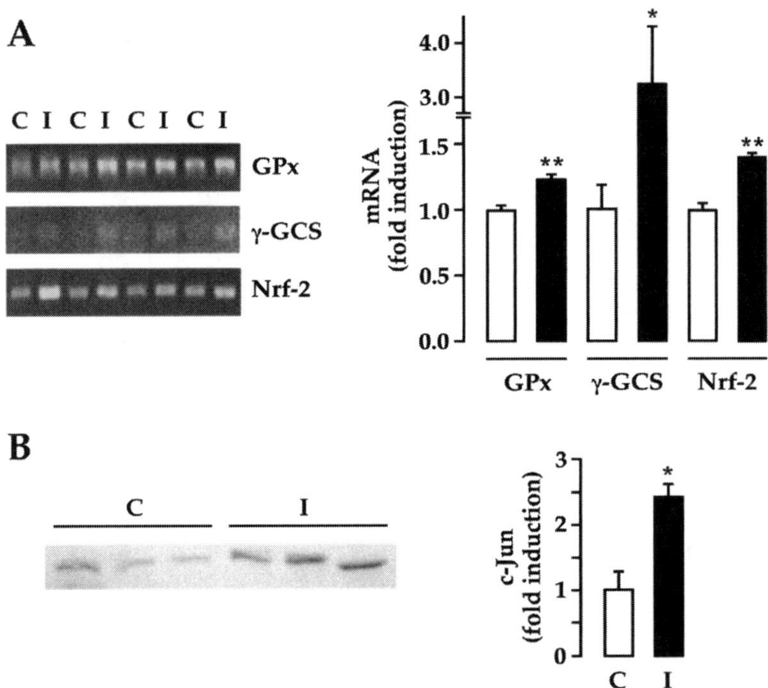

FIGURE 2. (**A**) Representative RT-PCR showing GPx, g-GCS and Nrf-2 mRNA levels in the contralateral (C, *open bars*) and ipsilateral (I, *closed bars*) gastrocnemius muscle of axotomized mice. A quantitative analysis is shown on the graph. The data represent means ± SE (N=5). (**B**) Representative Western blots show c-Jun protein levels in the gastrocnemius muscle of crushed mice as above. A quantitative analysis is shown on the graph. The data represent means ± SE (N=3). *P<.05, **P<.01 vs. corresponding Wt (Student's t test). Axotomized and crushed mice were prepared as follows: the right sciatic nerve of animals anesthetized with intraperitoneal administration of 250 mg/kg tribromoethanol was exposed at the mid-thigh level. Sciatic nerve was then dissected sharply with microscissors or crushed with a fine forceps. Muscles and skin incision were sutured, and animals were allowed to recover. Mice were sacrificed when no regeneration occurs (seven days after axotomy or three days after crush).

zymes is mostly under the control of key transcription factors such as c-Jun and the nuclear respiratory factor Nrf-2. In this regard, AP1 binding sites have been previously identified in the promoter region of the GPx gene.[4] In addition, recent studies have shown that Nrf-2 heterodimerizes with c-Jun to activate the antioxidant response element-mediated expression of γ-GCS.[5] In our hands, a concomitant overexpression of Nrf-2 and c-Jun was observed in both 90- and 105-day-old G86R mice, as determined by RT-PCR and Western blot analyses, respectively (FIG. 1A and B).

Denervation is one of the major features affecting skeletal muscle in ALS. To determine whether this process contributes by itself to activate the antioxidant defenses as observed in G86R mice, we induced denervation either by axotomy or by crush of sciatic nerve. As illustrated in FIGURE 2 (A), mRNA levels of GPx, γ-GCS, and Nrf-2 were significantly increased in the ipsilateral muscle of axotomized mice. Similarly, c-Jun protein levels were also higher after sciatic nerve crush (FIG. 2B). It should be stressed that the biological function of c-Jun seems to depend on the cell type and the physiological conditions considered. Thus, an important role in the regulation of neuronal programmed cell death,[6] but also protective actions associated with axonal regeneration[7] and cardiac myocyte survival[8] have been reported. Our present findings in skeletal muscle would rather support a protective role for c-Jun in response to oxidative stress.

Taken together, our results are strongly indicative of the early and long-lasting activation of a series of molecular effectors that are thought to act coordinately in preventing the increased oxidative stress characteristic of ALS. It should be mentioned that oxidative injury is observed in both familial and sporadic forms of the disease. Therefore, the antioxidant response found in SOD1 mutant mice as well as in animals submitted to experimentally induced denervation suggests that oxidative stress may represent a degenerative mechanism common to different denervation-related disorders. The precise molecular events leading to this common mechanism need further investigation.

ACKNOWLEDGMENTS

This work was supported by the Amyotrophic Lateral Sclerosis Association (ALSA, U.S.A). N.J. is recipient of a grant from the Ministère des Affaires Etrangères. Our thanks to Dr. J.W. Gordon for G86R mice.

REFERENCES

1. ROSEN, D., T. SIDDIQUE, D. PATTERSON, *et al.* 1993. Mutations in Cu/Zn superoxide dismutase are associated with familial amyotrophic lateral sclerosis. Nature **362:** 59–62.
2. ROBBERECHT, W. 2000. Oxidative stress in amyotrophic lateral sclerosis. J. Neurol. **247:** I/1–I/6.
3. DUPUIS, L., M. DE TAPIA, F. RENE, *et al.* 2000. Differential screening of mutated SOD1 transgenic mice reveals early up-regulation of a fast axonal transport component in spinal cord motor neurons. Neurobiol. Dis. **7:** 274–285.
4. JORNOT, L. & A.F. JUNOD. 1997. Hyperoxia, unlike phorbol ester, induces glutathione peroxidase through a protein kinase C-independent mechanism. Biochem. J. **326:** 117–123.

5. JEYAPAUL, J. & A.K. JAISWAL. 2000. Nrf-2 and c-Jun regulation of antioxidant response element (ARE)-mediated expression and induction of γ-glutamylcysteine synthetase heavy subunit gene. Biochem. Pharmacol. **59:** 1433–1439.
6. HAM, J., A. EILERS, J. WHITFIELD, *et al.* 2000. C-Jun and the transcriptional control of neuronal apoptosis. Biochem Parmacol. **60:** 1015–1021.
7. HERDEGEN, T., P. SKENE & M. BAHR. 1997. The c-Jun transcription factor—bipotential mediator of neuronal death, survival and regeneration. Trends Neurosci. **20:** 227–231.
8. DOUGHERTY, C.J., L.A. KUBASIAK, H. PRENTICE, *et al.* 2002. Activation of c-Jun N-terminal kinase promotes survival of cardiac myocytes after oxidative stress. Biochem. J. **362:** 561–571.

The Role of Apoptosis in Acetaminophen-Induced Injury

GEORGE E. N. KASS, PATRICIA MACANAS-PIRARD, PAULINE C. LEE,[a] AND RICHARD H. HINTON

School of Biomedical and Life Sciences, University of Surrey, Guildford, Surrey GU2 7XH, United Kingdom

[a]GlaxoSmithKline Research and Development, Ware, Herts SG12 0DP, United Kingdom

ABSTRACT: Apoptosis plays a critical role in acetaminophen (AAP)-induced hepatic injury, since inhibiting apoptosis also prevents the development of acute liver failure. In this study, the mechanism of apoptosis induction by AAP was investigated in the human hepatoblastoma cell line HuH7. AAP caused marked cytotoxicity in HuH7 cells as a result of apoptosis. Processing of execution caspases to their corresponding active fragments and cleavage of cytokeratin-18 were observed, supporting a role of caspases in AAP-induced apoptosis. The manifestation of apoptosis was preceded by a translocation of cytochrome *c* from mitochondria to the cytosol. In conclusion, AAP induces apoptosis in human hepatoblastoma HuH7 cells through mitochondrial cytochrome *c* release and caspase activation.

KEYWORDS: acetaminophen; liver failure; caspase activation; human hepatoblastoma cell line

Acetaminophen (paracetamol, AAP), a widely used analgesic drug, can induce fatal liver injury through a combination of apoptosis and necrosis, when taken in large doses. We have previously shown that apoptosis plays a critical role in AAP-induced hepatic injury since inhibiting apoptosis also prevents the development of acute liver failure.[1] In this study, the mechanism of apoptosis induction by AAP was investigated in the human hepatoblastoma cell line HuH7. AAP caused marked cytotoxicity in HuH7 cells as a result of apoptosis. This was evidenced by chromatin condensation. Processing of execution caspases to their corresponding active fragments and cleavage of cytokeratin-18 were observed, supporting a role of caspases in AAP-induced apoptosis. The manifestation of apoptosis was preceded by a translocation of cytochrome *c* from mitochondria to the cytosol. In conclusion, AAP induces apoptosis in human hepatoblastoma HuH7 cells through mitochondrial cytochrome *c* release and caspase activation.

Address for correspondence: Dr. George E.N. Kass, School of Biomedical and Molecular Sciences, University of Surrey, Guildford, Surrey GU2 7XH, United Kingdom. Voice: +44-1483-686449; fax: +44-1483-300374.
g.kass@surrey.ac.uk

Ann. N.Y. Acad. Sci. 1010: 557–559 (2003). © 2003 New York Academy of Sciences.
doi: 10.1196/annals.1299.103

TABLE 1. Acetaminophen-induced apoptosis in HuH7 cells

	Apoptosis (%)	
Time (h)	Control	AAP
0	2.1 ± 0.5	n.d.
8	n.d.	6.0 ± 0.5
24	n.d.	6.7 ± 2.0
30	n.d.	9.4 ± 0.6
48	3.9 ± 1.2	52.8 ± 7.9

NOTE: Cells were treated with acetaminophen (10 mM), harvested at the indicated time points, and fixed in ethanol. Following treatment with RNAase and staining with propidium iodide, the cells were analyzed by flow cytometry to quantitate the number of cells with <2N DNA.

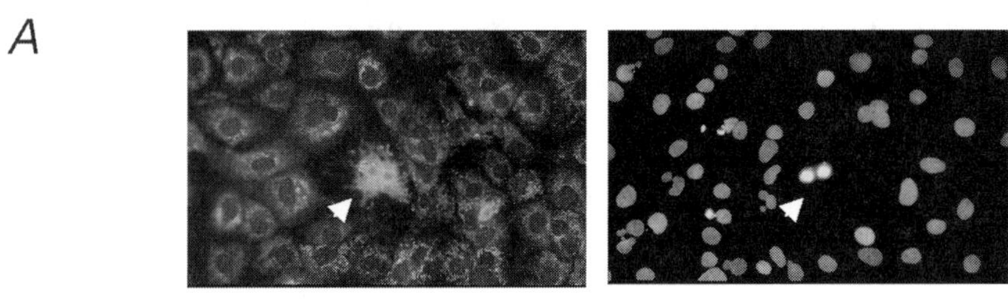

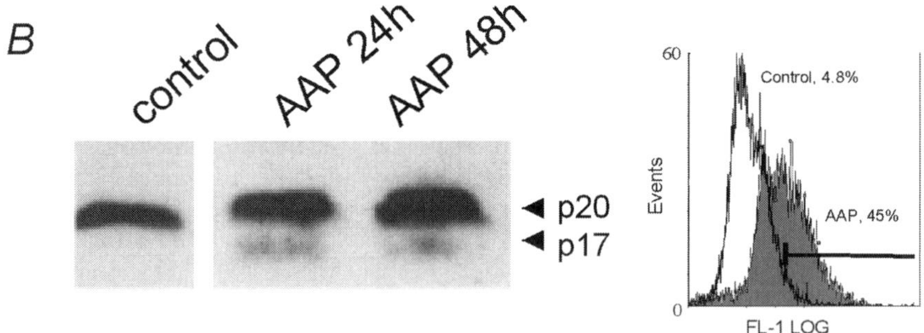

FIGURE 1. Induction of cytochrome *c* release from mitochondria and caspase activation. HuH7 cells were treated with acetaminophen (10 mM) for 24 h (**A**) or 48 h (**B**). In (**A**) the cells were immunostained for cytochrome *c* (*left*) and the chromatin labeled to show the nuclear morphology (*right*). The release of cytochrome *c* from mitochondria is seen as a generalized redistribution and accompanied by plasma membrane blebbing and chromatin condensation (*arrow*). In (**B**) the appearance of the active fragments of caspase-3 and the *in situ* cleavage of cytokeratin-18 by caspase-3 are shown.

MATERIALS AND METHODS

HuH7 cells were exposed to AAP or solvent control for the indicated time points, fixed, and the nuclei stained with HOE33258 followed by visualization using fluorescence microscopy. Alternatively, the cells were harvested, fixed in ethanol, stained with propidium iodide, and analyzed by flow cytometry. Activation of caspases was carried out as previously reported.[2] The release of cytochrome *c* from mitochondria was assessed by immunocytochemistry. Analysis of caspase processing was carried out by immunoblotting and cytokeratin-18 cleavage was detected by flow cytometric analysis of cells immunostained for cleaved cytokeratin-18 neoepitope detection (M30 mAb).

RESULTS AND DISCUSSION

HuH7 is an *in vitro* cell model for the study of the intrinsic mechanism of AAP-induced apoptosis. The induction of apoptosis (TABLE 1) involved the mitochondrial pathway of apoptosis with cytochrome *c* release (FIG. 1A). This was followed by the activation of execution caspases and proteolysis of their downstream substrates (FIG. 1B), which was necessary for the execution of apoptosis (data not shown).

REFERENCES

1. EL-HASSAN, H., K. ANWAR, P. MACANAS-PIRARD, *et al.* 2003. Toxicol. Appl. Pharmacol. **191:** 118–129.
2. JONES, R.A., V.L. JOHNSON, N. BUCK, *et al.* 1998. Hepatology **27:** 1632–1642.

CD4$^+$ Lymphocyte Increases in HIV Patients during Potent Antiretroviral Therapy Are Dependent on Inhibition of CD8$^+$ Cell Apoptosis

SANDRO GRELLI,[a] GABRIELLA D'ETTORE,[b] FILIPPO LAURIA,[c] FRANCESCO MONTELLA,[c] LOIDE DI TRAGLIA,[a] CARTESIO D'AGOSTINI,[a] MIRIAM LICHTNER,[b] VINCENZO VULLO,[b] CARTESIO FAVALLI,[a] STEFANO VELLA,[d] BEATRICE MACCHI,[e,f] AND ANTONIO MASTINO[g]

[a]*Department of Experimental Medicine and Biochemical Science, "Tor Vergata," University Hospital, 00133 Rome, Italy*

[b]*Department of Infectious and Tropical Diseases, University of Rome, "La Sapienza," viale del Policlinico 155, 00161 Rome, Italy*

[c]*S. Giovanni Hospital, via Amba Aradam 9, 00184, Rome, Italy*

[d]*Istituto Superiore Sanità, viale Regina Elena 299, Rome, Italy*

[e]*Department of Neuroscience, University of Rome, "Tor Vergata," via Montpellier 1, 00133 Rome, Italy*

[f]*IRCCS, S. Lucia, via Ardeatina 306, 00179 Rome, Italy*

[g]*Department of Microbiology, Genetics, and Molecular Science, University of Messina, salita Sperone 31, 98166 Messina, Italy*

Abstract: Although suppression of apoptosis contributes to immune-reconstitution during potent antiretroviral therapy, its relationship with the majors indicators of response to therapy, that is, changes in CD4$^+$ cell counts and in viral loads (VL), is still debated. We extended our previous study by collecting data on the relationships among apoptosis and immunological and virological parameters during a long-term follow-up of HIV patients with an overall positive response to potent antiretroviral therapy. We report results from 15 patients who completed two years of therapy. In a smaller group of patients, we focused our attention on investigating the specific contribution of the CD8$^+$ subset in the overall changes in lymphocyte apoptosis, which occur concomitantly with the response to the therapy. Our data, while again confirming that inhibition of PBMC apoptosis is a phenomenon strictly related to a positive response to potent antiretroviral therapy, suggest that CD4$^+$ cell rescue is not directly dependent on inhibition of CD4$^+$ cell apoptosis but rather on that of the CD8$^+$ subset.

Keywords: CD4$^+$ cells; CD8$^+$ cells; viral loads; antiretroviral therapy

Address for correspondence: Antonio Mastino, Department of Microbiology, Genetics, and Molecular Science, University of Messina, salita Sperone 31, 98166 Messina, Italy. Voice: +39-090-6765198; fax: +39-090-392733.
mastino@med.uniroma2.it

Ann. N.Y. Acad. Sci. 1010: 560–564 (2003).
doi: 10.1196/annals.1299.104

Although there is general agreement that suppression of apoptosis contributes to immune-reconstitution during potent antiretroviral therapy,[1] its relationship with the majors indicators of response to therapy, that is, changes in CD4^{+} cell counts and in viral loads (VL), is still highly debated. Recent studies reported that levels of apoptosis in peripheral blood lymphocytes from HIV patients undergoing therapy were correlated with CD4^{+} cell counts, but not with viral load levels.[1,2] Conversely, using all the data collected during the first six months of potent antiretroviral therapy in a selected cohort of HIV patients who were naïve at enrollment, we found that levels of spontaneous or anti–Fas-induced apoptosis in peripheral blood mononuclear cells (PBMC) were significantly correlated not only inversely with CD4^{+} cell counts, but also directly with plasma viral load.[3] This discrepancy could be due to the fact that data reported by other authors were based on cross-sectional observations or to the relatively high proportion of patients who did not achieve complete viral inhibition in the same studies. Another reason could be that the reported observations were limited to an early phase after potent antiretroviral therapy. One unresolved aspect of pathogenesis of HIV infection is whether CD4^{+} cell depletion is mainly due to direct destruction following viral infection or rather to other indirect mechanisms. Recent reports sustain that the latter mechanisms are very likely involved in CD4^{+} lymphopenia during infection.[4,5] In fact, studies of the dynamics of distinct PBMC subpopulations indicate that the turnover of the CD8^{+} subset is remarkably affected by infection and, as a consequence, in response to potent antiretroviral therapy.[6,7] Thus, the rescue of CD8^{+} cells could be involved in the process of immune-reconstitution during therapy. This could occur through activation of a specific CTL response[8] or of a nonspecific response, such as that of recently identified alpha-defensins.[9]

Based on the considerations described above, we decided to extend our previous study by collecting data on the relationships among apoptosis and immunological and virological parameters during a long-term follow-up of HIV patients with an overall positive response to potent antiretroviral therapy. Here we report results from fifteen patients who completed two years of therapy. Moreover, in a smaller group of patients, we focused our attention on investigating the specific contribution of the CD8^{+} subset in the overall changes in lymphocyte apoptosis, which occur concomitantly with the response to the therapy. Potent antiretroviral therapy administered to patients consisted of at least two reverse transcriptase inhibitors in combination with one protease inhibitor or with a non-nucleoside reverse transcriptase inhibitor. The following drugs and dosage were utilized: zidovudine (AZT, 150 mg 2/day), didanosine (ddI, 400 mg 1/day), stavudine (d4T, 40 mg 2/day), lamivudine (3TC, 150 mg 2/day), efavirenz (EFV, 600 mg 1/day), indinavir (IDV, 800 mg 3/day), ritonavir (RTV, 600 mg 2/day), and nelfinavir (NVF, 750 mg 3/day). Four of the patients changed from the initial therapy to a new combination of drugs or experienced brief interruptions of therapy, without compromise of the overall response. All patients were examined at the Department of Infectious and Tropical Disease, University of Rome "La Sapienza," or at the "AIDS Center," S. Giovanni Hospital, in Rome. None of the patients suffered opportunistic or severe infections, cancer, autoimmune disorders, or major vascular or neurological diseases. Spontaneous or Fas-induced apoptosis, viral load, CD4^{+} counts, and all other parameters were measured before (time 0) and regularly during potent antiretroviral therapy. PBMC, isolated from heparinized blood were cultured with 10% fetal calf serum for 66 h before detection of spontaneous or anti–Fas-induced apoptosis. Anti-human Fas, at 500 ng/mL, was

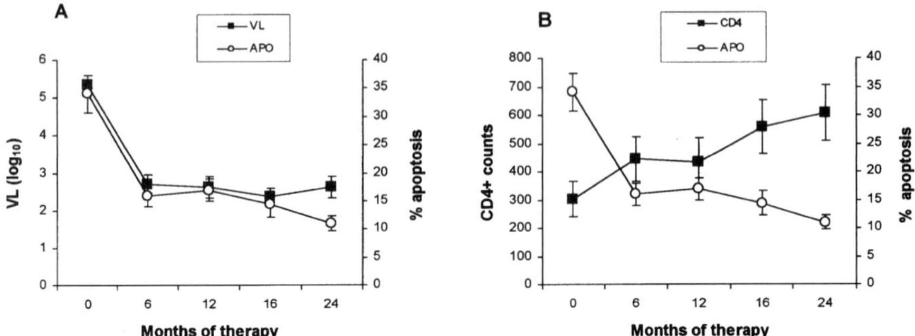

FIGURE 1. Two-year follow-up in a group of HIV-infected individuals with overall positive response to potent antiretroviral therapy. (**A**) Viral load expressed as $(\log_{10}) \times 10/\mu L$ (*solid square*, *y* axis, *left*) and susceptibility to spontaneous apoptosis (*open circle*, *y* axis, right). (**B**) $CD4^+$ cell counts (*solid square*, *y* axis, *left*) and susceptibility to spontaneous apoptosis (*open circle*, *y* axis, *right*). Apoptosis was detected in the PBMC of patients after 66 h of culture. Data from 15 patients, collected at time 0 and at 6, 12, 16, and 24 months of therapy, are presented as mean values ± SE.

added to some of the replicate cultures during the final 18 h of culture. Apoptosis was evaluated by flow cytometry analysis of isolated nuclei following detergent treatment and propidium iodide staining, as previously described by us.[10] Results, reported in FIGURE 1, refer to baseline (time 0) and 6, 12, 16 and 24 months after initiation of therapy, respectively. VL levels were permanently suppressed (FIG. 1A), except occasional, transitory rebounds at interruptions or change of therapy (not shown). In parallel, mean $CD4^+$ cell counts increased from 303.27 ± 241.13 (mean ± SD) cells/μL at baseline to 607.31 ± 348.51 cells/μL at the final observation time (FIG. 1B). $CD4^+$ cell counts were higher at the last observation time, in comparison with baseline, in all patients. Apoptosis was completely inhibited in all the 15 responder patients, with values overlapping those of normal controls (around 10% in our experimental conditions). Correlations among apoptosis, immunological, or virological parameters were investigated using values obtained before and after different time-points of potent antiretroviral therapy (1, 2, 4, 6, 8, 12, 20, 24, 36 months), by bivariate Spearman's analysis. For correlation, the number of HIV RNA copies/mL were transformed as decimal logarithms while $CD4^+$ and $CD8^+$ cell counts were transformed as corresponding natural logarithms. Total PBMC apoptosis was directly correlated with the level of viral RNA in plasma (r = .373 and P < .0001) and negatively with $CD4^+$ total cell counts (r = (–).386, P < .0001). Similar results were obtained when anti–Fas-induced apoptosis, instead of spontaneous apoptosis, was correlated with VL and with $CD4^+$ cell count (data not shown). These results, extending previous observations by us and other authors on apoptosis changes during potent antiretroviral therapy for a longer period of time, confirm that apoptosis levels even after three years of therapy remain remarkably decreased in patients showing positive response. Moreover, in a group of nine patients we evaluated whether susceptibility of total PBMC to apoptosis before and during potent antiretroviral therapy was selectively correlated with cell death of $CD4^+$ or $CD8^+$. For this

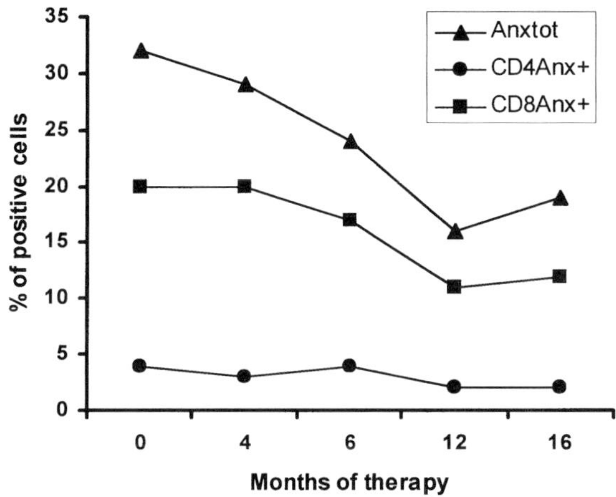

FIGURE 2. Variations in percentages of total annexin V–positive cells (Anxtot, *solid triangle*) and in percentages of distinct annexin V/CD8 double-positive (CD8Anx$^+$, *solid square*) or annexin V/CD4 double-positive (CD4Anx$^+$, *solid circle*) cells. Percentages, evaluated by flow cytometry in PBMC from nine HIV patients undergoing potent antiretroviral therapy at time 0 and at 4, 6, 12, and 16 months of therapy, are represented as mean values.

investigation, cell death of CD4$^+$ and CD8$^+$ was assessed by evaluating the percentage of annexin-positive cells in samples stained for double-fluorescence analysis. As shown in FIGURE 2, we found that variations in CD4$^+$ and CD8$^+$ cell apoptosis during therapy followed the same trend as those observed in total PBMC. In fact, analysis of correlations revealed that death of both CD4$^+$ and CD8$^+$ cells contributed to the percentage of total cells positive for annexin, in that the levels of total annexin-positive cells directly correlated with those of the single annexin-positive CD4$^+$ and CD8$^+$ subsets. Surprisingly, however, correlation with annexin-positive CD8$^+$ T cells was more significant than that with annexin-positive CD4$^+$ T cells (r = .934, $P < .0001$, and $r = .292$, $P = .018$, respectively). To assess the implications of CD4$^+$ and CD8$^+$ T cell death in the rescue of CD4$^+$ cells during potent antiretroviral therapy, we also analyzed the correlations between the absolute number of CD4$^+$ T cells and the percentage of annexin-positive CD8$^+$ and CD4$^+$ T cells, respectively. We found that the number of CD4$^+$ cells was inversely correlated with the percentage of annexin-positive CD8$^+$ cells ($r = -.506$, $P < .0001$). Conversely, it was directly correlated with the percentage of annexin-positive CD4$^+$ cells ($r = .280$, $P = .024$).

Results presented in this report contribute to clarify in which proportion the modulation of cell death of the two main T cell subsets contribute to the decrease in overall PBMC apoptosis in response to potent antiretroviral therapy. Interestingly, we provide direct evidence that the decrease in CD8$^+$ cell death contributed more significantly than the decrease in CD4$^+$ cell death in inhibiting total PBL apoptosis. Our results are in agreement with data reported in a cross-sectional observation.[1] In any case, a possible direct involvement of the CD8$^+$ subset in CD4$^+$ T cell rescue during potent antiretroviral therapy is also supported by our observation that in one case of virological failure, where a high level of VL rebound was accompanied by a remarkable decrease in $CD4^+$ cells, the parallel increase in apoptosis preferentially occurred in $CD8^+$ cells, rather than in $CD4^+$ T cells (data not shown). Taken together our data, while again confirming that inhibition of PBMC apoptosis is a phenomenon strictly related to a positive response to potent antiretroviral therapy, suggest that $CD4^+$ cell rescue is not directly dependent on inhibition of $CD4^+$ cell apoptosis but rather on that of the $CD8^+$ subset. Moreover, the persistence of $CD4^+$ cells that are prone to undergo apoptosis in the PBMC of individuals undergoing potent antiretroviral therapy could explain the difficulty in achieving a complete immune reconstitution. There is need for further investigation in order to confirm our results and to understand the mechanisms involved in the observed phenomena.

ACKNOWLEDGMENTS

We are grateful to Prof. Enrico Garaci for his support, advice, and helpful discussion throughout this work. We wish to thank Alison Inglis, B.A., for her linguistic assistance. This work was supported by grants from the Istituto Superiore di Sanità, AIDS Project, and from the University of Rome "Tor Vergata" to B. Macchi and by grants from the Italian Ministry of Education, University and Research, Research Projects of National Interest, and from the University of Messina to A. Mastino.

REFERENCES

1. De Oliveira Pinto, L.M., H. Lecoeur, E. Ledru, *et al.* 2002. Lack of control of T cell apoptosis under HAART. Influence of therapy regimen in vivo and in vitro. AIDS **16:** 329–339.
2. Roger, P.M., J.P. Breittmayer, C. Arlotto, *et al.* 1999. Highly active anti-retroviral therapy (HAART) is associated with a lower level of CD4+ T cell apoptosis in HIV-infected patients. Clin. Exp. Immunol. **118:** 412–416.
3. Grelli, S., S. Campagna, M. Lichtner, *et al.* 2000. Spontaneous and anti-Fas-induced apoptosis in lymphocytes from HIV-infected patients undergoing highly active antiretroviral therapy. AIDS **14:** 939–949.
4. Grossman, Z., M. Meier-Schellersheim, A.E. Sousa, *et al.* 2002. CD4+ T-cell depletion in HIV infection: are we closer to understanding the cause? Nat. Med. **8:** 319–323.
5. Feinberg, M.B., J.M. McCune, F. Miedema, *et al.* 2002. HIV tropism and CD4+ T-cell depletion. Nat. Med. **8:** 537–538.
6. Grossman, Z. & W.E. Paul. 2000. The impact of HIV on naïve T-cell homeostasis. Nat. Med. **6:** 976–977.
7. Kovacs, J.A., R.A. Lempicki, I.A. Sidorov, *et al.* 2001. Identification of dynamically distinct subpopulations of T lymphocytes that are differentially affected by HIV. J. Exp. Med. **194:** 1731–1741.
8. Oxenius, A., H.F. Gunthard, B. Hirschel, *et al.* 2001. Direct ex vivo analysis reveals distinct phenotypic patterns of HIV-specific CD8+ T lymphocyte activation in response to therapeutic manipulation of virus load. Eur. J. Immunol. **31:** 115–1121.
9. Zhang, L., W. Yu, T. He, *et al.* 2002. Contribution of human alpha-defensin 1, 2 and 3 to the anti-HIV-1 activity of CD8 antiviral factor. Science **298:** 995–1000.
10. Matteucci, C., S. Grelli, E. De Smaele, *et al.* 1999. Identification of nuclei from apoptotic, necrotic, and viable lymphoid cells by using multiparameter flow cytometry. Cytometry **35:** 145–153.

Characterization of Antiapoptotic Activities of *Chlamydia pneumoniae* in Infected Cells

S. F. FISCHER AND G. HÄCKER

Institute for Medical Microbiology, Immunology and Hygiene, Technische Universität München, D-81675 Munich, Germany

ABSTRACT: *C. pneumoniae* has the capacity to inhibit apoptosis. Here we provide further insight into how infection with *C. pneumoniae* affects the apoptotic signal transduction in infected human epitheloid cells. For assessing changes in the apoptotic response, intact cells and cell extracts were used. These results show that *C. pneumoniae* is able to interfere with the host cell's apoptotic apparatus at least at two steps in the signal transduction.

KEYWORDS: apoptosis; bacterium; caspase; *Chlamydia*

Apoptosis is a process by which mammalian cells undergo suicide upon a specific stimulus. Besides its importance for tissue homeostasis and embryonic development, apoptosis plays an important role in the defense of a complex host organism against infectious microorganisms. In infections with intracellular pathogens, like viruses or some bacteria, modulation of the cell death system by the pathogen is frequently observed. *C. pneumoniae* is an obligate intracellular bacterium that causes airway infections in humans and has been implicated in atherosclerosis. Because chlamydiae depend on host factors for replication, the death of the infected cell is very likely to interfere with the developmental cycle of *C. pneumoniae* and can probably be viewed as advantageous to the host cell. Mechanistically, cell death by apoptosis is the result of the activation of a signal transduction pathway, in which members of the caspase family of cysteine proteases play a central role. Caspase-activity is required to transmit cell death signals and to induce the morphological changes of apoptosis. Caspases are present as inactive pro-caspases and can be divided into initiator (caspase-8 and -9) and effector caspases (caspase-3 and probably others). Recent research has started to address the question of whether chlamydial infections affect the apoptotic response of an infected cell. *C. trachomatis* has been found to induce apoptosis *in vitro* and *in vivo*[3,5] and to inhibit experimentally induced apoptosis *in vitro*.[1] *C. psittaci* has only been described to induce apoptosis.[4] We and others have found that *C. pneumoniae* has the capacity to inhibit apoptosis.[2,6,7] Here we

Address for correspondence: S.F. Fischer, Institute for Medical Microbiology, Immunology and Hygiene, Technische Universität München, D-81675 Munich, Germany. Voice: +49-89-4140-4180; fax: +49-89-4140-4139.
silke.fischer@lrz.tum.de

**Ann. N.Y. Acad. Sci. 1010: 565–567 (2003). © 2003 New York Academy of Sciences.
doi: 10.1196/annals.1299.105**

provide further insight into how infection with *C. pneumoniae* affects the apoptotic signal transduction in infected human epitheloid cells. For assessing changes in the apoptotic response, intact cells and cell extracts were used.

INHIBITION OF APOPTOSIS UPON INFECTION WITH *C. PNEUMONIAE* IN INTACT CELLS

Chlamydial infection on its own did not induce detectable apoptosis in HeLa cells. When HeLa cells were infected, however, a strong protection against external apoptosis-inducing stimuli (staurosporine, fas-signaling) was seen by all parameters of apoptosis investigated. In order to map the block imposed by chlamydia, individual steps in the activation of the apoptotic pathway were investigated. In infected and staurosporine-treated cells, effector caspase activity, proteolytic cleavage/activation of caspase-3 and of caspase-9, and cytochrome *c* redistribution were all profoundly inhibited compared to mock-infected cells. When cells were treated with anti-fas mAb, caspase-8 processing was normal but effector caspase activity was reduced. All these results point to a block of the apoptotic pathway on or upstream of a mitochondrial change. In order to investigate whether this protection against apoptosis required bacterial metabolism, the bacterial RNA synthesis-blocking antibiotic rifampin was used. In the presence of rifampin cells could be infected but no inclusions developed and no protection against apoptosis was seen. Time course experiments showed that rifampin had to be present during the early phase of infection. This suggests that the antiapoptotic effect of *C. pneumoniae* needs early bacterial protein synthesis. It is a well-known phenomenon that infection with chlamydiae changes the gene expression pattern at least in some host cells. It is therefore important to distinguish between a cellular anti-apoptotic factor generated from the host genome upon infection and a factor provided by the bacteria. Especially proteins regulated by the activity of the family of transcription factors NF-κB have the potential to inhibit apoptosis. NF-κB-activation is, however, unlikely to partake in the protection against apoptosis in HeLa cells for two reasons: first, infection with *C. pneumoniae* did not increase the detectable NF-κB activity. Second, NF-κB activation did not protect HeLa cells against staurosporine-induced apoptosis. These chlamydial activities appear to be able to regulate the release of cytochrome *c*, a process which is thought to be of great importance in the induction of apoptosis.

INHIBITION OF APOPTOSIS UPON INFECTION WITH *C. PNEUMONIAE* IN A CELL-FREE SYSTEM

However, in another approach we found a second anti-apoptotic activity. When cytochrome *c* is added to cytosolic fractions from human cells *in vitro*, it leads to caspase activation. When cytosolic fractions were prepared from infected HeLa cells and used in this assay, the formation of the caspase-activating signaling complex and the activation of caspases were reduced. This inhibitory activity was shown to derive from a factor not present in normal HeLa cells. These data point to the generation of a cytosolic antiapoptotic factor in human cells upon infection by *C. pneumoniae*.

We could show that infection with *C. pneumoniae* inhibits externally induced apoptosis of the host cell and that this block lies upstream of caspase activation. The blockade was shown to extend to two different molecular steps: both cytochrome *c* release and cytochrome *c*–dependent caspase activation were inhibited. Early bacterial protein synthesis was required for this protection. *C. pneumoniae* has a special intracellular developmental cycle and requires cellular components. It is therefore likely that chlamydial growth is sensitive to the death of the host cell. A bacterial capability to inhibit apoptosis may therefore be advantageous (or even essential) either to prolong the normal life span of the infected cell or to counter apoptosis, which may be triggered upon chlamydial infection. As mentioned above, a number of studies have found that chlamydiae also have the potential to induce apoptosis; the outcome of the infection may therefore depend on a balance of apoptosis-inducing and -inhibiting activities. Here we demonstrate a clear apoptosis-inhibiting effect of *C. pneumoniae* in epithelioid cells. In a number of cases, bacterial inhibition of apoptosis relies on the induction of NF-κB transcriptional activity. This is especially the case in cells from the myeloid lineage and has in fact be demonstrated to occur for *C. pneumoniae.*[7] Our results show, however, that this mechanism very likely is not important in epithelial cells, a main target cell for chlamydial infection. The finding that bacterial protein synthesis was required for the *C. pneumoniae*–dependent antiapoptotic activity might mean that a bacterial protein has a direct antiapoptotic function. It has been shown that *C. pneumoniae* is able to inject proteins into the host cell's cytosol,[8] most likely via a type III secretion apparatus.

We cannot exclude the possibility that the inhibition of apoptosis requires cellular gene expression, but no direct evidence of such regulation could be detected. In summary these results show that *C. pneumoniae* is able to interfere with the host cell's apoptotic apparatus at least at two steps in the signal transduction. This behavior might explain the ability of these bacteria to cause chronic infections in humans.

REFERENCES

1. FAN, T. *et al.* 1998. Inhibition of apoptosis in chlamydia-infected cells: blockade of mitochondrial cytochrome c release and caspase activation. J. Exp. Med. **187**(4): 487–496.
2. FISCHER, S.F. *et al.* 2001. Characterization of antiapoptotic activities of *Chlamydia pneumoniae* in human cells. Infect. Immun. **69**(11): 7121–7129.
3. GIBELLINI, D. *et al.* 1998. Induction of apoptosis by *Chlamydia psittaci* and *Chlamydia trachomatis* infection in tissue culture cells. Zentralbl. Bakteriol. **288**(1): 35–43.
4. OJCIUS, D.M. *et al.* 1998. Apoptosis of epithelial cells and macrophages due to infection with the obligate intracellular pathogen *Chlamydia psittaci*. J. Immunol. **161**(8): 4220–4226.
5. PERFETTINI, J.L. *et al.* 2000. Effect of *Chlamydia trachomatis* infection and subsequent tumor necrosis factor alpha secretion on apoptosis in the murine genital tract. Infect. Immun. **68**(4): 2237–2244.
6. RAJALINGAM, K. *et al.* 2001. Epithelial cells infected with Chlamydophila pneumoniae (*Chlamydia pneumoniae*) are resistant to apoptosis. Infect. Immun. **69**(12): 7880–7888.
7. WAHL, C. *et al.* 2001. Survival of *Chlamydia pneumoniae*-infected Mono Mac 6 cells is dependent on NF-kappaB binding activity. Infect. Immun. **69**(11): 7039-7045.
8. ZHONG, G. *et al.* 2001. Identification of a chlamydial protease-like activity factor responsible for the degradation of host transcription factors. J. Exp. Med. **193**(8): 935–942.

NF-κB and Bcl-2 in *Helicobacter pylori*–Induced Apoptosis in Gastric Epithelial Cells

SANG HUI CHU, JOO WEON LIM, KYUNG HWAN KIM, AND HYEYOUNG KIM

Department of Pharmacology and Institute of Gastroenterology, Brain Korea 21 Project for Medical Science, Yonsei University College of Medicine, Seoul 120-752, Korea

ABSTRACT: *Helicobacter pylori* (*H. pylori*) has been considered as an important pathogen of gastroduodenal inflammation and gastric carcinogenesis. However, the pathogenic mechanisms including *H. pylori*–induced apoptosis and subsequent molecular mechanisms have not been clarified yet. The present study examined the role of Bcl-2 and its relation to NF-κB in *H. pylori*–induced apoptosis in human gastric epithelial AGS cells. AGS cells were cultured in the presence of *H. pylori*, at a bacterium/cell ratio of 300:1, for the determination of apoptosis, NF-κB activation with IκBα degradation, and Bcl-2 level. AGS cells were transfected with a control vector (pCMV cells) or a full-length human Bcl-2 expression vector (Bcl-2 cells). *H. pylori*–induced apoptosis and NF-κB activation were compared in the wild-type cells and the transfected cells. As a result, *H. pylori* increased apoptotic cells with chromatin condensation and reduced Bcl-2 levels, which were accompanied with NF-κB activation. *H. pylori* induced sixfold increase in the number of apoptotic cells in wild-type cells and pCMV cells. *H. pylori*–induced increment of apoptotic cells were relatively lower in Bcl-2 cells than pCMV cells. *H. pylori*–induced NF-κB activation and IκBα degradation were not different in the wild-type cells, pCMV cells, and Bcl-2 cells. In conclusion, the reduced gastric Bcl-2 level may be the main pathogenic mechanism of *H. pylori*–induced apoptosis in gastric epithelial cells.

KEYWORDS: NF-κB; Bcl-2; *Helicobacter pylori*; apoptosis; gastric epithelial cells

INTRODUCTION

H. pylori induces the activation of NF-κB, an oxidant-sensitive transcription factor in gastric epithelial cells.[1] Inhibition of NF-κB activation suppressed *H. pylori*–induced apoptosis in AGS cells.[2] This study suggests that the main pathogenic mechanism of *H. pylori*–induced apoptotic cell death may involve NF-κB activation. The mechanism by which NF-κB activation induces apoptosis has not been clarified, but it is assumed that the mechanism involves the regulation of one or more genes known to play a role in apoptosis. Among the potential downstream regulators, antiapoptotic protein Bcl-2 has been downregulated in various models of apoptosis including *H. pylori*–induced apoptosis.[3,4] We conducted the present study to deter-

Address for correspondence: Hyeyoung Kim, Department of Pharmacology, Yonsei University College of Medicine, Seoul 120-752, Korea. Voice: +82-2-361-5232; fax: +82-2-313-1894. kim626@yumc.yonsei.ac.kr

**Ann. N.Y. Acad. Sci. 1010: 568–572 (2003). © 2003 New York Academy of Sciences.
doi: 10.1196/annals.1299.106**

mine the role of Bcl-2 and its relation to NF-κB in *H. pylori*–induced apoptosis in human gastric epithelial cells.

METHODS

AGS cells (gastric adenocarcinoma, ATCC CRL 1739) were cultured in the presence of *H. pylori* (NCTC 11637 obtained from ATCC), at a bacterium/cell ratio of 300:1, for the determination of apoptosis (24-h culture), NF-κB activation (time course for 4 h), and Bcl-2 level (time course for 24 h). AGS cells were transfected with a control vector CMVneo (pCMV cells) or a full-length human Bcl-2 expres-

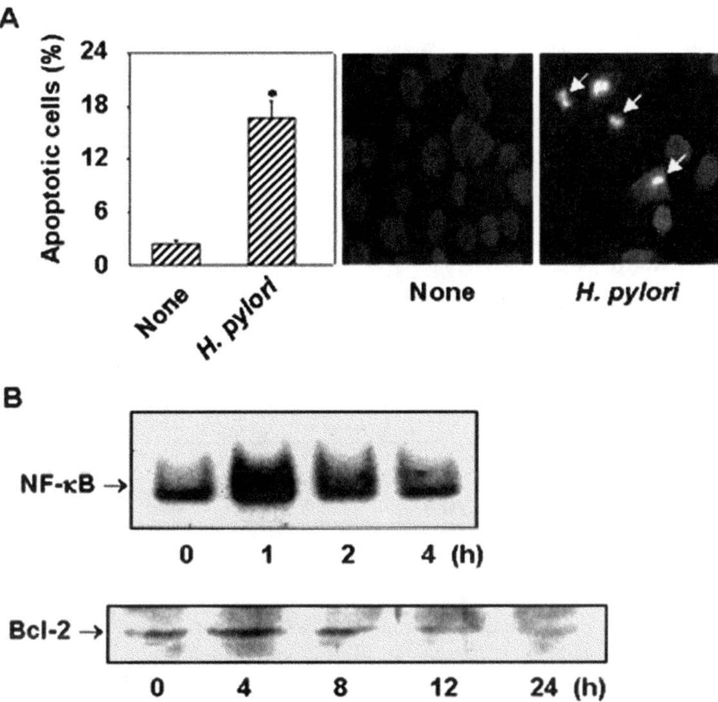

FIGURE 1. *H. pylori*–induced apoptosis, NF-κB activation, and reduced Bcl-2 protein level in the wild-type cells. (**A**) The wild-type cells were culture in the absence (None) or presence of *H. pylori* (*H. pylori*) for 24 h. Cells adhered as a monolayer onto coverslips were fixed with 4% paraformaldehyde and were stained with Hoechst 33258. Chromatin condensation (*arrows*) was observed in *H. pylori*–treated cells. The percentage of apoptotic cells was determined, based on total numbers of the cells. Data represent mean ± standard error of three separate experiments. $^{*}P<.05$ compared with None. (**B**) Time-course of NF-κB activation in the nuclear extracts of the cells cultured in the presence of *H. pylori* was determined by EMSA (*upper panel*). Time-course of Bcl-2 protein level in the whole cell extracts of the cells cultured in the presence of *H. pylori* was determined by Western blot analysis (*lower panel*).

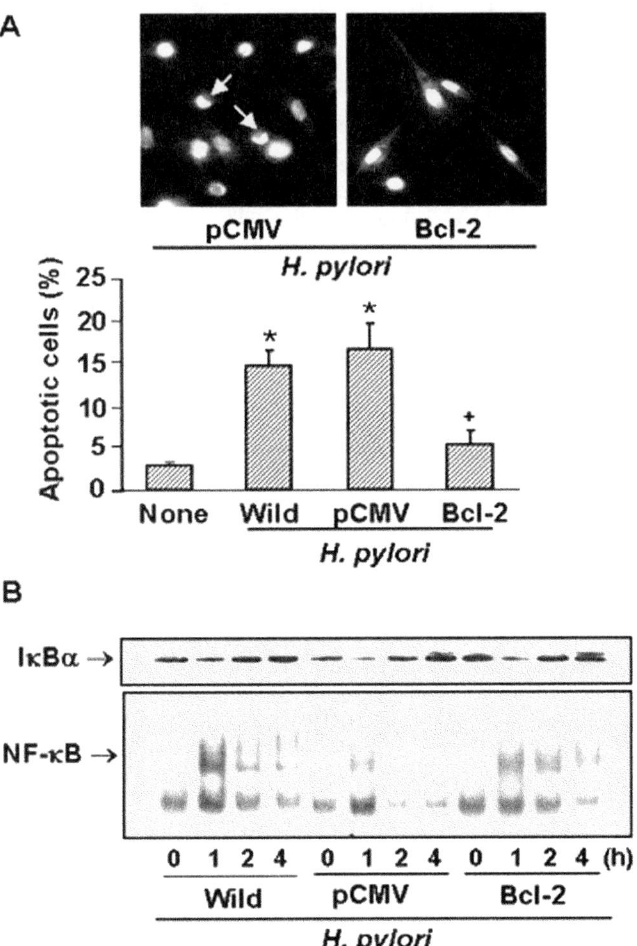

FIGURE 2. *H. pylori*–induced apoptosis, NF-κB activation, and IκBα degradation in the wild-type cells and the cells transfected with control vector or Bcl-2 overexpression gene. The cells were transfected with a control vector CMVneo (pCMV cells) or a full-length human Bcl-2 expression vector CMVbcl-2nl (Bcl-2 cells). The wild-type cells (Wild cells) and the transfected cells were cultured in the absence or presence of *H. pylori* (*H. pylori*) for 24 h (apoptosis) or 4 h (NF-κB activation, IκBα degradation). (**A**) Cells adhered as a monolayer onto coverslips were fixed with 4% paraformaldehyde and were stained with Hoechst 33258. Chromatin condensation (*arrows*) was observed in *H. pylori*–treated pCMV cells. Nuclear changes were not detected in Bcl-2 cells treated with *H. pylori* (*upper panel*). The percentage of apoptotic cells was determined, based on total numbers of the cells. Data represent mean±standard error of three separate experiments. $^{*}P<.05$ compared with None (Wild-type cells cultured in the absence of *H. pylori*). $^{+}P<.05$ compared with pCMV cells cultured in the presence of *H. pylori* (*lower panel*). (**B**) Time-course of IκBα degradation in the whole cell extracts of the cells cultured in the presence of *H. pylori* was determined by Western blot analysis (*upper panel*). Time-course of NF-κB activation in the nuclear extracts of the cells cultured in the presence of *H. pylori* was determined by EMSA (*lower panel*).

sion vector CMVbcl-2nl (Bcl-2 cells). The wild-type cells and the transfected cells were cultured in the absence or presence of *H. pylori* for 24 h (apoptosis) or 4 h (NF-κB activation, IκBα degradation). For the determination of apoptosis, the percentage of apoptotic cells, as assessed by staining with Hoechst 33258, was calculated based on total numbers of the cells. Chromation condensation stained with Hoechst 33258 was observed under a fluorescence microscope (×400). NF-κB activation in the nuclear extracts of the cells was determined by electrophoretic mobility shift assay (EMSA). Bcl-2 protein and IκBα levels in the whole cell extracts were determined by Western blot analysis.

RESULTS AND DISCUSSION

H. pylori induced apoptosis at 24-h culture, as determined by apoptotic cells stained with Hoechst 33258 and showing chromatin condensation (FIG. 1A). Time-course of NF-κB activation in the nuclear extracts of the cells cultured in the presence of *H. pylori* was determined by EMSA (FIG. 1B). NF-κB activation was detected at 1 h and decreased thereafter. Time-course of Bcl-2 protein level in the whole cell extracts of the cells cultured in the presence of *H. pylori* was determined by Western blot analysis (FIG. 1B). *H. pylori* induced a significant decrease in Bcl-2 level, an anti-apoptotic protein, at 8-h culture. Bcl-2 level was decreased by *H. pylori* in the cells in a time-dependent manner. Present results are supported by the report demonstrating that a subtle decrease in Bcl-2 level, through transcriptional repression by NF-κB and active degradation of Bcl-2 protein, resulted in a rapid apoptosis during early B lineage.[5] It is controversial for the relation between Bcl-2 and NF-κB. One report showed that Bcl-2 decreased NF-κB activity,[6] while others reported that Bcl-2 overexpression activates NF-κB by promoting degradation of inhibitory kB.[7] When Bcl-2 overexpression gene was transfected into AGS cells (Bcl-2 cells), *H. pylori*–induced apoptosis was significantly inhibited (FIG. 2A). *H. pylori* induced six-fold increase in the number of apoptotic cells in wild-type cells and pCMV cells, which were reduced twofold in Bcl-2 cells. However, both NF-κB activation and IκBα degradation induced by *H. pylori* were not different in wild-type cells, pCMV cells, and Bcl-2 cells (FIG. 2B). Our previous studies showed that NF-κB acted as a cell death signal at a transcriptional level in *H. pylori*–infected AGS cells.[1,2] Present study demonstrates that enhancement of Bcl-2 or inhibition of Bcl-2 degradation may prevent apoptotic cell death regardless of NF-κB activation. The main finding of the present study is that the reduced gastric Bcl-2 level may be the main pathogenic mechanism of *H. pylori*–induced apoptosis in gastric epithelial cells. Even though NF-κB is activated by *H. pylori*, if Bcl-2 level is maintained in the gastric epithelium, apoptotic cell death could be prevented. Present results suggest that enhancement of Bcl-2 may provide a therapeutic approach to the patients suffering with *H. pylori*–induced gastric diseases.

ACKNOWLEDGMENT

This study was supported by a grant from the Korea Ministry of Health and Welfare (H. Kim).

REFERENCES

1. Kim, H., J.Y. Seo & K.H. Kim. 2000. Inhibition of lipid peroxidation, NF-kappaB activation and IL-8 production by rebamipide in *Helicobacter pylori*-stimulated gastric epithelial cells. Dig. Dis. Sci. **45:** 621–628.
2. Lim, J.W., H. Kim & K.H. Kim. 2001. NF-kappaB, inducible nitric oxide synthase and apoptosis by *Helicobacter pylori* infection. Free Rad. Biol. Med. **31:** 355–366.
3. Konturek, P.C., P. Pierzchalski, S.J. Konturek, *et al.* 1999. *Helicobacter pylori* induces apoptosis in gastric mucosa through an upregultion of Bax expression in human. Scand. J. Gastroenterol. **34:** 375–383.
4. Reed, J.C. 1998. Bcl-2 family proteins. Oncogene **17:** 3225–3236.
5. Sohur, U.S., M.N. Dixit, C-L. Chen, *et al.* 1999. Rel/NF-κB represses *bcl-2* transcription in pro-B lymphocytes. Gene Expr. **8:** 219–222.
6. Grimm, S., M.K. Bauer, P.A. Baeuerle & K. Schulze-Osthoff. 1996. Bcl-2 down-regulates the activity of transcription factor NF-kappaB induced upon apoptosis. J. Cell Biol. **134:** 13–23.
7. De Moissac, D., H. Zheng & L.A. Kirshenbaum. 1999. Linkage of the BH4 domains of Bcl-2 and the nuclear factor kappaB signaling pathway for suppression of apoptosis. J. Biol. Chem. **274:** 29505–29509.

Role of Phospholipomannan in *Candida albicans* Escape from Macrophages and Induction of Cell Apoptosis through Regulation of Bad Phosphorylation

STELLA IBATA-OMBETTA, THIERRY IDZIOREK,[b] PIERRE-ANDRÉ TRINEL, DANIEL POULAIN, AND THIERRY JOUAULT

Laboratoire de Mycologie Fondamentale et Appliquée, Inserm EMI0360, Université de Lille II, 59037 Lille Cedex, France

[b]*Inserm U459, Faculté de Médecine H. Warembourg, Place Verdun, 59037 Lille Cedex, France*

ABSTRACT: *Candida albicans*, the most common opportunistic fungal pathogen of humans is a part of the normal microbial flora. To investigate host-parasite interaction related to the commensal-pathogen switch of this yeast we compared the response of macrophages to *C. albicans* and to the non-pathogenic yeast *Saccharomyces cerevisiae*. In contrast to *S. cerevisiae*, *C. albicans* survived within macrophages. This escape from macrophages was associated with qualitative differences in the sequential phosphorylation of MEK, ERK1/2, and p90RSK during phagocytosis. Decreased activation of this pathway was observed with *C. albicans* and was associated with a species-specific overexpression of the MEK phosphatase, MKP-1. Dysregulation of the ERK1/2/p90RSK signal transduction pathway by *C. albicans* was associated downstream with reduction in Bad phosphorylation, specifically at Ser-112, and disappearance of free Bcl-2. This ended at apoptosis of cells that have ingested *C. albicans,* as revealed by staining of phosphatidylserine exposure in the macrophage outer membrane. The role of phospholipomannan (PLM), a phylogenetically unique glycolipid with a phytoceramide moiety expressed at the surface of and shed by *C. albicans*, was examined. Addition of PLM to macrophages led to dysregulation similar to that observed with live *C. albicans* and promoted the survival of the sensitive *S. cerevisiae* within the cells. Evidence of externalization of membranous phosphatidylserine, loss of mitochondrial integrity, and DNA fragmentation after incubation of macrophages with PLM suggest that this molecule supported the activities observed with *C. albicans* yeast cells.

KEYWORDS: *C. albicans*; apoptosis; macrophage; signal transduction; escape

Address for correspondence: Thierry Jouault, Laboratoire de Mycologie Fondamentale et Appliquée, Université de Lille II, Faculté de Médecine H. Warembourg, Pôle Recherche, Place Verdun, 59037 Lille Cedex, France. Voice: +33-3-20-62-34-15; fax: +33-3-20-62-34-16.
tjouault@univ-lille2.fr

**Ann. N.Y. Acad. Sci. 1010: 573–576 (2003). © 2003 New York Academy of Sciences.
doi: 10.1196/annals.1299.107**

Macrophage effector functions, especially those induced in response to microbial stimuli, are known to be dependent on tyrosine phosphorylation processes and involve activation of members of the mitogen-activated protein kinase (MAPK) family. Mechanisms to counteract host defense by interfering with signal transduction pathways involved in endocytosis and phagocytosis have been developed by several pathogens. Intracellular pathogens have evolved diverse strategies to induce or inhibit host cell apoptosis, aiding dissemination within the host or facilitating intracellular survival. Host cell apoptosis is induced through different virulence mechanisms based on either surface glycolipids or type III secretion proteins.[1]

Candida albicans, a part of the normal microbial flora that colonizes mucocutaneous surfaces, is able to switch from a commensal to a pathogen capable of infecting a variety of tissues. Phagocytosis by mononuclear phagocytic cells is an important step in the control of the infection. However, in contrast with the nonpathogenic yeast *S. cerevisiae*, *C. albicans* is able to resist to lytic activity of these cells.[2] Endocytosis of *C. albicans* by J774 cells is associated with down-modulation of the MEK-ERK signal transduction pathway leading to decreased phosphorylation of ERK1/2 and its downstream product, p90rsk (FIG. 1). Phosphorylated p90rsk targets a Ser residue at position 112 of Bad, a pro-apoptotic member of the Bcl-2 family that plays an important role in mediating signal transduction pathways leading to apoptosis. As ERK1/2 and p90rsk phosphorylation were altered after ingestion of *C. albicans* by J774 cells, the effect of yeast engulfment on Bad phosphorylation was

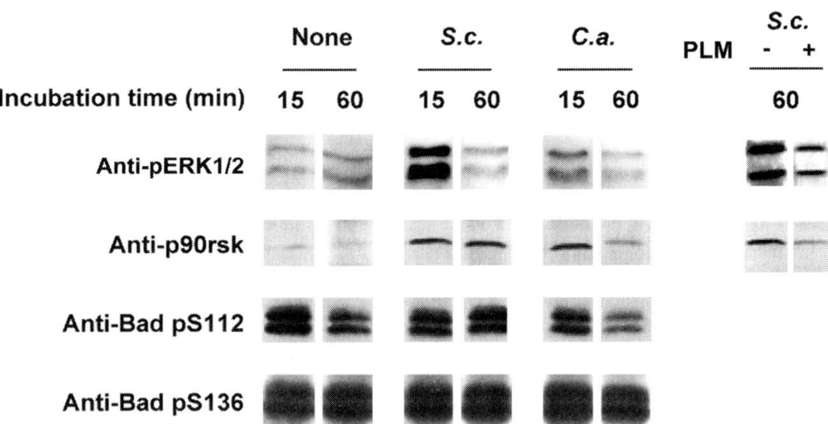

FIGURE 1. Signal transduction induced after endocytosis of yeasts by J774 macrophages. J774 cells were either untreated or incubated with *S. cerevisiae* (S.c.) or *C. albicans* (C.a.) blastoconidia. In some experiments (*right panel*), cells were incubated for 60 min with 50 μg/mL of PLM before addition of yeasts. After 15 or 60 min, cells were lysed. Whole cell lysates were separated by SDS-PAGE and transferred to nitrocellulose membranes. The blots were probed with antibodies specific for phosphorylated forms of either ERK1/2, p90rsk, Bad Ser-112, or Bad Ser-136. Blots were developed with ECL and the autoradiograms scanned. The data shown are representative of four independent experiments.

investigated. Incubation of J774 cells with *S. cerevisiae* blastoconidia was associated with high levels of phosphorylation of Bad at both Ser-112 and Ser-136 (FIG. 1). In cells incubated with *C. albicans*, a similar degree of phosphorylation of p90rsk and Bad at both Ser-112 and Ser-136 was observed at the early step of the endocytic process (15 min). However, after 60-min incubation with *C. albicans*, cells displayed a dramatic decrease in phosphorylation of p90rsk, and a simultaneous and significantly lower phosphorylation of Bad, specifically at Ser-112. Phosphorylation of Bad at Ser-112 has been shown to be critical for its binding to 14-3-3, leading to its sequestration (or inhibition) within the cytoplasma. The absence of such phosphorylation frees Bad, which may complex with the anti-apoptotic protein Bcl-2.[3] The effect of endocytosis of either *C. albicans* or *S. cerevisiae* on the distribution of Bcl-2, as investigated in an immunofluorescence assay, showed a similar distribution of Bcl-2 in untreated cells and in cells that had ingested *S. cerevisiae*. In contrast, a strong decrease in Bcl-2 staining was observed in cells that had ingested *C. albicans* yeasts, although analysis by Western blotting to detect both free and complexed protein, revealed similar levels of Bcl-2 expression in cell extracts. Together these data showed that the disappearance of Bcl-2 observed in the immunofluorescence assay could not be related to downmodulation of protein expression and suggested the formation of heterodimeric complexes between Bcl-2 and unphosphorylated Bad. Therefore, the effect of ingestion of *C. albicans* or *S. cerevisiae* on the expression of phosphatidylserine at the outer leaflet of macrophages was investigated. Staining with propidium iodide was similar irrespective of how the cells had been treated, showing that

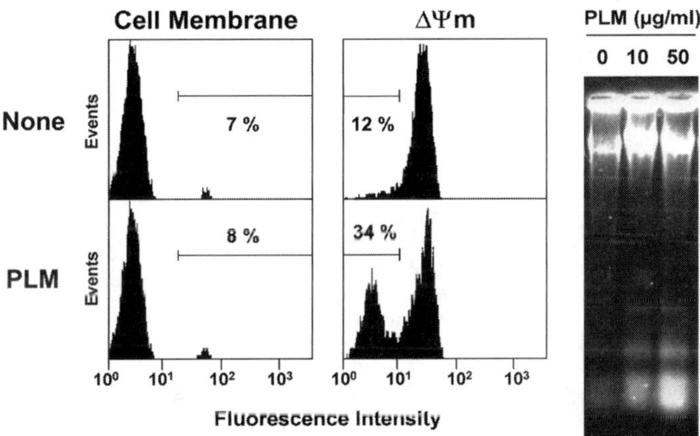

FIGURE 2. (*Left*) Cytometry analysis of plasma membrane and mitochondrial integrity of J774 cells after incubation with PLM. J774 cells were either left untreated (None) or treated with PLM (PLM) for 120 min at 37°C. After washing, the cells were recovered by trypsin treatment and incubated for 15 min with 10 μM YOPRO-1 to examine plasma membrane alterations (Cell Membrane), and 100 nM CMX-Ros to demonstrate the integrity of the mitochondrial transmembrane potential (Δψm). Staining was analyzed by flow cytometry. (*Right*) After incubation at 37°C for 16 h with different concentrations of PLM, cells were collected and DNA was extracted. Ten μL of purified DNA was applied to horizontal agarose gels (2%) and subjected to electrophoresis. Gels were stained with ethidium bromide and photographed under UV light. Data shown are representative of three independent experiments.

the cells were not affected by necrosis. No binding of annexin V was observed with control cells or with cells that had ingested *S. cerevisiae*. In contrast, binding of annexin V to cells that had ingested *C. albicans* yeast cells, revealed that phosphatidylserines were exposed at the plasma membrane of these cells.

Phospholipomannan (PLM), which is a characteristic of *C. albicans,* makes it different from other strains of *Candida* or from *S. cerevisiae*.[4] This glycolipid is a member of the mannose inositol phosphoceramide (MIPC) family, which is present at the surface of *C. albicans* and which is shed by the yeast on contact with macrophages.[5] Considering its structure, which is related to ceramide, we hypothesized that PLM could play a role in the alteration of the macrophage response observed with *C. albicans*. The effect of PLM on signal transduction was examined by incubating cells with PLM before addition of *S. cerevisiae*. The results obtained showed that addition of PLM to cells led to a decreased capability of the cells to phosphorylate ERK1/2 and p90RSK in response to ingestion of *S. cerevisiae* and was associated with an alteration of cell killing capability (28±2% vs. 5±5% of *S. cerevisiae* survived within cells treated with PLM). Moreover, this treatment led to reduced, if any, staining of Bcl-2 in cells having ingested *S. cerevisiae*, which was comparable to that observed after endocytosis of *C. albicans*. Finally, a direct effect of PLM on cell apoptosis was demonstrated by mitochondrial potential alteration, phosphatidylserine exposure at the plasma membrane and DNA fragmentation induced in cells that have been treated with PLM (FIG. 2).

The results of this study show that like most intracellular pathogens that invade macrophages, *C. albicans* is able to modulate cell response to resist to microbial activities and escape from the cell. Although participation of other yeast factors cannot be excluded, PLM, a specific MIPC expressed by *C. albicans*, may support the activities observed with the whole yeast cells.

REFERENCES

1. NAVARRE, W.W. & A. ZYCHLINSKY. 2000. Pathogen-induced apoptosis of macrophages: a common end for different pathogenic strategies. Cell Microbiol. **2:** 265.
2. IBATA-OMBETTA, S., T. JOUAULT, P.A. TRINEL & D. POULAIN. 2001. Role of extracellular signal-regulated protein kinase cascade in macrophage killing of *Candida albicans*. J. Leukoc. Biol. **70:** 149.
3. KORSMEYER, S.J. 1999. BCL-2 gene family and the regulation of programmed cell death. Cancer Res. **59:** 1693s.
4. TRINEL, P.A., E. MAES, J.P. ZANETTA, *et al.* 2002. *Candida albicans* phospholipomannan, A new member of the fungal mannose inositol phosphoceramide family. J. Biol. Chem. **277:** 37260.
5. JOUAULT, T.C. P.A. FRADIN, A. TRINEL, *et al.* 1998. Early signal transduction induced by *Candida albicans* in macrophages through shedding of a glycolipid. J. Infect. Dis. **178:** 792.

Glycoprotein of Nonpathogenic Rabies Viruses Is a Major Inducer of Apoptosis in Human Jurkat T Cells

STÉPHANIE LAY, CHRISTOPHE PRÉHAUD, BERNHARD DIETZSCHOLD,[a] AND MONIQUE LAFON

Unité de NeuroImmunologie Virale, Département de Neuroscience, Institut Pasteur, Paris, France

[a]Center for Neurology, Department of Microbiology and Immunology, Thomas Jefferson University, Philadelphia, Pennsylvania, USA

ABSTRACT: This study sought to identify the RV protein that causes apoptosis. For this purpose, we first compared the ability of G and N proteins of a pathogenic and a nonpathogenic strain to trigger apoptosis of Jurkat rtTA by using an inducible Tet-on expression system. Then we analyzed apoptosis induced by a reverse genetic-engineered recombinant rabies virus in which the G gene from a nonpathogenic strain was replaced by its pathogenic strain counterpart. No other virus proteins than G of nonpathogenic RV strains induce apoptosis, and the G polypeptide of RV is a critical determinant for apoptosis in human cells.

KEYWORDS: apoptosis; rabies virus; Rev-tet-on; Jurkat T cells

The identification of discrete proteins that directly trigger apoptosis is still a field of intense research interest. Besides the basic research knowledge in virus pathogenicity or virulence, the identification of "viral death proteins" may also have practical applications in human therapeutics. The induction of apoptosis by peptides or polypeptides could be valuable tools to eliminate undesirable cells, such as tumor, infected, and auto-immune cells or to trigger the formation of apoptotic bodies, which have been proposed to be powerful activator of immune responses. The rabies virus (RV) is an enveloped bullet-shaped virus belonging to the Rhabdoviridae family, genus Lyssavirus. The viral particle consists of a membrane composed of host lipids and two viral proteins, G and M, surrounding a helical nucleocapsid (NC). The NC is composed of a viral negative strand RNA molecule protected by N protein, P protein, and the RNA-dependent RNA polymerase—the protein L.

We reported previously that the attenuated live RV vaccine strain ERA triggers apoptosis in the human lymphoblastoid Jurkat T cell line involving the mitochondri-

Address for correspondence: Monique Lafon, Institut Pasteur, Unité de NeuroImmunologie Virale. 25, rue du Dr Roux, 75724 Paris cedex 15, France. Voice: +33 (0)1 45 68 87 52; fax: +33 (0)1 40 61 33 12.
mlafon@pasteur.fr

Ann. N.Y. Acad. Sci. 1010: 577–581 (2003). **© 2003 New York Academy of Sciences.**
doi: 10.1196/annals.1299.108

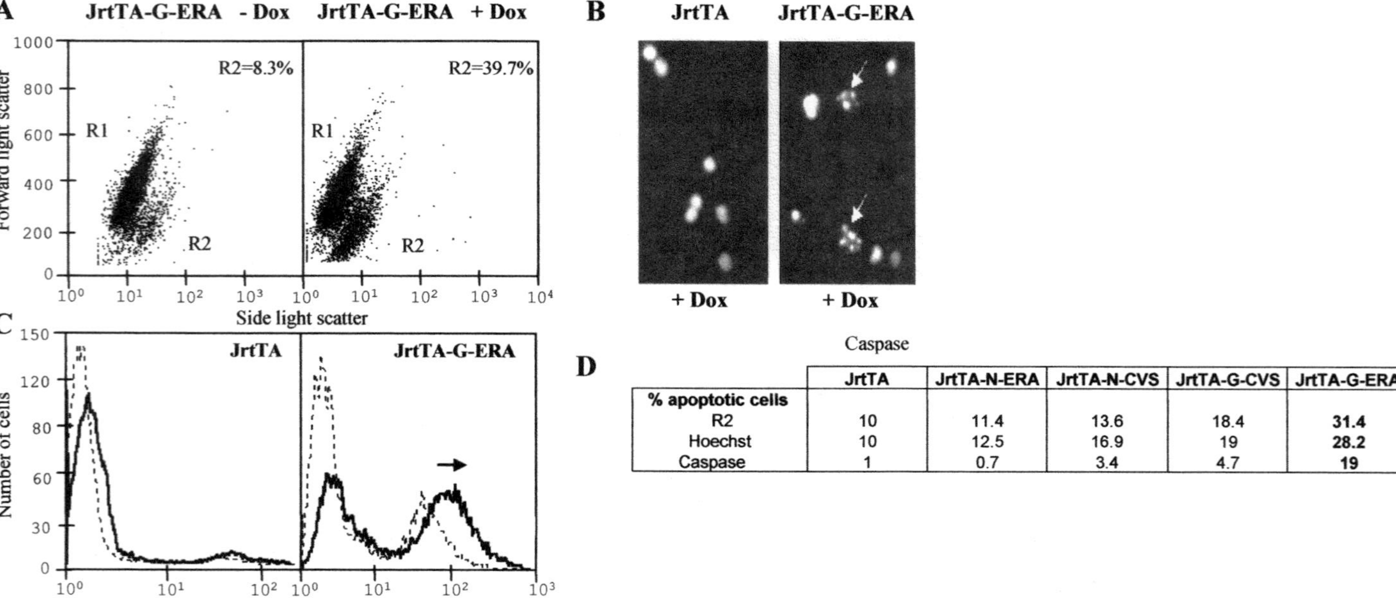

	JrtTA	JrtTA-N-ERA	JrtTA-N-CVS	JrtTA-G-CVS	JrtTA-G-ERA
% apoptotic cells					
R2	10	11.4	13.6	18.4	**31.4**
Hoechst	10	12.5	16.9	19	**28.2**
Caspase	1	0.7	3.4	4.7	**19**

FIGURE 1. Only G protein of ERA RV triggers apoptosis in Jurkat rtTA cells. (**A**) Jurkat rtTA containing the viral G protein gene of RV ERA were treated with Dox (*right*) or non-treated (*left*). Morphological changes, assessed by side and forward light scattering flow cytometry analysis, were used to identify apoptotic Jurkat T cells (R2). (**B**) Typical nuclei fragmentation after Hoechst 33342 (1 μg/mL) staining was observed in Dox-treated Jurkat rtTA-G-ERA (*arrows in the right panel*). (**C**) Caspase activation was compared in Jurkat rtTA (*left*) and in Jurkat rtTA-G-ERA cell lines (*right*) either treated with dox (*bold line*) or not (*dashed line*) using the FITC-conjugated analog of *N*-benzyloxycarbonyl-Leu-Glu-Thr-Asp-fluoromethylketone (FITC-LETD-FMK, Intergen). (**D**) Apoptosis was compared in control cells Jurkat rtTA and in Jurkat rtTA expressing the N protein of ERA or CVS, the G protein of CVS or ERA by determining the size of R2, the percentages of cells with fragmentized nuclei stained by Hoechst 33342 or expressing activated caspase. These different means of apoptosis were obtained simultaneously.

al pathway.[1,2] In contrast, the pathogenic RV strain CVS does not trigger apoptosis. Sequence analyses indicate these two RV strains are different by several amino acid substitutions, especially in both the G and the N protein, suggesting that some of the RV proteins can play a role in the apoptosis triggering. The goal of this study was to identify the RV protein that causes death. For this purpose, we firstly compared the ability of G and N proteins of the pathogenic strain (CVS) and nonpathogenic strain (ERA) to trigger apoptosis of Jurkat rtTA cells by using an inducible Tet-on expression system. In this system, measurements of apoptosis by light scatter analysis (FIG. 1A), nuclei fragmentation (FIG. 1B), and caspase activation indicate that expression of G ERA triggers a caspase-dependent apoptosis, leading to the fragmentation of the cell nucleus.

By using the same means, apoptosis was investigated for the other three transgenic Jurkat T cell lines expressing G-CVS, N-ERA or N-CVS. As illustrated on FIGURE 1D, in every case the G-ERA expression had the highest level of apoptosis, suggesting that G-ERA is a major inducer of apoptosis in human lymphoblastoid cells.

We then addressed the possibility that other proteins of the ERA strain, such as M, L, or P, were pro-apoptotic by using recombinant viruses obtained by reverse genetics. In order to test whether other proteins of the virus play a role in apoptosis, a recombinant virus, R-N2c, was constructed from the parental strain SN10, a virus closely related to ERA, that is non pathogenic. R-N2c results of the replacement of the SN10-G gene by its CVS-N2c-G counterpart (FIG. 2A). The recombinant virus R-N2c transcribes and replicates as efficiently as the SN10 strain as shown by its rate of virus production and its amount of N protein messenger RNAs.[3] As shown in FIGURE 2B, infection of Jurkat rtTA cells by the nonpathogenic strains of virus ERA or SN10 triggers apoptosis (determination percentage of cells of the R2, of cells with fragmentized nuclei with both Hoechst and TUNEL staining, and of cells harboring an activated caspase). In contrast, neither CVS nor the R-N2c can induce apoptosis. Absence of apoptosis in CVS- and R-2Nc-infected cultures was associated with an absence of caspase activation (FIG. 2C). Absence of apoptosis in CVS or R-N2c did not result in the lowest level of infection of the cells since infection range of Jurkat rtTA cells with the four viruses was 70–80% of infection (FIG. 2C).

Altogether, these data indicate that the switch of the G gene from the pro-apoptotic non pathogenic strain is sufficient to abrogate the pro-apoptotic property of this RV strain. It also shows that none of the N, P, M, or L proteins of the non–pathogenic viral strain is an apoptotic inducer. Some viral glycoproteins have been proposed to be inducers of apoptosis. It has been found that induction of apoptosis leads to the ability of some of them to disable receptor signaling. In particular the reovirus sigma1 is able to induce apoptosis upon binding to the cell surface.[4] HIV Gp120 and Gp41 induce apoptosis by interaction with CD4 and CxCR4, whereas Sindbis virus E2 protein targets an early step in the replication process that follows closely after entry. Similarly, the process of entry of vaccinia or HSV-1 can induce apoptosis.[5,6] In the case of RV, a direct interaction of virus particles with membrane is not sufficient to induce apoptosis since cells treated with inactivated, non-infectious virus particles do not encounter apoptosis.[1] This observation together with the delay required by RV infection to trigger apoptosis indicate that it is unlikely that the entry of the virus is a critical factor for apoptosis triggering. RV apoptosis requires that infectious cycle is launched, suggesting that newly produced viral G protein may

A **Parental rabies virus**

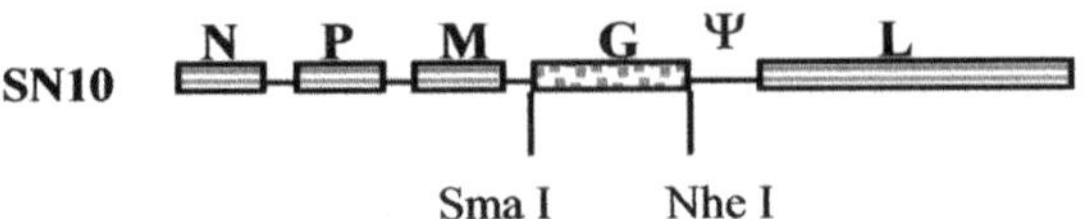

G-recombinant virus

B

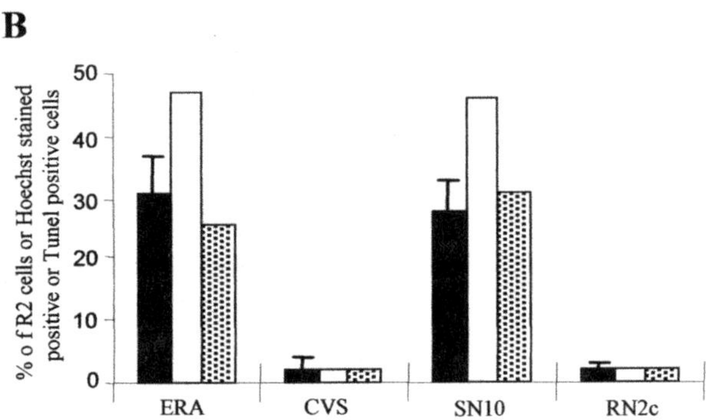

C

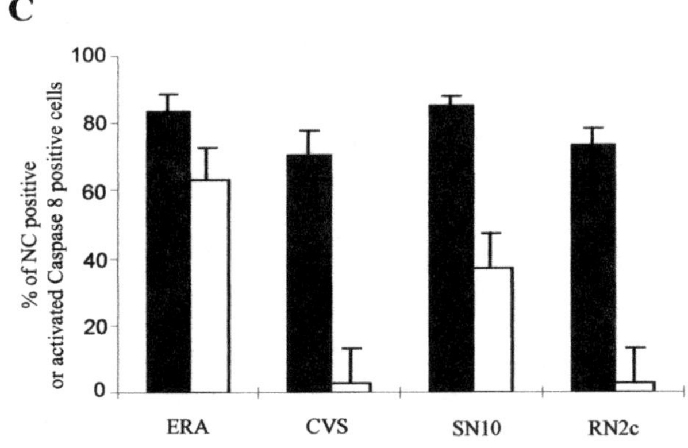

FIGURE 2. No other ERA RV proteins than G-ERA induces apoptosis in Jurkat T cell line. (**A**) Construction of the recombinant virus R-N2c by reverse genetics. The glycoprotein (G) gene of RV SN-10 (*stippled*) is replaced by the G gene of RV CVS-N2c (*diagonal stripes*) resulting in the G- recombinant virus R-N2c. (**B**) and (**C**) Comparison of apoptosis in two-day cultures of Jurkat-rtTA cells infected with ERA, CVS, SN10, or recombinant R-N2C as measured by determining (**B**) the percentage of R2 in the cell population (*black histograms*) of cells with fragmentized nuclei stained by Hoechst 33342 (*white histograms*, 500 cells counted) or by the TUNEL technique (*stippled histograms*, 500 cells counted). This experiment is representative of five others. (**C**) The percentage of cells expressing activated caspase (*white histograms*). Infection of cell culture is shown as black histograms. Results are presented as means ± standard deviation of three different experiments.

play a role in this process. Identification of the molecular mechanisms by which G protein from the ERA RV strain and not the CVS RV strain triggers apoptosis is under investigation.

ACKNOWLEDGMENTS

This work was supported by institutional grants from the Institut Pasteur. S.L. is supported by a Ministry of Research and Technology fellowship. We are indebted to J. Hiscott (McGill University, Canada) and A. Israel (Institut Pasteur, France) for the gift of the jurkat rtTA cell line. We would like to acknowledge the excellent technical assistance of M. Lafage for cell culture and virus production

REFERENCES

1. THOULOUZE, M.I., M. LAFAGE, J.A. MONTANO-HIROSE & M. LAFON. 1997. Rabies virus infects mouse and human lymphocytes and induces apoptosis. J. Virol. **71:** 7372–7380.
2. THOULOUZE, M.I., M. LAFAGE, V.J. YUSTE, *et al.* 2003. High level of Bcl-2 counteracts apoptosis mediated by a live rabies virus vaccine strain and induces long-term infection. Virology **314:** 549–561.
3. MORIMOTO, K., H.D. FOLEY, J.P. MCGETTIGAN, *et al.* 2000. Reinvestigation of the role of the rabies virus glycoprotein in viral pathogenesis using a reverse genetics approach. J. Neurovirol. **6:** 373–381.
4. TYLER, K.L., M.K. SQUIER, S.E. RODGERS, *et al.* 1995. Differences in the capacity of reovirus strains to induce apoptosis are determined by the viral attachment protein sigma 1. J. Virol. **69:** 6972–6979.
5. GALVAN, V. & B. ROIZMAN. 1998. Herpes simplex virus 1 induces and blocks apoptosis at multiple steps during infection and protects cells from exogenous inducers in a cell-type-dependent manner. Proc. Natl. Acad. Sci. USA **95:** 3931–3936.
6. RAMSEY-EWING, A. & B. MOSS. 1998. Apoptosis induced by a post-binding step of vaccinia virus entry into Chinese hamster ovary cells. Virology **242:** 138–149.

Redox-Dependent Apoptosis in Human Endothelial Cells after Adhesion of *Plasmodium falciparum*–Infected Erythrocytes

PACO PINO, IOANNIS VOULDOUKIS, NATHALIE DUGAS,[a] GERALDINE HASSANI-LOPPION, BERNARD DUGAS, AND DOMINIQUE MAZIER

INSERM U511, Immunobiologie Cellulaire et Moléculaire des Infections Parasitaires, CHU Pitié-Salpêtrière Paris VI, 75013 Paris, France

[a]*VigiCell sarl, 94240 Kremlin Bicêtre, France*

ABSTRACT: During *Plasmodium falciparum* infection leading to cerebral malaria, mechanisms such as cytokine generation and cytoadherence of parasitized red blood cells (PRBC) to post-capillary venules are clearly involved. We demonstrated that PRBC adhesion to human lung endothelial cells (HLEC) upregulated TNF-α superfamily genes and genes related to apoptosis and inflammation. Apoptosis was confirmed by standard techniques (annexin-V binding, genomic DNA fragmentation, and caspases activation). This apoptotic process involved the cytoplasmic pathway from a death receptor (DR-6, Fas, TNF-R1) through caspase 8, and the mitochondrial pathway though Bad and caspase 9 activation. Oxidative stress has been implicated in apoptosis induction in various pathological models. Superoxide anion ($O_2•^-$) is a key molecule in the oxidative stress pathway which can form peroxynitrites ($ONOO^-$) in association with nitric oxide (NO•). Even though the role of NO• in malaria physiopathology is still a matter of controversy, we demonstrated that PRBC-induced apoptosis in endothelial cells is mediated through an oxidative stress pathway. The inhibition of NO• synthesis protected the endothelial cells suggesting a deleterious role for NO•. In addition, the superoxide dismutase mimetic, MnTBAP, also protected the HLEC against PRBC-induced apoptosis, revealing the role of $O_2•^-$ and $ONOO^-$.

KEYWORDS: *Plasmodium falciparum*; endothelial cells; oxidative stress; nitric oxide

Cerebral malaria (CM), coupled with anemia, is one of the most tragic complications associated with an infection by *Plasmodium falciparum*. Although the physiopathology has been extensively investigated, the cellular and molecular bases of the neurological pathology are still unclear, particularly concerning the intricacy of the

Address for correspondence: Prof. Dominique Mazier, INSERM U511, Immunobiologie Cellulaire et Moléculaire des Infections Parasitaires, CHU Pitié-Salpêtrière Paris VI, 91 bd de l'Hôpital, 75013 Paris, France. Voice: +33-0-1-40-77-97-36; fax: +33-0-1-45-83-88-58.
mazier@ext.jussieu.fr

**Ann. N.Y. Acad. Sci. 1010: 582–586 (2003). © 2003 New York Academy of Sciences.
doi: 10.1196/annals.1299.109**

different factors described as being involved in the pathogenesis: sequestration of parasitized red blood cells (PRBC) leading to a mechanical blockade of microvessels; secretion of cytokines; modifications of the T cell repertoire; the immune status and genetic background of the host; and parasite factors.[1] Sequestration of PRBC to the surface of the microvasculature of various organs, including the brain and the lungs, is mediated by different endothelial cell surface receptors including thrombospondin (TSP), CD36, intercellular adhesion molecule-1 (ICAM-1), E-selectin, vascular cellular adhesion molecule-1 (VCAM-1), CD31, $\alpha_V\beta_3$ integrin, and hyaluronic acid.[1] A critical role for nitric oxide (NO•) has been proposed to explain malaria pathophysiology in humans, but it is still a matter of controversy. In some studies, increased NO• derivatives were found associated with coma and poor outcome, while other reports showed no correlation with disease progression. In contrast, other reports suggested a protective role for NO•.[2] We recently demonstrated that PRBC adhesion modulated human primary lung endothelial cell (HLEC) gene expression. TNF-α superfamily genes and genes related to apoptosis and inflammation (*Bad, Bax, Caspase-3, Fas, SARP 2, DFF45/ICAD, IFN-γ Receptor 2, Bcl-w, Bik,* and *iNOS*) appeared to be upregulated by PRBC adhesion. The induction of apoptosis was confirmed on the basis of morphology, with electron microscopy, annexin-binding, DNA fragmentation, and caspases activation. The apoptotic stimulus was a physical contact between HLEC and PRBC, and not parasite-secreted molecules. This effect was dependent on the time of co-culture, parasitemia, and hematocrit.

We used an ELISA assay quantification of the internucleosomal DNA cleavage for pharmacological evaluation of various drugs. As caspases are known to be the effectors of the terminal stage of apoptosis, we evaluated the pharmacological effect of different caspase inhibitors, Z-VAD-fmk (general caspase inhibitor), Z-IETD-fmk (caspase 8 inhibitor), and Z-LEHD-fmk (caspase 9 inhibitor). The cytoplasmic pathway from a death receptor (DR-6, Fas, TNF-R1) through caspase 8, and the mitochondrial pathway through caspase 9 activation were involved in this apoptotic process.[3]

Molecular oxygen is an important environmental and developmental signal that regulates cellular energetics, growth, and differentiation. However, oxygen plays various roles: while oxygen is indispensable for the cell to obtain the essential chemical energy as a form of ATP, it is often transformed into highly reactive forms, radical oxygen species (ROS), which are often toxic for the cell. Peroxynitrite ($ONOO^-$) is a potent oxidant formed from the non enzymatic reaction between superoxide anion ($O_2{\bullet}^-$) and nitric oxide (NO•) to form $ONOO^-$. Peroxynitrite can oxidize lipids, proteins and nucleic acids. It is then sensible to hypothesize that endothelial dysfunction may coincide with an enhanced NO• synthase expression and O_2.[4] We investigated the role of oxidative stress in this apoptosis induction.

HLEC were used for these experiments as previously described.[3] As shown in FIGURE 1 (A), a pretreatment of the HLEC with the L-arginine analogue L-NMMA, reduced the enrichment in cytoplasmic nucleosomes from 2.9±0.4 to 1.0±0.06 at 1.5 mM, concentrations that completely inhibited apoptosis in this model. The use of L-NMMA inhibited the formation of NO• and protected the endothelial cells against PRBC-induced apoptosis. We can hypotheses that the inhibition of NO• formation reduced peroxynitrite synthesis and then protected the cells against oxidative damages.

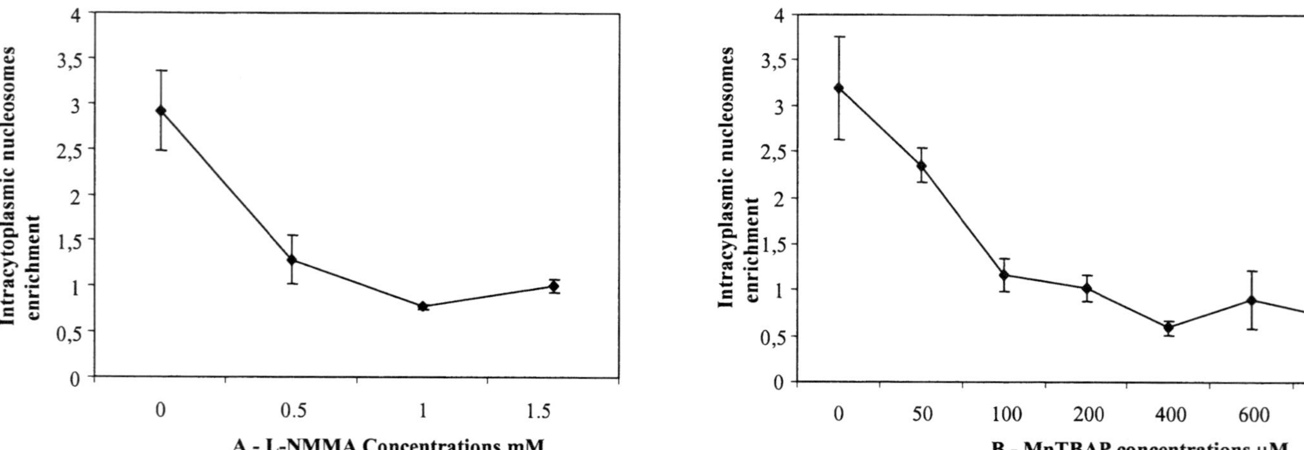

FIGURE 1. Inhibition of PRBC-induced apoptosis by antioxidant molecules. HLEC were pretreated with L-NMMA, for 2 h, before the addition of PRBC at a parasitemia of 50%, and a hematocrit of 5%. MnTBAP, which can diffuse freely in the cells, was added to the HLEC at the same time at PRBC. After 6-h co-culture, apoptotic cells were counted using the Cell Death Detection ELISA (Roche). In these representations, HLEC co-cultivated with PRBC are compared with HLEC co-cultivated with control RBC. The resulting apoptosis ratios are plotted against drugs concentrations.

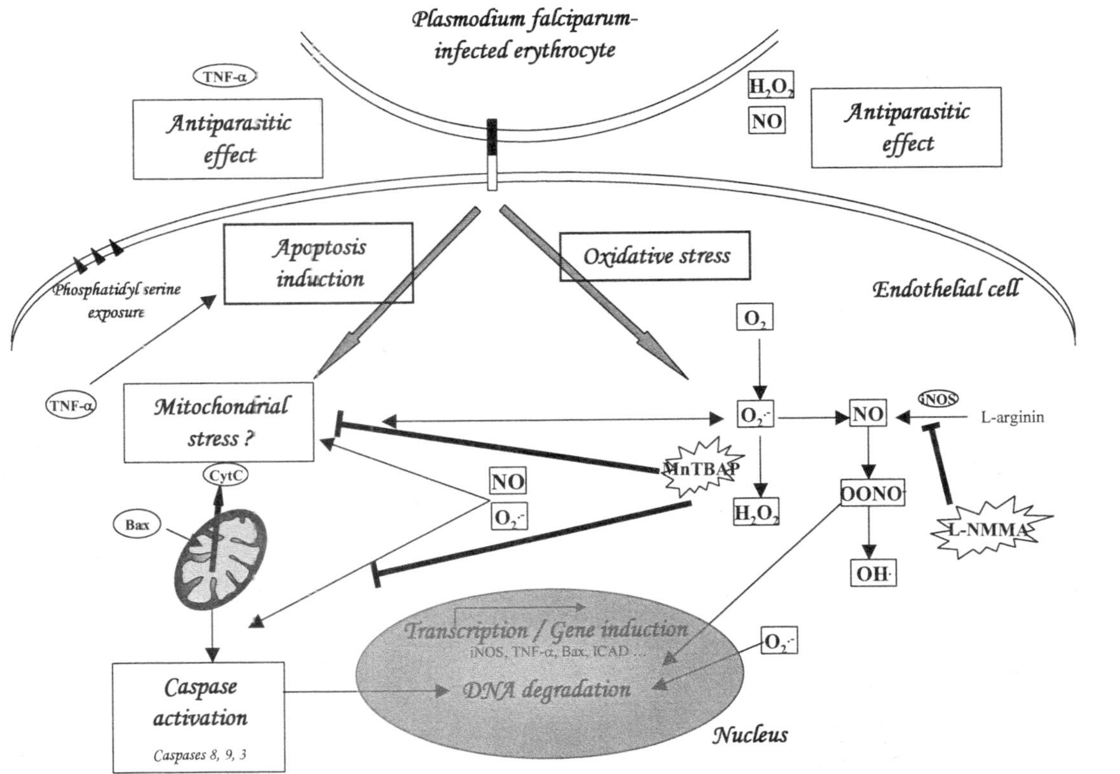

FIGURE 2. Schematic representation of consequences of PRBC adhesion on the endothelial cell. PRBC adhesion induced an oxidative stress and apoptosis in endothelial cells. Possible roles for NO• and $O_2{\bullet}^-$ are represented. Potential actions of antioxidant molecules SOD and L-NMMA are included.

In the same way, Cu/Zn-superoxide dismutase (SOD), an antioxidative enzyme that dismutes superoxide anion, was also tested, but no significant protection of HLEC was observed (data not shown). Free SOD in the culture medium is not sufficient to inhibit significantly apoptosis in this model. The important molecular weight of SOD and its poor diffusion ability can explain this weak activity. As described in FIGURE 1 (B), the use of the chemical-permeable SOD mimetic MnTBAP resulted in good protection of HLEC against PRBC-induced apoptosis, suggesting that the apoptotic process was mediated by O_2•$^-$ and NO•. In addition, MnTBAP also exhibits a peroxidase activity that detoxifies peroxynitrites resulting from reaction of NO• and O_2•$^-$.

In conclusion, *Plasmodium falciparum*–infected erythrocyte adhesion induces apoptosis in primary human endothelial cells. This apoptotic process involves caspases 8, 9, and terminal caspase 3; O_2•$^-$/NO•/ONOO$^-$ pathway mediated the endothelial cells death through an oxidative stress. FIGURE 2 is a schematic representation of the consequences of PRBC adhesion on endothelial cells: (1) an induction of apoptosis where caspases are terminal effectors and (2) an oxidative stress mediated by O_2•$^-$/NO•/ONOO$^-$. The protective role of L-NMMA and MnTBAP, by reducing the production or detoxifying the ROS are represented. Even though fine mechanisms and signal pathways involved in this process remain to be further defined, these data describe a new mechanism involved in the physiopathology of cerebral malaria and potential use of antioxidant molecules as complement therapy for the treatment of malaria.

REFERENCES

1. MAZIER, D., J. NITCHEU & M. IDRISSA-BOUBOU. 2000. Cerebral malaria and immunogenetics. Parasite Immunol. **22:** 613–623.
2. HOBBS, M.R. *et al.* 2002. A new NOS2 promoter polymorphism associated with increased nitric oxide production and protection from severe malaria in Tanzanian and Kenyan children. Lancet **360:** 1468–1475.
3. PINO, P., I. VOULDOUKIS, J.P. KOLB, *et al.* 2003. *Plasmodium falciparum*–infected erythrocyte adhesion induces caspase activation and apoptosis in human endothelial cells. J. Infect. Dis. **187:** 1283–1290.
4. CONNER, E.M. & M.B. GRISHAM. 1996. Inflammation, free radicals, and antioxidants. Nutrition **12:** 274–277.

Apoptosis in Cultured Cells Infected with Feline Calicivirus

LISA O. ROBERTS, NAEMA AL-MOLAWI, MICHAEL J. CARTER, AND GEORGE E. N. KASS

School of Biomedical and Life Sciences, University of Surrey, Guildford, GU2 7XH, United Kingdom

ABSTRACT: Caliciviruses are important pathogens of man and animals; feline calicivirus (FCV) is responsible for an acute upper respiratory tract disease in cats. To date, little is known about the mechanism of cell damage induced by these viruses. We set out to determine if apoptosis played any role in cell death in FCV infection of cultured cells. We demonstrate that caspase-2, -3, and -7 were activated during FCV infection, as evidenced by pro-form processing and an increase in acetyl-Asp-Glu-Val-Asp-7-amido-4-trifluoromethylcoumarin cleavage activity, as well as cleavage of poly(ADP-ribose)polymerase. Caspase activation coincided with the condensation of chromatin. At about 8 h post infection we also detected cleavage of the FCV capsid protein; this was prevented by caspase inhibitors. Taken together these results suggest that FCV triggers apoptosis within infected cells and that caspases are involved in the cleavage of the capsid protein.

KEYWORDS: feline calicivirus; caspase activation; capsid protein

INTRODUCTION

The Calicivirus family includes many viruses that are pathogenic for man and animals. Feline calicivirus (FCV) causes an upper respiratory tract disease in cats. The FCV genome is a single strand of positive sense RNA of about 7.5 kb and contains three open reading frames (ORFs). ORF1 encodes the virus non-structural proteins, ORF2 encodes the capsid protein, and ORF3 a minor structural protein.[1] Apoptosis is a process of cell death used by organisms to eliminate superfluous, cancerous, or virus-infected cells.[2] Viruses can either promote or inhibit this process, or even do both, in infected cells. Apoptosis progresses through a series of morphological and biochemical changes, such as chromatin condensation, DNA fragmentation, plasma membrane blebbing and cell shrinkage. Most of these changes are effected by caspases, a family of cysteine proteases. Initiator caspases activate a caspase cascade that culminates in the activation of downstream effector caspases. These in turn cleave a number of cellular target proteins and this is responsible for the destruction

Address for correspondence: George E.N. Kass, School of Biomedical and Life Sciences, University of Surrey, Guildford, GU2 7XH, United Kingdom. Voice: +44-1483-686449; fax: +44-1483-300374.
kass@surrey.ac.uk

Ann. N.Y. Acad. Sci. 1010: 587–590 (2003). © 2003 New York Academy of Sciences.
doi: 10.1196/annals.1299.110

of the cell. To date there have been few reports on the effects of caliciviruses on cells. Apoptosis has been demonstrated in rat liver following infection with rabbit hemorrhagic disease virus (RHDV).[3] Here we set out to look at potential apoptotic changes within FCV-infected cells. We report that activation of caspases is responsible for cellular apoptosis and that caspases also play a role in cleavage of the virus capsid protein.

MATERIAL AND METHODS

Crandell Reese Feline Kidney (CRFK) cells were infected with FCV and at various time points post infection (pi) were stained with Hoechst 33358 and visualized by fluorescence microscopy. Apoptotic proteins from infected cells were assayed by Western blot analysis. Infections were also carried out in the presence of caspase inhibitors (z-VAD-FMK and Ac-DEVD-CHO) to study the role of caspases in FCV capsid cleavage.

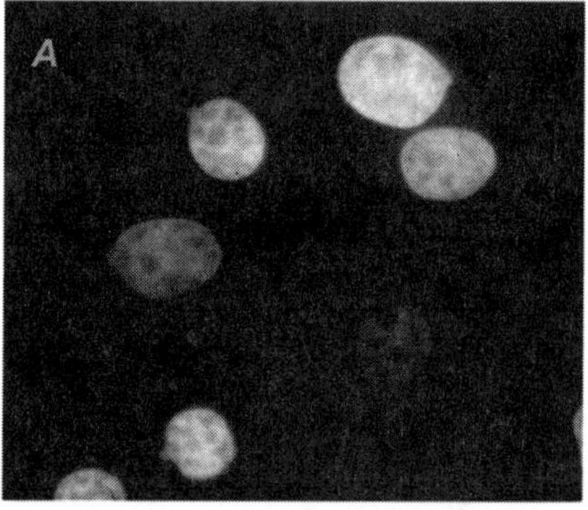

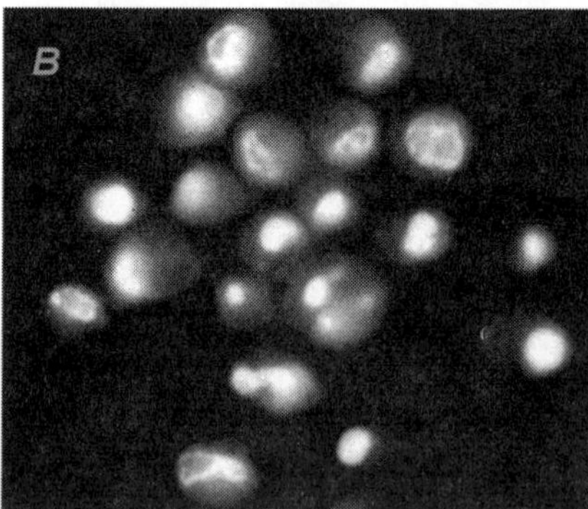

FIGURE 1. Induction of apoptosis during FCV infection. CRFK cells were mock-infected (**A**) or infected with FCV F9 (**B**) for 9 h, fixed with 4% formaldehyde, and stained with Hoechst 33358. The nuclei were visualized by fluorescence microscopy and photographed. The induction of apoptosis is evidenced by the appearance of highly condensed chromatin and fragmented nuclei.

RESULTS

FCV Infection Triggers Chromatin Condensation in Infected CRFK Cells

Nuclear integrity in infected cells was examined by staining cells with Hoechst 33358 which penetrates nuclei and binds to DNA. Condensation of chromatin, typical of apoptosis, was observed in FCV-infected cells by about 9 h pi (FIG. 1). Similar changes were seen in cells treated with the apoptosis-inducing agent staurosporine (1 μM; data not shown).

Caspase Activation in FCV-Infected Cells

We examined the activation of a number of caspases by Western blot analysis on infected cell extracts. Activation of caspase-3 was observed from about 4 h pi (FIG. 2) as evidenced by the appearance of the p20 and p17 fragments, which correspond to the large subunit of the pro-caspase. We also observed activation of caspase-7 and caspase-2 (data not shown). The activation of caspase-3 and -7 was confirmed by the increase in caspase-3-like activity measured as DEVDase activity (data not shown).

The activation of caspases in FCV infection was further confirmed by analysis of poly(ADP-ribose) polymerase (PARP) cleavage. This protein is normally involved in DNA repair but during apoptosis it is cleaved from a p116 protein into a p85 inactive form. Cleavage of PARP was detected in FCV-infected cells from about 4 h pi (FIG. 2).

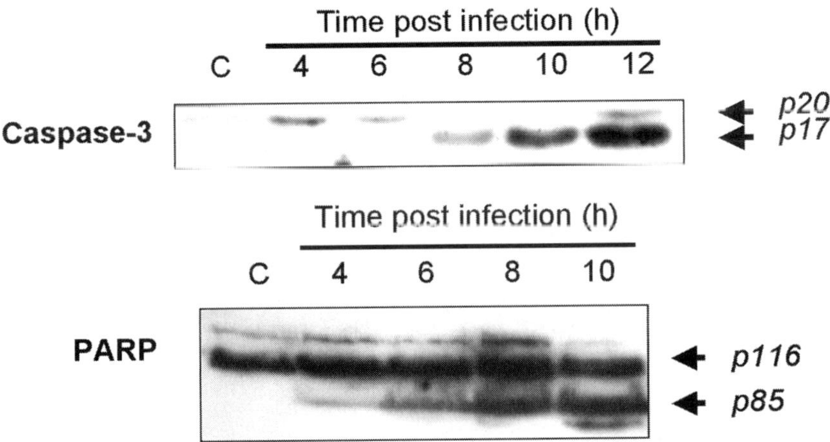

FIGURE 2. Caspase activation and PARP cleavage in FCV-infected CRFK cells. CRFK cells were infected with FCV for the times indicated or treated with 1 μM staurosporine (St) for 20 h or mock-infected. Cell extracts were subjected to SDS-PAGE (10%) and Western blotting as described under Materials and Methods. The 32 kDa pro-form (p32) of caspase-3 was processed to yield p20/p17 fragments that correspond to the (active) large subunit. Cleavage of PARP from the 116 kDa native protein to the 85 kDa fragment is shown.

DISCUSSION

FCV induces apoptosis in cultured cells and this may be a common feature of calicivirus infections as RHDV infection has also been shown to induce apoptotic changes.[3] In addition to the activation of caspases and chromatin condensation in FCV-infected cells, we have also observed cleavage of the viral capsid protein from a p62 form to a p40 form.[4] Furthermore, we have demonstrated that the capsid protein is cleaved by caspases. The impact of this capsid cleavage in unclear—is it a host cell defense mechanism to interfere with virus assembly or is the cleaved protein involved in virus pathogenesis?

REFERENCES

1. Clarke, I.N. & P.R. Lambden. 1997. The molecular biology of caliciviruses. J. Gen. Virol. **78:** 291–301.
2. Kaufmann, S.H. & M.O. Hengartner. 2001. Programmed cell death: alive and well in the new millennium. Trends Cell Biol. **11:** 526–534.
3. Alonso, C., J.M. Oviedo, J.M. Martin-Alonso, *et al.* 1999. Programmed cell death in the pathogenesis of rabbit hemorrhagic disease. Arch. Virol. **143:** 321–332.
4. Al-Molawi, N., V.A. Beardmore, M.J. Carter, *et al.* 2003. Caspase-mediated cleavage of the feline calicivirus capsid protein. J. Gen. Virol. **84:** 1237–1244.

Inhibition of Apoptosis by Human T-Lymphotropic Virus Type-1 Tax Protein

DANIELA SAGGIORO,[a,b] LIDIA ACQUASALIENTE,[a] LAURA DAPRAI,[a] AND LUIGI CHIECO-BIANCHI[a]

[a]*Department of Oncology and Surgical Sciences, Oncology Section, University of Padova, 35128 Padova, Italy*

[b]*IST, Viral and Molecular Oncology Section, University of Padova, 35128 Padova, Italy*

ABSTRACT: Deregulation of the apoptotic process can lead to pathophysiological changes that result in either degenerative diseases or cancer. Although the transactivator Tax has been established as an essential effector of human T-lymphotropic virus type–1 (HTLV-1)–mediated diseases, which include both a neurodegenerative pathology and leukemia/lymphoma, its exact role(s) in the pathogenesis remains to be clarified. Because the apoptotic potential of Tax is still being debated, we addressed this question by testing the susceptibility of Tax(–) and Tax(+) cells to apoptosis under conditions of growth factor withdrawal and Bax overexpression. Results showed that Tax(+) cells are protected from apoptosis triggered by depletion of growth factors. This protective effect seems to be the result of a block in the apoptotic program regulated by mitochondria. Furthermore, in an attempt to elucidate which transcriptional pathway activated by Tax is important in the observed Tax-induced resistance, we found that CREB/ATF activity plays a relevant role in protecting cells from apoptosis induced by Bax overexpression. All together, these data might suggest that the ability of Tax to inhibit certain apoptotic stimuli could be important in its role as a viral transforming protein.

KEYWORDS: HTLV-1; Tax; apoptosis; cytochrome c; serum starvation; Bax; CREB.

INTRODUCTION

Human T-lymphotropic virus type 1 (HTLV-1) is the etiological agent of leukemia/lymphoma (ATLL) and the inflammatory neurodegenerative disorder called tropical spastic paraparesis/HTLV-1–associated myelopathy (TSP/HAM). HTLV-1 encodes a 40-kDa phosphoprotein called Tax that has been shown to play an important role in viral pathogenesis. Tax acts as a transcriptional modulator by altering the expression of selected cellular genes, many of which are involved in cell cycle regulation. Indeed, Tax can interfere with the expression of genes responsive to NF-κB, serum response factor (SRF), and CREB/ATF (cAMP-responsive ele-

Address for correspondence: Daniela Saggioro, Department of Oncology and Surgical Sciences-Oncology Section, Via Gattamelata 64, 35128 Padova. Voice: +39-049-8215884; fax: +39-049-8072854.
d.saggioro@unipd.it

Ann. N.Y. Acad. Sci. 1010: 591–597 (2003). © 2003 New York Academy of Sciences.
doi: 10.1196/annals.1299.111

ment-binding protein/activating transcription factor). It has been suggested that it is through this ability to modulate cellular genes expression that Tax mediates HTLV-1 transformation.[1,2] Because cell cycle regulation and programmed cell death are closely related, Tax represents an attractive candidate for the deregulation of the normal apoptotic pathway. To date, contradictory data have been obtained concerning the apoptotic activity of Tax as the protein has been found to either induce or inhibit apoptotic cell death.[3,4] The discordant reports concerning the influence of Tax on apoptosis might reflect differences in cell type or methods used to induce apoptosis. In this study, we found that Tax(+) murine fibroblasts are resistant to serum deprivation–induced apoptosis, and that this resistance is the result of a block of the cytochrome c release from the mitochondrial intermembrane space. Furthermore, analysis of the role played in apoptosis resistance by the diverse transcriptional domains of Tax revealed that a functional CREB/ATF domain is necessary to protect cells from apoptosis.

MATERIALS AND METHODS

Cell Lines and Transfection

BC cell line was established from a C57BL/6 mouse. BC and 293 pools selected for Tax expression were established by transfecting cells with vectors expressing Tax-wt or the Tax mutants M22 and M47, together with a plasmid expressing the neomycin-resistance gene (pexo-neo) and then selecting for G418 resistance. Control BC and 293 cells were transfected with the neomycin-resistance gene together with a Tax-unrelated plasmid and selected as described above. Resistant cells were pooled and expanded. All the cells were cultured in Dulbecco's modified Eagle's medium with 10% fetal bovine serum (FBS) and 2 mM glutamine.

Western Blotting Analysis

After transfection with different amounts of the Bax-expressing vector, total proteins of 293-BS, 293-M22, and 293-M47 cells were extracted in Laemli buffer; a cocktail of protease inhibitors (Roche) was included in every extraction. Equal volumes of cell lysates, corresponding to 4×10^4 cells were subjected to SDS-PAGE in 10% gels. The proteins were transferred to PVDF membrane and probed with an anti-PARP–specific antibody (Cell Signaling Technology), or with a polyclonal anti-Tax antibody (AIDS Research Reagents Collection).

RESULTS

Susceptibility to Apoptosis Induced by Serum Deprivation in Tax(–) Murine BC Fibroblast and in Its Tax(+) Derivatives

Tax(–) BC fibroblasts, originally derived from a tail biopsy of a C57BL/6 mouse, and some Tax(+) nonclonal derivatives (BC-T01, BC-T03, and BC-T11) obtained by transfection with the pLcXL Tax-expressing vector were analyzed for their susceptibility to apoptosis when grown in 0.1% FBS. To minimize the variability due to in

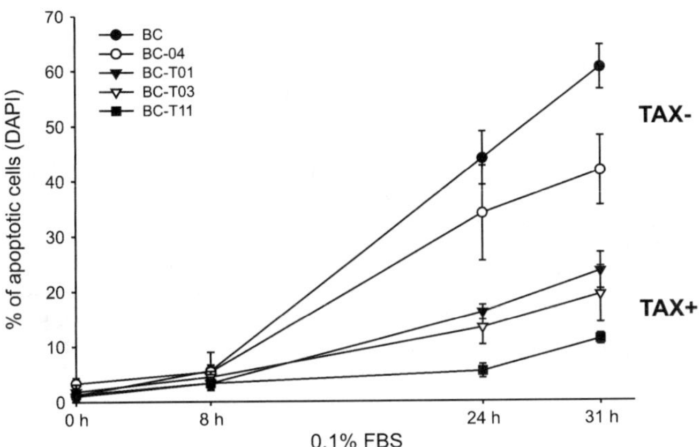

FIGURE 1. Percentage of Tax(+) and Tax(–) apoptotic cells after serum deprivation. Exponentially growing cells were cultured in 0.1% FBS for different time periods and then fixed and stained with DAPI. Nuclei showing an apoptotic morphology (chromatin condensation) were counted in randomly selected fields, and the percentage of apoptotic cells was calculated as the ratio between the number of apoptotic nuclei and the total number of nuclei counted.

vitro selection of transfected cells, the BC line was also transfected with a Tax-unrelated vector and a nonclonal cell population (BC-04) was included in these studies. In comparing the sensitivity of Tax(–) and Tax(+) cells to apoptosis induced by serum deprivation, we found that incubation of the cells in 0.1% FBS led to a rapid decrease in the viability of the Tax(–) cells. Morphological analysis showed that the Tax(–) BC and BC-04 cells exhibited an altered morphology resembling that reported to be associated with cell death; that is, the cells rounded up and detached from the tissue culture plate. These morphological alterations were absent or less evident in Tax(+) BC-T01, BC-T03, and BC-T11 cells, which in contrast maintained a normal morphology and viability after prolonged starvation. To confirm that the cell death induced by serum deprivation was caused by apoptosis, we stained the nuclei with DAPI to visualize chromatin condensation. We found that a consistent number of BC and BC-04 cells exhibited condensed chromatin within 24 h after culture in 0.1% FBS, with the number of apoptotic cells increasing with the length of treatment (FIG. 1). In contrast, very few Tax(+) cells showed an altered chromatin structure even after more than 30 h of serum deprivation (FIG. 1). Although these data might underestimate Tax(–) cell death due to cell loss from the substrata during prolonged treatment, they provide firm evidence that Tax(+) cells are more resistant than Tax(–) cells to apoptosis induced by growth factor withdrawal.

Block of Cytochrome c Release in Tax(+) Cells Grown in 0.1% FBS

Numerous proapoptotic signals and stress pathways converge on mitochondria, that therefore play a pivotal role in apoptosis. In particular, these cellular organelles coordinate caspase activation through the release of proapoptotic factors into the cytoplasm. One well-known mitochondrial proapoptotic factor is cytochrome c, whose

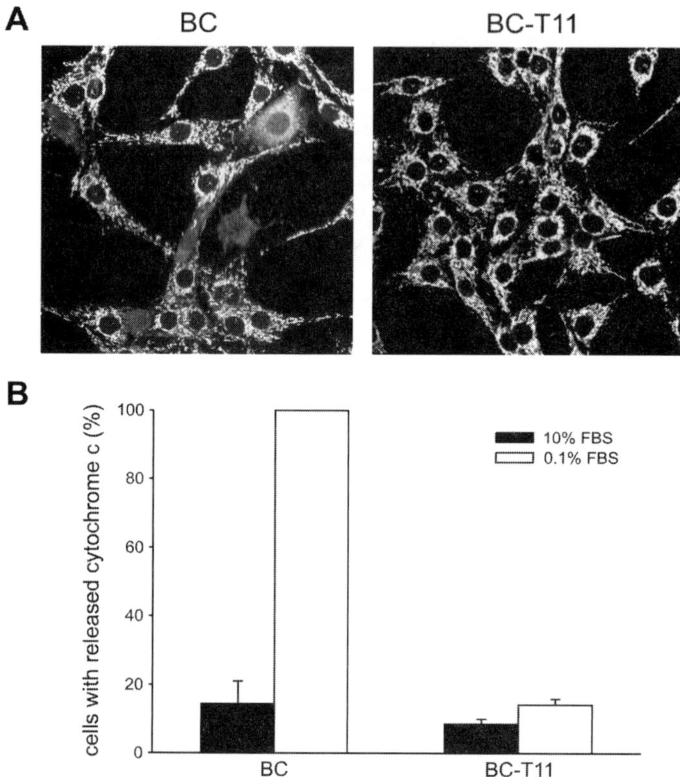

FIGURE 2. Immunofluorescence analysis of cytochrome c distribution during serum depletion-induced apoptosis. Tax(–) BC and Tax(+) BC-T11 cells were grown in 0.1% FBS for 8 h; cells grown in 10% FBS were included as a control. (**A**) After fixation and permeabilization, the cells were stained with anti–cytochrome c monoclonal antibody. (**B**) Bars represent the fraction of cells, treated as described in (**A**), with released cytochrome c calculated by counting at least 500 cells for each sample and considering 100% the release observed in the BC cells grown in 0.1% FBS.

release is considered a key initial step in the apoptotic process, although the precise mechanisms regulating this event remain elusive.[5]

We analyzed the subcellular distribution of cytochrome c in Tax(+) BC-T11 and Tax(–) BC cells grown in 0.1% FBS. Because release of cytochrome c from mitochondria is an early event in apoptosis that primarily involves the mitochondrial pathway, the cells were grown in the presence of low serum for 7–8 h. Then, the cells were fixed and analyzed by indirect immunofluorescence using a monoclonal anti–cytochrome c antibody. The fraction of Tax(+) cells showing cytochrome c release was determined by counting as positive the cells exhibiting a diffuse cytosolic cytochrome (see FIG. 2A) in randomly selected fields. We found only a few Tax(+) cells with released cytochrome c, whereas numerous Tax(–) cells, grown for the same period of time in 0.1 % FBS, exhibited cytocrome c release (FIG. 2B). These results suggest that Tax-expression inhibits apoptosis induced by serum deprivation by blocking cytochrome c release and thus interfering with the activation of a mitochondrial apoptotic pathway.

Analysis of the Functional Domain of Tax Responsible for the Apoptosis Resistance

Tax has been shown to activate promoters containing cAMP response elements (CREs), NF-κB response elements, or serum response elements (SREs) through interaction with CRE-binding protein (CREB/ATF), NF-κB and serum response factor (SRF), respectively. A series of Tax point mutations have been previously characterized for their ability to activate transcription.[6] We used two of these mutants, the M22 (which is unable to activate NF-κB pathway but is still CREB/ATF competent) and the M47 (which instead is CREB/ATF defective but NF-κB competent) to examine which of these two *trans*-activation pathways is important for the observed apoptosis protection. Evidence implicates the Bax protein as an essential regulator of apoptosis induced by growth factor withdrawal; furthermore, in the absence of any apoptotic stimulus its overexpression is sufficient to induce cytochrome *c* release.[7]

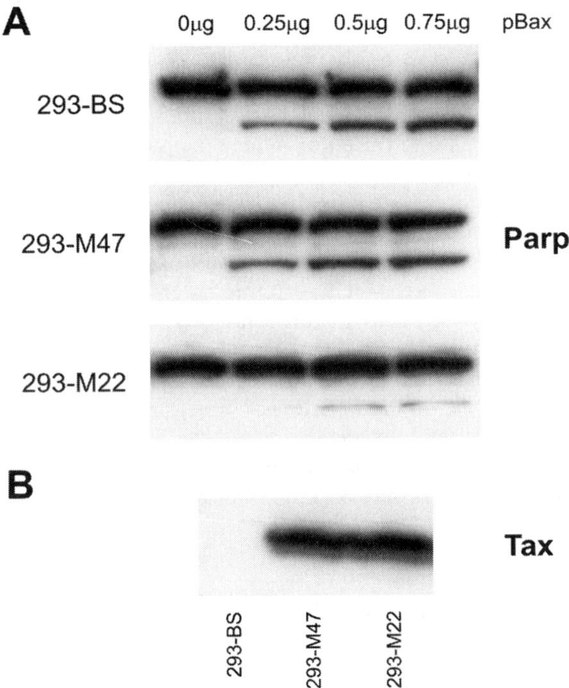

FIGURE 3. Bax-induced apoptosis in 293 cells expressing Tax mutants. Cells were transiently transfected with different amounts of a Bax-expressing plasmid. After 18 h, total proteins were extracted and aliquots, corresponding to total proteins of 6×10^4 cells, were subjected to Western blot analysis to detect PARP cleavage (**A**). Tax expression was also monitored, as control (**B**).

Thus, we investigated whether the M22 and M47 Tax mutants could counteract the apoptotic activity of Bax overexpression in 293 cells. To this end, we generated by transfection nonclonal populations of 293 cells, that either express M22 or M47 mutant (293-M22 and 293-M47, respectively); nonclonal 293 cells transfected with a Tax-unrelated plasmid (293-BS) were also generated, as control. The cells were transiently transfected with different amounts of a Bax-expressing vector and, after 18 h, the apoptosis was detected by Western blot, as PARP cleavage. PARP is a DNA-binding protein that is a specific substrate for caspase during the apoptotic process. We found that 293-M47 cells are susceptible to apoptosis induced by Bax overexpression to the same extent as the control 293-BS; in contrast, the 293-M22 cells exhibited a significantly reduced PARP cleavage (FIG. 3). These data seem to indicate that although activation of NF-κB pathway is irrelevant, the CREB/ATF activity of Tax plays a role in apoptosis protection.

DISCUSSION

Although it is known that Tax is a key regulator of HTLV-1–mediated transformation, the mechanism by which it transforms cells is not well understood. Because deregulation of the apoptotic program is thought to play an important role in transformation, the studies reported here were aimed at an understanding of Tax function in apoptosis. Several groups with different results have performed studies on the effects of Tax on cell death; indeed, it has been reported that Tax can favor or inhibit apoptosis. These seemingly contradictory data could be caused by the utilization of different cell systems or stimuli used to induce apoptosis. However, these results might also suggest that Tax affects the apoptotic process by multiple mechanisms. In this study, we found that Tax expression protects cells from apoptosis triggered by growth factor withdrawal. In an attempt to understand the involved mechanism, we found that this protection seems to be the result of a block of the mitochondrial apoptotic pathway. In fact, Tax(+) cells, after being grown in 0.1% FBS, do not undergo release of cytochrome *c* from the mitochondrial intermembrane space, a critical phenomenon to the mitochondrial apoptotic pathway. Furthermore, experiments with Tax mutants defective for only one of the transcriptional activities, although not directly addressing the role of Tax CREB/ATF function, show that the CREB/ATF-activating domain is important for preventing apoptosis triggered by overexpression of Bax. All together, these data allow for the hypothesis that the ability of Tax to inhibit certain apoptotic stimuli may be important in its role as a viral transforming protein.

ACKNOWLEDGMENTS

We thank Colette Case for assistance in preparing the manuscript and Pierantonio Gallo for the artwork. This work was supported in part by grants from the Italian Ministry of University and Research (MIUR), 40%, the Istituto Superiore di Sanità (Progetto AIDS), the Associazione Italiana per la Ricerca sul Cancro (AIRC), and the Fondazione Italiana per la Ricerca sul Cancro (FIRC). L.A. was supported by a fellowship from FIRC.

REFERENCES

1. YOSHIDA, M. 2001. Multiple viral strategies of HTLV-1 for dysregulation of cell growth control. Annu. Rev. Immunol. **19:** 475–496.
2. JOHNSON, J.M., R. HARROD & G. FRANCHINI. 2001. Molecular biology and pathogenesis of the human T-cell leukemia/lymphotropic virus type-1 (HTLV-1). J. Exp. Pathol. **82:** 135–147.
3. KAO, S.-Y., F.J. LEMOINE & S.J. MARRIOTT. 2000. HTLV-1 Tax protein sensitizes cells to apoptotic cell death induced by DNA damaging agents. Oncogene **19:** 2240–2248.
4. MORI, N., Y. YAMADA, S. IKEDA, *et al.* 2002. Bay 11-7082 inhibits transcription factor NF-κB and induces apoptosis of HTLV-I-infected T-cell lines and primary adult T-cell leukemia cells. Blood **100:** 1828–1834.
5. WANG, X. 2001. The expanding role of mitochondria in apoptosis. Genes Dev. **15:** 2922–2933.
6. SMITH, M.R. & W.C. GREEN. 1990. Identification of HTLV-I Tax trans-activator mutants exhibiting novel transcriptional phenotypes. Genes Dev. **4:** 1875–1885.
7. ESKES, R., B. ANTONSSON, A. OSEN-SAND, *et al.* 1998. Bax-induced cytochrome c release from mitochondria is independent of the permeability transition pore but highly dependent on Mg^{2+} ions. J. Cell Biol. **143:** 217–224.

Apoptosis Inversely Correlates with Rabies Virus Neurotropism

MARIA-ISABEL THOULOUZE,[a,e] MIREILLE LAFAGE,[a] VICTOR J. YUSTE,[b] GUIDO KROEMER,[c] SANTOS A. SUSIN,[b] NICOLE ISRAEL,[d] AND MONIQUE LAFON[a]

[a]*NeuroImmunologie Virale, Institut Pasteur, Paris 75724, France*

[b]*Apoptose et Système Immunitaire, Institut Pasteur, Paris 75724, France*

[c]*CNRS-UMR1599, Institut Gustave Roussy, Villejuif, France*

[d]*Biologie des Rétrovirus, Institut Pasteur, Paris 75724, France*

ABSTRACT: We report that non-neurotropic rabies virus (RV) strains, currently used to immunize wildlife against rabies, induces not only a caspase-dependent apoptosis in the human lymphoblastoid Jurkat T cell line (Jurkat-vect), but also a caspase-independent pathway. Cell redistribution of the apoptosis-inducing factor (AIF) was observed in Jurkat-vect infected with RV vaccine strain. Bcl-2 overproduction in Jurkat T cells (Jurkat-Bcl-2) abolished both caspase activation and AIF distribution. In contrast, strain of neurotropic RV did not induce apoptosis. The inverse correlation of the induction of apoptosis and the capacity of a virus strain to invade the brain suggests that blockage of apoptosis could be a strategy selected by neurotropic virus to favor its progression through the nervous system.

KEYWORDS: apoptosis; caspase; AIF; Bcl-2; rabies virus; Jurkat T cells; live virus vaccine

Viruses are useful tools to study the molecular pathways involved in apoptosis. Most types of virus-induced apoptosis follow caspase-dependent pathways. However, alternate pathways have been described, involving either noncaspase proteases such as cathepsins or calpain[1] or the mitochondrial apoptosis-inducing factor protein (AIF).[2] The AIF is a flavoprotein oxidoreductase present in the mitochondrial intermembrane space. When cells are primed to undergo apoptosis, AIF is released from mitochondria into the cytosol and then is translocated to the nucleus, where it induces partial chromatin condensation and large-scale DNA fragmentation into units of approximately 50 kb. Caspase inhibitors such as z-VAD do not prevent the translocation of AIF to the nucleus.[2] Apoptosis is controlled by a family of genes related to

Address for correspondence: Monique Lafon, Institut Pasteur, Unité de NeuroImmunologie Virale, 25, rue du Dr Roux, 75724 Paris cedex 15, France. Voice: +33-(0)1-45-68-87-52; fax: +33-(0)1-40-61-33-12.

mlafon@pasteur.fr

[e]Present address: Unité de Biologie des Interactions Cellulaires, Institut Pasteur, Paris 75724, France.

Ann. N.Y. Acad. Sci. 1010: 598–603 (2003). © 2003 New York Academy of Sciences.
doi: 10.1196/annals.1299.112

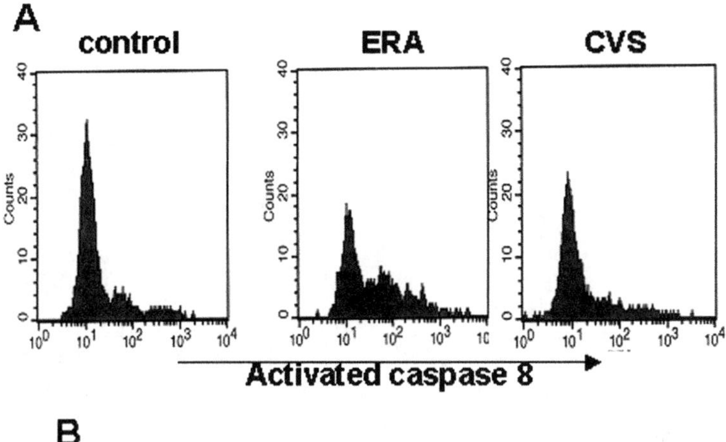

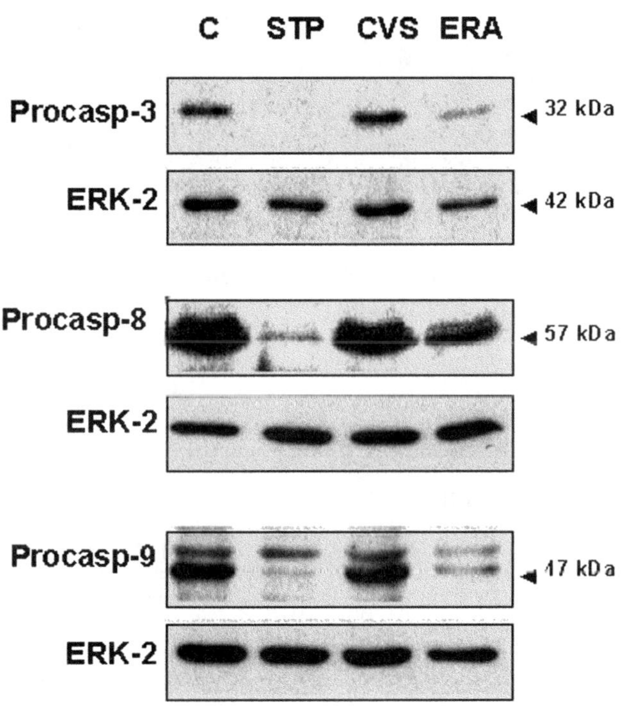

FIGURE 1. Caspase-dependent and caspase-independent (including AIF translocation) pathways are activated during the infection of Jurkat T cells by ERA but not by CVS. Jurkat T cells were infected with ERA or CVS RV strains (multiplicity of infection [MOI] = 3) or mock-infected (control) as previously described.[6] (**A**) Detection of activated caspase was performed with an FITC-conjugated analog of *N*-benzyloxycarbonyl-Leu-Glu-Thr-Asp-fluoromethylketone (LETD-FMK) for caspase 8. Fluorescence was analyzed in 10,000 gated cells.

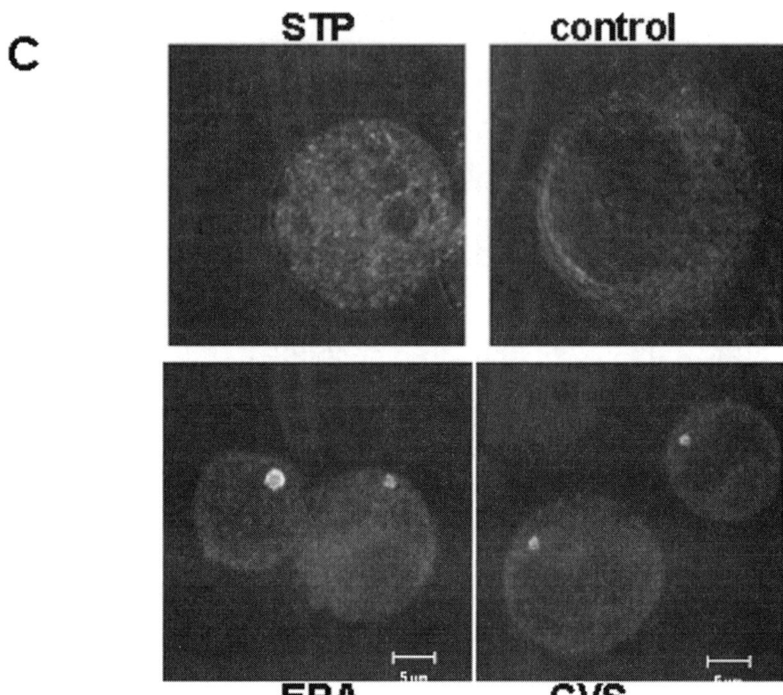

FIGURE 1. *Continued.*

Results were expressed as the percentage of gated cells emitting a fluorescent signal equal or higher than 150 fluorescence channel. (**B**) Protein extraction and Western blotting. After 72 h of infection or 4-h treatment with staurosporine (STP), 1×10^6 cells were lysed and loaded onto SDS-polyacrylamide gels. The proteins were electrophoresed and electrotransferred to polyvinylidene difluoride (PVDF) Immobilon filters. Then, filters were reacted with the appropriate specific primary Ab (anti-caspase 3, 8, 9 or anti-ERK) and incubated with the adequate secondary Ab conjugated with peroxidase. Finally, immunoblots were developed with the ECL Detection System. (**C**) AIF is translocated after STP treatment to the nucleus and during ERA-induced apoptosis in Jurkat T cells. STP-treated or CVS, ERA-infected, and mock-infected (control) cells were fixed in 4% paraformaldehyde (PFA) for 20 min at 4°C and incubated for 30 min at 4°C with FITC-conjugated anti–AIF rabbit Ab diluted in permeabilization buffer (1% heat-inactivated FCS, 0.1% [w/v] sodium azide, and 0.1% [w/v] saponin in PBS). Cells were further incubated with biotinylated anti–rabbit Ig and then with PE-conjugated streptavidin. Confocal microscopy was performed with a Zeiss LSM 510 confocal microscope which uses an helium-neon laser (1 mV). Data were analyzed with Image Browser version 2.8. Images were treated with Adobe-Photoshop 5.5.

the *Bcl-2* proto-oncogene. In contrast with the caspase-dependent pathway, which may or may not be abolished by *Bcl-2* expression, AIF translocation has been reported to be strictly abolished by *Bcl-2* expression.[3] This is consistent with the capacity of the *Bcl-2* to stabilize mitochondrial membrane function.

The rabies virus (RV) is an enveloped bullet-shaped virus of the *Rhabdoviridae* family, genus *Lyssavirus*. RV strain CVS (challenge virus standard) is a highly neurotropic virus strain that causes fatal encephaltomyelitis in mice. In contrast, ERA (Evelyn Rotnycki Abelseth) has lost pathogenicity after several passages on nonneuronal cultures. This strain rarely reaches the nervous system and is the prototype of live rabies vaccine distributed orally to immunize wildlife reservoir foxes in Western Europe.

We report here the triggering by live RV vaccine strains of caspase-dependent apoptosis in the human lymphoblastoid Jurkat T-cell line (Jurkat-vect). This process involves caspases 3, 8, and 9, according to staining with FITC-conjugated caspase substrate LEDT which reacts mainly with activated caspase 8 and disappearance of procaspases 3, 8, and 9 (FIG. 1, A and B). Treatment with the pan-caspase inhibitor ZVAD-fmk reduced but did not abolish apoptosis (data not shown), suggesting that a caspase-independent pathway also is induced by RV infection. Indeed, we detected translocation of the apoptosis-inducing factor (AIF) in ERA-infected Jurkat T cells (FIG. 1C). Bcl-2 overproduction in Jurkat T cells (Jurkat-Bcl-2)[4] (FIG. 1A) protects Jurkat T cells from natural apoptosis (FIG. 2B) and abolishes both caspase 3 and 8 activation and AIF translocation (FIG. 2C), suggesting that Bcl-2 overproduction blocks apoptosis by interacting upstream in both the caspase-dependent and the caspase-independent apoptotic pathway. RV infection and production were similar in Jurkat-vect and in Jurkat-Bcl-2 cells (data not shown), indicating that caspase activation is not a prerequisite for RV maturation or budding. Our data indicate also that Bcl-2 has no direct antiviral effects against rabies virus. Involvement of AIF, activation of caspase 8 and blockage of RV-induced apoptosis by Bcl-2 overexpression strongly supports that RV apoptosis follows an apoptosis signaling of "type II cell."[5] The type II cells are characterized by low levels of caspase 8 not sufficient to directly activate effector caspases as this occurs in type I cells. In type II cells, caspase 3 activation is obtained through the participation of the mitochondrial pathway and caspase 9 activation. The mitochondrial involvement is triggered by a truncated form of Bid, tBid, that cleaved itself by a minute amount of caspase 8. Because of the mitochondria involvement, the apoptosis of type II cells can involve the release of AIF and can be blocked by Bcl-2 overexpression.

Apoptotic signaling via FADD/caspase 8 can be obtained through extrinsic or intrinsic death stimuli. The extrinsic death stimuli results in ligation of membrane receptors TNFα-R and Fas/CD95, whereas the intrinsic pathway is activated by intracellular stress. Fas/CD95 probably is not involved in the extrinsic pathway in RV-induced apoptosis, because the treatment of RV-infected Jurkat cells with an antibody that mimicked FasL did not increase the rate of RV-induced apoptosis (data not shown). In virus-mediated apoptosis, extrinsic pathways also may involve the interaction of surface receptors lacking death domains with viral products. However, membrane receptors are unlikely to be involved in RV-mediated apoptosis because the treatment of cells with UV-inactivated virus did not trigger cell death.[6] The requirement for later steps in viral replication for RV-induced apoptosis[6] and the observation that apoptosis and the accumulation of RV proteins occur simultaneous-

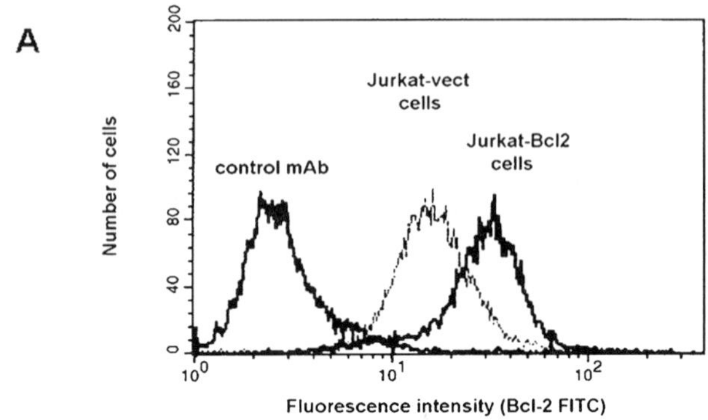

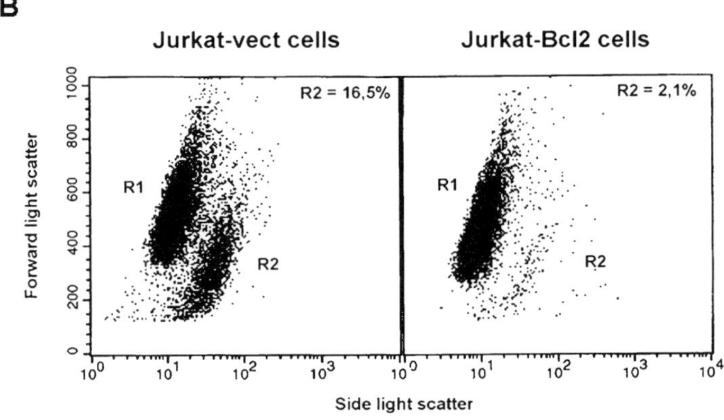

C

	Jurk-vect ERA	Jurk-Bcl-2 ERA	Jurk-vect staurosporine	Jurk-Bcl-2 staurosporine
Caspase 3 24H	30 +/-13.2	0.4 +/- 0.4	72.45 +/- 14.7	12.3 +/- 9.4
Caspase 8 24h	17.3 +/- 9.5	2.9 +/- 3.0	67.25 +/- 23	6.2+/- 3.8
AIF 24h	11 +/- 2.0	3.2 +/- 1.0	18.5 +/- 8.0	2.1 +/-0.35

FIGURE 2. Bcl-2 overexpression blocks both caspase activation and AIF translocation The human lymphoblastoid Jurkat Vb8-T cell line was stably transfected with the human *Bcl-2* gene under control of the cytomegalovirus early promoter (Jurkat-Bcl-2 cells) or with the insertless plasmid used as a control (Jurkat-vect cells).[4] Jurkat-Bcl-2 (seven clones) and Jurkat-vect (five clones) cells were analyzed. By flow cytometry (**A**) and Western blotting with an FITC-conjugated mouse mAb specific for the human Bcl-2 protein (data not shown), we found that Bcl-2 was overproduced in the seven Jurkat-Bcl-2 clones. For example, for clone 7, the Bcl-2 levels were higher in the Jurkat-Bcl-2 clone (fluorescence mean = 32) than in the Jurkat-vect clone (fluorescence mean = 15). (**B**) The natural background level of cell death was 16.5% of the total cell culture after 3 days in Jurk-vect and 2.1% in Jurk-Bcl-2,

ly support the hypothesis that the intrinsic pathway involving caspases 8 and 9 activation can be directly controlled by RV proteins.

In contrast with what was observed with the attenuated virus strain ERA, CVS, a neurotropic strain of rabies virus that causes fatal invasion of the nervous system, did not activate caspases nor translocate AIF (FIG. 1). RV is transmitted by the bite of a rabid animal. It enters the nervous system via a motor neuron through the neuromuscular junction, or via a sensory nerve through nerves spindles. It then travels from one neuron to the next, along the spinal cord up to the brain and the salivary glands. The virus particles are excreted in the saliva of the animal and are transmitted to another host by biting. Thus, to be transmitted to another animal, RV has to preserve the integrity of the neuronal network. The inverse correlation of the induction of apoptosis and the capacity of a virus strain to invade the brain suggests that blockage of apoptosis could be a strategy selected by neurotropic virus to favor their progression through the nervous system.

ACKNOWLEDGMENTS

This work was supported by institutional grants from the Institut Pasteur. M.-I.T. was a recipient of CANAM and Weizmann fellowships. We are grateful to Pascal Roux for the help with confocal microscopy.

REFERENCES

1. LEIST, M. & M. JAATTELA. 2001. Four deaths and a funeral: from caspases to alternative mechanisms. Nat. Rev. Mol. Cell Biol. **2:** 589–598.
2. SUSIN, S.A., H.K. LORENZO, N. ZAMZAMI, *et al.* 1999. Molecular characterization of mitochondrial apoptosis-inducing factor. Nature **397:** 441–446.
3. SUSIN, S.A., N. ZAMZAMI, M. CASTEDO, *et al.* 1996. Bcl-2 inhibits the mitochondrial release of an apoptogenic protease. J. Exp. Med. **184:** 1331–1341.
4. AILLET, F., H. MASUTANI, C. ELBIM, *et al.* 1998. Human immunodeficiency virus induces a dual regulation of Bcl-2, resulting in persistent infection of CD4(+) T- or monocytic cell lines. J. Virol. **72:** 9698–9705.
5. KRAMER, P.H. 2000. CD95's deadly mission in the immune system. Nature **407:** 789–795.
6. THOULOUZE, M.I., M. LAFAGE, J.A. MONTANO-HIROSE & M. LAFON. 1997. Rabies virus infects mouse and human lymphocytes and induces apoptosis J. Virol. **71:** 7372–7380.

indicating that Bcl-2 overexpression protected Jurkat T cells from spontaneous apoptosis. Despite their resistance to natural apoptosis, Jurkat-Bcl-2 cells did not grow any faster than Jurkat-vect cells. (**C**) Effect of Bcl-2 overexpression on caspases 3 and 8 activation. AIF translocation was measured in 4-h staurosporine-treated or 24-h ERA-infected Jurk-vect and Jurk-Bcl-2 cells. Cells were equally infected by ERA: 80% of Jurk-vect and 85% of Jurk-Bcl-2 were infected at 24 h.

Increased Apoptotic Cell Death in Sporadic and Genetic Alzheimer's Disease

ANNE ECKERT, CELIO A. MARQUES, UTA KEIL, KATRIN SCHÜSSEL, AND WALTER E. MÜLLER

Department of Pharmacology, Biocenter, University of Frankfurt, 60439 Frankfurt am Main, Germany

ABSTRACT: Mounting evidence indicates increased susceptibility to cell death and increased oxidative damage as common features in neurons from sporadic Alzheimer's disease (AD) patients but also from familial AD (FAD) cases. Autosomal dominant forms of FAD are caused by mutations of the amyloid precursor protein (APP) gene and by mutations of the genes encoding for presenilin 1 or presenilin 2 (PS1/2). We investigated the effect of the Swedish APP double mutation (APPsw) on oxidative stress–induced cell death mechanisms in PC12 cells. This mutation results in from three- to sixfold increased β-amyloid (Aβ) production compared with wild-type APP (APPwt). Because APPsw cells secrete low Aß levels similar to the situation in FAD brains, our cell model represents a very suitable approach to elucidate the AD-specific cell death pathways under more likely physiological conditions. We found that APPsw-bearing cells show decreased mitochondrial membrane potential after exposure to hydrogen peroxide. In addition, activity of the executor caspase 3 after treatment with hydrogen peroxide was elevated in APPsw cells, which seems to be the result of an enhanced activation of both intrinsic and extrinsic apoptosis pathways. Our findings provide evidence that the massive neurodegeneration in early age of FAD patients could be a consequence of an increased vulnerability of neurons by mitochondrial abnormalities resulting in activation of different apoptotic pathways as a consequence to elevated oxidative stress levels. Finally, we propose a hypothetical sequence of the pathogenic steps linking sporadic AD, FAD, Aβ production, mitochondrial dysfunction with caspase pathway, and neuronal loss.

KEYWORDS: APP mutation; β amyloid; caspases; oxidative stress; mitochondria; cell death

INTRODUCTION

Alzheimer's disease (AD) is a neurodegenerative disorder marked by progressive loss of memory and impairment of cognitive ability. AD can be classified into two forms: sporadic AD, which accounts for the vast majority of AD cases, and a familial form of AD (FAD), in which rare gene mutations have been identified. The latter

Address for correspondence: A. Eckert, Department of Pharmacology, Biocenter, University of Frankfurt, Marie-Curie-Strasse 9, 60439 Frankfurt am Main, Germany. Voice: +49-69-798-29377; fax: +49-798-29374.

a.eckert@em.uni-frankfurt.de

Ann. N.Y. Acad. Sci. 1010: 604–609 (2003). © 2003 New York Academy of Sciences. doi: 10.1196/annals.1299.113

patients suffer from an autosomal dominant inheritable variant of AD. Notably, patients with either sporadic AD or FAD share common clinical and neuropathological features including β-amyloid (Aβ) plaques, accumulation of intracellular neurofibrillary tangles, and pronounced neuronal cell loss. The amyloid plaque is composed of Aβ peptide, which is derived from the amyloid precursor protein (APP) through an initial β-secretase cleavage followed by an intramembraneous cut of γ-secretase. Three genes are known to be causatively linked with the pathogenesis of these early-onset FAD forms. Besides the genes encoding for presenilin 1 (PS1) on chromosome 14 and presenilin 2 (PS2) on chromosome 1, mutations in the amyloid precursor protein gene on chromosome 21 account for these FAD cases. Remarkably, more than 60 FAD mutations located in these three different genes apparently result in the overproduction of Aβ providing substantial evidence that Aβ plays a central role in the pathogenesis of AD. Moreover, a growing body of evidence points to a direct neurotoxic effect of Aβ on neurons. Although Aβ-mediated cell death is observed both *in vitro* and *in vivo* and may be apoptotic in nature, the underlying mechanisms triggering neuronal death are still not completely understood. Therefore, it is likely that knowledge of these mechanisms will help to identify potential molecular targets for development of new therapeutic strategies.

EVIDENCE FOR PATHWAYS ASSOCIATED WITH APOPTOSIS IN AD

Aβ-induced cell death has been studied *in vitro*. Aβ and its fragments induce cell death in neuronal cell cultures by exhibiting classic features of apoptotic cell death.[1–4] Notably, one possible mechanism initiating apoptosis is the generation of free radicals by the Aβ peptide leading to lipid peroxidation and oxidative stress.[5] In most of these studies, synthetic Aβ peptides generally have been applied at micromolar levels that are in contrast with the low nanomolar levels of natural Aβ found in the brain and cerebrospinal fluid. Furthermore, they can aggregate into different assembly forms which may mediate cellular effects unlike those found *in vivo*. Thus, the relevance of these findings to mirror cellular *in vivo* processes is not clear. Therefore, studies of Aβ-induced neurotoxic effects under more likely physiological conditions mimicking the situation in AD brain are needed.[6] Moreover, Aβ is produced intracellularly and can accumulate within cells.[7–9] Intracellular accumulation of Aβ, soluble or insoluble, might impair cellular functions and may represent the primary event within the neurotoxic Aβ cascade. Also importantly, the effects of intracellular Aß cannot be covered by studies using experimental settings that expose cells to extracellular amounts of Aβ.

Examination of postmortem tissue has implicated caspases (caspase 3, 8, and 9) in sporadic AD.[10–14] In addition, oxidative damage is a key feature in AD brain.[10,15–17] It is possible that neurons accumulating extensive oxidative insults may selectively undergo apoptosis. Mitochondria are particularly vulnerable to oxidative damage. Mitochondrial dysfunction has been observed in AD brain.[18] Neurons containing both oxidatively damaged DNA and the activation of caspase 9 suggest a link between these two events. Limitations of studies on postmortem tissue concern the extent of the disease process at the time of death, because each brain sample represents just one shot of the whole disease progress at a distinct time point and the lack of ability to study functional aspects of the underlying cell death pathways.[19] Consequently, the best approach to define these mechanisms is the use of cell culture models that attempt to mimic certain aspects of AD thereby overcoming many of the limitations of postmortem brain tissue. Interestingly, enhanced vulnerability to cell death also has been described in peripheral lymphocytes from AD patients.[20–23] Specifically oxidative stress mechanisms seem to be involved in this process. Therefore, peripheral cells represent a valuable cell model to study the effects of sporadic risk factors of AD in cell death mechanisms. To investigate specifically the toxic mechanisms of Aß and to clarify its final relevance for AD pathogenesis, the use of cell culture models which produce low levels of Aß seems to be the best experimental approach.

A CELL MODEL EXPRESSING ALZHEIMER'S APP MUTATION TO ELUCIDATE THE SPECIFIC CELL DEATH PATHWAYS

A double mutation preceding the N terminus of the Aβ domain (KM670/671NL) within the APP sequence (APPsw) has been identified in a Swedish family and results in from three- to six-fold increased Aβ production, which involves Aβ(1-40) as well as Aβ(1-42). The β-secretase cleavage site is mutated resulting in a very strong β-secretase activity. Cleavage occurs during passage through the late secretory pathway in Golgi-derived secretory vesicles, whereas wild-type APP (APPwt) must be reinternalized before β-secretase cleavage. APPsw and APPwt PC12 cells showed a moderate expression of human APP leading to low levels of Aβ(1-40) production (APPsw: 90 pg/mL; APPwt: 20 pg/mL) similar to the situation *in vivo*.[24] These two cell lines with different production levels of Aβ additionally may allow study of dose-dependent effects of Aβ. Expression of APPsw rendered PC12 cells vulnerable to the induction of cell death after exposure to oxidative stress (hydrogen peroxide).[24] One potential mechanism by which APP mutations enhance this process could be the increased production of Aβ, which is able to induce apoptotic cell death. However, in the absence of apoptosis-inducing treatments, we found no evidence for increased apoptosis per se.[24,25] Therefore, it seems rather likely that increased production of Aβ at physiological levels somehow primes APPsw cells to undergo cell death, leading to increased cell death only after additional stress, for example, oxidative stress, a scenario which also has been suggested to occur in AD brain. Moreover, we could show that, in PC12 cells bearing APPsw, caspase 3 activity was significantly enhanced compared with APPwt and vector-transfected control cells.[24] Interestingly, APPwt cells also showed an increased activity of caspase 3 compared with empty vector control cells even though to a lesser extent than APPsw cells did. Our data suggest that already very low Aβ levels in APPwt cells are probably sufficient to prime cells to undergo apoptosis and increasing Aβ levels, as in the case of APPsw cells, even strengthen the effects on caspase 3 activation. This is confirmed by findings that caspase 9 is increased and at the same time ATP levels and mitochondrial membrane potential are lowered in APPwt as well as in APPsw cells compared with vector-transfected control cells. In addition, we found that in APPsw PC12 cells activation of caspase 2 and caspase 8 is enhanced after exposure to oxidative stress.[26]

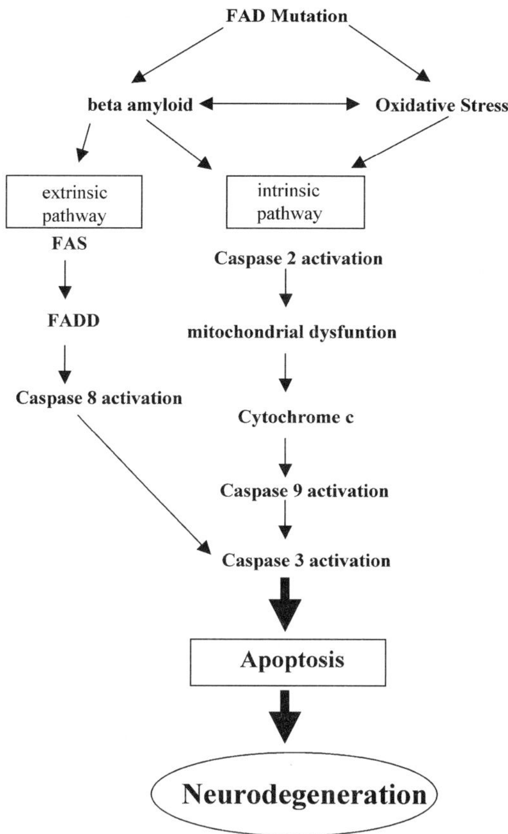

FIGURE 1. A hypothetical sequence of the pathogenetic steps linking sporadic AD, FAD, Aβ production, and mitochondrial dysfunction with caspase pathway and neuronal cell loss.

A HYPOTHETICAL SEQUENCE OF THE PATHOGENETIC STEPS OF AD

In accordance with findings from Rohn *et al.*,[13] we propose a hypothetical sequence of events linking sporadic AD, FAD, Aβ production, mitochondrial dysfunction with caspase pathway, and neuronal cell loss (FIG. 1). Two parallel pathways may occur in APP-transfected PC12 cells under oxidative stress conditions. Already at low Aβ levels, the intrinsic pathway is activated leading to mitochondrial dysfunction, for example, decrease in mitochondrial membrane potential and depletion in ATP. Cytochrome *c* is released by dysfunctional mitochondria and activates caspase 9. Interestingly, it was discovered recently that intrinsic pathways, such as those initiated by cell stress, induce activation of caspase 2, which is required for permeabilization of mitochondria.[27] Caspase 2 activation was strongly and specifically increased in APPsw PC12 cells, suggesting that this caspase is acting upstream of mitochondria and enforces the intrinsic pathway when Aß levels increase. Moreover, it has been reported that Aβ may lead to the cross-linking and activation of all death receptors resulting in caspase 8 activation. Indeed, in our cell model, caspase 8 activation was increased in the presence of high Aβ concentration produced by APPsw cells. Whether this increase is caused only by effects of extracellular Aβ on cell death receptors or if it can also be mediated by intracellularly accumulated Aβ is not clear. Both pathways then may converge by activating the effector enzyme, caspase 3, and the execution of cell death. Nondominant forms of AD including sporadic AD, in which Aβ and oxidative stress levels gradually increase with age, and dominantly inherited forms of AD, in which Aβ production increases throughout life, converge at a final common pathway of an increased vulnerability to apoptotic cell death.

Mitochondria may act as amplifiers rather than initiators of caspase activity. Efforts to block the activity of postmitochondrial caspase 9 and caspase 3 possibly may inhibit the amplification loop only, but not prevent cell death. Based on this assumption, strategies involving efforts to protect cells at the mitochondrial level or to target events upstream of mitochondria appear to be promising.

ACKNOWLEDGMENTS

This work was supported by grants from the Alzheimer Forschung Initiative e.V., from the Hirnliga e.V., and Dr. Robert Pfleger-Stiftung.

REFERENCES

1. Loo, D.T., A. Copani, C.J. Pike, *et al.* 1993. Apoptosis is induced by beta-amyloid in cultured central nervous system neurons. Proc. Natl. Acad. Sci. USA **90:** 7951–7955.
2. Pike, C.J., A.J. Walencewicz, C.G. Glabe & C.W. Cotman. 1991. In vitro aging of beta-amyloid protein causes peptide aggregation and neurotoxicity. Brain Res. **563:** 311–314.
3. Watt, J.A., C.J. Pike, A.J. Walencewicz Wasserman & C.W. Cotman. 1994. Ultrastructural analysis of beta-amyloid-induced apoptosis in cultured hippocampal neurons. Brain Res. **661:** 147–156.
4. Troy, C.M., S.A. Rabacchi, W.J. Friedman, *et al.* 2000. Caspase-2 mediates neuronal cell death induced by beta-amyloid. J. Neurosci. **20:** 1386–1392.
5. Hensley, K., J.M. Carney, M.P. Mattson, *et al.* 1994. A model for beta-amyloid aggregation and neurotoxicity based on free radical generation by the peptide: relevance to Alzheimer's disease. Proc. Natl. Acad. Sci. USA **91:** 3270–3274.
6. Selkoe, D.J. 2002. Alzheimer's disease is a synaptic failure. Science **298:** 789–791.
7. Hartmann, T. 1999. Intracellular biology of Alzheimer's disease amyloid beta peptide. Eur. Arch. Psychiatry Clin. Neurosci. **249:** 291–298.
8. Wirths, O., G. Multhaup, C. Czech, *et al.* 2001. Intraneuronal Abeta accumulation precedes plaque formation in beta-amyloid precursor protein and presenilin-1 double-transgenic mice. Neurosci. Lett. **306:** 116–120.
9. Hartmann, T., S.C. Bieger, B. Bruhl, *et al.* 1997. Distinct sites of intracellular production for Alzheimer's disease A beta40/42 amyloid peptides. Nat. Med. **3:** 1016–1020.
10. Su, J.H., K.E. Nichol, T. Sitch, *et al.* 2000. DNA damage and activated caspase-3 expression in neurons and astrocytes: evidence for apoptosis in frontotemporal dementia. Exp. Neurol. **163:** 9–19.

11. SU, J.H., T. SATOU, A.J. ANDERSON & C.W. COTMAN. 1996. Up-regulation of Bcl-2 is associated with neuronal DNA damage in Alzheimer's disease. Neuroreport **7:** 437–440.
12. ROHN, T.T., E. HEAD, W.H. NESSE, *et al.* 2001. Activation of caspase-8 in the Alzheimer's disease brain. Neurobiol. Dis. **8:** 1006–1016.
13. ROHN, T.T., R.A. RISSMAN, M.C. DAVIS, *et al.* 2002. Caspase-9 activation and caspase cleavage of tau in the Alzheimer's disease brain. Neurobiol. Dis. **11:** 341–354.
14. YANG, F., X. SUN, W. BEECH, *et al.* 1998. Antibody to caspase-cleaved actin detects apoptosis in differentiated neuroblastoma and plaque-associated neurons and microglia in Alzheimer's disease. Am. J. Pathol. **152:** 379–389.
15. LOVELL, M.A., S.P. GABBITA & W.R. MARKESBERY. 1999. Increased DNA oxidation and decreased levels of repair products in Alzheimer's disease ventricular CSF. J. Neurochem. **72:** 771–776.
16. LOVELL, M.A., W.D. EHMANN, M.P. MATTSON & W.R. MARKESBERY. 1997. Elevated 4-hydroxynonenal in ventricular fluid in Alzheimer's disease. Neurobiol. Aging **18:** 457–461.
17. LEUTNER, S., C. CZECH, K. SCHINDOWSKI, *et al.* 2000. Reduced antioxidant enzyme activity in brains of mice transgenic for human presenilin-1 with single or multiple mutations. Neurosci. Lett. **292:** 87–90.
18. HIRAI, K., G. ALIEV, A. NUNOMURA, *et al.* 2001. Mitochondrial abnormalities in Alzheimer's disease. J. Neurosci. **21**: 3017–3023.
19. TROY, C.M. & G.S. SALVESEN. 2002. Caspases on the brain. J. Neurosci. Res. **69:** 145–150.
20. ECKERT, A., K. SCHINDOWSKI, S. LEUTNER, *et al.* 2001. Alzheimer's disease-like alterations in peripheral cells from presenilin-1 transgenic mice. Neurobiol. Dis. **8:** 331–342.
21. ECKERT, A., C.W. COTMAN, R. ZERFASS, *et al.* 1998. Enhanced vulnerability to apoptotic cell death in sporadic Alzheimer's disease. Neuroreport **9:** 2443–2446.
22. ECKERT, A., M. OSTER, R. ZERFASS, *et al.* 2001. Elevated levels of fragmented DNA nucleosomes in native and activated lymphocytes indicate an enhanced sensitivity to apoptosis in sporadic Alzheimer's disease. Specific differences to vascular dementia. Dement. Geriatr. Cogn. Disord. **12:** 98–105.
23. PARSHAD, R.P., K.K. SANFORD, F.M. PRICE, *et al.* 1996. Fluorescent light-induced chromatid breaks distinguish Alzheimer disease cells from normal cells in tissue culture. Proc. Natl. Acad. Sci. USA **93:** 5146–5150.
24. ECKERT, A., B. STEINER, C. MARQUES, *et al.* 2001. Elevated vulnerability to oxidative stress-induced cell death and activation of caspase-3 by the Swedish amyloid precursor protein mutation. J. Neurosci. Res. **64:** 183–192.
25. LEUTZ, S., B. STEINER, M.A. MARQUES, *et al.* 2002. Reduction of trophic support enhances apoptosis in PC12 cells expressing Alzheimer's APP mutation and sensitizes cells to staurosporine-induced cell death. J. Mol. Neurosci. **18:** 189–201.
26. MARQUES, C., B. STEINER, C. HAASS, *et al.* 2002. Increased caspase-2, -3, and –8 activation in PC12 cells expressing Alzheimer mutations after exposure to oxidative stress. Naunyn-Schmiedeberg's Arch. Pharmacol. **365** (Suppl.1): 328.
27. KUMAR, S. & D.L. VAUX. 2002. A cinderella caspase takes center stage. Science **297:** 1290–1291.

Prion Protein Fragment 106-126 Induces a p38 MAP Kinase–Dependent Apoptosis in SH-SY5Y Neuroblastoma Cells Independently from the Amyloid Fibril Formation

A. CORSARO,[a] S. THELLUNG,[a] V. VILLA,[a] D. ROSSI PRINCIPE,[b] D. PALUDI,[b] S. ARENA,[a] E. MILLO,[c] D. SCHETTINI,[a,d] G. DAMONTE,[c] A. ACETO,[b] G. SCHETTINI,[a,d] AND T. FLORIO[a,d]

[a]*Section Pharmacology, Department Oncology, Biology and Genetics University of Genova, Genova, Italy*

[b]*Department Biomedical Sciences, University G. D'Annunzio, Chieti, Italy*

[c]*Department Experimental Medicine, University of Genova, Genova, Italy*

[d]*Pharmacology and Neuroscience IST Genova, Genova, Italy*

ABSTRACT: Prion diseases are neurodegenerative disorders of the central nervous system of humans and animals, characterized by spongiform degeneration of the central nervous system, astrogliosis, and deposition of amyloid into the brain. The conversion of a cellular glycoprotein (prion protein, PrP^C) into an altered isoform (PrP^{Sc}) has been proposed to represent the causative event responsible for these diseases. The peptide corresponding to the residues 106-126 of PrP sequence (PrP106-126) is largely used to explore the neurotoxic mechanisms underlying the prion diseases. We investigated the intracellular signaling responsible for PrP106-126–dependent cell death in the SH-SY5Y human neuroblastoma cell line. In these cells, PrP106-126 treatment induced apoptotic cell death and the activation of caspase-3. The p38 MAP-kinase blockers (SB203580 and PD169316) prevented the apoptotic cell death evoked by PrP106-126 and Western blot analysis revealed that the exposure of the cells to the peptide induced p38 activation. However, whether the neuronal toxicity of PrP106-126 is caused by a soluble or fibrillar form of this peptide is still unknown. In this study, we correlated the structural state of this peptide with its neurotoxicity. We show that the two conserved glycines in position 114 and 119 prevent the peptide to assume a structured conformation, favoring its aggregation in amyloid fibrils. The substitution of both glycines with alanine residues (PrP106-126AA) generates a soluble nonamyloidogenic peptide, that retained its toxic properties when incubated with neuroblastoma cells. These data show that the amyloid aggregation is not necessary for the induction of the toxic effects of PrP106-126.

KEYWORDS: prion protein; apoptosis; p38 MAP kinase; caspases; amyloid aggregation

Address for correspondence: Professor Tullio Florio, Department of Oncology, Biology and Genetics, University of Genova, c/o Advanced Biotechnology Center (CBA), Largo R. Benzi, 10, 16132 Genova, Italy. Voice: +39-10-5737255; fax: +39-10-5737257.
florio@cba.unige.it

**Ann. N.Y. Acad. Sci. 1010: 610–622 (2003). © 2003 New York Academy of Sciences.
doi: 10.1196/annals.1299.114**

INTRODUCTION

Transmissible spongiform encephalopathies (TSEs) or prion diseases are fatal neurodegenerative disorders of the central nervous system (CNS) of humans (Kuru; Creutzfeldt–Jakob disease [CJD]; fatal familial insomnia [FFI]; Gerstmann Straussler Sheinker syndrome [GSS]) and animals (bovine spongiform encephalopathy in cattle, and scrapie in sheep or goats).[1] TSEs are characterized by spongiform degeneration of the CNS, astrogliosis, and deposition of amyloid fibrils into the brain.[1]

Prion diseases may be inherited in an autosomal dominant fashion, may be sporadic, or may be transmitted through an infective mechanism.[1] The key event in the pathogenesis of all the TSEs is the conformational conversion of a normal cell surface glycoprotein (the cellular isoform of the prion protein, PrP^{C}) into a pathogenic isoform (named PrP^{Sc}, after the scrapie) that is characterized by a high content of β-sheet structure.[1] PrP^{Sc} accumulates into the brain of affected individuals in a detergent-insoluble and protease-resistant form, named PrP27-30. It has been demonstrated that the gene encoding for PrP^{C} (*Prn-P*) is essential for the development of spongiform encephalopathies. PrP^{c} is abundantly expressed in the CNS of all mammalian species, but its function is still largely unknown.

The neuronal death characterizing TSE, as well as other neurological disorders, is triggered by the activation of the apoptotic program.[2] Although in the past few years much data have been produced to better characterize the molecular mechanisms involved in the conversion of PrP^{C} in PrP^{Sc},[1] the intracellular signaling responsible for the PrP^{Sc}-induced cell death is much less studied. A peptide corresponding to the residues 106-126 of the PrP sequence (PrP 106-126) has been largely used to explore the neurotoxic mechanisms underlying the prion diseases.[3] PrP106-126 shows a high tendency to form amyloid fibrils *in vitro*,[4,5] and is partially resistant to proteolysis.[6] Moreover, this peptide was reported to induce apoptotic neuronal death in primary cultures of cortical, hippocampal, and cerebellar neurons[3,7–10] and to exert a trophic action on glial cells.[11] Thus, PrP106-126 represents a suitable tool to study the pathogenic effects of PrP^{Sc}. Programmed cell death or apoptosis is a process of pivotal importance in several physiological phenomena, such as the correct development and function of organs, the maturation/function of the immune system, and the development of the nervous system.[12] The apoptotic process is tightly regulated: a failure of the cell death program can lead to autoimmune diseases or cancer, whereas an excessive execution of apoptosis is linked to ischemic and neurodegenerative diseases.[2] Among the intracellular factors that contribute to the activation and execution of apoptosis, a central role is played by caspases.[13] Caspases are cysteine proteases that cleave different substrates with high specificity and take part to the apoptotic process at various levels.[12] Beside caspases, a role for the mitogen-activated protein (MAP) kinases in the control of several degenerative processes is now emerging.[14] In this study, we investigated the toxic effect of PrP106-126 in the SH-SY5Y human neuroblastoma cell line[15] and the involvement of caspases and MAP kinases in this phenomenon. We report that PrP106-126 induces apoptosis in SH-SY5Y cells through the activation of caspase-3 and p38 MAP kinase. Moreover, we analyzed the relationship between the structural state of PrP106-126 and its neurotoxicity. Indeed, it has not been determined yet whether the neuronal toxicity of PrP106-126 (and PrP^{Sc}) is caused by soluble or fibrillar insoluble forms of the peptide and which are the molecular components responsible for its tendency to aggregate into amyloid fibrils. In particular, although there is evidence for the toxicity of mature fibrils in some amyloidogenic pathologies, other investigations indicate that protofibrillar aggregates may be the primary toxic species.[16]

MATERIALS AND METHODS

Peptides

PrP 106-126 (sequence: KTNMKMAGAAAAGAVVGGLG) PrP 106-126AA (sequence: KTNMKHMAAAAAAAAVVGGLG) and PrP 106-126 scrambled (SCR) (sequence: NGAKALMGGHGATKVMVGAAA) were manually synthesized using the standard method of solid-phase peptide synthesis which follows the 9-fluorenylmethoxycarbonyl (Fmoc) strategy[17] with minor modifications. Peptides were dissolved in phosphate-buffered saline at a concentration of 10 mM and stored at –20°C.

Chemicals

The caspase-3 fluorogenic substrate Ac-DEVD-AMC, the caspase blockers (Z-VAD-FMK and Z-DEVD-FMK) and the MAP kinases inhibitors (PD98059, SB203580, and PD169316) were purchased from Calbiochem.

Cell Culture and Treatments

The human neuroblastoma cells SH-SY5Y were cultured in Dulbecco's modified Eagle's medium (DMEM) (Gibco) supplemented with 10% fetal bovine serum (FBS) (Gibco), 2 mM glutamine, 100 U/mL penicillin and 100 μg/mL streptomycin; cells were maintained at 37°C in a humidified incubator under 95% air and 5% CO_2. All the treatments were performed by single administration, directly into the wells from the stock solutions to reach the desired concentration.

Survival Assays

MTT Assay

The mitochondrial function, as an index of cell viability, was evaluated by measuring the reduction of 3-(4,5-dimethylthiazol-2-yl)-2,5,diphenyl-tetrazolium bromide (MTT). The cleavage of MTT to a purple formazan product by mitochondrial dehydrogenase was quantified spectrophotometrically as absorbance at 570 nm.

Detection of Apoptosis

DNA Ladder Detection

Cells were lysed in 10 mM Tris-HCl, pH 7.4, 5 mM EDTA, 0.5% Triton X-100 for 60 min at 4°C and centrifuged at 12,000 × *g* for 10 minutes. Supernatants were treated overnight with 200 μg/mL proteinase K at 50°C; DNA was phenol-extracted and ethanol-precipitated. The DNA was suspended in Tris-HCl 10 mM, pH 8, 1 mM

EDTA and treated with 10 μg/mL RNase for 30 min at 37°C. The samples were separated on 1.5% agarose gel stained with 5 μg/mL ethidium bromide and visualized under ultraviolet light.

Cell Death Detection ELISA Kit

The assay was performed following the manufacturer's instructions (Roche).

Caspase-3 Activity Detection

Caspase-3 activity induced by PrP106-126 was measured through the cleavage of its specific fluorogenic substrate Ac-DEVD-AMC.[18] Values have been expressed as fluorescence unit per microgram of protein.

Immunoblotting

Cells were lysed in a buffer containing 20 mM Tris-HCl, pH 7.4, 140 mM NaCl, 2 mM EDTA, 2 mM EGTA, 10% glycerol, 1% NP-40, 1 mM dithiothreitol, 1 mM sodium orthovanadate, 1 mM phenylmethylsulfonyl fluoride and the "Complete" protease inhibitor cocktail (Roche). Twenty-five micrograms of proteins from each sample were size-fractionated by 12% SDS-PAGE transferred to a polyvinylidene difluoride membrane (Bio-Rad) and probed with rabbit antibodies raised against the phosphorylated or total forms of p38 and ERK1/2 (New England Biolabs). The secondary antibody was a horseradish peroxidase–linked anti–rabbit IgG antiserum (Amersham). The antibody-reactive bands were visualized by ECL (Amersham).

Congo Red Staining

Amyloid fibril formation was analyzed by Congo red binding as reported.[7,19] Fifty micrograms of the peptides were incubated with 10 μM Congo red (Fluka; in 0.1 mM KH_2PO_4, 0.1 M NaCl, pH 7.4) for 8 h at room temperature. The amount of Congo red bound (CRB) was quantified by CRB (M) = (A540 ÷ 25,295) – (A488 ÷ 46,306), CRB (M) is the molar concentration of Congo red bound, and 25,295 and 46,306 are the molar extinction coefficients of bound and unbound Congo red, respectively.

Statistical Analysis

The experiments were performed, unless specified, in quadruplicate and repeated three times. Statistical analysis was performed by means of the analysis of variance. A *P* value less than .05 was considered statistically significant.

RESULTS

Using the MTT test, we quantified the time and dose dependency of PrP106-126 toxicity. SH-SY5Y viability was significantly affected after 5 days of exposure to PrP106-126 (100 μM) and was reduced to less than 45% of the control value after 7 days (TABLE 1). The effect was dose dependent in that the cell death was significantly detectable after treatment with 30 μM PrP106-126 and increased proportionally with the concentration of the peptide up to 100 μM (TABLE 1).

TABLE 1. Dose response and time course of SH-SY5Y cell death induced by PrP106-126

PrP106-126	**10 μM**	**30 μM**	**50 μM**	**100 μM**
Control	100	100	100	100
3 days	100 ± 1.4	111 ± 3.2	103 ± 2.1	90 ± 3.6*
5 days	103 ± 2.3	92 ± 2*	87.5 ± 4.5*	81 ± 0.5**
7 days	93.7 ± 3.5	87.5 ± 3.5*	71.2 ± 4.6**	46.4 ± 4.1**

NOTE: Cells were treated with PrP106-126 (10, 30, 50, and 100 μM), and cell viability was assessed by MTT after 3, 5, and 7 days. Data are expressed as a percentage of vehicle-treated samples, and each point represents the average of three experiments, performed in quadruplicate. $^{*}P < .05$ and $^{**}P < .01$ versus control values.

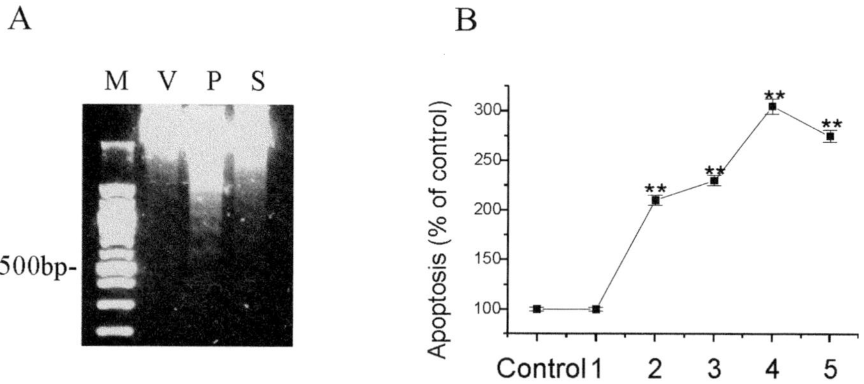

FIGURE 1. (A) Chromatographic analysis of SH-SY5Y DNA after PrP106-126 treatment. Cells were placed in 2% FBS-containing DMEM and treated with vehicle (*lane* 2), PrP106-126 100 μM (*lane* 3), or the scrambled peptide (SCR) 100 μM (*lane* 4). After 3 days of treatment, the DNA was extracted and separated onto a 1.5% agarose gel. *Lane* 1 contains a 100-bp DNA standard. (B) Time course of SH-SY5Y apoptosis induced by PrP106-126. Cells were treated with PrP106-126 (50 μM) for 1 to 5 days. The ELISA cell death detection kit was used to detect cytosolic mono- and oligonucleosomes. The values were expressed as a percentage of the vehicle-treated control. Each point represents the average of three experiments in quadruplicate. $^{**}P < .01$ versus control values.

The SH-SY5Y cell death, induced by the prion peptide, showed the typical characteristics of apoptosis as demonstrated by the detection of the characteristic electrophoretic ladder, after 3 days of treatment with PrP106-126 (100 μM), indicating the occurrence of DNA fragmentation in oligonucleosomes (FIG. 1A). Conversely, no signs of apoptosis were detectable in vehicle- or SCR- treated cells (FIG. 1A).

Apoptosis was quantified using an ELISA that measures the amount of mono- and oligonucleosomes released in the cytosol by the apoptotic endonucleases; PrP106-126 (50 μM) caused a time-dependent DNA cleavage significantly detectable after 2 days of treatment and reaching the highest value after 4 days (approximately 200% over control; FIG. 1A).

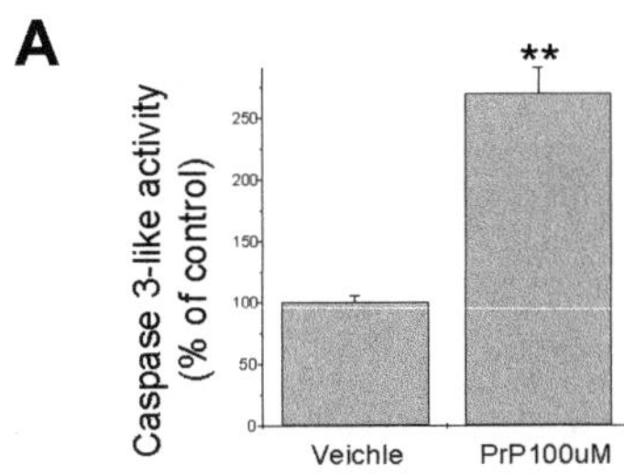

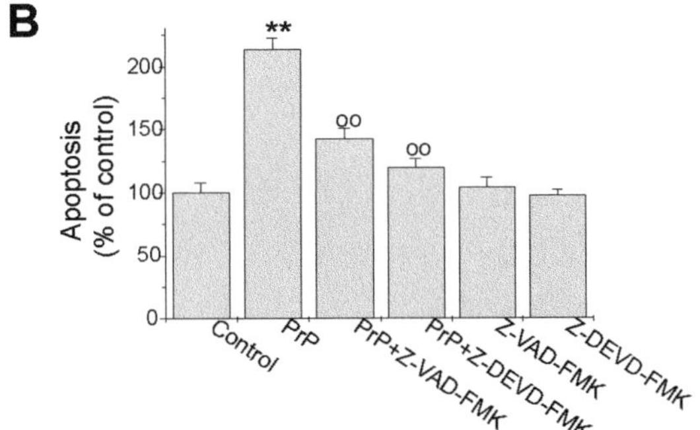

FIGURE 2. Caspase-3 activity induced by PrP106-126, in SH-SY5Y cells. (**A**) Cells were placed in 2% FBS-containing DMEM and treated with PrP106-126 (100 μM) for 5 h. Caspase-3 activity was measured evaluating the cleavage of the fluorogenic substrate Ac-DEVD-AMC, as described in MATERIALS AND METHODS. Data represent the average of three experiments in quadruplicate. $^{**}P < .01$ versus control values. (**B**) Effect of Z-VAD-FMK and Z-DEVD-FMK on SH-SY5Y apoptosis induced by PrP106-126. Cells were treated with PrP106-126 in the presence or absence of Z-VAD-FMK and Z-DEVD-FMK. Apoptosis induction was assayed by the ELISA test after 3 days of treatment. Data are expressed as percentage of control, and each value is the mean of three experiments in quadruplicate. $^{**}P < .01$ versus control and $^{oo}P < .01$ versus PrP106-126 values.

Because PrP106-126–induced SH-SY5Y cell death showed some features of apoptosis, we analyzed the involvement of caspase-3 in the toxic effects of the peptide. For this purpose, caspase-3 activation was assessed measuring the cleavage of the fluorogenic substrate Ac-DEVD-AMC. Caspase-3 proteolytic activity increased significantly after 2 h of treatment with 100 μM PrP106-126 (data not shown) and reached a maximal value (170% over the control value) after 5 h (FIG. 2A). To correlate caspase activation with the apoptotic activity of the prion fragment, we measured the effects of PrP106-126 on SH-SY5Y DNA cleavage in the presence of Z-VAD-FMK, a general caspase blocker, and Z-DEVD-FMK a more selective caspase-3 inhibitor.[20] The coincubation with the two caspase blockers inhibited the DNA fragmentation caused by 3 days of exposure to PrP106-126 (FIG. 2B).

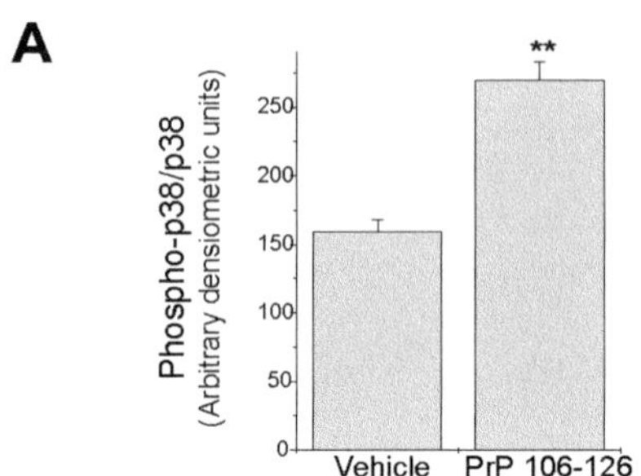

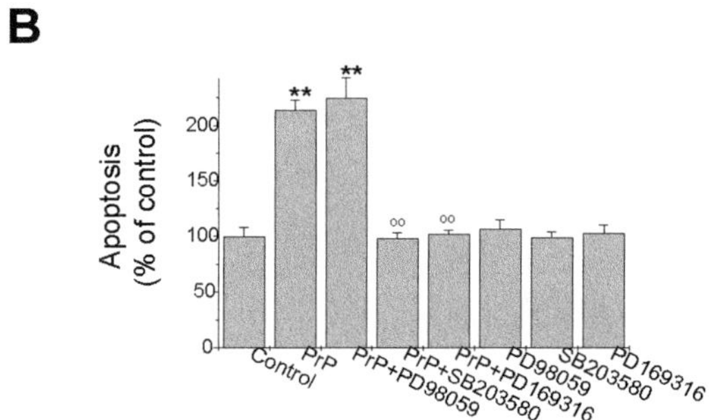

FIGURE 3. (**A**) Densitometric analysis of the phospho-p38/p38 ratio in vehicle- or PrP106-126–treated cells. Densitometric data were obtained as optical density (OD) × mm^2, and the ratios were expressed as a percentage of control (untreated) value. Each point represents the average of the results obtained from three independent Western blot experiments. (**B**) Effect of ERK1/2 and p38 inhibitors on SH-SY5Y apoptosis induced by PrP106-126. Cells were treated with PrP106-126 in the presence or absence of PD98059, SB203580, and PD169316. The induction of apoptosis was assayed by the ELISA after 3 days of treatment. Data are expressed as percentage of control, and each value is the mean of three experiments in quadruplicate. $^{**}P < .01$ versus vehicle and $^{oo}P < .01$ versus PrP106-126 values.

We also focused our study on the role of MAP kinases, and of p38 and ERK1/2, in particular, in the proapoptotic effects of PrP106-126. The p38 MAP kinase, beyond its role in immune response, represents an intracellular effector transducing different cell insults, including trophic factors withdrawal, oxidative stress, and glutamate toxicity.[21–25]

The activation/phosphorylation of p38 MAP kinase by PrP106-126 was directly assessed by Western blot. FIGURE 3A reports the densitometric analysis of the results obtained from three independent experiments expressed as ratio between the thr180/tyr182-phosphorylated enzyme and the total p38 levels. The treatment for 3 days with PrP106-126 (50 μM) caused a significant activation of p38 (FIG. 3A), although the phosphorylation of p38 started as early as 12 h of treatment (data not shown).

Conversely, PrP106-126 did not change ERK1/2 phosphorylation (data not shown). We used the selective MEK inhibitor PD98059 and the p38 MAP kinase blockers PD169316 and SB203580 to better investigate the involvement of MAP kinases in SH-SY5Y apoptotic death induced by PrP106-126. These compounds were added together with the peptide and apoptosis was assessed by ELISA after 3 days. Although p38 inhibitors (SB203580 and PD169316) totally abolished DNA cleavage induced by PrP106-126, PD98059 was unable to prevent the apoptotic process (FIG. 3B). These data confirm that the activation of p38 MAP kinase pathway is a critical step for the execution of apoptosis in our cell model, whereas ERK1/2 is not involved in such an effect.

Then, we analyzed whether the neuronal toxicity induced by PrP106-126 is caused by soluble or fibrillar isoforms of this peptide. To correlate the structural state of PrP106-126 with its neurotoxicity, we synthesized a mutated peptide (PrP106-126AA) in which the Gly114 and Gly119 were substituted with alanine residues. We choose to synthesize this mutated peptide because using the helix prediction algorithm AGADIR (data not shown) we identified the glycine in position 114 and 119 as key determinants for the three-dimensional structuration of the peptide. The choice of alanine to replace the glycines has been taken because it was considered the most conservative replacement. The mutated PrP106-126AA, in contrast with the *wild-type* peptide that rapidly aggregated and precipitated, resulted completely soluble in water. Interestingly, although the aggregates formed by PrP106-126 after 24 h of incubation at 37°C showed, in accordance with a previous report,[4] the characteristics of amyloid fibrils, no fibrils were detected in extensive electron microscopy analysis using PrP106-126AA preparations (data not shown). These observations were also confirmed using the Congo red assay that more specifically identifies amyloid structures.[7,19] As shown in FIGURE 4, PrP106-126 had the largest amount of Congo red bound, with the PrP106-126AA showing a binding less than 30% than the *wild-type* peptide.

Thus, we identified the Gly114 and 119 as major determinants of the amyloidogenic properties of PrP106-126. Subsequently, we analyzed whether the PrP106-126AA, despite being unable to aggregate in amyloid fibrils, retains the neu-

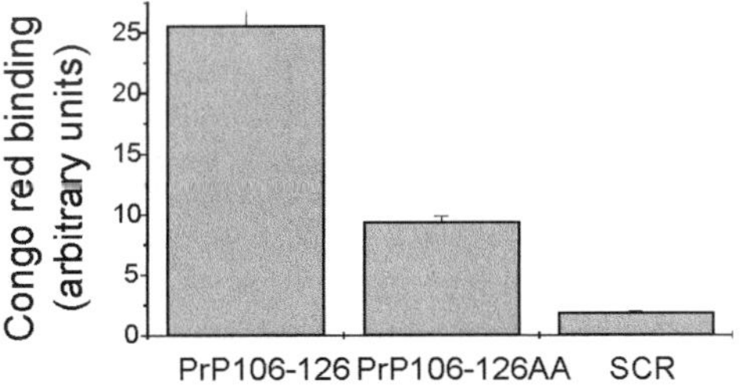

FIGURE 4. Quantitation of Congo red binding by PrP106-126 and mutated peptides. Fifty micrograms of the peptides were incubated with 10 μM Congo red for 8 h at room temperature. The amount of Congo red bound was quantified by CRB (M) = (A540 ÷ 25,295) – (A488 ÷ 46,306), where CRB (M) is the molar concentration of Congo red bound, and 25,295 and 46,306 are the molar extinction coefficients of bound and unbound Congo red, respectively. Data are shown as arbitrary units, as previously reported.[7]

TABLE 2. Time course of SH-SY5Y cell death induced by PrP106-126 and PrP106-126AA

Days	PrP106-126 wt	PrP106-126AA	SCR
0	100	100	100
1	98 ± 4.9	90 ± 6.3	97 ± 48
3	89 ± 4.6*	68 ± 4.7**	95 ± 4.7
4	77 ± 3.8**	62 ± 4.3**	97 ± 4.8
5	57 ± 2.8**	44 ± 3.1**	92 ± 4.6
7	52 ± 2.6**	31 ± 2.2**	91 ± 4.5

NOTE: Cells were treated with PrP106-126 PrP106-126AA or the scrambled peptide SCR (100 μM), and cell viability was assessed by MTT test after 1, 3, 4, 5, and 7 days. Data are expressed as a percentage of vehicle-treated samples, and each point represents the average of three experiments, performed in quadruplicate. $^{**}P < .01$ versus control values.

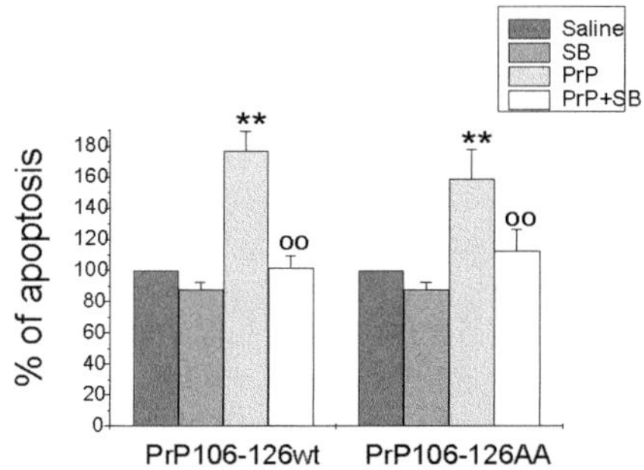

FIGURE 5. Effect of p38 MAP kinase inhibition on PrP106-126 and PrP106-126AA–induced apoptosis of SH-SY5Y cells. Cells were treated for 24 h with PrP106-126 and PrP106-126AA (100 μM) in the presence or absence of the p38 inhibitor SB203580 (10 mM) and then evaluated with the ELISA to detect cytosolic mono- and oligonucleosomes. The values were expressed as a percentage of the vehicle-treated control. The average of three independent experiments performed in triplicate is represented. $^{**}P < .01$ versus respective saline-treated values.

rotoxic properties of the *wild-type* peptide. For this purpose, we evaluated the cell survival of SH-SY5Y after prolonged treatment with both PrP106-126 *wild-type* and PrP106-126AA. PrP scrambled peptide was used as internal control of the specificity of the observed effects. As showed in TABLE 2, PrP106-126 caused a significant cell death starting after 4 days of treatment and reaching a maximum after 7 days (–48% of cell viability). Interestingly, PrP106-126AA showed, compared with *the wild-type* peptide, an increased toxicity, significantly impairing cell survival after only 3 days of treatment and reaching a maximal cell death effect after 7 days (approximately 69% of reduction of cell viability). As expected, the scrambled peptide did not affect cell survival even after prolonged treatment (7 days; TABLE 2).

To analyze whether PrP106-126AA retains the capability to activate the same cytotoxic intracellular pathways than the *wild-type* peptide also in the absence of amyloid fibril aggregation, we tested whether the cell death induced by the mutated peptide showed the characteristic of apoptosis and the effect of the pretreatment with the p38 MAP kinase inhibitor SB203580.

As shown in FIGURE 5, PrP106-126AA caused a significant induction of apoptosis (+60% after 24 h of treatment with 100 μM peptide), that was comparable to that induced by the *wild-type* peptide for the same treatment. Moreover, apoptotic effect of both PrP106-126 and its AA mutant were completely reverted in the presence of the P38 inhibitor SB203580 (10 μM), measured by quantification of the cytosolic oligonucleosomes (FIG. 5).

DISCUSSION

The aim of this work was to investigate the molecular components of the cell death machinery triggered by PrP106-126 and to correlate the structural state of PrP106-126 with its neurotoxicity. In particular, we studied the role of caspases and MAP kinases in this process, using the neuronal-like SH-SY5Y clonal cell line.[15] The SH-SY5Y cells are widely used to study intracellular events linked to neurotoxic insults such as trophic factors withdrawal, oxidative stress, DNA damage, and neurodegenerative disorders.[23,26,27] A prolonged exposure of SH-SY5Y cells to PrP106-126 significantly impaired the cell viability in a time- and concentration-dependent manner through the induction of the apoptotic DNA cleavage.

Because the cell death showed hallmarks of apoptosis, we studied the involvement of the caspases in the SH-SY5Y cell death induced by PrP106-126. To address this issue, we investigated whether PrP106-126 treatment caused a direct activation of caspase-3 and if the inhibition of these enzymes could prevent the cell death induced by the peptide. Our data show that PrP106-126 significantly increased caspase-3 activity and the inhibition of this enzyme reverted the apoptotic effects of PrP106-126. We also focused our study on the role of the MAP kinase enzymatic cascades. The p38 MAP kinase, beyond its role in immune response,[28] represents an intracellular effector that transduce different cell insults, including trophic factors withdrawal, oxidative stress, and glutamate toxicity.[21–25] Western blot analysis showed an increase of the immunoreactivity for the phosphorylated/activated p38 MAP kinase when SH-SY5Y cells were exposed to PrP106-126. Moreover, the cotreatment of the SH-SY5Y with PrP106-126 in the presence of the p38 inhibitors SB203580 and PD169316 fully blocked the apoptosis induced by 3 days of treatment with PrP106-126. These results supported the hypothesis of the p38 MAP kinase involvement in the SH-SY5Y death and indicated that the increase of p38 phosphorylation plays a critical role in the cell death triggered by the peptide rather than being a consequence of cellular stresses induced through other pathways. Conversely, the MEK inhibitor PD98059 did not prevent DNA cleavage, enforcing the evidence that the activation of ERK1/2 is not involved in the PrP106-126 death signaling in SH-SY5Y cells. Interestingly, it has been reported that ERK1/2 and p38 pathways exert opposite effects in cell proliferation and degeneration.[29,30] The prion-related peptide PrP106-126 has been shown to trigger several different molecular events and cellular responses depending on the experimental system and cell type. There is growing evidence that both PrP^{Sc} and PrP106-126 can bind different cellular structures and modulate cell response to extracellular stimuli. PrP^{Sc} and PrP106-126 bind the native prion protein and increase the cell sensitivity to the oxidative stress[31,32]; in addition, the PrP^{Sc} has been reported to impair the receptor-mediated calcium response in neuroblastoma cells.[33] We previously have demonstrated that PrP106-126 impairs the maintenance of a correct calcium homeostasis in excitable cells.[8,11] According to these reports, it is conceivable that neuronal apoptosis and glial proliferation induced, *in vitro*, by PrP106-126, are consequences of its capability to interact with a number of membrane-bound structures and intracellular effectors. How PrP106-126 treatment causes the activation of the p38 MAP kinase is still to be determined, and further studies will be required to address this issue. In agreement with our work, it was reported that, in SH-SY5Y cells, PrP106-126 induces caspase-3 activity and apoptosis.[34] Moreover, these researchers show that PrP106-126 alters mitochondrial membrane potential, as main mechanism to induce cell death. Interestingly, in SH-SY5Y cells, different apoptotic stimuli (i.e., NO and dopamine) caused mitochondrial activity impairment through the p38-dependent Bax translocation to mitochondria.[23,35] Thus, we propose that, upon exposure to PrP106-126, the activation of p38 MAP kinase is one of the early intracellular events that causes or allows the mitochondrial depolarization and the apoptotic cascade.

In this work, we also correlate the structural state of this prion fragment with its neurotoxicity. We synthesized a mutated form of PrP106-126, substituting both Gly114 and Gly119 residues with alanines. The mutated PrP106-126AA fragment results completely soluble in aqueous solution and, in contrast with the *wild-type* peptide, does not form a significant amount of amyloid fibrils as indicated by both electron microscopy and Congo red binding. These findings allow the identification of Gly114 and Gly119 as main determinants for the fibrillogenesis of the PrP106-126. Furthermore, we show that the soluble PrP106-126AA is even more neurotoxic than the fibrillogenic *wild-type* peptide. Moreover, both peptides induced toxic effects through the activation of the same mechanism (p38-dependent induction of apoptosis), confirming that the mutations that we introduced in the peptide did not alter its interaction with the cell membrane but only the amyloid fibril formation and that the two phenomena are not interrelated. These results, in line with earlier reports,[36,37] further demonstrate that nonfibrillar variants of this prion fragment retain the toxic properties. Note that a recent report[16] suggests that nonfibrillar aggregates, which precede the formation of mature amyloid fibrils, may be the primary toxic species. Here, we show, in addition to these findings, that also in a nonamyloidogenic state PrP106-126AA peptide causes cell death. Moreover, because both peptides induced apoptosis through the activation of the p38 MAP kinase cascade, it is conceivable that PrP106-126AA behaves similarly to a toxic protofibrillar intermediate of the *wild-type* PrP106-126. In summary, we show that a prolonged exposure of SH-SY5Y cells to PrP106-126 induces apoptotic DNA cleavage and cell death via the activation of both caspase-3 and p38 MAP kinase. Moreover, we identified the glycines in position 114 and 119 of the PrP106-126 sequence as major determinants of the fibrillogenesis of the peptide, and, in line with previous works, we show that the amyloid formation is not necessary for the toxic activity of the peptide.

ACKNOWLEDGMENTS

The financial support provided by CNR-MURST 5% 95/95, by Italian Ministry of Health (2000) "Basi molecolari della degenerazione neuronale da prioni e della neuroprotezione farmacologica" and by MIUR PRIN (2001) to G. Schettini, by Telethon Italy (Grant E.0975), by MIUR PRIN (2001) and MIUR FIRB (2001) to T. Florio, by MIUR PRIN (2001) and MISAN (2001) "Fattori genetici, patogenetici e biochimici responsabili della sensibilità/resistenza alle EST" to A. Aceto is gratefully acknowledged.

REFERENCES

1. PRUSINER, S.B. 1998. Prions. Proc. Natl. Acad. Sci. USA **95:** 13363–13383.
2. STEFANIS, L., R.E. BURKE & L.A. GREEN. 1997. Apoptosis in neurodegenerative disorders. Curr. Opin. Neurol. **10:** 299–305.
3. FORLONI, G., N. ANGERETTI, R. CHIESA, *et al.* 1993. Neurotoxicity of a prion protein fragment. Nature **362:** 543–546.
4. TAGLIAVINI, F., F. PRELLI, L. VERGA, *et al.* 1993. Synthetic peptides homologous to prion protein residues 106-147 form amyloid-like fibrils in vitro. Proc. Natl. Acad. Sci. USA **90:** 9678–9682.
5. DE GIOIA, L., C. SELVAGGINI, E. GHIBAUDI, *et al.* 1994. Conformational polymorphism of the amyloidogenic and neurotoxic peptide homologues to residues 106-126 of the prion protein. J. Biol. Chem. **269:** 7859–7862.
6. SELVAGGINI, C., L. DE GIOIA, L. CANTÙ, *et al.* 1993. Molecular characteristics of a protease resistant amyloidogenic and neurotoxic peptide homologous to residue 106–126 of the prion protein. Biochem. Biophys. Res. Commun. **194:** 1380–1386.
7. JOBLING, M.F., L.R. STEWART, A.R. WHITE, *et al.* 1999. The hydrophobic core sequence modulates the neurotoxic and secondary structure properties of the prion peptide 106–126. J. Neurochem. **73:** 1557–1565.
8. THELLUNG, S., T. FLORIO, V. VILLA, *et al.* 2000. Apoptotic cell death and impairment of L-type voltage-sensitive calcium channel activity in rat cerebellar granule cells treated with the prion protein fragment 106-126. Neurobiol. Dis. **7:** 299–309.
9. THELLUNG, S., T. FLORIO, A. CORSARO, *et al.* 2000. Intracellular mechanisms mediating the neuronal death and astrogliosis induced by the prion protein fragment 106-126. Int. J. Dev. Neurosci. **18:** 481–492.
10. THELLUNG, S., V. VILLA, A. CORSARO, *et al.* 2002. p38 MAP kinase mediates the cell death induced by PrP106-126 in the SH-SY5Y neuroblastoma cells. Neurobiol. Dis. **9:** 69–81.
11. FLORIO, T., M. GRIMALDI, A. SCORZIELLO, *et al.* 1996. Intracellular calcium rise through L-type calcium channels, as molecular mechanism for prion protein fragment 106-126-induced astroglial proliferation. Biochem. Biophys. Res. Commun. **228:** 397–405.
12. HENGARTNER, M.O. 2000. The biochemistry of apoptosis. Science **407:** 770–776.
13. GRUTTER, M.G. 2000. Caspases: key players in programmed cell death. Curr. Opin. Struct. Biol. **10:** 649–655.
14. Mielke, K. & T. Herdegen. 2000. JNK and p38 stress kinases-degenerative effectors of signal-transduction-cascades in the nervous system. Prog. Neurobiol. **6:** 45–60.
15. PAHLMAN, S., J.C. HOEHNER, E. NANBERG, *et al.* 1995. Differentiation and survival influences of growth factors in human neuroblastoma. Eur. J. Cancer. **31A:** 453–458.
16. BUCCIANTINI, M., E. GIANNONI, F. CHITI, *et al.* 2002. Inherent toxicity of aggregates implies a common mechanism for protein misfolding diseases. Nature **416:** 507–511.
17. WELLINGS, D.A. & E. ATHERTON. 1997. Standard Fmoc protocols. Methods Enzymol. **289:** 44–67.
18. STEFANIS, L., D.S. PARK, C. YAN, *et al.* 1996. Induction of CPP32-like activity in PC12 cells by withdrawal of trophic support. J. Biol. Chem. **29:** 30663–30671.
19. KLUNK, W.E., J.W. PETTEGREW & D.J. ABRAHAM. 1989. Quantitative evaluation of Congo red binding to amyloid-like proteins with a beta-pleated sheet conformation. J. Histochem. Cytochem. **37:** 1273–1281.
20. GARCIA-CALVO, M., E.P. PETERSON, B. LEITING, *et al.* 1998. Inhibition of human caspases by peptide-based and macromolecular inhibitors. J. Biol. Chem. **273:** 32608–32613.

21. KAWASAKI, H., T. MOROOKA, S. SHIMOHAMA, *et al.* 1997. Activation and involvement of p38 mitogen-activated protein kinases in glutamate-induced apoptosis in rat cerebellar granule cells. J. Biol. Chem. **272:** 18518–18521.
22. KUMMER, J.L., P.K. RAO & K.A. HEINDENREICH. 1997. Apoptosis induced by withdrawal of trophic factors is mediated by p38 mitogen-activated protein kinase. J. Biol. Chem. **272:** 20490–20494.
23. GHATAN, S., S. LARNER, Y. KINOSHITA, *et al.* 2000. p38 MAP kinase mediates Bax translocation in nitric oxide-induced apoptosis in neurons. J. Cell. Biol. **150:** 335–347.
24. KURATA, S. 2000. Selective activation of p38 MAPK cascade and mitotic arrest caused by low level oxidative stress. J. Biol. Chem. **275:** 23413–23416.
25. Ono, K. & J. Han. 2000. The p38 signal transduction pathway. Activation and function. Cell Signal. **12:** 1–13.
26. LI, Y.P., A.F. BUSHNELL, C.M. LEE, *et al.* 1996. Beta-amyloid induces apoptosis in human-derived neurotypic SH-SY5Y cells. Brain Res. **738:** 196–204.
27. MORIYA, R., T. UEHARA & Y. NOMURA. 2000. Mechanism of nitric oxide-induced apoptosis in human neuroblastoma SH-SY5Y. FEBS Lett. **484:** 253–260.
28. HAN, J., J.D. LEE, L. BIBBS & R.J. ULEVITCH. 1994. A MAP kinase targeted by endotoxin and hyperosmolarity in mammalian cells. Science **265:** 808–811.
29. XIA, Z., M. DICKENS, J. RAINGEAUD, *et al.* 1995. Opposite effects of ERK and JNK-p38 MAP kinases on apoptosis. Science **270:** 1326–1331.
30. ZHANG, H., X. SHI, M. HAMPONG, *et al.* 2001. Stress-induced inhibition of ERK1 and ERK2 by direct interaction with p38 MAP kinase. J. Biol. Chem. **276:** 6905–6908.
31. BROWN, D.R. 2000. Prion protein peptides: optimal toxicity and peptide blockade of toxicity. Mol. Cell. Neurosci. **15:** 66–78.
32. WONG, B.S., T. PAN, T. LIU, *et al.* 2000. Prion disease: a loss of antioxidants function? Biochem. Biophys. Res. Comm. **28:** 249–252.
33. KRISTENSSON, K., B. FEUERSTEIN, A. TARABULOS, *et al.* 1993. Scrapie prions alter receptor-mediated calcium responses in cultured cells. Neurology **43:** 2335–2341.
34. O'DONOVAN, C., D. TOBIN & T.G. COTTER. 2001. Prion protein fragment PrP106-126 induces apoptosis via mitochondrial disruption in human neuronal SH-SY5Y cells. J. Biol. Chem. **276:** 43516–43523.
35. JUNN, E. & M.M. MOURADIAN. 2001. Apoptotic signaling in dopamine-induced cell death: the role of oxidative stress, p38 mitogen-activated protein kinase, cytochrome c and caspases. J. Neurochem. **78:** 374–383.
36. FORLONI, G., O. BUGIANI, F. TAGLIAVINI & M. SALMONA. 1996. Apoptosis-mediated neurotoxicity induced by beta-amyloid and PrP fragments. Mol. Chem. Neuropathol. **28:** 163–171.
37. BROWN, D.R. 2000. PrPSc-like prion protein peptide inhibits the function of cellular prion protein. Biochem. J. **352:** 511–518.

Stable-Tau Overexpression in Human Neuroblastoma Cells

An Open Door for Explaining Neuronal Death in Tauopathies

PATRICE DELOBEL,[a] CHRISTEL MAILLIOT,[a] MALIKA HAMDANE,[a] ANNE-VÉRONIQUE SAMBO, SÉVERINE BÉGARD, ANNE VIOLLEAU, ANDRÉ DELACOURTE, AND LUC BUÉE

INSERM U422, Institut de Médecine Prédictive et Recherche Thérapeutique, Place de Verdun, F-59045 Lille cedex, France

ABSTRACT: Many neurodegenerative disorders referred to as "tauopathies" are characterized by the accumulation and aggregation of Tau proteins into filaments. In these pathologies, Tau proteins are hyperphosphorylated and also abnormally phosphorylated. Moreover, they differ from each other by the preferential aggregation of isoforms exhibiting either three microtubule-binding repeats (3R) or four repeats (4R) Tau. To investigate the effects of an intracellular accumulation of Tau, we stably transfected neuroblastoma cell line SY5Y with either 3R or 4R Tau. Our data showed that an increase in intracellular Tau expression has led to their hyperphosphorylation. Conversely, an abnormal Tau phosphorylation and/or aggregation were never observed. Furthermore, SY5Y cells transfected with 4R Tau showed an increased susceptibility to cell death. Finally, in apoptotic conditions, Tau proteins were degraded at their carboxy terminus by caspase, leading to an apparent decrease in Tau phosphorylation in this region. Because truncated Tau generated during apoptosis are not commonly found in Tau aggregates, apoptotic processes may not be of interest in neurofibrillary degeneration.

KEYWORDS: Tau proteins; FTDP-17 (frontotemporal dementia with parkinsonism linked to chromosome 17); hyperphosphorylation; abnormal phosphorylation; aggregation; apoptosis

INTRODUCTION

Intracellular aggregates of highly insoluble filamentous materials are a common feature of numerous neurodegenerative disorders (for recent review, see Lee *et al.*[1]) Among them, Tauopathies represent a subtype of neurodegenerative diseases in which these intracellular aggregates are made of microtubule-associated Tau proteins. In these disorders, Tau proteins are found abnormally phosphorylated and aggregated into intracellular filaments, leading to neuronal death.[2,3]

Address for correspondence and reprint requests: Dr. Luc Buée, INSERM U422, Place de Verdun, F-59045 Lille cedex, France. Voice: +33-0-320-622074; fax: +33-0-320-622079. Luc.Buee@lille.inserm.fr

[a]Patrice Delobel, Christel Mailliot, and Malika Hamdane contributed equally to this study.

Ann. N.Y. Acad. Sci. 1010: 623–634 (2003). © 2003 New York Academy of Sciences.
doi: 10.1196/annals.1299.115

Tau proteins are primarily expressed in neurons. They range from 45 to 74 kDa when separated by polyacrylamide gel electrophoresis in presence of sodium dodecyl sulfate (SDS-PAGE). Within human brain, six developmentally regulated isoforms are generated by alternative splicing.[4,5] They differ from each other by the presence of three (3R) or four microtubule-binding domains (4R) of 31 or 32 amino acids each in the carboxy-terminal half of the molecule, and the insertion or not of one or two amino-terminal sequences (29 or 58 amino acids).[6]

Recently, different teams have reported mutations in the *tau* gene as the main cause of frontotemporal dementia with parkinsonism linked to chromosome 17 (FTDP-17) (see references for reviews[1–3,7,8]). Approximately 20 mutations have been described in the *tau* gene and can be classified in two groups. Mutations of the first group affect the alternative splicing of exon 10 leading to a deregulation in the ratio of 4R to 3R *tau* isoforms. The second group includes mutations that are located in the vicinity of microtubule binding domains, and that decrease Tau-microtubule interactions.[9] In both groups, there is an increase in cytosolic Tau protein and/or a change in 3R/4R ratio (for reviews, see Lee *et al.*[1]; Buée *et al.*[2]).

This change in the balance 3R and 4R Tau isoforms is also encountered in other Tauopathies. In progressive supranuclear palsy (PSP) and corticobasal degeneration (CBD), there is a preferential aggregation of 4R Tau proteins,[10] whereas, in Pick's disease filaments, mostly 3R Tau isoforms are found.[11–14] Finally, mice overexpressing the 4R Tau protein showed axonopathy.[15] Taken together, all these findings suggest that Tau overexpression, and especially 4R Tau overexpression, may be deleterious for the neurons.

In this study, we addressed the question of whether a specific neuronal Tau content, for instance, 3R versus 4R isoforms, could account for a differential neuronal vulnerability to neurodegeneration. To answer this question, we have established stably transfected human neuroblastoma SKNSH-SY5Y cells with either 3R or 4R *tau* cDNAs. These cells then were submitted to apoptotic stimuli to test their resistance toward apoptosis in function of their Tau content. The phosphorylation state of Tau proteins was also investigated as well as the relationships between apoptosis and Tau protein phosphorylation.

MATERIALS AND METHODS

Antibodies

Several anti–Tau antibodies spanning the entire Tau molecule (numbering is given according to the longest human Tau isoform, 441 amino acids) were used for biochemical experiments. AD2 recognizes phosphorylated Ser-396 and Ser-404 in Tau.[16] AT270 is directed against phosphorylated Thr-181.[17] M19G and Tau Cter are two well-characterized immune sera recognizing the first 19 amino acids of the Tau sequence encoded by the exon 1[16] and the carboxy-terminal fragment of Tau (last 15 amino acids),[18] respectively.

During staurosporine (SSP) treatment, apoptosis was investigated regarding poly-(ADP-Ribose)-polymerase (PARP) cleavage using an anti–PARP antibody (Roche Molecular Biochemicals, Germany).

Cell Culture and Stable Tau Transfections

SY5Y human neuroblastoma cells were grown in 25-cm^2 flasks in Dulbecco's modified Eagle medium supplemented with 10% fetal calf serum, 2 mM L-glutamine, 1 mM nonessential amino acids, and penicillin/streptomycin (Life Technologies) in a 5% CO_2 humidified incubator at 37°C. cDNAs of 2+3–10– (3R) and 2+3–10+ (4R) Tau isoforms cloned in pSG5 vector (Stratagene) were a kind gift from Dr. Michel Goedert (Cambridge, UK). They were subcloned into stably eucaryotic expression vector pcDNA3.1Neo (Invitrogen), allowing for a eucaryotic G418 (Life Technologies) selection of stable clones. These 3R and 4R *tau* cDNAs were independently stably transfected into SY5Y cells using the ethyleneimine polymer ExGen500 (Euromedex, France) according to manufacturer's instructions. A mock cell line was stably transfected with the pcDNA3.1Neo vector alone, without any *tau* cDNA insert.

mRNA Isolation and RT-PCR Analysis

Total cellular RNA was extracted with the RNeasy kit (Qiagen) according to manufacturer's instructions. For first-strand cDNA synthesis, 20 pM of reverse primer, 0.5 mM of each dNTP (provided in the kit), 4 units of Omniscript Reverse Transcriptase (Qiagen Omniscript Reverse Transcriptase kit) and RNase inhibitor (RNaseOUT, Life Technologies) were added to 1 μg of total RNA in the convenient RT Buffer, as recommended by manufacturer's instructions. The reaction mixture without the RT enzyme was first incubated at 80°C for 5 min to abolish all secondary structures. Then, after addition of the enzyme, the cDNA first strand synthesis was performed for 60 min at 37°C. For the 100-μL PCR mixture, 0.3 unit of Expand High Fidelity polymerase enzyme mix (Roche), 20 pM of forward primer, and the provided enzyme buffer with 15 mM $MgCl_2$ were added to 15 μL of reverse transcription reaction mixture. Different couples of primers were used to amplify either the exon 2 and 3 or the exon 10–encoded sequences. PCR was performed in a Perkin-Elmer thermal cycler as previously described.[18] The double-strand DNA thus obtained was characterized by 2% gel agarose electrophoresis, and digestion was allowed with restriction endonucleases.

Tau Protein or Total Cell Homogenates Preparation and Immunoblotting

For biochemical analyses of the different cell lines Tau content, Tau proteins were prepared as follows: cells were harvested in Tris-EDTA solution at 4°C and centrifuged. Cell pellets were disrupted, homogenized, and sonicated in 0.3 M NaCl-Tris-Tween buffer in presence of a cocktail of protease inhibitors (Complete, Roche). After 30 min centrifugation at 13,000 *g*, supernatants were collected, diluted in Laemmli sample buffer ×2, and boiled again for an additional 5 minutes. For total protein content analyses, total cell homogenates were generated as previously described.[19] After convenient preparation of the samples, total protein concentrations were determined by using the BCA protein assay kit (Pierce). Electrophoresis and immunoblotting then were performed as previously described.[16,19] In brief, samples were loaded onto 10 or 15% SDS-PAGE (BioRad mini gel Protean III system). After transfer and blocking, membranes were incubated with primary antibodies at 4°C overnight. Horseradish peroxidase–conjugated antibodies (Sigma) were used as secondary antibodies and reaction product was detected using the Amersham ECL Western blotting system.

Analyses of Cell Lines Metabolism and Resistance to Apoptosis

To analyze cell metabolism, we seeded cells into 24-well plates and allowed them to attach for 24 h. They then were washed two times in PBS and grown in serum-free medium containing 0.2 % (w/v) BSA for 24 h. After that, cells were replaced in normal culture conditions as described above. A kinetic was performed for 36 h, and cells were fixed using 4% paraformaldehyde in PBS buffer. Nuclei were visualized after Hoechst 33258 staining allowing the detection of apoptosis and the different mitosis steps. Cell death also was measured using cytotoxic kits (see below).

We measured the mitochondrial activities of the different cell lines by using the CellTiter 96 AQ_{ueous} kit (Promega, France). For this purpose, we split the cells at the same low number (5×10^4 cells per 48 wells) and let them grow until the first cell line reached confluence. Then, we performed the CellTiter 96 AQ_{ueous} assay on all cell lines, whether or not they had reached confluence. This colorimetric method measures the ability of mitochondrial enzymes from metabolically active cells to transform MTS tetrazolium substrate into a soluble formazan product. Absorbance then is measured at 490 nm, and DO value is directly proportional to the number of viable cells in the wells.

For apoptosis experiments, we induced programmed cell death with 1 μM staurosporin (SSP) according to Chakravarthy *et al.*[20] We then measured the quantity of apoptosis either through a direct evaluation of the number of dead cells with the CytoTox 96 Non-Radioactive Cytotoxicity kit (Promega, France), which directly quantifies the quantity of lactate dehydrogenase (LDH) released in the culture medium by dead or dying cells, or through the indirect measure of mitochondria activity of still living cells with the CellTiter 96 AQ_{ueous} kit. The CytoTox 96 Non-Radioactive Cytotoxicity kit (Promega, France) also was used to measure the spontaneous cell death in each condition. Apoptosis also was assessed using the Apoptosis detection System Fluorescein (Promega).

RESULTS

Molecular and Biochemical Characterization of 3R and 4R Tau Cell Lines

Several different stable clones were generated through G418 selection. The first step in their characterization was to determine their *tau* content. RT-PCR was performed on total cell RNAs to amplify either exon 2– or exon 10–encoded sequences. In both cases, the primers also allowed DNA amplification of the endogenous *tau* isoform, which does not contain any of the alternatively spliced sequences. Indeed, the endogenous major *tau* isoform expressed by SY5Y cells corresponds to the smallest human Tau, also called the fetal one.[21] The PCR cycles were kept low to coamplify both isoforms, the endogenous as well as the exogenous one, and to give a true representation of the relative amounts of the different isoforms. Mock lines (transfection with empty vector) were also made. These lines did not show any amplification of alternatively spliced sequence. Regarding 3R and 4R clones, coampli-

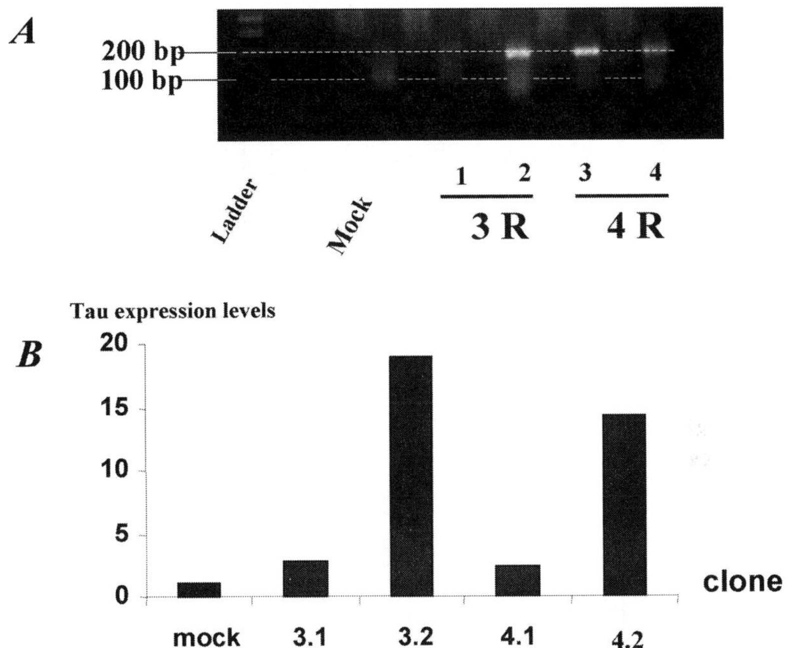

FIGURE 1. Cell line characterization. (**A**) Transgene expression in different cell lines. Transgene expressions were estimated by RT-PCR after mRNA isolation. Exogenous and endogenous *tau,* which were amplified using primers corresponding to exon 1 and 4, migrate at 199 and 112 bp when compared with the ladders. The mock cell line without transgene expression showed the endogenous *tau* expression. 3R or 4R *tau* clones also were characterized. Low or high degrees of transgene expression could be detected. (**B**) Endogenous or exogenous Tau protein content quantification. Endogenous and exogenous Tau protein expressions then were quantified using the Imagemaster program after M19G Western blotting. The mock cell line represents the endogenous Tau protein quantity. Exogenous 3R (panels 3.1 and 3.2) or 4R (panels 4.1 and 4.2) were chosen to have similar (3.1 and 4.1) or higher (3.2 and 4.2) expression than endogenous Tau.

fication of the endogenous *tau* DNA by PCR (25 cycles) allowed us to classify all clones: those expressing the transgene in a low degree with similar expression than endogenous *tau* isoform and others overexpressing the transgene compared with the endogenous *tau* isoform (FIG. 1A). All biochemical analyses confirmed these observations (see FIGS. 1B and 2): the polyclonal antibody M19G, which recognizes all Tau proteins independently of their phosphorylation state, labeled a weak band around 55 kDa in all cell lines. This latter corresponded to the endogenous Tau isoform (2–3–10–) (see FIG. 2, panel M19G). In some clones in which the RT-PCR experiment did not allow any amplification, there was no labeling of the exogenous Tau isoforms, whereas in other clones, a weak or more intense labeling was obtained with this polyclonal antibody. 3R and 4R Tau proteins migrated with an apparent molecular weight around 65 and 69 kDa respectively. For all further experiments, to

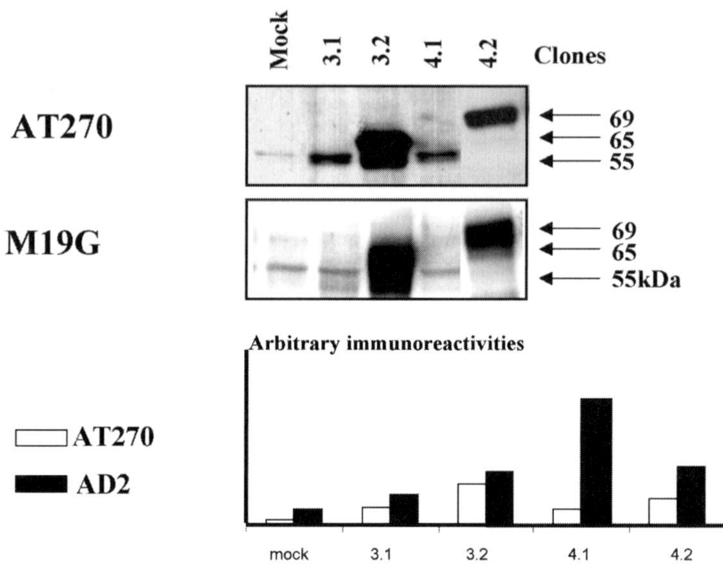

FIGURE 2. Tau phosphorylation in 3R and 4R cell lines. Immunoblot analysis of the phosphorylation state of Tau protein in naive neuroblastoma cell lines. All Tau isoforms (endogenous, 3R or 4R exogenous) were phosphorylated at AT270 (*top panel*) epitope. Note that AT270 immunoreactivity was more important for 3R and 4R clones than for the mock cell line. Total Tau proteins were estimated using M19G polyclonal antibody (*bottom panel*). Schematic representation of the Tau protein phosphorylation in 3R and 4R cell lines. Immunoreactivities with AD2 and AT270 were quantified and reported as a ratio on total Tau protein quantities observed with M19G. Note that overexpression of Tau proteins (3R and 4R) in neuroblastoma cells resulted in an increase of Tau protein phosphorylation for both AD2 and AT270. Moreover, the increase observed was correlated with the expression of 3R Tau but not with 4R Tau, which decreased in the line 4.2 when these 4R Tau proteins were overexpressed, compared with line 4.1 where the same expression of exogenous and endogenous was observed. This apparent decrease in phosphorylation at AD2 epitope in 4.2 clone is related to the loss of the carboxy-terminal Tau fragment during spontaneous apoptosis.

determine if overexpression of 3R or 4R Tau leads to neurodegeneration, two types of clones expressing either 3R or 4R Tau proteins were chosen. The first ones exhibited the same exogenous and endogenous Tau amounts. The second overexpressed Tau 10-fold more than the endogenous Tau quantities (FIG. 1, panel B).

Tau Phosphorylation in 3R and 4R Cell Lines

Further investigations were performed on these clones to analyze the phosphorylation state of exogenous Tau proteins. Antibodies AD2 and AT270, as shown in FIGURE 2, were used. Both antibodies labeled an immunoreactive band around 65 or 69 kDa for 3R or 4R clones, respectively, indicating that exogenous Tau isoforms were phosphorylated. Interestingly, the amount of expressed exogenous 3R Tau isoforms was correlated to the levels of phosphorylation, whereas it was not observed with 4R Tau isoforms. This could be visualized with M19G labeling after AT270 and AD2 (FIG. 2). In fact, even if an increase in the 4R Tau phosphorylation was observed, we also showed a differential phosphorylation on AD2 and AT270 epitopes (FIG. 2). However, we were unable to obtain any labeling either in the 3R or in the 4R cells with Alzheimer's disease–specific antibodies, such as AT100 or 988 (data not shown) suggesting that the exogenous Tau isoforms expressed in these cells are phosphorylated in a physiological but not in a pathological state.

Morphological Changes and Differential Metabolism in Function of Tau Content

From a morphological point of view, some differences between 3R and 4R cell lines could be highlighted. Considering nondifferentiated cells (i.e., cells growing in 15% FCS, or cells not submitted to a differentiation protocol), 3R cell lines, independently of the exogenous gene expression, exhibited a more differentiated morphology than 4R cell lines (data not shown). In fact, in these normal culture conditions, 3R cell lines had a high number of neurites and were grown slower than 4R cell lines. We have assessed the cell cycle of the different nondifferentiated cell lines. Control and 3R-Tau cell lines with a cell cycle of 26 h grew slower than 4R-Tau cell line with a 20-h cell cycle (data no shown). Surprisingly, using the Promega CellTiter 96 AQ_{ueous} kit, a higher mitochondrial activity was detected in 3R Tau cell lines compared with 4R Tau cell lines (FIG. 3A, left panel). This was confirmed using the CytoTox 96 Non-Radioactive Cytotoxicity kit (Promega, France) showing that more cells were dead in 4R Tau cell line versus 3R Tau cell line (FIG. 3A, right panel). Moreover, the number of cells showing DNA fragmentation after DAPI was correlated to the expression of 4R Tau (FIG. 3B). This correlation was not observed in cells overexpressing 3R Tau. These data suggested that 4R Tau overexpression was more deleterious for SY5Y cells than 3R Tau, probably by enhancing apoptosis. Similar data were obtained in differentiated cell lines in serum-free medium conditions with either 125 nM NGF or 10^{-8} M retinoic acid (RA). By immunoblotting, expression of high molecular mass neurofilaments was observed as early as 1 week in RA or 3 days in NGF (data not shown). 3R cell lines differentiated faster than 4R ones. In fact, 4R Tau cell line appeared to die upon differentiation. These data were consistent with those observed in undifferentiated cell lines. Taken together, these results suggested that 4R overexpression weakened cells.

Differential Cell Viability According to the Tau Content

Regarding these results on a differential cell death between 3R and 4R cells, we propose that overexpression of 4R Tau isoforms within a neuronal cell line could predispose it to death. To test this hypothesis, we submitted the different Tau-overexpressing cell lines (i.e., 3.2 and 4.2 clones with bigger Tau expression) to 1 μM SSP-induced apoptosis and evaluated the resistance of the different cell lines toward SSP-induced cell death (FIG. 3C). Twelve and 16 h after induction of apoptosis by 1 μM SSP, the amount of dying or still living cells for each cell line were assayed by CellTiter 96 AQ_{ueous} and CytoTox 96 kits. In both cases, we observed that 4R cell lines exhibited a significant higher mortality than 3R cells when treated for 12 or 16 h with 1 μM SSP (FIG. 3C). These data reinforced our hypothesis that 4R Tau overexpression renders cells more vulnerable to apoptotic death.

Link between Tau Protein Degradation and Caspase Activation

The link between Tau protein phosphorylation and degradation during apoptosis was also investigated. For this purpose, a time course experiment during 6 h after 1 μM SSP treatment of 3.2 clone was performed. The choice for this clone was evident because 4R Tau clones were too sensitive to apoptosis. Both phosphorylation and degradation of Tau proteins were analyzed during SSP-induced apoptosis (FIG. 4). The PARP cleavage indicated that caspases were activated rapidly after SSP

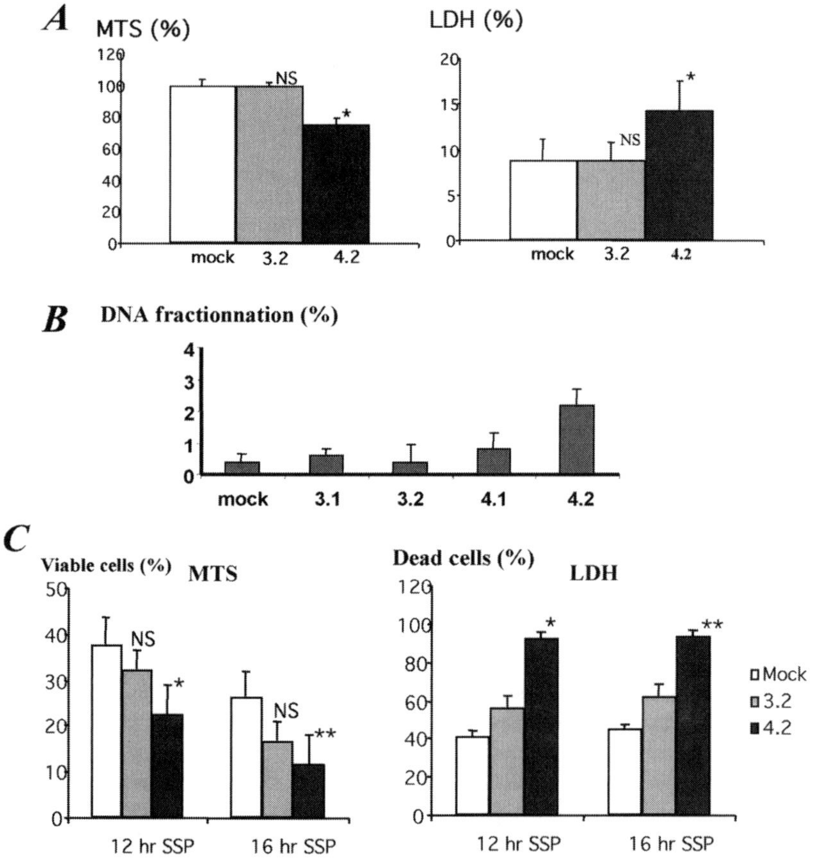

FIGURE 3. Metabolism analysis of 3R and 4R cell lines. (**A**) Cell viability in different clones. Cell viability was quantified by both MTS and LDH tests (see MATERIALS AND METHODS). Cell viability or death was quantified for clones overexpressing either 3R (3.2) or 4R (4.2) Tau proteins and compared with the mock cell line. (**B**) Spontaneous apoptosis quantification. Nuclear fragmentation was used to quantify spontaneous apoptosis in the different cell lines, after DAPI staining. The schematic representation was done after more than 1,000 cells were counted and reported in percentage of apoptotic cells on the totality of viable cells. (**C**) Induced apoptosis measurement in different cell lines. Apoptosis was induced by 1 μM staurosporine (SSP) and quantified after 12 and 16 h. Resultant cell viability for 3R (3.2) or 4R (4.2) Tau clones was compared with the mock cell line.

treatment. Indeed, just 1 h after apoptosis induction, the 113-kDa PARP, a caspase substrate, was cleaved into an 89-kDa product, demonstrating that there was a caspase activation during SSP treatment.[22,23] During the same time, Tau proteins were degraded at their carboxy-terminal extremity. Indeed, AD2- and anti-Tau Cter immunoreactivities decreased rapidly and finally disappeared after 6 h of SSP treatment. In parallel to these data, M19G immunoreactivity (anti–amino-terminal Tau

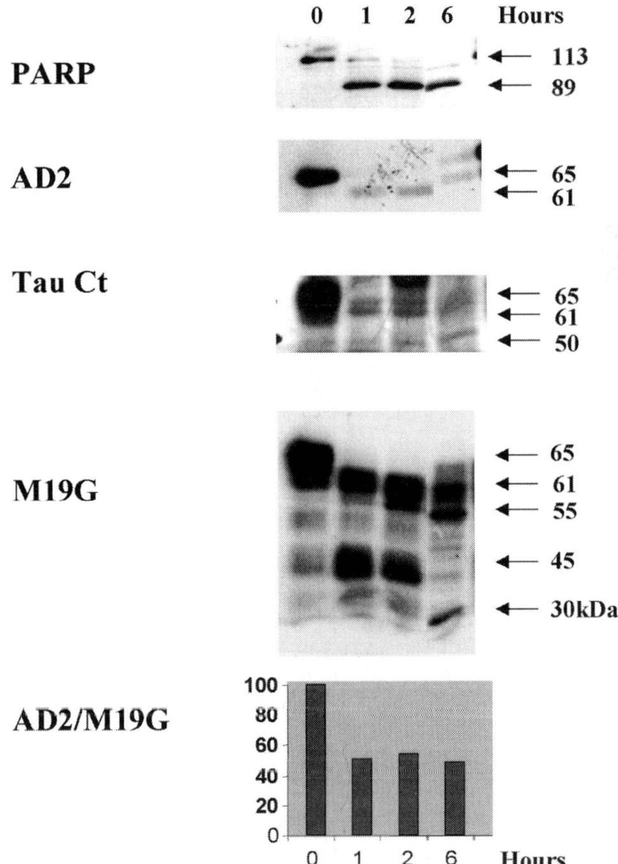

FIGURE 4. Tau protein degradation during apoptosis. Apoptosis was induced on 3.2 clone using 1 μM SSP during 6 h. Western blotting was performed with anti–PARP antibody. The appearance of cleaved PARP (89 kDa) after 1-h SSP treatment confirmed the caspase activation during the apoptosis induction. Tau was detected using AD2 phospho-dependent monoclonal antibody, Tau Cter, and M19G polyclonal antibodies. Phosphorylation-dependent or independent antibodies AD2 and Tau Cter were used to analyze the phosphorylation or the degradation of Tau proteins at carboxy-terminal part. M19G was also used to determine Tau content after apoptosis induction. Note that molecular weight is noted at the right. Histogram showing the ratio of Cter phosphorylation of Tau (AD2) on total Tau content (M19G) during SSP treatment. Loss of AD2 immunoreactivity may be explained by the carboxy-terminal degradation of Tau.

antibody) provided evidence that during apoptosis induction Tau proteins were effectively degraded more rapidly at their carboxy-terminal extremity, and that this proteolysis occurred at two levels: the first one generates high molecular weight Tau (61 and 55 kDa) and the second one low molecular weight (45 and 30 kDa). Furthermore, Tau proteins were also degraded at their amino-terminal extremity, but less rapidly (as showed by catabolism with Tau carboxy-terminal antibody [Cter]).

DISCUSSION

We have generated different cell lines expressing either 3R or 4R Tau isoforms, without differences in their N-terminal part. The different clones were characterized from molecular and biochemical points of view. Different levels of Tau expression were observed, from no expression to overexpression of Tau. We first reported a close correlation between Tau expression and levels of Tau phosphorylation within the cells. This could be assessed by the progressive increase of Tau molecular mass in Tau-overexpressing cells as compared with cell lines with lower Tau expression. Tau overexpression may simply lead to its hyperphosphorylation. In fact, SSP treatment indicated that Tau proteins were degraded during apoptosis and especially at their C-terminal extremity. This result was in agreement with recent findings, which show that Tau protein could be cleaved at their C-terminal extremity by caspases and become itself an effector of apoptosis.[24,25] Moreover, in these studies, it was demonstrated that 4R Tau proteins had a bigger proapoptotic effect than 3R Tau. Because spontaneous apoptosis was observed for the clone overexpressing 4R Tau isoform, we could assume that the loss of AD2 immunoreactivity for this clone is correlated to caspase activation and apoptosis. Consequently, levels of phosphorylation were also correlated to Tau overexpression in this clone (See AT270, FIG. 2), but this correlation was not visualized for phosphorylation sites (i.e., AD2) located at the carboxy-terminal Tau part.

Finally, Alzheimer-type phosphorylation of Tau proteins was never observed in this cell model. For instance, Alzheimer's disease–specific antibodies such as AT100, 988, or TG3 did not label Tau in 3R and 4R cells in physiological conditions. Conversely, when cells were treated by drugs such as either okadaic acid,[19] which inhibits Tau phosphatases, or mitotic inducers such as nocodazole or taxotere,[26] abnormal Tau phosphorylation was observed. Moreover, recent experiments also show that aggregation of Tau proteins occurred after abnormal phosphorylation and oxidative stress.[27] Our results suggest that even if overexpression and cell accumulation of Tau proteins leads to their hyperphosphorylation, it was not sufficient to their abnormal phosphorylation or aggregation.

Because several Tauopathies known currently are characterized by Tau overexpression, we studied Tau-overexpressing cell lines. In this work, we demonstrated that 4R Tau isoform overexpression generates a higher susceptibility to death with an increasing sensitivity for apoptosis. In other words, 4R Tau–overexpressing cells are less resistant to apoptosis than 3R cells. This result may be correlated to microtubule network dynamic. Indeed, it was recently shown that 4R Tau isoforms bind to microtubules four times more than 3R Tau isoforms.[28] So, the sensitivity of the 4R-overexpressing clones for neuronal death may be related to loss of microtubule dynamics. These observations are of particular interest regarding the hypothesis that

Tau overexpression would lead to neurodegeneration. They strengthen the conclusions of previous works from our laboratory,[2,13,19,29] indicating that different subsets of neurons can be distinguished by their Tau content and exhibit a differential vulnerability according to a given neurodegenerative disorder.

ACKNOWLEDGMENTS

We thank our colleagues from U422 INSERM laboratory for helpful discussion, Dr Michel Goedert (MRC Cambridge UK) for providing *tau* cDNA. P.D. is a recipient of a MRT (French Research ministry) Fellowship. These studies were supported by Centre National de la Recherche Scientifique (CNRS), Institut National de la Santé et de la Recherche Médicale (INSERM), grants from the Institute for the Study of Aging (New York), the "Région Guadeloupe," the "Région Nord Pas-de-Calais" (Génopole de Lille), and the F.E.D.E.R.

REFERENCES

1. Lee, V.M., M. Goedert & J.Q. Trojanowski. 2001. Neurodegenerative tauopathies. Annu. Rev. Neurosci. **224:** 1121–1159.
2. Buée, L., T. Bussière, V. Buée-Scherrer, *et al.* 2000. Tau protein isoforms, phosphorylation and role in neurodegenerative disorders. Brain Res. Rev. **33:** 95–130.
3. Goedert, M. 2001. The significance of tau and alpha-synuclein inclusions in neurodegenerative diseases. Curr. Opin. Genet. Dev. **11:** 343–351.
4. Goedert, M., M.G. Spillantini, M.C. Potier, *et al.* 1989. Cloning and sequencing of the cDNA encoding an isoform of microtubule-associated protein tau containing four tandem repeats: differential expression of tau protein mRNAs in human brain. EMBO J. **8:** 393–399.
5. Goedert, M., M.G. Spillantini, R. Jakes, *et al.* 1989. Multiple isoforms of human microtubule-associated protein tau: sequences and localization in neurofibrillary tangles of Alzheimer's disease. Neuron **3:** 519–526.
6. Lee, G., R.L. Neve & K.S. Kosik. 1989. The microtubule binding domain of tau protein. Neuron **2:** 1615–1624.
7. Forman, M.S., V.M. Lee & J.Q. Trojanowski. 2000. New insights into genetic and molecular mechanisms of brain degeneration in tauopathies. J. Chem. Neuroanat. **20:** 225–244.
8. Goedert, M. & M.G. Spillantini. 2001. Tau gene mutations and neurodegeneration. Biochem. Soc. Symp. **67:** 59–71.
9. Delobel, P., S. Flament, M. Hamdane, *et al.* 2002. Functional characterization of FTDP-17 tau gene mutations through their effects on *Xenopus oocyte* maturation. J. Biol. Chem. **277:** 9199–9205.
10. Sergeant, N., A. Wattez & A. Delacourte. 1999. Neurofibrillary degeneration in progressive supranuclear palsy and corticobasal degeneration: tau pathologies with exclusively "exon 10" isoforms. J. Neurochem. **72:** 1243–1249.
11. Delacourte, A., Y. Robitaille, N. Sergeant, *et al.* 1996. J. Neuropathol. Exp. Neurol. **552:** 159–168.
12. Delacourte, A. & L. Buée. 1997. Normal and pathological Tau proteins as factors for microtubule assembly. Int. Rev. Cytol. **171:** 167–224.
13. Sergeant, N., J.P. David, D. Lefranc, *et al.* 1997. Different distribution of phosphorylated tau protein isoforms in Alzheimer's and Pick's diseases. FEBS Lett. **412:** 578–582.
14. Chambers, C.B., J.M. Lee, J.C. Troncoso, *et al.* 1999. Overexpression of four-repeat tau mRNA isoforms in progressive supranuclear palsy but not in Alzheimer's disease. Ann. Neurol. **46:** 325–332.

15. Probst, A., J. Gotz, K.H. Wiederhold, *et al.* 2000. Axonopathy and amyotrophy in mice transgenic for human four-repeat tau protein. Acta Neuropathol. **99:** 469–481.
16. Buée-Scherrer, V., O. Condamines, C. Mourton-Gilles, *et al.* 1996. AD2, a phosphorylation-dependent monoclonal antibody directed against tau proteins found in Alzheimer's disease. Brain Res. Mol. Brain Res. **39:** 79–88.
17. Goedert, M., R. Jakes, R.A. Crowther, *et al.* 1994. Epitope mapping of monoclonal antibodies to the paired helical filaments of Alzheimer's disease: identification of phosphorylation sites in tau protein. Biochem. J. **301:** 871–877.
18. Sergeant, N., B. Sablonniere, S. Schraen-Maschke, *et al.* 2001. Dysregulation of human brain microtubule-associated tau mRNA maturation in myotonic dystrophy type 1. Hum. Mol. Genet. **10:** 2143–2155.
19. Mailliot, C., N. Sergeant, T. Bussiere, *et al.* 1998. Phosphorylation of specific sets of tau isoforms reflects different neurofibrillary degeneration processes. FEBS Lett. **433:** 201–204.
20. Chakravarthy, B.R., T. Walker, I. Rasquinha, *et al.* 1999. Activation of DNA-dependent protein kinase may play a role in apoptosis of human neuroblastoma cells. J. Neurochem. **72:** 933–942.
21. Dupont-Wallois, L., P.E. Sautiere, C. Cocquerelle, *et al.* 1995. Shift from fetal-type to Alzheimer-type phosphorylated Tau proteins in SKNSH-SY 5Y cells treated with okadaic acid. FEBS Lett. **357:** 197–201.
22. Lippke, J.A., Y. Gu, C. Sarnecki, *et al.* 1996. Identification and characterization of CPP32/Mch2 homolog 1, a novel cysteine protease similar to CPP32. J. Biol. Chem. **271:** 1825–1828.
23. Gu, Y., C. Sarnecki, R.A. Aldape, *et al.* 1995. Cleavage of poly(ADP-ribose) polymerase by interleukin-1 beta converting enzyme and its homologs TX and Nedd-2. J. Biol. Chem. **270:** 18715–18718.
24. Canu, N., L. Dus, C. Barbato, *et al.* 1998. Tau cleavage and dephosphorylation in cerebellar granule neurons undergoing apoptosis. J. Neurosci. **18:** 7061–7074.
25. Fasulo, L., G. Ugolini, M. Visintin, *et al.* 2000. The neuronal microtubule-associated protein tau is a substrate for caspase-3 and an effector of apoptosis. J. Neurochem. **75:** 624–633.
26. Delobel, P., S. Flament, M. Hamdane, *et al.* 2002. Abnormal Tau phosphorylation of the Alzheimer-type also occurs during mitosis. J. Neurochem. **83:** 412–420.
27. Perez, M., F. Hernandez, A. Gomez-Ramos, *et al.* 2002. Formation of aberrant phosphotau fibrillar polymers in neural cultured cells. Eur. J. Biochem. **269:** 1484–1489.
28. Goode, B.L., M. Chau, P.E. Denis & S.C. Feinstein. 2000. Structural and functional differences between 3-repeat and 4-repeat tau isoforms. Implications for normal tau function and the onset of neurodegenetative disease. J. Biol. Chem. **275:** 38182–38189.
29. Sergeant, N., J.P. David, D. Lefranc, *et al.* 1997. Different distribution of phosphorylated tau protein isoforms in Alzheimer's and Pick's diseases. FEBS Lett. **412:** 578–582.

Neuronal Death versus Synaptic Pathology in Alzheimer's Disease

CARLO BERTONI-FREDDARI,[a] PATRIZIA FATTORETTI,[a] MORENO SOLAZZI,[a] BELINDA GIORGETTI,[a] GIUSEPPINA DI STEFANO,[a] TIZIANA CASOLI,[a] AND WILLIAM MEIER-RUGE[b]

[a]*Neurobiology of Aging Laboratory, INRCA Research Department, 60121 Ancona, Italy*

[b]*Institute of Pathology, University Medical School Basel, CH-4002 Basel, Switzerland*

ABSTRACT: Neuron and synapse numeric densities as well as the average size and surface density of the synaptic junctional areas were measured in the hippocampus and cerebellum of adult, old, and demented (Alzheimer's disease) patients. Our findings support the notion that synaptic loss represents *per se* a prominent and early damage affecting zones of the central nervous system reported to show a different vulnerability to age- and pathology-related changes.

KEYWORDS: Alzheimer's disease; apoptosis; neuronal death; neuron density; synaptic density; synapse-to-neuron ratio; synaptic plasticity; hippocampus; cerebellum

Alzheimer's disease (AD) is a devastating, age-related, neurodegenerative pathology leading to dementia.[1] In addition to an abnormal number of senile plaques and neurofibrillary tangles, considered as typical AD neuropathological hallmarks, a significant loss of neurons in discrete zones of the central nervous system (CNS) characterizes the progression of the pathology of this disease.[2] Neuron death is a very specific pathological feature of consistent functional significance: it represents the final outcome of different adverse circumstances; thus, a decrease in neuron density may reliably report on the physiopathological conditions of selected CNS zones. To assess the extent of neuron loss in physiological aging and AD, we conducted a computer-assisted morphometric study on the neuronal numeric density ($Nv_{(Neur)}$) in autopsy samples of the hippocampus (dentate gyrus granular layer) and the cerebellum (granular layer) from adult, old, and old-AD patients. According to a routinely used methodology set up in our laboratory,[3,4] in the same samples, we measured the numeric density ($Nv_{(Syn)}$), average size (S), and overall area of the synaptic contact zones (surface density: Sv) to correlate changes of the synaptic ultrastructural features with the number of neurons present in the same discrete areas of the CNS. In the hippocampal dentate gyrus, $Nv_{(Neur)}$ decreased by 32% and 25% in old and AD patients versus the adult group, respectively, whereas a 10% loss was found be-

Address for correspondence: Dr. Carlo Bertoni-Freddari, Neurobiology of Aging Laboratory, INRCA Research Department, Via Birarelli 8, 60121 Ancona, Italy. Voice: +39-071-8004153; fax: +39-071-206791.

c.bertoni@inrca.it

Ann. N.Y. Acad. Sci. 1010: 635–638 (2003). © 2003 New York Academy of Sciences. doi: 10.1196/annals.1299.116

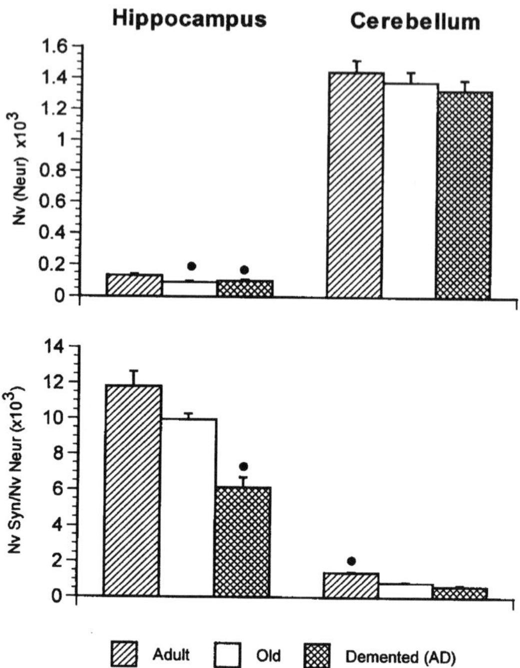

FIGURE 1. (*top histograms*) Neuron numeric density: $Nv_{(Neur)}$. Hippocampus: we found a decrease of 32% and 25% in old and demented patients versus the adult ones. No significant difference was found between old and AD subjects. Cerebellum: no significant change was found among the three groups taken into account. •$P < .05$ versus the adult value. (*bottom histograms*). Number of synapses/neuron: $Nv_{(Syn)}/Nv_{(Neur)}$. The synaptic and neuron densities measured in the same CNS zone are unaffected by alterations caused by age or tissue processing (e.g., shrinkage); therefore, this parameter provides a reliable quantitative evaluation directly related to morphologic aspects of synaptic efficacy. Hippocampus: we found a decrease of 15.6% and 48% in old and AD patients versus the adult ones, respectively. Cerebellum: there is a decrease of 41% and 56% in old and AD patients versus the adult ones, respectively. •$P < .01$ versus the other groups.

tween old and AD subjects (FIG. 1). Synaptic Nv and Sv significantly decreased between adult and old patients as well as between old and AD ones. No significant difference of S was found between old and AD groups, whereas it resulted significantly lower in adult versus old and AD patients (FIG. 2). The synapse-to-neuron ratio $Nv_{(Syn)}/Nv_{(Neur)}$ decreased by 15.6% and 48% in old and AD patients, respectively, when compared with the adult value (FIG. 1). In the cerebellar granular layer, the value of $Nv_{(Neur)}$ did not change among the three groups taken into account, although an age- and pathology-related decreasing trend also was seen in these samples (FIG. 1).

With reference to synaptic ultrastructure, $Nv_{(Syn)}$ significantly decreased both in the old and AD patients versus the adult ones. Sv was the same in adult and old patients, but it decreased significantly in the AD group. S was the same in old and AD

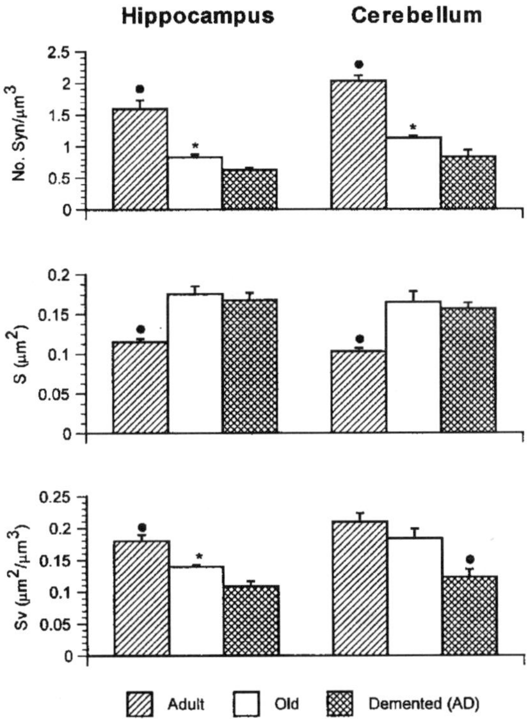

FIGURE 2. (*top histograms*) Synaptic numeric density: $Nv_{(Syn)}$. There is a significant decrease of this parameter between adult and old patients as well as between old and AD ones both in the hippocampus and cerebellum. •$P < .01$ versus the other groups; *$P < .01$ versus AD patients. (*middle histograms*) Synaptic average size: S. Both in the hippocampus and cerebellum, no significant difference of synaptic size was found between old and AD patients, whereas S resulted in significantly large versus the adult group. •$P < .01$ versus the other groups. (*bottom histograms*) Synaptic surface density: Sv. Hippocampus: we found a significant decrease between adult and old as well as between old and AD patients, respectively. Cerebellum: Sv was the same in adult and old patients, but it was significantly decreased in the AD group. •$P < .01$ versus the other groups; *$P < .01$ versus AD patients.

patients, but it was significantly higher versus the adult patients (FIG. 2). The cerebellar $Nv_{(Syn)}/Nv_{(Neur)}$ significantly decreased by 41% and 56%, in old and AD patients, respectively, versus the adult group (FIG. 1). The current findings document that in physiological aging and AD, whereas neuronal death occurs at a higher extent in the hippocampus than in the cerebellum (FIG. 1), a severe reduction of the number of synapses occurs in both the CNS zones of old and AD patients (FIG. 2).

Despite the significant enlargement of the average area (S) of the surviving contacts (FIG. 2), the outcome of this compensating reaction is a significant decrease of the overall synaptic surface/μm^3 of tissue (FIG. 2). We interpret these data to support that, regardless of the variable vulnerability of various CNS zones to aging and AD, synaptic pathology appears to represent a prominent and ubiquitous alteration of the old and demented CNS. Our assumption is supported by recent data from Martin *et al.* (1994)[5] reporting that in nonhuman primates synaptic pathology and glial responses to neuronal injury are early events preceding the formation of senile plaques, amyloid deposition, and progressive deafferentiation. Significant impairments in cell-to-cell communication, involving synaptic degeneration, appear to play a critical role in the consistent neuronal death associated with physiological aging and AD. The number of synapses/neuron was more severe in the CNS of AD patients (–48% hippocampus and –56% cerebellum, respectively), whereas in normal old subjects it decreased by 15.6% in the hippocampus and by 41% in the cerebellum (FIG. 1). In looking for tenable explanations of $Nv_{(Syn)}/Nv_{(Neur)}$ alterations, it must be taken into account the well-known dynamic condition of the synaptic contact areas which are capable of consistent plastic response in various CNS zones according to the environmental and experiential framework of each individual. As a reasonable explanation of our data we propose that the hippocampus, which is involved in memory retrieval processes, has a higher potential for recovery than the cerebellum, and this may result in a stronger resistance to changes due to physiological aging, but not to a severe pathological condition, such as AD. Conceivably, the 15.6% of the hippocampal number of synapses/neuron versus the 41% found in the cerebellum of normal old patients may be considered as peculiar reactive responses to physiological aging of different CNS zones.

REFERENCES

1. CRYSTAL, H.A., D. DICKSON, P. DAVIES, *et al.* 2000. The relative frequency of dementia of unknown etiology increases with age and is nearly 50% in nonagenarians. Arch. Neurol. **57:** 713–719.
2. BALL, M.J. 1988. Topographic distribution of neurofibrillary tangles and granulovacuolar degeneration in hippocampal cortex of ageing and demented patients. A quantitative study. Interdiscipl. Top. Gerontol. **25:** 16–37.
3. BERTONI-FREDDARI, C., P. FATTORETTI, R. PAOLONI, *et al.* 1996. Synaptic structural dynamics and aging. Gerontology **42:** 170–180.
4. BERTONI-FREDDARI, C., P. FATTORETTI, R. RICCIUTI, *et al.* 2002. Morphometry of E-PTA stained synapses at the periphery of pathological lesions. Micron **33:** 447–451.
5. MARTIN, L.J., C.A. PARDO, L.C. CORK & D.L. PRICE. 1994. Synaptic pathology and glial responses to neuronal injury precede the formation of senile plaques and amyloid deposits in the aging cerebral cortex. Am. J. Pathol. **145:** 1358–1381.

Does Nitric Oxide Synthase Contribute to the Pathogenesis of Alzheimer's Disease?

Effects of β-Amyloid Deposition on NOS in Transgenic Mouse Brain with AD Pathology

D. K. LAHIRI,[a] D. CHEN,[a] Y.-W. GE,[a] M. FARLOW,[a] G. KOTWAL,[b] A. KANTHASAMY,[c] D. K. INGRAM,[d] AND N. H. GREIG[d]

[a]*Department of Psychiatry, Institute of Psychiatric Research, Indiana University School of Medicine, Indianapolis, Indiana 46202, USA*

[b]*University of Louisville, Louisville, Kentucky 40202, USA*

[c]*Iowa State University, Ames, Iowa 50011, USA*

[d]*National Institute on Aging, GRC/NIH, Baltimore, Maryland 21224, USA*

ABSTRACT: Oxidative stress is a risk factor for Alzheimer's disease (AD) whose major hallmark includes brain depositions of the amyloid beta peptide (Aβ) derived from the β-amyloid precursor protein (APP). Our aim was to determine whether or not excessive Aβ deposition would alter nitric oxide synthase (NOS) activity, and thereby affect NOS-mediated superoxide formation. We compared NOS activity in brain extracts between Tg mice (expressing APP Swedish double mutation plus presenilin [PS-1] and nontransgenic [nTg] mice. Five brain regions, including cerebral cortex, hippocampus, cerebellum, and striatum from both nTg and Tg mice showed a detectable level of neuronal (n) NOS activity. Cerebellar extracts from both nTg and Tg mice displayed the highest level of nNOS activity, which was fourfold higher than cortical extracts. Although there was an increase in nNOS activity in Tg brain extracts, this did not attain statistical significance. A similar result was obtained for inducible NOS levels. Our results suggest that excess levels of Aβ failed to both trigger NOS activity and change NOS levels.

KEYWORDS: aging; beta-protein; dementia; neurodegeneration; nitric oxide; free radicals; oxidative stress; ROS

INTRODUCTION

The major neuropathological characteristics of Alzheimer's disease (AD) are cerebrovascular amyloid depositions, tau phosphorylation, cholinergic deficits, and oxidative stress.[1,2] Current research suggests that the neuronal cell death observed in AD may be attributed to apoptosis promoted by the amyloid beta peptide (Aβ),

Address for correspondence: D. K. Lahiri, Department of Psychiatry, Institute of Psychiatric Research, PR 313, 791 Union Dr., Indiana University School of Medicine, Indianapolis, IN 46202. Voice: 317-274-2706; fax: 317-274-1365.
dlahiri@iupui.edu

**Ann. N.Y. Acad. Sci. 1010: 639–642 (2003). © 2003 New York Academy of Sciences.
doi: 10.1196/annals.1299.117**

TABLE 1. Levels of nNOS and iNOS proteins in brain extracts from nontransgenic mice[a]

Brain Region	nNOS (band density)	iNOS (band density)
Cortex	2.95	6.51
Hippocampus	4.97	9.61
Cerebellum	12.08	10.51
Striatum	4.92	8.83
Rest of the brain	4.55	5.24

[a]Expressed as relative arbitrary densitometric units from the scanned Western blots.

which is derived from β-amyloid precursor protein (APP). Familial cases of AD are caused by mutations in APP and presenilin-1 (PS1) genes. It has been shown that PS1 mutations perturb neuronal calcium homeostasis, increase Aβ production, and enhance the vulnerability of neurons to synaptic dysfunction, excitotoxicity, and apoptosis.[3] Notably, astrocytes with elevated levels of NOS have been seen in direct association with Aβ deposits in AD and in APP transgenic (Tg) mice.[2] However, it remains to be determined whether Aβ generation directly induces apoptotic cell death or triggers an alternative pathway, NOS, that then leads to neurodegeneration. There is no quantitative and comparative study of NOS activity between APP-Tg and nontransgenic (nTg) mice to elucidate cause from effect. Our hypothesis is that (1) the aberrant expression of NOS isoforms is a primary event in the pathogenic cascade of AD and thus might be a potential therapeutic target, or (2) it reflects a secondary effect that occurs at more advanced stages of the disease process.

EXPERIMENTAL DESIGN

We studied NOS by three paradigms: (1) enzymatic activity of NOS by measuring the rate of conversion of ^{3}H-L-arginine to ^{3}H-L-citrulline,[4] (2) levels of neuronal NOS (nNOS), and (3) levels of inducible NOS (iNOS) by Western blotting with specific antibodies against nNOS and iNOS (Santa Cruz, CA). Proteins were analyzed by SDS-PAGE followed by Western immunoblotting and densitometry.[5] All were studied in brain extracts from both APP/PS-1 double transgenic and nTg mice of similar age and sex (5-month-old males). The brain of Tg mice showed a robust level of Aβ deposition and associated neuropathology compared with those from nTg mice.[6]

RESULTS

Five brain regions (cortex, hippocampus, cerebellum, striatum, and the remaining brain) from both Tg and nTg mice were homogenized and assayed for NOS enzymatic activity within the linear range of its detection. All brain extracts showed a quantifiable level of NOS enzymatic activity, with the cerebellum from nTg mice showing the greatest activity that was fourfold higher than cortical extracts. In ac-

TABLE 2. Comparison of NOS enzymatic activity in non-Tg and Tg mice brain extracts[a]

Brain Regions	Nontransgenic Mice	Transgenic Mice	Significance
Cortex	12%	19%	0.985
Cerebellum	47%	56%	0.954

[a]Expressed as percentage of conversion of labeled L-arginine to L-citrulline.

cord with this, cerebellar nNOS levels, quantified by Western blot analysis of the 155-kDa nNOS protein using a nNOS-specific antibody, were correspondingly higher that other measured brain regions in nTg mice (TABLE 1). Likewise, higher cerebellar iNOS levels were obtained when the 130-kDa iNOS protein was measured in different nTg brain extracts by Western blot using an iNOS-specific antibody. Cerebellum extracts were similarly prepared from the Tg mice and assayed for NOS enzymatic activity. Like the nTg brain, Tg cerebellum extracts showed the highest level of NOS activity (TABLE 2). However, although there was an increased trend of NOS activity in Tg cortical and cerebellum extracts, there was no significant difference of NOS activity between nTg and Tg brain extracts (TABLE 2).

DISCUSSION

The etiology of AD is complex; it is important to study other nongeneric risk factors, such as oxidative stress and inflammatory processes, to develop noval strategies for drug development in AD.[7] Our aim was to determine whether or not excessive Aβ deposition would alter NOS activity, and thereby affect NOS-mediated superoxide formation leading to neurodegeneration. To address this question, we compared both neuronal and inducible NOS activity in brain extracts between Tg mice (expressing human APP with the Swedish double mutation plus PS-1) and nTg mice. Our results suggest that excess levels of Aβ failed to both trigger NOS activity and change NOS levels. Thus, Aβ deposition likely does not contribute to the NOS-mediated nitric oxide release and subsequent generation of the superoxide radicals. Similarly, treatment with an Aβ-lowering cholinesterase inhibitor has no significant effect on NOS (data not shown). Hence, whereas NOS isoforms may be part of AD pathology, they likely are secondary to the amyloid pathology, a view consistent with previous reports.[2] Taken together, our results from transgenic mice suggest that NOS is a doubtful target for therapeutic intervention in AD. Although here we have checked NOS levels in the Tg animal model of AD, it will be interesting to carefully compare its levels in aged Tg and nTg mice as well as in brain regions from AD and age-matched control subjects. Similarly, the roles of NOS, caspase-3, and certain protein kinases vis-à-vis apoptosis and oxidative stress should be examined in other similar neurodegenerative and neurotrauma disease models, such as Parkinson's disease and spinal cord injury, and is a focus of our ongoing studies.

ACKNOWLEDGMENTS

We are sincerely thankful for support from the Alzheimer's Association and the National Institute on Aging, National Institutes of Health, USA.

REFERENCES

1. LAHIRI, D.K. *et al.* 2002. Current drug targets for Alzheimer's disease treatment. Drug Dev. Res. **56:** 267–281.
2. LUTH, H.J. *et al.* 2001. Expression of endothelial and inducible NOS-isoforms is increased in Alzheimer's disease in APP23 transgenic mice and after experimental brain lesion in rat: evidence for an induction by amyloid pathology. Brain Res. **913:** 57–67.
3. LEE, J. *et al.* 2002. Adverse effect of a presenilin-1 mutation in microglia results in enhanced nitric oxide and inflammatory cytokine responses to immune challenge in the brain. Neuromol. Med. **2:** 29–45
4. BONDY, S.C. *et al.* 2002. Dietary modulation of age-related changes in cerebral pro-oxidant status. Neurochem. Int. **40:** 123–130.
5. LAHIRI, D.K. *et al.* 1998. The secretion of amyloid beta-peptides is inhibited in the tacrine-treated human neuroblastoma cells. Mol. Brain Res. **62:** 131–140.
6. BORCHELT, D.R. *et al.* 1997. Accelerated amyloid deposition in the brains of transgenic mice coexpressing mutant presenilin 1 and amyloid precursor proteins. Neuron **19:** 939–945.
7. LAHIRI, D.K. *et al.* 2003. A critical analysis of new prolecular targets and strategies for drug developments in Alzheimer's disease. Curr. Drug Targets **4:** 97–112.

A Proximal Gene Promoter Region for the β-Amyloid Precursor Protein Provides a Link between Development, Apoptosis, and Alzheimer's Disease

D. K. LAHIRI, C. GHOSH, AND Y.-W. GE

Department of Psychiatry, Institute of Psychiatric Research, Indiana University School of Medicine, Indianapolis, Indiana 46202, USA

ABSTRACT: Abnormalities in regulation of the β-amyloid precursor protein (APP) gene might be a crucial factor in Alzheimer's disease (AD). Our aim is to study the role of a specific proximal APP promoter element under the apoptotic condition. Our transfection studies with APP promoter deletion constructs indicate that each cell type differently regulates promoter activity. The minimum region that was sufficient to drive basal promoter activity in neuronal PC12 and neuroblastoma SK-N-SH cells was –75/+104 and –47/+104 bp, respectively. In SK-N-SH cells, the –47/+104 construct displayed the highest promoter activity, and the –75/–46 region acted as a negative regulatory element. Results from the gel electrophoretic mobility shift assay (EMSA) indicate that the –75/–46 region binds to a distinct DNA–protein complex with nuclear protein(s) from HeLa, PC12, NIH-3T3, and neuroblastoma cells. EMSA results from HeLa cells, which were stimulated by serum starvation (SR), indicate a significant induction in the signal of the DNA–protein complex from controls. EMSA results from PC12 cells, which were subjected to hypoxia, indicate a significant reduction in the signal. Our results suggest that the –75/–46 region binds to a protein that is upregulated in serum starvation, and downregulated in hypoxia. Because serum starvation contributes to the induction of apoptosis, these results suggest a role of the 30-bp proximal APP promoter element in enhanced apoptotic neuronal cell death.

KEYWORDS: aging; apoptosis; dementia; gene regulation; gel shift assay; neurons; neurodegeneration; neuronal proteins; promoter; transcription

INTRODUCTION

One of the prominent features of Alzheimer's disease (AD) is a substantial loss of neurons in the hippocampus and cerebral cortex, and several lines of evidence suggest that this loss be attributed to apoptotic neuronal cell death promoted by amyloid beta peptide (Aβ), which is derived from the β-amyloid precursor proteins (APP).[1,2] Aging and the accumulation of Aβ are important risk factors for the impair-

Address for correspondence: D. K. Lahiri, Department of Psychiatry, Institute of Psychiatric Research, PR 313, 791 Union Dr., Indiana University School of Medicine, Indianapolis, IN 46202. Voice: 317-274-2706; fax: 317-274-1365.
dlahiri@iupui.edu

Ann. N.Y. Acad. Sci. 1010: 643–647 (2003). © 2003 New York Academy of Sciences.
doi: 10.1196/annals.1299.118

ment of memory and development of dementia. There is an age-dependent increase in plasma levels of Aβ in control and pre-AD subjects, and those age-dependent changes also occur in brain and plasma Aβ levels in the TG2576 model of AD. This change in AB levels is probably due to alterations in APP gene expression. Abnormalities in APP gene regulation might be a crucial factor in AD neuropathology.[1,3] To study the APP gene regulation, we have cloned and characterized the genetic regulatory region and APP promoter from different species.[4] Our hypothesis is that an overexpression of the APP gene is one of the factors leading to the accumulation of Aβ in the affected areas of the brain, and that APP gene promoter plays an important role in aging, development, and apoptosis. Our aim was to study the role of a specific proximal APP promoter element (–75 to –46) in APP gene expression in different cell types and under the apoptotic condition. We report that a minimum region of –75 to +104 bp was essential for basal level of promoter activity in a cell type–dependent manner, and that this region of the APP promoter may participate in apoptosis.

EXPERIMENTAL DESIGN

We prepared a series of recombinant plasmids containing progressive deletions from the upstream region to the transcription start site (+1) of the APP gene cloned to a reporter gene, chloramphenicol acetyl transferase (CAT).[4] Promoter activity of deletion constructs was tested by transient transfection in different cell types and by assaying CAT enzymatic activity.[4,5] Specific DNA–protein interaction was studied by the gel electrophoretic mobility shift assay (EMSA) as described previously.[4]

RESULTS

DNA Transfection Studies Suggest a Cell Type–Specific Promoter Element

Our transfection studies with APP promoter deletion constructs indicate that each cell type regulates promoter activity differently. The minimum region that was sufficient to drive basal promoter activity in neuronal PC12 and neuroblastoma SK-N-SH cells was –75/+104 and –47/+104 bp, respectively.

DNA Transfection Studies Reveal a Negative Regulatory Element

A serial deletion analysis from –3416 to –1 bp of the APP promoter reveals that promoter constructs (in bp) corresponding to –3416/+104, –2542/+104, –2105/+104, –1131/+104, –490/+104, –309/+104, –204/+104, –143/+104, and –75/+104 displayed a significantly lower promoter activity than –47/+104 in SK-N-SH cells. Notably, the –47 to +104 construct displayed the highest promoter activity than others tested (from –3416 to –75). Thus, the –75 to –46 region (30 bp) acts as a negative regulatory element in SK-N-SH cells.

Gel Shift Analysis to Study DNA–Protein Interactions under Serum Starvation

To examine whether this region binds to a specific nuclear protein, we performed EMSA with the radiolabeled 30-bp region in different nuclear extracts. We observed a distinct DNA–protein complex with a nuclear protein from HeLa, PC12, NIH-3T3, and neuroblastoma cells. When EMSA was performed with nuclear extracts from HeLa cells that were stimulated by serum starvation (SR), we observed a significant induction in the signal of the DNA–protein complex versus controls. A similar EMSA result was obtained with nuclear extracts from the serum-starved NIH3T3 cells (FIG. 1).

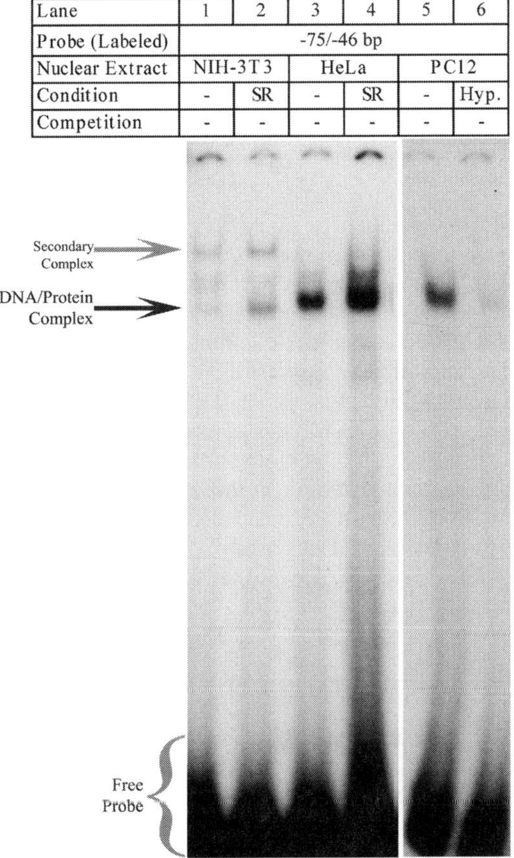

FIGURE 1. Electrophoretic mobility shift assay to detect nuclear protein(s) interacting with the –75/–46-bp region of APP promoter under different conditions. EMSA was performed with the [^{32}P]-labeled –75/–46-bp probe and with different nuclear extracts and analyzed in a 5% native polyacrylamide gel as indicated. EMSA was carried with the radiolabeled probe in the presence of nuclear extracts from either NIH-3T3 (*lanes 1, 2*), HeLa (*lanes 3,4*), or PC12 (*lanes 5, 6*) cell lines. Nuclear extracts were prepared in normal condition (*lanes 1, 3, 5*), subjected to SR (*lanes 2, 4*), or hypoxia (*lane 6*). The *top arrows* indicate the position of the DNA–protein complex, and the *bottom arrow* (in *broken line*) shows the position of unbound probe.

Gel Shift Analysis to Study DNA–Protein Interactions under Hypoxia

On the contrary, when EMSA was performed with nuclear extracts from PC12 cells that were prepared after hypoxia, we observed a significant reduction in the signal of the DNA–protein complex versus control cells (FIG. 1).

DISCUSSION

Apoptosis and DNA damage have been reported in Alzheimer's disease.[2] To study the transcriptional control, we have cloned and characterized a 5′ flanking region of the APP gene from the human and a rhesus primate.[4,5] Here, we have studied the proximal region of the APP promoter under apoptotic conditions. In this context, evidence suggests that the 30-bp region of APP promoter plays a significant biological role. First, the sequence homology and "BLAST computer search" reveal a near perfect homology of the CGAGTG motif with that of *zeste* binding domain, a developmental protein which was characterized in the fruit fly. Recently, human homologs of the *Drosophila* enhancer of zeste gene (EZH1 and EZH2) have been cloned,[6] one of which maps to chromosome 21q22.2 which is very close to the *AD* locus. APP gene can be developmentally regulated by the action of the zeste protein.[7] Second, the 30-bp region also contains binding sites for *Puf* which is a human c-*myc* factor. It is identified as nucleoside diphosphate kinase, a potential regulator of human apoptosis. Activation of *Puf* results in apoptosis and the 30-bp region can promote apoptosis by activating these genes. Third, this region also contains the SP1 binding domain, whose role in TATA-less promoters such as APP is not clearly understood. Apoptosis is also promoted by the c-*myc* gene.[8] Thus, the binding protein for the 30-bp region can be an activator of the APP promoter and is also a regulator of c-*myc* gene expression. Recently, it has been shown that APP is regulated at both transcriptional and post-transcriptional levels.[9] Taken together, these results suggest that the APP regulatory region, including the promoter, responds to various apoptotic and inflammatory processes which are relevant for AD pathogenesis.

CONCLUSION

Our results suggest that a 30-bp APP promoter region (–75/–46) binds to a protein that was induced in SR and downregulated in hypoxia. Because serum starvation contributes to the induction of apoptosis, these results suggest that the apoptotic protein binds to the 30-bp APP promoter element, which upregulates APP promoter activity, resulting in enhanced Aβ-mediated apoptotic neuronal cell death. The transcriptional regulation of both the APP and c-*myc* genes by a 30-bp binding protein provides a link between development, apoptosis, and AD.

ACKNOWLEDGMENTS

We sincerely thank the National Institute on Aging, National Institutes of Health, USA for support.

REFERENCES

1. Lahiri, D.K. *et al.* 2002. Current drug targets for Alzheimer's disease treatment. Drug Dev. Res. **56:** 267–281.
2. Anderson, A.J. *et al.* 1996. DNA damage and apoptosis in Alzheimer's disease: colocalization with c-Jun immunoreactivity, relationship to brain area, and effect of postmortem delay. J. Neurosci. **16:** 1710–1719.
3. Wavrant DeVrieze, F. *et al.* 1999. Genetic variability at the amyloid-beta precursor protein locus may contribute to the risk of late-onset Alzheimer's disease. Neurosci. Lett. **268:** 1–4.
4. Song, W. & D.K. Lahiri. 1998. Functional identification of the promoter of the gene encoding the Rhesus monkey beta-amyloid precursor protein. Gene **217:** 165–176.
5. Lahiri, D.K. *et al.* 2000. Analysis of the 5′-flanking region of the beta-amyloid precursor protein gene that contributes to increased promoter activity in differentiated neuronal cells. Mol. Brain Res. **77:** 185–198.
6. Abel, K.J. *et al.* 1996. Characterization of EZH1, a human homolog of *Drosophila* enhancer of zeste near BRCA1. Genomics **37:** 161–171.
7. Chen, H. *et al.* 1996. Cloning of a human homolog of the *Drosophila* enhancer of zeste gene EZH2 that maps to chromosome 21q22.2. Genomics **38:** 30–37.
8. Berberich, S.J. & E.H. Postel. 1995. PuF/NM23-H2/NDPK-B transactivates a human c-myc promoter-CAT gene via a functional nuclease hypersensitive element. Oncogene **10:** 2343–2347.
9. Lahiri, D.K. *et al.* 2003. Role of cytokines in gene expression of amyloid β-protein precursor: identification of a 5′-UTR-binding nuclear factor and its implication in Alzheimer's disease. J. Alzheiner's Dis. **5:** 81–90.
10. Lahiri, D.K. *et al.* 2003. A critical analysis of new molecular tartgets and strategies for drug developments in Alzheimer's disease. Curr. Drug Targets **4:** 97–112.

Induction of Cyclooxygenase-2 and Peroxisome Proliferator–Activated Receptor-γ during Nitric Oxide–Induced Apoptotic PC12 Cell Death

SO-YOUNG LIM, JUNG-HEE JANG, AND YOUNG-JOON SURH

Laboratory of Biochemistry and Molecular Toxicology, College of Pharmacy, Seoul National University, Seoul 151-742, Korea

ABSTRACT: Inappropriate expression of inducible nitric oxide synthase (iNOS) and unregulated production of nitric oxide (NO) may contribute to neuronal cell death implicated in neurological disorders such as Alzheimer's disease. In this study, we have investigated the molecular mechanisms underlying nitrosative cell death induced by NO in cultured rat pheochromocytoma (PC12) cells. Incubation of PC12 cells with the NO donor sodium nitroprusside (SNP) resulted in apoptotic death as revealed by the decrease of mitochondrial transmembrane potential ($\Delta\psi_m$), cleavage of poly(ADP-ribose) polymerase (PARP), and induction of p21$^{Waf1/Cip1}$. It has been reported that the expression of cyclooxygenase-2 (COX-2) and peroxisome proliferator–activated receptor-γ (PPARγ) is elevated in Alzheimer's disease, and certain nonsteroidal anti-inflammatory drugs (NSAIDs) can reduce the risk and delay the onset of Alzheimer's disease. Treatment of PC12 cells with a proapoptotic dose of SNP induced expression of both COX-2 and PPARγ. Addition of the PPARγ antagonist GW9662 to the media augmented the NO-induced cytotoxicity. Although cotreatment of PGE$_2$ (50 μM) and SNP (0.4 mM) aggravated the NO-induced cytotoxicity, preincubation of the same concentration of PGE$_2$ was cytoprotective. Taken together, the above findings suggest that the proinflammatory mediators such as PGE$_2$ and PPARγ may regulate the nitrosative stress–induced apoptotic cell death.

KEYWORDS: apoptosis; cyclooxygenase-2 (COX-2); nitric oxide; nitrosative stress; PC12 cells; peroxisome proliferator-activated receptor (PPARγ); prostaglandin E$_2$

INTRODUCTION

Neuroinflammation is a central feature of neurodegenerative disorders such as Alzheimer's disease, stroke, and neurovascular disease.[1,2] An excessive or chronic innate immune reaction results in neuronal toxicity and cell death. Nitric oxide (NO) production derived from inducible nitric oxide synthase (iNOS) is considered as one

Address for correspondence: Young-Joon Surh, College of Pharmacy, Seoul National University, Shinlim-dong, Kwanak-ku, Seoul 151-742, Korea. Voice: 82-2-877-3730; fax: 82-2-874-9775.

surh@plaza.snu.ac.kr

Ann. N.Y. Acad. Sci. 1010: 648–658 (2003).
doi: 10.1196/annals.1299.119

of main causes of neuronal cell death.[3] Several studies suggest that the proapoptotic effect of NO is mediated via peroxynitrite formed by the reaction between NO and $O^-{}_2$.[4,5] There are multiple lines of compelling evidence that nitrosative damage is implicated in both aging and the pathogenesis of neurodegenerative disorders. Thus, the level of 3-nitrotyrosine, a relatively specific marker of nitrosative damage mediated by peroxynitrite, has been reported to increase in Alzheimer's disease, Parkinson's disease, and amyotrophic lateral sclerosis as well as in normal aging process.[6] The process of reactive nitrogen species (RNS)-induced cell death involves mitochondria as a primary target. There are three main roles of mitochondria in RNS-induced cell death: (1) inhibition of respiration, (2) induction of mitochondrial permeability transition (MPT), and (3) triggering of intracellular signal transduction or DNA damage leading to activation of the p53 pathway.[7] It is well established that p53 is responsible for G_1 cell cycle arrest and apoptosis induction. NO could induce p53 activation and subsequent expression of $p21^{Waf1/Cip1}$, a potent inhibitor of the cell cycle kinases.[8] $p21^{Waf1/Cip1}$ has been known as a critical checkpoint target for cell cycle arrest and apoptosis. In cells that are prone to apoptosis, $p21^{Waf1/Cip1}$ is rapidly induced by DNA damage but is selectively cleaved by caspases.[9]

Cyclooxygenase (COX) is a rate-limiting enzyme responsible for the synthesis of a series of prostaglandins and thromboxanes that have multiple physiologic functions. Two isoforms of COX have been identified: COX-1 and COX-2. COX-1 is ubiquitously expressed in most tissues and produces prostanoids that are involved in maintaining homeostatic functions. COX-2 is induced under inflammatory conditions and can exert deleterious effects in the neurodegenerative diseases.[10,11] Prolonged intake of nonsteroidal anti-inflammatory drugs (NSAIDs) with COX inhibitory effects has been reported to reduce the risk of Alzheimer's disease and often delays its onset.[12,13] PGE_2, a major COX-2 product, could modulate oxidative damage and neurodegeneration.[14–19] Peroxisome proliferator-activated receptor γ (PPARγ) is a member of the nuclear hormone receptor superfamily and involved in several important physiological processes, such as adipocyte differentiation, insulin sensitivity, and inflammatory responses.[20,21] Recently, it has been recognized that the therapeutic actions of NSAIDs against Alzheimer's disease are mediated via PPARγ activation, independently of COX inhibition.[22] The anti-inflammatory actions of PPARγ may be mediated in part by antagonizing the activities of the transcription factors such as AP-1, STAT, and NF-κB.[23] The finding that PPARγ is increased in Alzheimer's disease implies its involvement in the regulation of inflammatory responses.[24]

In the current study, we evaluated the proapoptotic potential of NO in PC12 cells and examined the possible involvement of inflammatory mediators such as PGE_2 and PPARγ in the NO-induced cell death.

MATERIALS AND METHODS

Materials

Poly-D-lysine, MTT [3-(4,5-dimethylthiazol-2-yl)-2,5-diphenyltetrazolium bromide], sodium nitroprusside (SNP), and N-(1-naphthyl)ethylenediamine were purchased from Sigma Chemical Co. (St. Louis, MO). PGE_2 and GW9662 were products of Cayman Chemical (Ann Arbor, MI). Sulfanilamide was purchased from Junsei Chemical (Tokyo, Japan). Sodium nitrite was provided from Yakuri Pure Chemicals (Osaka, Japan). Tetraethylbenzimidazolcarbocyanine iodide (JC-1) was obtained from Molecular Probes, Inc. (Eugene, OR). Dulbecco's modified Eagle's medium (DMEM), fetal bovine serum, horse serum, F-12, and N-2 supplement were provided from Gibco BRL (Grand Island, NY).

Cell Culture

PC12 cells were maintained routinely in DMEM supplemented with 10% heat-inactivated horse serum and 5% fetal bovine serum at 37°C in a humidified atmosphere of 10% CO_2/90% air. All cells were cultured in poly-D-lysine–coated culture dishes. The medium was changed every other day, and cells were plated at an appropriate density according to each experimental scale. After 24 h subculture, cells were switched to serum-free N-2 defined medium for treatment.

Nitrite Assay

Production of nitrites released from SNP in culture medium was quantitated by the Griess reaction. Absorbance at 550 nm of the reaction mixture containing 100 μL cell culture media and the same volume of Griess reagent (1% sufanilamide, 0.1% naphtylethylene diamine in 5% phosphoric acid) was measured with an ELISA reader. A standard curve was prepared with known concentrations of sodium nitrite for the quantitation of NO released into media.

Determination of Cell Viability

PC12 cells were plated at a density of 5×10^4 cells/300 μL in 48-well plates, and the cell viability was determined by the conventional MTT reduction assay. The MTT assay relies primarily on the mitochondrial metabolic capacity of viable cells and hence reflects the intracellular redox state. After incubation, cells were treated with the MTT solution (final concentration, 1 mg/mL) for 2 h. The dark-blue formazan crystals formed in intact cells were solubilized with DMSO, and absorbance at 570 nm was measured with an ELISA reader. Results were expressed as the percentage of MTT reduction, assuming that the absorbance of control cells was 100%.

Measurement of Mitochondirial Transmembrane Potential ($\Delta\psi_m$)

PC12 cells were plated at a density of 10^5 cells/600 μL in 4-well chamber slide. After incubation with SNP for 5 h, cells were treated with JC-1 (4 μg/mL) in serum-free N-2 defined medium for 30 min at 37°C, washed twice with phosphate-buffered saline (PBS), and examined under a confocal laser scanning microscope. JC-1 fluorescence was monitored either as green fluorescent monomer at depolarized membrane potentials or as red fluorescent J-aggregate at hyperpolarized membrane potentials. The relative intensity of red to green fluorescence determined as the indicator of $\Delta\psi_m$ was quantified using a LEICA TCS NT software.

Western Blot Analysis

After treatment, cells (3×10^6 cells/6 mL in 60 ϕ dish) were collected and washed with PBS. After centrifugation, cell lysis was conducted at 4°C by vigorous shaking for 15 min in RIPA buffer (150 mM NaCl, 1% NP-40, 0.5% sodium deoxycholate, 0.1% SDS, 50 mM Tris-HCl, pH 7.4, 50 mM glycerphosphate, 20 mM NaF, 20 mM EGTA, 1 mM DTT, 1 mM Na_3VO_4, and protease inhibitors). After centrifugation at 15,000 rpm for 15 min, supernatant was separated and stored at –70°C until use. The protein concentration was determined by using the BCA protein assay kit (Pierce, Rockford, IL). After addition of sample loading buffer, protein samples were electrophoresed on a 12.5% SDS–polyacrylamide gel. Proteins were transferred to polyvinylidene difluoride blots at 300 mA for 3 h. The blots were blocked for 1 h at room temperature in fresh blocking buffer (0.1% Tween-20 in Tris-buffered saline, pH 7.4, containing 5% nonfat dried milk). Dilutions (1:1,000) of primary anti-PARP, anti-$p21^{Waf1/Cip1}$, anti-PPARγ, and anti-COX-2 were made in PBS with 3% nonfat dry milk. After three washes with PBST (PBS and 0.1% Tween-20), the blots were incubated with horseradish peroxidase–conjugated secondary antibodies in PBS with 3% nonfat dry milk for 1 h at room temperature. The blots were washed again three times in PBST buffer, and transferred proteins were incubated with ECL substrate solution (Amersham Pharmacia Biotech, Piscataway, NJ) for 1 min according to the manufacturer's instructions and visualized with x-ray film.

Measurement of PGE_2 Levels

PC12 cells were plated at a density of 5×10^4 cells/300 μL in 48-well plates and treated with SNP. The production of PGE_2 was measured by using an enzyme-immunoassay kit (Amersham Pharmacia Biotech, Arlington Heights, IL). Fifty-μL of culture medium previously centrifuged at 2,000 rpm for 10 min was mixed with 50 μL PGE_2 antibody solution and incubated on the shaker in a cold room for 3 h. After 50 μL conjugate was added to the reaction mixture, and the 96-well plate was washed four times. After incubation with 100 μL streptavidine-horseradish peroxidase solution for 30 min at room temperature followed by four times washing, 100-μL TMB substrate solution was applied to the 96-well plate and incubated for 30 min at room temperature in the dark. The reaction was terminated by addition of 100 μL stop solution, and absorbance at 450 nm was measured by the ELISA reader. PGE_2 was quantitated and represented as a pg/mL unit using a standard curve constructed with known concentrations of PGE_2.

RESULTS

SNP, a Nitric Oxide Donor, Decreased the Viability of PC12 Cells

To evaluate the cytotoxicity of NO, we treated PC12 cells with varying concentrations of the NO-releasing compound SNP for 24 hours. NO release by SNP was confirmed by the Griess assay (FIG. 1A). Treatment of PC12 cells with SNP decreased the cell viability in a concentration-dependent manner as determined by the MTT reduction assay (FIG. 1B). The SNP-treated cells exhibited proapoptotic features such as cell shrinkage (data not shown).

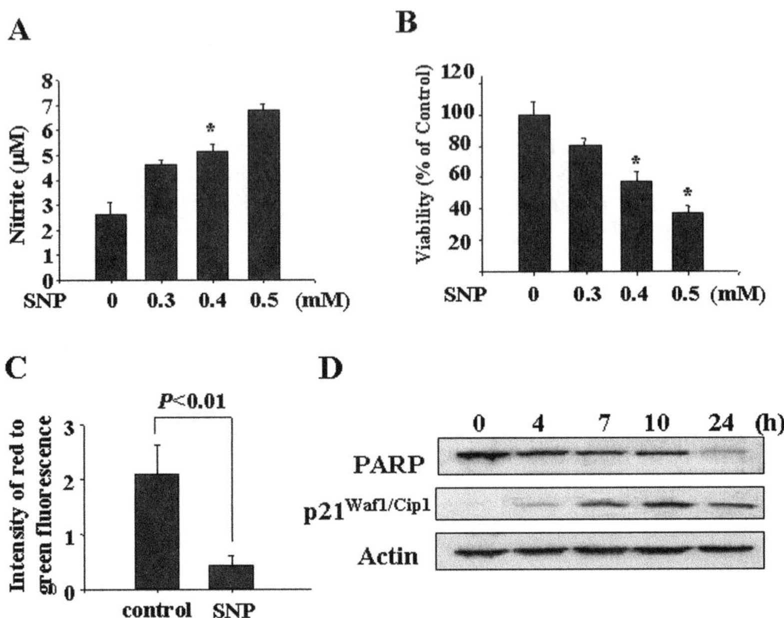

FIGURE 1. NO-induced apoptotic PC12 cell death. (**A**) Nitrite released from SNP in cell culture medium. PC12 cells were treated with indicated concentrations of SNP for 24 h at 37°C, and the nitrite concentrations in the culture supernatant were measured by the Griess reaction. *Significantly different from the untreated control ($P < .01$). (**B**) Cytotoxic effect of SNP in PC12 cells. Viable cells were determined by using the MTT reduction assay as described in MATERIALS AND METHODS. Data are presented as means ± SD ($n = 3$). (**C**) Mitochondrial depolarization induced by SNP. Mitochondrial depolarization in the cells treated with 0.4 mM SNP for 5 h was indicated by a decrease in the red/green fluorescence intensity ratio. (**D**) SNP-induced cleavage of PARP and upregulation of $p21^{Waf1/Cip1}$. Cells were treated with 0.4 mM SNP for indicated times and harvested for Western blot analysis as described in MATERIALS AND METHODS. Actin levels were measured to confirm equal amounts of protein loading.

Nitric Oxide–Induced PC12 Cell Death through Apoptosis

To examine the sequence of events associated with the SNP toxicity, we used some of the molecular markers of apoptotic cell death. Mitochondrial depolarization occurred in the cells treated with 0.4 mM SNP for 5 hours. JC-1 as monomer shows green fluorescence, and inside the intact mitochondrial membrane it aggregates to be detected as red fluorescence. The untreated control cells with intact mitochondria were selectively marked by red fluorescent staining. In contrast, in cells treated with 0.4 mM SNP, the red staining was replaced by diffused green monomer fluorescence indicating mitochondrial depolarization. The quantitative analysis of the intensity of red to green fluorescence was conducted (FIG. 1C). The failure in maintenance of the mitochondrial transmembrane potential has been known as one of the characteristics of the early stage of apoptosis. Another representative biological marker of apoptosis is PARP cleavage. PARP is a 116-kDa nuclear protein that is specifically cleaved

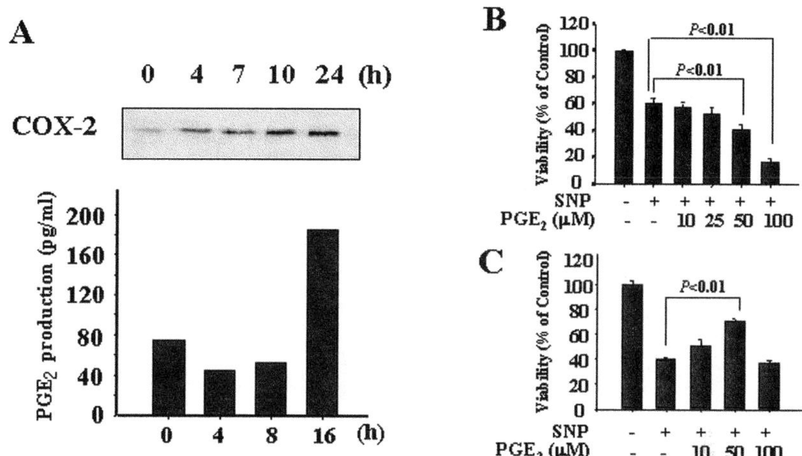

FIGURE 2. Induction of COX-2 expression by SNP and effects of PGE_2 on NO-induced cytotoxicity in PC12 cells. (**A**) NO-induced COX-2 expression and subsequent PGE_2 production. Cells were treated with 0.4 mM SNP for indicated times and harvested for Western blot analysis. Actin levels were measured to confirm equal amounts of protein loading. PGE_2 was measured by the enzyme immunoassay kit as described in MATERIALS AND METHODS. (**B**) Aggravation of SNP-induced cytotoxicty by coincubation of PGE_2. PC12 cells were cotreated with 0.3 mM SNP and indicated concentrations of PGE_2 for 24 h. (**C**) Attenuation of SNP-induced cytotoxicity by PGE_2 preincubation. After PC12 cells were treated with indicated concentrations of PGE_2 for 24 h, SNP was added to the media and incubated for additional 24 h. Viable cells were determined by using the MTT reduction assay as described in MATERIALS AND METHODS. Data are presented as means ± SD (n = 3).

by an active caspase to an 85-kDa apoptotic fragment. Treatment with 0.4 mM SNP caused cleavage of PARP in a time-related manner (FIG. 1D). We have also found that $p21^{Waf1/Cip1}$ accumulated upon SNP treatment (FIG. 1D).

Expression of COX-2 and Its Role in Nitric Oxide–Induced Cell Death

To investigate the possible involvement of the inflammatory pathway in the nirosative stress–induced PC12 cell death, a COX-2 level in the SNP-treated PC12 cells was analyzed by Western blotting. FIGURE 2A shows that COX-2 expression and subsequent PGE_2 production are increased by NO. To elucidate the role of COX-2 during nitrosative cell death, we coincubated PGE_2, a major product of COX-2, with 0.4 mM SNP for 24 h. PGE_2 aggravated SNP-induced cytotoxicty in a concentration-dependent manner (FIG. 2B), although PGE_2 alone did not elicit any toxicity in PC12 cells at concentrations up to 100 μM (data not shown). This finding suggests that COX-2 induction may contribute to NO-induced PC12 cell death. Interestingly, when cells were preincubated with 50 μM PGE_2 for 24 h, PC12 cells became resistant to SNP-induced cytotoxicity (FIG. 2C). However, a higher concentration of PGE_2 failed to protect PC12 cells from NO-induced cell death (FIG. 2C).

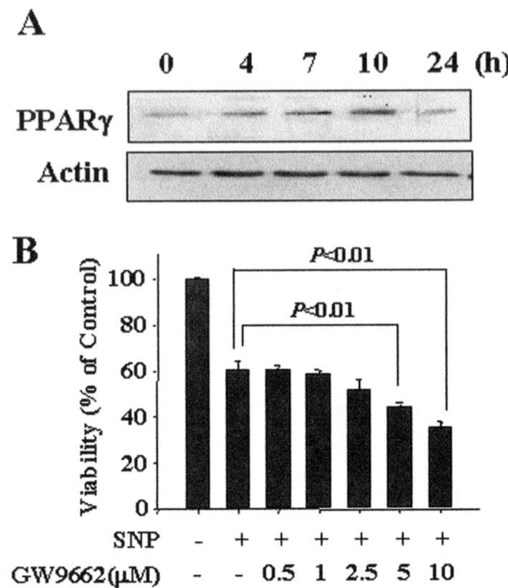

FIGURE 3. Expression of PPARγ and its role in NO-induced cell death. (**A**) NO-induced PPARγ expression in PC12 cells. Cells were treated with 0.4 mM SNP for indicated times and harvested for Western blot analysis as described in MATERIALS AND METHODS. (**B**) Enhancement of SNP-induced cytotoxicity by coincubation with GW9662, a PPARγ antagonist. PC12 cells were treated with 0.3 mM SNP and indicated concentrations of GW9662 for 24 h. Cell viability was determined by using the MTT reduction assay. Data are presented as means ± SD (n = 3).

Expression of PPARγ and Its Protective Effect on Nitric Oxide–Induced Cell Death

In another experiment, PPARγ expression was increased by SNP at the concentration (0.4 mM) that could cause apoptosis in PC12 cells (FIG. 3A). To determine the role of PPARγ during the nitrosative PC12 cell death, we added GW9662, a PPARγ antagonist, to the media together with SNP. As illustrated in FIGURE 3B, GW9662 enhanced SNP-induced cytotoxicity, suggesting that PPARγ could contribute to the defensive mechanism against the nitrosative stress.

DISCUSSION

Neuroinflammation characterized by improper induction of iNOS expression, subsequent NO production, and excessive COX-2 expression has been considered to be associated with certain neurological disorders.[3,11] This study demonstrates that NO released from SNP induces apoptotic death in PC12 cells, and the proinflammatory cascade is implicated in the nitrosative stress–induced cell death. SNP-induced apoptosis was evident from mitochondrial depolarization and PARP cleavage. Nuclear apoptosis is preceded by the disruption of the mitochondrial transmembrane potential which, in turn, facilitates the release of proapoptotic proteins such as cytochrome *c* through the opening of the mitochondrial permeability transition pore. Opening of this pore is known to be controlled by the mitochondrial redox potential. Released cytochrome *c* initiates formation of the apoptosome, and activates executioner caspases which cleave PARP and finally induce the cell disassembly. The ectopic expression of Bcl-2, a prototype antiapoptotic mitochondrial protein, rescued PC12 cells from NO-induced cytotoxicity, further supporting that NO induced cell death via apoptosis (J.-H. Jang and Y.-J Surh, unpublished observation). Although, mitochondrial depolarization could result in either apoptosis or necrosis, and PARP cleavage does not unequivocally distinguish between apoptotic versus necrotic pathways,[7,25] NO-induced PC12 cell death is likely to be mediated through an apoptotic process. According to previous studies by other investigators, both types of cell death can be simultaneously induced by nitrosative stress.[25] Interestingly, SNP-induced cell death was characterized by an induction of $p21^{Waf1/Cip1}$ protein capable of inducing G_1 arrest in the cell cycle. It has been known that NO can also induce cell cycle arrest which is accompanied by expression of $p21^{Waf1/Cip1}$.[8] The failure to arrest the cell cycle at the right time in the proper stage may trigger the apoptotic program, and the abrogation of G_1 cell cycle arrest can sensitize cells toward NO-induced apoptosis.[9] In this study, $p21^{Waf1/Cip1}$ began to disappear in the late stage of cell death when approximately 60% of NO-treated cells were under severe cytotoxicity. This observation coincides with the fact that $p21^{Waf1/Cip1}$ is cleaved by the caspase-like activity before the onset of extensive apoptosis and that the insufficient expression of $p21^{Waf1/Cip1}$ may trigger the cellular apoptotic program.[9]

We have observed that COX-2 expression and subsequent PGE_2 production were enhanced in PC12 cells undergoing apoptotic death. COX-2 expression may be regarded as part of a cellular defense mechanism against nitrosative stress. It has been reported that COX-2 expression may modulate apoptosis by regulation of proteins related to cell survival and death.[26–29] In many cell types including both normal and tumor cells, overexpression of COX-2 induced Bcl-2, thereby conferring resistance to apoptosis.[26,27] Upregulation of Bcl-2 by COX-2 overexpression reduced apoptotic susceptibility through inhibition of cytochrome *c*–dependent apoptotic pathway.[26] In COX-2–transfected human colon cancer cells, death receptor 5 (DR5) was repressed and attained resistance to TRAIL-induced apoptosis.[27] Furthermore, COX-2 overexpression promoted the survival of differentiated PC12 cells through upregulation of the endogenous protein inhibitor of neuronal NOS (PIN).[28] According to a proposed model for regulation of apoptosis by COX-2, PGE_2 released from COX-2 stimulates the prostaglandin receptors on plasma membrane of PC12 cells, leading to an increase in PIN expression. Stimulation of PIN expression results in an increase in association of PIN with nNOS, causing inactivation of this enzyme and thus reducing the production of NO. Therefore, in cells under nitrosative stress, PGE_2 production through COX-2 induction may provide a negative feedback to prevent production of NO from nNOS. In another experiment using Raw 264.7 macrophages, PGE_2 was found to induce hemeoxygenase-1 (HO-1), which is an important enzyme in the cellular defense against oxidative stress and has potent anti-inflammatory properties. The protective effect of PGE_2 on UV-induced cell death was blocked by a HO-1 inhibitor SnPP, suggesting that PGE_2 can exert a cytoprotective effects through HO-1 induction.[29] In agreement with this notion, we also found that the preincubation of PC12 cells with PGE_2 conferred the protection against the nitrosative stress through possible upregulation of apoptosis-related proteins, such as Bcl-2 and HO-1. In contrast, the coincubation of PGE_2 with SNP failed to rescue PC12 cells from nitrosative cell death, probably because of lack of enough time or capacity to execute *de novo* protein expression. The concentration of prostanoids is an important factor to determine their physiological functions. Submicromolar concentrations of PGE_2 have been reported to be neuroprotective against hypoxia and glutamate-induced excitotoxicity,[30] whereas higher amounts of prostanoids could be toxic to neurons.[14–17] Consistent with this notion, we found that 100 μM PGE_2 did not show any protective effect against SNP-induced cytotoxicity. In agreement with our findings, exogenous PGE_2 sensitized human oeseoarthritis chondrocytes to nitric oxide, whereas PGE_2 alone did not induce chondrocyte death.[31] The molecular mechanisms underlying such a dual effect of PGE_2 have not been fully elucidated yet because of complex physiology of PGE_2 and the limitation of appropriate experimental systems. PGE_2 has multiple receptor subtypes, that is, EP1, EP2, EP3, and EP4, and their expression is dependent on cell types. Therefore, PGE_2 commonly has versatile and often opposing actions in many tissues and cells.[32]

From another point of view, COX-2 expression may protect cells from nitrosative stress through the production of prostaglandin D_2 (PGD_2). PGD_2 is the second major COX-2 product which often can be converted to 15-deoxy-$\Delta^{12,14}$-prostaglandin J_2 (15d-PGJ_2). In the resolution phase of inflammatory tissues damage, COX-2 upregulation and elevated 15d-PGJ_2 production are observed, and the COX-2 inhibitor even aggravates the inflammatory damages.[33] 15d-PGJ_2 binds to PPARγ, a member of the nuclear hormone receptor superfamily with a potent anti-inflammatory property. PPARγ has been reported to inhibit the expression of iNOS and COX-2 in part by antagonizing the activities of the transcription factors such as AP-1, STAT, and NF-κB, and 15d-PGJ_2 itself has been known to interfere with the NF-κB signaling, resulting in the deceased expression of proinflammatory molecules such as COX-2, iNOS, and cytokines.[23] Recently, beneficial effects of NSAIDs for patients with Alzheimer's disease have been attributed to PPARγ activation, independent of COX inhibition, because NSAIDs showed their therapeutic effects at concentrations higher than those required for inhibition of COX-2.[22,34] The increased PPARγ levels in patients with Alzheimer's disease suggest that PPARγ could play a pivotal role in the regulation of pathophysiology of this neurodegenerative disorder. Based on our observation that NO upregulates PPARγ expression, and GW9662, the PPARγ antagonist, sensitizes cells to NO-induced cell death, it is likely that PPARγ has a defensive function against nitrosative PC12 cell death. In macrophage/monocyte cell systems, noncytotoxic NO treatment switched the cells from a pro- to an anti-inflammatory phenotype by augmenting anti-inflammatory properties of the PPARγ. PPARγ mediated downregulation of p47 phagocyte oxidase ($p47^{phox}$), a component of the NAD(P)H oxidase system, causing inhibition of superoxide anion formation.[35]

In conclusion, NO causes apoptotic events such as mitochondrial depolarization, PARP cleavage, and $p21^{Waf1/Cip1}$ induction which were accompanied by COX-2 expression and PPARγ expression and/or activation. Additional studies will be necessary to elucidate the molecular events controlling defensive roles of COX-2, PGE_2, and PPARγ against nitrosative cell death.

ACKNOWLEDGMENTS

This work was supported by a grant (2002-2-20800-003-5) from the Basic Research Program of the Korea Science and Engineering Foundation (KOSEF).

REFERENCES

1. BREITNER, J.C. 1996. Inflammatory processes and antiinflammatory drugs in Alzheimer's disease: a current appraisal. Neurobiol. Aging **17:** 789–794.
2. MCGEER, E.G. & P.L. MCGEER. 1998. The importance of inflammatory mechanisms in Alzheimer disease. Exp. Gerontol. **33:** 371–378.
3. HEALES, S.J., J.P. BOLANOS., V.C. STEWART, *et al.* 1999. Nitric oxide, mitochondria and neurological disease. Biochim. Biophys. Acta **1410:** 215–228.
4. BRUNE, B., C. GOTZ, U.K. MESSMER, *et al.* 1997. Superoxide formation and macrophage resistance to nitric oxide-mediated apoptosis. J. Biol. Chem. **272:** 7253–7258.
5. SZABO, C., B.J. DAY & A.L. SALZMAN. 1996. Evaluation of the relative contribution of nitric oxide and peroxynitrite to the suppression of mitochondrial respiration in immunostimulated macrophages using a manganese mesoporphyrin superoxide dismutase mimetic and peroxynitrite scavenger. FEBS Lett. **381:** 82–86.
6. BEAL, M.F. 2002. Oxidatively modified proteins in aging and disease. Free Radic. Biol. Med. **32:** 797–803.
7. BROWN, G.C. & V. BORUTAITE. 2001. Nitric oxide, mitochondria, and cell death. IUBMB Life **52:** 189–195.
8. YANG, F., A. VON KNETHEN & B. BRUNE. 2000. Modulation of nitric oxide-evoked apoptosis by the p53-downstream target $p21^{WAF1/CIP1}$. J. Leukoc. Biol. **68:** 916–922.
9. GERVAIS, J.L., P. SETH & H. ZHANG. 1998. Cleavage of CDK inhibitor $p21^{WAF1/CIP1}$ by caspases is an early event during DNA damage-induced apoptosis. J. Biol. Chem. **273:** 19207–19212.
10. KITAMURA, Y., S. SHIMOHAMA, H. KOIKE, *et al.* 1999. Increased expression of cyclooxygenases and peroxisome proliferator activated receptor-gamma in Alzheimer's disease brains. Biochem. Biophys. Res. Commun. **254:** 582–586.
11. NOGAWA, S., F. ZHANG, M.E. ROSS, *et al.* 1997. Cyclo-oxygenase-2 gene expression in neurons contributes to ischemic brain damage. J. Neurosci. **17:** 2746–2755.
12. MCGEER, E.G. & P.L. MCGEER. 1998. The importance of inflammatory mechanisms in Alzheimer disease. Exp. Gerontol. **33:** 371–378.
13. STEWART, W.F., C. KAWAS, M. CORRADA, *et al.* 1997. Risk of Alzheimer's disease and duration of NSAID use. Neurology **48:** 626–632.
14. TAKADERA, T., H. YUMOTO, Y. TOZUKA, *et al.* 2002. Prostaglandin E_2 induces caspase-dependent apoptosis in rat cortical cells. Neurosci. Lett. **317:** 61–64.
15. PRASAD, K.N., A.R. HOVLAND, F.G. LA ROSA, *et al.* 1998. Prostaglandins as putative neurotoxins in Alzheimer's disease. Proc. Soc. Exp. Biol. Med. **219:** 120–125.
16. MONTINE, T.J., D. MILATOVIC, R.C. GUPTA, *et al.* 2002. Neuronal oxidative damage from activated innate immunity is EP2 receptor-dependent. J. Neurochem. **83:** 463–470.
17. BEZZI, P., G. CARMIGNOTO, L. PASTI, *et al.* 1998. Prostaglandins stimulate calcium-dependent glutamate release in astrocytes. Nature **391:** 281–285.
18. KIM, E.J., K.J. KWON, J.Y. PARK, *et al.* 2002. Neuroprotective effects of prostaglandin E_2 or cAMP against microglial and neuronal free radical mediated toxicity associated with inflammation. J. Neurosci. Res. **70:** 97–107.
19. CAGGIANO, A.O. & R.P. KRAIG. 1999. Prostaglandin E receptor subtypes in cultured rat microglia and their role in reducing lipopolysaccharide-induced interleukin-1beta production. J. Neurochem. **72:** 565–575.
20. LEMBERGER, T., O. BRAISSANT, C. JUGE-AUBRY, *et al.* 1996. PPAR tissue distribution and interactions with other hormone-signaling pathways. Ann. N.Y. Acad. Sci. **804:** 231–251.
21. VAMECQ, J. & N. LATRUFFE. 1999. Medical significance of peroxisome proliferator-activated receptors. Lancet **354:** 141–148.

22. LEHMANN, J.M., J.M. LENHARD, B.B. OLIVER, *et al.* 1997. Peroxisome proliferator-activated receptors alpha and gamma are activated by indomethacin and other non-steroidal anti-inflammatory drugs. J. Biol. Chem. **272:** 3406–3410.
23. RICOTE, M., A.C. LI, T.M. WILLSON, *et al.* 1998. The peroxisome proliferator-activated receptor-gamma is a negative regulator of macrophage activation. Nature **391:** 79–82.
24. KITAMURA, Y., S. SHIMOHAMA, H. KOIKE, *et al.* 1999. Increased expression of cyclooxygenases and peroxisome proliferator-activated receptor-gamma in Alzheimer's disease brains. Biochem. Biophys. Res. Commun. **254:** 582–586.
25. UCHIYAMA, T., H. OTANI, T. OKADA, *et al.* 2002. Nitric oxide induces caspase-dependent apoptosis and necrosis in neonatal rat cardiomyocytes. J. Mol. Cell. Cardiol. **34:** 1049–1061.
26. SUN, Y., X.M. TANG, E. HALF, *et al.* 2002. Cyclooxygenase-2 overexpression reduces apoptotic susceptibility by inhibiting the cytochrome c-dependent apoptotic pathway in human colon cancer cells. Cancer Res. **62:** 6323–6328.
27. TANG, X., Y.J. SUN, E. HALF, *et al.* 2002. Cyclooxygenase-2 overexpression inhibits death receptor 5 expression and confers resistance to tumor necrosis factor-related apoptosis-inducing ligand-induced apoptosis in human colon cancer cells. Cancer Res. **62:** 4903–4908.
28. CHANG, Y.W., R. JAKOBI, A. MCGINTY, *et al.* 2000. Cyclooxygenase 2 promotes cell survival by stimulation of dynein light chain expression and inhibition of neuronal nitric oxide synthase activity. Mol. Cell. Biol. **20:** 8571–8579.
29. CHEN, Y.C., SHEN, S.C., W.R., LEE, *et al.* 2002. Nitric oxide and prostaglandin E_2 participate in lipopolysaccharide/interferon-gamma-induced heme oxygenase 1 and prevent RAW264.7 macrophages from UV-irradiation-induced cell death. J. Cell. Biochem. **86:** 331–339.
30. AKAIKE, A., S. KANEKO, Y. TAMURA, *et al.* 1994. Prostaglandin E_2 protects cultured cortical neurons against *N*-methyl-D-aspartate receptor-mediated glutamate cytotoxicity. Brain Res. **663:** 237–243.
31. NOTOYA, K., D.V. JOVANOVIC, P. REBOUL, *et al.* 2000. The induction of cell death in human osteoarthritis chondrocytes by nitric oxide is related to the production of prostaglandin E_2 via the induction of cyclooxygenase-2. J. Immunol. **165:** 3402–3410.
32. BREYER, R.M., C.K. BAGDASSARIAN, S.A. MYERS, *et al.* 2001. Prostanoid receptors: subtypes and signaling. Annu. Rev. Pharmacol. Toxicol. **41:** 661–690.
33. COLVILLE-NASH, P.R. & D.W. GILROY. 2000. COX-2 and the cyclopentenone prostaglandins—a new chapter in the book of inflammation. Prostaglandins Other Lipid Mediat. **62:** 33–43.
34. JIANG, C., A.T. TING & B. SEED. 1998. PPAR-gamma agonists inhibit production of monocyte inflammatory cytokines. Nature **391:** 82–86.
35. VON KNETHEN, A. & B. BRUNE. 2002. Activation of peroxisome proliferator-activated receptor gamma by nitric oxide in monocytes/macrophages down-regulates $p47^{phox}$ and attenuates the respiratory burst. J. Immunol. **169:** 2619–2626.

Tellurium Compound AS101 Induces PC12 Differentiation and Rescues the Neurons from Apoptotic Death

D. MAKAROVSKY,[a] Y. KALECHMAN,[a] T. SONINO,[a] I. FREIDKIN,[a] S. TEITZ,[a] M. ALBECK,[a] M. WEIL,[b] R. GEFFEN-ARICHA,[a] G. YADID, [a] AND B. SREDNI[a]

[a]*C.A.I.R. Institute, Faculty of Life Science, Bar Ilan University, Ramat Gan, 59200 Israel*

[b]*Department of Cell Research and Immunology, Faculty of Life Sciences, Tel Aviv University, Ramat Aviv 69978, Israel*

ABSTRACT: Parkinson's disease is characterized by the loss of dopaminergic neurons in the substantia nigra (SN). Studies show that anti-apoptotic and neurotrophic agents are suitable candidates to prevent delayed cell death and/or restore neural function. Here we present the nontoxic immunomodulating compound AS101, which has the ability to induce neurite outgrowth and neural differentiation in PC12 cells. The present study shows that components of the ras signaling pathway are crucial for AS101-induced PC12 differentiation. These include p21ras and its downstream effectors, c-raf-1 and MEK, as well as PI3K. Moreover, these components mediate AS101-induced upregulation of p21waf, which is obligatory for AS101-induced PC12 differentiation. Furthermore, nitric oxide plays a significant role in these AS101 activities. Finally, we show that AS101 prevents apoptosis of NGF-differentiated PC12 cells after NGF withdrawal. Taken together, these results suggest that AS101 induces PC12 cell differentiation and survival by activating the ras-ERK1/2 and ras-PI3K signal transduction pathways, as well as inducing NO production. Our findings may be important in understanding the regulation of survival/apoptosis of neurons deprived of neurotropic support. Futhermore the data propose that AS101 may have clinical potential in the treatment of neurodegenerative disorders like Parkinson's disease.

KEYWORDS: NGF; apoptosis; survival; p21ras; neurodegeneration

INTRODUCTION

During mammalian development and in the mature organism, various extracellular stimulli act to promote survival and suppress cellular suicide programs. Under pathological conditions, as well as during deficiency of target-derived survival factors, a lot of neurons in flawed tissues die during the development of the disease. The signaling pathways that lead to apoptosis are beginning to be defined, and a number of proteins have been identified that either induce or prevent apoptosis. For example,

Address for correspondence: Dr. B. Sredni, C.A.I.R. Institute, Faculty of Life Science, Bar Ilan University, Ramat Gan, 59200 Israel. Voice: +972-3 531-8250; fax: +972-3 635-6041. srednib@mail.biu.ac.il

Ann. N.Y. Acad. Sci. 1010: 659–666 (2003).
doi: 10.1196/annals.1299.120

the activation of the p21ras by different neurotrophic factors, like NGF, is essential for neural differentiation and proliferation of PC12 cells, as well as for neural survival. P21ras phosphorylation leads to activation of c-raf-1 and PI3K.[1] On the one hand, the p21ras-PI3K/Akt pathway leads to deactivation of different pro-apoptotic proteins, like Bad and caspase 9, and to the activation of, for example, atypical PKC and NF-κB. This pathway thus plays a crucial role in cell proliferation and survival.[2] On the other hand, the p21ras-raf-ERK1/2 pathway participates in cell differentiation and nitric oxide (NO) production.[3] NO is a unique molecule that plays an important role in both neural development and degeneration. Studies show that the outcome of nitric oxide's effect depends on its concentration, the cell type, and its accumulation in a specific stage in the cell cycle. Small amounts of NO can prevent apoptosis of NGF-differentiated PC12 cells after NGF withdrawal. Survival is mediated by NO through the p21ras-PI3K signal transduction pathway, or through the direct inhibition of caspase activity by S-nitrosylation.[4]

The nontoxic immunomodulator AS101 has been shown to possess neurotrophic properties in a hemiparkinsonean model. We have demonstrated that treatment of parkinsonean rats by AS101 leads to reduction of IL-10 levels and increased GDNF, IL-1, and IL-6 protein levels in the substantia nigra (SN). These activities are associated with a significant improvement in behavior parameters that characterize Parkinson's disease (PD).[5] The aim of the present study is to investigate the influence of AS101 on the differentiation of PC12 cells and their survival/apoptosis under conditions of deprivation of trophic support. In addition, determination of the signal-transduction pathways involved in these AS101 activities will be assessed.

RESULTS AND DISCUSSION

The Ability of AS101 to Induce PC12 Differentiation and Neurite Outgrowth

PC12 differentiation by NGF is accompanied by morphological changes, neurite outgrowth, and expression of different neural markers. FIGURE 1a illustrates that addition of AS101 to PC12 cells results in their differentiation. Differentiation of PC12 cells was microscopically observed following addition of AS101 at 0.5–1.5 μg/mL concentrations (data not shown). AS101's ability to induce PC12 differentiation was confirmed by immunohistochemistry studies showing positive staining to the neuronal marker neurofilament M, following AS101 treatment (FIG. 1a).

FIGURE 1. The role of p21ras-ERK1/2, p21ras-PI3K, and NO in AS101 activity on PC12 differentiation. (**a**) Photographs (magnification × 100) of differentiated PC12 cells induced by AS101 (AS), which are positively stained for neurofilament M. (**b**) Inhibitors of p21ras, c-raf-1, MEK, PI3K, and NOS abrogate AS101 activity. (**c**) Western blot analysis for determination of p21waf expression in different treatments. (**d**) The role of p21waf upregulation by AS101 in PC12 cells. Differentiation was assessed by showing that transfection of anti-sense p21waf (as) to these cells abrogated the differentiating capabilities of AS101. (**e**) Western blot analysis for determination phosphorylation of MAP kinase and PI3K activity. The phosphorylation of ERK1/2 was determinated two times simultaneously. L-NAME abrogated the late but not early phosphorylation, while both activations of MAP kinase were abolished by farnesyltransferase inhibitor (FTi). The PI3K activity was abolished by the MEK inhibitor (PD), ras inhibitor (FTi), and L-NAME within 24 hours.

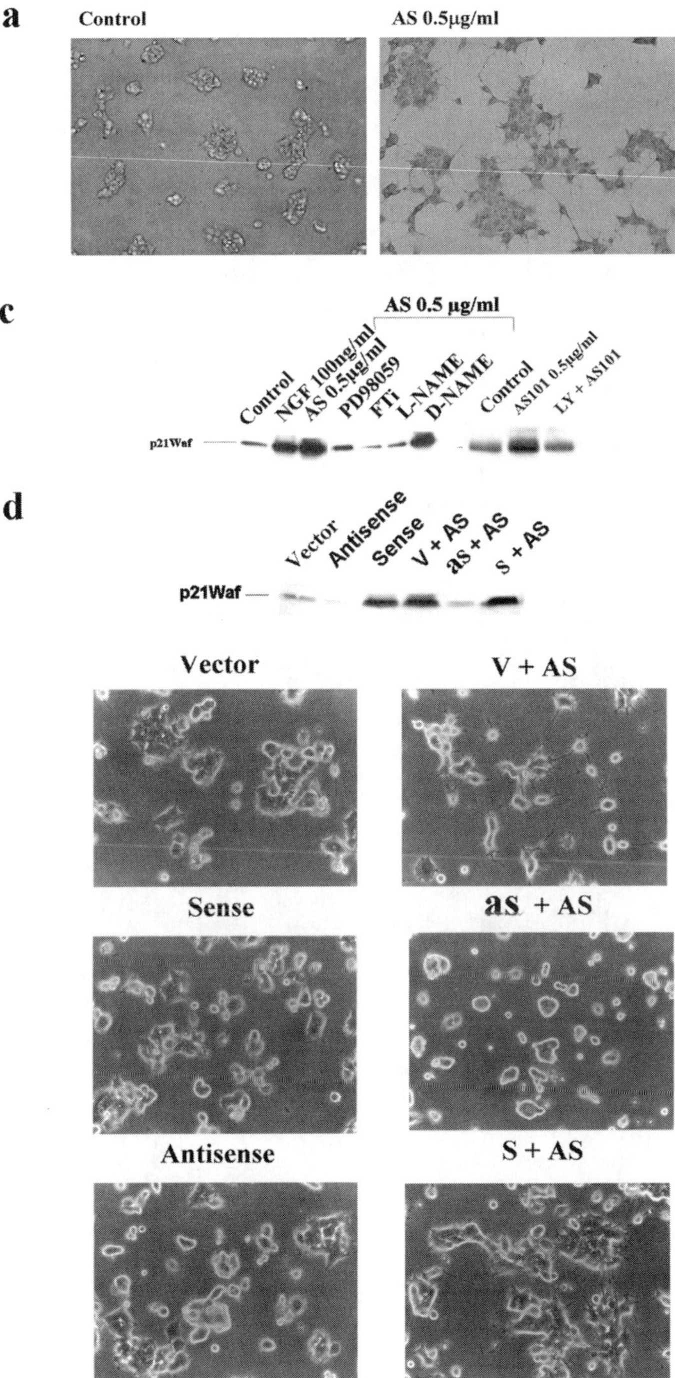

FIGURE 1. *See previous page for legend.*

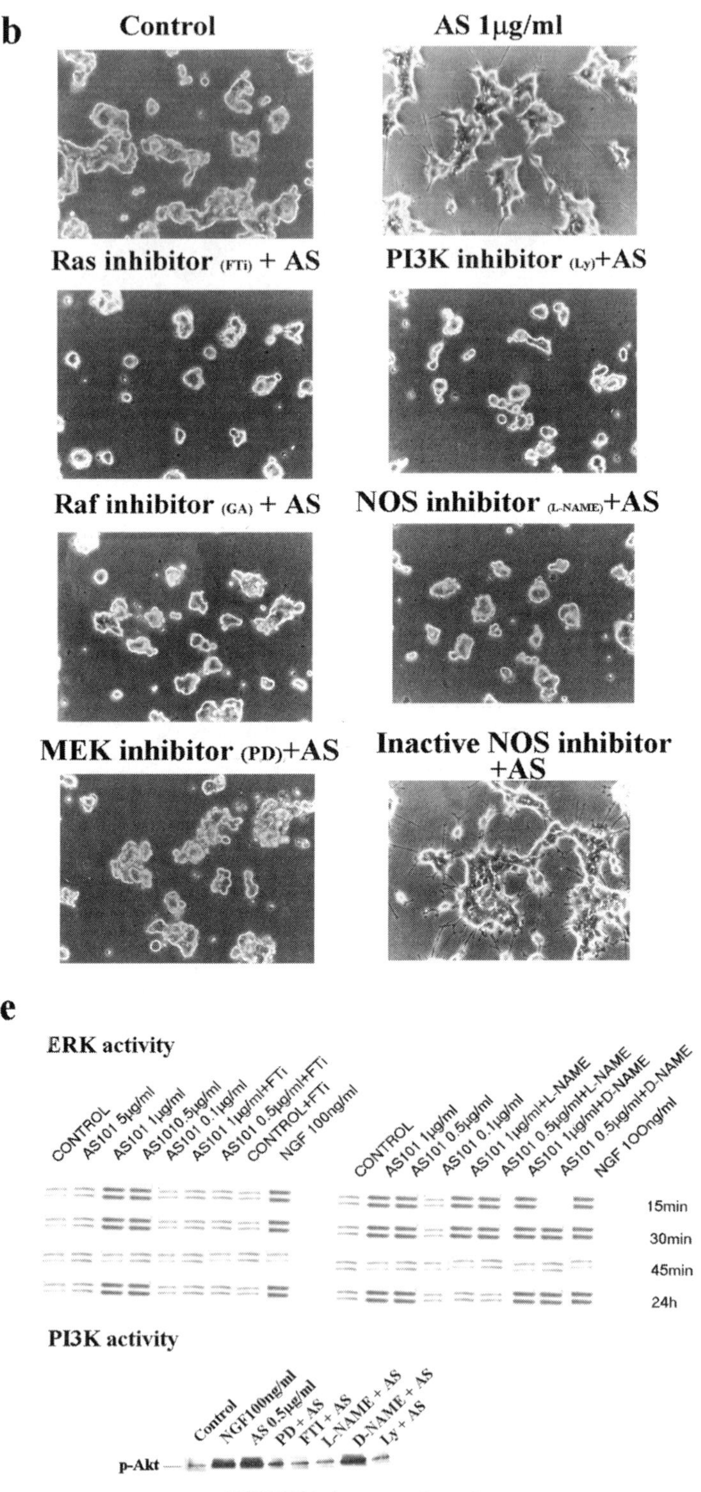

FIGURE 1 — *continued.*

AS101 Induces Activation of Two Different p21 Ras-Dependent Cascades

p21ras is a small GTP-protein molecule that initiates certain intracellular signaling cascades following induction with, for example, neurotrophins, growth factors, and interlukins. Two important pathways, which can be activated by p21ras, are the c-raf-1/ERK and PI3K/Akt pathways. It is known that neuronal differentiation and survival requires the activation of those cascades. Since AS101 alone leads to the differentiation of PC12 cells, we attempted to elucidate the mechanism of its activity. For this purpose we used the following imhibitors: farnesyl transferase inhibitor (Fti), which inhibits p21ras activity; geldanamycin (inhibitor of c-raf-1); PD98059 (MEK inhibitor); and LY294002 (PI3K inhibitor). FIGURE 1b shows that incubation of PC12 cells in the presence of either one of the inhibitors detailed below abrogates the neural PC12 differentiation induced by AS101. These results indicate that both the p21ras-ERK1/2 and p21ras-PI3K pathways are necessary for PC12 differentiation.

Role of Nitric Oxide and p21waf in AS101-Induced PC12 Cells Differentiation

It is known that the activation of the p21ras-c-raf-1-ERK1/2 cascade induces NO production, and that NO has a potential role in the regulation of p21waf protein expression and PC12 differentiation. AS101 has been previously shown to induce the production of NO in mouse peritoneal macrophages.[6] We therefore analyzed the role of NO as a second messenger in the increase of p21waf by AS101 and in its differentiating ability. As shown in FIGURE 1b, addition of AS101 at 0.5 μg/mL with an active NOS inhibitor to PC12 cells abolishes their differentiation. FIGURE 1c demonstrates that AS101 treatment at 0.5 μg/mL upregulates p21waf expression compared to levels in untreated cells. L-NAME, but not D-NAME, inhibited the accumulation of p21waf, and treatment of PC12 cells with either p21ras inhibitor or MEK inhibitor abrogates AS101-induced PC12 differentiation. Thus, NO, at physiological levels, acts as a second messenger, following stimulation of PC12 cells by AS101, to enhance expression of p21waf. The role of p21waf upregulation by AS101 in PC12 cell differentiation was assessed by showing that transfection of anti-sense p21waf to these cells abrogated the differentiating capabilities of AS101 (FIG. 1d). Suprisingly, we found that activation of MAPK by AS101 could be detected at both early (15–30 min) and late (24 h) stimulation times. NO was found to mediate the late, but not early, ERK1/2 activation, for only the latter was inhibited by L-NAME, while both activations of MAPK were abolished by farnesyltransferase inhibitor (FIG. 1e). Moreover, AS101-induced PI3K activity, as reflected by phosphorylation of Akt at 24h, was also abrogated by p21ras, MEK, and NOS inhibition (FIG. 1e).

The Ability of AS101 to Rescue the NGF-Differentiated PC12 from Neurotrophic Deprivation

On the basis of the ability of AS101 to activate the p21ras-ERK1/2 pathway, and to upregulate p21waf and the endogenous production of NO, all of which effects have been shown to mediate the survival of PC12 cells, we analyzed its ability to prevent apoptotic cell death of differentiated PC12 cells following withdrawal of trophic support. As shown in FIGURE 2a, treatment of PC12 cells with AS101 resulted in the induction of G1 arrest. Following incubation of the cells with AS101 for 24 h, 83.1% of the cells stimulated with 0.5 μg/mL AS101 accumulated in G1, and

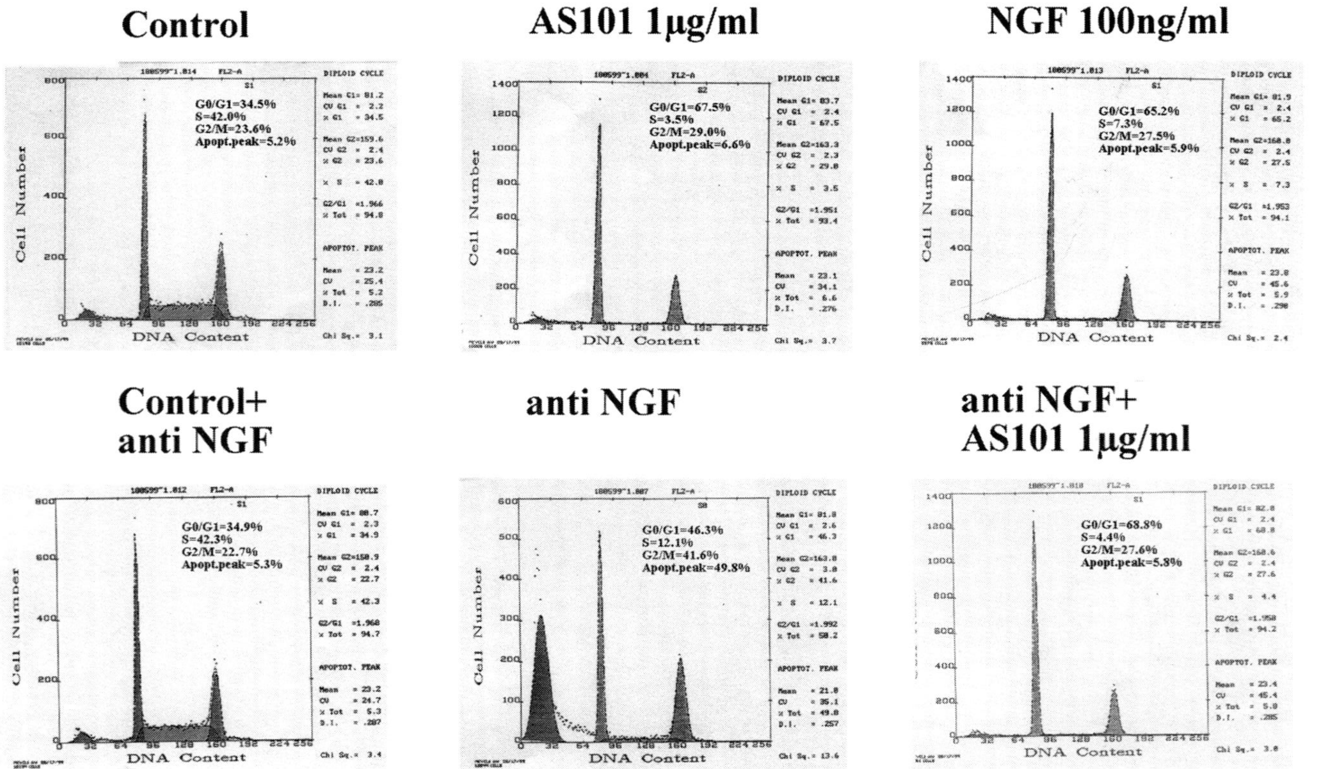

FIGURE 2. AS101 rescued the NGF-differentiated PC12 from NGF withdrawal. **(a)** The cell cycle analysis for determination of AS101 ability to rescue the NGF-differentiated PC12 cells.

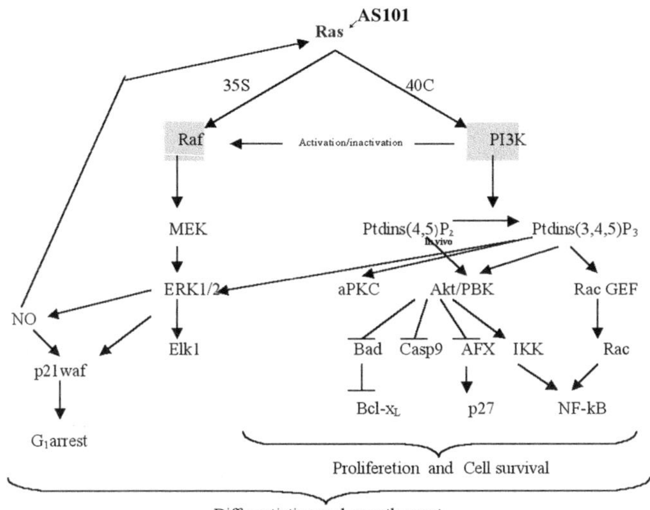

FIGURE 2 — *continued.* AS101 rescued the NGF-differentiated PC12 from NGF withdrawal. (**b**) The summary of the ras multiple signal pathway, which plays an important role in AS101 activity.

only 3.5% in S-phase compared to 81.2% in G1 and 42.8% in S-phase of untreated cells. More importantly, treatment of PC12 cells with anti-NGF antibodies 5 days following incubation of the cells with NGF resulted in 50% apoptosis 24 h later. Addition of 0.5 μg/mL AS101 with anti-NGF antibodies significantly decreased the rate of apoptosis occurring one day later, when it did not significantly differ from that of control cells incubated with AS101 (FIG. 2a).

Parkinson's disease (PD) is characterized by neural abnormalities that result largely from progressive degeneration of dopaminergic neurons in the SN. The causes of neuronal degeneration in PD are not yet known, but studies of the last few years suggest important roles of oxidative stress, mitochondrial dysfunction, and apoptosis in the progression of this disease. The neurons appear to be in a pre-apoptotic state for an extended period of time prior to being irreversibly damaged. Thus, both the anti-apoptotic and neurotrophic agents are suitable candidates for preventing delayed cell death and/or restoring neural function. Regulation of the balance by different extracellular stimuli between JNK-p38 and ERK in NGF-differentiated PC12 cells will determine what the future of the cell is—death or life.[7] The present study shows that AS101 induces the ras-raf-ERK1/2 and the ras-PI3K pathways in PC12 cells, as well as nitric oxide induction (FIG. 2b). We suggest that the nontoxic immunomodulating compound AS101, currently used in Phase II/III clinical trials in cancer patients, may prevent further neural degeneration and thus restore the functional capacity of diseased neurons in PD.

ACKNOWLEDGMENTS

This work was partly supported by the Dr. Tovi Comet-Wallerstein Cancer Research Chair and by the Dave and Florence Muskovitz Chair in Cancer Research.

REFERENCES

1. XUE, L. *et al.* 2000. The ras/phosphatidylinositol 3-kinase and ras/ER pathways function as independent survival modules each of which inhibits a distinct apoptotic signaling pathway in sympathetic neurons. J. Biol. Chem. **275:** 8817–8824.
2. BRUNET, A. *et al.* 2001. Transcription-dependent and -independent control of neuronal survival by the PI3K-Akt signaling pathway. Curr. Opin. Neurobiol. **11:** 297–305.
3. SCHONHOFF, C.M. *et al.* 2001. The ras-ERK pathway is required for the induction of neuronal nitric oxide synthase in differentiated PC12 cells. J. Neurochem. **78:** 631–639.
4. DEORA, A.A. *et al.* 2000. Recruitment and activation of Raf-1 kinase by nitric oxide-activated Ras. Biochemistry **39:** 9901–9908.
5. SREDNI, B. *et al.* Anti-apoptotic immunomodulating tellurium compound (AS101) as apotential treatment of Parkinson's disease. Data submitted.
6. ROSENBLATT-BIN, H. *et al.* 1996. Antibabesial effect of the immunomodulator AS101 in mice: role of increased production of nitric oxide. Parasite Immunol. **18**(6)**:** 297–306.
7. XIA, Z. *et al.* 1995. Opposing effect of ERK and JNK-p38 MAP kinases on apoptosis. Science **270:** 1326–1331.

Drugs of Abuse Induce Apoptotic Features in PC12 Cells

M. T. OLIVEIRA,[a] A. C. REGO,[a] T. R. A. MACEDO,[b] AND C. R. OLIVEIRA[a]

[a]*Institute of Biochemistry, Faculty of Medicine and Center for Neuroscience and Cell Biology of Coimbra,* [b]*Institute of Pharmacology and Experimental Therapeutics, Faculty of Medicine, University of Coimbra, 3004-504 Coimbra, Portugal*

ABSTRACT: Drugs of abuse induce the release of dopamine in the central nervous system, particularly in the mesolimbic–mesocortical pathway. As dopamine may act as a neurotoxin, in this study, we analyzed the effects of the drugs of abuse, cocaine, heroin, and amphetamine, on the neurodegeneration of PC12 cells, a dopaminergic cell line, by evaluating the activity of caspase-3 and mitochondrial cytochrome c release. All the drugs were shown to induce caspase-3 activation, similarly to staurosporine, a classical inducer of apoptotic cell death. Furthermore, like staurosporine, the drugs of abuse induced a decrease in mitochondrial cytochrome c content, suggesting the involvement of the mitochondrial apoptotic pathway.

KEYWORDS: amphetamine; apoptosis; cocaine; dopamine; heroin; mitochondria; PC12 cells

INTRODUCTION

The repeated abuse of drugs is sustained by the activation of the mesolimbic/ mesocortical circuit, in which dopamine is the main neurotransmitter. Dopamine bears potential neurotoxicity due to its oxidative metabolism, which produces reactive oxygen species, namely hydrogen peroxide. Drug abuse is also associated with changes in brain function and neurodegenerative processes, which, for some drugs, have been shown to be associated with the induction of apoptotic cell death.[1–3] Therefore, the study of the mechanisms involved in the neurodegeneration induced by these drugs may be useful to prevent further neural demise.

Apoptosis is a programmed form of cell death, executed by a family of proteases named caspases. This form of cell suicide can be triggered by several external or internal stimuli. When cellular homeostasis is affected, the mitochondrial apoptotic pathway is activated, inducing the release of cytochrome c, frequently associated with a loss of mitochondrial membrane potential. Caspase-9, classically considered as an initiator caspase, is further activated in the presence of Apaf-1 and dATP, and is responsible for activating the effector caspase-3. This caspase activates nuclear DNAses, inducing DNA fragmentation.

Address for correspondence: Prof. Catarina Oliveira, M.D., Ph.D., Institute of Biochemistry, Faculty of Medicine, Center for Neuroscience and Cell Biology of Coimbra, University of Coimbra, 3004-504 Coimbra, Portugal. Voice: +351-239-820190; fax: +351-239-822776.
catarina@cnc.cj.uc.pt

Ann. N.Y. Acad. Sci. 1010: 667–670 (2003). © 2003 New York Academy of Sciences.
doi: 10.1196/annals.1299.121

In this study, using a dopaminergic cell line (PC12 cells), we analyzed the activation of a central executioner caspase—caspase-3—and the content of cytochrome c within the mitochondria, in cells exposed to cocaine, heroin, or amphetamine, as compared to staurosporine, a classical inducer of apoptotic cell death.

METHODS

Culture of Undifferentiated PC12 Cells

PC12 cells (ATCC) were cultured in 75 cm^2 flasks, in RPMI 1640 medium (Sigma) supplemented with 10% (v/v) horse serum, 5% (v/v) bovine serum, 50 U/mL penicillin, and 50 mg/mL streptomycin. Cultures were maintained at 37° C in a humidified incubator containing 95% air and 5% CO_2. The cells were plated on poly-L-lysine–coated multiwells at a density of 160,000 cells/cm.[2] The cells were further incubated for 5 h in RPMI medium, containing 4% bovine and horse serums (in the same proportion as for culturing the cells), simple or supplemented with staurosporine (1 μM) or the drugs of abuse, cocaine (300 μM), heroin (30 μM), and amphetamine (300 μM). These concentrations were chosen based on our previous studies, in which we investigated the toxic effects of the drugs of abuse.[4] In the present work lower concentrations of the drugs were tested to avoid induction of necrotic cell death. The use of culture medium with 4% serum concentration was previously shown not to affect the viability PC12 cells.[5]

Caspase-3 Activity Assay

The cells were lysed in lysis buffer (in mM: 25 HEPES, 2 $MgCl_2$, 1 EDTA, 1 EGTA, 2 DTT, 0.1 PMSF and 1:1000 of protease inhibitor cocktail [chymostatin, leupeptin, antipain, and pepstatin A] 1 mg/mL). The resulting extracts were frozen and thawed three times and further centrifuged at 15,000 *g* for 10 min (4°C). The supernatant (25 μg of protein) was tested for the activity of caspase-3 by reaction with Ac-Asp-Glu-Val-Asp-pNA (Ac-DEVD-pNA, Calbiochem) for 2 h (37°C), and the absorbance was measured at 405 nm.

Western Blotting Analysis of Mitochondrial Cytochrome C Content

The cells were homogenized in sucrose buffer (in mM: 250 sucrose, 20 HEPES, 10 KCl, 1.5 $MgCl_2$, 1 EDTA, 1 DTT, 0.1 PMSF, and 1:1000 of protease inhibitor cocktail), and the mitochondrial fraction (P2) was obtained by subcellular fractionation.[6] Cytochrome c was detected by Western blotting (15% SDS-PAGE), following the incubation with an antibody against the denatured form of cytochrome c (Pharmingen, 1:500). The secondary detection was made with an alkaline-phosphatase–bound antibody. The bands were revealed with ECF (Amersham) and quantified using a STORM 860 device and the software Image Quant 5.0.

RESULTS AND DISCUSSION

Caspase-3 is a central executioner of the apoptotic program that is activated by several types of apoptotic stimuli. Exposure of PC12 cells to 1 μM staurosporine in-

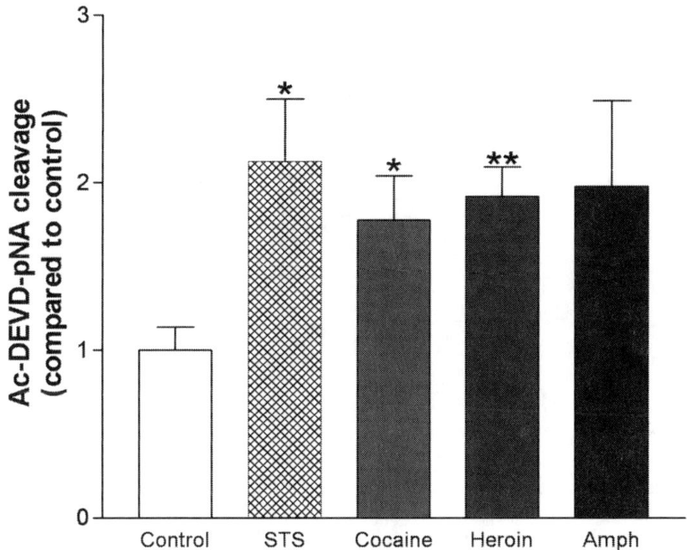

FIGURE 1. Caspase-3–like activity induced by the drugs of abuse. PC12 cells were incubated for 5 h with cocaine (300 μM), amphetamine (Amph, 300 μM), or heroin (30 μM), as compared to staurosporine (STS, 1 μM) in medium containing 4% serum. The extent of cleavage of Ac-DEVD-pNA was determined at 405 nm, in control cells and in cells incubated with the drugs for 5 h. The results are the means ± SEM of at least 3 experiments performed in duplicate. Statistical analysis (*t* test) $^{*}P < .05$, $^{**}P < .01$, as compared to the control.

creased caspase-3 activity by about 2-fold (FIG. 1). Moreover, exposure to cocaine (300 μM), heroin (30 μM), or amphetamine (300 μM) similarly increased the activity of this caspase (about 2-fold, FIG. 1). In order to determine if the mitochondrial apoptotic pathway was involved in caspase-3 activation, we evaluated the content in mitochondrial cytochrome c (FIG. 2). When incubated with staurosporine, the mitochondrial cytochrome c content was decreased by about 30%, in agreement with previous reports showing induction of cytochrome c release by staurosporine.[6] PC12 cells incubated with the drugs of abuse also showed a decrease in the levels of mitochondrial cytochrome c, when compared to control conditions. Heroin and amphetamine induced a decrease of about 40%, whereas cocaine induced a 15% decrease in cytochrome c (FIG. 2). These differences, however, were not reflected by significant changes in caspase-3 activation between the drugs tested (FIG. 1).

In previous studies,[4] we have shown that heroin and amphetamine were more toxic to PC12 cells than cocaine and were able to induce intracellular dopamine depletion, associated with chromatin condensation, another feature of cell death. Our previous findings also suggest that heroin's effects may be mediated by oxidative stress, since PC12 cells incubated with this drug showed higher levels of intracellular peroxide production and increased dopamine metabolization, when compared to control cells.[4] Furthermore, the effects of amphetamine may be due to the induction of metabolic dysfunction in these cells, since this drug was shown to decrease the

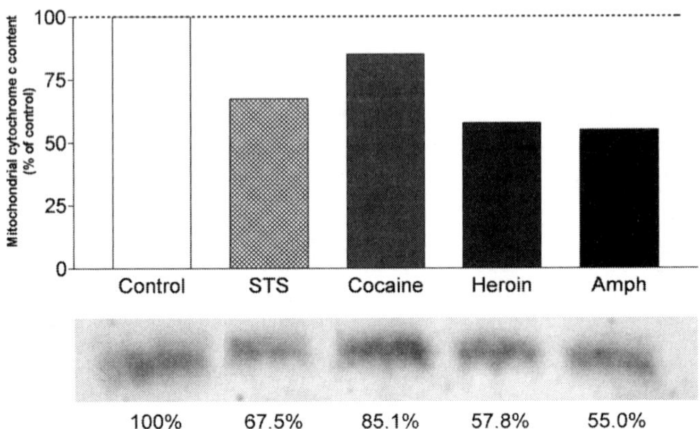

FIGURE 2. Mitochondrial cytochrome c levels upon exposure to the drugs of abuse. PC12 cells were incubated for 5 h with cocaine (300 μM), amphetamine (Amph, 300 μM), or heroin (30 μM), as compared to staurosporine (STS, 1 μM) in culture medium containing 4% serum, and the levels of cytochrome c in the mitochondrial fraction were determined by Western blot. The results show a representative experiment, expressed as a percentage of the levels of cytochrome c in the control.

ATP/ADP ratio,[4] probably due to alkalinization of the mitochondrial matrix.[1] The present results suggest that heroin and amphetamine activate the mitochondrial apoptotic pathway in PC12 cells. These results are in agreement with reports showing induction of apoptosis by amphetamine[1] and heroin.[3] Similar to our previous findings,[4] cocaine seems to be the less toxic drug, because it decreased mitochondrial cytochrome c levels only by a certain extent, although it had significantly activated caspase-3. These results seem to be in agreement with previous studies showing that this drug of abuse induces apoptosis in several cell types (reviewed in Ref. 2).

REFERENCES

1. Davidson, C., A.J. Gow, T.H. Lee & E.H. Ellinwood. 2001. Methamphetamine neurotoxicity: necrotic and apoptotic mechanisms and relevance to human abuse and treatment. Brain Res. Brain Res. Rev. **36:** 1–22.
2. Zhang, L., Y. Xiao & J. He. 1999. Cocaine and apoptosis in myocardial cells. Anat. Rec. **257:** 208–216.
3. Fecho, K. & D.T. Lysle. 2000. Heroin-induced alterations in leukocyte numbers and apoptosis in the rat spleen. Cell. Immunol. **202:** 113–123.
4. Oliveira, M.T., A.C. Rego, M.T. Morgadinho, *et al.* 2002. Toxic effects of opioid and stimulant drugs on undifferentiated PC12 cells. Ann. N. Y. Acad. Sci. **965:** 487–496.
5. Oberdoerster, J., A.R. Kamer & R.A. Rabin. 1998. Differential effect of ethanol on PC12 cell death. J. Pharmacol. Exp. Ther. **287:** 359–365.
6. Rego, A.C., S. Vesce & D.G. Nicholls. 2001. The mechanism of mitochondrial membrane potential retention following release of cytochrome c in apoptotic GT1-7 neural cells. Cell Death Differ. **8:** 995–1003.

Inhibition of CDKs: A Strategy for Preventing Kainic Acid–Induced Apoptosis in Neurons

E. VERDAGUER,[a] ELVIRA G. JORDÀ,[a] ALESSANDRA STRANGES,[b] ANNA M. CANUDAS,[a] ANDRÉS JIMÉNEZ,[a] FRANCESC X. SUREDA,[c] MERCÈ PALLÀS,[a] AND ANTONI CAMINS[a]

[a]*Unitat de Farmacologia i Farmacognòsia, Facultat de Farmàcia, Universitat de Barcelona, Barcelona, Spain*

[b]*Societat Italiana de Farmacologia (SIF) i Societat Italiana de Bioquímica i Biologia Molecular (SIB), Cosenza, Italy*

[c]*Unitat de Farmacologia, Facultat de Medicina i Ciències de la Salut, Universitat Rovira i Virgili, Reus, Tarragona, Spain*

ABSTRACT: Stimulation of ionotropic glutamate receptors are implicated in neurodegenerative diseases such as Alzheimer's and Parkinson's diseases. Recently this has been demonstrated in the expression of cell cycle proteins in vulnerable neurons in Alzheimer's disease. Thus, the aim of the present study was to evaluate the expression of cell cycle proteins in cerebellar granule cells after stimulation of AMPA/KA receptors and likewise to study the neuroprotective effects of CDK inhibitors. Our results demonstrated that after a treatment with CDK inhibitors, a significant decrease in apoptotic nuclei induced by kainic acid was found in the presence of flavopiridol and 3-ATA. We concluded that CDK activation is involved, at least, in part, in the pro-apoptotic effects of kainic acid.

KEYWORDS: 3-ATA; flavopiridol; CDK; cerebellar granule cells; kainic acid

Cyclin-dependent kinases (CDKs) are a family of serine/threonine protein kinases that regulate cell-cycle progression. In this process, cyclins D and E, associated with CDK4/6 and CDK2, respectively, are involved in the G1/S transition, a key step in cell proliferation. CDKs are able to phosphorylate the retinoblastoma protein (pRb) and thus release the transcription factor E2F-1, which promotes the G1 to S-phase transition. In neurons, amount data suggest that the CDK/pRb/E2F pathway plays a prominent role in promoting neuronal cell death. An increase in the expression of cell-cycle proteins has been determined in regions of the brain from Alzheimer's disease patients.[1–3]

Glutamate, the main excitatory amino acid in the brain is likely to be involved in numerous neurological disorders, including Alzheimer's and Huntington's disease.

Address for correspondence: Ester Verdaguer, Unitat de Farmacologia i Farmacognòsia, Facultat de Farmàcia, Universitat de Barcelona, Nucli Universitari de Pedralbes, E-08028 Barcelona, Spain. Voice: +34-934024531; fax: +34-934035982.
verdague@farmacia.far.ub.es

Ann. N.Y. Acad. Sci. 1010: 671–674 (2003). © 2003 New York Academy of Sciences.
doi: 10.1196/annals.1299.122

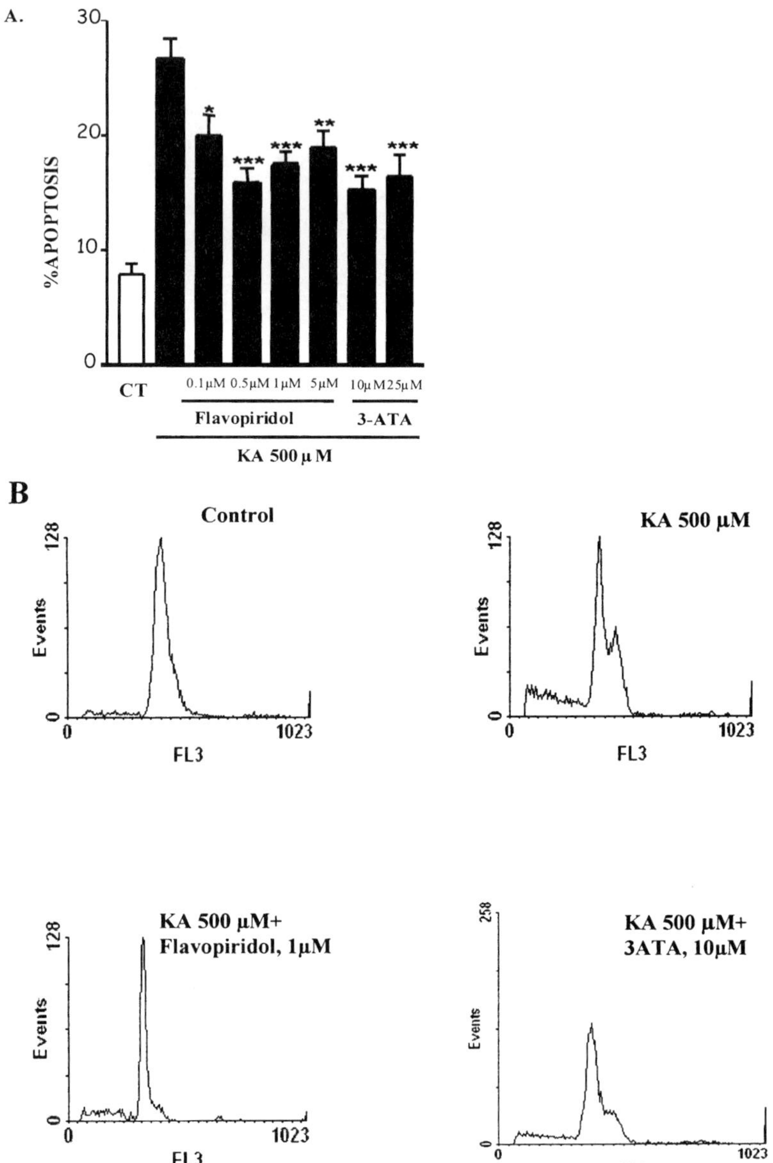

FIGURE 1. Effect of CDK inhibitors on KA-induced apoptosis in CGCs from rat. Methods: flow cytometry with PI staining. (**A**) Percentage of apoptotic nuclei to cells. An increase in the percentage of apoptotic nuclei in cells treated with KA 500 μM (7.9±0.8%, n=8 control cells versus 26±1.7%, n=9 treated cells) was reduced by flavopiridol (1 μM) and 3-ATA (10 μM) (17±1.1%, n=8; 17.4±1.3%, n=8) (**B**) Representative flow cytometry histograms. Distribution of fluorescent frequency nuclei. $^{*}P<.05$ vs. KA; $^{**}P<.01$ vs. KA; $^{***}P<.001$ vs. KA. ANOVA post-test to Tukey-Kramer. Each point is the mean±SEM of 4 experiments, carried out in duplicate.

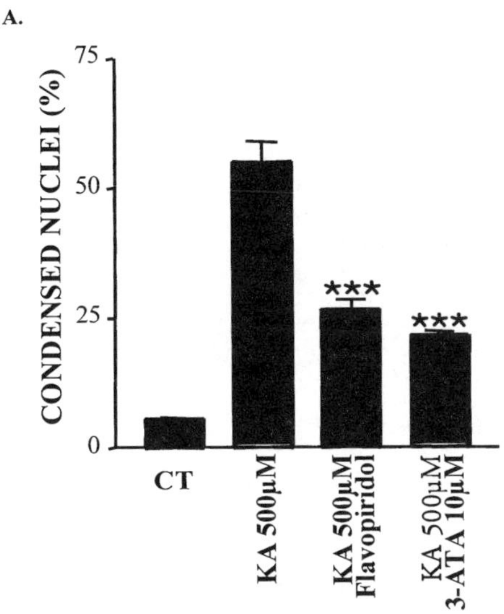

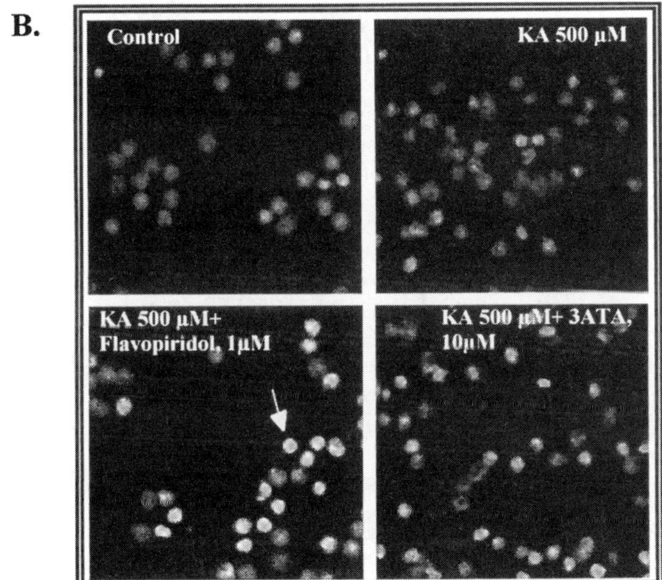

FIGURE 2. Quantification of apoptotic nuclei by staining with propidium iodide followed by fluorescence microscopy in CGCs. (**A**) In control culture the percentage is 5.57±0.2. This is increased in the presence of KA (500 μM) (55.16±3.7) and reduced by 3-ATA (10 μM) (21.81±0.5) and flavopiridol (1 μM) (26.79±1.7) (**B**) After treatment, with flavopiridol (1 μM) and 3-ATA (10 μM), the morphology of the nuclei was normal and the apoptotic features disappeared. The values represent the average of five to seven independent experiments, and errors bars represent the SEM.

Therefore, drugs that protect against excitotoxicity may be useful in the treatment of neurodegenerative disorders.

Apoptosis was evaluated in primary culture cerebellar granular cells (CGCs)[4] after adding kainic acid (KA) in the absence or presence of 3-aminothioacridone (3-ATA), a specific inhibitor of CDK4, and flavopiridol, a pan inhibitor of CDKs. When CGCs were preincubated with 3-ATA (10 μM) or flavopiridol (1 μM) plus KA (500 μM), these inhibitors significantly reduced the percentage of apoptotic cells measured by flow cytometry (FIG. 1). Furthermore, when the cells were stained with propidium iodide and observed under fluorescence, changes in the morphology of the nuclei were seen. Apoptotic nuclei showing high fluorescence and condensed chromatin were also observed in comparison to control cultures. When 3-ATA (10 μM) or flavopiridol (1 μM) were coincubated with KA (500 μM), the morphology of the nuclei was normal and the apoptotic features disappeared (FIG. 2).

It is now widely accepted that the activation of caspases causes neuronal apoptosis. In particular, caspase-3 and caspase-6 are the major executioners of apoptotic signals in neurons. In previous studies we demonstrated that KA only induces a 20% increase in caspase-3 activity. Addition of these CDK inhibitors to the medium did not block KA-induced activation in caspase-3. We concluded, therefore, that the anti-apoptotic effect of 3-ATA is not mediated through inhibition of the caspase pathway.

The data from the present study, showing that CDK activation was prevented by 3-ATA and flavopiridol, support the role of reentry into the cell cycle in KA-mediated apoptosis. We do not, however, rule out the possibility that other pathways might also be involved in KA-induced apoptosis in CGCs. Among these, free radical production has been reported to play a key role in the neurotoxic effects of KA, and several antioxidants, such as melatonin or lazaroids, have shown neuroprotective effects. Although 3-ATA and flavopiridol have been developed as drugs for cancer treatment,[5] the present study suggests that they may also be useful in the prevention of excitotoxic processes.

ACKNOWLEDGMENTS

This study was supported by SAF2002-00790 to A.M.C. E.V. is the recipient of a fellowship from the University of Barcelona. E.G.J. was a fellow of the Fundació Agustí Pedro Pons.

REFERENCES

1. McShea, A., A.F. Wahl & M.A. Smith. 1999. Re-entry into the cell cycle: a mechanism for neurodegeneration in Alzheimer's disease. Med. Hypotheses **52:** 525–527.
2. Yang, Y., D.S. Geldmacher & K. Herrup. 2001. DNA replication precedes neuronal cell death in Alzheimer's disease. J. Neurosci. **21:** 2661–2668.
3. Jordan-Sciutto, K., L.M. Malaiyandi & R. Bowser. 2002. Altered distribution of cell cycle transcriptional regulators during Alzheimer disease. J. Neuropathol. Exp. Neurol. **61:** 358–367.
4. Verdaguer, E., E. Garcia-Jordà, A. Jiménez, *et al.* 2002. Kainic acid-induced neuronal cell death in cerebellar granule cell is not prevented by caspase inhibitors. Br. J. Pharmacol. **135:** 1297–1307.
5. Kubo, A., K. Nakagawa, R.K. Varma, *et al.* 1999. The p16 status of tumor cell lines identifies small molecule inhibitors specific for cyclin-dependent kinase 4. Clin. Cancer Res. **5:** 4279–4286.

Peripheral Markers of Apoptosis in Parkinson's Disease

The Effect of Dopaminergic Drugs

F. BLANDINI,[a] A. MANGIAGALLI,[a] M. COSENTINO,[b] F. MARINO,[b] A. SAMUELE,[a] E. RASINI,[b] R. FANCELLU,[a,b] E. MARTIGNONI,[c] G. RIBOLDAZZI,[b] D. CALANDRELLA,[b] G. M. FRIGO,[d] AND G. NAPPI[a,e]

[a]*IRCCS "C. Mondino," Pavia, Italy*

[b]*University of Insubria, Varese, Italy*

[c]*University of Piemonte Orientale "A. Avogadro," Novara, Italy*

[d]*University of Pavia, Pavia, Italy*

[e]*University "La Sapienza," Rome, Italy*

ABSTRACT: In this study, we measured the lymphocyte levels of proteins involved in apoptosis regulation, such as Bcl-2, the peripheral benzodiazepine receptor (PBR), caspase-3, and Cu/Zn superoxide dismutase (Cu/Zn SOD), in patients with Parkinson's disease (PD), either untreated or under therapy with dopaminergic agents (L-Dopa alone or L-dopa + dopamine agonists) and in healthy volunteers. All PD groups showed increased activity of caspase-3, compared to controls, particularly those under treatment only with L-Dopa. In this latter group, the increase in caspase-3 activity was also paralleled by an increase in the concentration of Cu/Zn SOD. In addition, patients taking L-Dopa + dopamine agonists showed marked decrease in Bcl-2 levels and increased PBR expression, which seems in keeping with the hypothesis that PBR may be functionally related to Bcl-2. In conclusion, we found clear modifications in the levels of proteins involved in the control of apoptosis in lymphocytes of PD patients. These changes were disease related but also modulated by the pharmacological treatment, which confirms the potential role of apoptosis in PD pathogenesis and the modulatory influence of dopaminergic agents.

KEYWORDS: cell death; L-dopa; lymphocytes; caspase-3; Bcl-2; Cu/Zn SOD; PBR

INTRODUCTION

Defective regulation in the control of apoptosis (programmed cell death) may play a primary role in the pathogenic mechanisms of Parkinson's disease (PD). Apoptotic cell death and increased levels of pro-apoptotic molecules are present in the substantia nigra of PD patients.[1,2] In addition, the toxins used to replicate PD in animals induce apoptosis.[3] Another important issue is the relationship between dopaminergic stimulation and apoptosis. L-Dopa, still the most used drug for PD therapy, may pro-

Address for corrrespondence: Fabio Blandini, IRCCS "C. Mondino," Pavia, Italy. Voice: +39-0382-380333; fax: +39-0382-380286.
fabio.blandini@mondino.it

Ann. N.Y. Acad. Sci. 1010: 675–678 (2003). © 2003 New York Academy of Sciences.
doi: 10.1196/annals.1299.123

TABLE 1. Clinical and demographic characteristics of subjects participating in the study

	PD			Controls
N	10	21	38	29
Therapy	—	L-dopa	L-dopa+DA agonists	—
Gender (M/F)	8/5	14/7	18/20	18/11
Age (yrs)	59.8 ± 11.5	67.9 ± 8	63 ± 5.5	60 ± 6.1
Disease duration (yrs)	2.6 ± 1.8	6.1 ± 4.2	8.8 ± 5	—
UPDRS score (on)	13.8 ± 6.2	23.1 ± 9.8	23.6 ± 8.6	—
H & Y stage (on)	1.4 ± 0.5	1.8 ± 0.7	2.5 ± 0.6	—

NOTE: UPDRS: Unified Parkinson's Disease Rating Scale; H & Y: Hoehn & Yahr; M± SD.

mote oxidative stress and trigger apoptosis under experimental conditions,[4] while dopamine agonists may exert the opposite effect.[5] Because of the inherent difficulty of studying PD pathogenic mechanisms in the central nervous system of patients, peripheral blood cells—particularly lymphocytes—have been used as an accessible source of information on biomolecular aspects relevant to PD pathogenesis.[6] The aim of our study was to determine the lymphocyte levels of proteins involved in apoptosis regulation, such as major anti-apoptotic protein Bcl-2, the peripheral benzodiazepine receptor (PBR), which participates in a key step of the process, such as the opening of the mitochondrial permeability transition pore, and caspase-3, which is a major effector of the apoptotic cascade. We also determined the lymphocyte levels of enzyme Cu/Zn superoxide dismutase (Cu/Zn SOD), which intervenes in the regulation of the apoptotic process by scavenging the superoxide anion.

SUBJECTS AND METHODS

Three groups of PD patients, untreated, treated with L-dopa, or with L-dopa + dopamine (DA) agonists, were enrolled; healthy volunteers, matched for age and sex, served as controls (TABLE 1).

Sample Preparation

All subjects underwent a venous blood sampling (25 mL). Lymphocytes were isolated on Lympholite®-H, through serial centrifugations; PBS was then added, and we counted lymphocytes on a blood cell counter (Coulter Corp. USA). Samples were then split in two aliquots. The first aliquot—for the Bcl-2/caspase/SOD study—was pelleted by centrifugation and stored at −80°C. The second aliquot—for the PBR study—was pelleted, resuspended in saline solution containing 10% dymethylsulfoxide, as cryoprotectant, and stored at −80°C.

Caspase-3, Bcl-2, and Cu/Zn SOD Assays

Lymphocyte pellets were resuspended in ice-cold PBS and homogenized by ultrasound. Homogenates were ultracentrifuged, and resulting supernatants were used for the assays. Bcl-2 and Cu/Zn *SOD* concentrations were measured using ELISA assays (Bender MedSystem, Austria), while a fluorometric assay was used for the determination of caspase-3 protease activity (Molecular Probes, USA).

PBR Expression

PBR expression in lymphocytes was evaluated by flow cytometry (Coulter Corp., USA).

RESULTS

As shown in the FIGURE, PD patients treated with L-dopa+DA agonists showed the lowest concentrations of Bcl-2 and the highest expression of PBR receptors, with respect to the other groups. Both untreated PD patients and patients taking L-dopa alone showed Bcl-2 and PBR similar to those found in the controls. All PD groups

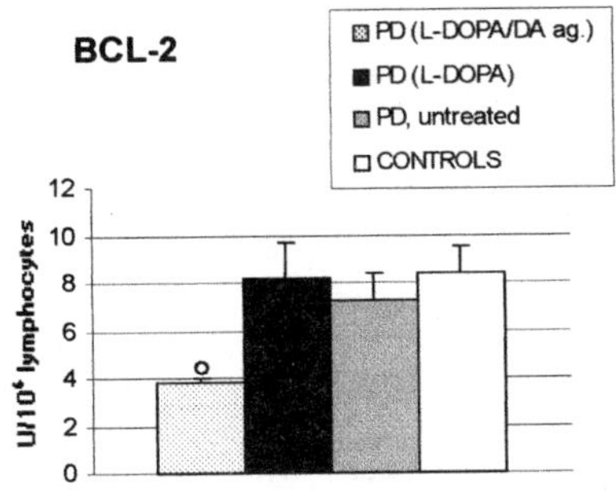

ANOVA: F=6.45; p<0.001
° p<0.001 vs. untr. PD, PD (L-Dopa), Controls

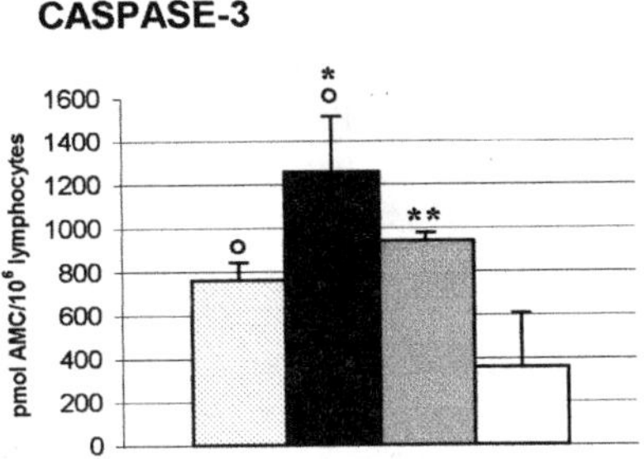

ANOVA: F=7.07; p<0.001
**p<0.01, ° p<0.001 vs. Controls; *p<0.05 vs. PD (L-Dopa/DA ag.)

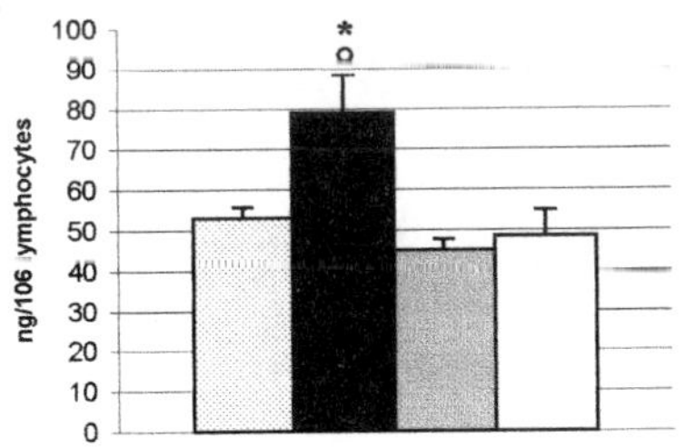

ANOVA: F=6.53; p<0.001
°p<0.005 vs. PD (L-Dopa/DA ag.) and Controls; *p<0.05 vs. untr. PD

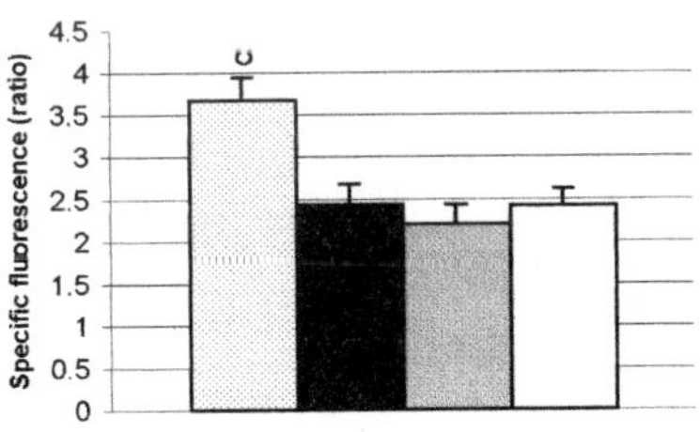

ANOVA: F=7.85; p<0.001
°p<0.005 vs. untr PD, PD (L-Dopa), Controls

FIGURE 1. Mean (± SEM) lymphocyte levels of Bcl-2, caspase-3, Cu/Zn SOD, and PBR ratio in the four subject groups (PD patients treated with L-dopa+DA agonist, patients treated with L-dopa alone, untreated PD patients, and controls).

showed increased activity of caspase-3, compared to controls, the highest values being found in patients taking L-dopa alone. In patients treated with L-dopa+DA agonists, caspase-3 levels were higher than controls, but significantly lower than the values found in patients treated with L-dopa alone. Patients taking L-dopa showed a significant increase in lymphocyte concentrations of Cu/Zn SOD, compared to all the other groups, while both untreated patients and patients treated with L-dopa+DA agonist had Cu/Zn SOD levels similar to those observed in the control group.

DISCUSSION

The main finding of our study is the increased activity of caspase-3 in lymphocytes of all PD groups. Pharmacological treatment seemed to modulate these changes. Indeed, untreated PD patients showed increased caspase-3 activity, which was further elevated in patients taking L-dopa, but tended to decrease in patients treated with L-dopa+DA agonists. Increased caspase-3 activity of patients taking L-dopa was paralleled by an increase in the concentration of Cu/Zn SOD, which counteracts the pro-apoptotic effect of the superoxide anion. Therefore, the increase in Cu/Zn SOD levels may have a compensatory meaning, in the setting of an enhanced pro-apoptotic status. Patients taking L-dopa+DA agonists showed marked decrease in Bcl-2 levels and increased PBR expression. In general, this seems in keeping with the hypothesis that PBR is functionally related to Bcl-2, the decrease in Bcl-2 levels being associated with upregulation of PBR. In the context of PD pathogenesis, however, the meaning of this finding is unclear. Bcl-2 is a major anti-apoptotic protein, and previous experimental studies have shown that dopamine agonists tend to enhance its expression. Considering the caspase-3 data, it may be hypothesized that adding a DA agonist in the therapeutic schedule of PD patients may "downsize" the expression of the apoptotic phenomenon, acting on both the anti- and pro-apoptotic sides.

In conclusion, we found clear modifications in the levels of proteins involved in the control of apoptosis in lymphocytes of PD patients, which confirms the role played by apoptosis in the pathogenesis of the disease. Changes were disease related and tended to be modulated by treatment with dopaminergic agents.

REFERENCES

1. Turmel, H., A. Hartmann, K. Parain, *et al.* 2001. Caspase-3 activation in MPTP-treated mice. Mov. Disord. **16:** 185–189.
2. Ziv, I., R. Zilkha-Falb, D. Offen, *et al.* 1997. Levodopa induces apoptosis in cultured neuronal cells—a possible accelerator of nigrostriatal degeneration in Parkinson's disease? Mov. Disord. **12:** 17–23.
3. Corona-Morales, A.A., A. Castell & L. Zhang. 2000. L-Dopa-induced neurotoxic and apoptotic changes on cultured chromaffin cells. Neuroreport **11:** 503–506.
4. Uberti, D., L. Piccioni, A. Colzi, *et al.* 2002. Pergolide protects SH-SH5Y cells against neurodegeneration induced by H_2O_2. Eur. J. Pharmacol. **434:** 17–20.
5. Takata, K., Y. Kitamura, J. Kakimura, *et al.* 2000. Increase of bcl-2 protein in neuronal dendritis processes of cerebral cortex and hippocampus by the antiparkinsonian drugs, Talipexole and Pramipexole. Brain Res. **872:** 236–241.
6. Migliore, L., L. Petrozzi, C. Lucetti, *et al.* 2002. Oxidative damage and cytogenetic analysis in leukocytes of Parkinson's disease patients. Neurology **58:** 1809–1815.

Dopaminergic Modulation of Apoptosis in Human Peripheral Blood Mononuclear Cells

Possible Relevance for Parkinson's Disease

CRISTINA COLOMBO,[a] MARCO COSENTINO,[a] FRANCA MARINO,[a] EMANUELA RASINI,[a] MARIA OSSOLA,[a] FABIO BLANDINI,[a,b] ANNA MANGIAGALLI,[b] ALBERTA SAMUELE,[b] MARCO FERRARI,[a] RAFFAELLA BOMBELLI,[a] SERGIO LECCHINI,[a] GIUSEPPE NAPPI,[b,c] AND GIANMARIO FRIGO[a]

[a]*Laboratory of Pharmacology, Faculty of Medicine, University of Insubria, Varese, Italy*

[b]*Laboratory of Functional Neurochemistry, Neurological Institute 'C. Mondino,' Pavia, Italy*

[c]*University 'La Sapienza,' Rome, Italy*

ABSTRACT: Dopamine (DA) modulates apoptosis in neuronal and non-neuronal cells, and dopaminergic pathways contribute to neurodegenerative disease. Human lymphocytes express dopaminergic receptors and DA transporters, and synthesize endogenous catecholamines, which may modulate apoptosis in these cells. In the present study, dopaminergic modulation of apoptosis was investigated in human peripheral blood mononuclear cells (PBMCs) obtained from healthy donors. Twenty-four–hour DA reduced at 0.1–5 $\times 10^{-6}$ M and enhanced at 1–5 $\times 10^{-4}$ M spontaneous apoptosis. DA 1 $\times 10^{-6}$ M was inhibited by the D1-like receptor antagonist SCH 23390 1 $\times 10^{-6}$ M, but not by the D2-like receptor antagonists domperidone 1 $\times 10^{-6}$ M or haloperidol 1 $\times 10^{-6}$ M, while the effect of DA 5 $\times 10^{-4}$ M was prevented by the antioxidants glutathione 5–10 mM or *N*-acetyl-L-cysteine 1–10 mM. Intracellular reactive oxygen species were respectively reduced and increased by 1–3 h incubation with DA 0.1–10 $\times 10^{-6}$ M and 1–5 $\times 10^{-4}$ M. Twenty-four–hour DA 1 $\times 10^{-6}$ M or 5 $\times 10^{-4}$ M had no effect on PBMC expression of Cu/Zn superoxide dismutase or Bcl-2; however, DA 5 $\times 10^{-4}$ M decreased caspase-3 activity. In human PBMCs, DA seems to promote apoptosis through oxidative mechanisms but may also result in cell rescue from apoptotic death possibly through activation of D1-like receptors. The dual effect of DA on human PBMCs closely resembles that on striatal neurons. Lymphocytes of patients with Parkinson's disease may show reduced DA content and impaired DA transporter immunoreactivity. Human PBMCs may thus represent a simple and readily accessible model to study DA-related mechanisms relevant for neurodegenerative disease.

KEYWORDS: dopamine; neurodegenerative disease; Parkinson's disease

Address for correspondence: Marco Cosentino, M.D., Ph.D., Laboratory of Pharmacology, Faculty of Medicine, University of Insubria, Via Ottorino Rossi n. 9, 21100 Varese VA, Italy. Voice and fax: +39-0332-811601.

marco.cosentino@uninsubria.it; lab.pharm@uninsubria.it

Ann. N.Y. Acad. Sci. 1010: 679–682 (2003). © 2003 New York Academy of Sciences. doi: 10.1196/annals.1299.124

INTRODUCTION

Dopamine (DA) is involved in the modulation of apoptosis in both neuronal and non-neuronal cells, and there is evidence that dopaminergic mechanisms may contribute to neurodegeneration in Parkinson's disease (PD). In striatal neurons high-concentration DA is proapoptotic; however, low-concentration DA can prevent cell death, possibly due to the ability of DA to affect intracellular oxidative processes.[1] Better understanding of the mechanisms involved in the dual role of DA are likely to improve the pharmacological approaches to neurodegeneration.

Human peripheral blood mononuclear cells (PBMCs) express an intrinsic dopaminergic system that includes DA receptors,[2] DA membrane and monoamine vesicular transporters, as well as catecholaminergic, tyrosine hydroxylase (TH)–dependent biosynthetic pathways.[3] Interestingly, this system may be altered in PBMCs from PD patients.[4]

In order to assess whether PBMCs may be taken as a model to study the dopaminergic modulation of cell death processes, we have examined the ability of DA and of dopaminergic agents to affect apoptosis and intracellular levels of reactive oxygen species (ROS) in these cells. A preliminary investigation of the effects of DA on intracellular levels of Bcl-2, Cu/Zn-dependent superoxide dismutase (Cu/Zn SOD), and on caspase-3 activity was also performed.

MATERIALS AND METHODS

Experiments were performed on human PBMCs obtained from venous blood of healthy donors, separated with density-gradient centrifugation and subsequently cultured under standard conditions. Apoptotic cells were identified by double-fluorescence labeling with Annexin V FITC kit (Bender MedSystems, Vienna, Austria) using a Coulter Epics Elite ESP flow cytometer (Beckman Coulter Corp., Miami, FL, USA). Intracellular ROS levels were assessed by use of the redox-sensitive dye C-DCDHF-DA (Molecular Probes, Eugene, OR, USA) using a Perkin-Elmer LS-50B spectrofluorimeter (excitation wavelenght 488 nm). Cu/Zn SOD and Bcl-2 concentrations were determined using an enzyme-linked immunosorbent assay (ELISA) kit (Bender MedSystems Diagnostics GmbH, Austria) and a colorimetric microplate reader (Bio-Rad, Hercules, CA, USA). Caspase-3 protease activity was determined by using a fluorometric assay (Molecular Probes, Inc., USA). Readings were performed using a fluorometric microplate reader (Spectramax Gemini XS, Molecular Devices, USA).

RESULTS

Spontaneous apoptosis was reduced by incubation with DA 1 μM and was increased with DA 500 μM. The effect of DA 1 μM was suppressed by the D1-like antagonist SCH 23390, but not by the D2-like antagonists haloperidol and domperidone, while the effect of DA 500 μM was suppressed by the antioxidants glutathione (GSH) and *N*-acetyl L-cysteine (NAC), but not by SCH 23390, haloperidol, or domperidone (FIG. 1). DA had a similar dual effect also on intracellular ROS content

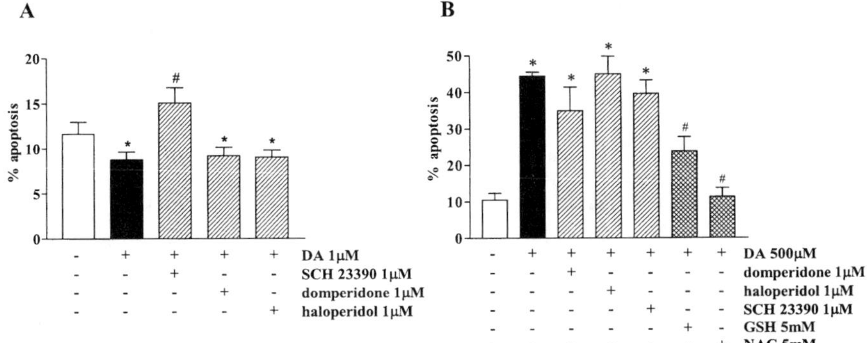

FIGURE 1. Effect of DA 1 μM (*panel A*) and 500 μM (*panel B*) on apoptosis levels in human PBMCs. Each column is the mean ± SEM of at least 6 different observations. * $P<.05$ vs. control; # $P<.05$ vs. DA alone.

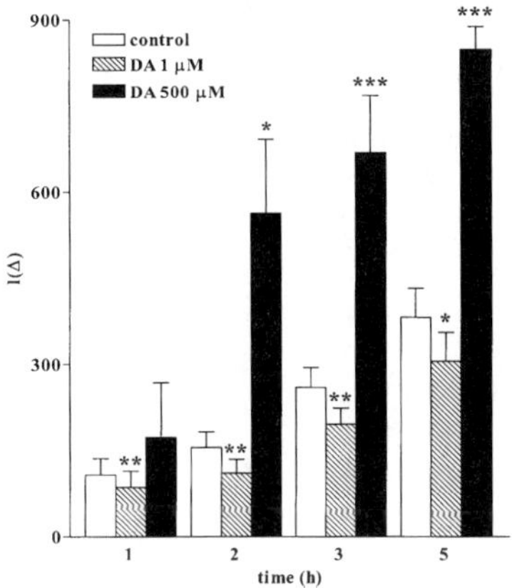

FIGURE 2. Effect of DA on intracellular ROS content in human PBMCs. Each column is the mean ± SEM of at least 6 different observations. * $P<.05$, ** $P<.01$, *** $P<.005$ vs. control.

(FIG. 2). The expression of Cu/Zn SOD and Bcl-2 was not modified by 24-h incubation with DA (1–500 μM) (data not shown). However, caspase-3 activity, which was 571.3 ± 444.6 pmol AMC/10^6 cells in control samples and 714.7 ± 403.6 pmol AMC/10^6 cells with DA 1 μM, was reduced by DA 500 μM to 44.5 ± 97.4 pmol AMC/10^6 cells ($P<.01$ vs. control).

CONCLUSIONS

In human PBMCs DA is able to modulate, in a complex way, oxidative metabolism and apoptotic processes through both receptor-dependent and -independent mechanisms. In particular, pharmacological evidence suggests that DA at low concentrations may have a (D1-like?) receptor-mediated antiapoptotic effect, which, to our best knowledge, has never been reported previously. On the other side, DA-induced apoptosis seems to involve receptor-independent, oxidative mechanisms. DA-induced reduction of caspase-3 activity was quite unexpected, since this caspase is thought to represent the final effector during the apoptotic cascade. However, recent evidence has been provided that caspase-3 may be required for neuroprotection following sublethal stimuli.[5] Unraveling the functional significance of DA-induced reduction of caspase-3 activity in PBMCs may thus provide further insight into the paradoxical protective role of mediators classically associated with cell death. The existence of a dopaminergic system in PBMCs and the ability of DA and of dopaminergic agents to affect PBMC function suggest the possibile use of these cells as a peripheral model for *in vitro/ex vivo* studies of the pathogenetic mechanisms and as pharmacological responses in neurodegenerative diseases such as PD.

REFERENCES

1. Blum, D., S. Torch., N. Lambeng, *et al.* 2001. Molecular pathways involved in the neurotoxicity of 6-OHDA, dopamine and MPTP: contribution to the apoptotic theory in Parkinson's disease. Prog. Neurobiol. **65:** 135–172.
2. Mckenna, P., P.J. McLaughlin, B.J. Lewis, *et al.* 2002. Dopamine receptor expression on human T- and B-lymphocytes, monocytes, neutrophils, eosinophiles and NK cells: a flow cytometric study. J. Neuroimmunol. **132:** 34–40.
3. Cosentino, M., F. Marino, R. Bombelli, *et al.* 2002. Stimulation with phytohaemagglutinin induces the synthesis of catecholamines in human peripheral blood mononuclear cells: role of protein kinase C and contribution of intracellular calcium. J. Neuroimmunol. **125:** 125–133.
4. Caronti, B., G. Antonini, C. Calderaro, *et al.* 2001. Dopamine transporter immunoreactivity in peripheral blood lymphocytes in Parkinson's disease. J. Neural Transm. **108:** 803–807.
5. McLaughlin, B., K. Hartnett, J. Erhardt, *et al.* 2003. Caspase 3 activation is essential for neuroprotection in preconditioning. Proc. Natl. Acad. Sci. USA **100:** 715–720.

Proteolytic Activation of Proapoptotic Kinase PKCδ Is Regulated by Overexpression of Bcl-2

Implications for Oxidative Stress and Environmental Factors in Parkinson's Disease

A. G. KANTHASAMY,[a] M. KITAZAWA,[a] S. KAUL,[a] Y. YANG,[a] D. K. LAHIRI,[b] V. ANANTHARAM,[a] AND A. KANTHASAMY[a]

[a]*Department of Biomedical Sciences, Iowa State University, Ames, Iowa 50011, USA*

[b]*Department of Psychiatry , Indiana University School of Medicine, Indianapolis, Indiana 46202, USA*

ABSTRACT: We previously demonstrated that the organochlorine pesticide dieldrin, a potential chemical risk factor for development of Parkinson's disease (PD), impairs mitochondrial function and promotes apoptosis in dopaminergic PC12 cells. We further demonstrated that caspase-3–dependent proteolytic activation of a member of the novel PKC family, protein kinase Cδ (PKCδ), contributes to apoptotic cell death in dopaminergic cells. In the present study, we report that the proapoptotic function of PKCδ can be regulated by overexpression of the mitochondrial anti-apoptotic protein Bcl2 in dieldrin-treated dopaminergic cells. Exposure to dieldrin (30 or 100 μM) for 3 h produced a dose-dependent increase in caspase-3 activation and DNA fragmentation in vector-transfected PC12 cells. Overexpression of human Bcl-2 in PC12 cells completely suppressed dieldrin-induced caspase-3 activation and DNA fragmentation. Furthermore, dieldrin-induced proteolytic activation of PKCδ was also remarkably reduced in Bcl-2–overexpressed cells. Together, these results suggest that the proapoptotic function of PKCδ can be regulated by mitochondrial redox modulators during neurodegenerative processes.

KEYWORDS: apoptosis; environmental factors; oxidative stress; protein kinase C; dieldrin; pesticide; dopaminergic degeneration

INTRODUCTION

Oxidative stress–mediated apoptotic cell death is recognized as one of the major cellular events in several neurodegenerative disorders, including Parkinson's disease (PD) and Alzheimer's disease (AD). Numbers of studies have implicated pesticide exposure with a significantly increased risk of PD.[2,3] Previously, we characterized that the organochlorine pesticide dieldrin, a potential chemical risk factor of PD,

Address correspondence: Dr. A.G. Kanthasamy, Parkinson Disorders Research Laboratory, Department of Biomedical Sciences, 2062 Veterinary Medicine Building, Iowa State University, Ames, IA 50011. Voice: 515-294-2516; fax: 515-294-2315.
akanthas@iastate.edu

Ann. N.Y. Acad. Sci. 1010: 683–686 (2003).
doi: 10.1196/annals.1299.125

promotes oxidative stress–dependent apoptosis in dopaminergic neuronal cells.[1] Also, we uncovered that activation of protein kinase Cδ (PKCδ) by caspase-3 plays a critical role both upstream and downstream in the dieldrin-induced apoptotic process. Herein, we have demonstrated that dieldrin-induced apoptotic cell death and proteolytic activation of PKCδ were significantly attenuated by overexpression of the mitochondrial anti-apoptotic protein Bcl-2, indicating that increased expression of the mitochondrial anti-apoptotic factor can negatively modulate the proapoptotic function of PKCδ.

MATERIALS AND METHODS

PC12V4 (Vector) and PC12HB2-3 (human Bcl2 overexpression) cells were grown and exposed to dieldrin, as previously described.[1] Caspase activity and DNA

(A)

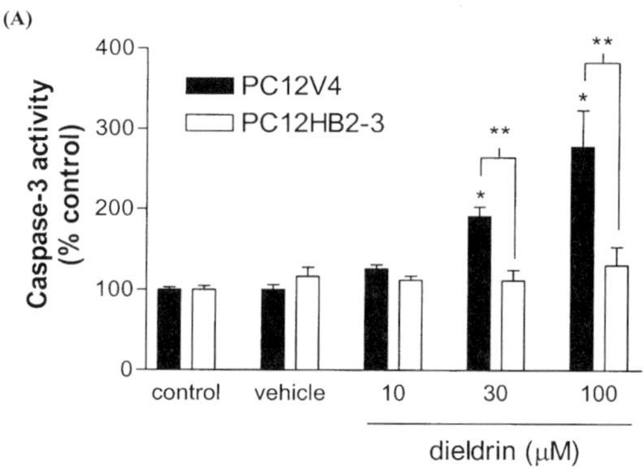

(B)

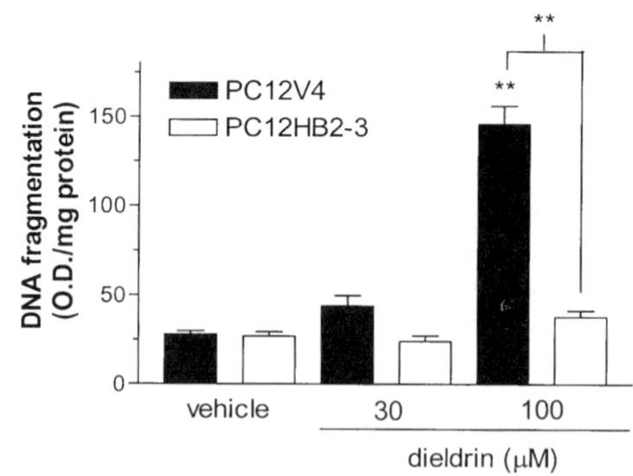

FIGURE 1. Dieldrin-induced caspase-3 activation and DNA fragmentation Vector PC12 cells (PC12V4) and Bcl-2–overexpressing PC12 cells (PC12HB2-3) were exposed to dieldrin (0–100 μM) for 3 h. (**A**) Caspase-3 activity and (**B**) DNA fragmentation were measured using Ac-DEVD-AMC and ELISA, respectively. Each bar represents the mean ± SEM for two separate experiments in triplicate. ($^{*}P < .05$ or $^{**}P < .01$).

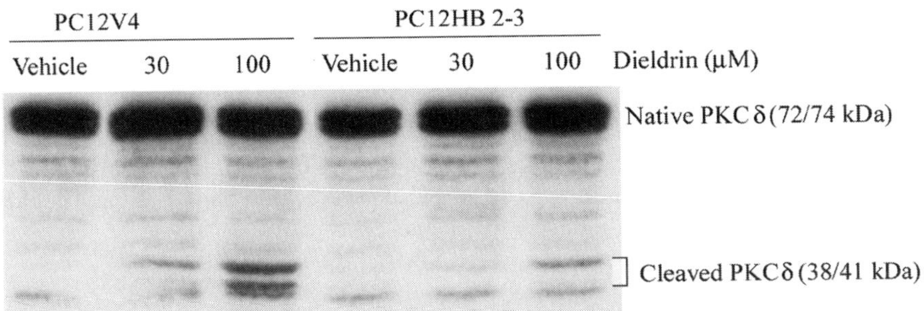

FIGURE 2. Proteolytic cleavage of PKCδ during dieldrin treatment. Vector PC12 cells (PC12V4) and Bcl-2–overexpressing PC12 cells (PC12HB2-3) were exposed to 30 μM or 100 μM dieldrin for 3 h at 37°C. Proteolytic cleavage of PKCδ was determined by 10% SDS-polyacrylamide gel electrophoresis and then immunoblotting with a PKCδ antibody.

fragmentation were measured, as described in our recent publication.[4] PKCδ cleavage was detected using 10% SDS-PAGE with rabbit polyclonal PKCδ primary antibody (Santa Cruz Biotechnology) and anti-rabbit secondary antibody. Protein-bound antibody was detected with an enhanced chemiluminescence detection kit (Amersham).

RESULTS

Vector-transfected PC12 cells (PC12V4) showed a dose-dependent increase in caspase-3 activity (FIG. 1A) and DNA fragmentation (FIG. 1B). Bcl-2–overexpressed PC12 cells (PC12HB2-3) significantly (P<.01) attenuated dieldrin-induced caspase-3 activation and DNA fragmentation, indicating that dieldrin activates cell death–signaling processes via mitochondrial dysfunction.

Activation of caspase-3 results in proteolytic cleavage and activation of PKCδ. We examined formation of the catalytically active fragment of PKCδ following dieldrin treatment in both PC12V4 and PC12HB2-3 cells (FIG. 2). The catalytically active PKCδ fragment appeared in the 30–100 μM dieldrin-treated PC12V4 cells in a dose-dependent manner. On the other hand, PC12HB2-3 cells almost completely suppressed formation of the catalytically active PKCδ fragment during treatment.

DISCUSSION

The dieldrin-induced apoptotic cell death pathway may share a common initiation process that is regulated by mitochondria, because Bcl-2–overexpressed PC12 cells significantly attenuated dieldrin-induced apoptosis. Recently, PKCδ, a member of the novel PKC family, was believed to play a critical role in the caspase-mediated apoptotic cell death. PKCδ (72–74 kDa) is proteolytically cleaved into 38 kDa and 41 kDa subunits by caspase-3.[4,5] We and other groups have demonstrated that proteolytic activation of PKCδ is necessary for not only the execution of apoptosis, but also for upstream amplification of the caspase cascade during chemical-induced apoptosis in different cell models.[4,5] Bcl-2 is a member of the anti-apoptotic Bcl-2 protein family that may exert its anti-apoptotic function by preventing the activation of various proapoptotic factors, including Bad, Bax, or Bid.

Our present study demonstrates for the first time that overexpression of Bcl-2 is sufficient to attenuate the proapoptotic function of PKCδ during chemical-induced apoptotic cell death in dopaminergic cells. Furthermore, mitochondria may be the central organelle for dieldrin-induced apoptosis.

ACKNOWLEDGMENT

This study is supported by National Institutes of Health (NIH) Grant ES10586 and NS 451 33.

REFERENCES

1. Kitazawa, M., V. Anantharam & A.G. Kanthasamy. 2001. Dieldrin-induced oxidative stress and neurochemical changes contribute to apoptopic cell death in dopaminergic cells. Free Radic. Biol. Med. **31:** 1473–1485.
2. Tanner, C.M. 1989. The role of environmental toxins in the etiology of Parkinson's disease. Trends Neurosci. **12:** 49–54.
3. Gorell, J.M., C.C. Johnson, B.A. Rybicki, *et al.* 1998. The risk of Parkinson's disease with exposure to pesticides, farming, well water, and rural living. Neurology **50:** 1346–1350.
4. Anantharam, V., M. Kitazawa, J. Wagner, *et al.* 2002. Caspase-3-dependent proteolytic cleavage of protein kinase Cdelta is essential for oxidative stress-mediated dopaminergic cell death after exposure to methylcyclopentadienyl manganese tricarbonyl. J. Neurosci. **22:** 1738–1751.
5. Reyland, M.E., S.M. Anderson, A.A. Matassa, *et al.* 1999. Protein kinase C delta is essential for etoposide-induced apoptosis in salivary gland acinar cells. J. Biol. Chem. **274:** 19115–19123.

Molecular Mechanisms for Cardiovascular Stem Cell Apoptosis and Growth in the Hearts with Atherosclerotic Coronary Disease and Ischemic Heart Failure

YONG-JIAN GENG

Center for Cardiovascular Biology and Atherosclerosis, Department of Internal Medicine, The University of Texas, Health Science Center at Houston, Medical School, and Heart Failure Research Laboratory, Texas Heart Institute, Houston, Texas, USA

ABSTRACT: In the heart with atherosclerotic coronary disease, chronic ischemia causes progressive loss of cardiovascular cells and ultimately triggers myocardial dysfunctions or heart failure. Various types of stem cells from embryonic and adult tissues have potentials for regenerating functional cardiovascular cells in the heart undergoing ischemic injury. However, native or exogenous stem cells in the ischemic hearts are exposed to various proapoptotic or cytotoxic factors. Furthermore, during repopulation and differentiation, certain numbers of newly produced cells may die by apoptosis during neocardiovascular tissue remodeling and morphogenesis. Embryonic and adult stem cells may have different life spans, as being programmed genetically to apoptosis. The endogenous and environmental factors play important roles in regulation of stem cells, including inflammatory cytokines, growth factors, surface receptors, proteolytic enzymes, mitochondrial respiration, nuclear proteins, telomerase activities, hypoxia-responding proteins, and stem cell–host cell interaction. Clarification of the molecular mechanisms may help us understand and design stem cell therapies.

KEYWORDS: apoptosis; atherosclerosis; heart failure; myocytes; stem cells

INTRODUCTION

Disregulated cardiovascular cell death by apoptosis occurs in atherosclerosis[1–4] and heart failure[5,6] as well as many other types of cardiovascular disease.[7] In the blood vessel system, vascular smooth muscle cells and endothelial cells have been found to undergo apoptosis when exposed to inflammatory cytokines produced by activated immune cells, for example, T lymphocytes and macrophages, in atherosclerotic plaques[8] and in culture.[9–11] The Fas/caspase death pathway may operate in

Address for correspondence: Yong-Jian Geng, M.D., Ph.D., Director, Center for Cardiovascular Biology and Atherosclerosis, Department of Internal Medicine, The University of Texas, Health Science Center at Houston, Medical School, Houston, TX 77030. Voice: 713-500-6607; fax: 713-500-0658.

yong-jian.geng@uth.tmc.edu

Ann. N.Y. Acad. Sci. 1010: 687–697 (2003). © 2003 New York Academy of Sciences.
doi: 10.1196/annals.1299.126

the vascular cells and trigger the apoptotic cell death.[10,11] Cardiac myocytes may also die by apoptosis when they are exposed to altered microenvironment intracellularly, extracellularly, or both. For example, enhanced activation of beta-adrenergic signaling by overexpression of the stimulatory G-protein alpha-subunit in the hearts induces apoptosis of cardiac myocytes as well as cardiomyopathy or heart failure.[12] Hence, progressive loss of functional cardiovascular cells represents a key cellular basis in the pathogenesis of atherosclerosis and chronic heart failure.

The adult, mature cardiac myocytes are terminally differentiated and usually do not proliferate. Hence, when myocardial infarction occurs, no new myocytes will be produced to replace dead muscle cells. To compensate the loss of cardiac myocytes, remaining myocytres become hypertrophied, and nonmuscle cells, particularly fibroblasts, proliferate to form scar tissue. Stem cell therapy has shown great promise in regenerating and repopulating the damaged myocardium, restoring its function.[13–16] Several types of stem cells from fetal and adult tissues are capable of generating mature cardiovascular cells, including embryonic stem cells, fetal cardiac myoblasts, adult skeletal myoblasts, and adult bone marrow stem cells. Rapidly accumulating evidence indicates that stem cells from embryonic and adult tissues may differentiate into cardiac myocytes in the heart with infarction.[17,18] Transplantation of fetal cardiac myoblasts, adult skeletal myoblasts, and bone marrow stem cells into the heart with experimental infarction reportedly leads to development of neomyocardium and neovascular tissues as well as improvement of heart function. These observations raise new hope for patients with ischemic heart failure who desperately need heart transplantation. A major challenge to stem cell therapy is that transplanted stem cells undergo apoptosis at an accelerated rate, as they are exposed to an extremely harsh, proapoptotic microenvironment in the infarcted heart. In addition to poor blood supply and low oxygen tension, inflammation triggered by tissue damage may generate oxidative stress, cytotoxic radicals, and proteins, all of which may cause death of the transplanted cells in many different ways, especially by an apoptotic mechanism. Thus, protection of stem cells against apoptosis is critical for successful cellular therapy. This review is aimed at exploiting the molecular mechanisms by which stem cells survive apoptosis and develop into mature and functional cardiovascular cells.

CASPASES AND MITOCHONDRIA

Apoptosis, a form of genetically programmed cell death, represents a major mechanism by which tissue removes unwanted, aged, or damaged cells under both physiological and pathological conditions. Caspases are a group of aspartate-specific cysteinyl proteases that function as effectors in apoptosis. Thus far, more than 14 members of the caspase family have been identified in noncardiovascular tissues, particularly in the immune and nervous systems. All members of the caspase family show a similar substrate cleavage at an aspartate residue, and they are synthesized as proenzymes. This protease family participates in downstream events of death signaling by the members of the TNF receptor family (e.g., CD95 or Fas and TNF receptor-1), which share a related intracellular "death domain" of about 70 amino acid residues. Activation of caspase-8 by CD95 or Fas death signaling may have two effects. First, it may directly trigger caspase-3 activation, leading to mitochondrial-independent apoptosis; second, activated caspase-8 may promote cytochome-C (Cyt-C) release from mitochondria by cleavage of Bid, a Bcl-2 family member that is a substrate for caspase-8. After release from mitochondria, Cyt-C forms a complex with Apaf-1 and caspase-9, leading to caspase-9 activation. The Bcl-2 protein family represents another group of cellular proteins that regulate apoptosis in mammalian cells, which has more than 15 members. On the basis of their differences in regulation of apoptosis, members of this family fall into two subgroups. Each of them contains five or more function- and structure-related proteins. The first group includes the anti-apoptotic proteins, such as Bcl-2, Bcl-X, Mcl-1, Bcl-w, and A1; members of the second group promote apoptosis and include Bax, Bak, Bad, Bik, Hrk, Bid, and Bcl-xs. However, whether caspases play a role in regulation of stem cell apoptosis remains unclear. Using a quantitative real time PCR approach to assess long-term cell survival and development, Muller-Ehmsen *et al.* found that transplanted male donor stem cells from neonatal Fischer 344 rats (1–2 days) were undergoing apoptosis in a time-dependent fashion.[16,19] They administered the caspase inhibitor AcYVADcmk but failed to improve transplanted cell survival at 24 hours. Histology revealed that transplanted cells became more elongated over time, developed cross-striations, and that their nuclei increased in size. However, at 12 weeks, transplanted cells and their nuclei were still smaller than those of the host myocardium. There was significant loss of cells within 24 h, and 12 weeks after transplantation, 15% of stem cells or stem cell–derived cells remained in the hearts, which carried the male-specific gene Sry. Those cells that did survive underwent differentiation and developed visible sarcomeres, suggesting a potential contribution toward ventricular function.

IMMUNE TOLERANCE OR PRIVILEGE IN STEM CELLS

Because of a limited availability of autologous stem cells, many investigators have been using allogeneic stem cells for studies of cardiac cellular therapy. It has been reported that both allogeneic ESC and MSC can survive for a fairly long period of time (if not for ever), suggesting that the ESC and MSC have, to some degree, escaped immune surveillance. Recent studies have demonstrated that certain tissues or organs can develop an immune privilege by triggering apoptosis of certain T lymphocyte subsets through activation of the Fas/FasL death signaling pathway.[20,21] Such tissues and organs with immune privilege include the eyes, brain, testes, and some tumors, which constitutively express high levels of FasL, resulting in apoptosis of invading T cells. Successful gene transfer with FasL can also confer immune privilege on other tissues. FasL may interact with other cytokines or growth factors, such as TGF-β. This interaction is important for maintaining immune privilege.

Immune tolerance may develop in allogeneic stem cell transplantation. Saito *et al.*[22] recently performed studies on bone marrow stromal cells engrafted into xenogeneic fetal recipients. They examined the fate of xenogeneic marrow stromal cells after their systemic transplantation into fully immunocompetent adult recipients without immunosuppression. Bone marrow stromal cells were isolated from C57BL/6 mice and retrovirally transduced with the LacZ reporter gene for cell labeling, and then injected into immunocompetent adult Lewis rats. One week later, the recipient animals underwent coronary artery ligation and were sacrificed at various time points ranging from 1 day to 12 weeks after ligation. They observed that labeled mice cells engrafted into the bone marrow cavities of the recipient rats for at least 13 weeks after transplantation without any immunosuppression. On the other hand, circulating cells were positive only for the animals with one-day-old myocardial infarction. At various time points, numerous murine cells could be found in the infarcted myocardium seen before coronary ligation. Some of these cells subsequently showed positive staining for cardiomyocyte-specific proteins, while other labeled cells participated in angiogenesis in the infarcted area. Thus, these results suggest that a unique immunologic tolerance might have developed in the animals, allowing the cell engraftment into a xenogeneic environment, while preserving their ability to be recruited to an injured myocardium by way of the bloodstream and to undergo differentiation to form a stable cardiac chimera.

TELOMERASE IN CARDIAC STEM CELL DIFFERENTIATION

Telomerase is a ribonucleoprotein that acts as a reverse transcriptase and participates in maintaining telomere length in stem cells and immortal and actively dividing cells. Borges and Liew report that telomerase is developmentally regulated in the normal rat heart. Comparison of rat hearts at different developmental stages revealed that telomerase activity decreased to 20% of the fetal level by 5 days after birth and was undetectable by 20 days after birth. These results indicate that the rate of cardiomyocyte proliferation decreases dramatically soon after birth in the rat. Of several noncardiac tissues examined, telomerase activity was highest in fetal and adult rat liver, suggesting that there is an active mechanism for maintaining long telomeres in liver tissue at all stages. The disappearance of telomerase activity in the rat heart at the time that cardiomyocytes become terminally differentiated suggests that telomerase downregulation is important in the permanent withdrawal of cardiomyocytes from the cell cycle. Telomerase appears to play an important role in maintaining genomic integrity. Recent studies point to the involvement of telomerase in the regulation of stem cell life in hearts. Most mammalian cells, but not germ cells, tumor cells, and stem cells, possess a finite replicative life span, manifested by the eventual cessation of cell proliferation. Clinically, this is germane not just to the overt derangements of cell growth in cancer, but also to organs such as the heart, in which the capacity for cell replacement and repair is insufficient to maintain organ function following cell death. Among the intrinsic mechanisms that control a conserved program of replicative senescence is the enzyme telomerase, which synthesizes the telomeric repeat for end-capping of each chromosome. The implications of telomerase for cardiac growth have recently begun to be defined. Other functions of telomerase, in maintaining genome integrity, also hold importance for cardiac muscle as a novel means to suppress apoptosis and, thus, salvage the myocardium following ischemic injury.

CYTOKINES AND GROWTH FACTORS

Very recently, autologous stem cells have been used for cellular therapies in patients with myocardial infarction or heart failure. The traditional therapeutic armamentarium involves medical treatment and revascularization either by coronary artery bypass grafting (surgical approach) or through coronary angioplasty and stent placement (percutaneous approach). Despite many successful technological breakthroughs, the limitations of this strategy have become quite clear. Once the presence of extensive myocardial scar tissue and/or an unsuitable coronary anatomy, the opportunity to perform revascularization procedures is limited. In addition, coronary arterial disease (CAD) is a leading cause of heart failure, posing significant morbidity and mortality along with increasing health costs within a rapidly enlarging patient population. Many of those patients with ischemic cardiomyopathy have no further treatment options. Therapeutic angiogenesis has been proposed for patients not amenable to revascularization procedures, and several trials were conducted to evaluate this new therapy. Until recently angiogenesis trials basically employed two families of growth factors: fibroblast growth factors (FGF) and vascular endothelial cell growth factor (VEGF).

Vascular smooth muscle cell (VSMC) differentiation is important in understanding vascular disease. Drab *et al.*[23] used totipotent mouse embryonic stem (ES) cells to establish a model of VSMC differentiation. They tested whether the ES cell–derived cells express VSMC-specific properties by (1) morphological analysis, (2) gene expression, (3) immunostaining for VSMC-specific proteins, (4) expression of characteristic VSMC ion channels, and (5) formation of $[Ca^{2+}]i$ transients in response to VSMC-specific agonists. Treatment of embryonic stem cell–derived embryoid bodies with retinoic acid and dibutyryl-cyclic adenosine monophosphate (db-cAMP) can enhance differentiation of spontaneously contracting cell clusters in 67% of embryoid bodies compared with 10% of untreated controls. The highest differentiation rate occurs when retinoic acid and db-cAMP are applied to the embryoid bodies between days 7 and 11 in combination with frequent changes of culture medium. Other protocols with retinoic acid and db-cAMP, as well as single or combined treatment with VEGF, ECGF, bFGF, aFGF, fibronectin, matrigel, or hypoxia show no major influence on the differentiation rates. Single-cell RT-PCR and sequencing of the PCR products demonstrates the myosin heavy chain (MHC) splice variants distinguishing between gut and VSMC isoforms. RT-PCR with VSMC-specific MHC primers and immunostaining can also confirm the presence of VSMC transcripts and MHC protein. Furthermore, VSMC expressing MHC contain typical ion channels and respond to specific agonists with an increased $[Ca^{2+}]i$. ES cell–derived cells expressing VSMC-specific MHC and functional VSMC properties may be a suitable system for studying mechanisms of VSMC differentiation.

Strategies to apply growth factors to stem cells have been proposed and tested, including both protein and gene transfer. Different delivery routes have been reported with varying success. Recent studies have demonstrated that one VEGF mechanism of action is the mobilization of endothelial precursors cells originating in the bone marrow. Bone marrow has been intensively studied during the last three years as the most promising source of cells that could repopulate an injured area of the heart. Bone marrow–derived stem cells (MSC) are one of the so-called adult stem cells, defined by their multipotential and self-renewable cell lineages. MSC have been shown to differentiate into neural cells, hepatocytes, endothelium, myocytes, and cardiomyocytes. Preliminary evidence recently accumulated has demonstrated that bone marrow cells can differentiate into endothelial cells associated with angiogenesis or more notably into cardiomyocytes. Endothelial precursor cells (EPCs) can be identified as CD34 positive, which frequently have VEGF receptors. Human mesenchymal stem cells are thought to be multipotent cells, which are present in adult marrow, that can replicate as undifferentiated cells and that have the potential to differentiate to lineages of mesenchymal tissues, including bone, cartilage, fat, tendon, muscle, and marrow stroma. Cells that have the characteristics of human mesenchymal stem cells can be isolated from marrow aspirates of volunteer donors. These cells display a stable phenotype, remained as a monolayer *in vitro*, and proliferate when treated with growth factors.

Recently, Xu *et al.*[24] reported an interesting protein, Cripto-1(Cr1), encoded by the tdgfl gene, as a secreted growth factor that is expressed early in embryonic development as well as in some tumors of the breast and colon. During embryonic development, Cr1 is expressed in inner cell mass cells and the primitive streak, and later is restricted to the developing heart. To investigate the role of Cr1 during mouse development, mice were generated that contain a null mutation of both Cr1 genes, derived from homologous recombination in embryonic stem cells. No homozygous $Cr1^{-/-}$ mice were born, indicating that Cr1 is necessary for embryonic development. Embryos initiated gastrulation, and some embryos produced mesoderm up to day E7.5. Increasingly aberrant morphogenesis gives rise to disordered neuroepithelium that does not produce a recognizable neural tube, or head fold. Although some biochemical markers of differentiating ectoderm, mesoderm, and endoderm were expressed, all the cardiac-specific markers are absent from day E8.7 embryos: α-MHC, β-MHC, MLC2A, MLC2V, and ANF, whereas they are expressed in wild-type embryos. The yolk sac and placental tissues continue development in the absence of the embryo until day E9.5 but lacked large yolk sac blood vessels. In chimeric mice constructed by microinjection of double-targeted $Cr1^{(-/-)}$ embryonic stem cells into normal C57BL/6 blastocysts, the Cr1 produced by the normal C57BL/6 cells fully rescued the phenotype of $Cr1^{(-/-)}$ cells, indicating that Cr1 protein acts in a paracrine manner. Cells derived from the embryo show lower levels of proliferation and poor migration, and they develop different adhesion properties, when compared to wild type 24. Therefore, lethality in the absence of Cr1 likely results largely from defective precardiac mesoderm that is unable to differentiate into functional cardiomyocytes.

TUMOR SUPPRESSOR GENE p53

Targeted expression of the SV40 large T-antigen oncoprotein (T-Ag) induces cardiomyocyte proliferation in the atria and ventricles of transgenic mice. Previous studies have identified the p53 tumor suppressor, p107 (a homologue of the retinoblastoma tumor suppressor), and p193 (a novel BH3-only proapoptosis protein) as prominent TAg binding proteins in cardiomyocyte cell lines derived from these transgenic mice. To further explore the significance of these protein–protein interactions in the regulation of cardiomyocyte proliferation, Huh *et al.*[25] has generated a transgene comprising the human atrial natriuretic factor (ANF) promoter and sequences encoding a mutant T-Ag lacking the p53 binding domain. Repeated microinjection of this DNA gave rise to genetically mosaic animals with minimal transgene content, suggesting that widespread cardiac expression of mutant T-Ag is deleterious. This notion is supported by the observation that the transgene is selectively removed from the cardiac myocytes (but not the cardiac fibroblasts) in the mosaic animals. Crosses between the mosaic mice and animals expressing a cardiac-restricted dominant negative p53 resulted in transgene transmission with ensuing overt cardiac tumorigenesis. Transfection of the mutant T-Ag in embryonic stem (ES) cell–derived cardiomyocytes resulted in wide-spread cell death with characteristics typical of apoptosis. Cotransfection with a dominant negative p53 transgene can rescue mutant T-Ag–induced cell death in the ES-derived cardiomyocyte cultures, resulting in a marked proliferative response similar to that seen *in vivo* with the rescued transgenic mouse study. These results indicate that T-Ag expression in the absence of p53 functional abrogation results in cardiomyocyte death.

p53 may cooperate with other apoptosis-regulating proteins for controlling survival of cardiovascular stem cells. It is known that expression of adenoviral E1A in cardiomyocytes results in the activation of DNA synthesis followed by apoptosis. In contrast, expression of simian virus 40 large T antigen induces sustained cardiomyocyte proliferation. T antigen binds to the two proapoptotic proteins, p53 and p193, in cardiomyocytes. p193 represents a new member of the BH3-only proapoptosis subfamily. Structure–function analyses has shown that a p193 C-terminal truncation mutant encodes prosurvival activity. E1A induced apoptosis in cardiomyocytes derived from differentiating embryonic stem cells. Expression of the prosurvival p193 mutant alone or a mutant p53 alone cannot block E1A-induced apoptosis. In contrast, combinatorial expression of mutant p193 and mutant p53 attenuates E1A-induced apoptosis, resulting in a proliferative response indistinguishable from that seen with T antigen.[26] These results confirm the hypothesis that there are two proapoptotic pathways, encoded by p53 and p193, respectively, that restrict cardiomyocyte cell cycle activity in differentiating embryonic stem cell cultures. Furthermore, these results explain in molecular terms the phenotypic differences of E1A versus T-antigen gene transfer in cardiomyocytes.

RXR NUCLEAR RECEPTORS

Nuclear receptors play important roles for stem cell survival and differentiation. Tran *et al.*[27] recently reported that germline mutation in mice of the retinoic acid receptor gene RXRalpha results in a proliferative failure of cardiomyocytes, which leads to an underdeveloped ventricular chamber and midgestation lethality. Mutation of the cell cycle regulator N-myc gene also leads to an apparently identical phenotype. In this study, they demonstrate by chimera analysis that the cardiomyocyte phenotype in RXRalpha$^{-/-}$ embryos is a non-cell-autonomous phenotype. In chimeric embryos made with embryonic stem cells lacking RXRalpha, cardiomyocytes deficient in RXRalpha develop normally and contribute to the ventricular chamber wall in a normal manner. Because the ventricular hypoplastic phenotype reemerges in highly chimeric embryos, we conclude that RXRalpha functions in a nonmyocyte lineage of the heart to induce cardiomyocyte proliferation and accumulation, in a manner that is quantitatively sensitive. The studies also show that RXRalpha is not epistatic to N-myc, and that RXRalpha and N-myc regulate convergent obligate pathways of cardiomyocyte maturation.[27]

HYPOXIA, HIF, AND ERYTHROPOIETIN

Hypoxia is an essential developmental and physiological stimulus that plays a key role in the pathophysiology of cancer, heart attack, stroke, and other major causes of mortality.[28,29] Hypoxia-inducible factor 1 (HIF-1) is the only known mammalian transcription factor expressed uniquely in response to physiologically relevant levels of hypoxia. Iyer *et al.* have shown in Hif1a$^{-/-}$ embryonic stem cells lacking the O_2-regulated HIF-1alpha subunit that levels of mRNAs encoding glucose transporters and glycolytic enzymes are lower and cellular proliferation impaired. Vascular endothelial growth factor mRNA expression was also markedly decreased in hypoxic Hif1a$^{-/-}$ embryonic stem cells and cystic embryoid bodies. Complete deficiency of HIF-1alpha results in developmental arrest and lethality by E11 of Hif1a$^{-/-}$ embryos that manifested neural tube defects, cardiovascular malformations, and marked cell death within the cephalic mesenchyme. In Hif1a$^{+/+}$ embryos, HIF-1alpha expression increases between E8.5 and E9.5, coincident with the onset of developmental defects and cell death in Hif1a$^{-/-}$ embryos.

These results demonstrate that HIF-1alpha is a master regulator of cellular and developmental O_2 homeostasis. Another mechanism of hypoxia effects on stem cell survival may be through the erytheropoietin signaling. Erythropoietin is an essential growth factor that promotes survival, proliferation, and differentiation of mammalian erythroid progenitor cells. Erythropoietin$^{(-/-)}$ and erythropoietin receptor$^{(-/-)}$ mouse embryos die around embryonic day 13.5 due, in part, to failure of erythropoiesis in the fetal liver. Recently, Wu, *et al.*[30] demonstrated a novel role of erythropoietin and erythropoietin receptor in cardiac development *in vivo*. The erythropoietin receptor is expressed in the developing murine heart in a temporal and cell type–specific manner: it is initially detected by embryonic day 10.5 and persists until day 14.5. Both erythropoietin$^{(-/-)}$ and erythropoietin receptor$^{(-/-)}$ embryos suffered from ventricular hypoplasia at day 12–13 of gestation. This defect appears to be independent from the general state of hypoxia and is likely due to a reduction in the number of proliferating cardiac myocytes in the ventricular myocardium. Cell proliferation assays revealed that erythropoietin acts as a mitogen in cells isolated from erythropoietin$^{(-/-)}$ mice, while it has no effect in hearts from erythropoietin receptor$^{(-/-)}$ animals. Erythropoietin$^{(-/-)}$ and erythropoietin receptor($^{(-/-)}$ embryos also suffered from epicardium detachment and abnormalities in the vascular network. Finally, through a series of chimeric analysis, they provided evidence that erythropoietin acts in a manner that is non-cell-autonomous. These results elucidate a novel role of erythropoietin in cardiac morphogenesis and suggest a combination of anemia and cardiac failure as the cause of embryonic lethality in the erythropoietin$^{(-/-)}$ and erythropoietin receptor$^{(-/-)}$ animals.

CONCLUSIONS

Stem cell–mediated replacement of apoptotic cardiovascular cells, in particular cardiomyocytes, smooth muscle cells, and endothelial cells is essential for repair of damaged myocardium and restoration of heart function. However, the stem cells transplanted into cardiovascular tissues with atherosclerosis and ischemia are themselves exposed to a harsh, proapoptotic environment. Therefore, controlling apopto-

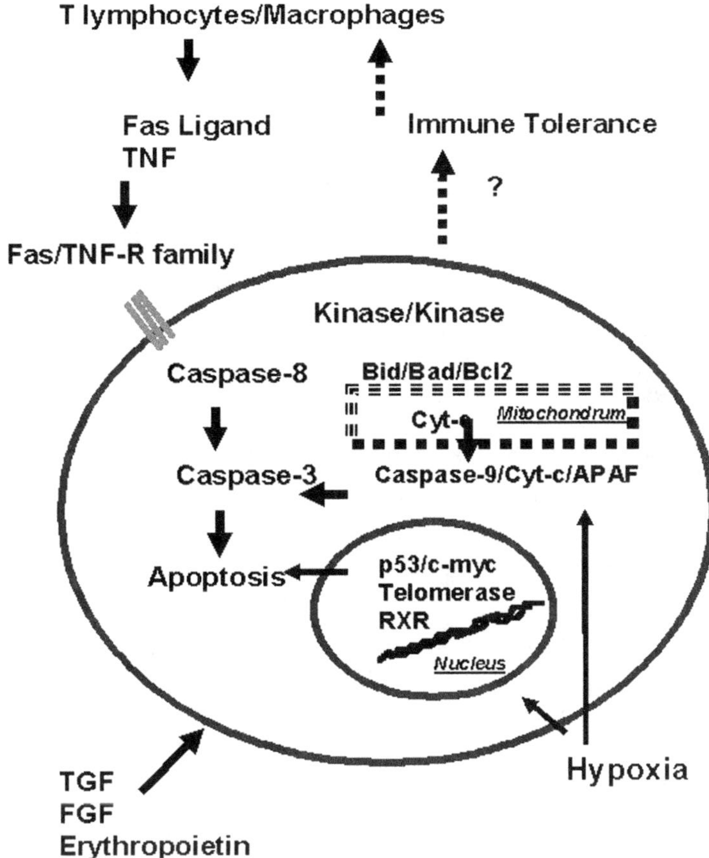

FIGURE 1. Schematic presentation of potential molecular pathways that regulate apoptosis of cardiovascular stem cells in hearts with atherosclerotic choronary arterial disease and ischemic failure. Abbreviations: APAF: apoptosis protein–associated factor; Cyt-C: cytochrome-C; FGF: fibroblast growth factor; IFN: interferon; TGF: transforming growth factors; TNF: tumor necrosis factor; TNF-R: tumor necrosis factor receptor.

sis is extremely important for stem cell survival and differentiation in the cardiovascular tissues. Many factors may influence stem cell apoptosis, including stimulation with cytokines and growth factors, expression of endogenous apoptosis-regulating genes (e.g., Fas, caspases, and p53), mitochondrial dysfunction, immune tolerance, telomerase activities, and hypoxia-responding proteins (FIG. 1). The information on stem cell apoptosis can help develop new stem cell therapies for patients with atherosclerotic coronary disease or heart failure.

ACKNOWLEDGMENTS

This study was supported by grants from the National Institutes of Health (R01HL59249 and R01HL69509), the U.S. Army DREAMS (17-98-1-8002), the Fourjay Foundation, Park-Davis/Pfizer ARA Research Awards, and the Texas State High Education Board.

REFERENCES

1. GENG, Y.J. 1997. Regulation of programmed cell death or apoptosis in atherosclerosis. Heart Vessels Suppl. **12:** 76–80.
2. GENG, Y.J. 2001. Molecular signal transduction in vascular cell apoptosis. Cell Res. **11:** 253–64.
3. GENG, Y.J. 2001. Biologic effect and molecular regulation of vascular apoptosis in atherosclerosis. Curr. Atheroscler. Rep. **3:** 234–242.
4. GENG, Y.J. & P. LIBBY. 2002. Progression of atheroma: a struggle between death and procreation. Arterioscler. Thromb. Vasc. Biol. **22:** 1370–1380.
5. BISHOPRIC, N.H., P. ANDREKA, T. SLEPAK & K.A. WEBSTER. 2001. Molecular mechanisms of apoptosis in the cardiac myocyte. Curr. Opin. Pharmacol. **1:** 141–150.
6. DISPERSYN, G.D. & M. BORGERS. 2001. Apoptosis in the heart: about programmed cell death and survival. News Physiol. Sci. **16:** 41–47.
7. PAULUS, W.J. 2001. The role of nitric oxide in the failing heart. Heart Fail Rev. **6:** 105–118.
8. GENG, Y.J. & P. LIBBY. 1995. Evidence for apoptosis in advanced human atheroma. Colocalization with interleukin-1 beta-converting enzyme. Am. J. Pathol. **147:** 251–266.
9. GENG, Y.J., Q. WU, M. MUSZYNSKI, *et al.* 1996. Apoptosis of vascular smooth muscle cells induced by in vitro stimulation with interferon-gamma, tumor necrosis factor-alpha, and interleukin-1 beta. Arterioscler. Thromb. Vasc. Biol. **16:** 19–27.
10. GENG, Y.J., L.E. HENDERSON, E.B. LEVESQUE, *et al.* 1997. Fas is expressed in human atherosclerotic intima and promotes apoptosis of cytokine-primed human vascular smooth muscle cells. Arterioscler. Thromb. Vasc. Biol. **17:** 2200–2208.
11. GENG, Y.J., T. AZUMA, J.X. TANG, *et al.* 1998. Caspase-3-induced gelsolin fragmentation contributes to actin cytoskeletal collapse, nucleolysis, and apoptosis of vascular smooth muscle cells exposed to proinflammatory cytokines. Eur. J. Cell Biol. **77:** 294–302.
12. GENG, Y.J., Y. ISHIKAWA, D.E. VATNER, *et al.* 1999. Apoptosis of cardiac myocytes in Gsalpha transgenic mice. Circ. Res. **84:** 34–42.
13. HUGHES, S. 2002. Cardiac stem cells. J. Pathol. **197:** 468–478.
14. MENASCHE, P. 2002. Cellular transplantation for the treatment of heart failure. Ernst Schering Res. Found. Workshop: 67–72.
15. TAYLOR, D.A. 2001. Cellular cardiomyoplasty with autologous skeletal myoblasts for ischemic heart disease and heart failure. Curr. Control Trials Cardiovasc. Med. **2:** 208–210.
16. MULLER-EHMSEN, J., L.H. KEDES, R.H. SCHWINGER & R.A. KLONER. 2002. Cellular cardiomyoplasty—a novel approach to treat heart disease. Congest. Heart Fail. **8:** 220–227.
17. MENASCHE, P. 2002. Cell transplantation for the treatment of heart failure. Semin. Thorac. Cardiovasc. Surg. **14:** 157–166.
18. ANVERSA, P., A. LERI, J. KAJSTURA & B. NADAL-GINARD. 2002. Myocyte growth and cardiac repair. J. Mol. Cell. Cardiol. **34:** 91–105.
19. MULLER-EHMSEN, J., P. WHITTAKER, R.A. KLONER, *et al.* 2002. Survival and development of neonatal rat cardiomyocytes transplanted into adult myocardium. J. Mol. Cell. Cardiol. **34:** 107–116.
20. GREEN, D.R. & T.A. FERGUSON. 2001. The role of Fas ligand in immune privilege. Nat. Rev. Mol. Cell Biol. **2:** 917–924.

21. LI, J.H., D. ROSEN, P. SONDEL & G. BERKE. 2002. Immune privilege and FasL: two ways to inactivate effector cytotoxic T lymphocytes by FasL-expressing cells. Immunology **105:** 267–277.
22. SAITO, T., J.Q. KUANG, B. BITTIRA, *et al.* 2002. Xenotransplant cardiac chimera: immune tolerance of adult stem cells. Ann. Thorac. Surg. **74:** 19–24; discussion 24.
23. DRAB, M., H. HALLER, R. BYCHKOV, *et al.* 1997. From totipotent embryonic stem cells to spontaneously contracting smooth muscle cells: a retinoic acid and db-cAMP in vitro differentiation model. FASEB J. **11:** 905–915.
24. XU, C., G. LIGUORI, M.G. PERSICO & E.D. ADAMSON. 1999. Abrogation of the Cripto gene in mouse leads to failure of postgastrulation morphogenesis and lack of differentiation of cardiomyocytes. Development **126:** 483–494.
25. HUH, N.E., K.B. PASUMARTHI, M.H. SOONPAA, *et al.* 2001. Functional abrogation of p53 is required for T-Ag induced proliferation in cardiomyocytes. J. Mol. Cell. Cardiol. **233:** 1405–1419.
26. PASUMARTHI, K.B., S.C. TSAI & L.J. FIELD. 2001. Coexpression of mutant p53 and p193 renders embryonic stem cell-derived cardiomyocytes responsive to the growth-promoting activities of adenoviral E1A. Circ. Res. **88:** 1004–1011.
27. TRAN, C.M. & H.M. SUCOV. 1988. The RXRalpha gene functions in a non-cell-autonomous manner during mouse cardiac morphogenesis. Development **125:** 1951–1956.
28. IYER, N.V., L.E. KOTCH, F. AGANI, *et al.* 1998. Cellular and developmental control of O_2 homeostasis by hypoxia-inducible factor 1 alpha. Genes Dev. **12:** 149–162.
29. ADELMAN, D.M. & M.C. SIMON. 2002. Hypoxic gene regulation in differentiating ES cells. Methods Mol. Biol. **185:** 55–62.
30. WU, H., S.H. LEE, J. GAO, *et al.* 1999. Inactivation of erythropoietin leads to defects in cardiac morphogenesis. Development **126:** 3597–3605.

Oxidation of LDL, Atherogenesis, and Apoptosis

CLAUDIO NAPOLI

Departments of Medicine and Clinical Pathology, University of Naples, Naples, Italy

ABSTRACT: A plethora of studies in cultured cells have established that oxidized low-density lipoprotein (oxLDL) may enhance arterial apoptosis that involves both mitochondrial and death receptor pathways (Fas/FasL, TNF receptors I and II), thereby activating caspase cascade and other proteases. When apoptosis is inhibited by Bcl-2 overexpression, oxLDL may trigger necrosis through a calcium-dependent pathway. Despite this effort, the pathophysiological relevance of apoptosis *in vivo* remains to be elucidated. In principle, apoptosis occurring in atherosclerotic areas could be involved in endothelial cell lining defects, necrotic core formation, and plaque rupture or fissuring. This complex pathogenic framework may favor coronary atherothrombotic events. To date, the pathogenic role of apoptosis in thrombosis is attractive, but a solid evidence is still needed. When the precise role of oxLDL in vascular programmed cell death occurring *in vivo* is clarified, this may aid in the development of novel therapeutic approaches to adverse atherogenesis and its clinical sequelae.

KEYWORDS: lipoproteins; lipoperoxidation; apoptosis; atherosclerosis

INTRODUCTION

Early human atherosclerotic lesions develop in the arterial intima, in which accumulation of lipids, oxidative by-products, and extracellular matrix yields a fatty streak that thickens the arterial wall.[1–4] By counterbalancing proliferation, vascular apoptosis may limit cell buildup in the intima. On the other hand, decreased apoptosis may increase the tissue cellularity and promote vascular smooth muscle cell (VSMC) proliferation. An antiapoptotic mechanism could favor lipid-laden foam macrophage (foam cell) accumulation in lesions.[5,6] Apoptosis can occur during different stages of the disease.[6] Apoptosis causes cell loss as atherosclerotic lesion progresses, often yielding a late or mature lesion containing a dense extracellular matrix with a relatively sparse cell population. In addition to lipids and connective tissue, the core of the typical plaque contains many dead cells or cell debris. Quantification of cell death showed that many cells in the lesions bear the markers of apoptosis, even though necrotic cells may also occur in the lipid core area.[7]

There is a high percentage of apoptotic cells in advanced atherosclerotic plaques. Because the tissue volume might decrease when the rate of apoptosis exceeds that of

Address for correspondence: Claudio Napoli, M.D., Ph.D., Medicine-UN, P.O. Box, Naples 80131, Italy.
claunap@tin.it

Ann. N.Y. Acad. Sci. 1010: 698–709 (2003). © 2003 New York Academy of Sciences.
doi: 10.1196/annals.1299.127

TABLE 1. Receptors, caspases, and genes involved in vascular cell apoptosis during atherosclerosis

Receptors, caspases, and genes	Receptor ligand and function
Death receptors (other names)	
DR3 (Apo3, Wsl1, TRAMP, LARD)	TWEAK
DR4 (Apo2, Trail-R1)	Trail
DR5 (Trail-R2, TRICK2)	Trail
Fas (CD95, Apo1)	FasL (CD95L)
TNFR1 and TNFR2	TNF
Caspases (other names)	
Caspase-1 (ICE)	Proinflammatory
Caspase-2 (ICH-1)	Proapoptotic
Caspase-3 (CPP32, Yama)	Proapoptotic
Caspase-4 (ICErel II, TX, ICH-2)	Proinflammatory
Caspase-5 (ICErel III, TY)	Proinflammatory
Caspase-6 (Mch2)	Proapoptotic
Caspase-7 (Mch3, ICE-LAP3, CMH-1)	Proapoptotic
Caspase-8 (FLICE, MACH, Mch5)	Proapoptotic
Caspase-9 (ICE-LAP6, Mch6, APAF-3)	Proapoptotic
Caspase-10 (Mch4)	Proapoptotic
Caspase-11	Proinflammatory
Caspase-12	Proinflammatory
Caspase-13	Proinflammatory
Caspase-14 (MICE)	Proinflammatory/proapoptotic
Genes	
Proapoptotic members of the Bcl-2 family	Bad, Bak, Bax, Bid, Bik, Bcl-Xs, Nip3, Mtd, DP5
Antiapoptotic members of the Bcl 2 family	Bcl-2, Bcl-w, Mcl-1, Bfl-1, BHRF-1

cell proliferation, this finding poses a paradox. Nevertheless, the presence of vascular apoptosis may not always be linked with reduction of atherosclerotic tissue. Indeed, accumulation of apoptotic cells in the plaque suggests that the scavenging system for the dead cells may operate inefficiently in advanced atheroma. Some apoptotic cells may remain "mummified" rather than undergo removal by phagocytosis.[7] The following mechanisms might lead to retention of apoptotic cells or bodies in atherosclerotic lesions: (a) phagocytosis; (b) intracellular accumulation of native and oxidized lipids may reduce the ability of macrophages and VSMCs to digest apoptotic cells; and (c) cross-linking of macromolecules, such as proteins, nucleic acids, and carbohydrates, may stabilize apoptotic cells in the artery. Many cellular receptors and intracellular proteins or enzyme systems participate in the regulation of apoptosis. TABLE 1 lists some of these elements; the specific role of each factor has been reviewed elsewhere.[6,8]

APOPTOSIS AND ATHEROSCLEROTIC LESION PROGRESSION

VSMCs and endothelial cells (ECs) encounter a broad variety of environmental factors, such as mechanical force,[9–11] oxidative stress,[1,3,12] radiation,[13,14] reactive oxygen or nitrogen species,[15,16] cholesterol and its oxides,[17–19] viruses,[20,21] bacterial products,[22] and inflammatory cytokines.[23,24] These extracellular environmental and immunologic factors can induce apoptosis in vascular cells (TABLE 1). Fas/apo-1/CD95, a cell surface–borne protein, belongs to the TNF receptor (TNFR) superfamily. Several lines of evidence implicate the Fas ligand (FasL)/Fas–caspase death pathway in the induction of apoptosis in atherosclerotic plaques.[25–28] Overexpression of FasL in arteries of rabbits with hyperlipidemia promotes the development of atherosclerosis as a result of intimal VSMC accumulation.[29] Moreover, Fas-associated signaling induces the expression of monocyte-chemoattractant protein-1 and IL-8 and causes massive immigration of macrophages *in vivo*.[30]

Early human atherosclerotic lesions of fetuses and children contain modified lipoproteins, in particular, oxidized low-density lipoproteins (oxLDL).[2,31–33] Intrinsically, cholesterol and its esters have proapoptotic action, but they become more cytotoxic after oxidation. High levels of cholesterol oxides in the cell membrane can trigger foam cell death by apoptosis.[34] Oxysterol-mediated apoptosis may foster the formation of a necrotic lipid core.[35] There is a plethora of signal transduction and nuclear events that are involved in the phenomenon, including TNFR1 and TNFR2 activation, the complex cascade of kinases, regulatory events of the Bcl-2 family of genes, activation of death-effector caspases and sphingomyelinase, and regulatory action of the nuclear transcription factor-κB and other transcription factors, which may contribute to oxLDL-induced apoptosis.[17,18,35–40] Coadministration of oxygen radical scavengers or antioxidants reduces the rate of oxLDL-induced apoptosis (see below).[40,41]

Plaque disruption causes the thrombotic complications of atherosclerosis that underlie clinical events such as unstable angina pectoris, myocardial infarction, and other acute syndromes.[42,43] Production of apoptosis-related mediators by activated inflammatory or immune cells may lead to the death of VSMCs, which can weaken and destabilize the plaques. Either apoptotic ECs[44] or VSMCs[45] can promote coagulation. Rapid exposure of membrane phosphatidylserine and loss of the anticoagulant membrane components in apoptotic ECs can occur, which may promote a procoagulant environment. Thus, apoptotic cells, if not promptly removed, become thrombogenic or proinflammatory through local thrombin activation, thereby contributing to lesion progression.[45] Experimental models of plaque instability are difficult to establish. This is the reason why many pathophysiological mechanisms are difficult to understand. Now, murine models can provide a useful approach in this experimental setting.[46]

Although apoptosis in lesions usually occurs in intimal cells, under certain pathophysiological conditions, such as the development of aortic aneurysms, this type of cell death may attack medial VSMCs. The medial VSMCs of arteries with aneurysms exhibit considerable levels of apoptosis and *p53* expression.[47] Moreover, macrophages and T lymphocytes migrate into the arterial wall in atherosclerotic aneurysms and produce death-promoting proteins (perforin, Fas, and FasL).[48] Beyond lipid accumulation, atherosclerosis involves innate and acquired immunity.[12,49,50] The antigenic substances may include oxLDL and some stress proteins, such as heat shock protein-

60. The effector of acquired immunity, the CD8-positive T lymphocyte, can kill cells by apoptosis via the production of perforin, granzyme, and FasL.[38,48] Mediators of innate immunity may also contribute to this process. Indeed, exposure of human VSMCs to IL-1β, TNF-α, and IFN-γ can promote apoptosis.[23] Thus, the effector limbs of innate and acquired immunity may have synergistic effects in causing the demise of cells within lesions. The cytokines produced by activated macrophages and T cells could induce activation of the sphingomyelin-ceramide signaling pathway[51] and the L-arginine–nitric oxide (NO) pathway.[26,52–55] At high concentrations, NO can attack several important iron-containing enzymes involved in DNA synthesis and mitochondrial respiration, leading to apoptosis of the target cells. NO can react with superoxide anion to form the cytotoxic species, peroxynitrite, augmenting the proapoptotic action of NO.[54–56] The proapoptotic effect of NO highly depends on its concentrations and interactions with other reactive species. Low physiological amounts of NO show no harm and, under certain circumstances of normal blood flow, demonstrate protective effects against vascular cell apoptosis and oxidation-sensitive gene expression.[10,55,57,58] More important, pharmacologic modulation of NO[59] can lead to reduced lesion formation in hypercholesterolemic mice.[57,60]

APOPTOSIS AND THROMBOGENICITY OF THE RUPTURED ATHEROSCLEROTIC PLAQUE

Tissue factor (TF) is a 47-kDa transmembrane glycoprotein that initiates blood coagulation by binding the coagulation factor VII and its activated form (factor VIIa) to form a high-affinity complex.[61] This binary complex proteolytically activates factors IX and X, which in turn leads to thrombin generation.[61] Pathological and functional studies have identified cellular and extracellular TF as one of the determinants of the thrombogenicity of the lipid core in humans.[62,63] Intervention with recombinant TF pathway inhibitor is associated with a significant reduction of acute thrombus formation in lipid-rich plaques.[63] TF is operational on the surface of cell membranes and it is known that its activity is highly dependent on the presence of anionic phospholipids, chiefly PS.[64] Because apoptosis is associated with significant PS exposure on the cell surface and leads to the shedding of PS-containing membrane microparticles, apoptosis may be directly responsible for TF activation within the plaque environment. There is a colocalization between TF expression (cellular and extracellular) and apoptotic death, particularly in the lipid core, suggesting that TF may be released in apoptotic microparticles.[65] Most of the microparticles originated from macrophages and lymphocytes that are known to be abundant at sites of plaque rupture.[66] Thus, shed membrane apoptotic microparticles may play a role in the initiation of the coagulation cascade after plaque rupture and exposure of the lipid core to the circulating blood. Consistently, macrophage apoptotic death is significantly increased at sites of plaque rupture and thrombosis in patients with sudden coronary death,[67] and apoptosis is significantly increased in unstable versus stable human plaques.[68]

Plaque rupture of a thin fibrous cap overlying a lipid core is not necessarily the only final common pathway in the formation of coronary thrombi. It was reported in several studies that plaque erosion without rupture is an important predisposing substrate for acute coronary syndromes and sudden cardiac death.[69,70] Although risk factors predisposing to plaque erosion have been identified, the cellular and molecular mechanisms responsible for this process remain largely unknown. Apoptosis of luminal ECs may be one of the mechanisms leading to erosion and thrombosis. Indeed, ECs are continuously exposed to hemodynamic forces that have a great impact on their cellular structure and function. ECs cultured under static conditions undergo a basal low level of apoptosis, whereas exposure to flow inhibits the apoptotic process.[9] Using carotid human atherosclerotic plaques, it was demonstrated that blood flow exerts a direct influence on EC survival in human atherosclerosis.[71] Analysis of longitudinal plaque sections revealed the presence of luminal EC apoptosis in the majority of plaques examined. Furthermore, luminal EC apoptosis in these nonruptured plaques occurred preferentially in the downstream parts of the plaques where low shear prevails in comparison with the high shear stress regions.[71] Given the high procoagulant and proadhesive potentials of apoptotic ECs and the propensity of denuded vessel segments to increased vasospasm and platelet aggregation, EC apoptosis may lead to lumen thrombosis favoring plaque progression or occurrence of coronary syndromes. Interestingly, steep outflow angle of a stenosis, which is a feature characteristic of disturbed flow downstream from the stenosis, is an independent predictor of infarction at 3-year follow-up.[72] Moreover, the amount of fibrin and platelets within thrombi formed on the contact of eroded plaques was similar to that in thrombi formed on the contact of ruptured plaques.[69] Thrombi formed on eroded plaques were predominantly composed of fibrin, which means that early fibrin deposition is present in the absence of plaque rupture. Hence, early deposition of a significant amount of fibrin in superficially eroded plaques could involve the TF-dependent extrinsic pathway of coagulation.

Despite this large body of evidence for a prothrombogenic role of apoptosis in vascular disease, recent experimental models in which massive apoptosis was induced within the vessel did not mention the occurrence of thrombosis.[73,74] Apoptotic death was specifically induced in neointimal smooth muscle cells of nonruptured plaques through seeding of VSMCs overexpressing the Fas-associated death domain or through downregulation of intimal cell bcl-x_L expression with the use of antisense oligonucleotides.[73,74] ECs were not reported to be apoptotic in these models.[73,74] Nevertheless, there was no direct contact between apoptotic cells and the circulating blood; thus, the absence of thrombosis is not particularly surprising.

PATHOGENIC MECHANISMS OF VASCULAR APOPTOSIS

The regulation of apoptosis involves negative as well as positive signaling, summarized in TABLE 2. Flow disturbances or lack of normal shear stress in regions characteristic of sites of predilection to atheroma may induce the apoptosis of ECs.[75] The inhibitory effect of shear stress on the apoptosis of ECs may be related in part to the increased levels of NO produced by the low-capacity endothelial isoform of NO synthase (eNOS), a shear stress–activated gene.[54,55] In contrast to the cytokine-inducible form of NO synthase (iNOS), eNOS generates relatively small amounts of NO that may inhibit caspase activities and protect ECs from apoptosis.[10,54,55] Analysis of human atherosclerotic lesions shows more apoptosis in the downstream parts of plaques, where disturbed flow occurs.[3,5,6,9] Macrophages deprived of macrophage colony stimulating factor (CSF) undergo apoptosis induced by denatured aggregated LDL,[76] whereas platelet-derived growth factor (PDGF) protects VSMCs from

TABLE 2. Vascular cell apoptosis: proapoptotic and antiapoptotic stimuli

Stimuli	References
Proapoptotic stimuli	
OxLDL	35–38, 40, 41
Oxysterols and modified fatty acids	17, 94
Reactive oxygen species (ROS)	16, 35, 40
Reactive nitrogen species (NO at high levels)	16, 54, 55
Viral and bacterial products	96
Inflammatory cytokines	23, 26
Radiations	13, 95
C-reactive protein	90
Antiapoptotic stimuli	
NO at low level	54, 55, 97
Shear stress	9–11
PDGF	98
bFGF	79, 100
VEGF	78, 99
IGF-I	101
Cowpox virus CrmA	102
Baculovirus protein *p35*	103
IAP protein family	104
Vitamins C and E	22, 41, 81, 105
SOD and catalase	40, 83

apoptosis.[77] EC growth factor can inhibit the apoptosis of bone marrow cells induced by radiation.[78] Basic fibroblast growth factor (FGF) can inhibit the apoptosis of ECs induced by TNF-α. Attenuation of basic FGF expression by an antisense strategy can trigger the apoptosis of VSMCs.[79] Protein kinase C may mediate the protective effect of FGF on apoptosis induced by serum starvation.[80]

Because oxidative stress can injure vascular cells and promote apoptosis as stated above, antioxidants might mitigate this process. Inhibitors of apoptosis relevant to atherosclerosis include oxygen radical scavenger enzymes (e.g., superoxide dismutase [SOD]), catalase (CAT),[35,40] and pharmacological doses of certain antioxidant vitamins.[41,81] For instance, during 7-ketocholesterol-induced apoptosis of U937 cells, the content of cellular antioxidants falls rapidly.[82] The administration of antioxidants, the aminothiols glutathione and *N*-acetylcysteine, can protect the monocytic cells from 7-ketocholesterol-induced apoptotic cell death.[82] CAT can prevent VSMC apoptosis triggered by hydrogen peroxide.[40,83] Thus, different reactive oxygen species may exert different effects on different cell types.

The proto-oncogene c-*myc* can mediate either cell death or proliferation depending on its level of expression. It functions as a nuclear phosphoprotein with certain properties of a transcription factor.[84,85] Recent data indicate an important oxidation-sensitive regulation of c-*myc* activity in experimental atherosclerosis.[86,87] In serum-deprived cultures, cells overexpressing c-*myc* readily undergo apoptosis. Deregulation of c-*myc* causes apoptosis of the VSMCs deprived of growth factors or treated with cytokines such as IFN-γ.[88] *p53* may mediate the proapoptotic effect of c-*myc* in various cell lines. Further studies are warranted to establish the causal role of c-*myc* in vascular apoptosis and atherosclerotic lesion progression. Interestingly, the presence of the tumor-suppressor gene *p53* in advanced atherosclerotic plaques and the sensitivity to *p53*-induced cell death of VSMCs isolated from these plaques have fueled speculation about the role of *p53* in lesion destabilization and plaque rupture.[89] Carotid atherogenesis was initiated in apolipoprotein E knockout mice by placement of a perivascular Silastic collar, and the resulting plaques were incubated transluminally with recombinant adenovirus carrying either a *p53* or β-galactosidase (lacZ) transgene.[89] *p53* overexpression resulted in a marked decrease in the cellular and extracellular content of the cap, reflected by a markedly reduced cap/intima ratio ($P < 0.001$). The latter is one of the characteristic features of plaque vulnerability to rupture; in fact, whereas spontaneous rupture of *p53*-treated lesions was rare, it was found in 40% of cases after treatment with the vasopressor compound, phenylephrine ($P = 0.003$).[89] Another link between vascular inflammation and apoptosis comes from the recent study of Chang *et al.*[90] Indeed, C-reactive protein binds to both oxidized LDL and apoptotic cells through recognition of a common ligand: phosphorycoline of oxidized phospholipids.[90] Apoptosis of VSMCs induced by Bcl-x is also associated with initial hyperplasia after heart transplantation.[91] Antisense Bcl-x ODN inhibits VSMC proliferation by inducing apoptosis in graft coronary arteries.[91]

CONCLUSIONS

OxLDL exhibit a wide range of biological effects on cultured arterial cells including apoptosis. OxLDL contain various toxic oxidized lipids (such as lipid peroxides, oxysterols, aldehydes), but the respective role of each compound in apoptosis remains to be fully understood. Oxidized lipids from oxLDL induce modifications of cell protein structure, elicit ROS generation and lipid peroxidation of cellular lipids, and alter the regulation of various signaling pathways and gene expression. An intense, delayed, and sustained calcium signal is apparently a trigger of cell death, but the initial targets and the subsequent sequence of events leading to cell death are only partly understood. Programmed cell death also may occur in atherosclerotic areas, leading to defects in the endothelial cell lining, necrotic core formation, and plaque rupture or fissuring.[106] Thus, apoptosis could participate in progression or regression of lesions, vascular remodeling, and plaque instability. The pathogenic role of apoptosis in thrombosis is attractive, but a solid evidence is still needed. Many environmental and endogenous factors can influence apoptosis through various signaling and nuclear coregulation events in the arterial wall.[92,93] Pharmacological modulation of these events could provide new therapeutic targets in the treatment of atherothrombosis-related diseases.

ACKNOWLEDGMENTS

This paper is dedicated to the memory of Adolfina Concha Molinari, who died on January 23, 2002. Work by the author described in this review was supported by National Institutes of Health Grant No. HL-56989.

REFERENCES

1. Ross, R. 1999. Atherosclerosis—an inflammatory disease. N. Engl. J. Med. **340:** 115–126.
2. Napoli, C. & W. Palinski. 2001. Clinical and pathogenic implications of maternal hypercholesterolemia during pregnancy for the later development of atherosclerosis. Eur. Heart J. **22:** 4–9.
3. De Nigris, F., L.O. Lerman, M. Condorelli *et al.* 2001. Oxidation-sensitive transcription factors and molecular mechanisms in the arterial wall. Antioxidant Redox Signal. **3:** 1119–1130.
4. Palinski, W. & C. Napoli. 2002. The fetal origins of atherosclerosis: maternal hypercholesterolemia and cholesterol-lowering or antioxidant treatment during pregnancy influence *in utero* programming and post-natal susceptibility to the disease. FASEB J. **16:** 1348–1360.
5. Geng, Y.J. 2001. Biologic effect and molecular regulation of vascular apoptosis in atherosclerosis. Curr. Atheroscler. Rep. **3:** 234–242.
6. Kockx, M.M., G.R. De Meyer, J. Muhring *et al.* 1998. Apoptosis and related proteins in different stages of human atherosclerotic plaques. Circulation **97:** 2307–2315.
7. Geng, Y.J. & P. Libby. 1995. Evidence for apoptosis in advanced human atheroma: colocalization with interleukin-1β-converting enzyme. Am. J. Pathol. **147:** 251–266.
8. Suh, Y. 2002. Cell signaling in aging and apoptosis. Mech. Ageing Dev. **123:** 881–890.
9. Dimmeler, S., J. Haendeler, V. Rippmann *et al.* 1996. Shear stress inhibits apoptosis of human endothelial cells. FEBS Lett. **399:** 71–74.
10. Dimmeler, S., J. Haendeler, M. Nehls & A.M. Zeiher. 1997. Suppression of apoptosis by nitric oxide via inhibition of interleukin-1β-converting enzyme (ICE)–like and cysteine protease protein (CPP)-32–like proteases. J. Exp. Med. **185:** 601–607.
11. Urbich, C., D.H. Walter, A.M. Zeiher & S. Dimmeler. 2000. Laminar shear stress upregulates integrin expression: role in endothelial cell adhesion and apoptosis. Circ. Res. **87:** 683–689.
12. Glass, C.K. & J.L. Witztum. 2001. Atherosclerosis: the road ahead. Cell **104:** 503–516.
13. Langley, R.E., E.A. Bump, S.G. Quartuccio *et al.* 1997. Radiation-induced apoptosis in microvascular endothelial cells. Br. J. Cancer **75:** 666–672.
14. Lopez-Candales, A., D.R. Holmes, M.J. Scott *et al.* 1996. Effects of ultraviolet light in vascular cells *in vitro* and in intact atherosclerotic explants: potential role of apoptosis in vascular biology. Biochem. Cell Biol. **74:** 333–345.
15. Johnson, T.M., Z.X. Yu, V.J. Ferrans *et al.* 1996. Reactive oxygen species are downstream mediators of p53-dependent apoptosis. Proc. Natl. Acad. Sci. USA **93:** 11848–11852.
16. Patel, R.P., D. Moellering, J. Murphy-Ullrich *et al.* 2000. Cell signaling by reactive nitrogen and oxygen species in atherosclerosis. Free Radical Biol. Med. **28:** 1780–1794.
17. Nishio, E. & Y. Watanabe. 1996. Oxysterols induced apoptosis in cultured smooth muscle cells through CPP32 protease activation and bcl-2 protein downregulation. Biochem. Biophys. Res. Commun. **226:** 928–934.
18. Harada-Shiba, M., M. Kinoshita, H. Kamido & K. Shimokado. 1998. Oxidized low density lipoprotein induces apoptosis in cultured human umbilical vein endothelial cells by common and unique mechanisms. J. Biol. Chem. **273:** 9681–9687.
19. Liao, H.S., T. Kodama & Y.J. Geng. 2000. Expression of class A scavenger receptor inhibits apoptosis of macrophages triggered by oxidized low density lipoprotein and oxysterol. Arterioscler. Thromb. Vasc. Biol. **20:** 1968–1975.
20. Kovacs, A., M.L. Weber, L.J. Burns *et al.* 1996. Cytoplasmic sequestration of p53 in cytomegalovirus-infected human endothelial cells. Am. J. Pathol. **149:** 1531–1539.
21. Sela, D.D., M. Korner, M. Pick *et al.* 1996. Programmed endothelial cell death induced by an avian hemangioma retrovirus is density dependent. Virology **223:** 233–237.
22. Haendeler, J., A.M. Zeiher & S. Dimmeler. 1996. Vitamin C and E prevent lipopolysaccharide-induced apoptosis in human endothelial cells by modulation of Bcl-2 and Bax. Eur. J. Pharmacol. **317:** 407–411.
23. Geng, Y.J., Q. Wu, M. Muszynski *et al.* 1996. Apoptosis of vascular smooth muscle cells induced by *in vitro* stimulation with interferon-gamma, tumor necrosis factor-alpha, and interleukin-1-beta. Arterioscler. Thromb. Vasc. Biol. **16:** 19–27.

24. Henderson, E.L., Y.J. Geng, G.K. Sukhova *et al.* 1999. Death of smooth muscle cells and expression of mediators of apoptosis by T lymphocytes in human abdominal aortic aneurysms. Circulation **99:** 96–104.
25. Ashkenazi, A. & V.M. Dixit. 1998. Death receptors: signaling and modulation. Science **281:** 1305–1308.
26. Geng, Y.J., L.E. Henderson, E.B. Levesque *et al.* 1997. Fas is expressed in human atherosclerotic intima and promotes apoptosis of cytokine-primed human vascular smooth muscle cells. Arterioscler. Thromb. Vasc. Biol. **17:** 2200–2208.
27. Cai, W., B. Devaux, W. Schaper & J. Schaper. 1997. The role of Fas/APO 1 and apoptosis in the development of human atherosclerotic lesions. Atherosclerosis **131:** 177–186.
28. Walsh, K. & M. Sata. 1999. Is extravasation a Fas-regulated process? Mol. Med. Today **5:** 61–67.
29. Schneider, D.B., G. Vassalli, S. Wen *et al.* 2000. Expression of Fas ligand in arteries of hypercholesterolemic rabbits accelerates atherosclerotic lesion formation. Arterioscler. Thromb. Vasc. Biol. **20:** 298–308.
30. Schaub, F.J., D.K. Han, W.C. Liles *et al.* 2000. Fas/FADD-mediated activation of a specific program of inflammatory gene expression in vascular smooth muscle cells. Nat. Med. **6:** 790–796.
31. Napoli, C., F.P. D'Armiento, F.P. Mancini *et al.* 1997. Fatty streak formation occurs in human fetal aortas and is greatly enhanced by maternal hypercholesterolemia: intimal accumulation of low density lipoprotein and its oxidation precede monocyte recruitment into early atherosclerotic lesions. J. Clin. Invest. **100:** 2680–2690.
32. Napoli, C., J.L. Witztum, F. De Nigris *et al.* 1999. Intracranial arteries of human fetuses are more resistant to hypercholesterolemia-induced fatty streak formation than extracranial arteries. Circulation **99:** 2003–2010.
33. Napoli, C., C.K. Glass, J.L. Witztum *et al.* 1999. Influence of maternal hypercholesterolemia during pregnancy on progression of early atherosclerotic lesions in childhood: Fate of Early Lesions in Children (FELIC) Study. Lancet **354:** 1234–1241.
34. Kellner-Weibel, G., Y.J. Geng & G.H. Rothblat. 1999. Cytotoxic cholesterol is generated by the hydrolysis of cytoplasmic cholesteryl ester and transported to the plasma membrane. Atherosclerosis **146:** 309–319.
35. Heermeier, K., R. Schneider, A. Heinloth *et al.* 1999. Oxidative stress mediates apoptosis induced by oxidized low-density lipoprotein and oxidized lipoprotein(a). Kidney Int. **56:** 1310–1312.
36. Escargueil-Blanc, I., N. Andrieu-Abadie, S. Caspar-Bauguil *et al.* 1998. Apoptosis and activation of the sphingomyelin-ceramide pathway induced by oxidized low density lipoproteins are not causally related in ECV-304 endothelial cells. J. Biol. Chem. **273:** 27389–27395.
37. Dimmeler, S., J. Haendeler, J. Galle & A.M. Zeiher. 1997. Oxidized low-density lipoprotein induces apoptosis of human endothelial cells by activation of CPP32-like proteases: a mechanistic clue to the "response to injury" hypothesis. Circulation **95:** 1760–1763.
38. Sata, M. & K. Walsh. 1998. Oxidized LDL activates Fas-mediated endothelial cell apoptosis. J. Clin. Invest. **102:** 1682–1689.
39. Garrington, T.P. & G.L. Johnson. 1999. Organization and regulation of mitogen-activated protein kinase signaling pathways. Curr. Opin. Cell Biol. **11:** 211–218.
40. Napoli, C., O. Quehenberger, F. De Nigris *et al.* 2000. Mildly oxidized low-density lipoprotein activates multiple apoptotic signaling pathways in human coronary cells. FASEB J. **14:** 1996–2007.
41. De Nigris, F., F. Franconi, I. Maida *et al.* 2000. Modulation by α- and γ-tocopherol of apoptotic signaling induced by oxidized low-density lipoprotein in human coronary smooth muscle cells. Biochem. Pharmacol. **59:** 1477–1487.
42. Fuster, V., B. Stein, J.A. Ambrose *et al.* 1990. Atherosclerotic plaque rupture and thrombosis: evolving concepts. Circulation **82**(suppl. II)**:** II47–II59.
43. Libby, P. 1995. Molecular bases of the acute coronary syndromes. Circulation **91:** 2844–2850.
44. Bombeli, T., A. Karsan, J.F. Tait & J.M. Harlan. 1997. Apoptotic vascular endothelial cells become procoagulant. Blood **89:** 2429–2442.

45. FLYNN, P.D., C.D. BYRNE, T.P. BAGLIN *et al.* 1997. Thrombin generation by apoptotic vascular smooth muscle cells. Blood **89:** 4378–4384.
46. PALINSKI, W. & C. NAPOLI. 2002 Unraveling pleiotropic effects of statins on plaque rupture. Arterioscler. Thromb. Vasc. Biol. **22:** 1745–1750.
47. HOLMES, D.R., C.A. LOPEZ, S. LIAO & R.W. THOMPSON. 1996. Smooth muscle cell apoptosis and p53 expression in human abdominal aortic aneurysms. Ann. N.Y. Acad. Sci. **800:** 286–287.
48. HENDERSON, E.L., Y.J. GENG, G.K. SUKHOVA *et al.* 1999. Death of smooth muscle cells and expression of mediators of apoptosis by T lymphocytes in human abdominal aortic aneurysms. Circulation **99:** 96–104.
49. PALINSKI, W. & J.L. WITZTUM. 2000. Immune responses to oxidative neoepitopes on LDL and phospholipids modulate the development of atherosclerosis. J. Intern. Med. **247:** 371–380.
50. BINDER, C.J., M.K. CHANG, P.X. SHAW *et al.* 2002. Innate and acquired immunity in atherogenesis. Nat. Med. **8:** 1218–1226.
51. SALVAYRE, R., N. AUGE, H. BENOIST & A. NEGRE-SALVAYRE. 2002. Oxidized low-density lipoprotein–induced apoptosis. Biochim. Biophys. Acta **1585:** 213–221.
52. HANSSON, G.K., Y.J. GENG, J. HOLM *et al.* 1994. Arterial smooth muscle cells express nitric oxide synthase in response to endothelial injury. J. Exp. Med. **180:** 733–738.
53. GENG, Y.J., A.S. PETERSSON, A. WENNMALM & G.K. HANSSON. 1994. Cytokine-induced expression of nitric oxide synthase results in nitrosylation of heme and nonheme iron proteins in vascular smooth muscle cells. Exp. Cell Res. **214:** 418–428.
54. IGNARRO, L.J., G. CIRINO, A. CASINI & C. NAPOLI. 1999. Nitric oxide as a signaling molecule in the vascular system: an overview. J. Cardiovasc. Pharmacol. **34:** 876–884.
55. NAPOLI, C. & L.J. IGNARRO. 2001. Nitric oxide and atherosclerosis. Nitric Oxide **5:** 88–97.
56. PATEL, R.P., D. MOELLERING, J. MURPHY-ULLRICH *et al.* 2000. Cell signaling by reactive nitrogen and oxygen species in atherosclerosis. Free Radical Biol. Med. **28:** 1780–1794.
57. DE NIGRIS, F., L.O. LERMAN, S. WILLIAMS-IGNARRO *et al.* 2003. Beneficial effects of antioxidants and L-arginine on oxidation-sensitive gene expression and endothelial nitric oxide synthase activity at the sites of disturbed shear stress. Proc. Natl. Acad. Sci. USA **100:** 1420–1425.
58. DIMMELER, S., V. RIPPMANN, U. WEILAND *et al.* 1997. Angiotensin II induces apoptosis of human endothelial cells: protective effect of nitric oxide. Circ. Res. **81:** 970–976.
59. NAPOLI, C. & L.J. IGNARRO. 2003. Nitric oxide–releasing drugs. Annu. Rev. Pharmacol. Toxicol. **47:** 97–123.
60. NAPOLI, C., E. ACKAH, F. DE NIGRIS *et al.* 2002. Chronic treatment with nitric oxide–releasing aspirin inhibits plasma LDL oxidation and oxidative stress, oxidation-specific epitopes in the arterial wall, and atherogenesis in hypercholesterolemic mice. Proc. Natl. Acad. Sci. USA **99:** 12467–12470.
61. PIKE, A.C., A.M. BRZOZOWSKI, S.M. ROBERTS *et al.* 1999. Structure of human factor VIIa and its implications for the triggering of blood coagulation. Proc. Natl. Acad. Sci. USA **96:** 8925–8930.
62. TOSCHI, V., G. GALLO, M. LETTINO *et al.* 1997. Tissue factor modulates thrombogenicity of human atherosclerotic plaques. Circulation **95:** 594–599.
63. BADIMON, J.J., M. LETTINO, V. TOSCHI *et al.* 1999. Local inhibition of tissue factor reduces the thrombogenicity of disrupted human atherosclerotic plaques: effects of tissue factor pathway inhibitor on plaque thrombogenicity under flow conditions. Circulation **99:** 1780–1787.
64. BACH, R. & D.B. RIFKIN. 1990. Expression of tissue factor procoagulant activity: regulation by cytosolic calcium. Proc. Natl. Acad. Sci. USA **87:** 6995–6999.
65. MALLAT, Z., B. HUGEL, J. OHAN *et al.* 1999. Shed membrane microparticles with procoagulant potential in human atherosclerotic plaques: a role for apoptosis in plaque thrombogenicity. Circulation **99:** 348–353.
66. VAN DER WAL, A.C., A.E. BECKER, C.M. VAN DER LOOS & P.K. DAS. 1994. Site of intimal rupture or erosion of thrombosed coronary atherosclerotic plaques is characterized by an inflammatory process irrespective of the dominant plaque morphology. Circulation **89:** 36–44.

67. KOLODGIE, F.D., J. NARULA, A.P. BURKE *et al.* 2000. Localization of apoptotic macrophages at the site of plaque rupture in sudden coronary death. Am. J. Pathol. **157:** 1259–1268.
68. BAURIEDEL, G., R. HUTTER, U. WELSCH *et al.* 1999. Role of smooth muscle cell death in advanced coronary primary lesions: implications for plaque instability. Cardiovasc. Res. **41:** 480–488.
69. FARB, A., A.P. BURKE, A.L. TANG *et al.* 1996. Coronary plaque erosion without rupture into a lipid core: a frequent cause of coronary thrombosis in sudden coronary death. Circulation **93:** 1354–1363.
70. BURKE, A.P., A. FARB, G.T. MALCOM *et al.* 1997. Coronary risk factors and plaque morphology in men with coronary disease who died suddenly. N. Engl. J. Med. **336:** 1276–1288.
71. TRICOT, O., Z. MALLAT, C. HEYMES *et al.* 2000. Relation between endothelial cell apoptosis and blood flow direction in human atherosclerotic plaques. Circulation **101:** 2450–2453.
72. LEDRU, F., P. THEROUX, J. LESPERANCE *et al.* 1999. Geometric features of coronary artery lesions favoring acute occlusion and myocardial infarction: a quantitative angiographic study. J. Am. Coll. Cardiol. **33:** 1353–1361.
73. SCHAUB, F.J., D.K. HAN, W.C. LILES *et al.* 2000. Fas/FADD-mediated activation of a specific program of inflammatory gene expression in vascular smooth muscle cells. Nat. Med. **6:** 790–796.
74. POLLMAN, M.J., J.L. HALL, M.J. MANN *et al.* 1998. Inhibition of neointimal cell bcl-x expression induces apoptosis and regression of vascular disease. Nat. Med. **4:** 222–227.
75. KAISER, D., M.A. FREYBERG & P. FRIEDL. 1997. Lack of hemodynamic forces triggers apoptosis in vascular endothelial cells. Biochem. Biophys. Res. Commun. **231:** 586–590.
76. KUBO, N., J. KIKUCHI, Y. FURUKAWA *et al.* 1997. Regulatory effects of aggregated LDL on apoptosis during foam cell formation of human peripheral blood monocytes. FEBS Lett. **409:** 177–182.
77. BENNETT, M.R., G.I. EVAN & S.M. SCHWARTZ. 1995. Apoptosis of human vascular smooth muscle cells derived from normal vessels and coronary atherosclerotic plaques. J. Clin. Invest. **95:** 2266–2274.
78. KATOH, O., H. TAUCHI, K. KAWAISHI *et al.* 1995. Expression of the vascular endothelial growth factor (VEGF): effect of VEGF on apoptotic cell death caused by ionizing radiation. Cancer Res. **55:** 5687–5692.
79. FOX, J.C. & J.R. SHANLEY. 1996. Antisense inhibition of basic fibroblast growth factor induces apoptosis in vascular smooth muscle cells. J. Biol. Chem. **271:** 12578–12584.
80. ARAKI, S., Y. SIMADA, K. KAJI & H. HAYASHI. 1990. Role of protein kinase C in the inhibition by fibroblast growth factor of apoptosis in serum-depleted endothelial cells. Biochem. Biophys. Res. Commun. **172:** 1081–1085.
81. SIOW, R.C., H. SATO, D.S. LEAKE *et al.* 1999. Induction of antioxidant stress proteins in vascular endothelial and smooth muscle cells: protective action of vitamin C against atherogenic lipoproteins. Free Radical Res. **31:** 309–318.
82. LIZARD, G., S. GUELDRY, O. SORDET *et al.* 1998. Glutathione is implied in the control of 7-ketocholesterol-induced apoptosis, which is associated with radical oxygen species production. FASEB J. **12:** 1651–1663.
83. LI, P.F., R. DIETZ & R. VON HARSDORF. 1997. Differential effect of hydrogen peroxide and superoxide anion on apoptosis and proliferation of vascular smooth muscle cells. Circulation **96:** 3602–3609.
84. EVAN, G. & T. LITTLEWOOD. 1998. A matter of life and cell death. Science **281:** 1317–1322.
85. NAPOLI, C., L.O. LERMAN, F. DE NIGRIS & V. SICA. 2002. c-Myc oncoprotein: a dual pathogenic role in neoplasia and cardiovascular diseases? Neoplasia **4:** 185–190.
86. DE NIGRIS, F., T. YOUSSEF, S. CIAFRE *et al.* 2000. Evidence for oxidative activation of c-Myc-dependent nuclear signaling in human coronary smooth muscle cells and in early lesions of Watanabe heritable hyperlipidemic rabbits: protective effects of vitamin E. Circulation **102:** 2111–2117.
87. DE NIGRIS, F., L.O. LERMAN, M. RODRIGUEZ-PORCEL *et al.* 2001. c-Myc activation in early coronary lesions in experimental hypercholesterolemia. Biochem. Biophys. Res. Commun. **241:** 945–950.

88. BENNETT, M.R., G.I. EVAN & A.C. NEWBY. 1994. Deregulated expression of the c-myc oncogene abolishes inhibition of proliferation of rat vascular smooth muscle cells by serum reduction, interferon-gamma, heparin, and cyclic nucleotide analogues and induces apoptosis. Circ. Res. **74:** 525–536.
89. VON DER THUSEN, J.H., B.J. VAN VLIJMEN, R.C. HOEBEN *et al.* 2002. Induction of atherosclerotic plaque rupture in apolipoprotein E–/– mice after adenovirus-mediated transfer of p53. Circulation **105:** 2064–2070.
90. CHANG, M.K., C.J. BINDER, M. TORZEWSKI & J.L. WITZTUM. 2002. C-reactive protein binds to both oxidized LDL and apoptotic cells through recognition of a common ligand: phosphorycoline of oxidized phospholipids. Proc. Natl. Acad. Sci. USA **99:** 13043–13048.
91. SUZUKI, J., M. ISOBE, R. MORISHITA *et al.* 2000. Antisense Bcl-x oligonucleotide induces apoptosis and prevents arterial neointimal formation in murine cardiac allografts. Cardiovasc. Res. **45:** 783–787.
92. NAPOLI, C., F. DE NIGRIS & W. PALINSKI. 2001. Multiple role of reactive oxygen species in the arterial wall. J. Cell. Biochem. **82:** 674–682.
93. DE NIGRIS, F., L.O. LERMAN & C. NAPOLI. 2002. New insights in the transcriptional activity and coregulator molecules in the arterial wall. Int. J. Cardiol. **86:** 153–168.
94. SCHROEPFER, G.J., JR. 2000. Oxysterols: modulators of cholesterol metabolism and other processes. Physiol. Rev. **80:** 361–354.
95. KITADA, S., S. KRAJEWSKI, T. MIYASHITA *et al.* 1996. Gamma-radiation induces upregulation of Bax protein and apoptosis in radiosensitive cells *in vivo*. Oncogene **12:** 187–192.
96. MATTURRI, L., A. CAZZULLO, P. TURCONI *et al.* 2000. Inflammatory cells, apoptosis, and *Chlamydia pneumoniae* infection in atherosclerosis. Int. J. Cardiol. **75:** 23–33.
97. JEREMY, J.Y., D. ROWE, A.M. EMSLEY & A.C. NEWBY. 1999. Nitric oxide and the proliferation of vascular smooth muscle cells. Cardiovasc. Res. **43:** 580–594.
98. NEWBY, A.C. & S.J. GEORGE. 1996. Proliferation, migration, matrix turnover, and death of smooth muscle cells in native coronary and vein graft atherosclerosis. Curr. Opin. Cardiol. **11:** 574–582.
99. REINMUTH, N., W. LIU, Y.D. JUNG *et al.* 2001. Induction of VEGF in perivascular cells defines a potential paracrine mechanism for endothelial cell survival. FASEB J. **15:** 1239–1241.
100. KONDO, S., Y. KONDO, D. YIN *et al.* 1996. Involvement of interleukin-1β-converting enzyme in apoptosis of bFGF-deprived murine aortic endothelial cells. FASEB J. **10:** 1192–1197.
101. BUERKE, M., T. MUROHARA, C. SKURK *et al.* 1995. Cardioprotective effect of insulin-like growth factor I in myocardial ischemia followed by reperfusion. Proc. Natl. Acad. Sci. USA **92:** 8031–8035.
102. TEWARI, M., L.T. QUAN, K. O'ROURKE *et al.* 1995. Yama/CPP32 beta, a mammalian homolog of CED-3, is a CrmA-inhibitable protease that cleaves the death substrate poly(ADP-ribose) polymerase. Cell **81:** 801–809.
103. DATTA, R., D. BANACH, H. KOJIMA *et al.* 1996. Activation of the CPP32 protease in apoptosis induced by 1-beta-D-arabinofuranosylcytosine and other DNA-damaging agents. Blood **88:** 1936–1943.
104. WANG, C.Y., M.W. MAYO, R.G. KORNELUK *et al.* 1998. NF-kappaB antiapoptosis: induction of TRAF1 and TRAF2 and c-IAP1 and c-IAP2 to suppress caspase-8 activation. Science **281:** 1680–1683.
105. SIOW, R.C., J.P. RICHARDS, K.C. PEDLEY *et al.* 1999. Vitamin C protects human vascular smooth muscle cells against apoptosis induced by moderately oxidized LDL containing high levels of lipid hydroperoxides. Arterioscler. Thromb. Vasc. Biol. **19:** 2387–2394.
106. DE NIGRIS, F., A. LERMAN, L.J. IGNARRO *et al.* 2003. Oxidation-sensitive mechanisms, vascular apoptosis, and atherosclerosis. Trends Mol. Med. **9:** 351–359.

Glycoxidation of Low-Density Lipoprotein Increases TUNEL Positivity and CPP32 Activation in Human Coronary Cells

FILOMENA DE NIGRIS,[a] GIANFRANCO TAJANA,[a] MARIO CONDORELLI,[b] FRANCESCO P. D'ARMIENTO,[b] GIACOMO SICA,[b] LILACH O. LERMAN,[c] AND CLAUDIO NAPOLI[b]

[a]*Department of Pharmacological Sciences, Chair of Anatomy, Faculty of Pharmacy, University of Salerno, Fisciano-Salerno, Italy*

[b]*Departments of Medicine and Human Pathology and Clinical Pathology, University of Naples, Naples, Italy*

[d]*Department of Medicine, Division of Hypertension, Mayo Clinic, Rochester, Minnesota, USA*

ABSTRACT: Apoptosis of arterial cells induced by oxidized low-density lipoprotein (oxLDL) is thought to contribute to the progression of vascular dysfunction and atherogenesis. It is well established that diabetes mellitus is accompanied by both glycosylation and oxidation of LDL (glc-oxLDL), but the biological effects of these modified lipoproteins are poorly understood. We demonstrate here for the first time that glc-oxLDL increases TUNEL positivity and caspase-3 activation (by Western blot and immunocytochemistry) of human coronary smooth muscle cells. Overall, these effects induced by glc-oxLDL were greater than those achieved with oxLDL. Thus, glc-oxLDL activated downstream apoptotic signaling. This may influence the evolution of atherogenesis and vascular complications in diabetes.

KEYWORDS: atherosclerosis; glc-oxLDL; oxLDL; caspase

INTRODUCTION

Oxidation of low-density lipoprotein (oxLDL) is a complex pathophysiological process (reviewed in ref. 1). The presence of oxLDL is detected in early human atherogenesis.[2,3] A plethora of experiments demonstrated that oxLDL induces apoptosis in arterial cells (reviewed in ref. 4). The molecular mechanism of apoptosis in arterial cells involves the activation of endogenous proteases and several programmed death signals.[5–8]

Early studies have well established that proteins exposed to elevated glucose concentration may undergo glycosylation in diabetes (reviewed in ref. 9). There is

Address for correspondence: F. de Nigris, Ph.D., Department of Pharmacological Sciences, Chair of Anatomy, Faculty of Pharmacy, University of Salerno, 84084 Fisciano-Salerno, Italy. fnigris@yahoo.com

Ann. N.Y. Acad. Sci. 1010: 710–715 (2003).
doi: 10.1196/annals.1299.128

now evidence that glycosylated LDL (glc-LDL) and glycoxidized-oxidized LDL (glc-oxLDL) are present in human atherosclerotic plaques.[10] Moreover, glc-oxLDL downregulates eNOS in coronary cells even more potently than glc-LDL,[11] suggesting that endothelial dysfunction in diabetic patients may be related to glc-oxLDL-mediated alteration of signaling. However, the possible role of glc-oxLDL in arterial cell apoptosis remains incompletely understood. Therefore, the goal of the present study was to explore whether glc-oxLDL can promote apoptosis and activate caspase-3 (CPP32) in human coronary vascular smooth muscle cells (VSMC) and whether these effects are amplified compared to glc-LDL alone.

MATERIALS AND METHODS

Preparation of Human Glc-LDL and Oxidation

LDL was isolated by ultracentrifugation as described.[12] Protein content was assayed by Lowry assay.[13] Glc-LDL was prepared by incubating native LDL (n-LDL) with 80 mM glucose in sterile phosphate buffer containing 2 mM EDTA, aprotinin (1000 IU/100 mL), 100 mM deferoxamine, and 0.009% gentamycin for 10 days at 37°C as described previously.[11,14] The extent of glycosylation achieved was typically 10–12% measured immunoenzymatically.[14] Native and glc-LDL (300 μg/mL) were incubated for 18 h at 37°C with 1 μM of the oxidant copper sulfate.[14]

Cell Culture

VSMCs from human coronary arteries were cultured using standard procedures.[5–8] Cells were incubated at 37°C in a humidified atmosphere of 95% air and 5% CO_2 for 24 h for each treatment.

TUNEL Assay and Immunocytochemistry for CPP32 Protein

Apoptosis in cultured cells was assessed by the "*In Situ* Cell Death Detection Kit" (Boehringer) as described in detail.[5–8] Immunocytochemistry for the caspase-3 (CPP32) protein was done using a monoclonal antibody against CPP32 (H-277) (Santa Cruz Biotechnology, San Diego, CA) as previously described.[8]

Western Blot Analysis of CPP32

Whole-cell extracts were prepared by a modification of the standard procedure[5–8] and hybridized with antibodies against CPP32 (H-277). Membranes were normalized with a polyclonal antibody against γ-tubulin protein (Sigma). Semiquantitative densitometry of blots was done using a Scan LKB (Pharmacia, Sweden).[5–8]

Flow Cytometric Analysis

Cells were analyzed by a FACScan flow cytometer (FACS Vantage™, Becton Dickinson, San Jose, CA) interfaced with a Hewlett Packard computer as described in detail.[5–8]

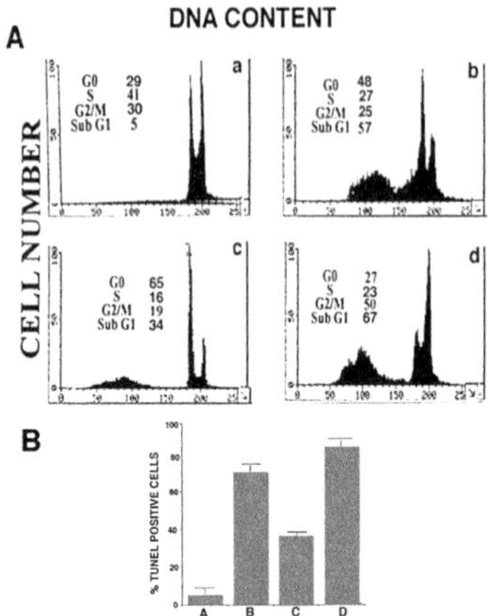

FIGURE 1. **(A)** A representative experiment on the apoptotic effects of 24-h exposure to oxLDL, glc-LDL, and glc-oxLDL on human coronary VSMCs tested by FACScan analysis: a, native LDL; b, oxLDL; c, glc-LDL; d, glc-oxLDL. **(B)** Percentage of TUNEL-positive cells after treatment with native LDL (A), ox-LDL (B), glc-LDL (C), and glc-oxLDL (D). C vs. A, $P < 0.001$; D vs. C, $P < 0.01$. Data are the mean ± SD of 4 different experiments.

Statistical Analysis

Data are expressed as the mean ± SEM or SD. Differences among the groups were tested by one-way ANOVA, followed by Bonferroni's corrected *t* test, with significance defined at a value of $P < 0.05$. Correlations were evaluated by the least-squares linear regression analysis.

RESULTS

Effects of Glc-LDL and Glc-oxLDL on TUNEL Assay and FACS Analysis

Human VSMCs were incubated for 24 h with glc-LDL, glc-oxLDL, n-LDL, or oxLDL and subjected to FACS analysis, which allows detection of cellular content of DNA (FIG. 1A). After incubation with glc-LDL, about 34% of the cells were present in the sub-G1 phase, showing fragmented DNA (FIG. 1A, panel c), whereas glc-oxLDL induced a significantly greater increase in the sub-G1 fraction to 67% (FIG. 1A, panel d). Similar results were obtained with TUNEL assay (FIG. 1B). The TUNEL-positive cells were approximately 38% for glc-LDL and 88% for glc-oxLDL, significantly higher than the results obtained with n-LDL and ox-LDL.

Activation of CPP32

In order to verify whether glc-oxLDL stimulated also the activation of CPP32, Western blots were performed on total protein extracts from cells treated with oxLDL, glc-LDL, and glc-oxLDL (FIG. 2A, top). Densitometry (FIG. 2A, bottom)

TABLE 1. Percentage of immunocytochemical CPP32-positive human coronary VSMCs after exposure to different forms of lipoproteins

Lipoprotein	%
n-LDL	1.1 ± 0.9
glc-LDL	3.8 ± 1.8[a]
oxLDL	23.6 ± 4.9[b,c]
glc-oxLDL	38.5 ± 8.5[b,c,d]

NOTE: Mean ± SD of 6 experiments. Terms: n-LDL, native LDL; glc-LDL, glycosylated LDL; oxLDL, oxidized LDL; glc-oxLDL, glycosylated and oxidized LDL. Statistics: [a]$P < 0.05$ vs. n-LDL; [b]$P < 0.0001$ vs. n-LDL; [c]$P < 0.001$ vs. glc-LDL; [d]$P < 0.05$ vs. oxLDL.

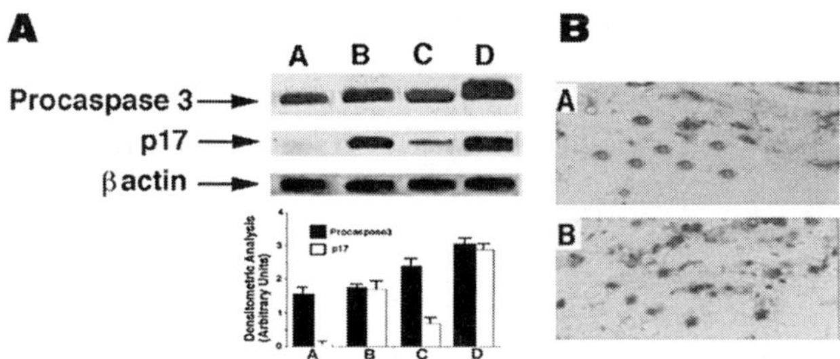

FIGURE 2. **(A)** *Top*: Procaspase-3 and active caspase subunit p17 generated by proteolytic cleavage were determined by Western blot as described in MATERIALS AND METHODS. Lanes: A, native LDL; B, oxLDL; C, glc-LDL; D, glc-oxLDL. β-Actin served as reference protein. *Bottom*: Densitometric analysis. Bars represent the mean ± SD of 3 separate experiments. A: native LDL; B: oxLDL; C: glc-LDL; D: glc-oxLDL. B vs. A, $P < 0.0001$; D vs. C, $P < 0.01$; D vs. B, $P < 0.05$; C vs. A, $P < 0.05$. **(B)** Immunocytochemical results using antibody against CPP32 subunit p17 after treatment with oxLDL (A) and glc-oxLDL (B).

showed that glc-oxLDL stimulated the activation of catalytic subunit p17 of CPP32 more potently than oxLDL.

We did not detect p17 activation by n-LDL in protein extracts, whereas glc-LDL per se did induce p17 activation (FIG. 2A, lane C). These results were comparable to those obtained by immunocytochemistry (FIG. 2B). TABLE 1 shows the percentage of immunocytochemical CPP32-positive cells after exposure to different forms of lipoproteins and demonstrates a higher degree of CPP32 expression elicited by glc-oxLDL.

DISCUSSION

The present study demonstrates that glc-oxLDL potently stimulates the fragmentation of DNA, as assessed by TUNEL and fluorescent DNA scanning, and potentiates the activation of CPP32 in human coronary VSMCs. TUNEL staining and increased protein expression of the active subunit of CPP32 are commonly used as indicators of programmed cell death. These measurements obtained in our study indicated that glc-oxLDL induces apoptosis in VSMCs more potently than oxLDL. Indeed, recent studies suggest that apoptosis is increased in several diseases, including diabetes,[4] and TUNEL-positive cells were observed in diabetic rat aortas.[15] Thus, our *in vitro* findings are supported by similar *in vivo* findings in a model of diabetes. Consistent with previous observations,[8] after cell exposure to oxLDL, the percentage of CPP32-positive cells that we observed was lower than TUNEL-positive cells. *In vivo* apoptosis seems to occur only in transitional and more advanced lesions rather than in fatty streaks, and the susceptibility of arterial cells to apoptosis is variable.[3]

Glycosylation and oxidation of LDL may have a profound impact on multiple regulatory apoptotic pathways in human VSMCs. Increasing evidence shows a complex network of signaling events in the arterial wall.[16–18] Thus, additional experiments are necessary to understand which are the molecular pathways and genes that are involved in glc-oxLDL-induced VSMC apoptosis. Furthermore, oxLDL can also induce cellular lysis and necrosis independently of apoptosis and CPP32,[19] and additional studies will be needed to explore whether glc-oxLDL also potentiates this deleterious pathway. If these mechanisms play a similar pathogenic role *in vivo*, this cascade of events may in turn influence vascular diabetic complications and progression of atherosclerosis.

REFERENCES

1. STEINBERG, D. 1997. Low density lipoprotein oxidation and its pathobiological significance. J. Biol. Chem. **272:** 20963–20966.
2. NAPOLI, C., F.P. D'ARMIENTO, F.P. MANCINI *et al.* 1997. Fatty streak formation occurs in human fetal aortas and is greatly enhanced by maternal hypercholesterolemia: intimal accumulation of LDL and its oxidation precede monocyte recruitment into early atherosclerotic lesions. J. Clin. Invest. **100:** 2680–2690.
3. NAPOLI, C., C.K. GLASS, J.L. WITZTUM *et al.* 1999. Influence of maternal hypercholesterolemia during pregnancy on progression of early atherosclerotic lesions in childhood: Fate of Early Lesions in Children (FELIC) Study. Lancet **354:** 1234–1241.
4. SALVAYRE, R., N. AUGE, N. BENOIST & A.N. SALVAYRE. 2002. Oxidized low-density lipoprotein-induced apoptosis. Biochim. Biophys. Acta **1585:** 213–221.
5. NAPOLI, C., O. QUEHENBERGER, F. DE NIGRIS *et al.* 2000. Mildly oxidized low-density lipoprotein activates multiple apoptotic signaling pathways in human coronary cells. FASEB J. **14:** 1996–2007.
6. BJORKERUD, B. & S. BJORKERUD. 1996. Contrary effects of lightly and strongly oxidized LDL with potent promotion of growth versus apoptosis on arterial smooth muscle cells and fibroblasts. Arterioscler. Thromb. Vasc. Biol. **16:** 416–424.
7. JOVINGE, S., M. CRISBY, J. THYBERGAND & J. NILSSON. 1997. DNA fragmentation and ultrastructural changes of degenerating cells in atherosclerotic lesions on smooth muscle cells exposed to oxidized LDL *in vitro*. Arterioscler. Thromb. Vasc. Biol. **17:** 2225–2231.
8. DE NIGRIS, F., F. FRANCONI, I. MAIDA *et al.* 2000. Modulation by α- and γ-tocopherol of the signaling induced by copper oxidized LDL in human coronary smooth muscle cells. Biochem. Pharmacol. **59:** 1477–1487.
9. KENNEDY, L. & J.W. BAYNES. 1984. Nonenzymatic glycosylation and the chronic complications of diabetes: an overview. Diabetologia **26:** 93–98.
10. IMANAGA, Y., N. SAKATA, S. TAKEBAYASHI *et al.* 2000. *In vivo* and *in vitro* evidence for the glycoxidation of low density lipoprotein in human atherosclerotic plaques. Atherosclerosis **150:** 343–355.
11. NAPOLI, C., L.O. LERMAN, F. DE NIGRIS *et al.* 2002. Glycoxidized low-density lipoprotein downregulates endothelial nitric oxide synthase in human coronary cells. J. Am. Coll. Cardiol. **40:** 1516–1521.

12. NAPOLI, C., F.P. MANCINI, G. CORSO *et al.* 1997. A simple and rapid purification procedure minimizes spontaneous oxidative modifications of low density lipoprotein and lipoprotein (a). J. Biochem. **121:** 1096–1101.
13. LOWRY, O.H., H.J. ROSEBROUGH, A.L. FARR & R.J. RANDALL. 1951. Protein measurement with the folin-phenol reagent. J. Biol. Chem. **193:** 265–275.
14. NAPOLI, C., M. TRIGGIANI, G. PALUMBO *et al.* 1997. Glycosylation enhances oxygen radical–induced modifications and decreases acetylhydrolase activity of human low density lipoprotein. Basic Res. Cardiol. **92:** 96–105.
15. CHU, Y., F.M. FARACI, H. OOBOSHI & D.D. HEISTAD. 1997. Increase in TUNEL positive cells in aorta from diabetic rats. Endothelium **5:** 241–250.
16. NAPOLI, C., F. DE NIGRIS & W. PALINSKI. 2001. Multiple role of reactive oxygen species in the arterial wall. J. Cell. Biochem. **82:** 674–682.
17. DE NIGRIS, F., L.O. LERMAN, M. CONDORELLI *et al.* 2001. Oxidation-sensitive transcription factors and molecular mechanisms in the arterial wall. Antioxidant Redox Signal. **3:** 1119–1130.
18. DE NIGRIS, F., A. LERMAN, L.J. IGNARRO *et al.* 2003. Oxidation-sensitive mechanisms, vascular apoptosis, and atherosclerosis. Trends Mol. Med. **9:** 351–359.
19. ASMIS, R. & J.G. BEGLEY. 2003. Oxidized LDL promotes peroxide-mediated mitochondrial dysfunction and cell death in human macrophages: a caspase-3-independent pathway. Circ. Res. **92:** E20–E29.

TNFα Stimulated ATP-Sensitive Potassium Channels and Attenuated Deoxyglucose and Ca Uptake of H9c2 Cardiomyocytes

D. EL-ANI AND R. ZIMLICHMAN

Cardiovascular and Hypertension Research Laboratory and Department of Medicine, Institute of Physiologic Hygiene, Wolfson Medical Center, Holon and Tel-Aviv University School of Medicine, Tel-Aviv, Israel

ABSTRACT: Tumor necrosis factor (TNFα) is an inflammatory cytokine that induces programmed cell death in a variety of tissue types, including the heart. Recent experimental data suggest that the TNF expressed within the myocardium in response to environmental injury plays an important role in initiating homeostatic response. The effect of TNFα (10–50 ng/mL) was studied on ^{86}Rb efflux, ^{3}H-deoxyglucose uptake, or ^{45}Ca uptake in H9c2 cardiomyocytes. TNFα stimulated ^{86}Rb efflux from cultures, while 2 μM glibenclamide blocker of ATP-sensitive potassium channels or 20 μM zvad-fmk caspase inhibitor attenuated this effect. TNFα also depressed the stimulatory effect of 80 mM KCl on ^{45}Ca uptake in the cardiomyocytes. TNFα inhibited the stimulatory effect of 100 nM insulin on ^{3}H-deoxyglucose uptake. Our findings further suggest that TNFα mediated adaptive and protective effects in the heart during a brief environmental injury.

KEYWORDS: TNF; H9c2; cardiomyocytes; ATP-sensitive potassium channels; glibenclamide; zvad-fmk; Ca uptake; deoxyglucose uptake; preconditioning

INTRODUCTION

Tumor necrosis factor (TNFα) is an inflammatory cytokine that induces programmed cell death in a variety of tissue types, including the heart. Recent experimental data suggest that the TNF expressed within the myocardium in response to environmental injury plays an important role in initiating homeostatic response.

TNFα biosynthesis within the myocardium following stress (like ischemia or hypoxia) occurs within 60 min of the stressful stimulus.[1] TNFα stimulation has been shown to reduce LDH (lactate dehydrogenase) release in Langendorff-perfused rat hearts that had been subjected to global ischemia or hypoxia.[1,2]

TNF depressed myocardial contractile function by disrupting calcium release by the sarcoplasmic reticulum. TNF disrupts L-type channel–induced calcium influx and thereby depresses calcium transients.[1]

Address for correspondence: D. El-Ani, NVR Labs, 17 Haharash Street, Ness Ziona 74031, Israel. Voice: +972-89301099 ext. 204; fax: +972-89302662.
daliae@nvrlab.com

Ann. N.Y. Acad. Sci. 1010: 716–720 (2003). © 2003 New York Academy of Sciences.
doi: 10.1196/annals.1299.129

ATP-sensitive K+ channels (K_{ATP}) and mitogen-activated protein kinases (MAPK) are involved in regulation of coronary blood flow. Activation of K_{ATP} channels in smooth muscle during ischemia results in marked coronary vasodilation, and endothelial K_{ATP} channels might contribute to the regulation of vascular tone via NO release. K_{ATP} channels reduce action potential duration, decrease contractility, and conserve energy during periods of ischemia.[1]

The present study was performed in H9c2 cardiomyocytes. H9c2 is a cardiac cell line that is characterized biochemically and electrophysiologically by membrane currents under voltage-clamp conditions and signal transduction G proteins.[3] The effect of TNF (10–50 ng/mL) was studied on ^{86}Rb efflux, ^{3}H-deoxyglucose uptake, or ^{45}Ca uptake in H9c2 cardiomyocytes.

MATERIALS AND METHODS

Materials

H9c2 cells were purchased from the American Tissue Culture Collection (ATCC, Rockville, MD). Cell culture reagents and phosphate-buffered saline (PBS) were obtained from GIBCO (PBS with 0.1 mM $CaCl_2$, 0.1 mM $MgCl_2$, and 10 mM HEPES, pH 7). Zvad-fmk and human TNFα were obtained from Calbiochem. Insulin and glibenclamide were obtained from Sigma.

2-Deoxyglucose, ^{86}Rb, and ^{45}Ca were purchased from Amersham Biosciences.

Cell Culture

H9c2 cells were grown in DMEM containing 10% fetal bovine serum (FBS) and pencillin-streptomycin in 5% CO_2 at 37°C. Experiments were carried out in cells at passage 30 or lower. Cells were seeded at 10^6 cells/well in 6 wells for ^{86}Rb efflux or deoxyglucose uptake; 10×10^6 cells were seeded in 100-mm dishes for ^{45}Ca uptake experiments.

Uptake of 2-Deoxy-(^{3}H)glucose

Cell cultures were incubated in PBS containing 1 μCi/mL 2-deoxy-(^{3}H)glucose in the presence or absence of 100 nM insulin or 10 ng/mL TNFα for 20–25 min at 37°C.

Uptake of ^{45}Ca

Cell cultures were preincubated for 10 min in histidine sucrose buffer (300 mM sucrose, 30 mM histidine, 4 mM KCl, 1 mM $CaCl_2$, pH 7.4). Then, medium was replaced with fresh buffer containing 2 μCi/mL ^{45}Ca. Cells were exposed to 80 mM KCl for depolarization or 20 ng/mL TNFα (or both). ^{45}Ca uptake was terminated after 0.5 min.

At the end of the incubation in ^{45}Ca or 2-deoxy-(^{3}H)glucose uptake experiments, cultures were rinsed 5 times with cold PBS (4°C). Nonspecific binding of ^{45}Ca or 2-deoxy-(^{3}H)glucose was detected in cultures that were preincubated on ice for 10 min in cold buffer (4°C).

^{86}*Rb Efflux*

Cultures were preincubated in cell medium containing 2 μCi/mL ^{86}Rb at 37°C in 5% CO_2 for 21–24 h. Efflux experiments were performed at room temperature. Cultures were rinsed 7 times in PBS to remove nonspecific binding of ^{86}Rb. Then, 1.5 mL PBS containing 50 ng/mL TNFα, 2 μM glibenclamide, 20 μM zvad-fmk, or various combinations of the drugs was added to cultures for 30 min of efflux. The efflux liquids were collected from each well and transferred to liquid scintillation tubes. For determination of the total influx to the cells, cell cultures were collected to scintillation tubes. Both influx or efflux tubes were detected for Cherenkov radioactivity in a liquid scintillation counter.

^{86}Rb effluxes were calculated for each well as percentages from the total (influx + efflux).

RESULTS

^{86}Rb efflux from H9c2 cardiomyocytes increased from 30.77% ± 0.75% to 38.51% ± 0.183% following 30 min of 50 ng/mL TNF application (t test, $P < 0.05$) (FIG. 1). Glibenclamide, an ATP potassium channel blocker, inhibited the effect of TNF: 29.25% ± 0.56% ^{86}Rb efflux was detected in H9c2 cells treated with TNF and glibenclamide (t test, $P < 0.05$) (FIG. 1). Caspase inhibitor zvad-fmk also inhibited ^{86}Rb efflux from 38.51% ± 0.183% in TNF treatment to 30.11% ± 0.2% (t test, $P < 0.05$) (FIG. 1). TNF accelerated ^{86}Rb efflux from 67.58% ± 0.44% to 71.23% ± 0.41% (t test, $P < 0.01$) following 2.5 h of exposure to 50 ng/mL TNFα (data not shown).

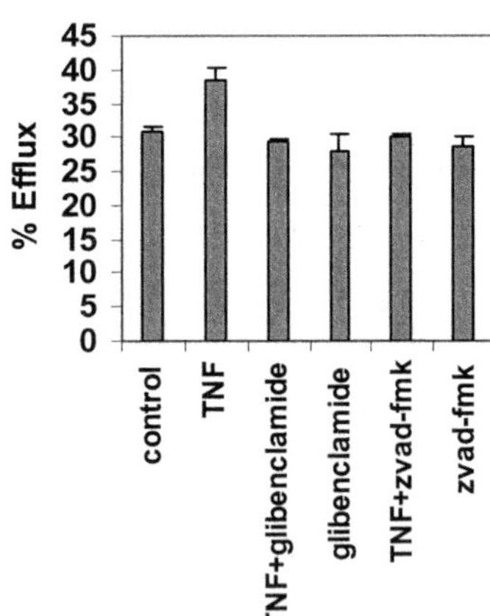

FIGURE 1. The effect of TNFα on ^{86}Rb efflux in H9c2 cardiomyocytes. Cultures were preincubated in medium containing 2 μCi/mL ^{86}Rb for 21–24 h. Cells were rinsed 7 times in PBS to remove nonspecific binding of ^{86}Rb. Then, 1.5 mL PBS containing 50 ng/mL TNFα, 2 μM glibenclamide, 20 μM zvad-fmk, or various combinations of the drugs was added to cultures for 30 min of efflux. ^{86}Rb efflux was calculated for each well as a percentage from the total (influx + efflux). Data represent the mean ± SD.

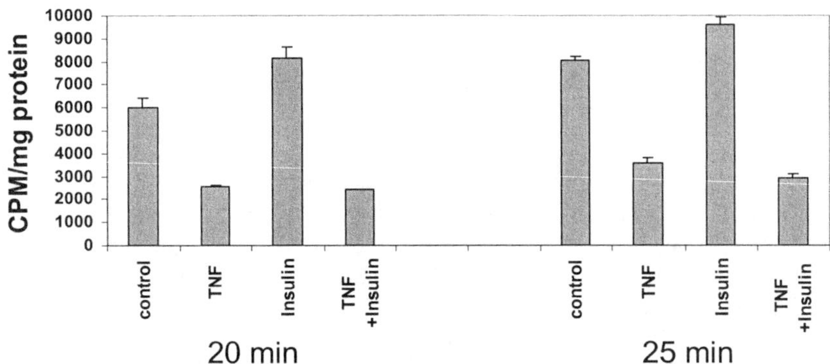

FIGURE 2. The effect of TNFα on ^{3}H-deoxyglucose uptake in H9c2 cardiomyocytes. Cell cultures were incubated in PBS containing 1 μC/mL 2-deoxy-(^{3}H)glucose in the presence or absence of 100 nM insulin or 10 ng/mL TNFα for 20–25 min at 37°C. Data represent the mean ± SD.

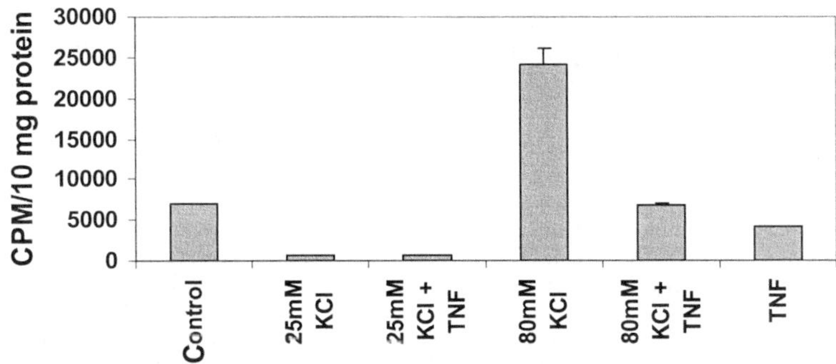

FIGURE 3. The effect of TNFα on ^{45}Ca uptake. Cell cultures were preincubated for 10 min in histidine sucrose buffer. Then, medium was replaced with fresh buffer containing 2 μCi/mL ^{45}Ca. Cells were exposed to 80 mM KCl for depolarization or 20 ng/mL TNFα (or both). ^{45}Ca uptake was terminated after 0.5 min. Data represent the mean ± SD.

In H9c2 cultures, 2-deoxy-(^{3}H)glucose uptake decreased from 5996 ± 410 to 2549 ± 79 cpm/mg protein (*t* test, $P < 0.01$) following 20-min exposure to 10 ng/mL TNFα (FIG. 2). During 20 min of 2-deoxy-(^{3}H)glucose uptake, TNF attenuated the stimulatory effect of 100 nM insulin: from 8178 ± 477 to 2412 ± 10 cpm/mg protein (*t* test, $P < 0.01$) (FIG. 2).

^{45}Ca uptake decreased in H9c2 cardiomyocytes from 6991 ± 84 to 4117 ± 116 cpm/10 mg protein (*t* test, $P < 0.01$) following 20 ng/mL exposure to TNFα (FIG. 3). TNF (20 ng/mL) inhibited ^{45}Ca uptake from 24,127 ± 2009 to 6821 ± 138 cpm/10 mg protein (*t* test, $P < 0.01$) during depolarization of 80 mM KCl (FIG. 3).

DISCUSSION

Exposing the heart to brief episodes of ischemia is known as ischemic preconditioning (IP). IP protects the myocardium against functional damage and cell death caused by subsequent prolonged ischemia. IP of the myocardium reduced infarct size produced by sustained ischemia[4] in several animal species and in humans. TNFα is involved in IP.[2,4–6] Following IP in null (TNF–/–) and wild-type mice, infarct size was significantly reduced by 43% in wild-type mice; in contrast, infarct size was not attenuated by IP in the TNF–/– versus ischemic reperfusion controls.[6]

ATP-sensitive K+ channels (K_{ATP}) are involved in regulation of coronary blood flow.[5] K_{ATP} channels reduce action potential duration, decrease contractility, and conserve energy during periods of ischemia.[5]

The present study indicated a physiological protective effect of TNF in H9c2 cardiomyocytes due to the following: (i) TNF activated K_{ATP} channels—this effect was blocked by glibenclamide and caspase inhibitor zvad-fmk (FIG. 1); (ii) TNF depressed the depolarization effect of 80 mM KCl on ^{45}Ca uptake in H9c2 cardiomyocytes (FIG. 2)—this effect can be attributed to the negative ionotropic effect of TNF that reduced cardiac contractility and attenuated Ca influx and Ca release by the sarcoplasmic reticulum;[1] and (iii) TNF restricted the stimulatory effect of 100 nM insulin on ^{3}H-deoxyglucose uptake (FIG. 3).

Our findings further suggest that TNF mediated adaptive and protective effects in the heart during a brief environmental injury.

REFERENCES

1. MELDRUM, D.R. 1998. Tumor necrosis factor in the heart. Am. J. Physiol. **274:** R577–R595.
2. EDDY, L.J., D.V. GODDEL & G.H.W. WONG. 1992. TNFα pretreatment is protective in a rat model of myocardial ischemia-reperfusion injury. Biochem. Biophys. Res. Commun. **184:** 1056–1059.
3. HESCHLER, J., R. MEYER, S. PLANT *et al.* 1991. Morphological, biochemical, and electrophysiological characterization of a clonal cell (H9c2) line from rat. Circ. Res. **69:** 1476–1486.
4. RUBINO, A. & D. YELLON. 2000. Ischemic preconditioning of the vasculature: an overlooked phenomenon for protecting the heart? TIPS **21:** 225–230.
5. LECOUR, S., R.M. SMITH, B. WOODWARD *et al.* 2002. Identification of a novel role for sphingolipid signaling in TNFα and ischemic preconditioning mediated cardioprotection. J. Mol. Cell. Cardiol. **34**(5)**:** 509–518.
6. SMITH, R.M., N. SULEMAN, J. MCCARTHY & M.M. SACK. 2002. Classic ischemic but not pharmacologic preconditioning is abrogated following genetic ablation of the TNFα gene. Cardiovasc. Res. **55**(3)**:** 553–560.

Preconditioning with Cardioplegia Is More Effective in Reducing Apoptosis Than Is Preconditioning with Ischemia in the Human Myocardium

HUNAID A. VOHRA, ALAN G. FOWLER, AND MANUEL GALIÑANES

Department of Integrative Human Cardiovascular Physiology and Cardiac Surgery, Glenfield Hospital, Leicester LE3 9QP, UK

Keywords: preconditioning; cardioplegia; apoptosis; ischemia; myocardium

INTRODUCTION

Strategies to minimize myocardial cell death by apoptosis and necrosis following ischemia and reperfusion continue to generate an ongoing investigation and are of great clinical importance. This is particularly true in the context of cardiac surgery, in which surgeons continue to seek the ideal myocardial protective strategy to minimize ischemic damage while achieving a bloodless field during open-heart surgery. Postoperatively, the aim is to restore cardiac contractility to facilitate weaning from cardiopulmonary bypass, especially in high-risk patients. Inability to achieve this may result in low cardiac output syndrome.

Apoptosis and necrosis are two distinct forms of cell death. Both have been shown to be important in reperfusion injury in cardiomyocytes.[1] Preconditioning by a short ischemic insult or pharmacologic means has been demonstrated as strong cardioprotective intervention. However, the role of apoptosis and necrosis in the protection afforded by this phenomenon in the heart is undefined. In the present study, the precise contribution of preconditioning by ischemia and by cardioplegia to reduce both types of cell death in the human myocardium was investigated.

MATERIAL AND METHODS

Preparation of Atrial Tissue and Solutions. The human right atrial appendage[2] from patients undergoing elective coronary artery bypass graft surgery was harvested after local ethical approval and patients' informed consent were obtained. Tissue from patients with atrial fibrillation, those with a poor ejection fraction (EF <30%),

Address for correspondence: Professor Manuel Galiñanes, University Clinical Sciences, Department of Integrative Human Cardiovascular Physiology and Cardiac Surgery, Glenfield Hospital, Groby Road, Leicester LE3 9QP, UK. Voice: +44 1162563031; fax: +44 1162502449. mg50@le.ac.uk

Ann. N.Y. Acad. Sci. 1010: 721–727 (2003). © 2003 New York Academy of Sciences.
doi: 10.1196/annals.1299.130

or those with diabetes receiving oral hypoglycemic drugs, insulin, or potassium channel activators (nicorandil or diazoxide) was excluded. Free hand sections were cut from atrial muscle trabeculae of approximately 300–500 μm in thickness and of 30–50 mg wet weight. These sections were equilibrated for 30 minutes in oxygenated (95% O_2 and 5%CO_2) Krebs Henseleit Hepes (KHH) buffer containing (in mM): NaCl (118), KCl (4.8), $NaHCO_3$ (27.2), $MgCl_2$ (1.2), KH_2PO_4 (1.0), $CaCl_2$ (1.25), glucose.H_2O (10), and HEPES (20) at a pH of 7.4 and a temperature of 37°C. The buffer was supplemented with 10% fetal calf serum (FCS; Harlanseralabs #S-0001A). Ischemia was simulated by bubbling the media with 95% N_2 and 5% CO_2 at pH 6.8 in the absence of glucose. To induce preconditioning with cardioplegia, we employed St. Thomas' cardioplegic solution no. 2, with the following components (in mM): NaCl (110), $NaHCO_3$ (10), KCl (16), $MgCl_2$ (16), and $CaCl_2$ (1.2) at a pH of 7.8.

Experimental Protocols. Sections of atrial muscle ($n = 8$/group) were equilibrated for 30 minutes and then subjected to the following protocols: Group I: time-matched aerobic controls incubated in the medium for 210 minutes; Group II: 90 minutes of simulated ischemia (SI); Group III: 90 minutes of SI followed by 120 minutes of reoxygenation (R); Group IV: ischemic preconditioning with 5 minutes of ischemia plus 5 minutes of reoxygenation followed by 90 minutes of SI/120 minutes R; Group V: preconditioning with cardioplegia by incubation for 5 minutes with the cardioplegic solution followed by 5 minutes of washout (oxygenated KHH) before the 90 minutes of SI/120 minutes of R.

Measurement of Tissue Injury by Creatine Kinase Release. Tissue injury was measured by myocyte-specific creatine kinase (CK) release[2] into the perfusate. The enzyme was measured at 25°C by a kinetic assay (IU/gr wet weight) at the end of the experimental protocol, and a commercial assay kit was employed (DG147-K; Sigma Chemicals).

Assessment of Apoptosis and Necrosis. Necrotic nuclei were identified by incubating the section of muscle for 10 minutes on ice with 5 μM propidium iodide in 0.1M glycine and 20 mM phosphate-buffered saline solution (PBS), pH 7.4, prior to fixation. Sections were fixed twice, once for 30 minutes, then overnight on ice, with 4% paraformaldehyde in 30% sucrose and 20 mM PBS at a pH of 7.4. Serial sections of 10 μm were then cut with a Bright cryomicrotome (model OTF) at –25°C in tissue-embedding matrix (Tissue Tek® OCT compound). The cryopreserved tissue sections were briefly washed with 20 mM PBS, pH 7.4, for 2 minutes, permeabilized in 0.02 mM proteinease-K for 10 minutes at 37°C, and then presensitized for 1 minute in a microwave oven at 800 watts in 0.1% Triton X-100, 0.1M trisodium citrate, pH 6. Terminal deoxynucleotidyl transferase was used to incorporate fluorescein (FITC)-labeled dUTP oligonucleotides to DNA strand breaks at the 3′-OH termini in a template-dependent manner (TUNEL technique) using a commercially available kit (Roche; #1684795).

Fluorescein (FITC) label via terminal-deoxynucleotidyl transferase-mediated dUTP-biotin nick-end labeling (TUNEL) incorporated in nucleotide polymers was detected by laser confocal epifluorescence microscopy (courtesy of D. Ciantar, UCL Confocal Unit, London University) with a ×10 oil immersion objective and with the FITC fluorescence emission (range 600–630 nm) using argon-ion fluorescence excitation at 488 nm. Fluorescence from propidium iodide-labeled nuclei excited with krypton-ion excitation at 546 nm was detected using the propidium iodide emission (range 680–730 nm) in order to abolish fluorescence "bleed-through" from FITC-labeled nuclei. Analysis was done using NIH Image software (Scion Corp.) with the Cavalieri-3 macro (G. MacDonald, University of Washington). Nuclei with areas less than 16 μm^2 were not included in the count. Absolute numbers of green fluorescent apoptotic (A) and necrotic (N) red fluorescent nuclei in any image field were determined by dividing by the total number of propidium iodide-labeled nuclei (M) in the next serial section (that is, the mirror image section). Then, the absolute percentage of apoptotic cells is given by A/M *100% and the percentage of necrotic cells is N/M *100%.

Statistical Analysis. Data were expressed as mean ± standard error of the mean (SEM). Analysis of variance (ANOVA) was used for comparisons of means. A P value of less than 0.05 was considered statistically significant.

RESULTS

Creatine Kinase Leakage. As shown in FIGURE 1A, SI/R resulted in a significant increase in creatine kinase release (3.99 ± 0.28) compared to that of the time-matched aerobic controls (2.32 ± 0.19 IU/gr wet weight, respectively; $P < 0.05$). Preconditioning with both ischemia and with cardioplegia significantly reduced creatine kinase leakage to mean values that were similar to those seen in the aerobic control group (2.47 ± 0.18 IU/gr wet weight and 2.56 ± 0.25 IU/gr wet weight, respectively; $P < 0.05$ versus SI/R group; P = NS versus aerobic control).

Tissue Necrosis and Apoptosis. FIGURE 1B shows that when compared to that in the aerobic control group, 90 minutes of simulated ischemia alone resulted in small increases in the levels of both apoptosis and necrosis (not statistically different). However, reoxygenation for 120 minutes after 90 minutes of ischemia led to significant increases in both apoptosis and necrosis that were greater in the former (29.5 + 2.9%) than in the latter (12.6 ± 1.6%; $P < 0.05$ versus apoptosis). Preconditioning with ischemia and with cardioplegia equally reduced necrosis from 12.6 ± 1.6% to 7.9 ± 0.9% and 7.9 ± 1.8%, respectively ($P < 0.05$ versus the SI/R alone group). Interestingly, preconditioning with ischemia reduced apoptotic cell death to 14.2 ± 3.04% ($P < 0.05$ versus the SI/R group), whereas preconditioning with cardioplegia resulted in a further reduction to 7.2 ± 2.1% ($P < 0.05$ versus preconditioning with ischemia).

DISCUSSION

It is well documented that progressive ischemia can lead to the rapid development of necrosis. Recent studies in animal cardiomyocytes have shown that ischemia may also lead to cellular apoptosis.[3] Here we show that both apoptosis and necrosis increase in human myocardium following ischemia and reoxygenation, with apoptosis playing a more important role in inducing cellular injury. As far as we are aware, this report is the first to demonstrate that apoptosis is the more important form of ischemia/reperfusion injury in the human myocardium and equates to values reported for the rat.[4,5] Our observations are consistent with the presumption that in the

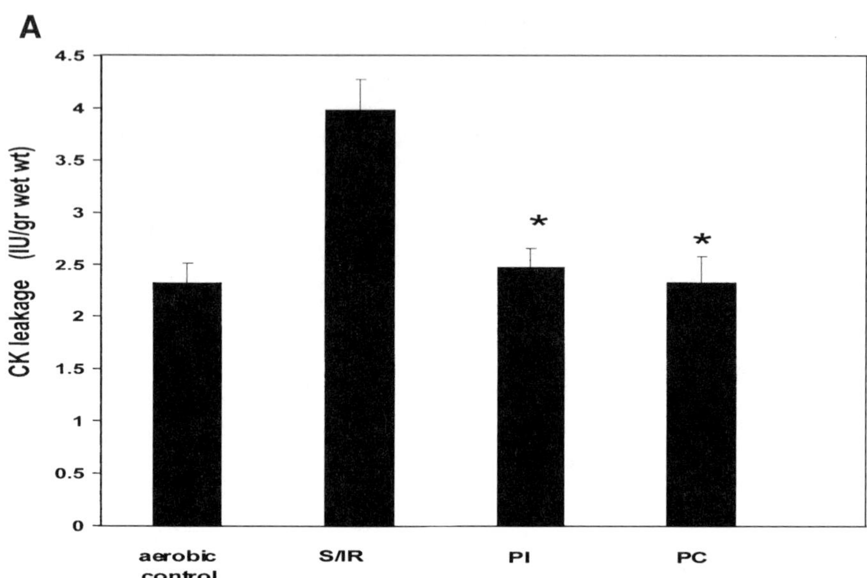

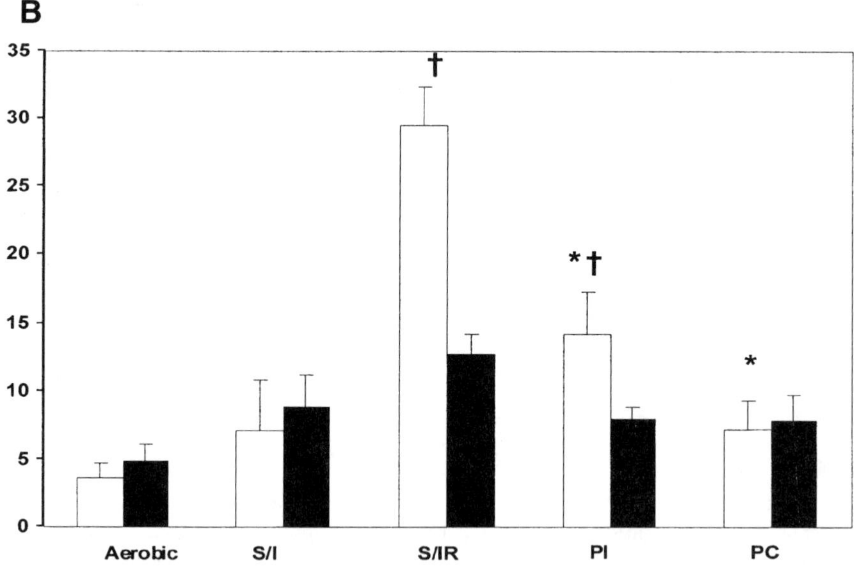

FIGURE 1. (**A**) Creating kinase (CK) leakage at the end of a 120-minute reoxygenation period. (**B**) TUNEL (*white bar*) and propidium iodide staining (*black bar*) at the end of each protocol ($n = 8$). *P <0.05 versus the S/IR group; † P <0.05 versus necrosis.

highly oxygenated human cardiomyocyte, apoptosis primarily occurs as a result of reperfusion-induced oxidative stress, which may follow the burst of reactive oxygen species (ROS) released on reperfusion.[6,7]

A second important finding of this study is that preconditioning with cardioplegia is a more potent intervention than is preconditioning with an ischemic stimulus. Although the cardioprotective role of cardioplegia has been investigated intensively in experimental/animal models and in the human heart, we believe the present report is the first to demonstrate a beneficial effect of preconditioning.

Ischemic preconditioning, in which a brief period of ischemia is followed by an equally short period of reperfusion, is well known to confer protection to myocardium that is subsequently subjected to longer periods of lethal ischemia. Whereas several studies have shown that apoptosis contributes to cardiac cell death after ischemia, emerging evidence indicates that preconditioning may suppress apoptosis in addition to necrosis in the intact heart.[8] The exact mechanism(s) by which preconditioning leads to cardioprotection by inhibition of apoptosis is still unknown, but it may include the opening of channels in the mitochondrion.

Briefly, apoptosis, or "programmed cell death," is mediated by the caspase cascade following activation of one or both of two signal transduction pathways. Signals trigger these pathways either externally through the Fas membrane receptor family, leading to activation of procaspase 8, or internally via the release of cytochrome c into the cytoplasm and the activation of procaspase 9, through a compromised mitochondrial membrane.

The opening of mitochondrial ATP-sensitive potassium (mito-K_{ATP}) channels can lead to either the enhancement or the attenuation of cardioprotection with the preconditioning mechanism highly dependent on the phasing of channel opening. Thus, mito-K_{ATP} channels may signal protection through (1) inhibition of cytochrome c release, (2) the optimization of energy production through depolarization of the mitochondrial membrane potential (Ψ_m) with alterations in mitochondrial Ca^{2+} handling, and (3) the modulation of ROS production, including oxygen-free radicals[6,7] and nitric oxide,[9] during ischemia and reoxygenation. In the present study, we have evidence that preconditioning by ischemia results in the inhibition of apoptosis.

Previously, Akao *et al.*[10] showed that the opening of mito-K_{ATP} channels inhibits apoptosis induced by oxidative stress in rat cardiomyocytes. Using ischemic and pharmacologic preconditioning of human tissue, our group has shown that opening of mitochondrial K_{ATP} channels signals the activation of both PKC and p38MAPK. Both downstream elements are shown to be inhibited by the antagonists 5-hydroxydecanoate (5-HD, a presumed specific mito-K_{ATP} channel inhibitor) or chelerythrine and SB203580, respectively.[11] By inducing the sublethal production of ROS, the short ischemic stimulus associated with preconditioning by ischemia may confer cardioprotection through opening of mito-K_{ATP} channels[12] or activation of protein kinase C isozymes, leading to attenuation of ROS[13] and the inhibition of caspase-8–induced apoptosis.[14] In the present study, we have evidence that preconditioning with cardioplegia is superior to preconditioning with ischemia through the inhibition of proapoptotic and/or activation of antiapoptotic pathways.

A recent comparative study of the cardioprotective role of hypothermic cardioplegic arrest versus ischemia/reoxygenation on the regulation of mRNA gene expression in isolated rabbit heart, using cDNA microarrays, has revealed cardioplegic arrest to downregulate the proapoptotic p53 pathway and enhancement of antiapoptotic heterodimer Bcl-2 family members, with upregulation of Bcl-2 and downregulation of bak.[15] Furthermore, preservation of mitochondrial ATP synthase mRNA expression is accompanied by improved indices of functional recovery after reoxygenation/reperfusion when hypothermic cardioplegia is employed. More interestingly, it has also been demonstrated that the high extracellular K^+ concentration results in inhibition of cytochrome c release, with decreased caspase 3 activation and endonuclease activity, leading to reduced internucleosomal DNA laddering and inhibition of apoptosis.[16]

Thus, our observations on the human myocardium are consistent with preconditioning by cardioplegia to provide cardioprotection, with hyperkalemic arrest resulting in inhibition of calcium overload and the reduction of apoptosis,[17] and maintenance of Na/K pump activity,[18] with higher energy and nutrient reserves and transcriptional capacity.[15] This effect of cardioplegia on cell membrane function combined with upregulation of the pathways of apoptosis inhibition (this study) and with mito-K_{ATP} channel activation[13] may provide for an enhanced antiapoptotic myocardial protection through preconditioning with cardioplegia compared to preconditioning with ischemia for the early/classic phase of ischemic preconditioning.

In the context of preconditioning in human atrial tissue, the results suggest that the overall cardioprotective effect in the first window of cardioprotection is mediated, to a greater extent, by a reduction in apoptotic cell death. Finally, unpublished data from our recent work indicate that necrosis may dominate pathways to cell death following the longer periods (>24 hours) of reoxygenation.

Present techniques of intraoperative myocardial protection are constantly evolving as current strategies have proved to be suboptimal in high-risk patients. We have shown that preconditioning with cardioplegia is as effective as ischemic preconditioning in reducing necrosis but more efficacious in reducing apoptosis in the human atrial appendage model. Although we observed that both interventions reduce apoptotic cell death, more studies are needed to delineate the mechanisms underlying this antiapoptotic effect.

REFERENCES

1. GOTTLIEB, R.A., K.O. BURLESON, R.A. KLONER, *et al.* 1994. Reperfusion injury induces apoptosis in rabbit cardiomyocytes. J. Clin. Invest. **94:** 1621–1628.
2. ZHANG, J.G., S. GHOSH, C. OCKLEFORD & M. GALINANES. 2000. Characterisation of an in-vitro model for the study of the short and prolonged effects of myocardial ischemia and reperfusion in man. Clin. Sci. **99:** 339–350.
3. BORUTAITE, V., A. BUDRIUNAITE, R. MORKUNIENE & G.C. BROWN. 2001. Release of mitochondrial cytochrome c and activation of cytosolic caspases induced by myocardial ischaemia. Biochim. Biophys. Acta **1537:** 101–109.
4. KAJSTURA, J., W. CHENG, K. REISS, *et al.* 1996. Apoptotic and necrotic myocyte cell deaths are independent contributing variables of infarct size in rats. Lab. Invest. **74:** 86–107.
5. OKAMURA, T., T. MIURA, G. TAKEMURA, *et al.* 2000. Effect of caspase inhibitors on myocardial infarct size and myocyte DNA fragmentation in the ischemia-reperfused rat heart. Cardiovasc. Res. **45:** 642–650.
6. ZWEIER, J.L., J.T. FLAHERTY & M.L. WEISFELDT. 1987. Direct measurement of free radical generation following reperfusion of ischemic myocardium. Proc. Natl. Acad. Sci. USA **84:** 1404–1407.

7. Blasig, I.E., S. Shuter, P. Garlick & T. Slater. 1994. Relative time-profiles for free radical trapping, coronary flow, enzyme leakage, arrhythmias, and function during myocardial reperfusion. Free Radical Biol. Med. **16:** 35–41.
8. Zhao, Z.Q. & J. Vinten-Johansen. 2002. Myocardial apoptosis and ischemic preconditioning. Cardiovasc. Res. **55:** 438–455.
9. Joyeux, M., D. Godin-Ribuot & C. Ribuot. 1998. Resistance to myocardial infarction induced by heat stress and the effect of ATP-sensitive potassium channel blockade in the rat isolated heart. Br. J. Pharmacol. **123:** 1085–1088.
10. Akao, M., A. Ohler & M.E. O'Rourke. 2001. Mitochondrial ATP-sensitive potassium channels inhibit apoptosis induced by oxidative stress in cardiac cells. Circ. Res. **88:** 1267–1275.
11. Loubani, M. & M. Galiñanes. 2002. Pharmacological and ischemic preconditioning of the human myocardium: mito-KATP channels are upstream and p38MAPK is downstream of PKC. BioMed. Central Physiol. **18:** 10.
12. Lui, H., B. McPherson & Z. Yao. 2001. Preconditioning attenuates apoptosis and necrosis: role of protein kinaseC∈ and –δ isoforms. Am. J. Physiol. Heart Circ. Physiol. **281:** H404–H410.
13. Wang, Y. & M. Ashraf. 1999. Role of protein kinase C in mitochondrial KATP channel-mediated protection against Ca21 overload injury in rat myocardium. Circ. Res. **84:** 1156–1165.
14. Scaffidi, C., I. Schmitz, J. Zha, *et al.* 1999. Differential modulation of apoptosis sensitivity in CD95 type I and type II cells. J. Biol. Chem. **274:** 22532–22538.
15. Ning, X.H., S.H. Chen, C.S. Xu, *et al.* 2002. Hypothermic protection of the ischemic heart via alterations in apoptotic pathways as assessed by gene array analysis. J. Appl. Physiol. **92:** 2200–2207.
16. Yu, S.P. 2003. K^+ channels and K^+ homeostasis in apoptosis. Presented at Apoptosis 2003: From Signalling Pathways to Therapeutic Tools, Luxembourg.
17. Ju, H. 1992. Protective effects of high potassium administered after ischemic arrest against reperfusion injury in isolated rat hearts. Zhonghua Xin Xue Guan Bing Za Zhi **20:** 182–184.
18. Ko, T., H. Otani, H. Imamura, *et al.* 1995. Role of sodium pump activity in warm induction of cardioplegia combined with reperfusion of oxygenated cardioplegic solution. J. Thorac. Cardiovasc. Surg. **110:** 103–110.

Hypoxia and Ischemia Induce Nuclear Condensation and Caspase Activation in Cardiomyocytes

MARTIN E. O'MAHONEY, SUSAN LOGUE, EVA SZEGEZDI, CATHERINE STENSON-COX, UNA FITZGERALD, AND AFSHIN SAMALI

Cell Stress and Apoptosis Research Group, Department of Biochemistry, National Centre for Biomedical Engineering Science, National University of Ireland, Galway, Ireland

KEYWORDS: **cardiomyocytes; ischemia; hypoxia; caspases**

INTRODUCTION

Cardiovascular disease is characterized by the progressive loss of cardiomyocytes. Programmed cell death, or apoptosis, is thought to play a role in this loss and has been linked specifically to ischemia-induced myocardial injury.[1,2] However, the apoptotic pathways functional within ischemic/hypoxic cardiomyocytes remain largely unknown. Caspases play a central role in apoptosis, and classically, two pathways lead to their activation. These are the death receptor–mediated/extrinsic and mitochondrial/intrinsic pathways. Recently, a novel endoplasmic reticulum (ER) stress-mediated pathway has been reported.[3] Its relevance to ischemia *in vivo* was emphasized when ischemic/hypoxic conditions *in vitro* were shown to interfere primarily with calcium homeostasis in the ER, leading to the accumulation of malfolded proteins and the triggering of an ER stress response.[4] Induction of the activity of caspases-8 or -9 is a hallmark of the death receptor or mitochondrial pathways, whereas the processing of caspase-12 has been linked specifically to ER stress-mediated apoptosis in embryonic fibroblast, PC12 cells.[4] In an expansion of existing information regarding ER stress-mediated apoptosis, we have compared the pattern of caspase activation in primary ischemic neonatal rat cardiomyocytes with that seen in hypoxic H9c2 cells.

MATERIAL AND METHODS

Cell Culture

The embryonic rat cardiomyocyte-derived cell line H9c2 was grown in Dulbecco's modified Eagle's medium supplemented with 10% fetal bovine serum, penicillin 50 U/mL, and streptomycin 5 mg/mL. Primary cultures were derived from 1–4-

Address for correspondence: Afshin Samali, Cell Stress and Apoptosis Research Group, Department of Biochemistry, NUI, Galway, Eire. Voice: 353-91-750393; fax: 35391-512504. afshin.samali@nuigalway.ie

Ann. N.Y. Acad. Sci. 1010: 728–732 (2003). © 2003 New York Academy of Sciences.
doi: 10.1196/annals.1299.131

day-old Sprague-Dawley rats according to a modified method of Simpson *et al.*[5] All reagents were from Sigma unless otherwise stated.

Analysis of Caspase Activity

Cell lysates and 50 μM of the appropriate caspase substrate (Enzyme System Products) suspended in reaction buffer (100 mM HEPES, 10% sucrose, 5 mM DTT, 0.1% NP-40, and 0.1% 3-[(3-cholamidopropyl) dimethylammonio] propane-1-sulphonic acid (CHAPS), pH 7.25] were added in triplicate to a microtiter plate. Caspase-liberated AMC was measured over time and the data analyzed by linear regression. To allow comparisons to be made between the two cell types, enzyme activity was expressed as a fold induction relative to untreated controls.

Western Blot Analysis

Cells were lysed in a buffer containing 20 mM HEPES, pH 7.5, 350 mM NaC-inhibitors. Proteins were separated and transferred onto nitrocellulose membranes using standard procedures. Blots were probed with caspase-12 antibody (rat polyclonal antibody diluted 1:50, a kind gift from Dr. J. Yuan) and bound protein visualized using horseradish peroxidase (HRP)-conjugated secondary antibody (Pierce, 1:10,000 dilution). Protein bands were detected with Super Signal Ultra Chemiluminescent Substrate (Pierce) on X-ray film (Agfa).

Hoechst Nuclear Staining

Control and treated cells were fixed at room temperature in 3% formaldehyde (Sigma) and were permeabilized using methanol. Following a phosphate-buffered saline (PBS) wash, coverslips were incubated for 15 minutes at room temperature with 100 mg/ml Hoechst 3342 stain. Excess stain was removed by washing three times with PBS. Stained nuclei were visualized using an LSM510 laser scanning confocal microscope.

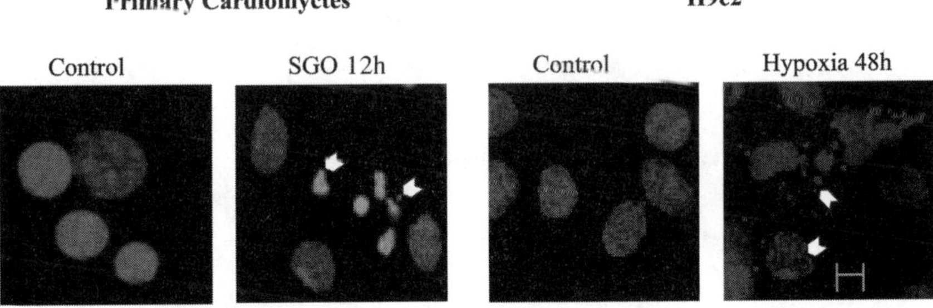

FIGURE 1. Nuclear condensation and fragmentation in ischemic/hypoxic cardiomyocytes. Twelve hours of serum, glucose, and oxygen (SGO) deprivation in primary cardiomyocytes caused condensation of nuclei (*arrowheads*), whereas pyknotic nuclei are present in H9c2 cells cultured for 48 hours under hypoxic conditions. Cells were subjected to hypoxia (O2/N2/CO_2, 0.5:94.5:5), using a hypoxia gas chamber (Analytica). Hypoxia with glucose/serum deprivation (SGO) was simulated using glucose-free DMEM (Life Technologies, Inc.) supplemented with the above and 1 mM/L of 2-deoxy-D-glucose, for the indicated time periods. *Bar* represents 20 μm.

RESULTS AND DISCUSSION

The extent of apoptotic cell death induced in hypoxic/ischemic cardiomyocytes was determined by staining with Hoechst 33342. Cells displaying condensed or pyknotic nuclei were evident in H9c2 cells after 48 hours of hypoxia and were present in primary cardiomyocytes deprived of serum, glucose, and oxygen (SGO) for 12 hours (FIG. 1). Peptide substrates specific for caspase-3–like, caspase-8, and caspase-9 activity (DEVD, IETD, and LEHD, respectively) were used to detect

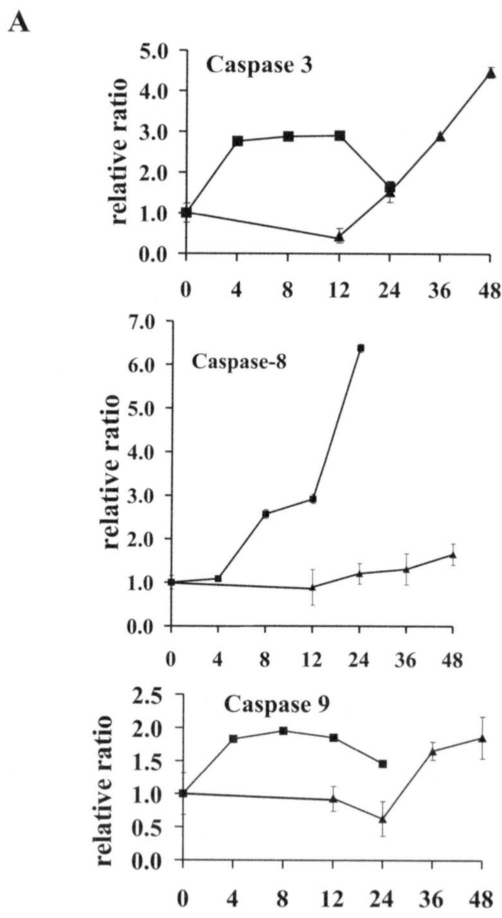

FIGURE 2. Ischemia/hypoxia in cardiomyocytes activates caspases-3, -8, -9, and -12. **(A)** Primary cardiomyocytes under serum, glucose, and oxygen (SGO) conditions (*squares*) show a more rapid induction of caspases-3, -8 and -9, when compared to hypoxic H9c2 cells (*triangles*).

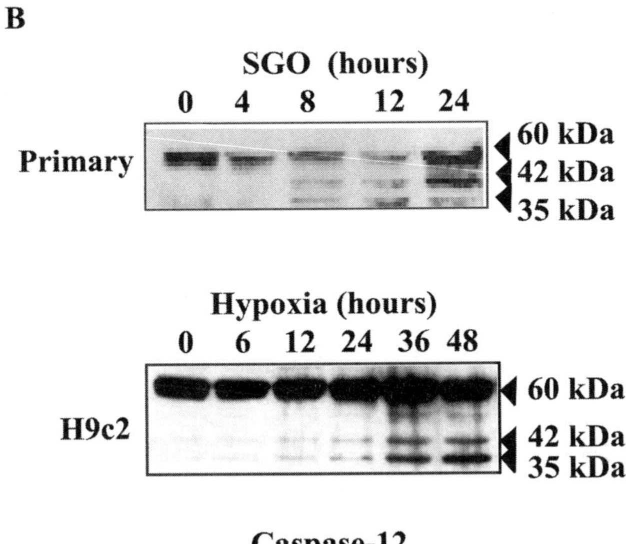

FIGURE 2—*Continued*. **(B)** Processing of pro-caspase-12 into 42- and 35-kDa subunits occurs within 8 hours in SGO conditions, while the same effect is seen in hypoxic H9c2 cells at 24 hours. Cells were subjected to hypoxia/SGO, as described in FIGURE 1.

enzyme activation. There was a fourfold increase in caspase-3–like activity after 48 hours of hypoxia. A slightly smaller (threefold) increase in caspase-3–like activity occurred in primary SGO cardiomyocytes within 3 hours, which was maintained for 24 hours. A slight increase in caspase-8 activity in hypoxic H9c2 cells was noted by 48 hours, whereas a significant increase in casapse-8 activity was detected in SGO primary cardiomyocytes, within 8 hours, that further increased by 24 hours. Caspase-9 activity doubled within 4 hours in SGO primary cardiomyocytes, whereas 36 hours of hypoxia were required to produce the same effect in H9c2 cells (FIG. 2A). Due to the lack of availability of a caspase-12–specific fluorescent substrate, detection (by Western blot analysis) of the cleavage of 60-kDa pro-caspase-12 into its 42 and 35 kDa processed forms, was used to indicate caspase-12 enzyme activity. Cleavage products were detected in SGO primary cells within 8 hours, whereas 12 hours of hypoxia led to pro-caspase-12 cleavage in H9c2 cells (FIG. 2B). In summary, simulated ischemia causes apoptosis and caspase activation in primary cardiomyocytes. A similar, albeit delayed, pattern is seen in hypoxic H9c2 cells. While caspase-3, -8, and -9 induction indicates the activation of classical apoptotic pathways in response to ischemia, processing of caspase-12 points to a possible role for this caspase in ER stress-mediated cardiomyocyte apoptosis. Thus, signaling molecules and enzymes associated with caspase-12 merit further study, in order to develop novel heart disease-targeted therapies.

ACKNOWLEDGMENTS

This work was supported by the Higher Education Authority of Ireland (PRTLI), the Irish Heart Foundation, Enterprise Ireland, and the Millenium Research Fund of NUI, Galway, Ireland.

REFERENCES

1. YAOITA, H. *et al.* 2000. Apoptosis in relevant clinical situations: contribution of apoptosis in myocardial infarction. Cardiovasc. Res. **45:** 630–641.
2. GILL, C. *et al.* 2002. Losing heart: the role of apoptosis in heart disease: a novel therapeutic target? FASEB J. **16:** 135–146.
3. KAUFMAN, R.J. 1999. Stress signalling from the lumen of the endoplasmic reticulum: coordination of gene transcriptional and translational controls. Genes Dev. **13:** 1211–1233.
4. NAKAGAWA, T. *et al.* 2000. Caspase-12 mediates endoplasmic-reticulum-specific apoptosis and cytotoxicity by amyloid-beta. Nature **403:** 98–103.
5. SIMPSON, P. & S. SAVION. 1982. Differentiation of rat myocytes in single cell cultures with and without proliferating nonmyocardial cells: cross-striations, ultrastructure, and chronotropic response to isoproterenol. Circ. Res. **50:** 101–116.

Apoptosis Induced in Vascular Smooth Muscle Cells by Oxidative Stress Is Partly Prevented by Pretreatment with CGRP

C. SCHAEFFER, L. THOMASSIN, L. ROCHETTE, AND J. L. CONNAT

Université de Bourgogne, Biologie Animale Cellulaire et Moléculaire, LPPCE, IFR Santé 100, 21000 Dijon, France

KEYWORDS: vascular smooth muscle cell; neuropeptide; hydrogen peroxide; glucose oxidase; MAPK; ERK

INTRODUCTION

Calcitonin gene–related peptide (CGRP) is a 37–amino acid neuropeptide largely distributed around the cardiovascular system. Its first described biological actions were potent vasodilatory effects together with a positive inotropic and chronotropic action on the heart. Recently, studies of cell cultures revealed that this peptide could influence proliferation and differentiation of the different types of vascular cells.[1,2] Ischemia/reperfusion were shown to generate free radicals[3] that cause apoptosis in myocardial cells.[4] In addition, it was demonstrated that CGRP was responsible for the cardioprotection induced after ischemic preconditioning in the rat heart.[5] We therefore hypothesized that CGRP could reduce the oxidative stress damages via inhibition of apoptosis due to the ischemia/reperfusion sequence. In this paper, we tested this possibility on cultured vascular smooth muscle cells (VSMCs) submitted to oxidative stress.

METHODS

Smooth muscle cells were isolated from the aorta of Wistar rats and cultured until fifth passage. An oxidative stress was generated on 80% confluent cells using 0.2 IU/ml glucose oxidase (EC 1.1.3.4, Sigma) for 1 hour. Cells were used 16 hours later to estimate metabolic activity with MTT, number, morphology, and proportion of apoptotic nuclei with Hoechst 33354, DNA laddering with agarose electrophoresis and expression/activation of ERK1/2 MAPK with Western blotting. Cells were pretreated with 10^{-8} M alpha human CGRP (BACHEM) 6 and 3 hours before the oxidative stress.

Address for correspondence: Jean-Louis Connat, Université de Bourgogne, Biologie Animale Cellulaire et Moléculaire, 6 Bd. Gabriel, 21000 Dijon, France. Voice: 0033 380 39 62 16; fax: 0033 380 39 38 25.
Connatjl@u-bourgogne.fr

Ann. N.Y. Acad. Sci. 1010: 733–737 (2003). © 2003 New York Academy of Sciences.
doi: 10.1196/annals.1299.132

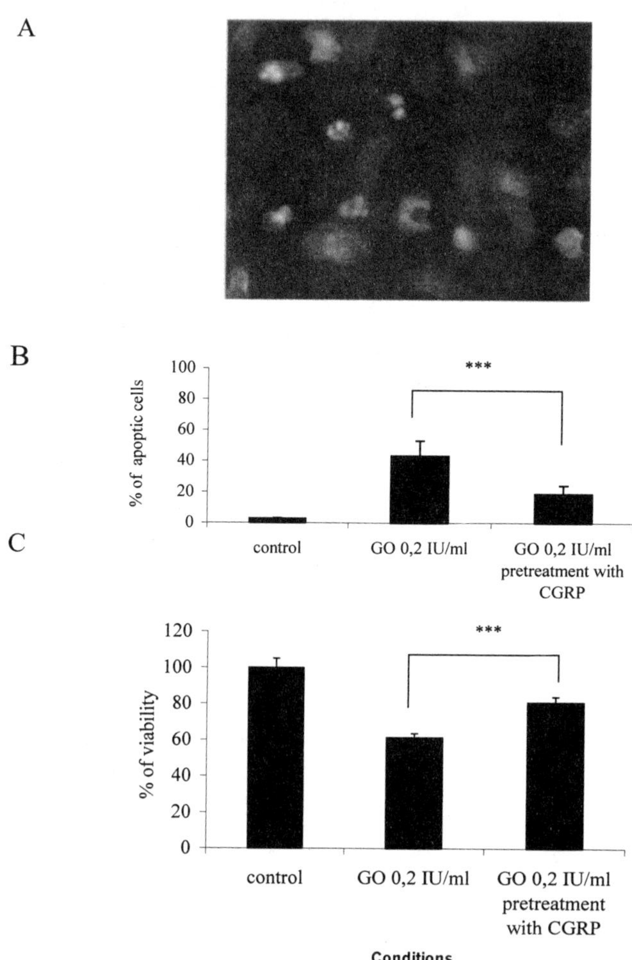

FIGURE 1. (**A**) Visualization of apoptosis on nuclei from cells stressed with glucose oxidase using Hoechst 33342 staining. (**B**) Influence of calcitonin gene–related peptide (CGRP) on the proportion of apoptotic cells induced by the oxidative stress. The total number of nuclei and the number of apoptotic cells presenting chromatin condensation were determined after staining with Hoechst. Percentage of apoptotic cells was then calculated. **(C)** Influence of CGRP on the metabolic activity of the surviving cells. MTT bioreduction was estimated spectrophotometrically and compared to controls as a percentage.

RESULTS

Effects of Glucose Oxidase on Cultured Vascular Smooth Muscle Cells

Oxidative stress generated with 0.2 IU/ml glucose oxidase added to the culture medium decreases the viability of cells by 30–40% when it was estimated with MTT 16 hours after the oxidative stress (data not shown). Hoechst staining revealed that numerous cells contained nuclei with abnormal condensed chromatin typical of apoptosis (FIG. 1A). By using DNA electrophoresis from nucleic extracts from detached cells collected into the culture medium, we observed a ladder characteristic of DNA fragmentation (data not shown).

Protective Effect of CGRP Treatment

When cells were pretreated with CGRP before oxidative stress, the number of apoptotic nuclei detected in the attached cells was significantly decreased to about half the proportion in nontreated cells (FIG. 1B). This was correlated with the global rate of NADH production estimated by MTT reduction (mainly due to glycolysis) of all the cells remaining in the dishes. Metabolic activity of cells pretreated with CGRP was significantly higher than that of stressed cells without neuropeptide.

CGRP Modulation of ERK 1/2 Activation

In the absence of CGRP pretreatment, glucose oxidase treatment induced a dose-dependent activation of ERK1/2 (FIG. 2). When cells were pretreated with the peptide, the signal intensity was approximately two times greater.

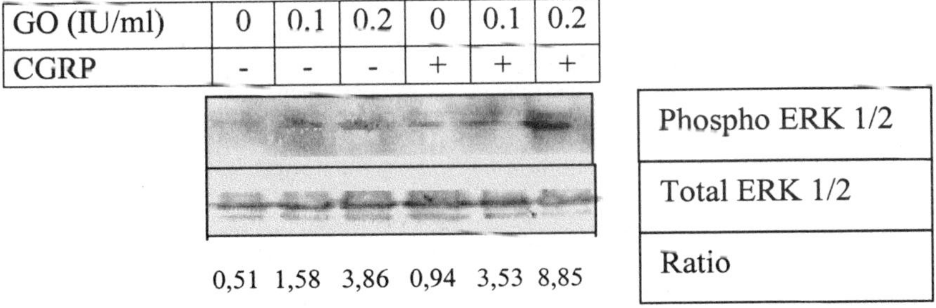

FIGURE 2. Expression and activation of ERK 1/2 MAPK in vascular smooth muscle cells in different experimental conditions. Western blotting realized with 10 μg proteins per track and using antibodies that recognize either the activated (phosphorylated) form of the kinases or whole (activated and inactivated) kinases. Expression of Total ERK was similar in the different cases. Glucose oxidase (GO) induced slight phosphorylation of the kinases. Pretreatment with CGRP increased the signal, especially at the Phospho-ERK2 level (p42). Ratio (Total ERK/Phospho ERK) corresponds to the activation level of the kinases and was estimated by densitometric quantification of the signals.

DISCUSSION

It was previously demonstrated that ERK 1/2 kinases are implicated in oxidant injury.[6,7] They can regulate expression of pro- and antiapoptotic genes.[8,9] Our work demonstrates that ERK1/2 kinases are only slightly activated by the generated oxidative stress. By contrast, pretreatment by CGRP significantly increased the level of activation. The neuropeptide could thus protect VSMCs from oxidative stress-induced apoptosis via this way. In pathophysiologic conditions, free radicals are produced by macrophages in the atheromatous plaque. Apoptosis induced in VSMCs destabilized the fibrous cap, leading to plaque rupture and thrombosis. The presence of a strong CGRP innervation around vessels should thus be of great importance for cellular homeostasis of the media (proliferation versus apoptosis). These studies could probably be extended to aging in which oxidant stress appears to be increased[10] and vascular CGRP to be diminished. In this situation, parts of cells from the media are lost following apoptosis.[11]

ACKNOWLEDGMENTS

We gratefully acknowledge Laboratoires Fournier (21121 Daix, France) for supplying the antibodies and Pharmacies Perrin (89430 Tanlay) and Pharmacie du Fort, C. Avit (21490 Varois & Chaignot) for their support.

REFERENCES

1. HAEGERSTRAND, A., C.J. DALSGAARD, B. JONZON, *et al.* 1990. Calcitonin gene-related peptide stimulates proliferation of human endothelial cells. Proc. Natl. Acad. Sci. USA **87:** 3299–3303.
2. CONNAT, J., V. SCHNURIGER, R. ZANONE, *et al.* 2001. The neuropeptide calcitonin gene-related peptide differently modulates proliferation and differentiation of smooth muscle cells in culture depending on the cell type. Regul. Pept. **101:** 169–178.
3. VERGELY, C., A. TABARD, V. MAUPOIL & L. ROCHETTE. 2001. Isolated perfused rat hearts release secondary free radicals during ischemia reperfusion injury: cardiovascular effects of the spin trap alpha-phenyl N-tert-butylnitrone. Free Radical Res. **35:** 475–489.
4. SCHMITT, J.P., J. SCHRODER, H. SCHUNKERT, *et al.* 2002. Role of apoptosis in myocardial stunning after open heart surgery. Ann. Thorac. Surg. **73:** 1229–1235.
5. LU, R., Y.J. LI & H.W. DENG. 1999. Evidence for calcitonin gene-related peptide-mediated ischemic preconditioning in the rat heart. Regul. Pept. **82:** 53–57.
6. GUYTON, K.Z., Y. LIU, M. GOROSPE, *et al.* 1996. Activation of mitogen-activated protein kinase by H2O2. Role in cell survival following oxidant injury. J. Biol. Chem. **271:** 4138–4142.
7. ZHANG, J., N. JIN, Y. LIU & R.A. RHOADES. 1998. Hydrogen peroxide stimulates extracellular signal-regulated protein kinases in pulmonary arterial smooth muscle cells. Am. J. Respir. Cell Mol. Biol. 19: 324–332.
8. BOUCHER, M.J., J. MORISSET, P.H. VACHON, *et al.* 2000. MEK/ERK signaling pathway regulates the expression of Bcl-2, Bcl-X(L), and Mcl-1 and promotes survival of human pancreatic cancer cells. J. Cell Biochem. **79:** 355–369.
9. SCHEID, M.P., K.M. SCHUBERT & V. DURONIO. 1999. Regulation of bad phosphorylation and association with Bcl-x(L) by the MAPK/Erk kinase. J. Biol. Chem. **274:** 31108–31111.

10. POLLACK, M., S. PHANEUF, A. DIRKS & C. LEEUWENBURGH. 2002. The role of apoptosis in the normal aging brain, skeletal muscle, and heart. Ann. N.Y. Acad. Sci. **959:** 93–107.
11. CONNAT, J.L., D. BUSSEUIL, S. GAMBERT, *et al.* 2001. Modification of the rat aortic wall during aging; possible relation with decrease of peptidergic innervation. Anat. Embryol. **204:** 455–468.

Upregulation and Formation of SDS-Resistant Oligomers of the Proapoptotic Factor Bax in Experimental Atherosclerosis

CHEIKH I. SEYE,[a] GUIDO RY DE MEYER,[a] WIM MARTINET,[a] MICHIEL W. M. KNAAPEN,[b] HIDDE BULT,[a] ARNOLD G. HERMAN,[a] AND MARK M. KOCKX[a]

[a]*Division of Pharmacology, University of Antwerp, Wilrijk, Belgium*

[b]*HistoGeneX, Edegem, Belgium*

Keywords: apoptosis; smooth muscle cells; Bax oligomerization; atherosclerosis; plaque destabilization

Apoptosis of vascular smooth muscle cells has been identified as an important process in the destabilization of atherosclerotic plaques.[1] Members of the Bcl-2 family, including Bcl-2, Bcl-x_L, and Bax, represent some of the most well-known upstream regulators of apoptosis. We previously reported increased Bax immuno-reactivity in cholesterol-induced atherosclerotic plaques of the rabbit aorta.[2] In the present study, this animal model for atherosclerosis was used to investigate apoptotic cell death and the changes in the expression of Bax before and after cholesterol withdrawal. Male New Zealand white rabbits were fed a diet supplemented with a low-dose cholesterol (0.3%) for 26 weeks. The diet induced a pronounced hypercholesterolemia (total cholesterol >700 mg/dL) as well as increased apoptotic cell death (1–2% TUNEL-positive nuclei) in the induced plaques of the thoracic aorta. According to a competitive RT-PCR assay, Bax mRNA was increased 3.5-fold in plaques of cholesterol-fed rabbits at week 26 in comparison with control animals fed a normal diet. With linear regression curves (FIG. 1), 100 ng of poly A^+ RNA corresponded to 13,619 ± 451 Bax mRNA molecules ($n = 3$) in cholesterol-fed rabbits and 3,564 ± 224 molecules in controls ($n = 3$, $P < 0.01$ versus cholesterol, one-way ANOVA). After 26 weeks of cholesterol withdrawal, Bax mRNA normalized to control levels (2,518 ± 212 molecules ($n = 3$), which were statistically different from those of the cholesterol group ($P < 0.01$) but not from the controls ($P < 0.17$).

Immunohistochemical staining of plaque tissue demonstrated that Bax was mainly expressed in smooth muscle cells, as previously reported.[2] Bax expression was further examined at the protein level by SDS-PAGE and Western blotting. The aorta

Address for correspondence: Dr. Mark M. Kockx, Department of Pathology, A.Z. Middelheim, Lindendreef 1, B-2020 Antwerp, Belgium. Voice: +32 3 280 48 16; fax: +32 3 280 48 16. mark.kockx@ua.ac.be

Ann. N.Y. Acad. Sci. 1010: 738–741 (2003). © 2003 New York Academy of Sciences. doi: 10.1196/annals.1299.133

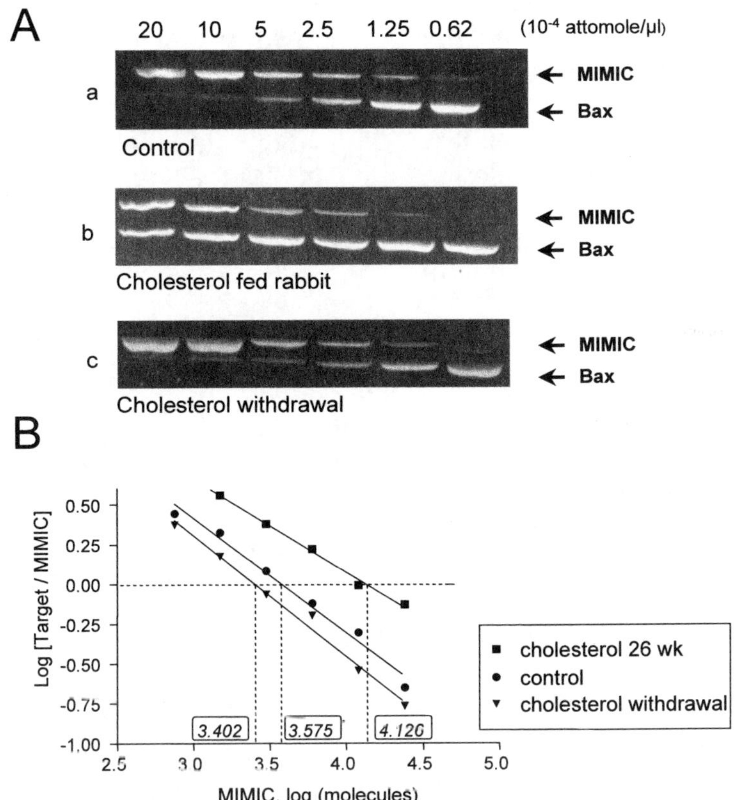

FIGURE 1. Competitive RT-PCR analysis of Bax mRNA in aortas of control and cholesterol-fed rabbits. The procedure used one set of primers to amplify both target cDNA and an externally added heterologous DNA called competitor or MIMIC of known concentration. MIMIC for Bax quantitation was constructed with the use of a PCR-construction MIMIC kit (Clontech). The primer pairs used for amplification of Bax mRNA and construction of the MIMIC fragment were as follows: Bax mRNA forward, 5′-cagctctgagcagatcatgaagaca-3′; reverse, 5′-gcccatcttcttccagatggtgacc-3′ and Bax MIMIC forward, 5′-cagctctgagcagatcatgaagacacgcaagtgaaatcttccg-3′, reverse 5′-gcccatcttcttccagatggtgagcttgagtcctggggagcttt-3′. Poly A^+ RNA (100 ng) was used as a template for cDNA synthesis. Ten-percent portions of the cDNA were then amplified in the presence of twofold serial dilution of the Bax MIMIC. Ten-percent portions of the PCR products were resolved on agarose gel. (**A**) PCR products derived from poly A^+ RNA isolated from the normal rabbit aorta (**a**) and the aorta of 26-week cholesterol-fed rabbits before (**b**) and after (**c**) cholesterol withdrawal. *Lanes* contain $0.625–20\times10^{-4}$ attomol/μl of the MIMIC. (**B**) Quantitative analysis of the competitive PCR shown in **A**. Results are representative of three independent experiments.

from control animals showed a 21-kDa band, which is compatible with the Bax-α isoform (FIG. 2A, *lane 1*). By contrast, positive signals located at 42 (dimers) and 63 kDa (trimers) were detected in the aorta of cholesterol-fed rabbits at 26 weeks (FIG. 2A, *lane 2*). Because trifluoroacetic acid (TFA) has been used successfully to dis-aggregate vesicular monoamine transporter,[3] we investigated the effect of this powerful solvent to dissociate Bax oligomers. As shown in FIGURE 2B, TFA treatment was able to dissociate Bax oligomers in a time-dependent manner, resulting in

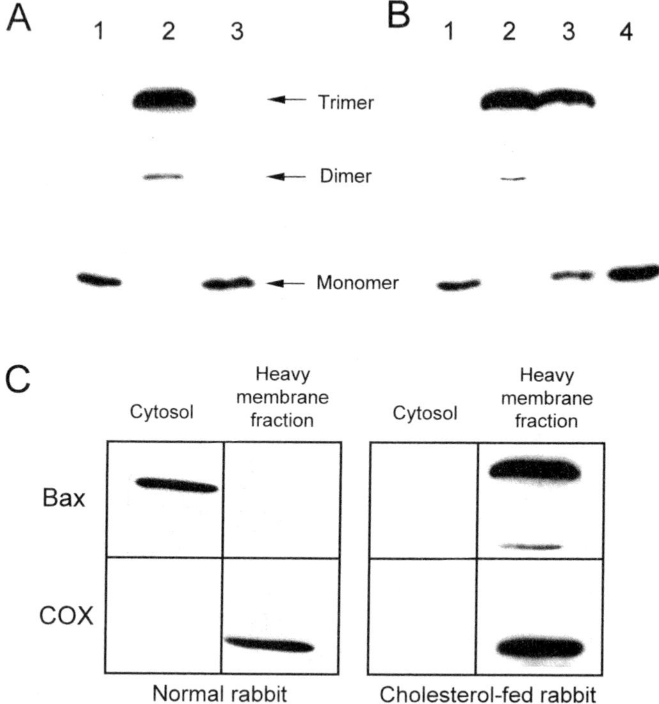

FIGURE 2. Immunoblot analysis of Bax in the aorta of normal and cholesterol-fed rabbits. **(A)** Total protein samples were separated by SDS-PAGE and analyzed with Bax 6A7 monoclonal antibody (mAb, PharMingen). A single 21-kDa band (monomer) was detected in the aorta of normocholesterolemic rabbits (*lane 1*). However, in the aorta of cholesterol-fed rabbits, this antibody detects two bands located at 42 (dimer) and 63 kDa (trimer) (*lane 2*). Bax monomers reappeared at the expense of oligomers after 26 weeks of cholesterol withdrawal (*lane 3*). **(B)** Reversal of Bax oligomers by treatment with trifluoroacetic acid (TFA). Protein samples were incubated with 200 μl of the anhydrous TFA and then analyzed by SDS-PAGE using Bax 6A7 mAb. *Lane 1*, control rabbit; *lane 2*, cholesterol-fed rabbit; *lane 3-4*, protein samples of aorta from cholesterol-fed rabbits incubated for 2 and 15 hours in TFA, respectively. **(C)** Cytosolic and heavy membrane fractions were prepared by differential centrifugation and analyzed with Bax 6A7 mAb and a monoclonal antibody for bovine cytochrome oxydase (subunit V, COX, Molecular Probes). In the normal aorta, Bax was present in the cytosolic fraction but not in the heavy membrane fraction. In contrast, in the aorta of cholesterol-fed rabbits, Bax oligomers were associated with the heavy membrane fraction, but were absent from the cytosol.

the accumulation of a single 21-kDa band. No additional bands appeared after this treatment, which excludes aspecific cleavage. Moreover, treatment of protein samples from thymus with TFA for up to 15 hours did not change the apparent molecular mass of Bax, confirming that this treatment did not promote intrachain breakage of the Bax polypeptide. Heavy membrane and cytosolic fractions were isolated from aortas and characterized by immunoblot analysis with antibodies specific for proteins with distinct known subcellular distributions. Antibodies used were directed against cytochrome oxydase (COX, specific for mitochondrial inner membrane) or poly(ADP-ribose) polymerase (PARP, marker for nuclei). As shown in FIGURE 2C, the mitochondrial marker was found in the heavy membrane fraction, but not in the cytosolic fraction. The nuclear marker, PARP, was not present in the heavy membrane or cytosolic fraction (not shown), demonstrating the absence of nuclei in both fractions. In the control aorta, Bax was present in the soluble fraction but not in the membrane fraction (FIG. 2C). By contrast, Bax oligomers in the aorta of cholesterol-fed rabbits were present only in the heavy membrane fraction (FIG. 2C). After 26 weeks of cholesterol withdrawal, the percentage of TUNEL-positive cells was almost comparable to that of control aortas (<0.1%), and this was associated with the disappearance of Bax oligomers (FIG. 2A, *lane 3*). Our findings confirm *in vitro* data showing that Bax oligomerizes and then integrates in the outer mitochrondrial membrane, where it triggers cytochrome *c* release and apoptotic cell death.[4] We believe that Bax oligomerization plays a key role in smooth muscle cell apoptosis and therefore could help to clarify the mechanisms of atherosclerotic plaque destabilization.

REFERENCES

1. GENG, Y.-J. & P. LIBBY. 2002. Progression of atheroma. A struggle between death and procreation. Arterioscler. Thromb. Vasc. Biol. **22:** 1370–1380.
2. KOCKX, M.M., G.R.Y. DE MEYER, N. BUYSSENS, *et al.* 1998. Cell composition, replication and apoptosis in atherosclerotic plaques after 6 months of cholesterol withdrawal. Circ. Res. **83:** 378–387.
3. SAGNÉ, C., M.-F. ISAMBERT, J.P. HENRY & B. GASNIER. 1996. SDS-resistant aggregation of membrane proteins: application to the purification of the vesicular monoamine transporter. Biochem. J. **316:** 825–831.
4. ESKES, R., S. DESAGHER, B. ANTONSSON & J.-C. MARTINOU. 2000. Bid induces the oligomerization and insertion of Bax into the outer mitochondrial membrane. Mol. Cell. Biol. **20:** 929–935.

Transient Leukocytosis, Granulocyte Colony-Stimulating Factor Plasma Concentrations, and Apoptosis Determined by Binding of Annexin V by Peripheral Leukocytes in Patients with Severe Sepsis

MANFRED WEISS,[a] MOHAMMED ELSHARKAWI,[a] KATRIN WELT,[a] AND ELISABETH-MARION SCHNEIDER[b]

[a]Department of Anesthesiology and [b]Department of Experimental Anesthesiology, Universitätsklinikum Ulm, D-89075 Ulm, Germany

ABSTRACT: This study was undertaken to clarify the relation between transient increases in the numbers of leukocytes and granulocyte colony-stimulating factor (G-CSF) plasma concentrations as well as the degree of apoptosis, as determined by binding of annexin V by these cells in patients with severe sepsis and septic shock. Over a 6-month period, annexin V binding by leukocytes was determined daily using flow cytometry and FITC-labeled annexin V in 33 post-operative patients with severe sepsis or septic shock during their intensive care unit stay. The percentage of annexin V binding neutrophils, monocytes, and lymphocytes was significantly lower, and G-CSF plasma concentrations were higher in patients than in controls on most days. Nine, 19, and 18 peaks in neutrophil, monocyte, and lymphocyte counts (increase of ≥30% within 2 days, followed by a decrease of ≥30% within 2 days) occurred in 6, 11, and 16 patients. During leukocyte peaks, the absolute numbers of annexin V binding neutrophils, monocytes, and lymphocytes paralleled those of neutrophil, monocyte, and lymphocyte numbers. However, the percentage of annexin V binding neutrophils, monocytes, and lymphocytes did not differ significantly from one day to another. Increased G-CSF serum concentrations preceded neutrophil and monocyte peaks. In conclusion, apoptosis of leukocytes is lowered during severe sepsis and septic shock in critically ill patients. Moreover, the degree of apoptosis does not increase during transient leukocytosis. G-CSF might contribute to the low degree of apoptosis of neutrophils and monocytes in those patients.

KEYWORDS: granulocyte colony-stimulating factor; annexin V; leukocytes; sepsis

Address for correspondence: Manfred Weiss, Department of Anesthesiology, Universitätsklinikum Ulm, Steinhoevelstr. 9, D-89075 Ulm, Germany. Voice: ++49-731-50027931; fax: ++49-731-50027917.

manfred.weiss@medizin.uni-ulm.de

Ann. N.Y. Acad. Sci. 1010: 742–747 (2003). © 2003 New York Academy of Sciences.
doi: 10.1196/annals.1299.134

INTRODUCTION

In the cecal ligation and puncture sepsis model in mice, differential expression of apoptotic frequency in neutrophils, monocytes/macrophages, and lymphocytes has been reported from different sites in the body.[1,2] Immediate autopsy in critically ill patients revealed that caspase 3–mediated apoptosis caused extensive lymphocyte apoptosis in sepsis and may contribute to the impaired immune response that characterizes the disorder.[3] However, the proportion of apoptotic neutrophils in bronchoalveolar lavage fluid (BAL) was low throughout the course of the acute respiratory distress syndrome (ARDS), and the appearance of DNA laddering paralleled the increase in annexin V binding and the appearance of morphologic criteria for apoptosis.[4] Interestingly, BAL from patients with ARDS prolonged survival of normal human neutrophils *in vitro*, and this effect was partially mediated by granulocyte colony-stimulating factor (G-CSF) and granulocyte macrophage colony-stimulating factor (GM-CSF).[4,5]

The present study was undertaken to clarify the relation between transient increases in the numbers of neutrophils, monocytes, and lymphocytes and the degree of apoptosis, as determined by binding of annexin V by these cells in patients with severe sepsis and septic shock. The study also examined whether an association exists between the presence of G-CSF in the circulation and leukocyte numbers.

MATERIAL AND METHODS

After approval by the Human Studies Committee at the University of Ulm, over a 6-month period, leukocyte counts and annexin V binding were determined daily in 33 postoperative patients with severe sepsis or septic shock during their intensive care unit stay. Twenty-four patients with newly documented and confirmed infections, severe sepsis, and septic shock lasting at least 3 days, as defined by the criteria of the ACCP/SCCM Consensus Conference,[6] without immunomodulatory drugs, such as G-CSF or cytostatic drugs, were available for final evaluation. Thirteen healthy volunteers (aged 22–50 years, median 24), seven female and six male, served as controls. Annexin V binding by leukocytes was determined daily using flow cytometry (Epics XL-MCL, Beckman Coulter Electronics, Krefeld, Germany) and FITC-labeled annexin V (Annexin V-Fluos, Boehringer Mannheim, Germany). Annexin V was used in conjunction with the vital dye propidium iodide (PI) to distinguish apoptotic cells (annexin V positive, PI negative) from necrotic or late-stage apoptotic cells (annexin V positive, PI positive).[7] PI-positive cells were excluded from analysis. Fifty thousand events were measured per test. Neutrophils were discriminated from lymphocytes and monocytes by their characteristic forward/sideward light scatter. Results are expressed as the absolute numbers and the percentage of annexin V positive neutrophils, lymphocytes, and monocytes. Transient leukocytosis, or peaks in leukocyte counts, were defined as individual increases of at least 30% within 2 days, followed by a decrease of at least 30% within the following 2 days. Leukocyte subpopulation counts were regarded as stable (plateau) if they did not change more than ± 10% from one day to the next. Serum concentrations of G-CSF were determined by sandwich-type ELISA (R & D Systems, Minneapolis, MN, USA).

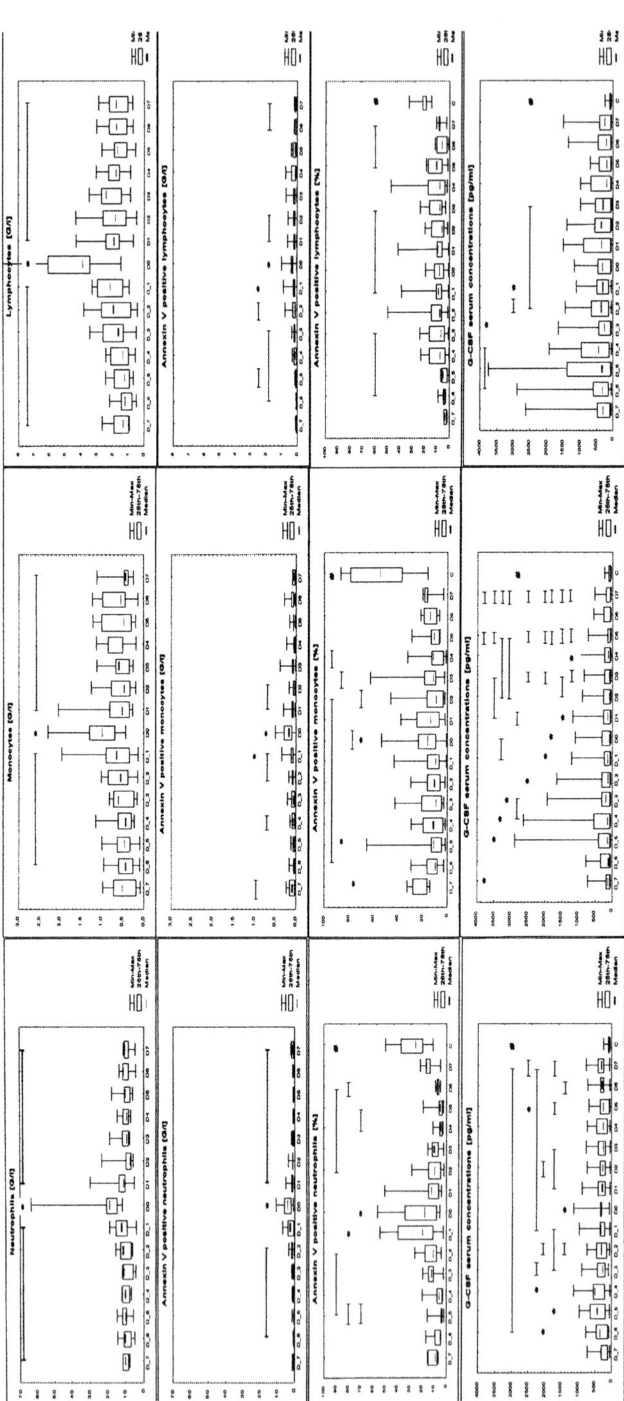

FIGURE 1. Peaks in neutrophil, monocyte, and lymphocyte counts (day 0), corresponding absolute numbers and percentage of annexin V binding cells and G-CSF plasma concentrations, respectively, in patients with severe sepsis and septic shock. Neutrophil, monocyte, and lymphocyte counts, annexin V binding, and G-CSF plasma concentrations are presented as *box plots* demonstrating median, 25th, and 75th percentiles, minimal and maximal values. **P* <0.05, significant differences between the day indicated by an *asterisk* (*) and the days denoted by a *line* over the respective days (I-I). #*P* <0.05, significant differences between healthy volunteers (C) and patients denoted by a *line* over the respective days of the patients (I-I). D, day.

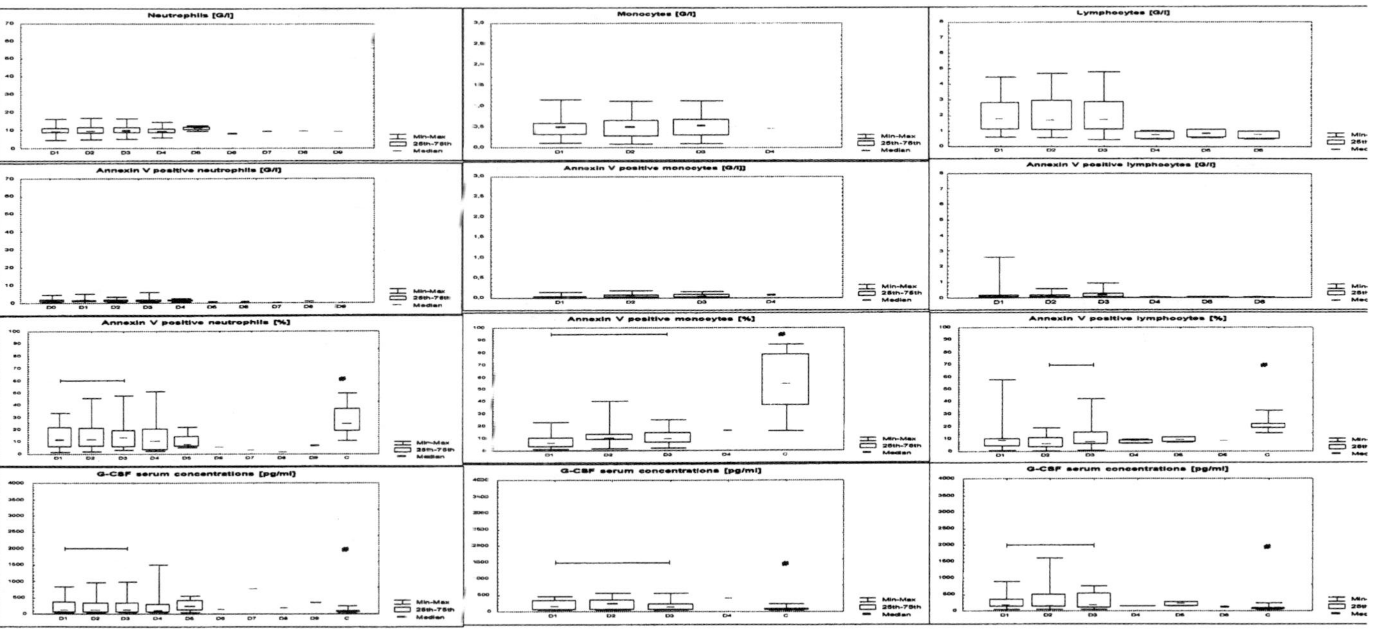

FIGURE 2. Plateaus in neutrophil, monocyte, and lymphocyte counts, corresponding absolute numbers and percentage of annexin V binding cells, and G-CSF plasma concentrations, respectively, in patients with severe sepsis and septic shock. #*P* <0.05, significant differences between healthy volunteers (C) and patients denoted by a *line* over the respective days of the patients (I-I). D, day.

RESULTS AND DISCUSSION

The percentage of annexin V binding neutrophils, monocytes, and lymphocytes was significantly lower and G-CSF plasma concentrations were higher in patients than in controls on most days (FIG. 1). Nine, 19, and 18 peaks in neutrophil, monocyte, and lymphocyte counts occurred in 6, 11, and 16 patients. During leukocyte peaks, the absolute numbers of annexin V binding neutrophils, monocytes, and lymphocytes paralleled those of neutrophil, monocyte, and lymphocyte numbers. However, the percentage of annexin V binding neutrophils, monocytes, and lymphocytes did not differ significantly from one day to another. Increased G-CSF plasma concentrations preceded neutrophil (5 and 4 days before) and monocyte (4 days before) peaks.

Twenty-eight, 13, and 13 plateaus in neutrophil, monocyte and lymphocyte counts were observed in 19, 9, and 12 patients, respectively (FIG. 2). During the neutrophil, monocyte, and lymphocyte plateau phases, absolute numbers and the percentage of annexin V binding cells and G-CSF serum concentrations did not differ significantly from one day to another. During the granulocyte, monocyte, and lymphocyte plateau phases in patients, the percentage of annexin V binding cells was lower and G-CSF serum concentrations were higher than in controls.

The present study demonstrates that apoptosis of leukocytes is decreased during severe sepsis and septic shock in critically ill postoperative patients. Moreover, the degree of apoptosis did not increase during transient leukocytosis. This might contribute to preserving leukocyte function in these patients. The antiapoptotic effects of G-CSF may play a major role in the low degree of apoptosis and the elevations and peaks in neutrophils and monocytes, but not in lymphocytes in these patients. This is in agreement with the clear detection, by flow cytometry, of G-CSF receptors on adult human peripheral granulocytes and monocytes, but not on lymphocytes.[8] *In vitro*, G-CSF significantly decreased apoptosis for neutrophils from AIDS patients, in whom accelerated apoptosis contributed to impaired neutrophil function and bacterial infections.[9,10] A variety of inflammatory mediators, such as LPS, C5a, G-CSF, GM-CSF, gamma interferon (IFN-γ), interleukin (IL)-2, IL-6, and LTB_4, inhibit apoptosis and prolong neutrophil survival *in vitro* (as reviewed in Ref. 4). However, *in vivo*, not all of these cytokines appear to be involved in inhibition of neutrophil apoptosis. In the acute respiratory distress syndrome, depleting the bronchoalveolar lavage fluid of G-CSF or GM-CSF increased the number of apoptotic neutrophils, but depleting IFN-γ and IL-6 had no significant effect.[4] Thus, major antiapoptotic effects in patients with severe sepsis and septic shock may be due to G-CSF or GM-CSF. Moreover, elevated G-CSF and GM-CSF levels might have a dual effect on neutrophils, prolonging their survival by delaying apoptosis, while at the same time enhancing phagocytosis of those neutrophils that have become apoptotic, as demonstrated in the bronchoalveolar lavage fluid of patients with acute respiratory distress syndrome.[4]

REFERENCES

1. AYALA, A., C.D. HERDON, D.L. LEHMAN, *et al.* 1996. Differential induction of apoptosis in lymphoid tissues during sepsis: variation in onset, frequency, and the nature of the mediators. Blood **15:** 4261–4275.

2. AYALA, A., S.M. KARR, T.A. EVANS, *et al.* 1997. Factors responsible for peritoneal granulocyte apoptosis during sepsis. J. Surg. Res. **69:** 67–75.
3. HOTCHKISS, R.S., P.E. SWANSON, B.D. FREEMAN, *et al.* 1999. Apoptotic cell death in patients with sepsis, shock, and multiple organ dysfunction. Crit. Care Med. **27:** 1230–1251.
4. MATUTE-BELLO, G., W.C. LILES, F.N. RADELLA, *et al.* 1997. Neutrophil apoptosis in the acute respiratory distress syndrome. Am. J. Respir. Crit. Care Med. **156:** 1969–1977.
5. MATUTE-BELLO, G., W.C. LILES, F. RADELLA, 2nd, *et al.* 2000. Modulation of neutrophil apoptosis by granulocyte colony-stimulating factor and granulocyte/macrophage colony-stimulating factor during the course of acute respiratory distress syndrome. Crit. Care Med. **28:** 1–7.
6. AMERICAN COLLEGE OF CHEST PHYSICIANS/SOCIETY OF CRITICAL CARE MEDICINE CONSENSUS CONFERENCE COMMITTEE. 1992. American College of Chest Physicians/Society of Critical Care Medicine Consensus Conference: definitions for sepsis and organ failure and guidelines for the use of innovative therapies in sepsis. Crit. Care Med. **20:** 864–874.
7. VERMES, I., C. HAANEN, N.H. STEFFENS, *et al.* 1995. A novel assay for apoptosis. Flow cytometric detection of phosphatidylserine expression on early apoptotic cells using fluorescein labelled Annexin V. J. Immunol. Methods **184:** 39–51.
8. SHIMODA, K., S. OKAMURA, N. HARADA, *et al.* 1992. Detection of the granulocyte colony-stimulating factor receptor using biotinylated granulocyte colony-stimulating factor: presence of granulocyte colony-stimulating factor receptor on CD34-positive hematopoietic progenitor cells. Res. Exp. Med. Berl. **192:** 245–255.
9. RAS, G.J., R. ANDERSON, G. LECATSAS, *et al.* 1984. Acquired neutrophil dysfunction in male homosexuals with the acquired immunodeficiency syndrome. S. Afr. Med. J. **65:** 873–884.
10. PITRAK, D.L., H.C. TSAI, K.M. MULLANE, *et al.* 1996. Accelerated neutrophil apoptosis in the acquired immunodeficiency syndrome. J. Clin. Invest. **98:** 2714–2719.

Evaluation of Apoptosis Markers in Conjunctival and Eyelid Benign and Malignant Tumors

JOANNA RESZEC,[a] MARIOLA SULKOWSKA,[a] LUIZA KANCZUGA-KODA,[a] JERZY JANICA,[b] MALGORZATA SKAWRONSKA,[b] WITOLD PEPINSKI,[b] AND STANISLAW SULKOWSKI[a]

Department of [a]Clinical Pathology and [b]Forensic Medicine, Medical Academy of Bialystok, Bialystok, Poland

ABSTRACT: The balance between cell proliferation and programmed cell death plays a crucial role in malignant development. Bcl-2 family proteins, including proapoptosis protein Bak and antiapoptosis protein Bcl-2, regulate the apoptotic process. Mutation of the p53 gene, which results in P53 protein accumulation, was observed in many types of human cancer. The aim of our study was to evaluate immunohistochemical Bcl-2, Bak, and P53 protein expression and the relation between these proteins in conjunctival and eyelid benign and malignant tumors. We examined a series of 42 papillomas (CEP), 12 squamous cell cancers (SCC), and 19 cases of basal cell cancer (BCC). The age in the CEP group ranged from 18–94 years, and in the SCC and BCC groups from 42–87 years. Staining patterns were correlated with sex, age, and tumor localization. P53 protein-positive immunostaining was observed in 71% of cases, Bcl-2 in 83.9%, and Bak in 74.2 cases in the SCC and BCC groups. In the CEP group, P53 overexpression was observed in 90.5% of cases, Bcl-2 in 71.4%, and Bak in 76.2%. No statistically significant correlation was found between examined protein expression and sex, age, and tumor localization. An inverse correlation was observed between P53 and Bak protein expression in the CEP group. No statistically significance correlation was noted between Bcl-2 and P53 and Bcl-2 and Bak protein expression in both examined groups. The obtained data suggests that P53 and Bcl-2 protein expression coupled with decreasing Bak expression are associated with apoptosis and proliferation as well as malignant progression in conjunctival and eyelid tumors.

KEYWORDS: papilloma; squamous cell cancer; basal cell cancer; Bcl-2, P53; Bak protein expression

INTRODUCTION

Squamous papilloma and squamous cell carcinoma of the eye are common epithelial tumors of the conjunctiva and eyelid.[1] Apoptosis is a cell death process in

Address for correspondence: Joanna Reszec, M.D., Department of Clinical Pathology, Medical Academy of Bialystok, Ul. Waszyngtona 13, 15-269 Bialystok, Poland. Voice: +48 (085) 742-02-80; fax: +48 (085) 742-34-38.
joasia@zpk.amb.edu.pl

Ann. N.Y. Acad. Sci. 1010: 748–751 (2003).
doi: 10.1196/annals.1299.135

which one of the most important factors is the BCL-2 protein family. It plays a role in the regulation of apoptosis, in the modulation of the cell cycle–regulating proteins, and in either the inhibition (Bcl-2 and Bcl-Xl) or the promotion of cell death (Bax, Bak). The Bak protein binds Bcl-XL and Bcl-2 and is thought to induce apoptosis by counteracting the protective functions of Bcl-2 and Bcl-XL. Bcl-2 protein inhibits programmed cell death and thus may be considered to allow malignant cells to proliferate. Abnormalities of the p53 gene in the nucleus of P53 protein accumulation is one of the most important proteins associated with carcinogenesis.

The aim of our study was to evaluate the immunohistochemical expression of Bcl-2, Bak, and P53 protein expression and to establish the relation between these proteins and benign and malignant lesions of the conjunctiva and eyelid.

MATERIAL AND METHODS

In a series of 73 tumors, there were 42 cases of squamous cell conjunctival and eyelid papillomas (CEP), including 17 keratopapillomas, and a series of 11 cases of squamous cell carcinomas (SCC) and 19 cases of basal cell carcinomas (BCC). The ageof the patients ranged from 18–94 years in the papilloma group and from 42–87 years in the cancer group. All specimens were fixed in 10% buffered formalin and routinely processed for paraffin embedding at 56°C. Samples were immunostained with monoclonal antibody for P53 protein (DAKO/P53 No M7001), polyclonal antibody for Bcl-2 protein (DAKO/Bcl-2 No M0887), and polyclonal antibody for Bak protein (Santa Cruz Biochemicals, No. sc-832). The reactions were performed using the labeled streptavidin avidin biotin (LSAB) technique. Evaluation of immunostaining was calculated as the percentage of epithelial cells with positive reaction to the total number of cells encountered in 10 representative fields by light microscopy using a ×20 objective lens. Scores were based on the following scale: (+++), over 80% of the field showing positive staining; (++), 40%–79%; (+), 10%–39%; and (–), below 10% for P53, Bcl-2, and Bak protein. The values obtained were subjected to statistical analysis with the use of the Fisher or chi-square test. Values at <0.05 were considered significant.

RESULTS

In the conjunctival and eyelid papilloma group (CEP), P53 overexpression was seen in 38 of 42 cases (90.4%), Bcl-2 immunoexpression was observed in 30 (71.4%) cases, and Bak expression in 32 (76.2%) (see TABLE 1). P53, Bcl-2,and Bak staining did not correlate with sex, age, and tumor site. Decreased Bak protein expression and increased P53 protein expression were noted in the CEP group. We found no statistically significant correlation between Bcl-2 protein expression and Bak and P53 protein expression, although overexpression of Bcl-2 protein was noted in papillomas with low expression of Bak. Similarly, overexpression of Bcl-2 protein was observed with decreasing P53 protein expression.

In squamous (SCC) and basal cell (BCC) cancers, P53 positive immunoreaction was recorded in 22 cases (71%), Bcl-2 in 26 cases (83.9%), and Bak in 23 cases (71.4%). No correlation was noted between P53, Bcl-2, and Bak protein expression

TABLE 1. Correlation between immunohistochemical expression of P53, Bcl-2, and Bak as well as some of the clinical features of papilloma and cancer in the tumors examined

Variable		Group (SCC, BCC)		Group (CEP)		*P* value
		n	%	*n*	%	
Sex	F	10	32.2	18	42.8	NS
	M	21	67.8	24	61.2	
Age (yr)	≤49	0	0	9	21.4	NS
	41-49	5	16.1	7	16.6	
	≥50	26	83.9	26	62.0	
Bcl-2	–	5	16.1	12	28.6	$P = 0.08$
	+	26	83.9	30	71.4	
P53	–	9	29.0	4	9.5	$P = 0.005$
	+	22	71.0	38	90.5	
Bak	–	8	25.8	10	23.8	NS
	+	23	74.2	32	76.2	

and sex, age, and tumor site. Also, there was no statistically significant correlation between Bcl-2, P53, and Bak expression in both cancer groups. However, increased Bcl-2 expression accompanied increased P53 protein expression.

DISCUSSION

Carcinogenesis is associated with the dysregulation of cellular growth, differentiation, and cell death. In this study, we observed P53 overexpression as a more frequent event in the papilloma group than in the cancer groups although Bcl-2 expression occurred more frequently in SCC and BCC groups. Similarly, Mahomed *et al.*[2] showed that expression of P53 is a common finding in squamous neoplasia, whereas Bcl-2 immunoexpression occurs in a minority of cases. In our study, Bak protein expression was comparable in both groups. In SCC and BCC groups, the correlation between Bcl-2, Bak, and P53 protein expression was not statistically significant, although an increase in Bcl-2 expression was found concomitant with P53 expression. Likewise, Klatka *et al.*[3] observed no correlation between P53 and Bcl-2 expression in laryngeal squamous cell cancer. Niu *et al.*[4] showed Bcl-2 overexpression in all cases of basal cell cancer, suggesting that suppression of apoptosis might play a role in tumorigenesis of eyelid BCC. In the present study, no correlation was observed between P53 and Bcl-2 protein expression and between Bak and Bcl-2 protein expression in conjunctival and eyelid papillomas, although a significant inverse correlation was found between P53 and Bak protein expression. Schoelch *et al.*[5] showed no association between P53 and Bak protein expression, but they observed increasing P53 and Bak expression during oral malignant development. Our study showed that dysregulation of apoptosis might play a crucial role in conjunctival and eyelid malignant progression. In benign lesions such as papilloma prevailing processes are associated with alterations in P53 protein function, which results in an increase in P53 and a decrease in Bak protein expression. In squamous and basal cell cancers, inhibition of the apoptotic process was more important, which occurred in Bcl-2 protein overexpression. The obtained data suggest that P53 and Bcl-2 protein expression coupled with decreasing Bak expression is associated with apoptosis and proliferation alterations as well as malignant progression in conjunctival and eyelid tumors.

REFERENCES

1. Sjo, N. *et al.* 2000. Conjunctival papilloma. Acta Ophtalmol. Scand. **78:** 663–665.
2. Mahomed, A. *et al.* 2002. Human immunodeficiency virus infection, Bcl-2, p53 protein, and Ki-67 analysis in ocular surface squamous neoplasia. Arch. Ophthalmol. **120:** 554–558.
3. Klatka, J. 2001. Prognostic value of the expression of p53 and bcl-2 in patients with laryngeal carcinoma. Eur. Arch. Otorhinolaryngol. **258:** 537–541.
4. Niu Y. *et al.* 2002. Expression of P16 and Bcl-2 protein in malignant eyelid tumours. Chin. Med. J. (Engl.) **115:** 21–25.
5. Schoelch, M.L. *et al.* 1999. Apoptosis-associated proteins and the development of oral squamous cell carcinoma. Oral Oncol. **35:** 75–85.

Immunohistochemical Status of p53, MDM2, bcl2, bax, and ER in Invasive Ductal Breast Carcinoma in Tunisian Patients

SAMI BACCOUCHE,[a] JAMEL DAOUD,[b] MOUNIR FRIKHA,[b]
RAJA MOKDAD-GARGOURI,[a] ALI GARGOURI,[a] AND RACHID JLIDI[b]

[a]*Centre de Biotechnologie de Sfax, Tunisia*

[b]*CHU Habib Bourguiba Sfax, Tunisia*

ABSTRACT: TP53 gene alterations have been associated with sporadic breast cancer. To assess the role of p53 in invasive ductal carcinoma (IDC) of the breast among Tunisian patients, p53 protein status was studied by immunohistochemical analysis. The p53 protein was expressed in 41 of 70 (58%) tumors. Study of the status of its target gene expression showed that MDM2 was overexpressed in 43 tumors (61%), bcl2 in 29 (41%), and bax in only 9 (12%). Estrogen receptor (ER) was detected in 38 tumor tissues (54%). The accumulated p53 was significantly associated with MDM2-positive, bcl2-negative, and ER-negative tumors (P = 0.024, P = 0.000027, and P = 0.000008, respectively), whereas with bax the correlaton was not significant. Bcl2 immunostaining displayed a positive correlation with ER (P = 0.001). A significantly higher fraction of p53-positive cells was observed in ER-negative SBRII-SBRIII tumors than in ER-positive SBRI-SBRII tumors (P = 0.000066). bcl2-positive tumors were significantly correlated with ER-positive/SBRI-SBRII tumors (P = 0.007), but negatively correlated with p53/bax (P = 0000004). MDM2 immunostaining displayed the same phenotype as p53 in the correlation with bcl2 and ER (P = 0.003), strengthened by significant associations between MDM2-positive/p53-positive and bcl2-negative or ER-negative, respectively (P = 0.00005 and P = 0.000001, respectively). MDM2-positive cells were significantly correlated with the p53-positive/bax-negative phenotype (P = 0.04). These results suggest that p53 accumulated in these tumor tissues is associated with bad prognostic markers (ER-negative, SBRIII) of IDC. MDM2 overexpression might be responsible for the accumulated p53 value in IDC. Regulation of the apoptotic process is involved in IDC; bcl2 is associated with a good prognostic marker (ER-positive and SBRI-II), whereas the regulation of bax is complex and does not necessarily correlate with the overexpression of p53.

KEYWORDS: p53; MDM2; bcl2; bax; estrogen receptor (ER); invasive ductal carcinoma of the breast (IDC); prognostic factors; immunohistochemistry (IHC)

Address for correspondence: Ali Gargouri, Laboratoire de Génétique Moléculaire des Eucaryotes, Centre de biotechnologie de Sfax B.P. "K" 3038 Sfax, Tunisia. Fax: +216-4-275970. Faouzi.Gargouri@cbs.rnrt.tn

Ann. N.Y. Acad. Sci. 1010: 752–763 (2003). © 2003 New York Academy of Sciences. doi: 10.1196/annals.1299.136

INTRODUCTION

It has long been known that tumorigenesis is a multistage event involving a number of genetic modifications whose accumulation over time leads to modified cells with transformed potential that become increasingly more aggressive, leading to clinically detectable tumor and eventual metastasis. Rigorous control of the stability of genetic information is a necessary condition for the survival of cells and multicellular organisms. The p53 gene is a tumor suppressor gene involved in the control of genome integrity.[1,2]

Cellular homeostasis is regulated by several active processes, the most important of which are cell proliferation, differentiation, and apoptosis. P53 is in the center of this dynamic and multifaceted control network.[3–9]

About half the tumors produce aberrant p53 protein. The level and activity of p53 have been shown to increase in response to irradiation and other DNA-damaging agents.[10] Activation of the p53 protein can induce either growth arrest at G1/S or G2/M cell cycle checkpoints or an apoptotic cell death.[3,5] P53-dependent apoptosis can modulate the cytotoxicity of chemotherapetic agents and can inhibit tumor growth and progression.[5] In the absence of functional wild-type p53, cells fail to arrest in G1 in response to DNA damage and to undergo apoptosis.[8]

Growth suppression by p53, through the induction of either growth arrest or apoptosis, can be modulated by viral proteins such as the SV40 T antigen[11] and cellular proteins such as MDM2.[10,12] The MDM2 protein has a very important role in the regulation of activity and the level of the p53 protein.[1,12] Indeed, by concealing the activation domain of p53, MDM2 can inhibit p53 functionality. MDM2 protein is at the same time responsible for the ubiquitin-dependent degradation of p53 protein[13] and is a target gene for p53 transcriptional transactivation.[1,14] Interestingly, MDM2 does not act on mutated or phosphorylated p53.[9,12] MDM2 protein was later found to be significantly overexpressed (30–40%) in many human tumors.[15] Among other genes regulated by p53, bax and bcl2 are, respectively, activated and repressed[9,16] by p53 at a transcriptional level.[17] Bax gene is homologous to bcl2 and its encoded protein coimmunoprecipitates with bcl2 both *in vitro* and in mitochondria. Indeed, bcl2 possesses a putative transmembrane domain at its carboxyl terminus and is found associated with mitochondria.[9] Bax protein is an activator of apoptosis, whereas bcl2 inhibits it; therefore, bcl2 has the unusual property of extending cell survival rather than stimulating cell proliferation.[9,18] In fact, protein-protein interaction and the balance of both proteins influence the decision of a cell to either live or die.[3] Upregulation of bax transcription is the means by which p53 induces apoptosis.[10] In functional assays, bax suppresses the ability of bcl2 to block apoptosis.[8,9] Moreover, bcl2 gene expression is critical in regulating apoptosis in breast carcinomas.[7,19]

Estrogen, by interacting with its receptor (ER), plays a central role in regulating the proliferation and differentiation of normal breast epithelium. During the last 10 years, many groups have used immunohistochemistry (IHC) to measure ER in breast cancers. These studies showed a 20–30% average reduction in recurrence/mortality in ER-positive patients receiving adjuvant endocrine therapy and an approximately 60% overall clinical response rate in patients with ER-positive advanced breast cancer treated with endocrine therapy. The primary reason for measuring ER in breast cancer today is this ability to predict response to endocrine therapy.[20,21]

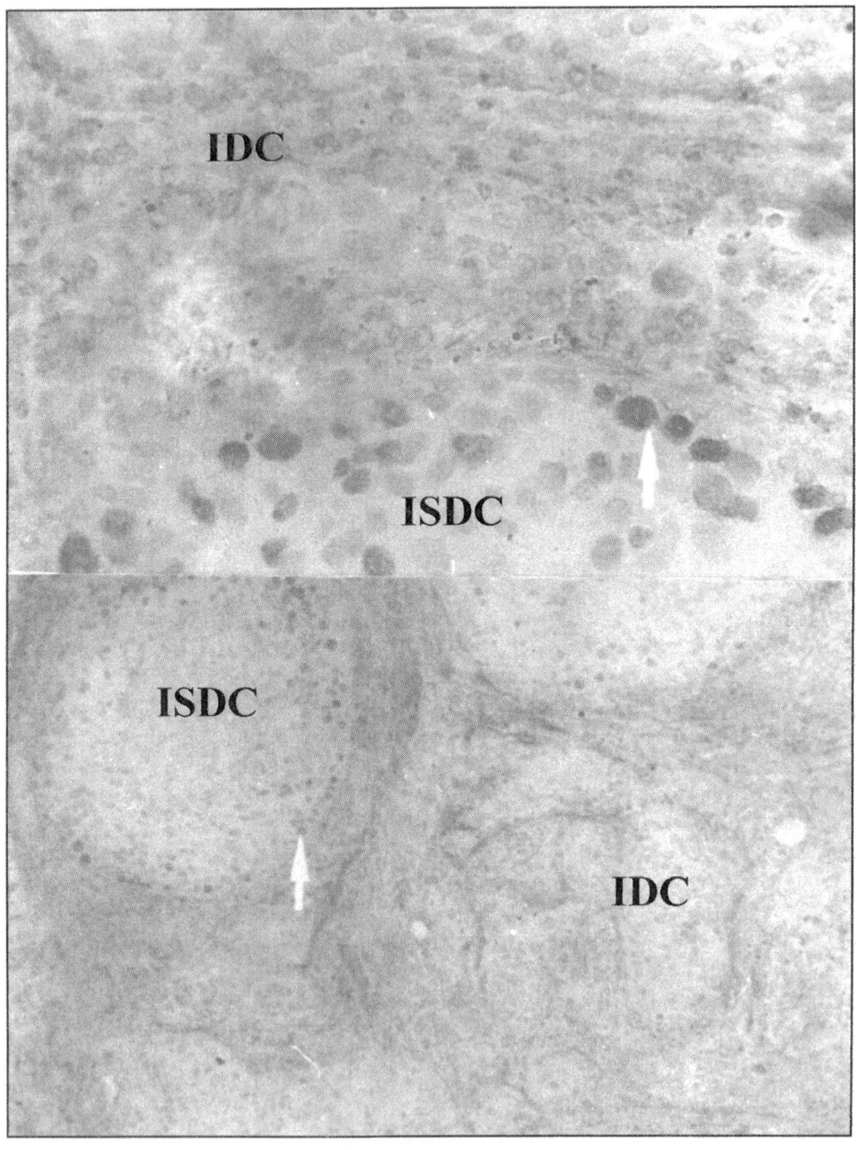

FIGURE 1. Difference in immunostaining of p53 between *in situ* ductal carcinoma (ISDC) and invasive ductal carcinoma (IDC) cells; high p53 immunoreactivity (*arrow*) was observed in tumor cells next to the basal membrane of ISDC.

The prognostic significance of p53 in breast cancer is still controversial. Several studies reported that TP53 gene alteration (overexpression and mutations) is an indicator of increased malignant potential and worse prognosis for breast cancer.[22] However, such a negative prognostic value has not been confirmed in other studies.[23]

Concerning invasive ductal carcinoma of the breast, several studies reported the bad prognostic significance of p53 protein.[24] In the present study, the significant value of p53 accumulated in breast invasive ductal carcinoma (IDC) among Tunisian patients was assessed, and an association between biological and prognostic markers was studied.

MATERIAL AND METHODS

Tissue Samples

Samples from 70 primary breast carcinomas, diagnosed as IDC of the breast, were obtained from the Department of Anatomic-Cytologic-Pathology, CHU (Sfax, Tunisia). Patients (66 women and 4 men) ranged in age from 25–74 years (mean age 53.1 years). Tumors were graded using the Scarff-Bloom-Richardson scale (well differentiated, SBRI; intermediately differentiated, SBRII; and poorly differentiated, SBRIII). Clinicopathologic patient information was obtained from patients' files and pathologic reports from the Department of Radiation and Clinical Oncology of CHU (Sfax, Tunisia).

Immunohistochemistry

Tumors, fresh frozen sections 4 μm thick, were treated with 0.3% hydrogen peroxide-methanol to remove endogenous peroxidase activity. After blocking with 2% nonfat milk, sections were incubated with appropriate monoclonal antibody. The mouse-raised antibodies were: p53 (DO7; Dako), MDM2 (Clone IF2, monoclonal antibody; Calbiochem), bcl2 (clone 124 monoclonal antibody; Dako), and ER (clone1D5, monoclonal antibody; Dako). Goat-antimouse antibody (Dako) was used as a secondary antibody and was revealed by peroxidase-labeled streptavidin (Dako). For bax immunostaining, a polyclonal antibody (Calbiochem) followed by the same procedure just outlined was used.

Slides were incubated in the presence of diaminobenzinidine or hydrogen peroxide chromogen substrate and 1% osmium tetroxide to develop the chromogen signal.

We considered nuclear immunoreactivity in more than 20% of cases as tumor p53 and ER positive. Cytoplasmic immunostaining of bcl2 in more than 20% of cells and bax in more than 30% was considered positive. For MDM2, cytoplasmic or nuclear immunostaining in more than 20% was considered positive.

Clinical Biological Correlations

In only 50 female patients could we obtain useful information about conventional prognostic factors (histologic grade, ER status, tumor size, and nodule metastasis). Therefore, in these 50 cases, we looked for the eventual association between biological markers (p53, bcl2, MDM2, bax, and ER), age, and prognostic factors. To

TABLE 1. Summary of statistically significant associations between biological markers (p53, MDM2, bcl2, bax, and ER) and prognostic markers (ER and SBR) obtained in univariate and multivariate analysis of 70 cases of invasive ductal carcinoma (IDC)

Analysis	Combination	Probability
Univariate	p53+/MDM2+	0.025
	p53+/ER	0.000008
	p53+/bcl2	0.00002
	Bcl2+/ER+	0.001
p53 correlations	P53+/ER-,SBRII-SBRIII	0.000066
	p53+/MDM2+,bax–	0.04
	p53+/bcl2–, ER-	0.000004
	p53+/MDM2+,ER– and MDM2–,ER–	0.000001
	p53+/ER–,bax–	0.000004
bcl2 correlations	bcl2+/ER+,SBRI-SBRII	0.007
	bcl2+/p53–,bax– and bcl2–/p53+,bax–	0.000004
	bcl2+/ER+,bax–	0.002
MDM2 correlations	MDM2+/bcl2–,ER–	0.003
	MDM2+/p53+,bcl2–	0.00005
	MDM2+/p53+,ER–	0.000001

exclude heterogeneity in our study, we did not investigate associations in patients 40 years or older.

Statistical Analysis

The $\chi 2$ test was used to statistically evaluate either the relations between the status of different proteins (p53, bcl2, MDM2, bax, and ER) or the associations between these protein expressions and conventional prognostic factors (histologic grade, ER status, tumor size, and nodal metastasis). The results of this test were usually associated with Yate's Correction (Program CONTING version 2.80) and confirmed by Fisher's test in multivariate analysis.

RESULTS

Status of Different Proteins in Breast IDC. The p53 protein was expressed in 41 of 70 tumors (58%). Study of the status of its target gene expression showed that the *mdm2* gene was expressed in 43 cases (61%), Bcl2 in 29 (41%), and Bax in only 9 (12%). Estrogen receptor was accumulated in 38 tumor tissues (54%).

Relation between Biological Markers. TABLE 1 summarizes the statistically significant association between biological markers (p53, MDM2, bcl2, bax, and ER) and prognostic markers (ER and SBR) obtained in univariate and multivariate analyses. The accumulated p53 was significantly associated with MDM2-positive, bcl2-negative, and ER-negative cells ($P = 0.024$, $P = 0.000027$, and $P = 0.000008$, respectively), whereas with bax there was no significant correlation. Bcl2 immunostaining displayed a positive correlation with ER ($P = 0.001$) and, by multivariate analysis, a negative one with the couple p53/bax ($P = 0.000004$). This analysis also showed a significantly higher fraction of p53-positive cells in ER-negative/SBRII-III tumors than in ER-positive/SBRI-II tumors ($P = 0.000066$). On the contrary, bcl2-positive staining was significantly correlated with ER-positive/SBRI-II tumors ($P = 0.007$). Mdm2 overexpression displayed the same phenotype as p53 in the correlation with bcl2 and ER ($P = 0.003$); indeed, significant associations were observed between "MDM2-positive/p53-positive" and bcl2-negative or ER-negative ($P = 0.00005$ and 0.000001, respectively). MDM2-positive was also slightly correlated with the phenotype "p53-positive/bax-negative" ($P = 0.04$).

Associations with Prognostic Factors (TABLES 2 and 4). In 50 women with useful data files of prognostic factors (histologic grade, ER status, tumor size, and nodal metastasis), we investigated the eventual associations between biological markers (p53, bcl2, MDM2, bax, and ER), age, and these prognostic factors. The results clearly showed a different pattern of p53 status between two groups based on age (TABLE 2): the association "older than 40 years and p53+" is statistically significant ($P = 0.03$). To exclude the heterogeneity study, we limited our investigation of associations to groups older than 40 years of age.

The associations between the biological markers (p53, bcl2, bax, mdm2, and ER), just mentioned (for all 70 cases, TABLE 1) were also present in this homogeneous study of women older than 40 years (TABLE 3). For instance, p53+/ER– ($P = 0.001$) and bcl2+/ER+ ($P = 0.027$) was found (TABLE 4A). A significant association was also found between accumulated p53 and SBRIII ($P = 0.03$) (TABLE 4A).

DISCUSSION

This work is the first report in Tunisia of the significant value of p53 breast IDC, which constitutes the most frequent type of carcinoma in our region (89.5% of all breast cancer diagnosed in 2000).[25]

p53 was frequently accumulated in IDC (58%, 41 of 70 cases), and this accumulation constituted a poorer prognosis, because a significantly higher fraction of p53-positive cells was observed in ER-negative and SBRIII tumors ($P = 0.000066$), suggesting that p53 function was altered in these tumors.

The study of p53 target gene expression, through which p53 exercises its essential function, showed various expression profiles. The *mdm2* gene was expressed in 58.8% of the 70 cases, bax in 12.8%, and the bcl2 gene in 41.4%. The frequent overexpression of *mdm2* in invasive breast cancer was already reported and taken as an indicator of breast cancer prognosis.[15]

Statistical study of different relations between the different biological markers showed that the accumulated p53 was significantly associated with MDM2-positive, bcl2-negative, and ER-negative cases ($P = 0.025$, $P = 0.0001$, and $P = 0.000006$, respectively), whereas no correlation was found with bax. The association of p53+

TABLE 2. Statistical evaluations of relation between biological markers p53, MDM2, bcl2, bax, and ER with age and all prognostic factors in 50 cases of IDC

	p53+	p53−	P^c	bcl2+	bcl2−	P^c	MDM2+	MDM2−	P^c	bax+	Bax−	P^c	ER+	ER−	P^c
Age: ≤40	3	10	**<u>0.03</u>**	9	4	**0.1**	9	4	**0.5**	3	10	**0.1**	19	18	**0.2**
≥40	22	16		16	21		22	15		3	35		9	4	
N+[a]	15	15	**0.8**	12	18	**0.1**	18	12	**0.7**	4	26	**1.0**	17	13	**0.4**
N−[a]	8	9		11	6		11	6		2	15		8	9	
T2[b]	14	14	**0.5**	18	13	**0.06**	19	10	**0.5**	4	25	**0.5**	14	15	**0.4**
T3–T4[b]	10	12		6	13		11	10		2	19		9	6	
ER+	7	21	**<u>0.0002</u>**	19	9	**<u>0.001</u>**	18	10	**0.5**	4	23	**0.5**	xxx	xxx	
ER−	17	5		5	17		14	8		2	21		xxx	xxx	
SBRI−	0	3	**0.09**	3	0	**<u>0.041</u>**	2	1	**0.9**	1	2	**0.2**	3	0	**<u>0.016</u>**
SBRII−	18	19		18	19		23	14		3	34		23	14	
SBRIII−	7	3		2	8		6	4		2	8		2	8	

[a] N+/N−, presence or absence of nodal metastasis.
[b] T2, T3, and T4, tumor size according to WHO classification.
[c] *P* values less than 0.05 are underlined.

TABLE 3. Statistical study of different biological markers (p53, MDM2, bcl2, bax, and ER) in IDC in a group of women older than 40 years

A. Univariate analysis

	bcl2+	bcl2−	*P*	MDM2+	MDM2−	*P*	bax+	bax−	*P*	ER+	ER−	*P*
p53+	4	18	<u>**0.002**</u>	16	6	<u>**0.04**</u>	3	19	NS	6	16	<u>**0.001**</u>
p53−	11	4		6	9		0	15		13	2	

B. Multivariate analysis

	MDM2+/ ER+	MDM2+/ ER−	MDM2/ ER+	MDM2/ ER−	*P*	bcl2+/ MDM2+	Bcl2+/ MDM2−	bcl2−/ MDM2+	Bcl2−/ Mdm2−	*P*
p53+	5	10	0	6	<u>**0.0003**</u>	4	0	12	6	<u>**0.001**</u>
p53−	5	1	8	1		5	6	1	3	
	MDM2+/ bax+	MDM2+/ bax−	MDM2/ bax+	MDM2/ bax−		bcl2+/ bax+	bcl2+/ bax−	bcl2−/ bax+	bcl2−/ bax−	
p53+	3	13	0	6	<u>**0.08**</u>	2	2	1	17	<u>**0.0008**</u>
p53−	0	6	0	9		0	11	0	4	
	bcl2+/ ER+	bcl2+/ ER−	bcl2−/ ER+	bcl2−/ ER−		ER+/ bax+	ER+/ bax−	ER−/ bax+	ER−/ bax−	
p53+	2	2	4	14	<u>**0.0005**</u>	2	4	1	15	<u>**0.0007**</u>
P53−	9	2	4	0		0	13	0	2	

NOTE: *P* values less than 0.05 are underlined.

TABLE 4. Statistical study of different biological markers (p53, MDM2, bcl2, bax, and ER) and prognostic factors in invasive ductal carcinoma (IDC) in a group of women older than 40 years

A. Univariate analysis

	p53+	p53–	*P*	bcl2+	bcl2–	*P*	MDM2+	MDM2–	*P*	ER+	ER–	*P*
N+	13	11	<u>0.9</u>	9	15	**0.6**	14	10	**1.0**	15	9	**0.09**
N–	8	4		6	6		7	5		4	8	
T2	13	8	<u>0.5</u>	9	12	**0.4**	13	8	**0.5**	9	12	**0.3**
T3–T4	7	5		4	8		6	6		7	5	
SBRI–	0	1	<u>0.03</u>	1	0	**0.2**	1	0	**0.1**	1	0	**0.1**
SBRII–	14	16		13	17		16	14		17	13	
SBRIII–	6	0		1	5		5	1		1	5	
ER+	6	13	<u>0.001</u>	11	8	<u>0.027</u>	10	8	**0.7**	xx	xx	xx
ER–	16	2		4	14		11	7		xx	xx	xx

B. Multivariate analysis

	p53+/ bcl2+	p53+/ bcl2–	p53–/ bcl2+	p53–/ bcl2–	*P*	p53+/ ER+	p53+/ ER–	p53–/ ER+	p53–/ ER–	*P*
N+	2	11	7	4	**0.4**	5	8	10	1	**0.4**
N–	2	6	4	0		1	7	3	1	
T2	4	10	6	2	**0.4**	3	11	6	2	**0.6**
T3–T4	0	7	4	1		2	5	5	0	
SBRI–	0	0	1	0	**0.3**	0	0	1	0	**0.3**
SBRII–	3	13	10	4		5	11	10	4	
SBRIII–	1	5	0	0		1	5	0	0	

NOTE: *P* values less than 0.05 are underlined.

with bcl2− or ER− cases was also reported in several studies of breast cancer,[20,24] but its positive correlation with MDM2 in IDC was never reported, to our knowledge, and its statistical significance is doubtful ($P = 0.02$). In fact, according to its inactivation potential of p53-mediated transactivation, MDM2 is thought to induce oncogenesis by inactivating the tumor suppressor p53.[15] Therefore, overexpression of MDM2 is frequent in cancers with wild-type p53,[15] but other studies revealed alteration in the expression of both proteins.[15]

Bcl2 immunostaining displayed a positive correlation with ER ($P = 0.003$) and, by multivariate analysis, a negative one with p53/bax ($P = 0.000004$). A significantly higher fraction of p53-positive cells was observed in ER-negative/SBRII-SBRIII tumors than in ER-positive/SBRI-SBRII tumors ($P = 0.000066$), but bcl2-positive cells were significantly correlated with ER-positive/SBRI-SBRII tumors ($P = 0.007$) and with ER+/bax− tumors ($P = 0.002$), which suggests that bcl2 accumulated in tumor cell inhibited the apoptotic process but maintained the histologic grade of surviving tumor cells.

The relation of p53+/bcl2−,bax− with P = 0.000004 indicated that the accumulation of p53 in the absence of the pro-apoptotic factor (bax) and the pro-survival factor (bcl2) might be responsible for the loss of the differentiation process. The association of p53+/ER−,bax− ($P = 0.000004$) strengthened this conclusion, because ER could be a marker of epithelial breast cell differentiation.

Our results show a significant association between bcl2+/ER+,SBRI-SBRII (TABLE 1). In this context, a significant association of bcl2+/ER+ was already described in univariate analysis, but only a weak significant association was reported for bcl2+/SBRI-SBRII.[7] p53 was known to regulate cell differentiation;[4–6] therefore, these results of statistical analysis, either univariate or multivariate, confirmed that p53 accumulation is correlated with a poorer prognosis in IDC, as described in various breast cancer studies.[21,22,24] Interestingly, these results also showed that p53 accumulated in our group of cases of breast IDC was associated with the absence of pro-apoptotic factor (bax) and pro-survival factor (bcl2), which might be responsible for the prognostic value of p53 in IDC.

MDM2 immunostaining displayed the same correlations as p53 with bcl2 and ER. Moreover, significant associations were observed between MDM2-positive/p53-positive, bcl2-negative, and ER-negative tumors, respectively ($P = 0.00005$ and 0.000001, respectively). MDM2-positive cells were significantly correlated with the p53-positive/bax-negative phenotype ($P = 0.04$), but no significant associations were noted between MDM2, ER, and bax ($P = 0.4$) or between MDM2+/bcl2−,bax− ($P = 0.07$), suggesting that MDM2 was not the sole inactivation factor of p53 function.

Statistical study of different relations between different markers (p53, bcl2, MDM2, bax, and ER) with age and all conventional prognostic factors showed a significantly different pattern of p53 status between two age groups (TABLE 2). Interestingly, p53 was not associated with sporadic breast cancer oncogenesis in Tunisian women younger than 40 years, suggesting that other genetic factors, such as BRCA1 and BRCA2, could be responsible for these cancers.

The associations between the biological markers presented in TABLE 1 (for the 70 cases) were also present in the homogeneous study group of only cases more than 40 years of age (TABLE 3), which confirmed the suggestions just outlined. Moreover, a significant association was noted between the overexpression of p53 and SBR-III histological grade of tumor cells (TABLE 4A).

CONCLUSIONS

This work, focused on the relation between p53 and MDM2, bcl2, bax, and ER, has highlighted the importance of considering p53 as part of a dynamic and multifaceted control network rather than its isolation. The combined evaluation of ER/p53 and ER/bcl2 improved prognostic/predictive capability and allowed the separation of ER-positive and ER-negative cases into subgroups with different prognoses. A difference was noted between the poorly differentiated IDC, found to be associated with the p53+/ER– phenotype, and the well differentiated IDC, which was associated with the bcl2+/ER+ phenotype. The p53 accumulated in these tumor tissues is associated with bad prognostic markers (ER– and SBRIII), and the potential prognostic significance of p53 was significantly manifested in sporadic breast IDC of women older than 40 years. The *mdm2* gene expression might be responsible for the p53 poorest prognosis. Regulation of the apoptotic process is involved in IDC; therefore, bcl2 overexpression is associated with good prognostic markers (ER+ and SBRI-II), whereas the regulation of bax is complex and does not necessarily correlate with the overexpression of p53.

ACKNOWLEDGMENTS

We warmly thank Leila Sellami Hakim, Semia Borchéni, Raoudha Abdennather, and Mohammed Zeghal for their expert technical assistance in immunohistochemistry. We also deeply thank Drs. Wissem Siala, Olfa Hdiji, and Jlaïl Mohammed for all clinicopathologic patient data. We would like to thank Dr. Ahmed Rebaï for providing the Utility Program for Analysis of Genetic Linkage: Program CONTING version 2.80. We thank Hafedh Belghith for discussion. This work was partially supported by the "Contrat-Programme" governmental funds provided by the "Secrétariat d'Etat à la Recherche Scientifique et à la Technologie" of Tunisia.

REFERENCES

1. LEVINE, J.A. p53, the cellular gatekeeper for growth and division. 1997. Cell **88:** 323–331.
2. LEVINE, A., J. MOMAND & A.C. FINLY. 1991. The p53 tumour suppresser gene. Nature **351:** 453–456.
3. GOTTLIEB, T. & M. OREN. 1996. p53 in growth control and neoplasia. Biochim. Biophys. Acta Rev. Cancer **1287:** 77–102.
4. HERNANDEZ, B.G., Q. ZHAN, R. WONG & S.Y. CHENG. 1998. Thyroid hormone receptor is a negative regulator in p53-mediated signalling pathways. DNA Cell Biol. **17:** 743–750.
5. HOLLY, S., K. LEONARD, R. LEE, ***et al.*** 1994. P53-dependent apoptosis suppresses tumour growth and progression in vivo. Cell **78 :** 703–711.
6. LI, Z.R., R. HROMCHAK, A. MUDIPALLI & A. BLOCH. 1998. Tumour suppressor protein as regulator of cell differentiation. Cancer Res. **58:** 4282–4287.
7. REHMAN, S., J. CROW & P.A. REVELL. 2000. Bax protein expression in DCIS of the breast in relation to invasive ductal carcinoma and other molecular markers. Path. Oncol. Res. **6:** 256–263.
8. KUERBITZ, S.J., B.S. PLUNKETT, W.V. WALSH & M.B. KASTAN. 1992. Wild-type p53 is a cell cycle checkpoint determinant following irradiation. Proc. Natl. Acad. Sci. USA **82:** 1092–1096.

9. WHITE, E. 1996. Life, death and the pursuit of apoptosis. Genes Dev. **10:** 1–15.
10. HALL, P.A., P.H. MCKEE, H.D.P. MENAGE, ***et al.*** 1996. High levels of p53 protein in UV irradiated human skin. Oncogene **8:** 203–208.
11. MIERTZ, J.A., T. UNGER, J.M. HUBERGETSE & P.M. HOWLEY. 1992. The transcriptiona transactivation function of wtp53 is inhibited by SV40 large T-antigen and by HPV-16 E6 oncoprotein. EMBO J. **11:** 5013–5020.
12. HAUPT, Y., B. YAACOV & M. OREN. 1996. Cell type-specific inhibition of p53 mediated apoptosis by MDM2. EMBO J. **15:** 1596–1606.
13. HONDA, R., H. TANAKA & H. YASUDA. 1997. Oncoprotein MDM2 is a ubiquitn ligase E3 for tumour suppressor p53. FEBS Lett. **420:** 25–27.
14. KUBBUTAT, H.G.M., N.J. STEPHEN & H.V. KAREN. 1997. Regulation of p53 stability by MDM2. Nature **387:** 299–230.
15. DEB, S.P. 2002. Function and dysfunction of the human oncoprotein MDM2. Front. Biosci. **7:** 235–245.
16. FRIEDLANDER, P., H. HAUPT, C. PROIVES & M. OREN. 1996. A mutated p53-responsive genes cannot induce apoptosis. Molec. Cell. Biol. **16:** 4961–4971.
17. JOHNSON, A., R. INCE, A. TAN & K.W. SCOTTO. 2001. Transcription repression by p53 through direct binding to a novel DNA element. J. Biol. Chem. **276:** 27716–27727.
18. STRASSER, A., W.A. HARRIS, J. TYLER & Z. CORY. 1994. DNA damage can induce apoptosis in proliferating lymphoid cells via p53-independent mechanism inhibitable by bcl2. Cell **79:** 329–339.
19. ROCHAIX, P., S. KRAJEWSKI, J.C. REED, ***et al.*** 1999. *In vivo* patterns of Bcl-2 family protein expression in breast carcinomas in relation to apoptosis. J. Pathol. **187:** 410–415.
20. CRAIG, D., M.H. JENNET, B. MELORA & M.C. GRAY. 1998. Prognostic and predictive factors in breast cancer by immunihistochemical analysis. Mod. Pathol. **11:** 155–168.
21. EARLY BREAST CANCER TRIALISTS' COLLABORATIVE GROUP. 1992. Systemic treatment of early breast cancer by hormonal, cytotoxic, or immune therapy: 133 randomised trials involving 31,000 recurrences and 24,000 deaths among 75,000 women. Lancet **339:** 71–85.
22. GRETARSDOTTIR, S., L. TRYGGVADOTTIR, J.G. JONASSON, ***et al.*** 1996. TP53 mutation analysis on breast carcinomas: a study of paraffin-embedded archival material. Br. J. Cancer **74:** 555–561.
23. VARLEY, J.M., W.J. BRAMMAR, D.P. LANE, ***et al.*** 1991. Loss of chromosome 17p13 sequences and mutation of p53 in human breast carcinomas. Oncogene **6:** 413–421.
24. GURSAN, N., M. KAAKOK, I. SARI & M.S. GURSAN. 2001. The relationship between expression of p53/bcl2 and histopathological criteria in breast invasive ductal carcinomas. Int. J. Clin. Pract. **55:** 589–590.
25. JLIDI, R., A. SELLAMI, M. HSAÏRI & N. ACHOUR. 2002. Incidence des cancers années 1997–2000. Registre du cancer du sud Tunisien.

No Evidence of Correlation between p53 Codon 72 Polymorphism and Risk of Bladder or Breast Carcinoma in Tunisian Patients

IMED MABROUK,*[a,c] SAMI BACCOUCHE,*[a] RYM EL-ABED,[c] RAJA MOKDAD-GARGOURI,[a] ALI MOSBAH,[d] SALEM SAÏD,[c] JAMEL DAOUD,[b] MOUNIR FRIKHA,[b] RACHID JLIDI,[b] AND ALI GARGOURI[a]

[a]*Centre de Biotechnologie de Sfax, BP "K" 3038, Sfax, Tunisia*

[b]*CHU Habib Bourguiba, Sfax, Tunisia*

[c]*Faculté de Médecine de Sousse, Tunisia*

[d]*CHU Sahloul-Sousse, Tunisia*

ABSTRACT: The TP53 gene, frequently mutated in human cancers, carries several polymorphisms. The one most informative and studied concerns codon 72; a single base changes the CGC (arginine) to CCC (proline). The arginine form was considered to be a significant risk factor in the development of cancer. However, various reports on this polymorphism are controversial. We carried out the same investigation in two groups of patients, a group with bladder cancer and another with breast cancer, and in healthy controls in two regions of our country, using an improved PCR-RFLP method. The number of Arg/Arg, Arg/Pro, and Pro/Pro genotypes was as follows: 21, 23, 3 and 13, 19, 2 for patients (total 47) and controls (34), respectively, in the first group; 18, 9, 3 and 19, 26, 4 for patients (30) and controls (49), respectively, in the second group. Statistical analysis of the genotype and allele frequencies did not reveal any difference between patients and controls in both groups except for a weak difference between the homozygotes to heterozygotes in the second group with a chi square of 4.1 ($P = 0.045$); the number of breast cancer patients is actually low (30) and should be increased in order to assess such a conclusion. Our overall results are therefore not consistent with a high risk associated with *TP53* codon 72 polymorphism in breast and in bladder cancers.

KEYWORDS: *TP53*; codon 72 polymorphism; PCR; statistics; breast cancer; bladder cancer

INTRODUCTION

Breast cancer is the most frequently encountered cancer in women throughout the world. Its incidence in our region is increasing and reached 27.6 per 100,000 women,

*Contributed equally to this work.

Address for correspondence: Ali Gargouri, Laboratory Génétique Moléculaire des Eucaryotes, Centre de Biotechnologie de Sfax, BP "K" 3038, Sfax, Tunisia. Voice: 216-74 27 41 10; fax: 216-74 27 59 70.

Faouzi.gargouri@cbs.rnrt.tn

Ann. N.Y. Acad. Sci. 1010: 764–770 (2003).
doi: 10.1196/annals.1299.137

in 1999.[1] Bladder cancer is also frequent among men and is usually ranked just after lung carcinoma. In our region, its incidence is about 17.2 per 100,000 men in 1999.[1]

Epidemiologic studies have suggested a number of risk factors, including genetic and environmental ones. The *TP53* tumor suppressor gene is the most involved genetic factor for breast cancer, in addition to BRCA1 and BRCA2, and, to a lesser extent, for bladder cancer.

The human tumor suppressor gene *TP53*, located on chromosome 17p13, encodes for a transcription factor that regulates the expression of different genes related to cell cycle control, apoptosis,[2] and antiproliferative effects.[3] The human *P53* protein is mutated and accumulated in more than 50% of cancers.[4] The *TP53* gene mutation can damage its DNA-binding or transactivation functions, thereby inhibiting its key role in cell cycle control.

The *TP53* gene exhibits several restriction site polymorphisms, but the most informative is the *Acc*II (or *BstU* I), located at codon 72 of exon 4, with a frequency of *Acc*II in about 65%.[5,6] This polymorphism involves a single base change that substitutes arginine (CGC) with proline residue (CCC) within the *P53* transactivation domain.

The involvement of this polymorphism in carcinogenesis is well studied but is a matter of controversy. Indeed, an association between the codon 72 polymorphism and human cancer susceptibility has been reported in bladder,[7,8] breast,[9] and lung cancer.[10,11] However, in other studies, this polymorphism does not seem to affect the risk of breast[12] and cervical cancer.[13] The codon 72 polymorphism might also be a risk factor in the development of human virus-associated cancer such as human papillomavirus (HPV)-associated skin cancer,[14] HPV-associated cervical cancer,[15,16] and esophageal cancer.[17] This polymorphism causes a functional difference between the two *P53* versions; in fact, a *P53* Arg/Arg genotype induces apoptosis and suppresses transformation more efficiently than does a Pro/Pro genotype.[18] However, the presence of an arginine residue significantly increases the susceptibility of *P53* protein to degradation mediated by the HPV E6 oncoprotein.[19]

In most studies, the *TP53* gene polymorphism has been identified by a simple method. It consists of the amplification of the exon 4 fragment followed by digestion with the *Acc*II restriction enzyme. Such RFLP analysis allowed the identification of genotypes at codon 72: Arg/Arg, Pro/Pro, or Arg/Pro. However, with the previous method, partial digestion of the amplified fragment (exon 4) could cause false conclusions. Therefore, we introduced a sensitivity improvement to this method, which is based on amplification of the exon 4-6 fragment that contains an additional *Acc*II site.[20]

In the present study, we examined the relation between the codon 72 polymorphism in *TP53* and the risk for tumorigenesis of the breast or the bladder in Tunisian patients.

MATERIAL AND METHODS

Cancer Cases and Healthy Subjects. Two groups of patients/controls were collected from two regions of Tunisia. One consisted of 47 cases of bladder carcinoma and 34 healthy persons from regions surrounding Sousse town, the center of Tunisia.

The second group, originating from Sfax town and the Tunisian south, was made up of 30 breast cancer cases and 49 controls.

Genomic DNA Extraction. Genomic DNA was extracted from the blood of patients and controls according to the method of Sambrook *et al.*[21]

PCR and AccII Digestion Conditions. The primers used for the amplification "exons 4-6" were: sense: 5′-TCC CCC TTG CCG TCC CAA GC - 3′; antisense: 5′-GTC AGG CGG CTC ATA GG –3′.

The target sequence was amplified in a 50-μl reaction volume containing 100 ng of genomic DNA, 0.2 mM dNTPs, 1 × Taq buffer, 2 mM $MgCl_2$, 100 ng of each primer, and 1 unit of Taq DNA polymerase *(Amersham Pharmacia Biotech)*. The amplification program (30 seconds at 94°C, 30 seconds at 58°C, and 2 minutes at 72°C) was performed for 40 cycles, followed by one extending cycle at 72°C for 10 minutes. Fifteen microliters from the PCR product were digested overnight by 10 units of *Acc*II restriction enzyme *(Amersham Pharmacia Biotech)*. Such a large amount of restriction enzyme and long incubation time ensured complete digestion of DNA. DNA fragments were separated by electrophoresis on 2% agarose gel, along with a 100-bp ladder marker *(Amersham Pharmacia Biotech)* and visualized under UV after staining with ethidium bromide.

Statistical Analysis. Statistical analysis was performed using the chi-square test with Yates' corrections. Statistical significance was regarded at a P value <0.05.

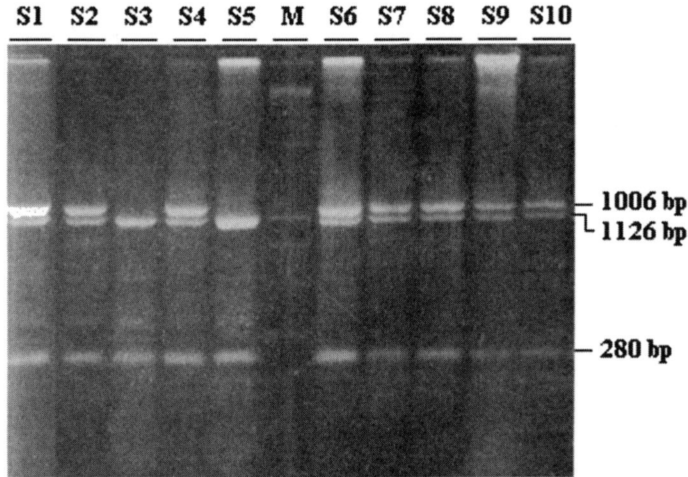

FIGURE 1. Detection of codon 72 polymorphism by *Acc*II digestion of PCR fragment ex4-6 of the *TP53* gene in 10 blood samples (S1-S10) from independent subjects. We focused on the doublet 1006-1126 bp fragments of *Acc*II digestion, because it was sufficient to conclude for the codon 72 genotype.[20] The 100-bp fragment was not shown. S1, S2, S4, and S6–S10, heterozygotes (Arg/Pro). S3 and S5, homozygotes (Arg/Arg). M, 100-bp marker ladder.

RESULTS

The typical pattern of codon 72 evaluated by restriction analysis of exon 4-6 fragment was presented in FIGURE 1. The presence of a 1126–1006-bp doublet was sufficient to verify a heterozygote case (Arg/Pro). For the homozygote cases, the presence of either 1126 bp or 1006 bp indicates a Pro/Pro or Arg/Arg genotype, respectively. The 280-bp fragment is present in all cases and represents internal control of the total DNA digestion, as reported in our previous work.[20]

Blood specimens from 47 patients with bladder cancer, 34 healthy subjects (control population), 30 patients with breast cancer (invasive ductal carcinoma, IDC), and 49 healthy subjects were analyzed for codon 72 polymorphism of the *TP53* gene. Analysis was performed on blood samples and not on the corresponding tumor samples, because of the loss of the heterozygosity (LOH) phenomenon. TABLES 1 and 2 compare the distribution of TP53 genotype and alleles frequencies at codon 72 in healthy subjects and patients with bladder and breast cancers, respectively. For both groups and for all subjects, the Arg/Arg genotype was more frequent than the Pro/Pro genotype, and the Arg allele is roughly three times more represented than the Pro one. In view of these results, no significant difference was observed in the distribu-

TABLE 1. Distribution of TP53 genotypes and allele frequencies of codon 72 polymorphism in bladder cancer cases and in a control population

	Genotypes			Alleles	
	Arg/Arg	Arg/Pro	Pro/Pro	Arg	Pro
Bladder cancer					
Number (T = 47)	21	23	03	65	29
Percentage	44.5%	49%	6.5%	69%	31%
Control populations					
Number (T = 34)	13	19	02	45	23
Percentage	38%	56%	6%	66%	34%
Qui2 = 0.39, P = 0.8				*Qui2 = 0.16, P = 0.68*	

TABLE 2. Distribution of TP53 genotypes and alleles frequencies of codon 72 polymorphism in breast cancer and in a control population

	Genotypes			Alleles	
	Arg/Arg	Arg/Pro	Pro/Pro	Arg	Pro
Breast cancer					
Number (T = 30)	18	09	03	45	15
Percentage	60%	30%	10%	75%	25%
Controls population					
Number (T = 9)	19	26	04	64	34
Percentage	39%	53%	8%	65.3%	34.7%
Qui2 = 4.09, P = 0.1				*Qui2 = 1.63, P = 0.1*	

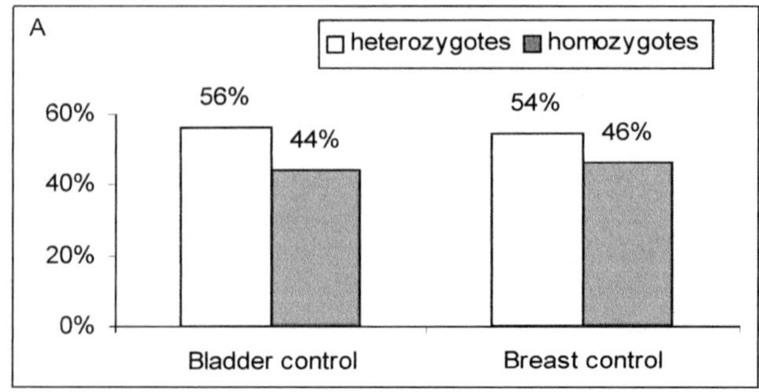

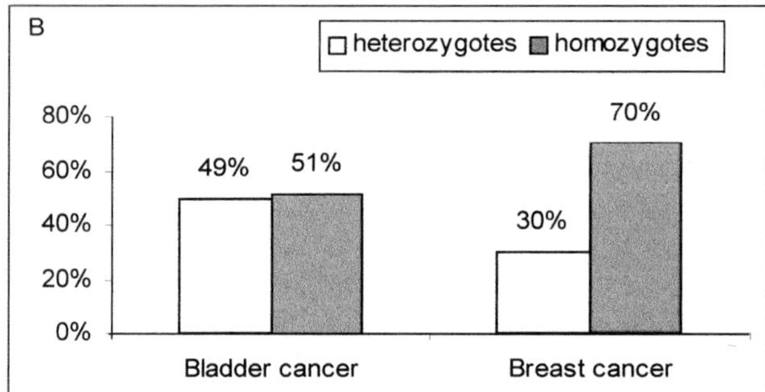

FIGURE 2. Distribution of heterozygote versus homozygote states at TP53 codon 72. (**A**) Comparison between bladder and breast controls. (**B**) Comparison between bladder and breast cancers.

tion of the three genotypes (Arg/Arg, Arg/Pro, and Pro/Pro) between cases of either bladder or breast cancer and controls. The same result was observed in the distribution of allele frequencies.

The distribution of homozygotes versus heterozygotes (FIG. 2) shows a slightly significant statistical association of the homozygote state with the risk of developing breast cancer ($P = 0.04$). On the contrary, no significant difference was observed between cases of bladder cancer and controls.

DISCUSSION

The results presented here indicate that the most common allele in our two populations is the one encoding arginine at position 72, as in the majority of populations tested to date, except some Chinese ones.[16] These authors suggest that the frequencies of codon 72 are ethnically related.

A large controversy already exists about the relation between the risk of tumorigenesis and the codon 72 polymorphism.[13,16,22–28] In this study, we examined the prevalence of the *TP53* codon polymorphism in Tunisian patients with bladder or breast cancer and controls. We found that the frequency distribution of genotypes and alleles did not show any significant difference between patients and healthy controls in both groups. This result is in agreement with similar studies on bladder[29] and breast cancers,[12] in which the frequency distributions were comparable, and on other cancers such as head and neck squamous cell carcinoma[23] and cervical cancer.[13,22] By contrast, it is controversial with several groups detecting an association between the *p53* Arg/Arg genotype and the increased risk of developing bladder,[8] breast,[24] skin,[25] laryngeal,[26] and cervical cancer.[27,28] These studies confirm the original finding of Story *et al.*[19] in HPV-associated cervical cancer in which the *p53Arg* homozygous genotype indicates a significantly higher risk of developing cancer.

Nevertheless, if the ratio of heterozygotes and homozygotes were roughly similar in bladder cancer cases and both control groups, homozygotes were more frequent than heterozygotes among breast cancer patients. Because the p53 protein acts in a tetrameric form, the heterotetramer (in heterozygotes) could possibly be a disfavor for the action of the p53 mutation. Meanwhile, the statistical significance of this difference is not too great ($P = 0.045$), and the number of breast cancer cases studied is low (30) and must be increased in order to verify this result.

Obviously, great care was taken in our methodology, because the method of detection of the p53 codon 72 polymorphism is known to be a source of biased results. A more accurate method of p53 codon 72 polymorphism detection was applied[20] to circumvent the recurrent problem of partial enzyme digestion.

In conclusion, our overall results are not concordant with an association between the polymorphism at codon 72 and the risk of developing bladder or breast cancer. Moreover, an eventual association between the p53 codon 72 polymorphism and pathologic data or genetic alteration in biopsies, point mutations, and loss of heterozygosity in the *TP53* gene should be tested.

ACKNOWLEDGMENTS

We warmly thank Prof. Jalel Gargouri for providing the blood samples. This work was partially supported by the Contrat-Programme governmental funds provided by the Secrétariat d'Etat à la Recherche Scientifique et à la Technologie of Tunisia.

REFERENCES

1. JLIDI, R., A. SELLAMI, M. HSAIRI & N. ACHOUR. 2002. Incidence des cancers année 1997-2000. Register du cancer du sud Tunisien.
2. RAYCROFT, L., H. WU & G. LOZANO. 1990. Transcriptional activation by wild-type but not transforming mutants of the p53 anti-oncogene. Science **249:** 1049–1051.
3. MARX, J., J. MOMAND & C.A. FINLAY. 1993. How p53 suppresses cell growth. Science **262:** 1644–1645.
4. NIGRO, J.M., S.J. BARKER, A.C. PREISINGER, *et al.* 1989. Mutations in the p53 gene occur in diverse human tumour types. Nature **342:** 705–708.
5. ARA, S., P.S.Y. LEE, M.F. HANSEN & H. SAYA. 1990. Codon 72 polymorphism of the TP53 gene. Nucl. Acids Res. **18:** 4961.
6. DELACALLE-MARTIN, O., V. FABREGAT, M. ROMERO, *et al.* 1990. *Acc*II polymorphism of the p53 gene. Nucl. Acids Res. **18:** 4963.
7. OLSCHWANG, S., P. LAURENTPUIG, A. VASSAL, *et al.* 1991. Characterization of a frequent polymorphism in the coding sequence of the Tp53 gene in colonic cancer patients and a control population. Hum. Genet. **86:** 369–370.

8. Soulitzis, N., G. Sourvinos, D. Dokianakis & D.A. Spandidos. 2002. p53 codon 72 polymorphism and its association with bladder cancer. Cancer Lett. **179:** 175–183.
9. Sjalander A., R. Birgander, G. Hallmans, *et al.* 1996. p53 polymorphisms and haplotypes in breast cancer. Carcinogenesis **17:** 1313–1316.
10. Rong, F., W. Ming-Tsang, M. David, *et al.* 2000. The p53 Codon 72 polymorphism and lung cancer risk. Cancer Epidemiol. Biomark. Prev. **9:** 1037–1042.
11. Pierce, M.L., L. Sivaraman, W. Chang, *et al.* 2000. Relationships of TP53 codon 72 and HRAS1 polymorphisms with lung cancer risk in an ethnically diverse population. Cancer Epidemiol. Biomark. Prev. **9:** 199–1204.
12. Weston, A. & J.H. Godbold. 1997. Polymorphisms of H-ras-1 and p53 in breast cancer and lung cancer: a meta-analysis. Environ Health Perspect. Suppl **4:** 919–926.
13. Tenti, P., N. Vesentini, M.R. Spaudo, *et al.* 2000. p53 Codon 72 polymorphism does not affect the risk of cervical cancer in patients from Northern Italy. Cancer Epidemiol. Biomark. Prev. **9:** 435–438.
14. O'Connor, D.P., E.W. Kay, M. Leader, *et al.* 2001. p53 codon 72 polymorphism and human papillomavirus associated skin cancer. J. Clin. Pathol. **54:** 539–542.
15. Madeleine, M.M., K. Shera, S.M. Schwartz, *et al.* 2000. The p53 Arg72Pro polymorphism, human papillomavirus, and invasive squamous cell cervical cancer. Cancer Epidemiol. Biomark. Prev. **9:** 225–227.
16. Guimaraes, D.P., S.H. Lu, P. Snijder, *et al.* 2001. Absence of association between HPV DNA, TP53 codon 72 polymorphism and risk of oesophageal cancer in a high-risk area of China. Cancer Lett. **162:** 231–235.
17. Klug, S.J., R. Wilmotte, C. Santos, *et al.* 2001. TP53 polymorphism, HPV infection, and risk of cervical cancer. Cancer Epidemiol. Biomark. Prev. **10:** 1009–1012.
18. Thomas, M., A. Kalita, S. Labrecque, *et al.* 1999. Two polymorphic variants of wild-type p53 differ biochemically and biologically. Mol. Cell. Biol. **19:** 1092–1100.
19. Storey, A., M. Thomas, A. Kalita, *et al.* 1998. Role of a p53 polymorphism in the development of human papillomavirus-associated cancer. Nature **393:** 229–234.
20. Baccouche, S., I. Mabrouk, S. Said, *et al.* 2003. A more accurate detection of codon 72 polymorphism and LOH of the TP53 gene. Cancer Lett. **189:** 91–96.
21. Sambrook, J., E.F. Fritsch & T. Maniatis. 1989. Isolation of High-Molecular-Weight DNA from Mammalian Cells in Molecular Cloning. A Laboratory Manual. 2nd edit. **3:** 9–19. Cold Spring Harbor Laboratory Press. Cold Spring Harbor, NY.
22. Hildesheim, A., M. Schiffman, L.A. Brinton, *et al.* 1998. *p53* polymorphism and risk of cervical cancer. Nature **396:** 531–532.
23. Hamel, N., M.J. Black, P. Ghadirian & W.D. Foulkes. 2000. No association between *p53* codon 72 polymorphism and risk of squamous cell carcinoma of the head and neck. Br. J. Cancer **82:** 757–759.
24. Papadakis, E.N., D.N. Dokianakis & D.A. Spandidos. 2000. *p53* codon 72 polymorphism as a risk factor in the development of breast cancer. Mol. Cell. Biol. Res. Commun. **3:** 389–392.
25. Dokianakis, D.N., E. Koumantaki, K. Billiri & D.A. Spandidos. 2000. *p53* codon 72 polymorphism as a risk factor in the development of HPV-associated non-melanoma skin cancers in immunocompetent hosts. Int. J. Mol. Med. **5:** 405–409.
26. Sourvinos, G., E. Rizos & D.A. Spandidos. 2001. *p53* codon 72 polymorphism is linked to the development and not the progression of benign and malignant laryngeal tumours. Oral Oncol. **37:** 572–578.
27. Dokianakis, D.N. & D.A. Spandidos. 2000. *p53* codon 72 polymorphism as a risk factor in the development of HPV-associated cervical cancer. Mol. Cell. Biol. Res. Commun. **3:** 111–114.
28. Agorastos, T., A.F. Lambropoulos, T.C. Constantinidis, *et al.* 2000. *p53* codon 72 polymorphism and risk of intra-epithelial and invasive cervical neoplasia in Greek women. Eur. J. Cancer Prev. **9:** 113–118.
29. Toruner, G.A., A. Ucar, M. Tez, *et al.* 2001. p53codon 72 polymorphism in bladder cancer: no evidence of association with increased risk or invasiveness. Urol. Res. **29:** 393–395.

p53, Bcl-2, PCNA Expression, and Apoptotic Rates during Cervical Tumorigenesis

HORACIO ASTUDILLO,[a,b] TEZCATLIPOCA LOPEZ,[a] SEBASTIAN CASTILLO,[a] PATRICIO GARIGLIO,[c] AND LUIS BENITEZ[a]

[a]*Unidad de Investigacion Medica en Enfermedades Oncologicas, Hospital de Oncologia, CMN Siglo XXI, IMSS, CP 06720, Mexico*

[b]*Programa de Biomedicina Molecular ENMyH-IPN, and* [c]*Departamento de Genetica y Biologia Molecular, CINVESTAV-IPN, Mexico*

KEYWORDS: apoptosis; p53; Bcl-2; PCNA; TUNEL; cervical cancer

Among all women's malignancies, cervical cancer is the first cause of death in most underdeveloped countries, including Mexico, and the second most common cancer in women worldwide. An association between human papillomavirus (HPV) infection and the development of precancerous and malignant lesions in the uterine cervix has been proposed.

It is known that deregulation of genes involved in apoptosis, such as Bcl-2 and p53, plays a crucial role in tumor formation. p53 functions as a tumor suppressor by arresting the cell cycle and by triggering apoptosis, and Bcl-2 has been proposed to be able to inhibit apoptosis. Recent studies have focused on the interaction between these two apoptotic regulatory proteins in carcinogenesis.

Cervical dysplasia is an ideal model in which to study tumorigenesis, considering that carcinoma arises from pre-malignant lesions. In order to estimate the apoptotic rates during the development of cervical cancer, we investigated apoptosis by TUNEL assay and morphological criteria, as well as by the expression of proteins related to this process such as p53, Bcl-2 and PCNA in all grades of cervical tumor progression. Additionally, the presence of HPV-DNA was also examined by polymerase chain reaction (PCR) and DNA sequence to type HPV-positive samples.

A total of 105 paraffin-embedded human cervical tissues were selected for this study as follows: 31 were of normal cervical epithelium; 10 of cervical intraepithelial neoplasia grade I (CIN I), 5 of CIN II, 23 of CIN III; and 36 of invasive squamous cell carcinoma (ISCC). For identification of p53, Bcl-2, and PCNA protein expression, the streptavidin–biotin peroxidase complex method and three polyclonal antibodies—clone (FL-393)-G anti-human p53 primary antibody (work dilution 1:100),

Address for correspondence: Dr. Horacio Astudillo, Unidad de Investigacion Medica en Enfermedades Oncologicas, Hospital de Oncologia, CMN Siglo XXI, IMSS, CP 06720, Mexico. Voice: 525-56276900, ext. 4318.
hastud2@aol.com

**Ann. N.Y. Acad. Sci. 1010: 771–774 (2003). © 2003 New York Academy of Sciences.
doi: 10.1196/annals.1299.138**

clone (C21 anti-human Bcl-2 primary antibody (1:100), and clone (FL-261) anti-human PCNA primary antibody (1:100)—were used.

Immunoreactivity was quantified with a 12-point weighted score, previously reported by Sinicrope *et al.*[1] (average weighted score). Apoptotic activity in epithelial cells was detected by *in situ* apoptosis detection system (TUNEL assay). Average apoptotic index, which represented the average of apoptotic cells in each sample, was identified in histological sections stained with H&E using established morphological criteria. At least two 10-μm-thick sections were cut for DNA extraction from paraffin-embedded tissues. The GP+PCR for HPV was performed as described by Walboomers and colleagues[2] to detect HPV infection followed by DNA sequence analysis.

Bivariate comparisons between the weighted scores of p53, Bcl-2, and PCNA were made using the Pearson's product moment correlation coefficient and significance was defined at $P<0.05$. Surprisingly, the average weighted score (AWS) for p53 immunoreactivity in normal cervical epithelium and different types of cervical lesions increased in the following order: normal (1.48±0.36) < CIN I (2.45±0.34) < CIN II/III (2.51±0.48) < ISCC (3.8±0.47). Bcl-2 protein expression increased from normal epithelium (0.92±0.19) to CIN I (3.20±0.48), but decreased in CIN II/III (2.94±0.48), and was highest in ISCC (3.7±0.67). We determined quantitative differences in PCNA expression between the various histological groups, with a high expression directly associated to dysplasia grade (FIG. 1). Statistical analysis reveals correlation between p53 and PCNA expression in CIN I (r=0.378, P=0.016, n=10),

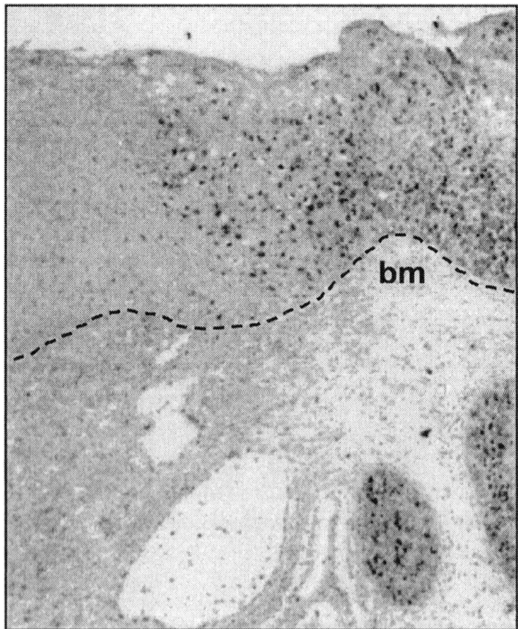

FIGURE 1. Photomicrograph of section taken from CIN II/III showing a strong nuclear PCNA immunoreactivity (bm: basal membrane).

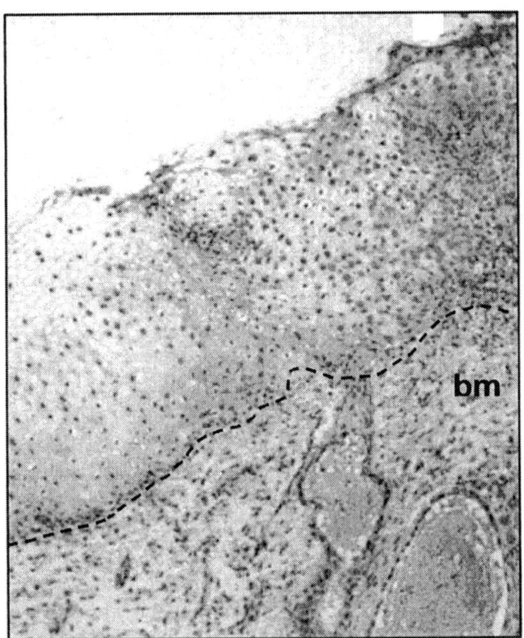

FIGURE 2. *In situ* detection of apoptotic cells in a CIN II/III section by TUNEL assay, showing an increased number of TUNEL-positive cells (bm: basal membrane).

indicating that CIN I lesions were always associated with a strong PCNA and weak p53 expression. A stronger p53 expression than Bcl-2 in CIN II/III was detected, giving a significant correlation (r−0.385, P−0.023, n=28). An increased apoptosis index by TUNEL assay was found toward malignity. TUNEL-positive cell averages were low for epithelial cells of normal epithelium, and increased in the following order: normal < CIN I < CIN II/III < ISCC (FIG. 2). On the other hand, apoptosis by morphological criteria decreased from normal cervical epithelium to CIN I, but no differences were found between CIN I and CIN II/III. ISCC showed the lowest apoptotic indexes.

A total of 76 of 105 (71.4%) cervical samples were HPV-positive, with HPV 16 being the most common high-risk type detected (72.4%) and increasing according to the degree of histological abnormality. Anti-human p53 used herein does not allow wild and mutant types of this protein to be distinguished. Wild-type p53 protein has a short half-life, resulting in immunohistochemically undetectable protein levels, but the intracellular level can increase either due to p53 mutation, cellular response to double-strand DNA damage, or because of posttranslational stabilization by complex-binding to cellular proteins.[3] Interestingly a recent report suggests that a high level of p53 could be used as shorter overall survival indicator in invasive cervical cancer.[4] Moreover gadd45 pathway induces functional p53 in HPV-positive cancer cells, providing evidence that p53 protein is competent to activate a cellular target gene, despite coexpression of the viral E6 oncogene.[5] The Bcl-2 early increase

(CIN I) raises the possibility that this may be closely related to the loss of cell differentiation at the beginning of the dysplasia. PCNA protein expression was significantly increased as the grade of cervical lesion became higher, suggesting that it may be a useful adjunct to histological diagnosis of dysplasia grades in cervical neoplasia independent of HPV type associated with the lesion.

In conclusion TUNEL assays and the expression of apoptosis-related proteins (p53, Bcl-2), as well as PCNA protein expression, could be useful markers in uterine cervical cancer development.

ACKNOWLEDGMENTS

This work was supported by grants from the Mexican Council of Science and Technology (CONACYT) and the Medical Research Council of Mexican Social Security Institute (IMSS), Mexico. We thank M. Sc. Iris A. Feria for statistical analysis and Dr. Jaime Berumen (Hospital Central Militar, Mexico) for technical advice in HPV detection and typification.

REFERENCES

1. SINICROPE, F.A., S.B. RUAN, K.R. CLEARY, *et al.* 1995. bcl-2 and p53 oncoprotein expression during colorectal tumorigenesis. Cancer Res. **55:** 237–241.
2. DE RODA HUSMAN, A.M., J.M. WALBOOMERS, A.J. VAN DEN BRULE, *et al.* 1995. The use of general primers GP5 and GP6 elongated at their 3' ends with adjacent highly conserved sequences improves human papillomavirus detection by PCR. J. Gen. Virol. **76:** 1057–1062.
3. ASHCROFT, M. & K.H. VOUSDEN. 1999. Regulation of p53 stability. Oncogene **18:** 7637–7643.
4. TJALMA, W.A., J.J. WEYLER, J.J. BOGERS, *et al.* 2001. The importance of biological factors (bcl-2, bax, p53, PCNA, MI, HPV and angiogenesis) in invasive cervical cancer. Eur. J. Obstet. Gynecol. Reprod. Biol. **97:** 223–230.
5. BUTZ, K., N. WHITAKER, C. DENK, *et al.* 1999. Induction of the p53-target gene GADD45 in HPV-positive cancer cells. Oncogene **18:** 2381–2386.

Assessment of *In Vivo* Chemotherapy-Induced DNA Damage in a p53-Mutated Rat Tumor by Micronuclei Assay

GREGORY DRIESSENS,[a] LAURA HARSAN,[b] PATRICK BROWAEYS,[a] XENOFON GIANNAKOPOULOS,[a] THIERRY VELU,[a,c] AND CATHERINE BRUYNS[a]

[a]*Interdisciplinary Research Institute (IRIBHM), School of Medicine, Université Libre de Bruxelles, 1070 Brussels, Belgium*

[b]*Faculty of Physics, Cluj-Napoca University, Cluj-Napoca, Romania*

[c]*Department of Oncology, Erasme Hospital, Université Libre de Bruxelles, 1070 Brussels, Belgium*

ABSTRACT: Production of DNA damage is the basis of cancer treatments such as chemo- and radiotherapy. Such treatments induce mitotic catastrophe, a form of cell death resulting from abnormal mitosis and leading to the formation of interphase cells with multiple micronuclei. In this study, we compared apoptosis induction and micronuclei formation to assess the DNA damage provoked *in vivo* by cytotoxic agents in established 9L rat gliosarcoma tumors expressing a mutated p53 gene. Results from TUNEL assays revealed the efficiency of local γ-irradiation at the tumor site to induce apoptosis within 9L tumor mass. However, little or no apoptosis was detected after systemic (ip) injection of cisplatin (1 mg/kg). Interestingly, the micronuclei assays showed that not only γ-irradiation but also cisplatin treatment led to an increase in the emergence of binucleated cells with micronuclei. Apoptosis induction and micronuclei emergence are thus not absolutely correlated. However, micronuclei assays, rarely performed on solid tumors, appear more sensitive than apoptosis assays in evaluating DNA damage linked to chemotherapy.

KEYWORDS: tumor; apoptosis; micronuclei; chemotherapy; γ-irradiation

INTRODUCTION

New efficient treatments of human cancers may emerge from combining different therapies. The association of dendritic cell vaccination with treatments like chemotherapy or radiotherapy that induce apoptosis and damage in tumor cells could be one of them.[1] As an index for DNA damage, micronuclei (MN) are formed in dividing cells that contain chromosome breaks lacking centromeres (acentric fragments) and/or whole chromosomes that are unable to travel to the spindle poles during mitosis.[2]

Address for correspondence: Gregory Driessens, Route de Lennik 808 (bâtiment C), 1070 Bruxelles, Belgium. Voice: +0032/2.555.41.04; fax: +0032/2.555.46.55.
gregory.driessens@ulb.ac.be

**Ann. N.Y. Acad. Sci. 1010: 775–779 (2003). © 2003 New York Academy of Sciences.
doi: 10.1196/annals.1299.139**

Most studies on MN quantification were done to assess the *in vivo* genotoxicity of potential carcinogens, especially in lymphocytes. Only a few recent publications reported MN formation in cell lines derived from solid tumors.[3] However, simultaneous evaluation of apoptosis and MN formation in tumor cells should help in improving the efficacy and in decreasing side effects of radio- and/or chemotherapy, given alone or in combination with other treatments. This study thus aimed to compare, after *in vivo* chemo- or radiotherapy, the induction of apoptosis and MN formation in established 9L rat gliosarcoma tumors expressing a mutated p53 gene.

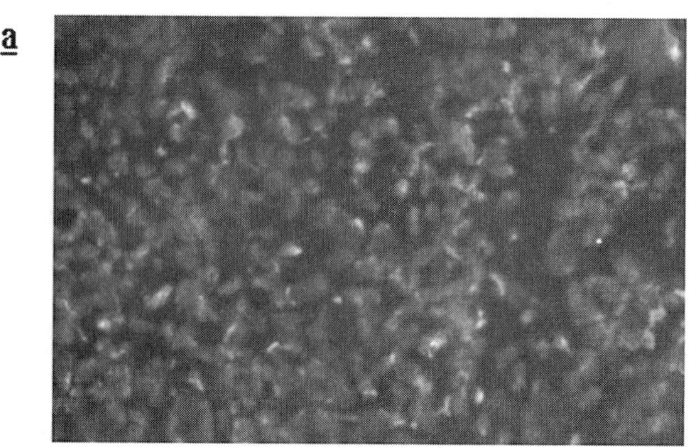

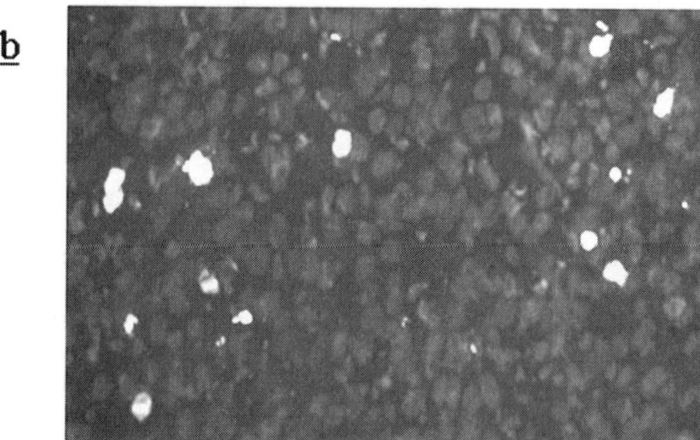

FIGURE 1. Apoptosis induced *in vivo* in established 9L tumors after **(a)** systemic injection of the rats with cisplatin (1 mg/kg) or **(b)** local γ-irradiation (20 Gy). Data from TUNEL assays revealed the presence of positive cells only after irradiation, but not in control cells or after cisplatin treatment. See text.

RESULTS

In Vivo *γ-Irradiation, but Not Cisplatin Treatment, Induces Detectable Apoptosis in 9L Tumors*

We evaluated the apoptosis induced *in vivo* in established 9L tumors after local γ-irradiation (20 Gy) or systemic injection of the rats with cisplatin (1 mg/kg). Tumor samples were excised from rats, directly cryopreserved in OCT, and cut into 10-μm-thick sections. Before immunostaining with *In Situ* Cell Death Detection kit, frozen tissue sections were fixed in PAF and permeabilized. Data from TUNEL assays revealed the presence of positive cells only after irradiation (FIG. 1b), but not in control cells or after cisplatin treatment (FIG. 1a).

Both In Vivo *γ-Irradiation and Cisplatin Treatment Induce DNA Damage in 9L Tumors*

We also analyzed in parallel, by the MN assay, the DNA damage induced *in vivo* in preimplanted 9L tumors following local γ-irradiation or systemic injection of cisplatin at 1 mg/kg. Cells were first recovered in suspension by enzymatic dissociation of resected tumor mass in DNase/collagenase solution, followed by vigorous pipetting and reculture overnight in complete medium. 9L cells were then treated with cytochalasin-B for a further 24 hours before being exposed to a mild hypotonic shock and fixed in methanol/acetic acid solution. Cells were cytospined on slides, stained with Hoechst 33342, and examined with a fluorescence microscope under UV light. Then, 1500 BN cells were scored for each rat and each agent. FIGURE 2a illustrates a photomicrograph of a binucleated (BN) cell with one MN. As shown in FIGURE 2b, MN formation was increased in tumors submitted to γ-irradiation or cisplatin treatment as compared to untreated controls. The percentages of BN cells with MN were similar for both cytotoxic treatments, but the number of BN cells bearing three or more MN per cell was higher in the case of γ-irradiation (not shown).

DISCUSSION

In this study, we compared the effects of cytotoxic agents on the induction of apoptosis and DNA damage within *in vivo* established 9L rat glioma cells. If data confirmed the efficiency of local γ-irradiation to induce apoptosis within 9L tumor mass *in vivo*, no or little apoptosis was however detected after cisplatin systemic injection. The lower efficiency of cisplatin to induce apoptosis may be linked to the moderate dose (1 mg/kg) used to avoid its general toxicity. Moreover, due to the *in vivo* clearance of apoptotic cells by neighbor macrophages, there was no means to quantify the time-dependent accumulation of apoptotic cells.[4] Finally, by expressing a mutated p53 protein, 9L cells could be also less sensitive to cisplatin-induced apoptosis (shown by Dorigo *et al.*).[5] Interestingly, MN assays that we also performed on *ex vivo* resected 9L tumors showed that both treatments led to an increase in the emergence of BN cells with MN. This result stressed that *in vivo* cisplatin injection could also cause important DNA damage in established 9L tumors.

a

b

control	21 +/- 1 %
cisplatin 1µg/ml	33 +/- 2 % *
irradiation	36 +/- 2 % *

* p<0.01 versus control

FIGURE 2. **(a)** Photomicrograph of a binucleated cell with one MN. **(b)** MN formation was increased in tumors submitted to γ-irradiation or cisplatin treatment as compared to untreated controls. See text.

In conclusion, we have successfully applied the MN assay, generally used to determine the genotoxicity of chemicals in normal cells, but rarely performed on solid tumors, to evaluate the *in vivo* efficacy of radio- or chemotherapy. Moreover, our data demonstrate that there is no direct relationship between induced apoptosis and MN formation and suggest that a little dose of chemotherapy could be sufficient to induce damage within a p53-mutated tumor, even without inducing apoptosis. Limiting the amount of cytotoxic agents administered along clinical protocols might be important to avoid toxicity and secondary deleterious effects of chemo- and radiotherapy, allowing optimization of anticancer treatments. Evaluating the efficacy of a cytotoxic treatment of the tumor mass by correlating apoptosis induction and DNA damage by MN assay should help to develop successfully one of the most promising strategies, the association of dendritic cell vaccinations with radio- or chemotherapy.

ACKNOWLEDGMENTS

This work was supported by the Belgian State, Prime Minister's office, Service for Science, Technology, and Culture; the Fonds National de la Recherche Scientifique; the Fonds de la Recherche Scientifique Médicale; the Actions de Recherche Concertées of the Communauté Française de Belgique; and Télévie.

G. Driessens is a fellow of the Fonds pour la Formation à la Recherche dans l'Industrie et dans l'Agriculture.

G. Driessens and L. Harsan contributed equally to this work.

REFERENCES

1. OKADA, H., I. POLLACK, F. LIEBERMAN *et al.* 2001. Gene therapy of malignant gliomas: a pilot study of vaccination with irradiated autologous glioma and dendritic cells admixed with IL-4 transduced fibroblasts to elicit an immune response. Hum. Gene Ther. **12:** 575–595.
2. FENECH, M. 2000. The *in vitro* micronucleus technique. Mutat. Res. **455:** 81–95.
3. TRUTER, E.J., A.S. SANTOS & W.J. ELS. 2002. Correlation between cell survival, clonogenic activity, and micronuclei induction in DMBA-OC-1R cells treated with immunospecific albumin microspheres containing cisplatin and 5-fluorouracil. Cell Biol. Int. **26:** 505–516.
4. SAVILL, J. & V. FADOK. 2000. Corpse clearance defines the meaning of cell death. Nature **407:** 784–788.
5. DORIGO, O., S.T. TURLA, S. LEBEDEVA *et al.* 1998. Sensitization of rat glioblastoma multiform to cisplatin *in vivo* following restoration of wild-type p53 function. J. Neurosurg. **88**(3)**:** 535–540.

Changes in Apoptotic and Mitotic Index, bcl_2, and p53 Expression in Rectum Carcinomas after Short-Term Cytostatic Treatment

E. FARCZÁDI,[a] I. KASZÁS,[b] M. BAKI,[a] AND B. SZENDE[c]

[a]*Oncology Department, St. Margit Hospital, Budapest, Hungary*

[b]*Pathology Department, St. Margit Hospital, Budapest, Hungary*

[c]*First Institute of Pathology and Experimental Cancer Research, Semmelweis University, and Research Group of Molecular Pathology, Hungarian Academy of Sciences and Semmelweis University, Budapest, Hungary*

ABSTRACT: Twenty patients with rectal adenocarcinoma were endoscopically biopsied and given short-term cytostatic therapy [5-fluorouracil (5-FU) (600 mg/m^2) and Ca-folinate (60 mg/m^2) for 2 days]. Seven days later, the tumor was resected or a second biopsy was performed. Apoptotic and mitotic indices as well as (mutant) p53 and bcl_2 expression were determined in the tumor tissue before and after the short-term chemotherapy. The patients were treated thereafter with long-term, intermittent 5-FU administration and followed up clinically for 7–26 months. Six patients showed progression of the disease and died, whereas 14 improved or showed no tumor progression. Significant increase of the apoptotic index and nonsignificant decrease of the mitotic index after the short-term cytostatic treatment were seen in the tumor tissue of responder cases. Nonresponders showed no change in both mitotic activity and apoptotic activity. Both survivors and deceased showed high mutant p53 expression, and the changes after short-term 5-FU treatment were not significant. Expression of bcl_2 was present in only 5 cases, with the postchemotherapy changes being not significant. These findings suggest that apoptotic response to short-term cytostatic therapy may be an additional predictive factor in rectal adenocarcinoma.

KEYWORDS: apoptosis; mitosis; bcl_2; p53; carcinoma; treatment; therapy

INTRODUCTION

The high incidence of colorectal carcinomas justifies the search for new prognostic and predictive factors.[1–5] Apoptotic and mitotic activity as well as expression of bcl_2 and p53 in untreated colorectal tumors were investigated for their prognostic value in colorectal tumors.

Address for correspondence: Dr. Béla Szende, First Department of Pathology and Experimental Cancer Research, Semmelweis University, Üllői út 26, H-1085 Budapest, Hungary. Voice/fax: +36-1-266-0451.

bszende@korb1.sote.hu

Ann. N.Y. Acad. Sci. 1010: 780–783 (2003). © 2003 New York Academy of Sciences.
doi: 10.1196/annals.1299.140

The present study deals with the possible predictive value of changes in apoptosis, mitosis, bcl_2, and p53 expression as a result of short-term cytostatic treatment.

MATERIALS AND METHODS

Twenty patients with clinically advanced (Dukes B2, C, and D) carcinoma of the rectum were selected for these studies. The site, stenosing, or exulcerative nature of the tumor was determined by rectoscopy, and a biopsy was taken. The histologic diagnosis of carcinoma was followed by 2-day cytostatic therapy including 5-fluorouracil (5-FU) (600 mg/m^2) and Ca-folinate (60 mg/m^2). Seven days later, a second histological specimen was taken.

Afterwards, chemotherapy was administered to all patients according to a recent protocol [5-FU (425 mg/m^2) and Ca-folinate (20 mg/m^2) for 5 subsequent days in each subsequent month]. Prospective radiotherapy (30 Gy) was also applied to the patients whose tumors were operable. The patients have been followed up regularly, including physical examination, tumor marker (CEA), and search for tumor recurrence (repeated rectoscopy and search for metastases by ultrasound and/or CT).

Histologic evaluation included classification and grading of the tumors according to differentiation and nuclear grade, as well as determination of the mitotic and apoptotic index and the proportion of p53 and bcl_2 positive tumor cells (2000 tumor cells in each sample).

In most instances, apoptotic figures were clearly recognized in HE-stained sections according to established morphological criteria. The TUNEL reaction confirmed the presence of apoptosis and was decisive in doubtful cases.

Informed consent was obtained from each patient, according to the policy of the local research ethics committee.

RESULTS

Fourteen patients are still alive and have improved clinically or are in stable condition at the average of 20.8 months after the first biopsy. Six patients died because of progression of the disease, proven by autopsy. The average survival time of these 6 patients was 11.5 months (TABLE 1).

The most important result was observed with apoptotic activity. There was a slight, although statistically nonsignificant, difference in the pretreatment samples between the apoptotic index of the patients who improved clinically and that of the patients who died, the latter being higher. Posttreatment apoptosis index increased in both groups, but the increase in the survivor group was highly significant, whereas with the deceased group the increase was not significant. The mitotic activity slightly and not significantly decreased in both groups. Similarly, mutant p53 expression did not change significantly in either group.

Initial p53 expression of various degrees was present in 13 survivors and 5 deceased. Expression of p53 disappeared in 1 survivor and appeared in the deceased who previously showed no p53 expression, by the time of the second biopsy. Expression of bcl_2 was detected in 1 deceased, in both pretreatment and posttreatment samples, and in 6 cases of survivors in the pretreatment sample, out of which bcl_2

TABLE 1. Age, sex, TNM and Dukes stage histological grade, and survival of patients with rectum adenocarcinoma

No.	Age, sex	TNM stage	Dukes stage	Histological grade	Survival (months)
1	73, M	$T_xN_xM_0$	D	I	24
2	58, F	$T_2N_0M_0$	B1	II	18
3	81, M	$T_2N_0M_0$	B1	II	26
4	73, F	$T_2N_1M_0$	C1	II	16
5	78, M	$T_1N_0M_0$	B1	I	26
6	55, F	$T_2N_2M_0$	C1	I	20
7	73, M	$T_2N_1M_0$	C1	I	24
8	53, M	$T_2N_1M_0$	C1	II	26
9	77, F	$T_2N_1M_0$	C1	II	16
10	66, M	$T_3N_0M_0$	B2	III	18
11	64, F	$T_2N_1M_6$	C1	I	18
12	67, M	$T_2N_1M_0$	C1	I	22
13	72, F	$T_2N_0M_0$	B1	II	16
14	59, M	$T_3N_0M_0$	B2	I	22
15	66, M	$T_xN_xM_0$	D	I	20+
16	60, F	$T_xN_xM_1$	D	I	13+
17	72, M	$T_2N_1M_0$	C1	II	9+
18	73, F	$T_xN_xM_1$	D	II	7+
19	63, F	$T_4N_2M_0$	C2	III	8+
20	73, M	$T_xN_xM_1$	D	I	12+

NOTE: x = uncertain; + = deceased.

TABLE 2. Apoptotic, mitotic, p53, and bcl_2 index of tumor cells in rectal adenocarcinomas in pretreatment and posttreatment (5-FU) samples

	Apoptosis %		Mitosis %		p53 %		bcl_2 %	
	Pre	Post	Pre	Post	Pre	Post	Pre	Post
Survivors (N = 14)	0.7 ± 5.6	1.76 ± 1.17*	1.2 ± 0.64	0.97 ± 0.52	58.9 ± 38.74	66 ± 38.0	10.7 ± 16.1	5.36 ± 10.6
Deceased (N = 6)	1.08 ± 0.45	1.25 ± 0.48	1.33 ± 0.5	1.05 ± 0.9	70 ± 4.58	79 ± 29.2	8.33 ± 20.4	8.33 ± 20.4

NOTE: Pre = pretreatment; post = posttreatment; $*p < 0.001$.

expression disappeared in 2 cases by the end of the short-term treatment. Owing to high values of standard deviation, the difference is nonsignificant (TABLE 2).

CONCLUSIONS

In conclusion, in this prospective study utilizing histologically detectable data of two subsequent samples, only the changes in the apoptotic index after short-term 5-FU treatment may be considered to have a predictive value.

REFERENCES

1. SCHWANDNER, O. *et al.* 2000. Apoptosis in rectal cancer: prognostic significance in comparison with clinical, histopathologic, and immunohistochemical variables. Dis. Colon Rectum **43:** 1227–1236.
2. WATSON, D.S. *et al.* 1999. Growth dysregulation and p53 accumulation in human primary colorectal cancer. Br. J. Cancer **80:** 1062–1068.
3. SINICROPE, F.A. *et al.* 1999. Apoptotic and mitotic indices predict survival rates in lymph node–negative colon carcinomas. Clin. Cancer Res. **5:** 1793–1804.
4. LANGLOIS, N.E. *et al.* 1997. Apoptosis in colorectal carcinoma occurring in patients aged 45 years and under: relationship to prognosis, mitosis, and immunohistochemical demonstration of p53, c-myc, and bcl-2 protein products. J. Pathol. **182:** 392–397.
5. SCHNEIDER, H.J. *et al.* 1997. Bcl-2 expression and response to chemotherapy in colorectal adenocarcinomas. Br. J. Cancer **73:** 427–437.

Apoptosis in Prostate Carcinomas after Short-Term Treatment with Decapeptyl

B. SZENDE,[a] S. LOVÁSZ,[b] P. FARID,[a] AND I. ROMICS[b]

[a]*First Department of Pathology and Experimental Cancer Research, Semmelweis University, and Research Group of Molecular Pathology, Hungarian Academy of Sciences and Semmelweis University, Budapest, Hungary*

[b]*Department of Urology, Semmelweis University, Budapest, Hungary*

ABSTRACT: Altogether, 40 patients with advanced prostate cancer received the LH-RH analog, Decapeptyl (D), as monotherapy for 1 month. The dose of D was 3.75 mg/month, intramuscularly in microcapsules. Needle biopsy was taken from the prostate of 20 patients at 24 hours, of 10 patients at 7 days, and of 10 patients at 30 days after the injection. The biopsy samples were investigated for mitotic and apoptotic indices as well as expression of p53 and Ki67 in the tumor cells. The effect of the LH-RH analog was manifested in very low and even absent mitotic activity in all points in time. The expression of mutant p53 was high (above 80%) in all 1-day and 7-day samples and in 8 of 10 samples taken 30 days after the start of the therapy. In 2 cases, the ratio of p53-expressing tumor cells was 2% and 5%. At day 30, the previously high ratio of Ki67-positive tumor cells decreased to 2–10% in 8 cases and showed focally higher values (70–90%) in 2 cases. Extremely high (90–100%) apoptosis could be registered in scattered foci of the tumor tissue in all samples taken 1 and 7 days after D injection. The other parts of the tumor tissue showed lower (1–5%) apoptotic activity. At day 30 after the injection, the same phenomenon was observed in 4 cases, but in 6 cases the apoptotic ratio became diffusely low (1–2%). Focal increase of apoptosis observed in the 1- and 7-day samples may be attributed to the direct effect of the LH-RH analog on tumor cells. The findings at day 30 indicate a considerable decrease in proliferating activity and, in the majority of cases, also in apoptotic activity, reflecting the effect of D on the pituitary-gonadal axis.

KEYWORDS: LH-RH; Decapeptyl; prostate carcinoma; apoptotic/mitotic index

INTRODUCTION

In our previous studies, an increase in apoptotic index and a decrease in mitotic index was demonstrated in repeated biopsies of prostate carcinomas after total androgen blockade (TAB), 3 to 4 months after the initiation of the therapy. Untreated tumors showed evenly distributed, very low (1–2%) apoptosis in this series and the increase was significant, but moderate.[1,2] Postulating direct antitumor capacity for

Address for correspondence: Dr. Béla Szende, First Institute of Pathology and Experimental Cancer Research, Semmelweis University, Üllői út. 26, H-1085 Budapest, Hungary. Voice/fax: +36-1-266-0451.

bszende@korb1.sote.hu

Ann. N.Y. Acad. Sci. 1010: 784–788 (2003).
doi: 10.1196/annals.1299.141

LH-RH analogs, apoptotic as well as proliferative activities were studied in samples taken from prostate carcinomas at 24 hours and 7 days after the first injection of the LH-RH agonist, Decapeptyl, in the present study.

MATERIALS AND METHODS

Altogether, 40 patients with advanced prostate cancer received the LH-RH analog, Decapeptyl (D), as monotherapy for 1 month. The dose of D was 3.75 mg/month, intramuscularly in microcapsules. Needle biopsy was taken from the prostate of 20 patients at 24 hours, of 10 patients at 7 days, and of 10 patients at 30 days after the injection. The biopsy samples were investigated for mitotic and apoptotic indices as well as expression of p53 and Ki67 in the tumor cells.

TABLE 1. Age, Gleason's grade, mitotic index, and apoptotic index in prostate cancers of patients treated with Decapeptyl for 24 hours

				Apoptosis %	
No.	Age	Gleason's grade	Mitosis %	Pattern 1	Pattern 2
1	77	7	–	1	50
2	68	7	–	1	80
3	68	8	–	1	60
4	74	7	–	90	90
5	62	9	1	10	10
6	73	7	2	5	5
7	67	5	–	1	90
8	46	9	–	1	90
9	73	5	–	1	60
10	76	9	–	1	90
11	79	5	–	1	70
12	69	9	–	1	70
13	69	5	–	1	50
14	57	3	–	1	60
15	79	7	–	1	60
16	65	7	–	1	60
17	71	7	–	1	50
18	85	5	–	1	70
19	70	8	–	1	60
20	66	7	–	1	50

TABLE 2a. Age, Gleason's grade, mitotic index, and apoptotic index in prostate cancers of patients treated with Decapeptyl for 7 days

				Apoptosis %	
No.	Age	Gleason's grade	Mitosis %	Pattern 1	Pattern 2
1	61	n.m.	–	50	50
2	71	5	–	5	50
3	88	7	2	4	60
4	81	7	–	2	60
5	74	4	1	5	60
6	53	4	–	2	60
7	80	4	–	2	60
8	68	8	2	2	2
9	60	7	–	4	60
10	86	7	–	4	50

TABLE 2b. Age, Gleason's grade, mitotic index, and apoptotic index in prostate cancers of patients treated with Decapeptyl for 30 days

				Apoptosis %	
No.	Age	Gleason's grade	Mitosis %	Pattern 1	Pattern 2
1	69	8	0	1	80
2	82	8	0	2	30
3	63	6	0	1	2
4	85	8	2	2	2
5	75	8	0	1	80
6	54	7	1	1	1
7	80	8	1	1	1
8	82	8	0	1	8
9	64	6	0	80	80
10	71	9	2	2	1

RESULTS

The effect of the LH-RH analog was manifested in very low and even absent mitotic activity in all points in time. The expression of mutant p53 was high (above 80%) in all 1-day and 7-day samples and in 8 of 10 samples taken 30 days after the start of the therapy. In 2 cases, the ratio of p53-expressing tumor cells was 2% and 5%. At day 30, the previously high ratio of Ki67-positive tumor cells decreased to 2–10% in 8 cases and showed focally higher values (70–90%) in 2 cases.

Extremely high (90–100%) apoptosis could be registered in scattered foci of the tumor tissue in all samples taken 1 and 7 days after D injection. The other parts of the tumor tissue showed lower (1–5%) apoptotic activity. At day 30 after the injection, the same phenomenon was observed in 4 cases, but in 6 cases the apoptotic ratio became diffusely low (1–2%). Focal increase of apoptosis observed in the 1-day and 7-day samples may be attributed to the direct effect of the LH-RH analog on tumor cells (TABLES 1, 2a, and 2b).

DISCUSSION

Our study was based on the postulation that at least a proportion of tumor cells express LH-RH receptors,[3–5] and the apoptosis-inducing effect of an LH-RH agonist may be observed even 24 hours after starting the therapy. It should be mentioned that in our previous studies the ratio of apoptotic tumor cells in untreated prostate cancers was low and apoptotic tumor cells were evenly distributed within the tumor tissue, without focal arrangement.[1,2]

The results obtained in the 20 patients biopsied 24 hours after the first D injection showed that a high proportion of tumor cells in focal distribution responded to the LH-RH analog with active cell death. However, other foci of tumor cells did not react. Basically, the same phenomenon was observed 7 days after initiation of the D treatment.

The findings at day 30 indicate a considerable decrease in proliferating activity and, in the majority of cases, also in apoptotic activity, reflecting the effect of D on the pituitary-gonadal axis.

CONCLUSIONS

Short-term treatment (24 hours, 7 days) with the LH-RH agonist, Decapeptyl, of patients with advanced prostate carcinoma causes focal regressive changes in the tumor tissue consistent with apoptosis. The nonapoptotic tumor cells showed Ki67 positivity pointing to proliferative activity. These findings indicate that a proportion of tumor cells express LH-RH receptors, and direct effect of the LH-RH analog may be exerted by binding the analog to these binding sites. The focal nature of the regressive changes means that a number of tumor cells failed to react and may give rise to further progression of the disease. Combined hormonal and cytostatic therapy may target such viable tumor cells.

ACKNOWLEDGMENTS

We are grateful to Ferring (Germany) for providing Decapeptyl.

REFERENCES

1. SZENDE, B. *et al.* 1999. Apoptotis, mitosis, p53, bcl_2, Ki-67, and clinical outcome in prostate carcinoma treated by androgen ablation. Urol. Int. **63:** 115–119.

2. SZENDE, B. *et al.* 2001. Repeated biopsies in evaluation of therapeutic effects in prostate carcinoma. Prostate **49:** 93–100.
3. REDDING, T.W. & A.V. SCHALLY. 1981. Inhibition of prostate tumor growth in two rat models by chronic administration of D-Rrp-6-LH-RH. Proc. Natl. Acad. Sci. USA **78:** 6509–6512.
4. FEKETE, M. *et al.* 1989. Receptors for luteinizing hormone–releasing hormone, somatostatin, prolactin, and epidermal growth factor in rat and human prostate cancers and in benign prostate hyperplasia. Prostate **14:** 191–200.
5. DAMYANOV, C. *et al.* 2001. Treatment of an orchiectomized patient with hormone-refractory prostate cancer with LH-RH agonists. Eur. Urol. **40:** 474–477.

Apoptosis in Prostate Carcinoma before and after Neoadjuvant Therapy with LH-RH Analog

STEFANIA BENCINI,[a] MARIO MIGALDI,[a] GIOVANNI CASTAGNETTI,[b] CARLO PINCELLI,[c] PAOLO FERRARI,[a] CARMELA DE GAETANI,[a] AND GIAN PAOLO TRENTINI[a]

[a]*Department of Pathologic Anatomy and Forensic Medicine, Section of Pathologic Anatomy, University of Modena and Reggio Emilia, 41100 Modena, Italy*

[b]*Division of Urology, AUSL, 41100 Modena, Italy*

[c]*Department of Medicine and Medical Specialties, Section of Dermatology, University of Modena and Reggio Emilia, 41100 Modena, Italy*

ABSTRACT: This study investigated apoptosis in prostate cancer before and after neoadjuvant treatment with LH-RH analog, demonstrating that this therapy induced high AI and bax and survivin overexpression, thus acting as a pro-apoptotic agent in prostate cancer. A statistically significant correlation was found between high AI and overexpression of bax protein after therapy. It is probable that this kind of therapy is deeply implicated in promoting the apoptotic intrinsic pathway.

KEYWORDS: apoptosis; prostate cancer; bax; survivin; AI; LH-RH analog

INTRODUCTION

Some recent evidence suggests a role for apoptosis in the clinical evaluation of prostate cancer and its modifications in response to androgen-depletion treatment.[1] The aim of this study was to evaluate any modifications in apoptosis induced by treatment with LH-RH analog in 105 cases of prostate cancer and to investigate the apoptotic pathway involved in this process. Apoptosis was studied by TUNEL assay, which marks nuclei undergoing apoptosis, and by immunohistochemistry (IHC) to evaluate bax and survivin expression. Bax is a proapoptotic member of the Bcl-2 family of proteins, involved in the "intrinsic" pathway of apoptosis,[2] as well as p53 and bcl-2, which we previously reported by IHC in the same series.[3] Survivin is an inhibitor of apoptosis protein (IAP), also involved in cell proliferation. It has been found overexpressed in most prostate, breast, and lung malignancies.[4]

Address for correspondence: Prof. Gian Paolo Trentini, Department of Pathologic Anatomy and Forensic Medicine, Section of Pathologic Anatomy, University of Modena and Reggio Emilia, via del Pozzo 71, 41100 Modena, Italy. Voice: +39-059-4224808; fax: +39-059-4224820. trentini.gianpaolo@unimore.it

Ann. N.Y. Acad. Sci. 1010: 789–792 (2003).
doi: 10.1196/annals.1299.142

TABLE 1. Follow-up (range: 20–90 months; mean: 50.12 months) and clinicopathological data in 105 prostate carcinomas

	Cases	Alive, disease-free	Alive, biochemical progression	Died of disease	*P*
Gleason grade					
1–6	30	18	11	1	
7	66	26	33	7	0.001
>7	9	1	3	5	
Tumor stage					
PT2	58	34	20	4	
PT3N0	30	11	17	2	0.001
PT3N+	17	0	10	7	

NOTE: Tumor stage was determined according to the UICC staging system, and grade according to Gleason's grading system.

MATERIALS AND METHODS

One hundred five cases of clinically localized prostate cancer were retrieved from the files of the Department of Pathologic Anatomy, University of Modena (Italy). In each case, IHC evaluation of bax and survivin expression on needle biopsy of the naïve prostate cancer and the relevant surgical specimens was performed. Moreover, apoptotic features were detected by TUNEL assay on both samples. All cases were treated for three months with LH-RH analogs before undergoing prostatectomy. Tumor stage and grade (TABLE 1) were determined by histological examination, while follow-up was obtained from clinical charts in all cases (TABLE 1). For IHC, we retrieved a representative block from routinely formalin-fixed and paraffin-embedded tissues using the streptavidin-biotin complex method. The following polyclonal antibodies were applied: bax protein (BioGenex, San Ramon, CA; 1:500 dilution) and survivin protein (Novus Biologicals, Littleton, CO; 1:2000 dilution). TUNEL assay was performed according to the Apoptotic Kit protocol (Intergen Co., Resnova, Rome, Italy). AI was evaluated as the % of TUNEL-positive nuclei on at least 100 neoplastic cells in biopsies and 1000 neoplastic cells in prostatectomies using a semiautomatic system of cell analysis (Menarini Diagnostics, Florence, Italy). AI, bax, and survivin expression were then compared with tumor grade/stage and patient's follow-up. Also, the above IHC results were compared with p53 and bcl-2 expression as previously investigated in the same case series.[3] Statistical analysis was carried out using *t* test and ANOVA. Survival curves were calculated using Kaplan-Meier method and log-rank test. Values ≤ 0.05 were considered significant.

RESULTS

AI mean values (±SD) were 0.88 ± 1.16 and 2.10 ± 1.76 before and after treatment, respectively, with a highly significant increase ($P < 0.0001$). When the AI values

TABLE 2. IHC results of bax and survivin

	Bax			Survivin		
	Biopsy	Operatory	*P*	Biopsy	Operatory	*P*
++	24	42		37	64	
+	49	52	0.0004	39	32	0.00005
–	32	11		29	8	

NOTE: A case was considered positive when at least 50% of cancer cells were DAB-positive; ++, positive; +, weakly positive; –, negative.

recorded on biopsies were compared with those of prostatectomies, an increase was documented in 74 cases (70.4%), whereas 23 cases (21.9%) showed a decrease and 8 had no changes (7.6%). The IHC cytoplasmatic positivity was scored on a three-tiered scheme. Reaction intensity was regarded as positive (++), weakly positive (+), and negative (–), considering "++" as a case with not less than 50% of tumor cells positive. Both bax and survivin expression were significantly increased after therapy (TABLE 2). A significant correlation ($P < 0.01$) was found between AI and bax expression, both increased after therapy; in contrast, survivin overexpression did not correlate with AI. Grade and stage were significantly associated with follow-up, confirming their role as prognostic indicators in prostate cancer (TABLE 1). AI, bax, and survivin expression did not correlate with grade, stage, and follow-up. When bax expression was compared to p53 expression, a significant correlation was recorded before and after therapy ($P < 0.002$ and $P < 0.003$, respectively). Survivin expression correlated with p53 expression after therapy only ($P < 0.0002$).

DISCUSSION

In the current study, we investigated apoptosis in prostate cancer before and after neoadjuvant treatment with LH-RH analog, demonstrating that this therapy induced high AI and bax and survivin overexpression, thus acting as a proapoptotic agent in prostate cancer. A statistically significant correlation was found between high AI and overexpression of bax protein after therapy. It is probable that this kind of therapy is deeply implicated in promoting the apoptotic intrinsic pathway.[1,2] A significant correlation was also found between increased bax expression and p53 overexpression, confirming the control role of bax protein with regard to the Bcl-2 family.[2] Conversely, survivin overexpression did not correlate with AI. However, we found a significant association between survivin and p53 overexpression, suggesting a possible involvement of p53 in control of the survivin gene.[5] As expected, tumor grade and stage correlated significantly with follow-up, resulting in the best prognostic markers in our series, whereas AI and bax and survivin expression did not appear to significantly predict patients' prognosis. However, it is noteworthy that a better, although not significant, overall survival was observed in prostate cancer presenting high AI (median, 157 vs. 139 months), bax protein overexpression (153 vs. 128 months), and lack of bcl-2 protein immunostaining (145 vs. 127 months).

REFERENCES

1. BRUCKHEIMER, E.M. & N. KYPRIANOU. 2000. Apoptosis in prostate carcinogenesis: a growth regulator and a therapeutic target. Cell Tissue Res. **301:** 153–162.
2. ANTOSSON, B. 2001. Bax and other pro-apoptotic Bcl-2 family "killer-proteins" and their victim, the mitochondrion. Cell Tissue Res. **306:** 347–361.
3. CESINARO, A.M., M. MIGALDI, G. FERRARI *et al.* 2000. Expression of p53 and bcl-2 in clinically localised prostate cancer before and after neo-adjuvant hormonal therapy. Oncol. Res. **12:** 43–49.
4. ALTIERI, D.C. 2001. The molecular basis and potential role of survivin in cancer diagnosis and therapy. Trends Mol. Med. **17:** 542–547.
5. LU, C., D.C. ALTIERI & N. TANIGAWA. 1998. Expression of a novel antiapoptosis gene, survivin, correlated with tumor cell apoptosis and p53 accumulation in gastric carcinoma. Cancer Res. **58:** 1808–1812.

Index of Contributors